Proceedings of the IX International Palynological Congress Houston, Texas, U.S.A., 1996

Edited by:

David K. Goodman
the irf group, inc.
6721 Round Tree Drive
Anchorage, Alaska 99516
U.S.A.
e-mail: flyfish@alaska.net

Robert T. Clarke
725 Sam Hill St.
Irving, Texas 75062-7548
U.S.A.
e-mail: rtclarke1@aol.com

Published by the American Association of Stratigraphic Palynologists Foundation
Dallas, Texas, U.S.A.

2001

ISBN 0–931871–06–9
Library of Congress Catalog Card Number 00-108213

Proceedings of the IX International Palynological Congress
Houston, Texas, U.S.A.
June 23-28, 1996

Editors: David K. Goodman and Robert T. Clarke

CONTENTS

Preface

Part 1: Paleozoic Palynology

Part 2: Mesozoic Palynology

Part 3: Tertiary Palynology

Part 4: Quaternary/Pleistocene Palynology

Part 5: Recent Palynology (Melissopalynology, Vegetational Palynology, Archaeopalynology, Pollen Spectra, Ecological Palynology, Aeropalynology, Pollen Morphology, and Forensic Palynology)

Preface

The Ninth International Palynological Congress was held in Houston, Texas, June 23–28, 1996 and was attended by approximately 415 palynologists from around the world. Participants were able to select from 421 oral papers and 141 posters and attend the topics of their choice during the five days of technical sessions. This volume contains 65 articles, which represent slightly more than 10% of the papers presented. The program was divided into 30 symposia and topical sessions, and this proceedings volume certainly reflects the diversity, if perhaps not the sheer volume, of subject matter presented and discussed during the Congress.

The field of palynology spans an incredibly wide and diverse variety of technical disciplines and scientific applications, from melissopalynology and forensic palynology in the modern world to the deliniation of paleoclimates and the reconstruction of fossil floras in early geologic time. We trust that the variety of papers contained in this volume provides an accurate measure of the impact that palynology has made in the past, as well as an indication of its expanding role in the future.

This volume could not have been published without the efforts of the many people who served as technical coordinators for the numerous subject groups. The science is simply too big for a pair of (almost) ancient industrial palynologists to master, therefore we relied on a group of people whose collective expertise is as wide as the science itself. This group includes Vaughn Bryant, Jr., David Pocknall, John Wrenn, Gordon Wood, Francisca Oboh-Ikuenobe, Paul Strother, Lucy Edwards, Pierre Zippi, Robert Van Pelt, Joyce Lucas-Clark, Javier Helenes, Owen Davis, Jan Jansonius, Merrell Miller, and Martin Farley. The work of these individuals reduced our task load immensely, for which we are truly thankful. However, from a broader perspective it has also resulted in a well-constrained technical volume of scientific papers that will be used by the palynological community for many years.

Finally, what would there be without the dedicated work of the authors? To put it bluntly ... not much. This volume exists only because more than 100 scientists made it happen. We have enjoyed working through the editorial process with each author, and thank them for their extended efforts involved in seeing their work to the final publication. It has been a long journey—much longer than originally planned. Thank you all for your patience, your persistence, and above all for your dedication to this complex and exciting field of scientific endeavour.

David K. Goodman (Anchorage, Alaska)
Robert T. Clarke (Irving, Texas)

Part 1: Paleozoic Palynology

NEW EARLY AND MIDDLE PALEOZOIC REPRESENTATIVES OF THE HYDRODICTYACEAE (CHLOROPHYTA)

Merrell A. Miller and Gordon D. Wood

Abstract

A Middle Ordovician to Late Devonian planar coenobium, *Tapetisphaerites formale* gen. et sp. nov., and a Llandovery ring-shaped coenobium, *Kahfia circularis* sp. nov., are described and illustrated. They are assigned to the Order Chlorococcales, Family Hydrodictyaceae, of the Division Chlorophyta based on their coenobial habit. Extant Hydrodictyaceae are found only in fresh or brackish water; however, the new Paleozoic species described herein are present in sediments deposited in marine settings and may be allochthonous.

INTRODUCTION

Paleozoic multicellular palynomorphs occur either in irregular or regular aggregations. Irregular aggregations can either be clusters of spheromorphs lacking regular configurations (e.g., *Synsphaeridium* Eisenack 1965) or monospecific clusters of netromorphic, acanthomorphic, diacromorphic, and galeate acritarchs that otherwise occur as isolated specimens (see Combaz et al., 1967; Cramer, 1970; Downie, 1973; Rasul, 1974). Monospecific clusters are relatively rare, not widely reported, and classified as acritarchs. Downie et al. (1963, p. 7) and Fensome et al. (1990) excluded coenobial palynomorphs from the Acritarcha.

Geometrically regular aggregations of cells that form coenobia are becoming more widely recognized in the Early and Middle Paleozoic. *Deflandrastrum* Combaz 1962 is among the first coenobial species to be classified with the Hydrodictyaceae (Combaz, 1962). Subsequently, coenobial palynomorphs have been recovered from the Ordovician (Cramer and Díez, 1977; Miller, 1988; Miller and Wood, 1996; Vavrdová, 1990; Wood and Miller, 1996, 1997), Silurian (Miller, 1988; Miller and Wood, 1996; Le Hérissé et al., 1995; Colbath and Grenfell, 1995; Wood and Miller, 1996, 1997) and Devonian (Oliveira and Burjack, 1996; Wood and Miller, 1996, 1997; Wood and Turnau, 2001). The regular geometric arrangement of the cells (coenocytes sensu Batten, 1996) in planar or spherical (concentric) configurations and addition of cells in systematic multiples within a consistent geometric configuration (Batten, 1996) allies them with the Hydrodictyaceae. These characters separate the taxa described herein from the irregular, monotypic clusters mentioned above. Based on morphological comparisons, *Tapetisphaerites formale* gen. et sp. nov. and *Kahfia circularis* sp. nov. are considered to be related to the Hydrodictyaceae (Miller, 1988; Miller and Wood, 1997, p. 705–706).

Although morphologically similar to extant members of the Hydrodictyaceae, and placed in that family, the Early and Middle Paleozoic species described herein were recovered from sediments deposited in marine environments. Extant representatives of the Hydrodictyaceae live only in fresh or brackish water but may be transported and deposited into marine settings (Evitt, 1963; Foerster, 1973; Wood and Miller, 1998). It is possible that these two forms are nonmarine and were transported from freshwater environments (see Zippi, 1998, p. 9–10). However there is no direct evidence to support this supposition in the case of the Paleozoic species described herein.

Miller, Merrell A., and Wood, Gordon D.
2001 New Early and Middle Paleozoic representatives of the Hydrodictyaceae (Chlorophyta); *In:* Goodman, D.K., and Clarke, R.T. (eds.), *Proceedings of the IX International Palynological Congress, Houston, Texas, U.S.A., 1996;* American Association of Stratigraphic Palynologists Foundation, p. 3–10.

ISBN 0–931871–06–9

SAMPLE, PROCESSING AND REPOSITORY INFORMATION

Tapetisphaerites formale gen. et sp. nov. and *Kahfia circularis* sp. nov. were recovered from outcrop samples, borehole cuttings and cores. Their stratigraphic and geographic occurrences are summarized in the Locality Register and on Text-Figure 1. Palynomorphs were isolated from shales using standard HCL–HF–HCL acid maceration followed by zinc bromide separation (Barss and Williams, 1973; Wood et al., 1996). Screened palynological residues were embedded in Clearcol and cemented to slides with Elvacite for permanent palynological mounts. Specimens are located by England Finder coordinates listed in the plate explanations. Type specimens are housed at The Orton Geological Museum, The Ohio State University, 155 South Oval Mall, Columbus, Ohio 43210. Each type specimen is assigned an Orton Geological Museum repository number (OSU).

SYSTEMATIC PALEONTOLOGY

Division CHLOROPHYTA Pascher 1914
Class CHLOROPHYCEAE Kützing 1843
Order CHLOROCOCCALES Marchand 1895
orth. mut. Pascher 1915
Family HYDRODICTYACEAE (Gray 1821)
Dumortier 1829

Genus *Tapetisphaerites* gen. nov.

Derivation of Name. *Tapeti* + *sphaerites* (Latin), for the mat-like sheet of attached subspherical coenocytes.

Type Species. *Tapetisphaerites formale* sp. nov.

Text Figure 1. Occurrences of *Tapetisphaerites formale* gen. et sp. nov. (nos. 1–10) and *Kahfia circularis* sp. nov. (no. 11). 1) Cabot Head Shale, Hamilton, Ontario, Canada; 2) Neahga Shale, Lewiston, New York, U.S.A.; 3) Mountain Lake Member of the Bromide Formation, Arbuckle Mountains, Carter County, Oklahoma, U.S.A.; 4) Tuscarora Formation, Clay County, West Virginia, U.S.A.; 5) Akpatok Island, Ungava Bay, District of Franklin, Canada; 6) Undifferentiated Middle Ordovician, Ogemaw County, Michigan, U.S.A.; 7) Qalibah Formation, Saudi Arabia; 8) Amoco Jordan Wadi Sirhan No. 2 well, Jordan; 9) Jafar No. 1 well, Syria; 10) Geirud Formation, Central Alborz Range, Iran; 11) Osgood Member of the Salamonie Dolomite, Kentucky, U.S.A. Details are provided in the Locality Register.

Diagnosis. A geometrically regular arrangement of coenocytes connected laterally to form a planar sheet one coenocyte thick. Sheets quadrate to subcircular in outline and contain an even number of coenocytes. Ontogenetic development begins with an initial cruciform four coenocyte coenobium followed by eight and sixteen cell coenobia. Individual coenocytes circular to subcircular in outline, presumably subspherical when inflated, except area in contact with adjacent coenocytes. Individual coenocytes unilayered and surfaces, other than those contacting adjacent coenocytes, may be ornamented. A small central lumen may be present. No dehiscence structures recognized.

Remarks. The systematic addition of coenocytes is an ancient and successful evolutionary adaption in some Chlorophyta. In rare instances small bodies may be present in the cells. This may represent remains of cytoplasm or metabolic products (see Wood and Miller, 1997, pl. 1.fig. 5). Ontogenetically the four coenocyte cruciform morphology should have resulted from a two coenocyte form, however, the latter was not observed in the assemblages examined.

Comparison. Eisenack (1965, p. 268) established the genus *Synsphaeridium* for small organic-walled spheres connected closely with each other, and not occurring in a spherical mass. He remarked that most specimens in the adhering masses occur in one plane, but that overlap could

PLATE 1

Light photomicrographs of specimens of *Tapetisphaerites formale* gen. et sp. nov. Specimens illustrated in figs. 1, 2, 4–8, 10–13 are from the Shell Western Exploration and Production State Foster No. 1-28, Ogemaw County, Michigan, U.S.A.; those illustrated in figs. 3 and 9 are from the Neahga Shale, New York, U.S.A. (see Locality Register). All figures x975. Figs. 1, 4, 6, 7, 10, 12, 13 are phase contrast; figs. 2, 3, 5, 8, 9, 11 are Normarski differential interference contrast.

1–13 *Tapetisphaerites formale* gen. et sp. nov. 1, 2, Holotype. Complete 16 cell cenobium. Slide 29093 A02, U-40/1. OSU 49182. 3, Slide 61325 A01, R-46. 4, 5, Paratype. Slide 29092 A02, Q-38. OSU 49183. 6, Paratype. Slide 29092 A02, N-44. OSU 49184. 7, 8, Paratype. A central lumen is present in this specimen. Slide 29092 A02, V-43. OSU 49185. 9, Slide 61235 A02, N-37. 10, 11, Slide 29089 A03, K-31/1. 12, Four cell cenobium. Slide 29089 A02, N-13/4. 13, Eight cell cenobium. Slide 29091 A03, M-42/4.

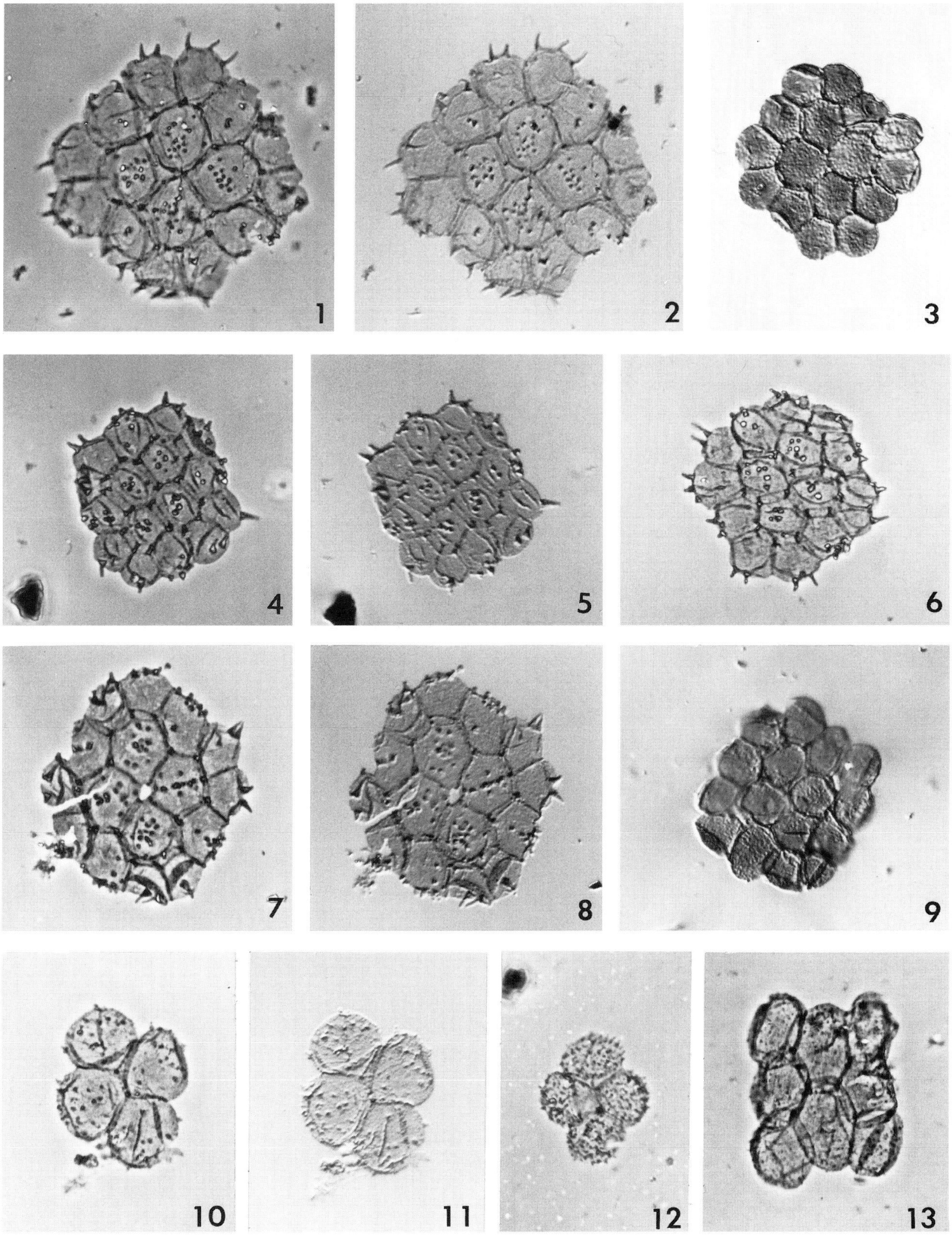
1
2
3
4
5
6
7
8
9
10
11
12
13

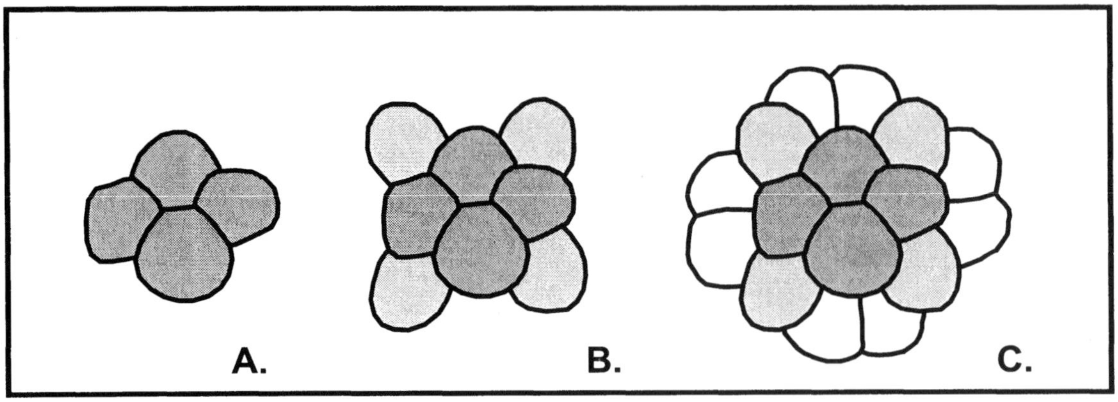

Text Figure 2. Ontogenetic sequence of coenocyte addition for *Tapetisphaerites formale* gen. et sp. nov. 2A is a representation of the specimens shown on Plate 1, figs. 10–12; 2B is an intermediate coenobium (Plate 1, fig. 13); 2C represents a 16 cell ceonobia (Plate 1, figs. 1–9).

occur. *Tapetisphaerites* gen. nov. differs from the holotype of *Synsphaeridium gotlandicum* Eisenack 1965 by having a geometric arrangement of a constant maximum number of coenocytes with a definable outline that do not overlap. Both Silurian species (*S. gotlandicum* and *S. tuberculata*) have variable numbers of coenocytes in given specimens, are more than one coenocyte thick, and have an irregular outline. Other species of *Synsphaeridium* have been reported from Precambrian strata but none possess geometrically arranged coenocytes (see Vidal, 1976, p. 32–33, for a brief discussion of some of these forms).

Moreau-Benoit (1974) and Playford (*in* Playford and Dring, 1981) assigned attached linear groups of spheromorphs to *Synsphaeridium*. Although Eisenack's diagnosis does not preclude this configuration, the linear forms are not comparable to the holotype of *Synsphaeridium*. *Ericantha* Cramer & Díez 1977 has seven 'cells'. Five are in one plane and two centrally situated on opposite sides of the group of five (Cramer and Díez, 1977, pl. 4, figs. 1–3).

Described four-celled coenobia include: *Bion* Eisenack 1951, *Deflandrastrum* Combaz 1962, and *Quadrisporites* Hennelly 1958 *ex* Potonié & Lele 1961. *Bion cruciferum* Eisenack 1951 (p. 200–201, pl. 3, figs. 1–4; neotype Eisenack 1959, p. 206, pl. 17, figs. 24, 25) is nearly twice as large and lacks a central cruciform group of four cells. *Deflandrastrum* is characterized by a prominent central lumen and angular coenobia that have long radiating processes (Combaz, 1962). *Quadrisporites* (Hennelly, 1959, p. 364–365, pl. 5, figs. 6, 7) has considerably more ornamentation than *Tapetisphaerites* and, to date, has only been found in four cell coenobia. *Tetraletes* (Cramer, 1966, p. 77–78, pl. 1, figs. 7–10,12) also only occurs as tetrads and has a thickened contact area between cells. Cramer and Díez de Cramer (1972, p. 171) suggested that *Tetraletes* is a junior synonym of *Quadrisporites*. Strother (1991) formally placed *Tetraletes* in synonomy with *Quadrisporites*.

Plaesiodictyon Wille 1970 (p. 285, figs. 9, 10; Brenner and Foster, 1994, p. 221–230) from the Triassic and Devonian coenobia (Wood and Miller, 1997, pl. 1, figs. 11–14; Wood and Turnau, 2001) have regularly-arranged quadrate, planar coenobia that may have considerably more coenocytes than *Tapetisphaerites*.

Tapetisphaerites formale sp. nov.
Plate 1, figs. 1–13; Text-Figure 2

Derivation of Name. *Formalis* (Latin) describing the set form of the individual coenocytes for this species.

Holotype. Plate 1, figs. 1, 2; OSU 49182.

Paratypes. Plate 1, figs. 4, 5; OSU 49183. Plate 1, fig. 6; OSU 49184. Plate 1, figs. 7, 8; OSU 49185.

Type Locality. Shell Western Exploration and Production State Foster No. 1–28, Ogemaw County, Michigan, U.S.A. (44.43792°N, 84.19204°W).

Type Stratum. Undifferentiated Middle Ordovician (probably Whiterockian), core depth 3,134.5–3,159.1 m (10,410,–10,425 ft).

Diagnosis. A geometrically arranged group of 16 individual hollow coenocytes attached to form a unilayered, planar sheet. Four coenocytes in cruciform arrangement are present in the coenobium center. These are surrounded by four or 12 coenocytes giving the sheet a crenate quadrate to subcircular outline. Individual cells are subcircular in outline except where in contact with adjacent coenocytes. Coenocyte walls unilayered except for contact areas with other coenocytes, < 0.5 µm thick. Ornamentation, scabrate, granulate to echinate (up to 3 µm in length). No excystment structure recognized.

Measurements. Maximum diameter 46.0(38.7)35.1 µm (holotype 46 µm); diameter of individual coenocytes 7.4–12.7 µm (holotype 11 µm). Six specimens measured. Diameter of four cell coenobia 25–29 µm; diameter of individual coenocytes 11–12 µm. Diameter of eight cell coenobia 31 µm; diameter of individual coenocytes 12.9 µm. Fourteen Silurian specimens and 40 Ordovician specimens observed.

Remarks. Four coenocytes form the central cruciform grouping (Text-Figure 2A). Two coenocytes of this central group meet to form a common boundry separating the remaining two coenocytes. Occasionally the cells in this central group are larger than those of the outer 'circlet'. Rarely there is a small pore or lumen in the center of the coenobium (Plate 1, figs. 7, 8). This arrangement maximizes packing and surface area. The outer 'circlet' of 12 coenocytes is also consistent in its packing arrangement with regard to the central cruciform group. Four cells are added at the junctions between the coenocytes of the central group to form an eight cell coenobium (Text-Figure 2B). For the complete 16 coenocyte coenobia, four pairs of cells that touch the central group coenocytes are added to the eight cell coenobia (see Text-figure 2C). Deviation from this pattern is probably the result of physiological factors or environmental conditions.

Ordovician representatives of *Tapetisphaerites formale* sp. nov. may have the most pronounced ornamentation. However, in a given population of Ordovician specimens some have granular ornamentation similar to that present on Silurian specimens (see Lower Silurian specimens in Wood and Miller, 1997, pl. 1, fig. 5; Le Hérissé, 1995, p. 72–73, pl. 4, fig.1). Thus we are not separating species or subspecies based on ornamentation development at this time but we recognize that the Silurian and younger specimens are generally not as robustly ornamented as are those from the Ordovician. Silurian representatives are typically 16 cell coenobia; eight cell coenobia were rarer as were the four cell forms.

Comparison. Specimens of *Tapetisphaerites formale* sp. nov., differ from those of *Synsphaeridium gotlandicum* Eisenack 1965 and *S. tuberculatum* Eisenack 1965 by their geometric planar arrangement of coenocytes, regular outline and one cell in thickness.

Occurrence. A specimen of *Tapetisphaerites formale* sp. nov. was figured by Tappan (as *Synsphaeridium* sp. cf. *S. gotlandicum;* 1980, p. 151, fig. 3.2 , no. 1) from the Middle Ordovician of Oklahoma. Le Hérissé *et al.* (1995; p. 72,73) reported *T. formale* (as *Synsphaeridium* sp.) from the Llandovery (Rhuddanian–Aeronian), Qalibah Formation of the RGHB-1 well, Saudi Arabia. Other specimens have been recovered from Tuscarora Sandstone in the United Fuel Gas Clay 513 well, Clay County, West Virginia, U.S.A.; the Cabot Head Shale, Flock Road, Hamilton, Ontario, Canada (Wood and Miller, 1997); and the Neahga Shale exposed along the access road to the Robert Moses Power Plant, Niagara County, New York, U.S.A. (see Locality Register, Wood and Miller, 1997; herein); Ghavidel-Syooki (1995, pl. 4, fig. 14) illustrated an unlabeled specimen along with the acritarch *Triangulina* from the Upper Devonian (Frasnian) Geirud Formation of Iran.

Genus *Kahfia* Le Hérissé et al. 1995

Type Species. *Kahfia arabica* Le Hérissé et al. 1995.

Kahfia circularis sp. nov.
Plate 2, figs. 1–4

Derivation of Name. *Circularis* (Latin) describing the ring-shaped coenobium.

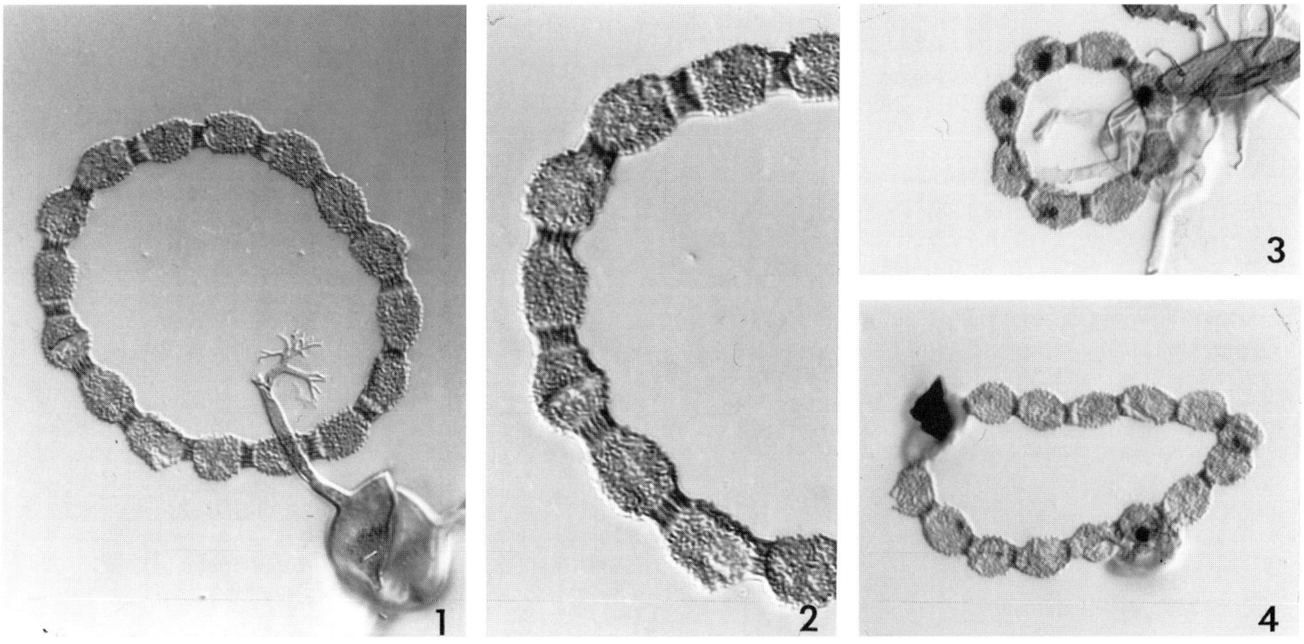

PLATE 2

Light photomicrographs of specimens of *Kahfia circularis* sp. nov. Specimens are from the Osgood Shale Member of the Salamonie Dolomite, Bardstown I Section, Nelson County, Kentucky, U.S.A. (see Locality Register; Noger and Kepferle, 1985).

1–4 *Kahfia circularis* n. sp. 1, 2, Holotype. 1: Complete 16 cell cenobium, x690. 2: Magnified view of the coenocyte junctions and ornamentation, x1130. Slide 30103 A02, M-8/1. OSU 49186. 3, Paratype. Complete eight cell cenobium. Slide 30103 A03, S-43/1, x690, OSU 49187. 4, Paratype. Complete 16 cell cenobium. Slide 30103 A03, P-37/4, x690, OSU 49188.

Holotype. Plate 2, figs. 1, 2; OSU number 49186.

Paratypes. Plate 2, fig. 3; OSU number 49187. Plate 2, fig. 4; OSU number 49188.

Type Locality. Bardstown I Section, Nelson County, Kentucky, U.S.A. (see Locality Register; Noger and Kepferle, 1985).

Type Stratum. Latest Llandovery Osgood Shale Member of the Salamonie Dolomite. *Gracilisphaeridium encantador* Cramer 1970 occurs with *Kahfia circularis* sp.nov.

Diagnosis. A planar, circular, coenobium consisting of eight or 16 individual coenocytes forming a large circular lumen. The coenocytes are elongate, biconvex, densely granulate to papillate, not communicating with adjacent cells, and connected at a constriction. The connection is thickened and has stria parallel to the longitudinal axis of the coenocytes. Dehiscence structure not observed.

Measurements. Cell width 7(9)11 µm; cell length 9(11)13 µm; constriction width 3.5(4.5)5.5 µm; overall diameter 8-cell coenobium 37 µm, 16-cell coenobium 74–77 µm (holotype 77 µm); wall thickness <0.5 µm (four specimens measured).

Comparison. *Kahfia circularis* sp. nov. differs from *K. arabia* Le Hérissé et al. (1995, p.72, pl. 4, figs. 2, 3) by having oval cenocytes. Additionally, *K. circularis* can occur in 16 cell coenobia.

Remarks. One specimen of *Kahfia circularis* (Plate 2, fig. 3) contains dark nearly opaque bodies in the cells. These bodies may represent condensed cytoplasm or metabolic remains.

Occurrence. Osgood Shale Member of the Salamonie Dolomite Bardstown I Section, Nelson County, Kentucky, U.S.A. (see Locality Register; Noger and Kepferle, 1985). The specimens illustrated in Plate 2 were first published as *Kahfia* sp. by Wood and Miller (1997, pl. 1, figs. 6–8).

LOCALITY REGISTER

1. Early Llandovery Cabot Head Shale, Medina Group, Flock roadcut (Sanford et al., 1972; Wood and Miller, 1997, pl. 1, fig. 5), ca. 9.1 m (30 ft) above the base of the Cabot Head Shale, City of Hamilton, Wentworth County, Ontario, Canada (43.26112°N, 79.85017°W).
2. Llandovery Neahga Shale, Clinton Group, exposed along the access road to the Robert Moses Power Plant (Berry and Boucot, 1970), 3.9 km (2.4 mi) south of the City of Lewiston, Niagara County, New York, U.S.A. (43°08'18"N, 79°02'30"W).
3. Middle Ordovician Simpson Group (Bromide Formation, Mountain Lake Member) of southern Oklahoma, Carter County, U.S.A. (Tappan, 1980; p. 151, fig. 3.2.1; Ross et al., 1982; Derby et al., 1991).
4. Llandovery Tuscarora Formation, United Fuel Gas Clay 513 well, core depth 2,266.4–2,271 m (7,479–7,494.5 ft) Clay County, West Virginia, U.S.A.
5. Middle Ordovician, Premium Homestead Akpatok Island F-26 well, core depth 308.1–308.1m (1,016.8–1,017 ft), Akpatok Island, Ungava Bay, District of Franklin, Canada (Barnes et al., 1981; Workum et al., 1976).
6. Undifferentiated Middle Ordovician (probably Whiterockian), Shell Western Exploration and Production State Foster No. 1-28, 3,134.5–3,159.1 m (10,410–10,425 ft), Ogemaw County, Michigan, U.S.A. (44.43792°N, 84.19204°W).
7. Llandovery (Rhuddanian–Aeronian), Qalibah Formation, RGHB-1 well, 2,468.2–2,468.8 m (8,145–8,147 ft), Saudi Arabia (Le Hérissé et al., 1995; p. 72,73, pl. 4, fig. 1).
8. Amoco Jordan Wadi Sirhan No. 2 well, core depth 954.6 m (3,150.2 ft), Jordan (31.11908°N, 37.2773°E).
9. Undifferentiated Llandovery, Jafar No. 1 well, cuttings depth 821.2 m (2,710 ft), Syria (30°06'15.6"N, 36°21'14.8"E).
10. Frasnian Geirud Formation, Hassanakdar area, Central Alborz Range, northern Iran (Ghavidel-Syooki, 1995; pl. 4, fig. 14).
11. Late Llandovery, Osgood Shale Member of the Salamonie Dolomite, Bardstown I Section (Noger and Kepferle, 1985; Wood, 1988, 1989), a road-cut exposure located on the south side of the Blue Grass Parkway, 1.6 km (ca. 1 mile) west of Kentucky Highway 31E interchange, just east of bridge across Beech Fork River, near Bardstown Kentucky, U.S.A. (37°46'5"N, 85°28'17"W).

ACKNOWLEDGMENTS

The authors gratefully acknowledge M. Schendel (Amoco Graphic Arts, Houston) for drafting text-figures and J.C. Lawson and H.M. Kuriger (Amoco Exploration and Production Group, Earth Science Lab) for processing the samples. Drs. D.K. Goodman (Arco Alaska, Inc.) and P.K. Strother (Boston College) provided excellent reviews. Dr. J. Jansonius assisted with etymology. Dr. C.B. Rexroad (Indiana Geological Survey) assisted in collecting the Osgood shale samples. We thank Amoco Exploration and Production Technology Group for technical support and permission to publish.

References Cited

BARNES, C.R., NORFORD, B.S., and SKEVINGTON, D.
1981 *The Ordovician System in Canada. Correlation chart and explanatory notes.* International Union of Geological Sciences. Publication No. 8, 27 pgs.

BARSS, S.M., and WILLIAMS, G.L.
1973 Palynology and nannofossil processing techniques. *Geological Survey of Canada, Paper*, 73–26, 25 pgs.

BATTEN, D.J.
1996 Colonial Chlorococcales. *In*: Jansonius, J., and McGregor, D.C. (eds.), *Palynology: principles and applications*. American Association of Stratigraphic Palynologists Foundation, Dallas, 1: 191–203.

BERRY, W.B.N., and BOUCOT, A.J.
1970 Correlation of the North American Silurian Rocks. *Geological Society of America, Special Paper*, 102, 289 pgs.

BRENNER, W., and FOSTER, C.B.
1994 Chlorophycean algae from the Triassic of Australia. *Review of Palaeobotany and Palynology*, 80: 209–234.

COLBATH, G.K., and GRENFELL, H.R.
1995 Review of biological affinities of Paleozoic acid-resistant, organic-walled eukaryotic algal microfossils (including "acritarchs"). *Review of Palaeobotany and Palynology*, 86: 287–314.

COMBAZ, A.
1962 Sur un nouveau type de microplanctonte cénobial fossile du Gothlandien de Libye, *Deflandrastrum* nov. gen. *Comptes rendus des séances de l'Académie des sciences*, 255: 1977–1979.

COMBAZ, A., LANGE, F.W., and PANSART, J.
1967 Les 'Leiofusidae' Eisenack, 1938. *Review of Palaeobotany and Palynology*, 1: 291–307.

CRAMER, F.H.

1966 Palynomorphs from the Siluro–Devonian boundary in NW Spain. *Instituto Geologico y Minero de España, notas y comunicaciones* no. 85: 71–82.

1970 Distribution of selected Silurian acritarchs. An account of the palynostratigraphy and paleogeography of selected Silurian acritarch taxa. *Revista Española de micropalaeontologia*, numero extradinario, 203 pgs.

CRAMER, F.H., and DÍEZ, M. del C.R.

1972 North American Silurian palynofacies and their spatial arrangement: acritarchs. *Palaeontographica*, Abteilung B, 138: 107–180.

1977 Late Arenigian (Ordovician) acritarchs from CIS-Saharan Morocco. *Micropaleontology*, 23: 339–360.

DERBY, J.R., BAUER, J.A., CREATH, W.B., DRESBACH, R.I., ETHINGTON, R.L., LOCH, J.D., STITT, J.H., McHARGUE, T.R., MILLER, J.F., MILLER, M.A., REPETSKI, J.E., SWEET, W.C., TAYLOR, J.F., and WILLIAMS, M.

1991 Biostratigraphy of the Timbered Hills, Arbuckle, and Simpson Groups, Cambrian and Ordovician, Oklahoma: A review of correlation tools and techniques available to the explorationist. *In*: Johnson, K.S. (ed.), Late Cambrian–Ordovician Geology of the Southern Midcontinent, 1989 Symposium. *Oklahoma Geological Survey, Circular*, 92: 15–41.

DOWNIE, C.

1973 Observations on the nature of the acritarchs. *Palaeontology*, 16: 239–259.

DOWNIE, C., EVITT, W.R., and SARJEANT, W.A.S.

1963 Dinoflagellates, hystrichospheres, and the classification of the acritarchs. *Stanford University Publications, Geological Sciences*, 7: 1–16.

EISENACK, A.

1951 Über Hystrichosphaerideen und andere Kleinformen aus baltischem Silur und Kambrium. *Senckenbergiana Lethaea*, 32: 187–204.

1959 Neotypen baltischer Silur-Hystrichosphären und neue Arten. *Palaeontographica*, Abteilung A, 112: 193–211.

1965 Mikrofossilien aus dem Silur Gotlands. Hystrichosphären, Problematika. *Neues Jahrbuch für Geologie und Paläontologie*, Abhandlungen, 122: 257–274.

EVITT, W.R.

1963 Occurrence of freshwater alga *Pediastrum* in Cretaceous marine sediments. *American Journal of Science*, 261: 890–893.

FENSOME. R., WILLIAMS, G.L., BARSS, M.S., FREEMAN, J.M., and HILL, J.M.

1990 Acritarchs and fossil prasinophytes: an index to genera, species and intraspecific taxa. *American Association of Stratigraphic Palynologists Contribution Series*, 25: 1–771.

FOERSTER, J.W.

1973 The fate of freshwater algae entering an estuary. *In*: Stevenson, L.H., and Colwell, R.R. (eds.), *Estustine Microbial Ecology*. University of South Carolina Press, Columbia, 387–420.

GHAVIDEL-SYOOKI, M.

1995 Palynostratigraphy and palaeogeography of a Palaeozoic sequence in the Hassanakdar area, Central Alborz Range, northern Iran. *Review of Palaeobotany and Palynology*, 86: 91–109.

HENNELLY, J.P.F.

1959 Spores and pollens from a Permian–Triassic transition, N.S.W. *Proceedings of the Linnean Society of New South Wales*, 83: 363–369 [1958].

LE HÉRISSÉ, A., AL-TAYYAR, H., and VAN DER EEM, H.

1995 Stratigraphic and paleogeographical significance of Silurian acritarchs from Saudi Arabia. *Review of Palaeobotany and Palynology*, 89: 49–74.

MILLER, M.A.

1988 An Early Paleozoic "acritarch" with morphology comparable to the Hydrodictyaceae. *Abstracts, 7th International Palynological Congress*, Brisbane, Australia, p. 112.

MILLER, M.A., and WOOD, G.D.

1996 Paleoecological significance of the Hydrodictyaceae, ?Scenedesmaceae and Zygnemataceae (Chlorophyta) in the Early Paleozoic. *Program and Abstracts, 9th International Palynological Congress*, Houston, U.S.A., p. 106.

MOREAU-BENOIT, A.

1974 Recherches de Palynologie et de Planctologie. Sur le Dévonien et quelques formations siluriennes dans le Sud-Est du Massif Armoricain. *Mémoires de la Société géologique et minéralogique de Bretagne*, no. 18, 248 pgs.

NOGER, M.C., and KEPFERLE, R.C.

1985 Stratigraphy along and adjacent to the Bluegrass Parkway. *Kentucky Geological Survey, Field Conference of the Geological Society of Kentucky*, 24 pgs.

OLIVEIRA, S. de F., and BURJACK, M.I.A.

1996 *Petrovina connata* gên. n., sp. n. Provavelmente relacionado à Chlorophyta, da Formação Ponta Grossa (Neodevoniano), Bacia do Paraná, Mt, Brasil. *Revista Universidade Guarulhos, Série Geociências*, I: 46–50.

PLAYFORD, G., and DRING, R.S.

1981 Late Devonian acritarchs from the Carnarvon Basin, Western Australia. *Special Papers in Palaeontology*, 27: 78 pgs.

RASUL, S.M.

1974 The Lower Palaeozoic acritarchs *Priscogalea* and *Cymatiogalea*. *Palaeontology*, 17: 41–63.

ROSS, R.J., Jr., ADLER, F.J., AMSDEN, T.W., BERGSTRÖM, D., BERGSTRÖM. S.M., CARTER, C., CHURKIN, M., CRESSMAN, E.A., DERBY, J.R., DUTRO, J.T., Jr., ETHINGTON, R.L., FINNEY, S.C., FISHER, D.W., FISHER, J.H., HARRIS, A.G., HINTZE, L.F., KETNER, K.B., KOLATA, D.L., LANDING, E., NEUMAN, R.B., SWEET, W.C., POJETA, J., Jr., POTTER, A.W., RADER, E.K., REPETSKI, J.E., SHAVER, R.H., THOMPSON, T.L., and WEBERS, G.F.

1982 *The Ordovician System in the United States: Correlation chart and explanatory notes*. International Union of Geological Sciences. Publication No. 12, 73 pgs.

SANFORD, J.T., MARTINI, I.P., and MOSHER, R.E.

1972 Road log. *In*: Segall, R.T., and Dunn, R.A. (eds.), *Niagaran Stratigraphy: Hamilton, Ontario*. Michigan Basin Geological Society Annual Field Excursion, p. 2–26.

STROTHER, P.K.

1991 A classification schema for the cryptospores. *Palynology*, 15: 219–236.

TAPPAN, H.

1980 *The Paleobiology of Plant Protists*. W.H. Freeman and Company (San Francisco), 1028 pgs.

VAVRDOVÁ, M.

1990 Coenobial acritarchs and other palynomorphs from the Arenig/Llanvirn boundary, Prague Basin. *Vestník Ústvredního ústavu geologického*, 65: 237–246.

VIDAL, G.

1976 Late Precambrian microfossils from the Visingsö beds in southern Sweden. *Fossils and Strata*, no. 9, 57 pgs.

WILLE, W.

1970 *Plaesiodictyon mosellanum* n.g., n.sp., eine mehrzellige Grünalge aus dem Unteren Keuper von Luxembourg. *Neues Jahrbuch für Geologie und Paläontologie*, Monatshefte, 1970: 283–310.

WOOD, G.D.

1988 Paleoecology and biostratigraphy of Lower Silurian (Llandovery–Wenlock) acritarchs from the Mid-continent United States. *Abstracts, 7th International Palynological Congress*, Brisbane, Australia, p. 182.

1989 Paleoecologic and biostratigraphic significance of chitinozoans and microplankton from the Lower Silurian Osgood Member of the Salamonie Dolomite, Indiana and Kentucky. *Palynology*, 13: 288.

WOOD, G.D., GABRIEL, A.M., and LAWSON, J.C.

1996 Palynological techniques—processing and microscopy. *In*: Jansonius, J., and McGregor, D.C. (eds.), *Palynology: principles and applications*. American Association of Stratigraphic Palynologists Foundation, Dallas, 1: 29–50.

WOOD, G.D., and MILLER, M.A.

1996 Pre–Carboniferous Chlorophyta: new occurrences of Hydrodictyaceae, ?Scenedesmaceae and Zygnemataceae. *Abstracts of the International Meeting and Workshop of the C.I.M.P. Acritarch Subcommission*, Prague, Czech Republic, p. 26.

1997 Pre–Carboniferous Chlorophyta: new reports of Hydrodictyaceae, ?Scenedesmaceae and Zygnemataceae. *In*: Fatka, O., and Servais, T. (eds.), Acritarchs in Praha, *Acta Universitatis Carolinae Geologica*, 40: 703–717 [1996].

1998 Stratigraphic, paleoecologic and petroleum generating significance of Chlorophyta (chlorococcalean algae) in the Cretaceous of Western Africa and South America. *Africa Geoscience Review*, 4: 499–510.

WOOD, G.D., and TURNAU, E.

2001 New Devonian coenobial Chloroccales (Hydrodictyaceae) from the Holy Cross Mountains and Radom–Lublin region of Poland: their paleoenvironmental and sequence stratigraphic implications. *Proceedings of the IX International Palynological Congress, Houston, Texas*, p. 53–63.

WORKUM, R.H., BOLTON, T.E., and BARNES, C.R.

1976 Ordovician geology of Akpatok Island, Ungava Bay, District of Franklin. *Canadian Journal of Earth Sciences*, 13: 157–178.

ZIPPI, P.A.

1998 Freshwater algae from the Mattagami Formation (Albian), Ontario: Paleoecology, botanical affinities and systematic taxonomy. *Micropaleontology*, 44, suppl. 1: 78 pgs.

Authors' addresses

MERRELL A. MILLER
Saudi Arabian Oil Co.
P.O. Box 12311
Dhahran 31311
Saudi Arabia

GORDON D. WOOD
the irf group, inc.
23222 Willow Pond Place
Katy, TX 77494
U.S.A.

EVOLUTIONARY HYPOTHESIS OF CRYPTOSPORE PRODUCING PLANTS BASED ON WALL ULTRASTRUCTURE

Wilson A. Taylor

Abstract

Based on a review of analyses completed to date, it is possible to recognize two basic types of cryptospores from near the Ordovician/Silurian boundary based on their wall ultrastructure—lamellated and non-lamellated. These are interpreted as separate lineages. The lamellated cryptospores include the dyads *Dyadospora murusdensa* and some *D. murusattenuata* (those with dark contact areas). These share ultrastructural features (two parted wall – inner lamellated) with modern sphaerocarpalean hepatics. The non-lamellated cryptospores include *Tetrahedraletes medinensis*, some *Dyadospora murusattenuata* (those without dark contact areas), *Pseudodyadospora* sp., *Velatitetras reticulata* and *Segestrespora membranifera*. The structural similarities between the non-lamellated types are not as strong. At present, it is not clear which of these lineages arose first, but the non-lamellated lineage may be ancestral to land plant miospores (trilete spores).

INTRODUCTION

Very little is known about cryptospore producing plants, especially prior to the Late Silurian. For the time being, phylogenetic hypotheses about such plants will have to be based on examination of the cryptospores themselves. Ultrastructural characters can be used to suggest affinities between different types of cryptospores, or between cryptospore producing and younger plants. These suggested hypotheses can then be tested against the growing database of fossil and extant spores. The cryptospores are an excellent research subject for ultrastructural analysis for several reasons: 1) there is a fairly small number of different types, making it possible to provide ultrastructural data for all, 2) the diversity begins at zero in the early Ordovician, and, 3) there are several recognizable ultrastructural types. In this paper, I will summarize what is known to date about the ultrastructure of cryptospores from near the Ordovician/Silurian boundary and propose one phylogenetic scenario based on that information.

DESCRIPTION OF BASIC ULTRASTRUCTURAL TYPES

Two basic types of wall ultrastructure are discernible in the cryptospores examined to date.

First Type

Wall two parted—outer wall homogeneous, surrounding the individual dyad members on non-contact faces and the edge of the contact face; inner wall at least partially lamellated. The inner walls of adjacent dyad members are in contact along the contact face.

Taylor, Wilson A.
2001 Evolutionary hypothesis of Cryptospore producing plants based on wall ultrastructure; *In*: Goodman, D.K., and Clarke, R.T. (eds.), *Proceedings of the IX International Palynological Congress, Houston, Texas, U.S.A., 1996*; American Association of Stratigraphic Palynologists Foundation, p. 11–15.
 ISBN 0–931871–06–9

Representative taxa. *Dyadospora murusdensa* (Strother and Traverse, 1979) Burgess and Richardson, 1991; some *Dyadospora murusattenuata* (Strother and Traverse, 1979) Burgess and Richardson, 1991.

Dyadospora murusdensa (Plate 1, fig. 1). These cryptospores tend to have a thick inner wall (approx. 1μm) formed from 10–15 discrete lamellae (Taylor, 1996). Some *Dyadospora murusattenuata* (Plate 1, fig. 2). These cryptospores have a thicker wall overall than the previous type, with an inner wall that consists of two sublayers—an outer sublayer of thick lamellae (b1) and an inner sublayer of globular units (b2). These thick-walled forms of *D. murusattenuata* are very dark over the contact face when viewed with the light microscope (Taylor, 1997).

Comments. Either of these cryptospores can have an outer envelope surrounding the entire dyad (i.e., be enclosed) or no envelope (i.e., be naked). Enclosed forms tend to have a very tightly adherent envelope that is probably not seen with the light microscope. There is no ultrastructural evidence of partial separation of these dyad members from near the Ordovician/Silurian boundary, but neither is there evidence of fusion that would prevent their separation. There is a rather substantial thickness of wall material (inner layer) that surrounds the individual dyad members that might serve to protect each if separation were to occur (preadaptation?). A higher incidence of wholly or partially dissociated dyads occurs in younger deposits (Wellman and Richardson, 1996) and may be the result of changes in the timing of cryptospore wall development (i.e., heterochrony—see below). Investigations of dispersed monads (e.g., *Laevolancis*) are in progress to determine if some monads are separated dyads as suggested by Burgess and Richardson (1991). The distinctive lamellated ultrastructure will allow monads and their putative parent dyads to be linked.

Second Type

Wall single layered, homogeneous (perhaps very densely spongy) with a single loose spongy layer usually detectable in the outer half of the spore wall.

Representative taxa. *Tetrahedraletes medinensis* Strother and Traverse, *emend.* Wellman and Richardson, 1993;

Pseudodyadospora sp. Johnson, 1985; and some *Dyadospora murusattenuata; Velatitetras reticulata* Burgess, 1991 and *Segestrespora membranifera* (Johnson) Burgess, 1991.

Tetrahedraletes medinensis (Plate 1, fig. 3). Contact face walls are about half the thickness of non-contact face walls (approx. 0.5 and 1.0 µm respectively). It has not been possible to elucidate the ultrastructural characteristics of any physical connection (e.g., proximal wall fusion), or proximal trilete scar, due to the extremely low contrast of these specimens (Taylor, 1995a).

Pseudodyadospora sp. (Plate 1, fig. 4) and some *Dyadospora murusattenuata* (Plate 1, fig. 5). Some cryptospores assignable to *D. murusattenuata* (in possessing two distinct crosswalls that are detectable with the light microscope) have a very similar ultrastructure to *Pseudodyadospora* sp., differing only with respect to the thickness of the contact face walls (250 nm in *Pseudodyadospora* sp. and 500 nm in *D. murusattenuata*), and possibly the extent of spongy "filler" wall material at the edges of the contact face. These members of *D. murusattenuata* tend to be more lightly colored and lack the dark contact face of the lamellelated *D. murusattenuata* mentioned above. It may be impractical or even impossible to use these coloration differences to distinguish these two subtypes of *D. murusattenuata* at the level of the light microscope. All dyads and pseudodyads examined thus far have two distinct crosswalls between the dyad members (Taylor, 1996, 1997).

Velatitetras reticulata (Plate 1, fig. 6) and *Segestrespora membranifera* (Plate 1, fig. 7). These forms tend to be somewhat more loosely constructed (i.e., more spongy). Both are enclosed by an envelope (e) that often stains more darkly than the underlying cryptospore wall. Ornamentation is formed strictly by the envelope which may be thickened, folded or both. A single lamella (l) occurs between the envelope and the cryptospore wall, and globules can sometimes be seen between the single lamella and the envelope.

Comments. These subtypes are united by a general similarity in wall thickness, and the presence of the less dense layer in the outer half of the wall. This layer can usually be traced, but is very difficult to illustrate in these low contrast specimens. They are also united in being "non-lamellated." While the absence of a characteristic is not the most rigorous criterion upon which to base a lineage, I have chosen to do so due to the lack of other characters.

PROPOSED PHYLOGENY OF CRYPTOSPORE TYPES BASED ON WALL ULTRASTRUCTURE

The following assumptions are built into the proposed phylogeny of cryptospore types.

Assumption #1 – Taxonomic identification using the light microscope accurately reflects ultrastructure regardless of geographic or stratigraphic position, e.g., the ultrastructure of *Dyadospora murusdensa* in Scotland is the same as the demonstrated ultrastructure of *D. murusdensa* in southern Ohio despite the geographic and/or stratigraphic separation. (This assumption reflects the infancy of ultrastructural analyses and will be tested in due time.)

Assumption #2 – Taxonomically distinct cryptospores with identical ultrastructure were probably produced by the same type of plant, e.g., the enclosed tetrad *Velatitetras* and the enclosed dyad *Segestrespora* (Taylor, 1996). This receives support from studies of ornamentation patterns by Wellman (1996), who, in fact extended the observation to include *Rugosphaera cerebra*.

The following phylogenetic hypothesis can be considered to explain the occurrence of cryptospores near the Ordovician/Silurian boundary in southern Ohio (Text-Figure 1).

No documented and repeatable occurrences of Ordovician trilete miospores are yet known (Wellman, 1996). Both permanent tetrads and (presumably) lamellate dyads appear in the earliest recognized cryptospore assemblages (Strother et al., 1996). Structural differences between the various cryptospore types suggest: a) an independent origin of these two lineages (perhaps from the same group of charophycean algae), and/or b) a much earlier origin than the early Middle Ordovician, and/or c) evolutionarily rapid modification of developmental patterns. Heterochronic change has already been suggested as a possible mechanism of evolutionary transformation with respect to cryptospores (specifically changes in the timing of the division phases of meiosis and deposition of sporopollenin; Hemsley, 1994; Strother, 1991; Strother et al., 1996).

The only extant plant group that produces spores with distinct lamellae that persist at maturity is the hepatics. Among the hepatics, the combination of spores with two parted walls and permanent tetrads is found only in the

PLATE 1

All figures are at x25,800; scale bar in fig. 1 represents 1µm.

1 *Dyadospora murusdensa* – wall cross-section of a naked dyad. Note the outer homogeneous layer (a) and the inner lamellated layer (b).

2 *Dyadospora murusattenuata*, lamellated variety – wall cross-section of a naked dyad. Note the outer homogeneous layer (a) and the inner two parted layer – b_1 with thick lamellae and b_2 with globular units.

3 *Tetrahedraletes medinensis* – wall cross-section of a naked tetrad. Note slightly less dense stratum in outer part of wall (marked by dashes).

4 *Pseudodyadospora* sp.– wall cross-section of a naked "pseudodyad."

5 *Dyadospora murusattenuata*, unlamellated variety – cross-section of two opposing walls. The lumen of the dyad member is at the level of the arrows. Less dense layer is marked by dashes.

6 *Velatitetras reticulata* – wall cross-section of an enclosed tetrad. The ornamentation here is formed by a fold in the envelope (arrow). Note structural similarity to fig. 7.

7 *Segestrespora membranifera* – wall cross-section of an enclosed dyad. Ornamentation here is formed from a thickening of the envelope (arrow). The single lamella is visible just beneath the envelope.

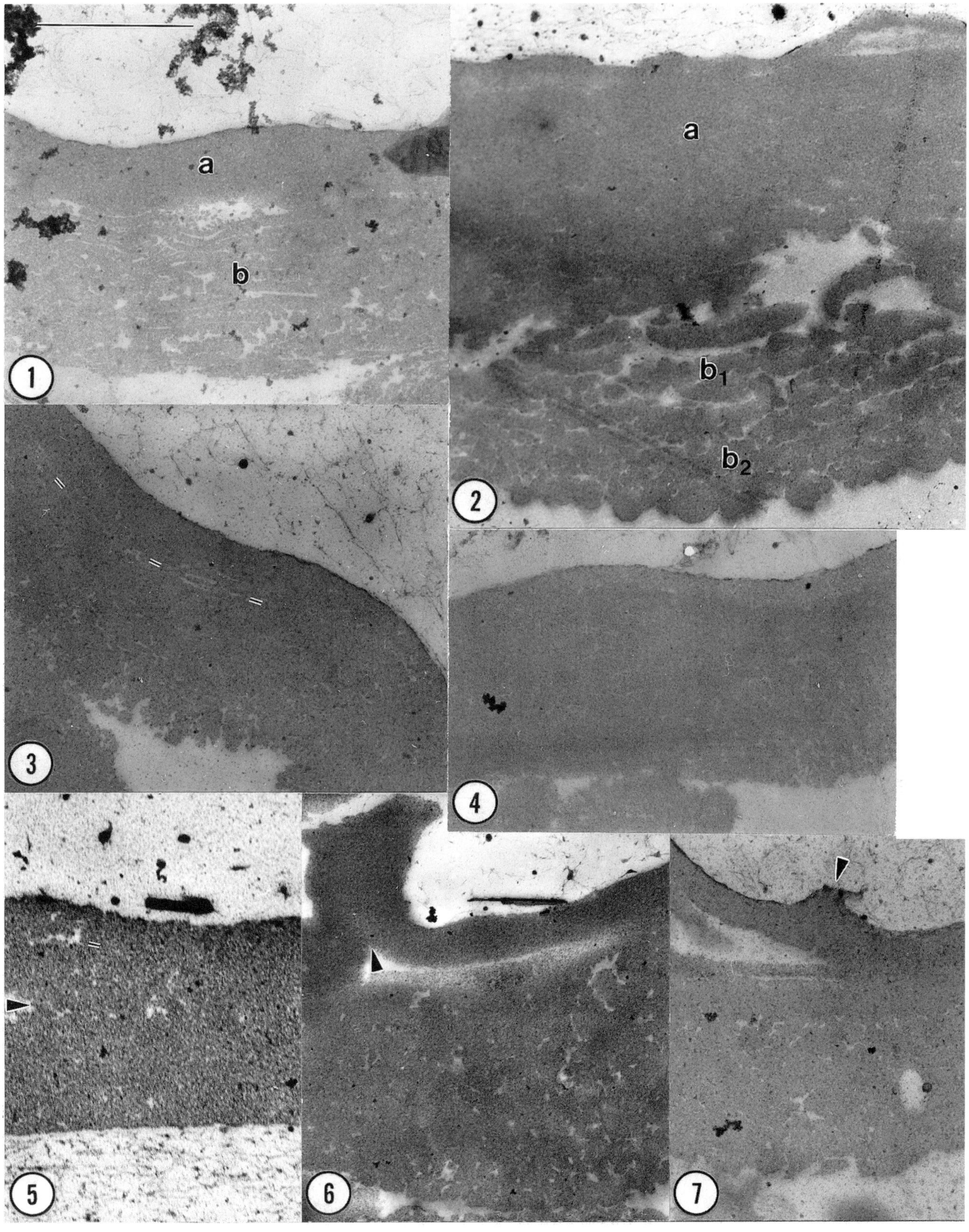
a
b
1
a
b1
b2
2
3
4
5
6
7

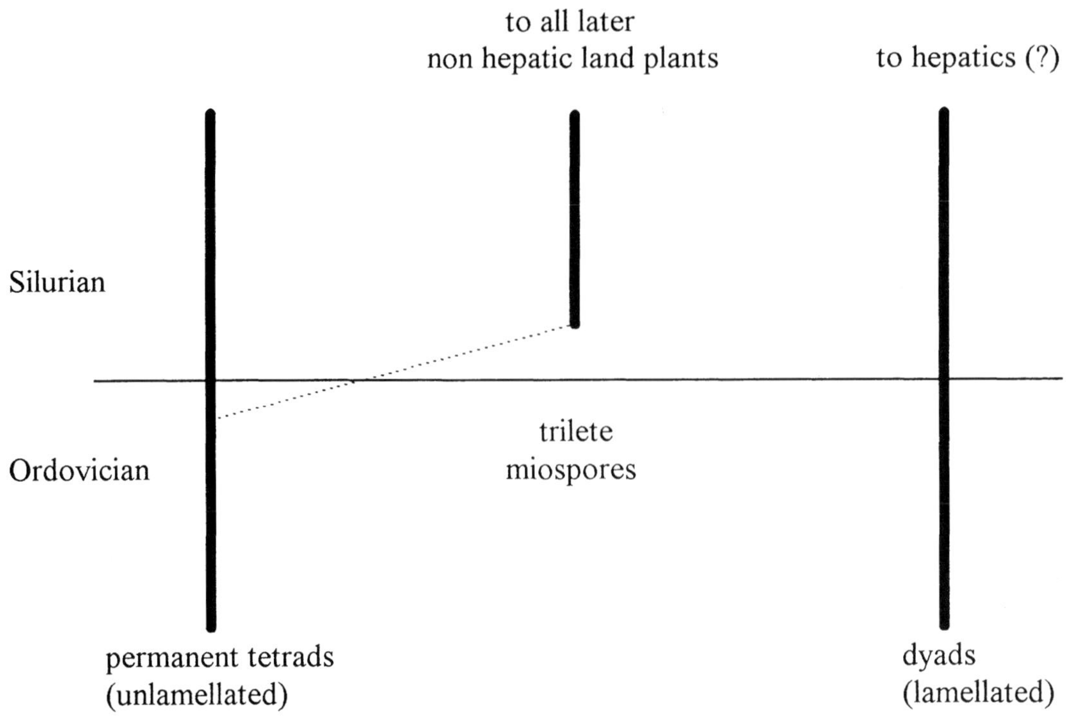

Text-Figure 1. Proposed phylogenetic hypothesis based on wall ultrastructure of cryptospores near the Ordovician/Silurian boundary.

Sphaerocarpales. Possible hepatic fossils that contain cryptospores (albeit permanent tetrads—no modern plants produce dyads as a normal product of development) have been described from the Lower Devonian (Edwards et al., 1995). Cladistic analyses of Mishler and Churchill (1985) assign liverworts to the most basal position among the bryophytes and the Sphaerocarpales to the most basal position within the hepatics. Ultrastructural information supports the suggestion that some cryptospores were produced by hepatics—possibly sphaerocarpalean hepatics (Gray, 1985; Taylor, 1995b, 1996).

The tentative link (dotted line) between the non-lamellated permanent tetrads and trilete miospores is based on tetrad geometry. Lower Silurian miospores have yet to be examined with the TEM. An additional link may exist between the two Ordovician lineages, but there is, as yet, no evidence to suggest which came first. Alternatively, they may both have arisen independently from charophycean algae in the early Ordovician or late Cambrian.

ACKNOWLEDGMENTS

Acknowledgment is made to the donors of The Petroleum Research Fund, administered by the ACS, for partial support of this research.

References Cited

BURGESS, N.D.
1991 Silurian cryptospores and miospores from the type Llandovery area, south-west Wales. *Palaeontology*, 34: 575–599.

BURGESS, N.D., and RICHARDSON, J.B.
1991 Silurian cryptospores and miospores from the type Wenlock area, Shropshire, England. *Palaeontology*, 34: 601–628.

EDWARDS, D., DUCKETT, J.G., and RICHARDSON, J.B.
1995 Hepatic characters in the earliest land plants. *Nature*, 374: 635–636.

GRAHAM, L., and TAYLOR, C.
1986 Occurrence and phylogenetic significance of "special walls" at meiosporogenesis in *Coleochaete*. *American Journal of Botany*, 73: 597–601.

GRAY, J.
1985 The microfossil record of early land plants: advances in understanding of early terrestrialization, 1970–1984. *Philosophical Transactions of the Royal Society of London B*, 309: 167–195.

HEMSLEY, A.R.
1994 The origin of the land plant sporophyte: an interpolational scenario. *Biological Review*, 69: 263–273.

JOHNSON, N.G.
1985 Early Silurian palynomorphs from the Tuscarora Formation in central Pennsylvania and their paleobotanical and geological significance. *Review of Palaeobotany and Palynology*, 45: 307–360.

MISHLER, B.D., and CHURCHILL, S.P.
1985 Transition to a land flora: phylogenetic relationships of the green algae and bryophytes. *Cladistics*, 1: 305–328.

STROTHER, P.K.
1991 A classification schema for the cryptospores. *Palynology*, 15: 219–236.

STROTHER, P.K., AL-HAJRI, S., and TRAVERSE, A.
1996 New evidence for land plants from the lower Middle Ordovician of Saudi Arabia. *Geology*, 24: 55–58.

STROTHER, P.K., and TRAVERSE, A.
1979 Plant microfossils from Llandoverian and Wenlockian rocks of Pennsylvania. *Palynology*, 3: 1–21.
TAYLOR, W.A.
1995a Ultrastructure of *Tetrahedraletes medinensis* (Strother and Traverse) Wellman and Richardson, from the Upper Ordovician of southern Ohio. *Review of Palaeobotany and Palynology*, 85: 183–187.
1995b Spores in earliest land plants. *Nature*, 373: 391–392.
1996 Ultrastructure of lower Paleozoic dyads from southern Ohio. *Review of Palaeobotany and Palynology*, 92: 269–279.
1997 Ultrastructure of lower Paleozoic dyads from southern Ohio II: *Dyadospora murusattenuata*, functional and evolutionary considerations. *Review of Palaeobotany and Palynology*, 97: 1–8.
WELLMAN, C.H.
1996 Cryptospores from the type area of the Caradoc series in southern Britain. *In*: Cleal, C.J. (ed.), Studies on early land plant spores from Britain. *Special Papers in Palaeontology*, 55: 103–136.
WELLMAN, C.H., and RICHARDSON, J.B.
1993 Terrestrial plant microfossils from Silurian inliers of the Midland Valley of Scotland. *Palaeontology*, 36: 155–193.
1996 Sporomorph assemblages from the "Lower Old Red Sandstone" of Lorne, Scotland. *In*: Cleal, C.J. (ed.), Studies on early land plant spores from Britain. *Special Papers in Palaeontology*, 55: 41–101.

Author's address

WILSON A. TAYLOR
Department of Biology
University of Wisconsin – Eau Claire
Eau Claire, Wisconsin 54701
U.S.A.

PALYNOSTRATIGRAPHY AND PALEOBIOGEOGRAPHY OF THE LOWER PALEOZOIC SEQUENCE IN THE NORTHEASTERN ALBORZ RANGE (KOPET-DAGH REGION) OF IRAN

Mohammad Ghavidel-Syooki

Abstract

Palynomorphs recovered from the Lashkarak, Ghelli and Niur formations of eastern part of Kuh-e-Saluk (northeastern Alborz Range, Kopet-Dagh region) were used to more precisely determine the age of these units and assess their paleogeographic importance in Southern and Northern Hemispheres. These formations yielded 56 acritarch taxa permitting the recognition of six stratigraphic zones. Zones I-III represent the Early Ordovician (Tremadoc–Arenigian), Zones IV-V indicate the Middle and Late Ordovician and Zone VI suggests the Early Silurian age. This is the first report of *Rhopaliophora* and *Peteinospheridium* from the Lower Ordovician strata of northeastern Alborz Range (Kopet-Dagh region). The presence of these two genera in the Arenigian acritarch assemblage of Lashkarak Formation suggests the Iranian Platform and southwestern China occupied similar paleolatitudes along the southern shore of the Tethys Ocean during the Early Ordovician.

Comparison of acritarch taxa from the Ghelli (Middle–Late Ordovician) and Niur (Early Llandovery) formations with those from other parts of the world indicate a broad similarity with those of the United States, southern Europe and northern Africa. This similarity suggests that the peri-Gondwanan Paleocontinent began moving northward during the Late Ordovician and by the Silurian formed the supercontinent of Pangea. The diverse acritarch taxa in the Early Ordovician (Lashkarak Formation), Middle and Late Ordovician (Ghelli Formation) and Early Llandovery (Niur Formation) indicate a marine depositional environment.

INTRODUCTION

Lower Paleozoic strata in the northeastern Alborz Range (Kopet-Dagh region), have yielded a diverse and well preserved acritarch assemblage. This study is directed toward developing palynological information from the Ordovician (Lashkarak and Ghelli formations) and Silurian strata (Niur Formation) of northeastern Albroz Range to aid in establishing the age relationships of the strata and resolve aspects of paleogeography and depositional environments.

PREVIOUS STUDIES

The Kuh-e-Saluk study area is approximately 32 km south of Bojnourd city (Text-Figure 1). The road from Bojnourd to Esfarayen is the main route into the region. The measured and sampled stratigraphic section was chosen along this road (Text-Figure 2). The total thickness of the Lower Paleozoic strata is 1200 m.

The Lower Paleozoic strata in Kuh-e-Saluk has been divided, in ascending order, into Mila, Lashkarak, Ghelli and Niur formations (Text-Figure 2) by Afshar-harb (1979). The study area is located in the Kopet-Dagh region (northeastern Alborz Range) where the Lower Paleozoic rock units extend towards the southern and eastern Caspian Sea.

Ghavidel-Syooki, Mohammad
2001 Palynostratigraphy and paleobiogeography of the Lower Paleozoic sequence in the northeastern Alborz Range (Kopet-Dagh Region) of Iran; *In*: Goodman, D.K., and Clarke, R.T. (eds.), *Proceedings of the IX International Palynological Congress, Houston, Texas, U.S.A., 1996*; American Association of Stratigraphic Palynologists Foundation, p. 17–35.

ISBN 0–931871–06–9

The Mila Formation consists mainly of medium to thin-bedded, cream-reddish brown limestones. Poorly preserved megafossils, such as brachiopods and trilobites, are not identifiable to the generic level. Based on stratigraphic position, the Mila Formation has been assigned to the Middle and Late Cambrian (Afshar-harb, 1979).

The Lashkarak Formation is a very distinctive sedimentary facies consisting of dark-gray shales, siltstones and fine-grained sandstones which are interbedded with thin limestones (Text-Figure 2). In the study area the Lashkarak Formation contains brachiopods and conodonts which have been assigned to the Tremadocian (Ahmadzadeh-heravi, 1983).

The Ghelli Formation consists mainly of olive-gray shales, micaceous siltstones, fine-grained sandstones and a few meters of limestones at the top. Near the upper part of the unit is a 44 m thick igneous sill. The Lower and Upper contact of Ghelli Formation are conformable with underlying and overlying formations (Text-Figure 2). Some intervals of this rock unit contain brachiopods, graptolites and conodonts. Ahmadzadeh-heravi (1983) recorded conodonts indicative of Late Ordovician age.

The Niur Formation, although well-developed in central Iran, was first described by Afshar-harb (1979) from the northeastern Alborz Range. In the study area, this formation consists of limestones, shales and sandstones. The fauna includes abundant corals, brachiopods and crinoids. Based on the brachiopod fauna Ahmadzadeh-heravi (1983) assigned a Silurian age to the Niur Formation.

LABORATORY TECHNIQUES

Palynological study was carried out on 140 samples from Lashkarak Formation, Ghelli Formation and Niur Formation. The field and laboratory description of the samples has been plotted on the stratigraphic section in

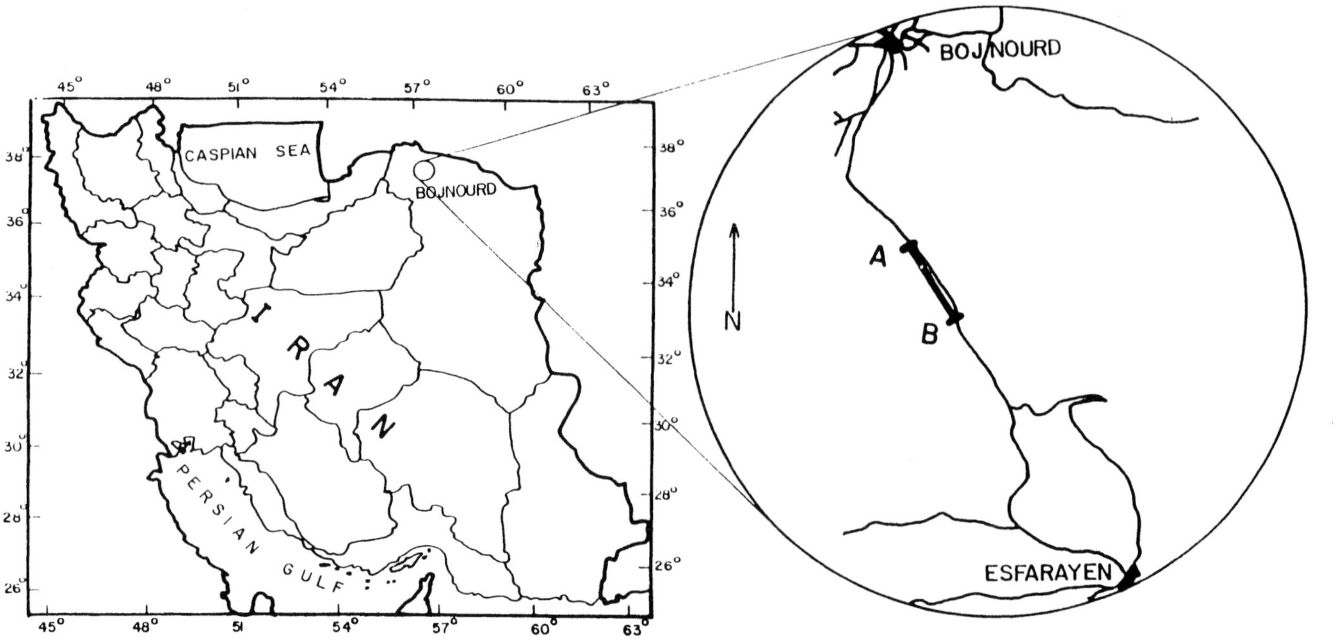

Text-Figure 1. Location map of study area.

Text-Figure 2. The code and number of each sample follow the policy of the National Iranian Oil Company. Fifty grams of rock were randomly selected from each sample and processed in the palynology laboratory of the Exploration Department of the National Iranian Oil Company.

Disaggregation of the rock samples was conducted using standardized techniques. All slides used in this study are housed in the Paleontology Section collections of the Exploration Department of the National Iranian Oil Company.

STRATIGRAPHIC PALYNOLOGY

The objectives of this study are to summarize the stratigraphic range of assemblages and species that occur in the Lashkarak, Ghelli and Niur formations and to compare these data with zonal assemblages that have been reported from other parts of the world. A total of 56 acritarch taxa were identified and their distribution is plotted on Text-Figure 2. Selected acritarch taxa are illustrated in Plates I-VIII. Six acritarch assemblages have been recognized and are discussed below in ascending stratigraphical order.

Text-Figure 2 (opposite page). Stratigraphic distribution of acritarch taxa in the lower Paleozoic strata of the northeastern Alborz Range (Kopet-Dagh region) of Iran. List of recorded taxa are noted below (numbers refer to the corresponding columns on Text-Figure 2). 1–*Multiplicisphaeridium cervinacornuum* Welsch; 2–*Acanthodiacrodium echinatum* (Timofeev) Deflandre & Deflandre-Rigaud; 3–*Acanthodiacrodium raia* (Deunff) Eisenack et al.; 4–*Cymatiogalea cuvillieri* (Deunff) Deunff; 5–*Saharidia downiei* Combaz; 6–*Cymatiogalea velifera* (Downie) Martin; 7– *Vulcanisphaera britannica* Rasul; 8– *Vulcanisphaera africana* Deunff; 9– *Vulcanisphaera cirrita* Rasul; 10–*Dactylofusa squama* (Deunff) Combaz et al.; 11– *Athabascaella playfordii* Martin; 12–*Goniosphaeridium piliferum* Martin; 13–*Lophosphaeridium torum* Rasul; 14– *Acanthodiacrodium angustum* (Downie) Combaz; 15–*Stelliferidium striatulum* (Vavrdovà) Deunff et al.; 16–*Stelliferidium stelligerum* (Gôrka) Deunff et al.; 17–*Acanthodiacrodium rectinerve* Burmann; 18–*Arbusculidium filamentosum* (Vavrdová) Vavrdovà; 19–*Arbusculidium sp.*; 20–*Marrocanium simplex* Cramer et al.; 21–*Striatotheca triangulata* (Cramer et al.) Eisenack et al.; 22–*Striatotheca principalis* Burmann; 23–*Striatotheca rugosa* Tongiorgi et al.; 24–*Pirea baculifera* Tongiorgi et al.; 25–*Coryphidium elegans* Cramer et al.; 26–*Coryphidium minutum* Cramer & Díez; 27–*Peteinosphaeridium tenuifilosum* Tongiorgi et al.; 28–*Peteinosphaeridium armatum* Tongiorigi et al.; 29–*Peteinosphaeridium robustiramosum* Tongiorgi et al.; 30–*Peteinosphaeridium angustilaminae* Playford et al.; 31–*Cymatiogalea deunffii* Jardiné, Combaz et al.; 32–*Rhopaliophora palmata* (Combaz & Peniguel) Playford & Martin; 33–*Rhopaliophora mamilliformis* Lu; 34–*Actinotodissus crassus* Loeblich & Tappan; 35– *Veryhachium reductum* (Deunff) Jekhowsky; 36–*Ordovicidium elegantulum* Tappan & Loeblich; 37– *Veryhachium oklahomense* Loeblich.; 38–*Baltisphaeridium perclarum* Loeblich Jr. & Tappan; 39–*Navifusa ancepsipuncta* Loeblich; 40–*Orthosphaeridium ternatum* (Burmann) Eisenack et al.; 41– *Villosacapsula setosapellicula* (Loeblich Jr.) Loeblich & Tappan; 42– *Veryhachium subglobosum* Jardiné et al.; 43–*Orthosphaeridium inflatum* Loeblich; 44–*Orthosphaeridium insculptum* Loeblich. 45–*Poikilofusa spinata* Staplin et al.; 46–*Rhiptosocherma improcera* (Loeblich) Loeblich & Tappan; 47–*Dactylofusa striata* (Staplin et al. 1965) Fensome et al.; 48–*Diexallophasis denticulata* (Stockmann & Williéré) Loeblich; 49–*Dactylofusa estillis* Cramer & Díez; 50–*Dactylofusa striatifera* (Cramer, 1964) Fensome et al.; 51–*Tunisphaeridium tentaculiferum* (Martin) Cramer; 52–*Helosphaeridium clavispinulosum* Lister; 53– *Multiplicisphaeridium neaghae* Cramer; 54– *Visbysphaera pirifera* (Eisenack) Lister; 55– *Visbysphaera oligofurcata* (Eisenack) Hill; 56– *Visbysphaera brevifurcata* (Eisenack) Le Hérissé;

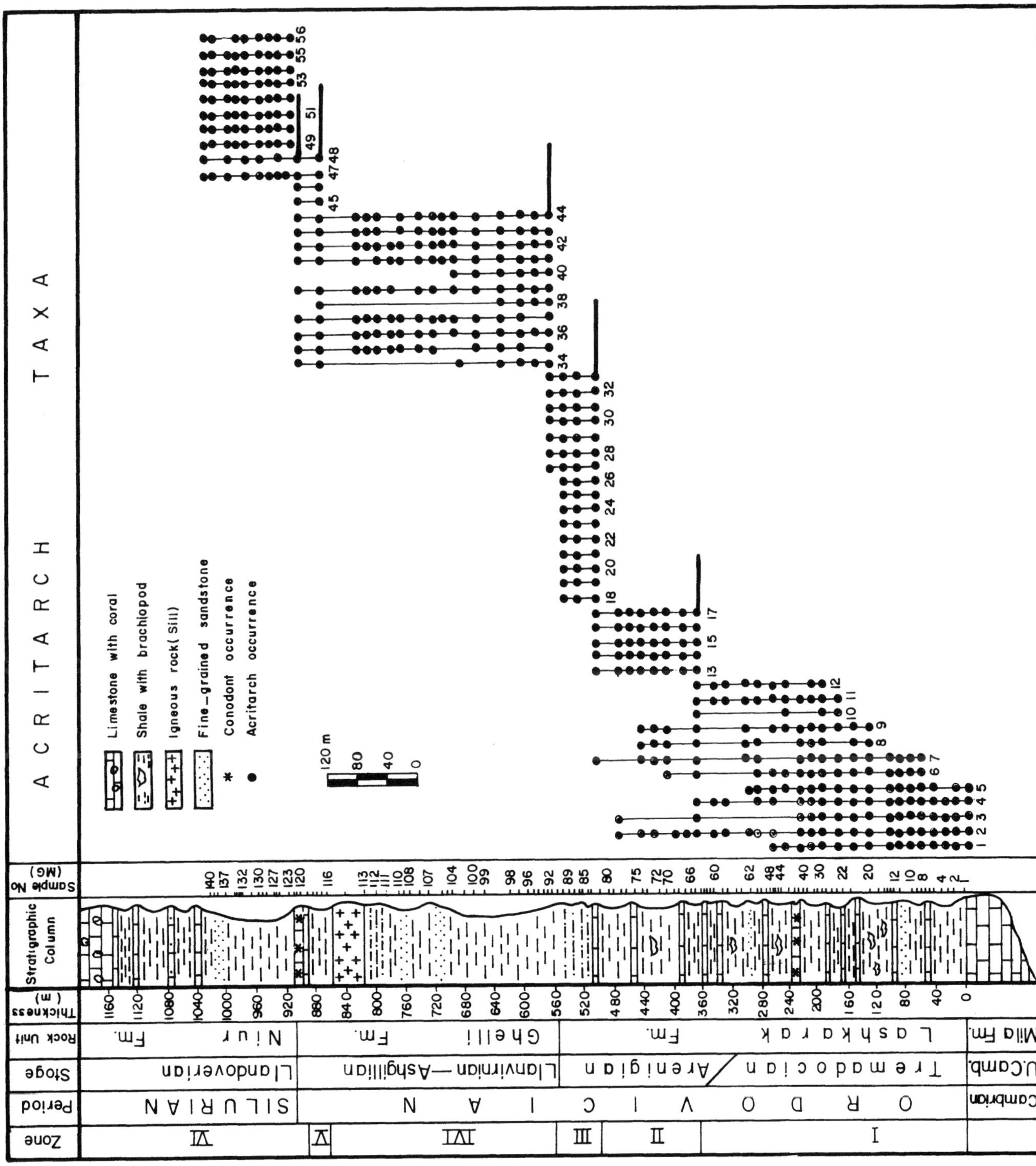

ACRITARCH TAXA
Limestone with coral
Shale with brachiopod
Igneous rock (Sill)
Fine-grained sandstone
Conodont occurrence
Acritarch occurrence
120 m
Zone
Period
Stage
Rock unit
Thickness (m)
Stratigraphic Column
Sample No (MG)
CAMBRIAN
ORDOVICIAN
SILURIAN
U.Camb.
Tremadocian
Arenigian
Llanvirnian—Ashgillian
Llandoverian
Mila Fm.
Lashkarak Fm.
Ghelli Fm.
Niur Fm.

Acritarch Assemblage Zone I

Assemblage Zone I begins at the lowermost part of the Lashkarak Formation and extends through a thickness of 368 m of the studied stratigraphic section (Text-Figure 2). This zone is characterized by presence of the acritarch taxa *Multiplicisphaeridium cervinacornuum, Acanthodiacrodium raia, Acanthodiacrodium echinatum, Cymatiogalea cuvillieri, Saharidia downiei,Cymatiogalea velifera, Vulcanisphaera africana, Vulcanisphaera britannica, Vulcanisphaera cirrita, Dactylofusa squama, Athabascaella playfordii* and *Goniosphaeridium piliferum*. This assemblage zone is considered to be Lower Ordovician (Early Tremadocian) based on the stratigraphic occurrence of acritarchs in England (Rasul, 1976, 1979; Downie, 1984), France (Rauscher, 1974), Belgium (Martin, 1965), Germany (Reitz, 1991), Norway (Welsch, 1986), Czech Republic (Vavrdová, 1974), Algeria (Combaz, 1967; Combaz et al., 1974; Jardiné et al., 1974), Sweden (Bagnoli et al., 1988). Ahmadzadeh-heravi (1983) has assigned an Early Ordovician (Tremadocian) age to this part of the Lashkarak Formation based conodont and brachiopod data (Text-Figure 2).

Acritarch Assemblage Zone II

This zone is marked by the first appearance of *Lophosphaeridium torum, Acanthodiacrodium angustum, Stelliferidium striatulum, Stelliferidium stelligerum* and *Acanthodiacrodium rectinerve*. This zone occurs in 132 m of the Lashkarak Formation (Text-Figure 2). The acritarch species are indicative of the Early Ordovician (Late Tremadocian) by comparison to their occurrence in strata from Algeria (Combaz, 1967; Combaz et al., 1974; Jardiné et al., 1974), southern Europe (Vavrdová, 1974) and southern and northern Iran (Ghavidel-Syooki, 1995).

Acritarch Assemblage Zone III

This zone comprises 60 m of the upper part of Lashkarak Formation. It is characterized by the presence of *Striatotheca triangulata, Striatotheca principalis, Striatotheca rugosa, Arbusculidium filamentosum, Arbusculidium sp., Marrocanium simplex, Pirea baculifera, Coryphidium elegans, Coryphidium minutum, Peteinosphaeridium tenuifilosum, Peteinosphaeridium armatum, Peteinosphaeridium velatum, Peteinosphaeridium robustiramosum, Peteinosphaeridium angustilaminae, Cymatiogalea deunffii, Rhopaliophora palmata* and *Rhopaliophora mamilliformis*.

Although some species of zones I and II extend into this zone there is a marked qualitative reduction in taxa (Text-Figure 2). Based on the above mentioned acritarchs, this zone is considered to belong to the uppermost part of the Early Ordovician (Arenigian). Acritarch taxa of this zone have been recorded in the Arenigian strata of southern Europe (Vavrdová, 1974; Rauscher, 1974; Burmann, 1968; Downie, 1984), Morocco (Cramer et al., 1974; Cramer and Díez, 1977), Algeria (Jardiné et al., 1974), southern and western China (Lu, 1987; Martin and Yin, 1988; Tongiorgi et al., 1995), southern and northern Iran (Ghavidel-syooki, 1990, 1993, 1995, 1996) and Argentina (Ottone and Toro, 1992). The assemblage contains typical Mediterranean taxa including *Corphidium, Arbusculidium* and *Striatotheca*, which have been reported from the Arenigian of peri-Gondwana (Mediterranean) acritarch paleo-province (Vavrdová, 1974). This is the first record of *Peteinosphaeridium* and *Rhopaliophora* from the lower Ordovician strata of Iran. These two genera have also been recovered from the Arenigian sediments of China (Tongiorgi et al., 1995). The occurrence of Mediterranean acritarch taxa and *Rhopaliophora* and *Peteinosphaeridium* indicates that Iranian Platform and southwestern China were in close proximity and about the same paleolatitude along the southern shore of the Tethys Ocean.

Acritarch Assemblage Zone IV

This assemblage zone begins in the lowermost part of Ghelli Formation and spans a thickness of 320 m (Text-Figure 2). This zone is marked by the absence of several lower Ordovician taxa and appearance of Middle and Late Ordovician acritarch species. The latter includes *Actinotodissus crassus, Veryhachium reductum, Ordovicidium elegantulum, Veryhachium oklahomense, Navifusa ancepsipuncta, Baltisphaeridium perclarum, Orthosphaeridium ternatum, Orthosphaeridium inflatum, Orthosphaeridium insculptum, Veryhachium subglobosum* and *Villosacapsula setosapellicula*. Many species present in this zone continue into the succeeding zone (Text-Figure 2). Acritarchs present in Zone IV indicate a Middle and Late Ordovician age based on comparisons with assemblages in the United States (Loeblich and Tappan, 1970, 1971 and 1978), Canada (Jacobson and Achab, 1985), England (Turner, 1984, 1985), southern Europe (Vavrdová, 1974), Sweden (Kjellström, 1971), Czech Republic (Dufka and Fatka, 1993), Libya (Molyneux et al., 1985), Algeria (Jardiné et al., 1974), Morocco (Elaouad-Debbaj, 1988), Jordan (Keegan et al., 1990), Saudi Arabia (Jachowicz, 1995) and Iran (Ghavidel-syooki, 1996; Ghavidel-syooki and Khosravi, 1995).

Acritarch Assemblage Zone V

This zone comprises 28 m of the Ghelli Formation (Text-Figure 2). The assemblage is characterized by occurrence of *Rhiptosocherma improcera, Poikilofusa spinata, Dactylofusa*

PLATE 1

All figures x1000.

1 *Stelliferidium striatulum* (Vavrdová 1966) Deunff, Górka & Rauscher 1974.

2 *Dactylofusa squama* (Deunff 1961) Combaz, Lange & Pansart 1967.

3 *Stelliferidium stelligerum* (Górka 1967) Deunff, Górka & Rauscher 1974.

4, 7 *Vulcanisphaera africana* Deunff 1961.

5 *Acanthodiacrodium angustum* (Downie 1958) Combaz 1967.

6 *Vulcanisphaera cirrita* Rasul 1976.

8 *Cymatiogalea cuvillieri* (Deunff 1961) Deunff 1964.

9 *Acanthodiacrodium raia* (Deunff 1961) Eisenack et al.

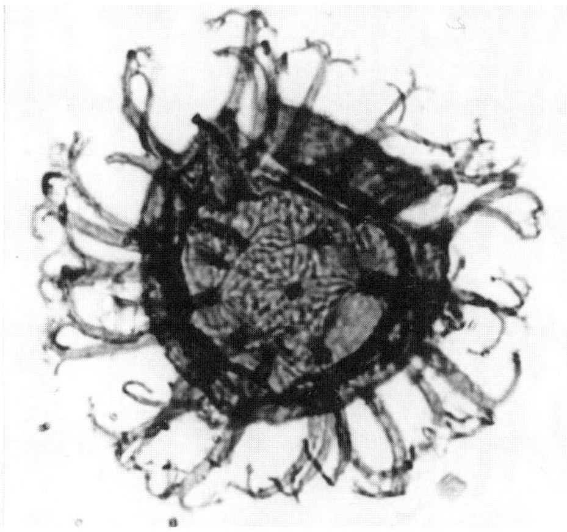

1

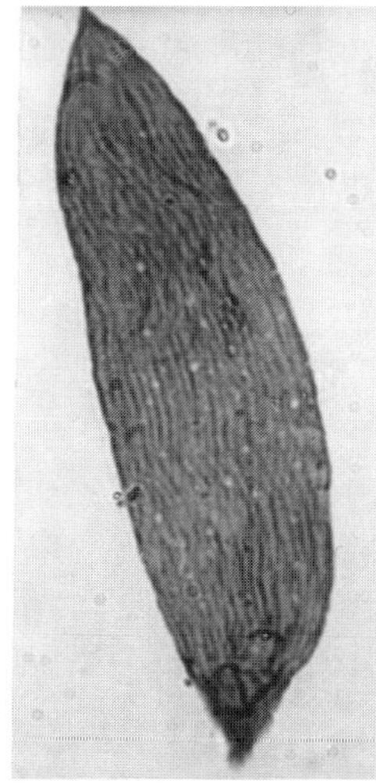

2

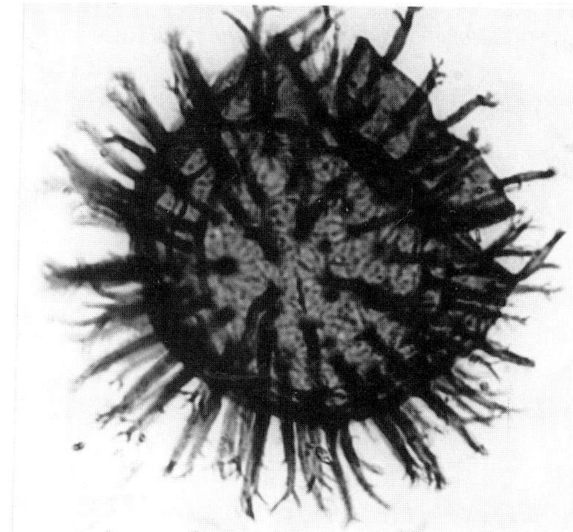

3

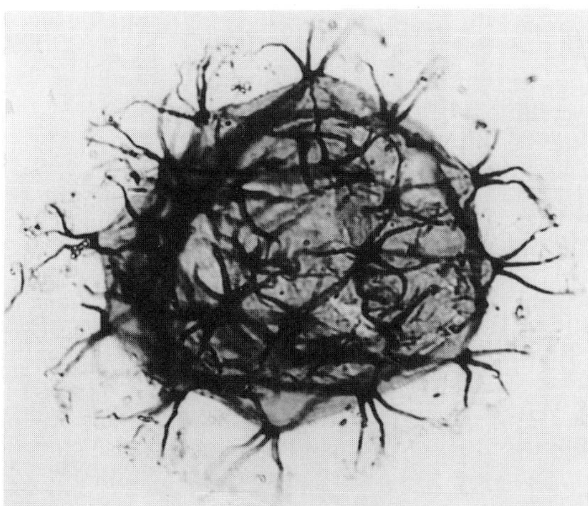

4

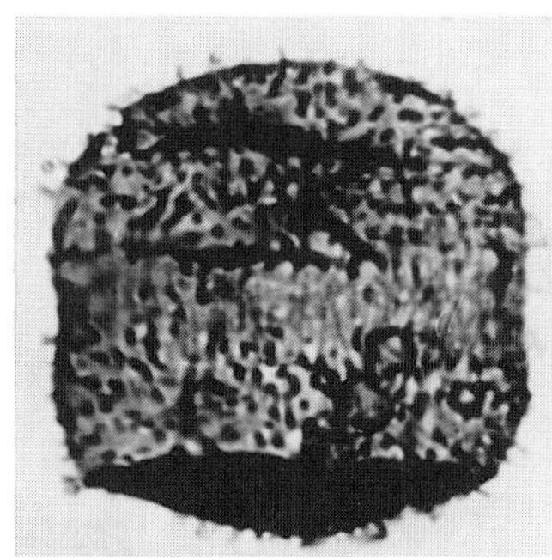

5

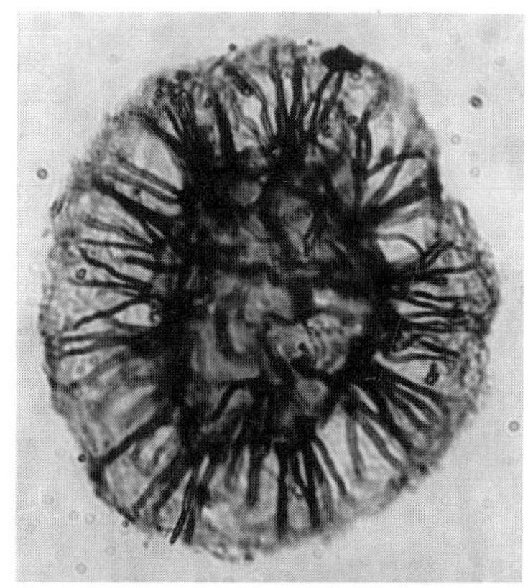

6

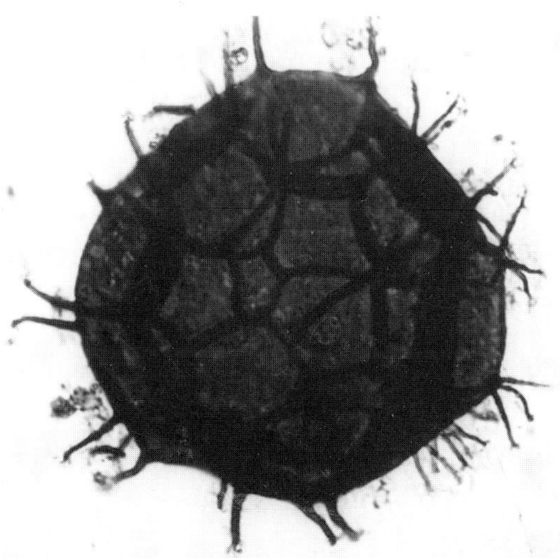

7

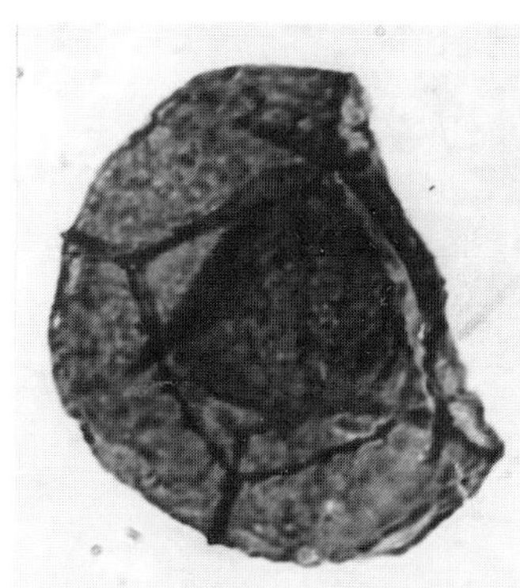

8

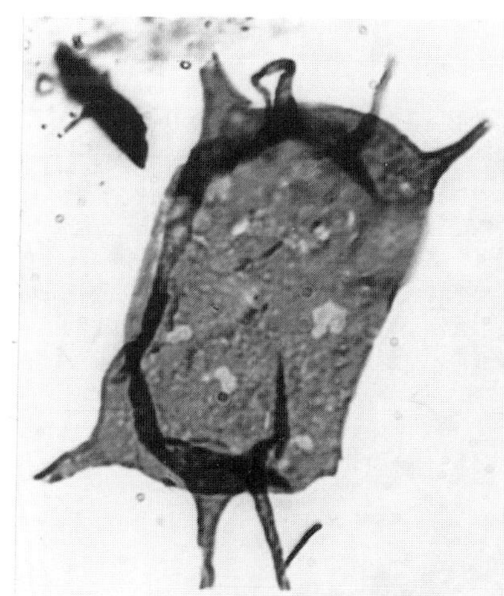

9

striata and *Diexallophasis denticulata*. This zone is assigned to the uppermost part of Late Ordovician (Ashgillian).

The acritarch taxa in this zone have been recorded from the Ashgillian strata in the United States (Loeblich and Tappan, 1978), Canada (Jacobson and Achab, 1985), Libya (Molyneux et al., 1985), Algeria (Jardiné et al., 1974) and Iran (Ghavidel-syooki, 1996). Ahmadzadeh-heravi (1983) suggested an Ashgillian age for this part of Ghelli Formation based on the presence of the conodonts *Zygeganthus plebia, Panderedus gracilis* and *Cordyledus flexus* (see Text-Figure 2; limestone horizon yielding conodonts is designated with an asterisk). A basaltic sill has intruded the upper part of Ghelli Formation and resulted in color changes in the acritarchs from orange to brown, reflecting contact metamorphism (Text-Figure 2).

Acritarch Assemblage Zone VI

This zone begins at the base of Niur Formation and extends through a thickness of 128 m (Text-Figure 2). The characteristic Silurian index acritarch species present include *Dactylofusa estillis, Helosphaeridium clavispinulosum, Multiplicisphaeridium neaghae, Dactylofusa straitifera, Tunisphaeridium tentaculiferum, Visbysphaera pirifera, Visbysphaera oligofurcata* and *Visbysphaera brevifurcata.*

This zone is considered lowermost Silurian (Rhuddanian) based on acritarch taxa recorded from the Early Llandovery strata in the United States (Cramer, 1971; Miller and Eames, 1982), England (Hill, 1974; Downie, 1984), Norway (Smelror, 1987), Sweden (Le Hérissé, 1989), Algeria (Jardiné et al., 1974), Libya (Hill et al., 1985), Jordan (Keegan et al., 1990), Saudi Arabia (Le Hérissé et al., 1995) and southern Iran (Ghavidel-syooki, 1990, 1996).

CONCLUSIONS

The Lashkarak, Ghelli and Niur formations yielded 56 acritarch taxa. The local stratigraphic distribution of these acritarchs is shown in Text-Figure 2. Zones I to III are present in the Lashkarak Formation and suggest a Early Ordovician (Tremadoc–Arenigian) age for this interval. Comparison of these Early Ordovician acritarch assemblages with acritarch taxa from other parts of the world indicates broad similarity with those from the Mediterranean Acritarch Province (Vavrdovà, 1974).

This acritarch province includes southern Europe, northern Africa, southwestern China, Saudi Arabia and Zagros basin of Iran. The presence of Mediterranean acritarch taxa in the Lashkarak Formation of northeastern Iran suggests that both southern and northern Iran were part of Peri-Gondwanan paleocontinent, possibly positioned along the southern shore of the Paleo-Tethys Ocean.

Zones IV to V occur in the Ghelli Formation and indicate a Middle and Late Ordovician age.

Acritarch species from Ghelli Formation suggest a similarity with those of recorded from Middle and Upper Ordovician strata in Morocco, Algeria, Libya, Jordan, Saudi Arabia, England, Germany, Czech Republic and United States of America.

This supports the opinion that the peri-Gondwanan paleo-continent and the United States have occupied a similar paleolatitude during the Middle and late Ordovician. Zone VI is present in the Niur Formation and the acritarch assemblage suggests Early Silurian (Rhuddanian) for the basal part. Comparison of this Early Silurian acritarch assemblage with those of other parts of the world indicates broad similarity with those from the United States, northern Africa and southern Europe.

This suggests that the Peri-Gondwanan or Mediterranean paleo-continent began to move up toward the Baltic paleo-continent in Middle and Late Ordovician and by the Silurian formed the supercontinent of Pangea.

ACKNOWLEDGMENTS

The author expresses his sincere appreciation to the management of the National Iranian Oil Company for permission to publish this paper. I am also grateful to Dr. M. Vavrdovà, Dr. G. D. Wood, Professor A. T. Cross and Professor R. Wicander for their helpful suggestions

References Cited

AFSHAR-HARB, A.
1979 *The Stratigraphy, Tectonics and Petroleum Geology of Kopet-Dagh region.* Unpubl. Ph.D. thesis, Imperial College of Science and Technology, University of London, England, 316 pgs.

AHMADZADEH-HERAVI, M.
1983 Brachiopods and conodonts from the Lower Palaeozoic sediments at southern Bojnourd city. *Journal of Technology and Science, Tehran University,* 45: 1–24.

BAGNOLI, G., STOUGE, S., and TONGIORGI, M.
1988 Acritarchs and conodonts from the Cambro–Ordovician Furuhäll (Köpingskint) section (Öland Sweden). *Rivista Italiana Paleontographia e Stratigraphia,* 94: 163–248.

PLATE 2

All figures x1000 except fig. 7.

1 *Acanthodiacrodium echinatum* (Timofeev 1959) Deflandre & Deflandre Rigaud 1962.
2 *Acanthodiacrodium rectinerve* Burmann 1968.
3 *Multiplicisphaeridium cervinacornuum* Welsch 1986.
4 *Cymatiogalea velifera* (Downie 1958) Martin 1968.
5 *Dactylofusa squama* (Deunff 1961) Combaz et al. 1967.
6 *Goniosphaeridium piliferum* (Martin 1966) Eisenack et al. 1973.
7 *Orthosphaeridium inflatum* Loeblich 1969 (x500).
8 *Saharidia downiei* Combaz 1967.
9 *Lophosphaeridium torum* Rasul 1979.

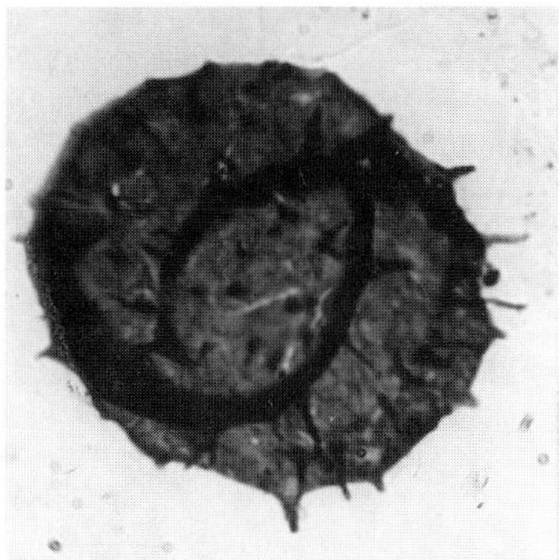

1

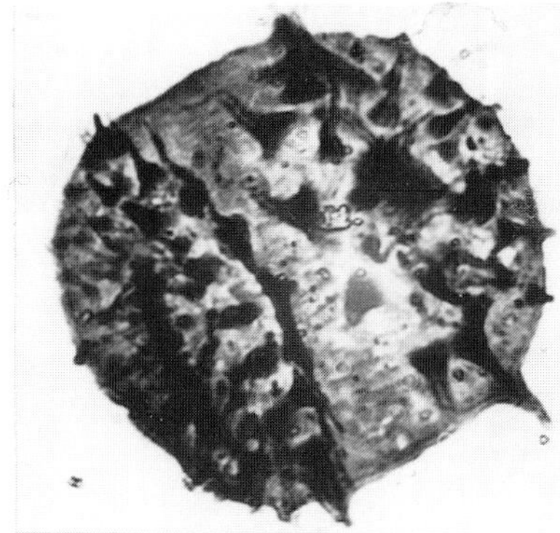

2

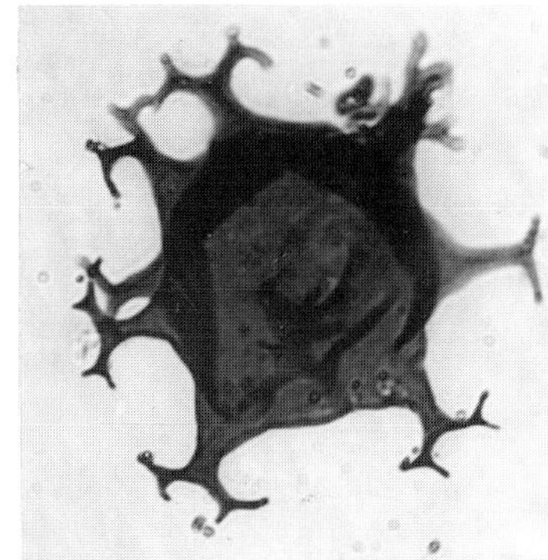

3

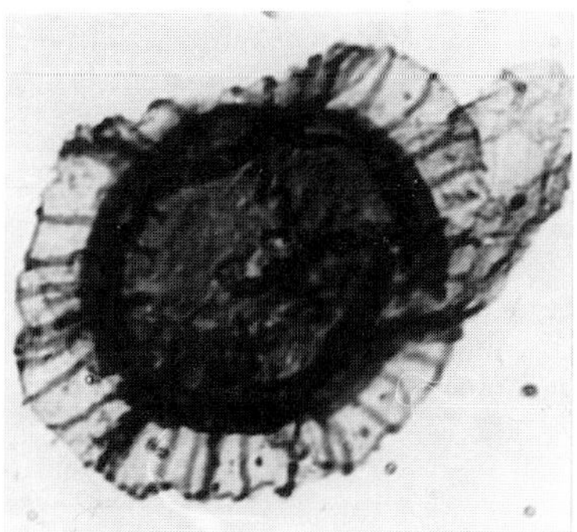

4

5

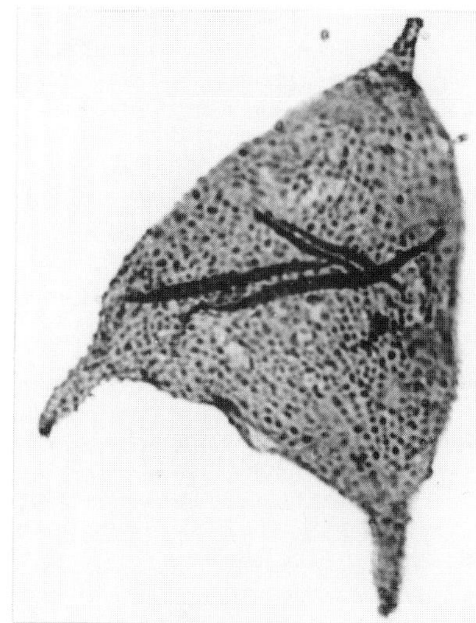

6

7

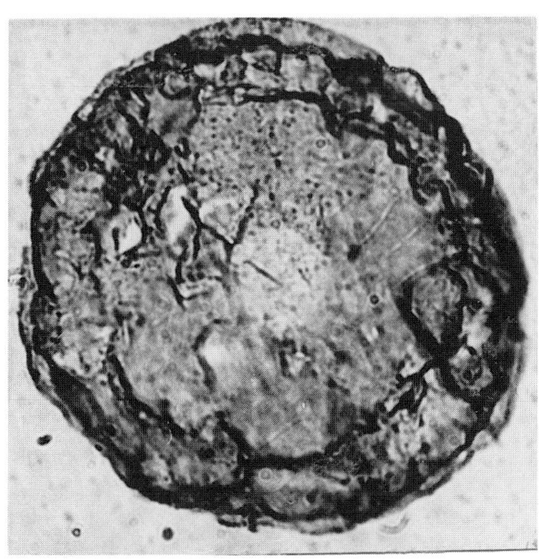

8

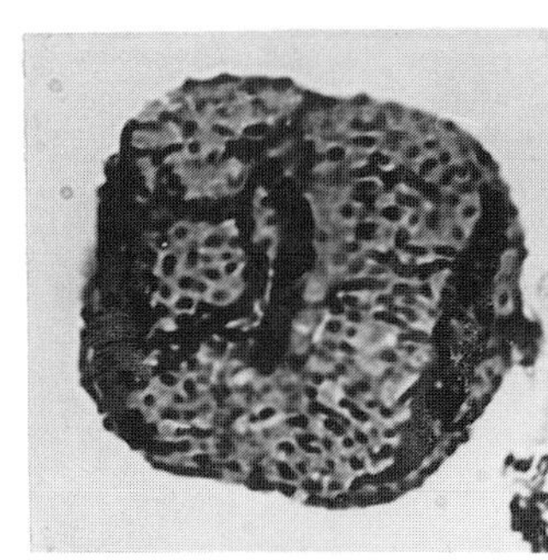

9

BURMANN, G.
1968 Diacrodien aus dem Unteren Ordovizium. *Paläontologisches Abhandlungen,* 2: 635–652.
COMBAZ, A.
1967 Un microbios du Trémadocien dans un sondage d'Hassi Massaoud. *Actes de la Société Linnéenne de Bordeaux,* 104 : 1–26.
COMBAZ, A., LANGE, F.W., and PANSART, J.
1967 Les "Leiofusidae" Eisenack (1938). *Review of Palaeobotany and Palynology,* 1: 291–307.
CRAMER, F.H.
1970 Distribution of selected Silurian acritarchs. An account of the palynostratigraphy and paleogeography of selected Silurian acritarch taxa. *Revista Española de Micropaleontología número extraordinario:* 1–203.
CRAMER, F.H., and DÍEZ, M.d.C.R.
1972 North American Silurian palynofacies and their spatial arrangement. *Palaeontographica,* Abteilung B, 138: 107–180.
1977 Late Arenigian (Ordovician) Acritarchs from Cis-Saharan Morocco. *Micropaleontology,* 23: 339–360.
CRAMER, F.H., KANES, W.H., DÍEZ, M.d.C.R., and CHRISTOPHER, R.A.
1974 Early Ordovician acritarchs from the Tadla Basin of Morocco. *Palaeontographica,* Abteilung B, 146: 57–64.
CRAMER, F.H., ALLAM, B., KANES, W.H., and DÍEZ, M.d.C.R.
1974 Upper Arenigian to Lower Llanvirnian acritarchs from the subsurface of the Tadla Basin in Morocco. *Palaeontographica,* Abteilung B, 145: 182–190.
DEUNFF, J.
1961 Un microplancton à hystrichosphères dans le Tremadoc du Sahara. *Revue de Micropaléontologie,* 1: 37–52.
1964 Systématique du microplancton fossile à Acritarches. Révision de deux genres de l' Ordovicien inférieur. *Revue de Micropaléontologie,* 7: 119–124.
DEUNFF, J., GÓRKA, H., and RAUSCHER, R.
1974 Observations nouvelles et précisions sur les Acritarches à large ouverture polaire du Paléozoïque inférieur. *Géobios,* 7: 5–18.
DOWNIE, C.
1984 Acritarchs in British stratigraphy. *Geological Society of London, Special Report 17:* 1–26.
DUFKA, P., and FATKA, O.
1993 Chitinozoans and acritarchs from the Ordovician–Silurian boundary of the Prague Basin. Czech Republic. *Special Papers in Palaeontology,* 48: 17–28.
EISENACK, A., CRAMER, F.H., and DÍEZ, M.d.C.R.
1976 *Katalog der fossilen Dinoflagellaten, Hystrichosphären und verwandten Mikrofossilien. Acritarcha, Band III, Teil 2;* E. Schweizerbart'sche Verlagsbuchhandlung, Stuttgart, 863 pgs.
1979 Katalog der fossilen Dinoflagellaten Hystrichosphären und verwandten Mikrofossilien *Acritarcha, Band VI, Teil 3,* E. Schweizerbart'sche Verlagsbuchhandlung Stuttgart, 531 pgs.
ELAOIUAD-DEBBAJ, Z.
1988 Acritarchs from the Upper Ordovician (Caradoc–Ashgill) of the Anti-Atlas Morocco. *Revue de Micropaléontologie,* 30: 232–248 .
FENSOME, R.A., WILLIAMS, G.L., BARSS, M.S., FREEMAN, J.M., and HILL, J.M.
1990 Acritarchs and fossil prasinophytes: an index to genera, species and infraspecific taxa. *American Association of Stratigraphic Palynologists Foundation, Contribution Series,* 25, 771 pgs.
GHAVIDEL-SYOOKI, M.
1990 The encountered acritarchs and chitinozoans from Mila, Ilebek and Zard-Kuh formations in Tang-e-Ilebek at Zard-Kuh region and their correlation with Palaeozoic sequence of Chal-i-Sheh area in Zagros Basin of Iran. *Symposium on Diaparism with Special Reference to Iran, Geological Survey of Iran,* 1: 141–218.
1993 Palynological study of Palaeozoic sediments of the Chal-i-Sheh area, southwestern Iran. *Iranian Journal of Science,* 4: 32–46.
1995 Palynobiostratigraphy and Palaeogeography of Palaeozoic sequence in the Hassanakdar area, Central Alborz Range, northern Iran. *Review of Palaeobotany and Palynology,* 89: 91–109
1997 Biostratigraphy of acritarchs in Palaeozoic rock units in the Zagros Basin of Iran. *In:* Fatka, O., and Servais, T. (eds.), *Acritarcha in Praha, Acta Universitatis Carolinae Geologica,* 40: 385–411.
GHAVIDAL-SYOOKI, and KHOSRAVI, M.E.
1995 Investigation of Palaeozoic sediments at Tang-e-Zakeen of Kuh-e- Faraghan and introduction to Seyahou and Sarchahan formations in the Zagros Basin. *Geological Survey of Iranian Sciences, Quarterly Journal,* 4: 1–21.
GÓRKA, H..
1987 Acritarches et prasinophyceae de l' Ordovicien moyen (Viruen) du Sondage de Smedsby Gård No. 1 (Gotland Suède). *Review of Palaeobotany and Palynology,* 52: 257–297.
HILL, P.J.
1974 Stratigraphic palynology of acritarchs from the type area of the Llandovery and the Welsh Borderland. *Review of Palaeobotany and Palynology,* 18: 11–23.
HILL, P.J., PARIS, F., and RICHARDSON, J.B.
1985 Silurian palynomorphs: *In:* Thusu, B., and Owens, B. (eds.), Palynostratigraphy of northeast Libya. *Journal of Micropalaeontology,* 4: 27–48.
JACHOWICZ, M.
1995 Ordovician acritarch assemblages from Central and northwestern Saudi Arabia. *Review of Palaeobotany and Palynology,* 89: 19–25.
JACOBSON, S.R., and ACHAB, A.
1985 Acritarch biostratigraphy of the *Dicellograptus complanatus* graptolite Zone from the Vaureal Formation (Ashgillian) Anticosti Island, Quebec Canada. *Palynology,* 9: 165–198.

PLATE 3

All figures x1000.

1 *Arbusculidium filamentosum* (Vavrdová 1965) Vavrdová 1972.
2 *Marrocanium simplex* Cramer et al. 1974.
3 *Arbusculidium* sp.
4 *Striatotheca triangulata* (Cramer et al. 1974) Eisenack et al. 1976.
5 *Pirea baculifera* Tongiorgi et al. 1995.
6 *Striatotheca principalis* Burmann 1970.
7 *Peteinosphaeridium tenuifilosum* Tongiorgi et al. 1995.
8 *Coryphidium minutum* Cramer & Déz 1976.
9 *Peteinosphaeridium armatum* Tongiorgi et al. 1995.

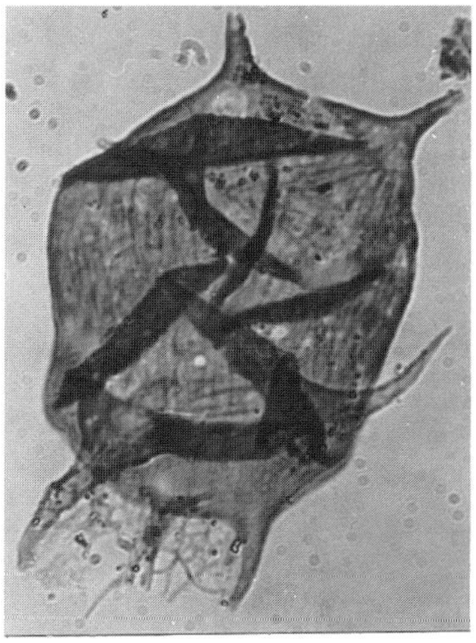

1

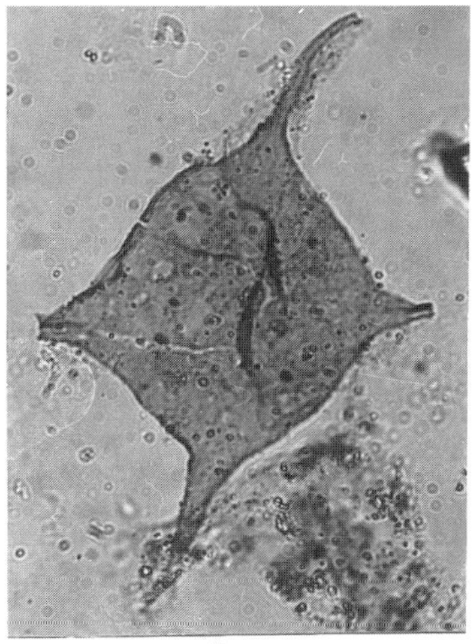

2

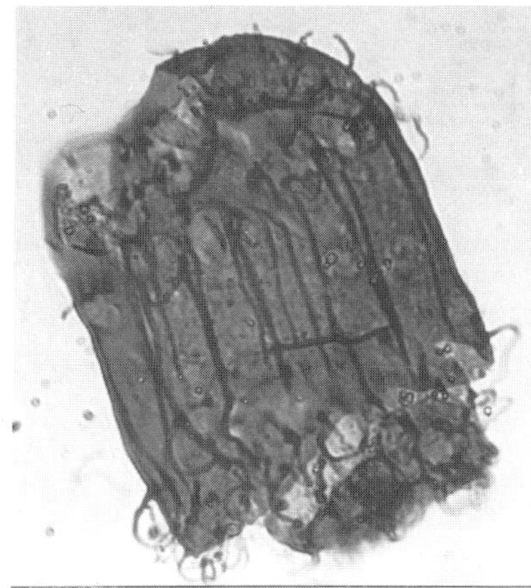

3

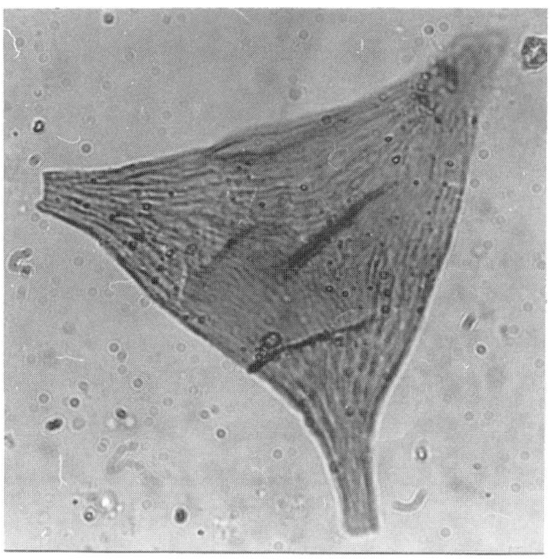

4

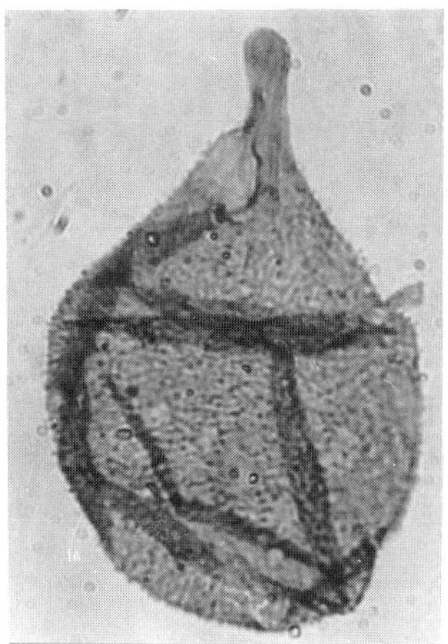

5

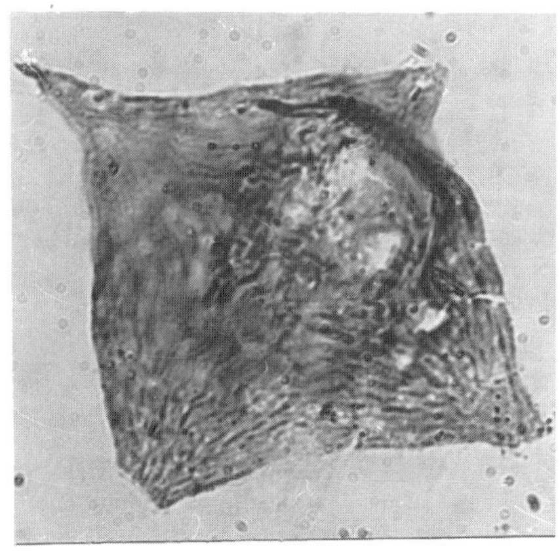

6

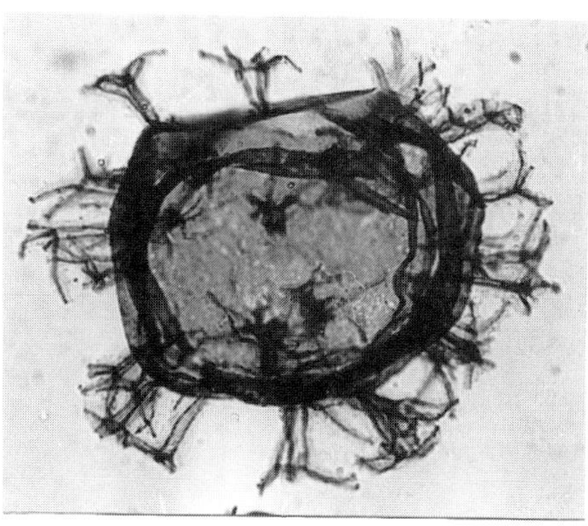

7

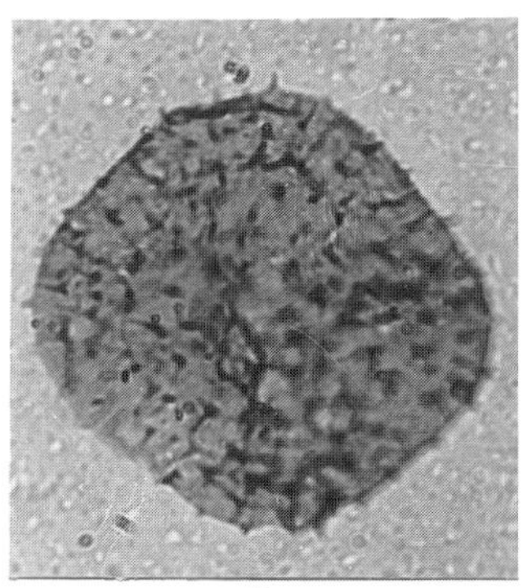

8

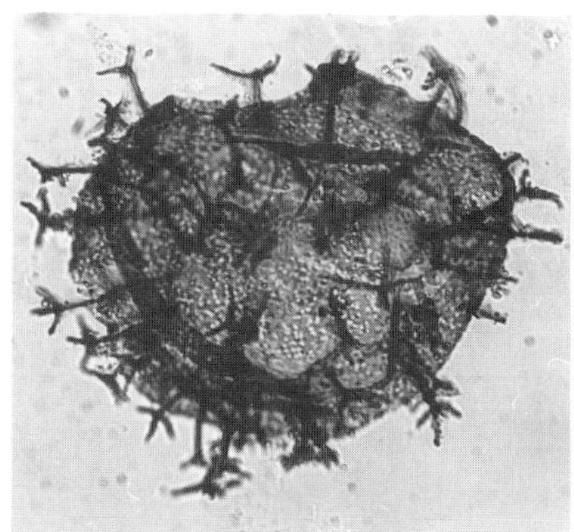

9

JARDINÉ, S., COMBAZ, A., MAGLOIRE, L., PENIGUEL, G., and VACHEY, G.
1974 Distribution stratigraphique des acritarches dans le Paléozoïque du Sahara algérien.*Review of Palaeobotany and Palynology*, 18: 99–129.
JEKHOWSKY, B.
1961 Sur quelques hystrichosphères d'Europe et d'Afrique. *Revue de Micropaléontologie*, 4: 207–212
KEEGEN, J.B., RASUL, S.M., and SHAHEEN, Y.
1990 Palynostratigraphy of the Lower Palaeozoic, Cambrian–Silurian sediments of the Hashemite Kingdom of Jordan. *Review of Palaeobotany and Palynology*, 66: 167–180.
KJELLSTRÖM, G.
1971 Ordovician microplankton (baltisphaerids) from the Grötlingbo borehole no. 1 in Gotland, Sweden. *Sveriges Geologiska Undersökning Avhandlingar och Uppsatser, Series C*, 65: 1–75.
Le HÉRRISÉ, A..
1989 Acritarches et Kystes d'algues prasinophycees du Silurien de Gotland Suede. *Paleontographica Italica*, 76: 57–302.
Le HÉRRISÉ, A., and GIOURVENNE, C.R.
1995 Biogeography of Upper Llandovery and Wenlock acritarchs. *Review of Palaeobotany and Palynology*, 89: 49–74
LI, J.
1987 Ordovician acritarchs from the Meitan Formation of Guizhou province, southwest China. *Palaeontology*, 30: 613–634.
LISTER, T.R.
1970 The acritarchs and chitinozoa from the Wenlock and Ludlow Serie of the Ludlow and Milichope area, Shropshire. *Palaeontographical Society, Monograph* 124: 1–100.
LOEBLICH, JR., A.R.
1970 Morphology, ultrastructure and distribution of Paleozoic acritarchs. *Proceedings of the North American Paleontological Convention*, Part G: 705–788.
LOEBLICH, JR., and TAPPAN, H.
1971 Two new *Orthosphaeridium* (acritarcha) from the Middle and Upper Ordovician. *Transactions of the American Microscopial Society*, 90: 182–188.
1978 Some Middle and Late Ordovician microplankton from North America. *Journal of Paleontology*, 52: 1233–1287.
LU, L.C.
1987 Acritarchs from the Dawan formation (Arenigian) of Huanghuachang in Yichang, western Hubei. *Acta Micropalaeontologica Sinica*, 4: 87–102.
MARTIN, F.
1965 Les acritarches de Sart-Bernard (Ordovicien Belge). *Bulletin de la Société de geologié Belge*, 74: 1–21.
1968 Les acritarches de l' Ordovicien et du Silurien Belges. Détermination et valeur stratigraphique. *Mémoire de l'Institut royal des Sciences naturelles de Belgique*, 160: 1–175.
MARTIN, F., and YIN, LEIMING
1988 Early Ordovician acritarchs from southern Jilin Province, northeast China. *Palaeontology*, 31: 109–127.
MILLER, M.A., and EAMES, L.E.
1982 Palynomorphs from the Silurian Medina Group (Lower Llandovery) of the Niagara Gorge, Lewiston, New York U.S.A. *Palynology*, 6: 221–254.
MOLYNEUX, S.G., and PARIS, F.
1985 Late Ordovician palynomorphs. *In*: Thusu, B., and Owens, B. (eds.), Palynostratigraphy of northeast Libya, *Journal of Micropalaeontology*, 4: 11–26.
MOLYNEUX, S.G., and RUSHTON, A.W.A.
1988 The age of the Watch Hill Grits (Ordovician), England, Lake district structural and palaeographical implication. *Transactions of the Royal Society of Edinburgh, Earth Sciences*, 79: 43–69.
OTTONE, E.G., TORO, B.A., and WAISFELD, B.G.
1992 Lower Ordovician palynomorphs from the Acoite Formation, northwestern Argentina. *Palynology*, 16 : 93–116.
PLAYFORD, G., RIBECAI, C., and TONGIORGI, M.
1995 Ordovician acritarch genera, *Petenosphaeridium*, *Liliosphaeridium* and *Cycloposphaeridium:* Morphology, Taxonomy, Biostratigraphy and Palaeogeographic significance. *Paleontolographica Italica*, 34: 3–54.
RASUL, S.M.
1976 New species of the genus *Vulcanisphaera* (Acritarcha) from the Tremadocian of England. *Micropaleontology*, 22: 479–484.
1979 Acritarch zonation of Tremadoc Series of Shineton Shales, Wrekin, Shrophshire. *Palynology*, 3: 53–72.
RAUSCHER, R.
1974 Les acritarches de l' Ordovicien en France. *Review of Palaeobotany and Palynology*, 18 : 83–97.
REITZ, E.
1991 Acritarchs of Early Tremadoc sediments in the western Frankenwald area, NE Bavaria. *Neues Jahrbuch für Geologie und Paläontologie, Abhandlungen*, 2: 97–104.
SMELROR, M.
1987 Early Silurian acritarchs and prasinophycean algae from the Ringerike district, Oslo region (Norway). *Review of Palaeobotany and Palynology*, 52 137–159.
STAPLIN, F.L., JANSONIUS, J., and POCOCK, S.A.J.
1965 Evaluation of some acritarchous hystrichosphere genera. *Neues Jahrbuch für Geologie und Paläontologie, Abhandlungen*, 123 : 167–201
TONGIORGI, M., YIN, L.M., and Di MILIA, A.
1995 Arenigian acritarchs from the Daping section (Yangtze Gorge area Hubei province, southern China) and their palaeogeographic significance. *Review of Palaeobotany and Palynology*, 86: 13–48.
TURNER, R.E.
1984 Acritarchs from the type area of the Ordovician Caradoc Series. Shropshire, England. *Palaeontographica*, Abteilung, B., 190 : 87–157.

PLATE 4

All figures x1000.

1 *Rhopaliophora palmata* (Combaz & Peniguel 1972) emend. Playford & Martin 1984.
2, 3 *Rhopaliophora mamilliformis* Lu 1987.
4, 6 *Cymatiogalea deunffii* Jardiné, Combaz et al. 1974.
5 *Vulcanisphaera britannica* Rasul 1976.
7 *Athabascaella playfordii* Martin 1984.
8 *Pirea baculifera* Tongiorgi et al. 1995.
9 *Rhiptosocherma improcera* (Loeblich 1970) Loeblich & Tappan 1978.

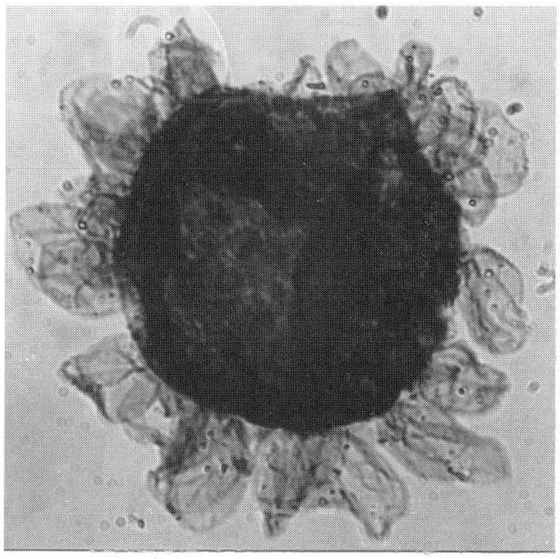

1

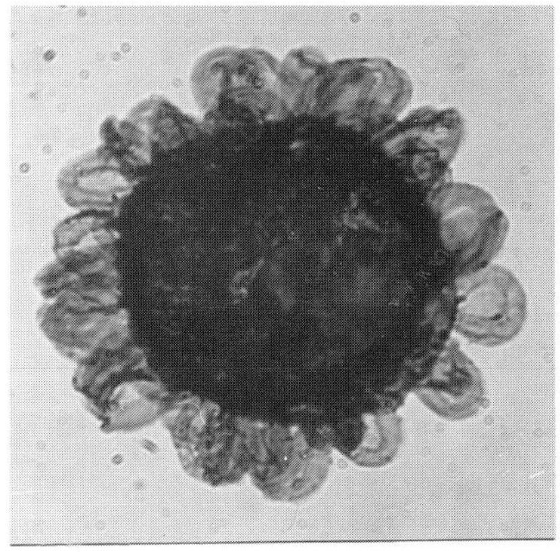

2

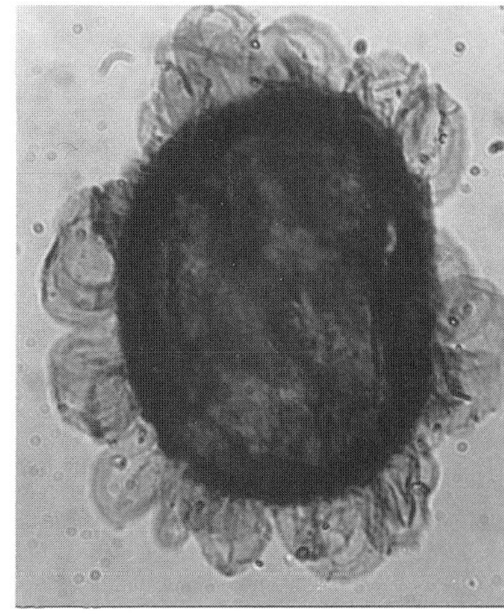

3

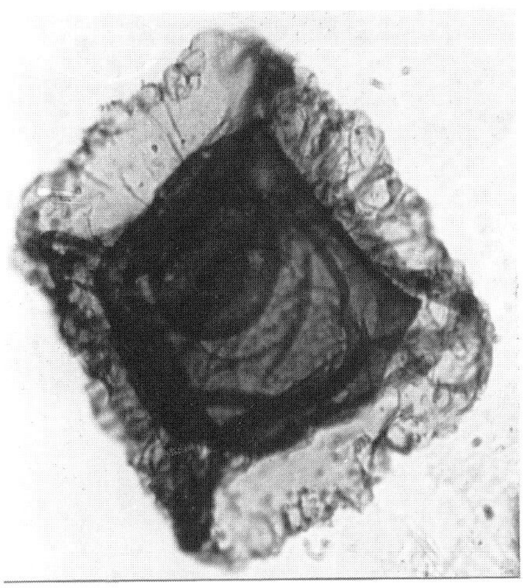

4

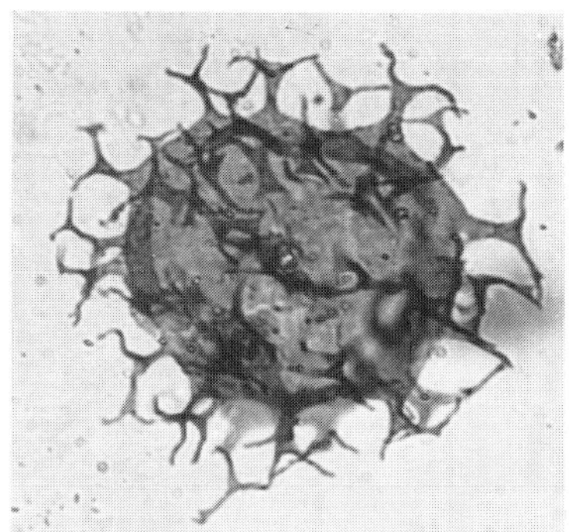

5

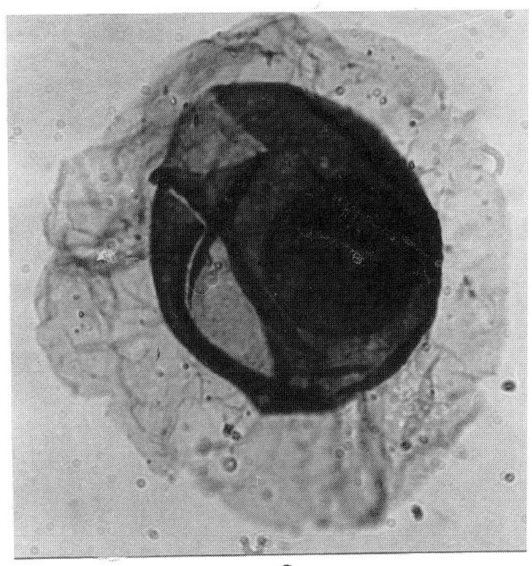

6

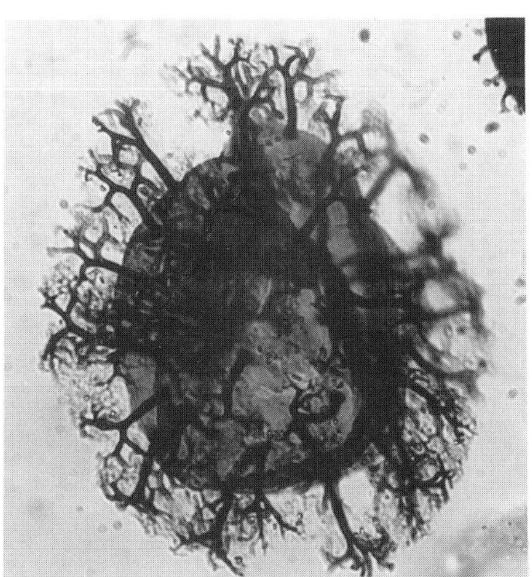

7

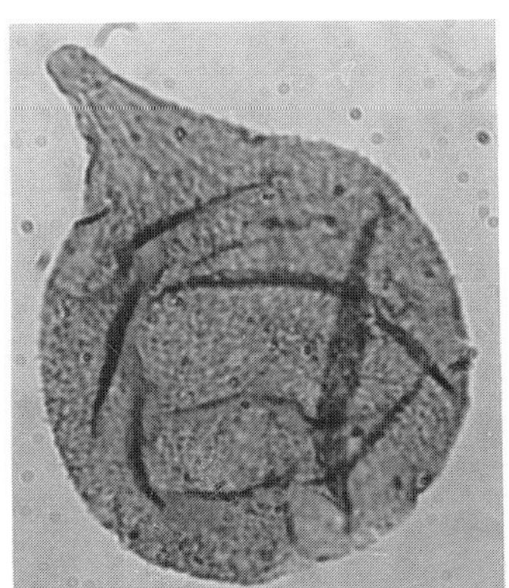

8

9

1985 Acritarchs from the type area of the Llandeilo Series, South Wales. *Palynology*, 9: 211–234.

VAVRDOVÁ, M.
1974 Geographical differentiation of Ordovician acritarch assemblage in Europe. *Review of Palaeobotany and Palynology*, 18: 171–175.

WELSCH, M.
1986 The acritarchs of the Upper Digermul group, Middle Cambrian–Tremadoc, eastern Finnmark, northern Norway. *Palaeontographica*, Abteilung B, 148: 2–43.

Author's address

MOHAMMAD GHAVIDEL-SYOOKI
Exploration Division
National Iranian Oil Company
P. O. Box 1065
Tehran
Iran

PLATE 5

All figures x1000.

1, 6 *Peteinosphaeridium robustiramosum* Tongiorgi et al. 1995.
2, 5 *Coryphidium elegans* Cramer et al. 1974.
3 *Peteinosphaeridium velatum* Kjellström 1971.
4, 7 *Peteinosphaeridium angustilaminae* Playford et al. 1995.
8 *Pirea baculifera* Tongiorgi, Yin & Di Milia 1995.
9 *Peteinosphaeridium angustilaminae* Playford et al. 1995.

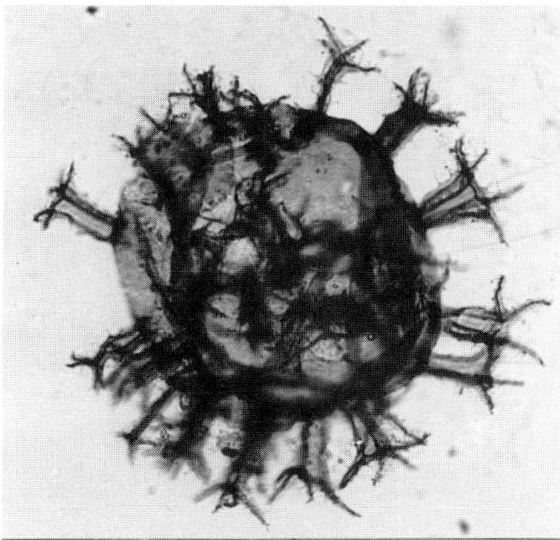

1

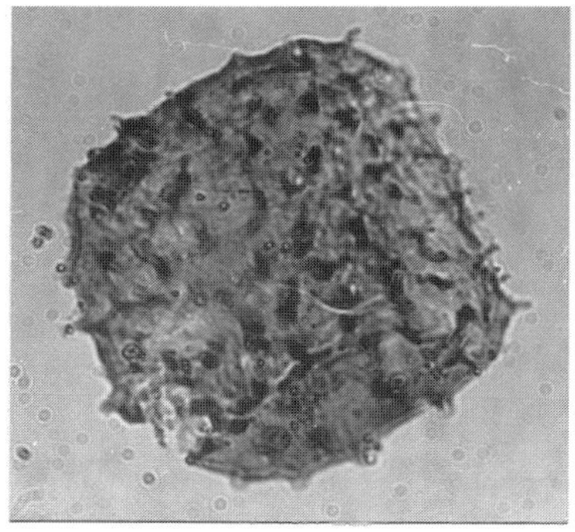

2

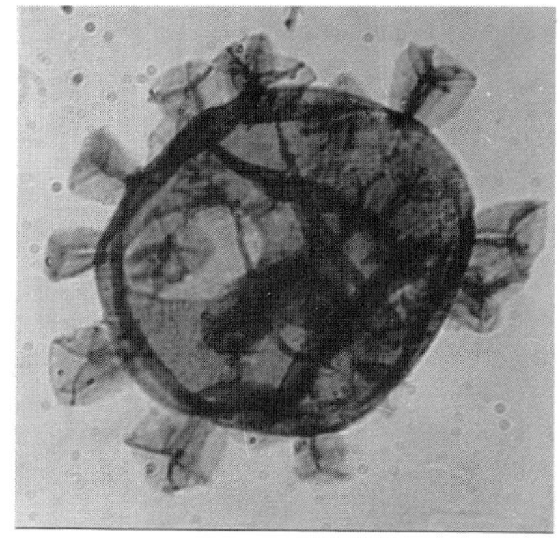

3

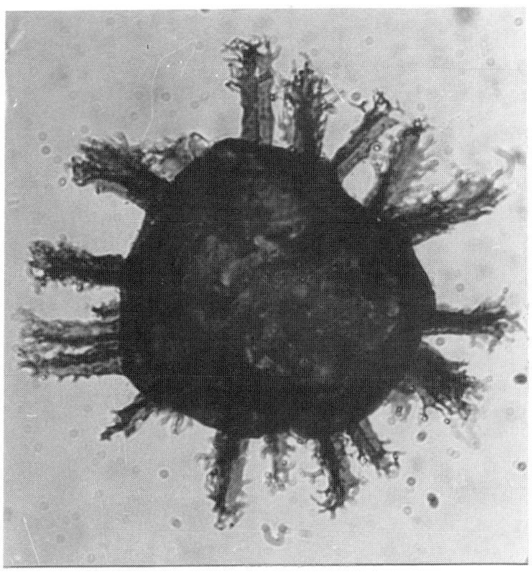

4

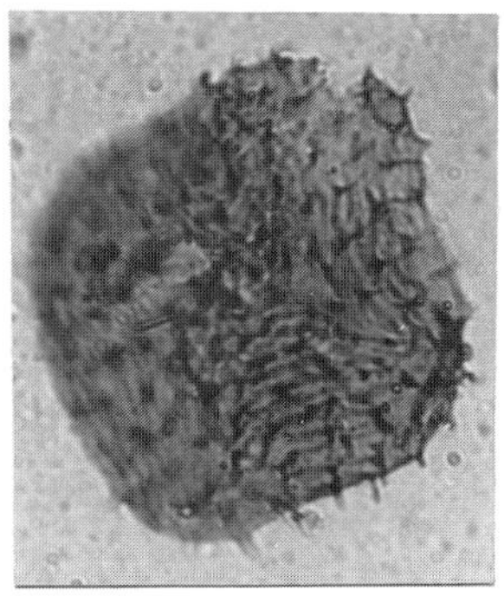

5

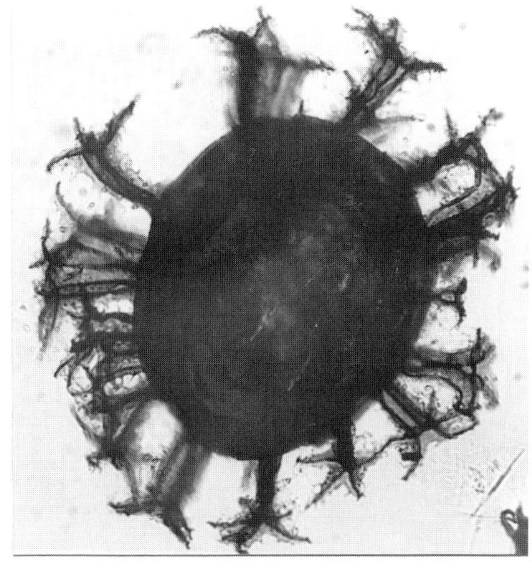

6

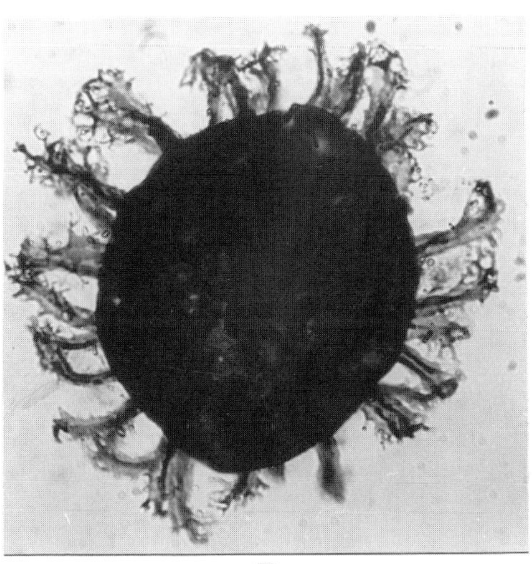

7

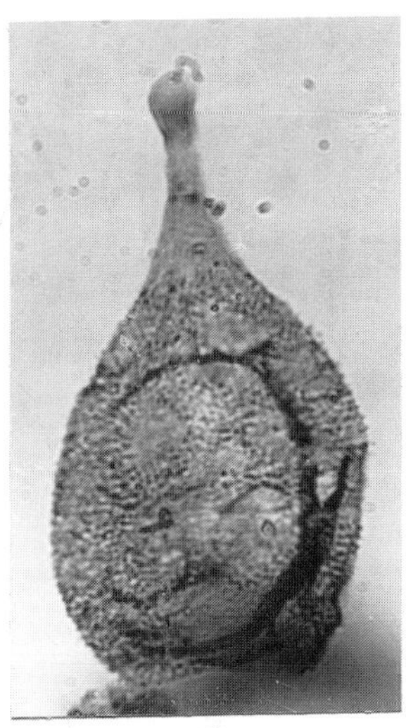

8

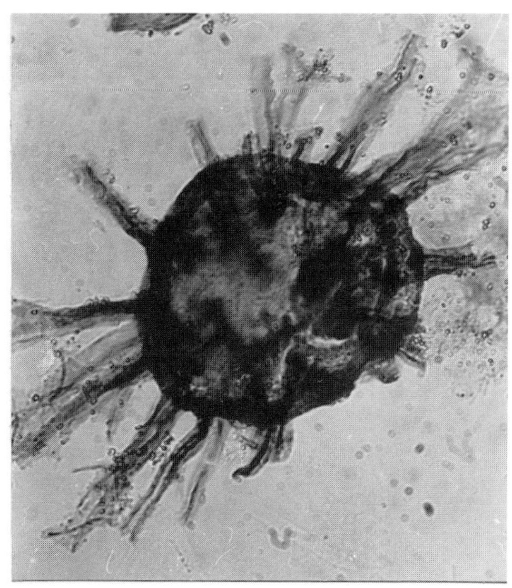

9

PLATE 6

All figures x1000 except number 2.

1 *Actinotodissus crassus* Loeblich & Tappan 1978.
2 *Dactylofusa striata* (Staplin et al. 1965) Fensome et al. 1990 (x500).
3 *Striatotheca rugosa* Tongiorgi et al. 1995.
4 *Villosacapsula setosapellicula* (Loeblich) Loeblich & Tappan 1976.
5 *Veryhachium reductum* (Deunff 1959) Jeknowsky 1961.
6 *Navifusa ancepsipuncta* Loeblich 1970.
7 *Veryhachium subglobosum* Jardiné et al. 1974.
8 *Veryhachium oklahomense* Loeblich 1969.

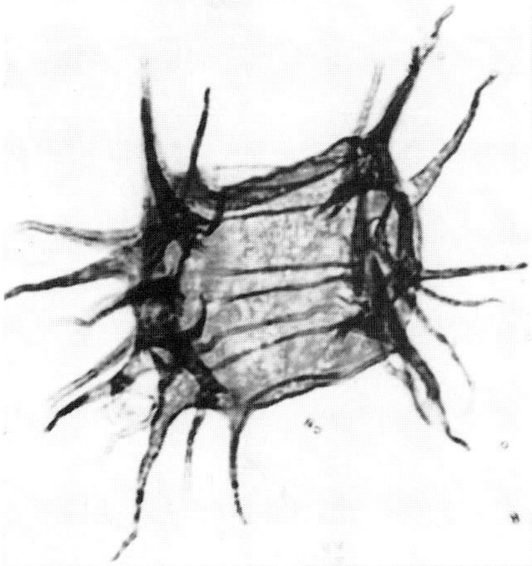

1

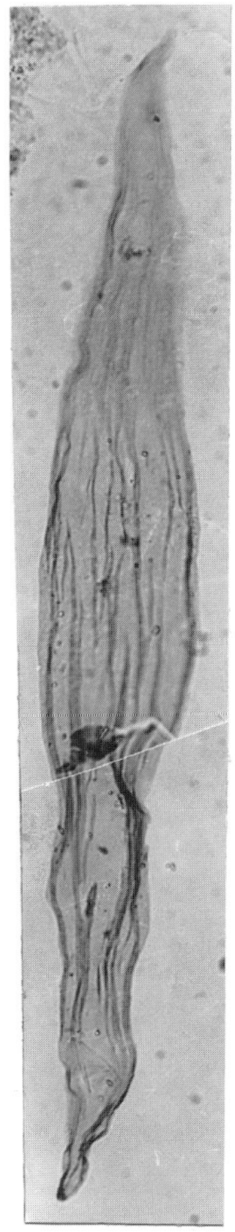

2

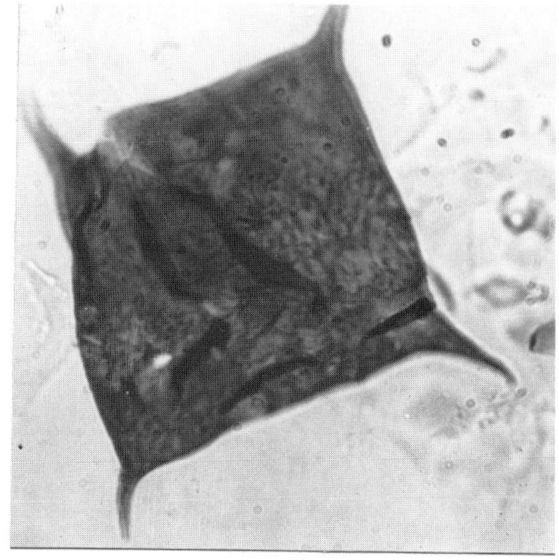

3

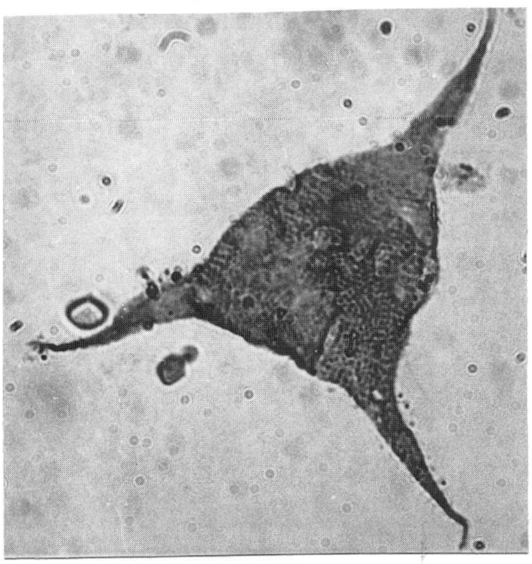

4

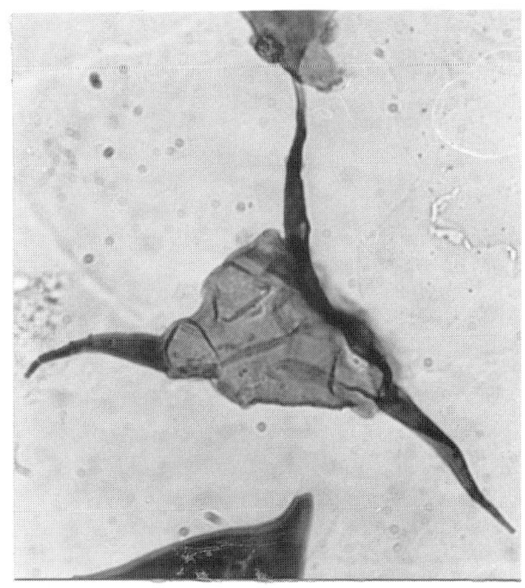

5

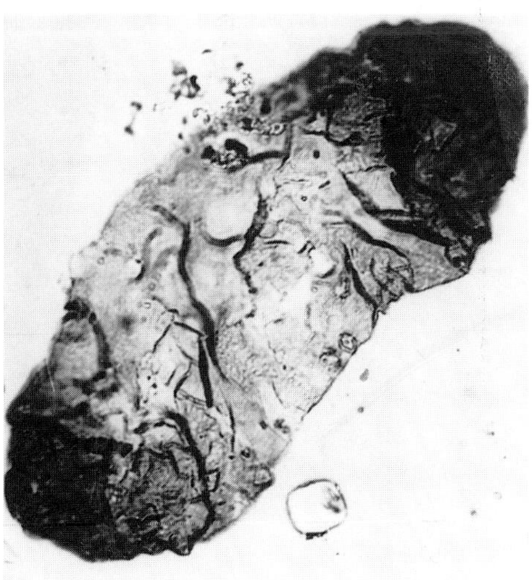

6

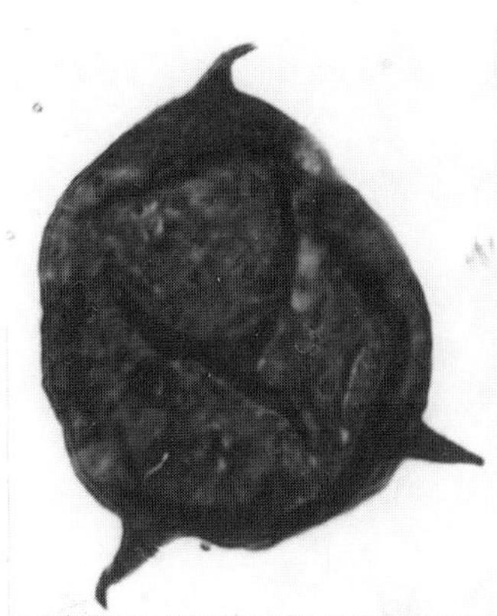

7

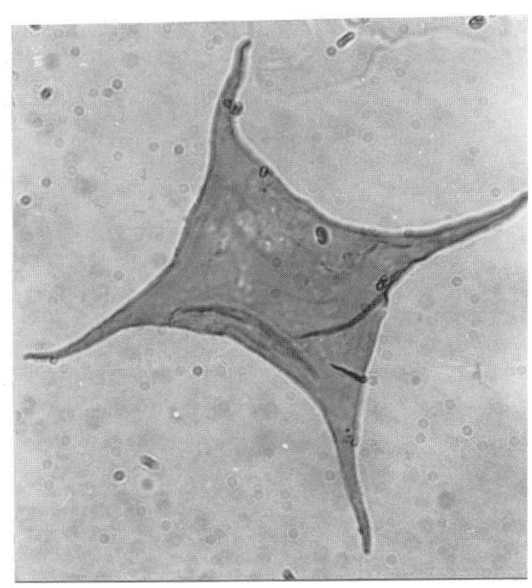

8

PLATE 7

All figures x500 with exception of figs. 2, 7, 8 and 9.

1, 4 *Orthosphaeridium insculptum* Loeblich 1969.

2 *Navifusa ancepsipuncta* Loeblich 1970 (x1000).

3 *Orthosphaeridium ternatum* (Burmann 1970) Eisenack et al. 1976.

5 *Orthosphaeridium inflatum* Loeblich 1969.

6 *Baltisphaeridium perclarum* Loeblich & Tappan 1978.

7 *Ordovicidium elegantulum* Tappan & Loeblich 1971 (x1000).

8 *Poikilofusa spinata* Staplin et al. 1965 (x1000).

9 *Diexallophasis denticulata* (Stockmanns & Williéré 1963) Loeblich 1970 (x1000).

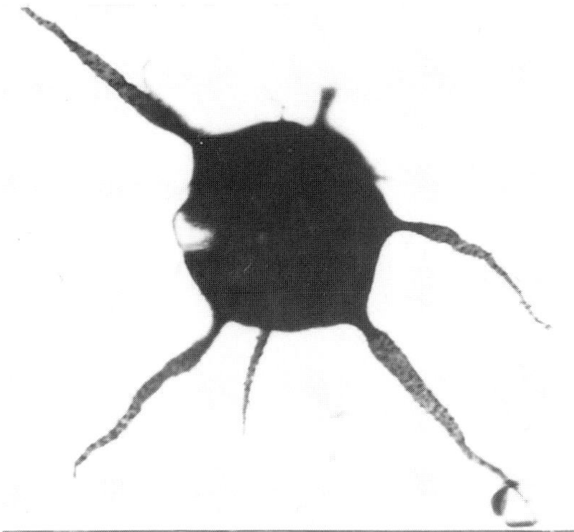

1

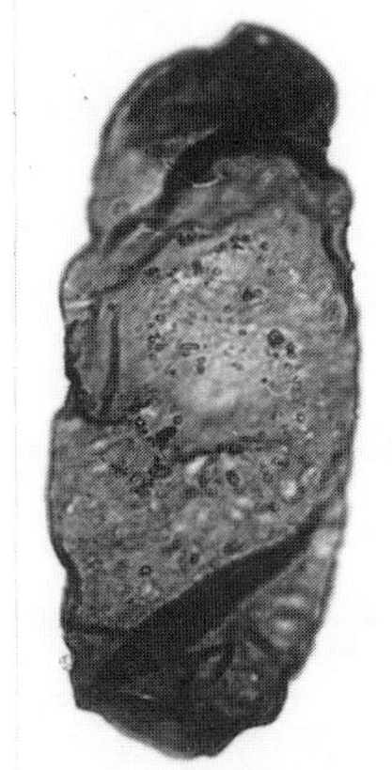

2

3

4

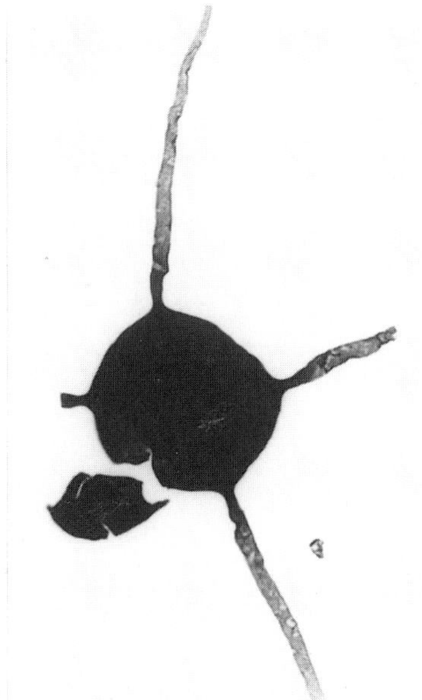

5

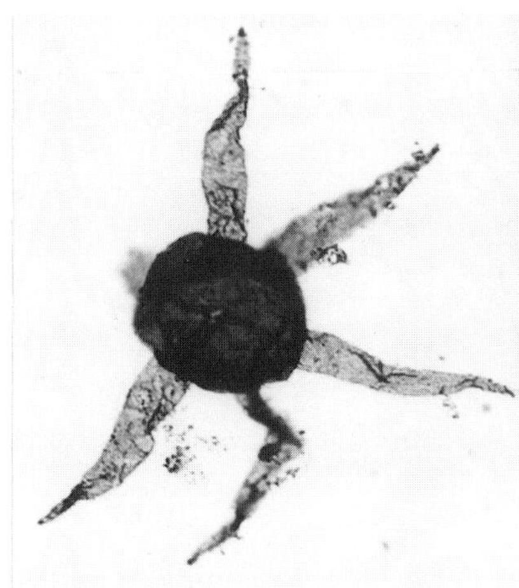

6

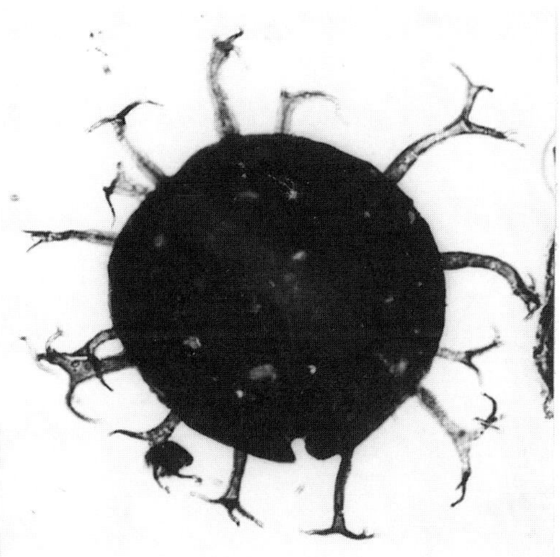

7

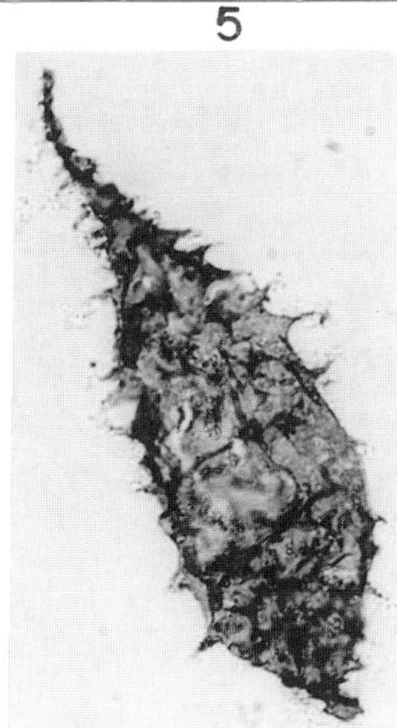

8

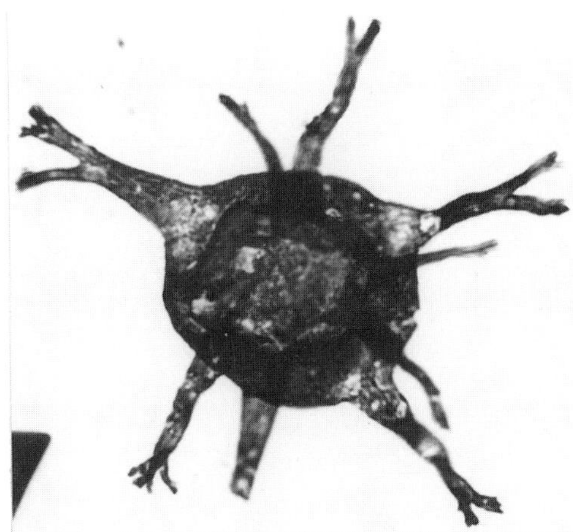

9

PLATE 8

All figure x1,000.

1 *Helosphaeridium clavispinulosum* Lister 1970.
2 *Dactylofusa estillis* Cramer & Díez 1972.
3 *Dactylofusa striatifera* (Cramer 1964) Fensome et al. 1990.
4 *Visbysphaera oligofurcata* (Eisenack 1959) Hill 1974.
5 *Tunisphaeridium tentaculiferum* (Martin 1967) Cramer 1970.
6 *Visbysphaera brevifurcata* (Eisenack 1954) Le Hérissé 1989.
7 *Visbysphaera pirifera* (Eisenack 1954) Lister 1970.
8 *Multiplicisphaeridium neaghae* Cramer 1970.

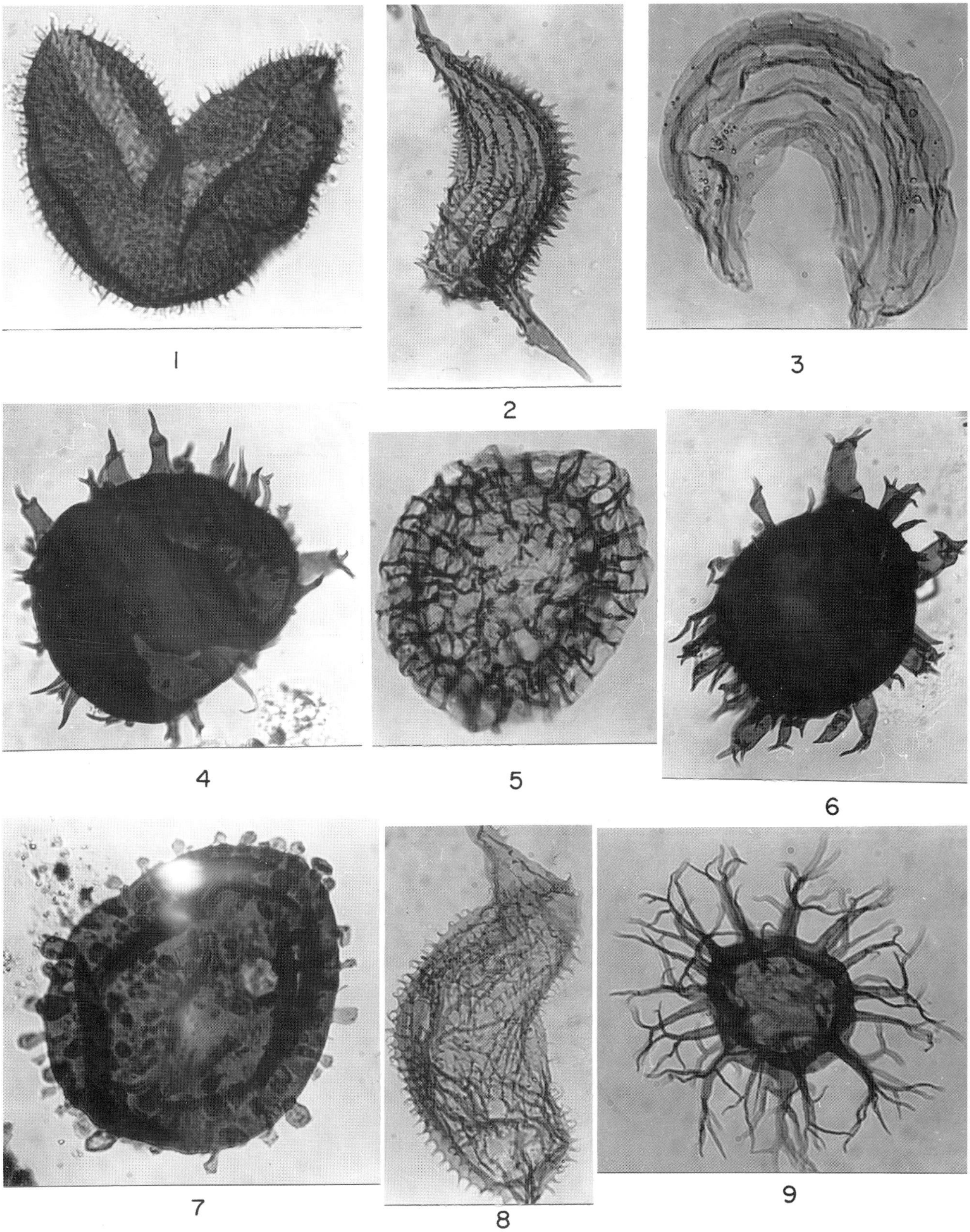
1
2
3
4
5
6
7
8
9

PALYNOSTRATIGRAPHY AND PALEOGEOGRAPHY OF THE LATE DEVONIAN IN NORTHEASTERN ESFAHAN CITY, CENTRAL IRAN

Mohammad Ghavidel-Syooki

Abstract

A palynological study of the Padeha and Bahram formations of the Chah-e-Riseh area, in northeastern Esfahan city, was initiated to more precisely determine the stratigraphic age of these rock units. These formations yield 44 diagnostic palynomorphs (23 miospore and 21 acritarch species) which are used to define five stratigraphic zones. Zone I is Early Frasnian and occurs in the Padeha Formation and the basal part of Bahram Formation. The Zones II through V are present in the Bahram Formation and palynomorphs suggest a Late Devonian (Middle Frasnian–Late Famennian) age. A hiatus between the Bahram Formation and overlying Jamal Formation spans essentially the entire Carboniferous and Early Permian, possibly time equivalent with the Hercynian Orogeny.

Few palynomorph taxa recorded in northeastern Esfahan city (Chah-e-Riseh area) are present in Europe and North America. However, many species, such as *Navifusa exilis, Deltotosoma intonsum, Papulogabata annulata, Melikeriopalla venulosa, Horologinella quadrispina, Crassiangulina tesselita, Somphophragma miscellum, Dictyotidium granulatum, Lophosphaeridium deminutum, Helosphaeridium microclavatum, Grandispora fibrilata, Cyclogranisporites isostictus, Dibolisporites turriculatus* and *Lophozonotriletes somphus* have been reported only from Late Devonian strata of western Australia, southern and northern Iran, and Algeria

Based on the mutual occurrence of palynomorph taxa it is reasonable to assume that the Iranian Platform, western Australia and Algeria were at a similar paleolatitude, along the southern shore of the Paleo-Tethys Ocean, during the Late Devonian. The presence of abundant acritarch taxa in the Padeha and Bahram formations suggests a shallow marine environment occupied the Central Iranian Basin during the Late Devonian. Moreover, the occurrence of abundant miospore taxa in these two formations indicates that a fairly robust terrestrial plant community existed in close proximity to the basin.

INTRODUCTION

A 644.0 m thick Devonian sequence is well-exposed at northeastern Esfahan city in central Iran (Text-Figure 1). This section has been divided into the Padeha and Bahram formations (Djafarian, 1977). These formations have received minimal interest for chronostratigraphy since the Padeha Formation lacks marine fauna.

The objectives of this paper is to demonstrate the biostratigraphic, paleoecologic and paleogeographic significance of the acritarchs and miospores in the Devonian of Iran. The paleogeography of the Devonian strata in northeastern Esfahan city is important in relation to the Iranian Alborz and Zagros mountain ranges and other regions of the world.

PREVIOUS STUDIES

The study locality, called the Chah-e-Riseh area, is approximately 55 km northeast of Esfahan city. The paved road from Esfahan to Ardestan city is the main connection to the Chah-e-Riseh area (Text-Figure 1). The Upper Paleozoic sequence in the study area has been divided, in ascending order, into the Padeha, Bahram and Jamal formations (Text-Figure 2). In this paper only the palynomorphs from the Padeha and Bahram formations are discussed since the Jamal Formation consists mainly of fusulinid bearing limestones that are not usually conducive to the preservation of spores and acritarchs.

Ghavidel-Syooki, Mohammad
2001 Palynostratigraphy and paleogeography of the Late Devonian in northeastern Esfahan City, central Iran; *In*: Goodman, D.K., and Clarke, R.T. (eds.), *Proceedings of the IX International Palynological Congress, Houston, Texas, U.S.A., 1996*; American Association of Stratigraphic Palynologists Foundation, p. 37–51.

ISBN 0–931871–06–9

The field work was carried out in 1993. The measured and sampled stratigraphic section was chosen near the Chah-e-Riseh area where the thickness of Padeha and Bahram formations is 644.0 m. The Padeha Formation is 236.0 m thick (Text-Figure 2) and consists mainly of dolomite, limestone, sandstone, gypsum and some dark-gray shale horizons. Although the formation lacks fauna, it has been assigned to the Lower Devonian based on stratigraphic position.

The Bahram Formation is 408.0 m thick and is composed of alternating fossiliferous limestones, sandstones and dark-gray shales. Djafarian and Brice (1973) and Djafarian (1977) have studied the brachiopods of Bahram Formation and assigned a Late Famennian–Lower Carboniferous age.

LABORATORY TECHNIQUES

A total of 100 surface samples from the Padeha and Bahram formations were selected for palynological study. The field and laboratory description of the samples are shown in Text-Figure 2. The code and number of each sample follow the policy of National Iranian Oil Company.

Thirty grams of sediments were randomly selected from each sample and processed in the palynology laboratory of the Exploration Department of the National Iranian Oil Company. Disaggregation of the rock samples was carried out using standard palynological procedures. All slides used in this study are on file in the Palaeontology Section of the Exploration Department of the National Iranian Oil Company.

STRATIGRAPHICAL PALYNOLOGY

In this study, a total of 44 palynomorph species (miospores and acritarchs) have been identified. Their distribution is plotted in Text-Figure 2 and selected

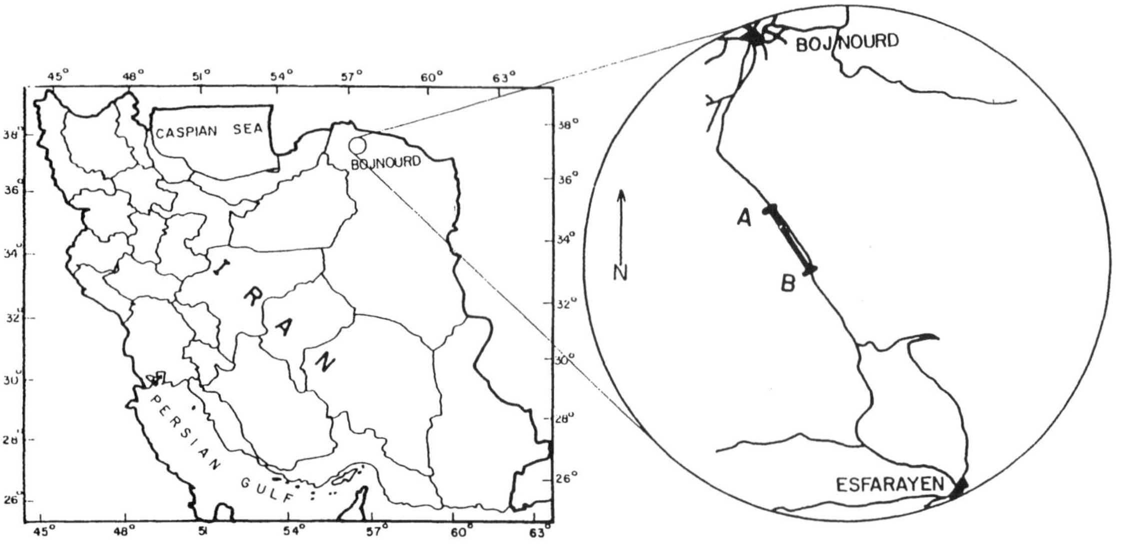

Text-Figure 1. Location map of study area.

palynomorphs are illustrated in Plates I-VI. Five palynomorph assemblages zones have been established and are discussed below in ascending stratigraphical order.

Miospore and Acritarch Assemblage Zone I

This assemblage zone occurs in a 310.0 m interval of the Padeha Formation and basal part of the Bahram Formation (Text-Figure 2). This assemblage is marked by presence of miospore taxa including *Retusotriletes distinctus, Retusotriletes rotundus, Emphanisporites rotatus, Archaeoperisaccus scabratus, Verrucosisporites scurrus, Samarisporites triangulatus, Grandispora fibrilata, Stenozonotriletes formosus, Cyclogranisporites isostictus, Retusotriletes pychovii, Lophozonotriletes somphus* and *Dibolisporites turriculatus.* Acritarchs occurring in this zone include *Stellinium micropolygonale, Stellinium comptum, Deltotosoma intonsum, Helosphaeridium microclavatum, Polyedryxium pharaonae, Chomotriletes vedugensis, Lophosphaeridium deminutum, Papulogabata annulata, Navifusa exilis, Somphophragma miscellum* and *Dictyotidium granulatum.* The miospore and acritarch species of this assemblage were compared to published information from the Devonian of Australia (Balme, 1960, 1988; Playford, 1981; Playford and Dring, 1981), Canada (Deunff, 1954; Owens, 1971; McGregor, 1961; McGregor and Camfield, 1982), the United States (Clendening et al., 1980; Wicander and Wood, 1981; Wicander, 1974; Wicander and Loeblich, Jr., 1977; Wicander and Playford, 1985), Argentina (Ottone, 1996), Russia (Naumova, 1953; Avchimovitch et al., 1993), France (Loboziak and Streel, 1980,1981; Rauscher, 1969), Belgium (Streel, 1967, 1986), Germany (Hartkopf-Fröder and Streel, 1994), Libya (Paris et al., 1985), Algeria (Jardiné et al., 1972; Moreau-Benoit et al., 1993) and Iran (Kimyai, 1972, 1979; Coquél et al., 1977; Ghavidel-syooki, 1988, 1994b, 1995). The presence of the miospores and acritarchs *Archaeoperisaccus*

Text-Figure 2 (opposite page). Stratigraphic distribution of acritarch and miospore taxa in the Padeha and Bahram formations in Chah-e-Riseh area, northeastern Esfahan city, central Iran (numbers correspond to the columns in Text-Figure 2). 1–*Retusotriletes distinctus* Richardson 1965; 2–*Retusotriletes rotundus* (Streel) Streel 1967; 3–*Emphanisporites rotatus* McGregor 1961; 4–*Archaeoperisaccus scabratus* Owens 1972; 5– *Verrucosisporites scurrus* McGregor & Camfield 1982; 6–*Samarisporites triangulatus* Allen 1965; 7–*Grandispora fibrilata* Balme 1988; 8–*Stenozonotriletes formosus* Naumova 1953; 9–*Cyclogranisporites isostictus* Balme 1988; 10–*Retusotriletes pychovii* Naumova 1953; 11–*Lophozonotriletes somphus* Balme 1988; 12–*Dibolisporites turriculatus* Balme 1988; 13–*Geminospora lemurata* Balme 1962; 14– *Verrucosisporites confertus* Owens 1971; 15–*Contagnisporites optivus* (Chibrikova) Owens 1971; 16–*Retusotriletes phillipsii* Clendening, Eames & Wood 1980; 17–*Dictyotriletes sphaericus* Kimyai 1979; 18–*Hymenozonotriletes perplexa* Balme & Hassel 1962; 19–*Diducites macronatus* (Kedo) emend. Van Veen 1981; 20–*Grandispora echinata* Hacquebard 1957; 21–*Grandispora famenensis* (Naumova) Streel 1986; 22– *Vallatisporites pusillites* (Kedo) Dolby & Neves 1970; 23–*Retispora lepidophyta* (Kedo) Playford 1976; 24–*Stellinium micropolygonale* (Stockmans & Williére) Playford 1977; 25–*Stellinium comptum* Wicander & Loeblich, Jr., 1977; 26–*Deltotosoma intonsum* Playford 1981; 27–*Helosphaeridium microclavatum* Playford 1981; 28–*Polyedryxium pharaonae* Deunff 1954; 29–*Chomotriletes vedugensis* Naumova 1953; 30–*Lophosphaeridium deminutum* Playford 1981; 31–*Papulogabata annulata* Playford 1981; 32–*Navifusa exilis* Playford 1981; 33–*Somphophragma miscellum* Playford 1981; 34–*Dictyotidium granulatum* Playford 1981; 35–*Unellium piriforme* Rauscher 1969; 36–*Maranhites perplexus* Wicander & Playford 1985; 37–*Multiplicisphaeridium amitum*; 38–*Ammonidium loriferum* (Deunff, 1965) Lister 1970; 39– *Veryhachium pannuceum* Wicander & Loeblich, Jr., 1974; 40–*Cymatiosphaera adaiochorata* Wicander 1974; 41–*Gorgonisphaeridium ohioense* (Winslow) Wicander 1974; 42–*Melikeriopalla venulosa* (Staplin) Jardiné et al. 1972; 43–*Crassiangulina tesselita* Jardiné et al. 1972; 44–*Hologinella quadrispina* Jardiné et al., 1972.

scabratus, Dibolisporites turriculatus, Cyclogranisporites isostictus, Lophozonotriletes somphus, Grandispora fibrilata, Chomotriletes vedugensis, Deltotosoma intonsum, Navifusa exilis, Helosphaeridium microclavatum, Papulogobata annulata, Somphophragma miscellum and *Dictyotidium granulatum* suggests an Early Frasnian age for this section in the Padeha Formation.

Miospore and Acritarch Assemblage Zone II

This assemblage comprises 122.0 m of the Bahram Formation (Text-Figure 2). This zone characterized by appearance of *Geminospora lemurata, Verrucosisporites confertus, Contagnisporites optivus, Maranhites perplexus, Multiplicisphaeridium amitum, Ammonidium loriferum* and

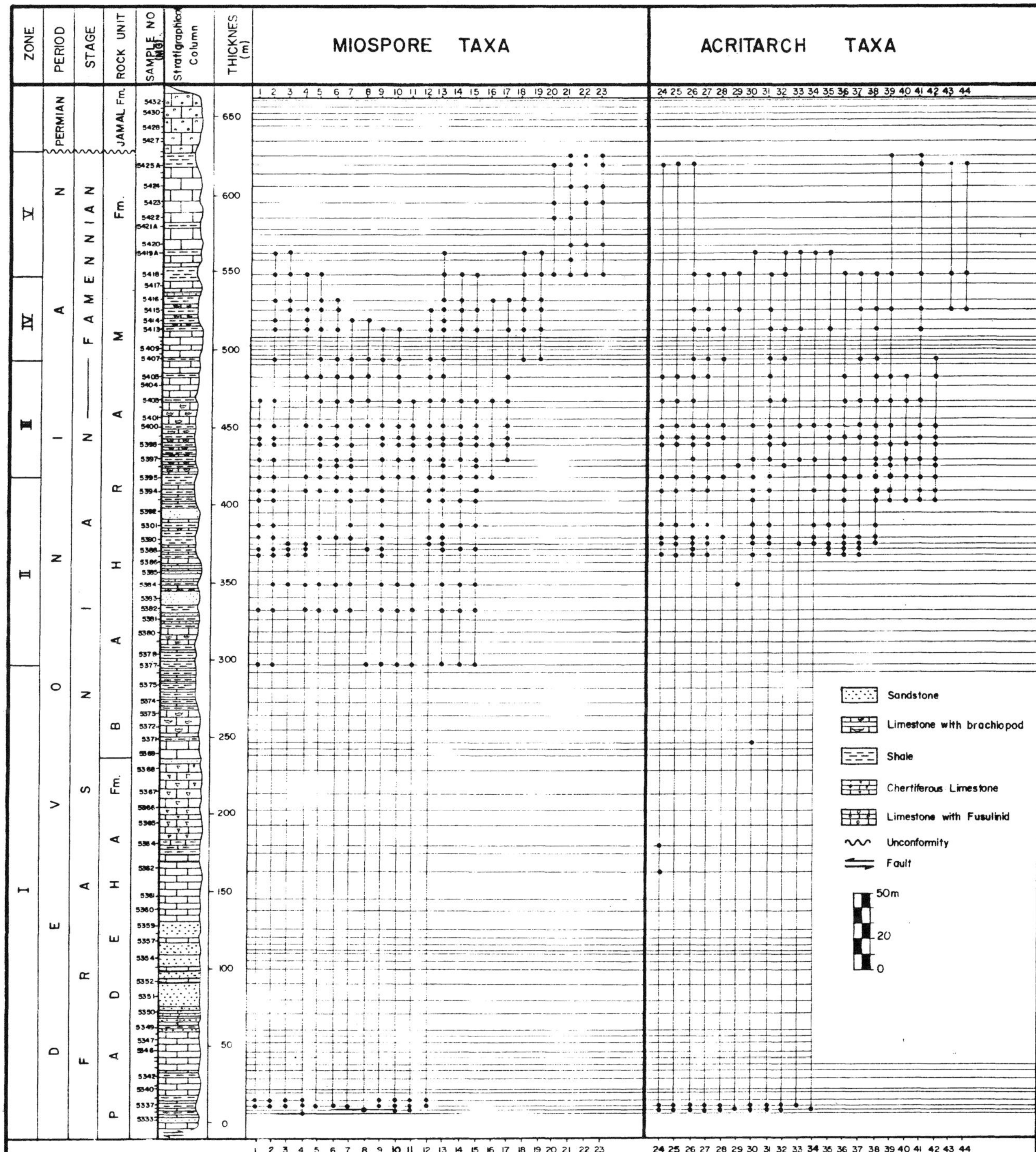

Text_Figure. 2. Stratigraphic distribution of miospore and acritarch taxa in the Padeha and Bahram formations in Chah_e_Riseh area, northeastern Esfahan city, Central Iran.

Unellium piriforme. The palynomorphs of this assemblage were compared to pertinent Late Devonian occurrences in the United States (Wicander and Loeblich, Jr., 1977; Wicander and Playford, 1985), Canada (Owens, 1971), Russia (Naumova, 1953; Avchimovitch et al., 1993), France (Loboziak and Streel, 1980, 1981), Libya (Paris et al., 1985), Algeria (Moreau-Benoit et al., 1993), Iran (Coquél et al., 1977; Kimyai, 1979; Ghavidel-syooki, 1988, 1994a, 1995). With the exception of the acritarch *Unellium piriforme* (Playford and Dring, 1981), a long-ranging form, the assemblage contains diagnostic Frasnian palynomorphs.

Miospore and Acritarch Assemblage Zone III

This assemblage spans 76.0 m of the Bahram Formation (Text-Figure 2). The zone is distinguished by first occurrences of *Retusotriletes phillipsii, Dictyotriletes sphaericus, Veryhachium pannuceum, Cymatiosphaera adaiochorata, Melikeriopalla venulosa* and *Gorgonisphaeridium ohioense.* This zone is considered to be Late Devonian (Late Frasnian) based on comparisons to the Late Devonian studies from The United States (Wicander and Loeblich, Jr., 1977; Wicander and Playford, 1985; Clendening et al., 1980), Australia (Balme, 1988; Playford, 1976, 1977, 1981), Algeria (Moreau-Benoit et al., 1993) and Iran (Coquél et al., 1977; Kimyai, 1979; Ghavidel-syooki, 1988, 1994a, 1994b, 1995).

Miospore and Acritarch Assemblage Zone IV

This assemblage is found in 54.0 m of the Bahram Formation (Text-Figure 2). This zone is characterized by the presence of *Hymenozonotriletes perplexa* and *Diducites macronatus.* These species have been recorded from Late Devonian strata in Russia (Naumova, 1953; Avchimovitch et al., 1993), Ireland (Van Veen, 1981), France (Loboziak and Streel, 1981), Belgium (Streel, 1986), Algeria (Moreau-Benoit et al., 1993), Libya (Paris et al., 1985) and Iran (Coquél et al., 1977; Kimyai, 1979; Ghavidel-syooki, 1988, 1994a, 1995). Comparison with these publications indicate this zone is Late Devonian (uppermost Frasnian–Early Famennian).

Miospore and Acritarch Assemblage Zone V

This zone appears in the upper 82.0 m of Bahram Formation. The most diagnostic species include *Grandispora famenensis, Grandispora echinata, Vallatisporites pusillites, Retispora lepidophyta, Horologinella quadrispina* and *Crassiangulina tesselita.* Some palynomorphs present in Zone IV extend into the basal part of this assemblage, however, there is a marked reduction in number of species (Text-Figure 2). Many of the acritarchs and miospores present in this zone have been recorded from the Famennian strata in Algeria (Moreau-Benoit et al., 1993), Libya (Paris et al., 1985), Russia (Naumova, 1953; Avchimovitch, et al., 1993), Belgium (Streel, 1986), Germany (Hartkopf-Fröder and Streel, 1994), France (Loboziak and Streel, 1981), Ireland (Van Veen, 1981), the United States (Wicander, 1974) and Iran (Coquel et al., 1977; Ghavidel-syooki, 1994a).

This comparison indicates that most of miospore taxa and some of acritarch species, such as *Chomotriletes vedugensis, Gorgonisphaeridium ohioense, Stellinium comptum, Stellinium micropolygonale, Ammonidium loriferum, Unellium piriforme, Maranhites perplexus, Cymatiosphaera adaiochorata, Multiplicisphaeridium amitum, Polyedryxium pharaonae* and *Veryhachium pannuceum* are also known from Europe and North America. However, many other acritarch species of the Padeha and Bahram formations including *Lophosphaeridium deminutum, Navifusa exilis, Deltotosoma intonsum, Papulogabata annulata, Dictyotidium granulatum, Melikeriopalla venulosa, Somphophragma miscellum, Helosphaeridium microclavatum, Horologinella quadrispina* and *Crassiangulina tesselita,* have been only recorded from the Late Devonian strata of Algeria (Moreau-Benoit et al., 1993), Australia (Playford and Dring, 1981; Playford, 1981), and southern and northern Iran (Ghavidel-syooki, 1994a, 1995). Based on these similarities, it is possible that the Iranian platform, Western Australia and Algeria were at a similar paleolatitude along the southern shore of the Tethys Ocean during the Late Devonian.

CONCLUSIONS

The Padeha and Bahram formations yield 44 diagnostic palynomorph taxa. The local stratigraphic distribution of each species permits division of the section into five assemblage zones. The Late Devonian (Early Frasnian) Zone I is present in a 310.0 m interval that includes the Padeha Formation and basal Bahram Formation. Zones II to V occur in the Bahram Formation and indicate Late Devonian (Middle Frasnian–Famennian). Comparison of these Late Devonian miospore and acritarch taxa with those from other parts of the world indicate broad similarity with those from western Australia, Algeria, Libya, and the Alborz and Zagros mountain ranges of Iran. This similarity suggests that the Iranian Platform, western Australia and north Africa have occupied the same tropical paleolatitude during the Late Devonian (Frasnian–Famennian).

Based on the palynological evidence, a hiatus is present between the Bahram Formation and Jamal Formation and extends from the Carboniferous through Early Permian. This hiatus possibly corresponds to the Hercynian Orogeny in Central Iranian Basin.

ACKNOWLEDGMENTS

The author expresses his sincere appreciation to the management of the National Iranian Oil Company for permission to publish this paper. I am also grateful to Dr. G.

PLATE 1

All figures x1000.

1, 2, *Crassiangulina tesselita* Jardiné et al. 1972.

3, 6

4 *Horologinella quadrispina* Jardiné et al. 1972.

5 *Deltotosoma intonsum* Playford 1981.

7, 8 *Chomotriletes vedugensis* Naumova 1953.

9 *Navifusa exilis* Playford 1981.

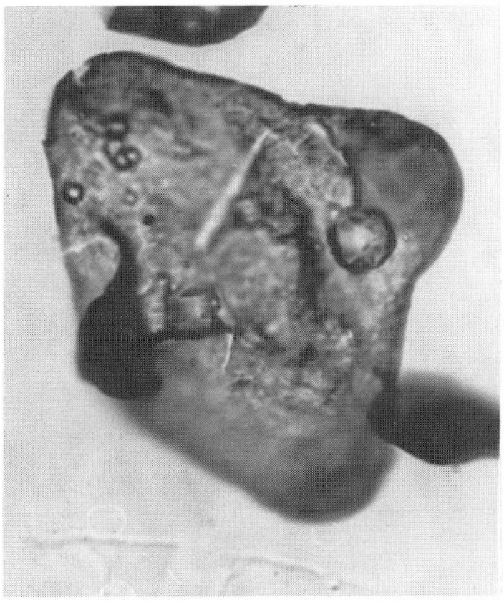
1

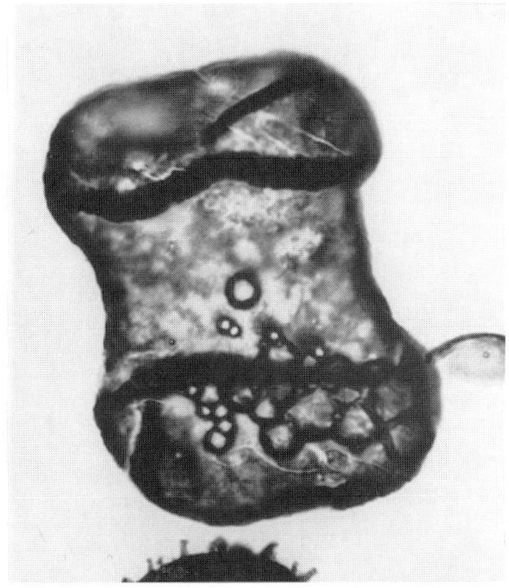
2

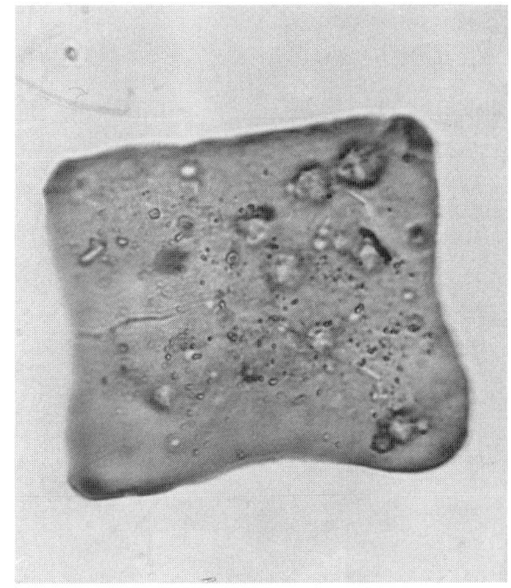
3

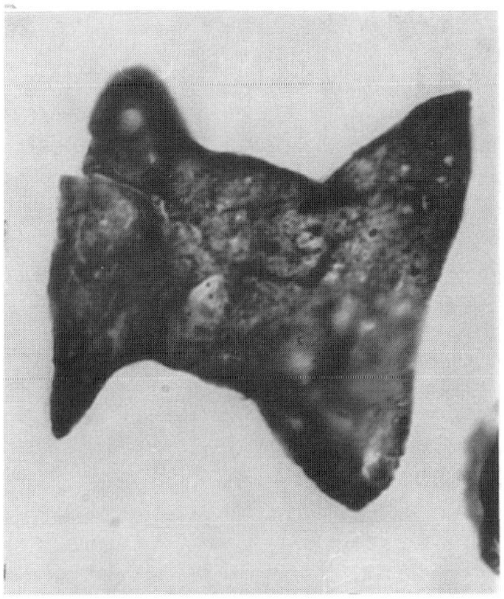
4

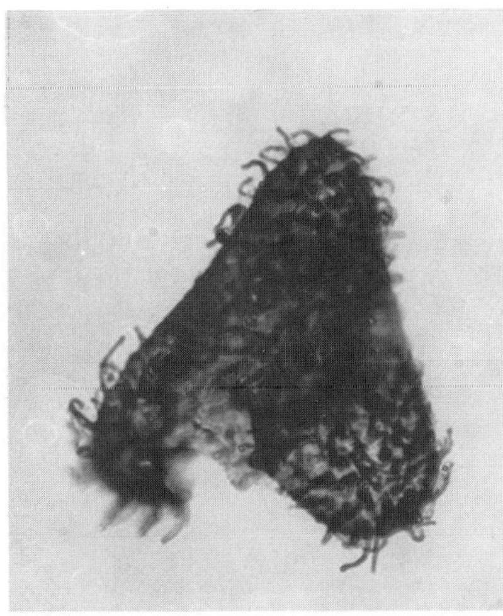
5

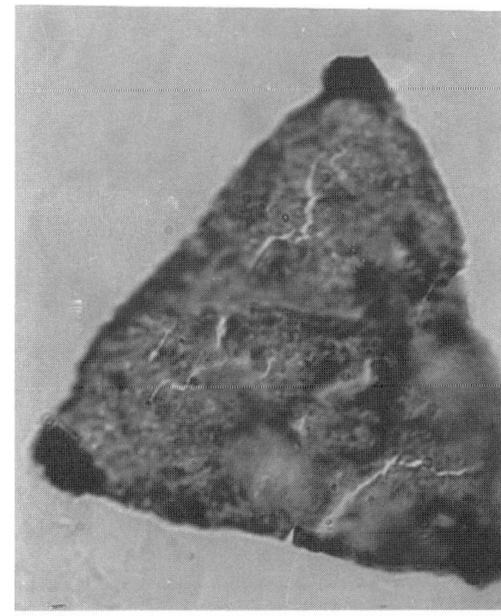
6

7

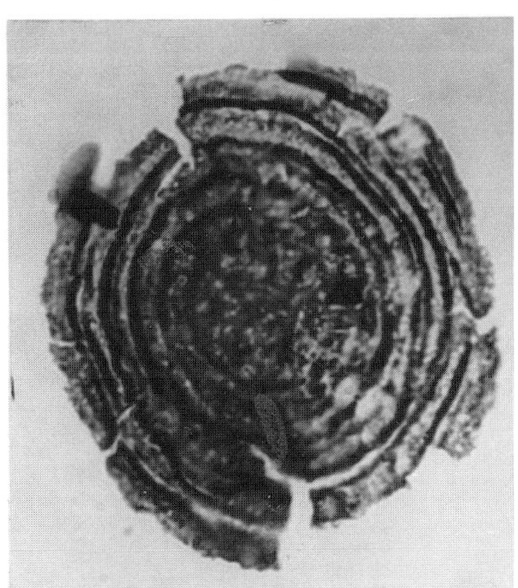
8

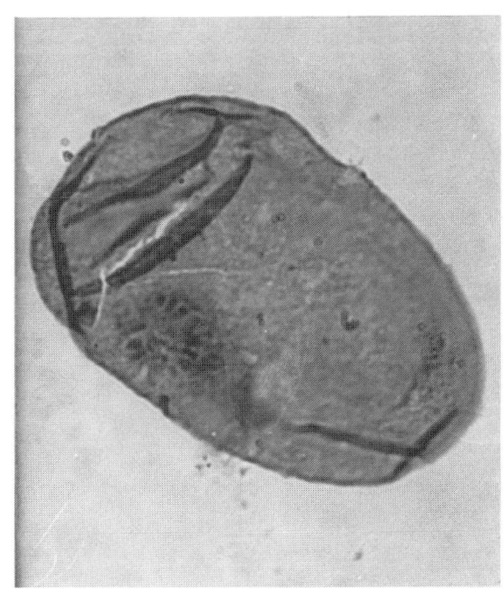
9

D. Wood and Professor R. Wicander for many helpful suggestions. Finally, I wish to acknowledge gratitude to my wife, Fatemeh, and my daughters Hedieh and Mona for many kindnesses.

References Cited

ALLEN, K.C.
1965 Lower and Middle Devonian spores of North and Central Vestspitsbergen. *Palaeontology*, 8: 687–748.

AVCHIMOVITCH, V.I., TCHIBRIKOVA, E.V., OBUKHOVSKAYA, T.G., NAZARENKO, A.M., UMNOVA, V.T., RASKATOVA, L.G., MANTSUROVA, V.N., LOBOZIAK, S., and STREEL, M.
1993 Middle and Upper Devonian miospore zonation of eastern Europe. *Bulletin des Centres de Recherches Exploration-Production Elf Aquitaine*, 17: 78–147.

BALME, B.E.
1962 Upper Devonian (Frasnian) spores from the Carnarvon Basin, western Australia. *The Palaeobotanist*, 9: 1–10.
1988 Miospores from Late Devonian (Early Frasnian) strata, Carnarvon Basin. *Palaeontographica*, Abteilung B, 209: 109–166.

CLENDENING, J.A., EAMES, L.E., and WOOD, G.D.
1980 *Retusotriletes phillipsii* nov. sp., a potential Upper Devonian guide palynomorph. *Palynology*, 14: 15–22.

COQUÉL, R., LOBOZIAK, S., STAMPFLI, G., and STAMPFLIVULLIE, B.
1977 Palynologie du Dévonien Supérieur et du Carbonifère inférieur dans l'Elburz oriental (Iran Nord-Est). *Revue de Micropaléontologie*, 20: 59–71.

DEUNFF, J.
1954 Sur un microplankton du Dévonien du Canada récélant des types nouveaux d'Hystrichosphaeridés. *Compte Rendu de l'Académie des Sciences de Paris*, 239: 1064–1066.

DJAFARIAN, M.A.
1977 Elements nouveaux concernont la geologie et les Models biostratigraphique de la region d' Esfahan. *Cahiers Geologique*, 93: 182–189.

DJAFARIAN, M.A., and BRICE, D.
1973 Biostratigraphie des brachiopodes dans le Famennien Supérieur de la region d' Esfahan (Central Iran). *Compte Rendu de l'Académie des Sciences de Paris*, 276: 2125–2128.

DOLBY, G., and NEVES, R.
1970 Palynological evidence concerning the Devonian–Carboniferous boundary in the Mendips, England. *Compte Rendu, Sixéme Congrés International de Stratigraphie et de Géologie du Carbonifére* (Sheffield, 1967), 2: 631–646.

GHAVIDEL-SYOOKI, M.
1988 *Palynostratigraphy and palaeogeography of the Faraghan Formation of southeastern Iran*. Unpubl. Ph.D. dissertation, Michigan State University, 279 pp.
1994a Upper Devonian acritarchs and miospores from the Geirud Formation in Central Alborz Range, northern Iran. *Iranian Journal of Science*, 5: 103–122.
1994b Biostratigraphy and palaeobiogeography of some Palaeozoic rocks at Zagros and Alborz Mountains. *Ministry of Mine and Metals, Geological Survey of Iran*, 18: 1–168.
1995 Palynobiostratigraphy and Palaeogeography of Palaeozoic sequence in the Hassanakdar area, Central Alborz Range, northern Iran. *Review of Palaeobotany and Palynology*, 89: 91–109.

HARTKOPF-FRÖDER, C., and STREEL, M.
1994 Late Famennian miospore assemblages from the Bergisch ladbach-Paffrath syncline, Rhenish Slate Mountains, Germany. *Annales de la Société géologique de Belgique*, 116: 333–357.

HIGGS, K.
1975 Upper Devonian and Lower Carboniferous miospore assemblage from Hook Head, County Wexford, Ireland. *Micropaleontology*, 21: 393–419.

HIGGS, K., CLAYTON, G., and KEEGAN, J.B.
1988 Stratigraphic and systematic palynology of the Tournasian rocks of Ireland. *Geological Survey of Ireland, Special Paper* 7: 1–93.

JARDINÉ, S., COMBAZ, A., MAGLOIRE, L., PENIGUEL, G., and VACHEY, G.
1972 Acritarchs du Silurien terminal, et du Dévonien du Sahara algérien. *Compte Rendu, Septième Congrés International de Stratigraphie et de Géologie du Carbonifère* (Krefeld, 1971), 1: 295–311.

KIMYAI, A.
1972 Devonian plant microfossils from the Central Elburz, Iran. *Pollen et Spores*, 14: 186–201.
1979 Devonian spores from the Hassanakdar area, Iran. *Pollen et Spores*, 21: 481–498.

LOBOZIAK, S., and STREEL, M.
1980 Miospores in Givetian to Lower Frasnian sediments dated by conodonts from Boulonnais, France. *Review of Palaeobotany and Palynology*, 29: 288–299.
1981 Miospores in Middle–Upper Frasnian to Famennian (Boulonnais, France). *Review of Palaeobotany and Palynology*, 58: 173–196.

McGREGOR, D.C.
1961 Spores with proximal radial pattern from the Devonian of Canada. *Geological Survey of Canada, Bulletin*, 76: 1–11.

McGREGOR, D.C., and CAMFIELD, M.
1982 Middle Devonian miospores from the Cape de Bray Weatherall and Hecla Bay formations of northeastern Melville Island, Canadian Arctic. *Geological Survey of Canada, Bulletin*, 384: 1–101.

MOREAU-BENOIT, A., COQUEL, R., and LATRECHE, S.E.
1993 Étude palynologique de Dévonien du bassin d'Illizi (Sahara oriental algérien). Approache biostratigraphy. *Géobios*, 26: 1–33.

NAUMOVA, S.N.
1953 Upper Devonian spore-pollen assemblages of the Russian Platform and their stratigraphic significance. *Trudy Instituta Geologischeskikh Nauk, Academiia Nauk SSSR*, 143, 204 pp.

PLATE 2

All figures x1000.

1, 4 *Papulogabata annulata* Playford 1981.
2 *Somphophragma miscellum* Playford 1981.
3 *Dictyotidium granulatum* Playford 1981.
5 *Helosphaeridium microclavatum* Playford 1981.
6 *Melikeriopalla venulosa* Playford 1981.
7 *Maranhites perplexus* Wicander & Playford 1985.
8 *Unellium piriforme* Rauscher 1969.
9 *Cymatiosphaera adaiochorata* Wicander 1974.

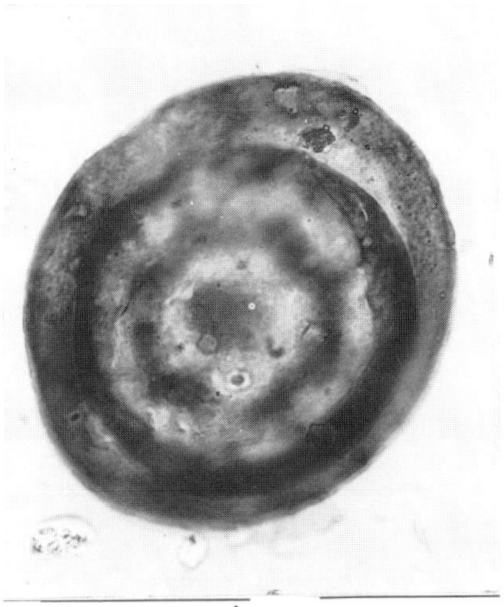
1

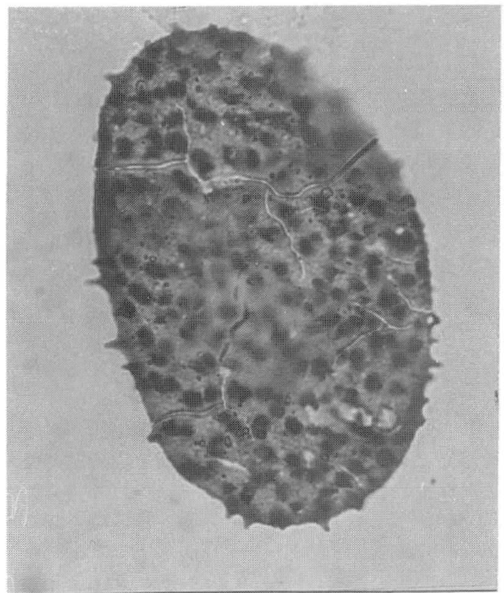
2

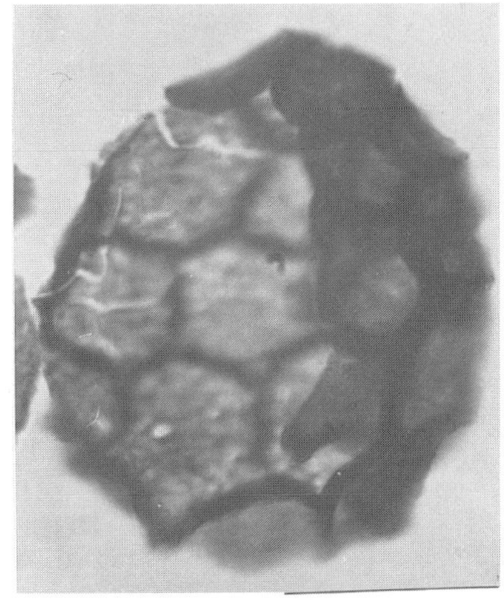
3

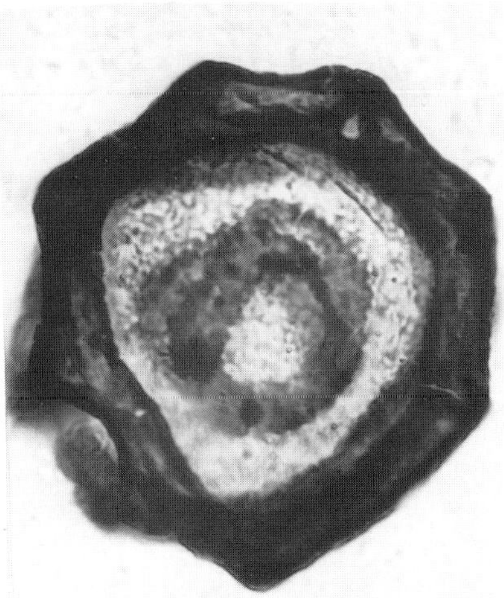
4

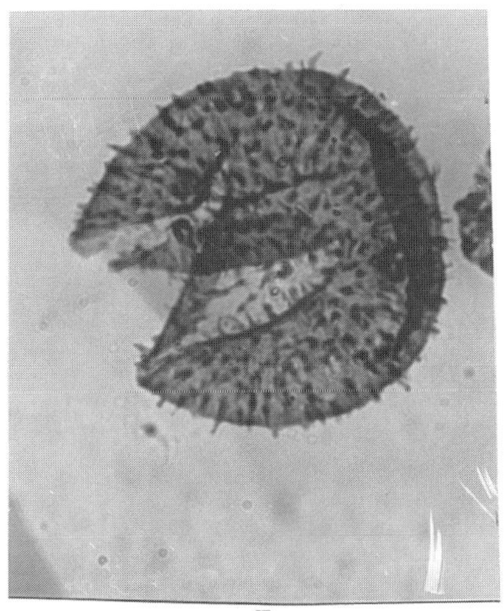
5

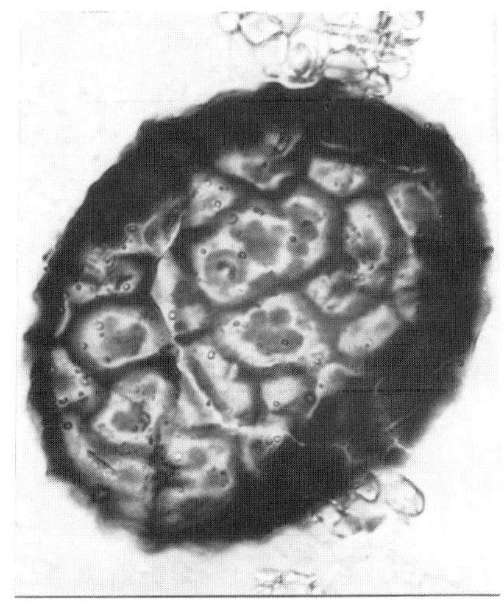
6

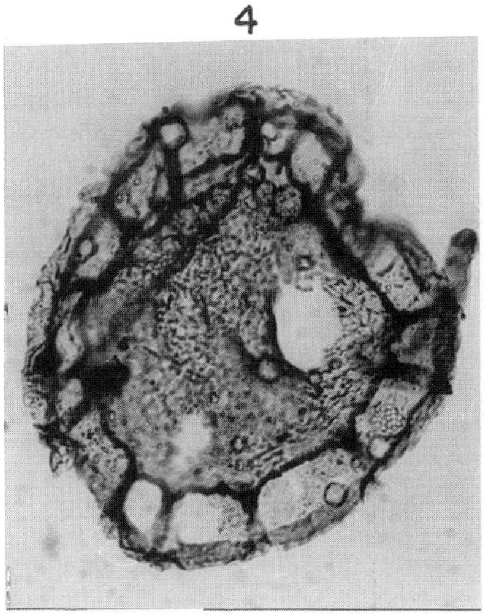
7

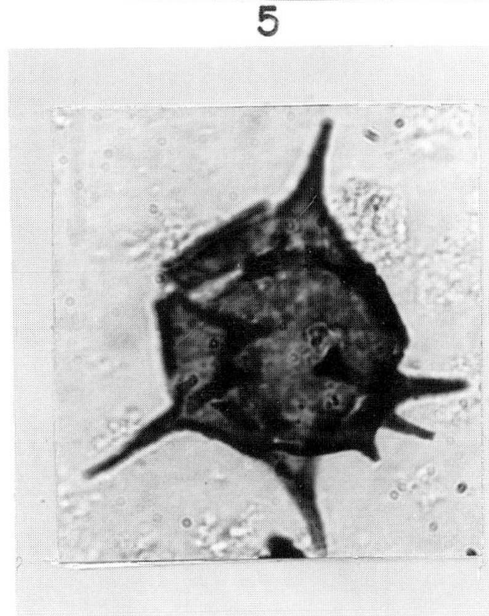
8

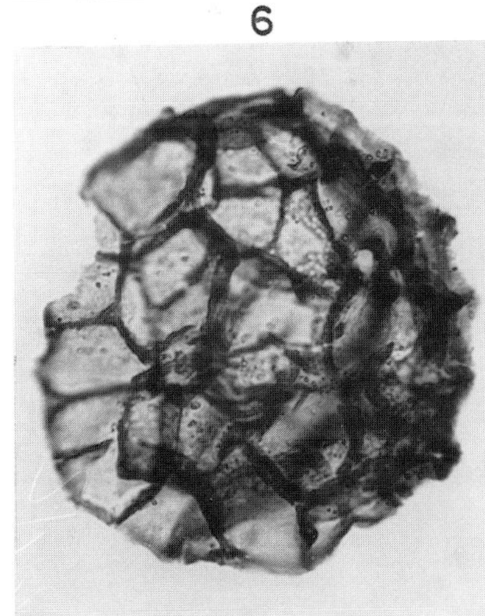
9

OTTONE, E.G.
1996 Devonian palynomorphs from the Los Monos Formation Tarija Basin, Argentina. *Palynology*, 20: 105–155.

OWENS, B.
1971 Miospores from the Middle and early Upper Devonian rocks of the western Queen Elizabeth Islands, Arctic Archipelago. *Geological Survey of Canada, Paper* 70–38: 1–157.

PARIS, F., RICHARDSON, J.B., RIEGEL, W., STREEL, M., and VANGUESTAINE, M.
1985 Devonian (Emsian–Famennian) palynomorphs. *In*: Thusu, B., and Owens, B. (eds.), Palynostratigraphy of northeast Libya. *Journal of Micropalaeontology*, 4: 49–82.

PLAYFORD, G.
1976 Plant microfossils from Upper Devonian and Lower Carboniferous of the Canning Basin, Western Australia. *Palaeontographica*, Abteilung B, 158: 1–171.
1977 Lower to Middle Devonian acritarchs of the Moose River Basin, Ontario. *Geological Survey of Canada, Bulletin*, 279: 1–87.
1981 Late Devonian acritarchs from the Gneudna Formation in the western Carnarvon Basin, Western Australia. *Géobios*, 14: 145–171.

PLAYFORD, G., and DRING, R.S.
1981 Late Devonian acritarchs from the Carnarvon Basin, Western Australia.*Palaeontological. Association. London, Special Paper in Palaeontology*, 27: 1–78.

RAUSCHER, R.
1969 Presence d'une forme nouvelle d' Acritarches dans le' Devonien de Normandie. *Compte Rendu de l'Academie des Sciences de Paris, Series D*, 268: 34–36.

RICHARDSON, J.B.,
1965 Middle Old Red sandstone spore assemblages from the Orcadian Basin north-east Scotland. *Palaeontology*, 7: 559–605.

STREEL, M.
1967 Association de spores du Devonien Inférieur Belge et, leur signification stratigraphique. *Annales de la Société géologique de Belgique*, 90: 11–53.
1986 Miospore distribution to the Upper Famennian–Strunian event stratigraphy. *Annales de la Société Géologique du Nord*, 105: 85–95.

VAN VEEN, P.M.
1981 Aspects of Late Devonian and Early Carboniferous palynology of southern Ireland. V. The change in composition of palynological assemblages at the Devonian Carboniferous boundary. *Review of Palaeobotany and Palynology*, 34: 67–97.

WICANDER, E.R.
1974 Upper Devonian–Lower Mississipian acritarchs and prasinophycean algae from Ohio, U.S.A. *Palaeontographica*, Abteilung B, 148 : 9–43.

WICANDER, E.R. and LOEBLICH, Jr., A.R.
1977 Organic-walled microplankton and stratigraphic significance from the Upper Devonian Antrim Shale, Indiana, U.S.A. *Palaeontographica*, Abteilung, 160: 129–165.

WICANDER, E.R., and WOOD, G.D.
1981 Systematics and biostratigraphy of the organic-walled microphytoplankton from the Middle Devonian (Givetian) Silica Formation Ohio, U.S.A. *American Association of Stratigraphic Palynologists, Contribution Series*, 8: 1–136.

WICANDER, E.R., and PLAYFORD, G.
1985 Acritarchs and spores from the Upper Devonian Lime Creek Formation Iowa, U.S.A. *Micropaleontology*, 31: 97–138.

Author's address

MOHAMMAD GHAVIDEL-SYOOKI
Exploration Division
National Iranian Oil Company
P.O. Box 1065
Tehran
Iran

PLATE 3

All figures x1000.

1 *Ammonidium loriferum* (Deunff 1965) Lister 1970.
2 *Lophosphaeridium deminutum* Playford 1981.
3 *Stellinium micropolygonale* (Stockmans & Williéré 1960) Playford 1977.
4 *Gorgonisphaeridium ohioense* Winslow 1962.
5 *Unellium piriforme* Rauscher 1969.
6 *Polyedryxium pharaonae* Deunff 1954.
7 *Stellinium comptum* Wicander & Loeblich, Jr. 1977.
8 *Veryhachium pannuceum* Wicander & Loeblich, Jr. 1977.
9 *Multiplicisphaeridium amitum* Wicander & Loeblich, Jr. 1977.

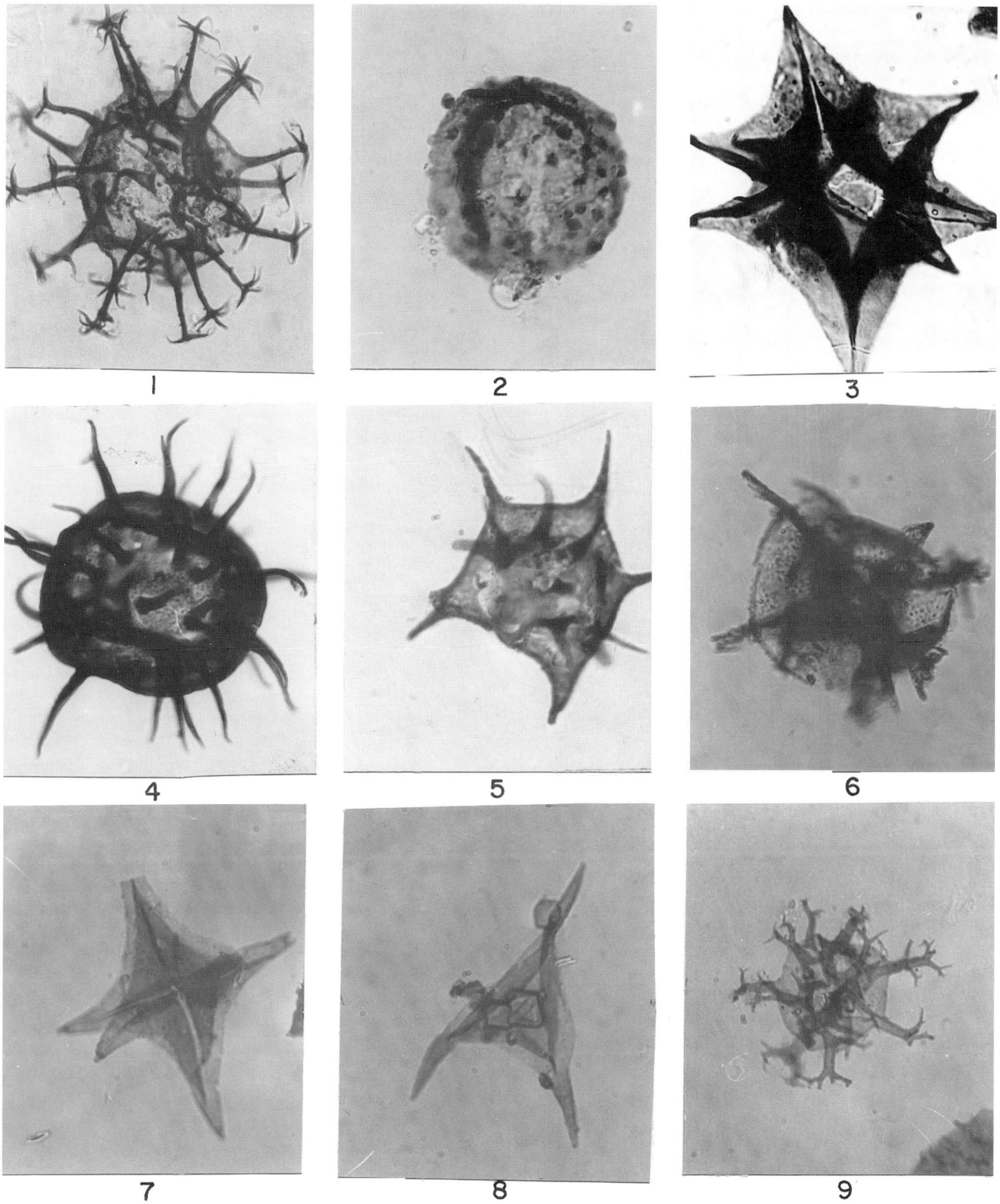
1
2
3
4
5
6
7
8
9

PLATE 4

All figures x1000.

1 *Retusotriletes distinctus* Richardson 1965.

2 *Geminospora lemurata* Balme 1962.

3 *Cyclogranisporites isostictus* Balme 1988.

4, 7 *Retusotriletes phillipsii* Clendening et al. 1980.

5 *Retusotriletes rotundus* Streel 1967.

6 *Dibolisporites turriculatus* Balme 1988.

8 *Geminospora lemurata* Balme 1962.

9 *Retusotriletes pychovii* Naumova 1953.

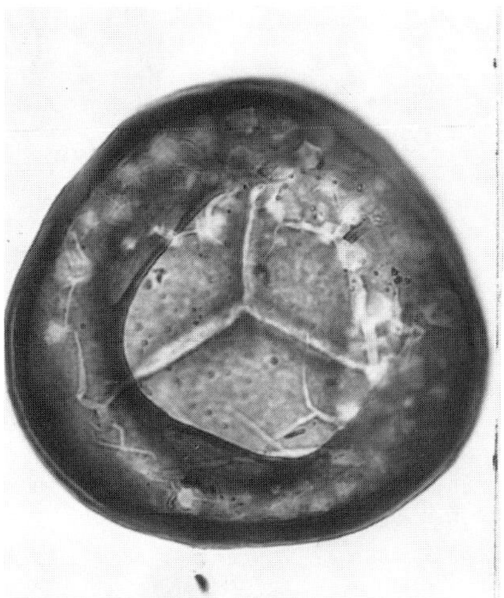
1

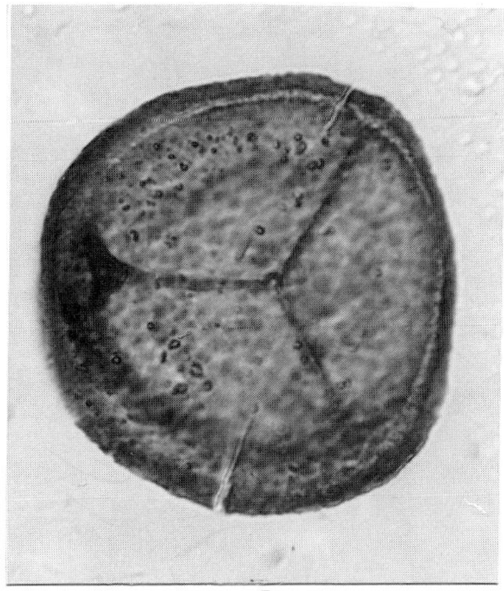
2

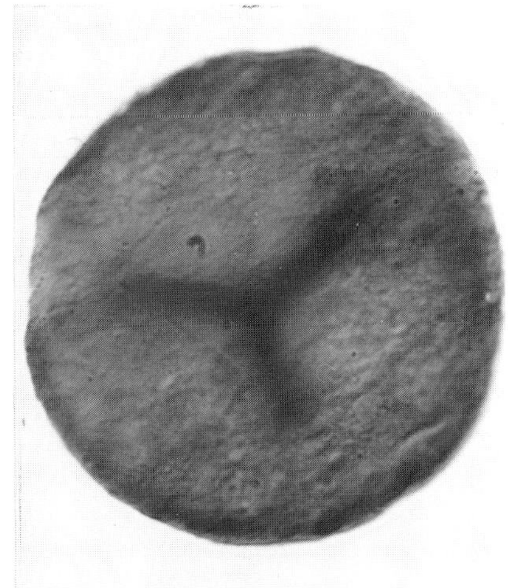
3

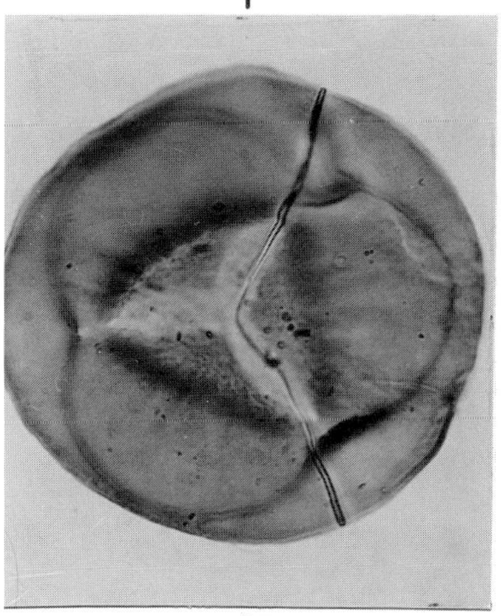
4

5

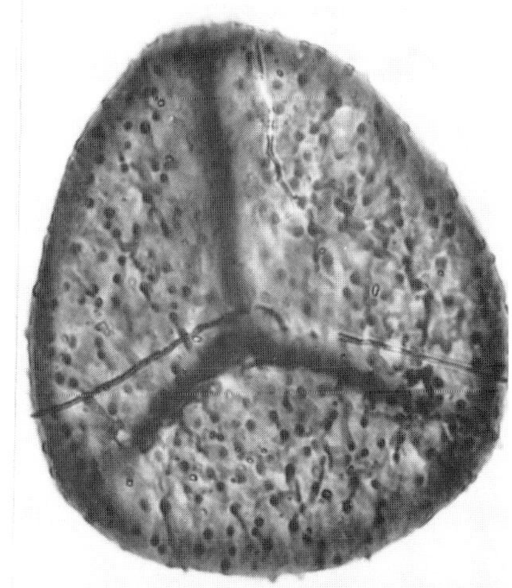
6

7

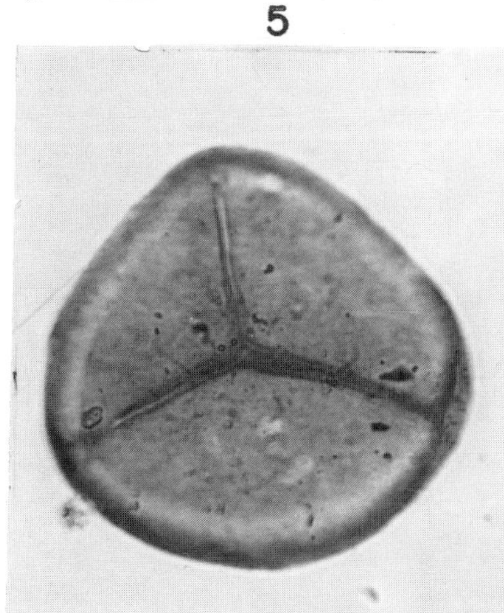
8

9

PLATE 5

All figures x1000.

1 *Lophozonotriletes somphus* Balme 1988.
2 *Grandispora famenensis* (Naumova 1953) Streel 1986.
3 *errucosisporites confertus* Owens 1971.
4 *Dictyotriletes sphaericus* Kimyai 1979.
5, 6 *Verrucosisporites scurrus* (Naumova 1953) McGregor & Camfield 1982.
7 *Stenozonotriletes formosus* Naumova 1953.
8 *Emphanisporites rotatus* McGregor 1961.
9 *Grandispora fibrilata* Balme 1988.

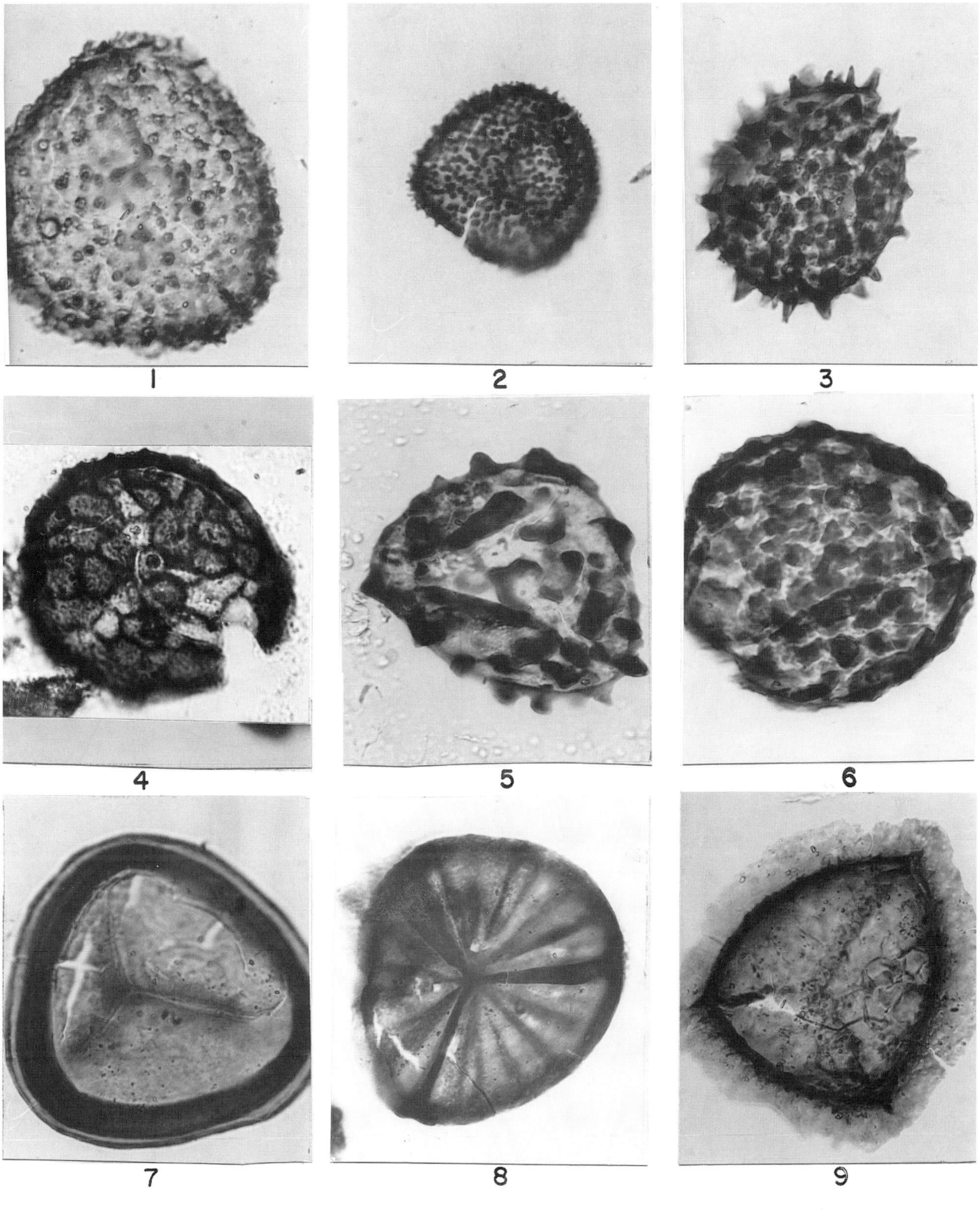
1
2
3
4
5
6
7
8
9

PLATE 6

All figures x1000.

1 *Grandispora echinata* Hacquebard 1957.

2 *Vallatisporites pusillites* (Kedo 1963) Dolby & Neves 1970.

3 *Hymenozonotriletes perplexa* Balme & Hassel 1962.

4 *Diducites macronatus* (Kedo 1963) emend. Van Veen 1981.

5, 7 *Archaeoperisaccus scabratus* Owens 1971.

6 *Retispora lepidophyta* (Kedo 1963) Playford 1976.

8 *Contagnisporites optivus* (Chibrikova) Owens 1971.

9 *Samarisporites triangulatus* Allen 1965.

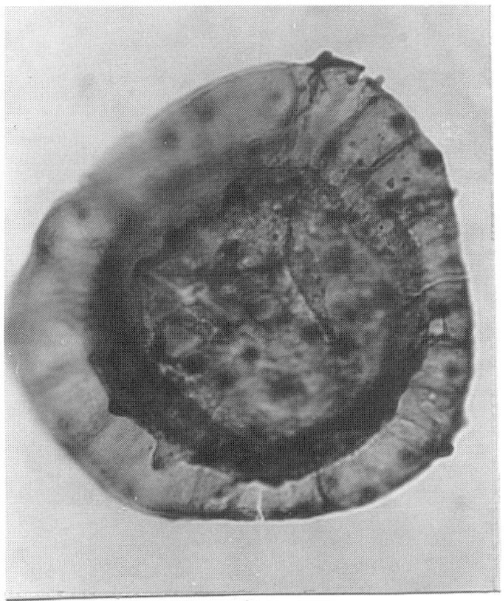

1

2

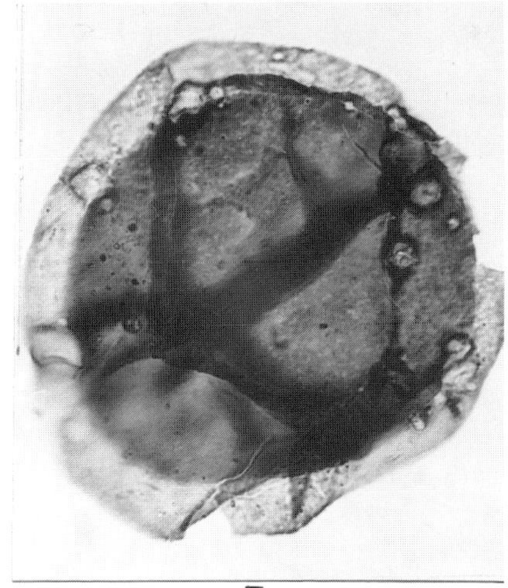

3

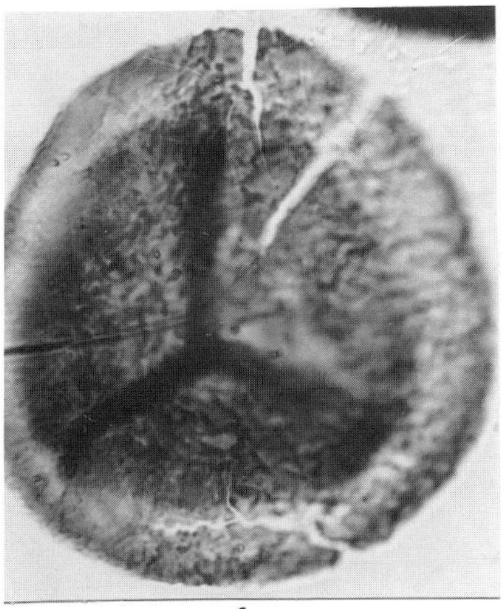

4

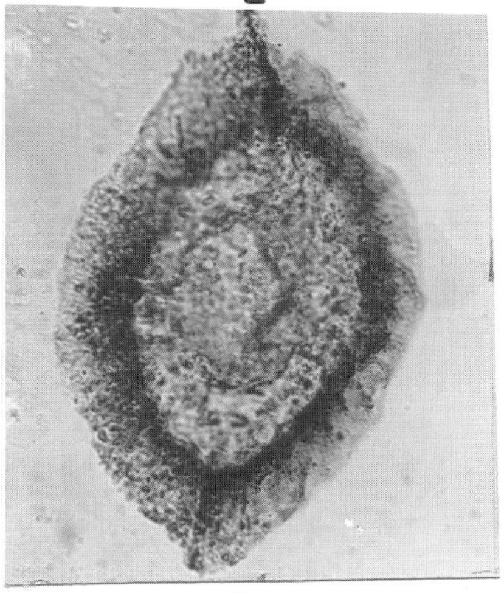

5

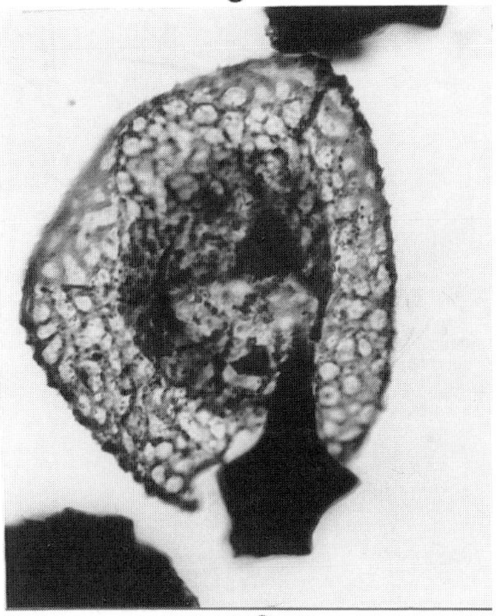

6

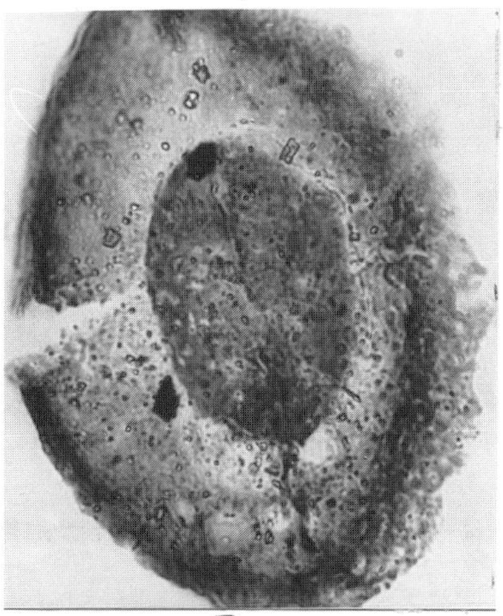

7

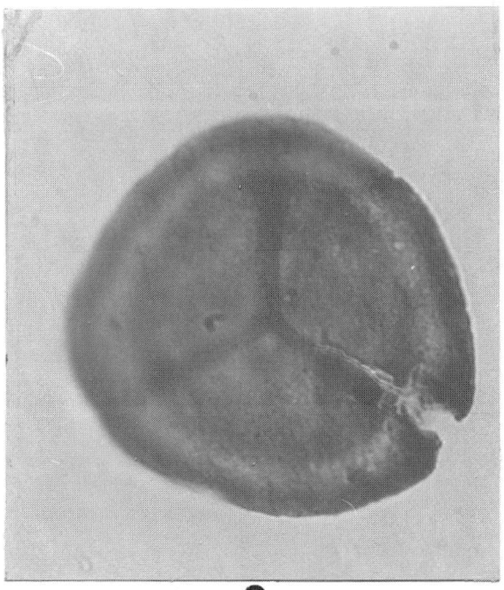

8

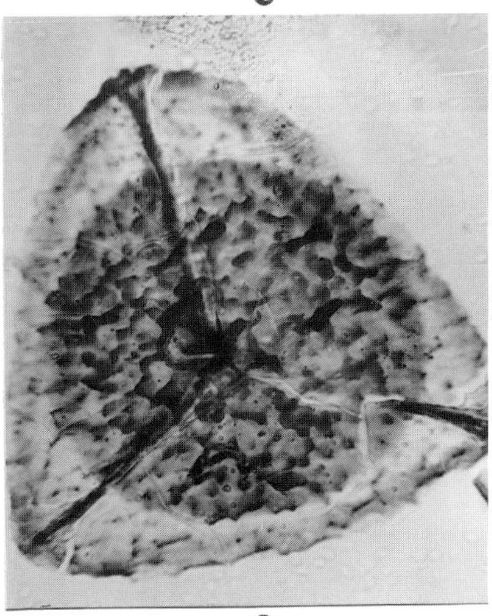

9

NEW DEVONIAN COENOBIAL CHLOROCOCCALES (HYDRODICTYACEAE) FROM THE HOLY CROSS MOUNTAINS AND RADOM–LUBLIN REGION OF POLAND: THEIR PALEOENVIRONMENTAL AND SEQUENCE STRATIGRAPHIC IMPLICATIONS

Gordon D. Wood and Elżbieta Turnau

Abstract

Chlorophyceaen algae have been recovered from Devonian core and outcrop samples of the Holy Cross Mountains (Góry Świętokrzyskie) and the Radom–Lublin region of Poland. *Musivum gradzinskii* gen. et sp. nov. and *Bijugum porebskii* gen. et sp. nov. are described and illustrated. They are assigned to the Order Chlorococcales, Family Hydrodictyaceae of the Chlorophyta based on their coenobial habit. Extant Hydrodictyaceae are found only in fresh to brackish water, however, the new species described here are present in samples whose depositional environments are interpreted as very nearshore (paralic-lagoonal) to offshore-shelf. Fluvial transport is believed to be the primary mechanism for dispersal of these algae into marine settings.

INTRODUCTION

Algae displaying a regular arrangement of cells (coenocytes) in a planar or spherical/concentric configuration (coenobium) are assignable to the Family Hydrodictyaceae, Order Chlorococcales, Class Chlorophyceae, Division Chlorophyta (Batten, 1996). Extant representatives of the Hydrodictyaceae, such as *Pediastrum*, have fossil records as old as the Cretaceous (see Batten and Lister, 1988a, 1988b; Brenner and Foster, 1994; Colbath and Grenfell, 1995; Evitt, 1963; Tappan, 1980; Wood and Aleman, 1997; Wood and Miller, 1997). Modern coenobial species live only in fresh to brackish water but may be transported and deposited into marine environments (Batten, 1996; Miller and Wood, 1996; Wood and Miller, 1996, 1997).

Reports of Pre-Carboniferous Hydrodictyaceae are rare in the palynological literature (Miller, 1988; Miller and Wood, 1996; Vavrdová, 1990; Wood and Miller, 1996, 1997). *Deflandrastrum* Combaz 1962 (p. 1977–1978; Fig. 1), *Ericantha* Cramer & Díez 1977 (p.346; Pl. 4, figs. 1-3), *Kahfia* Le Hérissé et al. 1995 (p. 72; Pl. 4, figs. 2-3) and *Petrovina* Oliveira & Burjack 1996 (p. 47; Pl. 1, figs. 1-6, Pl. 2, figs 1-2) have been reported from Devonian and older sediments. Possible coenobia, such as *Proteolobus* Wood 1997 (p. 327–328; Pls. 1-2), and others illustrated, but not discussed (see Ghavidel-Syooki 1995, pl. 4, fig. 14), are also evident. This paper describes and illustrates two new representatives of the Hydrodictyaceae from Devonian strata of the Holy Cross Mountains and Radom–Lublin region of Poland.

Wood, Gordon D., and Turnau, Elżbieta
2001 New Devonian Coenobial Chlorococcales (Hydrodictyaceae) from the Holy Cross Mountains and Radom–Lublin region of Poland: Their paleoenvironmental and sequence stratigraphic implications; *In*: Goodman, D.K., and Clarke, R.T. (eds.), *Proceedings of the IX International Palynological Congress, Houston, Texas, U.S.A., 1996*; American Association of Stratigraphic Palynologists Foundation, p. 53–63.
 ISBN 0–931871–06–9

GENERAL GEOLOGY

The Devonian epicontinental sediments of southern Poland were deposited in a shelf-basin ranging from 150 to 600 km in width, trending NW–SE, and stretching from the Ardenne–Rhenish regions to the Ukraine (Narkiewicz, 1988). To the south, the basin sloped into the Variscan geosyncline, and to the north and east it was bordered by the positive Belorussian High (East European Platform).

The Holy Cross Mountains are the area where Paleozoic rocks emerge at the surface (Text-Figure 1). The Devonian–Carboniferous in this region has been studied by many researchers but, in spite of this, the lithostratigraphy has not been formalized (Szulczewski, 1994, 1995). Differing tectonic histories allow the Holy Cross Mountains to be divided into the southern Kielce and a northern Łysogóry regions (Kłossowski, 1985; Racki, 1988, 1993).

The algae described here were recovered from two Devonian sites in the Łysogóry region. The area of study was subject to a slow transgressive phase immediately after the cessation of Lower Devonian clastic deposition (Old Red Sandstone facies). The Middle Devonian basinal sequence (the Skały, Pokrzywianka and Nieczulice beds) consists mainly of shales intercalated with marls and limestones. They usually contain abundant and diverse benthonic assemblages, but nectonic fossils (e.g., goniatites) may also be present. The Świętomarz beds are the exception, consisting primarily of fine grained clastics reflecting a global lowering of sea level.

The Radom–Lublin area is located north–east of the Holy Cross Mountains (Text-Figure 1), in proximity to the margin of the East European Platform, and both display similar depositional histories. The most comprehensive discussion concerning the Devonian in this area was published by Miłaczewski (1981). The uppermost Lower Devonian and Middle Devonian rocks yielding the algal coenobia described here were deposited in a shallow, near-shore marine setting dominated by clastic deposition. This interval was followed by an open marine, carbonate phase.

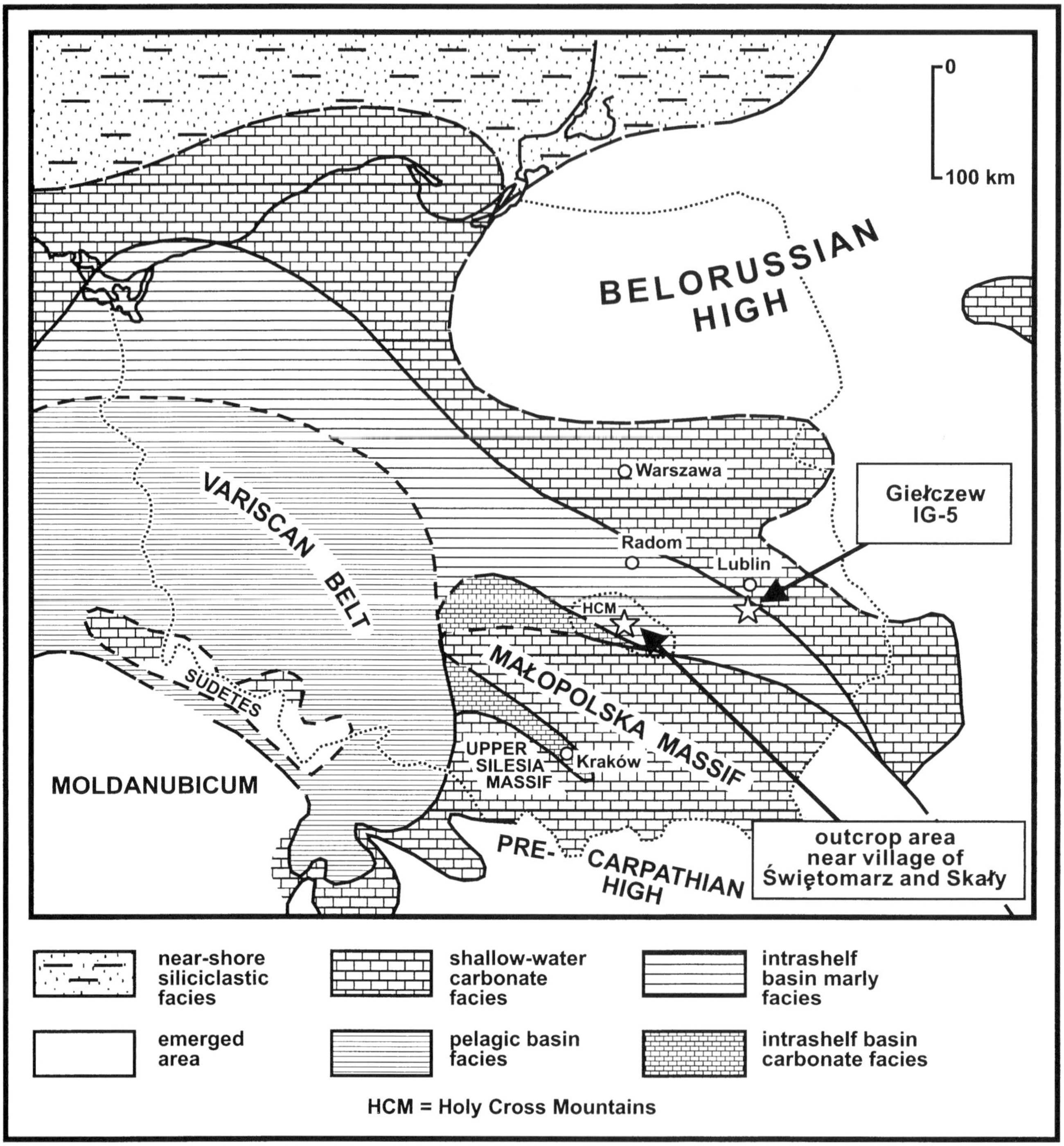

Text-Figure 1. Collecting localities plotted on a Givetian and Frasnian facies map (from Racki 1993; based on Narkiewicz (1988), Hladil (1988), Malec, et al. (1996), Stupnicka (1989), and Tsien (1988).

SAMPLE, MACERATION AND REPOSITORY INFORMATION

The palynomorphs described and illustrated in this paper were recovered from outcrop material from the Holy Cross Mountains and core from the Radom–Lublin area (Text-Figure 1). Their stratigraphical distribution is summarized in (Text-Figures 2 and 3). Algal coenobia were isolated from the shales using standard palynological techniques. This includes HCL–HF–HCL acidation sequence followed by heavy-liquid separation (Wood et al., 1996). Residues are embedded in Clearcol and cemented to a microscope slide with Elvacite for permanent palynological mounts. Type specimens are housed at the Orton Geo-

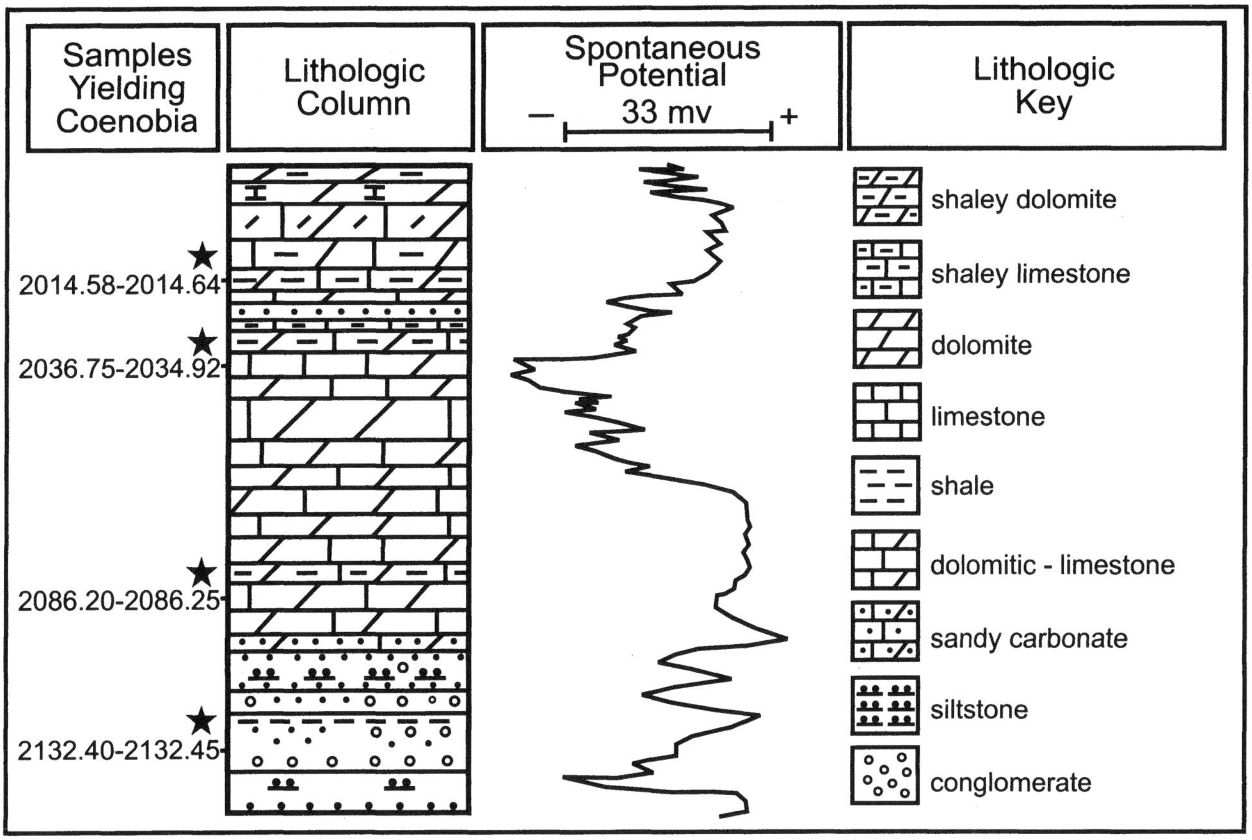

Text-Figure 2. Lithologies and gamma ray log from the P. I. G. Giełczew IG- 5 showing position of samples yielding coenobia (shallowest sample 2014.58-2014.64 m; deepest sample 2132.40-2132.45 m).

logical Museum, The Ohio State University, 155 South Oval Mall, Columbus, Ohio 43210. Each type specimen is assigned an Orton Geological Museum (OSU) repository number.

SUBSURFACE AND SURFACE SAMPLING SITES

Subsurface Samples

Coenobia were isolated between the 2014.58–2132.45 m interval of the Giełczew IG-5 borehole (Latitude 50°56'40"N; Longitude 20°40'41"E), located in the Radom–Lublin area (see Text Figures 1 and 2). The sampled intervals were assigned by Miłaczewski (personal communication, 1997) to the "Terrigenous-carbonate series" of the Telatyn Formation based on lithological comparison to the stratotype section penetrated by the Korczmin IG-1 borehole (Latitude 50°27'N; Longitude 23°56'E) approximately 70 km southeast of the Giełczew IG-5. The sample at 2132.40–2132.45 m (basal Telatyn Formation) can be assigned to the middle or early upper Emsian *foveolatus-dubia* miospore Zone of Streel et al. (1987). The sample at 2014.58–2014.64 m (uppermost Telatyn Formation) is assignable to the Givetian/Frasnian *triangulatus-concinna* miospore Zone of Streel et al. (1987). The miospore and the independent faunal data (Miłaczewski, personal information, 1977) indicate that the uppermost samples are not older than the Middle *varcus* conodont Subzone and not younger than the Givetian (see Wood, Turnau, et al., 1996).

Surface Samples

Outcrop samples which yield coenobia are from sections near the villages of Skały (Latitude 51°02'N, Longitude 21°04'E) and Świętomarz (Latitude 51°02'N, Longitude 21°05'E). The general location of these sections is shown in Text-Figure 1. The lithostratigraphic names in this region are largely informal. The Devonian succession is divided into "series," or "beds," some of which are subdivided into complexes (Pajchlowa, 1957). The material discussed here was collected from the Skały, Świętomarz and lower part of the Nieczulice beds (Text-Figure 3).

Skały Beds. Coenobial algae were recovered in the uppermost part of the Skały beds (complexes XXIV and XXV) near the village of Skały. The section consists of claystones, mudstones, and limestones, with abundant benthonic and nektonic fauna. This sequence yields a miospore assemblage assignable to the *triangulatus-ancyrea* Zone of Streel et al. (1987), and conodonts of the Lower *varcus* Subzone (Malec and Turnau, 1997).

Świętomarz Beds. Samples were collected from exposures of the so called Świętomarz-Śniadka profile, near the village Świętomarz and horizons transitional to the Nieczulice beds. Claystones, interbedded with fine grained quartz sandstones, were processed for palynomorphs. Fossils in these beds are rare, and consist

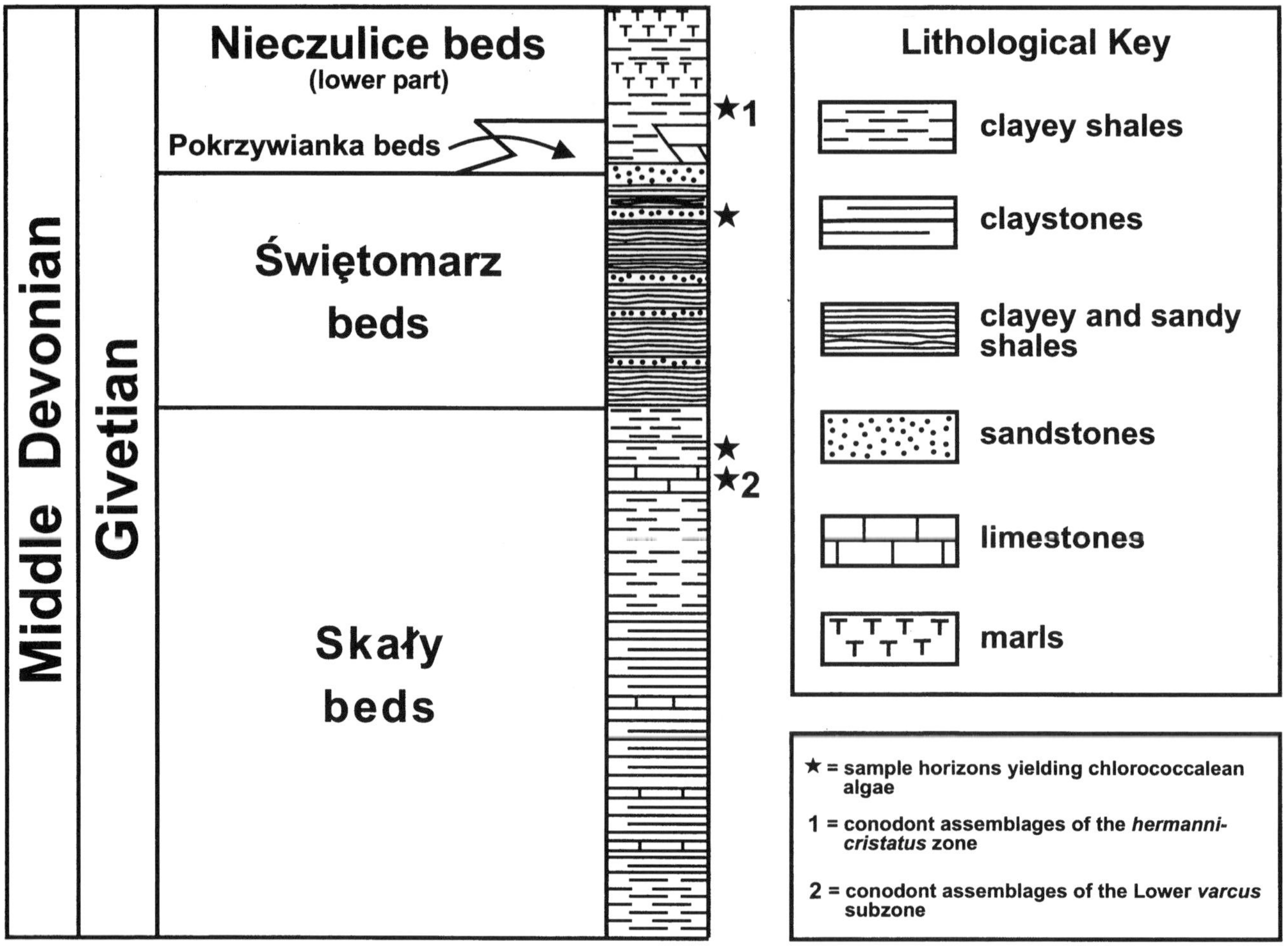

Text-Figure 3. Stratigraphical position of samples on a composite section as exposed near the villages of Świętomarz and Skały.

of brachiopods, crinoids, gastropods, lamellibranchs, tentaculitids and corals.

Nieczulice Beds. Samples yielding coenobia were collected from lower part of the Nieczulice beds in an exposure near the village Świętomarz and a trench at Pokrzywianka village, near Skały. The sequence near Świętomarz is clayey shale with a goniatite fauna and grades laterally into the Pokrzywianka beds, a organogenic and organodetrital limestone containing corals, ostracods, crinoids and brachiopods (Text-Figure 3). All samples from lower Nieczulice beds yielded miospore assemblages assignable to the *triangulatus-concinna* Zone of Streel et al. (1987). The goniatite fauna from the exposure at Świętomarz indicates the *Maenoceras terebratum* Zone (equivalent to a part of Middle

PLATE 1

Figures 1-16. *Musivum gradzinskii* gen. et sp. nov. All photomicrographs referenced with maceration number, slide number, Leitz Orthoplan coordinates, England Finder coordinates and Orton Museum specimen number (OSU). All figures magnified x690 unless otherwise indicated. Maceration number-subsurface/outcrop conversion information can be found in Appendix 1.

1 Paratype; 50430-A-10, 11.8/101.0, K12, OSU 49202.
2 Paratype; 50431-A-10, 33.9/101.0, K34/4, OSU 49203.
3 Paratype; 50431-A-10, 6.9/107.2, R7/2, OSU 49204.
4 Paratype; 50430-A-9, 15.9/100.9, K16, OSU 49205.
5 Paratype; 50431-A-10, 3.3/103.8, N4/1, OSU 49206.
6 Paratype; 41620-A-4, 51.8/94.0, C 52, OSU 49207.
7, 8 Paratype; 50431-A-9, 36.0/92.9, B36/2, OSU 49208. Fig. 7 is a phase contrast photomicrograph, x765.
9, 10 Holotype; 41620-A-4, 45.5/98.0, G46/3, OSU 49201. Fig. 9 is a phase contrast photomicrograph, x765.
11 Paratype; 41620-A-4, 52.2/107.0, R53/1, OSU49209.
12 Paratype; 50430-A-10, 39.8/97.8, F40, OSU49210.
13 Paratype; 50431-A-9, 17.1/103.9, N17/4, OSU 49211.
14 Paratype; 50430-A-10, 20.0/101.2, L10/2, OSU49213.
15 Paratype; 50430-A-11, 38.4/108.5, S39/1, OSU 49214.

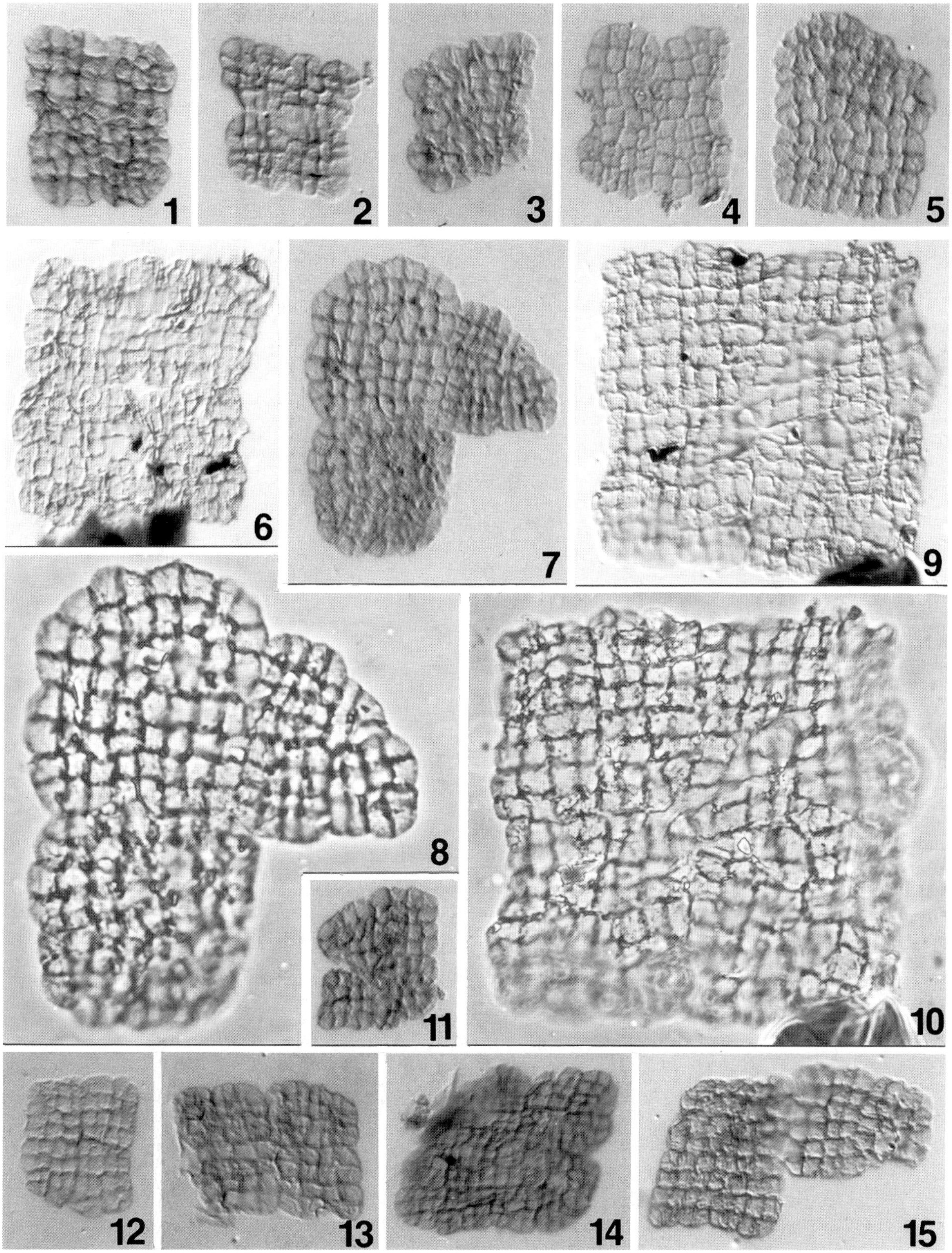
1
2
3
4
5
6
7
9
8
11
10
12
13
14
15

varcus Subzone). The higher part of the lower Nieczulice beds is dated as *hermanni-cristatus* Zone (Malec and Turnau, 1997; Turnau and Racki, 1999).

SYSTEMATIC PALEONTOLOGY

Division Chlorophyta Pasher 1914
Class Chlorophyceae Kützing 1843
Order Chlorococcales Marchard 1895
Family Hydrodictyaceae (Gray 1821) Dumoitier 1829

Genus *Musivum* gen. nov.

Type Species. *Musivum gradzinskii* gen. et sp. nov.

Derivation of Name. Latin, *Muses,* mosaic work, in reference to the overall morphologic pattern; from the "Muses," the gods presiding over the fine arts.

Diagnosis. Planar, equant, unilayered coenobial sheet with a straight to slightly undulatory outline. Individual coenocytes thin-walled, closely packed, square to slightly rectangular. The outer wall of the peripheral cells may be slightly bowed or rounded. Surface may be smooth or wrinkled. Perforations, ornament and dehiscence structure not observed.

Comparison. *Plaesiodictyon* Wille 1970 (p. 285, 287, 290–291; Figs. 9-10) has rectangular coenocytes, corner cells often exhibit elongations (horns), and perforations may be present.

Musivum gradzinskii gen. et sp. nov.
Plate 1, figs. 1-15

Derivation of Name. Named in honor of Prof. Richard Gradziński, Polish Academy of Sciences, Institute of Geological Sciences, Kraków, for his contributions to the study of the geology of Poland.

Holotype. Plate 1, figs. 9-10; slide 41620-A-4; Leitz Orthoplan coordinates 45.5/98.0; England Finder coordinates G46/3; OSU 49201.

Paratypes. Plate 1, fig. 1; slide 50430-A-10; Leitz Orthoplan coordinates 11.8/101.0; England Finder coordinates K12; OSU 49202.

Plate 1, fig. 2; slide 50431-A-10; Leitz Orthoplan coordinates 33.9/101.0; England Finder coordinates K34/4; OSU 49203.

Plate 1, fig. 3; slide 50431-A-10; Leitz Orthoplan coordinates 6.9/107.2; England Finder coordinates R7/2; OSU 49204.

Plate 1, fig. 4; slide 50430-A-9; Leitz Orthoplan coordinates 15.9/100.9; England Finder coordinates K16; OSU 49205.

Plate 1, fig. 5; slide 50431-A-10; Leitz Orthoplan coordinates 3.3/103.8; England Finder coordinates N4/1; OSU 49206.

Plate 1, fig. 6; slide 41620-A-4; Leitz Orthoplan coordinates 51.8/94.0; England Finder coordinates C52; OSU 49207.

Plate 1, figs. 7-8; slide 50431-A-9; Leitz Orthoplan coordinates 36.0/92.9; England Finder coordinates B36/2; OSU 49208.

Plate 1, fig. 11; slide 41620-A-4; Leitz Orthoplan coordinates 52.2/107.0; England Finder coordinates R53/1; OSU 49209.

Plate 1, fig. 12; slide 50430-A-10; Leitz Orthoplan coordinates 39.8/97.8; England Finder coordinates F40, OSU 49210.

Plate 1, fig. 13; slide 50431-A-9; Leitz Orthoplan coordinates 17.1/103.9; England Finder coordinates N17/4; OSU 49211.

Plate 1, fig. 14; slide 50430-A-10; Leitz Orthoplan coordinates 20.0/101.2; England Finder coordinates L10/2; OSU 49213.

Plate 1, fig. 15; slide 50430-A-11; Leitz Orthoplan coordinates 38.4/108.5; England finder coordinates S39/1; OSU 49214.

Type Stratum. The Holotype (Plate 1, figs. 9-10) was recovered from a core sample of the Telatyn Formation, 2132.40-2132.45 m interval of the Panstwowy Instytut Geologiczny Giełczew IG-5 (Latitude 50° 56'40"N, Longitude 20°40'41"E), Poland.

Diagnosis. Flat, square to slightly rectangular, single layered, equant, planar coenobium. Outline straight to slightly undulatory. Individual cells thin-walled, primarily square, rarely rectangular. Peripheral coenocytes may have rounded or bowed sides. Ornament and dehiscence structure not observed.

Measurements. Sheet size: 29 (61) 93 µm. Individual cell size: 3 (5) 7 µm. Maximum number of coenocytes observed in a coenobium: 256. (NOTE: caution is cited here because smaller coenobia may represent fragments of larger sheets

---

PLATE 2

1-13 *Bijugum porebskii* gen. et sp. nov. All photomicrographs referenced with maceration number, slide number, Leitz Orthoplan coordinates, England Finder coordinates and Orton Museum specimen number (OSU) All figures magnified x690 unless otherwise indicated. Maceration number-subsurface/outcrop conversion information can be found in Appendix 1.

1, 2 Holotype; 41620-A-4, 50.0/92.6, B50/2, OSU 49215. Fig. 2 is a phase contrast photomicrograph, x765.

3, 4 Paratype; 41620-A-4, 45.7/96.1, E46, OSU 49216. Fig. 4 is a phase contrast photomicrograph, x765.

5, 6 Paratype; 50430-A-10, 39.5/100.0, J40, OSU 49217. Fig. 6 is a phase contrast photomicrograph, x765.

7, 8 Paratype; specimen exhibiting dehisence slits (arrows) parallel to longitudinal axis of individual coenocyte; 50430-A-11, 25.9/96.0, E26, OSU 49218. Fig. 8 is a phase contrast photomicrogaph, x765.

9 Paratype; 50430-A-9, 7.0/112.2, W7/2, OSU 49219.

10, 11 Specimen showing probable coagulated cytoplasmic contents; 41620-A-4, 51.9/106.9, Q52. Figure 11 is a phase contrast photomicrograph, x765.

12 Paratype; 50431-A-9, 19.0/103.8, N19, OSU 49220.

13 Paratype; 50431-A-9, 20.1/101.0, L10/2, OSU 49212.

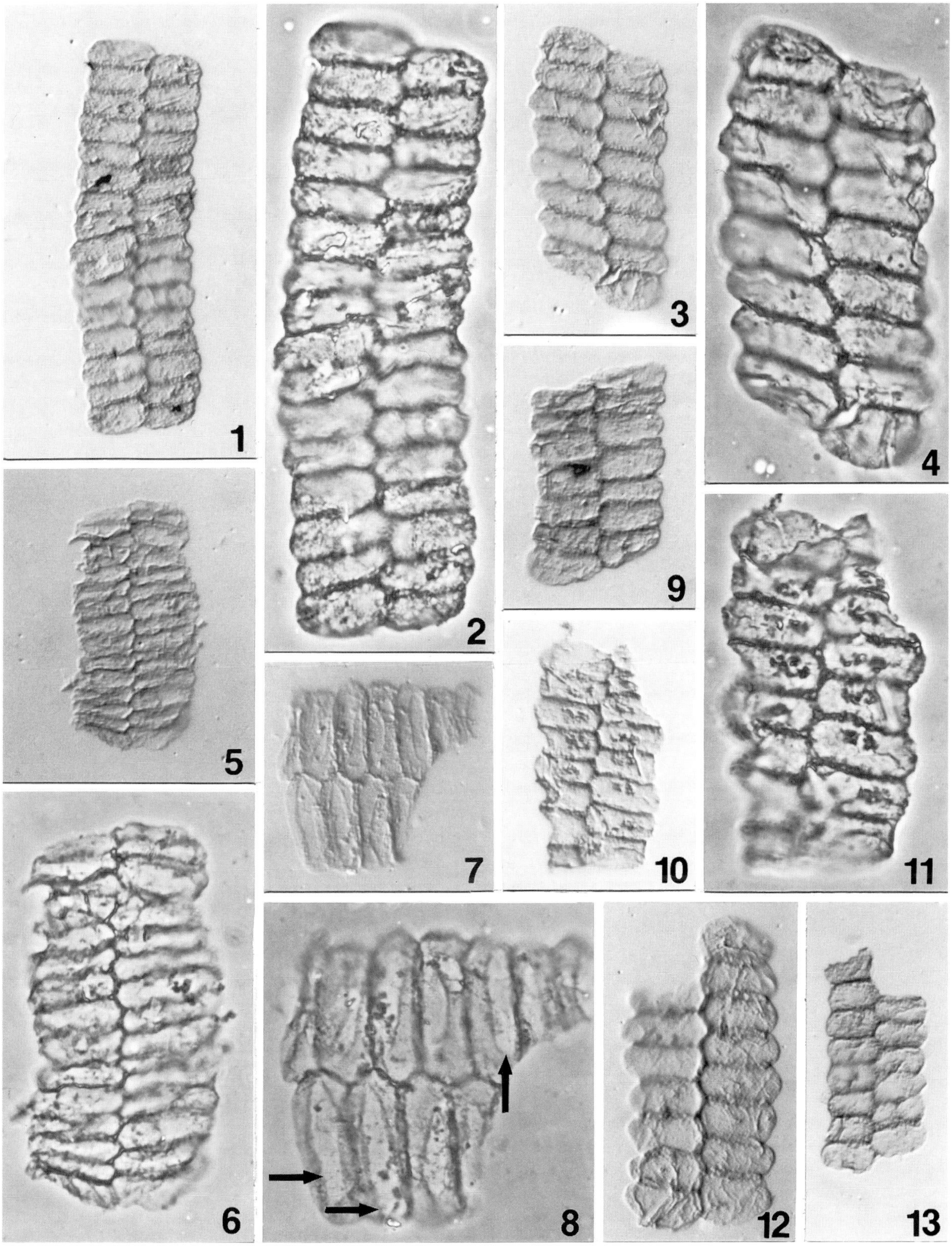
1
2
3
4
5
6
7
8
9
10
11
12
13

disaggregated by taphonomic mechanisms or separated during processing.)

Remarks. The majority of specimens are more or less equant in overall shape. Sheets of cells not exhibiting this morphology (see Plate 1, figs. 7-8, 15) were rare. Examination of numerous specimens, and comparison with published information suggests an ontogenetic sequence (see Text-Figure 4) comparable to those proposed for other chlorococcalean coenobia (e.g., Batten, 1996; Brenner and Foster, 1994). A four coenocyte cluster appears to be the primary element for a *Musivum gradzinskii* coenobium (see also Brenner and Foster, 1994, p. 221–223). This basic coenobial configuration would give rise to a 16 coenocyte coenobium (four coenocytes on a side) and forms with 64 coenocytes (8 coenocytes on a side; Plate 1, figs. 1, 2, 3, 12). The largest sheets observed, 256 total coenocytes (see Plate 1, figs. 6, 9-10), are apparently formed by the multiples of the 64 cell coenobium. The 256 cell coenobium may have a central lumen (see Plate 1, fig. 6). Specimens that do not exhibit this morphology could represent an artifact of preservation related to taphonomy, mechanical separation during processing or aberrant coenobium development (see Plate 1, figs. 7-8).

Genus *Bijugum* gen. nov.

Type Species. *Bijugum porebskii* gen. et sp. nov.

Derivation of Name. Latin, *bijugus*, yoked together; in reference to the distinctive "zigzag" contact between coenocytes.

Diagnosis. Planar, rectangular, unilayered coenobial sheet consisting of two linear rows of closely juxtaposed coenocytes. Individual coenocytes are elongate, straight sided to slightly bowed, and may be smooth or wrinkled. Ends of coenocytes are rounded to subrounded. Longitudinal axis of individual coenocytes offset resulting in an angular to subangular 'zig-zag', or 'zipper-like', arrangement connecting the two rows of cells. Dehiscence consists of linear slits parallel to long axis of coenocytes.

Bijugum porebskii gen. et sp. nov.
Plate 2, figs. 1-13

Derivation of Name. Named in honor of Dr. Szczepan J. Porębski, Polish Academy of Sciences, Institute of Geological Sciences, Kraków, for his geological research concerning Poland.

Holotype. Plate 2, figs. 1-2; slide 41620-A-4; Leitz Orthoplan coordinates 50.0/92.6; England Finder coordinates B50/2; OSU 42915.

Paratypes. Plate 2, figs. 3-4; slide 41620-A-4; Leitz Orthoplan coordinates 45.7/96.1; England Finder coordinates E46; OSU 42916.

Plate 2, figs. 5-6; slide 50430-A-10; Leitz Orthoplan coordinates 39.5/100.0; England Finder coordinates J40; OSU 42917.

Plate 2, figs. 7-8; slide 50430-A-11; Leitz Orthoplan coordinates 25.9/96.0; England Finder coordinates E26; OSU 42918.

Plate 2, fig. 9; slide 50430-A-9; Leitz Orthoplan coordinates 7.0/112.2; England Finder coordinates W7/2; OSU 49219.

Plate 2, figure 12; slide 50431-A-9; Leitz Orthoplan coordinates 19.0/103.8; England Finder coordinates N19; OSU 49220.

Plate 2, figure 13, slide 50431-A-9; Leitz Orthoplan coordinates 20.1/101.0; England Finder coordinates L10/2; OSU 49212.

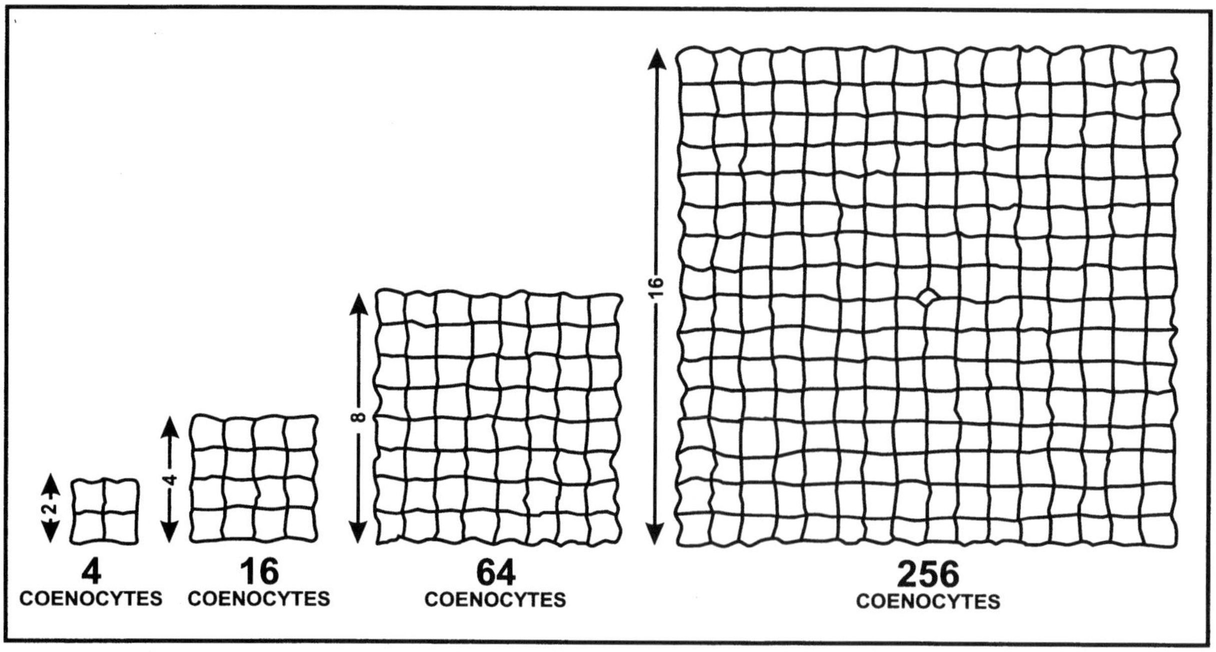

Text-Figure 4. Generalized depiction of the probable ontogenetic series associated with *Musivum gradzinskii* gen. et sp. nov.

Type Stratum. The Holotype (Plate 2, figs. 1-2) was recovered from a core sample of the Telatyn Formation, 2132.40–2132.45 m interval, of the Panstwowy Instytut Geologiczny Gieåczew IG- 5 (Latitude 50°56"40'N., Longitude 20°40'21"E), Poland.

Diagnosis. Planar, rectangular single layered coenobium. Coenocytes closely packed into two offset linear rows. Individual coenocytes elongate, straight sided on long axis, smooth to wrinkled. Distal ends of individual coenocytes are rounded to sub-rounded. Longitudinal axis of individual coenocytes in offset rows result in a "zigzag" or "zipper-like" appearance. Dehiscence mechanism is a linear slit parallel to the long axis of the coenocyte.

Measurements. Coenocyte width: 3 (5) 7 µm. Coenocyte length: 11 (14) 17 µm. Total maximum coenobium width 22 (28) 34 µm. Total maximum coenobium length: 33 (65) 97 µm. Maximum number of coenocytes observed in a coenobium: 32.

Remarks. There are no other Devonian algae similar to this form. *Scenedesmus* displays a single row of coenocytes (see Batten, 1996; Wood and Miller, 1998).

PALEOECOLOGICAL AND SEQUENCE STRATIGRAPHIC IMPLICATIONS

The maximum abundance of these coenobia in the samples examined from Telatyn Formation in the Giełczew IG-5 is 6.3% (2132.40–2132.45m). Spores/ terrestrial plant debris and acritarchs (primarily *Gorgonisphaeridium, Micrhystridium, Veryhachium*) range from 85–93% and 2.0–11.0 % of the total assemblage, respectively, in the four samples examined (Text-Figure 3). Coenobial abundances in the Skały, Swiętomarz and Nieczulice beds range from <1.0%–14.0%. All outcrop samples contain acritarchs, spores and terrestrial plant detritus. The samples from 2132.40–2132.45 m and 2014.58–2014.64 m in the Giełczew IG- 5 are shallow marine, possibly paralic and contain abundant land plant material (terrestrial C_{org}). On the basis of their macrofossil content the Skały and Nieczulice beds probably represent a highstand (HST) or transgressive systems tract (TST), whereas the intervening Swiętomarz horizon suggests a regressive, or lowstand systems tract (LST). Immediately above the uppermost Nieczulice sample containing coenobia there is an initiation of a major flooding surface represented palynologically by a reduction in terrestrial C_{org} and an absence of coenobia.

Generalized paleoenvironmental considerations can be derived from quantitative taphonomic information using abundance variations of non-marine (e.g., spores, terrestrial plant detritus, fresh/brackish water algae) and marine (e.g., scolecodonts, acritarchs) palynomorphs. Pasley et al. (1993) has shown that during an LST and HST a specific location will gradually become more proximal to a shoreline and the amount of terrestrial C_{org} increases. Conversely, a relative rise in sea level associated with a transgressive systems tract (TST) would result in a more distal proximity to a shoreline and a decrease in terrestrial C_{org} at a specific location. Since coenobia and terrestrial C_{org} can be transported to marine environments via a fluvial plume, and possibly redistributed by bottom currents, an in-depth regional study of the Devonian in the subject area is required to ascertain the direction of the shoreline and coarser grain clastic deposition. These parameters may be useful in palaeoenvironmental interpretations applicable to reservoir prediction. Chlorococcales have also been shown to display a distinctive hydrocarbon character. Geochemical studies of other Hydrodictyaceae (e.g., *Pediastrum*) indicate the presence of long chain (C_{22} and above) *n*-alkanes that can be associated with waxy hydrocarbons (He et al., 1997; Wang, F. et al., 1997; Wang, K. et al., 1997).

ACKNOWLEDGMENTS

The authors want to thank the following individuals, associated with the Panstwowy Instytut Geologiczny (P.I.G.), Warsaw for unselfishly giving their time and expertise concerning the Devonian of Poland: Drs. L. Miłaczewski, M. Narkiewicz, R. Wagner and A. Żelichowski. G. Buchart, also associated with P.I.G., assisted GDW in logistical efforts during sample collection at the Holowno Core Facility. Drs. G. Racki (Laboratoire of Palaeontology and Stratigraphy, Silesian University–Sosnowiec) kindly provided the outcrop samples discussed in this paper. M. A. Miller (Saudi Arabian Oil Co.) critically reviewed the manuscript and offered several excellent suggestions for improvement. Dr. J. Jansonius (Calgary, Canada) provided orthographic assistance for the derivation of new names and Prof. S. Bergström (The Ohio State University, Columbus) assisted in securing repository numbers for Holotypes and Paratypes. The authors also gratefully acknowledge M. Schendel (Amoco Graphic Arts, Houston) for drafting the text-figures. G. D. Wood thanks the former Amoco Exploration and Production Technology Group Production Company for technical support and permission to publish.

References Cited

BATTEN, D.J.
1996 Chapter 7C. Colonial Chlorococcales. *In*: Jansonius, J., and McGregor, D.C. (eds.), *Palynology: Principles and Applications*, American Association of Stratigraphic Palynologists Foundation: 191–203.

BATTEN, D.J., and LISTER, J.K.
1988a Evidence of freshwater dinoflagellates and other algae in the English Wealden (Early Cretaceous). *Cretaceous Research*, 9: 171–179.
1988b Early Cretaceous dinoflagellate cysts and chlorococcalean algae from freshwater and low salinity palynofacies in the English Wealden. *Cretaceous Research*, 9: 337–367.

BRENNER, W., and FOSTER, C.B.
1994 Chlorophycean algae from the Triassic of Australia. *Review of Paleobotany and Palynology*, 80: 209–234.

COLBATH, G.K., and GRENFELL, H.R.
1995 Review of biological affinities of Paleozoic acid-resistant, organic-walled eukaryotic algal microfossils (including "acritarchs"). *Review of Palaeobotany and Palynology*, 86: 287–314.

COMBAZ, A.
1962 Sur un nouveau type de microplanctonte cénobial fossile du Gothlandien de Libye, *Deflandrastrum* nov. gen. *Comptes rendus des séances de l' Academie des sciences*, 225: 1977–1979.

CRAMER, F.H., and DÍEZ, M.DEL C.R.
1977 Late Arenigian (Ordovician) acritarchs form CIS-Saharan Morocco. *Micropaleontology*, 23: 339–350.

EVITT, W.R.
1963 Occurrence of freshwater algae *Pediastrum* in Cretaceous marine sediments. *American Journal of Science*, 261: 890–893.

GHAVIDEL-SYOOKI, M.

1995 Palynostratigraphy and palaeogeography of a Palaeozoic sequence in the Hassanakdar area, Central Alborz Range, northern Iran. *Review of Palaeobotany and Palynology*, 86: 91–109.

HE CHENG-QUAN, WANG KAI-FA, ZHANG YU-LAN, WO GUO-XUAN, LI YI-YIN, and ZHANG HUI-ZHI

1997 Relationships between fossil *Pediastrum* and petroleum formation. *Palaeoworld*, 7: 21–27.

HLADIL, J.

1988 Structure and microfacies of Middle and Upper Devonian carbonate buildups in Moravia, Czechoslovakia. *Canadian Society of Petroleum Geologists, Memoir*, 14: 607–618.

KŁOSSOWSKI, J.R.

1985 Sedimentation of the Middle Devonian in the Łysogóry region (Swiętomarz-Sniadka section). *Przegląd geologiczny*, 33: 264–267.

LE HÉRISSÉ, A., AL-TYYAR, H., and VAN DER EEM, H.

1995 Stratigraphic and paleogeographical significance of Silurian acritarchs from Saudi Arabia. *Review of Palaeobotany and Palynology*, 89: 49–74.

MALEC, J., MIŁACZEWSKI, L., NARKIEWICZ, K., and NARKIEWICZ, M.

1996 Stratigraphy of the Devonian in the Szwejki IG-3 deep well, Central Poland. *Kwartalnik Geologiczny*, 40: 367–392.

MALEC, J., and TURNAU, E.

1997 Middle Devonian conodont, ostracod and miospore stratigraphy of the Grzegorzowice-Skały section, Holy Cross Mountains, Poland. *Bulletin of the Polish Academy of Sciences, Earth Sciences*, 45: 68–86.

MIŁACZEWSKI, L.

1981 The Devonian of the south–eastern part of the Radom–Lublin area (eastern Poland) (in Polish, English summary). *Prace Instytutu Geologicznego*, 101: 3–90.

MILLER, M.A.

1988 An early Paleozoic "acritarch" with morphology comparable to the Hydrodictyaceae. *VII International Palynological Congress Abstracts* (Brisbane, Australia): 112.

MILLER, M.A., and WOOD, G.D.

1996 Paleoecological significance of the Hydrodictyaceae, ? Scenedesmaceae and Zygnemataceae (Chlorophyta) in the Early Paleozoic. *IX International Palynological Congress Program and Abstracts*, (Houston, TX): 106.

NARKIEWICZ, M.

1988 Turning points of sedimentary development in the Late Devonian in Southern Poland. *Canadian Society of Petroleum Geologists, Memoir* 14: 619-636.

OLIVEIRA, S.DE F., and BURJACK, M.I.A.

1996 *Petrovina connata* gên. n., sp. n., provavelmente relacionado à Chlorophyta, da Formação Ponta Grossa (Neodevoniano), Bacia do Paraná, Mt., Brasil. *Revista Universidade Guarulhos, Série, Geosciências*, 1: 45–50.

PAJCHLOWA, M.

1957 The Devonian in the Grzegorzowice-Skały section. *Biuletyn Instytutu Geologiczhnego* 122: 145–254.

PASLEY, M.A., RILEY, G.W., and NUMMENDAL, D.

1993 Sequence stratigraphic significance of organic matter variations: Example from the Upper Cretaceous Mancos Shale of the San Juan Basin, New Mexico. *In*: Katz, B.J., and Pratt, L.M. (eds.), *Source Rocks in a Sequence Stratigraphic Famework*, American Association of Petroleum Geologists Studies in Geology, No. 37: 221–242.

RACKI, G.

1988 Middle to Upper Devonian boundary beds of the Holy Cross Mountains, Central Poland: introduction to ecostratigraphy. *Canadian Society of Petroleum Geologists, Memoir 14*: 119–131.

1993 Evolution of the bank to reef complex in the Devonian of the Holy Cross Mountains. *Acta Palaeontologica Polonica*, 37: 87–182.

STREEL, M., HIGGS, K., LOBOZIAK, S., RIEGEL, W., and STEEMANS, P.

1987 Spore stratigraphy and correlation with faunas and floras in the type marine Devonian of the Ardenne-Rhenish regions. *Review of Palaeobotany and Palynology*, 50: 211–229

STUPNICKA, E.

1989 Regional geology of Poland (in Polish). *Wydawnictwa Geologiczne* (Warszawa), 286 pp.

SZULCZEWSKI, M.

1994 The Holy Cross Mountains during the Devonian and the Carboniferous. Europrobe *Trans-European Suture Zone Workshop*, Excursion Guidebook, Holy Cross Mountains (Kielce, Poland).

1995 Depositional evolution of the Holy Cross Mts. (Poland) in the Devonian and Carboniferous—a review. *Geological Quarterly*, 4: 471–488.

TAPPAN, H.

1980 *The Paleobiology of Plant Protists.* W.H. Freeman and Company (San Francisco): 1–1028.

TSIEN, H.H.

1988 Devonian paleogeography and reef development of northwestern and central Europe. *Canadian Society of Petroleum Geologists, Memoir 14*: 341–358.

TURNAU, E., and RACKI, G.

1999 Givetian palynostratigraphy and palynofacies: new data from the Bodzentyn Syncline (Holy Cross Mountains, Poland). *Review of Palaeobotany and Palynology*, 106: 237–271.

VAVRDOVÁ, M.

1990 Coenobial acritarchs and other palynomorphs from the Arenig/Llanvirn boundary, Prague Basin. *Vestnik Ústredního ústavu geologického*, 65: 237–246.

WANG, F., HE, P., and HAO, S.

1997 Origin of Chinese lacustrine high-wax oils and organic petrology of its source rocks. *American Association of Petroleum Geologists/14th Society of Organic Petrology*, Abstracts & Programs: 97.

WANG, K., ZHANG, Y., WU, G., LI, Y., HE, C., and ZHANG, H.

1997 Study of the thermal maturation for oil generation by *Pediastrum*. *Journal of Tongji University*, 22: 184–191.

WILLE, W.

1970 *Plaesiodictyon mosellanum* n. g. n. sp., eine mehrzellige Grünalge aus dem Unteren Keuper von Luxembourg. *Neues Jahrbuch für Geologie und Paläontologie, Monatschefte*, 5: 283–310.

WOOD, G.D.

1997 The acritarch *Proteolobus wallii* gen. et sp. nov. from the Iquiri Formation of Bolivia: a possible coenobial alga. *Micropaleontology*, 43: 325–331.

WOOD, G.D., and ALEMAN, A.M.

1997 Cretaceous chlorococcalean algae (Chlorophyta) from Peru: paleoecologic, sequence stratigraphic and geochemical implications. *IX Congreso Peruano de Geología, Résumes Extendidos, Sociedad Geológica del Peru*, 1: 597–598.

WOOD, G.D., GABRIEL, A.M., and LAWSON, J.C.

1996 Chapter 3. Palynological techniques-processing and microscopy. *In*: Jansonius, J., and McGregor, D.C. (eds.), *Palynology: Principles and Applications*, American Association of Stratigraphic Palynologists, Foundation: 29–50.

WOOD, G.D., and MILLER, M.A.

1996 Pre–Carboniferous Chlorophyta: new occurrences of Hydrodictyaceae, ?Scenedesmaceae and Zygnemataceae. *Abstracts of the International Meeting and Workshop of the C.I.M.P. Acritarch Subcomission* (Prague): 26.

1997 Pre–Carboniferous Chlorophyta: new reports of Hydrodictyaceae, ?Scenedesmaceae and ?Zygnemataceae. *In*: Fatka, O., and Servais, T. (eds.), Acritarcha in Praha, *Acta Universitatis Carolinae*, 40: 703–717.

1998 Stratigraphic, paleoecologic and petroleum generating significance of Chlorophyta (Chlorococcalean algae) in the Cretaceous of Western Africa and South America. *Africa Geoscience Review*, 4: 499–510.

WOOD, G.D., TURNAU, E., MIŁCZEWSKI, L., and NARKIEWICZ, M.

1996 Middle and Upper Devonian palynology and palynofacies, Holy Cross Mountains Region, Poland. *IX International Palynological Congress, Abstracts and Programs* (Houston, USA): 175.

Authors' addresses

GORDON D. WOOD
Amoco Exploration and Production Technology Group
P.O. Box 3092
Houston, TX 77253
U.S.A.

Present address:
the irf group, inc.
23222 Willow Pond Place
Katy, TX 77494-3566
U.S.A.

ELŻBIETA TURNAU
Instytut Nauk Geologicznych
Polska Akademia Nauk
Ul. Senacka, 1
31-002 Kraków
Poland

APPENDIX 1. Maceration Number/Locality/Holotype and Paratype conversions.

Maceration Number 41620 = Państwowy Instytut Geologiczny Giełczew IG-5 (Latitude 50°56'40"N; Longitude 20°40'21"E) 2132.40-2132. 45 m interval.

Maceration Number 50430 = Świętomarz–Śniadka outcrop section, Exposure 2, Sample 4, Swiętomarz beds (Latitude 51°02"N, Longitude 21°05"E) near the village of Świętomarz.

Maceration Number 50431 = Świętomarz - Śniadka outcrop section, Exposure 3, Sample 11, Skały beds, (Latitude 51°02"N, Longitude 21°04"E), near the village of Skały.

LATE PALEOZOIC MIOSPORE NOMENCLATURE DATABASE (LPMN)

Mayya Oshurkova, Aleksandr Oleynikov,
Elena Paevskaya and Anna Suvorova

Abstract

The Late Paleozoic Miospore Nomenclature (LPMN) database consists of:

a) a modification of the Potonié and Kremp morphological classification for Devonian, Carboniferous and Permian miospores;

b) a revision and description of the dispersed miospore form-genera; and

c) a diagnostic key for dispersed miospore identification.

The database includes information concerning synonymy, stratigraphic range, geographic distribution, number of species, and bibliographic data for descriptions of dispersed miospore genera. Seventy-two supergeneric groups (12 of which are new) are treated in the database, as are 240 genera (13 of which are new and 16 are emended; 148 genera are entered as synonyms). The database contains information on 2092 species, and 494 new combinations are made as part of the project.

INTRODUCTION

In the Division of Stratigraphy and Paleontology at VSEGEI, work is progressing on the development of an automated system for the comprehensive processing of paleontological and stratigraphic information. The system allows storage, retrieval and processing of multidimensional data. A package of applied programs, termed the "Automated Workstation of a Paleontologist–Stratigrapher (ARM-PS)," has been developed using the Clipper database management system (SDBM) (Paevskaya et al., 1993). The system operates on an IBM-compatible PC.

One of the modules of the system, named "Paleontology," is designed to solve the traditional tasks of data organization, classification, diagnostics, construction of phylogenetic models, etc., and can be extended by application of additional programs to include specialized analysis of biostratigraphic information. We also plan to integrate separate smaller programs as subsystems into this module, each of which represents a routine oriented at solving special tasks for certain paleontological groups. Another module is "Paleopalynology," which addresses collection, storage, and formalized processing of the results of spore and pollen analysis.

Palynological research is widely used in geological practice to support detailed stratigraphic models during geological surveying. However, the palynological analyses in some laboratories are often based on outdated classifications of miospores and on subjective interpretation of spore and pollen analysis. There are virtually no documentation or storage capabilities for primary paleontological information in the modern system of geological exploration. It is quite clear to each geologist–stratigrapher, that detailed characterization and reliability of stratigraphic constructions depend on the precise definitions of paleontological species. The system "Paleopalynology", therefore, is based on a standardized classification of dispersed miospores and a unified terminology of the morphological features of miospores.

Oshurkova, Mayya, Oleynikov, Aleksandr, Paevskaya, Elena, and Suvorova, Anna
2001 Late Paleozoic miospore nomenclature database (LPMN); *In*: Goodman, D.K., and Clarke, R.T. (eds.), *Proceedings of the IX International Palynological Congress, Houston, Texas, U.S.A., 1996*; American Association of Stratigraphic Palynologists Foundation, p. 65–68.
 ISBN 0–931871–06–9

In Upper Paleozoic palynology, Russian specialists, until recently used the classification of miospores proposed by S.N. Naumova in 1937–1939. Foreign palynology uses the classification by R. Potonié and G. Kremp, developed in 1954–1956. More recently the morphological classification of R. Potonié and G. Kremp was modified by various researchers and now is widely used in the systematic treatment of dispersed miospores for both the Paleozoic and the Mesozoic.

A.A. Lyuber (1966) proposed a slightly modified version of this classification. However, up to now, there were no publications of generic diagnoses for Upper Paleozoic miospores in the Russian language. At the same time, the palynological publications in English are noted for an unlimited growth in the number of new genera of dispersed miospores, many of which we consider to be synonyms.

The systematic work of operational palynologists is hampered by the lack of a clearly and logically structured scheme for definition of dispersed miospore taxa. In this light, the primary task in the development of the system "Paleopalynology" was to produce a computerized database of the Late Paleozoic Miospore Nomenclature (LPMN), consisting of a glossary of standardized terms of morphological features of dispersed miospores, a model of diagnostic morphological classification, and a synopsis of generic diagnoses of the Upper Paleozoic miospores. This task required a revision of the terminology used and of the diagnoses of dispersed miospore genera from the Devonian, Carboniferous and Permian deposits known from the published literature. Palynologists from the Division of Coal Deposits Geology of VSEGEI participated in this work, including M.V. Oshurkova, I.Z. Faddeeva and Z.P. Prosviryakova.

DATABASE STRUCTURE

The database structure of the LPMN is composed of a number of units. The first unit is a glossary of standardized definitions of terms used in the morphological description of miospores. Its use will allow an unambiguous interpretation of morphological features observed in fossil miospores. During compilation of the database, all of the spore and pollen structures that are known in Upper Paleozoic rocks were analyzed. On the basis of the authors' experience, as well as analyses of available publications on morphological terminology of spores and pollen (Kremp, 1965; Kupriyanova and Aleshina, 1967; Grebe, 1971, Meyer-Melikyan, 1987; Punt et al., 1994), a standardization of the definitions of terms for morphological description was conducted. Definitions of these terms are accompanied by illustrations.

The second unit deals with the characterization and differentiation of generic diagnoses. A set of morphological features, which are necessary and sufficient for determination of the form-genus, was established for data input. A characteristic of each genus includes those features inherited by, and recurring in, miospore species belonging to it. Species radiation results in variations of morphological features; however, this variability should be confined to the limits of the Procrustean bed of genus diagnosis. The selected type of the genus, according to S.V. Meyen, represents its archetype, i.e. it contains the entire set of morphological elements characterizing the genus. For example, when determining the correct form-genus among miospores which have no equatorial growths of exine (i.e. subturma Azonotriletes), the distinguishing features are the outlines of miospores, the expressed character of the area, the thickness of exine, the type of ornamentation, and its presence on the distal and/or proximal surface of the spore (Oshurkova and Pashkevich, 1990). A change in the character of one of the above features seems sufficient ground for distinguishing a different form-genus. For instance, spores with granular ornamentation covering the entire surface of the spore, and differing only in outline of the spores, are classified under two genera: (1) the rounded *Cyclogranisporites* Potonié & Kremp, and (2) the triangular *Granulatisporites* Ibr. emend. Potonié & Kremp.

For the purpose of a reliable confirmation of each form-genus and its comparison with taxa having a similar miospore structure, the diagnoses of the form-genera are given in a standardized format containing the entire set of features that are necessary (and sufficient), in a standard succession. The diagnosis of a given genus should then also govern the assignment of particular species to that genus.

The third unit contains the file with descriptions or diagnoses of the form-genera for Devonian, Carboniferous and Permian dispersed miospores. The following data are included for each genus: name, author(s), year of publication, younger synonyms and place of their publication; the type (species) with author(s), year of publication; the unified diagnosis of the genus; reasons for inclusion of a taxon into the synonymy; changes made in the redescription of the given genus, as well as the grounds for distinguishing the genus (in case of a new genus); comparison with taxa having similar morphological features; stratigraphic range; geographic distribution; possible natural affinity, with indication of the place of publication of the given interpretation; and, finally, the species composition of the genus, including the following data for each species: name, author(s), year of publication, locality of the holotype, stratigraphic position, size of miospores.

Generally, the description of the genus in the system has the following format:

Genus *Cyclogranisporites* Potonié & Kremp 1954

Synonymy. *Granisporites* (pars) – Dybova & Jachowicz, 1957, p. 76.

Type of Genus. *C. leopoldi* (Kr.) Potonié & Kremp – Potonié & Kremp, 1955.

Diagnosis. Radial triradiate miospores of a rounded outline. Simple slit of dehiscence. Rays of slit straight, 1/2–2/3 of spore radius in length. Area not expressed. Exine of medium thickness. Granulate ornamentation. Sculptural elements in the form of grains (grana), always of a more or less rounded shape, predominantly of the same size (0.5–1.0 µm) and distributed over the spore surface.

Size. 20–150 µm.

Remarks. *Granisporites* (pars) Dybova & Jachowicz here is included as synonym, since the species of this genus with granulate ornamentation differ only in density and nonuniform arrangement of sculptural elements. These features do not seem to warrant a separate genus. Species having spores with punctate ornamentation should be referred to a genus *Punctatisporites* (Ibr.) Potonié & Kremp.

Comparison. Ornamentation is similar to *Granulatisporites* (Ibr.) Potonié & Kremp, but the genus differs from the latter by a rounded outline of spores. Among genera of the infraturma Apiculati which have miospores of a rounded outline, it is characterized by having a granular ornamentation.

Stratigraphic Range. Carboniferous, Permian; most common in the Middle Carboniferous.

Geographic Distribution. Global.

Natural Assignment. Pteridophytes (Potonié, 1962).

Species Composition of the Genus.

1. *Cyclogranisporites leopoldi* (Kremp) Potonié & Kremp 1954 – Potonié and Kremp, 1954, p. 126; holotype re-illustrated by Potonié and Kremp, 1955, p. 62, pl. 13, figs. 174–178. Germany, Ruhr Basin, Westphalian B (25–35 µm).

Basionym: *Granulatisporites leopoldi* Kremp 1952, p. 348, pl. 15b, fig. 15.

2. *C. lasius* (Waltz) Playford 1962 – Luber & Waltz, 1938, p. 11, pl. 1, fig. 4, pl. 4, fig. 4. Kizel Basin, Lower Carboniferous (50–95 µm).

Basionym: *Azonotriletes lasius* Luber & Val'ts 1938.

3. *C. punctulatus* (Waltz) Lub. 1941 – Lyuber & Waltz, 1941, p. 14, pl. 3, figs. 16 a, b. Kizel Basin, Lower Carboniferous (25–115 µm).

Basionym: *Azonotriletes punctulatus* Luber & Waltz 1941

4. *C. ovatus* (Knox) Potonié & Kremp 1955 – Knox, 1950, p. 316, pl. 17, fig. 222. Scotland, Carboniferous (80–150 µm).

Basionym: *Planisporites ovatus* Knox 1950.

5. *C. orbicularis* (Kosanke) Potonié & Kremp 1955 – Kosanke, 1950, p. 16, pl. 2, fig. 9. USA, Illinois Basin, Pennsylvanian (35–44 µm).

Basionym: *Punctatisporites* Kosanke 1950.

6. *C. major* (Dybova & Jachowicz) Oshurkova comb. nov. – Dybova & Jachowicz, 1957, p. 77, pl. 10, figs. 1–4. Poland, Upper Silesian Basin, Carboniferous, Westphalian B–Westphalian D (85–105 µm).

Basionym: *Granisporites major* Dybova & Jachowicz 1957.

The description of each genus is accompanied by a schematic representation which reflects the combination of features of a given taxon.

The fourth unit comprises a diagnostic system for Upper Paleozoic dispersed miospores. Experience shows that the most useful system for grouping the Paleozoic miospores in a morphological classification is the "turma" system, which includes five supergeneric ranks.

Taxa of the rank anteturma of dispersed miospores are established on the character of the germinal aperture of spores:

Anteturma PROXIMEGERMINANTES R. Potonié 1970 (= Sporites H. Potonié, 1893) comprises miospores that have a germinal aperture in the form of a tetrad scar (trilete or monolete aperture), located on the proximal side.

Anteturma VARIEGERMINANTES R. Potonié, 1970 (= Pollenites R. Potonié, 1931) comprises miospores with a germinal aperture of various type and location (furrow, colpus, sulcus, pores etc.).

Anteturma AGERMINANTES Oshurkova, ant. nov. comprises miospores that have no germinal aperture.

At each successive stage of morphological classification, a definitive feature of miospore structure is also used to distinguish taxa.

In the anteturma Proximegerminantes the next subordinate, that of turma, is established from the character of the laesura; the third rank, suprasubturma, is distinguished on the structure of the exine; the fourth rank, subturma, is differentiated using the presence or lack of equatorial growth of the exine; the fifth rank, infraturma, is defined using the character of ornamentation.

For diagnosis of the Upper Paleozoic disperses miospore form-genera, we propose using a polychotomous classification table (Oshurkova and Pashkevich, 1990) and dichotomous determination keys. The example below is a portion of the key to define form-genera of the infraturma Apiculati:

1.	Ornamentation elements in the form of verrucae	2
	Ornamentation elements of other forms	6
2(1)	Similar ornamentation over the entire miospore surface	3
	Different ornamentation on the proximal and distal sides	5
3(2)	Rounded or triangular-rounded outline of miospores	4
	Triangular or rounded-triangular outline of miospores	genus ***Converrucosisporites***
4(3)	Area not expressed	genus ***Verrucosisporites***
	Area expressed	genus ***Verruciretusispora***
5(2)	Rounded or triangular-rounded outline of miospores	genus ***Schopfites***
6(1)	Triangular or rounded-triangular outline of miospores	genus ***Interradispora***

Compilation of the determination keys is the logical completion of the compilation of a diagnostic classification system.

The last unit, the bibliographic file, includes a list of publications used in compilation of the LPMN base.

Data input for both stratigraphic position and geographic distribution of miospore taxa is effected in accordance with two auxiliary glossaries, "Stratigraphic scale" and "Geographic distribution." The structure of the stratigraphic glossary reflects hierarchical relations between stratigraphic units established by the Stratigraphic Code (Zhamoida, 1992); the geographic glossary contains geographic names, corresponding to four hierarchical levels.

LPMN SYSTEM OPERATION MODES

The LPMN system comprises six basic operational modes:

1. data input
2. reference information on taxa
3. glossaries
4. diagnostics
5. search
6. general information on the system

The "Data Input" mode enables input and editing of glossaries and reference information on taxa.

The "Glossary" mode comprises six glossaries. The glossary "Terms for Morphological Description of Miospores" contains the alphabetical list of terms (in Russian and English) used to describe miospores. If the information contains a term whose meaning seems unclear to the user, the address to the appropriate glossary provides the necessary definition. The glossary "Morphological Classification" reflects systematics of the Upper Paleozoic dispersed miospores. The glossary "Alphabetical Catalogue of Dispersed Miospores" comprises the alphabetical list of taxa whose descriptions are input into the database. The glossaries "Stratigraphic Scale" and "Geographic Distribution" ensure that a common set of terms are recorded in descriptions of taxa. The glossary "Bibliography" contains the alphabetical list of authors' names with citations of the works published by each author.

The "Diagnostics" mode contains the determination keys which make it possible to determine whether the studied object belongs to the specific form-genus.

The "Search" mode assesses the sample of objects based on the assigned criteria. The possibility of definition of the systematic position of a taxon of any level is envisaged, as well as the retrieval of taxa based on their stratigraphic position and geographic occurrence.

The "General Information on the System" mode comprises three sections:

1. Description of the System. The section contains information on the structure and purpose of IRS LPMN.

2. User's Guide. The section contains instructions for operating the system.
3. Instrumental Means. The section contains information on required hardware for the system, and on the function of directories in the LPMN catalogue.

CONCLUSIONS

The system enables users to supplement and correct the LPMN base and deliver information based on user's requests in accordance with the notions on nomenclature of the Late Paleozoic dispersed miospores accepted at the present time.

LPMN is a database containing the diagnoses of genera of Upper Paleozoic miospores. Species from the earlier published works are given in accordance with the Potonié and Kremp school of morphological classification, that provide identification of old definitions. This is the first time that such a generalized program of paleontological data-basing has been undertaken in Russia. Upon completion the program will provide more accessible palynological information for international applications.

The completed IRS system will be applicable to any paleontological group, and can thus be used as the prototype for creation of regional data banks based on different groups of fossils.

ACKNOWLEDGMENTS

We thank J. Jansonius for constructive criticism of the manuscript and his help in making the English more idiomatic.

References Cited

GREBE, H.
1971 A recommended terminology and descriptive method for spores. *Microfossiles Organiques du Paléozoïque C.I.M.P., Spores*, no. 4: 11–34.

KREMP, G.O.W.
1965 *Morphologic encyclopedia of palynology*. University Arizona Press, Tucson, 186 pgs.

KUPRIYANOVA, L.A., and ALESHINA, L.A.
1967 *Palynological terminology of angiosperms*. Nauka, Leningrad, 85 pgs.

LYUBER, A.A.
1966 Morphological classifications of spores and pollens and its unification. *In*: Pokrovskaya, I.M. (ed.), *Paleopalinologyi, v. 1*, Nedra, Leningrad, p. 104–111.

MEYER-MELIKYAN, N.R.
1987 Spore and pollen morphology. *In*: Nesterov, I.I. (ed.), *Methodological aspects of palynology*. Nedra, Moscow, p. 17–40.

OSHURKOVA, M.V., and PASHKEVICH, N.G.
1990 Morphological classification of dispersed fossil spores. *In*: Panova, L.A., Oshurkova, M.V. and Romanovskaya, G.M. (eds.), *Practical palynostratigraphy*. Nedra, Leningrad, p. 5–15.

PAEVSKAYA, E.B., OLEYNIKOV, A.N., VAKORINA, L.I., LEONT'EVA, E.N., SUVOROVA, A.G., SHIRINOVA, E.M., and ANDREEVSKIY, L.I.
1993 To develop the information-mathematical model for automated processing of paleontological–stratigraphic data. Annual of *VSEGEI*: 95–96.

PUNT, W., BLACKMORE, S., NILSSON, S., and LE THOMAS, A.
1994 *Glossary of pollen and spore terminology*. LPP Foundation, Utrecht, 71 pgs.

ZHAMOIDA, A.I. (ed.)
1992 *Stratigraphic Code.* Second edition, supplemented. St. Petersburg, 120 pgs.

Authors' addresses

MAYYA OSHURKOVA
ALEKSANDR OLEYNIKOV
ELENA PAEVSKAYA
ANNA SUVOROVA
Russian Geological Research Institute (VSEGEI)
Sredniy pr. 74
St. Petersburg, 199106
Russia

EARLY CHESTERIAN SPORES FROM NORTHWESTERN ALABAMA, U.S.A.

Yu-Xing Zhou, Anne Raymond and Cortland F. Eble

Abstract

Palynomorphs from the lower Chesterian (V3c) of northwestern Alabama are similar to those from the Hardinsburg Formation of Illinois and from the upper Visean of Britain. Sixty-five taxa are recognized, and two new species, *Raistrickia bangorii* and *Crassispora punctata*, are proposed.

INTRODUCTION

The Mississippian–Pennsylvanian boundary was a time of profound global climate change, and this climate change probably affected land plants. Nevertheless, few studies have been made since Kosanke (1982) assessed palynomorph changes across the Mississippian–Pennsylvanian boundary in the United States. The depositional hiatus, which exists throughout most of the United States, has hindered such studies. Nevertheless, recent work suggests that the Mississippian–Pennsylvanian boundary, which falls in the Parkwood Formation, may be continuous in northern Alabama (Pashin, 1993), providing a very good opportunity to study the floral response to Mississippian–Pennsylvanian climate change. This is the first report of such serial investigations, and focuses on the microfloras of the early Chesterian prior to the Mississippian–Pennsylvanian boundary.

GEOLOGICAL SETTINGS AND STRATIGRAPHY

The Black Warrior Basin, a late Paleozoic foreland basin formed by tectonic loading of the Alabama continental promontory during Appalachian–Ouchita orogenesis (Thomas, 1988), lies at the northwest corner of Alabama and northern Mississippi. Chesterian strata are widespread in northwest Alabama, and are divided into the Pride Mountain Formation, the Hartselle Sandstone, the Bangor Limestone, and the Parkwood Formation in ascending order. The Pride Mountain Formation consists of dark gray shales, but includes two shale–sandstone–limestone cycles: the Lewis and the Evans. Drahovzal (1967) considered the Pride Mountain Formation equivalent to strata from the Ste. Genevieve through the Haney formations of Illinois, i.e. Genevievian, Gasperian and lower Hombergian Stages of the lower Chesterian series. The Hartselle Sandstone consists of quartzarenite. Butts (1926) assigned the Hartselle Sandstone to the Gasperian and Hombergian Stages, whereas Drahovzal (1967) considered the Hartselle Sandstone correlative to the Hardinsburg Formation of Illinois, i.e. the middle of the Hombergian Stage. Gray shale and limestones comprise the lower part of the Bangor Limestone, which correlates to the Glen Dean Limestone of Illinois, i.e. the top of the Hombergian Stage (Butts, 1926; Drahovzal, 1967; Burdick and Strimple, 1982) (Text-Figure 1). In their revision of the Meramecian/Chesterian boundary, Maples and Waters (1987) correlate the lower Chesterian to the Brigantian stage of Britain based on marine fossils (Text-Figure 1).

At the Alabama Rock Asphalt Quarry, the Pride Mountain Formation contains the Lewis cycle. Slope deposits cover most of the lower part of the section, which is composed mainly of siltstones and mudstones. The upper parts of the section comprise the Lewis cycle, the lower portion of which contains some interbedded siltstones (Text-Figure 1) (Pashin and Rindsberg, 1993). At Fielder Ridge, the lower Hartselle Sandstone consists of medium-grained sandstone with trough cross beds; the middle Hartselle Sandstone consists of interbedded sandstone and shale with wavy, flaser, and lenticular bedding; and the upper Hartselle Sandstone consists of calcareous sandstone. The basal Bangor Limestone at Fielder Ridge consists of interbedded limestone and shales at the base, slightly fissile, calcareous gray shales with some thin limestone beds in the middle, and a thick bed of crossbedded limestone at the top (Text-Figure 1). The contact between the Hartselle Sandstone and the Bangor Limestone at Fielder Ridge is distinct but gradual (Rindsberg and Pashin, 1993).

SAMPLE LOCATION AND PREPARATION

Samples were taken during the 30th Annual Field trip of the Alabama Geological Society. Samples of the Pride Mountain Formation are from the Alabama Rock Asphalt Quarry, in 1/4SE sec. 1, 1/4NE sec. 12, T.4S., R.15W., Bishop 7.5 minute Quadrangle (Pashin and Rindsberg, 1993). Samples of the Hartselle Sandstone and of the lower part of the Bangor Limestone are from Fielder Ridge, on the west side of Alabama Highway 247, in 1/4SW1/4SE sec. 3, T.5S., R.13W., Barton 7.5-minute Quadrangle (Rindsberg and Pashin, 1993).

Ten grams of crushed sample, screened through a 710 μm sieve, were prepared by conventional techniques using hydrochloric and hydrofluoric acids to dissolve the rock matrix in the Palynology Laboratory at Texas A&M University. No subsequent oxidation was performed. Residue was screened through a 15 μm nylon sieve mesh in an ultrasonic device. Permanent mounts of the palynomorph preparations were made in loctite, an ultraviolet adhesive agent. Slides are stored in the Department of Geology and Geophysics at Texas A&M University. The types are housed in the Pollen Laboratory at Texas A&M University, College Station, Texas. The *Raistrickia bangori* holotype and paratype, and the *Crassispora ? punctata* holotype reference number is STOP 1-3(1); the *Crassispora ? punctata* paratype reference number is STOP 2-3(1).

Zhou, Yu-Xing, Raymond, Anne, and Eble, Cortland F.
2001 Early Chesterian spores from northwestern Alabama, U.S.A.; *In*: Goodman, D.K., and Clarke, R.T. (eds.), *Proceedings of the IX International Palynological Congress, Houston, Texas, U.S.A., 1996*; American Association of Stratigraphic Palynologists Foundation, p. 69–77.
 ISBN 0–931871–06–9

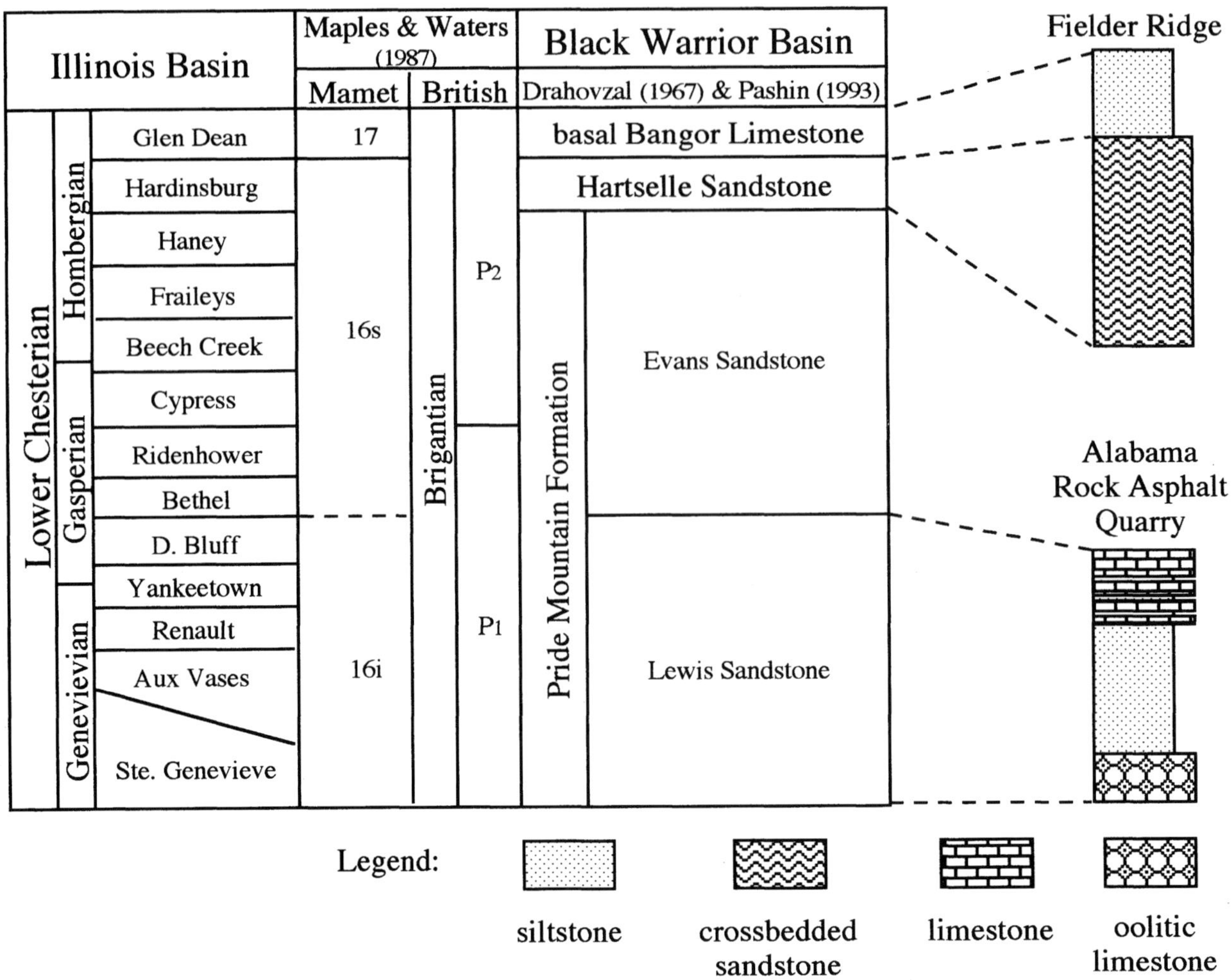

Text-Figure 1. Correlation of the Pride Mountain Formation, the Hartselle Sandstone, and the basal Bangor Limestone in northwestern Alabama with early Chesterian strata in Illinois Basin, and with strata in Great Britain. On the right side is a sketch of the sampled sections.

PLATE 1

All specimens magnified x600. All the stage coordination is recorded on a Leitz Ortholux microscope.

1 *Leiotriletes politus* (Hoffmeister et al.) Love 1960. Slide number STOP1–4(1), stage coordination 9.2-111.5.

2 *Anaplanisporites baccatus* (Hoffmeister et al.) Smith & Butterworth 1967. Slide number STOP2-1(1), stage coordination 15.5-119.

3 *Microreticulatisporites fundatus* Hoffmeister et al. 1955. Slide number STOP1-4(1), stage coordination 0-103.5.

4 *Comasphaeridium* sp. Slide number STOP1-2(1), stage coordination 9-107.2.

5 *Calamospora macer* William. Slide number STOP1-4(1), stage coordination 14.5-110.8.

6 *Laevigatosporites vulgaris* Ibrahim 1933. Slide number STOP1-4(1), stage coordination 16-102.5.

7 *Gorgonisphaeridium plerispinosum* Wicander 1974. Slide number STOP1-2(1), stage coordination 1.8-109.6.

8 *Gorgonisphaeridium winslowiae* Staplin et al. 1965. Slide number STOP1-2(1), stage coordination 16.3-112.2.

9 *Lycospora pusilla* (Ibrahim) Schopf et al. 1944. Slide number STOP1-4(1), stage coordination 15.3-115.

10 *Lycospora pellucida* (Wicher) Schopf et al. 1944. Slide number STOP2-1(1), stage coordination 10.2-106.3.

11 *Rotaspora fracta* Schemel 1950. Slide number STOP2-1(1), stage coordination 15-115.8.

12 *Rotaspora knoxi* Butterworth & William 1958. Slide number STOP1-3(1), stage coordination 14-109.

13, 14 *Raistrickia bangorii* n. sp. 13. Slide number STOP1-3(1), stage coordination 13-124.4; 14. Slide number STOP1-3(1), stage coordination 3-119.5.

15 *Veryhachium downiei* Stockmans & Williere 1962. Slide number STOP1-2(1), stage coordination 13.2-107.

16 *Raistrickia nigra* Love 1960. Slide number STOP2-1(1), stage coordination 17-117.

17 *Reticulatasporites* cf. *R. lacunosus* Felix & Burbridge 1967. Slide number STOP1-4(1), stage coordination 0-115.2.

18 *Auroraspora macra* Sullivan 1968. Slide number STOP1-4(1), stage coordination 3.3-111.

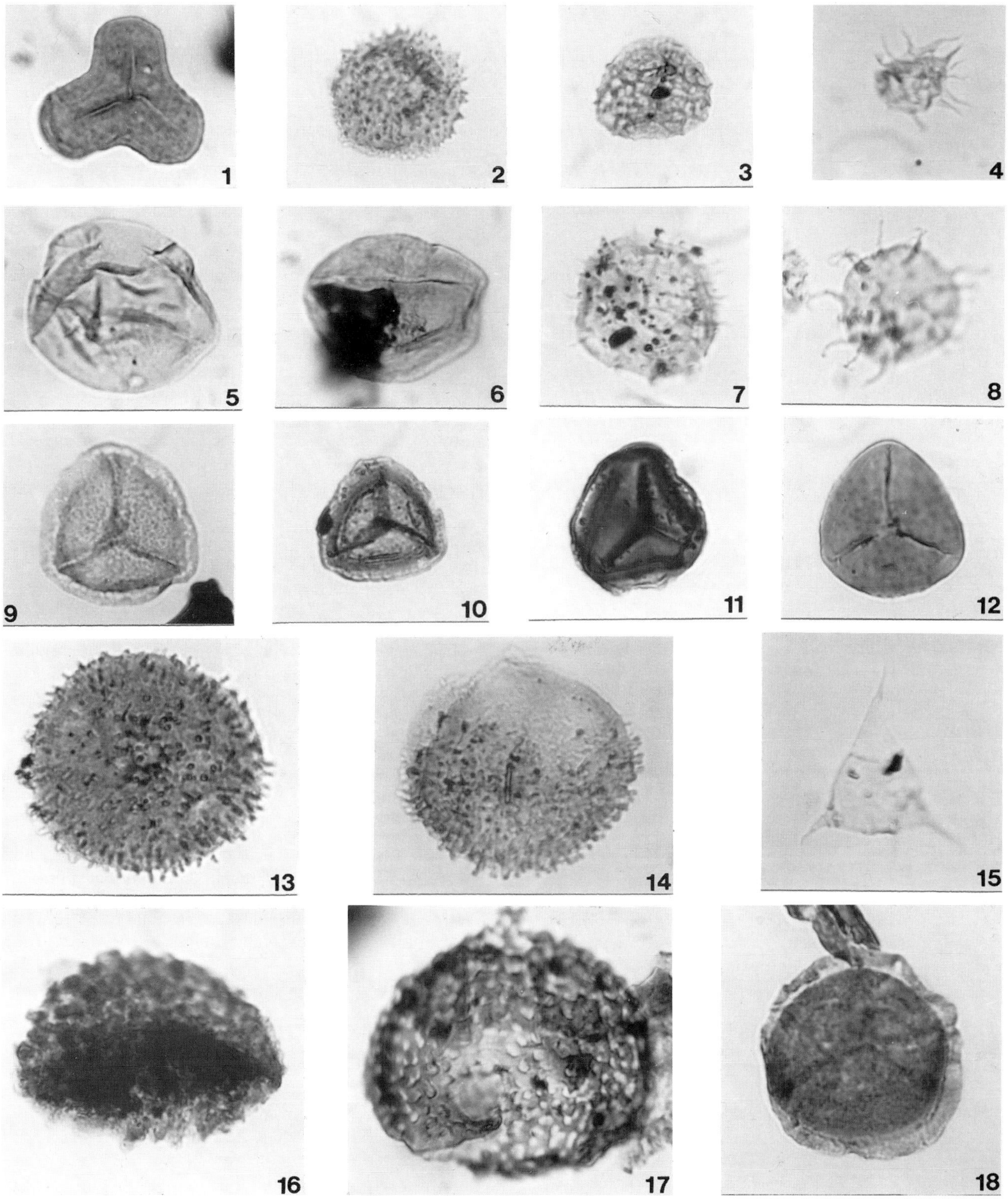
1
2
3
4
5
6
7
8
9
10
11
12
13
14
15
16
17
18

PALYNOMORPH COMPOSITION

Four samples from the Pride Mountain Formation and one sample from the Hartselle Sandstone were examined. The palynomorph assemblages in the Pride Mountain Formation and the Hartselle Sandstone are generally indistinguishable. A total of 65 taxa were recognized. Stratigraphically significant species include: *Anaplanisporites baccatus* (Hoffmeister et al.) Smith & Butterworth, *Auroraspora macra* Sullivan, *Cirratriradites granulatipunctatus* Hoffmeister et al., *Convolutispora florida* Hoffmeister et al., *C. ampla* Hoffmeister et al., *C. lessellata* Hoffmeister et al., *Crassispora kosankei* (Potonié & Kremp) Bharadwaj, *C. maculosa* (Knox) Sullivan, *C. trychera* Neves & Ioannides, *C. punctata* sp. n., *Knoxisporites rotatus* Hoffmeister et al., *Densosporites tenuis* Hoffmeister et al,, *Discernisporites irregularis* Neves, *Grandispora spinosa* Hoffmeister et al., *Leiotriletes politus* (Hoffmeister et al.) Love, *Microreticulatisporites fundatus* Hoffmeister et al., *Raistrickia nigra* , *Reticulatisporites macunosus* Felix & Burbridge, *Raistrickia nigra* Love, *Lycospora pusilla* (Ibrahim) Schopf et al., *L. pellucida* (Wicher) Schopf et al., *Rotaspora fracta* Schemel, *Schulzospora ocellata* (Horst) Potonié & Kremp, *Tripartites vetustus* Schemel, *T. marginatus* Hoffmeister et al., *Punctatisporites nitidus* Hoffmeister et al., and *Laevigatosporites vulgaris* Ibrahim.

Lycospora, with percentages generally over 45%, predominates in the assemblages of the Pride Mountain Formation and the Hartselle Sandstone. *Anaplanisporites baccatus* constitutes slightly over 20% of the assemblages. The percentage of *Anaplanisporites baccatus* decreases slightly to 11% in the Lewis cycle of the Pride Mountain Formation, possibly due to the change in depositional environment, which is indicated by the appearance of marine acritarchs. *Crassispora punctata* new species, *Discernisporites micromanifestus* (Hacquebard) Sabry & Neves, and *Punctatisporites nitidus* Hoffmeister et al. occur more commonly in some assemblages.

Three samples from the basal Bangor Limestone, distributed more or less evenly throughout the sequence, were examined. Twenty-five taxa were found, and the following are considered stratigraphically significant: *Punctatisporites nitidus* Hoffmeister et al., *Anaplanisporites baccatus* (Hoffmeister et al.) Smith & Butterworth, *Calamospora macer* William, *Rastrickia bangorii* sp. n., *Lycospora pusilla* (Ibrahim) Schopf et al., *L. pellucida* (Wicher) Schopf et al., *Rotaspora knoxi* Butterworth & William, *R. fracta* Schemel, *Grandispora spinosa* Hoffmeister et al., *Crassispora kosankei* (Potonié & Kremp) Bharadwaj, *C. punctata* sp. n., *Cirratriradites granulatipunctatus* Hoffmeister et al., *Schulzospora ocellata* (Horst) Potonié & Kremp, acritarch *Gorgonisphaeridium plerispinosum* Wicander, *G. winslowiae* Staplin et al. *Veryhachium downiei* Stockmans & Williere, and *Comasphaeridium* sp. The composition of spores and pollen grains is similar to that of the underlying Pride Mountain Formation and the Hartselle Sandstone. However, the frequency of *Lycospora* decreases dramatically to about 25%, from over 45% in the underlying strata, as does the frequency of *Anaplanisporites baccatus,* to about 5%, from over 20% in the underlying strata. Increased sea level, which is indicated by the high frequency of marine acritarchs (32%) and the low diversity of the palynomorph assemblages in the basal Bangor Limestone may account for these changes. Concomitant with the decrease of *Lycospora* and *Anaplanisporites baccatus,* the amount of *Calamospora macer* in the assemblages increases.

DISCUSSION

The Pride Mountain Formation, the Hartselle Sandstone, and the basal Bangor Limestone have been correlated to the Genevievian, the Gasperian and the Hombergian of Illinois based on marine fossils (Butts, 1926; Drahovzal, 1967), and the latter are equivalent to the British Brigantian (late Visean) (Maples and Waters, 1987). The Brigantian of Britain contains the VF and the lower part of the NC palynomorph zones. The base of the VF zone is marked by the first occurrence of several species: *Rotaspora fracta, Grandispora spinosa, Tripartites vetustus, Savitrisporites nux, Crassispora maculosa,* and *Rotaspora knoxi* (Clayton et al., 1977). The NC zone also contains these characteristic species, but the upper part of the NC zone, which is Namurian E1 in age, contains *Crassispora kosankei* and *Potonieisporites elegans* as well. In palynomorph assemblages of the Pride Mountain Formation, the Hartselle Sandstone and the basal Bangor Limestone of northwestern Alabama, *Rotaspora fracta, Grandispora spinosa, Tripartites vetustus, Crassispora maculosa,* and *Rotaspora knoxi* occur, which is consistent with an age assignment to the VF or lower NC zone. Nevertheless, the occurrence of

PLATE 2

All specimens magnified x600. All the stage coordination is recorded on a Leitz Ortholux microscope.

1 *Cirratriradites granulatipunctatus* Hoffmeister et al. 1955. Slide number STOP1-3(1), stage coordination 6.2-130.2.

2 *Densosporites annulatus* (Loose) Smith & Butterworth 1967. Slide number STOP1-4(1), stage coordination 18-101.8.

3 ?*Schulzospora* sp., Slide number STOP1-4(1), stage coordination 3-112.5.

4,5 *Crassispora kosankei* (Potonié & Kremp) Bharadwaj 1957. 4. Slide number STOP2-3(1), stage coordination 0.5-121; 5. Slide number STOP2-3(1), stage coordination 16-123.

6 *Knoxisporites rotatus* Hoffmeister et al. 1955. Slide number STOP1-4(1), stage coordination 6-100.

7 *Schulzospora ocellata* (Horst) Potonié & Kremp, Slide number STOP1-4(1), stage coordination 12.5-106.7

8 *Crassispora trychera* Neves & Ioannides 1974. Slide number STOP1-4(1), stage coordination 16-103.

9 *Crassispora maculosa* (Knox) Sullivan 1964. Slide number STOP1-4(1), stage coordination 17-113.

10 *Radiizonates tenuis* (Loose) Butterworth & Smith 1964. Slide number STOP1-4(1), stage coordination 1.8-108.5.

11 *Densosporites triangulatus* Kosanke 1950. Slide number STOP1-4(1), stage coordination 13.5-102.5.

12 *Discernisporites micromanifestus* (Hacquebard) Sabry & Neves 1971. Slide number STOP2-1(1), stage coordination 10-112.

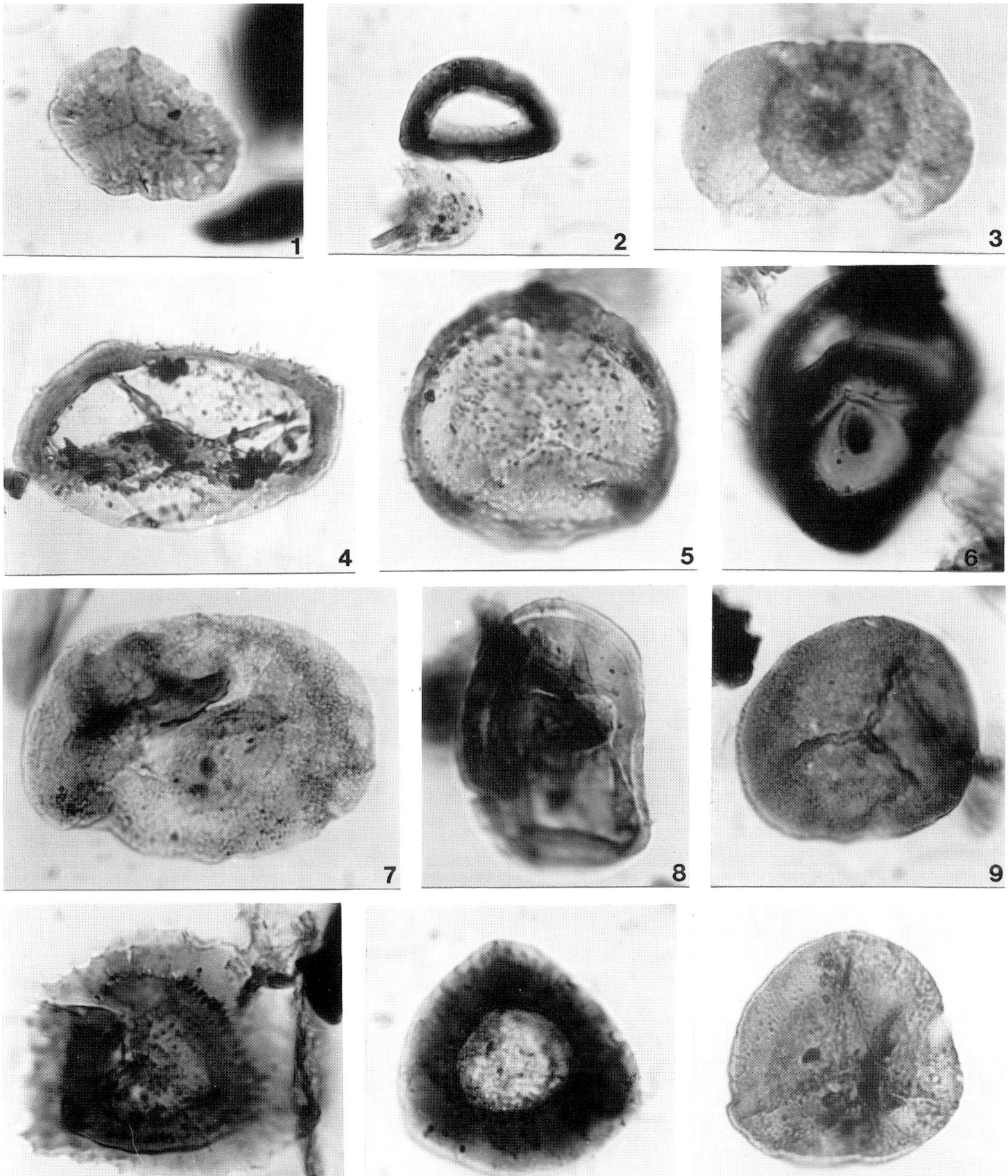
1
2
3
4
5
6
7
8
9
10
11
12

Crassispora kosankei in the Lewis cycle of the Pride Mountain Formation and the basal Bangor Limestone suggests that these Formations belong to the upper NC zone, which is Namurian (E1) in age. *Crassispora kosankei* makes its first occurrence at the base of the E1 goniatite zone in Europe (Clayton et al., 1977). Assuming that *Crassispora kosankei* appeared in Alabama and Europe at the same time, we would assign a Namurian (E1) age to the Pride Mountain Formation, the Hartselle Sandstone and the basal Bangor Limestone. Nevertheless, such an age assignment contradicts that of Butts (1926), Drahovzal (1967), and Maples and Waters (1987) based on marine fossils. On the other hand, if we were to accept the age assignment based on marine fossils, the species *Crassispora kosankei* appeared earlier in Alabama than in Europe. Further studies are necessary to clarify this discrepancy in age assignments based on palynomorphs and marine fossils. Nevertheless, *Crassispora kosankei* may not appear synchronously in Alabama and Europe.

The palynomorph assemblages from the Pride Mountain Formation, the Hartselle Sandstone, and the basal Bangor Limestone of northwest Alabama are very similar to assemblages from the Hardinsburg Formation of Illinois and Kentucky (Hoffmeister et al., 1955) and from the Upper Sedimentary Group and the Lower Limestone Group of the Midland Valley of Scotland (Sullivan and Marshall, 1966). Assemblages from these three localities have a high percentage of *Lycospora* and *Anaplanisporites baccatus* and share many species including *Rotaspora fracta*, *Tripartites vetustus*, *Grandispora spinosa*, *Raistrickia nigra*, *Leiotriletes politus*, *Convolutispora ampla*, *C. florida*, *Microreticulatisporites fundatus*, and *Knoxisporites rotatus*. *Cirratriradites granulatipunctatus* occurs in assemblages from northwest Alabama, Illinois and Kentucky, but not in assemblages from the Midland Valley of Scotland and other VF and NC zone assemblages of west Europe (Neves et al., 1972; Clayton et al., 1977).

Neville (1968) described assemblages of late Visean age from the East Fife section between St. Monance and Pittenweem, Scotland. This section is separated by 75 miles from that of Sullivan and Marshall (1966) and includes rocks older and probably younger than those of Sullivan and Marshall (1966). The St. Monance-Pittenweem assemblages also have a high percentage of *Anaplanisporites* and *Lycospora*, and contain *Rotaspora fracta*, *Tripartites vetustus*, *Grandispora spinosa*, *Raistrickia nigra*. They may share other species with assemblages of northwest Alabama; however, Neville (1968) did not include a complete species list for St. Monance-Pittenweem.

The late Mississippian palynomorph assemblage from Daggett County, Utah (Schemel, 1950) showed a high frequency of *Densosporites* (over 92%). Only the occurrence of *Tripartites vetustus* and *Rotaspora fracta* suggests that Dagget County assemblages are correlative with assemblages from northwest Alabama.

The dominant types of palynomorph assemblages from the Golata Formation of Alberta, Canada (Staplin, 1960), *Lycospora* and *Anaplanisporites baccatus*, are similar to those of northwest Alabama, and do not appear to be affected by facies changes within the Golata Formation. *Densosporites* contributes less than 1% to slightly over 15%. However, the Golata assemblages lack those characteristic species found in the Chesterian of United States and in the late Visean of western Europe, for example *Rotaspora fracta*, *Tripartites vetustus*, and *Grandispora spinosa*. Instead, they contain species that generally do not occur in the Chesterian of United States and in the late Visean of Europe, for example *Monilospora* and *Murospora*.

Utting (1987) recovered palynomorphs from the boundary between the Windsor and Canso Groups of Nova Scotia, Canada, and placed them in the SM zone. On the basis of first occurrence of *Grandispora spinosa* and *Raistrickia nigra* and the appearance of *Schulzospora*, Utting (1987) compared the SM zone with the VF zone (late Visean) of Western Europe. *Crassispora trychera* predominates the SM zone of Nova Scotia. A few species are common to the palynomorph assemblages of the Windsor-Canso boundary of Nova Scotia and the lower Chesterian of northwest Alabama: *Auroraspora macra*, *Crassispora trychera*, *Discernisporites micromanifestus*, *Grandispora spinosa*, *Raistrickia nigra*, *Retusosporites incohatus*, and *Convolutispora tessellata*.

PALEOPHYTOGEOGRAPHY

Sullivan (1965) divided Mississippian palynofloras of Euramerica into three suites: the *Monilospora*, the *Grandispora*, and the Kazakhstan suites. The *Grandispora* suite is found in the United States, Maritime Provinces of Canada, and Western Europe, and contains *Anaplanisporites baccatus*, *Tripartites vetustus*, *Rotaspora fracta*, *Rotaspora knoxi*, *Crassispora maculosa*, *Grandispora spinosa*, *Savitrisporites nux*, and *Bellisporites nitidus*. All but *Savitrisporites nux*, and *Bellisporites nitidus* occur in the lower Chesterian of northwest Alabama. Thus, the palynomorphs from the Pride Mountain Formation, the Hartselle Sandstone, and the basal Bangor Limestone of northwest Alabama further confirm Sullivan's proposal.

PLATE 3

All specimens magnified x600. All the stage coordination is recorded on a Leitz Ortholux microscope.

1 *Tripartites vetustus* Schemel 1950. Slide number STOP1-4(1), stage coordination 17.7-104.8.

2 *Tripartites marginatus* Hoffmeister et al. 1955. Slide number STOP1-4(1), stage coordination 7-113.

3 *Densosporites regalis* (Bharadwaj & Venkatachala) Smith & Butterworth 1967. Slide number STOP1-4(1), stage coordination 11-104.2.

4-6 *Crassispora? punctata* n. sp. 4. Slide number STOP1-3(1), stage coordination 17.3-126.8; 5. Slide number STOP1-3(1), stage coordination 9-109; 6. Slide number STOP2-1(1), stage coordination 11-116.4.

7 *Laevigatosporites* sp. Slide number STOP2-1(1), stage coordination 13-121.

8 Scoledontes. Slide number STOP1-3(1), stage coordination 18-104.

9 *Grandispora spinosa* Hoffmeister et al. 1955. Slide number STOP1-4(1), stage coordination 6.8-106.

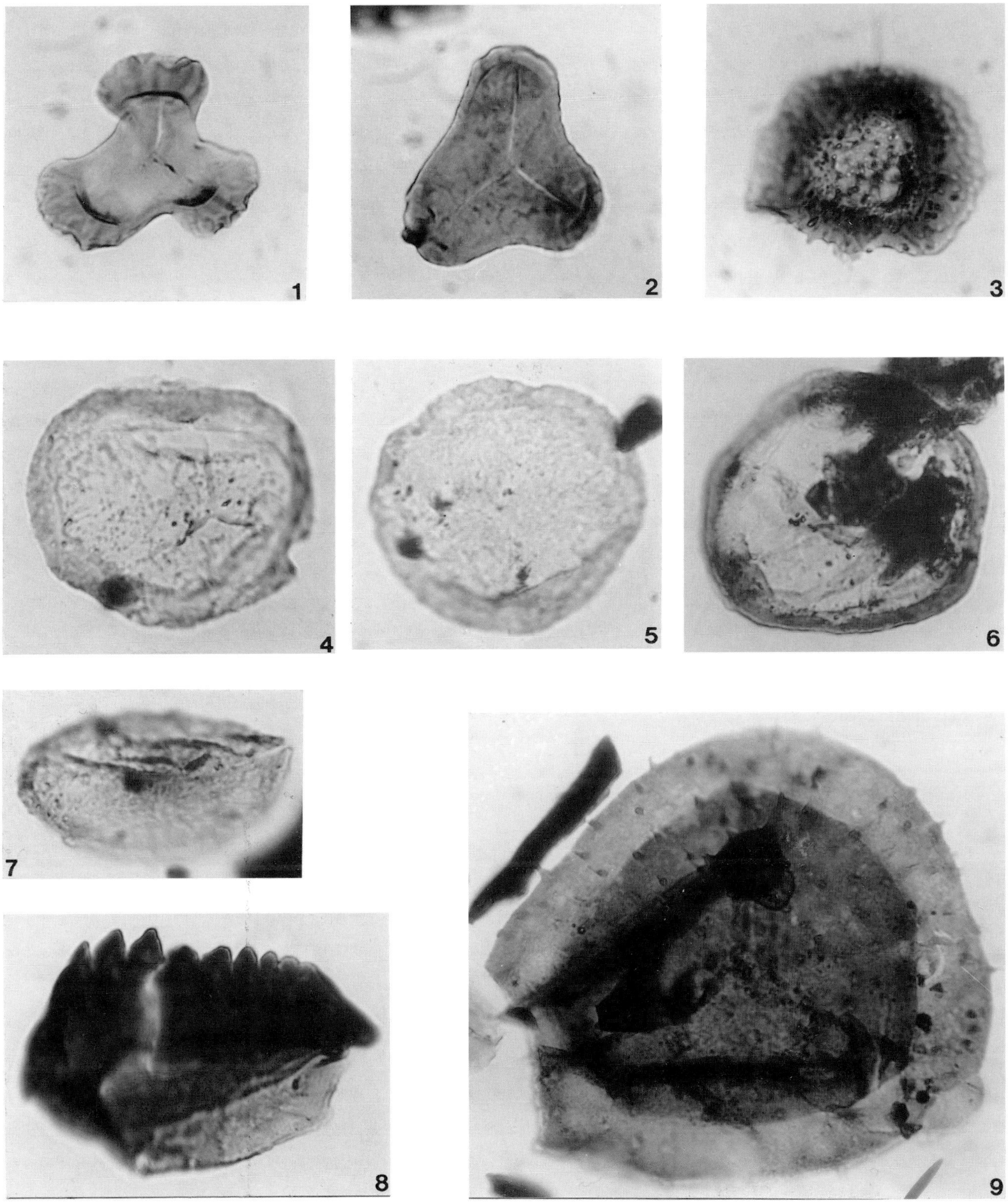

SYSTEMATIC PALEONTOLOGY

Infraturma APICULATI (Bennie & Kidston) Potonié 1956
Subinfraturma Baculati Dybova & Jachovicz 1957

Genus *Raistrickia* (Schopf et al. 1944)
Potonié & Kremp 1954

Raistrickia bangorii sp. nov.
Plate 1, figs. 13–14

Derivation of Name. The specific epithet, bangorii, is derived from the Bangor Limestone, referring to the first occurrence.

Types. Holotype: Plate 1, Figure 13, Paratype: Plate 1, Figure 14.

Holotype Locality and Age. Basal Bangor Limestone, Fielder Ridge, Colbert County, Alabama. Early Chesterian or late Visean (V3c), VF zone or lower NC zone.

Description. Amb circular to subcircular. Laesurae not always visible; when present, Laesurae extend at least half of the radius of spore. Equatorial and distal exine covered by bacula, closely spaced; the base of process is slightly broader; apices rounded, or truncate. Proximal exine laevigate. Exine thin.

Measurement. Size range: 61.7–71.8 μm; Holotype size: 61.7x66.7 μm; bacula: base width about 1.5 μm, length 2–4.5 μm.

Comparison. *Raistrickia bangorii* sp. nov. differs from other species assigned to *Raistrickia* by its densely spaced, slim bacula.

Occurrence. Basal Bangor Limestone, lower Chesterian.

Infraturma CRASSITI (Bharadwaj & Venkatachala 1957)
Smith & Butterworth 1967

Genus *Crassispora* (Bharadwaj 1957) Sullivan 1964

Crassispora punctata sp. nov.
Plate 3, figs. 4–7

Derivation of Name. The specific epithet, punctata, is derived from the Latin *punctum*, referring to the type of ornamentation.

Types. Holotype: Plate 3, Figure 4; Paratype: Plate 3, Figure 6.

Holotype Locality and Age. Basal Bangor Limestone, Fielder Ridge, Colbert County, Alabama. Early Chesterian or late Visean (V3c), VF zone or lower NC zone.

Description. Amb subcircular, often distorted into an irregular outline. Trilete mark obscure in most specimens, when present, it is simple and extends to inner margin of the thickening at equator. Inner margin of the crassitudinous thickening at the equator usually poorly defined. Exine is punctate, and sometimes folded.

Measurement. Size range: 73x80 μm; Holotype size: 51–87 μm; thickness of the crassitudinous thickening 3.5–10 μm.

Remarks. The current species is tentatively assigned to *Crassispora* on the basis of its crassitudinous thickening at the equator. *Crassispora* includes species with distal cones or spines.

Comparison. *C. punctata* differs from other species assigned to Crassispora in its lack of raised ornamentation, such as grains, cones, and spines.

Occurrence. The Pride Mountain Formation, the Hartselle Sandstone, and the basal Bangor Limestone; lower Chesterian.

ACKNOWLEDGMENT

This research is supported in part by funds from Sigma Xi to the senior author and from the National Science Foundation (EAR-9418424) to the second and third authors.

References Cited

BURDICK, D.W., and STRIMPLE, H.L.
1982 Genevievian and Chesterian crinoids of Alabama. *Alabama Geological Survey Bulletin* 121, 277 pgs.

BUTTS, C.
1926 The Paleozoic rocks. *In*: Adams, G.I., Butts, C., Stephenson, L.W., and Cooke, C.W. (eds.), *Geology of Alabama. Alabama Geological Survey Special Report* 14: 41–230.

CLAYTON, G., COQUEL, R., DOUBINGER, J., GUEINN, K.J., LOBOZIAK, S., OWENS, B., and STREEL, M.
1977 Carboniferous miospores of western Europe: illustration and zonation. *Mededelingen Rijks Geologische Dienst*, 29:1–71

DRAHOVZAL, J.A.
1967 The Biostratigraphy of Mississippian rocks in the Tennessee valley. *In*: Smith, W.E. (ed), *A field guide to Mississippian sediments in northern Alabama and south-central Tennessee. Guidebook for 5th Annual Field Trip of Alabama Geological Society*, pp. 10–24.

HOFFMEISTER, W.S., STAPLIN, F.L., and MALLOY, R.E.
1955 Mississippian plant spores from the Hardinsburg Formation of Illinois and Kentucky. *Journal of Paleontology*, 29(3): 372–399.

KOSANKE, R.M.
1982 Mississippian–Pennsylvanian boundary in the United States based on palynomorphs. *In*: Rainsbottom, W.H.C. and Owens, B. (eds.), *Biostratigraphic Data for a Mid-Carboniferous Boundary*, Leeds, pp. 27–35.

MAPLES, C.G., and WATERS, J.A.
1987 Redefinition of the Meramecian/Chesterian boundary (Mississippian). *Geology*, 15: 647–651.

NEVILLE, R.S.W.
1968 Ranges of selected spores in the upper Visean of the East Fife Coast section between St. Monance and Pittenweem. *Pollen et Spores*, 10(2): 431–462.

NEVES, R., GUEINN, K.F., CLAYTON, G., IOANNIDES, N.S., and NEVILLE, R.S.A.
1972 A scheme of miospore zones for the British Dinantian. *VII Congrés International de Stratigraphie et de Géologie du Carboniféře*, Krefeld, Compte Rendu 1, pp. 347–353.

PASHIN, J.C.
1993 *New perspectives on the Mississippian System of Alabama.* Guidebook for the 30th Annual Field Trip of the Alabama Geological Society, November 11–13, 1993, 151 pgs.

PASHIN, J.C., and RINDSBERG, A.F.
1993 Stop 2: Alabama Rock Asphalt Quarry. *In*: Pashin, J.C. (ed.), *New perspectives on the Mississippian System of Alabama, Guidebook for the 30th Annual Field Trip of the Alabama Geological Society*, November 11–13, 1993, pp. 119–123.

RINDSBERG, A.F., and PASHIN, J.C
1993 Stop 1: Ichnology of the Hartselle Sandstone at Fielder Ridge. *In*: Pashin, J.C. (ed.), *New perspectives on the Mississippian System of Alabama, Guidebook for the 30th Annual Field Trip of the Alabama Geological Society*, November 11–13, 1993, pp. 116–119.
SCHEMEL, M.P.
1950 Carboniferous plant spores from the Daggett County, Utah. *Journal of Paleontology*, 24(2): 232–244.
STAPLIN, F.L.
1960 Upper Mississippian plant spores from the Golata Formation, Alberta, Canada. *Palaeontographica*, Abteilung B, 107(1-3): 1–40.
SULLIVAN, H.J.
1965 Palynological evidence concerning the regional differentiation of Upper Mississippian floras. *Pollen et Spores*, 7(3): 539–561.
SULLIVAN, H.J., and MARSHALL, A.E.
1966 Visean spores from Scotland. *Micropaleontology*, 12(3): 265–285.
THOMAS, W.A.
1988 The Black Warrior Basin. *In*: Sloss, L.L. (ed.), *Sedimentary Cover-North American Craton: Geological Society of America, The Geology of North America*, D-2: 471–492.
UTTING, J.
1987 Palynology of the Lower Carboniferous Windsor Group and Windsor-Canso boundary beds of Nova Scotia, and their equivalents in Quebec, News Brunswick and Newfoundland. *Geological Survey of Canada Bulletin*, 374, 93 pgs.

Authors' addresses

YU-XING ZHOU
ANNE RAYMOND
Department of Geology and Geophysics
Texas A&M University
College Station, TX 77843-3115
U.S.A.

CORTLAND F. EBLE
Kentucky Geological Survey
University of Kentucky
Lexington, KY 40506
U.S.A.

PALYNOLOGY OF THE PERMIAN BOLZANO VOLCANIC COMPLEX, SOUTHERN ALPS, ITALY, PART 1: MIOSPORE PRESERVATION, QUANTITATIVE SPORE COLOR AND QUANTITATIVE FLUORESCENCE MICROSCOPY

Christoph Hartkopf-Fröder, Gordon D. Wood, and Karl Krainer

Abstract

Fluvial-lacustrine sediments, intercalated with lavas and ignimbrites of the Bolzano Volcanic Complex (Permian, Southern Alps), yield a rich, diverse and well-preserved miospore assemblage. Due to early diagenetic silica precipitation miospores from chert layers and lenses are three-dimensionally preserved in contrast to assemblages from mudstones, which are usually compressed. The Thermal Alteration Index (TAI) is assessed between 1+ and 2+. Quantitative spore color analysis and quantitative fluorescence microscopy indicate a very low thermal alteration in the Bolzano Volcanic Complex. Miospores from the base of the overlying siliciclastic Gröden Formation have a TAI of 2 to 3-. It is possible that the existence of dense chert layers in the lacustrine sediments suppressed the circulation of pore fluids and acted as an effective barrier against physical, chemical and biological alteration.

The presence of small numbers of dark brown and black spores, spore clusters and terrestrial plant detritus is interpreted to represent charring during wildfires or volcanic activity. Considering the thick volcanic succession of lavas and pyroclastics, and a semiarid to arid climate, wildfires were probably a widespread and recurrent phenomenon.

INTRODUCTION

This paper is a preliminary assessment of the palynological assemblage from the Bolzano Volcanic Complex (BVC), also known as the Atesino Platform Volcanics, from the Southern Alps, northern Italy (see Text-Figure 1). Palynomorphs were recovered from limestones, mudstones and cherts, the latter yielding three-dimensionally preserved specimens. The objectives of this paper are to outline the geology and introduce pertinent palynological information applicable to palynofacies and organic maturation. An extensive taxonomic treatment of the assemblage and its palynostratigraphic significance is in preparation.

GENERAL GEOLOGY

In the central Southern Alps (Alto Adige/South Tyrol; northern Italy) the Permian succession consists of two tectono-sedimentary cycles. The first involves infilling of small intermontane grabens, or half-grabens, and is dominated by alluvial fan and lacustrine sediments and volcanics. The second cycle, which begins above a marked unconformity, consists of fluvial, coastal sabkha and shallow marine deposits. These cover the sediments of the lower tectono-sedimentary cycle and the structural highs in between (Cassinis and Neri, 1992). The BVC belongs to the lower cycle and reaches a maximum thickness of approximately 3 km, and an areal extent of 4,000 km^2, including an area covered by Mesozoic sediments. In the study area (see Text-Figures 1 and 2), which is located in the Bolzano graben system, the BVC is superjacent to the Variscan basement ("Brixener Quarzphyllit"), or the locally developed Ponte Gardena Conglomerate (Waidbrucker Konglomerat, Conglomerato di Ponte Gardena; Krainer, 1989). It is subjacent to the Gröden Formation (Val Gardena Sandstone) which represents the base of the second tectono-sedimentary cycle (Conti et al., 1986; Conti et al., 1997; Cassinis and Neri, 1992). In the Tregiovo Basin the top of the BVC contains a distinctive unit called the Tregiovo Beds (Giannotti, 1962; Cassinis and Neri, 1992; Barth and Mohr, 1994).

Hartkopf-Fröder, Christoph, Wood, Gordon D., and Krainer, Karl

2001 Palynology of the Permian Bolzano Volcanic Complex, southern Alps, Italy, Part 1: Miospore preservation, quantitative spore color and quantitative fluorescence microscopy; *In*: Goodman, D.K., and Clarke, R.T. (eds.), *Proceedings of the IX International Palynological Congress, Houston, Texas, U.S.A., 1996*; American Association of Stratigraphic Palynologists Foundation, p. 79–97.

 ISBN 0–931871–06–9

The BVC sequence consists primarily of ignimbrites, pyroclastics and subordinate rhyodacitic lava flows and andesitic rocks. Although the complex lithostratigraphy of the BVC is still a matter of discussion, the succession can be roughly divided into a lower "group" dominated by andesitic to dacitic-rhyodacitic volcanic rocks, and an upper "group" consisting of rhyodacitic to rhyolithic volcanic rocks. Recent geological mapping (Hartkopf-Fröder and Krainer, 1990; Krainer and Spötl, 1998) has revealed intervolcanic sediments including silica layers intercalated with stromatolitic carbonates, sandstones, siltstones, and mudstones (see Text-Figures 3 and 4). These are interpreted as fluvial-lacustrine sediments deposited during volcanic quiescence. Clastic sediments containing weathered volcanic rock transported to small basins formed by volcano-tectonic processes within the BVC are the result of sheet floods and debris flows. There is no evidence of volcanic activity during the deposition of these sediments. Krainer and Spötl (1998) have interpreted chert deposition to be abiogenic precipitates analogous to modern volcanic lake settings. Physico-chemical weathering of rhyolithic and rhyodacitic volcanic rocks released the silica transported to lacustrine environments. The presence of soft-sediment deformation structures, shrinkage features and pristine three-dimensionally preserved palynomorphs suggests very early diagenetic precipitation and consolidation prior to mechanical compaction of sediment. Lakes are assumed to

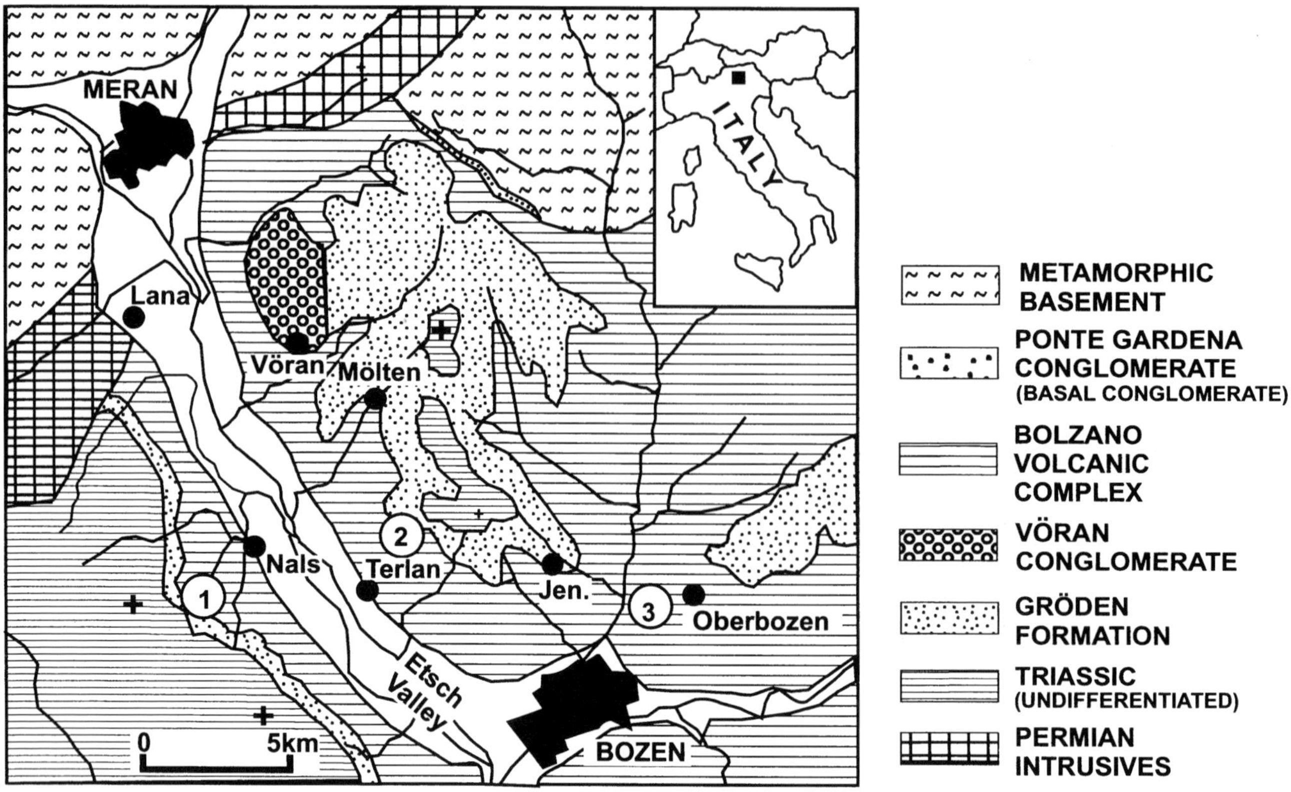

Text-Figure 1. Idealized geological map showing location of sections studied (1=Grissian Gorge; 2=Gorl I and II; 3=Oberbozen).

have been highly alkaline because of the elevated silica content and were probably perennial in nature., This is aupported by the domination of fine-grained laminated sediments and absence of desiccation cracks and evaporites.

Excluding scattered plant macrofossil remains, which do not allow precise age assignments, no biostratigraphic data are available from the BVC. Several papers concerning isotopic age determinations of the lower and upper "group" have been published, however, there is no agreement on the age of the emplacement of the BVC (Bargossi et al. 1993; Barth and Mohr, 1994; DAmico, 1986; D'Amico and Del Moro, 1988; D'Amico et al., 1980; Hess, 1990; Savelli and Balboni, 1984; Storzer, 1970). On the basis of $^{40}Ar/^{39}Ar$ ages of biotites Hess (1990) determined a minimum time span of volcanic activity from 298 to 286 Ma. Using Rb-Sr biotite isochrone ages from biotites D' Amico (1986), D'Amico et al. (1980) and D'Amico and Del Moro (1988) found that the age of the BVC ranged from 276±2 Ma to 267±2 Ma.

The BVC, and at least the lower part of the Gröden Formation, are characterized by reversed polarities and belong to the Carboniferous–Permian Reversed Superchron (Dachroth, 1976; Becke and Mauritsch, 1984). The Illawarra Reversal, which marks the base of a zone of mixed polarity (Permo–Triassic Mixed Superchron), is located within the Gröden Formation (Dachroth, 1976; Mauritsch and Becke, 1983).

The Gröden Formation, which overlies the BVC, has yielded a wealth of paleontological and biostratigraphical information. In general, palynomorphs (e.g., Massari et al., 1988; Massari et al., 1994; see below), plant remains (e.g., Clement-Westerhof, 1984), tetrapod footprints (e.g., Conti et al., 1977; Ceoloni et al., 1988; Massari et al., 1988), and a marine fauna which includes orthocone nautiloids, have been recovered. The nautiloids are restricted to rather thin horizons (e.g., Broglio Loriga et al., 1986; Broglio Loriga and Cassinis, 1992). The flora and fauna indicate a Late Permian age. Biostratigraphical correlation with Permian deposits of middle European intramontane basins and especially with the marine standard scale is still very difficult, and is probably related to regional climatic differences.

The Gröden Formation yields specimens of *Peltaspermum martinsii* and *Pseudovoltzia liebeana* which are strongly cutinized and densely covered by papillae (Poort and Kerp, 1990; Kerp, 1996). Cuticles of the same species from the Zechstein of Germany and England are thinner and weakly papillate. The Gröden Formation specimens exhibit morphological features typical of a drier and more arid (xeric) climate during the Late Permian in the Southern Alps. Recently, in a tentative chronostratigraphic scheme of the Permian deposits in the eastern Southern Alps, the BVC was regarded as Artinskian to Ufimian in age, the Gröden Formation was referred to the Midian to Dzhulfian/ Changxingian and the Bellerophon Formation to the Dzhulfian to Changxingian (Massari et al., 1994).

MATERIALS AND METHODS

Four sections were sampled in this preliminary study: Grissian Gorge, Gorl I, Gorl II, and Oberbozen (see Text-Figures 1-3). Palynomorphs were recovered from lime-

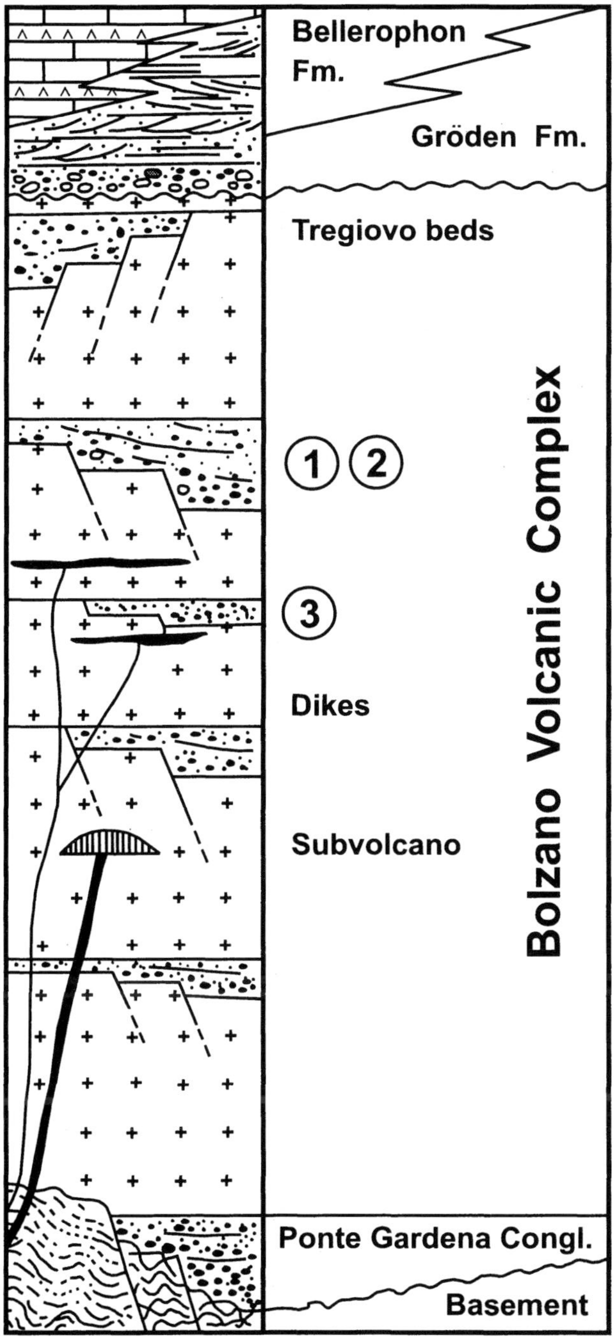

Text-Figure 2. Generalized stratigraphic column of Permian and associated rocks, Bolzano/Bozen area, Southern Alps (after Brandner and Mostler, 1982). Sampling locations are indicated by encircled numbers (1=Grissian Gorge; 2=Gorl I and II; 3=Oberbozen).

stones, mudstones and cherts using standard palynological processing techniques (Wood, Gabriel and Lawson, 1996). Selected chert samples were studied in thin section. However, only certain samples are discussed in this paper.

Elvacite 2044™ epoxy resin was used as a mounting medium for fluorescence study because glycerine jelly fluoresces intense dark blue when exposed to long-wave UV light. Palynomorphs embedded in rock were studied in polished rock samples using incident light fluorescence microscopy. The filter block incorporated an excitation filter with a bandwidth of 395–440 nm (blue-violet light), a suppression filter LP 470 nm and a beam splitting mirror of 460 nm. Methods for quantitative spore color analysis and quantitative fluorescence microscopy are described in detail under their respective subheadings.

Scanning Electron Microscopy (SEM) of three-dimensionally preserved miospores within the sediment follows the technique described by Munnecke and Servais (1996) with the modification that hydrochloric and hydrofluoric acids were used to etch the palynologically-rich chert layers.

Permanent repository for palynological slides, SEM stubs and material used in quantitative color and fluorescence analyses is the Geological Survey of North Rhine–Westphalia (Krefeld).

PERMIAN PALYNOLOGICAL STUDIES OF THE SOUTHERN ALPS

The Permian plant macrofossils of the Italian Alps have been the subject of several papers (e.g., Clement-Westerhof, 1974, 1984, 1987). There are few studies concerning miospore assemblages from West European upper Lower and lower Upper Permian ("Saxonian") sediments. This is primarily due to facies such as coarse-grained alluvial fan sediments, red beds and thick volcanic sequences that are not conducive to palynomorph preservation. However, several studies concerning the Southern Alps discuss palynomorph assemblages from the Gröden Formation, the Bellerophon Formation, and lacustrine sediments of the Tregiovo and Collio Basins. One of the first investigations was published by Klaus (1963) and the excellent taxonomic descriptions and photographic documentation mark this paper as a benchmark contribution to Permian palynology. A preliminary review of palynomorphs and palynofacies (see Plates 2 and 3) of the BVC was discussed by Wood et al. (1996) and Krainer et al. (1996).

Diverse and abundant miospores have been recovered from the Gröden and Bellerophon formations (Massari et al., 1988; 1994) which unconformably overlay the BVC. The major components of this assemblage are *Lueckisporites virkkiae, Nuskoisporites dulhuntyi, Lunatisporites* spp., *Jugasporites delasaucei, Klausipollenites schaubergeri, Alisporites* spp., *Falcisporites* spp., *Vesicaspora* spp. and *Paravesicaspora* spp. Rare, but consistently present taxa include *Gardenasporites* spp., *Gigantosporites* spp., *Inaperturopollenites dolomiticus, Jugasporites* spp., *Perisaccus granulosus, Scheuringipollenites* spp., *Striatoabieites* spp., *Striatopodocarpites* spp. and *Strotersporites* spp.

Lacustrine-fluviatile sediments of the Tregiovo Basin are exposed at the northwestern margin of the Bolzano Volcanic Complex. These sediments are approximately 200 m in thickness and intercalated between volcanic sequences. The miospores of the Tregiovo Formation have been studied by Cassinis and Doubinger (1991a, b) and Barth and Mohr (1991, 1994). Cassinis and Doubinger (1991a) differentiated two distinct assemblages. Both are dominated by taeniate and alete bisaccate pollen, however, many forms were not identified at the generic level. Characteristic species of the older assemblage are *Trizonaesporites grandis,*

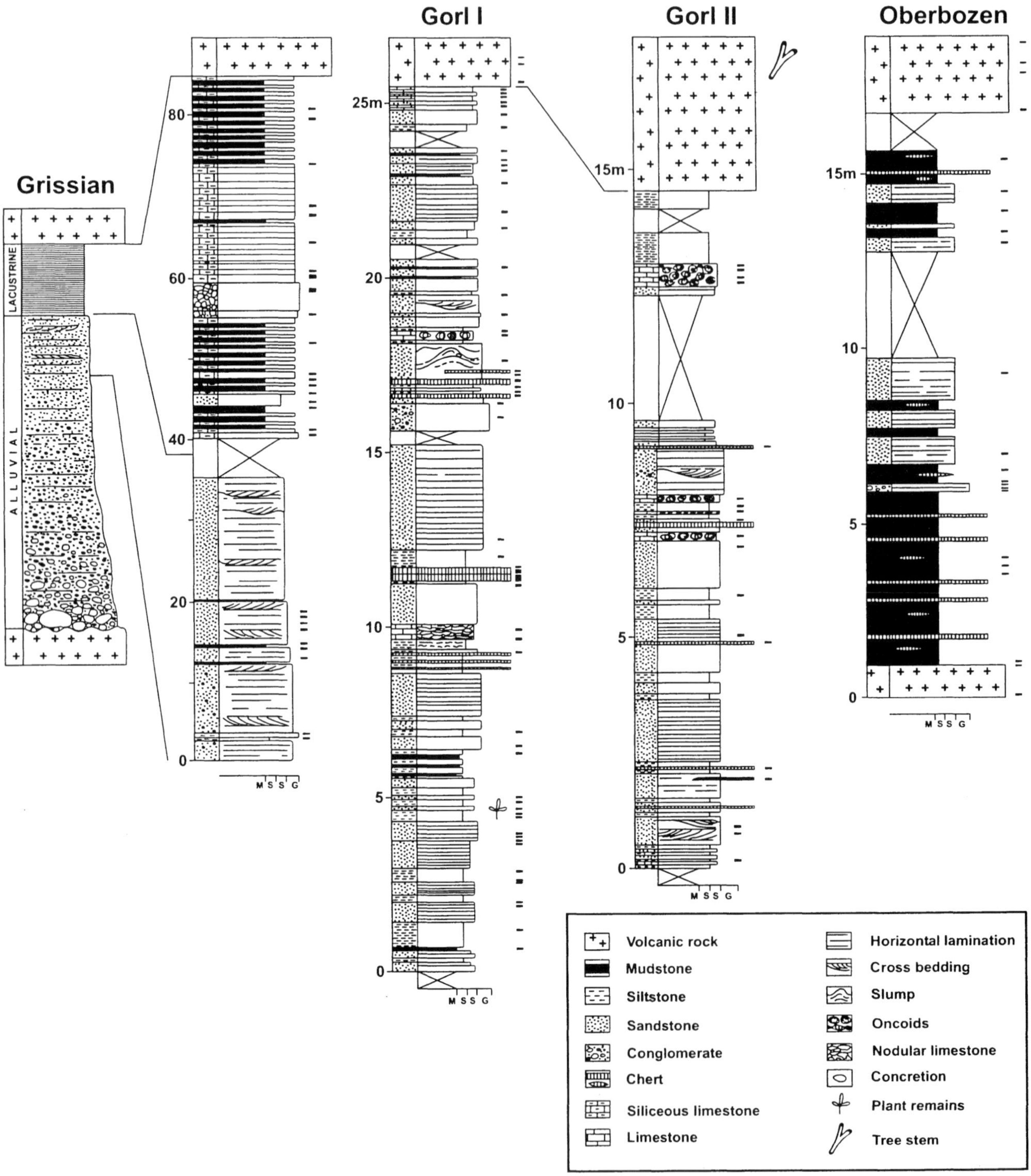

Text-Figure 3. Columnar section at Grissian Gorge (= locality 1 in Text-Figures 1 and 2), Gorl I and II (= locality 2 in Text Figures 1 and 2) and Oberbozen (= locality 3 in Text-Figures 1 and 2) showing alluvial fan and lacustrine sediments intercalated in the Bolzano Volcanic Complex (after Krainer and Spötl, 1998, fig. 2). Dashed lines to the right of sections indicate a horizon sampled for palynological analysis.

Lunatisporites cf. *L. labdacus*, and *Lueckisporites microgranulatus*. A second assemblage, present only in the two uppermost samples, also contains *Plicatipollenites* sp., *Densipollenites oviformis, Gigantosporites aletoides, Gardenasporites* sp., *Lueckisporites virkkiae*, rare specimens of *L. microgranulatus, Corisaccites alutas, Crucisaccites variosulcatus* and *Paravesicaspora splendens. Trizonaesporites grandis* and *Lunatisporites* cf. *L. labdacus* are absent from the assemblage. Based on palynostratigraphic evidence Cassinis and Doubinger (1991 a, b) and Barth and Mohr (1994) suggest a Kungurian–Ufimian age for the Tregiovo Formation.

The Collio Basin lies southwest of the Tregiovo Basin, in the Brescian Alps. Basin fill consists of alluvial to lacustrine sediments and volcanic rocks (Cassinis and Doubinger, 1991 b). At the type section the Collio Formation is approximately 750 m thick. Clement-Westerhof et al. (1974) first reported on the miospores from the Collio Formation. However, no stratigraphic information was given for the single sample yielding palynomorphs. The assemblage is dominated by forms assigned to *Cordaitina, Vittatina* and taeniate pollen. Clement-Westerhof et al. (1974) emphasized that the composition of the assemblage cannot be compared to others presently known from either Autunian or Thuringian deposits of western Europe and suggests a Late Artinskian to Early Ufimian age. Cassinis and Doubinger (1991a, b) studied three productive samples from the lower third of the Collio Formation. The most characteristic taxa are *Potonieisporites* cf. *P. enormis, Nuskoisporites* sp., *Lueckisporites microgranulatus, Vittatina costabilis, V. foveolata* and bisaccate pollen. Minor constituents include *Lundbladispora simoni, Cordaitina* sp., *Vestigisporites minutus, Playfordiaspora crenulata, Falcisporites* cf. *F. zapfei, Vittatina* sp. A *sensu* Hochuli (1985) and taeniate bisaccate pollen. In these samples *Cordaitina* is not abundant indicating that the assemblage described by Clement-Westerhof et al. (1974) reflects local ecological conditions. Cassinis and Doubinger (1991a, b) suggested a Late Artinskian–?Early Ufimian age based on palynomorph data.

PALYNOMORPH PRESERVATION IN THE BOLZANO VOLCANIC COMPLEX

Most samples, particularly those from the Grissian Gorge section, yielded exquisitely preserved miospores. Spore color ranges from primarily light yellow to light brown, with the exception of a few dark brown to black specimens (see Plate 5). Bacterial or fungal pitting of the exine and corrosion caused by pyrite formation and fragmentation (e.g. of bisaccates) is rare. Some samples yield three-dimensionally preserved miospores (see Plate 1, figs. 1-2; Plate 2, figs. 1-4; Plate 4, figs. 1-3). Three-dimensionally preserved palynomorphs are abundant in chert layers and stringers but are rarely found in the surrounding carbonate beds. The excellent miospore preservation in chert suggests that syndepositional silica precipitation and very early consolidation provided an efficient barrier to physical, chemical and biological degradation. In this respect, the miospores entombed in the BVC chert layers exhibit preservation comparable to that reported in Precambrian organic-walled microfossil assemblages found in early diagenetic cherts (Knoll, 1985). Obviously, early diagenetic cementation and the lack of compression is an important, but certainly not the only factor, contributing to the excellent state of palynomorph preservation (see Ferris et al., 1988; Hemsley and Scott, 1991). The sources of the silica are probably the rhyolithic to rhyodacitic volcanic rocks surrounding the lake and acidic volcanic glass (Krainer and Spötl, 1998).

The lake was probably alkaline with anoxic conditions at or near the sediment-water interface (Krainer and Spötl, 1998). Lake water salinity increased during dry seasons. These conditions inhibit early microbial degradation of organic matter (Southgate, 1986) but facilitate rapid chert formation. This protected the miospores from corrosive bottom water, abiotic oxidation, and distortion (e.g., flattening) during sediment compaction. Post-lithification oxidation from pore fluids or subsurface waters was reduced in the impermeable chert compared to more porous rocks like mudstones or poorly sorted sandstones. As Knoll and Ohta (1988) have demonstrated, acritarchs and organic-walled filaments of cyanobacterial origin can be well preserved even in chert nodules from greenschist facies metasediments because the chert reduces the effects of deformation, diagenesis and metamorphism.

Maximum burial temperatures of the Permian chert-bearing rocks in the study area are unknown, however, preliminary results from vitrinite reflectance measurements of wood fragments from the Oberbozen locality have 0.54% Ro. This value corresponds to a maximum paleotemperature of 76–87°C (using both low and fast heating rates as proposed by Barker and Pawlewicz, 1986). Higher paleotemperatures have been reported from similar rocks in the Tregiovo Basin based on Thermal Alteration Index, suggesting a rather complex thermal regime partially influenced by both Permian and mid–Triassic hydrothermal heating (Barth and Mohr, 1994). The occurrence of hydrothermal processes and low organic maturation in the BVC is perplexing. It is possible that zones of chert may have acted in some manner to dissipate heat flow in the BVC resulting in less than expected maturation. Thus, the non-porous character of the cherts in the BVC sedimentary sequence acted as an impermeable 'seal' and did not allow migration of hot fluids. Additional detailed analyses concerning the thermal conductivity of cherts are required to accurately determine the effect of these rocks in the section (see Carter et al., 1998).

QUANTITATIVE MIOSPORE COLOR

There are numerous organic maturation scales based on spore color. Most of these scales are subjective, requiring visual estimates by an individual researcher. Several attempts have been made to circumvent these shortcomings and to create quantitative spore color indices which are objective and considerably more accurate than the visual Thermal Alteration Index (TAI) estimates (see discussion and references in Marshall, 1991). The quantitative technique of Marshall (1990, 1991) and Milton (1993) has been applied to the miospores of the BVC. This approach uses transmitted light micro-spectrophotometry with the acquired spectra converted into the Commission Internationale de l'Éclairage (International Commission of Illumination or CIE) color system and expressed numerically. Complete description of a color requires the three dimensional representation of a*, b* (the two chromaticity coordinates), and L* (lightness). Color data are usually presented using a combination of two diagrams, the chromaticity diagram with a* versus b* and the color tone diagram with chroma (C^*_{ab}) versus L*. The chroma value can be calculated from a* and b* (Hunt, 1987).

Text-Figure 4. Diagrammatic representation of Grissian Gorge sample GRI 27 showing interval composed of intercalated algal laminites, silica layers and micritic limestone layers (after Krainer and Spötl, 1998, fig. 3).

Quantitative spore color measurements were taken using a Zeiss Universal Microspectrophotometer (USMP 50) in the School of Oceanography and Earth Science, University of Southampton, England. The microscope is equipped with a stabilized 12V 100W halogen bulb whose color temperature is modified by a filter to 2854 K required by Standard Illuminant A of the Commission Internationale de l'Éclairage color system. All measurements were made in transmitted light mode. After completion of approximately forty measurements (see Table 1) the calibration of the system was checked against the color of the illuminant. The system is controlled by a personal computer which calculates the color spectra (a*, b* and L*) for single spores, the mean and standard deviation, and plots the values graphically and in spreadsheet format (for more details see Marshall, 1990, 1991; Milton, 1993).

Bisaccate pollen were chosen for measurement because of the rarity of single layered trilete spores. The measurement spot was placed, without exception, on the saccus of a bisaccate pollen free of extraneous organic matter. Only non-corroded specimens were chosen with sacci of uniform thickness and free of folds or scars. This ensures that color measurements are free of variation caused by wall thickness. Eight samples (see Table 1) collected from the Grissian Gorge Section, were measured. Additionally, two samples (89317, 89318) from gray mudstones near the base of the overlying Gröden Formation in the Grissian Gorge have been measured for comparison.

As previously stated, the BVC is a ≤ 3 km sequence of rhyolithic and andesitic-dacitic volcanic rocks. Dikes related to a magmatic event of predominantly Ladinian age are also present (Hess, 1990). Due to these lengthy and intensive magmatic processes, a high heat-flow, and accordingly, a high thermal alteration would be expected. The area was also affected by the Alpine Orogeny, and a thick sequence of Triassic and ?Jurassic/Cretaceous sediments covers the BVC (Bosellini, 1991; Cassinis et al., 1997). For this reason, miospores from the BVC would be expected to exhibit colors of the higher end of spore color scales, i.e. brown or even black. However, the thermal alteration is surprisingly low. Pollen are light yellow to light brown with TAI values between 1+ and 2+ (color scale of Pearson, 1990). Few (less than 2%) dark brown to black spores and spore clusters (TAI 4- to 5) are present.

Chromaticity plots for selected samples are shown in Text-Figure 5. Samples GRI 0, GRI 15, and GRI 27 from Grissian Gorge (see Text-Figure 3) have values of a* and b* concentrated in a continuous and narrow band, with the exception of the very few dark spores. The diagrams indicate that the miospores all have low, positive values of both a* and b* which are characterized by low chroma (weak colors) in the yellow red quadrant. The variation in the a* parameter is relatively low but somewhat higher in the b* parameter. The color tone plot denotes a near transparent state with high lightness values (high chroma values are associated with lower lightness values). Samples GRI 1, GRI 5, GRI 16 and GRI 17 show that a low maturity pattern is still present but the spread of color is not as tightly defined in terms of a* and b* as in GRI 0, GRI 15 and GRI 27.

PLATE 1

1 Transmitted light photomicrograph of miospores in rock thin section, x90.

2 Transmitted light photomicrograph of miospores in rock thin section, x90.

3 Bisaccate pollen in rock thin section, x100.

4, 5 High and low focus light photomicrographs of cuticular material, slide 50439-A-5, 13.7/99.8, x590.

6 Light photomicrograph of cuticular material, slide 50439-A-2, 37.2/99.7, x590.

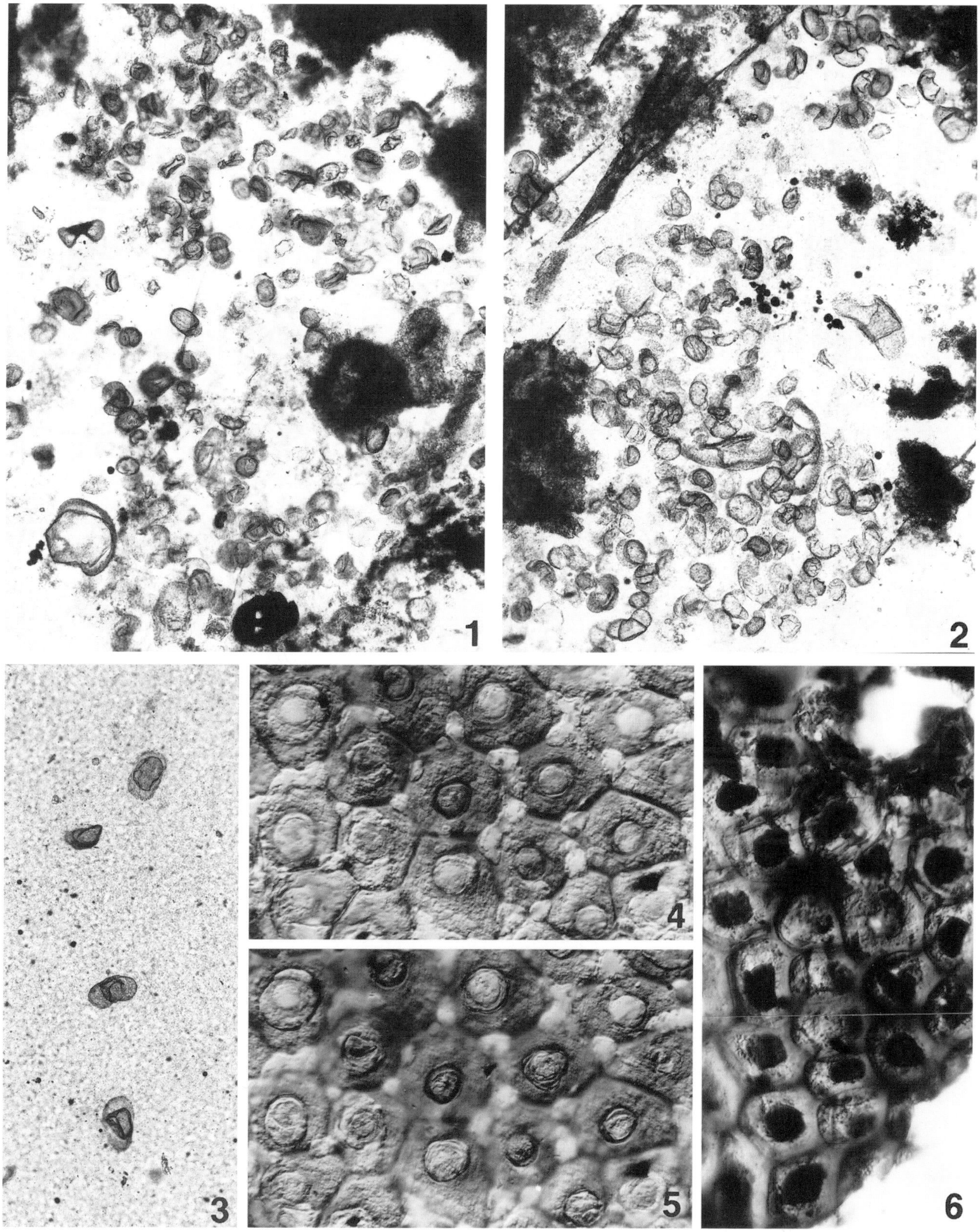
1
2
3
4
5
6

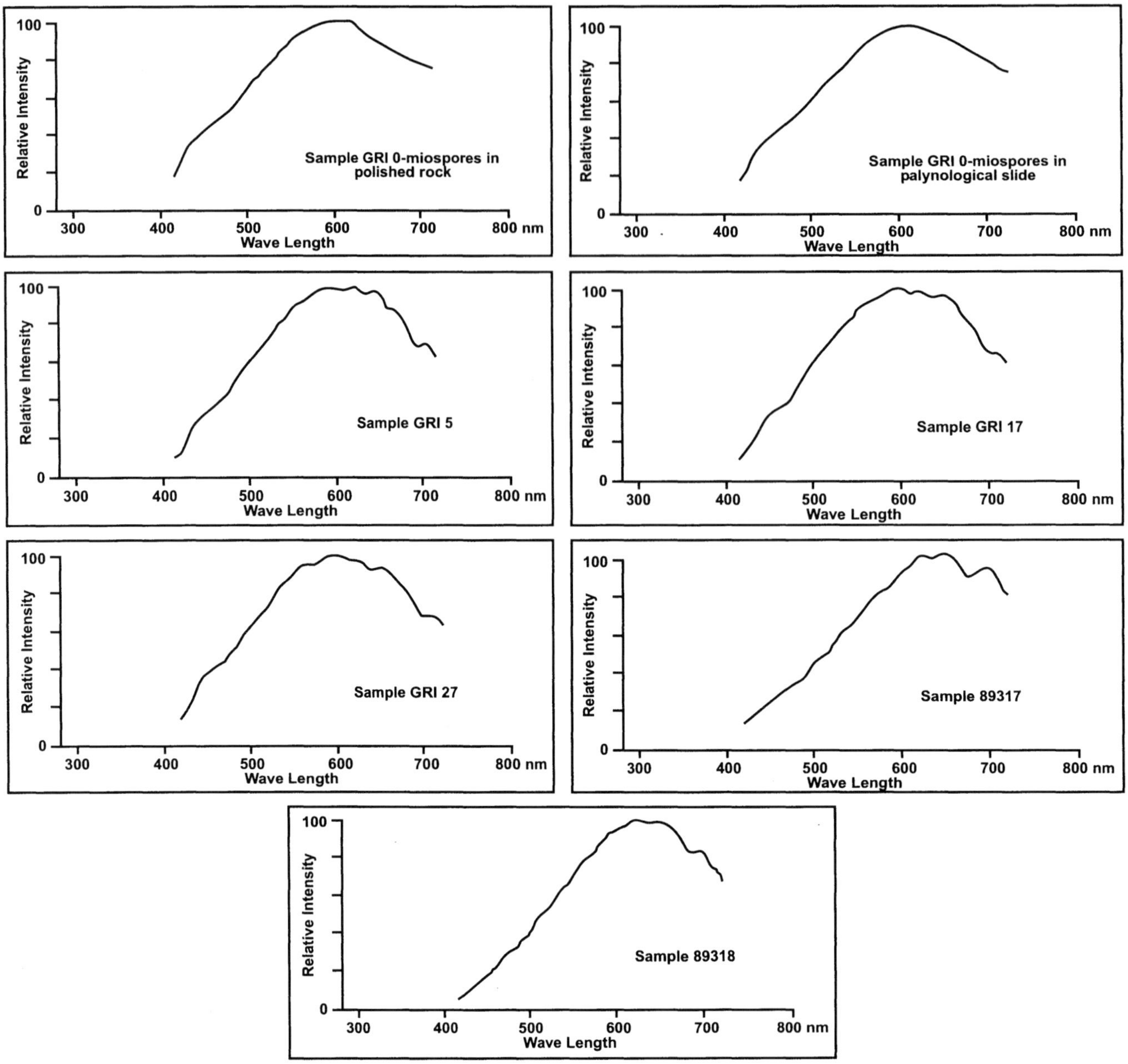

Text-Figure 5. Chromaticity and color tone plot of all measured miospores in selected samples.

PLATE 2

All light photomicrographs magnified x660.

1-4 Light photomicrograph at four focal levels of a three-dimensionally preserved monolete bisaccate, slide 50438-A-3, 26.4/104.0.

5 cf. *Scheuringipollenites* sp., slide 50447-A-4, 29.9/110.6.

6 cf. *Scheuringipollenites* sp., slide 50447-A-4, 46.0/95.7.

7 cf. *Jugasporites* sp., slide 50446-A-6, 5.8/114.1.

8 Monolete bisaccate, slide 50439-A-4, 41.9/106.1.

9 Monolete bisaccate, slide 50447-A-3, 4.6/106.0.

10, 11 Taeniate bisaccate, 50437-A-3, 39.8/101.7.

12 *Florinites* cf. *F. mediapudens*, slide 50437-A-3, 8.8/93.0.

13 Verrucate monolete, slide 50439-A-4, 29.8/99.2.

14 cf. *Striatopodocarpites* sp., 50466-A-6, 22.7/101.1.

15 Striate monolete bisaccate, slide 50446-A-6, 40.1/102.1.

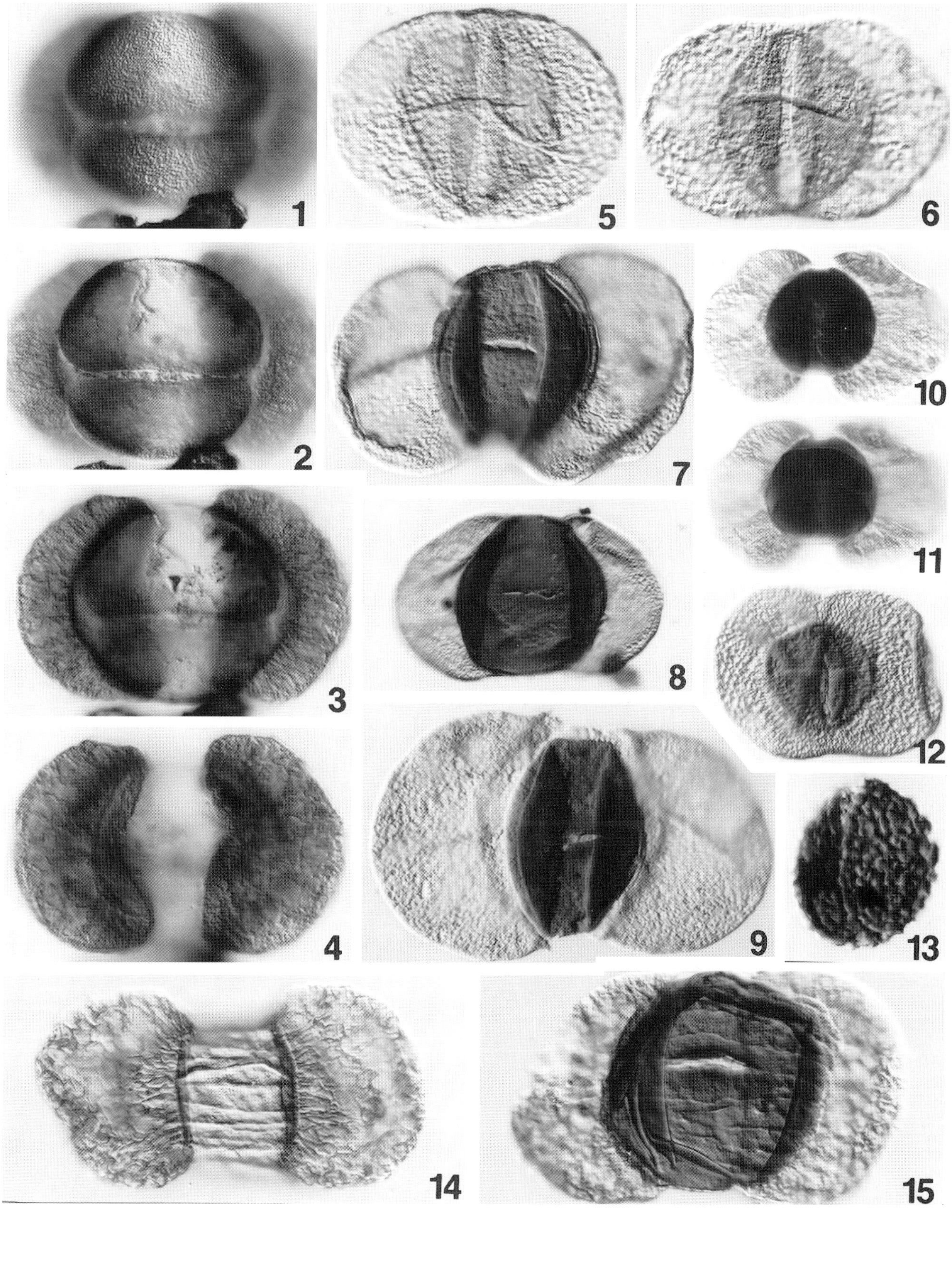
1
2
3
4
5
6
7
8
9
10
11
12
13
14
15

TABLE 1. Chromaticity and color information. Average (mean) color data (CIE 1976 quantities L*, a*, b* and CIE 1976 a, b chroma) of measured samples (l.c. = light colored specimens; d.c. = dark colored specimens).

SAMPLE	L*	a*	b*	C^*_{ab}	n
GRI 0	85.8	7.7	23.3	24.5	40
GRI 1 (l.c.)	74.0	13.6	33.3	36.0	37
GRI 1 (d.c.)	30.7	11.6	2.9	14.9	10
GRI 5	78.1	12.7	36.3	38.5	40
GRI 15	82.2	10.9	28.2	30.3	40
GRI 16	79.0	14.4	34.1	37.1	40
GRI 17	80.4	11.8	36.6	38.5	40
GRI 26 (l.c.)	81.6	9.4	26.1	27.8	40
GRI 26 (d.c.)	34.3	10.7	5.0	13.2	16
GRI 27 (l.c.)	83.2	9.8	26.8	28.3	39
GRI 27 (d.c.)	31.0	8.9	1.1	11.8	5
89317	75.8	17.3	44.4	47.7	40
89318	77.4	16.2	44.2	47.1	40

There is a concurrent increase in chroma and a* and b* values. The values of these parameters in the two samples from the base of the Gröden Formation (89317, 89318) are of higher maturity as shown by the greater b* and slightly higher a* values.

A conspicuous difference exists between the light and dark colored pollen (samples GRI 1, GRI 26, GRI 27) in all color parameters except a*. Most important is the lack of colors with b*>20 and very low chroma and lightness values. Additionally, very low b* values appear.

The inverted "V"-shaped maturity path which was described in detail by Marshall (1991) and Milton (1993) can be recognized in the chromaticity plots (see Text- Figure 5). The values of the light colored spores clearly fall in the range of the ascending limb with progressive increases in the a* and b* parameters. Correspondingly, the color tone diagram exhibits an increase in chroma and a reduction of L*. The dark colored spores, however, are located on the descending limb wherein the chromaticity plot a* and b* and in the color tone plot C^*_{ab} and L* are reducing. The values of a*, b* and L* do not show any peculiarities (e.g. a shift towards uncommon colors or chroma) which might be expected because of the unusual depositional environment and host sediment type. Text-Figure 5 shows that the cusp of the maturity path, where a wide variation of colors expands from the well-defined and narrow pattern, is not present as highly chromatic palynomorphs with similar high values of both a* and b* are absent. Instead, when compared to the normal maturity path the chromaticity curves produced by miospores from samples GRI 1, GRI 26 and GRI 27 abruptly terminate before reaching the cusp but continue only on the lower part of the descending limb. Hence, the color tone plots exhibit two well-separated, discrete sections of the maturity path.

Correlation of quantitative spore color with vitrinite reflectivity is still difficult. Beyond the cusp with its wide spread of colors, changes are initially progressive but from 1.3% reflectivity alteration of the color parameters gradually cease (Marshall, 1991; Milton, 1993). However, from both the quantitative spore color data and spore color scales a vitrinite reflectivity of approximately 0.5 to 0.6% can be deduced.

A single vitrinite reflectance value indicates a low thermal alteration (Ro= 0.54%; 76–87°C) for the BVC (Krainer and Spötl, 1998). Barth and Mohr (1994) reported highly degraded miospores from the Tregiovo sediment complex. These exhibit a TAI of 2+ to 3- according to the Pearson (1990) color chart (equivalent to a vitrinite reflectivity of 0.6–0.8%). The authors ascribe the maturation grade to a sedimentary overburden or a major hydrothermal event of Triassic age. Miospores from the Collio Formation and the Tregiovo Formation figured by Cassinis and Doubinger (1991a, b) are badly preserved. TAI values were not determined by Cassinis and Doubinger (1991a, b), however, the miospores from the Collio Formation are dark brown to black. The thermal alteration of the Gröden Formation which overlays the BVC is low. Massari et al. (1994) stated that miospores are yellow in color.

It is quite surprising that Gröden Formation miospores from Grissian Gorge (samples 89317, 89318 from the base of the unit) show a slightly higher thermal maturation. These samples come from a gray silty mudstone approxi-

PLATE 3

All light photomicrographs x660 unless otherwise stated.

1 Trilete monosaccate, slide 50446-A-6, 25.8/108.2, x500.

2 Trilete monosaccate, slide 50447-A-2, 44.9/105,7, x500.

3, 4 *Nuskoisporites* sp. exhibiting three dimensional preservation, slide 50437-A-4, 4.0/100.0, x525.

5, 6 ?Cavate trilete, slide 50439-A-3, 12.5/109.0.

7 *Alisporites* sp., slide 50437-A-3, 14.2/101.0.

8 *Alisporites* sp., slide 50447-A-4, 16.2/107.8.

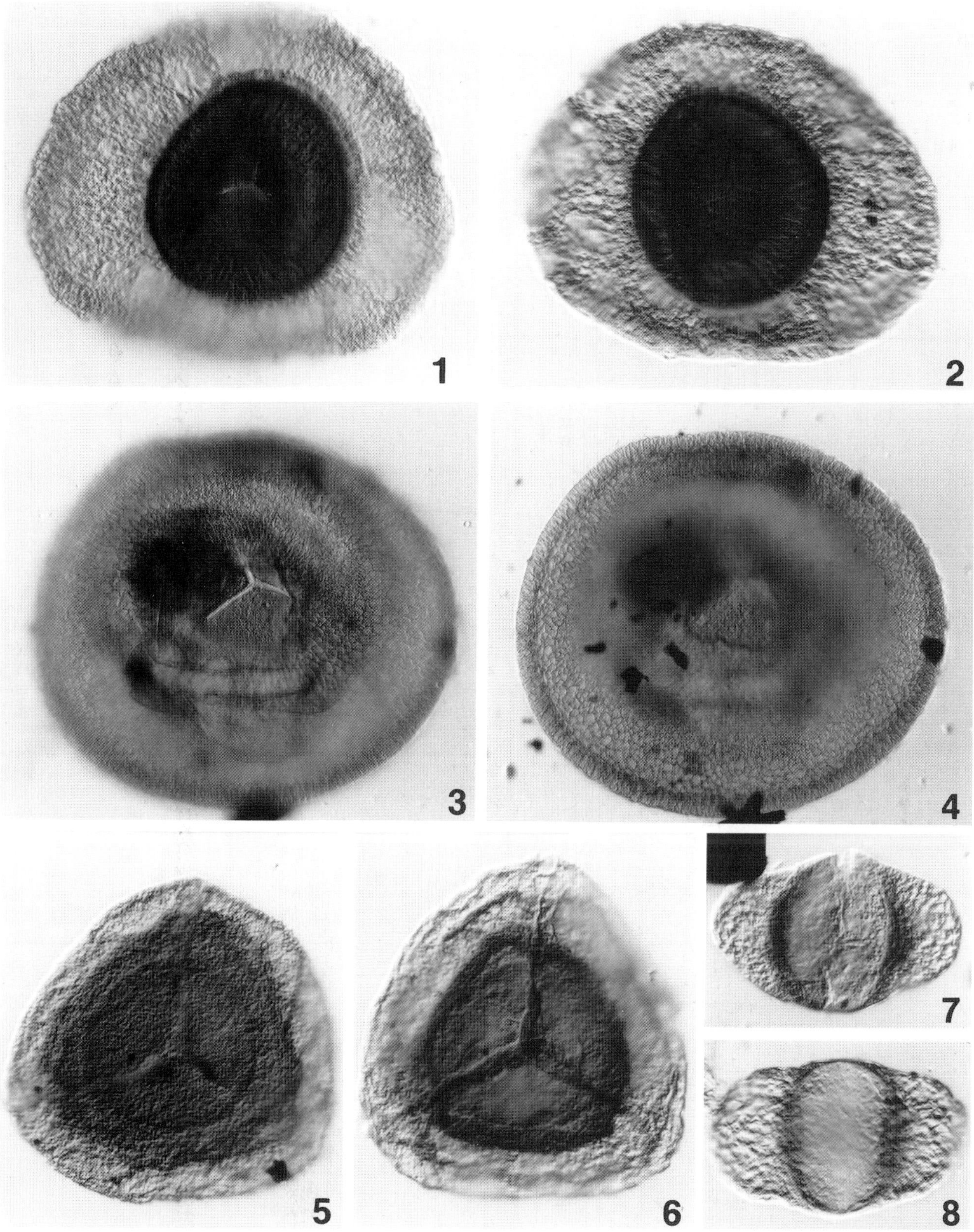
1
2
3
4
5
6
7
8

TABLE 2. Fluorescence spectral parameters and fluorescence intensity of miospores from the Bolzano Volcanic Complex (*=miospores after palynological preparation; **miospores in polished rock; all other measurements conducted on miospores subjected to palynological preparation techniques).

SAMPLE	λ_{max} (nm)	Q	Q_{max}	I_{546nm}	s	n
GRI 0*	605	1.49	1.60	0.43	0.11	50
GRI 0**	605	1.38	1.55	0.46	0.18	50
GRI 5	590	1.66	1.71	0.15	0.05	50
GRI 17	595	1.60	1.67	0.16	0.07	50
GRI 27	595	1.43	1.55	0.24	0.19	50
89317	640	2.36	2.40	0.07	0.02	50
89318	620	2.41	2.45	0.09	0.03	50

mately 18 m above the top of the BVC. In such sediments connate water can circulate much easier than in dense chert layers and increased water circulation may occur at or near the boundary between the siliciclastic and volcanic rocks. Differences in lithology (e.g., grain size), clay mineralogy and thermal conductivity or the presence of faults will also influence local maturation levels (Batten, 1981).

It is probable that the few dark brown to black spores, spore clusters, parts of sporangia with well-preserved spores, and cuticular fragments (samples GRI 1, GRI 26 and GRI 27) are not reworked because pale colored specimens of the same species are also present. However, it is likely that the dark colored spores and plant debris were charred during wildfires or volcanic activity. Skarby (1986, p. 237), in her study of the Upper Cretaceous of Sweden, suggested that 'partly burned anthers with blackened grains....as well as dispersed pollen grains' were damaged by fire. It is now well established that wildfires played a major role in shaping late Paleozoic ecosystems (Cope and Chaloner, 1985; Scott and Jones, 1994). These are usually ignited by lightning strikes or volcanic activity. Fusain and charcoal are profuse in Recent and Carboniferous volcanogenic sediments, ashes and lavas (e.g., Scott et al. 1985; Scott, 1990; Scott and Jones, 1994). Considering that ignimbrites and pyroclastics are abundant in the BVC the presence of charred plant material in perennial lake sediments is to be expected. In addition, the semiarid to arid climate with dry seasons (Krainer and Spötl, 1998) would have expedited the spreading of wildfires.

QUANTITATIVE FLUORESCENCE MICROSCOPY

Change of fluorescence colors has been used to recognize contamination by modern pollen grains, to identify caving or reworked palynomorphs (e.g., van Gijzel, 1967) and to facilitate classification of amorphous organic matter or detection of small algal phytoclasts in palynofacies analysis (McPhilemy, 1988). Most importantly, fluorescence properties can be used as an indicator of thermal maturation (see Plate 4, figs. 4–6). As maturity increases the fluorescence shifts from green through yellow, orange and red to dark brown. Simultaneously, the intensity of fluorescence decreases. Visible fluorescence of sporinite extinguishes at a vitrinite reflectance value of 1.2–1.3% which is equivalent to a medium volatile bituminous coal rank (Ottenjann et al., 1974). A vitrinite reflectance value of 1.2–1.5% coincides with a transition of the late mature to over-mature level where liquids are not usually generated. Therefore, visible fluorescence is of extreme importance in assessing hydrocarbon source potential. Sporinite fluorescence and the classification of fluorescence colors are an inexpensive and rapid method for estimating a rank below the end of catagenesis. However, color determination is highly subjec-

PLATE 4

1 Scanning electron photomicrograph of *in situ* monosaccate miospore in polished, slightly etched, rock section showing weakly folded nexine and alveolate infrastructure of the sexine. Note three-dimensional preservation and ?distal sexine-free area/sexine thinning (sample GRI 0; x400).

2 Detail of fig. 1 (x1,500).

3 Taeniate bisaccate miospores in rock. Monolete mark of pollen grain in the upper right corner clearly visible (sample GRI 0; x650).

4 Incident light fluorescence microscopy of chert lense with three-dimensionally preserved bisaccate miospores. Note very bright fluorescence (water immersion objectives not corrected for cover-slips). A cavum is developed between the folded nexine and the alveolate sexine on the large, brightly fluorescing bisaccate pollen grain. This cavum can be either a genuine feature (eusaccate condition) but can also form due to mechanical tearing during folding of the nexine (protosaccate/quasisaccate condition; see Foster and Balme, 1994). Black dots are pyrite framboids (polished rock section of sample GRI 0; x400).

5 Bisaccate miospores in chert lense. In the upper right corner proximal side of a taeniate pollen grain. Note intexine is slightly fluorescent and alveolate sexine is brightly fluorescent (sample GRI 0; x250).

6 Densely packed miospores in chert lense. Note section through the central part of a trilete cavate/zonate miospore showing laesura (sample GRI 0; x250).

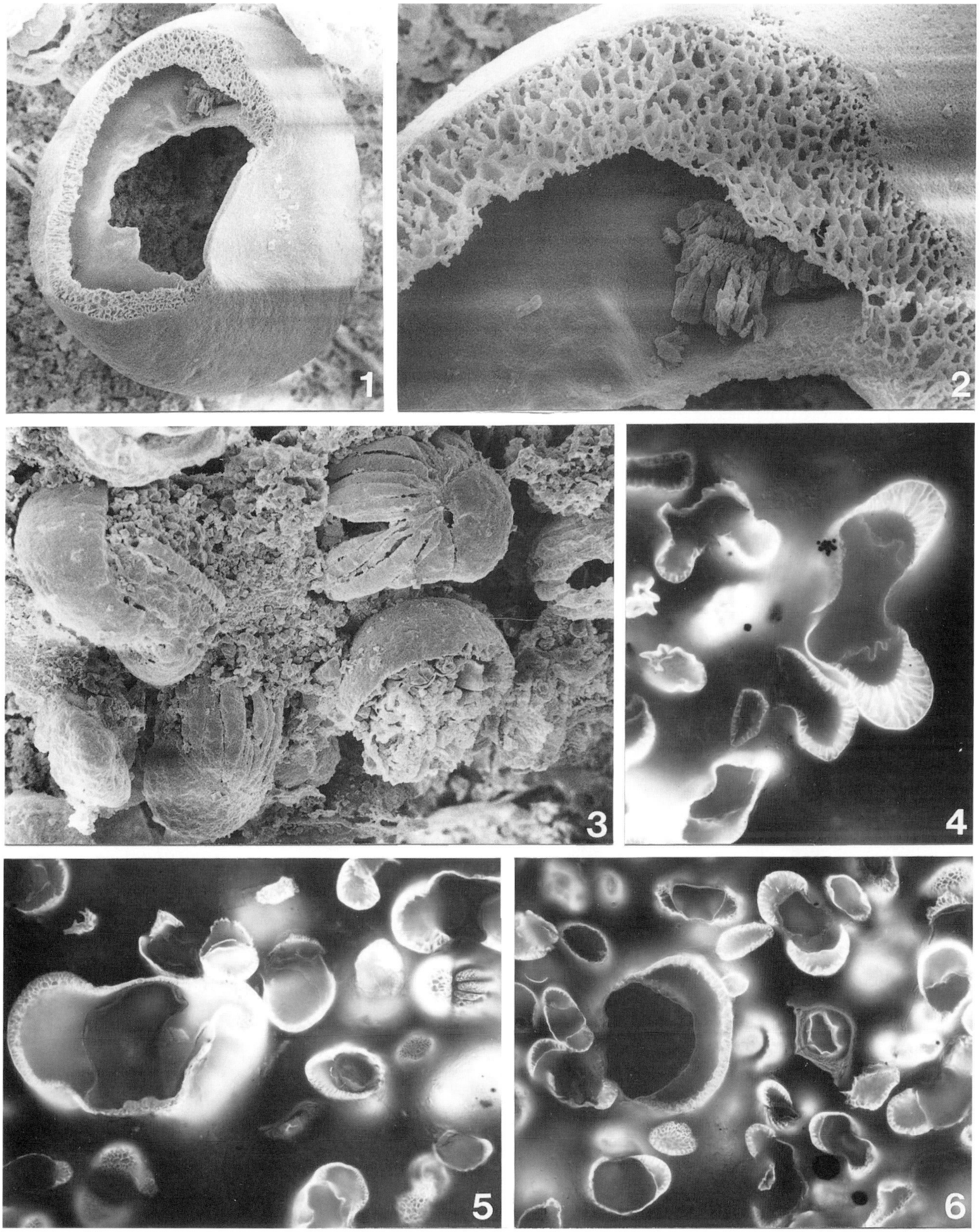
1
2
3
4
5
6

tive since the same color might be perceived differently depending on fluorescence intensities.

To circumvent these shortcomings the fluorescence spectra can be measured. Usually the spores are excited with long-wave UV light and the spectra of the emitted light, the fluorescence, are recorded in the visible range over a wave length of approximately 400 to 700 nm. Several parameters can be calculated using the spectral distribution of relative fluorescence intensity. The most important include the wavelength of the spectral peak (λ_{max}), the ratio of the spectral intensity at 650 nm and 500 nm (the red/green ratio Q), and the ratio of the spectral intensity at λ_{max} and $I_{546\ nm}$ (Q_{max}). In addition, the fluorescence intensity at 546 nm ($I_{546\ nm}$) and the fluorescence alteration during prolonged exposure to long-wave UV light have also been used (Ottenjann, 1980; 1988). Since the fluorescence intensity at $I_{546\ nm}$ also depends on the thickness of the palynomorph the measurements may show a wider range of variation than other fluorescence parameters. Although with increasing maturity the fluorescence intensity of all liptinites decreases and their spectral maximum shifts from green to red in unison, sporinites are particularly suitable for maturity studies. Fluorescence of cutinites and resinites can be highly variable, even within the same sample. Alginites and fluorinites are characterized by a bright and intense fluorescence but above a vitrinite reflectance of approximately 0.9% these organic types are inferior to sporinites for rank determinations (Teichmüller, 1974; 1982).

In order to exclude possible variations of fluorescence parameters dependent on different spore groups only the sacci of bisaccate miospores were considered in this study. Basically, the instrumentation used for measuring the spectra is the same as described by Ottenjann (1980; 1988). This includes a high-pressure mercury lamp, a long-wave UV light gained by excitation filter BG 38 (4 mm in thickness), a barrier filter of 410 nm, and a dichroic mirror of 400 nm.

Van Gizel (1967) suggested that acid preparation can have an adverse influence on the fluorescence of fossil sporinite. To evaluate this hypothesis, fluorescence spectra were measured on bisaccate spores both embedded in the host matrix (cut and polished chert) and in a palynological mount for sample GRI 0 (see Table 2). The similarity of the two spectra is shown in Text-Figure 6. The wavelength of the spectral peak is identical and the red/green ratio Q and Q_{max} are similar. Fluorescence intensity is very high in both measurements emphasizing a bright, intense yellow color (see examples in Plate 4, figs. 4–6). The application of acid digestion appears to result in a negligible difference in fluorescence character between the two preparation types.

In general, two types of fluorescence spectra can be distinguished in the samples examined (Text-Figure 6). Samples from siliceous limestones yield spores whose wavelength of spectral peak is located between 590–605 nm. The Q values range from 1.38 to 1.66 and Q_{max} from 1.55 to 1.71. The shape of the spectral curves for these five samples (GRI 0*, GRI 0**, GRI 5, GRI 17, GRI 27) varies only slightly. Only marked differences in regard to the relative intensity (I_{546}) exist. As indicated, the GRI 0 spores, from a cherty limestone containing many layers and lenses of silica, exhibit the highest intensity. Also the chromaticity diagram for this sample shows a narrow distribution of a* and b* values (see Text-Figure 5). The second type of fluorescence spectra is evidenced by measurements of samples 89317 and 89318 from the base of the Gröden Formation. The overall shape of the two spectra corresponds, although the curve is quite irregular beyond 600 nm (additional measurements would probably 'flatten' the peaks). The similarity of the two graphs is also manifested by only minor differences with regard to the fluorescence spectral parameters. Compared with the spectra of the aforementioned samples (GRI 0*, GRI 0**, GRI 5, GRI 17, GRI 27) the wavelength of the spectral maximum is clearly shifted from yellow towards the red. The Q and Q_{max} values have increased and the relative fluorescence intensity at $I_{546\ nm}$ is considerably reduced.

Compared to published fluorescence spectra (e.g. Ottenjann et al., 1974; Teichmüller, 1974) the curves of samples GRI 0, GRI 5, GRI 17, and GRI 27 are most similar to those of sporinites from a high volatile bituminous (B) coal, whereas the curves of samples 89317 and 89318 are similar to spectral distributions of a sporinite from a high volatile bituminous (A) coal. Teichmüller (1982) published a correlation diagram of the red/green ratio (Q) versus vitrinite reflectance. Following this diagram the samples from the siliceous limestones have a reflectance between 0.7% and 0.9% and those from the base of the Gröden Formation around 1.0%. Therefore, the fluorescence parameters indicate a slightly higher vitrinite reflectance equivalent than shown by the quantitative spore color data.

CONCLUSIONS

Quantitative spore color and quantitative fluorescence microscopy of the miospores from lacustrine sediments within the BVC indicate a remarkably low thermal maturity. Quantitative spore color data indicate a vitrinite reflectance equivalent of approximately 0.5–0.6% and quantitative fluorescence data suggest values of 0.7–0.9%. Krainer and Spötl (1998) have recorded a vitrinite reflectance value of 0.54% from a single sample from the Oberbozen section (locality 3 of this paper). This corresponds to the quantitative spore color data. These low maturity values are difficult to justify because the lacustrine and alluvial fan sediments are intercalated with thick rhyolithic and rhyodacitic lavas and ignimbrites and additional evidence indicates that a 3000–3200 m Triassic–Cretaceous section may have been present in the Bozen area. Approximately 30 km northeast of the sampling area Triassic andesitic dikes are known to penetrate the BVC. Therefore, a high geothermal gradient and high thermal alteration would initially be assumed.

The thermal effects of ignimbrite flows on the sediments below may have been limited (cf. Marshall, 1988, p. 13) and the Triassic volcanism is restricted to a small area. Effects of hydrothermal overprint (Brondi et al., 1973) may have been confined to regions where duration of exposure was the dominant parameter (Barth et al., 1993). Another factor which has influenced the maturity path of the organic material is the extraordinary depositional environment and host sediment. Rapid chert formation may have shielded the miospores from pore fluids and suppressed, or at least delayed, chemical and physical alteration. The significance of the early diagenetic chert is emphasized by the fact that miospores from the base of the overlying Gröden Formation reveal a higher thermal alteration than those from the BVC.

The miospore assemblage from the BVC is quite diverse compared to assemblages from sediments of similar age (e.g., the Collio and Tregiovo Formation) in western Europe. Charred spores, sporangia and cuticules suggest that the vegetation was effected by wildfires. Considering the

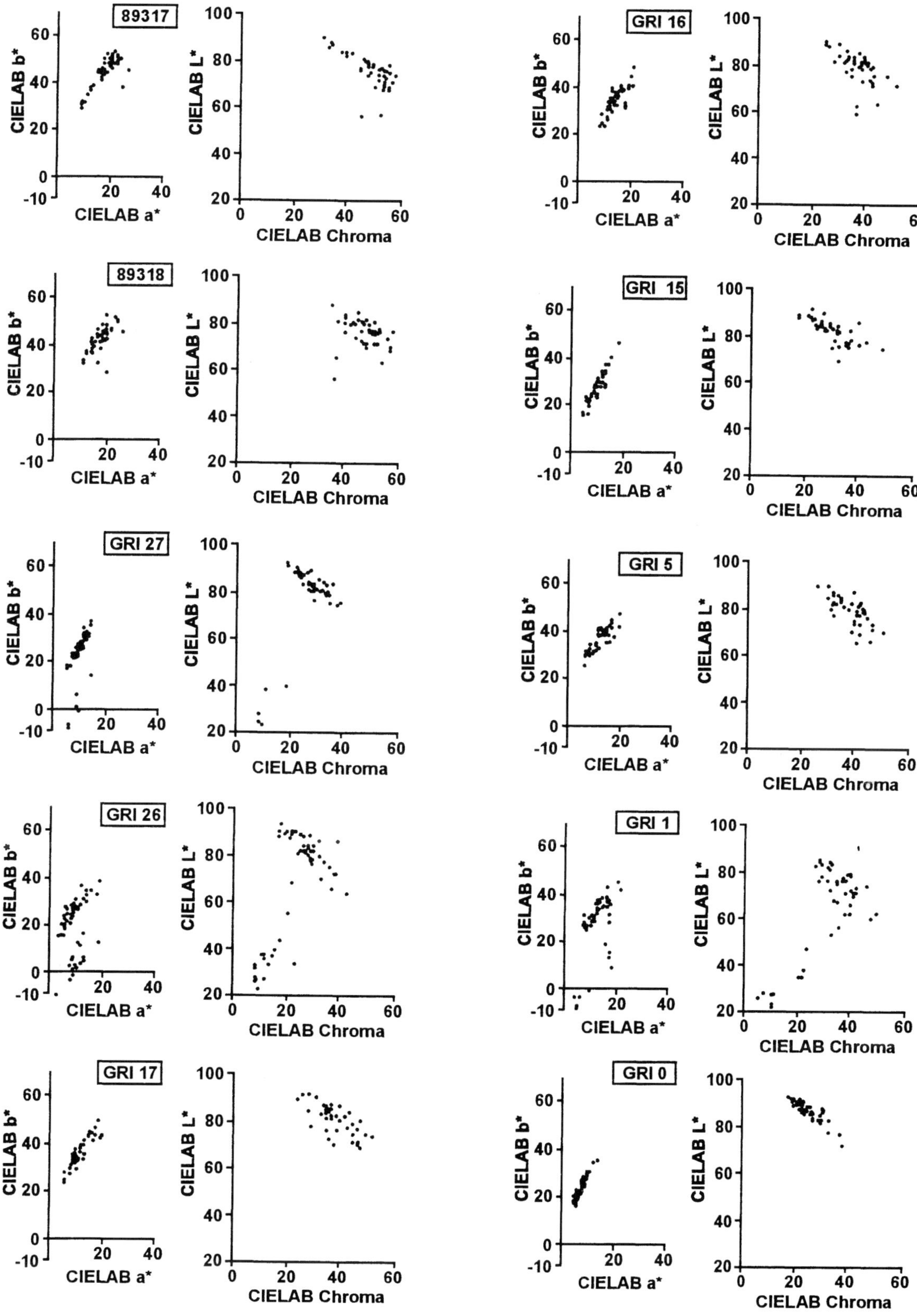

Text-Figure 6. Spectral distribution of fluorescence intensities (mean spectra) of miospores measured in selected samples.

abundance of ignimbrites and the semiarid to arid climate, wildfires must have been a recurrent phenomenon. It can be speculated that volcanic activity caused different pre- and post-eruptive vegetation and that wildfires produced mineral-rich soils colonized by pioneer plants (Scott 1990; Scott and Jones, 1994). Hence, the frequent changes in vegetational successions may be the reason for the diverse miospore assemblage.

The presence of excellently preserved palynomorphs in cherts and siliceous limestones has long been overlooked, at least in late Paleozoic palynology. Early diagenetic silicified limestones are frequently reported from terrestrial sediments of Late Carboniferous and Permian age as acidic volcanism and intense weathering provided high amounts of silica from volcanic glass or rhyolithic to rhyodacitic rocks. A carefully directed search of silicified limestones from the Rotliegend of western Europe resulted in several palynologically productive samples which yielded three-dimensionally preserved miospores (e.g. from the Erbendorf Basin, southern Germany; Hartkopf-Fröder, in prep.). For all intents and purposes, siliceous limestones appear visually worthless as a palynological repository, but may actually entomb a wealth of palynological information.

Frequently, miospore assemblages from uppermost Carboniferous and Permian ("Autunian" and "Saxonian") successions in intramontane basins of western Europe are badly preserved, or destroyed, due to high thermal alteration or adverse facies conditions such as red beds, coarse-grained alluvial fan sediments and widespread volcanic rocks. Nevertheless, besides conchostracans, insect and fish remains and vertebrate tracks, palynology plays a major role in dating these strata (e.g., Boy and Martens, 1991). The rich and diverse miospore assemblage from the BVC offers a unique opportunity to clarify the age of this thick sequence and to settle taxonomic problems which need excellently preserved material.

ACKNOWLEDGMENTS

The authors would like to thank K. Ottenjann, A. Vieth-Redemann (Geological Survey of North Rhine-Westphalia, Krefeld) and J. E. A. Marshall (School of Oceanography and Earth Science, Southampton) for their discussions concerning the maturation significance of palynomorph fluorescence. Dr. Marshall also graciously allowed access to optical equipment for spore color measurement and to his 'CP' program. He is also thanked for his hospitality when CHF visited the School of Oceanography and Earth Sciences and R. C. Tobin, B. L. Claxton and C. Skirius (Source, Migration and Reservoir Technology Team, Amoco-Houston) and C. A. Law (Basin Modeling Group, Amoco-Houston) are gratefully acknowledged for their discussions concerning vitrinite reflectance, fluorescence and heat flow gradients. Prof. A. C. Scott (Department of Geology, Royal Holloway, University of London) provided helpful discussions concerning the effects of fire on organic materials. The authors also express thanks to H. M. Kuriger (Amoco Earth Science Lab, Houston) and M. Kirchoff (Geological Survey of North Rhine-Westphalia, Krefeld) for processing the palynological samples and M. Schendel (Amoco Graphics Department, Houston) for drafting the text-figures. CHF would like to acknowledge Zeiss (Germany) for the loan of special objectives for fluorescence microscopy. KK acknowledges the government of the province of South Tyrol and the district of Mölten for financial support. CHF and GDW thank the Geological Survey of North Rhine-Westphalia (Krefeld) and Amoco Exploration & Production Technology Group (Houston) for technical support and permission to publish.

References Cited

BARGOSSI, G.M., Di BATTISTINI, G., GASPATOTTO, G., VERCSEI, P.L., and GUALTIERI, S.
1993 The Permian volcano-sedimentary sequence of the Tione Basin (Val Rendena, Trento, Northern Italy). *Mineralogica Petrographica Acta*, 36: 19–33.

BARKER, C.E., and PAWLEWICZ, M.J.
1986 The correlation of vitrinite reflectance with maximum temperature in humic organic matter. *In*: Buntebarth, G., and Stegena, L. (eds.), *Lecture notes in Earth Sciences 5. Palaeogeothermics*. Springer-Verlag, Berlin, pp. 79–93.

BARTH, S., and MOHR, B.A.R.
1991 Late Hercynian magmatism in the Southern Alps, N. Italy: Oxygen isotope data of the Atesino Volcanic Complex and Cima d'Asta intrusion and palynological age of lacustrine sediments from Tregiovo. *Terra Cognita*, 3: 209–210.
1994 Palynostratigraphically determined age of the Tregiovo sedimentary complex in relation to radiometric emplacement ages of the Atesina Volcanic Complex (Permian, Southern Alps, N. Italy). *Neues Jahrbuch für Geologie und Paläontologie, Abhandlungen*, 192: 273–292.

BARTH, S., OBERLI, F., MEIER, M., BLATTNER, P., BARGOSSI, G.M., and Di BATTISTINI, G.
1993 The evolution of calc-alkaline basic to silicic magma system: geochemical and Rb–Sr, Sm–Nd, and $^{18}O/^{16}O$ isotopic evidence from the Late Hercynian–Atesina–Cima d'Asta volcano-plutonic complex, northern Italy. *Geochimica et Cosmochimica Acta*, 57: 4285–4300.

BATTEN, D.J.
1981 Palynofacies, organic maturation and source potential for petroleum. *In*: Brooks, J. (ed.), *Organic maturation studies and fossil fuel exploration*. Academic Press, London, pp. 201–223.

BECKE, M., and MAURITSCH, H.J.
1984 Rock and paleomagnetic investigations on the Quartzporphyry Complex near Bolzano (Bozen), Italy. *International Geological Correlation Program No. 5 Newsletter*, 6: 11–19.

BOSELLINI, A.
1991 Geology of the Dolomites. An introduction. *Dolomieu Conference on Carbonate Platforms and Dolomitization, Ortisei/St. Ulrich, Val Gardena/Groedental, The Dolomites, Italy*, 43 pgs.

BOY, J.A., and MARTENS, T.
1991 Zur Problematik chronostratigraphischer Korrelationen im mitteleuropäischen Rotliegend (?oberstes Karbon–Perm). *Newsletters on Stratigraphy*, 25: 163–192.

BRANDNER, R., and MOSTLER, H.
1982 Der geologische Aufbau des Schlerngebietes und seiner weiteren Umgebung. *In*: Mostler, H. (ed.), *Seis am Schlern, Südtirol 1982, Exkursionsführer zur 4. Jahrestagung der Österreichischen Geologischen Gesellschaft*. Österreichische Geologische Gesellschaft, Innsbruck, p. 1–42.

BROGLIO LORIGA, C., and CASSINIS, G.
1992 The Permo-Triassic boundary in the Southern Alps (Italy) and in adjacent Periadriatic regions. *In*: Sweet, W.C., Yang Zunyi, Dickins, J.M., and Yin Hongfu (eds.), *Permo-Triassic events in the eastern Tethys: stratigraphy, classification, and relations with the western Tethys*. Cambridge University Press, Cambridge, p. 78–97.

BROGLIO LORIGA, C., NERI, C., PASINI, M., and POSENATO, R.
1986 Marine fossil assemblages from the Upper Permian to lowermost Triassic in the western Dolomites. *Memorie della Società Geologica Italiana*, 34: 5–44.

BRONDI, A., CARRARA, C., and POLIZZANO, C.
1973 Uranium and heavy metals in Permian sandstones near Bolzano (N. Italy). *In*: Amstutz, G.C., and Bernard, A.J., (eds.), *Ores in Sediments*, p. 65–77.

CARTER, L.S., KELLEY, S.A., BLACKWELL, D.D., and NAESER, N.D.
1998 Heat flow and thermal history of the Anadarko Basin, Oklahoma. *American Association of Petroleum Geologists, Bulletin*, 82: 291–316.

CASSINIS, G., and DOUBINGER, J.
1991a Artinskian to Ufimian palynomorph assemblages from the Central Southern Alps, Italy, and their regional stratigraphic implications. *In*: *Contributions to Eurasian Geology*, International Congress on the Permian System of the World, Occasional Publications of the Earth Science Research Institute, n.s., 8B: 9–18.
1991b On the geological time of the typical Collio and Tregiovo continental beds in the Southalpine Permian (Italy), and some additional observations. *Atti Ticinensi di Scienze della Terra*, 34: 1–20.

CASSINIS, G., and NERI, C.
1992 Sedimentary and palaeotectonic evolution of some Permian continental basins in the central Southern Alps, Italy. *Cuadernos de Geología Ibérica*, 16: 145–176.

CASSINIS, G., PEROTTI, C.R., and VENTURINI, C.
1997 Examples of the late Hercynian transtensional tectonics in the Southern Alps (Italy). *In*: Dickins, J.M., Yang Zunyi, Yin Hongfu, Lucas, S.G., and Acharyya, S.K. (eds.), *Late Palaeozoic and Early Mesozoic Circum-Pacific events and their global correlation*. Cambridge University Press, Cambridge, pp. 41–50.

CEOLONI, P., CONTI, M.A., MARIOTTI, N., MIETTO, P., and NICOSIA, U.
1988 Tetrapod footprint faunas from southern and central Europe. *Zeitschrift für Geologische Wissenschaften*, 16: 895–906.

CLEMENT-WESTERHOF, J.A.
1974 *In situ* pollen from gymnospermous cones from the Upper Permian of the Italian Alps, a preliminary account. *Review of Palaeobotany and Palynology*, 17: 63–73.
1984 Aspects of Permian palaeobotany and palynology. IV. The conifer *Ortiseia* Florin from the Val Gardena Formation of the Dolomites and the Vicentinian Alps (Italy) with special reference to a revised concept of the Walchiaceae (Göppert) Schimper. *Review of Palaeobotany and Palynology*, 41: 51–166.
1987 Aspects of Permian palaeobotany and palynology, VII. The Majonicaceae, a new family of Late Permian conifers. *Review of Palaeobotany and Palynology*, 52: 375–402.

CLEMENT-WESTERHOF, J.A., VAN DER EEM, J.G.L.A., VAN ERVE, A.W., KLASEN, J.J., SCHUURMAN, W.M.L., and VISSCHER, H.
1974 Aspects of Permian, Triassic and Early Jurassic palynology of western Europe — a research project. *Geologie en Mijnbouw*, 53: 329–341.

CONTI, M.A., FONTANA, D., MARIOTTI, N., MASSARI, F., NERI, C., NICOSIA, U., PASINI, M., and PITTAU, D.
1986 The Bletterbach-Butterloch Section (Val Gardena Sandstone and Bellerophon Formation). *In*: Field Conference on Permian and Permian–Triassic boundary in the South-Alpine segment of the Western Tethys. *Società Geologica Italiana, Excursion Guidebook*, p. 91–110.

CONTI, M.A., LEONARDI, G., MARIOTTI, N., and NICOSIA, U.
1977 Tetrapod footprints of the "Val Gardena Sandstone" (North Italy). Their paleontological, stratigraphic and paleoenvironmental meaning. *Palaeontographia Italica*, 70: 1–91.

CONTI, M.A., MARIOTTI, N., NICOSIA, U., and PITTAU, P.
1997 Succession of selected bioevents in the continental Permian of the Southern Alps (Italy): improvements in intrabasinal and interregional correlations. *In*: Dickins, J.M., Yang Zunyi, Yin Hongfu, Lucas, S.G., and Acharyya,

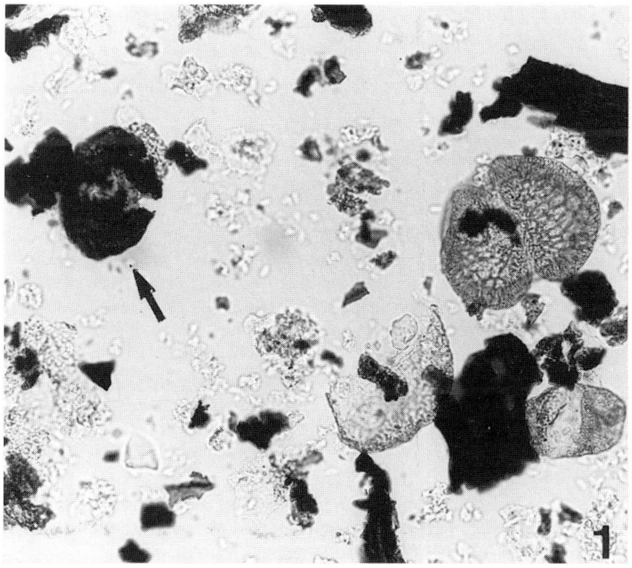

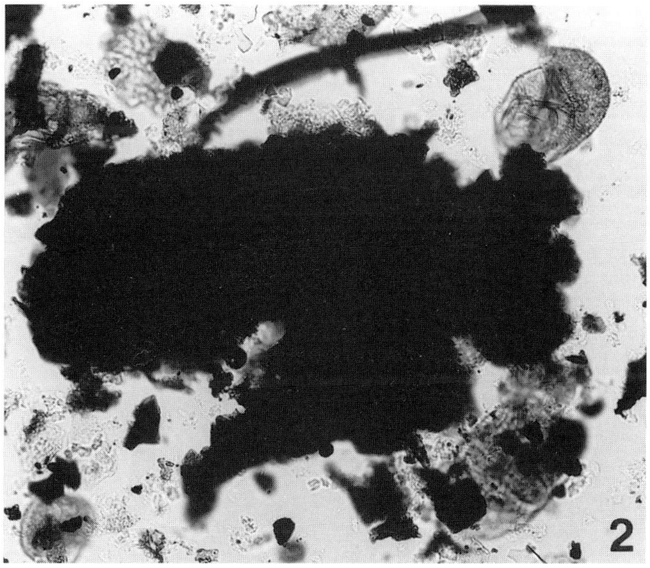

PLATE 5

1 Dark, nearly black bisaccate miospore (arrow). Note weak color and near transparent state of uncharred bisaccate miospores (sample GRI 0; x250).

2 Spore cluster composed of blunt bisaccate miospores. Alveolate structure of sacci is present on the outermost margins of the cluster (sample GRI 0; x250).

S.K. (eds.), *Late Palaeozoic and Early Mesozoic Circum-Pacific events and their global correlation.* Cambridge University Press, Cambridge, pp. 51–65.

COPE, M.J., and CHALONER, W.G.
1985 Wildfire: an interaction of biological and physical processes. *In*: Tiffney, B.H. (ed.), *Geological factors and the evolution of plants.* Yale University Press, New Haven, pp. 257–277.

DACHROTH, W.
1976 Gesteinsmagnetische Marken im Perm Mitteleuropas. *Geologisches Jahrbuch*, E10: 3–71.

D'AMICO, C.
1986 Volcanic sequence in Trentino-Alto Adige. *In*: Field Conference on Permian and Permian–Triassic boundary on the South Alpine segment of the Western Tethys. *Società Geologica Italiana, Excursion Guidebook*, pp. 16–22.

D'AMICO, C., and DEL MORO, A.
1988 Permian and Triassic Rb–Sr dating in the Permian rhyodacitic ignimbrites of Trentino (Southern Alps). *Rendiconti della Società Italiana di Mineralogia e Petrologia*, 43: 171–180.

D'AMICO, C., DEL MORO, A., FREDDO, A., and PARDINI,G.
1980 Studio radiometrico delle ignimbriti riolitiche atesine, Gruppo superiore. *Rendiconti della Società Italiana di Mineralogia e Petrologia*, 36: 703–716.

FELS, H., and PAUL-KOCH, G.
1985 Alluviale Schüttungsfächer innerhalb der unterpermischen Vulkanite Südtirols (Italien). *Zeitschrift der Deutschen Geologischen Gesellschaft*, 136: 167–179.

FERRIS, F.G., FYFE, W.S., and BEVERIDGE, T.J.
1988 Metallic ion binding by *Bacillus subtilis*: implications for the fossilization of microorganisms. *Geology*, 16: 149–152.

FOSTER, C.B., and BALME, B.E.
1994 Ultrastructure of *Teichertospora torquata* (Higgs) from the Late Devonian: oldest saccate palynomorph. *In*: Kurmann, M.H., and Doyle, J.A. (eds.), *Ultrastructure of fossil spores and pollen.* Royal Botanic Gardens, Kew, pp. 87–97.

GIANNOTTI, G.P.
1962 Intercalazioni lacustri entro le vulcaniti paleozoiche atesine. *Atti della Società Toscana Scienze di Naturali*, 69: 598–617.

HARTKOPF-FRÖDER, C., and KRAINER, K.
1990 Fluviatile und lakustrine Sedimente innerhalb der permischen Bozener Quarz porphyrabfolge (Südtirol/ Italien): Ihre biostratigraphische und palökologische Bedeutung. *Sediment '90 Conference*, Abstracts (Bonn): p. 3.

HEMSLEY, A., and SCOTT, A.C.
1991 Controls on the ultrastructural preservation of megaspore walls. *International Congress on Carboniferous and Permian Geology and Stratigraphy*, Abstracts (Buenos Aires, Argentina), pp. 45–46.

HESS, J.C.
1990 Numerische Stratigraphie permokarbonischer Vulkanite Zentraleuropas. Allgemeine Einführung und Teil I: Südtirol. *Zeitschrift der Deutschen Geologischen Gesellschaft*, 141: 1–11.

HOCHULI, P.A.
1985 Palynostratigraphische Gliederung und Korrelation des Permo–Karbon der Nordschweiz. *Eclogae Geologicae Helvetiae*, 78: 719–831.

HUNT, R.W.G.
1987 *Measuring colour.* Ellis Horwood, Chichester, 221 pgs.

KERP, H.
1996 Post–Variscan late Palaeozoic Northern Hemisphere gymnosperms: the onset to the Mesozoic. *Review of Palaeobotany and Palynology*, 90: 263–285.

KLAUS, W.
1963 Sporen aus dem südalpinen Perm. *Jahrbuch der Geologischen Bundesanstalt*, 106: 229–361.

KNOLL, A.H.
1985 Exceptional preservation of photosynthetic organisms in silicified carbonates and silicified peats. *Philosophical Transactions of the Royal Society, London*, B311: 111–122.

KNOLL, A.H., and OHTA, Y.
1988 Microfossils in metasediments from Karls Forland, western Svalbard. *Polar Research*, 6: 59–67.

KRAINER, K.
1989 Composition and evolution of Lower Permian molasse sediments (Ponte Gardena Conglomerate) at the base of the Bolzano Volcanic Complex, Southern Alps (N. Italy). *Neues Jahrbuch für Geologie und Paläontologie, Monatshefte*, 1989: 400–424.

KRAINER, K., and SPÖTL, C.
1998 Abiogenic silica layers within a fluvio-lacustrine succession, Bolzano Volcanic Complex, northern Italy: A Permian analogue for Magadi-type cherts? *Sedimentology*, 45: 489–505.

KRAINER, K., WOOD, G.D., and KOZUR, H.
1996 Geology, palynology and palynofacies of the Bolzano Volcanic Complex (Permian) of Northern Italy. *30th International Geological Congress, Abstracts* (Beijing, China), 2: 179.

MARSHALL, J.E.A.
1988 Devonian miospores from Papa Stour, Shetland. *Transactions of the Royal Society of Edinburgh, Earth Sciences*, 79: 13–18.
1990 Quantitative spore colour for the determination of thermal maturation. *Mededelingen Rijks Geologische Dienst*, 45: 111–114.
1991 Quantitative spore colour. *Journal of the Geological Society*, 148: 223–233.

MASSARI, F., CONTI, M.A., FONTANA, D., HELMOLD, K., MARIOTTI, N., NERI, C., NICOSIA, U., ORI, G.G., PASINI, M., and PITTAU, P.
1988 The Val Gardena Sandstone and *Bellerophon* Formation in the Bletterbach Gorge (Alto Adige, Italy): biostratigraphy and sedimentology. *Memorie di Scienze Geologiche*, 40: 229–273.

MASSARI, F., NERI, C., PITTAU, P., FONTANA, D., and STEFANI, C.
1994 Sedimentology, palynostratigraphy and sequence stratigraphy of a continental to shallow-marine rift-related succession: Upper Permian of the eastern Southern Alps (Italy). *Memorie di Scienze Geologiche*, 46: 119–243.

MAURITSCH, H.J., and BECKE, M.
1983 A magnetostratigraphic profile in the Permian (Gröden Beds, Val Gardena Formation) of the Southern Alps near Paularo (Carnic Alps, Friuli, Italy). *International Geological Correlation Program No. 5 Newsletter*, 5: 80–86.

McPHILEMY, B.
1988 The value of fluorescence microscopy in routine palynofacies analysis: Lower Carboniferous successions from the counties Armagh and Roscommon, Ireland. *Review of Palaeobotany and Palynology*, 56: 345–359.

MILTON, J.A.
1993 *The application of quantitative spore colour measurement to thermal maturity studies.* Unpublished Ph.D. Dissertation, University of Southampton, 322 pgs.

MUNNECKE, A., and SERVAIS, T.
1996 Scanning electron microscopy of polished, slightly etched rock surfaces: a method to observe palynomorphs *in situ*. *Palynology*, 20: 163–176.

OTTENJANN, K.
1980 Spektrale Fluoreszenz-Mikrophotometrie von Kohlen und Ölschiefern. *Leitz-Mitteilungen für Wissenschaft und Technik*, 7: 262–273.

1988 Fluorescence alteration and its value for studies of maturation and bituminization. *Organic Geochemistry*, 12: 309–321.

OTTENJANN, K., TEICHMÜLLER, M., and WOLF, M.
1974 Spektrale Fluoreszenz-Messungen an Sporiniten mit Auflicht-Anregung, eine mikroskopische Methode zur Bestimmung des Inkohlungsgrades gering inkohlter Kohlen. *Fortschritte in der Geologie von Rheinland und Westfalen*, 24: 1–36.

PEARSON, D.L.
1990 Pollen/spore color "standard." 2nd printing of version 2, Phillips Petroleum Company, Bartlesville U.S.A.

POORT, R.J., and KERP, J.H.F.
1990 Aspects of Permian palaeobotany and palynology. XI. On the recognition of true peltasperms in the Upper Permian of Western and Central Europe and a reclassification of species formerly included in *Peltaspermum* Harris. *Review of Palaeobotany and Palynology*, 63: 197–225.

SAVELLI, C, and BALBONI, S.
1984 Determinazioni di eta K/Ar de vulcaniti permiane sudalpine. *Mineralogica Petrographica Acta*, 28: 243–252.

SCOTT, A.C.
1990 Preservation, evolution, and extinction of plants in Lower Carboniferous volcanic sequences in Scotland. *In*: Lockley, M.G., and Rice, A. (eds.), Volcanism and fossil biotas. *Geological Society of America, Special Paper*, 244: 25–38.

SCOTT, A.C., GALTIER, J., and CLAYTON, G.
1985 A new late Tournaisian (Lower Carboniferous) flora from the Kilpatrick Hills, Scotland. *Review of Palaeobotany and Palynology*, 44: 81–99.

SCOTT, A.C., and JONES, T.P.
1994 The nature and influence of fire in Carboniferous ecosystems. *Palaeogeography, Palaeoclimatology, Palaeoecology*, 106: 91–112.

SKARBY, A.
1986 Normapolles anthers from the Upper Cretaceous of southern Sweden. *Review of Palaeobotany and Palynology*, 46: 235–256.

SOUTHGATE, P.N.
1986 Depositional environment and mechanism of preservation of microfossils, upper Proterozoic Bitter Springs Formation, Australia. *Geology*, 14: 683–686.

STORZER, D.
1970 Fission track dating of volcanic glass and the thermal history of rocks. *Earth and Planetary Science Letters*, 8: 55–60.

TEICHMÜLLER, M.
1974 Entstehung und Veränderung bituminöser Substanzen in Kohlen in Beziehung zur Entstehung und Umwandlung des Erdöls. *Fortschritte in der Geologie von Rheinland und Westfalen*, 24: 65–112.
1982 *Fluoreszenzmikroskopische Änderungen von Liptiniten und Vitriniten mit zunehmendem Inkohlungsgrad und ihre Beziehungen zu Bitumenbildung und Verkokungsverhalten*. Geologisches Landesamt Nordrhein-Westfalen, Krefeld: 119 pgs.

VAN GIJZEL, P.
1967 Autofluorescence of fossil pollen and spores with special reference to age determination and coalification. *Leidse Geologische Mededelingen*, 40: 263–317.

WOOD, G.D., GABRIEL, A.M., and LAWSON, J.C.
1996 Palynological techniques-processing and microscopy. *In*: Jansonius, J. and McGregor, D.C. (eds.), *Palynology: Principles and Applications*, American Association of Stratigraphic Palynologists, Foundation, pp. 29–50.

WOOD, G.D., HARTKOPF-FRÖDER, C., and KRAINER, K.
1996 Sedimentological and palynological characterization of the Permian Bolzano Volcanic Complex, Southern Alps, Italy. *IX International Palynological Congress* (Houston, U.S.A.), p. 175.

Authors' addresses

CHRISTOPH HARTKOPF-FRÖDER
Geologisches Landesamt NRW
De-Greiff-Strasse 195
D-47803 Krefeld
Germany

GORDON D. WOOD
BP AMOCO
Upstream Technology Group (UTG)
P. O. Box 3092
Houston, TX 77253-3092
U.S.A.

Present address:
the irf group, inc.
23222 Willow Pond Place
Katy, TX 77494
U.S.A.

KARL KRAINER
Institut für Geologie und Paläontologie
Universität Innsbruck
Innrain 52
A-6020 Innsbruck
Austria

ULTRASTRUCTURE OF SOME PERMIAN POLLEN GRAINS FROM THE RUSSIAN PLATFORM

N. E. Zavialova, N. R. Meyer-Melikian, and A. V. Gomankov

Abstract

Dispersed pollen grains from the Upper Tatarian of the Russian platform have been studied using LM, SEM and TEM. Saccate pollen grains assigned to *Vesicaspora* and *Protohaploxypinus* (of possible peltaspermous affinity) are characterized by a regular alveolate ectexine in the region of the cappa, by a thin homogeneous ectexine at the apertural area, and by a fine-grained endexine and protosaccus. The protosaccoid appearance of pollen grains may be the result of preservational flattening of the pollen. Protobisaccate pollen assigned to *Piceapollenites*, which could also be produced by peltasperms, is divided into two groups based on ultrastructure: (I) a group with granulous ectexine, and (II) a group with a faint alveolate ectexine. The sporoderm of asaccate monosulcate pollen assigned to *Cycadopites* (possibly Leptostrobales or Ginkgoales) consists of only ectexine which becomes thinner in the apertural region. Asaccate ovoid pollen with four furrows, assigned to *Mulinopollenites* (unknown affinity), has some similarities with *Eucommiites* and *Praecolpatites*. Basic structural elements of the sporoderm (attributed to different representatives of seed ferns) and their possible development within more specialized lines of the seed plants are discussed.

INTRODUCTION

The Upper Tatarian flora of the Russian platform has attracted the attention of palaeobotanists for a long time (Gomankov and Meyen, 1986). Dominated by the peltasperms, it formed an ecotone between tropical Euramevian and boreal Angara floras and was at the same time a phylogenetic source of many typical Mesozoic groups of plants (e.g., Ginkgoales). However, this key evolutionary position of the flora, revealed mostly from plant megafossils, needs to be elucidated in palynological terms. The pollen grains dominating the assemblages of dispersed miospores from the Upper Tatarian of the Russian platform were for a long time morphologically and taxonomically interpreted as the bisaccate pollen of conifers (e.g. Efremova, 1966; Sivertseva, 1966). Other authors attributed them to the pteridosperms (Gomankov and Meyen, 1980, 1986; Gomankov, 1986; Esaulova, 1986), which provoked a set of morphological questions which can be solved by means of electronic microscopy. The principal questions can be formulated as following:

1. Are these pollen grains protosaccate or eusaccate?
2. If they are protosaccate, then how significantly different were they from the asaccate *Vittatina*-like pollen grains, which were also produced by the Upper Tatarian peltasperms (Gomankov and Meyen, 1986; Gomankov, 1986)?
3. Are there any ultrastructural features that can be considered as common for the family Peltaspermaceae, for the order Peltaspermales, or for the class Ginkgoopsida (*sensu* Meyen, 1987)?

The last question focuses attention onto an interesting type of pollen grains from the Upper Tatarian of the Russian platform, i.e. the asaccate and colpate pollen. Usually referred to as "ginkgocycadophytes," it was most probably produced by arising "Mesozoic" taxa such as the Ginkgoales or the Leptostrobales (*sensu* Meyen, 1987) and certainly merits an ultrastructural investigation in comparison with the later representatives of these taxa (Zavialova, 1998).

Five genera were selected for the present study. Three of them (*Vesicaspora, Protohaploxypinus* and *Piceapollenites*) are saccate and belong to different groups of peltasperms (Cardiolepidaceae, Peltaspermaceae, and an uncertain family, respectively). Two other genera (*Cycadopites* and *Mulinopollenites*) possess colpi but are deprived of sacci. The first of these genera could be the initial candidate for being a pollen of "youthful" Ginkgoales or Leptostrobales (Gomankov, 1986), while the natural affinity of the second genus is enigmatic; the external morphology of *Mulinopollenites* resembles pollen grains of *Ephedra* or even the angiosperms (which is quite incredible in the Permian), but the need for the scrupulous investigation of its colpi was stated (Gomankov, 1996) in the first description of the genus.

All the material investigated comes from a single palynological sample collected on the right bank of the Vyatka River near the village of Mulino (Kirovskaya region, Russia). This outcrop is incorporated into the stratotype of the Vyatsky horizon of the Upper Tatarian (Upper Permian) and yields abundant, diverse and well-preserved palynomorphs, as well as plant megafossils and remains of vertebrates (Gomankov, 1997; Gomankov et al., 1998). The assemblages of both plant megafossils and palynomorphs (for which the age could be precisely determined due to the position of the locality in the type section of the Tatarian) were recently described by A. V. Gomankov (1996).

MATERIAL AND METHODS

The pollen grains were macerated by standard methods (Rybakova and Smirnova, 1988). Selected pollen grains were mounted in glycerine jelly for examination with transmitted light. The pollen grains were then mounted onto standard SEM stubs, coated with approximately 200 Å of gold–palladium and examined with a Hitachi S-405A SEM. For transmission electron microscopy, individual pollen grains were picked from the SEM stubs with a needle, placed in the capsule and embedded in epon (Meyer-

Zavialova, N. E., Meyer-Melikian, N. R., and Gomankov, A. V.
2001 Ultrastructure of some Permian pollen grains from the Russian Platform; *In*: Goodman, D.K., and Clarke, R.T. (eds.), *Proceedings of the IX International Palynological Congress, Houston, Texas, U.S.A., 1996*; American Association of Stratigraphic Palynologists Foundation, p. 99–114.
 ISBN 0–931871–06–9

Melikian and Telnova, 1991). The pollen grains were sectioned by ultramicrotome with a diamond knife and examined using a Hitachi H-600 transmission electron microscope.

DESCRIPTIONS AND COMPARISONS

Vesicaspora

The genus *Vesicaspora* was established by M. P. Schemel (1951) for monosaccate pollen grains from the Pennsylvanian of Iowa. The majority of representatives of the genus found in the Tatarian of the Russian platform are usually identified as *Vesicaspora ex gr. magnalis* (Andreyeva) Hart, which is interpreted as a plexus of three species — *V. magnalis* (Andreyeva) Hart, *V. ovata* (Balme & Hennelly) Hart, and *V. potoniei* (Lakhanpal, Sah & Dube) Hart — with the differences among the three species still remaining rather indefinite (Gomankov and Meyen, 1980). Besides dispersed pollen grains, similar forms were extracted from sporangia of *Permotheca vesicasporoides* S. Meyen, Esaulova & Gomankov, a synangiate organ of peltaspermaceous affinity (Gomankov and Meyen, 1986; Esaulova, 1986).

The dispersed pollen grains identified as *Vesicaspora ex gr. magnalis* are bilaterally symmetrical and monosulcate. In transmitted light they usually appear bisaccate but the examination of a number of specimens (Gomankov and Meyen, 1980) revealed that the grains are monosaccate with a central body surrounded by a bilobed saccus. This observation was confirmed by the present ultrastructural investigations (see below).

The outline of the pollen grain is almost square. In transmitted light the surface appears reticulate or fine-tubercular near the edge of the sulcus; the sulcus membrane appears smooth (Plate 1, fig. 6). The central body, when seen in polar view, is often surrounded by a dark ring which was referred to as "equatorial cingulum" and regarded as a specific character of *V. magnalis* by G. F. Hart (1965).

In SEM, the surface of the pollen grain appears irregularly pitted-tubercular (Plate 2, fig. 5). The sculpture is most obvious on the surface of the saccus. There is a narrow sulcus in the center of the distal side, and the sulcus often has lacerated margins. The ends are acuminate and poorly defined. The sulcus membrane is comparatively smooth and fine-grained. The breadth of the pollen grain measured by SEM is 20 μm, the length is 24 μm.

Ultra-thin transverse sections through the pollen grains reveal that they are strongly compressed (Text-Figure 1). The sporoderm is about 1.5 μm thick (Plate 2, figs. 1–3). Only rare and small fragments of endexine are preserved. These fragments can be observed in exceptional cases as

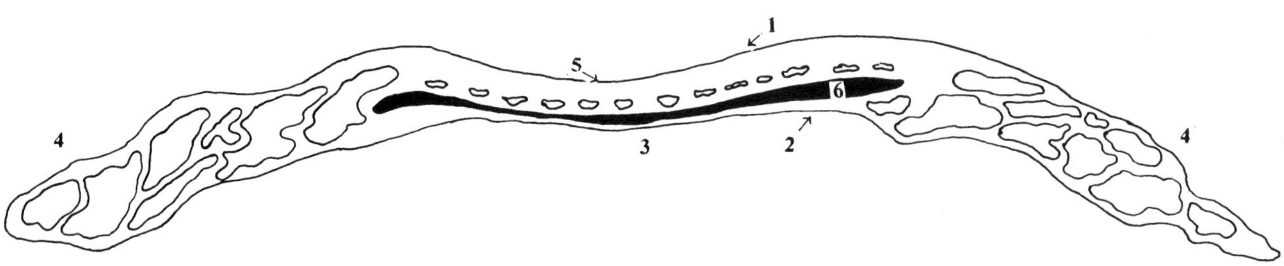

Text-Figure 1. Scheme of the section of a compressed *Vesicaspora* pollen grain (in the region of the sulcus). 1, Proximal side; 2, Distal side; 3, Area of sulcus; 4, Saccus with irregular-shaped alveolae; 5, Cappas region; 6, Compressed hollow of the pollen grain.

PLATE 1

Figures 1–8, 11, 12 are light microscope photographs, x500. Figures 9, 10, 13 are oblique sections of *Protohaploxypinus*, TEM. x10,000.

1 *Mulinopollenites* (see Plate 5, fig. 1. Surface of the pollen grain).

2 *Mulinopollenites* (see Plate 5, figs. 2, 3. Surface of the furrows).

3 *Cycadopites* (see Plate 5, figs. 5, 6. Transverse section of the pollen grain).

4 *Cycadopites* (see Plate 6, figs. 3, 5. Longitudinal section of the pollen grain).

5 *Cycadopites* (see Plate 5, fig. 4. Surface of the pollen grain).

6 *Vesicaspora* (see Plate 2, figs. 1–3, 6–7. Transverse section of the pollen grain). Polar view. Note the "equatorial cingulum" visible as a dark ring.

7 *Piceapollenites*, group 2 (see Plate 3, fig. 4 - surface of the pollen grain, Plate 4, fig. 6 - section of the pollen grain). Polar view. Note the fold at the region of corpus.

8 *Piceapollenites*, group 1 (see Plate 3, fig. 3 - surface of the pollen grain, Plate 4, figs. 1–5, 7 - longitudinal section of the pollen grain). Polar view.

9 The first part of the oblique section of *Protohaploxypinus* (see figs. 10, 13). Ultrastructure of the saccus.

10 The middle part of the section (see figs. 9, 13). Arrows indicate endexine.

11 Polar view of *Protohaploxypinus* (see Plate 3, fig. 2. Surface of the pollen grain). Note the striate corpus of the pollen grain.

12 Lateral view of *Protohaploxypinus* (see Plate 3, fig. 1. Surface of the pollen grain). Note the striae and reticulations of the saccus.

13 The last part of the section (see figs. 9, 10). Proximal side of the pollen grain in the region of cappa. Arrows indicate striae.

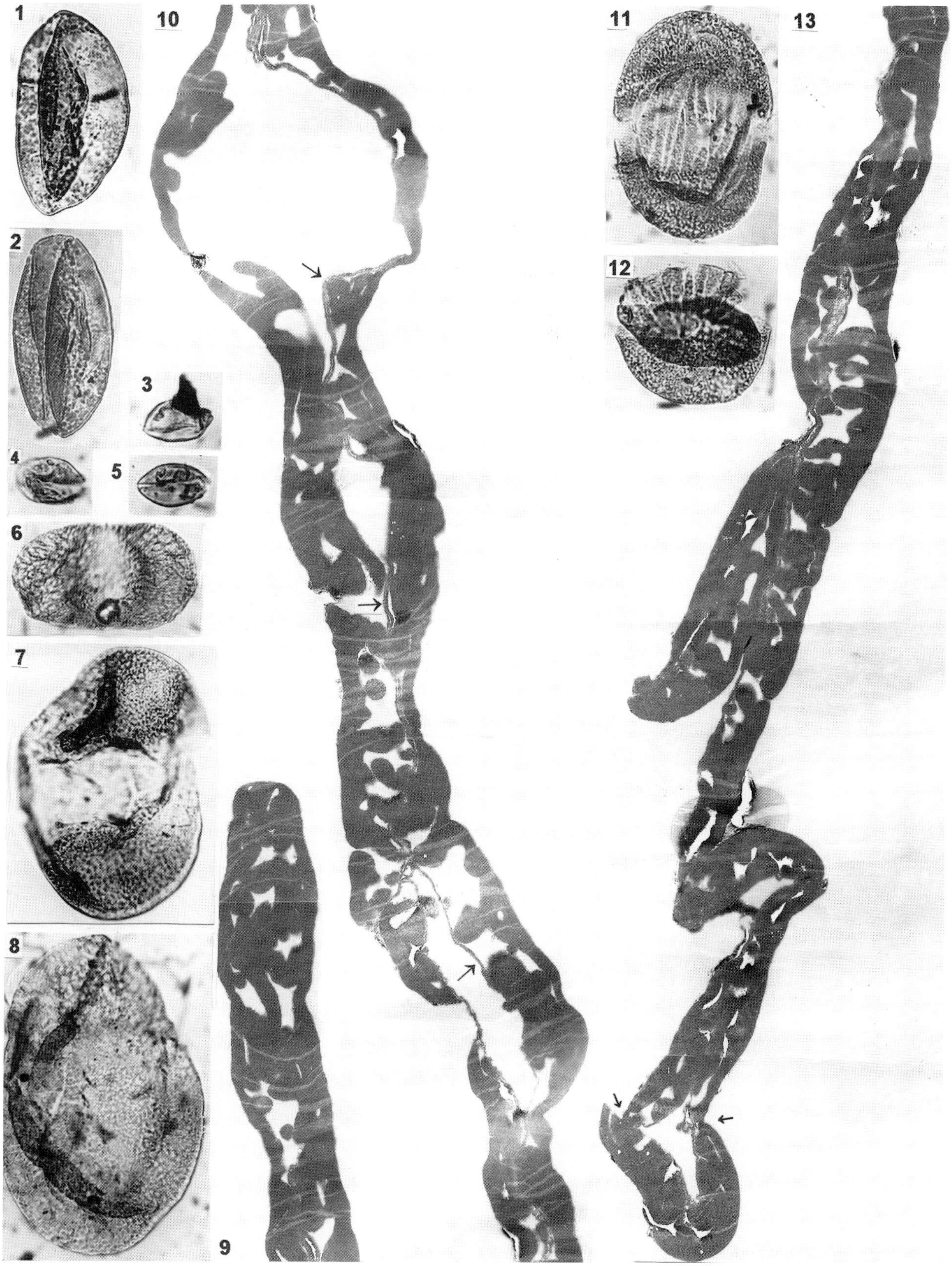
1
2
3
4
5
6
7
8
9
10
11
12
13

pieces of thin granular membrane (Plate 2, fig. 6, at right), while most of the visible patterns are formed by the ectexine. There is a distinctive cappa at the proximal side presented by two homogeneous layers with the intermediate layer of alveolae between them; the length of alveolae is greater than the height (Plate 2, fig. 3, above). The saccus possesses a protosaccate structure with alveolae of variable size and form that are divided by relatively thick partitions whose form is also highly diverse. Two denser areas (with thicker partitions, smaller alveolae, and more abundant remains of the endexine) are observed near the edges of the cappa (Plate 2, figs. 2, 6, at right). It is possible that they correspond to the dark ring seen in transmitted light around the central body. The ectexine gradually gets thinner towards the sulcus, where it is represented by a single and very thin layer (Plate 2, fig. 1, above, fig. 7, the arrow).

The pollen of *Vesicaspora* has been the subject of numerous investigations using electron microscopy, although these investigations were restricted by one species, *V. schaubergeri* (Klaus) Jizba from the Pennsylvanian of North America, which is attributed to the Callistophytales. Pollen grains of this type were initially identified in the Middle Pennsylvanian synangiate pollen organ *Idanothekion* (Millay and Eggert, 1970) and in the Upper Pennsylvanian genus *Callandrium* (Stidd and Hall, 1970). M. A. Millay and T. N. Taylor (1970) have examined *Vesicaspora*, using both light and scanning electron microscopy, as well as TEM (Millay and Taylor, 1974). The early ontogenetic stages of *V. schaubergeri* pollen have been detailed by Hall and Stidd (1971) including tetrad configurations and saccus development. G. W. Rothwell (1972, 1977) described *V. schaubergeri* pollen grains lodged in the nucellus of a *Callospermarion*-type ovule. A tube-like structure extended from the distal sulcus and divided twice to form three branches. Nothing is known about either the tube content or its function. Millay and Eggert (1974) investigated several developmental stages of *V. schaubergeri* extracted from the reproductive organs *Idanothekion* and *Callandrium*.

All of the above studies revealed a eusaccate organization of *V. schaubergeri* unlike the protosaccate organization of *V. ex gr. magnalis*, although the general similarity of both types of pollen grains within the single genus corroborates the close relationship between the Callistophytales and Peltaspermales suggested by S. V. Meyen (1984). It should be mentioned, however, that the described protosaccate structure of *V. ex gr. magnalis* differs strongly from that of other protosaccate pollen grains described, for example those studied by C. B. Foster (1979) or by Taylor and L. Grauvogel-Stamm (1995). The main differences between the two are that the pollen grains of *V. ex gr. magnalis* have larger alveolae and a general "single-story" appearance of the structure of the saccus due to the small number of partitions oriented either parallel or subparallel to the surface of the pollen grain. Such a structure may be the result of preservational flattening of the pollen grain, which was a eusaccate one in reality. The close positioning of some partitions to the cappa (Plate 2, figs. 2, 6) seems to be an additional argument supporting this assumption. Still, it seems most likely that the pollen grains of *V. ex gr. magnalis* did not possess true eusacci, but sacci with somewhat intermediate structure distinguishing itself by few and relatively large alveolae.

Protohaploxypinus

The genus *Protohaploxypinus* was established by Samoilovich (1953) and it has been widely adopted by modern palynologists since the emendation by Hart (1964). It is one of the most abundant genera of the dispersed miospores in the Permian on the Russian platform. The majority of representatives of the genus found in the Upper Tatarian can be assigned to the species *P. dvinensis* (Sedova) Hart (Plate 1, figs. 11, 12), although some authors use a number of different generic and specific names to reflect the variability within this pool of pollen grains (e.g. Molin et al., 1986).

The pollen grains assigned to *P. dvinensis* belong to the Peltaspermaceae using the methodology of Gomankov (1986) including extracting the pollen from the synangia of *Permotheca striatifera* Gomankov & Meyen, as well as from the micropyle of the seeds *Salpingocarpus bicornutus* Meyen and *S. variabilis* Meyen.

Pollen grains of *P. dvinensis* are bilaterally symmetrical in polar view, and they are striate (Plate 1, fig. 11). In transmitted light they appear to be bisaccate, however some SEM photomicrographs indicate that the saccus appears to girdle the entire corpus. There are small tubercles and foveolae irregularly distributed over the surface (Plate 3, figs. 1, 2). The sulcus is not clear but according to the TEM data the entire distal part of the corpus appears to be an apertural area (leptoma).

There are several similarities between the ultrastructure of *Protohaploxypinus* and that of *Vesicaspora*, namely the structure of saccus, aperture and endexine. The endexine is similarly preserved as small and indistinct fragments of a thin fibrous membrane. The proximal side of the ectexine

PLATE 2

1 The first part of the transverse section of monosaccate pollen grain *Vesicaspora* (see the polar view of the same pollen grain at the Plate 1, fig. 6, LM). Note a thin sporoderm layer at the area of sulcus at the distal side of pollen (a) and a distinctive cappa presented by two homogenous layers with alveolate layer between them (b). TEM. x10,000.

2 Same transverse section of *Vesicaspora* (continuing) as in fig. 1. Note the alveolate infrastructure of the saccus. TEM. x10,000.

3 Area of transverse section of *Vesicaspora*. A region of the cappa (above). Sporoderm becomes thinner at the distal side of the pollen (below), a thin membrane between proximal and distal sides indicates the locality of the compressed cavity of the pollen. TEM. x15,000.

4 Area of transverse section of *Protohaploxypinus*. Note the fine-grained endexine, two ribs with a single layer of regular alveolae, a stria between the ribs presented by a thin homogenous layer of ectexine. TEM. x10,000.

5 Distal view of *Vesicaspora*. SEM. x1,500.

6, 7 Area of transverse section of *Vesicaspora*. The arrow indicates the area of sulcus represented by a thin homogenous layer of ectexine. TEM. x15,000. a - distal side of pollen grain; b - proximal side of pollen grain.

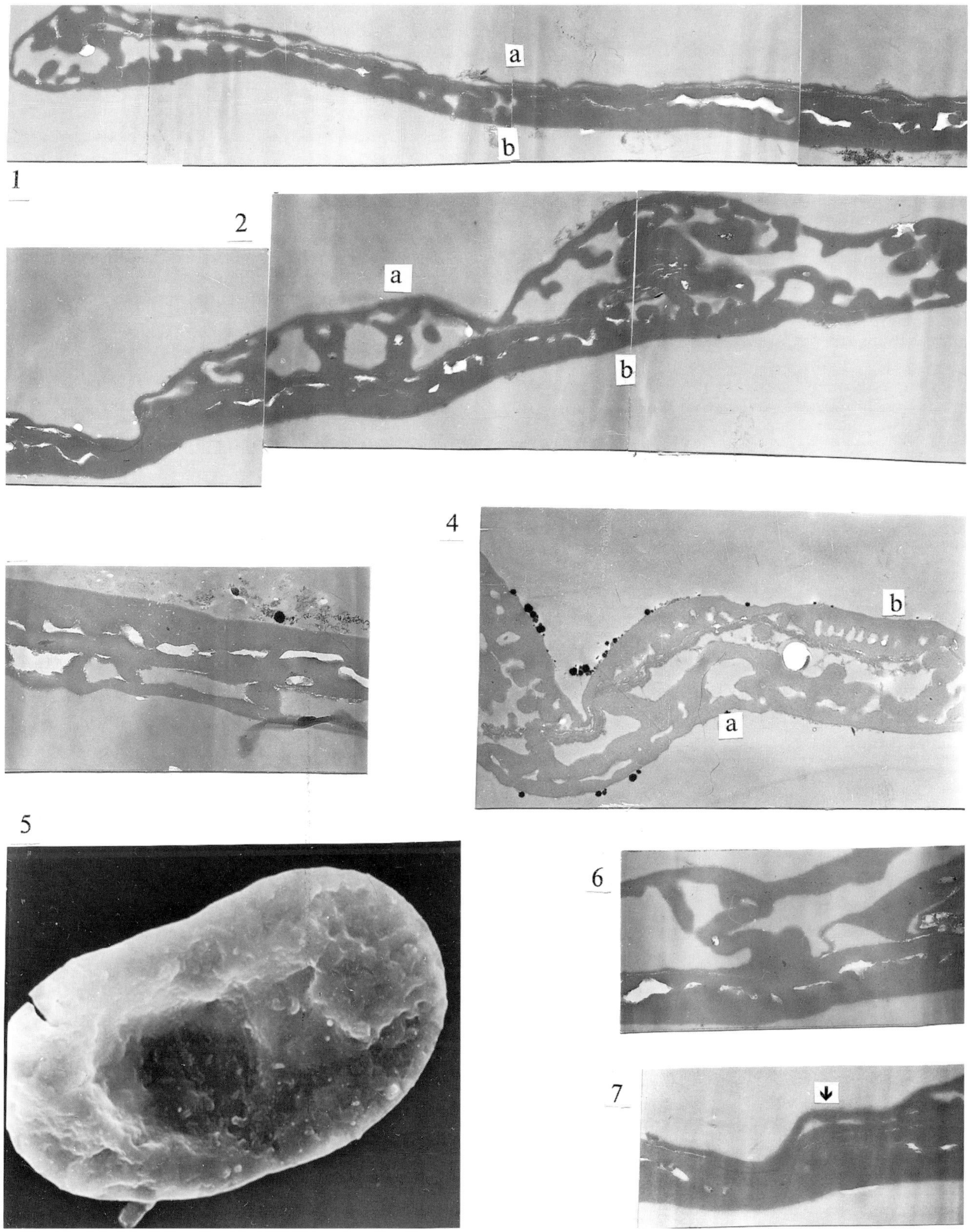
1
a
b
2
a
b
4
b
a
5
6
7

demonstrates a cappa with a single layer of rather regular alveolae whose height (unlike *Vesicaspora*) is usually larger than breadth (Text-Figure 2). The cappa is divided by narrow striae into several longitudinal or oblique ribs, whereas at the sites of striae the alveolae are not present; here the ectexine is represented by a single thin and homogeneous layer (Plate 1, fig. 13, arrows; Plate 2, fig. 4). A circular stria of the same construction surrounds the whole cappa.

The dimensions of the cappa in comparison to the breadth of the whole pollen grain suggest the latter is monosaccate, similar to *Vesicaspora*. Still, the distribution of the endexine fragments and the occurrence of the leptoma on the distal side show that the central body was sufficiently wider than the cappa dividing the saccus into two separate sections.

The saccate structure is similar to that of *Vesicaspora*. It exhibits large alveolae (some are so large to suggest a eusaccate structure of the whole saccus) divided by relatively thick partitions (Plate 2, fig. 4). Both alveolae and partitions have an irregular form. In some cases, a layer formed by more or less uniform small alveolae can be observed on the distal side of the saccus. The structure of this layer resembles that of the cappa, although the outlines of the alveolae are usually less regular. In the apertural area the ectexine is represented by a single thin homogeneous layer.

The pollen grains of the *Protohaploxypinus*-type found in the Gondwana countries are usually attributed to the Glossopteridales because they were repeatedly extracted from male fructifications of this order (Balme, 1995). The ultrastructure of dispersed pollen grains of *P. limpidus* (Balme & Hennely) Balme & Playford from Australia was described by Foster (1979). It differs strongly from that of

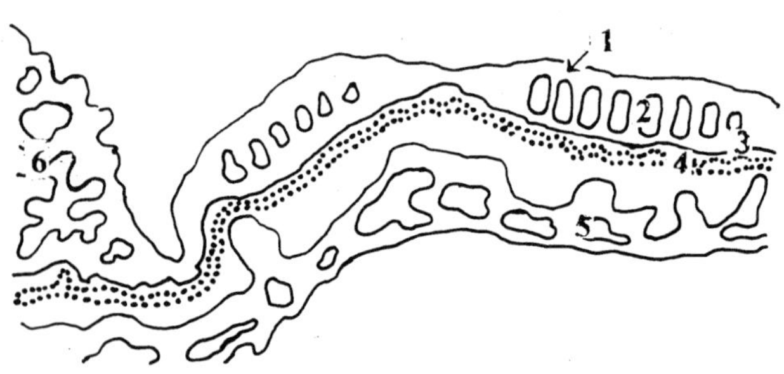

Text-Figure 2. Scheme of the section of a *Protohaploxypinus* pollen grain. 1, Homogenous tectum; 2, Alveolate layer; 3, Foot layer; 4, Possible endexines of distal and proximal sides pressed together; 1–3, Proximal side; 5, Distal side; 6, Part of the saccus.

P. dvinensis described above in exhibiting the extremely small alveolae and thick partitions between them both at the cappa and in the saccus (thus undoubtedly demonstrating the protosaccate nature of the saccus) as well as the irregular arrangement of alveolae on the ribs of cappa.

Another ultrastructural investigation of Gondwanan *Protohaploxypinus* was published by Zavada (1991), who extracted pollen grains of this type from the sporangia of *Arberiella*. The main difference between *P. dvinensis* and Gondwanan *Protohaploxypinus* is the well-defined homogeneous (with occasional lamellations) endexine of the latter. They are similar in the structure of the apertural region represented by a thinning of the ectexine. The ectexines of both taxa become thinner at the striae, but for the pollen in this study this trend is more distinct and is interpreted as a single thin and homogeneous layer (Plate 1, fig. 13, arrows; Zavada, 1991, fig. 13, arrows). Pollen assigned to Gondwanan *Protohaploxypinus* is protomonosaccate, however sometimes the structure is very close to those of the pollen in this study (Plate 1, fig. 9). It confirms a complicated influence of flattening of the pollen to possibly reveal the true nature of the saccus.

Piceapollenites

We use the name *Piceapollenites* in the sense proposed by Hart (1965). Most modern palynologists prefer to accommodate the pollen assignable to *Piceapollenites* under the name *Alisporites*, although Jansonius (1971) has clearly shown that the holotype of the type species of *Alisporites* possesses a distinct distal sulcus. We keep the pollen grains described here under the name *Piceapollenites* separately from the mentioned specimen because of this feature.

Pollen grains assignable to *Piceapollenites* are rather diverse in the Upper Tatarian on the Russian platform and may include several species, but a detailed systematic treatment of this group has not been completed to date. Gomankov (1986) has suggested the relationship between the pollen grains *Piceapollenites* and the leaves *Rhaphidopteris*, which could have been produced by peltasperms (Meyen, 1987), but such a connection remains equivocal.

Pollen grains of *Piceapollenites* are bilaterally symmetrical, protobisaccate and elliptical in polar view. They all appear to be of the same type in transmitted light and SEM (Plate 1, figs. 8, 7; Plate 3, figs. 3, 4), but the ultrastructural characteristics revealed by TEM permits separation into two groups which probably correspond to different species.

Group I

Group I is represented by 4 specimens (Plate 1, fig. 8 - LM, Plate 3, fig. 3 - SEM, Plate 4, figs. 1–5, 7). The endexine

PLATE 3

1 The lateral view of *Protohaploxypinus* showing striate corpus and the surface of the saccus (see this pollen grain in Plate 1, fig. 12, LM). SEM. x1,500.

2 Proximal view of *Protohaploxypinus* (see this pollen grain in Plate 1, fig. 11, LM). SEM. x1,000.

3 Proximal view of bisaccate pollen grain *Piceapollenites*, group I (see the same pollen grain in Plate 1, fig. 8, LM). Note the granulate surface in the area of central body. SEM. x1,000.

4 Proximal view of *Piceapollenites*, group II (see the same pollen grain in Plate 1, fig. 7, and ultrastructure of the corpus in Plate 4, fig. 6, TEM). SEM. x1,000.

1

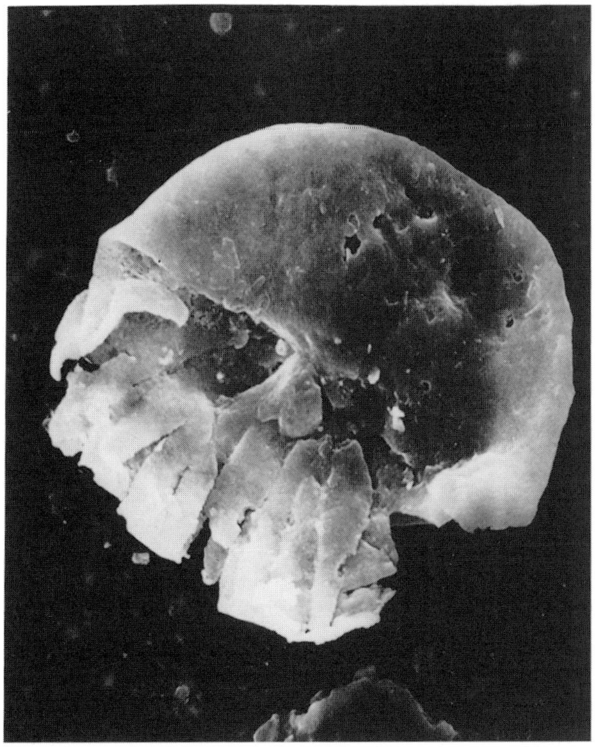

2

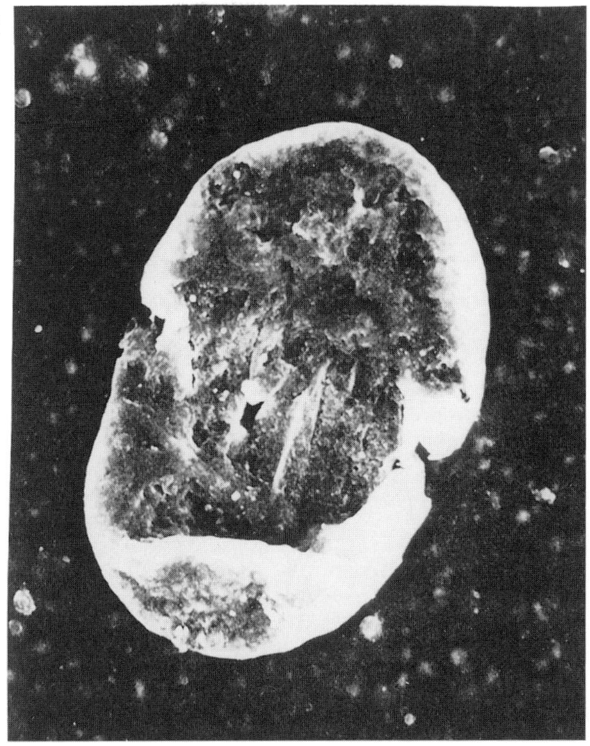

3

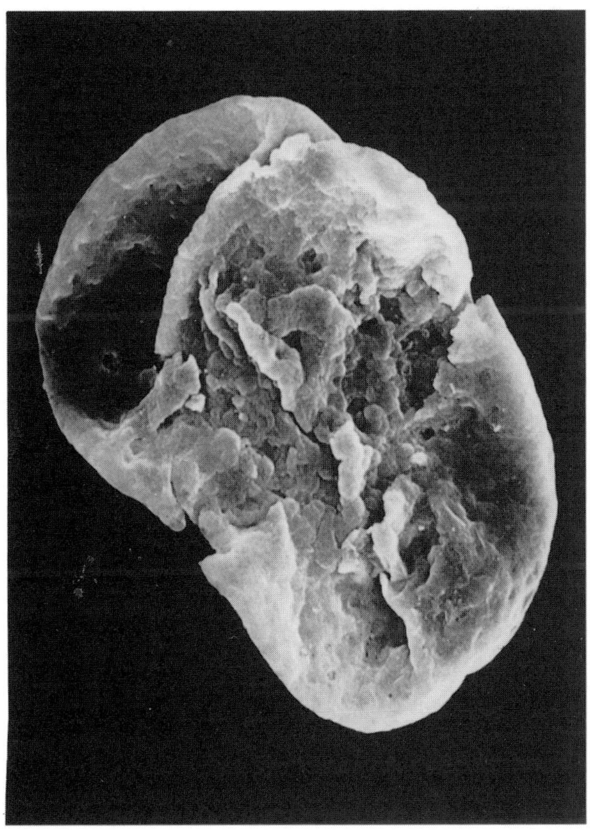

4

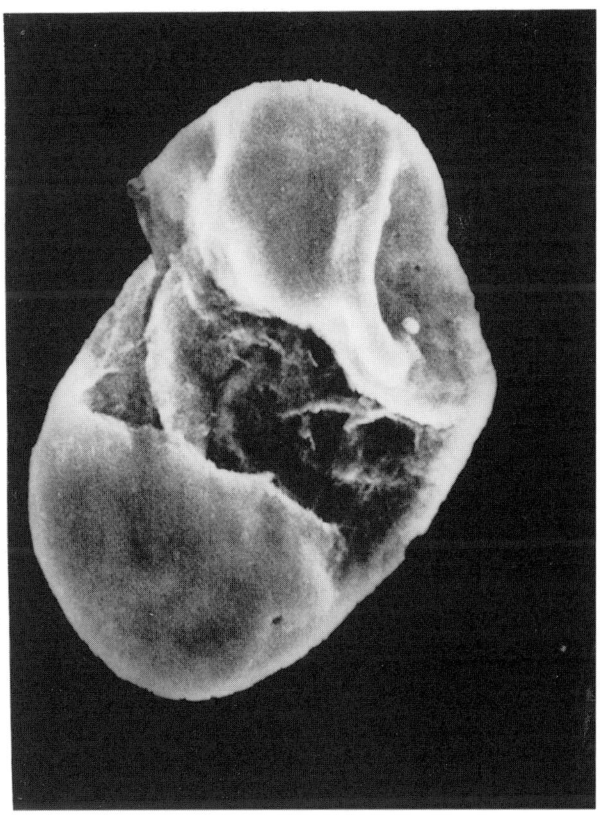

is not preserved. The ectexine in the area of the central body is similar on both the proximal and distal sides and has a "spongy" appearance with a large number of very small cavities (alveolae) (Plate 4, figs. 1–4). On one side (assumed to be distal) the ectexine is very thin and consists mainly of only one layer of alveolae (Plate 4, figs. 3, 4), whereas on the other side it can be up to 3 µm thick (Plate 4, figs. 1, 2). Apparently the ectexine on both sides becomes granular in the area of the transition to the sacci. The sacci possess a protosaccate structure similar to that described for *Vesicaspora* and *Protohaploxypinus,* which consists of relatively large alveolae divided by partitions of irregular shape and thickness (Plate 4, figs. 1, 3–5, 7).

Group II

Group II is represented by 3 specimens (Plate 1, fig. 7 - LM, Plate 3, fig. 4 - SEM, Plate 4, fig. 6 - TEM). The endexine is preserved as a thin fine granular membrane (Plate 4, fig. 6). The ectexine at the cappa is thick and contains numerous relatively large, irregularly-shaped lacunae usually located in one layer. The whole distal side of the corpus is covered by thin single-layered homogeneous ectexine and functioned apparently like a leptoma. Available information about the sacci is poor, we suggest that the sacci are of the same structure as the structures described above.

Cycadopites

This genus is usually considered as Mesozoic. A number of species was described from the Permian of Siberia (Hart, 1965), but there is doubt as to whether they represent true monocolpate pollen and are not simply rolled up monosaccate grains (Gomankov and Meyen, 1980; Dyupina, 1986). The pollen of "true" *Cycadopites* is rare in the Permian of the Russian platform. In the Upper Tatarian it is probably represented by a new, undescribed species. Gomankov (1986) has attributed these pollen grains to the Leptostrobales or Ginkgoales, the leaves and female fructifications of which are found in the same deposits.

The pollen grains examined herein (Plate 1, figs. 3–5) are ovoid and monosulcate. The surface of the pollen grains appears smooth using both transmitted and SEM microscopy (Plate 1, fig. 5, Plate 5, fig. 4). The breadth of the pollen measured by SEM is 11.7 µm, and the length is 21 µm.

The sporoderm consists of one layer, apparently representing the ectexine. It is dense and almost entirely homogeneous. In transverse section (Plate 5, figs. 5, 6) there are some morphological differences between the distal and proximal sides. The proximal ectexine contains a single layer of very small alveolae, and this layer divides the proximal ectexine into outer and inner homogeneous parts. The length of alveolae usually exceeds their height; distally they become more distinct, more circular in shape, and are rarely fused.

At the distal side the alveolate layer and inner part of homogeneous ectexine disappear. The outer homogeneous part thins out gradually (in some transverse sections becomes absent). Longitudinal sections show that the inner surface of the distal ectexine (formed by the outer homogeneous part) is uneven, unlike that of the proximal ectexine which is usually smooth (Plate 6, fig. 5).

Thus, the proximal ectexine is distinctly thicker than the distal ectexine (0.65–0.8 µm and 0.15 µm, respectively), and it has an alveolate layer unlike that at the distal side that is entirely homogeneous.

The ultra-thin sections revealed that all of the pollen grains were compressed (Text-Figure 3); we believe the pollen grains were originally almost circular in polar view (this idea agrees with the conclusions of Harris, 1974, about the initially spherical pollen grains of *Williamsoniella lignieri* Nathorst). Longitudinal sections show close association of the distal and proximal sides of the pollen grains. The hollow of each pollen grain is represented by a narrow space between the sporoderm layers. Transverse sections show not only the compression but a rolling up of the exine. The rolling occurred in the formation of two folds aligned more or less parallel to the sulcus with a narrow chink between them (Plate 5, fig. 5). In transmitted light this chink may be taken for a sulcus. In point of fact the chink usually indicates the position of sulcus, a large part of the thin distal side, which cannot be observed by means of light microscopy or SEM.

During the Mesozoic, the pollen grains assigned to *Cycadopites* were produced by a number of different gymnosperms such as the Peltaspermales, Ginkgoales, Leptostrobales, Pentoxylales, Cycadales, Bennettitales and Gnetales (Balme, 1995). Meyen (1987) noted that pollen grains of at least some of these taxa were not distinguishable by means of light microscopy, whereas the electron microscope investigations were carried out under only one or two representatives of each taxon (Table 1). Information detailing the ultrastructure of the pollen of the Ginkgoales, Leptostrobales and Gnetales (identified as Cycadopites*)* is still not available.

Cycadopites pollen from the Russian platform, as well as that of the Bennettitales and Pentoxylales, has a smooth surface in SEM, whereas fossil cycad pollen grains are characterized by different species-specific sculpture.

The sulcus is formed by a thinning of the ectexine and encompasses a large part of the distal side of pollen grains in all pollen discussed herein. The exine of all pollen (Table 1), except for one taxon, consists of two layers, the endexine

---

PLATE 4

1, 2, 4, 5, 7 Area of one and the same longitudinal section of *Piceapollenites*, group I (see the polar view of this pollen grain in Plate 1, fig. 8). Note the granules (arrow), alveolate structure of the saccus (fig. 5). TEM. x6,000.

3 Area of longitudinal section of *Piceapollenites*, group I. TEM. x15,000.

6 Area of transverse section of *Piceapollenites*, group II (see the polar view of this pollen grain in Plate 1, fig. 7) at the region of corpus. Note the thin layer of electron-dense endexine. A fold at the upper right part of the photo corresponds to the fold of exine of the corpus (Plate 3, fig. 4). TEM. x15,000.

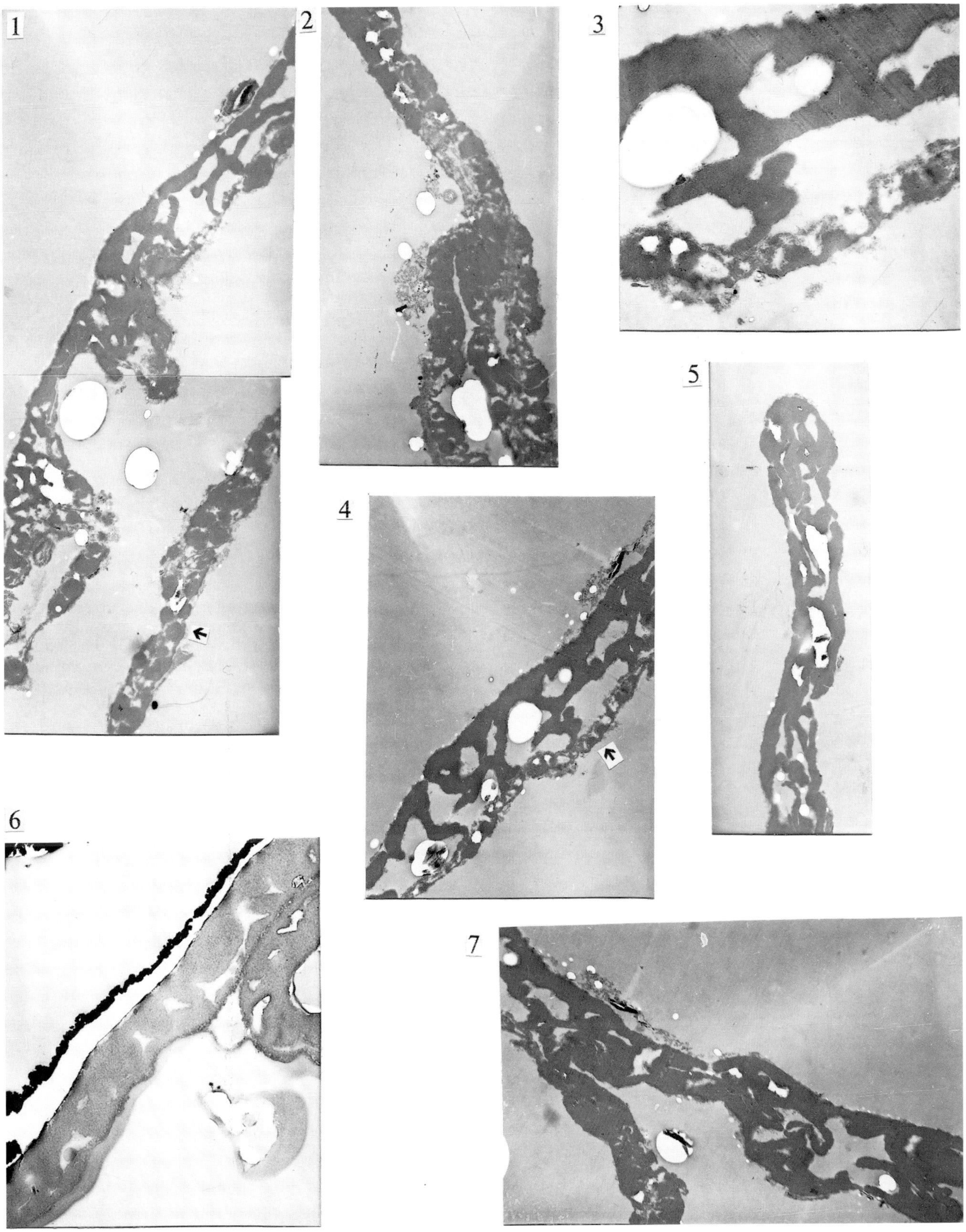
1
2
3
4
5
6
7

and ectexine. The endexine is homogeneous or faintly lamellate (the lamellae usually are difficult to see), whereas the Permian *Cycadopites* examined herein have no endexine. However, this difference may be caused by variable preservational conditions.

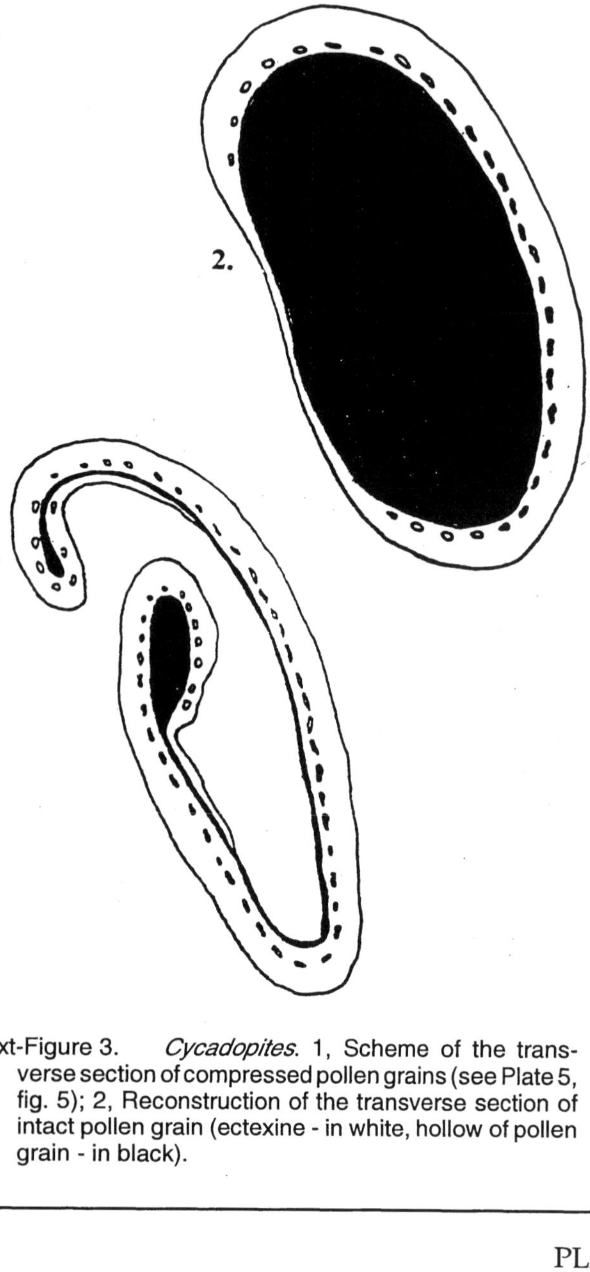

Text-Figure 3. *Cycadopites.* 1, Scheme of the transverse section of compressed pollen grains (see Plate 5, fig. 5); 2, Reconstruction of the transverse section of intact pollen grain (ectexine - in white, hollow of pollen grain - in black).

The ectexine of Permian *Cycadopites* clearly distinguishes it from the specific regular-alveolate ectexine of the Cycadales, and from *Cycadeoidea* pollen which possesses an inner ectexine layer consisting of very small granules.

Though the middle ectexine layer of *Leguminanthus* and the inner layer of *Sahnia* was interpreted by Ward et al. (1989) and Osborn et al. (1991) as granular, the ultra–thin sections of Permian *Cycadopites* seem to be more similar to sections of this pollen. The *Cycadopites* ectexine of the present study is also similar to that of the extant *Ginkgo biloba* (Audran, 1987; Audran and Masure, 1978), but there are some differences in sculpture.

For more definitive conclusions it will be necessary to investigate the pollen ultrastructure of fossil Ginkgoales and Leptostrobales.

Mulinopollenites

This genus of dispersed pollen grains recently described by Gomankov (1996) is still known only from the Upper Tatarian of the Russian platform and is represented by the single species *M. bonus* Gomankov.

Pollen grains (Plate 1, figs. 1, 2) are ovoid (40 µm in length and 23 µm in breadth), with four long, broad furrows. In SEM the surface of nonapertural regions is almost smooth; the ridges of furrows appear uneven and lacerated; the furrow ends are acute. The furrow membrane is tuberculate with rare pits (Plate 5, figs. 1, 2).

In ultra–thin sections, two layers of exine are revealed. The ectexine is thick, almost homogeneous, especially in its outer part. The inner part of the ectexine is perforated by numerous narrow canals with dark electron dense filling. The inner surface of the ectexine is tuberculate with variable spaces between the tuberculae. The spaces between the tuberculae often contain pieces of fine-grained endexine, which underlies the ectexine as a thin layer. Near the furrows the ectexine gradually becomes thinner. The furrow membrane is 0.1 µm thick and is composed solely of endexine, which does not become appreciable thinner (Plate 6, figs. 1, 2, 4; Text-Figure 4).

The natural affinity of *Mulinopollenites* pollen remains unknown. In the first description of the pollen (Gomankov, 1996) its morphological similarity with *Praecolpatites* was mentioned, but the ultrastructure of this genus (Foster and Price, 1981) differs from that of *Mulinopollenites* in having lamellate endexine and an "incipient-alveolate" intrastructure of the ectexine (which is not interrupted at the bottom of the furrows). On the contrary, the aforementioned similarity with some ancient angiosperms is strengthened by the fine-grained endexine and tuberculate eleva-

PLATE 5

1 The surface of *Mulinopollenites* (see the pollen grain at the Plate 1, fig. 1). Note the lacerated edge of the furrow. SEM. x4,000.

2 *Mulinopollenites* (see this pollen grain at the Plate 1, fig. 2). Note two furrows, the difference between granulate bottom of one of the furrows and smooth sculpture of the pollen at the nonapertural region. SEM. x1,500.

3 The region of fig. 2. *Mulinopollenites*. Area of furrow, note granulate sculpture. SEM. x3,800.

4 Smooth surface of monosulcate pollen grain *Cycadopites* (see this pollen grain in Plate 1, fig. 5), distal view. SEM. x3,000.

5 The transverse section of compressed pollen grain *Cycadopites* (see this pollen grain in Plate 1, fig. 3). TEM. x8,000.

6 Area of section 5. Note almost homogenous ectexine contained a single layer of very small alveolae. TEM. x15,000.

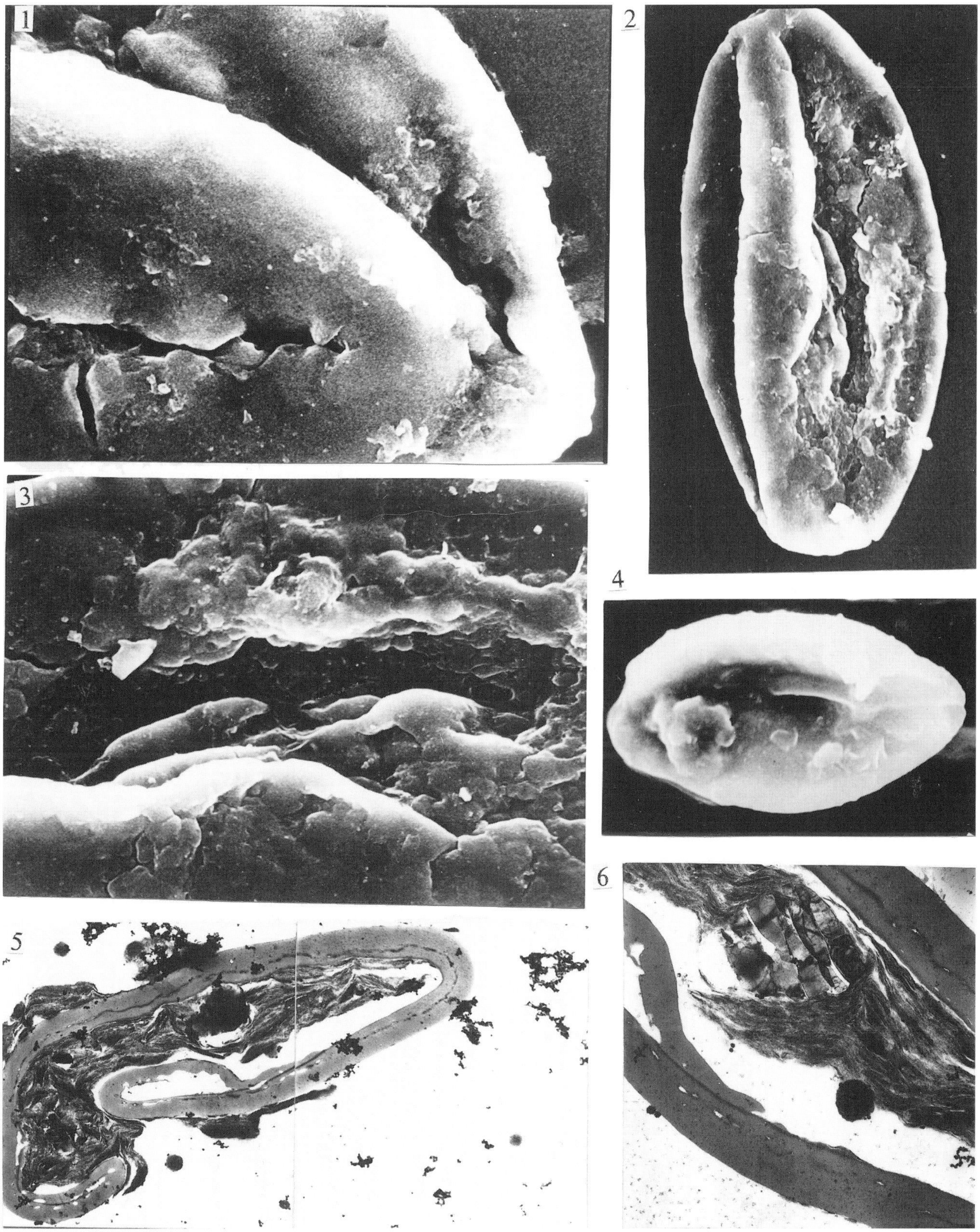
1
2
3
4
5
6

TABLE 1. Comparative ultrastructure of investigated disperse and in situ *Cycadopites* pollen grains.

Taxon	Size of pollen grain, (μm)	Organization of ectexine/ thickness, (μm)	Organization of endexine/ thickness, (μm)	References
Leptostrobales or Ginkgoales (?)	21x11.7	homogeneous with a single layer of very small alveolas/ 0.15–0.8	absent	present investigation
Cycadeoidea dacotensis/ Bennettitales	25x12	thin, homogeneous outer layer, thicker granular inner layer (sometimes appears homogeneous because of very small granules and little lacunae between them)/ 0.5	homogeneous/ 0.23	Osborn and Taylor, 1995
Legyminanthus siliquosus/ Bennettitales	23x12	thick homogeneous outer layer, intermediate layer of coarse, densely packed granules fuse basally into a thin homogeneous inner layer/ up to 0.6	faintly lamellate/ 0.14–0.3	Ward et al., 1989
Androstrobus balmei/ Cycadales	24.3x17.7	alveolate/ 0.25–0.68	faintly lamellate/ 0.27	Hill, 1990
Androstrobus szei/ Cycadales	24.3–31x 14.5–19.5	alveolate/ 0.55	homogeneous/ 0.1	Hill, 1990
Sahnia laxiphora/ Pentoxylales	26x23	homogeneous outer layer, inner layer presented by large granules separated by irregular lacunae/ 0.69	faintly lamellate/ 0.13	Osborn et al., 1991

tions at the inner surface of the ectexine, which have a morphological potential for developing into the baculate infrastructure typical for the angiosperms. Still the most ancient angiosperms (e.g., *Eupomatia* and *Degeneria*) have a homogeneous ectexine, which is occasionally (*Degeneria*) underlain with sporopollenin granules (Takhtajan and Meyer, 1976). The same features (fine-grained endexine and tuberculate elevations at the inner surface of the ectexine) unite *Mulinopollenites* with *Eucommiidites troedssonii* Erdtman, whose ultrastructure was described by Trevisan (1980). *Eucommiidites* is usually attributed to the Cycadales or Gnetales, although its natural affinity is not absolutely clear.

DISCUSSION

The diversity of pollen observed among the pteridosperms is extremely high (Millay et al., 1978; Millay and Taylor, 1970, 1976; Rothwell, 1972, 1977). While the most primitive representatives of the group, the Lagenostomales, produced prepollen, the more advanced pteridosperms possess a number of features both of gross morphology and exine ultrastructure, which are considered as characteristic for gymnosperms. Many of these features were probably a function of the necessity of regulation of microspore volume associated with the development of the male gametophyte inside the sporoderm. Plants such as *Archaeopteris*, which possessed spores with the elastic sporoderm (Telnova and Meyer-Melikian, 1993), could represent an ancestral stock which developed a variety of pteridosperm pollen. The most important features to develop in connection with the changes in volume are (1) harmomegathy, which later took the function of apertures on themselves, and (2) cavities in the exine which developed into sacci.

Therefore, asaccate pollen grains with one or two sulcae and a monolete suture (*Monoletes* Taylor, 1978), monosaccate (*Vesicaspora*) and bisaccate (*Protohaploxypinus*, *Piceapollenites*) pollen grains with sacci of varying com-

PLATE 6

1, 4 Montage of transverse section at the middle part of *Mulinopollenites*. The pollen grain is strongly compressed. Note ectexine transpierced by narrow canals, fine-grained endexine. Ectexine gets thinner toward the furrows and disappears at the furrow regions. TEM. x10,000.

2 Montage of transverse section of the same pollen grain *Mulinopollenites* (see section in figs. 1, 4). Section placed at the end of the grain. Note the thin fine-grained endexine. TEM. x8,000.

3 Montage of longitudinal section of *Cycadopites* (see the pollen grain in Plate 1, fig. 4). Endexine is absent. TEM. x8,000.

5 Area of section 3. Note thicker proximal side of the grain with narrow canals, thinner distal side represented by homogenous ectexine and compressed cavity between them. TEM. x25,000. a - distal side of pollen grain; b - proximal side of pollen grain.

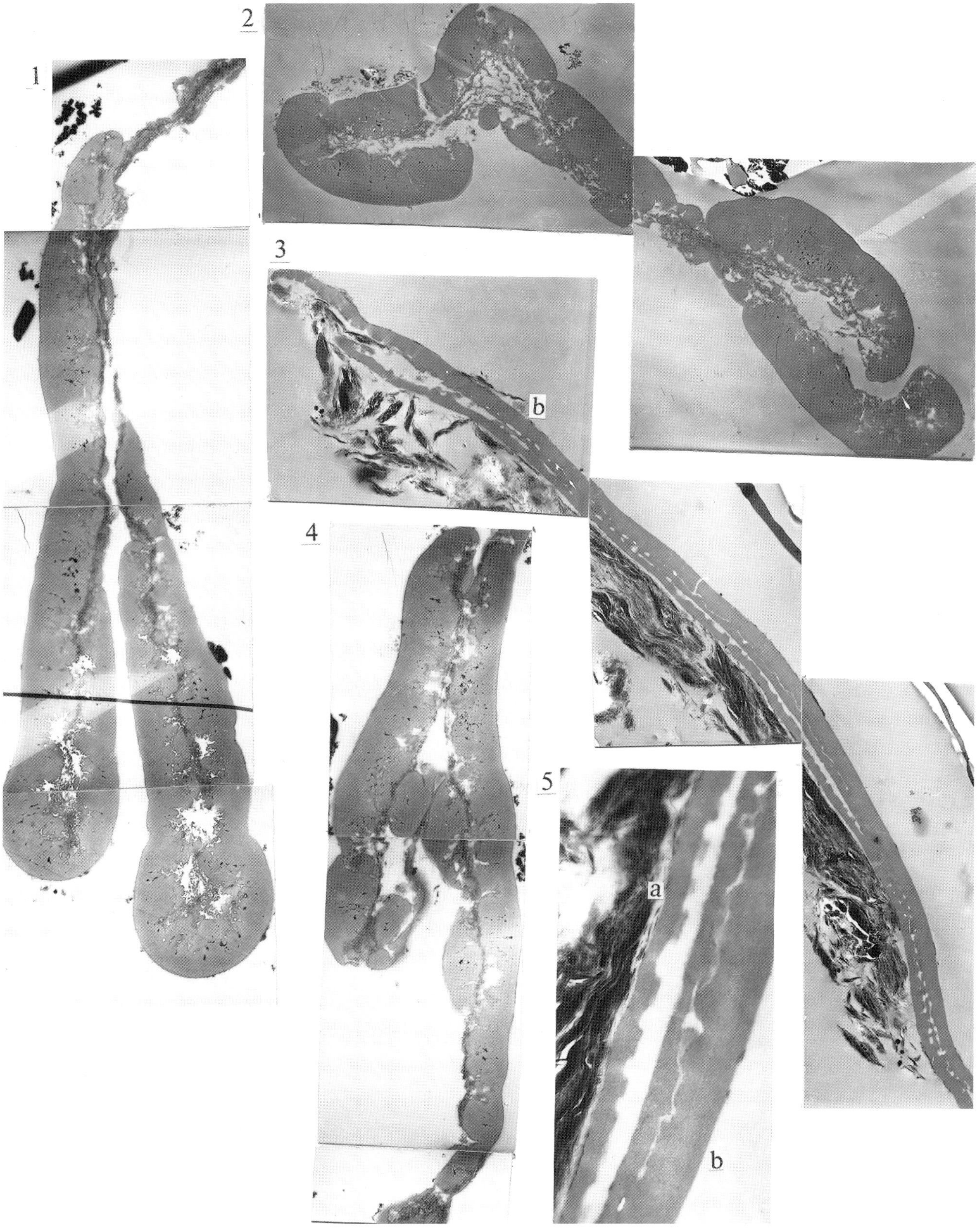
1
2
3
b
4
5
a
b

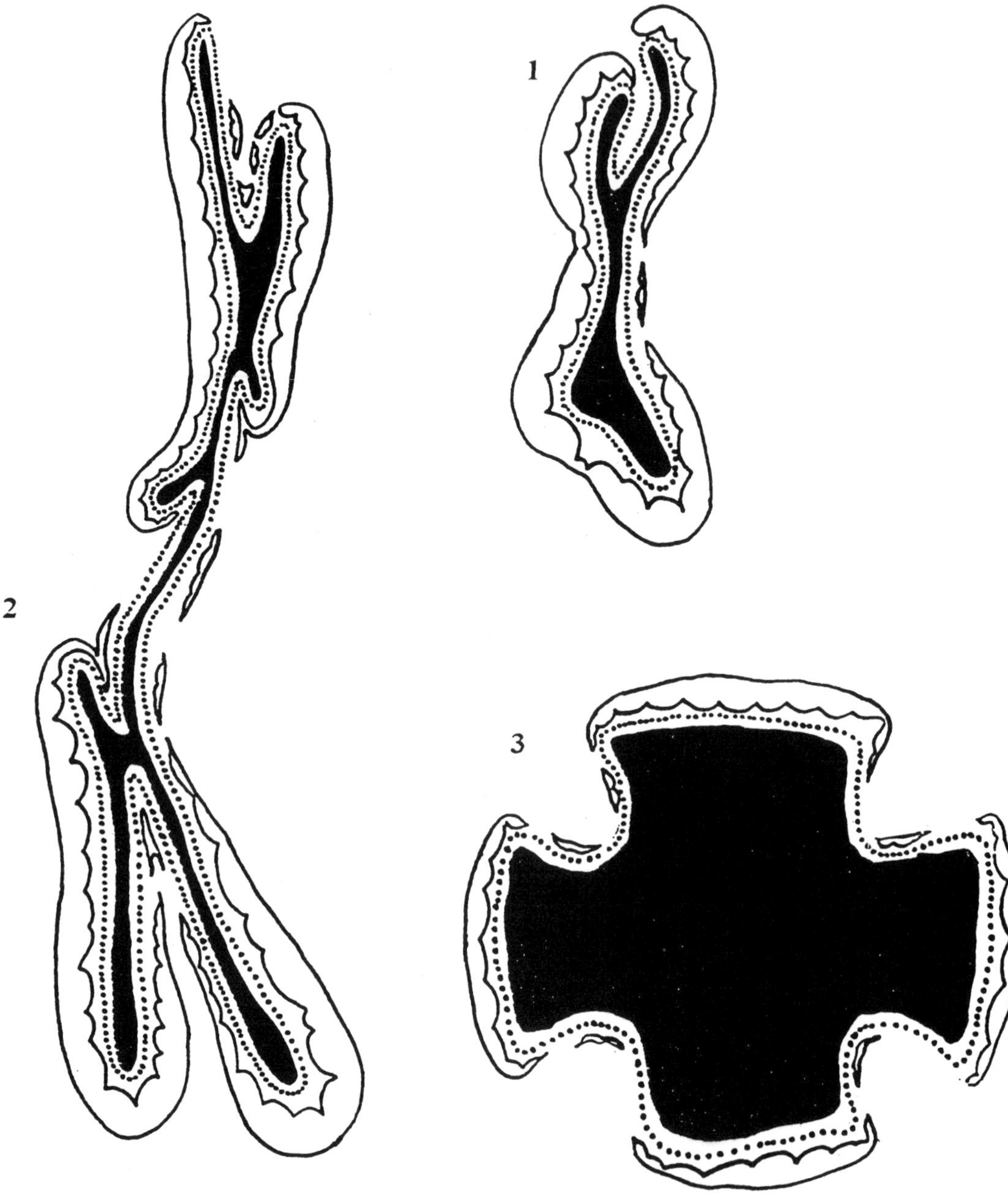

Text-Figure 4. Schemes of the transverse sections of a single pollen grain of *Mulinopollenites* (see Plate 6, figs. 1, 2, 4). 1, The section in the end of the pollen grain (see Plate 6, fig. 2); 2, The section in the middle part of the pollen grain (see Plate 6, figs. 1, 4); 3, Reconstruction of the transverse section of intact pollen grain (ectexine in white; endexine, points; hollow of pollen grain in black). Abbreviations: a - alveolae; f - furrow; g - granule; sa - saccus; su - sulcus; d.s. - distal side of pollen grain; p.s. - proximal side of pollen grain; f-g - fine-grained endexine.

plexity appeared among the pteridosperms. The infrastructure of the exine may be also highly variable. Some pollen grains possess a single layer of exine, i.e. the ectexine (e.g. pollen of *Crossotheca hughesiana* Taylor and Taylor, 1987). Others have two layers, i.e., ectexine and endexine. The endexine can be lamellate such as in *Potoniea* and *Monoletes* (Taylor, 1978). In other cases it can be granulate or fine-grained. The ectexine can be homogeneous as in the pollen of *Crossotheca hughesiana* (Taylor and Taylor, 1987); alternatively the layer can be almost homogeneous with a single layer of small alveolae, and with definite alveolae as in the Medullosales (Taylor and Taylor, 1987), or it can form a saccus with either proto- or eusaccate structure.

The pollen grains described in the present study belonged to the peltasperms or their nearest descendants. Having been dispersed, they are usually strongly flattened and the endexine is preserved as rare fragments, but the main features of their morphology can be still reconstructed. Although they demonstrate a high diversity in some characters, all fall within the generally accepted concept of the variability of the pteridosperm pollen reviewed above. All of them have a distal aperture and therefore they can be assigned to the true pollen and not to prepollen (although double germination, as for their modern descendant, *Ginkgo biloba,* cannot be excluded). The construction of the aperture is variable. It may be a simple thinning of the ectexine (leptoma) as in *Piceapollenites* or *Protohaploxypinus,* or it can be represented by a sulcus (as in *Vesicaspora*) or by highly specialized furrows (as in *Mulinopollenites*).

The endexine, when observed, is very thin and fine-grained. It can be considered as a common feature of the peltasperms and allied taxa (Class Ginkgoopsida *sensu* Meyen, 1987), since most of the other gymnospermous pollen grains possess thicker, lamellate endexine. The infrastructure of ectexine outside of the sacci is also similar among the genera examined herein, demonstrating small alveolae disposed usually in a single layer. *Mulinopollenites* represents an exception, but the inner surface of its ectexine is uneven as if bearing "opened" alveolae.

Three of the genera studied which are assigned to the Peltaspermales possess protosacci. However, the structure of these structures is rather peculiar and more "friable" than in other well-known protosaccate genera, possibly suggesting the imitation of the protosaccus by the eusaccus (which is highly flattened due to the preservation). It should be mentioned here that primitive conifers (*Triadispora*) possessed the true protosaccus (Taylor and Grauvogel-Stamm, 1995), and thus the development of eubisaccate pollen proceeded independently in both the Coniferopsida and Ginkgoopsida.

At the same time, the ultrastructure of asaccate *Vittatina*-like pollen, produced by the other representatives of the Peltaspermaceae (to be described in detail elsewhere), differs strongly from that of saccate genera *Vesicaspora, Protohaploxypinus,* and *Piceapollenites.*

This situation enlarges the diversity of pollen ultrastructure among the peltasperms and hampers the elucidation of characters common for all of the Peltaspermales and Peltaspermaceae. In summary, the different features of pollen grains arose once in the pteridosperms, and subsequently developed in parallel and convergent ways, often across phylogenetic lines, as it was initially proposed by Meyen (1984).

ACKNOWLEGMENTS

This research was in part funded by grants from the International Science Foundation (MG 6000).

References Cited

AUDRAN, J.C.
1987 Comparaison des ultrastructures exiniques et des modalités de l'ontogenése pollinique chez les Cycadales et Ginkgoales actuelles (Préspermaphytes). *Bulletin de la Societé botanique de France, Actualités Botaniques,* 134: 9–18.

AUDRAN, J.C., and MASURE, E.
1978 La sculpture et l'infrastructure du sporoderme de Ginkgo biloba comparées à celles des envelopes polliniques des Cycadales. *Review of Palaeobotany and Palynology,* 26: 363–387.

BALME, B.E.
1995 Fossil in situ spores and pollen grains: an annotated catalogue. *Review of Palaeobotany and Palynology,* 87: 81–323.

DYUPINA, G.V.
1986 The taxonomy, nomenclature and stratigraphical occurrence of pollen grains *Ginkgocycadophytus* Samoilovich. *In: The theory and practice of the palynological studies of Permian and Triassic of the USSR.* Manuscript deposited at VINITI, Syktyvkar, 4839-B86: 17–26 (in Russian).

EFREMOVA, G.D.
1966 On classification of dispersed pollen Striatiti Pant 1954. In: On the methods of the palaeopalynogical studies. *Materials for the 2nd International Palynological conference.* Leningrad, pp. 42–57 (in Russian).

ESAULOVA, N.K.
1986 Connection of pollen *Vesicaspora* with foliage of fossil plants. *In: The theory and practice of the palynological studies of Permian and Triassic of the USSR.* Manuscript deposited at VINITI, Syktyvkar, 4839-B86: 9–12 (in Russian).

FOSTER, C. B.
1979 Permian plant microfossils of the Blair Athol Coal Measures, Baralaba Coal Measures and basal Rewan Formation of Queensland. Brisbane. *Geological Survey of Queensland. Publication* 372, *Palaeontological Paper* 45, 244 pgs.

FOSTER, C.B., and PRICE, P.L.
1981 Exine ultrastructure of *Praecolpatites sinuosus* (Balme & Hennelly) Bharadwaj & Srivastava, 1969, and *Marsupipollenites triradiatus* Balme et Hennelly, 1956. *The Palaeobotanist,* 28–29: 177–187.

GOMANKOV, A.V.
1986 Systematic affinities of the Permian miospores. *In: The theory and practice of the palynological studies of Permian and Triassic of the USSR.* Manuscript deposited at VINITI, Syktyvkar, 4839-B86: 4–9 (in Russian).
1996 Palynological and floristical characteristics of the Tatarian, In: *Stratotypes and reference sections of the Upper Permian of regions of the Volga and Kama Rivers.* Kazan', Ecocentre press, pp. 365–380.
1997 The Permian (Tatarian) flora from the Kotel'nich vertebrate locality (Kirov Oblast). *Stratigraphy and Geological Correlation,* 3: 309–318.

GOMANKOV, A.V., BLOM, G.I., GUSEV, A.K., and LOZOVSKY, V.R.
1998 The Vyatka River Basin. *In:* Lozovsky, V.R., and Esaulova, N.K. (eds.), *Permian–Triassic boundary in the continental series of East Europe.* Moscow, "GEOS", pp. 31–37.

GOMANKOV, A.V., and MEYEN, S.V.
1980 On the relationships of plant mega- and microfossil assemblages in the Permian of Angaraland. *Paleontologicheskii zhurnal,* 4: 114–122 (in Russian).
1986 Tatarian flora (composition and distribution in the Late Permian of Eurasia). *Transactions of Geological Institute of Academy of Sciences USSR,* 401: 174 pgs. (in Russian).

HALL, J.W., and STIDD, B.M.
1971 Ontogeny of *Vesicaspora,* a late Pennsylvanian pollen grain. *Palaeontology,* 14: 431–436.

HARRIS, T.M.
1974 *Williamsoniella lignieri*: its pollen and the compression of spherical pollen grains. *Palaeontology,* 17: 125–148.

HART, G.F.
1964 A review of the classification and distribution of the Permian miospore: Disaccate Striatiti. *Compte Rendu, 5th International Congress de Stratigraphie et Geologie de Carboniferous,* pp. 1171–1199.
1965 The systematics and distribution of Permian miospores. Witwatersrand University Press, Johannesburg, 252 pgs.

HILL, C.R.
1990 Ultrastructure of *in situ* fossil cycad pollen from the English Jurassic, with a description of the male cone *Androstrobus balmei* sp. nov. *Review of Palaeobotany and Palynology,* 65: 165–173.

JANSONIUS, J.
1971 Emended diagnosis of *Alisporites* Daugherty 1941. *Pollen et Spores*, 13: 349–357.
MEYEN, S.V.
1984 Basic features of gymnosperm systematics and phylogeny as evidenced by the fossil record. *The Botanical Review*, 50: 1–111.
1987 Fundamentals of paleobotany. Chapman and Hall, London New York, 432 pgs.
MEYER-MELIKIAN, N.R., and TELNOVA, O.P.
1991 On the method of study of fossil spores and pollen using light, scan electron and transmission electron microscopes. In: *Palynological taxa in biostratigraphy, Materials of the 5th Soviet Union palynological conference*. Moscow, pp. 8–9 (in Russian).
MILLAY, M.A., and EGGERT, D.A.
1970 *Idanothekion* gen. n., a synangiate pollen organ with saccate pollen from the Middle Pennsylvanian of Illinois. *American Journal of Botany*, 57: 50–61.
1974 Microgametophyte development in the Paleozoic seed fern family Callistophytaceae. *American Journal of Botany*, 61: 1067–1075.
MILLAY, M.A., EGGERT, D.A., and DENNIS, R.L.
1978 Morphology and ultrastructure of four Pennsylvanian prepollen types. *Micropaleontology*, 24: 303–315.
MILLAY, M.A., and TAYLOR, T.N.
1970 Studies of living and fossil saccate pollen grains. *Micropaleontology*, 16: 463–470.
1974 Morphological studies of Paleozoic saccate pollen. *Palaeontographica*, Abteilung B, 147: 75–99.
1976 Evolutionary trends in fossil gymnosperm pollen. *Review of Palaeobotany and Palynology*, 21: 65–91.
MOLIN, V.A., BUDANOV, G.F., KOLODA, N.A., and PLOTNIKOV, M.A.
1986 Permian red-colored formations of the northern part of the Russian platform. Leningrad, "Nauka", 112 pgs. (in Russian).
OSBORN, J.M., and TAYLOR, T.N.
1995 Pollen morphology and ultrastructure of the Bennettitales: *in situ* pollen of *Cycadeoidea*. *American Journal of Botany*, 82: 1074–1081.
OSBORN, J.M., TAYLOR, T.N., and CRANE, P.R.
1991 The ultrastructure of *Sahnia* pollen (Pentoxylales). *American Journal of Botany*, 78: 1560–1569.
ROTHWELL, G.W.
1972 Evidence of pollen tubes in Paleozoic pteridosperms. *Science*, 175: 772–774.
1977 Evidence for a pollination-drop mechanism in Paleozoic pteridosperms. *Science*, 198: 1251–1252.
RYBAKOVA, H.O., SMIRNOVA, S.B.
1988 Elements of palynology. Moscow. MSU Press. 98 pgs. (in Russian).
SAMOILOVICH, S.R.
1953 Pollen and spores from the Permian deposits of the Cherdynian and Aktyubinian Cis-Urals. *In*: *Palaeobotanical collection*, GOSTOPTEKHIZDAT, Leningrad - Moscow, 5–91 (Transactions of VNIGRI, new series, iss. 75; in Russian).
SCHEMEL, M.P.
1951 Small spores of the Mystic coal of Iowa. *American Midland Naturalist*, 46: 743–750.
SIVERTSEVA, I.A.
1966 On Permian vegetation of the Soviet Union. *In*: On the methods of the palaeopalynogical studies. *Materials for the 2nd International Palynological conference*. Leningrad, pp. 105–117 (in Russian).
STIDD, B.M., and HALL, J.W.
1970 *Callandrium callistophytoides*, gen. et sp. nov., the probable pollen-bearing organ of the seed fern, *Callistophyton*. *American Journal of Botany*, 57: 394–403.
TAKHTAJAN, A.L., and MEYER, N.R.
1976 Some additional data an the morphology of pollen grains of *Degeneria vitiensis* (Degeneriaceae). *Botanicheskii zhurnal*, 61: 1531–1534. (In Russian).
TAYLOR, T.N.
1978 The ultrastructure and reproductive significance of *Monoletes* (Pteridospermales) pollen. *Canadian Journal of Botany*, 56: 3105–3118.
TAYLOR, T.N., and GRAUVOGEL-STAMM, L.
1995 The ultrastructure of voltzialean pollen. *Review of Palaeobotany and Palynology*, 84: 281–303.
TAYLOR, T.N., and TAYLOR, E.L.
1987 The ultrastructure of fossil gymnosperm pollen. *Bulletin de la Societé botanique de France*, 134, *Actualités Botaniques*: 134, 121–140.
TELNOVA, O.P., and MEYER-MELIKIAN, N.R.
1993 *Spores of Devonian–Carboniferous of Timano-Pechorschia region*. St. Petersburg, "Nauka", 126 pgs. (in Russian).
TRAVISAN, L.
1980 Ultrastructural notes and considerations of *Ephedripites*, *Eucommiidites* and *Monosulcites* pollen grains from Lower Cretaceous sediments of southern Tuscany (Italy). *Pollen et Spores*, 22: 85–132.
WARD, J.V., DOYLE, J.A., and HOTTON, C.L.
1989 Probable granular angiosperm magnoliid pollen from the Early Cretaceous. *Pollen et Spores*, 31: 113–132.
ZAVADA, M.S.
1991 The ultrastructure of pollen found in dispersed sporangia of *Arberiella* (Glossopteridaceae). *Botanical Gazette*, 152: 248–255.
ZAVIALOVA, N.E.
1998 Morphology and ultrastructure of fossil pollen grains (from the Upper Perm sediments of the Vyatka River and the Low Jurassic Deposits of West Siberia). Thesis for the degree of Doctor of Philosophy submitted to Russian Academy of Sciences. 280 pgs. (in Russian, unpubl.).

Authors' addresses

N. E. ZAVIALOVA
N. R. MEYER-MELIKIAN
Department of Higher Plants
Biological Faculty
Moscow State University
Vorobjevy Gory
119899, Moscow
Russia

A. V. GOMANKOV
Geological Institute
Russian Academy of Sciences
109017, Pyzhevsky per., 7
Moscow
Russia

USING ACRITARCH ASSEMBLAGES TO CORRELATE UPPER PROTEROZOIC AND LOWER PALEOZOIC METAMORPHIC SEQUENCES IN ROMANIA

Leonard Olaru

Abstract

For metamorphic Proterozoic and Paleozoic formations in Romania where generally body fossils are not preserved, assemblages of the acritarchs reflect paleoenvironmental factors associated with primary sedimentation. Acritarch assemblages not only are a reflection of primary sedimentation but also are useful for correlation of these older formations. We have analysed and interpreted acritarch assemblages from the Bretila Group (Riphean), Rebra Group (Vendian) and Tulgheş Group (Lower Cambrian to Middle Cambrian) from the East Carpathians, and these have been correlated with other zones from Romania.

In Dobrogea County (south-eastern Romania), the older metamorphic formations are represented by the Orliga Group (Riphean–Vendian) and the Greenschist Group (Lower Cambrian).

INTRODUCTION

The regions of Romania studied for paleobiologic correlation, are the East Carpathians and Dobrogea County (Text-Figure 1). The formations analysed have a Proterozoic and Lower Paleozoic age and have been affected by different phases of tectonics and metamorphism. Biostratigraphic correlation is based on acritarch assemblages, the only paleobiological witness to the primary sedimentary conditions in the basin, before the beginning of metamorphic and tectonic events. The acritarchs are very resistant to all thermal and dynamic processes that affected the sedimentary rocks with standing temperatures greater than 180°C, but exhibiting progressive browner color and thermic alteration with increasing temperature. Moczydlowska and Vidal (1992) suggest that TAI discrepancies may be caused by irradiation from natural radioactive decay. Thus acritarchs are considered the geothermometers of the rocks for temperature interval between 60°C and 150°C (Martin, 1993). The TAI evolution of the spread of organic matter in the metamorphic rocks in the East Carpathians and in the North Dobrogea County has not been analysed. However, acritarchs and amorphous organic matter fragments, are discolored or slightly colored to a pale yellow. This suggests, possibly that the initial sedimentary rocks, with acritarch included, were immature and they didn't undergo subsequent thermic alteration (small TAI).

The metamorphism, which later affected both, the rocks and the enclosed organic matter was a regional dynamic typeof metamorphism. Temperature was secondar, but was an important factor in the sdiscoloration of organic matter. The mineralization fluids completed the fossilization process and preserved the morphology and structure of acritarchs. The major adaptations in body's morphology of acritarchs took place during their life in different euphotic zone conditions (Mendelson, 1987; Martin, 1993). After the tectonic and metamorphic event, small changes took place (the body became folded and shrunken, or smaller modifications to body outline occurred).

Almost all acritarchs were preserved in marine strata such as marls, clays, silts, fine sandstones and carbonate rocks. Of these, the mud rocks are probably the best source for acritarchs because they behave, hydrodynamically, like clay sized particles (Mendelson, 1987). During metamorphism these rocks were converted into shales, slates, chlorite–sericite schists, graphite schists, black quartzites, micaschists, paragneisses, crystalline limestones and dolomites. All of these were encountered in the East Carpathians and the Dobrogea County, and all have been analysed for acritarchs.

Olaru, Leonard
2001 Using acritarch assemblages to correlate Upper Proterozoic and Lower Paleozoic metamorphic sequences in Romania; *In*: Goodman, D.K., and Clarke, R.T. (eds.), *Proceedings of the IX International Palynological Congress, Houston, Texas, U.S.A., 1996*; American Association of Stratigraphic Palynologists Foundation, p. 115–127.

 ISBN 0–931871–06–9

BIOSTRATIGRAPHIC CORRELATION AMONG ANALYSED METAMORPHIC FORMATIONS

The biostratigraphic correlation among different geological formations is the final result of the interpretation of some comparative studies of organic fossil tests. The correlation of metamorphic formations is difficult because macrofossils are not preserved. Thus, the acritarchs represent the only means of correlation, if they are originally in the primary sedimentary rocks and not are reworked. In the metamorphic rocks it is difficult to establish biozones, because, in most instances, some of the biozone taxa are missing, or not well preserved. In this case, for correlation with similar metamorphic rocks cenozone assemblages can be used. Palynological age can be determined, in part, from isotopic age data (Mânzatu et al., 1975) and also by correlation with "acritarch assemblages" from similar sedimentary rocks and with other fossils. Although the tectonic evolution of both regions of Romania, the East Carpathians and the North Dobrogea County, is not the same, the presence of the same acritarch assemblages in both initial sedimentary basins, provides a possible means for their correlation.

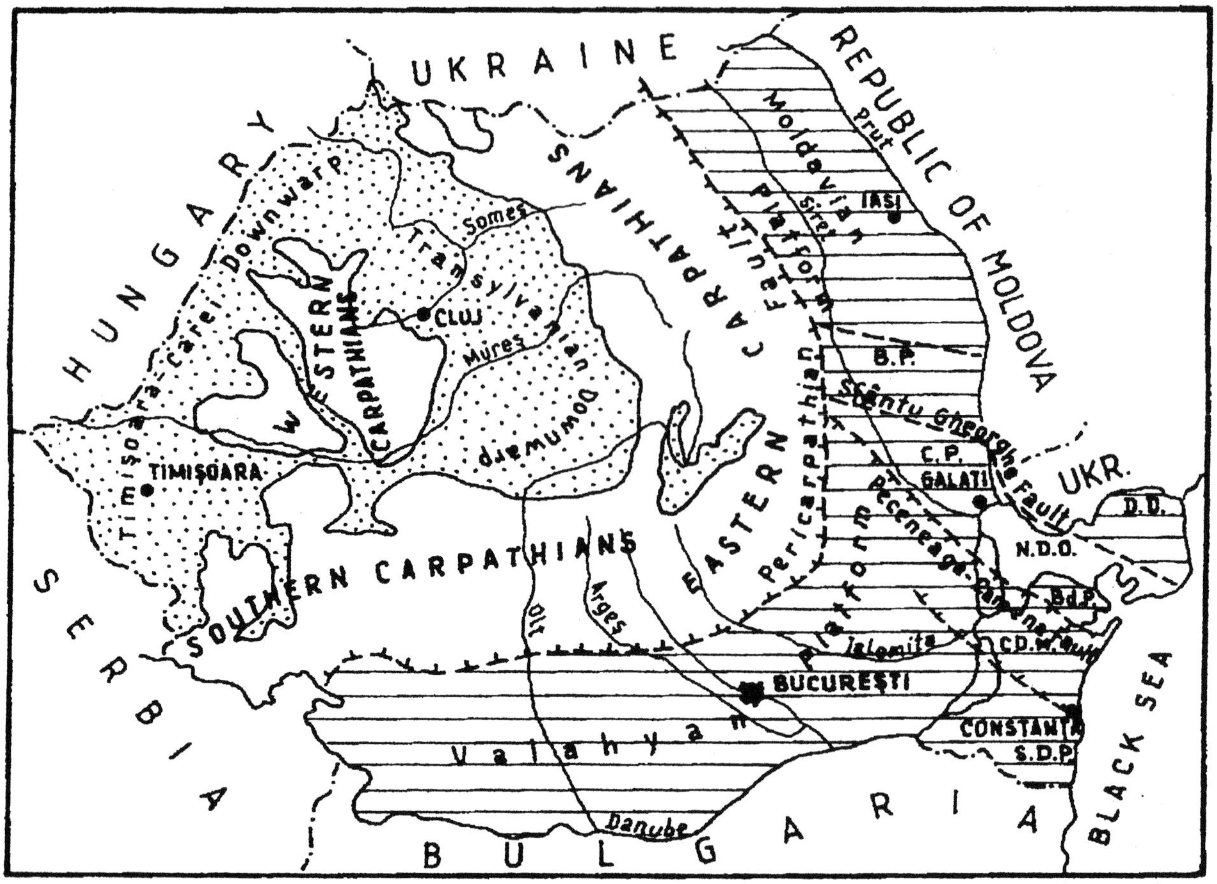

Text-Figure 1. Major structural units of Romania (after Ionesi, 1994): Bârlad Platform (B.P.); Covurlui Platform (C.P.); Delta Danube Platform (D.D.); North Dobrogean Orogene (N.D.O.); Babadag Platform (Bd.P.); Central Dobrogean Massif (C.D.M.); South Dobrogean Platform (S.D.P.).

The Bretila Group

The first rock sequence analysed is the Bretila Group, the oldest metamorphics in the East Carpathians (850 ± 50 Ma). The rocks outcrop in the Bretila anticline and in other zones (Text-Figure 7) in the Bistrita Valley basin (Balintoni, 1984). The Bretila Group was metamorphosed in the first cycle of metamorphism, the Barrovian facies, corresponding to the Dalslandian event (Kräutner, 1980). Only the micaschists and paragneisses of the upper part of the lithological column (Text-Figure 2) were analysed. A Riphean (Cryogenian) acritarch assemblage (Oniceanu et al., 1974; Olaru and Oniceanu, 1985; Olaru, 1991) was recovered.

The characteristic taxa from this acritarch assemblage are the following: *Leiosphaeridia ternata, L. crassa, L. tenuissima, Margominuscula rugosa, Globiferum deminatum, Kirbia multipartita* and other species of acritarchs (Table 1). This assemblage is the oldest from the East Carpathians, yielded from rocks isotopically dated as 850 ± 50 Ma. A part of these acritarchs are synonymous to types of smooth sphaeromorphs belonging to the genera *Kildinella, Favosphaeridium, Protosphaeridium, Trachysphaeridium or Turuchanica,* which have been reported from the Riphean of the Baltic Region (Timofeev, 1959; Jankauskas et al., 1989), Chuar Group of Utah, U.S.A. (Vidal and Ford, 1985), which has been assigned an isotopic age of 850–790 Ma, and from the Southern Urals (Mikhailova and Potkovirov, 1987). Also, the presence in the assemblage of the taxa *Satka undosa, S.elongata* and *S. colonialica* (I regret, all of these taxa are badly preserved), characteristic of a Riphean age (Jankauskas et al., 1989), makes this assemblage important for biostratigraphical correlation. Together the characteristic assemblage and isotopic age of 850 ± 50 Ma established for rocks analysed from the Bretila Group, define a Riphean age for this group. The correlation can be extended to other regions, such as the East European Platform, Southern Sweden and Canada using acritarch data (Timofeev, 1959; Volkova, 1969; Volkova et al., 1979; Vidal, 1976; Jankauskas et al., 1989; Hofman, and Jackson, 1994; Vidal and Moczydlowska, 1995).

It is true that species of *Leiosphaeridia* and *Satka* are dominant in the Late Proterozoic and Mesoproterozoic (1600–1000 Ma) and are common in the Tonian (Knoll, 1996). It is also reality that Vidal and Ford (1985) gave a range for *Satka colonialica* to Upper Riphean and Lower Vendian. But the authors mentioned referred only to sedimentary analysed formations. The metamorphic rocks here are affected by Caledonian, Hercynian and Alpine tectonic phases and different metamorphic facies. The limits between metamorphic formations changed many times, because they were

TABLE 1. Range chart showing stratigraphic distribution of acritarchs in the Bretila and Rebra Groups (East Carpathians) and the Orliga Group (North Dobrogea), Romania.

Taxonomic Units	Chronostratigraphy			Bretila Group	Rebra Group			Orliga Group
	Riph.	Vnd.	L. Cb.		Rb.1	Rb.2	Rb.3	
Leiosphaeridia ternata (Timofeev) Mikhailova & Jankauskas	—			•	•	•		
Margominuscula rugosa (Naumova) Jankauskas	—			•	•	•		
Leiosphaeridia holtedahlii (Timofeev) Jankauskas	—			•	•			
Globiferum deminatum German & Timofeev	—			•				
Satka undosa (Jankauskas) Jankauskas	—			•				•
Satka elongata Jankauskas	—			•	•			•
Satka colonialica Jankauskas	—			•				
Satka squamifera Pjatiletov	—							•
Melanocyrillium sp.	—							•
Eomycetopsis robusta (Schopf) Knoll & Golubic	—							•
Siphonophycus sp.	—							•
Kirbia multipatita Mikhailova & Volkova	—			•				•
Flabelleforma compacta German	—			•				•
Stictosphaeridium sinapticuliferum (Timofeev) Mikhailova & Jankauskas	—	—		•	•	•	•	•
Germinosphaera guttaformis Mikhailova	—	—						•
Polytrichoides lineatus (German) Jankauskas	—	—						•
Brevitrichoides tiulmenicus Jankauskas & Kozlova	—	—						•
Leiotrichoites typicus (German) Jankauskas	—	—		•				
Leiosphaeridia crassa (Naumova) Jankauskas	—	—	—	•	•	•	•	•
Leiosphaeridia obsuleta (Naumova) Jankauskas	—	—	—	•	•	•	•	•
Leiosphaeridia tenuissima Eisenack	—	—	—	•	•	•	•	
Retiforma tolparica Mikhailova		—				•		
Podoliella irregulare (Timofeev) Mikhailova		—				•		•
Spumosina rubiginosa (Andreeva) Jankauskas & Medvedeva		—				•		•
Bavlinella foveolata Schepeleva		—				•		•
Octoedryxium truncatum (Rudavskaya) Rudavskaya		—				•		
Leiosphaeridia atava (Naumova) Jankauskas		—	—			•		
Leiosphaeridia laminarita (Timofeev) Jankauskas		—				•		•
Granomarginata notata Naumova		—				•		
Vendotaenia sp.		—				•	•	
Omalophyma solida Golub		—						•
Leiosphaeridia minutissima (Naumova) Jankauskas		—	—				•	
Botuobia wernadskii (Schepeleva) Kolosov		—	—				•	
Asteridium tornatum (Volkova) Moczydlowska		—	—				•	
Asteridium pallidum (Volkova) Moczydlowska			—				•	
Globosphaeridium cerinum (Volkova) Moczydlowska			—				•	
Comasphaeridium mollicum Moczydlowska & Vidal			—				•	
Dictyotidium birvetense Paškevičiene			—				•	
Synsphaeridium switjassium Kirjanov			—				•	
Fimbriaglomerella minuta (Jankauskas) Moczydlowska & Vidal			—				•	

overthrusted. The only geologic control for the palynological data is isotopic age. This isotopic age, for the analysed rocks, e.g., paragneisses, quartzite schists and micaschists, from the upper part of the Bretila Group (Text-Figure 2) are 850 ± 50 Ma. These rocks are dominated by *Leiosphaeridia* spp., *Satka* is present as are *Globiferum deminatum, Stictosphaeridium sinapticuliferum, Kirbia multipartita, Leiotrichoides typicus* (Table 1). The lower part of the Bretila Group (unanalyzed), is possibly Tonian in age. The upper part of the Bretila Group is Cryogenian (Riphean) in age.

The Rebra Group

The second group studied from the Proterozoic of the East Carpathians is the Rebra Group. Lithologically this group is more complex than Bretila Group (Text-Figure 3) and more widespread in the Rodna and Bistrita mountains of the East Carpathians (Text-Figure 7). It was metamorphosed in the second cycle of metamorphism, the Cadomian type, corresponding to the Baikalian or Assyntian event. This was later affected by the third cycle of retrograde metamorphism,

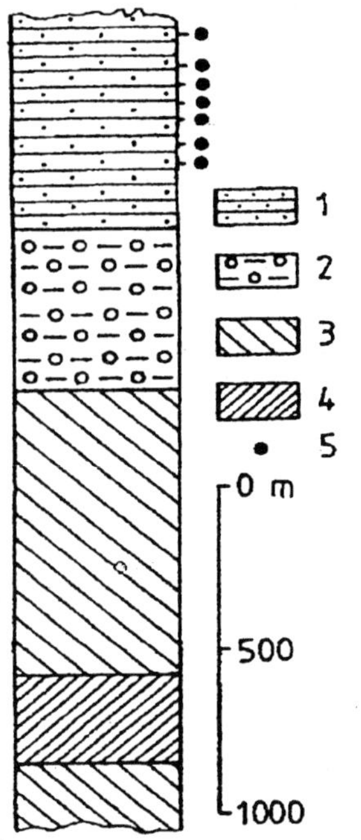

Text-Figure 2. Lithologic column of the Bretila Group (after Olaru and Oniceanu, 1985). 1: micaschists, quartzite schists and paragneisses; 2: gneiss with feldspar eyes; 3: micaschists with garnets and orthoamphibolites; 4: amphibolites; 5: position of the samples analysed.

the Hercynian type. The Rebra Group is divided into three lithological formations: Lower (Rb. 1), Middle (Rb. 2) and Upper (Rb.3).

The tectonic relationship between the Rebra and Bretila Groups is not well established. The Rebra Group has a higher lithostratigraphical position to the Bretila Group, but direct stratigraphic relationship between the two these groups have not been found (Bercia et al., 1976).

The Lower Formation (Rb. 1) is a 1000 m thick succession of micaschists, paragneisses and quartzite schists. It is lithologically equivalent to the upper part of the Bretila Group.

The Middle Carbonatic Formation (Rb. 2) is 1500 m thick and has yielded a rich acritarch assemblage (Table 1) which comes from crystalline limestones and dolomites (Text-Figure 3). The assemblage includes the following representative species: *Podoliella irregulare, Bavlinella foveolata, Octoedryxium truncatum, Stictosphaeridium sinapticuliferum, Granomargi-nata notata, Retiforma tolparica, Leiosphaeridia crassa, L. obsuleta, L. laminarita, L. atava* (Table 1). This assemblage reflects a Vendian age (Olaru and Oniceanu, 1983, 1985; Olaru, 1991), and has been found in other regions. These include the Sudete Mountains (Olaru and Gunia, 1987), South Ural Mountains (Mikhailova and Podkovirov, 1987), Southern Sweden (Vidal, 1976), Baltic Sea Region (Paškievičiene, 1980), and the East European Platform (Jankauskas et al., 1989). An isotopic age of 650 ± 50 Ma was established for the carbonatic rocks (Mânzatu et al., 1975; Bercia et al., 1976; Kräutner, 1980). The palynostratigraphic position of this formation is above that of the Bretila Group that is at the beginning of Vendian Period (Younger Neoproterozoic A, Knoll, 1996).

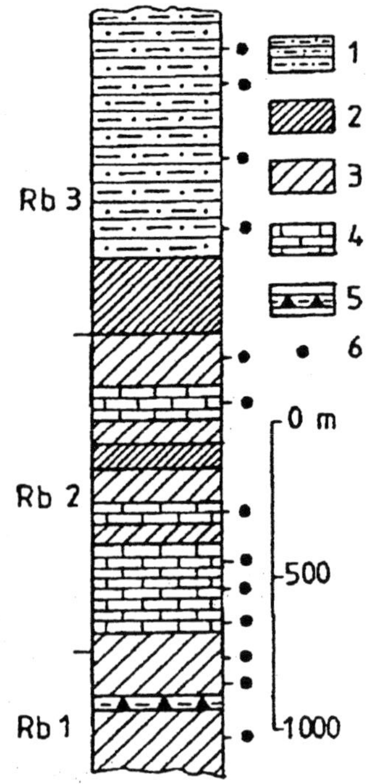

Text-Figure 3. Lithological column of the Rebra Group (after Olaru and Oniceanu, 1985). 1: terrigenous rocks, micaschists with biotite, schists with muscovite, chlorite and sericite, acid metatuffs; 2: amphibolites; 3: micaschists with biotite, paragneisses, quartzites and garnets; 4: crystalline limestones and dolomites with tremolite or talc; 5: gneiss; 6: position of the samples analysed.

The Upper Rebra Formation (Rb. 3) is younger that the Middle Rebra Formation (Rb. 2). This formation is predominantly terrigenous with interbedded basic and carbonatic rocks. Lithologically it is represented by micaschists with biotite, chlorite–sericite schists, acid metatuffs with sequences of amphibolites and crystalline limestones, especially in the lower part (Text-Figure 3). The acritarchs recovered from this formation were *Botuobia wernadskii, Leiosphaeridia obsuleta, L.* crassa, *L. minutissima* and rarely *Asteridium tornatum, Globosphaeridium cerinum, Fimbriaglomerella minuta, Dictyotidium birvetense* and *Synsphaeridium switjassium* (Table 1). The age established by this assemblage is Late Vendian, although the palynological assemblage is composed of Vendian and some Lower Cambrian species. The presence of some rare Lower Cambrian species, e.g. *Asteridium tornatum* (Moczydlowska, 1991) with typical Vendian acritarchs, represents the effect of slow changes of the Vendian to the Lower Cambrian microflora. It is possible that in this mixture of palynomorphs is involved in sedimentological and tectonic factors. Distinguishing between the Upper Rebra Formation (Late Vendian) and the Lower Tulgheş Formation (Tg.1) of Lower Cambrian is difficult to do based on lithology (Olaru, 1991). It can be observed the persistence of the same rocks, than in the Upper Rebra Formation, such as chlorite–sericite schists alternating with quartzite schists, graphite schists, acid metatuffs and crystalline limestone interbedded, in the lower part.

Lithologically, the Upper Rebra Formation is missing amphibolite intercalations. Palynologically, without the rare occurrences of some Vendian species (Table 2) in the Lower Tulgheş Formation (Tg.1), the prevailing acritarchs are those with thin walls, pylomes, small thorns and membranaceous ridges. The stratigraphic Vendian–Lower Cambrian boundary is marked by the acritarch species *Teophypolia lacerata, Vendotaenia* sp., *Ceratophyton vernicosum, Leiosphaeridia dehisca, Leiosphaeridia pylomifera, Asteridium* sp., *Globosphaeridium cerinum, Fimbriaglomerella minuta.* If these species of acritarchs extend into the Lower Cambrian, they are either absent or rare in the Upper Rebra Formation (Rb.3), which is Vendian age (Olaru and Gunia, 1987; Olaru, 1991). The isotopic age of this boundary is 550–570 Ma (Bercia et al., 1976; Kräutner, 1980). Therefore the Vendian–Lower Cambrian boundary is defined by the acritarch radiation above the more typical Lower Cambrian acritarchs (Table 2). This boundary is positioned within the Rovno horizon (Moczydlowska, 1989, 1991), because the transition from Vendian to Lower Cambrian microflora is gradational, with a mixed palynomorph assemblage until the new Lower Cambrian acritarch associations. This boundary section is present in several regions (Volkova, 1969; Konzalowa, 1974; Paškievičiene, 1980; Moczydlowska and Vidal, 1986; Mikhailova and Potkovirov, 1987; Moczydlowska, 1989, 1991; Jankauskas and Lendzion, 1992, 1994; Vidal and Moczydlowska, 1995; Knoll, 1996), but there are different opinions among these authors concerning the precise position of this boundary.

The Orliga Group

Moving from the East Carpathians to the Proterozoic formations of the North Dobrogea County, one group that is interesting for correlation is the Orliga Group. This group has undergone low-grade metamorphism and is less widespread than the Bretila Group and the Rebra Group from the East Carpathians (Text-Figure 8). In the Orliga Group are prevailing metaterrigenous rocks, with amphibolites interbedded in the lower part and carbonaceous rocks in the upper part (Seghedi, 1980). These rocks were metamorphosed in the amphibolite with almandine facies. The presence of basic metamorphic rocks in the Orliga Group, denote rifting process by opened of the first sedimentation basin proper to Dalslandian orogene cycle. These rocks were reworked during the Cadomian, Caledonian, Hercynian and Alpine orogenesis (Seghedi, 1980; Săndulescu, 1984). The absolute age of this group is 711 Ma, established by comparison with the equivalent Ceamurlia–Altân Tepe Group from Central Dobrogea County (Text-Figure 8). The second area of sedimentation opened in the time of the Assyntian orogenesis of Upper Proterozoic and Lower Cambrian age, when the first crystalline schists were folded and retrogressed (Kräutner et al., 1988).

The first assemblage of acritarchs comes from micaschists and paragneisses (Text-Figure 4), like those of Bretila Group from the East Carpathians, but is poorer. This first acritarch assemblage reflects the first cycle of sedimentation and is Riphean (Cryogenian) in age (Table 1). It has yielded the acritarchs *Satka undosa, S. squamifera, S.elongata, Kirbia multipartita* and *Polytrichoides lineatus*. The presence of *Satka undosa* and *Satka elongata* is evidence for a sedimentation cycle beginning in the Middle Riphean which can also be seen in the Bretila Group. The second assemblage from the Orliga Group is characteristic of the second sedimentation cycle. This assemblage is younger and comes from the upper quartzites assemblage is younger and comes from the upper quartzites that are interbedded carbonatic rocks (Text-Figure 4). Characteristic acritarchs include *Bavlinella foveolata, Podoliella irregulare, Omalophyma solida, Leiosphaeridia laminarita, L. obsuleta, L. crassa* (Table 1). This assemblage is equivalent to that from the Middle Carbonatic Formation (Rb. 2) of the Rebra Group, is thus of Vendian age. Thus, when we consider the Proterozoic acritarch assemblages from the two regions of Romania, it is evident that the Orliga Group from the North Dobrogea County can be correlated with both the Bretila Group and the Lower and Middle Formations (Rb. 1 and Rb. 2) of the Rebra Group, from the East Carpathians. From this we can deduce the age of Orliga Group as Riphean to Vendian (Middle to Late Neoproterozoic).

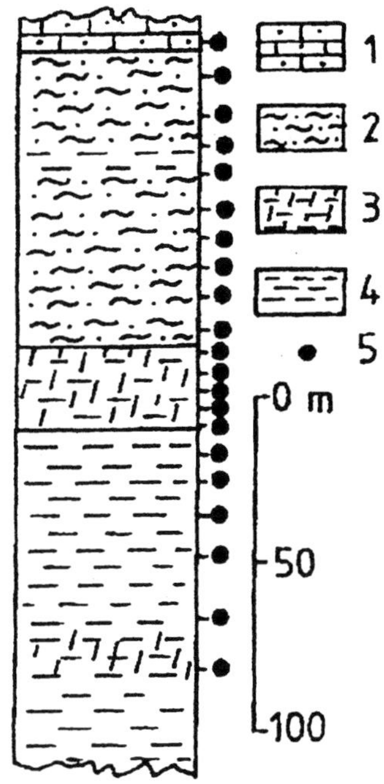

Text-Figure 4. Lithological column of the Orliga Group (partially after Kräutner, Mureşan and Seghedi, 1988). 1: calc–silicate gneisses and crystalline limestones, interbedded; 2: mica-quartzites and micaschists, interbedded; 3: paragneisses with biotite and garnets; 4: micaschists with paragneisses interbedded; 5: position of the samples.

The Tulgheş Group

Returning to the correlation of the Lower Paleozoic formations, we begin with the Tulgheş Group from the East Carpathians (Text-Figure 7). This group transgressively overlies the metamorphic basement of the Bretila or Rebra Group, and they are included in the same Bucovinical nappe System (Bercia et al., 1976; Kräutner, 1980). This group is metamorphosed to a very low grade and lithologically is

TABLE 2. Range chart showing stratigraphic distribution of acritarchs in the Tulgheş Group (East Carpathians) and the Greenschist Group (Central Dobrogea), Romania.

Taxonomic Units	**Chronostratigraphy**			**Tulgheş Group**			**Greenschist Group**		
	Vnd.	L. Cb.	M. Cb	Tg. 1	Tg. 2	Tg. 3	I. Grw.	L. Grw.	U. Grw.
Bavlinella faveolata Schepeleva	—			•			•		
Asteridium tornatum (Volkova) Moczydlowska	—	—		•	•				
Leiosphaeridia minutissima (Naumova) Jankauskas	—	—		•			•		
Botuobia wernadskii (Schepeleva) Kolosov	—	—		•			•		
Leiosphaeridia atava (Naumova) Jankauskas	—	—		•			•		
Leiosphaeridia crassa (Naumova) Jankauskas	—	—		•			•		
Leiosphaeridia obsuleta (Naumova) Jankauskas	—	—		•			•	•	
Leiosphaeridia tenuissima Eisenack	—	—		•					
Granomarginata squamacea Volkova		—			•				
Granomarginata prima Naumova		—			•				
Tasmanites tenellus Volkova		—		•	•		•	•	•
Globosphaeridium cerinum (Volkova) Moczydlowska		—		•	•		•		
Comasphaeridium brachyspinosum (Kirjanov) Moczydlowska & Vidal		—			•				
Skiagia ornata (Volkova) Downie		—			•				
Lophosphaeridium dubium (Volkova) Moczydlowska		—		•	•			•	
Asteridium pallidum (Volkova) Moczydlowska		—		•	•				
Micrhystridium brevicornum Jankauskas		—		•	•				
Micrhystridium villosum Kirjanov		—			•				
Leiosphaeridia pylomifera Paškevičiene		—		•	•		•	•	•
Fimbriaglomerella membranacea (Kirjanov) Moczydlowska & Vidal		—		•	•				
Dictyotidium birvetense Paškevičiene		—		•	•			•	
Teophypolia lacerata Kirjanov		—		•	•			•	
Brestovia annulata Paškevičiene		—			•			•	
Navifusa sp.		—			•			•	
Ceratophyton sp.		—		•	•				
Dictyotidium priscum Kirjanov & Volkova		—			•			•	•
Lophosphaeridium tentativum Volkova		—			•		•		•
Fimbriaglomerella minuta (Jankauskas) Moczydlowska & Vidal		—		•	•				
Pulvinosphaeridium antiquum Paškevičiene		—			•			•	
Skiagia compressa (Volkova) Downie		—			•	•		•	
Leiovalia striatella Paškevičiene		—			•			•	
Archaeodiscina umbonulata Volkova		—			•			•	
Cymatiosphaera nerisica Jankauskas		—		•	•		•	•	
Cymatiosphaera favosa Jankauskas		—		•	•	•	•		
Cymatiosphaera cristata Jankauskas		—			•		•		
Retisphaeridium densum Paškevičiene		—			•	•			•
Baltisphaeridium acerosum Jankauskas & Posti		—			•		•		•
Heliosphaeridium dissimilare (Volkova) Moczydlowska		—	—			•			
Tasmanites volkovae Kirjanov		—	—			•			
Alliumella baltica Vanderflit		—	—			•			
Cymatiosphaera postii (Jankauskas) Jankauskas		—	—			•			
Lophosphaeridium truncatum Volkova		—	—			•			
Ovullum saccatum Jankauskas		—	—			•			•
Ovullum lanceolatum Jankauskas		—	—			•			•
Baltisphaeridium pseudofaveolatum Fridrichsone			—			•			
Leiosphaeridia dehisca Paškevičiene		—		•	•		•	•	
Ceratophyton vernicosum Kirjanov		—		•					
Vendotaenia sp.	—	—		•					
Asteridium sp.		—		•					

represented by a succession of chlorite, sericite, graphite schists, black quartzites and interbedded metatuffs (Text-Figure 5). It was affected by Caledonian greenschist facies metamorphism and later retrogressed during the Hercynian orogeny. The thickness of the Tulgheş Group is more than 4,500 m and it is the most widespread group in the East Carpathians (Bercia et al., 1976). The Tulgheş Group has been divided into four lithological formations (Tg. 1 to Tg. 4), from which we palynologically analysed only three formations (Tg. 1, Tg. 2, Tg. 3), which are widespread in the studied region. This group has been approximately dated at 600–472 Ma (Olaru and Gunia, 1988).

The palynostratigraphical boundary between the Vendian and the Lower Cambrian is established in the chlorite–sericite schists interbedded with grey limestones and quartzite schists from the lower part of the Lower Cambrian (Tg. 1). The isotopical age level, is at the 550–570 Ma (Bercia et al., 1976; Kräutner, 1980). From this level we recovered an assemblage with *Globosphaeridium cerinum, Asteridium pallidum, Fimbriaglomerella membranacea, Leiosphaeridia pylomifera, Teophypolia lacerata, Vendotaenia sp. and Ceratophyton* sp. (Olaru and Gunia, 1987,1988; Olaru, 1991).

A typically Lower Cambrian palynological assemblage of acritarchs (Text-Figure 5) is found in the Tg. 2 Formation, the black quartzites and graphite schists and accompanying chlorite–sericite schists that represent a lithological level mark for the Tg. 2 Formation of the Lower Cambrian age. This assemblage (Table 2) is represented by following taxa: *Globosphaeridium cerinum, Comasphaeridium brachyspinosum, Asteridium pallidum, A.tornatum, Micrhystridium villosum, Leiosphaeridia pylomifera, L. dehisca, Fimbriaglomerella membranacea, Teophypolia lacerata* and *Tasmanites tenellus* (Olaru and Gunia, 1987, 1988; Olaru,1991).This assemblage represents the real period of radiation of such acritarch genera as *Asteridium (=Micrhystridium), Globosphaeridium, Comasphaeridium (= Baltisphaeridium), Fimbriaglomerella (= Cymatiosphaera)* and *Tasmanites* (Knoll, 1996) and is representative of the Lower Cambrian Period of the East Carpathians.

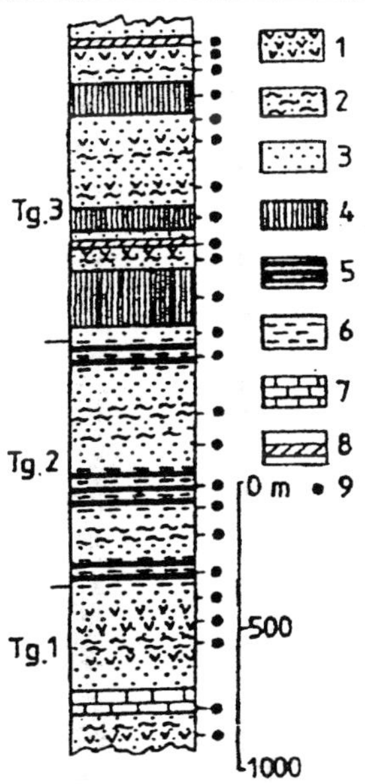

Text-Figure 5. Lithological column of the Tulgheş Group (after Bercia, Kräutner, and Mureşan, 1976; Olaru, 1991). 1: quartzite schists; 2. chlorite–sericite schists; 3: sericite–chlorite schists; 4: rhyolite metatuffs; 5: black quartzites; 6: graphite schists; 7: gray limestones; 8: stratiform pyrites and base metal sulfides; 9: position of samples analysed.

The Middle Formation (Tg.2) of the Tulgheş Group (Text-Figure 5), with a typical Lower Cambrian acritarch assemblage, can be correlated with the Lontova horizon of the East European Platform (Volkova, 1969), and also possibly with Mazowsze Formation of the Poland (Moczydlowska, 1991).

The last palynological assemblage comes from the Upper Formation (Tg. 3) of the Tulgheş Group, which is composed of the sericite–quartzite schists interbedded with rhyolite metatuffs and graphite schists. The palynological assemblage has acritarch taxa typical of the Lower Cambrian (Table 2) and some, typical of the Middle Cambrian, such as *Baltisphaeridium pseudofoveolatum, Heliosphaeridium dissimilare, Ovulum lanceolatum, O.saccatum, Lophosphaeridium truncatum* and *Tasmanites volkovae*. Based on the palynological data the Tulgheş Group can be considered Lower to earliest Middle Cambrian age (Iliescu et al., 1983; Olaru and Gunia, 1988; Olaru, 1991) and corresponds to an isotopic age of 570–472 Ma. This group can be correlated with Niemcza–Kamienetz Series of the Sudete Mountains of the Poland (Olaru and Gunia, 1988), with the East European Platform Series, and Baltic Sea region (Volkova, 1969; Volkova et al.,1979; Moczydlowska and Vidal, 1986; Moczydlowska, 1991; Jankauskas, 1991; Jankauskas and Lendzion, 1992, 1994).

The Greenschist Group

The last of the Paleozoic formations investigated are from Central Dobrogea County — the Greenschist Group (Text-Figure 8). This group overlies the older basement of the Ceamurlia–Altân Tepe Group (Riphean), has a terrigenous character, and is 5,000 m thick. It is divided (Text-Figure 6) into four lithological formations (Mirăuţa, 1969). Only the first three formations were palynologically analysed (Olaru et al., 1990).

This group begins with the Infragraywacke Formation (I.Grw.), or the Dorobanţu Formation at its base (Ionesi, 1994). It contains the first Lower Cambrian assemblage with *Globosphaeridium cerinum, Baltisphaeridium acerosum, Leiosphaeridia pylomifera, Tasmanites tenellus* and *Lophosphaeridium dubium* (Table 2). Vendian age species, such as *Leiosphaeridia crassa, L. obsuleta, L. atava, Bavlinella foveolata* and *Botuobia vernadskii* were also recovered (Table 2). These last species probably were reworked from typical Vendian basement formation. This palynological mixture is the same with the Lower Tulgheş Formation (Tg. 1) of the East Carpathians. These two lithological formations, the Lower Tulgheş Formations (Tg.1) from the East Carpathians, and the Infragraywacke Formation of the Greenschist Group (Central Dobrogea County), can be correlated and have a Lower Cambrian age.

The Lower Graywacke Formation (L.Grw.), or the Istria Formation (Ionesi, 1994), has a descriptive acritarch assemblage that included the Vendian forms, *Leiosphaeridia crassa, L. obsuleta,* and the Lower Cambrian acritarchs species *Pulvinosphaeridium antiquum, Dictyotidium birvetense, Leiovalia striatella, Leiosphaeridia pylomifera* and *Tasmanites tenellus* (Table 2). The age of this formation is certainly Lower Cambrian.

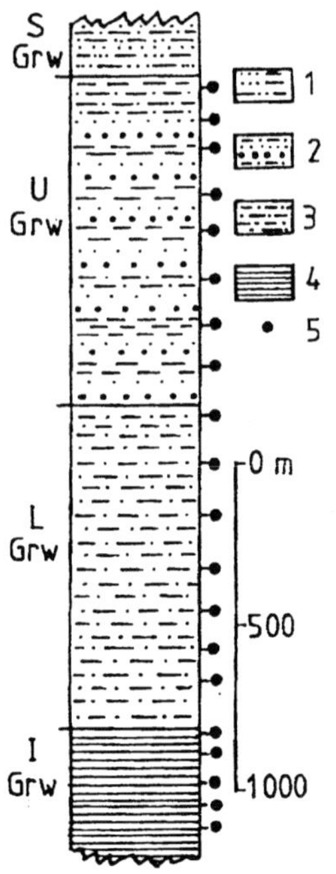

Text-Figure 6. Lithological column of the Greenschist Group (after Mirăuța, 1969; Olaru, Erhan and Antoche, 1990). 1: greenschists and green–violet siltites; 2: silicic compact rocks, quartz microconglomerates, quartzites and violet pelite schists; 3: rhythmical succession of graywackes and green pelites; 4: chlorite-green quartzites and quartz-green sandstones with sericite–chlorite green phyllites, interbedded; 5: position of the samples analysed.

The third formation, the Upper Graywacke Formation (U.Grw.), or the Măgurele Formation (Ionesi, 1994) has a predominantly Lower Cambrian assemblage, with such species *Leiosphaeridia pylomifera, Retisphaeridium densum, Tasmanites tenellus* and *Lophosphaeridium tentativum* (Table 2). This Upper Graywacke Formation (U.Grw.) and possible Lower Graywacke Formation (L.Grw.) can be correlated with the Middle Tulgheş Formation (Tg.2) from the East Carpathians.

The fourth formation, the Supragraywacke Formation (S.Grw.), or the Băltăgesti Formation was not analysed.

In conclusion, the whole Greenschist Group, has yielded good Lower Cambrian palynological assemblages and can be correlated with the Lower and Middle Tulgheş Formations (Tg. 1 and Tg. 2) of the East Carpathians. Its isotopic age is 546–470 Ma.

CONCLUSIONS

For the correlation of metamorphic formations from the East Carpathians and Dobrogea County of Romania, the only fossils, which can be used are the acritarch assemblages (Table 3). These assemblages frequently constitute cenozones, because the biozone acritarch species are not well preserved in all formations analysed. Partial control of this paleobiological correlation can be done by isotopic data. The acritarchs, because of their composition, generally were resistant to all paleoenvironmental changes and tectonic and metamorphic events, and represent the only paleobiological evidence for correlation.

The oldest metamorphic formation of the East Carpathians, is the Bretila Group, of Riphean (Cryogenian) age (850 ± 50 Ma). The second oldest formation is the Rebra Group of Vendian (Younger Neoproterozoic A) age (650 ± 50 Ma). Both lithological groups have undergone low-grade metamorphism. The stratigraphic boundary between the

PLATE 1

All figures x 1,200 unless otherwise indicated.

1 *Spuminosa rubiginosa* (Andreeva) Jankauskas & Medvedeva, Vendian, Rebra Group (Rb.2).
2 *Leiosphaeridia obsuleta* (Naumova) Jankauskas, Upper Riphean–Vendian, Rebra Group (Rb. 2).
3 *Leiosphaeridia obsuleta* (Naumova) Jankauskas, Upper Riphean–Vendian, Orliga Group.
4 *Satka squamifera* Pjatielov, Upper Riphean, Orliga Group.
5 *Stictosphaeridium sinapticuliferum* (Timofeev) Mikhailova & Jankauskas, Upper Riphean–Vendian, Rebra Group (Rb. 2).
6 *Stictosphaeridium sinapticuliferum* (Timofeev) Mikhailova & Jankauskas, Upper Riphean, Bretila Group.
7 *Satka squamifera* Pjatiletov, Upper Riphean, Orliga Group.
8 *Dictyotidium birvetense* Paškievičiene, Lower Cambrian, Tulgheş Group (Tg. 1).
9 *Stictosphaeridium sinapticuliferum* (Timofeev) Mikhailova & Jankauskas, Upper Riphean–Vendian, Oriliga Group.
10 *Polytrichoides lineatus* Herman, Upper Riphean, Orliga Group, x480.
11 *Omalophyma solida* Golub, Vendian, Orliga Group.
12 *Syphonophycus* sp., Upper Riphean, Orliga Group.
13 *Brevitrichoides tiulmenicus* Jankauskas & Kozlova, Upper Riphean, Orliga Group.
14 *Omalophyma angusta* Golub, Vendian, Orliga Group, x480.
15 *Satka* cf. *elongata* Jankauskas, Upper Riphean, Bretila Group.
16 *Leiosphaeridia holtedahlii* (Timofeev) Mikhailova, Vendian, Bretila Group, x800.
17 *Podoliella irregulare* (Timofeev) Mikhailova, Vendian, Rebra Group (Rb 2).
18 *Leiosphaeridia dehisca* Paškievičiene, Lower Cambrian, Tulgheş Group (Tg. 1).
19 *Dictyotidium priscum* Kirjanov & Volkova, Lower Cambrian, Tulgheş Group (Tg. 2).
20 *Baltisphaeridium acerosum* Jankauskas & Posti, Lower Cambrian, Tulgheş Group (Tg. 1–Tg. 2), x800.
21 *Lophosphaeridium tentativum* Volkova, Lower Cambrian, Tulgheş Group, (Tg. 2).
22 *Leiosphaeridia pylomifera* Paškievičiene, Lower Cambrian, Tulgheş Group (Tg. 1).
23 *Teophypolia lacerata* Kirjanov, Lower Cambrian, Tulgheş Group (Tg. 2), x480.
24 *Ovullum saccatum* Jankauskas, Lower–Middle Cambrian, Tulgheş Group (Tg 3).
25 *Granomarginata squamacea* Volkova, Lower Cambrian, Tulgheş Group (Tg. 2).
26 *Leiosphaeridia crassa* (Naumova) Jankauskas, Upper Riphean, Bretila Group.
27 *Podoliella irregulare* (Timofeev) Mikhailova, Vendian, Rebra Group (Rb 2).

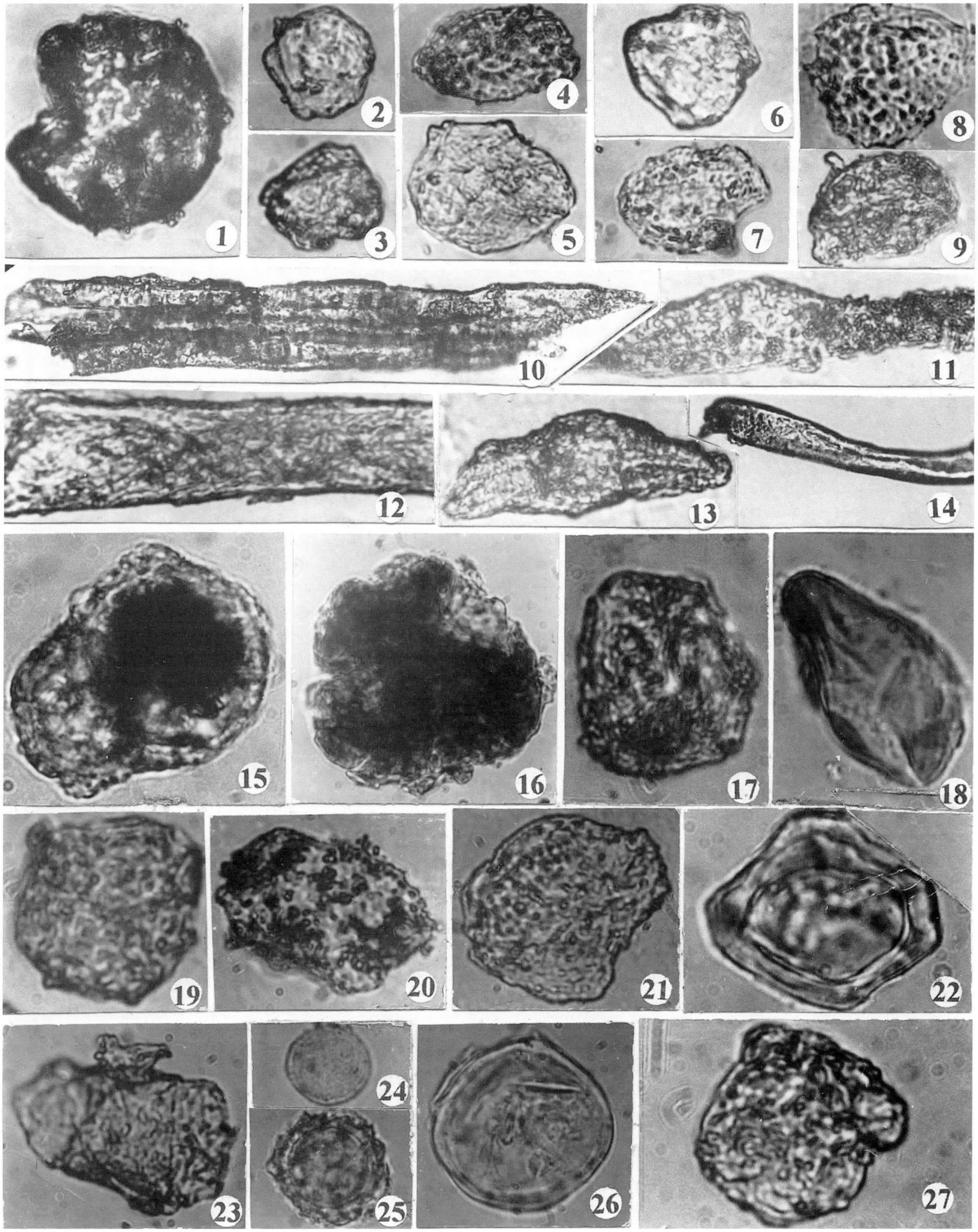
1
2
3
4
5
6
7
8
9
10
11
12
13
14
15
16
17
18
19
20
21
22
23
24
25
26
27

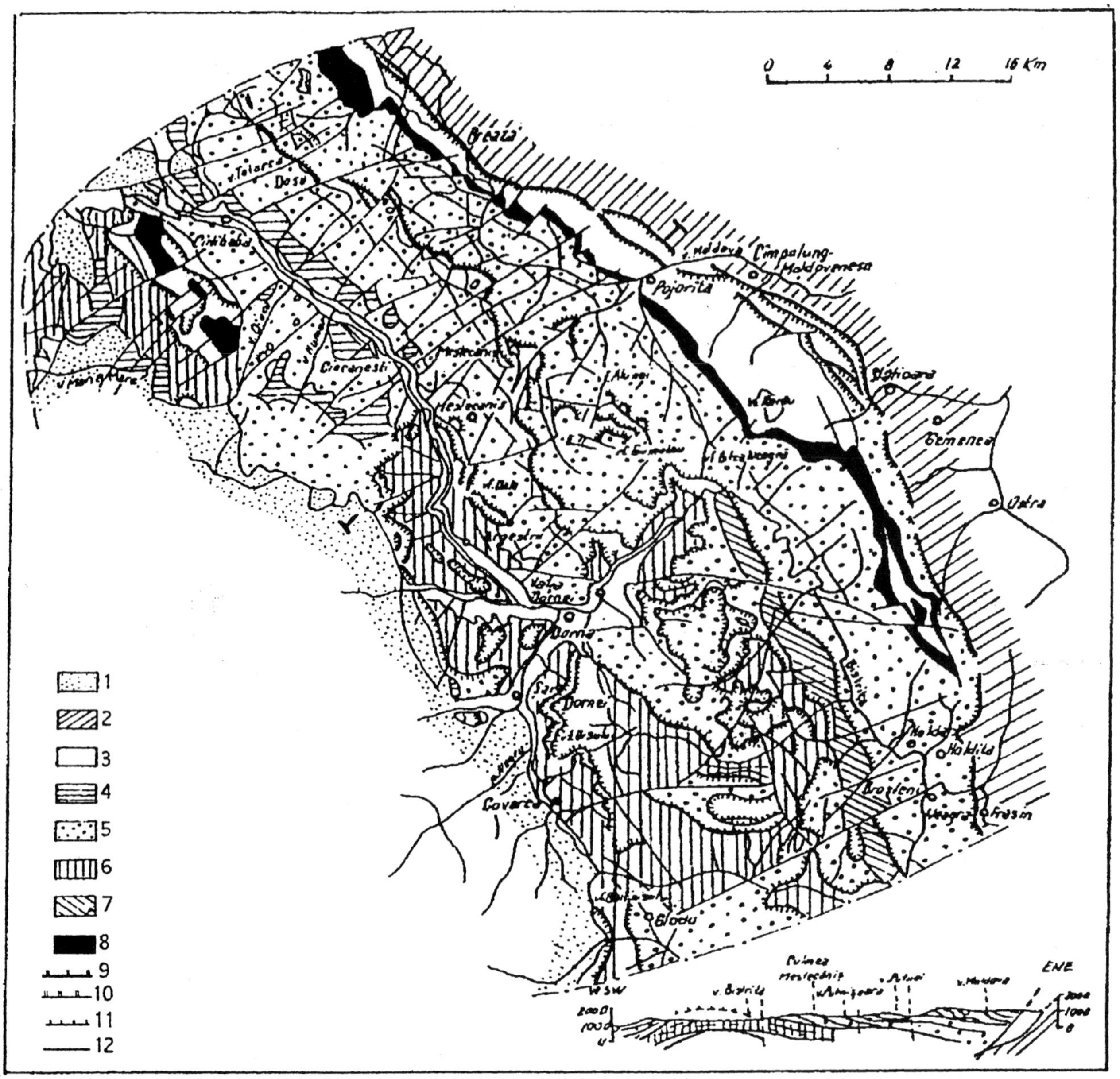

Text-Figure 7. Simplified geologic map of the Eastern Carpathians: the studied areas (after Bercia et al., 1976; Olaru and Oniceanu, 1985). 1: Paleogene sedimentary deposits; 2: flysch deposits; 3: Jurassic deposits; 4. Ţibău Formation; 5: Tulgheş Group; 6: Rebra Group; 7: Pietrosu porphyroide; 8: Bretila Group; 9: Austrian nappes; 10: Pre–Alpine nappes; 11: inverse faults; 12: normal faults.

Vendian and Lower Cambrian, was established at the level of initial radiation Lower Cambrian acritarch species, equivalent with the base of Lower Cambrian Formation (Tg.1) of the Tulgheş Group from the East Carpathians. The isotopical age of boundary is 550–570 Ma, and it is positioned within the Rovno horizon (Moczydlowska, 1989, 1991).

The Lower Paleozoic from the East Carpathians, is represented by the Tulgheş Group of Lower up to Middle Cambrian age (570–472 Ma). This group has undergone metamorphic change to a very low grade, in the greenschist facies.

Geological formations equivalent to those in the East Carpathians ones were analysed from Dobrogea County. The Orliga Group, has a Riphean to Vendian age and can be correlated with the Bretila Group and the Lower and Middle Rebra Formations (Rb. 1 and Rb. 2) from the East Carpathians. The isotopic age is 711 Ma, and it has undergone low-grade metamorphism. It also can be correlated with the Ceamurlia Group, from the basement of the Central Dobrogea County. The Greenschist Group, from the Central Dobrogea County, can be correlate with the Tulgheş Group of the East Carpathians, and has metamorphosed very low grade, and a Lower Cambrian (possible up to Middle Cambrian) age, corresponding to an isotopic age of 570–472 Ma.

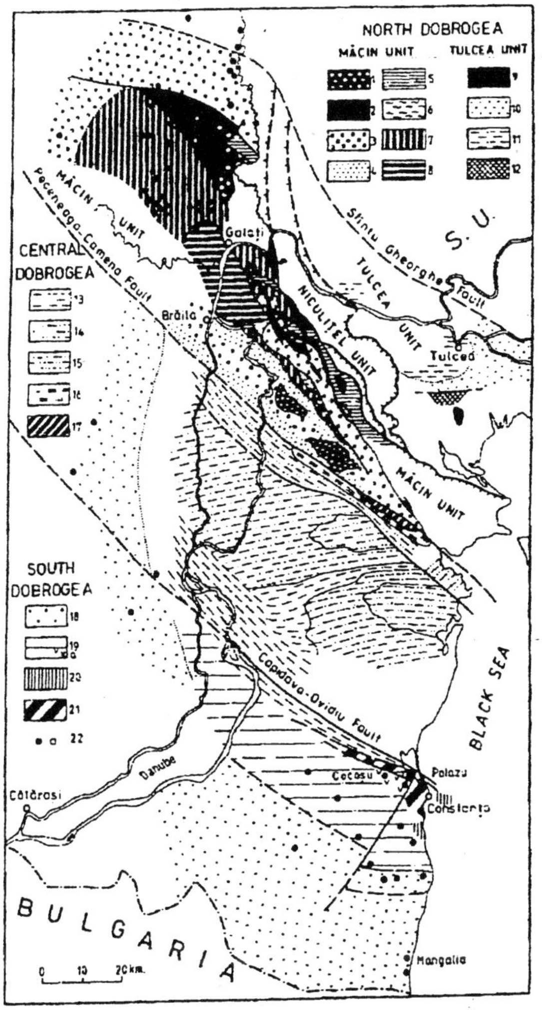

Text-Figure 8. Geological and structural map of Dobrogea, without Mesozoic and Cenozoic deposits (after Kräutner, Mureşan and Seghedi, 1988, with completions). 1: Variscan alkali granites; 2: Variscan and Caledonian granites; 3: Carapelit Formation (Lower Carboniferous); 4: Cerna and Bujoare Formation (Silurian and Lower Devonian); 5: Boclugea Group (Lower Paleozoic); 6: Megina Gneisses (?Vendian); 7: Megina Group (?Vendian); 8: Orliga Group (Riphean–Vendian); 9: Variscan granites; 10: Rediu and Beştepe Formation (Silurian and Devonian); 11: Dealul Horia Formation (Ordovician); 12: Uzum Bair Formation (?Riphean); 13–16: Dobrogea Greenschist Formation (Lower Cambrian); Băltăgeşti Member (S. Grw.); 14 Măgurele–Sibioara Member (U. Grw.); 15: Istria Member (L. Grw.); 16: Dorobanţu Member (U. Grw.); 17: Altân Tepe Group (Riphean–Vendian); 18: Paleozoic platform cover; 19: Cocoşu Group (Neo Proterozoic); a: spilites; 20: Palazul Mare banded iron ore Formation (Proterozoic); 21: Gneisses (?Archean–Lower Proterozoic); 22: drillings.

ACKNOWLEDGMENTS

I first gratefully acknowledge Paul K. Strother, from Weston Observatory, Massachusetts, who, with accuracy and generosity has reviewed this manuscript and has made very useful critical remarks and comments, for revision and publication of this work. Our thanks are also extended to David K. Goodman, Anchorage, Alaska, for his pertinent editorial interventions that contributed greatly to the final form of this work and to Robert T. Clarke, Dallas, Texas, for his helpful editorial work.

References Cited

BALINTONI, I.
1984 Structure of right side of the Bistrita River between Ciocăneşti and Vatra Dornei (East Carpathians). *Dări de seamă, Institutul de Geologie şi Geofizică, Bucureşti*, 68, (5): 5–21.

BERCIA, I., KRÄUTNER, H.G., and MUREŞAN, M.
1976 Pre–Mesozoic Metamorphites of the East Carpathians. *Anuarul Institutului de Geologie şi Geofizică, Bucureşti*, 50: 37–70.

EKLUND, C.
1990 Lower Cambrian acritarch stratigraphy of the Bårstad 2 core, Östergötland, Sweden. *Geologiska Föreningens I Stockholm Förhandlingar*, 112: 19–44.

HOFMANN, H.J., and JACKSON, G.D.
1994 Shale–facies microfossils from the Proterozoic Bylot Supergroup, Baffin Island, Canada. *Journal of Paleontology*, 37(4): 1–35.

ILIESCU, V., KRÄUTNER, H.G., KRÄUTNER, F. and HANN, H.
1983 New palynological proofs on the Cambrian age of the Tulgheş Group (East Carpathians). *Anuarul Institutului de Geologie şi Geofizică, Bucureşti*, 59: 7–17 (in Romanian).

IONESI, L.
1994 Geologia unităţilror de platformă şi a orogenului Nord-Dobrogean. *Editura Tehnică, Bucureşti*, 280 p.

JANKAUSKAS, T.
1991 Biostratigraphy of the Lower and Middle Cambrian boundary beds of Lithuania. *Proceedings of the Estonian Academy of Sciences, Geology*, 40(4): 152–159 (in Russian).

JANKAUSKAS, T., MIKHAILOVA, N.S., and GERMAN, T.N.
1989 Precambrian microfossils of the USSR. *Nauka, Leningrad*, 188 p. (in Russian).

JANKAUSKAS, T. and LENDZION, K.
1992 Lower and Middle Cambrian acritarchs based biozonation of the Baltic Syneclize and adjacent areas (East European Platform). *Przeglad Geologiczny*, 9: 519–524.
1994 Biostratigraphic correlation of Lower and Middle Cambrian section in the Baltic Syneclize and adjacent areas. *Przeglad Geologiczny*, 40(5): 365–370.

KNOLL, A.H.
1996 Archean and Proterozoic Paleontology. *In*: Jansonius, J. and McGregor D.C. (eds.): Palynology: Principles and Applications. *American Association of Stratigraphic Palynologists Foundation*, volume 1, p. 51–80.

KONZALOWA, M.
1994 Acritarchs from the Bohemian Precambrian (Upper Proterozoic) and Lower–Middle Cambrian. *Review of Palaeobotany and Palynology*, 18: 41–56.

KRÄUTNER, H.G.
1980 Lithostratigraphic correlation of Pre–Precambrian in the Romanian Carpathians. *Anuarul Institutului de Geologie şi Geofizică, Bucureşti*, 57: 229–296.

TABLE 3. Palynostratigraphical correlation of metamorphic formations from different regions of Romania.

AGE Ma	EON	PERIODS	EAST CARPATHIANS LITHOLOGICAL GROUPS	EAST CARPATHIANS MEMBERS	NORTH DOBROGEA LITHOLOGICAL GROUPS	CENTRAL DOBROGEA LITHOLOG. GROUPS	CENTRAL DOBROGEA MEMBERS
500				TG.4			
		UPPER CAMBRIAN	TULGHES	?			
	LOWER	MIDDLE CAMBRIAN		TG.3		?	U. GRW.
						GREEN-	
	PALEOZOIC	LOWER CAMBRIAN	GROUP	TG.2		SCHIST	L. GRW.
				TG.1		GROUP	I. GRW:
570				RB.3	? MEGINA GROUP	CEAMURLIA	
	NEO	VENDIAN	REBRA	RB.2		OR	
			GROUP				
650				RB.1	ORLIGA	ALTÂN TEPE	
		RIPHEAN	BRETILA	?	GROUP		
	PROTEROZOIC	OR	GROUP			GROUP	
		CRYOGENIAN			?	?	
850							

KRÄUTNER, H.G., MUREŞAN M., and SEGHEDI, A.
1988 Precambrian of Dobrogea. *In*: Zoubek, V. (ed): *Precambrian in younger fold belt, Wiley, Chichester*, p. 361–379.

MARTIN F.
1993 Acritarchs: A review. *Biological Review*, 68: 475–538.

MÂNZATU, S., LEMNE, M., VÂJDEA, E., TĂNĂSESCU, A., IONCICĂ, M., and TIEPAC, I.
1975 Geochronological data for crystalline formations and igneous massifs of Romania. *Dări de seamă ale Institutului de Geologie şi Geofizică, Bucureşti*, 61(5): 85–111 (in Romanian).

MENDELSON, C.
1987 Acritarchs. *In*: Lipps, J.H. (ed.), *Fossil Prokaryotes and Protists*. University of Tennessee, Department of Geological Sciences, 18: 62–67.

MIKHAILOVA, N.C., and PODKOVIROV, V.N.
1987 Characteristic microphytology of Riphean and Vendian boundary beds horizon of Southern Ural. *Izvestya Akademii Nauk SSSR, serya Geologicheskaya*, 9: 75–83 (in Russian).

MIRĂUŢĂ, O.
1969 The Upper Proterozoic tectonics of Central Dobrogea. *Anuarul Institutului Geologic, Bucureşti*, 37: 31–36 (in Romanian).

MOCZYDLOWSKA, M.
1989 Upper Proterozoic and Lower Cambrian acritarchs from Poland — micropalaeontology, biostratigraphy and thermal study. *Lund Publications in Geology*, 75:30.
1991 Acritarch biostratigraphy of the Lower Cambrian and Precambrian–Cambrian boundary in southeastern Poland. *Fossils and Strata*, 29: 1–127.

MOCZYDLOWSKA, M., and VIDAL, G.
1986 Lower Cambrian acritarch zonation in southern Scandinavia and southern Poland. *Geologiska Föreningens i Stockholm Förhandlingar*, 108: 201–223.
1992 Phytoplankton from the Lower Cambrian Læså formation on Bornholm, Denmark: biostratigraphy and palaeoennvironmental constraints. *Geological Magazine*, 129: 17–40.

OLARU, L.
1991 Stratigraphical evolution of the palynological assemblages from the crystalline formations of the Romanian East Carpathians and the correlation with other area from Europe. *Evolution and Adaptation*, 4: 77–86.

OLARU, L., ERHAN, V., and ANTOCHI, M.
1990 Palynological and petrographical observations about the Greenschist of Dobrogea. *Analele ştiinţifice ale Universităţii "Al. I. Cuza" Iaşi, Geologie*, 36: 5–16 (in Romanian).

OLARU, L., and GUNIA, T.
1987 Palynological correlation between the Pre–Cambrian Formations in the Sudetic Mountains (Poland) and the East Carpathians (Romania). *Analele ştiinţifice ale Universităţii "Al. I. Cuza" Iaşi, II b, Geologie-Geografie*, 33: 31–34.
1988 Palynological correlation between the Cambrian Formation in the Sudetic Mountains (Poland) and the East Carpathians (Romania). *Analele ştiinţifice ale Universităţii "Al. I. Cuza" Iaşi, II b, Geologie-Geografie*, 34: 23–31.

OLARU, L., and ONICEANU, M.
1983 Contribution to the sporo-protistological study of the formations with talc powder mineralization in the Drăgoiasa–Barnar (Eastern Carpathians). *Analele ştiinţifice ale Universităţii "Al. I. Cuza" Iaşi, II b, Geologie-Geografie*, 29: 25–30.
1985 Sur la présence de quelques associations d'acritarchs du Protérozoïque supérieur et du Paléozoïque inférieur dans les Carpates Orientales de Roumanie. *Science Géologique, Bulletin, Strasbourg*, 38(1): 131–136.

ONICEANU, M., OLARU, L, and ENCIU, P.
1974 New data about the age of some crystalline formations of Bretila anticline zone. *Analele științifice ale Universității "Al. I. Cuza" Iași, II b, Geologie*, 20: 23–27 (in Romanian).
PAŠKIEVIČIENE, L.T.
1980 Acritarchs from Vendian and Cambrian boundary deposits in western part of the East European Platform. *Akademii Nauk S.S.S.R., Paleontologitsheskiy Institut, Nauka, Moscow*, p. 75 (in Russian).
SĂNDULESCU, M.
1984 The Romanian geotectonics. *Technical Publishing House, Bucarest*, p. 334 (in Romanian).
SEGHEDI, A.
1980 Metamorphic Formations of North Dobrogea. *Anuarul Institutului de Geologie și Geofizică, București*, 41: 131–137.
TIMOFEEV, B.V.
1959 Drevneyshaya flora Pribaltiki i ee stratigraficeskoe znachenie. *Trudy. VNIGRI*, 129: 320 (in Russian).
VIDAL, G.
1976 Late Precambrian microfossil from the Visingosö Beds in southern Sweden. *Fossil and Strata*, 9: 57.
VIDAL, G., and FORD, T.D.
1985 Microbiotas from the Late Proterozic Chuar Group (Northern Arizona) and Uinta Mountain Group (Utah) and their Chronostratigraphic implications. *Precambrian Research*, 28: 349–389.
VIDAL, G., and MOCZYDLOWSKA, M.
1995 The Neoproterozoic of Baltica-stratigraphy, palaeobiology and general evolution. *Precambrian Research*, 73: 197–216.
VOLKOVA, N.A.
1969 Acritarchs of the north–western Russian Platform Tommotian Stage and the Cambrian lower boundary problem. *Trudy Akademii Nauk S.S.S.R.*, 206: 224–235 (in Russian).
VOLKOVA, N.A., KIRJANOV, V.V., PISKUN, L.V., PAŠKIEVIČIENE, L.T. and JANKAUSKAS, T.V.
1979 Plant Microfossils *In*: Keller, B.M., and Rozanov, A.Yu. (eds)., Upper Precambrian and Cambrian Palaeontology of the East European Platform. *Akademii Nauk S.S.S.R., Nauka, Moscow*, p. 4–38 (in Russian).

Author's address

LEONARD OLARU
Department of Geology
University "Al. I.Cuza" Iasi
Bd. Carol the 1st, 20 A
6600, Iasi
Romania
E:mail: lola@uaic.ro

Part 2: Mesozoic Palynology

PRELIMINARY REPORT ON THE PALYNOLOGY OF THE CRETACEOUS OF THE GOBI DESERT, MONGOLIA

D.J. Nichols, M. Watabe, N. Ichinnorov, Y. Ariunchimeg

Abstract

Palynological samples from units of Cretaceous age in the Gobi Desert, Mongolia, were collected as an aspect of the Japanese–Mongolian Joint Paleontological Expedition. Units sampled include those exposed in several badlands areas that are yielding dinosaurs and other vertebrate fossils. Palynomorphs were recovered from such outcrops at Khuren Dukh and Sainshand, and also from a drill core collected at Bulgan Uul.

Preliminary analyses of samples from Khuren Dukh reveal palynofloras dominated by gymnosperm pollen and including fern spores. With the exception of rare specimens of *Asteropollis*, angiosperm pollen is absent from these assemblages. These assemblages are indicative of Early Cretaceous (Albian) age. The Bulgan Uul core also yielded an assemblage of gymnosperm pollen that appears to be Early Cretaceous in age. Assemblages from strata of uncertain age near Sainshand are dominated by algal cysts indicative of lacustrine depositional environments. So far, samples collected from dinosaur-bearing formations of Late Cretaceous age have not yielded palynomorphs. Collection and analysis of palynological samples will continue in future field seasons.

INTRODUCTION

In 1995 the senior author joined the Japanese–Mongolian Joint Paleontological Expedition as a palynologist and conducted field work in the Gobi Desert of southern Mongolia. The expedition was in its third year of operation collecting vertebrate fossils and studying Cretaceous environments in the Gobi region for the Hayashibara Museum of Natural Sciences, a new facility being planned for construction in Okayama, Japan. The goal of the palynological research is to establish age relations between sites from which dinosaurs are being collected and to provide paleoecological data for environmental reconstructions. The 1995 field party included coauthor and expedition leader M. Watabe and other Japanese scientists; Mongolian scientists including Y. Ariunchimeg and N. Ichinnorov (who is Mongolia's sole palynologist); and one other American scientist, dinosaur paleontologist and sedimentologist D. E. Fastovsky of the University of Rhode Island.

Since the pioneering expeditions of Roy Chapman Andrews of the American Museum of Natural History in the 1920s, the Gobi Desert has been famous for its rich vertebrate fossil fauna. The world's first dinosaur eggs were discovered in the Gobi by Andrews at Bayn Dzak, which he dubbed the Flaming Cliffs. Those discoveries and other adventures were described especially well in two popular books by Andrews (1945, 1951). In subsequent years, other field parties have ventured into the desolate and inhospitable Gobi in search of dinosaurs, including several led by Russian and Polish paleontologists in the 1940s and 1960s, perhaps most notably those detailed by Kielan-Jaworowska (1969). More recent dinosaur hunts in the Gobi are described by Novacek (1996), and some of the recent work on Mongolian dinosaurs discussed in the present report is presented by Fastovsky and Weishampel (1996).

Nichols, D.J., Watabe, M., Ichinnorov, N., and Ariunchimeg, Y.
2001 Preliminary report on the palynology of the Cretaceous of the Gobi Desert, Mongolia; *In:* Goodman, D.K., and Clarke, R.T. (eds.), *Proceedings of the IX International Palynological Congress, Houston, Texas, U.S.A., 1996*; American Association of Stratigraphic Palynologists Foundation, p. 131–138.
 ISBN 0–931871–06–9

Until 1992 the former People's Republic of Mongolia was part of the Soviet bloc and was largely inaccessible to scientists from other countries. Since the breakup of the Soviet Union, Mongolia has become an independent country and has opened again to scientific exploration in various cooperative ventures that involve the Mongolian Institute of Geology and Mineral Resources and the Mongolian Paleontological Center. The Japanese–Mongolian Joint Paleontological Expedition is one such cooperative investigation. It is sponsored by the Hayashibara Museum of Natural Sciences and supported by Hayashibara Company Ltd. of Okayama, Japan.

Little recent information on the geology and fossils of the Gobi region is available in the English language. A comprehensive review was published by Jerzykiewicz and Russell (1991). Their study stemmed from their participation in a joint Canadian–Chinese expedition to the Mongolian Plateau in the People's Republic of China. That area, which is the Chinese province of Inner Mongolia, is distinct from the independent country of Mongolia referred to in the present report, which was sometimes previously called "Outer Mongolia." Jerzykiewicz and Russell provide a useful overview of the entire Gobi region, and their report includes an extensive bibliography; numerous reports listed therein are in Russian. The stratigraphy and depositional environments of the Cretaceous units of Mongolia were briefly summarized by Badamgarav et al. (1995). The most recent report on the geology and paleontology of the Gobi Desert in Mongolia is that by Fastovsky et al. (1997) on Tugrikin-shireh, a locality studied by the Japanese–Mongolian Joint Paleontological Expedition that is rich in fossils of the dinosaur *Protoceratops*.

GEOLOGIC SETTING

The country of Mongolia lies in the center of the Asian continent, between China to the south and Russia to the north (Text-Figure 1). The Gobi Desert occupies southern Mongolia and part of northern China. The Gobi region is located on a large plateau at the southern margin of the pre-Mesozoic Siberian Plate (Jerzykiewicz and Russell, 1991), between latitudes 42 and 46 degrees north and longitudes 96 and 112 degrees east. The area is underlain by fault-block mountains with intervening grabens and semigrabens that formed from Early Jurassic through Holocene time; Cretaceous deposits are preserved in grabens and semigrabens of

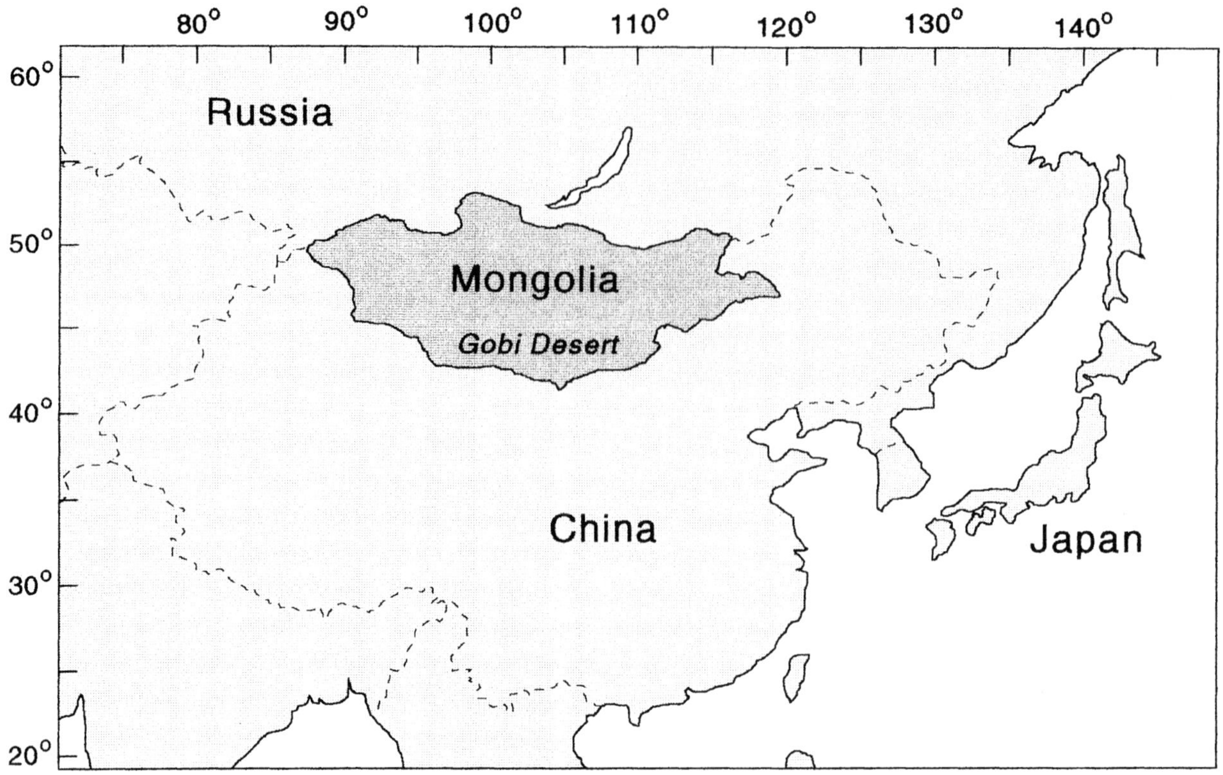

Text-Figure 1. Generalized map showing location of Mongolia in central Asia and the Gobi Desert in the southern part of Mongolia.

early Tertiary age (Jerzykiewicz and Russell, 1991). These rocks are exposed only in geographically isolated badlands areas called "tsavs" (Text-Figure 2). The vast areas between the badlands are covered with desert pavement or sparse vegetation and lack outcrops.

The isolated nature of exposures complicates correlation of the Cretaceous deposits. The stratigraphic nomenclature used in this report (Text-Figure 3) is based on lithostratigraphic correlations and limited data on age. Age determinations in terms of internationally accepted terminology for subdivision of the Cretaceous are few, and most of the existing information on ages tends to mix lithostratigraphic and chronostratigraphic concepts (Jerzykiewicz and Russell, 1991). Thus, the Upper Cretaceous Nemegt Formation is said to be of "Nemegtian" age, the underlying Barun Goyot Formation is said to be of "Barungoyotian" age, and so on. It is hoped that palynostratigraphy will help to resolve some uncertainties about age and correlation.

The Cretaceous rocks consist of sandstone, siltstone, and shale. They are of fluvial and lacustrine origin but include significant erg deposits that preserve ancient sand dunes, in some places in close proximity to modern dunes. Oxidation and coarse grain size of many of the deposits cause them to have low potential for preservation of palynomorphs. In fact, much of the Upper Cretaceous is composed of oxidized units that are discouragingly "ulaan" (red) in color. Palynomorphs have been recovered from more organic-rich outcrops at Khuren Dukh and Sainshand and from a drill core at Bulgan Uul.

FIELD WORK IN THE GOBI

Rigorous conditions of climate and terrain make geologic fieldwork in the Gobi difficult. There are no paved roads and few tracks to follow. Daytime temperatures in the summer routinely reach 30°C and may exceed 40°C. Windy conditions in the desert offer some relief from the heat but also tend to blow sand about, stinging the skin and filtering into tents and sleeping bags. At times the winds become so strong that sand storms develop. The desert pavement of gravel that covers the surface of the Gobi except in the tsavs is littered with ventifacts, many of classic three-sided pyramidal form, that attest to the effects of the wind. Water is quite scarce. There are very few springs; water for personal use and for making plaster to encase dinosaur bones must be obtained from scattered wells and carried along. Travel is by four-wheel-drive vehicles, which must also carry their own gasoline. Food for field parties is also carried in the vehicles, although it can be supplemented occasionally by purchase of a sheep or goat from nomads who keep herds in the better vegetated parts of the semi-arid desert region. The Mongolian nomads, who are very friendly people, usually offer camel cheese and fermented mare's milk ("airag") to visitors. Fermented milk of sheep, goats, and camels as well as of horses is distilled to make "arkhi," the local vodka, which tends to have its own hazards if consumed in excess. Conditions in this expedition's field camp were ameliorated by the presence of a "ger," the traditional felt-covered dwelling of Mongolian nomads (more widely known by the Russian word yurt), which served as a meeting place and for meals (Text-Figure 4).

Text-Figure 2. The landscape at Bugin Tsav, a typical badlands area ("tsav") in the Gobi Desert, Mongolia. The Nemegt Formation (Upper Cretaceous) is exposed here.

SYSTEMS	FORMATIONS	DINOSAURS
Tertiary (part)	Gashaat	
	Nemegt	*Tarbosaurus & Saurolophus*
Upper	Barun Goyot	
	Djadokhta (= Bayn Dzak)	*Protoceratops & Pinacosaurus*
Cretaceous	Bayn Shire	
	Sainshand (= Barun Bayan)	
Lower	Khukhtek (= Dzun Bayan)	*Iguanodon & Psittacosaurus*
Jurassic (part)	Tsagantsav	

Text-Figure 3. Cretaceous stratigraphy of the Gobi Desert, Mongolia. The Cretaceous is not subdivided into stages in this diagram because the positions of stage boundaries are uncertain.

Text-Figure 4. View of the field camp at Bugin Tsav, Gobi Desert, Mongolia, showing the ger (the traditional Mongolian building, more widely known as the Russian word "yurt"), the supply tents (at left), and two personal tents (at right).

The tsavs or badlands afford excellent local exposures of the Cretaceous section. One can prospect for vertebrate fossils and potentially palyniferous beds over wide areas, unimpeded by vegetational cover (Text-Figure 2). Quarrying dinosaur skeletons involves much more time and effort than collecting samples for palynological analysis (Text-Figure 5).

LOCALITIES SAMPLED

Bugin Tsav

Bugin Tsav is a badlands area located at lat 43° 51–53' N, long 100° 00–01' E, in the southern Gobi Desert (Text-Figure 6). The Nemegt Formation (Upper Cretaceous) is exposed here. These rocks have been said to be lacustrine in origin (cf. Badamgarav et al., 1995), but more recent investigations suggest that they are largely of fluvial origin (D. E. Fastovsky, oral communication, 1995). Their age is thought to be either Campanian–Maastrichtian (Badamgarav et al., 1995) or Maastrichtian (Jerzykiewicz and Russell, 1991). Remains of the dinosaurs *Tarbosaurus* and *Saurolophus* are found in these rocks at Bugin Tsav, among others, and dinosaur eggshell fragments are present, as are fossil turtles and mollusks.

Mudstone beds of gray, greenish gray, tan, and brown color were collected from various levels within the exposed section. All samples proved to be barren of palynomorphs.

Gurilin Tsav

Gurilin Tsav is a badlands area located at lat 43° 50–51' N, long 100° 08–10' E (Text-Figure 6). The Nemegt Formation is exposed here, as at nearby Bugin Tsav. Mudstone beds of gray, greenish gray, and brown color were collected, but they also proved to be barren. The presence of a well with potable water made the locality worth visiting, nonetheless.

Khure Tsav

Khure Tsav is a badlands area located at about lat 43° 59' N, long 101° 17' E (Text-Figure 6). The Barun Goyot Formation is exposed here. These rocks, which are stratigraphically below the Nemegt Formation, are mostly brownish red mudstone at this locality. Elsewhere the Barun Goyot Formation includes fine-grained sandstone and caliche (Badamgarav et al., 1995). Based on stratigraphic relations, the formation may be Campanian in age. Samples were collected although the palynological potential of these beds appeared to be poor; in addition to mudstone, a black, carbonaceous sandstone near the top of the formation was sampled. All samples proved to be barren of palynomorphs.

Khongil Tsav West

Khongil Tsav West is a badlands area located toward the western end of Udan Sayr at about lat 43° 10' N, long 102°

Text-Figure 5. View of the dinosaur quarry at Bugin Tsav. A skeleton of *Saurolophus*, a duck-billed dinosaur, is being excavated.

47' E (Text-Figure 6). Here the Djadokhta Formation is exposed. The Djadokhta underlies the Barun Goyot Formation and is stratigraphically equivalent to the unit exposed at Bayn Dzak, for which reason it is also known as the Bayn Dzak Formation (cf. Jerzykiewicz and Russell, 1991; Jerzykiewicz et al., 1993; and Badamgarav et al., 1995). At the Khongil Tsav West locality the unit is composed primarily of mudstone and sandstone red beds evidently of fluvial origin, although elsewhere this formation includes beds of both eolian and lacustrine origin. The Djadokhta is reported to contain a *Protoceratops* fauna although no such bones were observed at this locality; a skeleton of an anklyosaur was found, however. No palynomorphs were recovered.

Alag Teg

Alag Teg is located at about lat 44° 15' N, long 103° 18' E (Text-Figure 6), about 4 km northeast of the Tugrikin-shireh locality studied by Fastovsky et al. (1997). The Djadokhta Formation (= Bayn Dzak Formation) is exposed here. At Alag Teg the rocks are sandstone and mudstone, with mudstone of lacustrine origin dominant, in sharp contrast to the eolian erg paleoenvironment at the correlative Tugrikin-shireh locality. The dinosaur fauna differs as well as the sedimentary environment; whereas *Protoceratops* is common at Tugrikin-shireh, *Pinacosaurus* is common at Alag Teg. Palynological evidence of the vegetation that existed in the Alag Teg area would be quite useful, but unfortunately all samples collected during the 1995 expedition are barren.

Bulgan Uul core hole

The Bulgan Uul core hole drill site is located about 30 km west of Tsogt Ovoo at lat 44° 29' 10" N, long 104° 54' 22.4" E, (Text-Figure 6). A sample taken at a depth of 104 m in the core yielded palynomorphs. Available information indicates that the unit sampled underlies the Sainshand Formation. As noted, the Sainshand Formation evidently is Cenomanian in age, and thus the underlying unit, which was sampled in the core, may be Early Cretaceous (Albian) in age.

Khuren Dukh

Khuren Dukh is an extensive area located at lat 45° 49–51' N, long 108° 26' E, in the eastern Gobi (Text-Figure 6). Because of its great areal expanse, the locality is divided into three sub-localities, KD-1, KD-2, and KD-3, which correspond to a rough stratigraphic subdivision of lower, middle, and upper. The entire interval is in the Khukhtek Formation and thus is said to be "Khukhtekian" in age. The Khukhtek Formation may be equivalent to the Dushilin or Khulsyngol Formations as described by Jerzykiewicz and Russell (1991); those formations are Aptian–Albian in age. Several palynological samples collected at this locality yielded pollen and spores, and our palynological data (discussed below) indicate that the Khukhtek Formation is

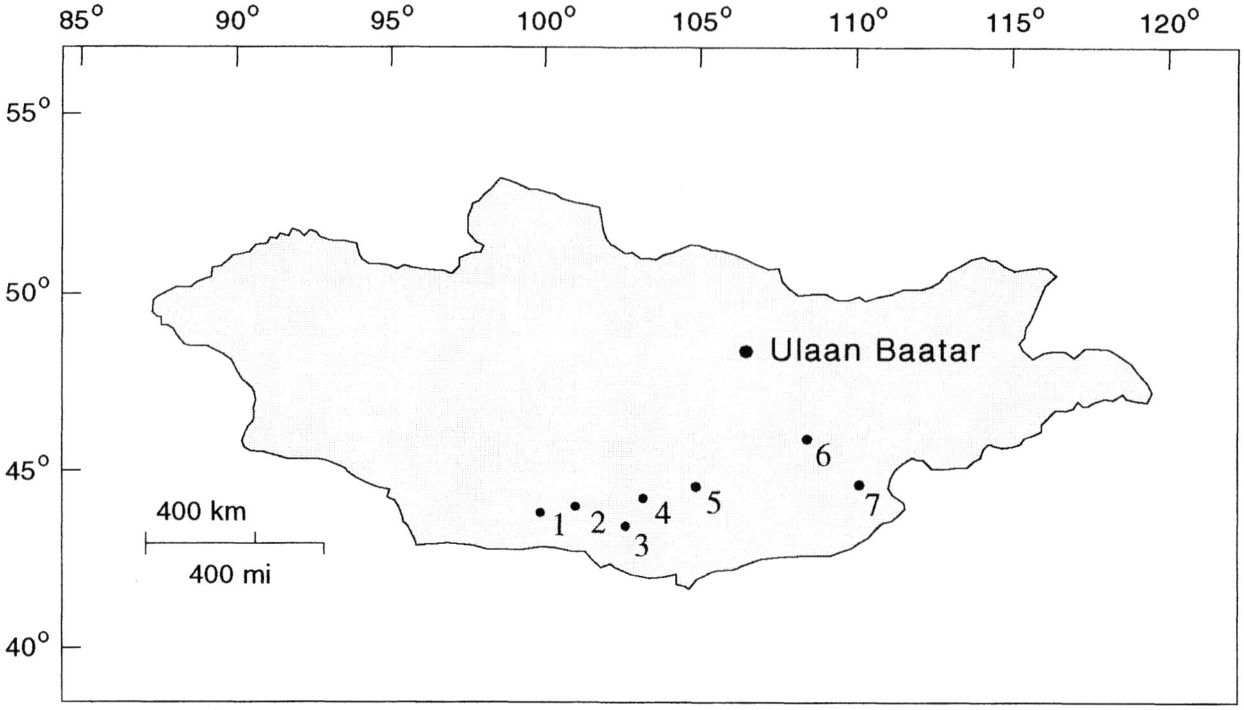

Text-Figure 6. Outline map of Mongolia showing locations of field areas and sample sites. Ulaan Baatar (sometimes spelled Ulan Bator) is the capital of Mongolia. The localities are: (1) Bugin Tsav and Gurilin Tsav, (2) Khure Tsav, (3) Khongil Tsav West, (4) Alag Teg, (5) Bulgan Uul, (6) Khuren Dukh, and (7) Southeast Sain Shand. The famous "Flaming Cliffs" locality (Bayn Dzak) is near Alag Teg.

Albian in age. The dinosaur *Iguanodon* has been found at the Khuren Dukh locality. A palynoflora from beds equivalent to KD-3 at Khuren Dukh was previously described by Brattseva and Novodvorskaya (1975), who referred to these deposits as the upper Dzun Bayan suite.

Southeast Sain Shand

The Southeast Sain Shand locality is located in the eastern part of the Gobi Desert at lat 44° 35' N, long 110° 13' 46" E (Text-Figure 6). It is about 45 km south of the town of Sain Shand and 10 km northeast of the town of Dzunbayan. It is somewhat uncertain which stratigraphic unit is exposed, but it may be the Sainshand Formation (formerly known as the Barun Bayan Formation), which is widely exposed in the area. The Sainshand Formation is Cenomanian in age (Jerzykiewicz and Russell, 1991). Several samples collected at the Southeast Sain Shand locality yielded palynomorphs.

PALYNOLOGICAL ASSEMBLAGES

Southeast Sain Shand

Assemblages from all productive samples collected at the Southeast Sain Shand locality are dominated by cysts of aquatic algae. These palynomorphs are strong evidence for a lacustrine depositional environment but provide no clear evidence on the age of the deposits. Analysis and interpretation of these assemblages in continuing. To date, species of the algal genera *Tetraporina*, *Schizophacus*, and *Schizosporis* have been identified. Rare unidentified bisaccate pollen and trilete spores are present in the assemblage also.

Bulgan Uul

A sample from the 104 m level of the Bulgan Uul core yielded an assemblage dominated by non-saccate gymnosperm pollen. Especially common are pollen attributable to *Ephedra* ("*Ephedripites*") and *Corollina*, and the assemblage includes a species of inaperturate, probably gymnospermous pollen. Bisaccates referable to *Pityosporites* and perhaps other genera are also present. Spores are less common but two trilete species have been found so far. This assemblage is consistent with, but not definitive of, the suggested Albian age of the sampled unit. Other samples from the Bulgan Uul core yet to be studied may provide more information on age.

Khuren Dukh

All samples from Khuren Dukh analyzed in the present study yielded generally similar assemblages of spores and gymnosperm pollen, although some differences are evident between assemblages from the lower (KD-1), middle (KD-2), and upper (KD-3) parts of the sampled interval. The assemblage from KD-1 is low in diversity and is dominated by bisaccate species referable to *Pityosporites*. Two species of monosaccate gymnosperm pollen and three species of monosulcate gymnosperm pollen referable to *Cycadopites* or *Entylissa* ("*Ginkgocycadophytus*") also are present; spores are rare in this assemblage. Probably the most interesting specimens present in the assemblage from KD-1 are pollen of the primitive angiosperm genus *Asteropollis*, probably *A. asteroides* (Text-Figure 7). *Asteropollis asteroides* is known from rocks of Albian age in North America (Hedlund and Norris, 1968; Wingate, 1980; Nichols and Jacobson, 1982). Thus, these specimens indicate an Albian age for at least the lower part of the interval at Khuren Dukh.

The assemblage from KD-2 is more diverse than that from KD-1, although it also is dominated by pollen of *Pityosporites* or related bisaccate genera. Monosaccate pollen and several species of monosulcate pollen are present. Spores are more numerous in the KD-2 assemblage; they include species of *Laevigatosporites*, *Gleicheniidites*, and others. Brattseva and Novodvorskaya (1975) reported that *Gleichenia* spores (= *Gleicheniidites*) are absent from Khuren Dukh; their samples were from the upper part of the interval (KD-3 of the present study).

Assemblages from KD-3 analyzed in the present study include considerably greater numbers of trilete spores than do those from the underlying parts of the interval. Brattseva and Novodvorskaya (1975) reported the angiosperm pollen species *Stephanocolpites* sp. aff. *fredericksburgensis* in their assemblage. This species also is known from the Albian of North America (Hedlund and Norris, 1968) and in conjunction with the record of *Asteropollis* in the lower part of the interval, it tends to confirm an Albian age determination for the Khukhtek Formation at Khuren Dukh.

SUBSEQUENT WORK

Palynological research on the Cretaceous of the Gobi Desert is ongoing. New collections were made in the southern Gobi in 1997 by the senior author in conjunction with studies conducted by other collaborating scientists who joined the Japanese–Mongolian Joint Paleontological Expedition during the 1997 field season. These related investigations and investigators are sedimentology (D.E. Fastovsky, University of Rhode Island), paleobotany (K.R. Johnson, Denver Museum of Natural History), and paleomagnetism (J.F. Hicks, Scripps Institution of Oceanography). All of this work is of an interdisciplinary nature and will be closely tied to the research on dinosaurs currently underway by the staff of the Hayashibara Museum in Okayama, Japan, and the Paleontological Center in Ulaan Baatar, Mongolia. The palynological studies will be conducted collaboratively by N. Ichinnorov of the Institute of Geology and Mineral Resources in Ulaan Baatar and D. J. Nichols.

NOTE ADDED IN PROOF

Since preparation of the manuscript for this paper, geologic investigations in the Gobi have continued. Palynological data refining the stratigraphy and age of the deposits at Khuren Dukh are especially important. See Hicks, J.F., Brinkman, D.L., Nichols, D.J., and Watabe, M., 1999, Paleomagnetic and palynologic analyses of Albian to Santonian strata at Bayn Shireh, Burkhant, and Khuren Dukh, eastern Gobi Desert, Mongolia. *Cretaceous Research*, 20: 829-850.

ACKNOWLEDGMENTS

We thank the Hayashibara Company, Ltd., for financial support. Special thanks go to Dr. K. Ishii, Director of the Hayashibara Museum of Natural Sciences, for guidance and support, and to Dr. R. Barsbold of the Institute of Geology and Mineral Resources of Mongolia and Mr. K. Tsogtbaatar of the Paleontological Museum in Ulaan Baatar for logistical assistance. This report is Hayashibara Museum of Natural Sciences Contribution Number 2.

References Cited

ANDREWS, R.C.
1945 *Under a lucky star*. Blue Ribbon Books, Garden City, New York, 300 p.
1951 *Heart of Asia*. Duell, Sloan and Pearce, New York, 224 p.
BADAMGARAV, D., KHAND, YO., and BARSBOLD, R.
1995 Nonmarine Cretaceous of Mongolia. *In*: The Cretaceous System in East and South Asia. *Newsletter Special Issue 2, IGCP*, 350: 17–23.
BRATTSEVA, G.M., and NOVODVORSKAYA, I.M.
1975 Spores and pollen from Lower Cretaceous deposits of a site Khuren Dukh, MNR (Mongolian People's Republic). *In*: Fossil fauna and flora of Mongolia. *Trudy Sovmestnaya Sovetsko-Mongoliskaya Paleontologicheskaya Ekspeditsiya*, 2: 205–209 [in Russian].

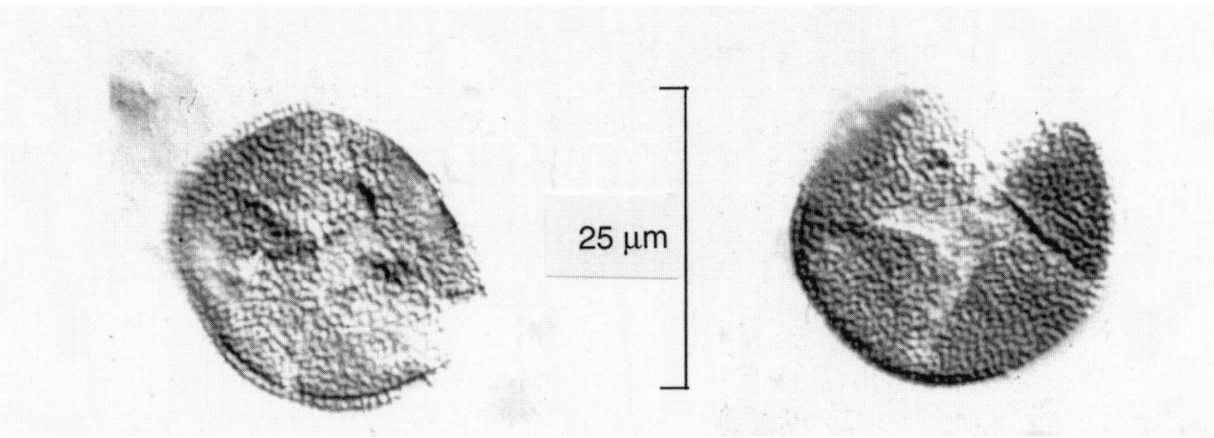

Text-Figure 7. Two specimens of *Asteropollis* sp. aff. *A. asteroides* from the Khuren Dukh locality. These fossils indicate an Albian age for the deposits. Scale bar is 25 µm.

FASTOVSKY, D.E., BADAMGARAV, D., ISHIMOTO, H., WATABE, M., and WEISHAMPEL, D.B.
1997 The paleoenvironments of Tugrikin-shireh (Gobi Desert, Mongolia) and aspects of the taphonomy and paleoecology of *Protoceratops* (Dinosauria, Ornithischia). *Palaios*, 12: 59–70.

FASTOVSKY, D.E., and WEISHAMPEL, D.B.
1996 *The evolution and extinction of the dinosaurs*. Cambridge University Press, Cambridge, U.K., 460 p.

HEDLUND, R.W., and NORRIS, G.
1968 Spores and pollen grains from Fredericksburgian (Albian) strata, Marshall County, Oklahoma. *Pollen et Spores*, 10: 129–159.

JERZYKIEWICZ, T., CURRIE, P.J., EBERTH, D.A., JOHNSTON, P.A., KOSTER, E.H., and ZHENG, J.-J.
1993 Djadokhta Formation correlative strata in Chinese Inner Mongolia: an overview of the stratigraphy, sedimentary geology, and paleontology and comparisons with the type locality in pre-Altai Gobi. *Canadian Journal of Earth Science*, 30: 2180–2195.

JERZYKIEWICZ, T., and RUSSELL, D.A.
1991 Late Mesozoic stratigraphy and vertebrates of the Gobi Basin. *Cretaceous Research*, 12: 345–377.

KIELAN-JAWOROWSKA, Z.
1969 *Hunting for dinosaurs*. The MIT Press, Cambridge, Mass., 177 p.

NICHOLS, D.J., and JACOBSON, S.R.
1982 Palynostratigraphic framework for the Cretaceous (Albian–Maestrichtian) of the overthrust belt of Utah and Wyoming. *Palynology*, 6: 119–147.

NOVACEK, M.J.
1996 *Dinosaurs of the Flaming Cliffs*. Doubleday, New York, 367 p.

WINGATE, F.H.
1980 Plant microfossils from the Denton Shale Member of the Bokchito Formation (Lower Cretaceous, Albian) in southern Oklahoma. *Oklahoma Geological Survey Bulletin* 130, 93 p.

Authors' addresses

D. J. NICHOLS
U.S. Geological Survey
M.S. 939, Box 25046
Denver, Colorado 80225
U.S.A.
nichols@usgs.gov

M. WATABE
Hayashibara Museum of Natural Sciences
Shimoishii 1-2-3
Okayama 700
Japan
wmahito@pa2.so-net.or.jp

N. ICHINNOROV
Institute of Geology & Mineral Resources
Tolgoit 49, P.O. Box 110
Ulaan Baatar 37
Mongolia

Y. ARIUNCHIMEG
Mongolian Paleontological Center
Ulaan Baatar 51
Mongolia

THE MAASTRICHTIAN FLORA AND DINOSAURS OF THE RUSSIAN FAR EAST

Valentina S. Markevich and Eugenia V. Bugdaeva

Abstract

Five localities with remains of reptiles, including dinosaurs, were studied in the Russian Far East. A study of fossil plants (pollen, spores, and megafossils) from bone-bearing strata permits determination of their geological age as middle Maastrichtian.

INTRODUCTION

The stratigraphy of continental beds has always been a principal point of discussion among paleontologists because it does not correspond to the general geochronological scale based on marine animals. As V.V. Menner noted in 1962, the only way marine and continental beds can be correlated is by palynological analysis. We attempt to divide Maastrichtian beds using this method. Deposits of this stage were distinguished in eastern Russia as either undivided or divided only into lower and upper parts, although the uppermost stage of the Cretaceous is most interesting. That was a time at the eve of a global crisis that influenced many regions of the Earth, which occurred at the Mesozoic–Cenozoic boundary. Although the Maastrichtian is divided into two parts, nevertheless some researchers dealing with of inocerams (Pokhialaynen, 1980; Pokhialaynen and Byalobzesky, 1984), vertebrates (Nessov and Golovneva, 1990; Nessov, 1995), plant megafossils (Golovneva, 1994b), and palynoflora (Markevich et al., 1994) believe that this stage has tripartite division. E.G.Kauffman (1986) held to the idea that the Maastrichtian of the Euro–American region and the Caribbean area could be divided into three parts.

Some of the Maastrichtian deposits in the Far East contain reptile remains. If one can define their exact age, environment, and changes of living conditions, it will shed light on the enigma of the extinction of dinosaurs, which played an important role in the Mesozoic ecosystems. The palynological method appears to be the same or even more efficient in comparison with traditional ammonite studies in detailed division of stratigraphic units, giving a more thorough knowledge about changes of flora, continental paleoecosystems, and paleogeography. Evolutionary changes of the Maastrichtian palynoflora allow us to distinguish three stages, probably reflecting ecosystem reorganizations, which are well-developed in the Far East deposits.

In eastern Russia, beds with fossil reptiles have been discovered in the Amur region (Priamurye), Koryak Upland, and Sakhalin Island (Text-Figure 1). More than 100 palynological samples were collected from these bone-bearing beds. Overlying and underlying deposits were also sampled for comparison. These data are shown in Tables 2–4.

Markevich, Valentina S., and Bugdaeva, Eugenia V.
2001 The Maastrichtian flora and dinosaurs of the Russian far east; *In*: Goodman, D.K., and Clarke, R.T. (eds.), *Proceedings of the IX International Palynological Congress, Houston, Texas, U.S.A., 1996*; American Association of Stratigraphic Palynologists Foundation, p. 139–148.

 ISBN 0–931871–06–9

PRIAMURYE (BLAGOVESHCHENSK AND KUNDUR LOCALITIES)

First notes about dinosaur discoveries in the region appeared in 1902, when A.Y. Gurov found bone fragments in the Beliye Kruchi locality, on the right bank of the Amur River. The duckbilled dinosaur *Mandshurosaurus amurensis* was first described from this section by A.N. Ryabinin (1930). Fossil vertebrates were further studied by A.K. Rozhdestvensky (1957), L.A. Nessov (Nessov, 1978, 1995; Nessov and Golovneva, 1990), and Yu.L. Bolotsky (Bolotsky, 1990; Bolotsky and Kurzanov, 1988; Bolotsky and Moiseenko, 1988; Markevich et al., 1994).

The richest dinosaur collection was excavated by Bolotsky from the correlative Blagoveschensk and Kundur localities in Priamurye. In that collection, the assemblage was dominated by bones and teeth of the hadrosaurid *Mandshurosaurus* cf. *amurensis* Riab. and the lambeosaurid *Amurosaurus riabinini* Bolotsky & Kursanov. Teeth of carnivorous dinosaurs, and the remains of crocodiles and turtles, were found as well (Table 1).

The most favorable place for paleontological study is extensive and nearly continuous outcrop near Kundur Village. A sufficient amount of well-preserved spores and pollen permits the definition of floral assemblages and the character of assemblage changes during deposition of this stratigraphic unit. Plant megafossils and a lacustrine fauna (conchostracans, ostracoda) from the lower part of the section extend our knowledge of the sedimentary basin paleoenvironment.

It should be noted that the well-known Tsagayan flora described by A.N. Kryshtofovich was obtained from the upper part of the Tsagayan Formation (Kryshtofovich and Baikovskaya, 1966). Later, Krassilov (1976) collected and studied the flora from this classic locality and from the upper middle part of this stratigraphic unit. He suggested the following structure of the "paleocatena" (a closely linked series, according to Krassilov, 1972):

1. water community;
2. coastal brush of *Carex, Phragmites, Arundo,* and uncertain grass;
3. marshy forest with *Taxodium* and *Nyssa;*
4. river valley forest with *Trochodendroides* and *Platanus;*
5. mixed mesophilous forest on mountain slopes with ginkgoaleans; conifers *Podocarpus, Araucarites, Androvettia, Metasequoia, Pinus;* dicots *Tiliaephyllum, Cyclocarya, Celtis;* various Menispermaceae and Leguminosae.

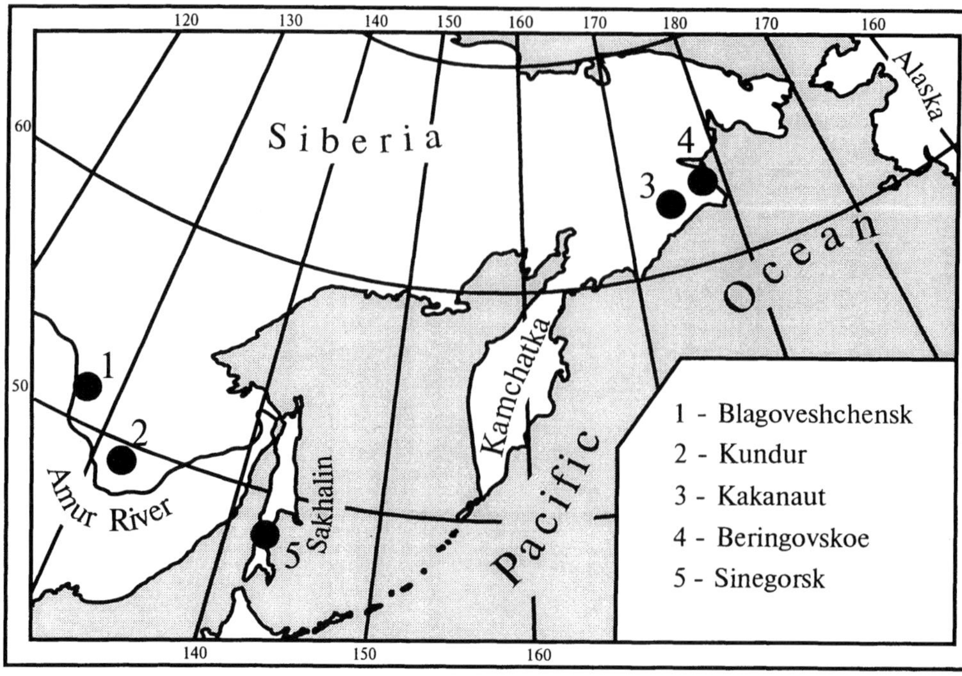

Text-Figure 1. Map of eastern Russia showing the recorded distribution of localities with fossil reptiles.

In the opinion of Krassilov, there was an increase in number of montane conifers and the angiosperm *Tiliaephyllum* during late Tsagayan time that indicates migration of these forms from hills down to lowlands. Thus, *Tiliaephyllum* became the dominant lowland plant. This event probably indicates a drop in temperature, resulting in movement of the paleocatena links down the mountain slopes.

Prevalence of leaves of dicots in the Tsagayan oryctocoenoses (according to I.A.Efremov, 1950, this is an assemblage of fossilized remains of animals or plants discoverd in a locality) correlates with the situation recently described by Wing et al. (1993) in the middle Maastrichtian deposits of Big Cedar Ridge, central Wyoming, USA. These authors point to the fact that dicots predominated in the river basin environments, while the *in situ* flora, buried under volcanic ash, is dominated by monocots and sporophytes, with the dicots comprising only 12% of the vegetative cover. This description provides an explanation for

TABLE 1. Fossil reptiles from the Maastrichtian of the Russian Far East.

FOSSIL REPTILES	Amur region		Magadan region		Sakhalin Island
	Blagoveshchensk locality	Kundur locality	Kakanaut locality	Beringovskoe locality	Sinegorsk
superorder Crocodilia	■	■			
family Dermatemydidae					
Mongolemys cf.*planicostatus*	■	■			
family Trionychidae					
Trionyx sp.	■	■			
family Tyrannosauridae	■	■			
family Ornithomimidae	■	■			
family Dromaeosauridae	■	■			
family Troodontidae	■	■	■		
family Titanosauridae	■				
family Hadrosauridae					
subfamily Hadrosauridae	■	■			
subfamily Lambeosaurinae	■	■	■		■
family Nodosauridae		■			
family Elasmosauridae				■	

TABLE 2. Angiosperm pollen from the Kundur locality.

Taxa \ System / Stage	CRETACEOUS		
	Upper Cretaceous		
	Maastrichtian		
	lower	middle	upper
Aquilapollenites subtilis	rare	abundant	rare
Kuprianipollis elegans	rare	rare	rare
K. santaloides	abundant	abundant	rare
Fothergilla gracilis	rare	rare	rare
Hamameliidaceae	rare	rare	abundant
Fagaceae	rare	rare	abundant
Yuglanspollenites	rare	rare	abundant
Cranwellia striata	abundant	rare	
C. aspera	abundant	rare	
C. sp.	abundant	rare	
Aquilapollenites conatus	abundant	abundant	
Orbiculapollis globosus	abundant	abundant	
O. lucidus	rare	rare	abundant
Proteacidites bellus	abundant	rare	
P. thalmanii	abundant	abundant	
Wodehouseia aspera	abundant	rare	
Aquilapollenites insignis	abundant	rare	
A. trialatus	abundant	rare	
A.striatus	abundant		
A. amygdaloides	abundant		
Parviprojectus dolium	abundant		
Mancicorpus anchoriforme	abundant		
Wodehouseia gracile	abundant		
Liliacidites variegatus		abundant	
Fibulapollis mirificus		abundant	
Aquilapollenites asper		abundant	
A. quadrilobus		abundant	
A. cruciformis		abundant	
A. amurensis		abundant	
Mancicorpus tenue		rare	
Wodehouseia spinata		abundant	
Triatriopollenites radiatostriatus		abundant	
Tricolpites gracilis		rare	
Erdmanipollenites albertensis		rare	
Ulmipollenites krempii	rare	abundant	abundant
U. tricostatus	rare	abundant	abundant
U. planaeriformis		abundant	abundant
Triatriopollenites aroboratus		rare	abundant
Triporopollenites plectosus		rare	abundant
Quercites sparsus	rare	rare	rare
Myricapollenites imperfectus			abundant
Comptonia sibirica			abundant
Tricolpites aff. *variegatus*			abundant
T. discus			abundant
T. coryloides			abundant

(grey) - rare fossil pollen

(black) - abundant fossil pollen

TABLE 3. Angiosperm pollen from the Beringovskoe and Kakanaut localities.

System / Taxa / Stage	CRETACEOUS		
	Upper Cretaceous		
	Maastrichtian		
	lower	middle	upper
Aquilapollenites subtilis		■	■
Kuprianipollis elegans	■	■	▒
K. santaloides	■	■	▒
Fothergilla gracilis			
Hamameliidaceae		■	■
Fagaceae	■	■	■
Yuglanspollenites	▒	■	■
Cranwellia striata	■	■	▒
C. aspera	▒	■	
C. sp.	▒	■	
Aquilapollenites conatus	■		
Orbiculapollis globosus	■	■	■
O. lucidus	■	■	■
Proteacidites bellus	■		
P. thalmanii	▒	■	
Wodehouseia aspera	▒	■	▒
Aquilapollenites insignis	■	■	
A. trialatus	■	■	
A.striatus	■		
A. amygdaloides	■		
Parviprojectus dolium	■	■	▒
Mancicorpus anchoriforme		▒	■
Wodehouseia gracile			
Liliacidites variegatus	▒	▒	■
Fibulapollis mirificus		■	
Aquilapollenites asper	▒	■	
A. quadrilobus	▒	■	
A. cruciformis	▒	■	
A. amurensis			
Mancicorpus tenue		■	
Wodehouseia spinata	■		
Triatriopollenites radiatostriatus			▒
Tricolpites gracilis	▒	▒	▒
Erdmanipollenites albertensis			
Ulmipollenites krempii	■	■	▒
U. tricostatus	■	■	▒
U. planaeriformis	▒	■	■
Triatriopollenites aroboratus	▒	▒	■
Triporopollenites plectosus		▒	■
Quercites sparsus	▒	▒	■
Myricapollenites imperfectus	▒	▒	■
Comptonia sibirica		▒	■
Tricolpites aff. *variegatus*			■
T. discus			■
T. coryloides		■	■

▒ - rare fossil pollen

■ - abundant fossil pollen

differences in plant megafossils (prevalence of dicots) and palynomorph assemblages (prevalence of sporophytes and gymnosperms). The latter reflects average vegetation of the region, whereas the fossil leaf assemblage was formed by vegetation of a river basin.

In the palynological assemblage in the lower part of the Tsagayan Formation, the thermophilic subtropical pteridophytes, gymnosperms (in general, forms close to Podocarpaceae, Araucariaceae, *Ginkgocycadophytus*), as well as flowering plants, predominate. Among the angiosperms, those close to Proteaceae, Santalaceae, Loranthaceae, Fagaceae (and some with unclear relationship to other plants, producing pollen of "unica" and "oculata" types), were abundant. Most probably, they formed rainforest-like vegetation during the late Campanian to early Maastrichtian when the climate was warm (subtropical) and humid (Markevich et al., 1994).

A palynological assemblage from the middle part of the section (with bone-bearing beds) differs from the lower one by an increased amount of deciduous warm-temperate species similar to Platanaceae, Ulmaceae, and Betulaceae. Plants close to Taxodiaceae, Cyatheaceae, and *Ginkgocycadophytus* are still numerous. The appearance of Gnetaceae and Cheirolepidiaceae, together with the sharp changes in species composition of the flowering plants with "unica" and "oculata" pollen, is evidence for the deterioration of climate—most likely a drop in temperature and humidity. Many plants such as the following thermophilic Campanian species have not been found: *Aquilapollenites striatus, A. conatus, A. amygdaloides, Parviprojectus dolium, Wodehousia asper, W. gracile, Mancicorpus solidus.* However these pollen prevailed in underlying deposits. *Aquilapollenites subtilis* N. Mtch., *A. insignis* N. Mtch., *A. cruciformis* N. Mtch., *A. quadrilobus* M. Mtch., *A. amurensis* Bratz., and *Wodehouseia spinata* Stanl., i.e., species typical of the Maastrichtian palynoflora, are dominant (Table 2). The diversity of the plants close to Proteaceae, Santalaceae, and among gymnosperms, the thermophilic plants close to Araucariaceae, Podocarpaceae, and *Ginkgocycadophytus,* decreased (Markevich et al., 1994). The palynoflora from these dinosaur-bearing beds is dated as the middle Maastrichtian. The palynological assemblage from the upper part of the section demonstrates a sharp reduction in "unica" and "oculata" pollen as well as pollen from other flowering species typical of the early and middle Maastrichtian. Plants close to the modern families Ulmaceae, Platanaceae, Betulaceae, Juglandaceae, and Fagaceae, as well as those bearing *Orbiculapollis* pollen and others, became dominant at that time. These data are evidence of a drop in temperature and a rise in humidity. Such a sharp change in the taxonomic composition is typical for the end of the Maastrichtian and early Danian stages in different regions of East Asia (Bratceva, 1965, 1969; Markevich, 1994). Thus, we have defined the successive changes of palynoflora diversity during the Maastrichtian.

THE KORYAK UPLAND (BERINGOVSKOE AND KAKANAUT LOCALITIES)

Within the Koryak Upland, two localities with reptile remains are known (Nessov and Golovneva, 1990). The first one is at Beringovskiy Village (Ugolnaya Bay, the upper Koryak subformation), where traces of crustaceans, teeth and bones of different fish, teeth and vertebrae of Elasmosauridae plesiosaurs, as well as abundant plant fossils, were found (Nessov et al., 1988). The Koryak floral assemblage is similar to the early Rarytkin from the middle unit of the Rarytkin Formation, differing only by a few species (Nessov and Golovneva, 1990; Golovneva and Herman, 1992; Herman, 1993). For the latter, *Nyssa* sp., *Celastrus septentrionalis* (Krysht.) Golovn., *Talowia* sp., *Arthollia rarytkensis* Golovn., *Haemanthophyllum cordatum* Golovn., *Vitis rarytkensis* Krysht., and *Quereuxia angulata* (Lesq.) Krysht. are typical.

Both the early and late Rarytkin assemblages are dominated by *Corylus anadyrensis* Budants., *Trochodendroides arctica* (Heer) Berry, *Metasequoia disticha* (Heer) Miki, *Glyptostrobus nordenskioldii* (Heer) R.W. Brown, and *Microconium beringianum* Golovn. (Nessov and Golovneva, 1990).

The taxonomic composition of the palynoflora from these beds is not diverse, but palynomorphs are well-preserved. The spores represent species related to the Cyatheaceae, Gleicheniaceae, and *Leiotriletes*. Among gymnosperms, the conifers, ginkgoaleans, cycadophytes, and a few plants close to Gnetaceae, are dominant. The angiosperms are much more diverse than the gymnosperms. They are commonly represented by species close to Platanaceae, Fagaceae, and Ericaceae, as well as species whose relation to the modern families is unclear (including *Aquillapollenites trialatus, A. cruciformis, A.* sp., *Parviprojectus dolium, Fibulapollis mirificus* and others). Taxonomic composition of the palynoflora is similar to that of the middle Tsagayan subformation from the Kundur dinosaur-bearing locality of the middle Maastrichtian age. This similarity is defined by the typical species of genus *Aquilapollenites,* which is geographically widespread, but occurs within a relatively narrow time interval.

The second locality of fossil reptiles is situated on the left bank of Kakanaut River, in the Pekulney Lake basin, Koryak Upland (the upper Kakanaut subformation). The remains of carnivorous and herbivorous dinosaurs, associated with the thin hollow bones probably belonging to some birds, were found there. The oryctocoenosis was formed by a mudflow or "quicksand" composed of volcanic ash (Nessov and Golovneva, 1990). Both the Kakanaut and Gornorechensk floral assemblages are similar, having common species and relatively small-leaved plants. Conifers are represented by numerous *"Cephalotaxopsis,"* while *Myrica heterophylla*, small-leaved *Trochodendroides, Celastrus* sp., as well as platanoids (two or three species) dominate among angiosperms. Cycadophyte *Encephalartopsis vassilevskajae* (Krassilov et al., 1990) forms the "leaf mat." Conifers *Araucarites, Metasequoia, Glyptostrobus,* as well as angiosperms *Corylus anadyrensis* Budants. and small-leaved birches, became more rare. *Pterospermites, Liriophyllum, Cissites, Tiliaephyllum* and the aquatic *Quereuxia angulata* (Lesq.) Krysht. are rare. This assemblage has much more *Ginkgo* and *Nilssonia* compared to the Gornorechensk assemblage (Golovneva and Herman, 1992).

In the assemblage from the bone-bearing beds of the Kakanaut locality, there are many spores and pollen close to Cyatheaceae, Polypodiaceae, Pinaceae, Podocarpaceae, *Ginkgocycadophytus*, and Taxodiaceae. Angiosperms are commonly represented by "unica" pollen, typical of the middle Maastrichtian palynoflora, and by plants close to Platanaceae, Fagaceae, Loranthaceae and Santalaceae (Table 3). Acritarchs, dinoflagellate cysts, and chitin fragments are abundant. This palynological assemblage is taxonomically comparable with that of dinosaur-bearing beds from

Ugolnaya Bay and Priamury. In our opinion, it makes sense to compare the taxonomic composition of assemblages from the fossiliferous bed and continental deposits of the lower and upper Rarytkin subformations and from Rarytkin Ridge, both of which contain abundant plant fossils (L.B. Golovneva's samples).

The palynoflora from the lower Rarytkin subformation (at the boundary with the Gornorechensk Formation) is characterized by relatively high diversity and predominance of flowering plants that is chiefly due to species with "unica" pollen. They are associated with the thermophilic plants closely related to the Santalaceae, Loranthaceae, Buxaceae, and Platanaceae, as well as to warm-temperate broad-leaved Ulmaceae. Among gymnosperms, Pinaceae and Taxodiaceae are common; the pteridophytes, however, are not so abundant. By taxonomic composition, this assemblage is very similar to those from the Campanian–lower Maastrichtian deposits of East Asia (Markevich, 1990, 1995; Markevich et al., 1994). Its age is most likely early Maastrichtian. However, the assemblage appears to be transitional between the Barykovski (Santonian–early Campanian) and the Gornorechensk floral stages (Herman, 1993; Nessov and Golovneva, 1990). Therefore, the palynological data appear to fill the gap for the late Campanian–early Maastrichtian interval.

The fossil flora from the upper part of lower Rarytkin subformation corresponds to the Gornorechensk stage of the Anadyr–Koryak region (Herman, 1993; Golovneva, 1994a,b). Typical representatives are angiosperms (genera *Peculnea, Trochodendroides, Celastrinites, Renea, Dyrana, Platanus, Viburnum, Quereuxia, Palaeotrapa, Corylus*). There are few conifers, represented by *Sequoia minuta, Cryptomerites, Taxites* and, less commonly, younger *Metasequoia* and *Glyptostrobus*. *Ginkgo* and *Nilssonia*, which could form the monodominant communities, were relatively abundant. Ferns are represented only by a single *Osmunda*. The floral assemblage is dated by L.B.Golovneva (1994 a, b) as middle Maastrichtian.

The palynoflora from the lower part of middle Rarytkin subformation is dominated by the flowering plants, mainly *Aquilapollenites asper, A. reductus, A. amygdaloides, A. insignis, A. cruciformis* with "unica" and "oculata" pollen, accompanied by different plants close to Santalaceae and Loranthaceae, as well as to Buxaceae, Menispermaceae, Ericaceae, Platanaceae, Fagaceae, and others. The age of the assemblage is considered to be the middle Maastrichtian because its taxonomic composition has similarity to the simultaneous assemblages of East Russia.

There are many conifers and angiosperms in the Rarytkin flora. *Corylus* and some species of *Trochodendroides, Metasequoia, Glyptostrobus* and *Microconium* dominate. The characteristic species are *Taxodium* sp., *Platanus raynoldsii* Newb., *Quercus groenlandica* Heer, *Arthollia* sp., *Rarytkinia* sp., *Celastrinites* sp., *Nyssa* sp., *Platimelis* sp., *Viburnum* sp., *Vitis* sp., *Quereuxia* sp., *Haemanthophyllum* sp. Rare specimens of *Ginkgo* sp. occur. Cycadophytes have not been found. Ferns are represented by *Onoclea* sp. and *Coniopteris* sp. (Golovneva, 1994a).

In the upper part of the Rarytkin Formation, we observe a sharp change in the taxonomic composition of the palynoflora. Flowering plants and ferns are more abundant and diverse. Commonly, monolete spores close to Polypodiaceae dominate together with *Leiotriletes* spp. Among gymnosperms, pollen close to Taxodiaceae and *Ginkgocycadophytus* is abundant. Flowering plants are chiefly represented by species close to the modern families Betulaceae, Juglandaceae, Fagaceae, Ericaceae. However, both the diversity and abundance of pollen of the "unica" and "oculata" types decrease in this interval. Taxonomic composition of the assemblage reflects the transitional stage of the Cretaceous palynoflora evolution from the late Maastrichtian to the early Danian in East Russia.

SOUTH SAKHALIN

In a locality near Sinegorsk City, South Sakhalin, an incomplete skeleton of *Nipponosaurus sachalinensis* (Nagao, 1936) was found in shallow marine deposits of the Ryugase Group. According to modern data, this finding belongs to the lower Krasnoyarka subformation (Nessov and Golovneva, 1990).

We examined spores and pollen from this subformation coeval to the bone-bearing beds of the Kawakami colliery (previously owned by the Mitsubishi Mining Company). The palynological assemblage from the lowermost Krasnoyarka Formation (unit I-II) containing a diverse and abundant fauna of ammonites, inocerams, bivalves, gastropods, and echinoids; it was comprehensively studied in the basins of the Naiba River (outcrop 0111/60) and Krasnoyarka River (outcrops 66, 67) (Bityutskaya et al., 1973; Budrin and Gromova, 1973; Kalishevich et al., 1981; Zonova, 1987; Markevich, 1995). In this assemblage, spores of pteridophytes, commonly Polypodiaceae, along with spores close to the Schizaeaceae and Gleicheniaceae, dominate. Among gymnosperms, pollen close to Pinaceae are abundant. Flowering plants are well-represented in number and diversity. The predominance of "unica" and "oculata" pollen is basically due to common representatives of the late Campanian and early Maastrichtian palynoflora characterized by *Aquilapollenites striatus, A. reductus, A. amygdaloides, A. insignis, Wodehouseia aspera, W. spinata*, and others. In addition, plants close to modern families Ulmaceae, Juglandaceae, Fagaceae, Buxaceae, Proteaceae, Santalaceae, and Loranthaceae were found. This palynoflora is dated as late Campanian–early Maastrichtian.

The next palynological assemblage characterizes the Krasnoyarka Formation (Krasnoyarka River Basin: outcrop 62, unit IV; outcrop 59, unit V), and contains a diverse fauna of inocerams, ammonites and brachiopods. *Leiotriletes* spp. (up to 16%) and pollen close to Pinaceae is abundant among the pteridophytes and gymnosperms, respectively. In the taxonomic composition of flowering plants (which comprise 25–30 % of the assemblage), the common Maastrichtian taxa with "unica" and "oculata" pollen decreased (Table 4). We suggest that this palynoflora correlates to the middle Maastrichtian in East Asia.

The palynological assemblage from the upper part of the lower Krasnoyarka subformation shows reduced amount of spores related to the Schizaeaceae and Gleicheniaceae. The diversity of the angiosperm pollen basically increases due to plants whose relation to the modern families is not determined, but their taxonomic composition changed greatly. The number of species typical of the Campanian–lower Maastrichtian were reduced. *Aquilapollenites asper, A. subtilis, A. conatus, A. quadrilobus, A. cruciferus* are common. Pollen of flowering plants, especially Proteaceae, Loranthaceae, Betulaceae and others, which are close to modern taxa, became more diverse.

Fossil spores and pollen from the mudstones in the upper part of the upper Krasnoyarka subformation were

TABLE 4. Angiosperm pollen from the dinosaur-bearing beds of Sakhalin Island.

System / Stage — Taxa	CRETACEOUS — Upper Cretaceous — Maastrichtian: lower	middle	upper
Aquilapollenites subtilis	rare	abundant	
Kuprianipollis elegans	abundant	abundant	
K. santaloides	abundant	rare	rare
Fothergilla gracilis		abundant	
Hamameliidaceae	rare	abundant	abundant
Fagaceae	rare	abundant	abundant
Yuglanspollenites	rare	abundant	abundant
Cranwellia striata	abundant	abundant	
C. aspera	abundant	abundant	
C. sp.	abundant	rare	rare
Aquilapollenites conatus			
Orbiculapollis globosus	rare	abundant	
O. lucidus	abundant	abundant	abundant
Proteacidites bellus	abundant	rare	rare
P. thalmanii	rare	rare	rare
Wodehouseia aspera	abundant	abundant	
Aquilapollenites insignis	abundant	rare	
A. trialatus			
A. striatus	abundant		
A. amygdaloides	abundant		
Parviprojectus dolium		abundant	
Mancicorpus anchoriforme		abundant	
Wodehouseia gracile		abundant	
Liliacidites variegatus		abundant	abundant
Fibulapollis mirificus		abundant	
Aquilapollenites asper		abundant	
A. quadrilobus		abundant	
A. cruciformis		abundant	
A. amurensis			
Mancicorpus tenue		abundant	
Wodehouseia spinata	rare	abundant	
Triatriopollenites radiatostriatus			
Tricolpites gracilis		abundant	
Erdmanipollenites albertensis			
Ulmipollenites krempii	abundant	abundant	abundant
U. tricostatus	abundant	abundant	abundant
U. planaeriformis	abundant	abundant	abundant
Triatriopollenites aroboratus		abundant	abundant
Triporopollenites plectosus			abundant
Quercites sparsus	rare	rare	
Myricapollenites imperfectus			abundant
Comptonia sibirica			abundant
Tricolpites aff. *variegatus*			rare
T. discus			rare
T. coryloides			rare

(grey) - rare fossil pollen

(black) - abundant fossil pollen

studied at outcrop 58, Razvedochny Creek in the Krasnoyarka River Basin. Angiosperms close to the Betulaceae, Ulmaceae, Juglandaceae, Ericaceae and porate pollen, i.e. broad-leaved, warm-temperate, climate-requiring species close to the modern ones, predominate. Plants common for the early–middle Maastrichtian, and producing pollen of the "unica" and "oculata" types, decreased in abundance. They are represented by *Orbiculapollis lucidius* Chlon., *O. globosus* Chlon., *Aquilapollenites asper*, *Pentapollenites normales*, *Mancicorpus tenue*, *Wodehouseia asper*, *W. spinata*. Rather diverse are pollen *Triatriopollenites aroboratus* Pfl., *T. plicoides*, *T.* sp., *T. confusus*, and *Anacolosidites insignis* Samoil., which are common to the late Maastrichtian–early Danian.

We suggest the unit where the dinosaur skeleton was found to be middle Maastrichtian in age, though there are some different opinions. Matsumoto and Obata (1979), and Matsukawa and Obata (1994), considered the age of the bone-bearing beds as late Santonian–early Campanian. They argued their statement by stratigraphic position of the horizon containing *Nipponosaurus*, which is situated in a bed 1500 m lower than the Paleogene basal beds; furthermore, it is stratigraphically lower than the *Inoceramus orientalis* zone. In addition, it should be noted that the ammonite *Pachydiscus subcompressus* Mats. and the foraminifer assemblage, both characteristic of the Maastrichtian, were found in the Krasnoyarka Formation (Kalishevich et al., 1981; Zonova and Turenko, 1986).

CONCLUSIONS

Taxonomic composition of palynofloras from the Maastrichtian dinosaur-bearing deposits of Priamurye, the Koryak Upland, and South Sakhalin show close similarities to each other and allow us to consider them to be coeval (Tables 2, 3, 4). Thus, the age of the lower part of Kundur locality, upper Koryak subformation, lower Rarytkin subformation and lower Krasnoyarka subformation is early Maastrichtian. More than 50 % of taxa from the above-mentioned palynological assemblages are common.

Comparison of the taxonomic composition of fossil spores and pollen from the middle part of Kundur locality (beds with dinosaur remains) with those of the upper Kakanaut subformation (fossils of dinosaurs, birds, and plants), the lower part of the middle Rarytkin subformation, and the upper part of the lower Krasnoyarka subformation (where the lambeosaurid specimen was collected), suggests that their age is the middle Maastrichtian.

In the same way, by taxonomic resemblance, close similarity between palynofloras of the upper part of the Kundur locality, the upper Rarytkin, and the Krasnoyarka has been established. The age of these stratigraphic units is late Maastrichtian–early Danian.

It is interesting that all known dinosaur-bearing strata in the Russian Far East are of Maastrichtian age (more exactly middle Maastrichtian), despite the different environments in which they occur. The Priamurye taphocoenosis was formed in the wide river valley under the influence of mudflows from the paleo–Hingang. The Kakanaut taphocoenosis was covered with "quicksand" in a river valley between volcanoes. *Nipponosaurus sachalinensis* on South Sakhalin was buried in the shallow marine deposits near the coast. In marine deposits of the Koryak Formation, teeth and vertebrae of Elasmosauridae were found also (Nessov and Golovneva, 1990). Nessov suggested that the hadrosaurs had intrazonal distribution. Matsukawa and Obata (1994) pointed out different sedimentary environments (of lake, river, eolian and marine genesis) in which this group of animals was distributed within the Asian continent. However, in spite of their adaptability to different conditions and environments, there have been no dinosaurs in the oryctocoenoses in the Far East since the late Maastrichtian. V.A. Krassilov suggested that the dinosaurs had been confined to non-forest biotopes and that their extinction had resulted from ecosystem reconstruction (Krassilov, 1981, 1985). Our data suggest the existence of vast savanna-like valleys with oasis vegetation along the banks of lakes and rivers during the middle Maastrichtian. According to the calculations of Beland and Russell (1978), who had worked at the Canadian burial places, the dinosaur biomass consisted of 2 tons per hectare. It has long been known that the Campanian–Maastrichtian dinosaur faunas from Asia and the western part of North America are similar (Nessov, 1995, and other). The fauna existed on the ancient continent Asiamerica (this land included eastern part of Asia and western part of North America [Nessov, 1995]) whose boundaries approximately coincided with the boundaries of the *Aquilapollenites* floral province. Allowing for resemblance of dinosaur faunas and floras of North America and East Asia, we assume that the calculations of Beland and Russell (1978) can be extended to eastern Asia paleoenvironments. Obviously such a large number of reptiles required luxuriant and abundant vegetation that probably regenerated well. Thus, the expansion of forest within these biotopes most likely began after the late Maastrichtian. This could conceivably lead to the local extinction of dinosaurs because the newly formed vegetation could not provide a forage reserve sufficient for the survival of the populations. The above-mentioned dinosaur-bearing beds of diverse origin (river, shallow marine deposits) are only of middle Maastrichtian age. In younger sediments, the fossil reptiles are absent. It can therefore be assumed that the extinction of these dinosaurs took place during the late Maastrichtian.

Our conclusion, based on study of the plant fossils, is in a contrast with the famous hypothesis of Alvarez et al. (1980) concerning the sudden extinction of dinosaurs as the result of an extraterrestrial impact. Unfortunately, the above mentioned localities have not been investigated for the iridium anomaly, but it seems likely that a detailed study of the phytooryctocoenosis along the sections, together with the sedimentological and paleogeographic data, could help in reconstructing the paleoenviroments and understanding the causes of dinosaur extinction. To the point, a fossil of turtle *Trionyx* was found in the Priamurye taphocoenoses (Table 1). This genus still exists in North America, Asia and Africa (Carrol, 1993). The reptile passed through the Cretaceous/Tertiary boundary probably because it was not closely connected with the terrestrial biotopes and ecosystems. The same reason may be suggested for the modern crocodiles whose ancestors lived in the paleo–Priamurye valleys. The dinosaurs strongly depended on the land vegetation and died out as a result of the radical reconstruction of the continental ecosystems. This is true for hadrosaurs too. These reptiles, though they lived in water (Ostrom, 1964), ate shoots, cones, and leaves of the conifers that grew near the water (Krausel, 1922).

ACKNOWLEDGMENTS

The authors thank Prof. Valentine A. Krassilov (Paleontological Institute of Russian Academy of Sciences, Moscow) for his helpful suggestions during the early stages of this study, Yury L. Bolotsky (Amur Scientific Center of Russian Academy of Sciences, Blagoveshchensk) for data on fossil reptiles of Russian Far East, and Drs. Lev A. Nessov (University of Sankt-Petersburg) and Lena B. Golovneva (Komarov Botanical Institute, Sankt-Petersburg) for their comments. We are deeply indebted to Dr. D.J.Nichols (US Geological Survey, Denver) for his critical review of the manuscript.

Our work was supported by the Russian Foundation for Basic Research (grant 98-05-65494) and partly by the International Science Foundation of George Soros.

References Cited

ALVAREZ, L.W., ALVAREZ, W., ASARO, F., and MICHEL, H.V.
1980 Extraterrestrial cause for the Cretaceous–Tertiary extinction. *Science*, 208: 1095–1108.

BELAND, P., and RUSSELL, D.A.
1978 Paleoecology of Dinosaur Provincial Park (Cretaceous), Alberta, interpreted from the distribution of articulated vertebrate remains. *Canadian Journal of Earth Sciences*, 15: 1012–1024.

BITYUTSKAYA, P.I., BRYZGALOVA, E.G., VERBITSKAYA, Z.I., BUDRIN, V.S., and MARKEVICH, V.S.
1973 Significance of palynological method for the stratigraphic dividing of the Cretaceous deposits in the northwest Pacific biostratigraphic area. *In*: Boyzova E.P., (ed.), *Palynologicheskiy metod v stratigraphii (Stratigraphy and palynological method)*. Leningrad. Trudy VSEGEI 195: 73–121 (In Russian).

BOLOTSKY, Y.L.
1990 The Blagoveshchensk locality of Cretaceous dinosaurs. *In*: Krassilov, V.A. (ed.), *Kontinental'niy mel SSSR (The continental Cretaceous of the USSR)*. Far Eastern Branch of the USSR. Academy of Sciences, Vladivostok, pp. 109–113 (In Russian).

BOLOTSKY, Y.L., and KURZANOV, S.M.
1988 The hadrosaurs of Priamurye. *In*: Moiseenko, V.G., and Zagruzina, I.A. (eds), *Geologiya Tikhookeanskogo obramleniya (Geology of Pacific frame)*. Part 3. AmurKNII DVO AN USSR, Blagoveshchensk, pp. 94–107 (In Russian).

BOLOTSKY, Y.L., and MOISEENKO, V.G.
1988 On the dinosaurs of the Amur river region. *Amurskaya Pravda*, Blagoveshchensk, 34 pgs. (In Russian).

BRATCEVA, G.M.
1965 Pollen and spores of the Maastrichtian deposits of Far East. *Nauka*, Moscow, 40 pgs. (In Russian).
1966 Palynological research of the Upper Cretaceous and Paleogene of Far East. *Nauka*, Moscow, 56 pgs. (In Russian).

BUDRIN, V.S., and GROMOVA, N.S.
1973 On the age of transitional deposits of the Cretaceous and Paleogene on Sakhalin. *In*: Boyzova, E.P. (ed.), *Palynologicheskiy metod v stratigraphii (Stratigraphy and palynological method)*. Leningrad. Trudy VSEGEI 195: 135–144 (In Russian).

CARROL, R.L.
1993 Vertebrate paleontology and evolution. *Mir*, Moscow, 2, 280 pgs. (In Russian).

EFREMOV, I.A.
1950 Taphonomy and geologic record. *Academy of Sciences of USSR*, Moscow–Leningrad, 177 pgs. (In Russian).

GOLOVNEVA, L.B.
1994a The Cretaceous/Paleogene boundary-related flora on the North-West of Russia. *Stratigraphiya. Geologicheskaya korrelyatsiya (Stratigraphy. Geological correlation)*, 2: 64–70 (In Russian).
1994b The flora of the Maastrichtian–Danian deposits of the Koryak Upland, Northeast Russia. *Cretaceous Research*, 15: 89–100.

GOLOVNEVA, L.B., and HERMAN, A.B.
1992 New data on composition and age of Koryak Formation flora (Koryak Upland, North East of Russia). *Botanicheskiy Zhurnal (Botanical Journal)*, 77: 60–71 (In Russian).

HERMAN, A.B.
1993 Evolutionary stages and recurrence of Late Cretaceous flora of the Anadyr-Koryak subarea. *Stratigraphiya. Geologicheskaya korrelyatsiya (Stratigraphy. Geological correlation)*, 1: 87–96 (In Russian).

KALISHEVICH, T.G., ZAKLINSKAYA, E.D., and SEROVA, M.Y.
1981 Development of the Pacific belt organic Nature at Mesozoic/Cenozoic boundary. *Nauka*, Moscow, 163 pgs. (In Russian).

KAUFFMAN, E.J.
1986 Extinction structure of marine biotas in the Cretaceous. In: Berggren, W.A., Van Couvering, J.A.(eds), Catastrophes and Earth History. The new Uniformitarianism. Mir, Moscow, p.156–254 (Translation from the English).

KRASSILOV, V.A.
1972 Paleoecology of terrestrial plants. *Far Eastern Scientific Center of Academy of Sciences of the USSR*, Vladivostok, 208 pgs. (In Russian).
1976 The Tsagayan flora of the Priamurye area. *Nauka*, Moscow, 92 pgs. (In Russian).
1981 Changes of Mesozoic vegetation and the extinction of dinosaurs. *Palaeogeography, Palaeoclimatology, Palaeoecology*, 34: 207–224.
1985 The Cretaceous. Earth crust and biosphere evolution. *Nauka*, Moscow, 239 pgs. (In Russian).

KRASSILOV, V.A., GOLOVNEVA, L.B., and NESSOV, L.A.
1990 Cycadophytes from the Late Cretaceous dinosaur locality in the North Koryakiya. *In*: Krassilov, V.A. (ed.), *Kontinental'niy mel SSSR (The continental Cretaceous of the U. S. S. R.)*. Far East Branch of the U.S.S.R. Academy of Sciences, Vladivostok, pp. 213–215 (In Russian).

KRAUSEL, R.
1922 Die Nahrung von *Trachodon*. *Palaontologische Zeitschrift*, 4, S. 80.

KRYSHTOFOVICH, A.N., and BAIKOVSKAYA, T.N.
1966 The Upper Cretaceous Tsagayan flora from Amur region. *In*: Baikovskaya, T.N. (ed.), A.N.Kryshtofovich. Selected publications. *Academy of Sciences of USSR*, Moscow-Leningrad, Part 3: 184–320 (In Russian).

MARKEVICH, V.S.
1990 Palynostratigraphy of Cretaceous deposits of the Northwest of the USSR. *In*: Krassilov, V.A. (ed.), *Kontinental'niy mel SSSR (The continental Cretaceous of the USSR)*. Far East Branch of the USSR Academy of Sciences, Vladivostok, pp. 132–143. (In Russian).

MARKEVICH, V.S.
1994 Palynological zonation of the continental Cretaceous and Lower Tertiary of eastern Russia. *Cretaceous Research*, 15: 165–177.

MARKEVICH, V.S.
1995 The Cretaceous palynoflora of North of East Asia. *Dalnauka*, Vladivostok, 172 pgs. (In Russian)

MARKEVICH, V.S., BOLOTSKY, Y.L., and BUGDAEVA, E.V.
1994 The Kundur dinosaur-bearing locality in the Priamurye (Amur River region). *Tikhookeanskaya geologiya (Pacific geology)*, 6: 96–107. (In Russian).

MATSUKAWA, M., and OBATA, I.
1994 Dinosaurs and sedimentary environments in the Japanese Cretaceous: a contribution to dinosaur facies in Asia based on molluscan palaeontology and stratigraphy. *Cretaceous Research*, 15: 101–125.

MATSUMOTO, T., and OBATA, I.
1979 Evaluation of ammonites and other fossils from the Cretaceous of Japan for inter-regional correlation. *Kaseki*, 29: 43–58. (In Japanese).

MENNER, V.V.
1962 Biostratigraphic principles of marine, lagoon and continental suites correlation. *USSR Academy of Sciences*, Moscow, 375 pgs. (In Russian).

NAGAO, T.
1936 *Nipponosaurus sachalinensis* a new genus and species of trachodont dinosaur from Japanese Saghalien. *Journal of the Faculty of Science, Hokkaido University*, 4: 187–220.

NESSOV, L.A.
1978 On the turtle family Dermatemydidae from the Cretaceous deposits of the Amur River basin. *In*: Bromley, G.F. (ed.), *Herpetofauna Dal'nego Vostoka I Sibiri (The Far East and Siberia herpetofauna)*. Far East Scientific Center, USSR Academy of Sciences, Vladivostok, pp. 21–22 (In Russian).
1995 Dinosaurs of Northern Eurasia: new data about assemblages, ecology and palaeobiogeography. *University of Sankt-Petersburg*, Sankt-Petersburg, 156 pgs. (In Russian).

NESSOV, L.A., and GOLOVNEVA, L.B.
1990 Stages of flora and vertebrate fauna development and climatic conditions in the Late Senonian on the northeast of Koryak Upland. *In*: Krassilov, V.A. (ed.), *Kontinental'niy mel SSSR (The continental Cretaceous of the USSR)*. Far East Branch of the USSR Academy of Sciences, Vladivostok, pp. 191–212 (In Russian).

NESSOV, L.A., IVANOV, A.O., and KHOZATSIY, L.I.
1988 On the find of ichthyosaur remains in USSR and problems of changes of fauna in the Middle Cretaceous. *Vestnik LGU. Biologiya (Leningrad State University (LSU) bulletin. Biology)*, 7: 15–25 (In Russian).

OSTROM, J.H.
1964 A reconsideration of the paleoecology of hadrosaurian dinosaurs. *American Journal of Sciences*, 262: 975–997.

POKHIALAYNEN, V.P.
1980 A new Maastrichrian inoceramids from Koryak Upland. *In*: Zopanov, O.Kh. (ed.), *Materialy po geologii i poleznym iscopaemym Severo-Vostoka SSSR (Materials on geology and mineral deposits of North-East of U.S.S.R)*. Magadan Book Press, Magadan, pp. 17–21. (In Russian).

POKHIALAYNEN, V.P., and BYALOBZESKY, S.G.
1984 New data on biostratigraphy of upper Maastrichtian of eastern Koryak Upland. *In*: Bely, V.Ph. (ed.), *Paleontologiya i biostratigraphiya Korykskogo nagorya (Paleontology and biostratigraphy of Koryak Upland)*. SVKNII DVNC AN USSR, Magadan, p.64–74. (In Russian).

ROZDESTVENSKIY, A.K.
1957 On the Upper Cretaceous dinosaur locality on the Amur River. *Vertebrata Palasiatica*, 1: 129–149 (In Russian).

RYABININ, A.N.
1930 *Mandschurosaurus amurensis* nov. gen. et nov. sp., the Upper Cretaceous dinosaur from the Amur River. *Russkoye paleontologicheskoye obschestvo (Russian Paleontological Society)*, Monograph II, Leningrad, 36 pgs. (In Russian).

WING, S.L., HICKEY, L.J., and SWISHER, C.C.
1993 Implications of an exceptional fossil flora for Late Cretaceous vegetation. *Nature*, 363: 342–344.

ZONOVA, T.D.
1987 The biostratigraphical analysis. *In*: Poyarkova, Z.N. (ed.), *Opornyi razrez melovykh otlozhenii Sakhalina (Naibinskyi razrez) (Key section of Sakhalin island Cretaceous deposits (the Naiba section)*. Nauka, Leningrad, pp. 61–103 (In Russian).

ZONOVA, T.D. and TURENKO, T.V.
1986 New data for the unified scheme of the Sakhalin island Cretaceous. *In*: Yanshin, A.L., and Dagis, A.S. (eds.), *Biostratigraphiya mezozoya Sibiri i Dal'nego Vostoka (Mesozoic Biostratigraphy of the Siberia and Far East)*. Nauka, Novosibirsk, pp. 159–171 (In Russian).

Authors' addresses

VALENTINA S. MARKEVICH
EUGENIA V. BUGDAEVA
Institute of Biology and Pedology
Far East Branch of Russian Academy of Sciences
Vladivostok, 690022
Russia

DINOFLAGELLATE CYSTS AND PALYNOFACIES VARIATIONS ACROSS THE CRETACEOUS–TERTIARY BOUNDARY IN THE UM SOHRYNGKEW RIVER SECTION, MEGHALAYA, INDIA

Arun Kumar, P. N. Kapoor, and J. Sharma

Abstract

Ten outcrop samples within a range of 80 cm above and 70 cm below the Iridium-rich limonitic (Ir) layer marking the Cretaceous–Tertiary boundary in the Um Sohryngkew River section, Meghalaya, in northeastern India, were studied to show the variations in dinoflagellate cyst assemblages and palynofacies. Three new dinoflagellate cyst biozones are defined. These include the late Maastrichtian *Dinogymnium* spp. Partial Range Zone and *Dinogymnium* spp. and *Selenopemphix maastrichta* Concurrent Range Zone, and the *Damassadinium californicum* Zone of Danian age. As observed elsewhere in the world, most of the late Maastrichtian species of dinoflagellate cysts occur above the Cretaceous–Tertiary boundary (KTB), although, dinoflagellate cyst population and species diversity is drastically reduced just above the Ir layer. The distribution of various organic matter constituents show a drastic change across this boundary and palynofacies change from predominantly sapropelic below the Ir layer to predominantly humic with marine influence just above it. Palynofacies changes have resulted due to a fall in the sea level of about 70 to 80 m. This sea level fall is estimated to have begun about 8.8 ka before the deposition of the Ir layer. This regressive phase has been observed globally at this boundary and is marked in the present section by both the relative and the absolute abundance of *Spiniferites* spp. and *Achomosphaera* spp. Late Maastrichtian sediments were deposited in water depths greater than 80 m and during the Cretaceous–Tertiary transition sea level fell to almost 10 to 20 m, or possibly less followed by gradual rise of sea level in the region.

INTRODUCTION

Jain et al. (1975) published on the stratigraphic distribution of dinoflagellate cysts across the KTB from Cherrapunji area of Shillong plateau of northeastern India. Misra et al. (1994) and Raju et al. (1995) used planktic foraminifera, dinoflagellate cysts, and calcareous nannoplankton to determine the location of KTB in subsurface well sections in the Krishna–Godavari Basin, on the east coast of peninsular India, and estimated the duration of the hiatus at the KTB in different well sections. Raju et al. (1996) published data on the duration of KTB hiatuses in large number of drill holes in this basin and produced a paleogeographic map for the late Maastrichtian and early Danian period.

A complete marine sedimentary section across the KTB is exposed in the Um Sohryngkew river section in Meghalaya, northeast India (Bhandari et al., 1994) (Text-Figure 1). A 1.5 cm thick Iridium (Ir) rich limonitic layer marks the KTB in this section (Text-Figure 2). Kumar et al. (1993) described a new dinoflagellate cyst species *Selenopemphix maastrichta* from the samples just below the KTB Ir layer in this section. This species is stratigraphically important since its extinction marks the top of the Cretaceous in this section.

The objective of this study is to demonstrate and explain changes in organic matter types, palynofacies, and dinoflagellate cyst assemblages across the KTB within a narrow range of only 70 cm below and 80 cm above the Ir layer. These events are then related to the global regression phenomenon observed during the KTB.

The known features of dinoflagellate cyst distribution across the KTB are: (1) Dinoflagellate cysts, unlike other oceanic phytoplankton, do not show accelerated rates of extinction at the KTB; (2) changes in the composition of the dinoflagellate cysts assemblages are most evident just below and above the KTB; (3) the last appearance datum (LAD) of *Dinogymnium* spp. and *Palynodinium grallator* marks the terminal Cretaceous, whereas first appearance datum (FAD) of *Damassadinium californicum* marks the base of the Danian; (4) the KTB is marked by a sudden global regression of the sea (this event is marked by increase in both absolute and relative abundance of *Spiniferites* spp. and *Achomosphaera* spp.); and (5) high diversity of dinoflagellate cysts is observed during transgressive phases in the late Maastrichtian and again in middle to late Danian.

STRATIGRAPHY

The Um Sohryngkew River cuts across Maastrichtian/Danian marine clastic sediments of the Mahadeo and Langpar formations in the vicinity of the village of Therria (Text-Figure 1). The stratigraphic section, thickness and corresponding lithologies are given in Table 1. The KTB defined on the basis of the Ir layer and planktic foraminifera is 10 m below the contact between the Mahadeo and Langpar formations (Pandey, 1981).

MATERIAL AND METHODS

The present study is based on ten samples (J-20 through J-29) collected in 1985 by Dr. J. Pandey of ONGC Ltd., India, from 70 cm below and 80 cm above the Ir layer (Text-Figure 2). Standard palynological techniques were used in maceration of samples and four slides of each sample were pre-

Kumar, Arun, Kapoor, P.N., and Sharma, J.
2001 Dinoflagellate cysts and palynofacies variations across the Cretaceous–Tertiary boundary in the Um Sohryngkew River section, Meghalaya, India; *In*: Goodman, D.K., and Clarke, R.T. (eds.), *Proceedings of the IX International Palynological Congress, Houston, Texas, U.S.A., 1996*; American Association of Stratigraphic Palynologists Foundation, p. 149–160.

ISBN 0–931871–06–9

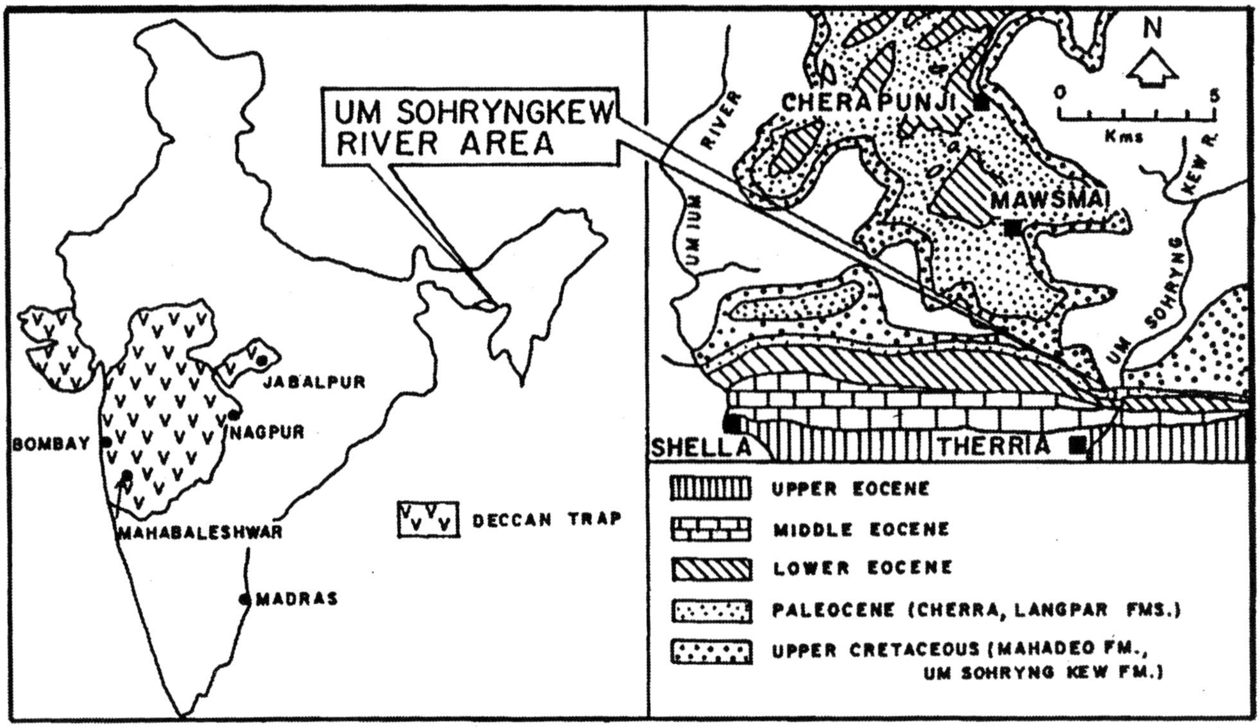

Text-Figure 1. Map of India showing the location and geology of the study area. (Modified after Pandey, 1981).

pared for the study of dinoflagellate cysts. In addition, two more slides were prepared for organic maturation studies and these were mounted before humification and sieving. The results are based on the examination of all six slides for each sample. The data on dinoflagellate cysts are presented in Text-Figures 3 and 4. The results of palynofacies studies are given in Text-Figure 5 and Table 2.

RESULTS

Text-Figure 3 shows the temporal variation in the percentages of chorate cysts, proximate cysts, and other palynomorphs; marine (dinoflagellate cysts and acritarchs) and terrestrial (pollen, spores) palynomorphs; and the absolute abundances of dinoflagellate cysts. All curves show similar trends. Samples J-20 through J-24 show higher percentages of both proximate and chorate dinoflagellate cysts than samples J-25 through J-27. The abundance of dinoflagellate cysts in samples J-20 through J-24 is high, exceeding over 200 individual specimens (all counts are only from two slides per sample). Sample J-25 (2 cm below the Ir layer) and J-26 (Ir layer) show a drastic reduction in the percentage of dinoflagellate cysts and a corresponding increase in terrestrial palynomorphs. The count of dinoflagellate cyst specimens is 40 and those present were poorly preserved and mostly unidentifiable. The reduction in dinoflagellate species diversity occurs just above the Iridium layer (only 3 species), but gradually recovers to higher diversity (7 species) 40 cm above the Ir layer. Sample J-27 (6 cm above the Ir layer) exhibits an increase in marine ele-

PLATE 1

All figures x 650 unless otherwise specified; coordinates are in parentheses.

1 *Cyclapophysis monmouthensis* Benson 1979. J23-1, (95x30.5), x900.

2 *Dinogymnium acuminatum* Evitt et al. 1967. J21-1, (106.4x38.6).

3 *Spiniferites ramosus* subsp. *ramosus* (Ehrenberg 1838) Davey & Williams 1966. J23-1, (107x69.2).

4 *Cerodinium speciosum* subsp. *elongatum* (Mao Shaozhi & Norris 1988) Lentin & Williams 1989. J21-3, (104.4x33.4), x 900.

5 *Cerodinium speciosum* (Alberti 1959) Lentin & Williams 1987. J20-1, (96.2x30.2).

6a, b *Andalusiella* sp. J20-1, (94.7x32).

7 *Andalusiella* sp. J20-1, (96.6x33).

8 *Cerodinium speciosum* subsp. *elongatum* (Mao Shaozhi & Norris 1988) Lentin & Williams 1989. J20-2, (111x53.4).

9 *Palaeocystodinium golzowense* Alberti 1961. J23-3, (95x71), x900.

10 *Andalusiella polymorpha* (Malloy 1972) Lentin & Williams 1977. J23-2,(96x61), x900.

11 *Andalusiella polymorpha* (Malloy 1972) Lentin & Williams 1977. J24-1,
(107.7x66.1), x900.

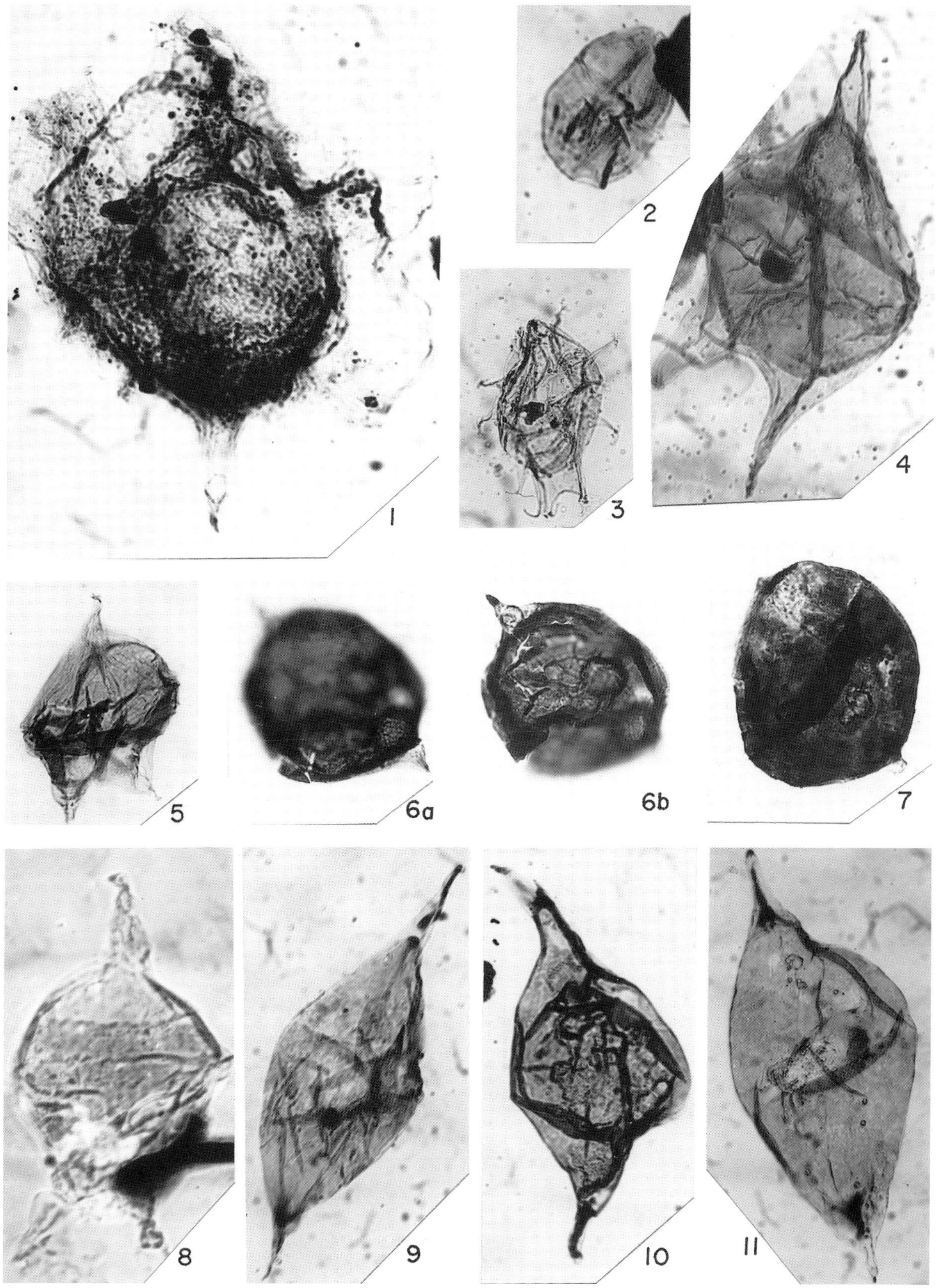
1
2
3
4
5
6a
6b
7
8
9
10
11

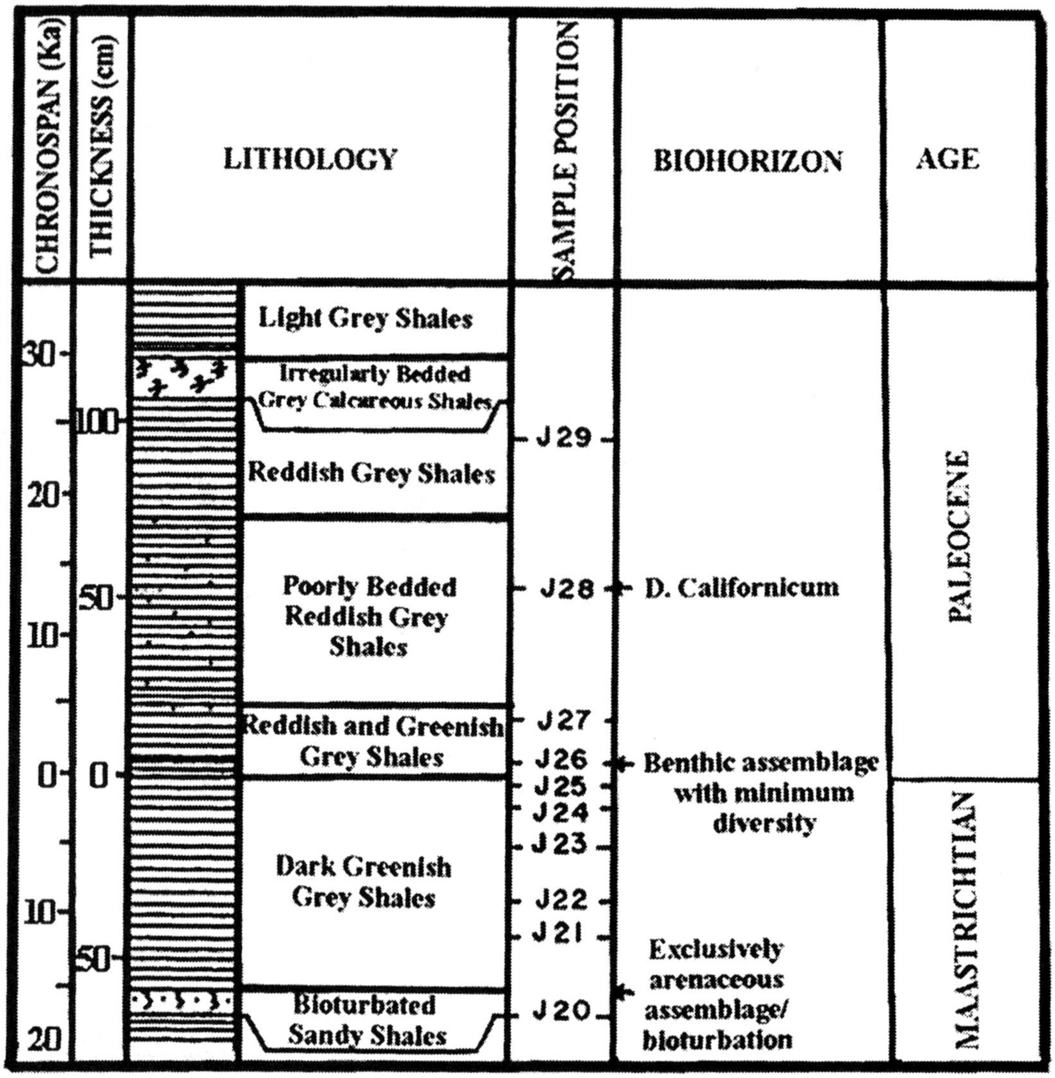

Text-Figure 2. Um Sohryngkew river section, Meghalaya showing lithostratigraphy, sample locations, biohorizons and age data (Modified after Pandey, 1981)

TABLE 1. Lithology, thickness, and age of the Mahadeo to Langpar Formation in Meghalaya (after Pandey, 1981).

FORMATION	MEMBER (Thickness m)	LITHOLOGY	AGE
LANGPAR	Upper (5.4)	Brownish gray silty shale with coaly matter	Late Danian to Thanetian
	Middle (15.1)	Brownish gray shale	
	Lower (28.5)	Dark gray carbonaceous shale and silty shale	
Unconformity/Hiatus			
MAHADEO	Upper (21)	Light to dark gray, greenish gray shale, greenish gray fine grained calcareous sandstone and thin sandy limestone	Early Danian and late Maastrichtian
	Lower (28)	Fine to medium grained greenish and brownish sandstone and sandy shale	Early Maastrichtian

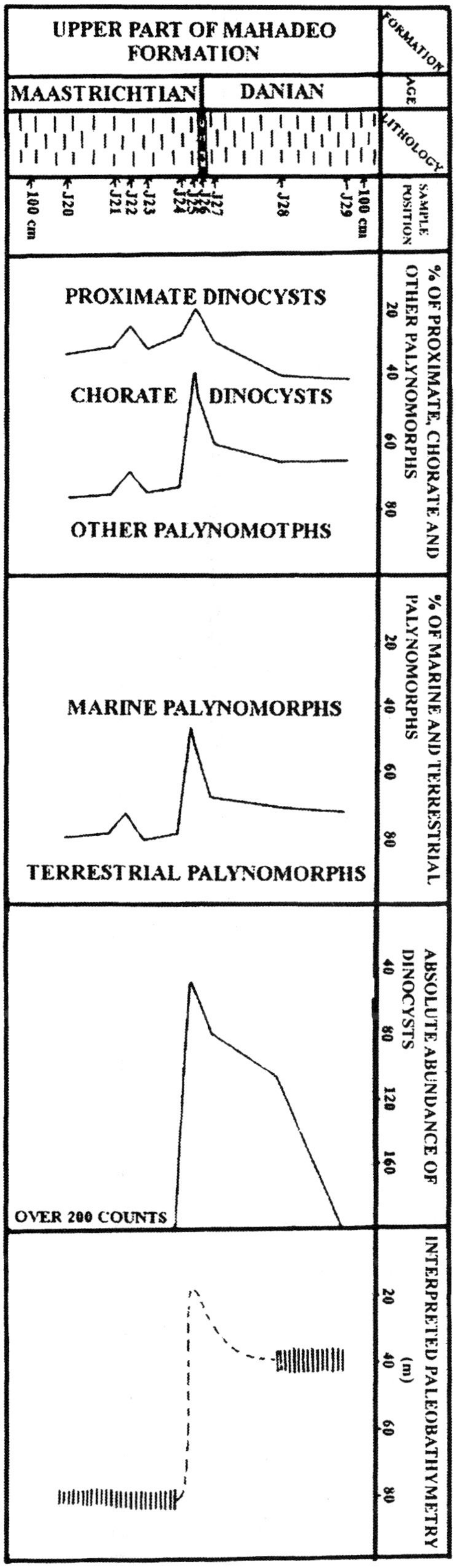

Text-Figure 3. Relative percentages of different groups of palynomorphs and interpreted paleobathymetry of the KTB interval in Meghalaya.

ments of the palynomorph assemblage, especially dinoflagellate cysts which increase to 80. Samples J-28 and J-29 also show an increase in number of dinoflagellates reaching 200 specimens in sample J-29.

The reduction in dinocysts in samples J-25 and J-26 are related to a sudden fall in sea level during the KTB transition. It is evident that the fall in sea level began prior to the geochemical KTB (i.e. the Ir layer). Samples J-20 through J-24 have high percentages of chorate dinoflagellate cysts indicating middle to outer shelf environments.

Bhandari et al. (1994) suggested a value of 0.45±0.02 cm ka^{-1} as the rate of sedimentation for Maastrichtian to Danian aged sediments in this section. This value can also be approximated to be 2.2 ka per cm. Since the fall in sea level is documented in the top 4 cm below the Ir layer (samples J-24 to J-26), it is suggested that the regression of the sea commenced approximately 8.8 ka before the KTB and the sudden drop in the sea level observed between samples J-25 and J-26 (2 cm apart) represents a time interval of 4.4 ka. According to Bhandari et al. (1994) the Ir layer is a sudden injection of clastic amorphous material devoid of common minerals (like quartz, mica, calcite and felspars) in an otherwise common sedimentation regime. Samples J-27 through J-29 represent the gradual transgression of the sea during early Danian time.

Marine palynomorph assemblages can be used to provide an estimate of relative paleobathymetry. Foraminiferal data is much better for making paleobathymetric interpretations. The presence of keeled *Globotruncana* (*G. rosetta* Carsey 1926, *G. elevata* Brotzen 1934, *G. stuarti* De Lapparent 1918 and *G. stuartiformis* Dalbiez 1955) in samples below J-25 and J-26 (Pandey, 1981) indicate a paleobathymetry of >80 m (Hart, 1980). The presence of small planktic foraminifera *Parvularugoglobigerina* (*Globigerina*) *eugubina* Luterbacher and Premoli Silva 1964, in samples J-27 through J-29 (Pandey and Ravindran, 1988) indicate evolution of a new foraminiferal community on the inner shelf at the beginning of a new transgressive phase (Hart, 1980), in water depths between between 40–50 m or even less. In this section we indicate a fall of sea level on the order of 70 to 80 m during the KTB regression from 80 to 100 m depth of water to almost 10 to 20 m at the KTB. Haq et al. (1987) have shown a fall of sea level on the order of 90 m towards the end of Maastrichtian. The latest Cretaceous regression appears to be widespread in the east coast basins of peninsular India where the sea regressed laterally over a distance of 50 km (Raju and Ravindran, 1990).

The increase in abundance of *Spiniferites* spp. and *Achomosphaera* spp. in the samples near the KTB represents the regressive phase that has also been recorded in various parts of the world (Davies et al., 1982; Firth, 1993). Edet and Nyong (1993) recorded >80 % dinoflagellate cysts (mainly *Spiniferites* spp.) representing inner neritic environments during latest Maastrichtian time in Nigeria. High relative abundances of *Spiniferites* spp. have been demonstrated to be controlled by an inshore-offshore gradient (Wall et al. 1977; Harland, 1982) with the largest concentrations of *Spiniferites* spp. observed in inner shelf environments (Morzadec-Kerfourn, 1979). In the present study it was noted that the percentages of *Spiniferites* spp./*Achomosphaera* spp. in the total dinoflagellate cyst assemblage was between 10% and 20% with the exception of samples J-25, J-26 and J-27 where this percentage was as high as 60 % to 70 %, indicating a regressive phase during the KTB interval.

DINOFLAGELLATE CYST BIOZONATION IN THE UM SOHRYNGKEW RIVER SECTION

The following biozones are proposed on the stratigraphic ranges of twelve species (Text-Figure 4).

Dinogymnium spp. Partial Range Zone

Definition. Sediments containing specimens of *Dinogymnium acuminatum* Evitt et al. 1967, *D. westralium* (Cookson & Eisenack) Evitt et al. 1967 emend. May 1977, *D. longicornis* (Vozzhennikova) Harland 1973, *D. sphaerocephalum* (Vozzhennikova) Lentin & Williams 1973 and *D. vozzhennikovae* Lentin & Williams 1973. The top of this zone is marked by the first appearance datum (FAD) of *Selenopemphix maastrichta* (Kumar et al. 1993) and the base of this zone is not yet defined because samples below J-20 were not available for further study.

Age. Late Maastrichtian.

Reference Section. Um Sohryngkew River Section, Meghalaya, India. This zone includes the upper part of the Mahadeo Formation. The top of this zone is 35 cm below the Ir layer.

Dinogymnium spp. and *Selenopemphix maastrichta* Concurrent Range Zone

Definition. Sediments containing specimens of *Dinogymnium* spp. and *Selenopemphix maastrichta* Kumar et al. (1993). The top of this zone is marked by the last appearance datum (LAD) of these two taxa and the base is marked by the first appearance datum (FAD) of *Selenopemphix maastrichta*.

Age. Latest Maastrichtian.

Reference Section. Um Sohryngkew River section, Meghalaya, India. This zone includes the upper part of the Mahadeo Formation. It is 35 cm thick and the top is just below the Ir layer.

Damassadinium californicum Range Zone

Definition. Sediments containing the total range of *Damassadinium californicum* (Drugg 1967) Fensome et al. (1993)

Age. Danian.

Reference Section. Um Sohryngkew River section, Meghalaya, India. This zone covers the uppermost part of Mahadeo Formation above the Ir layer.

Remarks. To refine the stratigraphic relationships of the proposed zones in the Um Sohryngkew River section, we plan to examine more closely spaced samples.

ORGANIC MATTER TYPES AND PALYNOFACIES

This study follows the classification and nomenclature of kerogen by Pocock (1982). The following six components in the kerogen have been identified in two broad groups for the interpretation of palynofacies. (A). Humic Organic Matter (HOM): which includes; 1. Woody/herbaceous and terrestrail kerogen like pollen, spores and cuticular membranes (Wood); 2. Biodegraded terrestrial organic matter (BDT), and 3. Charcoal, oxidized or metamorphosed carbon particles. (B). Sapropelic Organic Matter (SOM): which includes; 4. Gray amorphous and algal phytoplankton (dinoflagellate cysts, acritarchs and other algal remains); 5. Finely divided organic matter (FOM) is biodegraded, unrecognizable amorphous matter, and 6. Amorphous marine/terrestrial organic matter (AMTM). Palynofacies have been described as humic (H), sapropelic (S), humic-sapropelic (HS) and sapropelic-humic (SH) with dominance of either charcoal or wood. The variations in total organic matter (TOM), palynofacies, and in the percentages of various macerals across the KTB in the Um Sohryngkew river section are presented in Table 2. Changes in humic and sapropelic facies across the KTB are shown in Text-Figure 5. There are four discrete events observed from variations in palynofacies and percentages of different components of kerogen across the KTB. These events result from sea level changes and movement of paleoshoreline.

Event 1. This is represented in samples J-20 through J-24 and is marked by influx of aquatic derived organic matter which is moderate to rich in quantity. Palynofacies are mainly sapropelic and dominated by finely divided organic matter (FOM) indicating high energy conditions. This corresponds to the transgressive phase.

Event 2. This is represented in samples J-25 and J-26 and marked by a decrease in amorphous, finely divided organic matter (FOM) and amorphous terrestrial/marine organic matter (AMTM). Land derived organic matter increases. This marks the regressive phase.

PLATE 2

All figures x 650 unless otherwise specified; coordinates are in parentheses.

1. *Membranilarnacia? tenella* Morgenroth 1968. J20-2, (111.7x 30) Phase contrast, x900.
2. *Tanyosphaeridium magdalium* (Drugg 1967) Heisecke, 1970. J20-2, (95.5x58).
3. *Selenopemphix maastrichta* Kumar et al. 1993. J23-1, (106.4x69.3), x900.
4. *Selenopemphix maastrichta* Kumar et al. 1993. J23-1, (99x30).
5. *Selenopemphix maastrichta* Kumar et al. 1993. J22-1, (105x65).
6. *Selenopemphix maastrichta* Kumar et al. 1993. J22-1, (56x46).
7. *Phelodinium magnificum* (Stanley 1965) Stover & Evitt 1978. J21-2, (91.1x44).
8. *Cerodinium speciosum* (Alberti 1959) Lentin & Williams 1987. J24-2, (75x45).
9. *Andalusiella* sp. J20-1, (96x38.7)
10. *Glaphyrocysta assamica* (Jain et al. 1975) Jain 1982. J23-1, (95.5x37.8).
11. *Lejeunecysta hyalina* (Gerlach 1961) Sarjeant 1984. J23-1, (107.5x43.1).
12. *Cordosphaeridium* sp. J23-1, (100.4x69).

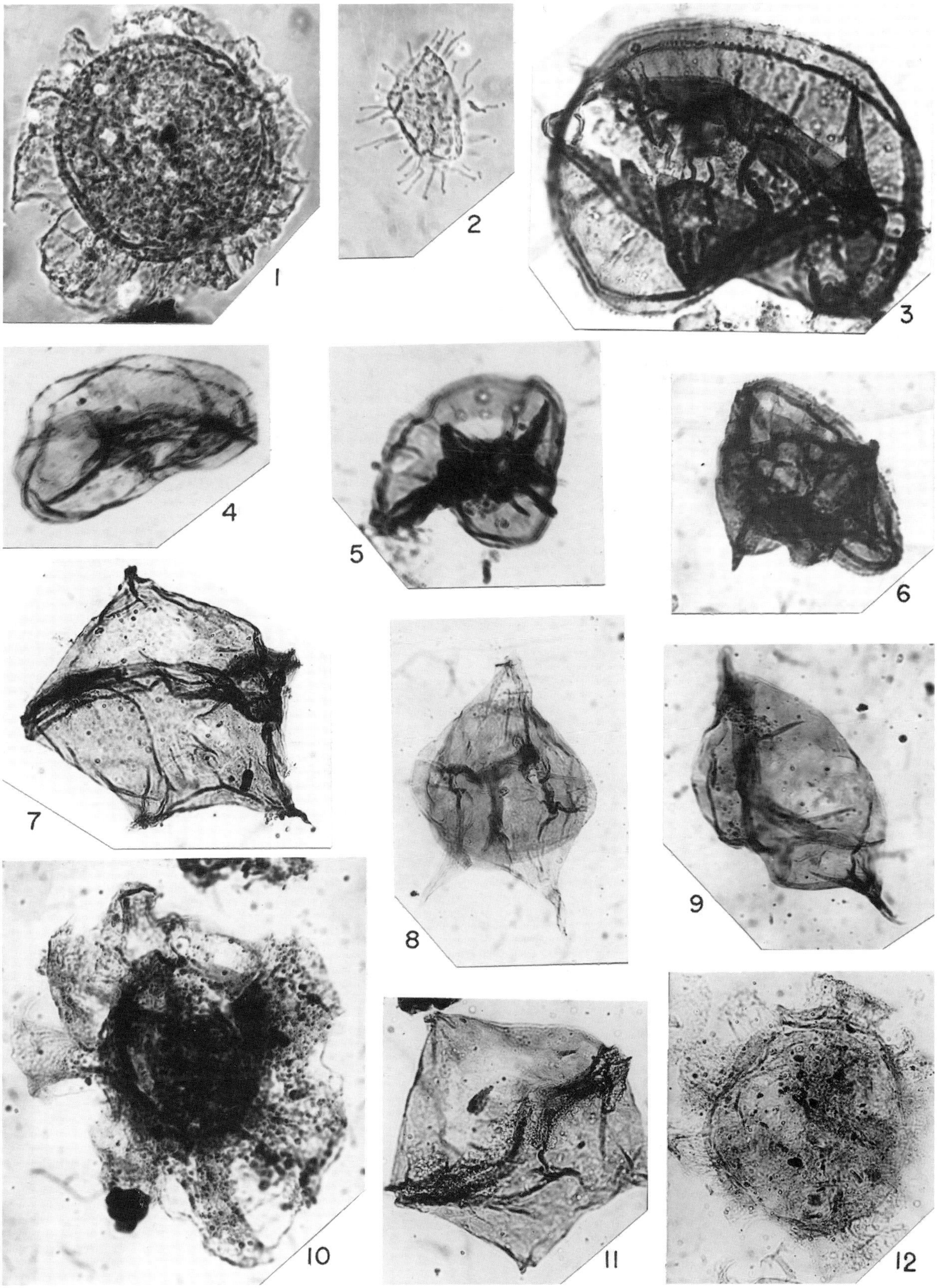
1
2
3
4
5
6
7
8
9
10
11
12

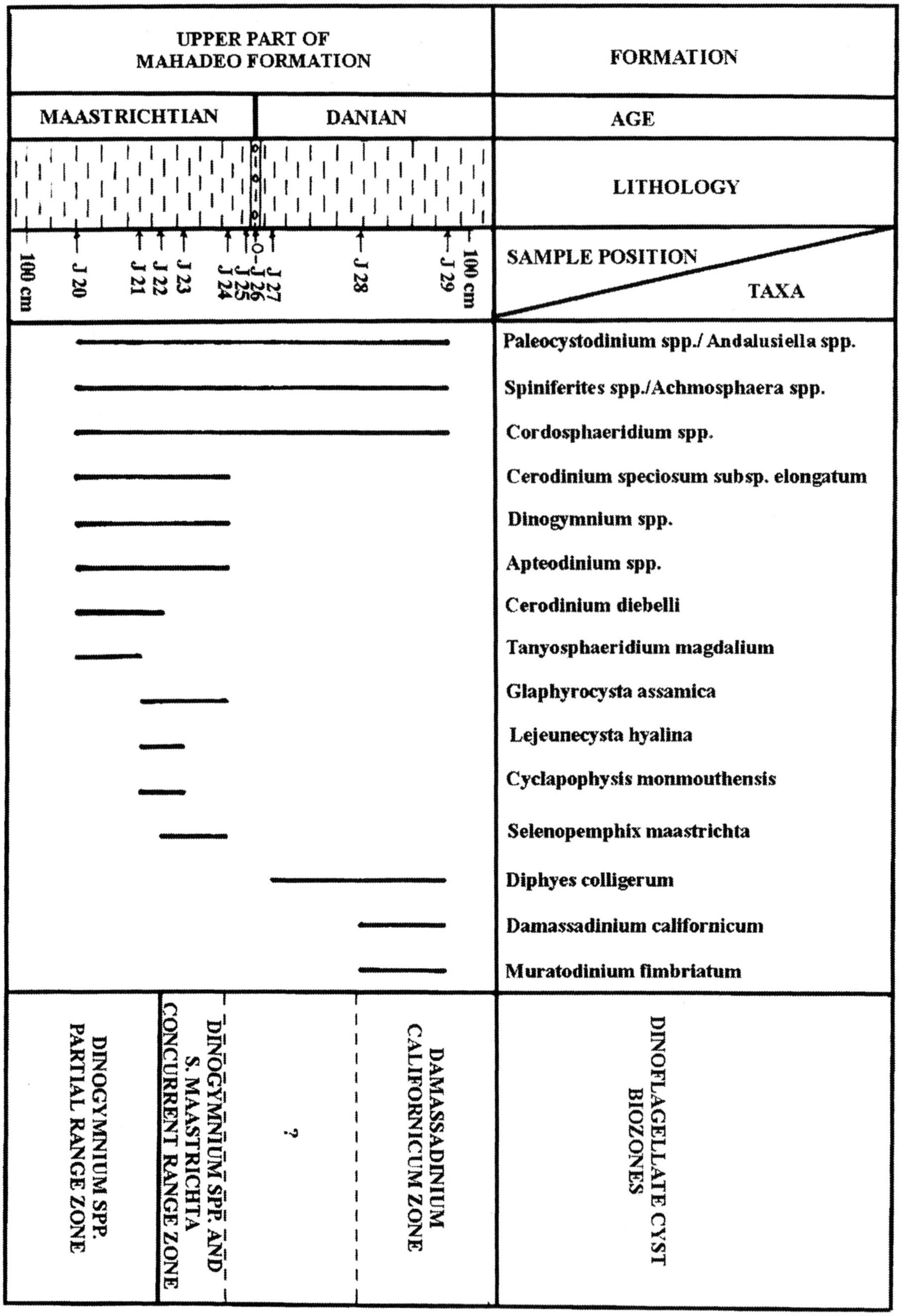

Text-Figure 4. Stratigraphic distribution of selected dinoflagellate cysts and biozones across the Cretaceous–Tertiary boundary in Meghalaya.

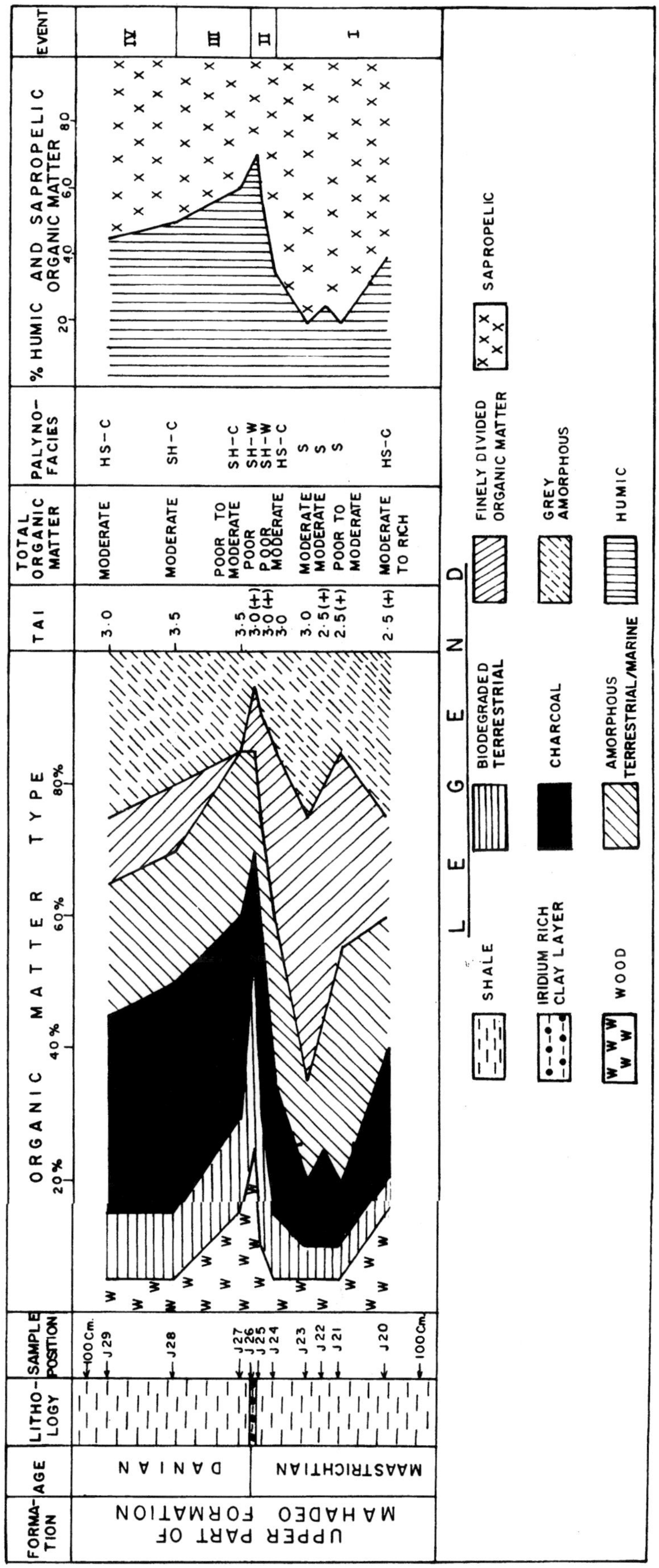

Text-Figure 5. Percentages of organic matter and variations and palynofacies across the KTB in Meghalaya.

Event 3. This event is recorded in sample J-27 where total organic matter is poor to moderate and there is an increase in percentage of amorphous marine/terrestrial organic matter (AMTM) and a decrease in terrestrially derived organic matter. The change in palynofacies between event 2 and 3 from dominantly humic to sapropelic suggests a relationship to a transgressive event

Event 4. This event is recorded in samples J-28 and J-29 and marks by the more stabilized state of a transgressive phase. It is a dominantly sapropelic facies, characterized by more than 60% aquatic derived organic matter. Total organic matter is moderate.

CONCLUSIONS

1. Ten samples within a range of 80 cm above and 70 cm below the Iridium rich (Ir) layer were studied to demonstrate the changes in palynofacies and dinoflagellate cyst assemblages close to the KTB.
2. A change occurs in dinoflagellate cyst assemblages from a *Dinogymnium* spp. dominated assemblage of late Maastrichtian age to a *Damassadinium californicum* and *Muratodinium fimbriatum* assemblage of early Danian age. The reduction in dinoflagellate species diversity occurs just above the Iridium layer (only 3 species), but gradually recovers to higher diversity (7 species) 40 cm above the Ir layer.
3. *Spiniferites* spp. and *Achomosphaera* spp. account for almost 60 to 70 % of all the dinoflagellates in the samples representing earliest Danian indicating a phase of regression and inner shelf deposition.
4. A fall of sea-level of the order of 70 to 80 m is estimated during the KTB in Meghalaya, which closely corresponds with the 90 m fall of sea-level at the KTB estimated by Haq et al. (1987).
5. The fall in sea level began approximately 8.8 ka prior to the deposition of the Ir layer which gradually rose again in Early Danian. The sea level fall was relatively sudden and rise was rather gradual.
6. The *Dinogymnium* spp. Partial Range Zone, *Dinogymnium* spp. and *S. maastrichta* Concurrent Range Zone and *D. californicum* Zone are defined.
7. The percentage of various OM constituents and palynofacies change dramatically near the KTB from predominantly sapropelic to almost humic with marine influence.

ACKNOWLEDGMENTS

We thank J. Pandey for providing us the samples, D.S.N. Raju and B.C. Jaiprakash for discussion. We also thank David Goodman, Robert S. Van Pelt and David Pocknall for helpful suggestions and to Anita Kumar for help in redoing figures on computer. We thank the Director, KDMIPE, ONGC Ltd. for permission to publish this paper. The opinions expressed in this paper are of the authors only and not necessarily those of ONGC Ltd.

References Cited

BHANDARI, N., GUPTA, M., PANDEY, J., and SHUKLA, P.N.
1994 Chemical profiles in K/T boundary section of Meghalaya, India: cometary, asteroidal or volcanic. *Chemical Geology*, 113:45–60.

DAVIES, E.H., BUJAK, J.P., and WILLIAMS, G.L.
1982 The application of dinoflagellates to paleoenvironmental problems. *Proceedings of the 3rd North American Paleontological Convention*, 1: 125–131.

EDET, J.J., and NYONG, E.E.
1993 Depositional environments, sea level history and paleobiogeography of the Late Campanian–Maastrichtian on the Calabar flank, S.E. Nigeria. *Palaeogeography, Paaleoclimatology, Palaeoecology*, 102: 161–175.

FIRTH, J.V.
1993 Dinoflagellate assemblages and sea level fluctuations in the Maastrichtian of Southwest Georgia. *Review of Palaeobotany and Palynology*, 79: 179–204.

HARLAND, R.
1982 Recent dinoflagellate cyst assemblages from the southern Barent Sea. *Palynology*, 6: 9–18.

HART, M.B.
1980 A water depth model for the evolution of the planktonic foraminiferida. *Nature*, 286: 252–254.

HAQ, B.U., HARDENBOL, J., and VAIL, P.R.
1987 Chronology of fluctuating sea levels since the Triassic. *Science*, 235: 1156–1166.

JAIN, K.P., SAH, S.C.D., and SINGH, R.Y.
1975 Fossil dinoflagellates across Maastrichtian–Danian boundary in Lower Assam, India. *The Palaeobotanist*, 22: 1–18.

PLATE 3

All figures x 650; coordinates are in parentheses.

1 *Apteodinium* sp. J23-1, (107.8x53).

2 cf. *Damassadinium californicum* (Drugg 1967) Fensome et al. 1993. J28-1,(101.2x34.3).

3 *Damassadinium californicum* (Drugg 1967) Fensome et al. 1993. J29-1, (100.7x40.3).

4 *Dinogymnium acuminatum* Evitt et al.1978. J20-5, (111.2x56.4).

5 *Achomosphaera* sp. J21-3, (104.9x63.3).

6 *Dinogymnium vozzhennikovae* Lentin & Williams 1973. J21-1,(107x38).

7 *Dinogymnium westralium* (Cookson & Eisenack 1958) Evitt et al. 1967 emend. May 1977. J24-3, (108x48).

8 *Dinogymnium westralium* (Cookson & Eisenack 1958) Evitt et al. 1967 emend. May 1977. J21-1, (96.4x33).

9 *Dinogymnium vozzhennikovae* Lentin & Williams 1973. J21-1, (103.2x39.5).

10 *Dinogymnium acuminatum* Evitt et al. 1961. J21-3, (92.5x58.2).

11 *Spiniferites ramosus* subsp. *ramosus* (Ehrenberg 1838) Davey & Williams 1966. J20-1, (95.3x56.5), phase contrast.

12 *Apteodinium* sp. J20-2, (95.5x32.7).

13 *Spiniferites ramosus* subsp. *ramosus* (Ehrenberg 1838) Davey & Williams 1966. J21-2, (95.5x38.2), phase contrast.

14 *Dinogymnium longicorne* (Vozzhennikova 1967) Harland 1973. J20-4, (107.5x41.3).

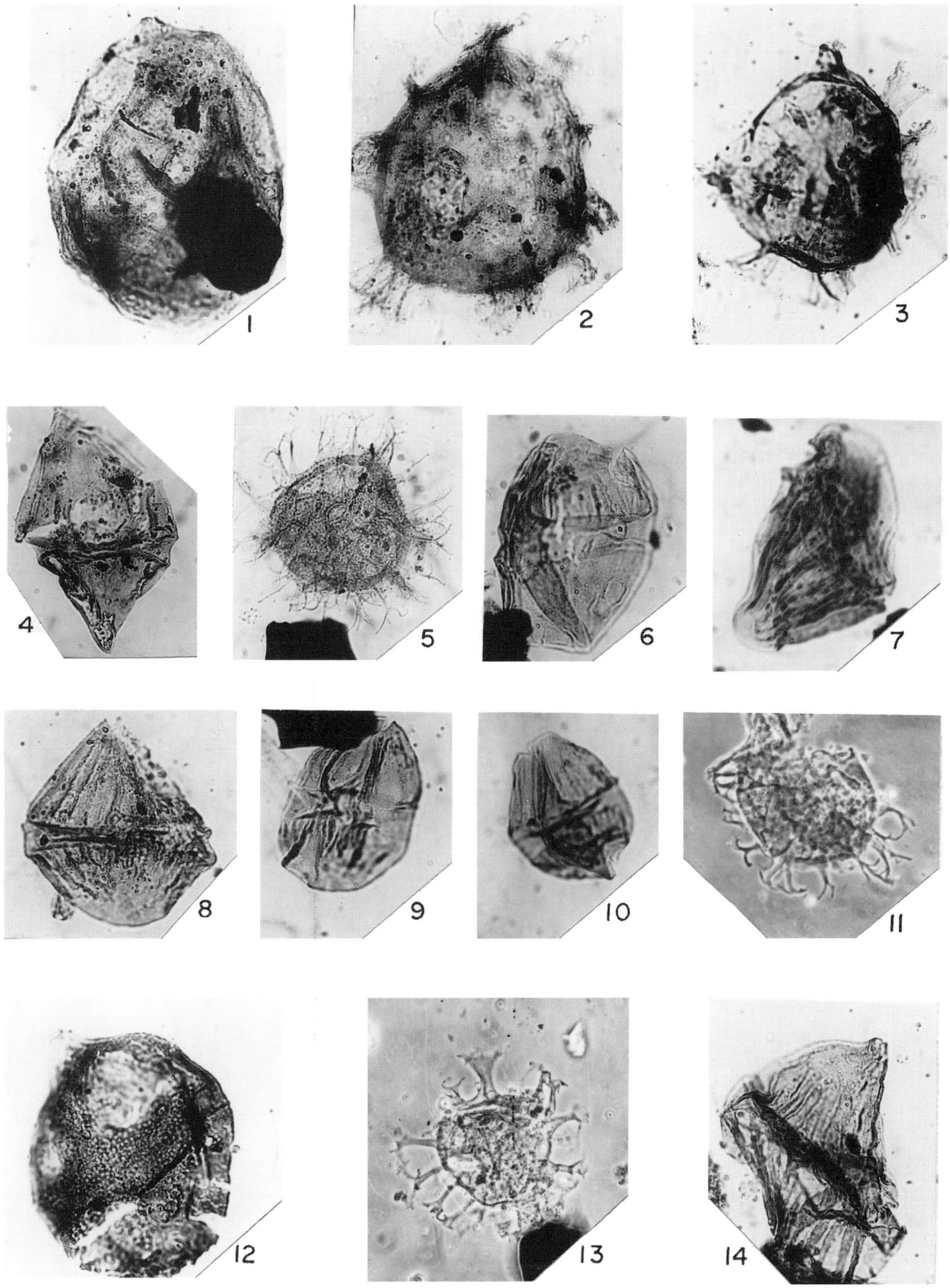
1
2
3
4
5
6
7
8
9
10
11
12
13
14

TABLE 2. Palynofacies and variation in organic matter types across Cretaceous–Tertiary boundary in Meghalaya. Abbreviations: OM – Organic Matter; BDT – Biodegraded Terrestrial Organic Matter; AMTM – Amorphous Marine/Terrestrial Organic Matter; FOM – Finely Divided Organic Matter; TAI – Thermal Alteration Index; TOM – Total Organic Matter

SAMPLE NUMBER	LITHOLOGY	HUMIC OM %			SAPROPELIC OM %			TAI	TOM	PALYNOFACIES
		WOOD	BDT	CHARCOAL	AMTM	FOM	GREY AMORPHOUS			
J-29	Grey Silty Shale	5	10	30	20	10	25	3	Moderate	Humic Sapropelic-Charcoal
J-28	Grey Silty Shale	5	10	35	20	10	20	3.5	Moderate	Sapropelic Humic-Charcoal
J-27	Grey Silty Shale	15	15	30	25	0	15	3.5	Poor–Moderate	Sapropelic Humic-Charcoal
J-26	Limonitic Shale	25	30	15	15	10	5	3(+)	Poor	Sapropelic Humic-Wood
J-25	Dark Grey Shale	10	20	25	20	15	10	3(+)	Poor	Sapropelic Humic-Wood
J-24	Dark Grey Shale	5	10	20	25	25	15	3	Moderate	Humic Sapropelic-Charcoal
J-23	Dark Grey Shale	5	5	10	15	40	25	3	Moderate	Sapropelic
J-22	Dark Grey Shale	5	5	15	20	35	20	2.5(+)	Moderate	Sapropelic
J-21	Dark Grey Shale	5	5	10	35	30	15	2.5(+)	Poor–Moderate	Sapropelic
J-20	Dark Grey Shale	15	5	20	20	15	25	2.5(+)	Moderate–Rich	Humic Sapropelic-Charcoal

KUMAR, A., SHARMA, J., and KAPOOR, P.N.
1993 *Selenopemphix maastrichta* sp. nov.: The first pre Tertiary record of the genus *Selenopemphix* (Benedek) Bujak in Bujak et al. 1990. *Bulletin of Oil & Natural Gas Corporation*, 30:139–146.

MISRA, C.M., KUMAR, A., PRASAD, B., PUNDIR, B.S., and SAXENA, R.K.
1994 A note on the bioevents across the K/T transition in the well Modi-A, Krishna–Godavari basin, India. *Geoscience Journal*, 15: 33–36.

MORZADEC-KERFOURN, M.T.
1979 Les kystes de dinoflagelles. *Geologic Mediterraneene, La Mer Pelagienne*, 6: 221–246.

PANDEY, J.
1981 Cretaceous foraminifera of Um Sohryngkew River section, Meghalaya. *Journal of the Palaeontological Society of India*, 25: 53–74.

PANDEY, J., and RAVINDRAN, C.N.
1988 Foraminiferal controls in the Indian Palaeocene. Palaeocene of India. *Indian Association of Palynostratigraphers*, Lucknow, p. 124–184.

POCOCK, S.A.J.
1982 Identification and recording of particulate sedimentary organic matter. *Society of Economic Paleontologists and Mineralogists* Short Course 7: 13–133.

RAJU, D.S.N., and RAVINDRAN, C.N.
1990 Cretaceous sea-level changes and transgressive/regressive phase in India; a review. *In:* Sahni, A., and Jolly, A. (eds.), *Contributions from the seminar cum workshop, International Geological Correlation Program* 216, p. 38–46.

RAJU, D.S.N., JAIPRAKASH, B.C., KUMAR, A., SAXENA, R.K., DAVE, A., CHATTERJEE, T.K., and MISRA, C.M.
1995 Age of Deccan volcanism across KTB in Krishna–Godavari Basin: New evidences. *Journal Geological Society of India*, 45:229–233.

RAJU, D.S.N., JAIPRAKASH, B.C., and KUMAR, A.
1996 Paleoenvironmental setup and age of basin floor just prior to the spread of Deccan volcanism in the Krishna–Godavari basin. *In:* Sahni, A. (ed.), B. Rama Rao memorial volume, *Geological Society of India Memoir* 37, pp. 285–295.

WALL, D., LOHMANN, G.P., and SMITH, W.K.
1977 The environmental and climatic distribution of dinoflagellate cysts in modern marine sediments from the regions in the North and South Atlantic Oceans and adjacent seas. *Marine Micropaleontology*, 2: 121–200.

ZACHOS, J.C., and ARTHUR, M.A.
1986 Paleoceanography of the Cretaceous/Tertiary boundary event: Inferences from stable isotopic and other data. *Paleoceanography*, 1: 5–26.

Authors' addresses

ARUN KUMAR
Ottawa-Carleton Geoscience Centre
and Department of Earth Sciences
Carleton University
1125 Colonel By Drive
Ottawa, Ontario K1S 5B6
Canada

P. N. KAPOOR
J. SHARMA
Palynology Laboratory
K.D.M. Institute of Petroleum Exploration
Oil & Natural Gas Corporation Ltd.
Kaulagarh Road, Dehradun 248 195
India

APPLICATION OF PALYNOLOGY TO ENVIRONMENTAL RESTORATION ACTIVITIES: AN EXAMPLE FROM THE SAVANNAH RIVER SITE, SOUTH CAROLINA, U.S.A.

Robert S. Van Pelt and Joyce Lucas-Clark

Abstract

Over the past eleven years, environmental issues have become a major focus of scientific research at the Savannah River Site (SRS) in an attempt to address air, water, and soil contamination from past industrial practices. Environmental restoration activities have required a thorough understanding of the surface and subsurface hydrogeological and geological conditions. Palynology has been and is currently being applied in combination with other geological tools to establish the subsurface hydrogeological framework. Palynology is given a critical role in identifying aquifers and confining units, and evaluating their lateral and vertical extent. Preferential pathways (e.g. faults, slump features, and facies changes) for ground water movement are also identified largely with the assistance of palynological research.

Cretaceous and Tertiary aquifers and confining units at SRS are described herein in terms of their age diagnostic and characteristic assemblages of dinoflagellates, and to a lesser extent, assemblages of pollen and spores. The units consist of the McQueen Branch, Crouch Branch, Gordon, Upper Three Runs, and Steed Pond aquifers, and their corresponding confining units, the Appleton Confining System, and the McQueen Branch, Crouch Branch, and Gordon confining units. A selection of diagnostic and characteristic fossil species is illustrated along with range data.

INTRODUCTION

The Savannah River Site (SRS) occupies 310 square kilometers within Aiken, Barnwell, and Allendale counties in southwestern South Carolina, U.S.A. (Text-Figure 1). Bedrock (Paleozoic metamorphic and Triassic clastic rocks) and overlying Coastal Plain sediments (Upper Cretaceous through Recent unconsolidated sediments) constitute the hydrologic system beneath the SRS and surrounding areas (Text-Figure 2). Direction of ground water flow in the aquifers at SRS is primarily towards the Savannah River and its tributaries. Aquifers in the Cretaceous age sediments are the primary source for drinking water for the SRS and surrounding communities.

The primary mission of the SRS is to produce nuclear materials for national defense. In the 1980s, concern about environmental issues spawned research into the hydrogeology of the SRS. Since 1990, the mission has expanded to include environmental restoration of over 40 years of waste-producing activities. Numerous waste-sites at SRS are undergoing environmental assessments and/or remediation under state and federal regulations.

An essential part of the environmental assessments is the characterization of the subsurface hydrogeology. Hydrogeological characterization involves establishing the hydrologic and geologic conditions (including the hydrostratigraphy) and incorporating this information into ground water flow and contaminant transport models. Results from the hydrogeology and ground water models are then applied in the selection of remediation strategies for a waste site. At the SRS, palynology has played a critical role in the establishment of the hydrogeologic framework, including the identification and understanding of hydrogeologic units and their relationships.

Distinguishing the sedimentary units that comprise the hydraulic system at SRS can be difficult based solely on lithology. Most of the units consist of sediments of near shore and deltaic origin, regardless of age, such that the similar depositional origin masks lithological distinctions. Many sedimentary units are lenticular, comprise several depositional facies, and transgress time. Palynology, particularly dinoflagellate stratigraphy, has been applied to the various sedimentary units to interpret age distinctions and paleoenvironmental facies.

Palynology has also played an important role in the detection of ground water preferential pathways such as structural features (e.g. faults, slump features) and facies changes. Prediction of such pathways is essential in modeling ground water contaminant flow and transport. The subsurface investigation at the SRS represents one of the frontier applications of palynology to hydrogeology and environmental restoration.

Previous Work At Savannah River Site

As early as the mid 1970s, the need for accurate information concerning hydrostratigraphy beneath the SRS became apparent to those concerned with potential contamination problems (Cahil, 1982; Faye and Prowell, 1982; Marine and Root, 1975). Palynology has been used from the mid 1980s as a basis with which to revise the formal stratigraphy (Siple 1967; Prowell et al., 1985a, 1985b; Gohn and Campbell, 1992; Nystrom et al., 1991; Fallaw and Price, 1992, 1995). For example, the Sawdust Landing Formation (hydrostratigraphically equivalent to the lower part of the Crouch Branch Confining Unit) was determined by dinoflagellate and pollen analysis to be Paleocene rather than Cretaceous in age (Prowell et al., 1985b). Subsequently, this confining unit was divided into three distinct units separated by unconformities, all based on palynological age interpretations (Prowell et al., 1985b). In general, age interpretations based on palynology allowed

Van Pelt, Robert S., and Lucas-Clark, Joyce
2001 Application of palynology to environmental restoration activities: An example from the Savannah River site, South Carolina, U.S.A.; *In*: Goodman, D.K., and Clarke, R.T. (eds.), *Proceedings of the IX International Palynological Congress, Houston, Texas, U.S.A., 1996*; American Association of Stratigraphic Palynologists Foundation, p. 161–170.
 ISBN 0–931871–06–9

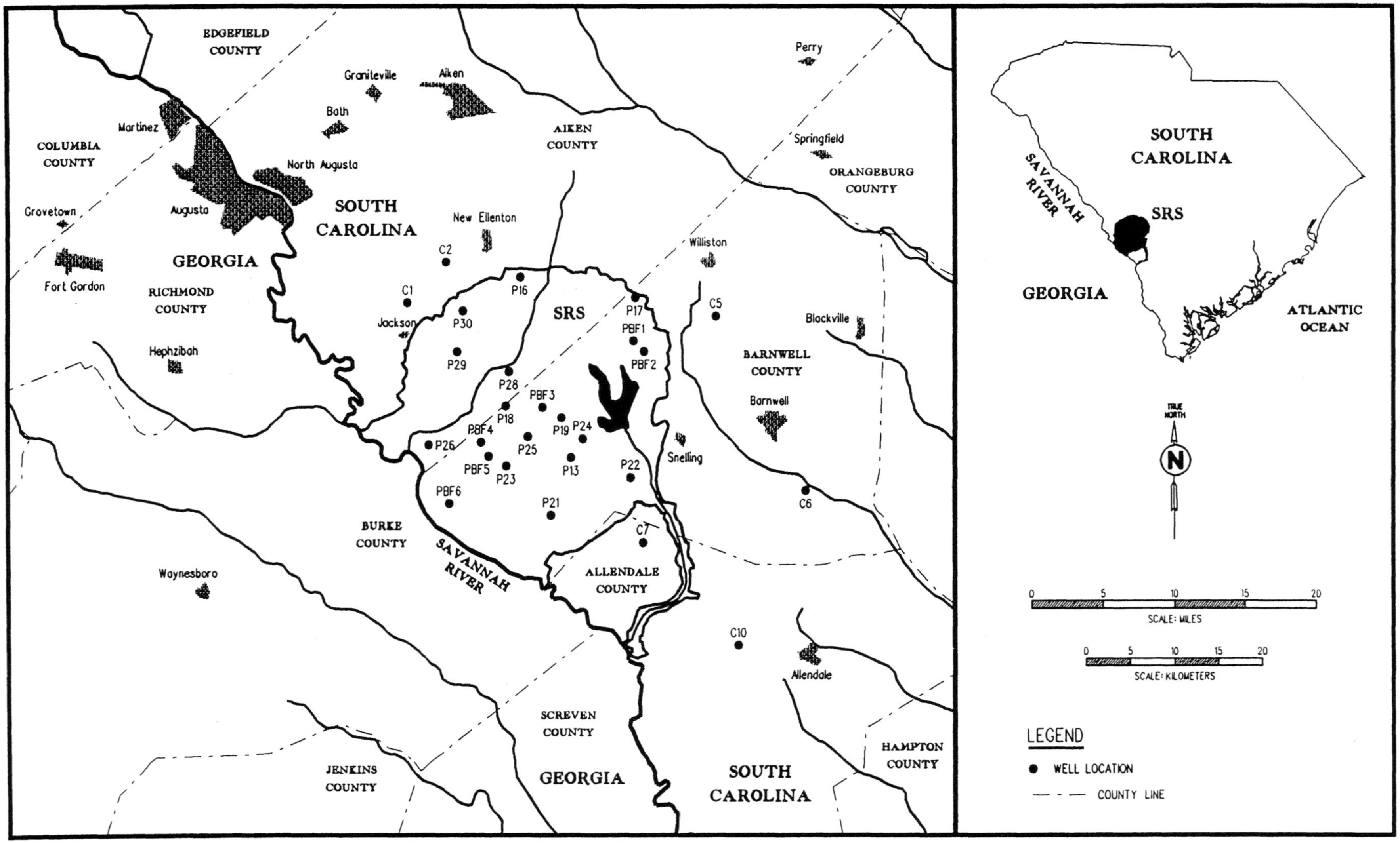

Text-Figure 1. Location of the study area, Savannah River Site. Palynological data presented herein corresponds to the P and PBF wells.

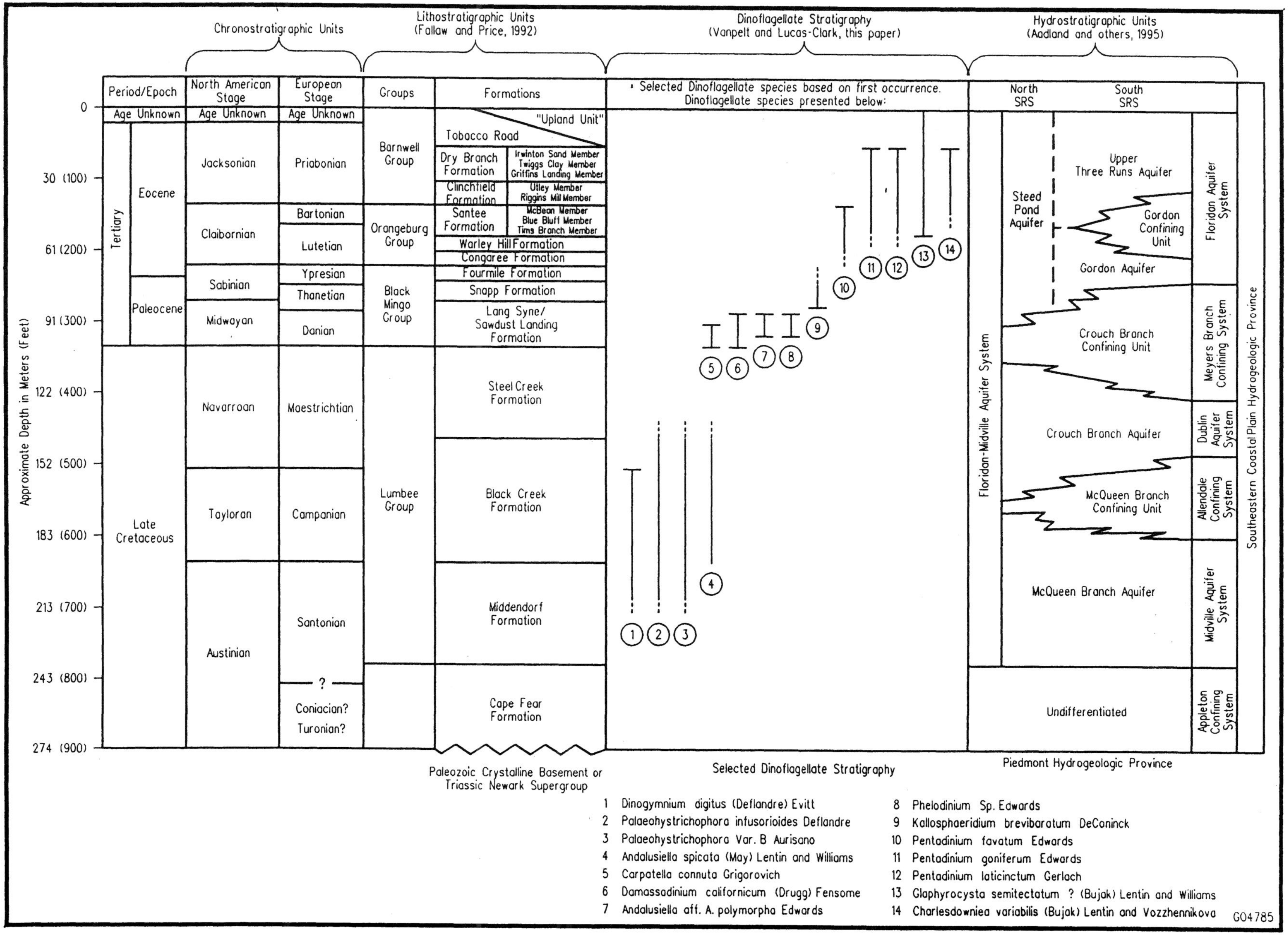

Text-Figure 2. Comparison of chronostratigraphic, lithostratigraphic, selected dinoflagellate stratigraphy and hydrostratigraphic units at the SRS.

for more accurate correlation to other stratigraphic units in the southeastern coastal plain, in what has come to be considered the Floridan Aquifer System. The revised lithostratigraphy and study of ground water transport led to a complete formal hydrostratigraphy for the SRS and surrounding vicinity (Aadland, 1990; Aadland and Bledsoe, 1992; Aadland et al., 1992; 1995); (Text-Figure 2).

Studies in the mid 1980s focused on Tertiary carbonate units due to concerns that these are potential conduits for contaminants, and because they are difficult to correlate due to their discontinuous, lenticular, and reef-like nature. Although palynomorphs are usually sparse within the carbonates, they are abundant in the surrounding and interbedded clay and mud. Identification of these otherwise unfossiliferous fine-grained units was determined to be more critical to the correlation of the carbonate units than the fossils recovered from the carbonates beds.

More recent emphasis on environmental restoration has required compliance with state and federal regulations, [e.g. Resource Conservation and Recovery Act (RCRA)/ Comprehensive Environmental Response Compensation and Liability Act (CERCLA)] which require a complete subsurface hydrogeologic/geologic characterization of waste sites, such as landfills, settling basins, storage tank areas, etc. (Van Pelt et. al., 1993; Lucas-Clark and Van Pelt, 1993; Van Pelt and Lucas-Clark, 1996). In virtually all cases, palynology has been utilized to aid in identification of hydrostratigraphic units, and to improve and confirm well log correlation.

Certain hydrostratigraphic units, specifically confining units, have characteristic dinoflagellate and pollen assemblages that can be interpreted for age and paleoenvironment. The interpretations allow for accurate correlation as well as prediction of pinch outs and leaks in the confining units. Some confining units, such as the lower part of the Crouch Branch Confining Unit, were deposited under marine conditions and represent a major transgressive sequence. Others units, such as the late Paleocene age upper part of the Crouch Branch Confining Unit, are characterized by dinoflagellates that occur abundantly in a deltaic and possibly brackish water environment. Overall, near shore to shelfal dinoflagellate assemblages appear to dominate confining units that are more reliable as continuous barriers to hydrological flow except where breached by unconformities; shallow water dinoflagellate assemblages generally comprise confining units that are not reliable barriers to vertical ground water flow. This is the case with the upper part of the Crouch Branch Confining Unit which grades into coarser

PLATE 1

Examples of dinoflagellate species used in hydrostratigraphic correlation at the Savannah River Site. Scale Bar = 20 um.

1, 2 *Glaphyrocysta semitectata* (Bujak in Bujak et al. 1980) Lentin & Williams 1981. Ventral view of dorsal surface (ventral surface has ectophram). University of South Carolina core USC 92-30, sample CGS 9210.310, Upper Three Runs Aquifer/Steed Pond Aquifer. Range: late middle to early late Eocene, Bartonian. Range at SRS: Gordon Confining Unit (upper part), Upper Three Runs/ Steed Pond aquifers (Santee, Clinchfield formations).

3 *Glaphyrocysta exuberans* (Deflandre & Cookson 1955) Stover & Evitt 1978. Ventral ? view. University of South Carolina core USC 92-30, sample CGS 9210.310, Upper Three Runs/Steed Pond aquifers. Range: late middle to early late Eocene, Bartonian. Range at SRS: Gordon Confining Unit (upper part), Upper Three Runs/Steed Pond aquifers (Santee, Clinchfield formations).

4 *Wetzeliella articulata* Eisenack 1938. Ventral view of dorsal surface. SRS core P23, sample CGS 8656.2001 Gordon Confining Unit (Santee Formation). Range: middle to late Eocene. Range at SRS: Gordon Confining Unit, Upper Three Runs/Steed Pond aquifers (Santee, Clinchfield, Dry Branch formations).

5 *Phelodinium* sp. of Edwards (1989). Dorsal view of ventral surface. SRS core P-20TA, sample CGS 8643.150 Crouch Branch Confining Unit (middle clay of Lang Syne Formation). Range: Danian to mid late Paleocene? Range at SRS: Crouch Branch Confining Unit (Sawdust Landing and Lang Syne formations).

6 *Spinidinium pulchrum* (Benson 1976) Lentin & Williams 1977. Two specimens, ventral and lateral views. SRS core P-20TA, sample CGS 8643.150. Range: Maastrichtian ? to mid Paleocene. Range at SRS: Crouch Branch Confining Unit (Sawdust Landing and Lang Syne formations).

7 *Wetzeliella symmetrica* Weiler 1956. Ventral view of dorsal surface. SRS core P23, sample CGS 8656.2001 Gordon Confining Unit (Santee Formation). Range: middle to late Eocene. Range at SRS: Gordon Confining Unit, Upper Three Runs/Steed Pond aquifers (Santee, Clinchfield, Dry Branch formations).

8 *Kallosphaeridium brevibarbatum* de Coninck 1969. Dorsal view of dorsal surface and operculum. SRS core LFW10SB, sample CGS 9103.088 Crouch Branch Confining Unit (upper clay of Fourmile Branch Formation). Range: late Paleocene–early Eocene. Range at SRS: upper part of Crouch Branch Confining Unit and Steed Pond/Gordon aquifers (Snapp and Fourmile Branch formations).

9, 12 *Pentadinium goniferum* Edwards 1982; (9) Dorsal view of dorsal surface, (12) Dorsal view, optical cross-section, SRS core P23, sample CGS 8656.2001 Gordon Confining Unit (Santee Formation). Range: middle to late Eocene, Clabornian to Jacksonian (upper Lisbon to Gosport at Little Stave Creek, Alabama). Range at SRS: Gordon Confining Unit, Upper Three Runs/Steed Pond aquifers (Santee, Clinchfield, Dry Branch formations).

10 *Damassadinium* aff. *D. californicum* (Drugg 1967) Fensome et al. 1993. Dorsal view, optical section. SRS core LFW 10SB, sample CGS lower clay of Crouch Branch Confining Unit (Sawdust Landing Formation). Range: K/T boundary to mid late Paleocene (NP7). Range at SRS: Crouch Branch Confining Unit (Sawdust Landing and Lang Syne formations).

11 *Damassadinium californicum* (Drugg 1967) Fensome et al.1993. Dorsal? view, mid-focus. SRS core FC-5A, sample CGS 8643.1318 Crouch Branch Confining Unit (Sawdust Landing Formation). Range: K/T boundary to mid late Paleocene (NP7). Range at SRS: Crouch Branch Confining Unit (Sawdust Landing and Lang Syne formations).

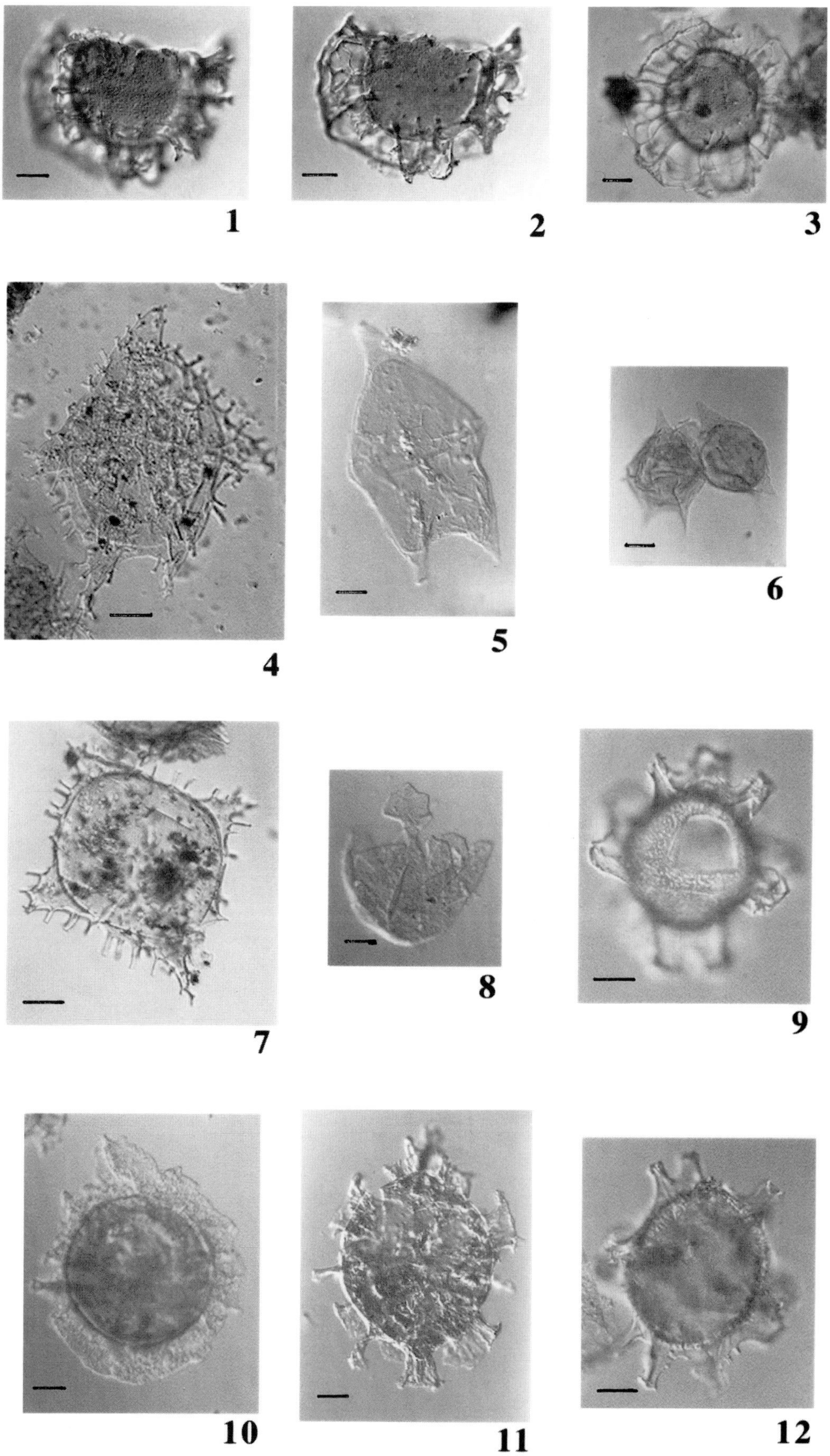
1
2
3
4
5
6
7
8
9
10
11
12

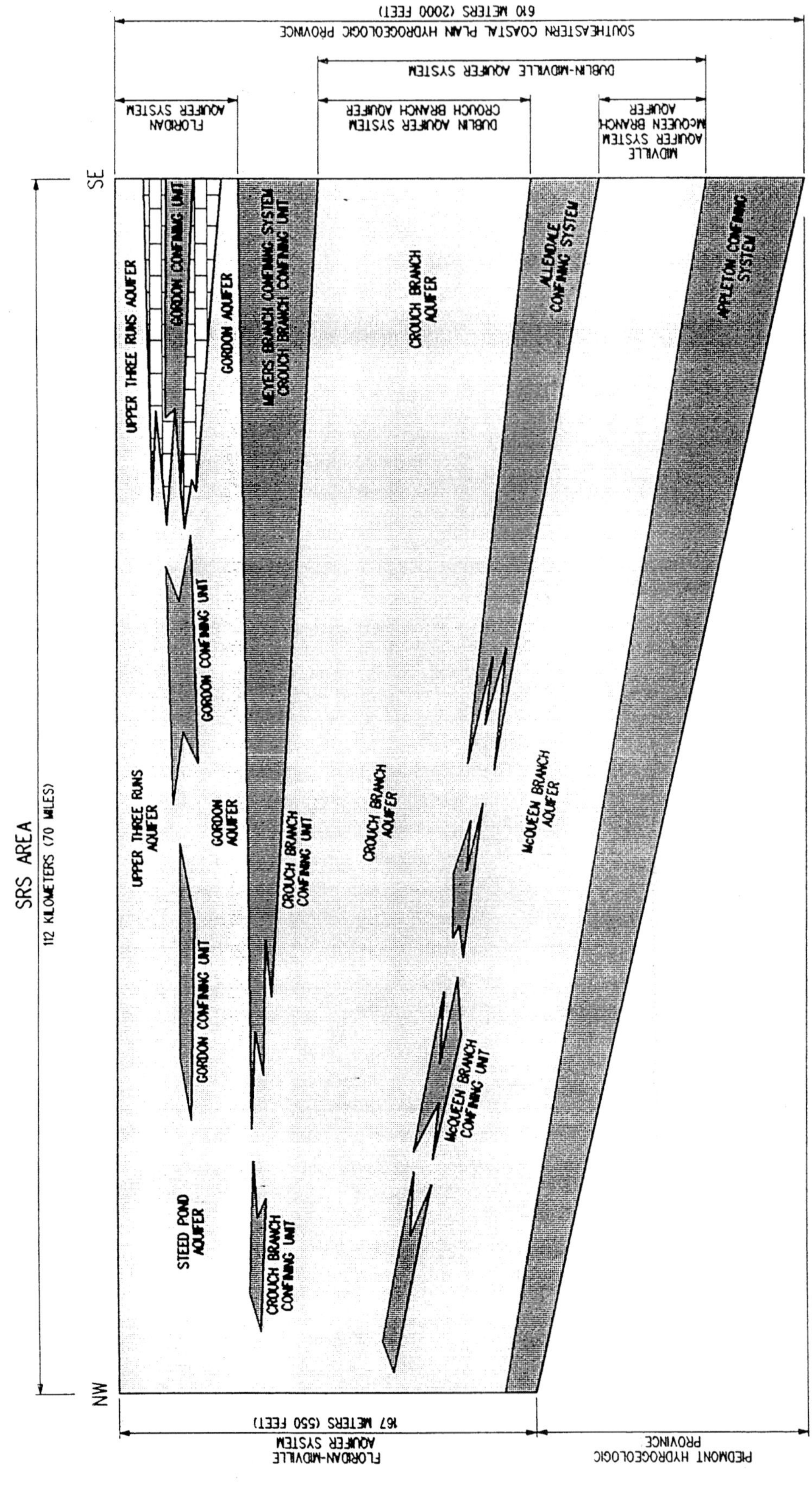

Text-Figure 3. Cross-section of Upper Cretaceous and Tertiary hydrostratigraphic sequence beneath the SRS.

TABLE 1. Alphabetical listing of pollen, spores, and dinoflagellates used in hydrostratigraphic correlation at the Savannah River Site, South Carolina, U.S.A.

Pollen/Spores
Brevicolporites Sp.B of Christopher 1978
Classopollis Pflug spp.
Complexipollis abdita Tschudy
Complexiopollis abditus Tschudy
Holkopollenites (Fairchild) Stover, Elsik, and Fairchild spp.
Parvisaccites rugulatus Brenner
Plicapollis Pflug spp.
Rugubivesiculites Pierce spp.
Dinoflagellates
Anadalusiella (Riegel) Riegel & Sarjeant spp.
Andalusiella aff. *A. polymorpha* Edwards
Andalusiella spicata (May) Lentin & Williams
Apectodinium homomorphum (Deflandre & Cookson) Lentin & Williams
Apectodinium quinquelatum (Williams & Downie) Lentin & Williams
Areoligera coronata (Wetzel) Lejeune-Carpenter
Areoligera senonensis Lejeune-Carpenter
Carpatella cornuta Grigorovich
Cerodinium Vozzennikova spp.
Cerodinium striatum (Drugg) Lentin & Williams
Charlesdowniea variabilus (Bujak) Lentin & Vozzhennikova
Cordosphaeridium Eisenack spp.
Cordosphaeridium cantharellus (Brosius) Gocht
Damassadinium californicum (Drugg) Fensome et al
Dinogymnium acuminatum Evitt et al.
Dinogymnium digitus (Deflandre) Evitt et al.
Dinogymnium euclaense Cookson & Eisenack
Eocladopyxis peniculata Morgenroth
Glaphrocysta Stover & Evitt spp.
Glaphyrocysta semitectatum ? (Bujak) Lentin & Williams
Kallosphaeridium brevibaratum DeConinck
Membranophoridium aspinatum Gerlach
Membranophoridium bilobatum Michoux
Palaeocystodinium Alberti spp.
Palaeocystodinium benjaminii Drugg
Palaeohystrichophora Deflandre spp.
Palaeohystrichophora infusorioides Deflandre
Palaeohystrichophora infusorioides var. A and B of Aurisano (1989)
Pentadinium favatum Edwards
Pentadinium goniferum Edwards
Pentadinium laticinctum Gerlach
Phelodinium sp. Edwards
Senegalinium ?dilwynense (Cookson & Eisenack) Stover & Evitt
Spinidinium Cookson & Eisenack spp.
Spinidinium pulchrum (Benson) Lentin & Williams
Spiniferites septatus (Cookson & Eisenack) McLean
Subtilisphaera Jain & Millepied spp.
Trithyrodinium Drugg spp.
Xenascus Cookson & Eisenack spp.
Xenascus ceratioides (Deflandre) Lentin & Williams
Xenascus sarjeantii (Corradini) Stover & Evitt
Wetzeliella articulata Eisenack

grain non marine sediments on the southeastern side of the Savannah River, and is absent on the northwest side of the river. This conclusion was particularly important to the "Trans River Flow Study" undertaken by the United States Geological Survey and the Georgia State Geological Survey in the mid 1990s. The focus of the "Trans River Flow Study" was to consider the possibility of contamination flow from the Savannah River Site under the Savannah River and into the aquifers on the Georgia side.

SAVANNAH RIVER SITE HYDROGEOLOGY AND DINOFLAGELLATE STRATIGRAPHY

Bedrock (Paleozoic metamorphic and Triassic clastic rocks) and overlying Coastal Plain sediments constitute the hydrologic system beneath the SRS and surrounding areas. Poor water quality and low permeability prevent usage of ground water within the bedrock. Overlying the bedrock system, however, the Coastal Plain sediments comprise a multilayer hydrologic system in which retarding clay and marl beds are interlayered with sand and limestone beds that transmit water more readily (Aadland and Bledsoe, 1992). The hydrologic system is divided, in ascending order, into four aquifers (McQueen Branch, Crouch Branch, Gordon, and Upper Three Runs aquifers), each separated by intervening confining units (Text-Figure 3). The Gordon and Upper Three Runs aquifers coalesce towards the north forming the Steed Pond Aquifer (Lewis and Aadland, 1992).

The relationship between the palynology and the hydrogeology is presented below. Table 1 lists the pollen, spores, and dinoflagellates discussed below. Selected dinoflagellate species useful in hydrostratigraphic correlation are presented in Plate 1. The hydrogeology is presented in ascending order:

Appelton Confining System

Separating the aquifers of the Coastal Plain sequence from the basement is the Appelton Confining System, which consists of saprolite overlying crystalline basement rock, along with the lithologic unit referred to as the Cape Fear Formation. The Cape Fear Formation is mainly nonmarine and dated by pollen and spores as Turonian to Santonian (Late Cretaceous) in age (Lucas-Clark, 1992a). The Appleton Confining System/Cape Fear Formation is characterized by pollen species such as *Rugubivesiculites* Pierce spp., *Classopollis* Pflug spp., *Plicapollis* Pflug spp. *Complexiopollis* Krutzsch spp., *Holkopollenites* (Fairchild) Stover, Elsik, & Fairchild spp., and various tricolpate, tricolporate, and porate pollen.

McQueen Branch Aquifer

The McQueen Branch Aquifer overlies the Appelton Confining System and is the lower of two Cretaceous aquifers at SRS. This aquifer represents the principal source for domestic use at SRS. Domestic wells in these sands commonly yield more than 1000 gallons per minute of good quality water.

The McQueen Branch Aquifer corresponds to the sands of the Middendorf Formation and the sand in the lower part of the Black Creek Formation at SRS. The Middendorf Formation is considered nonmarine to shallow marine in origin. Dinoflagellates in the shallow marine facies have yielded mainly Santonian to early Campanian (Late Cretaceous) ages. In most marine samples from the Middendorf Formation, dinoflagellates are sparse and assemblages consist mainly of spiny cysts and a few peridiniacean species of genera such as *Subtilisphaera* Jain & Millepied, *Spinidinium* Cookson & Eisenack, and *Palaeocystodinium* Alberti. Pollen species of the Middendorf Formation are similar to those in the Cape Fear Formation, with a greater abundance of *Rugubivesiculites* spp. and fewer *Classopollis* spp. Examples of some useful species in these nonmarine units include: *Parvisaccites rugulatus* Brenner, *Complexipollis abdita* Tschudy and *Brevicolporites* sp. B of Christopher (1978).

The lower part of the Black Creek Formation is marine to nonmarine in origin. Dinoflagellates in the shallow marine facies have yielded mainly Campanian ages. In most marine samples, dinoflagellates are abundant, and assemblages consist of abundant small peridiniacean species of *Subtilisphaera*, *Spinidinium*, *Palaeohystrichophora* Deflandre, and larger peridiniacean species of *Andalusiella* (Riegel) Riegel & Sarjeant, *Xenascus* Cookson & Eisenack, *Palaeocystodinium*, *Cerodinium* Vozzennikova, and *Trithyrodinium* Drugg. Gonyaulacacean dinoflagellates, other than ceratioid types, are sparse (Lucas-Clark, 1992a). Age diagnostic dinoflagellate species in the McQueen Branch Aquifer/Black Creek Formation include *Andalusiella spicata* (May) Lentin & Williams, *Xenascus sarjeantii* (Corradini) Stover & Evitt, and *Palaeohystrichophora infusorioides* Deflandre.

McQueen Branch Confining Unit

The McQueen Branch Confining Unit separates the two Cretaceous (McQueen Branch and Crouch Branch) aquifers. The unit thins and pinches out in the vicinity of the well cluster P 19 (Text-Figure 1) near the center of the SRS (where the two Cretaceous aquifers are in communication) but it persists throughout the remainder of the site region. The unit corresponds to the clays of the middle third of the Black Creek Formation.

Dinoflagellates from this unit are mainly late Campanian and early Maastrichtian in age. Characteristics of the dinoflagellate assemblages vary. Some assemblages are dominated by spiny cysts; others by *Areoligera coronata* (Wetzel) Lejeune-Carpenter and *A. senonensis* Lejeune-Carpenter; others by small peridiniacean cysts; and others by *Xenascus ceratioides* (Deflandre) Lentin & Williams and *X. sarjeantii*. Useful species of dinoflagellates include, for example: *Dinogymnium euclaense* Cookson & Eisenack, *D. digitus* (Deflandre) Evitt et al., and *Palaeohystrichophora infusorioides* var. A and B of Aurisano (1989). Resedimentation of early Cretaceous dinoflagellates is not uncommon (Lucas-Clark, 1992a).

Crouch Branch Aquifer

The Crouch Branch Aquifer represents the upper of the two Cretaceous aquifers at SRS. It consists of the sand layers in the uppermost part of the Black Creek Formation and the majority of the overlying Steel Creek Formation. Dinoflagellates from the fine-grained marine portions of this aquifer have yielded dates of latest Campanian through Maastrichtian. Unfortunately, the clean sand and highly altered kaolinitic clay comprising the upper part of this aquifer often do not contain well-preserved fossil material. However, generic level identification of dinoflagellates such as *Dinogymnium* spp. and species of *Rugubivesiculites* pollen are useful in distinguishing Cretaceous from overlying Tertiary units (Lucas-Clark and Van Pelt, 1993). Diagnostic dinoflagellate species include: *Dinogymnium acuminatum* Evitt et al, *Palaeocystodinium benjaminii* Drugg, and *Cerodinium striatum* (Drugg) Lentin & Williams.

Crouch Branch Confining Unit

The confining system that overlies the Crouch Branch Aquifer represents the principal confining unit for the SRS and vicinity. This confining unit prevents migration of contaminated water from the overlying Tertiary age Steed Pond and Gordon aquifers into the underlying Cretaceous Crouch Branch Aquifer. However, in some areas within the SRS, the confining unit is discontinuous resulting in local communication of the Tertiary and Cretaceous age aquifers. It is thus critical to identify the confining unit where it is present, and important to understand its relation to other units.

The Crouch Branch Confining Unit corresponds to early and late Paleocene and early Eocene lithologic units corresponding to the Sawdust Landing Formation, the Lang Syne Formation, the Snapp Formation, and the Fourmile Branch Formation. The base of the Crouch Branch Confining Unit/Sawdust Landing Formation is characterized by the first appearance of several Danian (Paleocene) dinoflagellate species: *Carpatella cornuta* Grigorovich, *Damassadinium californicum* (Drugg) Fensome et al, *Spiniferites septatus* (Cookson & Eisenack) McLean, *Senegalinium ?dilwynense* (Cookson & Eisenack) Stover & Evitt, and *Spinidinium pulchrum* (Benson) Lentin & Williams.

The upper part of the Crouch Branch Confining Unit (equivalent to the upper part of the Lang Syne, Snapp and Fourmile Branch formations) is characterized by dinoflagellate assemblages that are dominated by one or a few species. These assemblages may comprise small peridiniacean species along with *Areoligera* Lejeune-Carpenter spp. and/or *Cordosphaeridium* Eisenack spp. *Apectodinium homomorphum* (Deflandre & Cookson) Lentin & Williams, *A. quinquelatum* (Williams & Downie) Lentin & Williams, and *Eocladopyxis peniculata* Morgenroth make their first appearance in the upper part of the confining unit (Edwards, 1990; Lucas-Clark, 1992b).

Gordon/Steed Pond Aquifer and Gordon Confining Unit

The middle to late Eocene age aquifer (Steed Pond/Gordon) and confining unit (Gordon) system overlying the Crouch Branch Confining Unit is a complex set of strata consisting, in part, of carbonate (limestone) sediments that are lenticular and discontinuous, as well as alternating clay and sand layers that act as confining units. Understanding the potential complexity of groundwater flow requires considerable accuracy in the identification and correlation of the lithologic units.

Lithologic units that correspond to the Steed Pond and equivalent aquifers include: the Congaree Formation, Warley Hill Formation, McBean Member of the Santee Formation, Clinchfield Formation, Irwinton Sand and Griffins Landing members of the Dry Branch Formation, and the Tobacco Road Sand Formation. Sediments comprising the Tobacco Road Sand Formation are consistently barren of organic-walled microfossils. The other lithologic units have yielded dinoflagellate assemblages, usually from the clay/silt rich intervals.

Dinoflagellates useful in correlating the complex of middle to late Eocene units include: *Pentadinium favatum* Edwards, *P. goniferum* Edwards, *Glaphrocysta* Stover & Evitt spp., *Wetzelliella articulata* Eisenack group, *Membranophoridium aspinatum* Gerlach, *M. bilobatum* Michoux and *Charlesdowniea variabilis* (Bujak) Lentin & Vozzhennikova.

Throughout most of the SRS, confining conditions (Gordon Confining Unit) exist to separate the Steed Pond Aquifer into two distinct aquifer zones, the Upper Three Runs and underlying Gordon aquifer zones. The Gordon Confining Unit consists of sandy clay to clayey sand which represent parts of the Warley Hill Formation, and the Blue Bluff Unit Member. Dinoflagellates first appearing within this confining unit include: *Pentadinium laticinctum* Gerlach, *Cordosphaeridium cantharellus* (Brosius) Gocht, and *Charlesdowniea variabilis*. *Pentadinium goniferum* last appears within the upper part of this confining unit.

CONCLUSIONS

Dinoflagellate stratigraphy has been applied successfully as a geologic tool in the establishment of the subsurface hydrogeologic framework beneath the SRS. Because of the rapid vertical and lateral changes in up-dip coastal plain setting, dinoflagellate, pollen and spore biostratigraphy, and to some extent dinoflagellate paleoenvironmental interpretations have been critical in the correlation of aquifers and confining units and the identification of facies changes and structural features (e.g. faults, slump features). Recognition of these features is important as they serve as preferential pathways for groundwater movement and contaminant transport.

ACKNOWLEDGMENTS

We thank Lucy E. Edwards for helpful discussion on particular aspects of these palynofloras. We also thank Drs. John H. Wrenn, and Donald W. Engelhardt for their constructive reviews of the manuscript. Support for this project was provided by Department of Energy under contract No. DE-AC09-96-SR18500. We gratefully acknowledge Bechtel Savannah River Incorporated for permission to publish this information.

References Cited

AADLAND, R.K.
1990 Hydrogeologic Characterization of the Cretaceous–Tertiary Coastal Plain Sequence at the Savannah River Site, South Carolina. *In*: Zullo, V.A., Harris, W.B., and Price, V., (eds.) 1990, Savannah River Region: Transition between the Gulf and Atlantic Coastal Plains, *Proceedings of the Second Bald Head Island Conference on Coastal Plains Geology*, p. 57–60.

AADLAND, R.K., and BLEDSOE, H.W.
1992 Hydrogeologic Classification of the Cretaceous–Tertiary Coastal Plain Sequence at the Savannah River Site. *Geological Society of America Abstracts with Programs*, 24: 1.

AADLAND, R.K., GELLICI, J.A., and THAYER, P.A.
1995 Hydrogeologic Framework of West–Central South Carolina. *South Carolina Department of Natural Resources, Water Resources Division Report*, no. 5, 200 pgs.

AADLAND, R.K., THAYER, P.A., and SMITS, A.D.
1992 The Hydrogeological Atlas of West–Central South Carolina. *Westinghouse Savannah River Company, Westinghouse Savannah River Company Technical Report* 90–987.

AURISANO, R.W.
1989 Upper Cretaceous dinoflagellate biostratigraphy of the subsurface Atlantic Coastal Plain of New Jersey and Delaware, U.S.A. *Palynology*, 13: 143–180.

CAHIL, J.M.
1982 Hydrology of the low-level radioactive solid-waste burial site and vicinity near Barnwell, South Carolina. *U.S. Geological Survey Open File Report* 82–863, 101 pgs.

EDWARDS, L.E.
1990 Dinocysts from the Lower Tertiary Units in the Savannah River area, South Carolina and Georgia. *In*: Zullo, V.A., Harris, W.B., and Price, V., (eds.) 1990, Savannah River Region: Transition between the Gulf and Atlantic Coastal Plains, *Proceedings of the Second Bald Head Island Conference on Coastal Plains Geology*, pp. 89–90.

FALLAW, W.C., and PRICE, V.
1992 Outline of stratigraphy at the Savannah River Site. *In*: Fallaw, W.C., and Price, V., (eds.), 1992, *Carolina Geological Society Field Trip Guide Book, Nov. 13–15, 1992*, Geological Investigations of the central Savannah River area, South Carolina and Georgia: U.S. Department of Energy and S.C. Geological Survey, p. CGS-92-B-11-1–33.
1995 Stratigraphy of the Savannah River Site and Vicinity. *Southeastern Geology*, 35: 21–58.

FAYE, R.E., and PROWELL, D.C.
1982 Effects of Late Cretaceous and Cenozoic faulting on the geology and hydrology of the Coastal Plain near the Savannah River, Georgia, and South Carolina. *U.S. Geological Survey Open File Report* 82–156.

GOHN, G.S., and CAMPBELL, B.G.
1992 Recent revisions to the stratigraphy of subsurface Cretaceous sediments in the Charleston, South Carolina area. *South Carolina Geology*, 34: 25–38.

LEWIS, S., and AADLAND, R.K.
1992 Hydrogeologic Setting of the A/M Area: Framework for Groundwater Transport. *Westinghouse Savannah River Company Technical Report* 92-355, 80 pgs.

LUCAS-CLARK, J.
1992a Problems in Cretaceous palynostratigraphy (dinoflagellates and pollen) of the Savannah River Site Area. *In*: Zullo, V.A., Harris, W.B., and Price, V., (eds.) 1992, Savannah River Region: Transition between the Gulf and Atlantic Coastal Plains, *Proceedings of the Second Bald Head Island Conference on Coastal Plains Geology*, pp. 74.
1992b Some characteristic fossil dinoflagellate cysts of Eocene strata, Savannah River Site, South Carolina. *In*: Fallaw, W., and Price, V. 1992, Geological Investigations of the Central Savannah River Area, South Carolina and Georgia. *Carolina Geological Society Field Trip Guidebook 1992*, pp. 58–60.

LUCAS-CLARK, J., and VAN PELT, R.S.
1993 Application of dinoflagellate biostratigraphy to environmental restoration activities at the Savannah River Site, South Carolina, U.S.A., (abst.) *Abstracts with Programs: Fifth International Conference on Modern and Fossil Dinoflagellates, Zeist, The Netherlands, April 1993.*

MARINE, I.W., and ROOT, K.R.
1975 A ground water model of the Tuscaloosa Aquifer at the Savannah River Plant. *In*: Crawford, T.V. (ed.), *Savannah River Laboratory Environmental Transport and Effects Research Annual Report*, USERDA Report DP-1374, E.I. duPont de Nemours & Co., Savannah River Laboratory, Aiken, South Carolina.

NYSTROM, P.G., Jr., WILLOUGHBY, R.H., and PRICE, L.K.
1991 The Cretaceous and Tertiary stratigraphy of the South Carolina upper Coastal Plain. *In*: Horton, J.W., Jr., and Zullo, V.A., (eds.), *Geology of the Carolinas*. University of Tennessee Press, Knoxville, Tennessee, pp. 221–240.

PROWELL, D.C., CHRISTOPHER, R.A., EDWARDS, L.E., BYBELL, L.M., and GILL, H.E.
1985a Geologic section of the updip Coastal Plain from central Georgia to western South Carolina. *U.S. Geological Survey Miscellaneous Field Studies*, Map MF-1737.

PROWELL, D.C., EDWARDS, L.E., and FREDERIKSEN, N.O.
1985b The Ellenton Formation in South Carolina — A revised age designation from Cretaceous to Paleocene. *U.S. Geological Survey Bulletin*, 1605-A: A63–A69.

SIPLE, G.E.
1967 Geology and ground water of the Savannah River Plant and vicinity, South Carolina. *U.S. Geological Survey Water-Supply Paper* 1841, 113 pgs.

VAN PELT, R.S., WAANDERS, G.L., LUCAS-CLARK, J., and HARRIS, M.K.
1993 The application of palynology to waste-site characterization: an example from the Savannah River Site, South Carolina, (Abst.) *American Association of Stratigraphic Palynologists 1993 Annual Meeting Programs and Abstracts*, pp. 82.

VAN PELT, R.S., and LUCAS-CLARK, J.
1996 The application of palynology to environmental restoration activities: an example from the Savannah River Site, South Carolina, (Abst.) *Ninth International Palynological Congress Programs and Abstracts*, June 1996, pp. 82.

Authors' addresses

ROBERT S. VAN PELT
Bechtel Savannah River Inc.
Environmental Restoration Engineering & Technology
Savannah River Site, Building 730-2B
Aiken, SC 29808
U.S.A.

JOYCE LUCAS-CLARK
Clark Geological Services
1023 Old Canyon Road
Fremont, CA 94536
U.S.A.

THE PALYNOFLORAL SUCCESSION ACROSS THE CRETACEOUS TO PALEOCENE TRANSITION ZONE, MERIDA ANDES, WESTERN VENEZUELA

David T. Pocknall, Robert N. Erlich, Jeffrey A. Stein, and Maria Antonieta Lorente

Abstract

In northern South America the Cretaceous to Paleocene transition usually occurs at a change from predominantly marine depositional conditions of the Late Cretaceous to continental (fluvial/lacustrine) conditions in the Early Paleocene. Dating of this transitional interval has been largely based on terrestrially derived palynomorphs (spores, pollen) and calibration to the marine fossil record has not been possible. A measured section through the Mito Juan Formation (Late Cretaceous) to Los Cuervos Formation (Paleocene) at Río Lora in the Merida Andes of western Venezuela contains diverse palynomorph assemblages, including dinoflagellates. Graphic correlation of the dinoflagellate data against an extensive database indicates the presence of three unconformities, although it is possible that a more or less conformable Cretaceous–Tertiary boundary section is preserved within a covered interval of approximately 65 ft, assuming continuous deposition and preservation of sediments. Dinoflagellate data indicate that the sediments below the covered interval are latest Cretaceous and immediately above are very earliest Paleocene in age. Key dinoflagellates include (in ascending stratigraphic order) *Glaphyrocysta perforata, Dinogymnium pustulicostatum, Damassodinium californicum, Fibrocysta bipolaris, Kenleyia lophophora*, and *Carpatella cornuta*. Calibration to a global database containing marine fossils makes it possible to better refine the age ranges of some key spore and pollen taxa that have previously been used to define the Late Cretaceous *Proteacidites dehaani* and Early Paleocene *Spinizonocolpites baculatus* pollen zones in South America.

The rich palynomorph assemblages indicate environments of deposition ranging from nearshore/nonmarine to middle neritic throughout the interval studied. A predominant component of the palynofloras is an association of species from the dinoflagellate genera *Andalusiella, Palaeocystodinium*, and *Cerodinium*. The fungal fruiting body *Trichopeltinites*, which is known to disappear at or about the Cretaceous–Tertiary boundary in sections in North America is also recorded, although it ranges into the Paleocene section at Río Lora.

INTRODUCTION

Reconnaisance palynological studies of outcrop samples from western Venezuela in the early 1990s revealed the presence of well preserved and diverse palynofloras, essentially of terrestrial origin, that could be used to age date and determine the environments of deposition of the sections studied. During 1993–1995, three visits were made to rivers and road cuts along the northern flank of the Merida Andes, south of Lake Maracaibo, in order to study the Cretaceous and Paleogene sections in detail. In this paper we review only a small part of the overall study, focusing attention on the Upper Cretaceous to Lower Paleocene section exposed in Río Lora (Text-Figure 1), describing and illustrating the palynological assemblages, and highlighting the presence of a potential Cretaceous–Tertiary boundary in the section. The details of the boundary section and its significance to the topic of the terminal Cretaceous event (Alvarez et al., 1980; Hildebrand et al., 1991) have previously been discussed by Pocknall et al. (1997).

PREVIOUS WORK

The use of palynology in petroleum related investigations in Venezuela began in the 1930s, gained some momentum in the 1940s, and eventually during the 1950s developed as an essential tool in the study of sedimentary basins. In one of the earliest publications, Kuyl et al. (1955) used palynology to correlate some of the major Eocene-aged sand bodies in Lake Maracaibo, the Sierra de Perija (west of the lake), and the northern flank of the Merida Andes. In their classic work, Germeraad et al. (1968), defined pollen zones for use in correlation and they endeavoured to calibrate the age of the zones to the marine fossil record (foraminifera) where nonmarine sediments interfinger with fossiliferous marine sediments.

Over the past 30 years many zonal schemes have been defined for northern South American Cretaceous and Paleogene sediments (summarized in Text-Figure 2). In addition to Germeraad et al. (1968), the studies of Muller et al. (1987) and Sarmiento (1994) are relevant to the stratigraphic section discussed here. The age assignments used in this paper are not in the form of zonal designations; they are made through comparison of the Río Lora outcrop palynological assemblages with Amoco's corporate paleontological database. In Río Lora the palynological assemblages obtained across the Cretaceous to Paleocene "transition" contain both terrestrially-derived (spores and pollen) and marine-derived (dinoflagellates) palynomorphs. This has provided us with an opportunity to more accurately date the section in question and also calibrate some of the ages of the key spore and pollen types that have provided the basis for definition of some of the pollen zones shown in Text-Figure 2.

RÍO LORA SECTION

The section is located on the western bank of Río Lora, approximately 5 miles south of the main highway between the cities of El Vigia and La Fria (8°25.15'N, 71°51.63'W; Text-Figure 1). The basal unit in the measured

Pocknall, David T., Erlich, Robert N., Stein, Jeffrey A., and Lorente, Maria Antonieta
2001 The Palynofloral Succession Across the Cretaceous to Paleocene Transition Zone, Merida Andes, Western Venezuela; *In*: Goodman, D.K., and Clarke, R.T. (eds.), *Proceedings of the IX International Palynological Congress, Houston, Texas, U.S.A., 1996*; American Association of Stratigraphic Palynologists Foundation, p. 171–179.
 ISBN 0–931871–06–9

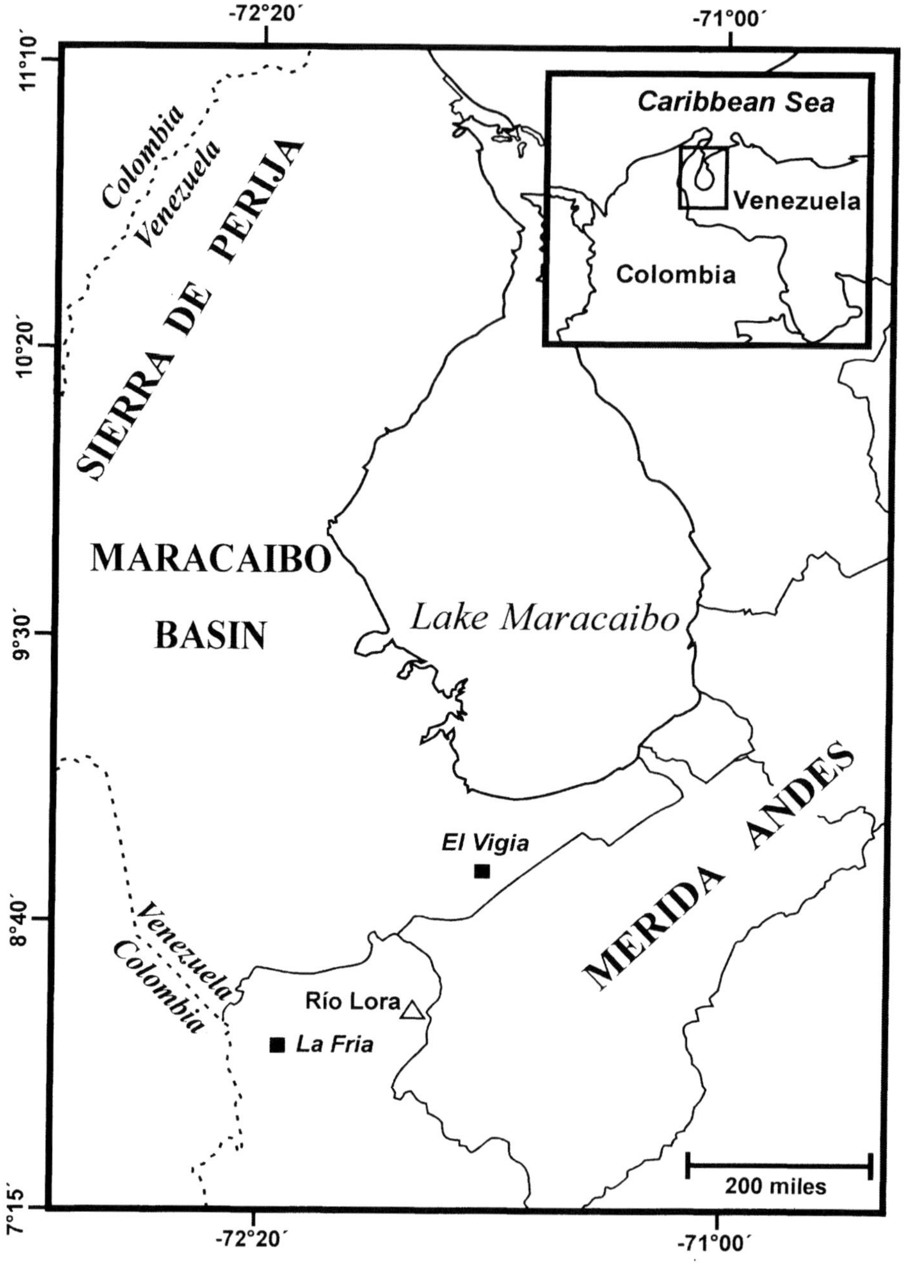

Text-Figure 1. Locality map showing the general location of the Río Lora outcrop (marked with a triangle) in the southwestern Merida Andes, western Venezuela.

section is the Mito Juan Formation, overlain successively by the Catatumbo, Barco, and Los Cuervos formations (Text-Figure 3). Other formations exposed in the Río Lora section include the Mirador (Eocene), Carbonera (Eocene–Oligocene) , Leon (Oligocene), and Palmar (Miocene); these are not discussed in this paper.

The Mito Juan Formation, originally defined by Garner (1926), consists of black shales, similar to the underlying Colon Formation, with occasional interbeds of thin sands and glauconite, although at Río Lora the predominant lithology is shale. The overlying Catatumbo Formation, defined by Notestein et al. (1944) is the basal unit of the Orocue Group and consists of interbedded fine sandstones, shales, and occasional limestones; it is often difficult to see a lithological distinction between the Catatumbo and the underlying Mito Juan. Kuyl et al. (1955) indicated that the Catatumbo Formation was of Paleocene age.

The Barco and Los Cuervos formations were defined by Notestein et al. (1944) from the Barco Concession in eastern Colombia. The Barco is the middle unit of the Orocue Group and consists of interbedded sands and shales; the sand in the Río Lora section between 4,010 and 4,020 ft (Text-Figure 3) is probably the basal sand of the formation. In some areas, including the type locality, the formation can be more than 400 ft thick (Gonzalez de Juana et al., 1980), but the thickness is considerably reduced in the region of Río Lora. The

AGE		TROPICAL ZONE: Germeraad, Hopping & Muller (1968) — Caribbean	TROPICAL ZONE: Germeraad, Hopping & Muller (1968) — Atlantic	TROPICAL ZONE: Germeraad, Hopping & Muller (1968) — Pantropical	NORTHERN SOUTH AMERICA: Muller, de Giacomo & Van Erve (1987)	COLOMBIA: Sarmiento (1994)
EOCENE	Late			*Verrucatosporites usmensis*	*Echiperiporites estelae*	
EOCENE	Middle	*Retitricolporites guianensis* *Psilatricolporites operculatus* *Psilatricolporites crassus*		*Monoporites annulatus*	*Janmulleripollis pentaradiatus* *Bombacacidites foveoreticulatus* *Retitricolpites guianensis* *Bombacacidites soleaformis* *Retitricolpites magnus* *Echitriporites trianguliformis Forma A*	
EOCENE	Early		*Retibrevitricolporites triangulatus*		*Rugutricolporites felix*	
PALEOCENE		*Foveotricolpites perforatus* *Ctenolophonidites lisamae* *Foveotriletes margaritae*	*Retidiporites magdalenensis*	*Proxapertites operculatus*	*Foveotricolpites perforatus* *Gemmastephanocolpites gemmatus* *Spinizonocolpites baculatus*	*Foveotriletes margaritae* (Zonas IIA & B)
LATE CRETACEOUS (MAASTRICHTIAN)			*Proteacidites dehaani*		*Proteacidites dehaani*	*Buttinia andreevi* (Zona I)

Text-Figure 2. Summary chart of Late Cretaceous to Eocene palynological zonal schemes used in northern South America (adapted from Lorente, 1986).

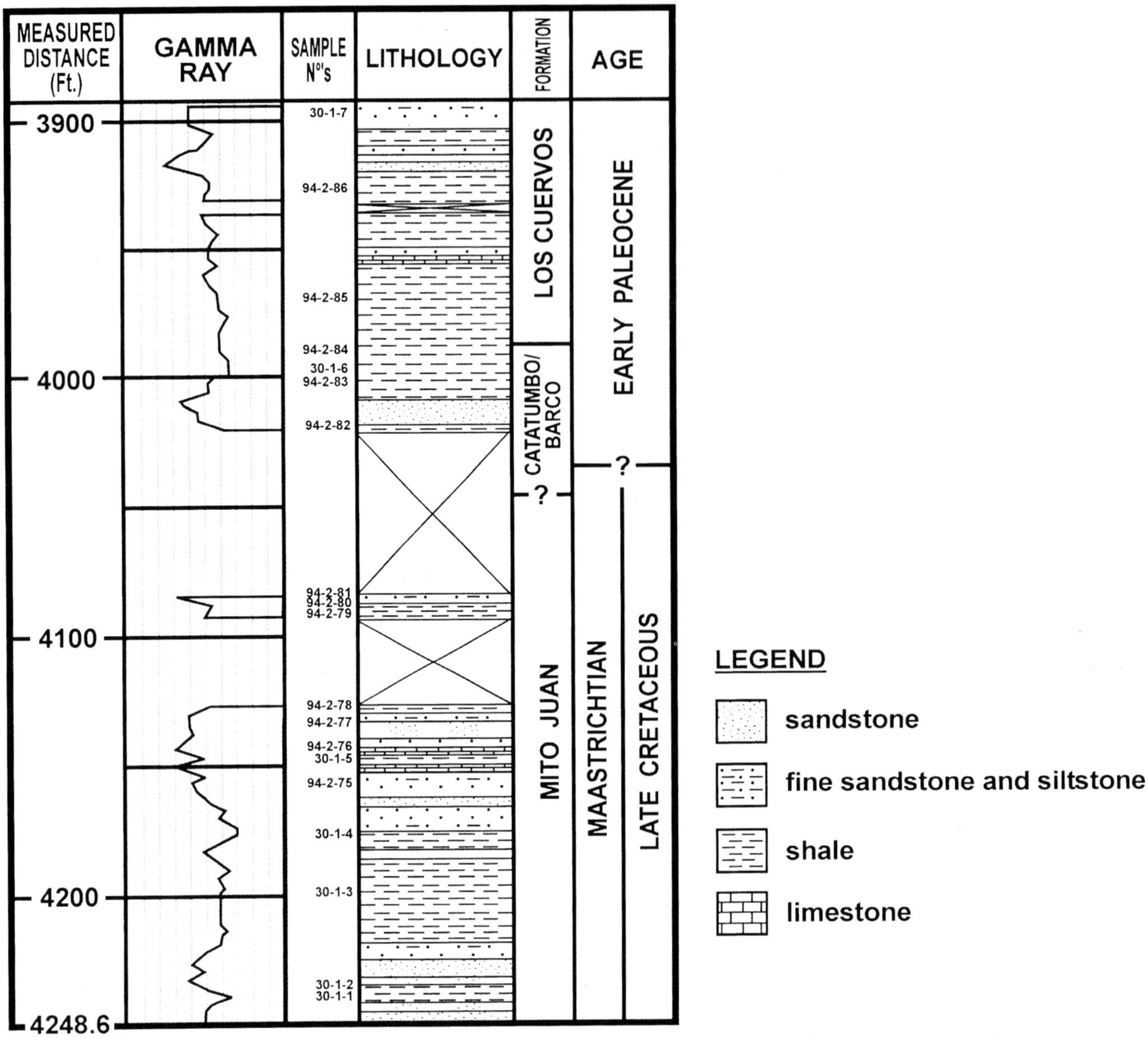

Text-Figure 3. Measured section for the Mito Juan to lower Los Cuervos interval exposed Río Lora. The columns show gamma ray, sample locations, lithology, formation designations, and age.

overlying Los Cuervos Formation represents the upper unit of the Orocue Group and consists of interbedded sands, carbonaceous shales, and thin coals, with occasional thin limestones near the basal contact with the Barco Formation.

RESULTS

Nineteen (19) samples were collected for palynology from the lower 350 ft of the Río Lora section (see Text-Figure 3 for locations). Samples were prepared using standard processing techniques (Wood et al., 1996). All samples, except one, contained well-preserved, and at times diverse assemblages of spores, pollen, freshwater algae, fungal remains, and dinoflagellates. A list of all specimens identified are recorded in Table 1.

A summary palynological log (Text-Figure 4) shows the abundance of some of the key palynomorph groups, and gives an indication of the relative fluctuations between land-derived palynomorphs and dinoflagellates. A graphic correlation plot of the Mito Juan to lower Los Cuervos interval graphed against Amoco's Maastrichtian–Danian composite standard database of global and cosmopolitan planktonic dinoflagellate species (Text-Figure 5) summarizes the age relationships of the section.

DISCUSSION

The interpreted graphic correlation plot (Text-Figure 5) indicates the presence of three unconformities (terraces) and four sequences (sloped lines) in the Río Lora section. The lower Maastrichtian sequence (consisting of 5 samples)

TABLE 1. List of abundance data for palynomorphs recovered from Cretaceous to Paleocene transition in Río Lora.

Taxon\Sample No.	30-1-7	94-2-86	94-2-85	94-2-84	30-1-6	94-2-83	94-2-82	94-2-81	94-2-80	94-2-79	94-2-78	94-2-77	94-2-76	30-1-5	94-2-75	30-1-4	30-1-3	30-1-2	30-1-1
Depth (feet)	3897	3927	3969	3989	3999	4002	4019	4084	4087	4091	4127	4133	4143	4146	4156	4176	4209	4235	4238
Spores and Pollen																			
Annutriporites iversenii	2	.	.	.	.	.	.	.	.	.	.	.	.	.	3	.	.	.	.
Araucariacites australis	.	.	1	.	1	1	.	.	.	.	.	.	.	.	.	.	.	.	.
Arecipites spp.	12	3	2	2	9	.	5	.	.	.	.	.	.	2	13	13	14	9	7
Azolla sp.	.	1	.	.	.	.	.	.	.	.	.	.	.	.	.	.	.	.	.
Colombipollis tropicalis	3	3	.	.	.	3	4	.	.	.	.	1	.	2	.	.	2	.	.
Crussafontites grandiosus	.	2	1	2	2	1	2	.	.	.	.	.	1	.	.	.	.	.	.
Cyathidites australis type	40	23	11	14	15	8	17	1	5	1	3	2	1	4	11	46	53	52	40
Droseridites senonicus	.	.	.	.	.	.	.	?	.	.	.	.	.	.	.	.	.	.	.
Echitriporites trianguliformis	9	.	.	.	13	.	1	.	.	.	.	.	.	2	8	2	7	3	.
Ephedripites spp.	.	.	1	.	.	.	2	.	.	.	.	.	1	.	.	.	.	.	5
Foveotricolpites perforatus	.	.	.	.	1	.	.	.	.	.	.	.	.	.	.	.	.	.	.
Foveotriletes margaritae	4	14	10	14	9	5	20	.	1	1	7	2	5	1	6	8	5	10	.
Laevigatosporites ovatus	.	.	.	1	.	.	1	.	2	.	.	.	.	.	.	.	.	.	.
Liliacidites spp.	6	.	.	.	4	.	1	.	.	.	.	.	.	1	1	3	5	5	3
Longapertites vaneendenburgi	1	.	.	.	.	.	.	.	.	.	.	.	.	.	.	.	.	.	.
Longapertites spp.	2	.	.	.	.	.	.	.	.	.	.	.	.	.	.	.	.	.	.
Mauritia crassibaculatus	10	7	16	8	27	5	5	1	3	1	2	.	.	.	18	24	25	18	27
Polypodiaceoisporites retirugatus	.	1	.	1	.	1	.	.	.	.	.	.	.	.	.	.	.	.	.
Proteacidites dehaani	1	2	8	2	5	6	3	.	3	.	1	1	1	.	.	1	2	.	.
Proxapertites cursus	.	7	.	2	2	.	1	.	1	.	.	.	.	.	.	2	.	.	.
Proxapertites maracaiboensis	2	3	1	9	2	3	5	1	6	1	2	3	4	5	1	1	7	.	.
Proxapertites operculatus	5	1	5	2	2	5	4	.	2	.	2	1	.	.	1	2	6	4	3
Retidiporites magdalenensis	3	.	.	.	3	.	.	.	1	.	.	.	.	.	2	.	1	.	.
Retisyncolporites giganteus	.	2	.	.	.	.	.	.	4	.	.	.	.	.	.	.	2	.	.
Spinizonocolpites baculatus	.	8	3	4	9	1	1	.	1	2	2	.	.	2	.	4	4	10	7
Spinizonocolpites echinatus	1	8	2	1	.	.	3	.	1	1	.	1	.	2	3	1	.	1	1
Stephanocolpites costatus	.	.	1	1	1	1	.	.	.	.	.	.	.	.	.	.	.	.	.
Tetradopollenites ericius	.	1	.	1	2	.	2	.	.	.	.	.	.	.	2	.	.	.	.
Tricolpites waiparaensis	.	.	.	.	2	.	.	.	.	.	.	.	.	.	2	1	5	3	1
Ulmoideipites krempii - 4 pores	3	.	.	1	1	.	.	.	.	.	.	.	.	.	2	1	.	.	1
Ulmoideipites krempii - 5 pores	1	.	.	.	.	.	.	.	.	.	.	.	.	.	.	1	.	.	.
Zlivisporis blanensis	1	.	.	.	.	1	1	.	.	.	.	.	.	.	.	.	.	.	.
	106	86	62	65	110	41	78	3	30	7	19	11	13	21	73	110	138	115	95
Others																			
Pediastrum spp.	.	3	.	2	1	.	1	.	.	.	.	.	.	.	.	1	2	1	.
Trichopeltinites sp.	1	4	3	2	3	2	6	.	.	.	.	.	.	.	.	1	.	.	.
Dinoflagellates																			
Achomosphaera ramulifera	.	.	.	.	3	.	.	.	.	.	.	.	.	.	.	.	.	.	.
Andalusiella spp.	.	.	.	1	3	.	2	5	.	2	.	.	1	1	1	15	6	3	.
Andalusiella polymorpha	.	.	.	.	.	.	.	15	.	.	.	.	.	.	.	.	.	.	.
Areoligera senonensis	.	.	.	.	3	1	.	.	.	.	.	.	.	.	.	.	.	.	.
Carpatella cornuta	.	1	.	1	.	.	.	.	.	.	.	.	.	.	.	.	.	.	.
Cassidium fragile	1	.	.	.	.	.	.	.	.	.	.	.	.	.	.	.	.	.	.
Ceratiopsis granulostriata	1	2	5	3	7	.	2	2	2	.	.	1	3	4	1	1	4	.	.
Cerodinium diebeli	.	.	.	.	.	.	.	.	.	.	1	.	.	.	.	.	.	.	.
Cerodinium striata	.	.	2	.	4	.	1	1	.	.	.	.	.	.	.	.	.	.	.
Cordosphaeridium fibrospinosum	.	.	.	6	.	1	1	1	.	1	1	.	1	.	.	.	.	.	.
Cyclopophysis sp.	.	.	.	.	.	.	.	?	.	?	.	.	.	.	.	.	.	.	.
Damassadinium californicum	1	2	.	1	1	.	2	.	.	.	.	.	.	.	.	.	.	.	.
Deflandrea spp.	.	.	.	.	1	.	.	1	1	.	.	?	.	.	.	.	.	.	.
Dinogymnium acuminatum	.	.	.	1	.	.	.	.	.	.	.	.	.	1	.	1	.	.	1
Dinogymnium pustulicostatum	.	.	.	.	.	.	.	.	2	.	.	.	.	.	.	.	.	.	.
Fibrocysta bipolaris	2	3	1	3	1	.	2	.	.	.	.	.	.	.	.	.	.	.	.
Glaphyrocysta perforata	.	2	.	.	2	1	1	.	.	50	.	.	.	.	.	.	.	.	.
Gymnodinium cf. *G.lanceolatum*	.	.	.	.	.	1	1	.	.	.	.	.	.	.	.	.	.	.	.
Kenleyia lophophora	.	.	.	.	1	1	.	.	.	.	.	.	.	.	.	.	.	.	.
Kleithriasphaeridium truncatum	1	.	.	.	.	.	.	.	1	.	1	1	.	3	15	.	.	.	.
Lejeunecysta hyalina	.	.	.	1	2	2	1	.	.	.	.	.	.	.	.	1	2	.	.
Odontochitina costata	.	.	.	.	.	.	.	.	.	.	.	.	.	.	.	1	.	.	.
Operculodinium sp.	.	.	.	3	.	.	.	.	.	.	.	.	.	.	.	.	.	.	.
Palaeocystodinium golzowense	.	.	46	2	6	5	1	23	41	2	30	4	3	10	5	36	21	6	4
Palaeoperidinium pyrophorum	.	.	.	1	.	.	1	.	.	.	.	.	.	.	.	.	.	.	.
Senegalinium dilwynense	.	.	.	.	.	2	.	.	1	1	.	.	.	.	.	.	.	.	.
Spiniferites spp.	4	18	2	19	10	39	15	100	.	.	.	.	1	.	.	.	.	.	3
Spiniferites ramosus granosus	.	.	.	1	.	.	.	100	30	20	81	100	100	100	7	.	5	11	.
Trithyrodinium evitti	.	.	.	1	.	.	.	.	.	.	.	.	.	.	.	.	.	.	.
Turbiosphaera filosa	.	2	.	1	1	.	.	.	1	.	.	.	.	.	.	.	.	.	.
	10	30	56	45	45	53	30	248	79	76	114	106	109	119	29	55	38	20	8

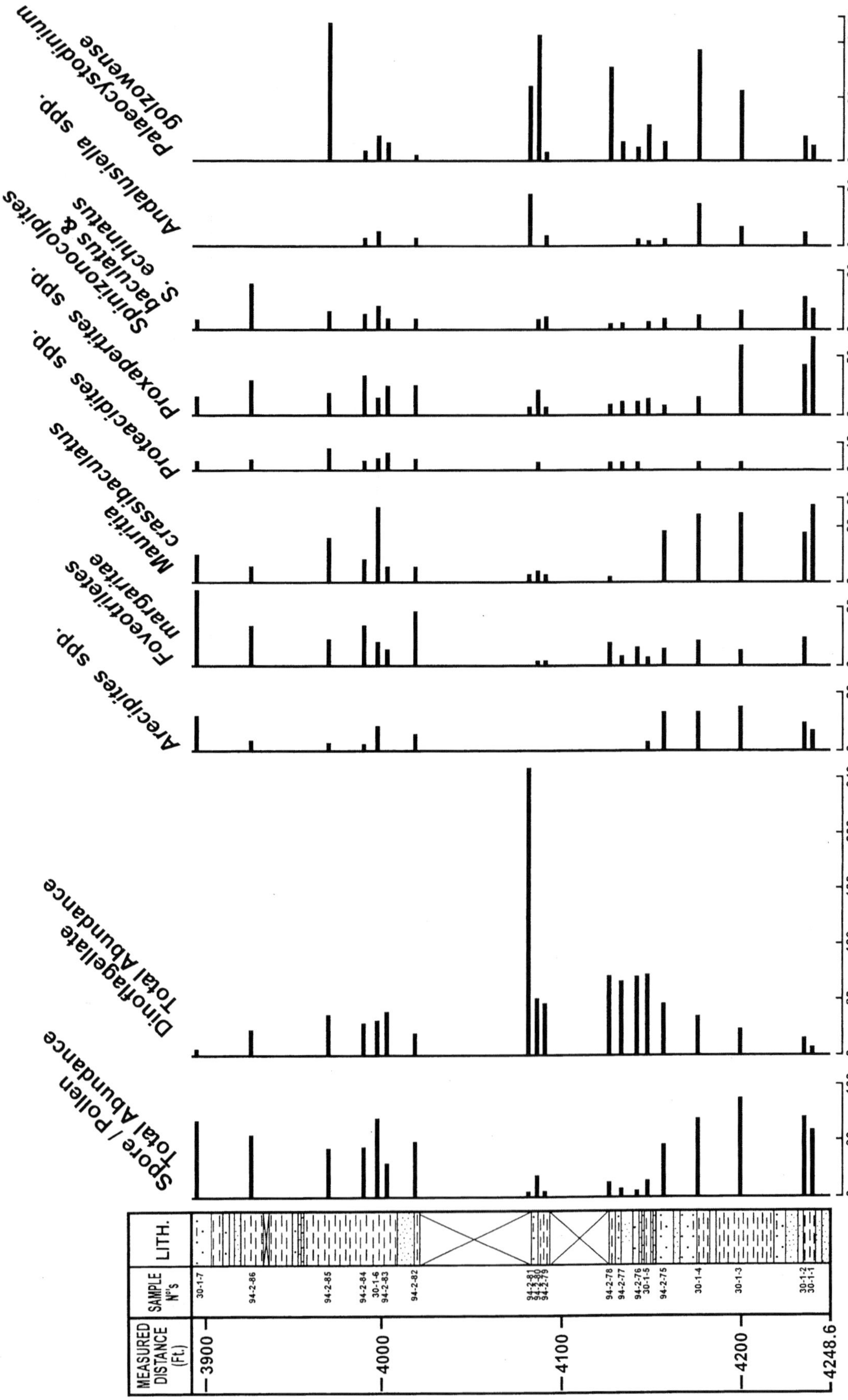

Text-Figure 4. Palynomorph abundance plot showing the relationships between the major palynomorph groups. For lithology see legend for Text-Figure 3.

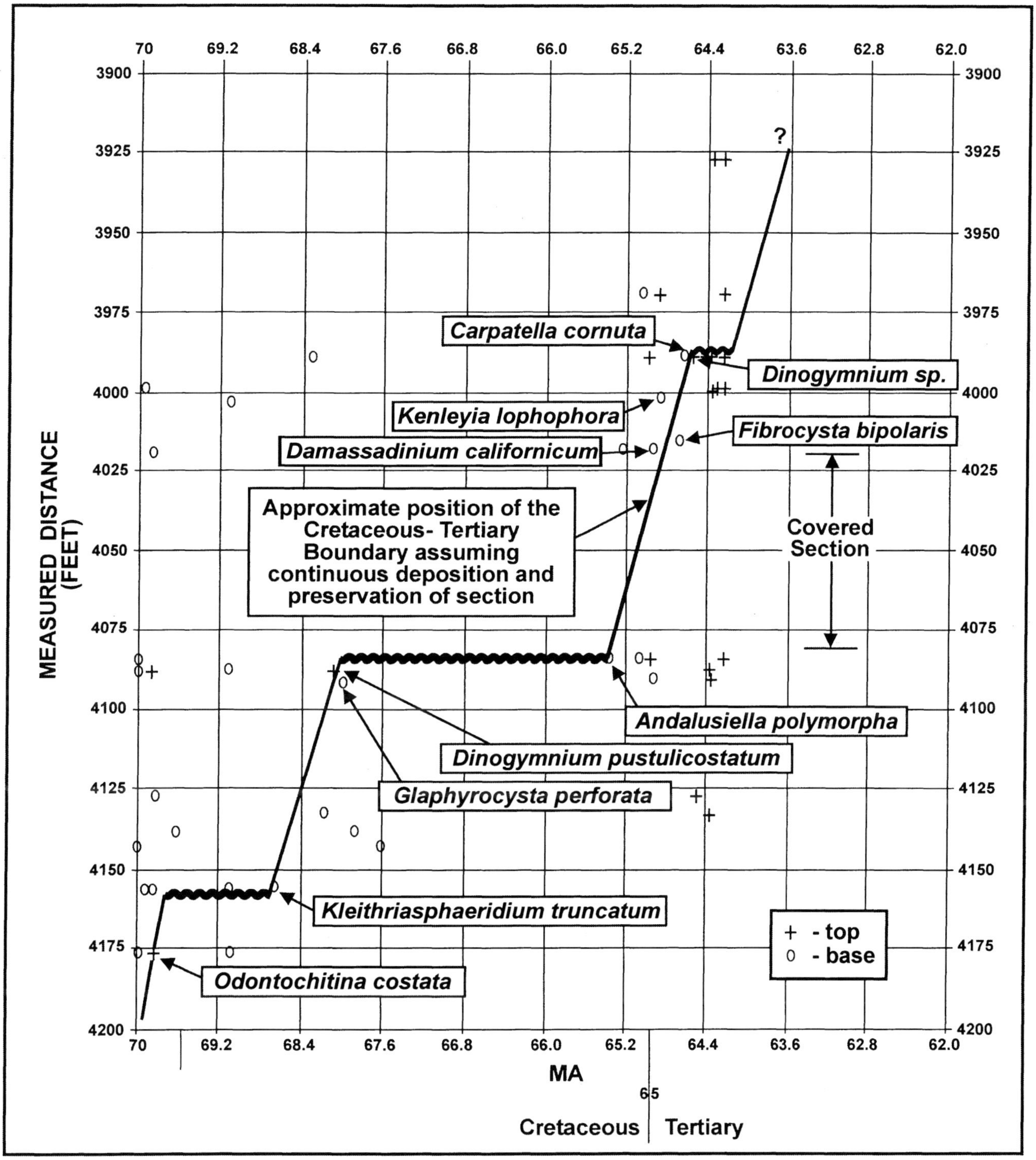

Text-Figure 5. Graphic correlation plot for Río Lora. Only dinoflagellates are plotted. The graph shows unconformities (missing section) as terraces and sequences of rock as solid, sloped lines. The approximate position of the Cretaceous–Tertiary boundary within the covered interval is indicated (after Pocknall et al., 1997).

in the interval 4156 to 4238 ft is characterized by a mixed assemblage of palynomorphs. Among the spores and pollen, *Cyathidites australis* type and *Mauritia crassibaculatus* are common, with minor abundances of *Spinizonocolpites baculatus, Echitriporites trianguliformis, Proxapertites* spp., *Arecipites* spp., and *Foveotriletes margaritae*. Dinoflagellates are dominated by *Palaeocystodinium golzowense* and *Andalusiella* species.

The line of correlation is constrained by the extinction of *Odontochitina costata* and the sequence is truncated at 4156 ft by the first (basal) appearance of *Kleithriasphaeridium truncatum*. Approximately 1.5 m.y. of time is missing at this unconformity which coincides with a lithological change from a sand dominated to a shale and limestone dominated sequence, and a shift in the gamma ray log signature. The first appearance of *Colombipollis tropicalis* is recorded within the lower sequence, representing the first published record of this species from Venezuela. *C. tropicalis* was previously reported only from Colombia by Sarmiento (1994).

A second sequence (7 samples) of Maastrichtian age extending from 4084 to 4146 ft overlies the lower sequence and consists of interbedded shales, limestones, and minor sandstones of the upper Mito Juan Formation. Palynomorph assemblages are dominated by dinoflagellates, especially *Palaeocystodinium golzowense*, but also by species of *Spiniferites*. Additionally, the common presence of *Glaphyrocysta perforata* and *Andalusiella polymorpha* in samples at 4091 and 4084 ft (Table 1), respectively, represents the greatest marine influence in this part of the Mito Juan Formation. *Proxapertites* spp., *Foveotriletes margaritae*, and *Cyathidites australis* are the most consistently recorded terrestrial palynomorphs, although they occur in very low numbers.

The top of the sequence is terminated by an unconformity at 4084 ft although the sequence could extend higher into the covered interval (Text-Figures 3–5). The terrace shown on Text-Figure 5 is constrained by the bases of *Glaphyrocysta perforata* and *Andalusiella polymorpha*, and the top of *Dinogymnium pustulicostatum*. The dramatic change in the assemblages (Text-Figure 4) in the next highest sample at 4019 ft indicates that at least a major facies change occurs between the samples, and this may coincide with an unconformity; only excavation of the covered interval and further sampling will resolve this question.

The covered section lies within the third sequence shown on Text-Figure 5. While the sequence probably extends from 3989 to 4084 ft, four of the samples are clustered in the uppermost 30 ft of the interval. The interval consists mostly of shale, although a 10 foot thick sandstone occurs at 4015 ft. The most common dinoflagellates are species of *Spiniferites*, however, dinoflagellates are generally not as common as they were in the previous sequence. Spores and pollen increase in abundance with *Arecipites* spp., *Cyathidites australis, Foveotriletes margaritae, Mauritia crassibaculatus, Proteacidites* spp., *Proxapertites* spp., *Spinizonocolpites baculatus* and *S. echinatus* commonly recorded (see Text-Figure 4).

Based on dinoflagellates, Pocknall et al. (1997) drew a line of correlation that inferred that a more-or-less conformable Cretaceous–Tertiary boundary sequence exists in Río Lora in a covered interval extending from 4019 and 4084 ft. Below the covered interval *Glaphyrocysta perforata* and *Dinogymnium pustulicostatum*, indicate a latest Maastrichtian age, and above the covered interval *Damassodinium californicum, Fibrocysta bipolaris, Kenleyia lophophora, Turbiosphaera filosa, Carpatella cornuta*, and *Cassidium fragile* indicate an earliest Danian (Paleocene) age. The sequence containing the Cretaceous–Tertiary boundary is thought to be truncated at another unconformity around 3989 ft, in part constrained by the top of *Andalusiella* spp. There are no major facies changes around this interval but a sample at 3969 ft contains the last occurrence of *Palaeocystodinium golzowense* indicating a significant shallowing between the Barco and the Los Cuervos formations. The overlying section is dominated by *Arecipites* spp., *Cyathidites australis, Foveotriletes margaritae, Mauritia crassibaculatus, Proxapertites* spp., *Spinizonocolpites baculatus* and *S. echinatus*. The palm elements (*Proxapertites, Arecipites*, and *Mauritia*) are indicative of a predominantly lower coastal plain or fluvial environment. Proximity of the shoreline during deposition of the Mito Juan Formation to the lower part of the Los Cuervos Formation sequence is indicated by the presence of *Spinizonocolpites baculatus* and *S. echinatus* which are considered similar to the pollen of the modern day mangrove palm *Nipa* (Fechner, 1988), together with floods of dinoflagellates. Ultimately the shoreline moved north and eastwards, and continental environments became established in the area of Río Lora and in many parts of the northern flank of the Merida Andes where the interbedded thin sandstones, siltstones, floodplain mudstones, and coals of the Los Cuervos Formation were deposited.

The dinoflagellate data from Río Lora enable us to calibrate the terrestrial palynological record to the marine record. In terms of spores and pollen, the transition from the Cretaceous to the Tertiary is generally considered to span the boundary between the *Proteacidites dehaani* and *Foveotriletes margaritae* zones of Germeraad et al. (1968), and the *Proteacidites dehaani* and *Spinizonocolpites baculatus* zones of Muller et al. (1987). In Colombia, the Late Cretaceous *Buttinia andreevi* Zone of Sarmiento (1994) is equivalent to the *Proteacidites dehaani* Zone (Text-Figure 2). The results of this study indicate that some of the spores and pollen that have been used to define the Late Cretaceous in this region appear to range into the Paleocene and some previously reported only from the Late Cretaceous (e.g., *Buttinia andreevi*) have not been found, perhaps as a result of the more marine nature of the section at Río Lora. Species long considered to be Late Cretaceous markers that range into the Early Paleocene at Río Lora include *Proteacidites dehaani* (or forms similar to it), *Crussafontites grandiosus, Periretisyncolpites giganteus*, and *Araucariacites australis*. Additionally, some taxa, such as *Spinizonocolpites baculatus*, that have been thought of as only Paleocene in distribution have been recorded in the Late Cretaceous section at Río Lora. An interesting occurrence noted in the Mito Juan and Catatumbo formations is the presence of the fungal fruiting body genus *Trichopeltinites*. This form is known to disappear at or about the Cretaceous–Tertiary boundary in sections in North America (Fleming, 1985) as a result of the extinction of its host plants at the end of the Cretaceous.

An argument could be made that many of these palynomorphs have been reworked into younger Tertiary sediments but there is very little evidence in support of this. It is our contention that these range extensions reflect the previous absence of calibration of spore-pollen bearing continental strata to fossiliferous marine strata. Further work on similar sections in other parts of northern South America will help to more clearly define the age ranges of specific, important spores and pollen.

ACKNOWLEDGMENTS

We wish to thank Amoco Venezuela and Maraven for permission to publish this paper. Throughout the study we are grateful of the help of many people. In Venezuela, during fieldwork, we received valuable assistance and guidance from Daniel Loureiro (formerly from Litos C.A, Caracas; now with Intevep, Los Teques) and Franklin Yoris Villasana (Litos C.A, Caracas). At Amoco in Houston we received assistance from Alvin Gabriel, Debra Wilkes, Hank Tran, J.C. Lawson, Harley Kuriger, Mary LaCoe, Elvira Kreger, and Mark Schendel. Without these valuable contributions this study could not have been completed.

References Cited

ALVAREZ, L.W., ALVAREZ, W., ASARO, F., and MICHEL, H.V.
1980 Extraterrestrial cause for the Cretaceous–Tertiary extinction. *Science*, 208: 1095–1108.

FECHNER, G.G.
1988 Selected palynomorphs from the Lower to Middle Eocene of the South Atlas Border Zone (Morocco) and their environmental significance. *Palaeogeography, Palaeoclimatology, Palaeoceology*, 65: 73–79.

FLEMING, R.F.
1985 Palynological observations of the Cretaceous–Tertiary boundary in the Raton Formation, New Mexico. *Palynology*, 9: 242.

GARNER, A.H.
1926 Suggested nomenclature and correlation of geological formations in Venezuela. *American Institute of Mining and Metallurgical Engineers Transactions*, pp. 677–684.

GERMERAAD, J.H., HOPPING, C.A., and MULLER, J.
1968 Palynology of Tertiary sediments from tropical areas. Review of *Palaeobotany and Palynology*, 6: 189–348.

GONZALEZ DE JUAN, A.C., AROZENA, J.A., and PICARD CADILLAT, X.
1980 *Geología de Venezuela y de sus cuencas petrolíferas*. Caracas, Venezuela, Ediciones Foninves, 1031 pgs.

HILDEBRAND, A.R., PENFIELD, G.T., KRING, D.A., PILKINGTON, M., CAMARGO. Z.A., JACOBSEN, S.B., and BOYNTON, W.V.
1991 Chicxulub crater: A possible Cretaceous/Tertiary boundary impact crater on the Yucatan Peninsula, Mexico. *Geology*, 19: 867–871.

KUYL, O.S., MULLER, J., and WATERBOLK, H.Th.
1955 The application of palynology to oil geology with reference to Western Venezuela. *Geologie en Mijnbouw*,17: 49–76.

LORENTE, M.A.
1986 Palynology and palynofacies of the Upper Tertiary in Venezuela. *Dissertationes Botanicae*, Band 99: 1–222.

MULLER, J., DE DI GIACOMO, E., and VAN ERVE, A.W.
1987 A palynological zonation for the Cretaceous, Tertiary, and Quaternary of Northern South America. *American Association of Stratigraphic Palynologists, Contributions Series*, 19: 7–76.

NOTESTEIN, F.B., HUBMAN, C.W., and BOWLER, J.W.
1944 Geology of the Barco concession, Republic of Colombia, South America. *Bulletin of the Geological Society of America*, 55: 1165–1216.

POCKNALL, D.T., ERLICH, R.N., STEIN, J.A., BERGEN, J.A., and LORENTE, M.A.
1997 A Cretaceous–Tertiary boundary section at Río Lora, Merida Andes, Western Venezuela. *Memorias del VI Simposio Bolivariano "Exploracion en las Cuencas Subandinas"*; Cartagena, Colombia, September 1997, Tomo I, pp. 552–564.

SARMIENTO, G.
1994 Estratigrafia, palinología y paleoecología de la formación Guaduas (Maastrichtiano–Paleoceno; Colombia). *Publicaciones Geológicas Especiales del Ingeominas*, 20: 1–192.

WOOD, G.D., GABRIEL, A.M., and LAWSON, J.C.
1996 Chapter 3. Palynological techniques — processing and microscopy. *In*: Jansonius, J., and McGregor, D.C. (eds), *Palynology: Principles and Applications.* American Association of Stratigraphic Palynologists Foundation, Vol. 1, p. 29–50.

Authors' addresses

DAVID T. POCKNALL
ROBERT N. ERLICH
JEFFREY A. STEIN
BP Amoco Corporation
P.O. Box 3092
Houston, TX 77253
U.S.A.

MARIA ANTONIETA LORENTE
Maraven
PDVSA
Caracas
Venezuela

SEDIMENTOLOGY, PALYNOLOGY, AND ROCK-EVAL PYROLYSIS OF THE UPPER CRETACEOUS KANGUK FORMATION, CANADIAN ARCTIC

Koldo Núñez-Betelu, and L. V. Hills

Abstract

Sedimentological, palynological, and Rock-Eval/TOC pyrolysis analyses of the uppermost Cenomanian to Campanian Kanguk Formation and contiguous strata are used to characterize a depositional sequence bounded by two major diachronous unconformities in the northeastern Canadian Arctic Archipelago. In this area, the Kanguk Formation can be traced from deposition in coastal areas (Mount Bridgman, 79°45'11"N and 82°39'12"W) westward and southward to deposition in offshore areas (Glacier Fiord, 78°38'N and 89°55'W). Marine and terrestrial palynomorphs are well-preserved and represented by 428 taxa, including 226 marine and 202 terrestrial species. Based on the palynological content, the Kanguk Formation can be subdivided into a lower marine dominated interval and an upper terrestrial dominated interval. Furthermore, the palynological assemblages can be used to subdivide the formation into four informal biozones. The upper three biozones roughly correlate with previously established biozones for the Kanguk Formation or equivalent units.

Combined sedimentological and Rock-Eval/Total Organic Carbon (TOC) pyrolysis data support the presence of three informal lithologic units in the Kanguk Formation, namely the "Mount Bridgman unit," the "bituminous unit," and the "Remus Creek unit." The "Mount Bridgman unit" and the "bituminous unit" are, in part, laterally equivalent and represent coastal and offshore deposition respectively. The "bituminous unit" was deposited in anoxic, offshore environments, and is correlative to the worldwide Cenomanian/Turonian anoxic event. This unit probably represents a good source-rock for oil and gas, whereas only gas could be generated from the deltaic "Remus Creek unit" and the coastal "Mount Bridgman unit." These last two units are organic-poor and terrigenous organic-matter is predominant.

The nature of the Kanguk Formation and the marked variations in thickness might be indicative of deposition in a tectonically active basin as suggested also by the abundance of bentonite beds. This tectonic activity is probably linked with the Eurekan Orogeny.

INTRODUCTION

The objective of this paper is to characterize the Kanguk Formation in the northeastern part of the Sverdrup Basin by combining sedimentology, palynology and Rock/Eval-TOC pyrolysis. Outcrops of this formation and equivalent units are known from a large area of the Canadian Arctic and nearby regions. However, the Kanguk Formation lacks outstanding lithological features and has no known economic prospects and thus, it has been studied with little detail. Therefore, the present publication will add to the scarce information previously published about this unit, which is representative of the final closure stages of the Sverdrup Basin (Núñez-Betelu et al., 1994).

The study area is in the north–eastern part of the Canadian Arctic Archipelago and includes part of Axel Heiberg and Ellesmere islands (Text-Figure 1). In the study area, the Upper Cretaceous strata include the Bastion Ridge and Kanguk formations, although the former is restricted to a single outcrop on southern Axel Heiberg Island (Glacier Fiord) and is not included in this work.

This study summarizes the results of detailed sedimentological, palynological and rock-eval pyrolysis analyses from 14 outcrop sections of the Kanguk Formation (Text-Figure 1). Of these, a representative of offshore to nearshore deposition, the Glacier Fiord section (78°38'N and 89°55'W), and a representative of nearshore and coastal deposition, the Mount Bridgman section (79°45'11"N and 82°39'12"W), were selected for detailed palynological analyses; whereas only the bases and tops of the rest of the sections were included in this part of the study to analyze all the environments of deposition present in the study area. A total of 190 outcrop samples was selected for this palynological study. These samples were processed following a standard processing technique (Barss and Williams, 1973). The rock-eval pyrolysis analyses include the entire 14 sections with the objectives of recognizing organic-rich strata and of assessing their maturity levels. A total of 599 samples was included in these rock-eval pyrolysis analyses. These samples were processed and studied following a standard technique, which is explained in detail in Núñez-Betelu and Baceta (1994).

Núñez-Betelu, Koldo, and Hills, L.V.
2001 Sedimentology, palynology, and Rock-Eval Pyrolosis of the Upper Cretaceous Kanguk Formation, Canadian Arctic; *In*: Goodman, D.K., and Clarke, R.T. (eds.), *Proceedings of the IX International Palynological Congress, Houston, Texas, U.S.A., 1996*; American Association of Stratigraphic Palynologists Foundation, p. 181–189.

ISBN 0–931871–06–9

REGIONAL GEOLOGICAL SETTING

The Sverdrup Basin includes most of the Canadian Arctic Archipelago and represents a depocentre in the Arctic Islands that was active from Carboniferous to Early Tertiary (Embry, 1991). It contains up to 13,000 m of strata (Balkwill, 1978). The Sverdrup Basin, a consequence of continental rifting and associated subsidence, has expanded and contracted several times, often following tectonic episodes. Presently it is approximately triangular in shape, almost 1,000 km long, and about 350 km at its widest point. Due to the tectonic activity, eustatic variations, and changes in regimes of sedimentation, numerous Carboniferous to Early Tertiary transgressive–regressive cycles or depositional sequences were deposited in the Sverdrup Basin (Embry and Podruski, 1988). These sequences are usually represented by a regressive succession of marine shale and siltstone that coarsens upward into shallow marine and

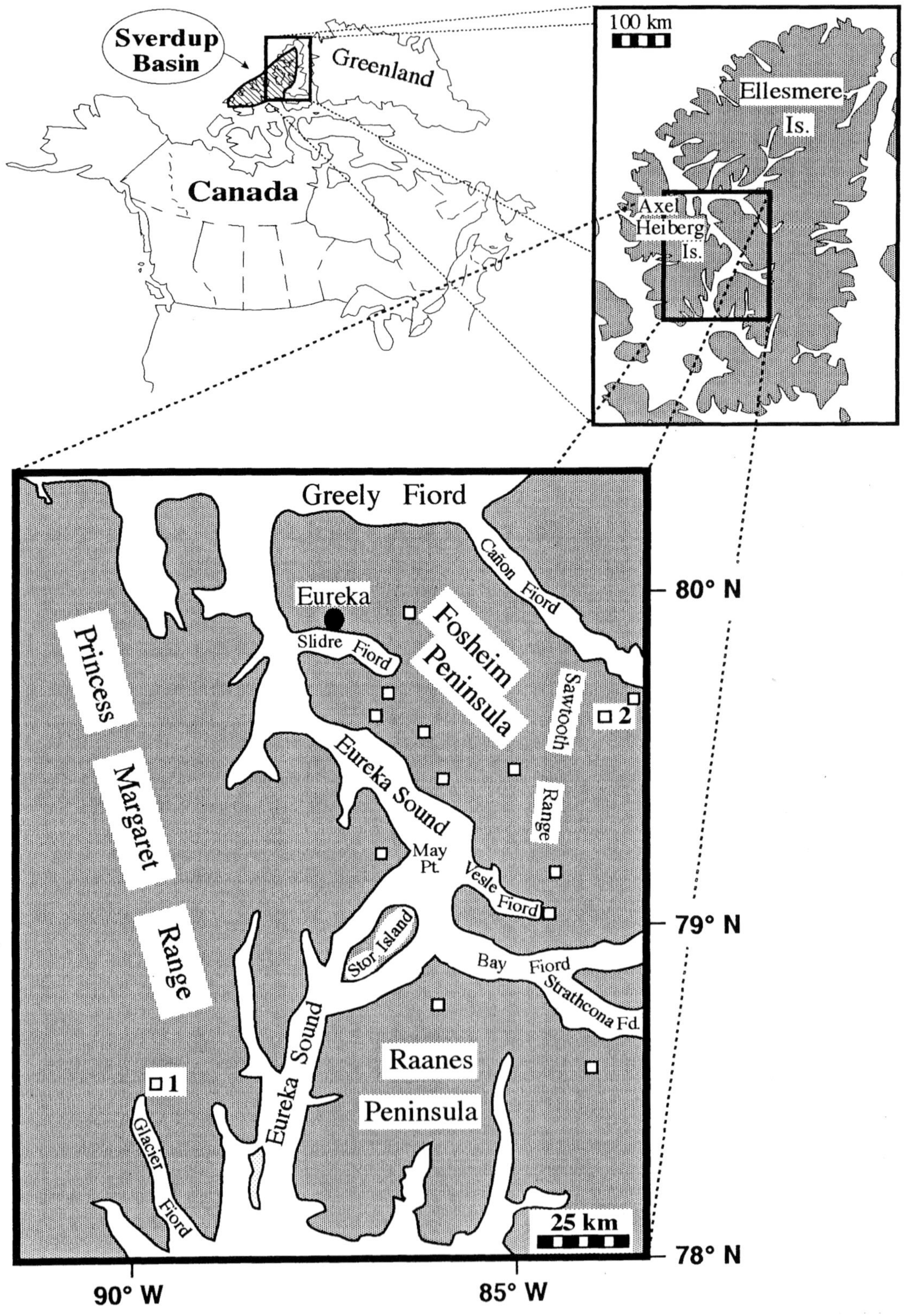

Text-Figure 1. Map of study area showing section location and geographic features. Each small square represents a studied section. (1) Glacier Fiord section, (2) Mount Bridgman section.

deltaic sandstones with subaerial unconformities on the basin margins. Unconformities mark the cycle boundaries (Embry, 1991).

The Sverdrup Rim, a linear high along the northwestern margin of the Sverdrup Basin, was intermittently exposed during the Mesozoic. During the Cretaceous, renewed uplift of the Sverdrup Rim produced marginal unconformities that become correlatable to conformable surfaces basinward. Renewed rifting in the nearby Amerasia Basin, in response to broad thermal subsidence, resulted in marine incursion of the Sverdrup Basin from the late Cenomanian/Turonian to the Campanian. This marine incursion resulted in deposition of the youngest major marine depositional sequence, which consists of the Kanguk Formation and, in some areas, the base of the overlying Expedition Formation of the Eureka Sound Group (i.e., Embry, 1991). This sequence was named as Sequence 1 by Ricketts and Stephenson (1994).

During the Turonian and Coniacian sedimentation rates were low. The reduced sedimentation, coupled with high productivity, resulted in widespread deposition of bituminous muds on offshore shelf areas and throughout most of the basin. Rates of sedimentation increased during the Santonian and culminated in the Campanian with tectonic pulses that disturbed the eastern region of the Sverdrup Basin and resulted in variable rates of sedimentation across the basin. Along the eastern margin deltaic systems prograded southward and westward, forming rapidly growing clastic wedges that signaled the initial episodes of the Eurekan Orogeny. On northern Ellesmere Island the Hansen Point volcanics occur in the upper part of the Kanguk depositional sequence. Uplifting and concomitant subaerial exposure eroded most deltaic sediments by the end of the Cretaceous producing a widespread unconformity on top of the Kanguk Formation. At the end of the Mesozoic and beginning of the Cenozoic, tectonic activity ended marine deposition in this basin and folded and faulted the units of the eastern Sverdrup Basin exposing the Kanguk Formation and providing excellent outcrops.

Presently, outcrops of the Kanguk Formation are widespread in the Canadian Arctic and occur not only in the Sverdrup Basin, but also in adjacent areas, such as Banks and Eglinton basins, the Franklinian Mobile Belt, and the Arctic Platform. Units laterally equivalent to the Kanguk Formation also occur in other areas around the Arctic including Baffin Bay, Lancaster Sound, West Greenland and the District of Mackenzie (McIntyre, 1974; Hea et al., 1980; Christiansen et al., 1996).

STRATIGRAPHY OF THE KANGUK DEPOSITIONAL SEQUENCE

In most of the Sverdrup Basin the Kanguk Formation rests unconformably on the Upper Albian Hassel Formation (Text-Figure 1). However, on southern Axel Heiberg Island the Bastion Ridge Formation (Upper Albian) occurs between the Hassel and Kanguk formations. In contrast, on central Axel Heiberg Island the Strand Fiord volcanics intertongue and overlie the Bastion Ridge Formation before coming in contact with the base of the Kanguk Formation (Núñez-Betelu et al., 1994).

Generally, the Kanguk Formation (sensu Souther, 1963) consists of dark, organic-rich bituminous mudstones, often referred as "papery mudstones, " in the lower part, and claystones, siltstones, and very fine-grained sandstones in the upper part (Núñez-Betelu, 1994). The bituminous mudstones and interbedded bentonites, which mostly occur in the lower part, were deposited in low "energy" environments. The quietness of the environment and high productivity rates in surficial waters resulted in the accumulation of abundant organic matter and a variety of marine palynomorph assemblages in sea bottom sediments (Núñez-Betelu, 1991, 1994; Núñez-Betelu and Hills, 1992a, 1992b, 1994).

In contrast to most areas, the easternmost sections of the Kanguk Formation include near their base, mudstone to fine-grained sandstones that are arranged into several coarsening upward cycles. Some of these cycles are capped by paleosols. These upward coarsening cycles are best developed at Mount Bridgman, central Ellesmere Island (Text-Figure 1). The silty claystones, siltstones, and sandstones, typical of the upper part in all sections, record shallow shelf and nearshore facies with abundant and diverse marine and terrestrially derived assemblages. These life-forms are represented among others by meter-sized inoceramids, large marine vertebrates, and marine and non-marine palynomorphs.

Bentonite beds are common in the Kanguk Formation in the northeastern Sverdrup Basin. Their presence in the upper part of some sections helps to define the Kanguk Formation/Eureka Sound Group boundary. Although bentonites are not present in the upper part of the formation in the type area, all bentonite beds should be included within the Kanguk Formation, because no bentonite beds are known to occur in the Eureka Sound Group (Núñez-Betelu, 1994). Inoceramids, on the other hand, are typical of the upper part of the Kanguk Formation both in the type section and in the sections studied herein.

An unconformity cuts into the Kanguk Formation and the overlying Expedition and Strand Bay formations of the Eureka Sound Group on west–central Ellesmere Island. The lower units of the Eureka Sound Group are missing in this area. This unconformity is usually present within the Eureka Sound Group, is widespread throughout the Sverdrup Basin, and extends to the south to include the Baffin Bay–Labrador–Scotian Shelf areas. This unconformity is time-trangressive. It extends from the Campanian to the Upper Paleocene in the Mount Bridgman locality of central Ellesmere Island. The two unconformities that bound the Upper Cretaceous depositional sequence have a tectonic origin that developed in response to changes in the adjacent Amerasia Basin as proposed by Embry (1991). Other less widespread unconformities may occur in the basin but are of limited extent.

SEDIMENTOLOGY

The Kanguk Formation ranges in thickness between 85 m and 500 m but most sections are about 180–200 m thick. Mudstone is the most typical lithology of the Kanguk Formation. Often, these mudstones are very finely laminated and are referred to as papery mudstones. They are very common in the lower part of the studied sections and readily recognized in the field due to their papery look. Coarser sediments, such as fine-grained sandstones, also occur in most sections. The dark, highly fissile mudstones, are interpreted as indicative of low-energy conditions. They are usually overlain, and even interbedded, by lighter, blocky mudstones, which are, in turn, overlain by brown and greyish mudstones, sandy siltstones, and silty sandstones. This general succession indicates a progressive shallowing of the basin and an increase in the energy of the environment. However, this shallowing trend is not uniform and is repeated, truncated, and altered through this formation, with intercalations of facies that at times repeat themselves. The Kanguk Formation represents deposition in a restricted, stressful environment that eventually gives way to the initiation of deltaic deposition. Fully deltaic sediments are represented by the overlying Expedition Formation of the Eureka Sound Group (Ricketts and Stephenson, 1994).

Good preservation of bentonites are also indicative of very low energy regimes, and, therefore, of restricted basins. Bentonites occur throughout the Kanguk Formation, even in the uppermost part of some sections, although they tend to be far more abundant in the lower part of the formation. Thus, a restricted, quiescent depositional setting for the Kanguk Formation would be supported by both the mudstone facies and the interbedded bentonite beds.

Shallow marine sandstones are also present in the lower part of the Kanguk Formation on the eastern sections of the study area. These sandstones represent a paleoshoreline position in the Sverdrup Basin. On other areas, thin sandstone beds appear randomly interbedded with offshore mudstones of the Kanguk Formation in the lower part of many sections. The base of the Kanguk Formation is representative of a basin-wide transgression and since transgressions provide a possible mechanism for placing sand in offshore settings, some of the lower Kanguk sandstones may be related to this mechanism. Yet, another mechanism that could supply sand from terrestrial environments to the shelf would be that of tidal river deltas where nearshore and inner shelf hydraulic regimes are in equilibrium.

The upper Kanguk siltstones and thin-bedded sandstones occur cyclically and represent the distal parts of the first stages of delta development. This was a low energy environment but probably stressful owing to relatively low oxygen levels in the Kanguk Sea as indicated by the low ichnofossil diversity. The upper sandstones are maintained within the Kanguk Formation following the original definition of the formation by Souther (1963). These upper sandstones represent the initiation of deltaic sedimentation but are still within the marine realm of the Kanguk Sea. By following Souther's definition, all bentonite beds and all inoceramids found in this study occur within the Kanguk Formation.

In summary, the Kanguk mudstones and sandstones represent deposition from coastal to outer shelf settings that grade upward into lower prodelta environments indicating a progressive shallowing of the basin. An unconformity is present at top of the Kanguk Formation at all sections except one (Text-Figure 2). This unconformity truncates the upper Kanguk beds and, is in part responsible for the difference in thickness from section to section.

PALYNOLOGICAL BIOSTRATIGRAPHY

A detailed palynological study (Núñez-Betelu, 1994) of the Kanguk Formation at Glacier Fiord and Mount Bridgman sections, and of the upper Hassel Formation and the lower Eureka Sound Group included the identification of 428 palynomorph taxa. Of these, 226 were marine and 202 were terrestrial.

Based on the palynological associations, the following four informal biozones (Núñez-Betelu, 1994) were erected for the Kanguk Formation and compared (Text-Figure 3) with previous studies that include biozonation schemes of the Kanguk Formation (Banks Island, Doerenkamp et al., 1976; Bylot and Devon Islands, Ioannides, 1986) or of Kanguk-equivalent units (Horton River, District of Mackenzie, McIntyre, 1974).

Biozone 1

This zone is characterized by stepwise introduction of many marine taxa and by the presence of several characteristic taxa that include common to abundant *Cribroperidinium exilicristatum, Kallosphaeridium ringnesiorum, Isabelidinium ? globosum, Apteodinium* spp., *Microdinium distinctum,* and *Trithyrodinium* spp. This zone is characterized in the lower sample by abundant reworked material from underlying formations including *Nyktericysta* spp. *Luxadinium primulum, Limbicysta* sp., *Rugubivesiculites multisaccus* and *Stephodinium* spp. No terrestrial taxa is specific to this interval. At Glacier Fiord the middle part of the biozone contains very low diversity assemblages, which may indicate changes in the paleoenvironmental conditions. By comparison with other studies this biozone is likely late Cenomanian and Turonian in age (Text-Figure 2) based on the presence of *Apteodinium* spp., *Cribroperidinium* spp. and *Isabelidinium ?globosum* (Núñez-Betelu and Hills, 1992a). This biozone appears to be unique to the study area and does not have many palynological features in common with any previously proposed palynological schemes for the Kanguk Formation. This may be due to the lack of previous detailed palynological studies on the Kanguk Formation and also to local characteristics such as a slightly older age and different paleoenvironmental conditions.

Biozone 2

The base of this biozone is characterized by the first appearances of *Areoligera* spp., *Chatangiella* spp., *Isabelidinium* spp., and *Laciniadinium williamsii.* The top of the biozone is marked by the base of biozone 3, which is represented by the introduction of new taxa. Species that are representative of biozone 2 are *Chatangiella ditissima, C. granulifera, C. spectabilis, C. williamsii, Laciniadinium arcticum,* and *Spinidinium* spp. Stepwise introduction of terrestrial palynomorph taxa occurs throughout this zone but no terrestrial forms are characteristic of this interval. This biozone roughly correlates (Text-Figure 3) with the lower part of other previously established biozones such as Division H1 of McIntyre (1974), biozone CV of Doerenkamp et al. (1976) and Interval I of Ioannides (1986). Biozone CV was defined for the lower part of the Kanguk Formation ("lower shale member") on Banks Island and was given a Santonian to Campanian age. Division H1 was proposed for the "bituminous zone" of the Horton River section, which was also assigned a Santonian to Campanian age. Interval I was defined after the study of six palynological samples from the lower half of the Kanguk Formation from Bylot Island and spans the Santonian and Campanian stages. However, a Coniacian age is most likely for biozone 2 of this study. This age is based on the presence of a late Coniacian ammonite, *Scaphites depressus* (Hills et al., 1994) at the base of next biozone. The lack of agreement with the age defined in previous studies may be due to the lack of any other independent age based on other fossil groups for those studies. In our

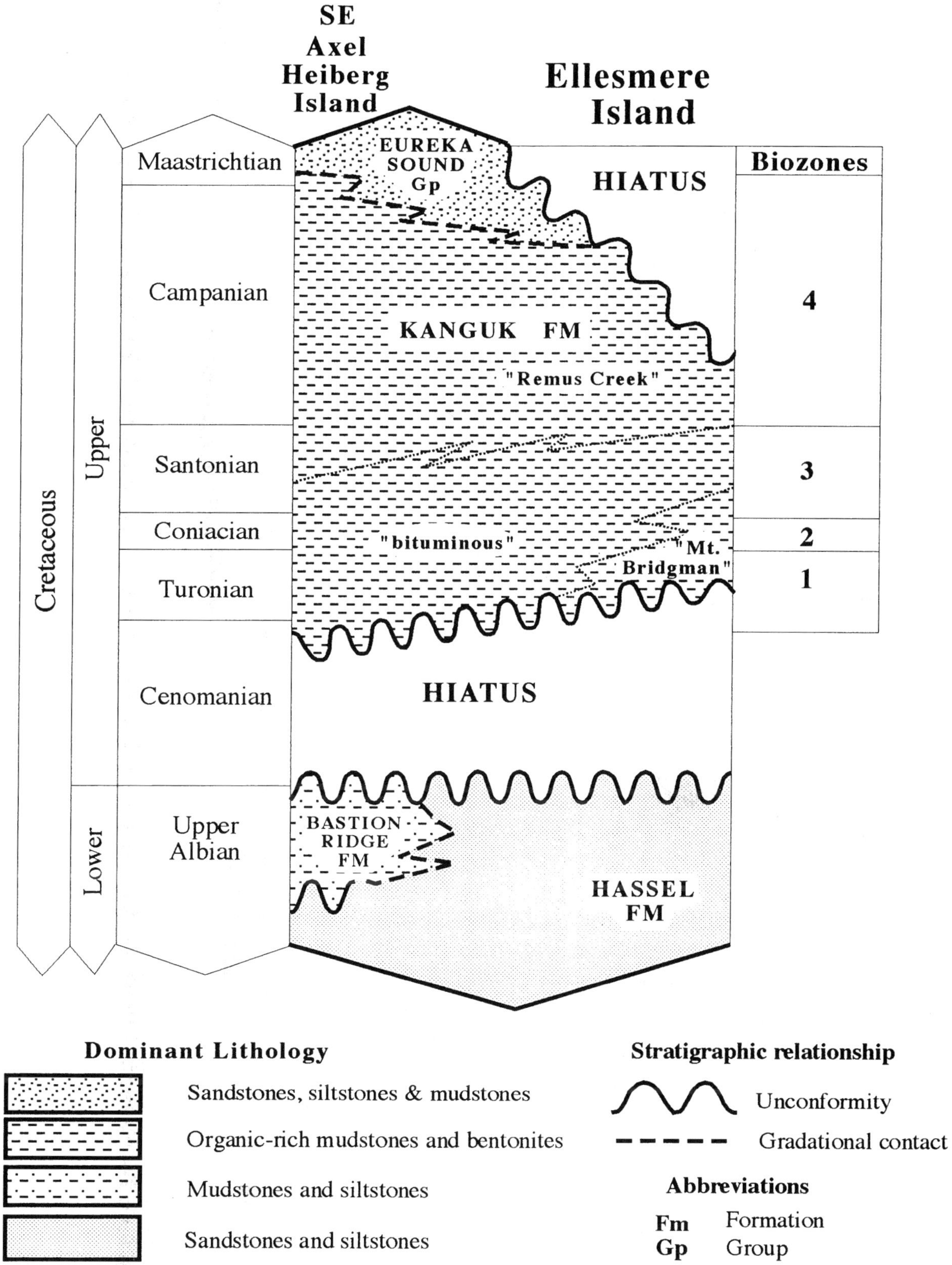

Text-Figure 2. Summary stratigraphic chart of the Kanguk Formation and adjacent units in the study area. The Bastion Ridge Formation is restricted to Axel Heiberg Island, Canadian Arctic.

AUTHORS / STRATIGRAPHY	**This Study**	McIntyre, 1974	Doerenkamp et al., 1976	Ioannides, 1986
Lower Maastrichtian		Division H2		Interval II
Campanian	Biozone 4	Division H1	CVIa	
Santonian	Biozone 3		CV	Interval I
Coniacian	Biozone 2			
Turonian	Biozone 1			
Upper Cenomanian				

Text-Figure 3. Comparative diagram of biozones defined in this study and biozones from previous studies.

study, the ammonite provides an argument for the age assignment.

Biozone 3

This biozone is transitional between the marine dominated lower Kanguk Formation and the terrestrially dominated upper Kanguk Formation. At the base of this zone many marine and terrestrial taxa reappear. The base of this biozone is marked by the presence of *Odontochitina* spp., *Paleoperidinium pyrophorum* and *Spongodinium delitiense*. The top of the zone is marked by the first appearance of several spore and angiosperm pollen taxa that mark the beginning of biozone 4. Species belonging to the genus *Spinidinium* are common to abundant in this zone and are the most representative. No terrestrial taxa are characteristic of this zone. This biozone approximately corresponds to the upper part of Division H1 of McIntyre (1974), biozone CV of Doerenkamp et al. (1976) and Interval I of Ioannides (1986). The base of the section is early Santonian in age whereas the top is late Santonian or early Campanian. The early Campanian age is questionable, but possible, based on the first appearances of species of the Triprojectatites group (*Aquilapollenites* spp., *Expressipollis* spp., etc.), which are traditionally considered as indicative of the Campanian in North America (Tschudy, 1969; Stanley, 1970; Norris et al., 1975; McIntyre, 1974; Doerenkamp et al., 1976; Ioannides and McIntyre, 1980; Ioannides, 1986).

Biozone 4

The base of this biozone is marked by a significant influx of terrestrial pollen and spores. Biozone 4 is characterized by the presence of common *Aquilapollenites* spp., *Expressipollis* spp., *Fibulapollis* spp., and other angiosperm pollen. The top of this zone is marked by a considerable decrease in marine palynomorphs and an important increase in the relative percentage of pollen grains. Despite many similarities between this biozone at Glacier Fiord and at Mount Bridgman, the upper part of the biozone seems to be missing from the Mount Bridgman section. Many species of *Aquilapollenites* are common in the upper part of the Kanguk Formation at the Glacier Fiord section, but they are not at Mount Bridgman. This zone is considered to be Campanian in age based on the presence of a variety of triprojectate pollen grains. Based on the palynological content this zone (Text-Figure 3) approximately corresponds to the lower part of Division H2 (the "Pale shale zone") of McIntyre (1974), biozone CVIa of Doerenkamp et al. (1976) and uppermost Interval I and lowermost Interval II of Ioannides (1986). All these zones are characterized by a rich terrestrial palynoflora that is, in part, similar to the Campanian boreal complex of the Ust-Yenissey Subprovince I and Middle Yenissey Subprovince II of Siberia (Russia) reported by Samoilovich (1966, 1967).

As previously stated, the Eureka Sound Group overlies the Kanguk Formation. The contact between the Kanguk

Formation and the Eureka Sound Group is very sharp in many parts of the study area, especially at the Mount Bridgman section. The Eureka Sound Group samples are very rich in gymnosperm pollen and some contain rare, reworked, poorly-preserved marine palynomorphs. These samples contain characteristic *Cycadopites* spp., *Juglanspollenites* spp., *Myricipites* spp., and abundant fungal spores and fruiting bodies. The palynomorph assemblage is similar to that from the TIb zone (Doerenkamp et al., 1976) of the Eureka Sound Formation on Banks Island that is Paleocene in age.

In summary, drastic change from terrestrially dominated assemblages to marine dominated assemblages occurs at the Hassel/Kanguk Formation boundary. The presence of a late Albian palynological assemblage in the upper Hassel samples and of a latest Cenomanian/Turonian assemblage within the Kanguk Formation documents the presence of a major unconformity between these two formations (Text-Figure 2). Similarly, the difference in age between the Campanian upper Kanguk beds and the Paleocene Eureka Sound Group also indicates the presence of a major unconformity between these two formations. Therefore, palynology indicates the presence of two major unconformities bounding the Kanguk depositional sequence.

INTEGRATED ROCK-EVAL/ TOC PYROLYSIS AND SEDIMENTOLOGICAL ANALYSIS

The Kanguk Formation is an organic-rich unit, as indicated by the high TOC values in all sections and with a maximum of 9.96 wt. %. Hydrogen Index (HI) values are relatively high too, up to 559 mg HC/g TOC, in the papery mudstones but usually are very variable and commonly low in other lithologies. In the Kanguk Formation, Tmax and PI values are relatively low and indicate the presence of only immature kerogen.

The samples from the Kanguk Formation contain Type II, Type III, and Type IV organic matter (Núñez-Betelu, 1994; Núñez-Betelu et al., 1995a, 1995b). These types are considered as derived from the maturation of organic matter of different origins (Núñez-Betelu and Baceta, 1994). For instance, Type II kerogen is derived from plurispecific phytoplankton and is the most common oil-prone kerogen type. Type III kerogen corresponds generally to terrestrially produced organic matter but could be oxidized marine organic matter as well. This type of kerogen is commonly regarded as a source for gas rather than oil. Finally, Type IV kerogen comprises inert carbon with no hydrocarbon potential. It represents highly oxidized and degraded organic matter. This type of kerogen could have originally been other types but was later oxidized.

In this study, Type III kerogen has been found in all sections and is more abundant than Type II, which is not present everywhere. Oxygen Index (OI) values are usually higher in the upper part of the sections, where Type III kerogen dominates. Type IV kerogen is also present in most sections but is minoritary.

The pyrolysis data indicate that organic matter types are distributed according to facies. The facies and the features of the organic matter seem to be closely interrelated and therefore, can be used to interpret the environments of deposition for the Kanguk Formation. Based on the specific features of the organic matter, the Kanguk Formation can be subdivided into three intervals, namely: (1) a lower interval with high terrestrial influence and present only at eastern sections, (2) a lower fully marine interval, and (3) an upper terrestrially influenced interval.

The first interval has been recognized only at eastern areas of this study, in sections such as Mount Bridgman. This interval is characterized by low TOC, low HI, and high OI values. Type III and Type IV kerogen occur in this interval. No cyclicity is observed in the organic matter distribution within this interval, which is best developed on Mount Bridgman and herein is informally called the "Mount Bridgman unit." This unit is characterized by sandy and silty lithologies. According to the interpretation of facies and the nature of the organic matter, this unit contains terrestrially derived organic matter that was deposited in nearshore and coastal environments.

The second interval is present in all the studied sections and corresponds to the informal "bituminous unit," which was already mentioned by Miall (1979). This interval is characterized by moderate to high TOC, moderate to high HI, and variable OI values. All Type II kerogen occurs in this interval but there is also Type III and some Type IV kerogen. Two to three TOC/HI cycles are observed in the organic matter distribution within this interval. If the "bituminous unit" starts from the base of the Kanguk Formation, three cycles are recognized by the combination of the distribution of TOC and HI values, but if the "Mount Bridgman unit" is present, then only two cycles are present in the "bituminous unit." Therefore, the "Mount Bridgman unit" is equivalent to the lowest cycle of the "bituminous unit." The papery mudstones constitute the main lithology of this interval. The "bituminous unit" is representative of the most offshore environments of the Kanguk Formation. It represents deposition during the most anoxic interval and has the best preservation of marine organic matter. This is supported by both, sedimentological and organic geochemical evidence.

The third interval is present in all sections and is informally referred to as the "Remus Creek unit" since it was first observed in this section. This unit is characterized by moderate to high TOC, very low HI, and high OI values. All samples from this unit contain mainly Type IV kerogen, but some Type III is also present. No cyclicity is reflected in the combined TOC and HI data in this interval. An intercalation of muddy, silty, and sandy lithologies is typical of this interval, which is characterized by the presence of common fossils, such as inoceramids. This interval represents deposition in nearshore and prodelta environments as indicated by sedimentological and organic matter features.

Therefore, by combining sedimentology and pyrolysis data the Kanguk Formation can be subdivided into three units that are representative of three major environments of deposition.

DISCUSSION AND CONCLUSIONS

The combination of palynological analyses, sedimentology and Rock-Eval/TOC pyrolysis from the Kanguk Formation appear to indicate that two types of deposition occurred: a marine-dominated, oxygen-deficient and low-energy environment for the lower part of the formation and a terrestrially-dominated, deltaic environment for the upper part. The uniformity, continuity, and synchroneity of the "papery mudstones" characteristic of the lower part of the Kanguk

Formation in the western study area, implyes a regional mechanism for their formation rather than upwelling. In this basin, shelf-productivity would be based on essentially autochthonous nutrient recycling due to relatively slight, seasonal, lowering of a regional pycnocline and the consequent turbulent entrainment of nutrients from the top of the bottom waters. These conditions have been recognized worldwide and have been defined as the E2 event for the North Atlantic ocean (de Graciansky et al., 1987).

In the marine-dominated part of the sections the palynological biozones do not closely match with the units recognized based on sedimentology combined with Rock-Eval/TOC pyrolysis (Text-Figure 2). In fact, the first biozone is not restricted to any sedimentological or organic facies and occurs in the "bituminous unit" and also in the "Mount Bridgman unit." If the marine floras were closely linked to the environment of deposition, it would imply that the anoxic conditions occurred in the entire water column. However, the different distribution of the biozones with respect to the units indicates that the upper part of the water-column where the dinoflagellates lived was relatively well-oxygenated, whereas the bottom of the ocean went through periods of anoxia. The organic-rich sediments of the "bituminous unit" may represent epeiric facies that formed by a combination of factors including a Turonian transgressive maximum, temperate warm climate that promoted watermass stratification and oxygen depletion, and lack of free water circulation between the Sverdrup Basin and the Atlantic Ocean (Gradstein and Srivastava, 1980). The ocean-bottom anoxic environments allowed preservation of organic matter and deposition of the papery mudstones. This is best understood in a basin where water masses are stratified with ascending and descending boundaries between the layers. This also explains the presence of papery mudstones interbedded with other lithologies in the lower part of the formation. When communication with surrounding basins was renewed, migrating faunas colonized the Arctic Sverdrup Basin. These faunas included marine reptiles, which were common during deposition of the upper part of the Kanguk Formation.

In all sections except Glacier Fiord, the base of the Kanguk Formation appears to be Turonian in age and the top Campanian. However, both the base and the top of the Kanguk may be diachronous, since the Glacier Fiord section is latest Cenomanian to middle Campanian in age whereas in the rest of the sections the Kanguk Formation appears to be Turonian to early Campanian in age. This diachroneity for the top is based on the comparison of assemblages that seem to correlate relatively well with the lower and middle part of zone 4 at Glacier Fiord but not with the upper part. The upper part of zone 4 may be restricted to the Glacier Fiord section. The Eureka Sound Group is characterized by Paleocene assemblages in all the studied samples. Therefore, the assemblages document the presence of major unconformities between the Hassel and Kanguk formations, and between the Kanguk Formation and the Eureka Sound Group.

ACKNOWLEDGMENTS

The authors wish to acknowledge the logistic support of Polar Continental Shelf Project. Financial support was provided by the American Association of Petroleum Geologists, American Association of Stratigraphic Palynologists, Geological Society of America, Northern Scientific Training Program (Department of Indian and Northern Affairs, Government of Canada), and the University of Calgary (Thesis Research Grants) to K. Núñez-Betelu, and Natural Sciences and Engineering Research Council Grants in Aid of Research to L. V. Hills. We want to warmly thank L. Bloom and B. Rhoades, who efficiently processed the palynological samples. We also want to thank Bob Cushman, Francisca Oboh-Ikuenobe and Javier Helenes, whose thorough reviews greatly improved the original manuscript.

References Cited

BALKWILL, H.R.

1978 Evolution of the Sverdrup Basin, Arctic Canada. *American Association of Petroleum Geologists Bulletin*, 62: 1004–1029.

BARSS, M.S., and WILLIAMS, G.L.

1973 Palynology and nannofossil processing techniques. *Geological Survey of Canada*, Paper 73–26, 25 pgs.

CHRISTIANSEN, F.G., BOJESEN-KOEFOED, J., DAM, G., NYTOFT, H.-P., LARSEN, L.M., PEDERSEN, A.K., and PULVERTAFT, T.C.R.

1996 The Marraat oil discovery on Nuussuaq, West Greenland: evidence for a latest Cretaceous–earliest Tertiary oil prone source rock in the Labrador Sea–Melville Bay region. *Bulletin of Canadian Petroleum Geology*, 44: 39–54.

DE GRACIANSKY, P.C., BROSSE, E., DEROO, E., HERBIN, J.-P., MONTADERT, L., MÜLLER, C., SIGAL, J., and SCHAAF, A.

1987 Organic-rich sediments and palaeoenvironmental reconstructions of the Cretaceous North Atlantic. *In*: Brooks, J., and Fleet, A.J. (eds.), *Marine Petroleum Source Rocks*. Geological Society Special Publication No. 26, pp. 317–344.

DOERENKAMP, A., JARDINE, S., and MOREAU, P.

1976 Cretaceous and Tertiary palynomorph assemblages from Banks Island and adjacent areas (N.W.T.). *Bulletin of Canadian Petroleum Geology*, 24: 372–417.

EMBRY, A.F.

1991 Mesozoic history of the Arctic Islands; Chapter 14. *In*: Trettin, H.P. (ed.), *Geology of the Innuitian Orogen and Arctic Platform of Canada and Greenland*. Geological Survey of Canada, Geology of Canada, No. 3; and Geological Society of America, The Geology of North America, v. E, pp. 369–433.

EMBRY, A.F., and PRODRUSKI, J.A.

1988 Third-order depositional sequences of the Mesozoic succession of Sverdrup Basin. *In*: James, D.P., and Leckie, D.A. (eds.), *Sequences, Stratigraphy, Sedimentology: Surface and Subsurface*. Canadian Society of Petroleum Geologists, Memoir 15, pp. 73–84.

GRADSTEIN, F.M., and SRIVASTAVA, S.P.

1980 Aspects of Cenozoic stratigraphy and paleooceanography of the Labrador Sea and Baffin Bay. *Paleogeography, Paleoclimatology, Paleoecology*, 30: 261–295.

HEA, J.P., ARCURI, J., CAMPBELL, G.R., FRASER, I., FUGLEM, M.O., O'BENOS, J.J., SMITH, D.R., and ZAGAT, M.

1980 Post–Ellesmerian basins of Arctic Canada: their depocentres, rates of sedimentation and petroleum potential. *In*: Miall, A.D. (ed.) *Facts and Principles of World Oil Occurrence*. Canadian Society of Petroleum Geologists, Memoir 6, pp. 447–488.

HILLS, L.V., BRAUNBERGER, W.F., NÚÑEZ-BETELU, K., and HALL, R.L.

1974 Paleogeographic significance of *Scaphites depressus* in the Kanguk Formation (Upper Cretaceous), Axel Heiberg Island, Canadian Arctic. *Canadian Journal of Earth Sciences*, 31: 733–736.

HOPKINS, W.S., Jr., and BALKWILL, H.R.
1973 Description, palynology, and paleoecology of the Hassel Formation (Cretaceous) on eastern Ellef Ringnes Island, District of Franklin. *Geological Survey of Canada*, Paper 72–37, 31 pgs.

IOANNIDES, N.S.
1986 Dinoflagellate cysts from Upper Cretaceous–Lower Tertiary sections, Bylot and Devon Islands, Arctic Archipelago. *Geological Survey of Canada, Bulletin*, 371: 1–99.

IOANNIDES, N.S., and McINTYRE, D.J.
1980 A preliminary palynological study of the Caribou Hills outcrop section along the Mackenzie River, District of Mackenzie. Current Research, *Geological Survey of Canada, Paper*, 80–1A, pp. 197–208.

McINTYRE, D.J.
1974 Palynology of an Upper Cretaceous Section, Horton River, District of Mackenzie, N.W.T. *Geological Survey of Canada, Paper*, 74–14, pp. 1–57.

MIALL, A.
1979 Mesozoic and Tertiary geology of Banks Island, Arctic Canada. The history of an unstable craton margin. *Geological Survey of Canada, Memoir*, 387, 235 pgs.

NORRIS, G., JARZEN, D.M., and AWAI-THORNE, B.V.
1975 Evolution of the Cretaceous terrestrial palynoflora in western Canada. *In*: Caldwell, W.G.E. (ed.), *The Cretaceous System in the western interior of North America*. The Geological Association of Canada, Special Paper No. 13, pp. 333–364.

NÚÑEZ-BETELU, K.
1991 *Palynology of the Kanguk Formation (Late Cretaceous). Remus Creek, Ellesmere Island, Canadian Arctic Archipelago.* Unpubl. M. Sc. thesis. The University of Calgary, 270 pgs.
1994 *Sequence stratigraphy of a coastal to offshore transition in the Kanguk Formation (Upper Cretaceous), Axel Heiberg and Ellesmere islands: a palynological, sedimentological, and Rock-Eval characterization of a depositional sequence in the Sverdrup Basin.* Unpubl. Ph. D. thesis, The University of Calgary, 569 pgs.

NÚÑEZ-BETELU, K., and BACETA, J.I.
1994 Basics and application of Rock-Eval/TOC pyrolysis: an example from the uppermost Paleocene/lowermost Eocene in the Basque Basin, Western Pyrenees. *Munibe*, 46: 43–62.

NÚÑEZ-BETELU, K., and HILLS, L.V.
1992a Preliminary paleopalynology of the Kanguk Formation (Upper Cretaceous), Remus Creek, Ellesmere Island. Canadian Arctic Archipelago: I. Marine Palynomorphs. *Revista Española de Paleontologia*, 7: 185–196.
1992b Preliminary paleopalynology of the Kanguk Formation (Upper Cretaceous), Remus Creek, Ellesmere Island. Canadian Arctic Archipelago: II. Terrestrial Palynomorphs. *Revista Española de Paleontologia*, 7: 197–206.
1994 Palynological data from the uppermost Hassel and Kanguk formations and the lowermost Eureka Sound Group (uppermost Lower Cretaceous–Paleocene), Axel Heiberg and Ellesmere islands, Canadian Arctic. Geological Survey of Canada, Open File 2489, 192 pgs.

NÚÑEZ-BETELU, K., MACRAE, R.A., HILLS, L.V., and MUECKE, G.K.
1994 Uppermost Albian–Campanian palynological biostratigraphy of Axel Heiberg and Ellesmere Islands (Canadian Arctic). *1992 International Conference on Arctic Margins, Proceedings Volume*, pp. 135–140.

NÚÑEZ-BETELU, K, RIEDEGER, C.L., and HILLS, L.V.
1995a Papery mudstones from the Upper Cretaceous Kanguk Formation as indicative of anoxic environments, Canadian Arctic Archipelago. *In*: Grimalt, J.O., and Dorronsoro, C. (eds.), *Organic geochemistry: developments and applications to energy, climate, environment and human history, Proceedings Volume*, pp. 526–527.
1995b Rock-Eval/TOC pyrolysis analysis of the Upper Cretaceous Kanguk Formation, Canadian Arctic Archipelago. *In*: Grimalt, J.O., and Dorronsoro, C. (eds.), *Organic geochemistry: developments and applications to energy, climate, environment and human history, Proceedings Volume*, pp. 528–529.

RICKETTS, B.D., and STEPHENSON, R.A.
1994 The demise of Sverdrup Basin: Late Cretaceous–Paleogene sequence stratigraphy and forward modelling. *Journal of Sedimentary Research*, B64: 516–530.

SAMOILOVICH, S.R.
1966 Tentative botanico-geographical subdivision of northern Asia in Late Cretaceous time. *In*: Methods of the paleopalynological investigations. *Materials to the second International Palynological Conference*, Utrecht, September, 1966, pp. 147–171.
1967 Tentative botanico-geographical subdivision of northern Asia in Late Cretaceous time. *Review of Palaeobotany and Palynology*, 2: 127–139.

SOUTHER, J.G.
1963 Geological traverse across Axel Heiberg Island from Buchanan Lake to Strand Fiord. *In*: Geology of the North-Central part of the Arctic Archipelago, Northwest Territories (Operation Franklin). Geological Survey of Canada, Memoir 320, pp. 426–448.

STANLEY, E.A.
1970 The stratigraphical, biogeographical, paleontological and evolutionary significance of the fossil pollen group Triprojectatites. *Bulletin, Georgia Academy of Sciences*, 28: 1–44.

TSCHUDY, B.D.
1969 Species of *Aquilapollenites* and *Fibulapollis* from two Upper Cretaceous localities in Alaska. U. S. Geological Survey, Professional Paper 643–A, pp. 1–17.

Authors' addresses

KOLDO NÚÑEZ-BETELU
Laboratory of Paleontology
Department of Stratigraphy and Paleontology
Faculty of Sciences
University of the Basque Country
P.O. Box 644
48080 Bilbao
Basque Country
e-mail: gpbnubek@lg.ehu.es

L. V. HILLS
University of Calgary
2500 University Dr. N.W.
Calgary, Alberta
T2N 1N4
Canada

PALYNOLOGY OF THE LOWER CRETACEOUS KACHAIKE FORMATION, SANTA CRUZ PROVINCE, ARGENTINA

Alicia M. Baldoni, Rosemary A. Askin, and Daniel Ragona

Abstract

Spores, pollen, megaspores and microplankton occur in the Kachaike Formation from Arroyo Caballo Muerto at Barrancas de Kachaike, Santa Cruz Province, southern Argentina. Of the 14 outcrop samples studied, 12 contain moderately well-preserved palynomorphs and a variety of kerogen types. Spores and pollen may be assigned to the bryophytes, lycophytes, ferns, gymnosperms (including *Classopollis* [the most abundant form], *Callialasporites, Cyclusphaera, Balmeiopsis* and saccate podocarps) and chloranthaceous angiosperms. In some samples megaspores with affinities to the families Marsileaceae, Selaginellaceae and Isoetaceae are common. The freshwater algal microflora includes colonial types and filamentous algae. Dinoflagellate cysts reflect marine incursions, particularly in the middle part of the section.

The Kachaike assemblages have many species in common with assemblages from the Río Mayer and Baqueró formations, Austral Basin (northwestern Santa Cruz and southwestern Chubut provinces); the Agrio and Huitrín formations, Neuquén Basin; and from Lower Cretaceous intervals in other basins in southern South America, the Antarctic Peninsula and Australia. Based on their stratigraphic records, certain spore, megaspore and dinoflagellate cyst species indicate a late Aptian–early Albian age for the Kachaike Formation. An age no younger than early Albian is consistent with the absence of tricolpate angiosperm pollen, which appear in the middle Albian in southern South America.

INTRODUCTION

Palynological study of sedimentary rocks of the Lago Viedma sheet, situated at latitude 49°50'S and longitude 72–73°30'W, was begun as part of the program of geologic map preparation, during the program of scientific activities in Argentina that included establishment of Dirección Nacional del Servicio Geológico. At this time, a stratigraphic profile was constructed with paleoenvironmental details to be based on lithostratigraphic and biostratigraphic characters. With this objective 14 outcrop samples were analyzed for palynomorphs, with rich assemblages recovered in 12 of these. The sampled sediments are assigned to the Kachaike Formation (Riccardi, 1971), which crops out in the gorges of Kachaike, Lago San Martín region of Santa Cruz Province. The Kachaike Formation, deposited within the Austral Basin, is underlain by the Río Mayer Formation and overlain by the Cardiel Formation.

The study area is approximately 3km southeast of Estancia Kach Aike (Text-Figure 1), with the described section (Text-Figure 2) located in Arroyo Caballo Muerto, between lakes Tar and San Martín. This location lies in the southwest of Santa Cruz Province on the northwestern border of the Austral (or Magallanes) Basin.

This work reaffirms the late Aptian–early Albian age previously assigned in a preliminary report based on the megaflora (studies by Archangelsky, Baldoni, and Romero, *in* Rebasa, 1982) and palynoflora (Gamerro, unpubl. report, 1982), and Baldoni and Batten (1991).

Baldoni, Alicia M., Askin, Rosemary A., and Ragona, Daniel
2001 Palynology of the Lower Cretaceous Kachaike Formation, Santa Cruz Province, Argentina; *In*: Goodman, D.K., and Clarke, R.T. (eds.), *Proceedings of the IX International Palynological Congress, Houston, Texas, U.S.A., 1996*; American Association of Stratigraphic Palynologists Foundation, p. 191–200.

 ISBN 0–931871–06–9

PREVIOUS WORK

Details of the geology of the Kachaike Formation within the study area are given by Rebasa (1982) as part of his Licenciatura (bachelor's) thesis. This report provides references to previous stratigraphic studies, the most important being Riccardi (1971), Riccardi and Rolleri (1980) and Nullo et al. (1978, 1981). Preliminary information on the megaflora by Archangelsky, Baldoni and Romero, and the palynoflora by Gamerro, are included in this unpublished report. The suggested late Aptian–early Albian age for the Kachaike Formation was based on the work of the various authors cited in the report (Rebasa, 1982).

Baldoni (1987) described two new species of megaspores (*Arcellites santacrucencis* and *Paxillitriletes kachaikensis*) from these sediments. More recently Baldoni and Batten (1991) extended the list of known megaspores with *Balmeisporites holodictyus, Bacutriletes triangularis, Minerisporites* sp. cf. *M. marginatus* and *Tenellisporites coronatus*. These authors followed the age of late Aptian–early Albian for their assemblages.

MATERIALS AND METHODS

Palynomorphs were recovered in 12 of the 14 outcrop samples analyzed. Palynological extraction followed conventional techniques, including sieving to yield > 25 μm and < 25 μm fractions. Quantitative analyses and most of the microscopic examination was carried out by Baldoni on a Leitz Laborlux at the Dirección Nacional del Servicio Geológico, which is the repository for the microslides.

The analyzed fossiliferous horizons contain a rich assemblage of moderately well-preserved spores (microspores and megaspores), pollen, algal remains, cuticle, woody tissue and microplankton. The criteria adopted for study of the recovered assemblage are general classification of the identified elements, including the most likely botanical affinity; quantitative determination of the most representative taxa in the different horizons of the section;

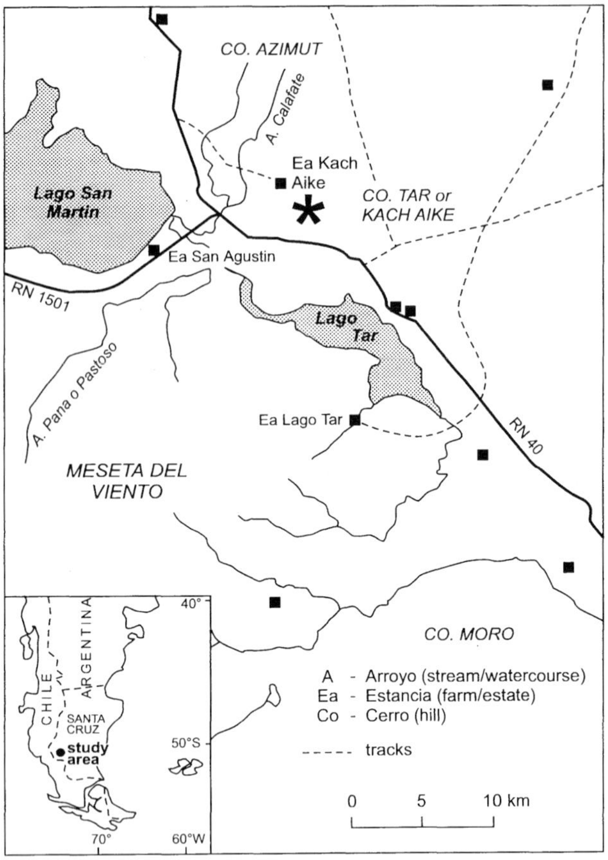

Text-Figure 1. Map showing location of the Arroyo Caballo Muerto (Dead Horse) section (*) in the vicinity of Tar and San Martín lakes, southwestern Santa Cruz Province (from Baldoni and Batten, 1991).

comparison of the general floristic composition with other assemblages of the same epoch from other localities, and paleoenvironmental information derived from previous studies.

CHARACTERISTICS OF THE ASSEMBLAGE

Among the recovered palynomorphs and kerogen are 48 species of trilete and monolete spores, 14 species of gymnosperm pollen, one species of angiosperm pollen, 20 dinoflagellate cyst taxa, plus other algae, a few fungal spores, and fragments of cuticle and woody tissue. The sampled succession is predominantly continental with some marine/marginal marine intervals. The spores suggest Bryophyta, Lycophyta and Filices derivation, and the gymnosperms are well represented by pollen of families Cheirolepidaceae and Podocarpaceae. Angiosperm pollen occur in very low numbers, and include only one species of Chloranthaceae. Identified palynomorph taxa are listed in the Appendix. Distribution of each of the identified elements in our samples is shown in Table 1, plotted as presence/absence data with taxa grouped by botanical affinity.

Quantitative Analysis

With the objective of characterizing the palynofloristic content, relative abundance counts were carried out on the 12 productive samples distributed through 190 meters of section. Preservation, abundance and diversity of palynomorphs were noted for each sample; counts were made of between 200 and 250 specimens.

The resulting data showed the existence of a few major representative taxonomic groupings. These are (with abbreviations used in Text-Figure 3): *Classopollis* (CL), *Callialasporites* (CA), saccates (SA), *Cyclusphaera* + *Balmeiopsis* (CY+BA), spores (ES, for "esporas"), and microplankton (MP). Monocolpate, inaperturate and angiosperm pollen are poorly represented and do not appear in the counts. Algal remains belonging to the genera *Botryococcus* and *Schizosporis* occur in all the samples, but are also rare enough to be excluded from the counted specimens. Text-Figure 3 shows relative frequencies in each of the productive horizons. Observations on distribution of each of the major categories are:

CL – (*Classopollis*). These pollen are very abundant in each of the continental samples, and also common in the marine samples where they are only surpassed by microplankton. The highest percentages, with values >50% to 65%, were observed in the upper part of the section (upper plant-bearing horizon, samples 2119, 2118, 2117)

CA – (*Callialasporites*). These are most common (6%) in the uppermost (plant) sample, and vary between 1 and 4% through the rest of the section.

ES – (Spores). Spores are common to abundant, dominating the lowest sample (2012, 49%) and reaching values of 39% and 38% in samples 2014 and 2117.

CY+BA – (*Cyclusphaera* + *Balmeiopsis*). These pollen are common to abundant, particularly in continental samples, reaching 29% and 36% in 2113 and 2114, and 20% and 30% in the upper samples, plant and 2119. The two genera are grouped together for counting purposes because it can be difficult to differentiate between them.

SA – (Saccates). These include bisaccate and trisaccate pollen. They are abundant (20%) in the lower marginally marine sample (2014), but are typically less than 5%, except for the upper continental sample 2118 (12%) and plant-bearing sample (14%).

MP – (Microplankton). Marine samples 2015 and 2112 contain 59% marine microplankton, and samples 2115 and 2116 contain 47%, while marginal marine sample 2014 contains 8% and 2117 less than 0.5%.

Degree of marine influence in the studied section does not significantly change the relative proportions of the important terrestrial taxa. The numerically most significant forms are gymnosperm pollen of the families Cheirolepidaceae and Podocarpaceae, which overwhelm other groups such as the cryptogam spores, in all but the lower two samples. The cryptogam spores represent several families typically important in the Lower Cretaceous: the Schizaceae, Gleicheniaceae, Lophosoriaceae, Lycopodiaceae, Selaginellaceae, Isoetaceae and Marsileaceae. Angiosperm, monocolpate and inaperturate pollen are always rare.

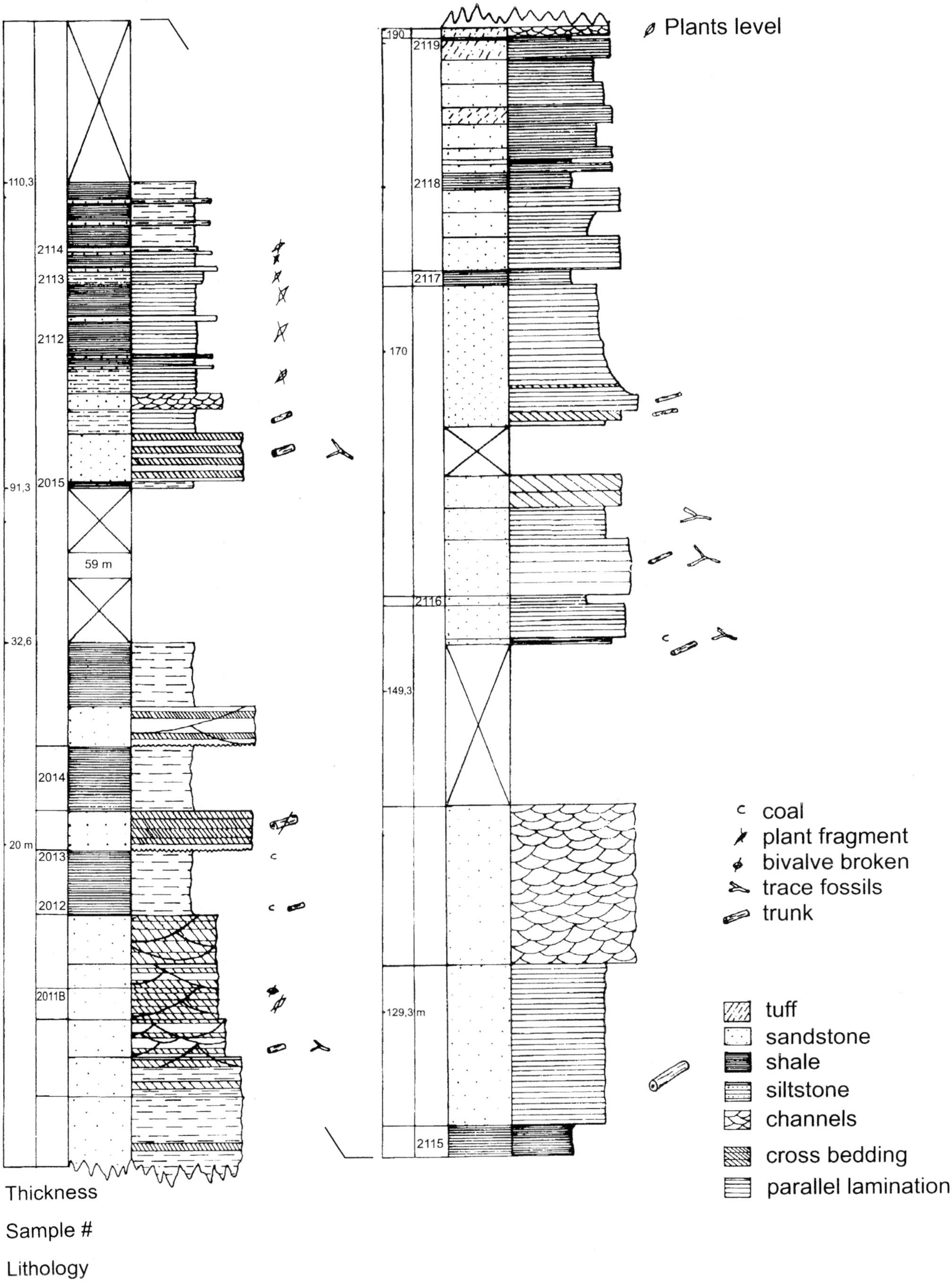

Text-Figure 2. Stratigraphic column of the Kachaike Formation at the Arroyo Caballo Muerto section to show lithology and sample positions.

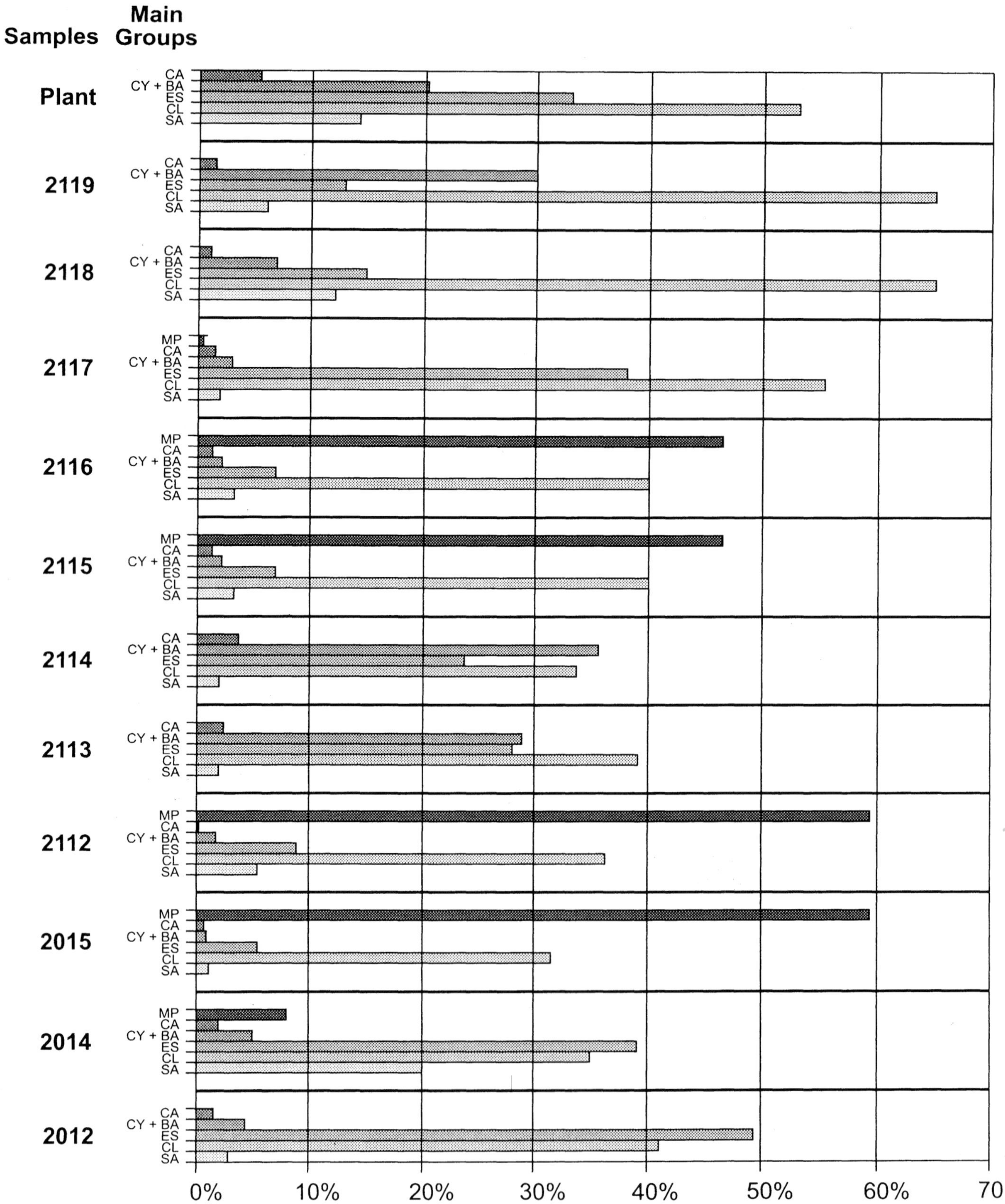

Text-Figure 3. Relative abundance histograms of major taxonomic groups found in the Kachaike Formation (see text for explanation of groups).

TABLE 1. Presence/absence data for palynomorphs in the Kachaike Formation.

	192	191	182.2	175.7	155.3	122.3	106.3	104.8	101.6	91.7	25.8	19.3	Depth m
	Plants	2119	2118	2117	2116	2115	2114	2113	2112	2015	2014	2012	Sample No.
PALYNOMORPH SPECIES													Botanical affinity
BRYOPHYTIC SPORES													
Aequitriradites verrucosus			+										Hepaticae
A. spinulosus	+			+	+		+	+	+				
Foraminisporis dailyi			+	+									
F. asymmetricus			+	+									Musci?
Polycingulatisporites sp.								+	+				
Triporoletes reticulatus								+	+				Hepaticae
Interulobites pseudoreticulatus	+	+						+	+	+			
Staplinisporites caminus		+										+	
LYCOPHYTIC SPORES													
Ceratosporites equalis		+	+			+	+		+	+		+	cf. Selaginella biformis, Lycopodium densum.
Leptolepidites verrucatus				+	+			+	+	+			cf. Selaginella
Neoraistrickia truncatus		+							+	+	+	+	Selaginella, cf. Lycopodium densum
Retritriletes austroclavatidites			+			+	+	+	+		+	+	Lycopodium
Densoisporites velatus			+		+			+	+	+	+	+	Selaginellaceae
D. corrugatus	+			+	+		+	+				+	Selaginellaceae
FILICEAN SPORES													
Baculatisporites comaumensis			+		+			+					cf. Osmundaceae
Cicatricosisporites spp.	+	+	+	+	+		+	+	+	+		+	Schizaceae
Cicatricosisporites australiensis		+		+			+				+	+	Schizaceae
C. hughesi					+		+		+			+	Schizaceae
C. venustus												+	Schizaceae
C. cf. ticoensis			+	+									Schizoceae
C. cf. baqueroensis			+						+			+	Schizoceae
Appendicisporites sp.				+		+		+				+	Schizaceae
Cyathidites australis		+		+	+				+		+		cf. Dicksoniaceae
C. minor		+		+	+				+	+			cf. Dicksonicaceae
Biretisporites spp.				+					+				
Gleichenidites circinidites	+	+	+	+	+	+	+	+		+	+		Gleicheniceae
G. sp.	+			+		+							Gleicheniceae
Klukisporites sp.	+				+	+		+				+	Schizaceae
Osmundacidites wellmanii			+	+		+	+						cf. Osmundaceae
Cyatheacidites annulatus	+	+	+				+						Lophosoriaceae
Contignisporites cooksoniae				+				+				+	
Taurocosporites segmentatus	+						+	+			+	+	
Trilobosporites purverulentus				+			+	+					
T. apiverrucatus			+	+								+	
Cribelosporites striatus			+	+	+		+	+	+	+	+		Marsileaceae
Rugulatisporites sp.		+		+				+					
Laevigatosporites ovatus	+			+						+		+	
MEGASPORES													
Arcellites santacrusensis	+		+	+							+	+	Marsileaceae
Balmeisporites holodictyus	+			+							+	+	Hydropteridineae
Paxillitriletes kachaikensis	+		+								+		Isoetaceae
Tenellisporites coronatus	+		+								+		Sellaginellaceae
Bacutriletes triangulatus											+		Sellaginelaceae
Minerisporites sp. cf. marginatus											+		Isoetaceae
GYMNOSPERMOUS POLLEN													
Callialasporites trilobatus forma trilobatus	+	+	+	+	+	+	+		+	+	+	+	Podocarpaceae
C. trilobatus forma dampieri	+		+	+				+	+	+		+	Podocarpaceae
C. trilobatus forma tenuisaccatus	+			+		+	+			+		+	Podocarpaceae
Podocarpidites spp.	+	+	+	+	+	+	+	+	+	+	+	+	Podocarpaceae
P. cf. elipticus	+	+						+				+	Podocarpaceae
Trisaccites microsacatum						+				+		+	Podocarpaceae
Microcachryidites antarcticus	+	+		+									Podocarpaceae
Vitreisporites pallidus	+	+											cf. Caytoniales
Araucariacites australis	+	+	+	+			+	+					Araucariaceae
Cyclusphaera psilata	+	+	+	+	+	+	+	+	+	+	+	+	Podocarpaceae
Balmeiopsis limbatus	+	+	+	+	+	+	+	+	+	+	+	+	Podocarpaceae
Cycadopites nitidus			+							+	+		Cycadophyta
Classopollis spp.	+	+	+	+	+	+	+	+	+	+	+	+	Cheirolepidaceae
ANGIOSPERMOUS POLLEN													
Clavatipollenites hughesii					+	+	+		+	+			Chloranthaceae
ALGAL PALYNOMORPHS													
cf. Botryococcus sp.	+	+	+	+	+	+	+	+	+	+	+	+	Chlorophyta
Schizosporis reticulatus				+								+	Chlorophyta
Kiokansium unituberculatum				+		+						+	Dinoflagellata
Oligosphaeridium complex					+	+	+			+	+	+	Dinoflagellata
Spiniferites spp.						+	+	+		+	+	+	Dinoflagellata
Chichaouadinium boydü										+			Dinoflagellata
Muderongia sp.										+			Dinoflagellata
Pseudoceratium turneri										+			Dinoflagellata
Eyrea nebulosa									+	+			Dinoflagellata
Florentinia sp.									+	+			Dinoflagellata
Muderongia tetracantha									+	+			Dinoflagellata
Dioxya armata						+				+			Dinoflagellata
Litosphaeridium arundum						+							Dinoflagellata
Florentinia cf. radiculata					+	+				+			Dinoflagellata
Odontochitina operculata					+	+			+	+			Dinoflagellata
Cribroperidinium sp.				+	+				+	+			Dinoflagellata
Subtilisphaera perlucida				+					+	+			Dinoflagellata
Cleistosphaeridium ancoriferum						+			+				Dinoflagellata
Oligosphaeridium pulcherrimum						+							Dinoflagellata
Hystrichodinium sp.					+	+							Dinoflagellata
Cannosphaeropsis sp.				+	+								Dinoflagellata
Cribroperidinium edwardsi				+	+	+							Dinoflagellata

CORRELATIONS AND AGE

The palynoflora offers a means of correlation and age interpretation of the Lower Cretaceous Kachaike Formation with other successions in Patagonia and the Southern Hemisphere, including Australia and Antarctica. Text-Figure 4 shows age distribution in various regions of 21 selected species of spores, pollen, megaspores and dinoflagellate cysts. Locality and stratigraphic details, with citations, for numbered areas in the right-hand column are as follows:

(1) Neuquén Basin, especially the Agrio and Huitrín formations which are dated as Hauterivian–Barremian and Aptian–Albian (Volkheimer et al., 1976; Volkheimer, 1980; Volkheimer and Quatrocchio, 1981); (2) Baqueró Formation, central Santa Cruz Province, Barremian to lower Aptian (Archangelsky and Gamerro, 1965, 1966a, b, c, 1967); (3) Austral Basin, southwest Chubut Province and northwest Santa Cruz Province, Cretaceous strata overlying an older succession, upper Berriasian–Albian (Archangelsky et al., 1981, 1983, 1984; Baldoni and Archangelsky, 1983); (4) San Jorge Basin, NE Santa Cruz Province, upper Albian–Cenomanian (Archangelsky et al., 1994); (4') San Jorge Gulf Basin, Albornoz Formation, Hauterivian–Barremian (Vallati, 1993); (5) Mayer River Beds, Santa Cruz Province, Hauterivian–Barremian (Pothe de Baldis and Ramos, 1980); (6) UN OIL OS-1, SO, Santa Cruz Province, Barremian–Aptian (Archangelsky and Seiler, 1980); (7) Lago San Martín, Santa Cruz Province, middle–upper Aptian (Pothe de Baldis and Ramos, 1983, 1988); (8) Various Australian basins, Lower Cretaceous sequences (e.g., Dettmann, 1963; Helby et al., 1987); (9) James Ross Island, Antarctic Peninsula, Kotick Point and Whisky Bay formations, Aptian–Cenomanian (Dettmann and Thomson, 1987; Keating et al., 1992).

(Note: For extent and stratigraphy of southern South American Cretaceous basins, the reader is referred to Riccardi, 1988).

As can be observed from Text-Figure 4, some of the selected species (*Contignisporites cooksonii, Trilobosporites purverulentus, Aequitriradites spinulosus, Foraminisporis dailyi, Cicatricosisporites hughesii*) have extensive stratigraphic ranges, appearing in the Jurassic and continuing to or through the Aptian–Albian interval. The rest are restricted to the Cretaceous, appearing during Berriasian to Barremian time. They are often important elements in the Aptian–Albian interval. *Trilobosporites purverulentus* and *Staplinisporites caminus* disappear during the Aptian–Albian. *Crybelosporites striatus* appears in the late Aptian in Australia and Albian in the Antarctic Peninsula. This species, which originated in the high southern latitude region, has not been previously recorded in published reports in Argentina, like other species in the Kachaike assemblage, although Gamerro (unpubl. report, 1982) notes its presence in the preliminary palynology study. We note, for age correlations, that first and last appearances of terrestrial species may be considerably diachronous across the Gondwana continents (e.g., Askin, 1989; Dettmann, 1989).

The dinoflagellate cyst species *Chichaouadinium boydii, Dioxya armata, Florentinia radiculata* and *Pseudoceratium turneri* appear during the Aptian. These species plus other taxa such as *Odontochitina operculata* and *Muderongia tetracantha* characterize the Aptian–Albian in southern basins. *Litosphaeridum arundum* appears at the base of the Albian in Australia, while *M. tetracantha* does not range into the upper Albian.

The age of the Kachaike Formation in Arroyo Caballo Muerto is thus late Aptian–early Albian (and perhaps early Albian), based on the total assemblage of spores, pollen and microplankton, and comparisons with data from other parts of Argentina, as well as the Antarctic Peninsula and Australia. This age is consistent with previous conclusions in megaspore reports by Baldoni (1987) and Baldoni and Batten (1991), based in part on the range of *Balmeisporites holodictyus*. *B. holodictyus* and the dinoflagellate cyst *Pseudoceratium turneri* indicate the age is no older than late Aptian, whereas possible *Litosphaeridium arundum* in one sample (2115) may restrict the age to Albian. Various spore and dinoflagellate cyst species, as discussed above, provide an upper age limit of Albian, and the presence of *Muderongia tetracantha* indicates the age is no younger than early Albian. The absence of tricolpate angiosperm pollen is consistent with an age no younger than early Albian, because these forms appear in the middle Albian in southern South America.

There is a rich fossil fauna nearby in Lago San Martín in the lower part of the Kachaike Formation. The most important fossil is the ammonite *Sanmartinoceras patagonicum.* Riccardi et al. (1987) suggested an age of middle Aptian to early Albian for this fauna. G. Blasco (verbal comm., 1996) suggests that the levels with fossils such as *S. patagonicum* can be assigned to the Aptian–Cenomanian interval.

PALEOENVIRONMENTAL AND PALEOPHYTOGEOGRAPHIC IMPLICATIONS

Spores and pollen, dispersed cuticle, and woody tissue and other plant remains dominate in the majority of samples in the studied section. Dinoflagellate cysts indicating marine influence are important constituents of several samples in the section, mainly within the middle part. The continental components are primarily conifer pollen of Cheirolepidaceae (*Classopollis*) and Podocarpaceae (*Callialasporites, Cyclusphaera–Balmeiopsis* and saccate forms), along with isosporous and heterosporous cryptogams. Abundance of *Classopollis,* and variations in presence and proportions of marine elements through the section, indicate coastal to nearshore marine environments. This is consistent with location of this Kachaike section near the edge of the Austral Basin. Preservation of the delicate structural ornaments of the megaspores indicates low energy deposition for much of the section. Depositional environment can be interpreted as coastal to nearshore marine reflecting relative sea-level fluctuations at the basin margin.

The Kachaike locality lies within the southern South American region that Volkheimer (1980) called the *Cyclusphaera psilata–Classopollis* subprovince of the Southern Gondwana Province of Brenner (1976). This palynofloristic subprovince, which existed during the early to middle Cretaceous, is characterized by *Cyclusphaera psilata* with high percentages of *Classopollis* plus *Callialasporites.* Plant communities in the Aptian–Albian Kachaike area are characterized by abundant cheirolepidacean conifers, in contrast to the high latitude southern Gondwana podocarp/araucarian communities of Antarctica–Australasia.

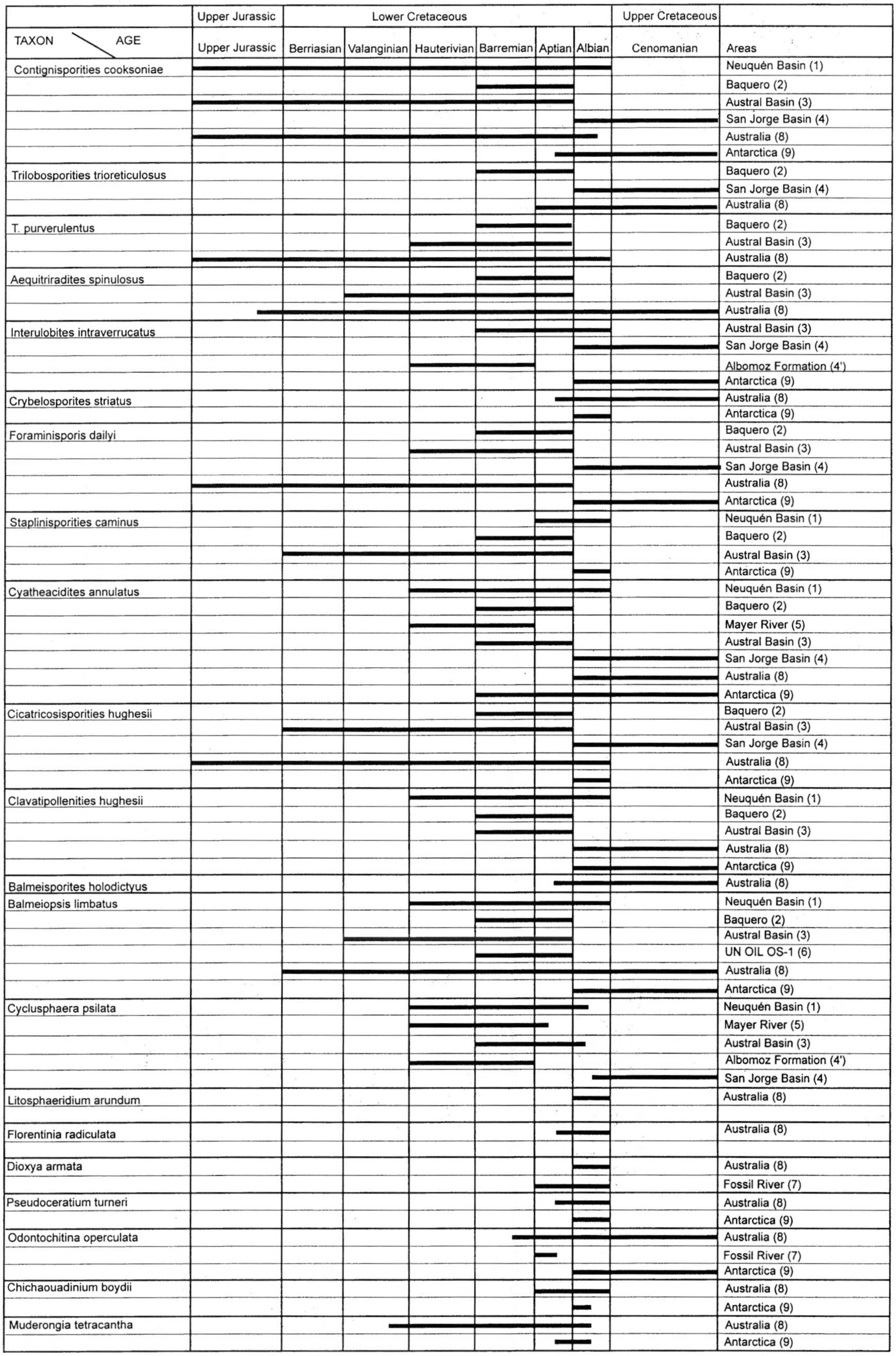

Text-Figure 4. Known ranges of selected taxa in the southern South American basins, Antarctic Peninsula and Australia (see text for explanation and references).

CONCLUSIONS

The twelve samples from the Kachaike Formation of Arroyo Caballo Muerto include 63 species of trilete and monolete cryptogam spores, gymnosperm and angiosperm pollen, 20 dinoflagellate cyst taxa, plus other algae and land plant debris. The numerically dominant elements in this association are gymnosperm pollen (mainly *Classopollis*, then *Cyclusphaera–Balmeiopsis*, saccates and *Callialasporites*), and fern spores. Angiosperm pollen (*Clavatipollenites hughesii*, no tricolpates) are rare.

The Kachaike plant communities were within the Southern Gondwana province, and their notable characteristics of abundant cheirolepidacean conifers plus *Cyclusphaera* (+*Balmeiopsis*) and *Callialasporites* make them part of the *Cyclusphaera psilata–Classopollis* subprovince.

Dinoflagellate cysts are relatively common in a few samples and reflect marine incursions, especially in the middle part of the section. Coastal to nearshore marine conditions with fluctuating relative sea-levels near the margin of the Austral Basin are envisaged.

A late Aptian–early Albian age is indicated for the Kachaike Formation in this area. Stratigraphic distribution of spores, pollen, megaspores and microplankton in other Lower Cretaceous successions in the Austral, San Jorge and Neuquén basins of South America, and from the Antarctic Peninsula and Australia were considered.

References Cited

ARCHANGELSKY, S., and GAMERRO, J.C.
1965 Estudio palinológico de la Formación Baqueró (Cretácico), Provincia de Santa Cruz, I. *Ameghiniana*, 4: 159–170.
1966a Estudio palinológico de la Formación Baqueró (Cretácico), Provincia de Santa Cruz, II. *Ameghiniana*, 4: 201–209.
1966b Estudio palinológico de la Formación Baqueró (Cretácico), Provincia de Santa Cruz, III. *Ameghiniana*, 4: 229–236.
1966c Estudio palinológico de la Formación Baqueró (Cretácico), Provincia de Santa Cruz, IV. *Ameghiniana*, 4: 363–372.
1967 Spores and pollen types of the Lower Cretaceous in Patagonia. *Review of Palaeobotany and Palynology*, 1: 211–217.

ARCHANGELSKY, S., BALDONI, A.M., GAMERRO, J.C., PALAMARCZUK, S., and SEILER, J.
1981 Palinología estratigráfica del Cretácico de Argentina Austral I.Diagramas de grupos polínicos del SO de Chubut y NO de Santa Cruz. *Actas VIII Congreso Geológico Argentino*. San Luis: 718–742.

ARCHANGELSKY, S., BALDONI, A.M., GAMERRO, J.C., and SEILER, J.
1983 Palinología estratigráfica del Cretácico de Argentina Austral II. Descripciones sistemáticas. *Ameghiniana*, 20: 190–226.
1984 Palinología estratigráfica del Cretácico de Argentina Austral III. Distribución de las especies y conclusiones. *Ameghiniana*, 21: 15–23.

ARCHANGELSKY, S., BELLOSI, E.S, JALFIN, G.A., and PERROT, C.
1994 Palynology and alluvial facies from the mid–Cretaceous of Patagonia, subsurface of San Jorge Basin, Argentina. *Cretaceous Research*, 15: 127–142.

ARCHANGELSKY, S., and SEILER, J.
1980 Algunos resultados palinológicos de la perforación UN OIL OS-1, del SO de la provincia del Chubut, Argentina. *II Congreso Argentino de Paleontología y Bioestratigrafía y I Congreso Latinoamericano de Paleontología*, 5: 216–222.

ASKIN, R.A.
1989 Endemism and heterochroneity in the Late Cretaceous (Campanian) to Paleocene palynofloras of Seymour Island, Antarctica: Implications for origins, dispersal and palaeoclimates of southern floras. *In*: Crame, J.A. (ed.), Origins and Evolution of the Antarctic Biota. *Geological Society (London) Special Publication*, 147: 107–119.

BALDONI, A.M.
1987 Dos nuevas especies de megasporas de la Formación Kachaike, Cretácico inferior de la Provincia de Santa Cruz, Argentina. *Anais do 10 Congreso Brasileiro de Paleontología*, Río de Janeiro: 669–689.

BALDONI, A.M., and ARCHANGELSKY, S.
1983 Palinología de la Formación Springhill (Cretácico Inferior) subsuelo de Argentina y Chile Austral. *Revista Española de Micropaleontología*, 15: 47–101.

BALDONI, A.M., and BATTEN, D.J.
1991 Megaspores from the Lower Cretaceous Kachaike Formation, Santa Cruz Province, Argentina. *Neues Jahrbuch für Geologie und Paläontologie, Abhandlungen*, 162: 377–393.

BRENNER, G.J.
1976 Middle Cretaceous floral provinces and early migrations of Angiosperms. *In*: Beck, C.B. (ed.), *Origin and Early Evolution of Angiosperms*. Columbia University Press, New York, p. 23–47.

DETTMANN, M.E.
1963 Upper Mesozoic microflora from south–eastern Australia. *Proceedings of the Royal Society of Victoria*, 77: 1–148.
1989 Antarctica: Cretaceous cradle of austral temperate rainforest? *In*: Crame, J.A. (ed.), Origins and Evolution of the Antarctic Biota. *Geological Society (London) Special Publication*, 147: 89–105.

DETTMANN, M.E., and THOMSON, M.R.A.
1987 Cretaceous palynomorphs from the James Ross Island area, Antarctica — a pilot study. *British Antarctic Survey Bulletin*, 77: 13–59.

GAMERRO, J.C.
1982 Informe palinológico del Perfil Arroyo Caballo Muerto, Santa Cruz. Y.P.F., Buenos Aires (Unpublished report).

HELBY, R.J., MORGAN, R., and PARTRIDGE, A.D.
1987 A palynological zonation of the Australian Mesozoic. *Association of Australasian Palaeontologists, Memoir*, 4: 1–94.

KEATING, J.M., SPENCER-JONES, M., and NEWHAM, S.
1992 The stratigraphical palynology of the Kotick Point and Whisky Bay formations, Gustav Group (Cretaceous), James Ross Island. *Antarctic Science*, 4: 279–292.

NULLO, F.E., PROSERPIO, C., and BLASCO, G. DE NULLO
1981 El Cretácico de la Cuenca Austral, entre el Lago San Martín y Río Turbio. *Comité sudamericano del Jurásico y Cretácico de América del Sur*, I, Buenos Aires: 181–220.

NULLO, F.E., PROSERPIO, C., and RAMOS, V.A.
1978 Estratigrafía y tectónica de la vertiente este del hielo continental patagónico Argentina–Chile. *VII Congreso Geológico Argentina, Actas* I, Buenos Aires: 455–470.

POTHE DE BALDIS, D., and RAMOS, V.A.
1980 Las microfloras de la Formación Río Mayer y su significado estratigráfico, Provincia de Santa Cruz, República Argentina. *II Congreso Argentino de Paleontología y Bioestratigrafía y I Congreso Latinoamericano de Paleontología*, 5: 201–213.

1983 Dinoflagelados del Aptiano inferior de Río Fósiles, Lago San Martín, Provincia de Santa Cruz, Argentina. *Revista Española de Micropaleontología,* 15: 427–446.

1988 Microplancton adicional del Aptiano de Río Fósiles, Lago San Martín, Provincia de Santa Cruz, Argentina y sus correlaciones con Australia. *IV Congreso Argentino de Paleontología y Bioestratigrafía, Actas* 3: 23–40.

REBASA, M.

1982 Análisis estratigráfico y Paleoambiental de la Formación Kachaike, aflorante en la barranca epónima, Provincia de Santa Cruz. Unpubl. licentiate (bachelor's) thesis, University of Buenos Aires, 52 p.

RICCARDI, A.C.

1971 Estratigrafía en el Oriente de la Bahía de la Lancha, Lago San Martín, Santa Cruz, Argentina. *Universidad Nacional La Plata, Museo La Plata, Geología, Revista,* 7: 245–318.

1988 The Cretaceous System of southern South America. *Geological Society of America, Memoir* 168, 161 p.

RICCARDI, A.C., and ROLLERI, E.O.

1980 Cordillera Patagónica Austral. *In:* Segundo Simposio de Geología Regional Argentina, 2: 1173–1306, Academia Nacional de Ciencias, Córdoba.

RICCARDI, C.A., AGUIRRE URRETA, M.B., and MEDINA, F.A.

1987 Aconeceratidae (Ammonitina) from the Hauterivian–Albian of Southern Patagonia. *Palaeontographica,* Abteilung A, 196: 105–185.

VALLATI, P.

1993 Palynology of the Albornoz Formation (Lower Cretaceous) in the San Jorge Gulf Basin, Patagonia, Argentina. *Neues Jahrbuch für Geologie und Paläontologie, Abhandlungen,* 187: 345–373.

VOLKHEIMER, W.

1980 Microfloras del Jurásico superior y Cretácico inferior de América Latina. II *Congreso Argentino de Paleontología y Bioestratigrafía y I Congreso Latinoamericano de Paleontología,* 5: 121–136.

VOLKHEIMER, W., and QUATROCCHIO, M.

1981 Distribución estratigráfica de los palinomorfos jurásicos y cretácicos de la faja andina y áreas adyacentes de América del Sur, con especial consideración de la Cuena Neuquina. *In: Cuencas Sedimentarias del Jurásico y Cretácico de América del Sur,* 2, Buenos Aires, p. 407–444.

VOLKHEIMER, W., QUATROCCHIO, M., SALAS, A., and SEPÚLVEDA, E.G.

1976 Caracterización palinológica de formaciones del Jurásico superior y Cretácico inferior de la Cuenca Neuquina (Argentina). *Actas VI Congreso Geológico Argentino,* Bahia Blanca, 1: 593–608.

Authors' addresses

ALICIA M. BALDONI
Dirección Nacional del Servicio Geológico
Julio A. Roca 651
Buenos Aires
Argentina

ROSEMARY A. ASKIN
Byrd Polar Research Center
The Ohio State University
Columbus, OH 43210
U.S.A.

DANIEL RAGONA
Dirección Nacional del Servicio Geológico
Julio A. Roca 651
Buenos Aires
Argentina

APPENDIX. Species list of palynomorphs.

BRYOPHYTIC SPORES

Aequitriradites verrucosus (Cookson & Dettmann) Cookson & Dettmann
Aequitriradites spinulosus (Cookson & Dettmann) Cookson & Dettmann
Foraminisporis dailyi (Cookson & Dettmann) Dettmann
Foraminisporis asymmetricus (Cookson & Dettmann) Dettmann
Interulobites pseudoreticulatus Archangelsky
Interulobites sp. of Archangelsky et al. 1983
Polycingulatisporites sp.
Triporoletes reticulatus (Pocock) Playford

LYCOPHYTIC SPORES

Ceratosporites equalis Cookson & Dettmann
Densoisporites velatus Weyland & Krieger
Densoisporites corrugatus Archangelsky & Gamerro
Leptolepidites verrucatus Couper
Neoraistrickia truncatus (Cookson) Potonié
Retitriletes austroclavatidites (Cookson) Döring et al.

FILICEAN SPORES

Appendicisporites sp. cf. *A. stylosus* (Thiegart) Déak
Appendicisporites sp.
Baculatisporites comaumensis (Cookson) Potonié
Biretisporites sp.
Cicatricosisporites australiensis (Cookson) Potonié
Cicatricosisporites hughesii Dettmann
Cicatricosisporites venustus Déak
Cicatricosisporites sp. cf. *C.baqueroensis* Archangelsky & Gamerro
Cicatricosisporites sp. 5 of Archangelsky et al. 1983
Cicatricosisporites sp. 1
Cicatricosisporites sp. 2
Contignisporites cooksoniae (Balme) Dettmann
Crybelosporites striatus (Cookson & Dettmann) Dettmann
Cyatheacidites annulatus Cookson *ex* Potonié
Cyathidites minor Couper
Cyathidites sp.
Gleicheniidites circinidites (Cookson) Dettmann
Gleicheniidites sp.
Ischyosporites sp.
Januasporites sp. cf. *J. spinulosus* Dettmann
Klukisporites pseudoreticulatus Couper
Laevigatosporites ovatus Wilson & Webster
Osmundacidites wellmanii Couper
Rugulatisporites sp. cf. *R. micraulaxus* Partridge
Taurocosporites segmentatus Stover
Trilobosporites purverulentus (Verbitskaya) Dettmann
Trilobosporites apiverrucatus Couper
Trilobosporites trioreticulosus Cookson & Dettmann

MEGASPORES

Arcellites santacrucencis Baldoni
Balmeisporites holodictyus Cookson & Dettmann
Bacutriletes triangulatus Taylor & Taylor
Minerisporites sp. cf. *M. marginatus* (Dijkstra) Potonié
Paxillitriletes kachaikense Baldoni
Tenellisporites coronatus Baldoni & Batten

GYMNOSPERM POLLEN

Araucariacites australis Cookson
Balmeiopsis limbatus (Balme) Archangelsky
Callialasporites trilobatus (Balme) Dev forma *trilobatus* Baldoni & Archangelsky
Callialasporites trilobatus (Balme) Dev forma *dampieri* Baldoni & Archangelsky
Callialasporites trilobatus (Balme) Dev forma *tenuisaccatus* Baldoni & Archangelsky
Classopollis sp.
Cycadopites nitidus (Balme) de Jersey
Cycadopites sp.
Cyclusphaera psilata Volkheimer & Sepúlveda
Cyclusphaera sp.
Microcachryidites sp.
Monosulcites sp.
Podocarpidites sp. cf. *P. ellipticus* (Cookson) Couper
Podocarpidites sp.
Trisaccites microsaccatum (Couper) Couper
Vitreisporites pallidus (Reissinger) Nilsson

ANGIOSPERM POLLEN

Clavatipollenites hughesii Couper

ALGAL PALYNOMORPHS

Botryococcus spp.
Schizosporis reticulatus Cookson & Dettmann
Cannosphaeropsis sp.
Chichaouadinium boydii (Morgan) Bujak & Davies
Cleistosphaeridium ancoriferum (Cookson & Eisenack) Davey et al.
Cribroperidinium edwardsii (Cookson & Eisenack) Davey
Cribroperidinium sp.
Dioxya armata Cookson & Eisenack
Eyrea nebulosa Cookson & Eisenack
Florentinia sp. cf. *F. radiculata* (Davey & Williams) Davey & Verdier
?*Florentinia* sp.
Hystrichodinium sp.
Kiokansium unituberculatum (Tasch) Stover & Evitt
Litosphaeridium arundum (Eisenack & Cookson) Davey
Muderongia tetracantha (Gocht) Alberti
Muderongia sp.
Odontochitina operculata (O.Wetzel) Deflandre & Cookson
Oligosphaeridium complex (White) Davey & Williams
Oligosphaeridum pulcherrimum (Deflandre & Cookson) Davey & Williams
Pseudoceratium turneri Cookson & Eisenack
Spiniferites spp.
Subtilisphaera perlucida (Alberti) Jain & Millipied

PALEOECOLOGICAL ASPECTS OF *AFROPOLLIS*/ ELATERATES PEAKS (ALBIAN–CENOMANIAN POLLEN) IN THE CRETACEOUS OF NORTHERN SUDAN AND EGYPT

Eckart Schrank

Abstract

High abundances of *Afropollis* and of elaterate pollen in marine sections in northern Egypt suggest that the parent plants were mainly inhabitants of humid coastal plains. Exceptional *Afropollis*/elaterates peaks in palynomorph assemblages from largely nonmarine sections of the Dongola area, northern Sudan, are probably related to a short-time transgressive event which allowed the southward spreading of humid conditions and of coastal plain vegetation during the late Albian to early Cenomanian (most probably early Cenomanian). The nonmarine assemblages from the Dongola area contain rare freshwater to low salinity dinoflagellates and higher percentages of *Araucariacites* and poly-aperturate angiosperm pollen than the northern, marine sections. This suggests that araucarians and early non-magnoliid dicots flourished especially in the relatively dry continental sites.

Two new combinations, *Spicadinium nenjiangense* (Gao et al. 1992) and *Clavatipollenites spheroidites* (Jardiné and Magloire 1965), are proposed.

INTRODUCTION

Among the most remarkable palynofloristic events in Cretaceous tropical areas is the early Albian appearance, and subsequent proliferation of elater-bearing pollen of probable ephedroid affinities (Crane, 1988). Their characteristic morphology and relatively short stratigraphic range, lower Albian to upper Cenomanian (e.g. Herngreen and Duenas Jiminez, 1990; Herngreen et al., 1996), attracted the attention of palynologists and contributed to their application in palynostratigraphy and paleobiogeography. Elaterate pollen were thus used to define several microfloral regions, most recently the Albian–Cenomanian Elaterates Province (Herngreen et al., 1996, and references therein).

In the present paper, *Afropollis*/elaterates assemblages from different paleogeographic settings in the Albian–Cenomanian of Egypt and Sudan are reviewed in order to understand the ecological factors which influence the distribution of these important pollen types.

DISTRIBUTION OF ELATERATES AND OF *AFROPOLLIS* IN EGYPT AND SUDAN

A comparison of Albian–Cenomanian palynomorph assemblages from northern Egypt and Sudan (Text-Figures 1 and 4) shows that highest percentages of elater-bearing pollen (up to 82.0%, El-Shamma, 1991) and of *Afropollis* (botanical affinity Winteraceae, Doyle et al., 1990) are generally attained in the paralic basins of northern Egypt where they are associated with abundant spores and marine dinoflagellates. This suggests that the parent plants of both groups flourished in relatively humid environments of the southern Tethyan coastal plains.

In the present paper, extraordinarily high abundances of *Afropollis* (78.0%, East Kodroka; 37.0%, El Bab; 35.0%, Kodroka) and of elaterates (12.0%, East Kodroka; 11.0%, El Bab; 15.0%, Kodroka) are reported from largely nonmarine subsurface equivalents of the Wadi Milk Formation (Wycisk, 1991) in the Dongola area, northern Sudan (Text-Figures 2-4). Possible explanations for these *Afropollis*/elaterates peaks, which are exceptional for the Sudan (see Kaska, 1989, and Text-Figure 1), should take into account the three following points: 1) The peak levels are short correlative stratigraphic intervals in three Dongola wells (El Bab, 105–110 ft; Kodroka/Pilot Farm, 280–285 ft; East Kodroka, 250–265 ft); 2) The associated palynological assemblages are predominantly terrestrially derived, but brackish water phytoplankton is also present and; 3) Local tectonic and sedimentological features suggest that Albian–Cenomanian transgressive events could have reached northern Sudan from the north (Mansour, 1990). The Misaha Trough, a distinct NNW–SSE trending tectonic structure in the Egyptian–Sudanese border area (Text-Figure 1), could have functioned as a gateway for such short-time transgressive events. Marine influence in presumably Albian (Misaha Trough) and Cenomanian strata (Gebel Abyad, about 200 km west of Dongola, Text-Figure 1) has been postulated (Mansour, 1990) on the basis of trace fossils and other sedimentological features, but age determinations and correlation with the Dongola wells remain uncertain.

The palynological assemblages of the Dongola wells (examples see Plates 1 and 2) contain rare (<1.5%) dinoflagellates (Plate 1, figs. 1, 2, 5, 6) related to forms previously described from freshwater or low salinity envrionments (Batten and Lister, 1988; Gao et al., 1992). The prasinophytes are represented by a single *Pterospermella* specimen (Plate 1, fig.7). *Pterospermella* is regarded as a marine alga, but rare brackish occurrences are known (Weiler, 1988). Freshwater algae include *Botryococcus* (Plate 1, fig. 14) and *Chomotriletes minor* (Plate 1, fig. 4). Although

Schrank, Eckart
2001 Paleoecological aspects of *Afropollis*/elaterates peaks (Albian–Cenomanian pollen) in the Cretaceous of northern Sudan and Egypt; *In*: Goodman, D.K., and Clarke, R.T. (eds.), *Proceedings of the IX International Palynological Congress, Houston, Texas, U.S.A., 1996*; American Association of Stratigraphic Palynologists Foundation, p. 201–210.

ISBN 0–931871–06–9

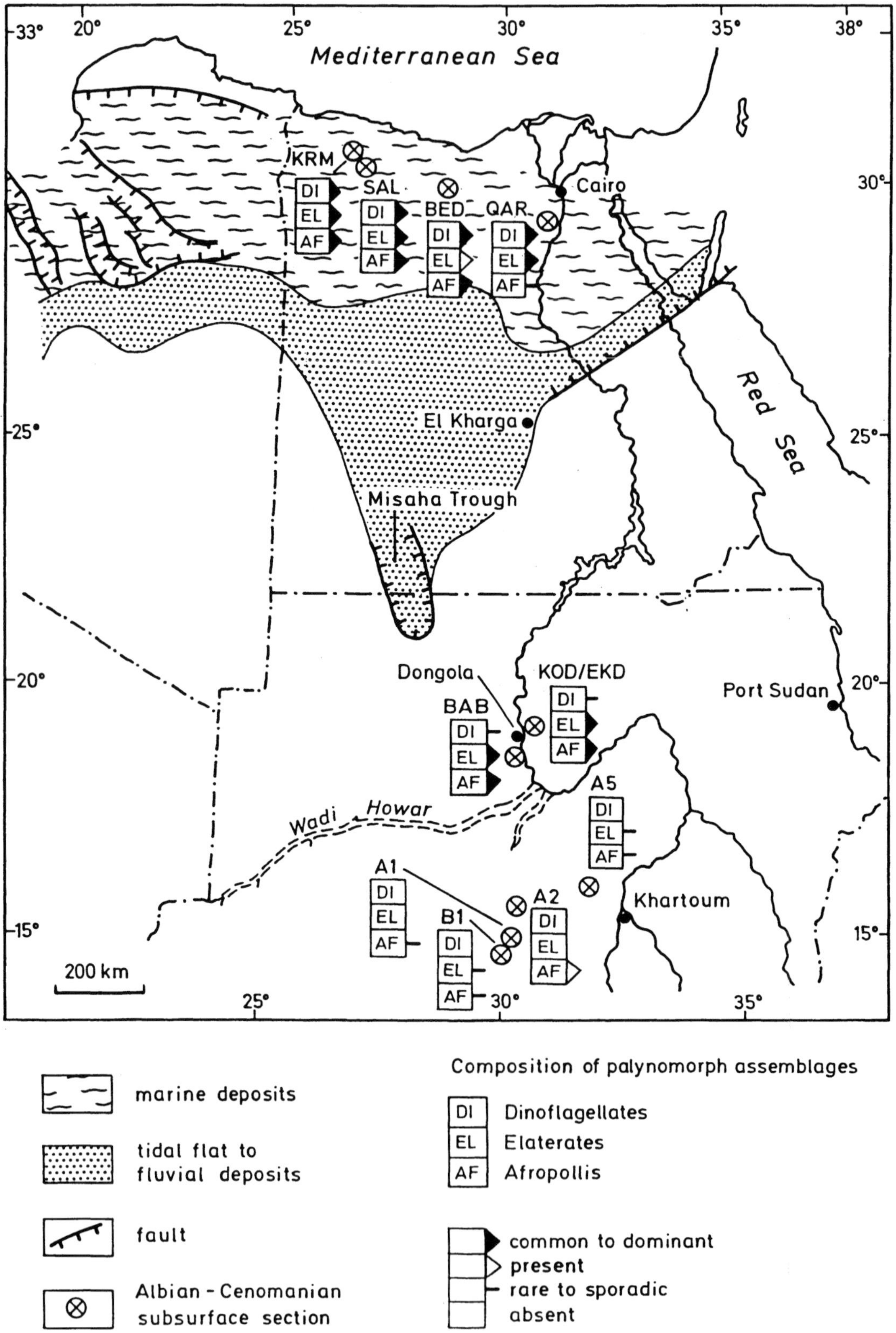

Text-Figure 1. Paleogeographic sketch of the Cenomanian plotted on a present day map of northeast Africa (partly simplified and modified from Klitzsch and Wycisk, 1987). The distribution of dinoflagellates, elaterate pollen and *Afropollis* in 11 Albian–Cenomanian well sections is also shown. The wells are Kahraman-1 (KRM, Schrank and Ibrahim, 1995), El-Salam 17 (SAL, El-Shamma, 1991), Bed 1-1 (BED, El-Beialy, 1994), Qarun 2-1 (QAR, El-Beialy, 1993); El Bab (BAB), Kodroka/Pilot Farm and East Kodroka (KOD/EKD)(Schrank, 1990); El Maghad (A1), Kirieb El Sanaya (A2), Gabrad Said (A5) and Kerkish (B1) after Schrank (1994).

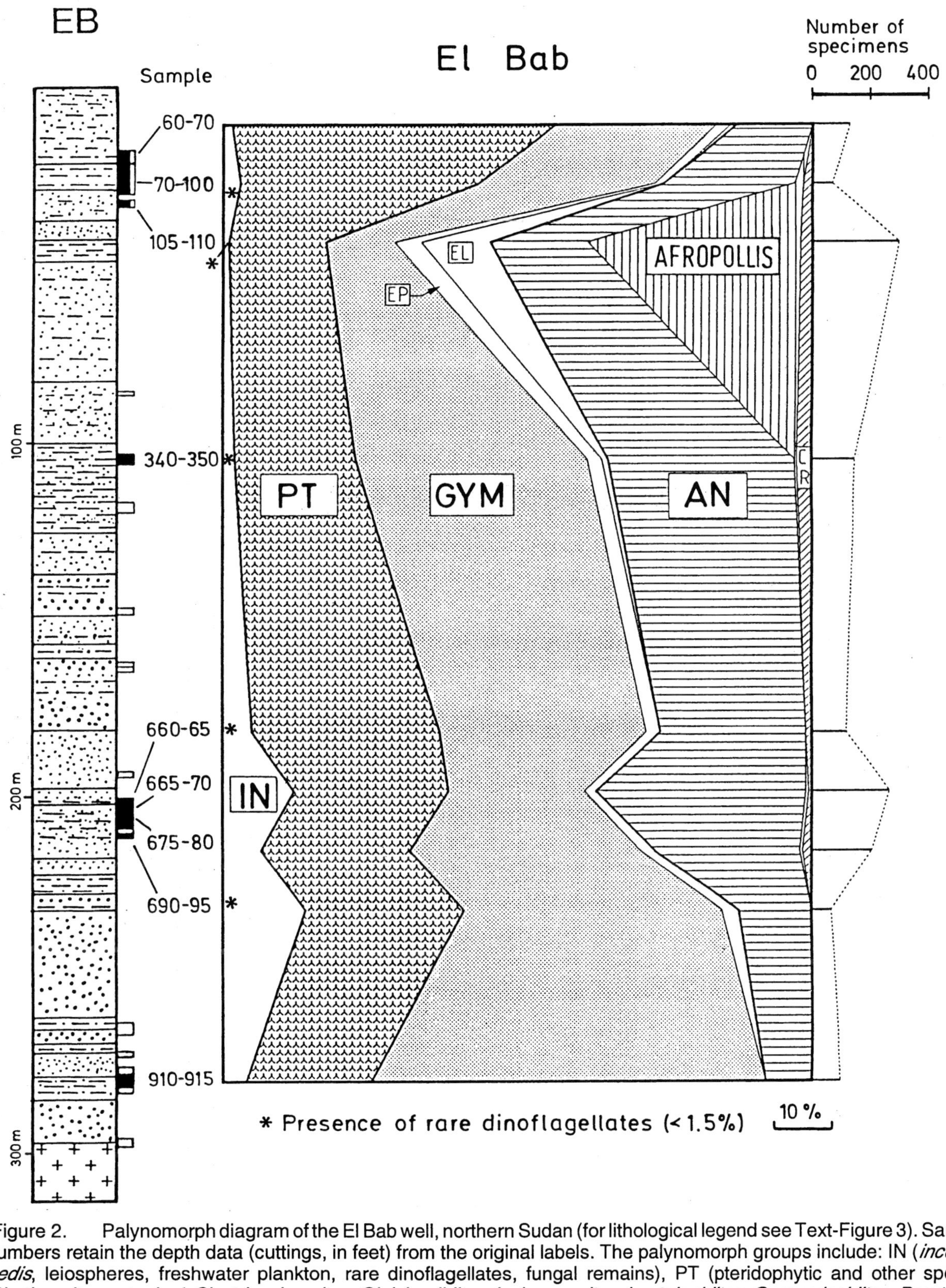

Text-Figure 2. Palynomorph diagram of the El Bab well, northern Sudan (for lithological legend see Text-Figure 3). Sample numbers retain the depth data (cuttings, in feet) from the original labels. The palynomorph groups include: IN (*incertae sedis*, leiospheres, freshwater plankton, rare dinoflagellates, fungal remains), PT (pteridophytic and other spores: *Biretisporites potoniaei*, *Cicatricosisporites*, *Gleicheniidites*, *Ischyosporites*, *Leptolepidites*, *Osmundacidites*, *Perotrilites* and others), GYM (gymnosperm pollen grains: mainly *Araucariacites*, also *Spheripollenites* and *Inaperturopollenites* types, rarely *Callialasporites*, *Cycadopites* and *Eucommiidites*), EP (mainly *Ephedripites*, also *Steevesipollenites binodosus*), EL (elaterate pollen grains: mainly *Elaterocolpites castelainii* and *Elateroplicites africaensis*, rarely *Elaterosporites klaszii*), AN (angiosperm pollen grains: mainly tricolpates and tricolporates, subordinate *Clavatipollenites*, *Brenneripollis*, *Retimonocolpites*, *Echimonocolpites*, *Tetracolpites* and others), *Afropollis* (mainly *A. jardinus* type), CR (polyporate angiosperm pollen: *Cretacaeiporites*).

rare, the zygnematacean spores *Brazilea parva* and *Lacunalites* sp. (Plate 1, fig. 8) are also present.

From the palynomorph curves of the relatively thick Albian–early Cenomanian succession of the El Bab well (Text-Figure 2) the following scenario can be tentatively reconstructed.

Araucarians (mainly *Araucariacites australis*, Plate 1, fig. 11) dominate the lower, predominantly sandy part of the section suggesting that araucarian conifers were important elements of the local flora in the early phase of basin formation. Younger in the section, the araucarian element decreases while angiosperm pollen gradually increases in abundance culminating in the characteristic *Afropollis*/ elaterates peak. This peak is immediately followed by a spore-dominated palynoflora in the uppermost part of the section, which is relatively rich in clay and silt. Thus the *Afropollis*/elaterates peak is associated with a late phase of basin development when the relief had lowered, conditions became more humid, and ferns, lycophytes and other spore producers became more abundant. This may have coincided with a short-time flooding event from the north, probably via the Misaha route, which brought partly brackish (?) and relatively humid conditions into the Dongola area and allowed coastal plain vegetation to spread southward.

The age range of the *Afropollis*/elaterates peak levels in the three Dongola wells (Text-Figures 2-4) is late Albian to early Cenomanian. Recent data from the foramimifera-dated KRM-1 well in northern Egypt (Schrank and Ibrahim, 1995), favours an early Cenomanian age because the two most common Dongola elaterates, *Elateroplicites africaensis* and *Elaterocolpites castelainii*, both appear in this interval of the well.

SYSTEMATIC PALEONTOLOGY

Division DINOFLAGELLATA (Bütschli 1885)
Fensome et al. 1993

Genus *Spicadinium* Batten and Lister 1988
Spicadinium nenjiangense (Gao et al. 1992) comb. nov.

Basionym. *Cleistosphaeridium nenjiangense* Gao et al., 1992, p. 18, 24, pl. 3, figs. 1–9; pl. 4, figs. 1–7.

Remarks. *Impletosphaeridium nenjiangense* (Gao et al. 1992) Williams et al. 1998 is transferred to *Spicadinium* mainly because of the similarity of its processes with those of the latter genus. The processes are relatively long and simple, which contrasts with the complex and distally branched processes common in *Impletosphaeridium.*

Spicadinium sp. cf. *S. nenjiangense*
(Gao et al. 1992) comb. nov.
Plate 1, figs. 1-2

Paleoecological Implications. *Spicadinium akidoton* Batten and Lister 1988, the type species, and *S. nenjiangense*, are from freshwater or low salinity envrionments. The former has been recovered from fresh to brackish palynofacies of the British Wealden (Batten and Lister, 1988) and the latter from Campanian brackish water, lacustrine environments of China (Gao et al., 1992).

Occurrences. El Bab, 70–100, 105–110, 660–665, 690–695 ft; Kodroka, 125–130 ft .

Pollen

Genus *Echimonocolpites* Van der Hammen and Garcia de Mutis 1965

Echimonocolpites sp. cf. *E. tersus* Ward 1986
Plate 1, figs. 15, 16

Remarks. *Echimonocolpites*-like pollen occurs in many monocotyledon families and in the Nymphaeaceae (Ward, 1986). It differs from *Spinizonocolpites* Muller 1968 (botanical affinity *Nypa*) in having a nearly smooth and dense wall in the spaces between the spines whereas *Spinizonocolpites* is reticulate and columellate (Muller, 1968). Therefore the presence of *Echimonocolpites* does not necessarily support the existence of *Nypa*-like mangrove palms in the Albian of

PLATE 1

Dinoflagellates (1, 2, 5, 6), acritarchs (3), prasinophytes (7), Zygnemataceae (4, 8) and other freshwater algae (14), spores (9, 10) and pollen grains (11-13, 15, 16) from the Albian–early Cenomanian of the Dongola area, northern Sudan. All photomicrographs are cited with a TU Berlin photo identification number (e.g. DQ1,3; EE17 etc.).

1, 2 *Spicadinium* sp. cf. *S. nenjiangense* (Gao et al. 1992) comb. nov., x800. El Bab, 70–100 ft, DQ1,3

3 *Leiosphaeridia*? sp. cf. *L. tenuissima* Eisenack 1958, x630. El Bab, 660–665 ft, EE17.

4 *Chomotriletes minor* (Kedves 1961) Pocock 1970, x1,000. El Bab, 665–670 ft, EL7.

5 *Cassiculosphaeridia* sp. cf. *C. parvula* Batten and Lister 1988, x630. El Bab, 340–350 ft, EA17.

6 *Spicadinium* sp., short-spined type, x1,000. El Bab, 675–680 ft, EN26.

7 *Pterospermella* sp., x800. El Bab, 890–895 ft, EQ34

8 *Lacunalites* sp., x1,250. El Bab, 890–895 ft, EQ33.

9, 10 *Perotrilites pannuceus* Brenner 1963, x1,000. Kodroka, 395–400 ft, FH26, 27.

11 *Araucariacites australis* Cookson 1947, x800. El Bab, 60–70 ft, DL27.

12 *Ephedripites multicostatus* Brenner 1963, x2,000. El Bab, 105–110 ft, AN28.

13 *Ephedripites multicostatus* Brenner 1963, x1,000. El Bab, 105–110 ft, DR26.

14 *Botryococcus* sp., x1,000. El Bab, 60–70 ft, DN29.

15, 16 *Echimonocolpites* sp. cf. *E. tersus* Ward 1986, x1,000. El Bab, 60–70 ft, DL13, 14.

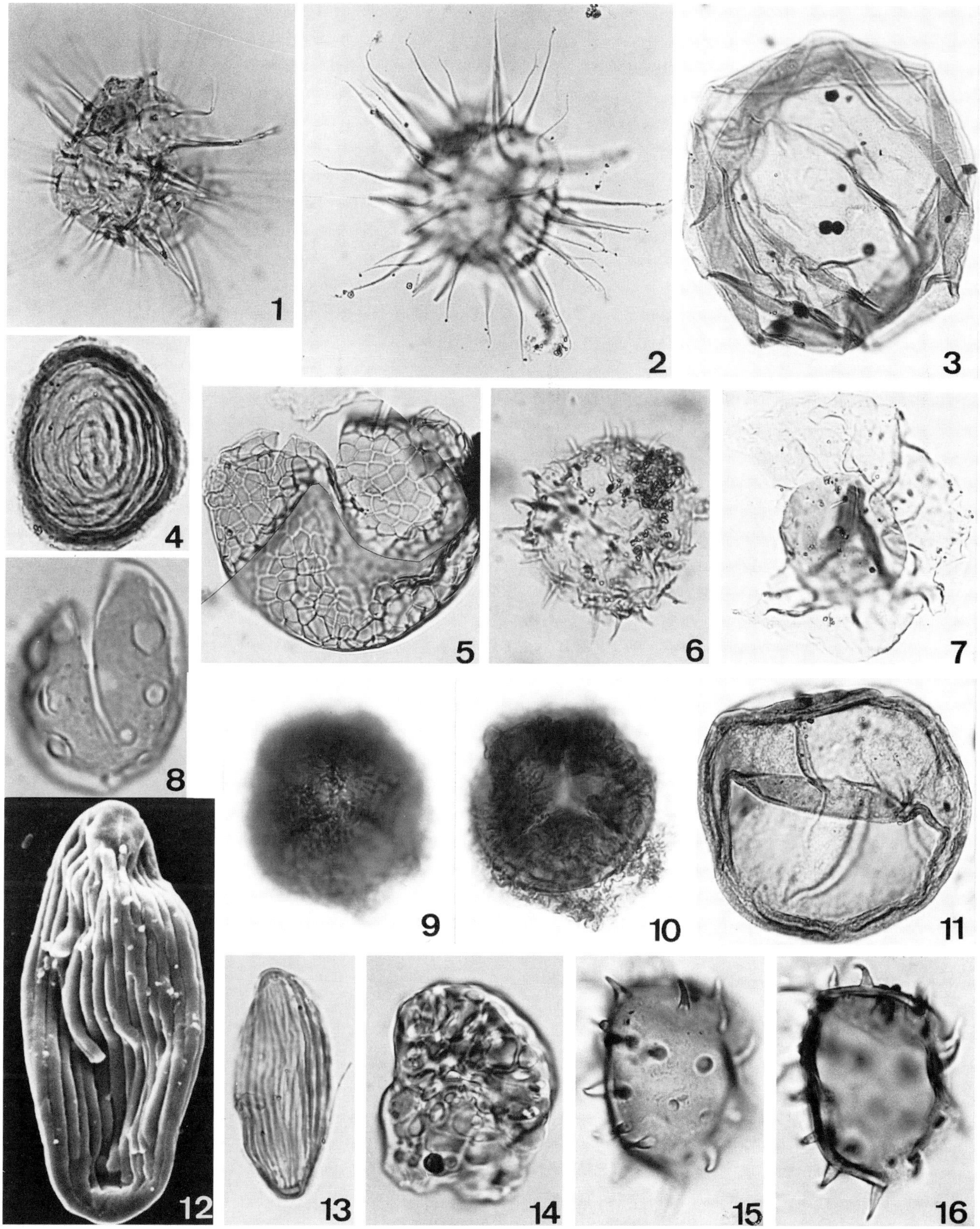
1
2
3
4
5
6
7
8
9
10
11
12
13
14
15
16

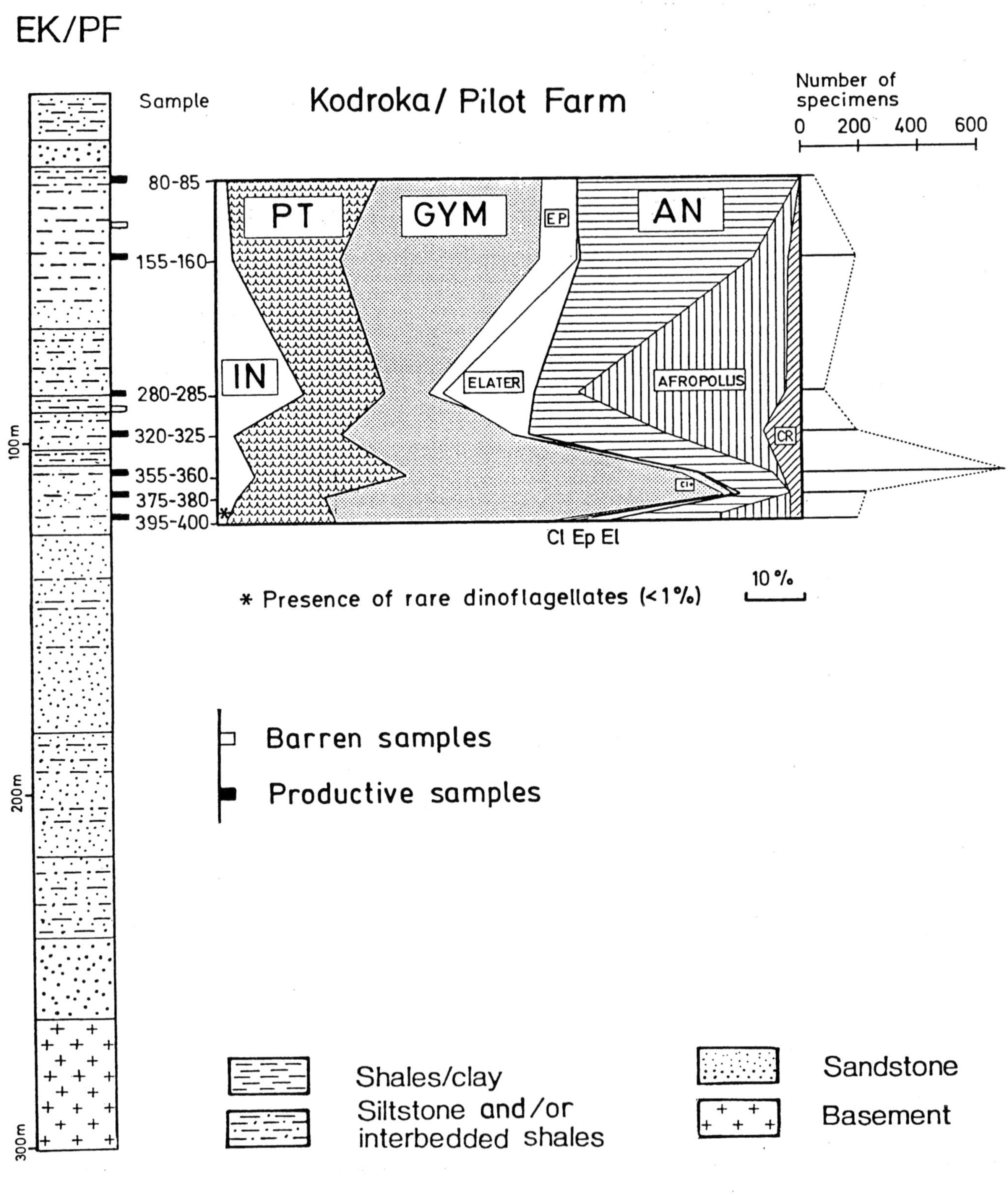

Text-Figure 3. Palynomorph diagram of the Kodroka/Pilot Farm well, northern Sudan, and generalized lithological log of the East Kodroka/Pilot Farm area. The palynomorph groups are the same as in Text-Figure 2, but in addition CL (*Classopollis*) is shown.

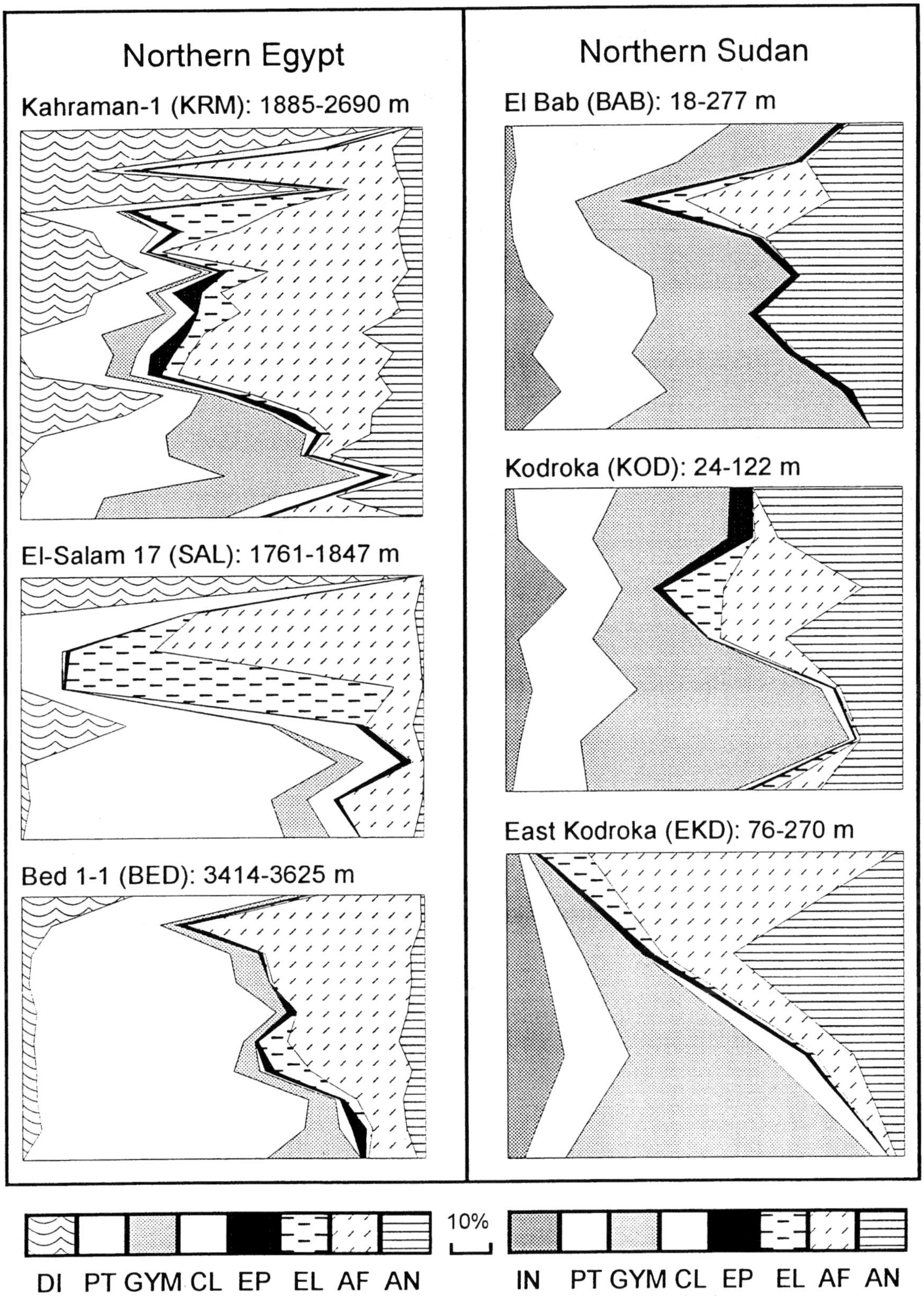

Text-Figure 4. Comparison of palynomorph diagrams from Albian–Cenomanian well sections in northern Egypt and northern Sudan (see Text-Figure 1 for the location of the wells and data sources). The palynomorph groups include: DI (marine dinoflagellates and organic-walled microforaminifera), PT (pteridophytic and other spores), GYM (mainly *Araucariacites*, *Inaperturopollenites*, *Spheripollenites*), CL (*Classopollis*), EP (*Ephedripites* and other ephedroids), EL (elaterate pollen grains: in northern Egypt mainly *Elaterosporites klaszii*, in northern Sudan mainly *Elaterocolpites castelainii* and *Elateroplicites africaensis*), AF (*Afropollis jardinus* group), AN (angiosperm pollen other than *Afropollis*: mainly tricolpates and tricolporates, but also monocolpates and polyporates), IN (*incertae sedis*, rare dinoflagellates and others, see Text-Figure 2).

the Sudan as suggested by Srivastava (1994) on the basis of earlier illustrations of this pollen type (Schrank, 1990, pl. 3, fig. 2). So far there are no unequivocal *Spinizonocolpites/ Nypa* records from pre–Campanian sediments. Some of the *Echimonocolpites* grains in the material from northern Sudan look similar to mainly Tertiary pollen of the *Nupharipollenites* and *Nupharipollis* type. This would indicate botanical affinities to the Nymphaeaceae, which have a well established macrofossil record from the Albian onward.

Occurrences. El Bab, 60–70, 340–350, 660–665 ft; Kodroka, 125–130, 155–160, 355–360 ft.

Genus *Clavatipollenites* Couper 1958

Clavatipollenites spheroidites (Jardiné and Magloire 1965) comb.nov.
Plate 2, fig. 11

Basionym. *Monocolpopollenites spheroidites* nov. spec. S.CI.145 Jardiné and Magloire 1965, p. 211, pl. 8, figs. 27–30.

Remarks. The transfer of this species from *Monocolpopollenites* to *Clavatipollenites* is done because of its general similarity with *Clavatipollenites rotundus* Kemp 1968 from the British Albian. In particular, the form of its colpus or sulcus which tends to taper towards the extremities is a feature known in *Clavatipollenites* (see Kemp, 1968). *Monocolpopollenites* is characterized by a colpus with flared or rounded ends (see Nichols et al., 1973). *C. spheroidites* differs from *C. rotundus* by its smaller size and finer sculpture of its outer exine layer. The stratigraphic range of *C. spheroidites* species is Aptian (Santana Formation, Brasil, Lima, 1990) to base of Maastrichtian (Senegal and Ivory Coast; see Jardiné and Magloire, 1965).

Occurrences. El Bab, 665–670 ft; Kodroka, 355–360, 395–400 ft .

CONCLUSIONS

Afropollis (aff. Winteraceae) and elaterates (at least partly ephedroids) producers are mainly inhabitants of paleotropical humid coastal plains. Therefore, they attain high abundances in shallow marine environments, for example, on both sides of the Proto-Atlantic (Herngreen and Duenas Jimenez, 1990) and on the southern Tethyan shores. Coastal plains bordering the Proto-Atlantic and the Cretaceous Tethyan Ocean (e.g. Libya, northern Egypt, southern Alps, Saudi Arabia, Qatar, China, Papua New Guinea) were major routes of migration for the elaterate pollen producing plants, and sites as distant from the Proto–Atlantic center of elaterate radiation as the western Tarim Basin, China, still have elaterate percentages of 7.2–20.0% (Song and Huang, 1997; and references therein).

Distinct *Afropollis*/elaterates peak levels in largely nonmarine sections of the Dongola area, northern Sudan, were formed when Tethyan coastal plain vegetation migrated southward during a short-time transgressive event, probably during the early Cenomanian (possible age range late Albian to early Cenomanian). Otherwise, the nonmarine palynofloras of the Dongola area are different from the mainly marine northern Egyptian assemblages (see Text-Figure 4) in the absence of undoubtedly marine dinoflagellates, in lower spore percentages and in the relative abundance of *Araucariacites* and of angiosperm pollen other than *Afropollis.* Among the latter, tricolpates and tricolporates are common, and polyporates (*Cretacaeiporites*) are present. This suggests that araucarians and early non-magnoliid dicots flourished in the relatively dry continental sites.

ACKNOWLEDGMENTS

Thanks go to Dr. M.I.A. Ibrahim, Alexandria University, for background information on the KRM-1 well, to Ms. B. Dunker, Technical University Berlin, for help with the

PLATE 2

Elaterate pollen (1, 2, 4, 5, 14-16), ephedroids (7) and angiosperm pollen (3, 6, 8-13, 17) from the Albian–early Cenomanian of the Dongola area, northern Sudan.

1 *Elateroplicites africaensis* Herngreen 1973, x1,704. East Kodroka, 250–265 ft, AF18.

2 *Elaterocolpites castelainii* Jardiné and Magloire 1965, x480. East Kodroka 250–265 ft, AF36.

3 *Afropollis jardinus* (Brenner 1968) Doyle et al. 1982, x4,917. East Kodroka 250–265 ft, AH6.

4, 5 *Elaterosporites klaszii* (Jardiné and Magloire 1965) Jardiné 1967, x800. El Bab, 105–110 ft, DU5, 6.

6 *Afropollis jardinus* (Brenner 1968) Doyle et al. 1982, x1,000. Kodroka, 280–285 ft, EX32.

7 *Ephedripites jansonii* (Pocock 1964) Muller 1968, x1,000. El Bab, 105–110 ft, DU20.

8-10 *Brenneripollis reticulatus* (Brenner 1963) Juhász and Góczán 1985, x1,250. El Bab, 665–670 ft, EK28-30.

11 *Clavatipollenites spheroidites* (Jardiné and Magloire 1965) comb. nov., x1,250. Kodroka, 355–360 ft, FC4.

12 *Tricolpites hians* Stanley 1965, x1,000. El Bab, 665–670 ft, DI26.

13 *Tricolporopollenites* sp., x1,250. El Bab, 70–100 ft, DO36.

14, 15 *Elateroplicites africaensis* Herngreen 1973, specimen with elaters rolled up, note the similarity with *Ephedripites* (7) x1,000. East Kodroka, 250–265 ft, FK37, FL1.

16 *Elateroplicites africaensis* Herngreen 1973, x1,000. El Bab, 105–110 ft, DW4.

17 *Cretacaeiporites polygonalis* (Jardiné and Magloire 1965) Herngreen 1973, x2,000. El Bab, 105–110 ft, AN23.

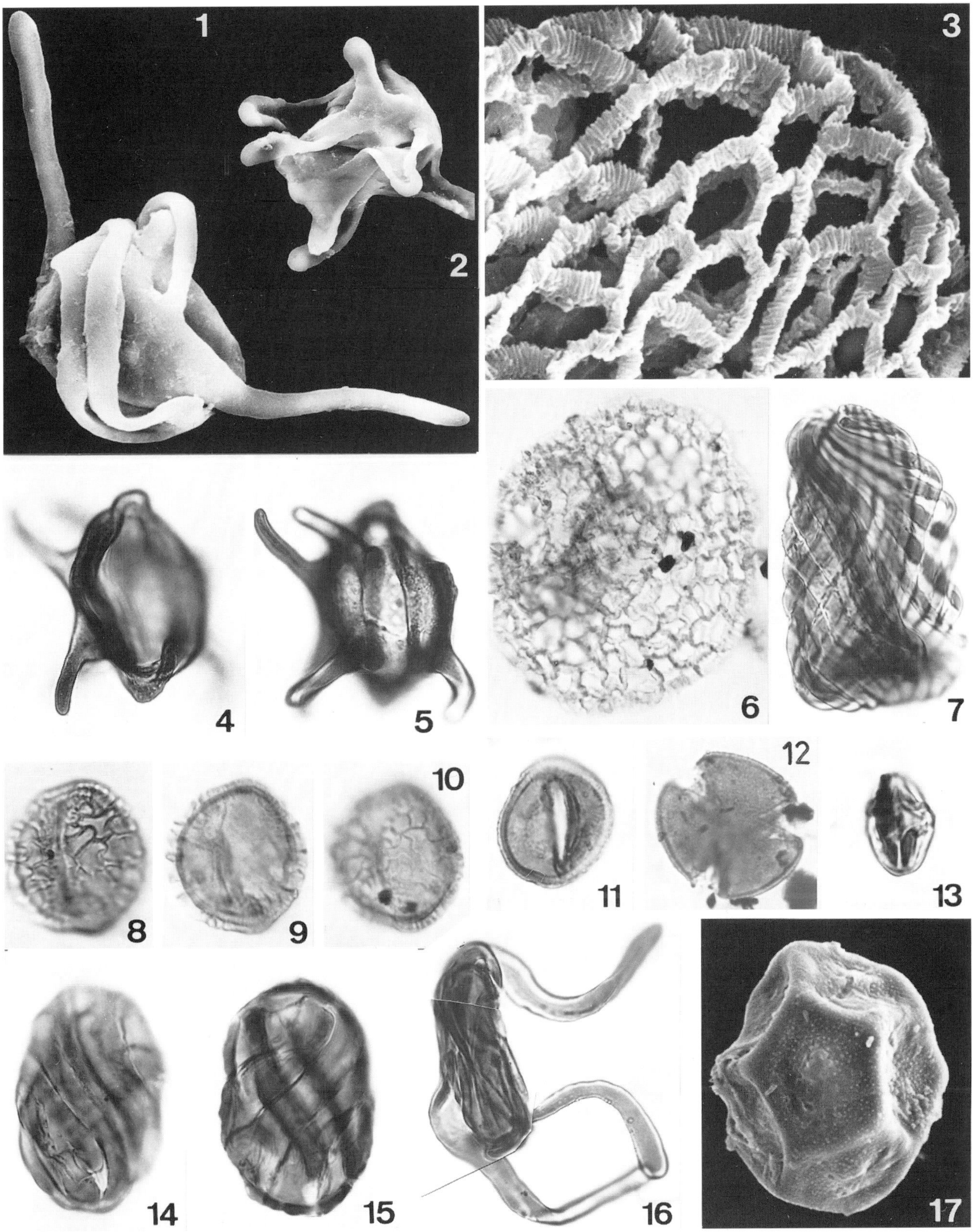
1
2
3
4
5
6
7
8
9
10
11
12
13
14
15
16
17

Text-Figures and last not least to the German Research Foundation (DFG) for a travelling grant which enabled participation in the Ninth IPC, Houston. I am grateful to the editor, Dr. D.K. Goodman, and tp Dr. G.D. Wood, for reviewing the manuscript.

References Cited

BATTEN, D.J., and LISTER, J.K.
1988 Early Cretaceous dinoflagellate cysts and chlorococcalean algae from freshwater and low salinity palynofacies in the English Wealden. *Cretaceous Research*, 9: 337–367.

CRANE, P.R.
1988 Major clades and relationships in the "higher" gymnosperms. *In*: Beck, C.B. (ed.), *Origin and evolution of gymnosperms*. Columbia University Press, New York, pp. 218–272.

DOYLE, J.A., HOTTON, C.L., and WARD, J.V.
1990 Early Cretaceous tetrads, zonasulculate pollen, and Winteraceae. II. Cladistic analysis and implications. *American Journal of Botany*, 77: 1558–1568.

EL-BEIALY, S.Y.
1993 Dinoflagellate cysts and miospores from the Albian–Cenomanian sequence of the Qarun 2-1 borehole, Western Desert, Egypt. *Qatar University Science Journal*, 13: 301–307.
1994 Palynostratigraphy and palynofacies analysis of some subsurface Cretaceous formations in the Badr El Dein (Bed 1-1) borehole, North Western Desert, Egypt. *Neues Jahrbuch für Geologie und Paläontologie Abhandlungen*, 192: 133–149.

EL-SHAMMA, A.E.
1991 Palynostratigraphic zonation of the Bahariya Formation in El-Salam no. 17 well, Western Desert, Egypt. *Egyptian Journal of Geology*, 35: 261–273.

GAO RUI-QI, HE CHENG-QUAN, and QIAO XIU-YUN
1992 A new genus and species of Cretaceous dinoflagellates from two transgressive beds in Songliao Basin, NE China. *Acta Palaeontologica Sinica*, 31: 17–29.

HERNGREEN, G.F.W., and DUENAS JIMINEZ, H.
1990 Dating of the Cretaceous Une Formation, Colombia and the relationship with the Albian–Cenomanian African–South American microfloral province. *Review of Palaeobotany and Palynology*, 66: 345–359.

HERNGREEN, G.F.W., KEDVES, M., ROVNINA, L.V., and SMIRNOVA, S.B.
1996 Cretaceous palynofloral provinces: a review. *In*: Jansonius, J., and McGregor, D.C. (eds.), *Palynology: Principles and Applications*. American Association of Stratigraphic Palynologists Foundation, 3, pp. 1157–1188.

JARDINÉ, S., and MAGLOIRE, L.
1965 Palynologie et stratigraphie du Crétacé des bassins du Sénégal et de Côte d'Ivoire. *Mémoires du Bureau de Recherches Géologiques et Minières*, 32: 187–245.

KASKA, H.V.
1989 A spore and pollen zonation of Early Cretaceous to Tertiary nonmarine sediments in central Sudan. *Palynology*, 13: 79–90.

KEMP, E.M.
1968 Probable angiosperm pollen from British Barremian to Albian strata. *Palaeontology*, 11: 421–434.

KLITZSCH, E., and WYCISK, P.
1987 Geology of the sedimentary basins of northern Sudan and bordering areas. *Berliner geowissenschaftliche Abhandlungen*, Reihe A, 75.1: 97–136.

LIMA, M.R. de
1990 Palinologia da Formação Santana (Cretáceo do nordest do Brasil). IV – Descrição sistemática dos polens das turmas Plicates e Poroses, esporos, *incertae sedis* e microplancton marinho. *Ameghiniana*, 26(1-2) (for 1989): 63–81.

MANSOUR, N.
1990 Geochemisch-sedimentologische und tektonische Untersuchungen am Uweinat-Safsaf-Aswan-Schwellensystem (Südägypten/Nordsudan). *Berliner geowissenschaftliche Abhandlungen*, Reihe A, 122: 1–115.

MULLER, J.
1968 Palynology of the Pedawan and Plateau Sandstone Formations (Cretaceous–Eocene) in Sarawak, Malaysia. *Micropaleontology*, 14: 1–37.

NICHOLS, D.J., AMES, H.T., and TRAVERSE, A.
1973 On *Arecipites* Wodehouse, *Monocolpopollenites* THOMSON & PFLUG, and the species "*Monocolpopollenites tranquillus*". *Taxon*, 22: 241–256.

SCHRANK, E.
1990 Palynology of the clastic Cretaceous sediments between Dongola and Wadi Muqaddam, northern Sduan. *Berliner geowissenschaftliche Abhandlungen*, Reihe A, 120.1: 149–168.
1994 Nonmarine Cretaceous palynology of northern Kordofan, Sudan, with notes on fossil Salviniales (water ferns). *Geologische Rundschau*, 83: 773–786.

SCHRANK, E., and IBRAHIM, M.I.A.
1995 Cretaceous (Aptian–Maastrichtian) palynology of foraminifera-dated wells (KRM-1, AG-18) in northwestern Egypt. *Berliner geowissenschaftliche Abhandlungen*, Reihe A, 177: 1–44.

SONG ZHICHEN; and HUANG FEI
1997 The boundaries between the Southern Laurasian and Northern Gondwana provinces and the *Aquilapollenites* and Normapolles palynofloras in East Asia. *Cretaceous Research*, 18: 1–15.

SRIVASTAVA, S.K.
1994 Evolution of Cretaceous phytogeoprovinces, continents and climates. *Review of Palaeobotany and Palynology*, 82: 197–224.

WARD, J.V.
1986 Early Cretaceous angiosperm pollen from the Cheyenne and Kiowa formations (Albian) of Kansas, U.S.A. *Palaeontographica*, Abteilung B, 202: 1–81.

WEILER, H.
1988 *Pterospermella* Eisenack (1972) (Prasinophyceae). Morphotypen aus mitteloligozänen Sedimenten Südwestdeutschlands. *Mainzer geowissenschaftliche Mitteilungen*, 17: 283–312.

WYCISK, P.
1991 Stratigraphic update of the nonmarine Cretaceous from SW Egypt and NW Sudan. *Cretaceous Research*, 12: 185–200.

Author's address

ECKART SCHRANK
Institut für Angewandte Geowissenschaften II
Technische Universität Berlin, Sekr. EB 10
Ernst-Reuter-Platz 1
D-10587 Berlin
Germany

TRIASSIC AND LOWER CRETACEOUS MIOSPORES OF THE MID–POLISH TROUGH (CENTRAL POLAND) AS A RECORD OF STRATIGRAPHY, PALEOECOLOGY, AND HISTORY OF THERMAL CHANGES

Marta Waksmundzka

Abstract

Triassic and Lower Cretaceous pollen spectra from sediments in the Polish Lowland have been used to define stratigraphy, paleoecology and maturity grade of organic matter (TAI index) in the sedimentary basin. Occurrence of biostratigraphic index forms in the sediments permitted some stratigraphical conclusion. Differentiated pollen spectra in the Triassic and Lower Cretaceous sediments, documented by quantitative analysis, might be applied together with miospore color analysis (TAI index), to develop a palynofacies reconstruction of the basin. Color variability also permitted preliminary estimates of thermal changes in the sedimentary basin.

INTRODUCTION

The Triassic and Lower Cretaceous palynological material analyzed was derived from boreholes of the Pomerania–Kujawy Rampart and adjacent depressions, Polish Lowlands (Text-Figure 1a). These structural units were formed in latest Cretaceous and earliest Tertiary time as a result of the Zechstein/Mesozoic inversion of the Mid–Polish Trough. The Mid–Polish Trough, elongated NW–SE, was an axial zone of an epicontinental sedimentary basin in Poland during the Permian and Mesozoic. Sea water of the two boreal seas, the Danish and North-German ones, flowed into this structure from the west and southwest and from the Tethys ocean from the south and southeast (Marek and Pajchlowa, 1997).

Analyzed miospores are substantially varied from the points of view of taxonomic content and color. This variability was discovered while using the material to study the sedimentary basin in different ways:

a. for stratigraphic analysis (after Orłowska-Zwolińska, 1983; Waksmundzka, 1992);
b. for quantitative analysis with paleoecological conclusions (after Tyson, 1995);
c. for color analysis with palaeothermal conclusions (after Batten, 1984; Gaupp and Batten, 1985).

The complex analysis permits correlation of results with results of bio- and lithostratigraphic studies and petrographic examination.

Waksmundzka, Marta
2001 Triassic and Lower Cretaceous miospores of the mid-Polish Trough (Central Poland) as a record of stratigraphy, paleoecology, and history of thermal changes; *In*: Goodman, D.K., and Clarke, R.T. (eds.), *Proceedings of the IX International Palynological Congress, Houston, Texas, U.S.A., 1996*; American Association of Stratigraphic Palynologists Foundation, p. 211–220.
 ISBN 0–931871–06–9

OBSERVATION RESULTS

Taxonomic Analysis

The Triassic material from the boreholes # 4, 5, 6, 7, 8, located in the Kujawy Central Segment of the Mid–Polish Trough (Text-Figure 1b), illustrates substantial predominance of bisaccate conifer pollen grains. The representatives of the following genera occur there:

a. *Angustisulcites* (Freudenthal) Visscher: distinct oval central body and distally inclined air sacs (Plate 1; fig. 3).
b. *Infernopollenites* Scheuring: wide taeniae ornamentation elements, closely jointed or divided with narrow ectexine fissures (Plate 1; fig. 1).
c. *Striatoabietites* Hart: narrow taeniae ornamentation elements on a central body and distinct network on air sac surface (Plate 1; fig. 2).
d. *Pinuspollenites* (Couper) Kemp: small (less than a central body) air sacs located subequatorially (Plate 1; fig. 4).
e. *Brachysaccus* Mädler: pollen grains of controversial morphology; they have been defined as bisaccate, where air sacs do not protrude or protrude only a little outside a grain corpus (Mädler, 1964), or non-bisaccate pollen grains with a sulcus (Potonié, 1970) (Plate 1, fig. 5).

The Lower Cretaceous palynological material was taken from boreholes #1, 2 and 3, located at the northwest slope of the Pomerania Segment of the Mid–Polish Trough (Text-Figure 1b). Substantial predominance of fern miospores of the families Gleicheniaceae, Schizeaceae and Cyathaceae–Dicksoniaceae characterizes this spectrum.

Among miospores of the family Gleicheniaceae, with characteristic equatorial exine thickenings and thickenings on the central part of the distal side, the following genera have been found:

a. *Gleicheniidites* Ross ex Delcourt & Sprumont emend. Bolkhovitina: smooth surface, equatorial exine thick-

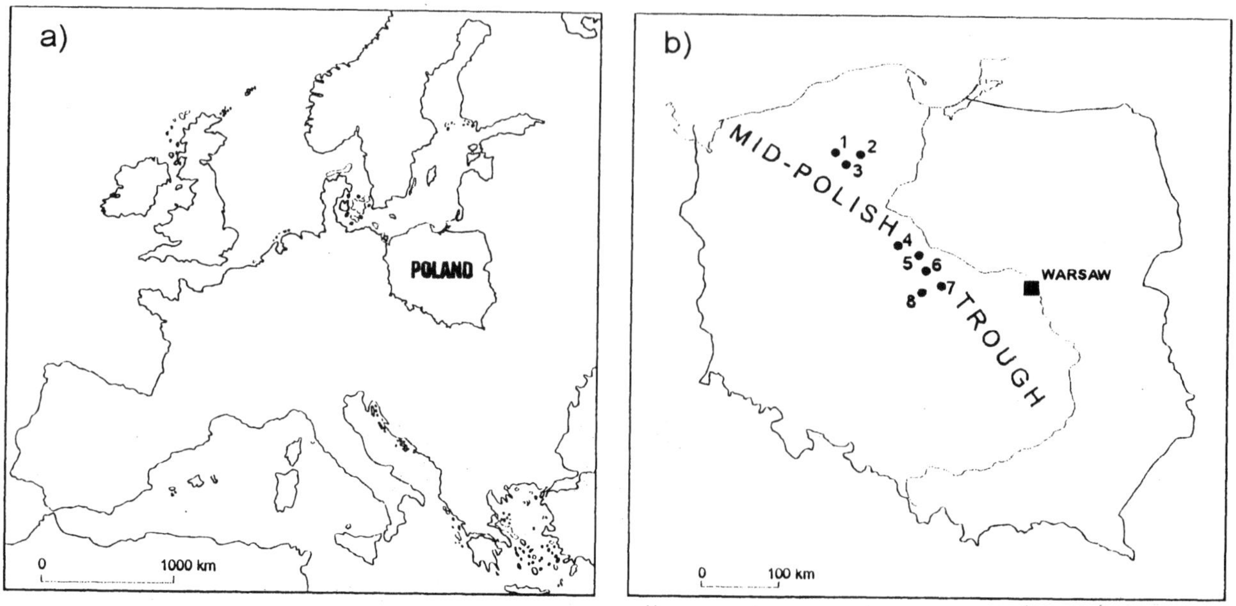

Text-Figure 1. 1a: Poland on a background of the Europe continent. 1b: The Mid–Polish Trough and location of the boreholes. 1 = boreholes # 1, 2, 3; Lower Cretaceous material. 2 = boreholes # 4, 5, 6, 7, 8; Triassic material.

ening in interradial area, exine thickenings on the central part of distal side (Plate 2; figs. 1, 2, 4).

b. *Clavifera* Bolkhovitina: smooth surface, equatorial exine thickening in radial and interradial areas, exine thickenings on the central part of distal side (Plate 2; fig. 3, 5, 7).

c. *Ornamentifera* Bolkhovitina: equatorial exine thickenings, exine thickenings on the central part of distal side, ornamentation on proximal and distal side (Plate 2; figs. 6, 8).

Among the miospores of the family Schizeaceae, with characteristic rib ornamentation, the following genera occur:

a. *Cicatricosisporites* Potonie & Gelletich: differentiated rib map (Plate 3; figs. 1-5).

b. *Appendicisporites* Weyland & Krieger: differentiated rib map, apical appendices (Plate 3; fig. 6).

The families Cyathaceae–Dicksoniaceae with smooth surface is represented by the genus *Deltoidospora* (Miner) Potonié (Plate 4; fig. 1).

Many fern miospores with undefined botanical position have been also recorded from the Lower Cretaceous samples. The following taxa belong to this assemblage:

a. *Impardecispora* Venkatachala et al.: characteristic verrucae elements of ornamentation with vari-diameter on apical area and on other areas (Plate 4; fig. 3).

b. *Verrucosisporites* (Ibrahim) Waksmundzka: characteristic verrucae elements of ornamentation with the same diameter on the whole spore surface area (Plate 4, fig. 4);

c. *Pilosisporites* Delcourt & Sprumont: papillae elements of ornamentation on the whole spore surface area (Plate 4, fig. 2).

Bisaccate pollen grains are represented only by single specimens belonging to the families Podocarpaceae, the genus *Callialasporites* Dev, and Caytoniaceae the genus *Vitreisporites* (Leschik) Jansonius (Waksmundzka, 1992). Also pollen grains belonging to other genera have been found in the Lower Cretaceous samples. There are pollen grains of the genus *Eucommiidites* (Erdtman) Hughes (Plate 4, fig. 5) with controversial botanical position (Bennettitinae, Gnetaceae, Pentoxylales or early Angiospermae?—Pedersen et al., 1989) and the species *Clavatipollenites hughesi* Couper (Plate 4, figs. 6', 6") assigned to primitive Angiospermae on the basis of aperture and exine structure (Jansonius and Mc Gregor, 1966).

Quantitative Analysis

Quantitative analysis of the material indicated different palynological spectra in every stratigraphic unit. Against a background of the total material, the Triassic spectra are characterized by substantial predominance of bisaccate conifer pollen grains, while the Lower Cretaceous spectra show important dominance of fern miospores belonging to the families Schizeaceae and Gleicheniaceae, with only the occasional appearance of bisaccate pollen grains.

Exine Color Analysis

Examination of exine color from the studied material recorded a wide color differentiation from light to very dark, controlled by different stages of organic matter maturity. The most mature miospores, with reflecting black and very dark brown-black colors (Plate 5; figs. 7, 8) occur in the

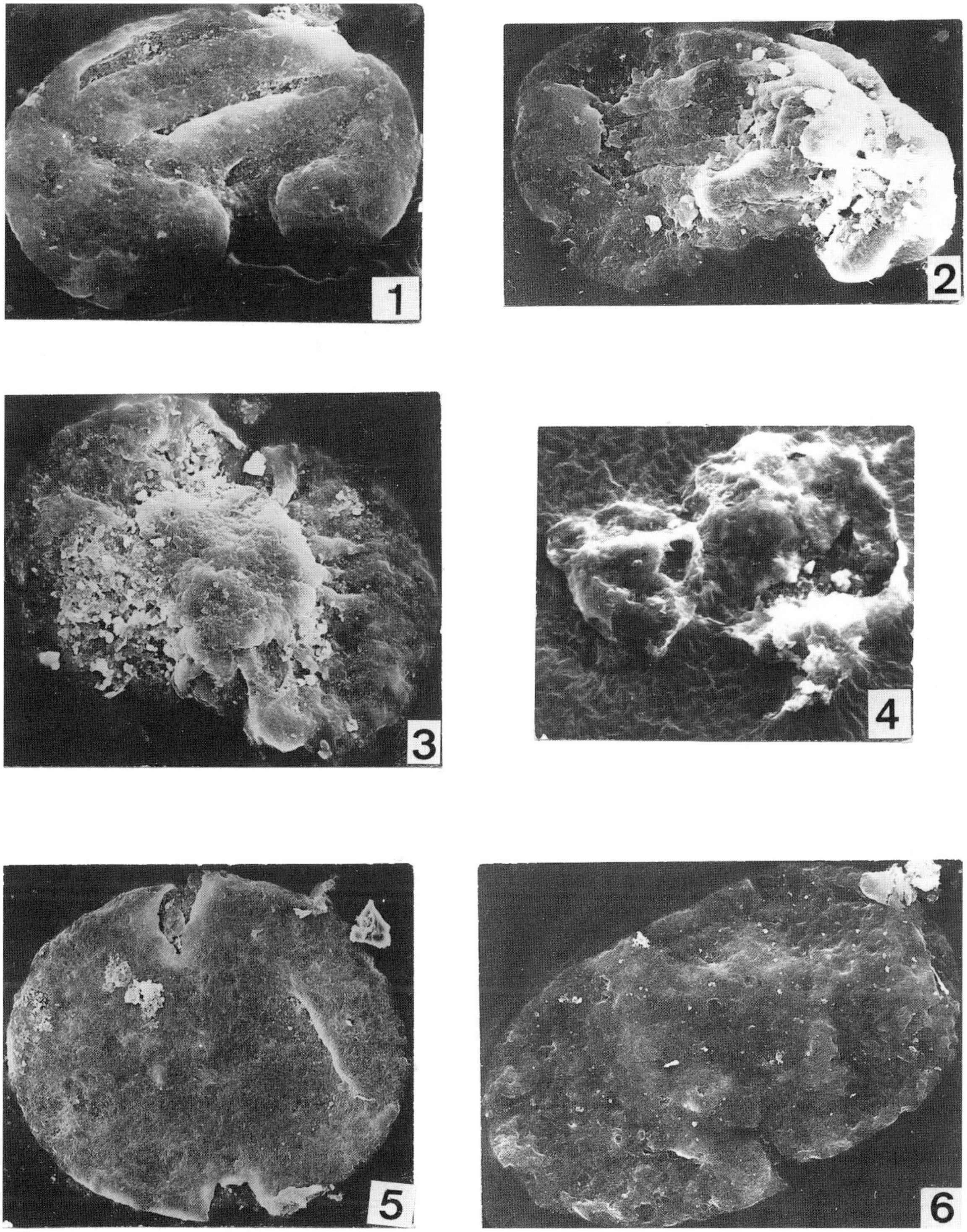

PLATE 1

All figures x1,500.

1 *Infernopollenites sulcatus* (Pautsch) Scheuring; SEM; proximal view.

2 *Striatoabietites aytugii* Visscher; SEM; proximal view.

3 *Angustisulcites klausii* Freudenthal; SEM; proximal view.

4 *Pinuspollenites minimus* (Couper) Kemp; SEM; equatorial view.

5 *Brachysaccus neomundanus* (Leschik) Mädler; distal view.

6 *Platysaccus niger* Mädler; SEM; proximal view.

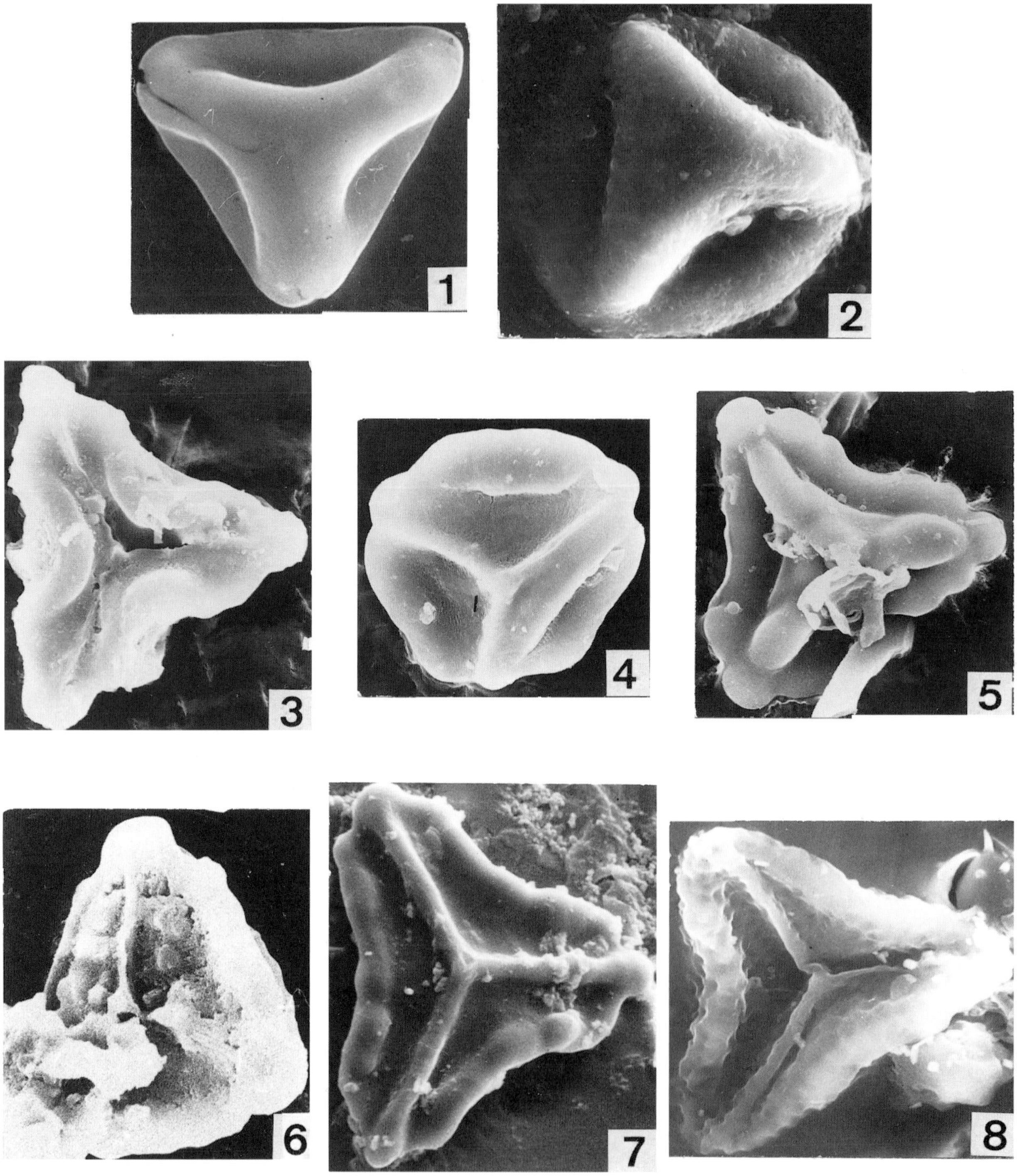

PLATE 2

Figures x1,500 unless otherwise noted.

1 *Gleicheniidites senonicus* Ross; SEM; distal view.

2 *G. carinatus* (Bolkhovitina) Bolkhovitina; SEM; distal view.

3 *Clavifera jachromiensis* Bolkhovitina; SEM; proximal view.

4 *Gleicheniidites rasilis* (Bolkhovitina) Bolkhovitina; SEM; x1,000; proximal view.

5 *Clavifera triplex* Bolkhovitina; SEM; distal view.

6 *Ornamentifera peregrina* (Grigorieve) Bolkhovitina; SEM; distal view.

7 *Clavifera* sp.; SEM; proximal view.

8 *Ornamentifera echinata* (Bolkhovitina) Bolkhovitina; SEM; proximal view.

PLATE 3

Figures x1,500 unless otherwise noted.

1 *Cicatricosisporites australiensis* (Cookson) Potonié; SEM; proximal view.

2 *C. annulatus* Archangelsky and Gamero; SEM; proximal view.

3 *C. spiralis* Singh; SEM; distal view.

4 *C. mediostratus* (Bolkhovitina) Pocock; SEM; x1,000; proximal view.

5 C. sp. cf. *Anemia exilioides* (Maljavkina) Bolkhovitina; SEM; distal view.

6 *Appendicisporites* sp.; SEM; proximal view.

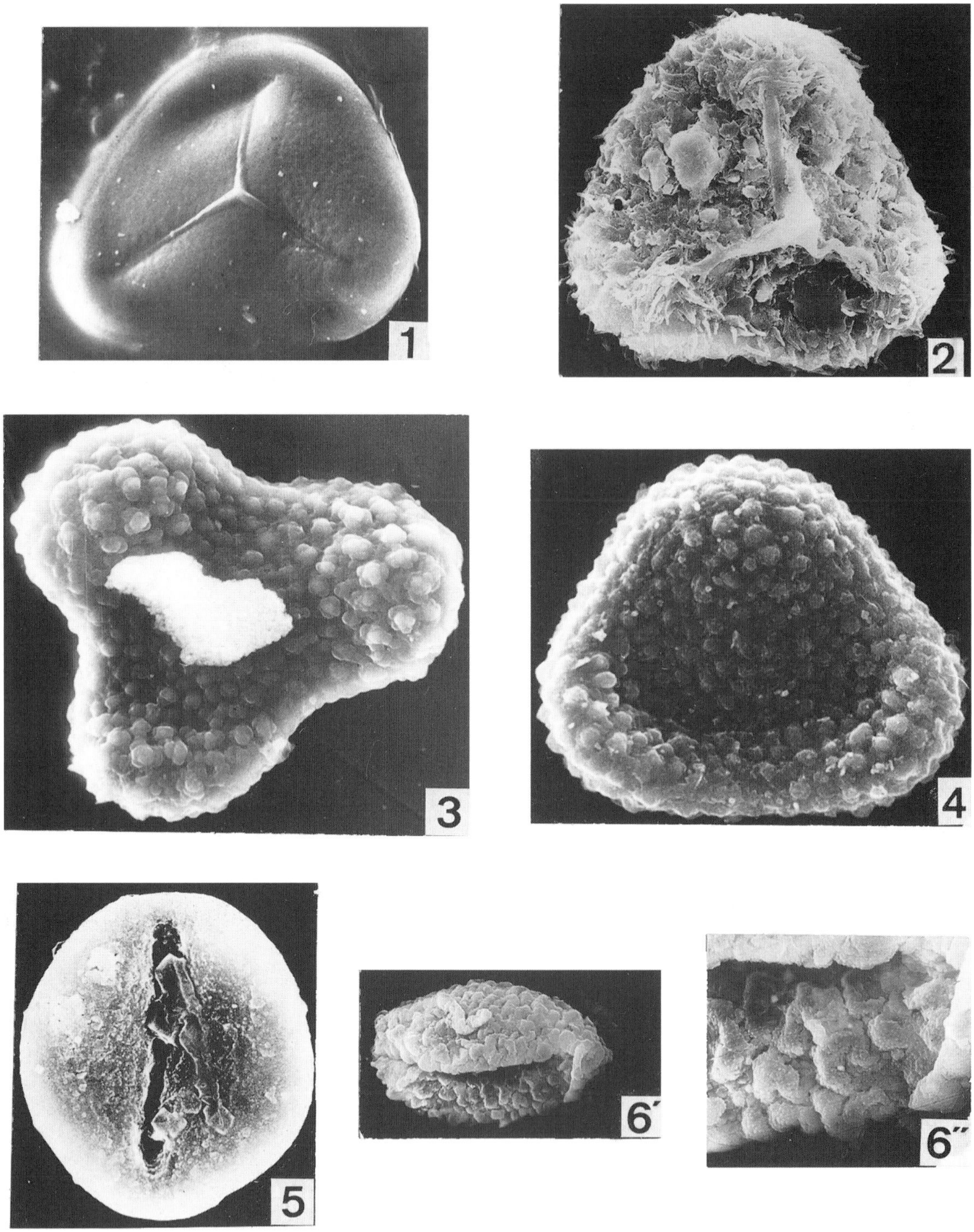

PLATE 4

Figures x1,500 unless otherwise noted.

1 *Deltoidospora australis* (Couper) Srivastava; SEM; proximal view.

2 *Pilosisporites verus* Delcourt & Sprumont; SEM; proximal view.

3 *Impardecispora apiverrucata* (Couper) Venkatachala et al.; SEM; distal view.

4 *Verrucosisporites microverrucatus* (Döring) Waksmundzka; SEM; distal view.

5 *Eucommiidites troedssonii* (Erdtman) Hughes; SEM; x2,000; front view.

6' *Clavatipollenites hughesi* Couper; SEM; x1,000; front view.

6'' *C. hughesi* — structure exine; SEM; x3,000.

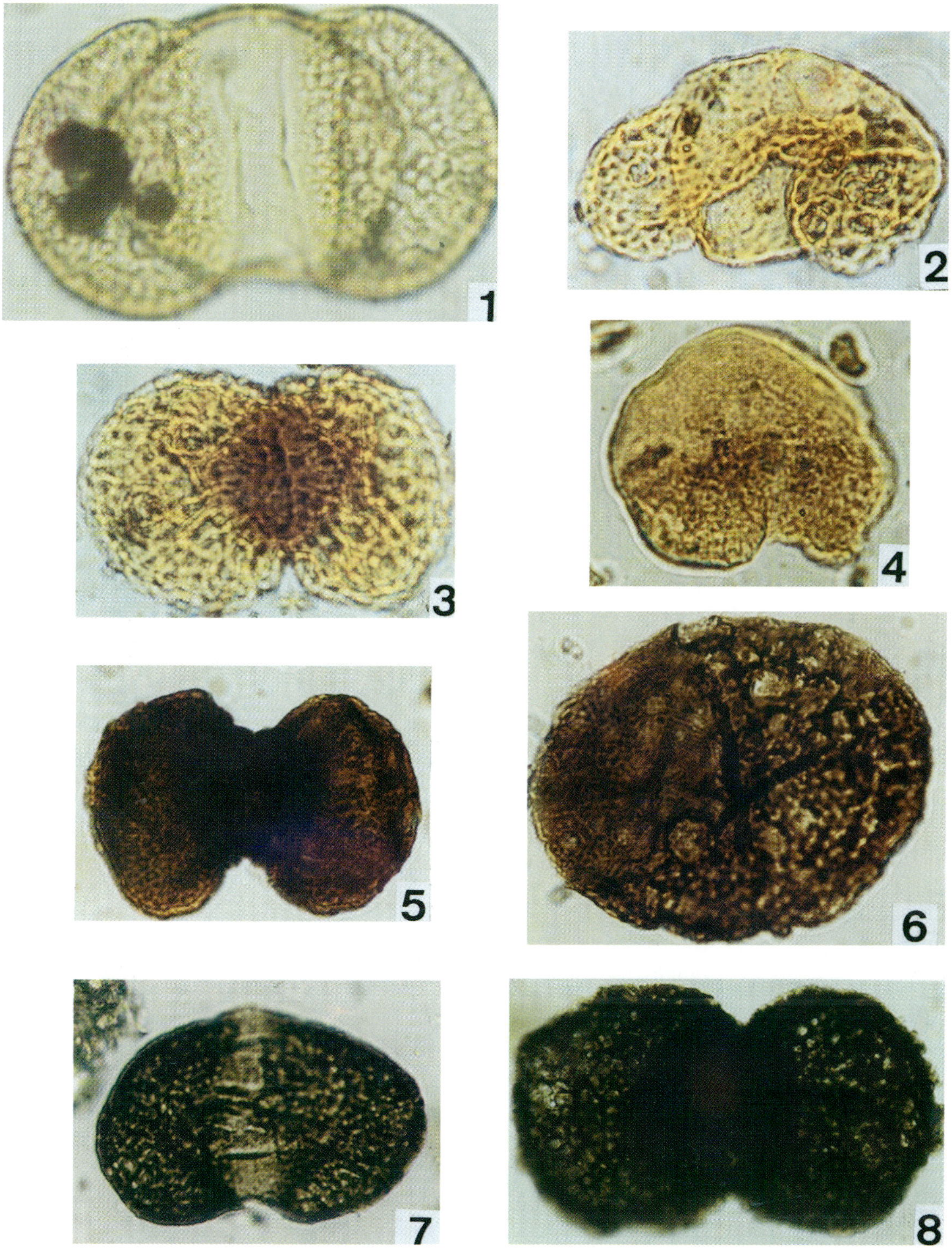

PLATE 5

Miospore color variability related to succeding phases of organic matter transformation (TAI); all miospores x1,000, transmitted light.

1 TAI 1/2
2 TAI 2/3
3, 4 TAI 3
5, 6 TAI 4
7 TAI 5
8 TAI 7

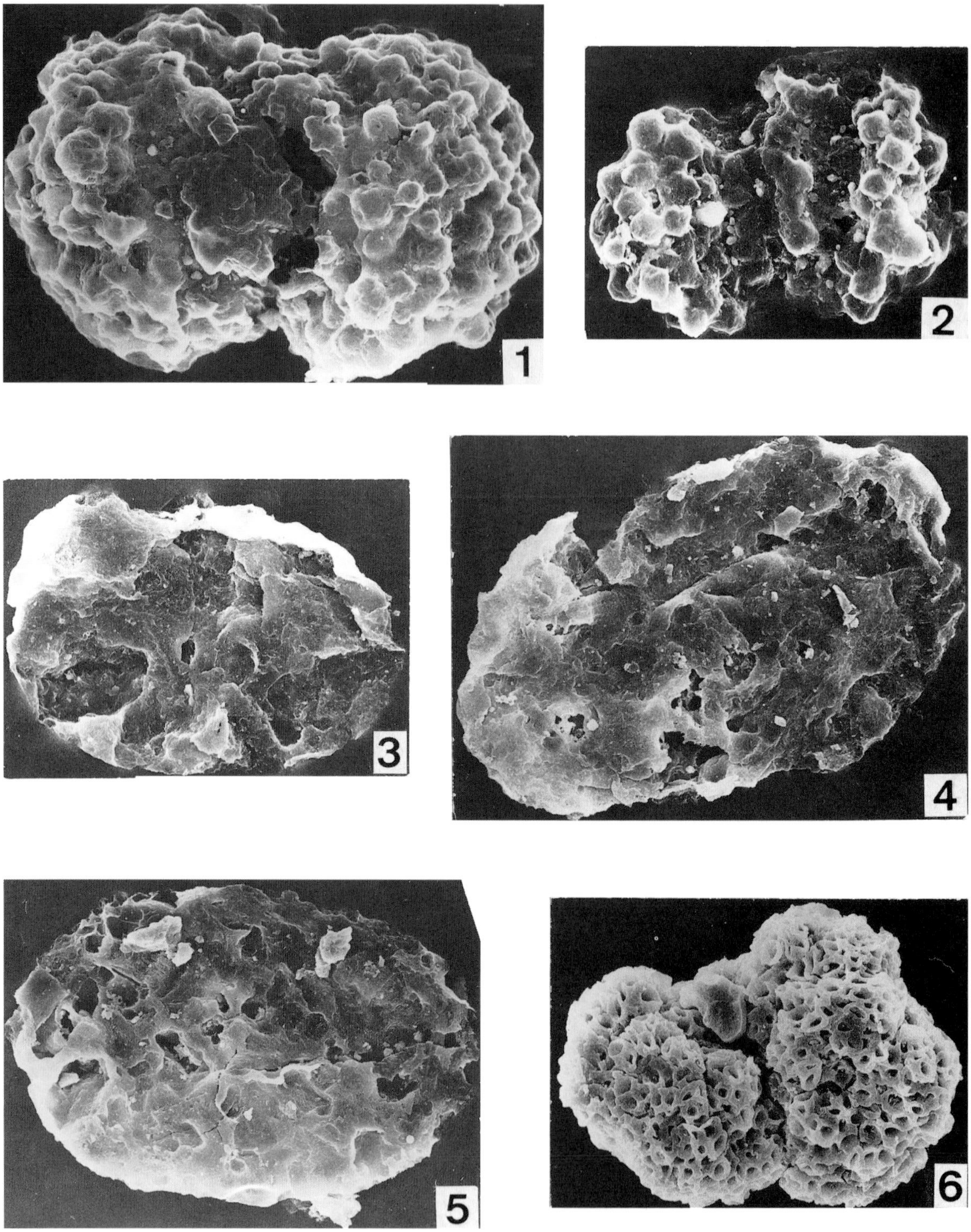

PLATE 6

Succeeding stages of transformation from exine to kukersive kerogene on surface of pollen grains in SEM:

1, 2 Spherical–oval structures: 1: x1,500, 2: x1,000.

3, 4, 5 Spherical–oval assemblages degrading miospore walls, x1,500.

6 Network ("reticulum"), x1,500.

Lower Triassic sediments from Kujawy (depth ca. 4600 m), in the central part of the Mid–Polish Trough (boreholes # 5 and 7). Pollen grains related to mature organic matter with dark-brown—(Plate 5; figs. 5, 6) through light medium-brown - (Plate 5, figs. 3, 4) to yellow–orange color (Plate 5; fig. 2) have been found in the shallower Upper Triassic sediments on the west slope of the Kujawy Segment of the trough (boreholes # 4 and 8) and in Lower Cretaceous sediments from the depth of 1200–1800 m. The lowest maturation stage, corresponding to light-brownish and yellow color (Plate 5; fig. 1) was observed in shallowest Upper Triassic sediments (depth 2879.5 and 3200 m) in boreholes # 4 and 8, and in the shallow Lower Cretaceous deposits (depth 927.9–1025.6 m) from the boreholes located at the northeast slope of the Mid–Polish Trough (boreholes # 1, 2 and 3).

Analysis of exine color permitted recording the additional surface structures characteristic for the pollen grains of the same color. Spheroidal-oval structures occur on the pollen grains with thermal alternation index (TAI) values of 3. Similar complexes degrading miospore walls are visible on the grains with TAI 4, and black reticulum resembling primary morphology is frequently characteristic for the grains with TAI - 6–7. Therefore, an analysis of the miospore surface morphology becomes an important part of the exine color examination.

DISCUSSION OF RESULTS

Detailed analysis of the published taxonomic data and intra-special variability of recent miospores (Waksmundzka, 1992) indicated a necessity to combine some taxa that were formerly separated, which consequently lowered the biostratigraphic value of these taxa. However, this did not exclude usage of palynological material for stratigraphic purposes. In both the stratigraphical units, the Triassic and Lower Cretaceous one, forms with short stratigraphical range still existed within palynological assemblages. *Cicatricosisporites spiralis* Singh, *Clavifera tuberosa* Bolkhovitina (characteristic of the Aptian of southern United States, Canada and Russia), *Cicatricosisporites hallei* Delcourt & Sprumont (characteristic of the Albian of Canada), *Platysaccus triassicus* Mädler (characteristic of the Rhaetian of Poland and Germany) and *Voltziaceaesporites nephrosaccus* Klaus (characteristic of the Middle Triassic of West Poland and Germany) may be good examples of these forms.

Miospore production with emphasis on great numbers of pollen grains deposited in the vicinity of parent floras is the basis of quantitative analysis for interpretation of palaeoecological conditions. In the Lower Cretaceous part of the profile, predominance of less-differentiated fern miospores transported mostly in fluvial environment and a scarce occurence of bisaccate pollen grains with absence of dinoflagellate cysts suggests nearshore/deltaic conditions of deposition. In contrast, substantial predominance of bisaccate pollen grains in the samples of Triassic deposits transported long-distance by air indicates a rather open-marine, sedimentary environment far from shore.

Variability of TAI values permits a maturity estimate of organic matter (Gaupp and Batten, 1985). Therefore, it may be used in the establishing of stages generation of liquid and gas hydrocarbons and enable a prediction of oil and gas resources. Moreover, it permits establishment of depth of burial time of the placement of the sediments at the defined depth, and the effective heating time (time the sediments existed at maximal temperature) on the basis of nomograms (Wright, 1980). Estimation of TAI permits correlation of palynological examination with stratigraphic and petrological data.

Detailed SEM-analysis of sculpture of Triassic pollen grains showed that the surficial structures are not elements of ornamentation but a transformed organic substance, secondarily replaced at the grain surface. Occurence of different kinds of structures related to succeeding stages of organic maturity are (1) spherical-oval structures related to TAI index 3 (Plate 6; figs. 1, 2); (2) spherical-oval structures degrading miospore walls, related to TAI index 4–5 (Plate 6; figs. 3-5); and (3) radial structures related to TAI index 6–7 (Plate 6; fig. 6). They confirm dependence of the discussed structures of different stages on organic matter maturation (Körts and Veski, 1994).

CONCLUSIONS

The Lower Cretaceous palynological assemblage in the studied deposits (Waksmundzka, 1992) indicates a tropical to subtropical climate on the Polish Lowland at that time.

Quantitative analysis confirmed differences between sedimentary environments of the Lower Cretaceous and Triassic sediments, reflected in the different character of palynological assemblages.

Color of spores and pollen grains is controlled by geological history of the basin. Dark miospores (black, very dark brown-black) indicate areas of extensive subsidence, resulting from deep burial of sedimentary series. Lighter miospores (dark-brown and brown ones) recorded areas of shallower burial of the sediments that occur in the borehole columns from marginal parts of the Mid-Polish Trough and outside.

Analysis of color and morphology of differentiated surficial structures of pollen grains indicated their relationships to succeeding stages of organic matter maturity and a transformation process from exine to kukersive kerogen (Körts and Veski, 1994), resulting from temperature and pressure increase.

References Cited

BATTEN, D.J.
1984 Pollen/spore color standard Munsell color standard. (Matte finish. Version 2).

GAUPP, R., and BATTEN, D.J.
1985 Maturation of organic matter in Cretaceous strata. *Neues Jahrbuch fur Geologie and Palaeontologie*, Mh. 3: 157–175.

JANSONIUS, J., and MC GREGOR (ed.)
1966 Palynology: Principles and Application. *American Association of Stratigraphic Palynologists Foundation*, vols.1–3, 1,324 pgs.

KÖRTS, A., and VESKI, R.
1994 Scaning electron microscopy of *Gleocapsomorpha* as producted from kerogen oxidation. *Oil Shale*, 11: 293–303.

MÄDLER, K.
1964 Die geologische Verbereitung von Sporen und Pollen in der Deutschen Trias. *Beihefte zum Geologischen Jahrbuch*, Heft 65: 147 pgs.

MAREK, S., and PAJCHLOWA, M.
1997 Epikontynentalny Perm i Mezozoik w Polsce. *Prace PIG*, 153, 432 pgs.

ORŁOWSKA-ZWOLIŃSKA, T.
1983 Palinostratygrafia epikontynentalnych osadów wyszego triasu w Polsce. *Prace IG*, 104, 82 pgs.
PEDERSEN, K.R; CRANE, P.R., and FRIIS, E.M.
1989 Pollen organs and seeds with *Eucommiidites* pollen. *Grana*, 30: 577–590.
POTONIÉ, R.
1970 *Synopsis of the Genera of the Sporae dispersae, Part V.*
TYSON, R.V.A.
1995 *Organic Facies and Palynofacies.* Chapman and Hall, 607 pgs.
WAKSMUNDZKA, M.
1992 Revision of taxonomy of the Lower Cretaceous miospores on the Polish Lowland. *Prace IG*, 136: 1–64.
WRIGHT, N.J.R.
1980 Time, temperature and organic maturation—The evolution of rank within a sedimentary pile. *Journal of Petroleum Geology*, 2,4: 411–425.

Author's address

MARTA WAKSMUNDZKA
Polish Geological Institute
Rakowiecka 4
Warsaw
Poland

TRIASSIC PALYNOFLORAS FROM THE TARIM BASIN OF SOUTHERN XINJIANG, CHINA

Lujin Zhang and Xinfu Wang

Abstract

Triassic palynofloras consisting of 48 genera and 66 species are reported from the Ehuobulake, Karamay and Huangshanjie formations in the Tarim Basin of China. In terms of paleophytogeography, the palynomorphs belong to the northern floristic region of China. The characteristics and age of the palynofloras and their paleophytogeographical significance, as well as the distribution of the genus *Cuneisporites*, are discussed.

INTRODUCTION

The Triassic fossil record in the Tarim Basin of Xinjiang is rather poorly known. The best outcrop of Triassic strata is in the Kuqa section along the northern margin of the Tarim Basin. In 1979, an expedition for biostratigraphy was established and a primary investigation on the palynology, palaeobotany and lamellibranchs in the Kuqa region was carried out by Zhang Lujin, Huang Baoyu, Huang Fengbao (Nanjing Institute of Geology and Palaeontology, Academia Sinica), and by Zhow Hanzhong, Luo Lianglu, and Zhong Xiaoyan (Xinjiang Corporation of Coal-field Geological Exploration). Preliminary palynological results from this investigation have previously been reported by Zhang (1980b). Some subsequent stratigraphic and palaeontological papers have also been published (e.g., Wu and Chen, 1992).

The Triassic sediments in the Kuqa section consist of five units. This paper deals with the lower three units (the Ehuobulake, Karamay and Huangshanjie formations). Eighteen samples were collected and four of them contain abundant spores and pollen (sample Sk2 from the Ehuobulake Formation, samples Sk12b, Sk12c from the Karamay Formation, and sample Sk20c from the Huangshanjie Formation).

STRATIGRAPHY

The Kuqa section is located in the upper and middle reaches of the Kuqa River, about 100 km north of Kuqa county (Text-Figure 1). The stratigraphic sequence is shown in Table 1, and a comparison of the Tarim Triassic sequence with other areas is shown in Text-Figure 2.

CHARACTERISTICS AND AGE OF THE PALYNOFLORAS

The Ehuobulake Palynoflora

The dominance of gymnospermous pollen (about 70% of the total palynological assemblage) is characteristic of the Ehuobulake Palynoflora. Most of the pollen is bisaccate. The chief representative genus is *Prototaeniaesporites* (12%). Secondary genera include *Pityosporites* (6%), *Pinuspollenites* (6%), *Angustisulcites* (5%), *Pseudoquadrisaccus* (5%), *Piceites* (4%), *Protopinus* (4%), and *Psophosphaera* (3%). A series of subsidiary forms also in this palynoflora; these include *Platysaccus*, *Vitreisporites*, *Protoconiferus*, *Podocarpidites*, and *Alisporites*, as well as a few specimens of *Taeniaesporites*, *Striatopodocarpites*, *Striatites*, *Ephedripites*, *Monosulcites*, and *Cycadopites*, among others.

Spores represent only about 30% of this palynoflora. The main genera are *Punctatisporites* (19%), *Todisporites* (5%), *Verrucosisporites* (5%), *Calamospora* (3%), *Aratrisporites* (1%), *Foveolatisporites* (1%), and *Ordosisporites* (1%). Other rare forms are *Monoletes*, *Marattisporites*, *Osmundacidites*, *Densoisporites*, *Asseretospora*, and *Apiculatisporites*.

Most of the above-mentioned genera appear in the Triassic. The Permian forms from Xinjiang (Zhang, 1983, 1990) become extinct in this formation; these include *Hamiapollenites*, *Crustaesporites*, *Potonieisporites*, and *Endosporites*. The difference between the Ehuobulake palynoflora and the assemblages from the younger formations is in the abundance of *Angustisulcites* and in the absence of *Heliosaccus dimorphus* and *Cuneisporites marginelis* (which are specifically characteristic of the Karamay and Huangshanjie formations).

The Ehuobulake palynoflora is virtually identical to that from the Upper Buntsandstein of The Netherlands (Visscher, 1966) in having *Angustisulcites klausii*, *Angustisulcites gorpii*, and some other forms. This palynoflora is also very similar to that from the Heshanggor Formation (upper part of Lower Triassic; see Ouyang and Norris, 1988) of Shannxi of China in having *Angustisulcites gorpii*, etc.

According to the palynological characters and stratigraphical sequence, the Ehuobulake palynoflora is considered to be late Early Triassic in age.

The Karamay Palynoflora

This palynoflora consists of approximately one-third gymnospermous pollen and two-thirds spores. The assemblage includes the primary genera *Heliosaccus* (4%) and *Vitreisporites* (4%), and the secondary genera *Pinuspollenites* (3%), *Pityosporites* (3%), *Psophosphaera* (3%), *Protoconiferus* (2%), and *Piceites* (2%). Other elements of the assemblage include *Ephedripites*, *Cycadopites*, *Monosulcites*, *Chordasporites*, *Pseudoquadrisaccus*, *Pseudopinus*, *Striatopodocarpites*, *Alisporites*, and *Amplexisaccites*.

Zhang, Lujin and Wang, Xinfu
2001 Triassic palynofloras from the Tarim Basin of southern Xinjiang, China; *In*: Goodman, D.K., and Clarke, R.T. (eds.), *Proceedings of the IX International Palynological Congress, Houston, Texas, U.S.A., 1996*; American Association of Stratigraphic Palynologists Foundation, p. 221–230.
 ISBN 0–931871–06–9

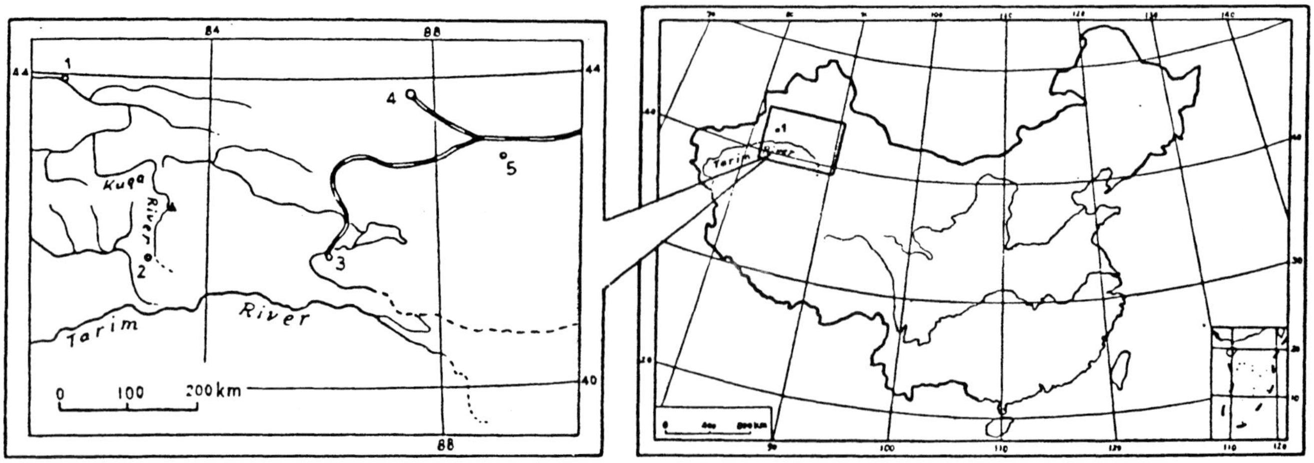

Text-Figure 1. Index map showing the geographic position of the Kuqa region of Xinjiang, China. 1 - Yining; 2 - Kuqa (▲ - Kuqa section); 3 - Kurla; 4 - Urumqi; 5 - Turpan.

The spores of the Karamay palynoflora are represented principally by *Punctatisporites* (18%), *Todisporites* (13%), *Aratrisporites* (10%), *Dictyophyllidites* (4%), *Calamospora* (4%), *Marattisporites* (3%), *Lygodiumsporites* (3%), *Limatulasporites* (2%), *Verrucosisporites* (2%), and *Osmundacidites* (2%).

The most characteristic feature of the Karamay palynoflora is the dominance of spores instead of gymnospermous pollen. The primary spores are *Punctatisporites fissus* and *Todisporites major*. *Aratrisporites* also occurs in abundance.

A significant number of spore species appeared at approximately the same time in this assemblage, including *Calamospora, Dictyophyllidites, Marattisporites, Lygodiumsporites, Verrucosisporites, Limatulasporites* and *Osmundacidites*. A secondary characteristic of this palynoflora is the presence of the large monosaccate pollen *Heliosaccus dimorphus*.

These characteristics of the Karamay palynoflora bear a striking resemblance to the palynological assemblages from the Ermaying Formation (lower Middle Triassic) of Shaanxi, China (Qu, 1980). It is possible that the Karamay and Ermaying palynofloras are coeval in age.

The Huangshanjie Palynoflora

Fifty-five percent of this palynoflora is represented by gymnospermous pollen. The primary genera are *Pityosporites* (11%), *Prototaeniaesporites* (9%), *Protoconiferus* (5%), *Psophosphaera* (5%), *Pinuspollenites* (4%), *Platysaccus* (4%), *Podocarpidites* (2%), *Vitreisporites* (2%), and *Sulcatisporites* (2%). Rare forms include *Monosulcites, Cycadopites, Angustisulcites, Piceites, Striatopodocarpites, Taeniaesporites, Chordasporites, Protohaploxypinus, Striatites,* and *Ephedripites*.

PLATE 1

Figures. 1-27, 29-30 = x600; fig. 28 = x1,000, fig. 31 = x2,400. Figs. 28 and 31 are SEM photomicrographs.

1 *Monoletes* sp. Slide SK12c-11.
2 *Aratrisporites* cf. *Parvispinosus* Leschik 1956. Slide SK2-5.
3 *Aratrisporites tongchuanensis* Zhang 1980a. Slide SK12c-10.
4 *Aratrisporites* sp. Slide SK2-14.
5, 6 *Aratrisporites exilis* Zhang 1980a. Slides SK12c-10, SK12c-6.
7 *Marattisporites* sp. Slide SK12c-6.
8 *Osmundacidites parvus* de Jersey 1962. Slide SK20a-7.
9 *Concavisporites* sp. Slide SK12c-4.
10 *Dictyophyllidites* sp. Slide SK12b-5.
11,12, & 16 *Limatulasporites limatulus (Playford)* Helby & Foster 1979. Slides SK12c-10, SK12c-10, SK12b-8.
13 *Calamospora mesozoica* Couper 1958. Slide SK20a-27.
14 *Punctatisporites fissus* Leschik 1955. Slide SK12c-7.
15 *Anguisporites* sp. Slide SK2-1.
17 *Ordosisporites kollybos* Zhang 1980a. Slide SK2-2.
18 *Psophosphaera* sp. Slide SK12b-4.
19 *Asseretospora gyrata* (Playford & Dettmann) Schuurman 1977. Slide SK20a-29.
20 *Densoisporites* cf. *Perinatus* Couper 1958. Slide SK2-5.
21, 22 *Foveolatitriletes tarimensis* Zhang & Wang 1999. Slides SK2-7. SK2-10.
23 *Todisporites dissemurus* Zhang & Want 1999. Slide SK2.
24 *Osmundacidites wellmanii* Couper 1953. Slide SK2-2.
25 *Cuneisporites marginelis* (Bharadwaj & Singh) Yang 1994. Slide SK20a-20.
26-28, 31 *Cuneisporites laevis* Zhang 1980a. Slide SK20a-20.
29, 30 *Verrucosisporites labrum* Zhang 1980a. Slides SK2-9, SK2-4.

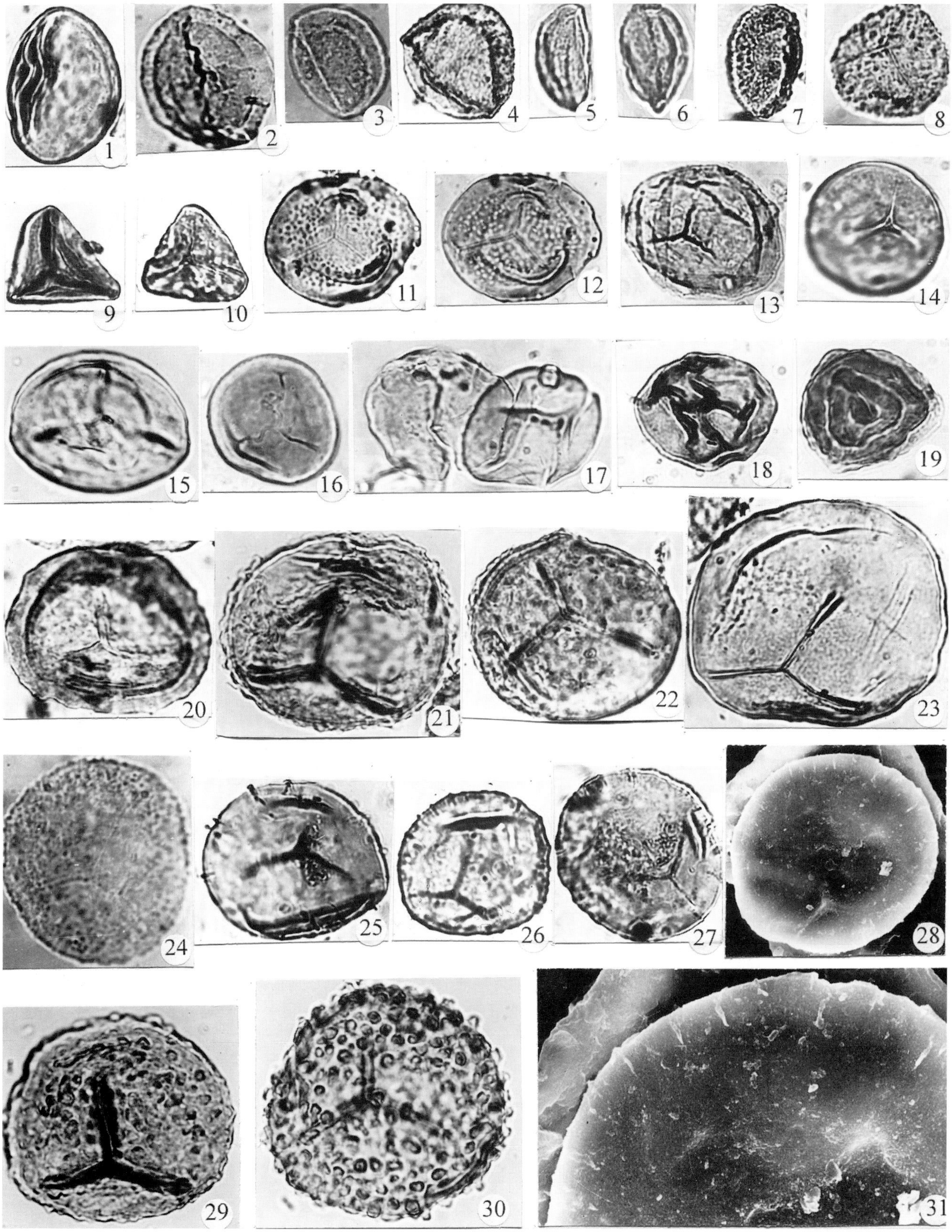
1
2
3
4
5
6
7
8
9
10
11
12
13
14
15
16
17
18
19
20
21
22
23
24
25
26
27
28
29
30
31

<table>
<tr><th colspan="2" rowspan="2">Region and Strata / Age</th><th colspan="3">Northern China</th><th rowspan="2">Europe</th></tr>
<tr><th>Tarim of Xinjiang (Kuqa Section)</th><th colspan="2">Shaanxi</th></tr>
<tr><td rowspan="7">Triassic</td><td rowspan="3">Upper</td><td>Ahe Fm</td><td rowspan="3">Yenchang Fm</td><td>Upper</td><td>Rhaetian</td></tr>
<tr><td>Tariqike Fm</td><td>Middle</td><td>Norian</td></tr>
<tr><td>Huangshanjie Fm</td><td>Lower</td><td>Carnian</td></tr>
<tr><td rowspan="2">Middle</td><td>?</td><td colspan="2">Tongchuan Fm</td><td rowspan="2">Muschelkalk</td></tr>
<tr><td>Karamay Fm</td><td colspan="2">Ermaying Fm</td></tr>
<tr><td rowspan="2">Lower</td><td>Ehuobulake Fm</td><td colspan="2">Heshanggou Fm</td><td rowspan="2">Buntsandstein</td></tr>
<tr><td></td><td colspan="2">Liujiagou Fm</td></tr>
</table>

Text-Figure 2. Comparison of the Tarim Triassic section with other regions.

Spores comprise about 45% of the Huangshanjie palynoflora. The most important characteristic is the presence of several species of the genus *Cuneisporites* (8%). Other genera include *Calamospora* (7%), *Punctatisporites* (6%), *Aratrisporites* (4%), *Osmundacidites* (3%), *Verrucosisporites* (3%), *Cyathidites* (2%), and *Schizosporis* (2%). *Marattisporites* and *Cingulatisporites* are rare.

The Huangshanjie palynoflora is similar to that of the lower part of the Upper Triassic Yenchang Formation of Shaanxi, and of the Upper Triassic Tanzhuang Formation of Henan, China, in containing the peculiar species *Cuneisporites marginelis* (Bharadwaj & Singh) Yang. Furthermore, these three formations share a relatively large number of common genera and main species, such as *Prototaeniaesporites pseudostriatus* (Kopytova), *Asseretospora gyrata* (Playford & Dettmann), *Angustisulcites klausii*, and the genera *Aratrisporites, Marattisporites, Punctatisporites, Verrucosisporites*, and *Calamospora*.

Both in general character, and in the occurrence of *Cuneisporites marginelis*, the present palynoflora is comparable with that from the Lunz Keuper of Austria (Bharadwaj

PLATE 2

1, 3, 4, 11 *Pinuspollenites* spp. Slides SK2-9, SK2-9, Sk2-7, Sk2-18.

2 *Quadraeculina* sp. Slide SK20a-21.

5, 6, 10 *Piceites* cf. *Podocarpoides* Bolkhovitina 1956. Slides SK12c-22, SK2-5, SK12c-25.

7 *Scopulisporites minor* Maedler 1964. Slide SK20a-27.

8 *Pityosporites devolvens* Leschik 1955. Slide SK20a-11.

9 *Platysaccus* sp. Slide SK20a-5.

12 *Protohaploxypinus perfectus* (Naumova) Samoilovich 1953. Slide SK12c-6.

13 *Taeniaesporites* sp. Slide SK2-10.

14, 15 *Monosulcites* spp. Slides SK12b-5, SK12c-12.

16 *Prototaeniaesporites pseudostriatus* (Kopytova) Zhang & Wang 1999. Slide SK2-7.

17 *Platysaccus kainarensis* Bolkhovitina 1953. Slide SK12c-22.

18 *Striatopodocarpites* sp. Slide SK12c-16.

19 *Cycadopites altilis* Zhang 1984. Slide SK2-14.

20 *Pseudoquadrisaccus irregularis* Liu 1981. Slide SK2-8.

21 *Chordasporites volziaformis* Visscher 1966. Slide SK12c-6.

22, 26 *Pseudoquadrisaccus minus* Zhang & Wang 1999. Slides SK2-1, SK12c-9.

23 *Ephedripites* cf. *notensis* Cookson 1957. Slide SK12c-15.

24, 25 *Pseudoquadrisaccus lobatus* Zhang & Wang 1999. Slides SK2-9, SK2-4.

27 *Ephedripites kuzaensis* Zhang & Wang 1999. Slide SK12c-25.

28 *Vitreisporites* sp. Slide SK12c-9.

29 *Vitreisporites pallidus* (Reissinger) Nilsson 1958. Slide SK12c-6.

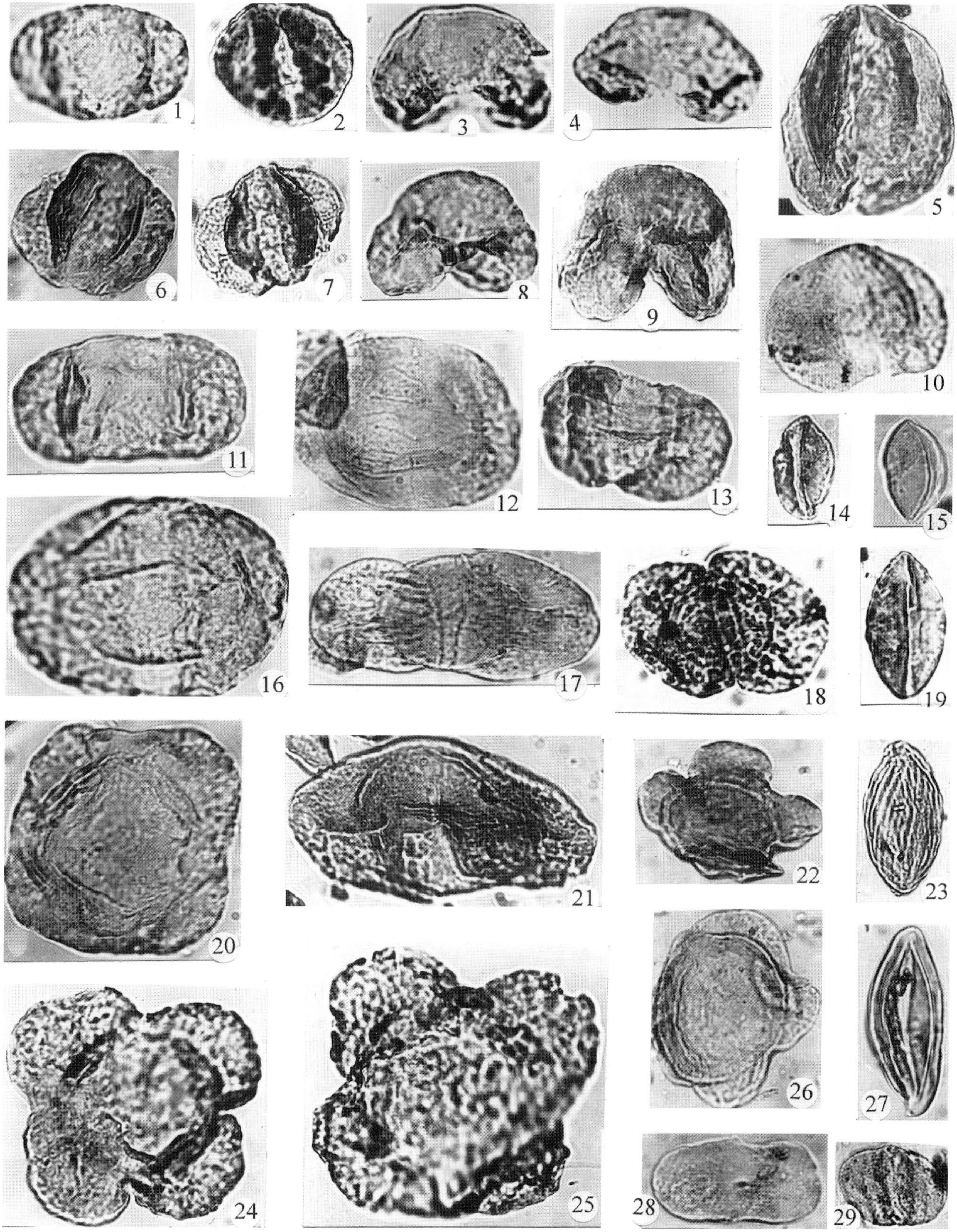
1
2
3
4
5
6
7
8
9
10
11
12
13
14
15
16
17
18
19
20
21
22
23
24
25
26
27
28
29

TABLE 1. Stratigraphic sequence of the Kuqu section.

21	Overlying strata: Tariqike Formation.	
	Conformity	
Huangshanjie Formation		
20	Dark-gray silty shale, yellow greenish siltstone and fine-grained sandstone intercalated with lenses of coal. (palynological samples Sk20a, Sk20b)	190m
19	Gray-white and yellow-greenish fine-grained sandstone with cross-bedding. (palynological samples SK19a, Sk19b)	104m
18	Light yellow chert and quartz conglomerate, dark-gray lenticular siltstone, coarse-fine-grained sandstone, gray-greenish and light-yellow siltstone intercalated with conglomerate and argillaceous siltstone, yielding plants, conchostracans and Kazacharthran fossils.(palynological samples Sk18a, Sk18.	96.9m
	?disconformity	
Karamay Formation		
17	Yellow greenish siltstone, fine-grained sandstone and dark purple siltstone.	19.5m
16	Yellow-greenish fine-coarse sandstone and gray-purple conglomerate.	22.3m
15	Yellow and purple siltstone, yellow fine-grained sandstone.	29.2m
14	Light-reddish fine conglomerate, gray-greenish and yellow-greenish fine grained sandstone and siltsone.	107m
13	Gray greenish fine grained sandstone.	3m
12	Yellow-greenish, gray-greenish sandstone, siltstone, occasionally intercalated with carbonaceous laminae, yielding plant fossils (palynological samples Sk12a, Sk12b, Sk12c, Sk12d).	83m
11	Purple siltstone, yellow greenish, gray-greenish fine-coarse sandstone and light-yellow conglomerate.	77.3m
10	Purple siltstone and sandstone, intercalated with light yellow coarse sandstone and conglomerate.	185.2m
9	Gray-greenish sandy mudstone, intercalated with carbonaceous laminae. (palynological samples Sk9a, Sk9b).	7m
8	Reddish siltstone intercalated with conglomerate.	41.5m
7	Gray greenish shale intercalated sandy mudstone, containing conchostracans. (palynological samples Sk7a, Sk7b).	29.5m
6	Dark-reddish, gray-greenish thick-bedded conglomerate.	17.7m
	Conformity or disconformity	
Ehuobulake Formation		
5	Gray greenish sandstone and siltstone, intercalated with purple siltstone. (palynological samples Sk5a, Sk5b).	18m
4	Purple, yellow-greenish siltstone.	17m
3	Gray-greenish sandstone interbedded with gray-greenish sandy mudstone. (palynological sample Sk3)	3.5m
2	Gray-greenish fine-grained sandstone and siltstone, intercalated with dark mudstone, yielding plant fossils.(palynological sample Sk2).	14m
	Unconformity	
1	Underlying strata: Volcanic clastic rocks.	

PLATE 3

Figs. 1-9, 11-13 = x600; figs. 10, 14 = x400.

1, 2 *Angustisulcites klausii* Freudenthal 1964. Slides SK2-5, SK12c-6.

3 *Podocarpus* sp. Slide SK2-8.

4, 5 *Angustisulcites strigatus* Zhang & Wang 1999. Slides SK2-14, SK2-5.

6 *Taeniaesporites rhaeticus* Schulz 1967. Slide SK2-10.

7 *Amplexisaccites notabilis* Zhang 1983. Slide SK2-1.

8 *Pseudopicea* sp. Slide SK2-15.

9 *Angustisulcites gorpii* Visscher 1966. Slide SK2-12.

10, 14 *Heliosaccus dimorphus* Maedler 1964. Slide SK12c-6.

11 *Todisporites* sp. Slide SK 12 c-6.

12 *Lygodiumsporites grandis* Zhang & Wang 1999. Slide SK12c-6.

13 *Schizosporis* sp. Slide SK12c-6.

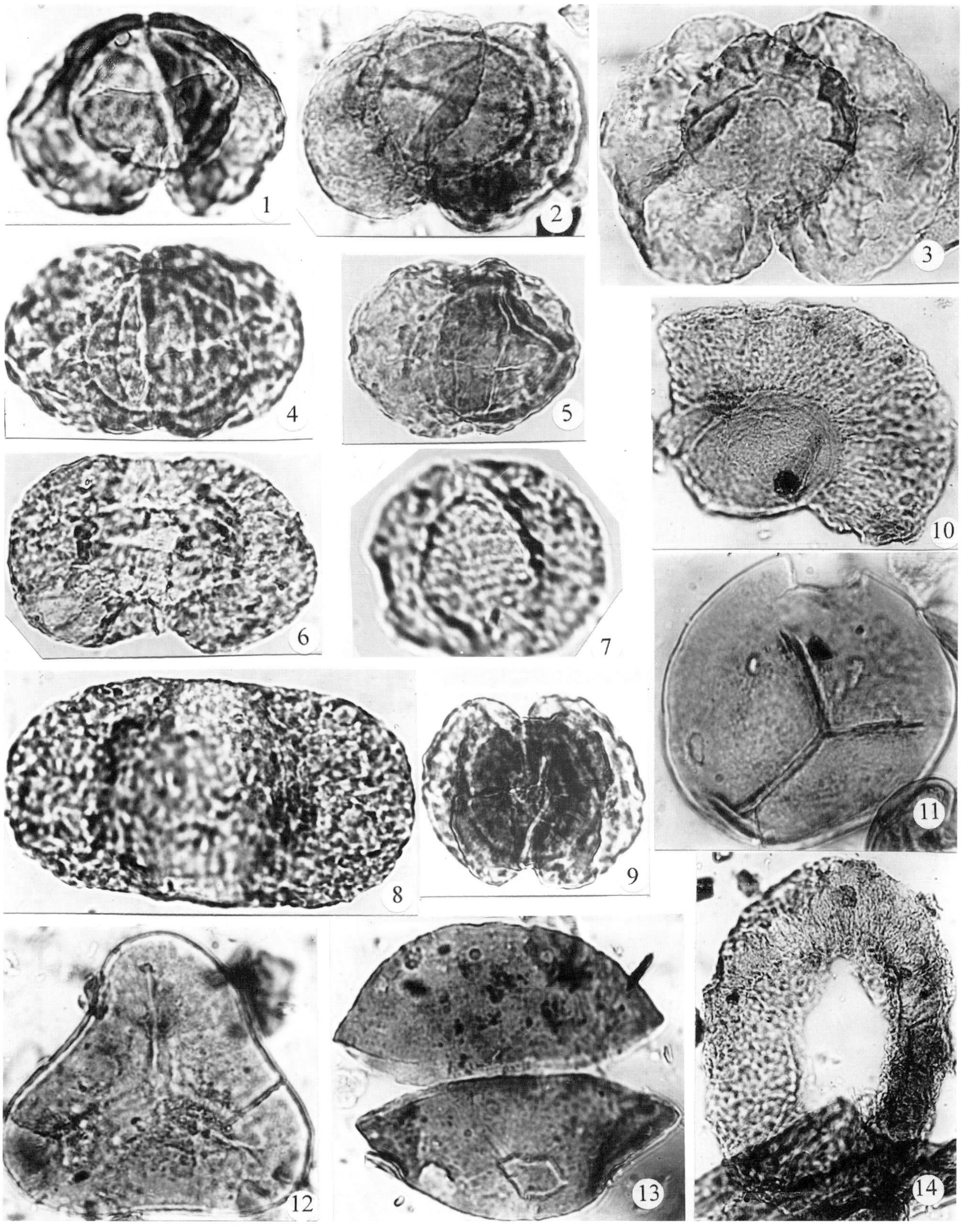
1
2
3
4
5
6
7
8
9
10
11
12
13
14

and Singh, 1964). The Huangshanjia Formation may be Carnian (Lower Keuper) in age.

THE STRATIGRAPHIC AND GEOGRAPHIC DISTRIBUTION OF THE GENUS *CUNEISPORITES* ZHANG 1980

The genus *Cuneisporites* was named by Zhang Lujin (1980a) to accommodate *Punctatisporites*-like or *Todisporites*-like spores with wedge-shaped ornaments, which are scattered prominently and radially along the equatorial margin. In the Tarim Basin, this genus makes its first appearance in the Huangshanjie Formation (this paper). Previous records of this spore are from the Upper Triassic Yenchang Formation of Shaanxi Province (Zhang, 1980a), the Upper Triassic Tanzhuang Formation of Henan Province (Yang, 1994) and the Upper Triassic Lunz coals of Austria (Bharadwaj and Singh, 1964); see Text-Figure 4. Therefore, the genus appears to be stratigraphically significant.

It is noteworthy that the genus *Cuneisporites* has not yet been reported in the region south of the Kunlun-Qinling Range (Text-Figure 3). Also notable is that the forms typical of the southern floristic region such as *Ovalipollis* and *Ricciisporites* do not occur in the localities that contain *Cuneisporites*. This fact seems to indicate that the genus *Cuneisporites* is a typical form of the northern floristic region of China during the Lower Keuper depositional cycle.

CONCLUSIONS

The Huangshanjie Palynoflora is Late Triassic (Carnian) in age. Representative taxa include *Cuneisporites, Punctatisporites, Verrucosisporites*, and *Prototaeniaesporites*.

The Karamay Palynoflora is regarded as early Middle Triassic in age. Representative taxa include *Heliosaccus, Punctatisporites, Todisporites, Aratrisporites*, and *Lygodiumsporites*.

The Ehuobulake Palynoflora is assigned an uppermost Early Triassic age. Representative taxa include *Angustisulcites, Prototaeniaesporites, Pseudoquadrisaccus, Punctatisporites, Todisporites*, and *Verrucosisporites*.

In terms of its palynofloral content, Upper Triassic (Carnian) deposits in China can be divided into southern and northern realms. The boundary between the two regions is herein proposed to range from the Kunlun Range to the Qinling Range.

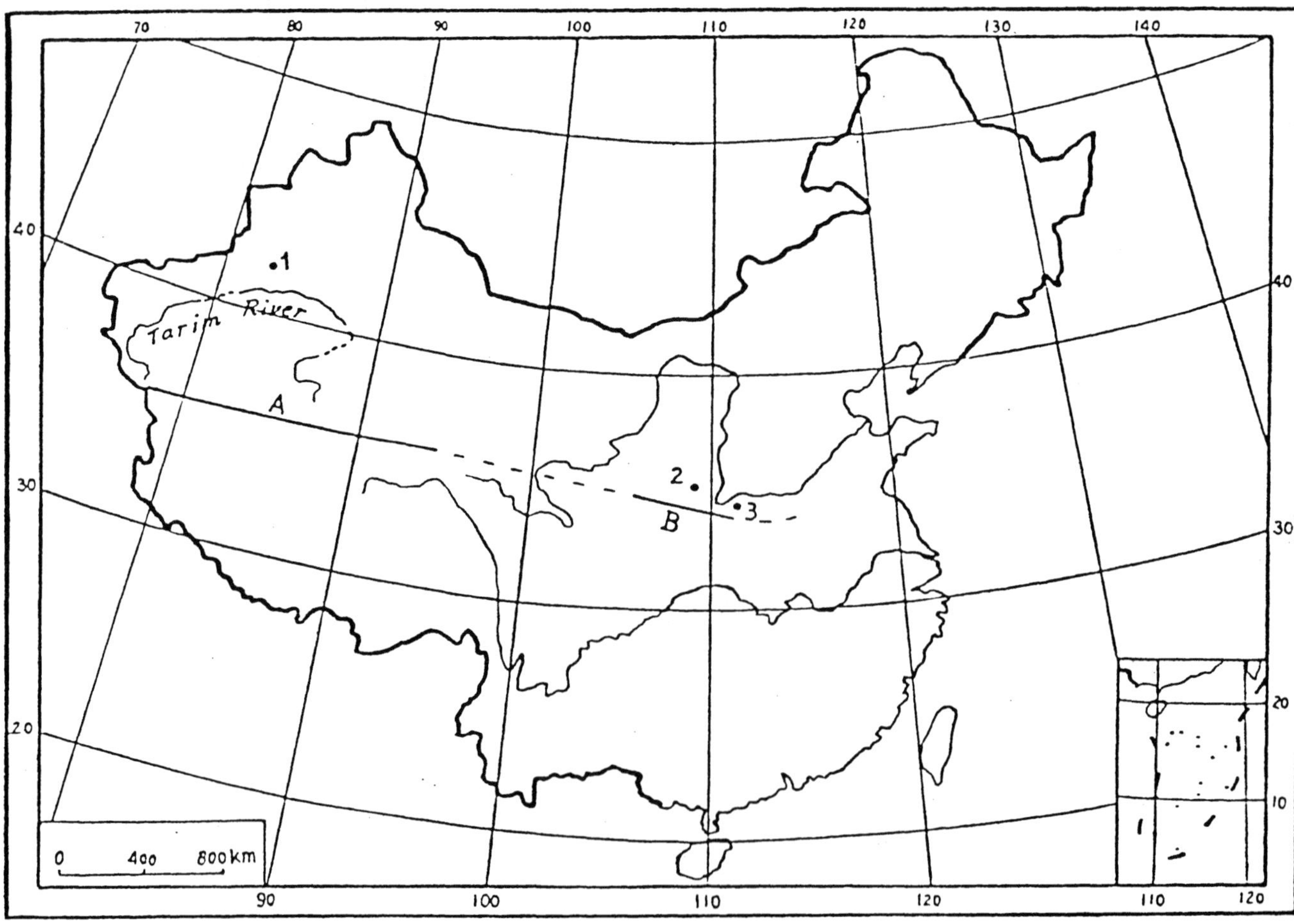

Text-Figure 3. Distribution of the genus *Cuneisporites* and the approximate position of the Upper Triassic boundary in the southern and northern regions of China. 1 - Kuqa of Xinjiang (this paper); 2 - Tongchuan of Shaanxi (Zhang, 1980a); Yima of Henan (Yang, 1994). A - Kunlun Range; B - Qinling Range.

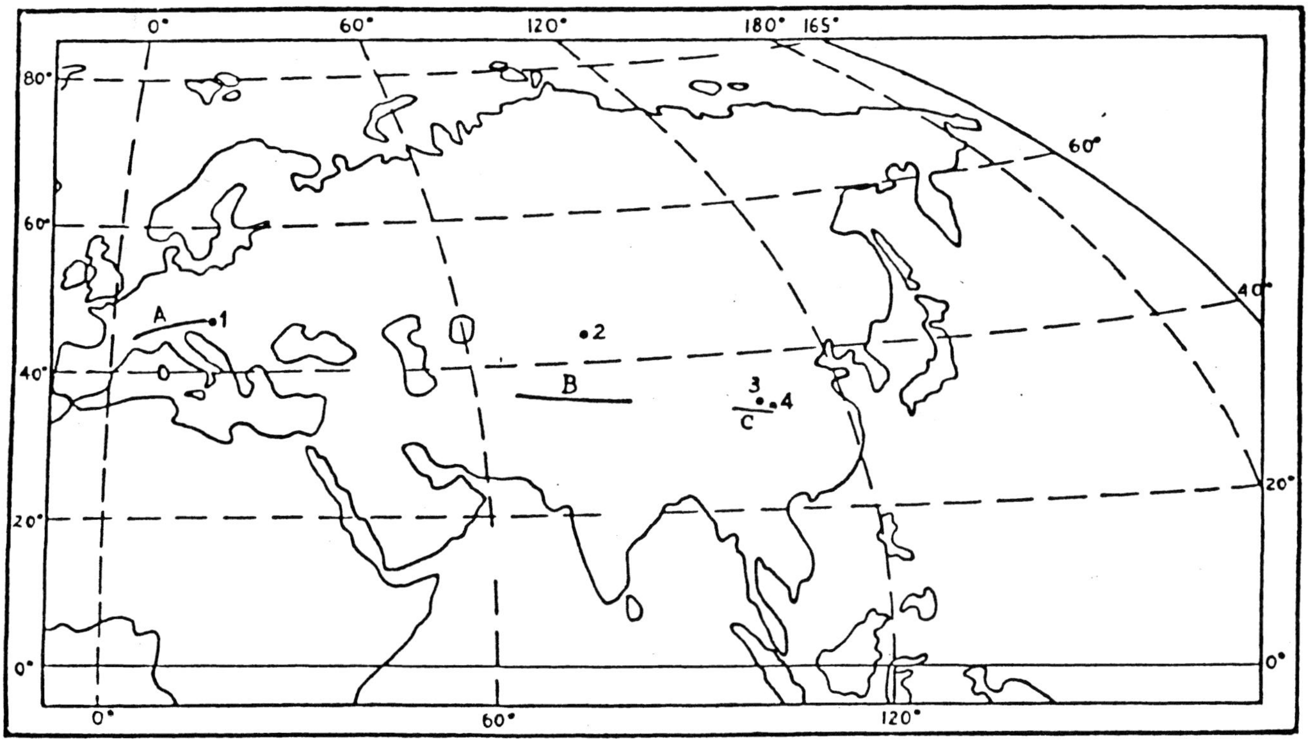

Text-Figure 4. Geographic distribution map showing the global occurrence of the genus *Cuneisporites*. 1 - Lunz, Austria (Bharadwaj and Singh, 1964); 2 - Kuqa of Xinjiang, China (this paper); 3 - Tongchuan of Shaanxi, China (Zhang, 1980a); 4 - Yima of Henan, China (Yang, 1994). A - Alps Range; B - Kunlun Range; C- Qinling Range.

NOTE ON PLATES

All palynological samples are from the Kuqa section at Xinjiang, China. Sample numbers SK2; SK12b, SK12c and SK20a were collected form the Ehuobulake, Karamaj, and Huangshanjia formations. All figures are x600, unless otherwise stated. The original specimens are kept in the Nanjing Institute of Geology and Palaeontology, Academia Sinica, China.

ACKNOWLEDGMENTS

The writers wish to express their deep gratitude to the Xinjiang Corporation of Coal-field Geological Exploration (including the attached team No. 156) and its chief engineer Yang Tongge, as well as engineers Zhou Hanzhong and Luo Lianglu. With their help, the field work in this region was accomplished. The palynological samples were macerated by Huang Fengbao and He Cuiling. Palynomorphs were photographed by Song Zhiyao and Fan Xiaoyi. The text-figures were modified by Ren Yugao. The writers express sincere thanks for their valuable support to the present study.

References Cited

BHARADWAJ, D.C., and SINGH, H.P.
1964 An Upper Triassic Miospore Assemblage from the Coals of Lunz, Austria. *Palaeobotanist*, 12: 28–44.

BOLKHOVITINA, N.A.
1953 Spore-pollen characters of Cretaceous deposits of the central provinces of the USSR. *Monograph of Geological Institute, Academia of Soviet Union*, 145 (Geologic Series 61): 1–184 (in Russian).
1956 Atlas of spores and pollen from Jurassic and Lower Cretaceous Deposits of the Vilyuisk Basin. *Monograph of Geological Institute, Academia of Soviet Union*, 2: 1–186 (in Russian).

COUPER, R.A.
1953 Upper Mesozoic and Cainozoic Spores and Pollen Grains from New Zealand. New Zealand Geological Survey. *Paleontological Bulletin*, 22: 1–77.
1958 British Mesozoic Microspores and Pollen grains. A Systematic and Stratigraphic Study. *Palaeontographica*, Abteilung B, 103: 75–179.

FREUDENTHAL, T.
1964 Palaeobotany of the Mesophytic. 1. Palynology of Lower Triassic rock salt Hengelo, the Netherlands. *Acta Botanica Neerlandica*, 13: 209–236.

KOPYTOVA, AI.A.
1963 New Genera of Spores and Pollen from Triassic Deposits of Western Kazakstan. *National Science* and *Technology Press (References of Geology and Mineralogy) Monograph* 37: 65–69 (in Russian).

LESCHIK, G.
1955 Die Keuper Flora von Neuewelt bei Basel: Die Iso- und Mikrosporen. *Schweizerische Palaeontologische, Abhandlungen*, 72: 1–70.
1956 Sporen aus den Salzton des Zechsteins von Neuhof (bei Fulda). *Palaeontographica*, Abteilung B, 100: 122–142.

LIU, Z.S., SHANG, Y.K., and LI, W.B.
1981 Triassic and Jurassic Sporo-pollen Assemblages from some Localities of Shaanxi and Gansu, North–West China. *Bulletin, Nanjing Institute of Geology* and *Palaeontology, Academia Sinica*, 3: 191–210. (in Chinese with English abstract.)

MAEDLER, K.
1964 Die geologische Verbreitung von Sporen und Pollen in der deutschen Trias. *Beihefte zum geologischen Jahrbuch*, 65: 1–147.

OUYANG, S., and NORRIS, G.
1988 Spores and Pollen from the Lower Triassic Heshanggou Formation, Shaanxi Province, China. *Review of Palaeobotany and Palynology*, 54: 187–231.

QU, L.F.
1980 Triassic spores and pollen. *In*: Mesozoic Stratigraphy and Palaeontology of the Shaanxi Gansu Ningxia Basin. (1). Geological publishing House, Beijing, pp. 115–143 (in Chinese).

SAMOILOVICH, S.R.
1953 Pollen and Spores from Permian Deposits of the Cherdynsk and Aktyubinsk Pre–Urals. *Trudy VNIGRI*, 75: 5–92 (in Russian).

SCHOPF, J.M., WILSON, L.R., and BENTALL, R.
1944 An Annotated Synopsis of Paleozoic Fossil Spores and the definition of generic groups. *Illinois Geological Survey Reports of Investigations*, 191: 1–73.

SCHULZ, E.
1965 Sporae Dispersae aus der Trias von Thueringen. *Zentrales Geologisches Institut*, 1: 257–287.
1967 Sporenpalaeontologische Untersuchungen raetoliassicher Schichten in Zentralteil des germanischen Beckens. *Palaeontographica*, Abteilung B, 11 (3): 427–633.

SCHUURMAN, W.M.L.
1977 Aspects of Late Triassic palynology. 2. Palynology of the 'Gres et schists a Avicula contorta' and 'Argiles de levallois' (Rhaetian) of northeastern France and southern Luxemburg. *Review of Palaeobotany and Palynology*, 23: 159–253.

VISSCHER, H.
1966 Palaeobotany of Mesophytic. Plant microfossils from the Upper Bunter of Hengelo, The Netherlands. *Acta Botanica Neerlandica*, 15: 316–375.

WU, S.Q,. and CHEN, P.J.
1992 Non marine Triassic of Tarim. In: Biostratigraphy and Geological Evolution of Tarim, pp. 278–292. *Science Press*, Beijing, China. (in Chinese).

YANG, S.R.
1994 Late Triassic–Jurassic sporopollen assemblages from Yima area, Henan. *Acta Palaeontologica Sinica*, 33(6): 765–779. (in Chinese with English abstract).

ZHANG, L.J.
1980a Late Triassic Spores and Pollen from Regions to the South and North of Qinling Range. *Paper for the 5th International Palynological Conference*, pp. 1–15.
1980b Palynological Contribution of the Chronology and Stratigraphy of Xinjiang (Part 1). *Paper for the 5th International Palynological Conference*, pp. 1–7.
1983 On the Characters of Permian Microflora in the Junggar Basin of Xinjiang. *Palaeontologia Cathayana*, 1: 327–365. Science Press, Beijing.
1984 Late Triassic Spores and Pollen from Certral Sichuan. *Palaeontologia Sinica (Monograph), New series A*, no. 8, whole no. 167: 1–100. Science Press, Beijing (in Chinese with English abstract).
1990 Permian spores and pollen from karamay Region of Xinjiang, *China*. *Palaeontologia Cathayana*, 5: 181–203. Science Press, Beijing.

ZHANG, L.J., and WANG, X.F.
1999 Important Triassic palynomorphs from the Tarim Basin of Xinjiang, China. *Acta Micropalaeontologica Sinica*, 16 (4): 393-404. Science Press, Beijing (in English with Chinese abstract).

Authors' addresses

ZHANG LUJIN
WANG XINFU
Nanjing Institute of Geology and Palaeontology
Academia Sinica
Nanjing, 210008
Peoples Republic of China

Part 3: Tertiary Palynology

PALYNOSTRATIGRAPHY OF THE UPPER PALEOCENE AND LOWER EOCENE WILCOX GROUP IN THE NORTHWESTERN GULF OF MEXICO BASIN

William C. Elsik and Jeff P. Crabaugh

Abstract

Palynostratigraphy is used to determine the times of prograding of four major delta systems in the upper Paleocene to lower Eocene Wilcox Group: one in the lower, one in the middle and two in the upper Wilcox. The *Thomsonipollis magnificus* Megazone spans the entire Wilcox deltaic complex. *Carya* spp. < 30 μm occur above the base of the lower Wilcox. The *Momipites dilatus* Top marks the top of the lower Wilcox. *Choanopollenites eximius* and *Maceopolipollenites granulatus* comb. nov. both have tops in the "Big Shale" in the lower middle Wilcox. The Paleocene–Eocene contact in the upper part of the middle Wilcox is marked by the *Ephedra voluta* and *Danea californica* tops below and the *Platycarya* spp. Base above. A narrow *Bagelopollis verrucatus* Zone occurs in some part of the uppermost middle Wilcox. The early Eocene *Platycarya* Abundance Zone characterizes the upper part of the middle and all of the upper Wilcox. The time of progradation of lower delta systems of the Carrizo Formation in the upper Wilcox is differentiated from the time of progradation of upper delta systems by the presence of an undescribed new species of *Callimothallus*, whereas *Annona? foveoreticulata* appears in the upper Carrizo. *Thomsonipollis magnificus* and *Muratodinium fimbriatum* tops are in the uppermost Wilcox strata. The *Annona? foveoreticulata* Zone characterizes transitional strata and the Reklaw Formation of the overlying lower Claiborne Group.

INTRODUCTION

Major sediment stacking in delta systems occurred during deposition of the late Paleocene and early Eocene Wilcox Group in Texas. Wilcox delta systems prograded in the northwestern Gulf of Mexico Basin during four major pulses of sedimentation (Crabaugh and McBride, 1993). The timing of those delta systems are the focus of this study.

Hargis (1986) assigned formations and subdivided the Wilcox Group into lower, middle and upper subgroups (Text-Figure 1). Delta outbuilding during the lower Wilcox, sedimentary pulse 1, is most pronounced in the Houston Embayment (Xue and Galloway, 1993). Delta outbuilding during the middle Wilcox, sedimentary pulse 2, records a period of shifting of the predominance of deltaic sedimentation from the Houston Embayment to the Rio Grande Embayment (Xue and Galloway, 1995). Delta outbuilding during the upper Wilcox, sedimentary pulses 3 and 4, is most pronounced in the Rio Grande Embayment (Edwards, 1981). The upper Wilcox consists of the Carrizo Formation (Hamlin, 1988). At least two delta systems have been recognized in the downdip Carrizo (Hargis, 1986).

An example of the extent of the lobes developed during a typical deltaic phase was described by Fisher and McGowen (1967). They noted that the geology of the Wilcox was complicated by the development of numerous growth faults. Whereas the Wilcox strata are measured in hundreds of feet at the surface, equivalent strata in the subsurface can be thousands of feet thick. Palynological control helps to determine stratigraphic position of strata and faulting (Elsik, 1978; Elsik and Crabaugh, 1996).

Elsik, William C., and Crabaugh, Jeff P.
2001 Palynostratigraphy of the Upper Paleocene and Lower Eocene Wilcox Group in the northwestern Gulf of Mexico Basin; *In*: Goodman, D.K., and Clarke, R.T. (eds.), *Proceedings of the IX International Palynological Congress, Houston, Texas, U.S.A., 1996*; American Association of Stratigraphic Palynologists Foundation, p. 233–237.

ISBN 0–931871–06–9

The Hooper, Simsboro, Calvert Bluff, Sabinetown and Carrizo formations received variable amounts of terrestrial and marine palynomorphs that reflect the age and paleoecology of the deltaic and associated strata. The diverse palynological elements are used to biostratigraphically zone the Wilcox Group. The eight palynological partitions of the five Wilcox formations (Text-Figure 1) are adequate to determine times of major deltaic progradations (Crabaugh and Elsik, 1996).

For this study, samples were selected to update the palynological model developed two decades ago by Fairchild and Elsik (1969; Elsik, 1968c) and revised by Elsik (1978). Approximately 115 new samples from seven cores and 16 surface exposures were processed for study. Palynomorphs typical of the late Paleocene and early Eocene Wilcox in the northwestern Gulf of Mexico Basin were described previously by Stover et al. (1966), Elsik (1968a, b; 1974) and Fairchild and Elsik (1969). A new species of *Callimothallus* and other new palynomorphs will be described in another taxonomic paper.

PALEOECOLOGY

The paleoenvironments varied from fluvial and shoreline lignite swamps to open marine. Several pollen taxa including *Taxodium* and *Thomsonipollis* (*Randia*) reflect fluvial to estuarine swamps. A small pollen morphotype similar to the black mangrove, *Rhizophorus*, is common in many samples. *Carexpollenites* is an indication of sedge marshes.

The presence of marine dinoflagellates in much of the Wilcox is unexpected. The upper deltaic sequence of the Carrizo in south Texas is dominantly marine with assemblages characterized by *Apectodinium* species.

An indeterminate fluctuation in climate is suggested by the palynosequences. Paleoclimates were subtropical to tropical with indications of temperate plants in highland or inland cooler areas. *Alnus* is one of the cooler indicators: never abundant but more likely to be seen in the shalier intervals. *Carya* spp. are common through the middle of the Calvert Bluff Formation.

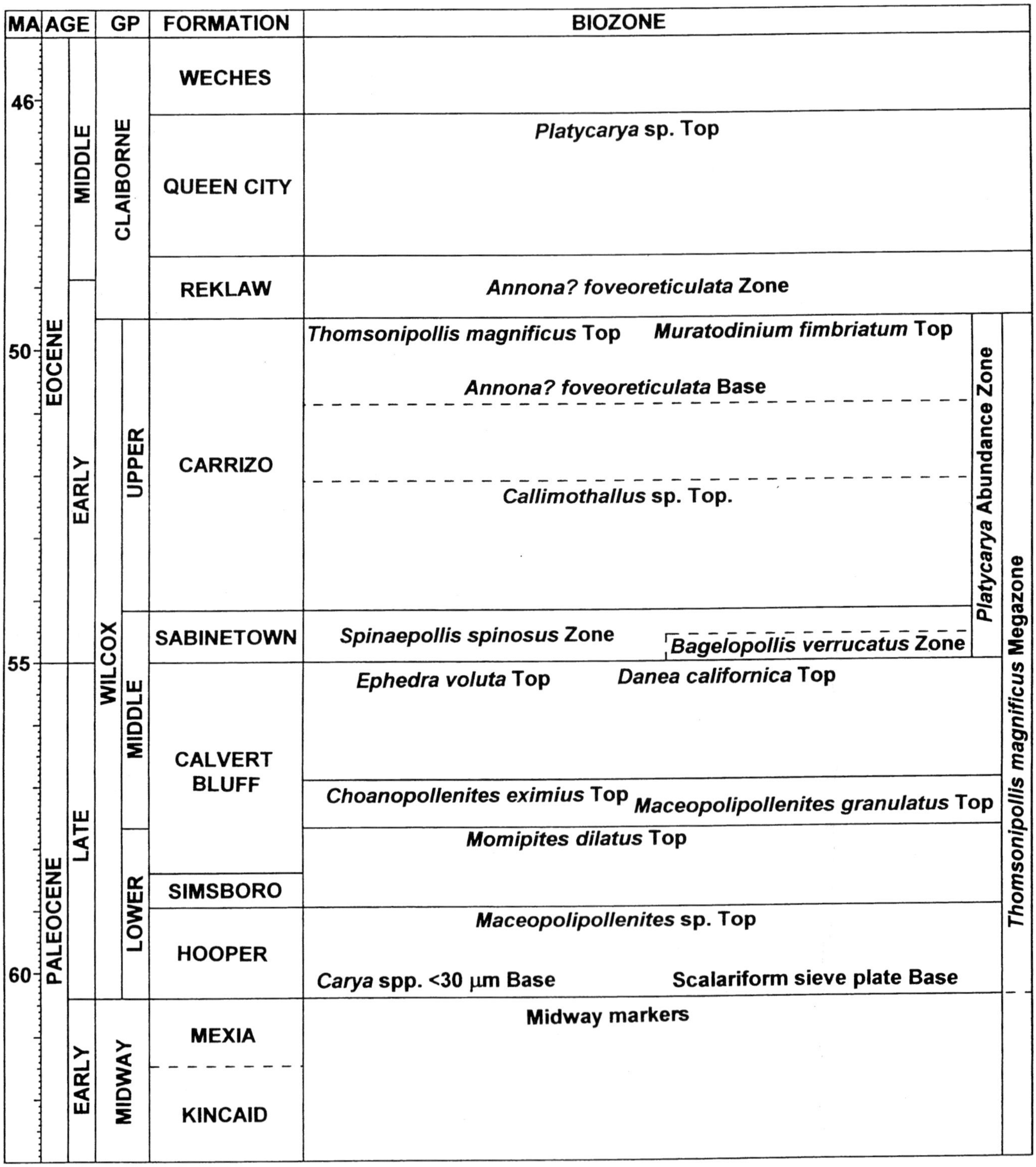

Text-Figure 1. Palynostratigraphy of the Wilcox Group. Absolute ages compiled by Crabaugh and McBride (1993).

BIOSTRATIGRAPHY

Wilcox biostratigraphy (Text-Figure 1) is based on a mixture of palynological markers including marine dinoflagellates, nonmarine gymnosperm and angiosperm pollen, plant tissue, and microfungi.

Upper Midway Group

At least two species of *Choanopollenites* and *Momipites* can be considered Midway markers; their last occurrences are at about the top of the Midway. The Midway–Wilcox contact can be recognized by common *Holkopollenites chemardensis* in the Midway and common *Taxodium* spp. in the Wilcox.

Wilcox Group

***Thomsonipollis magnificus* Megazone**. The *Thomsonipollis magnificus* Megazone encompasses all of the Wilcox Group: the *Thomsonipollis magnificus* Top marks the end of the early Eocene. *Thomsonipollis magnificus*, *T. magnificoides*, and new species of *Thomsonipollis* aff. and cf. *T. magnificus* occur throughout the Wilcox and appear to be limited almost exclusively to that group of strata. One or more of the new species may be restricted to shorter intervals within the Wilcox. The species complex does not occur above the top of the Carrizo. Its range of occurrence is designated the *Thomsonipollis magnificus* Megazone.

Lower Wilcox

***Carya* spp. < 30 μm Base; Scalariform sieve plate**. Pollen appearing in the lower lower Wilcox include *Carya* spp. < 30 μm. Rare scalariform sieve plates from dispersed lignified plant tissue, a type seen from the Late Paleocene to the Recent, also appear in the lower Wilcox.

***Maceopolipollenites* sp. Top**. An undescribed new species of *Maceopolipollenites* is limited to the Hooper Formation of the lower Wilcox. Early variants occur in samples from about the Midway–Wilcox contact.

***Brevicolporites colpella* Abundances**. The Hooper is also characterized by abundances of *Brevicolporites colpella* and occurrences of *Maceopolipollenites leboensis*. *Brevicolporites colpella* occurs up through the lower part of the Calvert Bluff Formation, but is most abundant in the Hooper and Simsboro.

***Momipites dilatus* Top**. *Momipites dilatus* occurs to the top of the lower Wilcox in the lower part of the Calvert Bluff Formation.

Middle Wilcox

***Maceopolipollenites leboensis* Occurrences**. *Maceopolipollenites leboensis* occurs through the Hooper, Simsboro and lower part of the Calvert Bluff to about the top of the "Big Shale" in the lower part of the middle Wilcox. The "Big Shale" is an interval of shales that is greatly expanded in some downdip sections (see for example Glawe, 1989).

***Maceopolipollenites granulatus* Top; *Choanopollenites eximius* Top**. *Maceopolipollenites granulatus* comb. nov. and *Choanopollenites eximius* apparently characterize the "Big Shale" through the lower part of the middle Wilcox in shorter sections. The tops of both *M. granulatus* and *C. eximius* are found at the top of the "Big Shale" in shorter sections. However, the two tops are generally found to be separated in greatly expanded sections in the subsurface: *C. eximius* persists into somewhat younger strata of the "Big Shale". *C. eximius* also is found in an updip equivalent of the "Big Shale" in a Calvert Bluff fluvial lignite (Elsik, 1968b).

***Ephedra voluta* Top; *Danea californica* Top**. *Ephedra voluta* and *Danea californica* apparently disappear at the end of the Paleocene. The last occurrences of *E. voluta* and *D. californica* at the top of the Calvert Bluff Formation and the first occurrence of *Platycarya* at the base of the Sabinetown Formation mark the Paleocene–Eocene contact.

***Bagelopollis verrucatus* Zone**. *Bagelopollis verrucatus* (Carroll, 1999), previously known as *Corsinipollenites? verrucatus* Frederiksen (Carroll, 1993; Carroll and Nunley, 1993), appears to be restricted to a short stratigraphic interval in the Sabinetown at about the Calvert Bluff–Sabinetown contact. Frederiksen (1979) found this distinctive angiosperm pollen restricted to a short interval in the Oak Grove core from northeastern Virginia. *Bagelopollis verrucatus* is rare in east Texas (Elsik, 1968d), but very common in the Tuscahoma Formation of Alabama (Carroll, 1999).

***Platycarya* spp. Base**. *Platycarya* spp. first occurs at the Calvert Bluff–Sabinetown contact. The earliest *Platycarya* morphotype evolved from the *Maceopolipollenites* morphotype in the latest Paleocene and was established by the earliest Eocene.

***Ilexpollenites* spp. Base; *Exesisporites* spp. Base**. *Ilexpollenites* spp. and *Exesisporites* spp. first appear in the earliest Eocene Sabinetown Formation.

***Spinaepollis spinosus* Zone; *Spinaepollis spinosus* Top**. The ranges of *Spinaepollis spinosus* and *Platycarya* species overlap in the Sabinetown at the top of the middle Wilcox. The *Spinaepollis spinosus* Zone is delimited by the lowermost occurrence of *Platycarya* at the base of the zone and the uppermost occurrence of *Spinaepollis spinosus* at the top, e.g. the *Spinaepollis spinosus* Top. The lowermost occurrences of *Spinaepollis spinosus* are in the upper Hooper at about the middle of the lower Wilcox. The species apparently does not occur above the top of the Sabinetown, the top of the middle Wilcox.

***Platycarya* Abundance Zone**. *Platycarya* species occur from the base of the Sabinetown, earliest Eocene, in the northwest Gulf of Mexico. Common to abundant *Platycarya* characterizes the Sabinetown and Carrizo formations of the uppermost middle and upper Wilcox.

The *Platycarya* Abundance Zone in the western Gulf Coast can be defined as that interval bounded at its base by the first occurrence of *Platycarya* in the section up through the last occurrence of *Thomsonipollis magnificus*, e.g. the *Thomsonipollis magnificus* Top. The *Platycarya* Abundance Zone can be subdivided into a lower portion equivalent to the *Spinaepollis spinosus* Zone in the uppermost middle Wilcox, and an upper portion in the upper Wilcox above the uppermost occurrence of *Spinaepollis spinosus*.

The *Platycarya* Abundance Zone is comparable at its base to the *Platycarya platycaryoides* Zone established by Pocknall (1987). The upper limit of the *Platycarya platycaryoides* Zone was not defined. *Platycarya* persists into the Oligocene and perhaps the Miocene in the Gulf Coast (Elsik, unpub. data).

Upper Wilcox

Occasional abundances of *Platycarya* species and occurrences of *Thomsonipollis magnificus* are seen to the top of the upper Wilcox.

***Callimothallus* sp. Top**. The two deltaic sequences within the upper Wilcox are differentiated on the presence of an undescribed new species of *Callimothallus* through the lower sequence to about the middle of the Carrizo Formation. The presence of *Annona? foveoreticulata* above the *Callimothallus* sp. Top also is useful in differentiating the uppermost sequence. The lowest occurrences of *Symplocoipollenites* spp. of the Claiborne type are between the *Callimothallus* sp. Top and the *Annona? foveoreticulata* Base in the upper part of the Carrizo.

***Annona? foveoreticulata* Base**. The first *Annona? foveoreticulata* and *Annona?* cf. *A.? foveoreticulata* are seen in the upper Carrizo Formation above the *Callimothallus* sp. Top.

***Thomsonipollis magnificus* Top; *Muratodinium fimbriatum* Top**. The top of the upper Wilcox is marked by the last occurrences of *Thomsonipollis magnificus s.l.* and *Muratodinium fimbriatum*. The presence of Wilcox palynofloras supports the placement of the Carrizo Formation in the Wilcox Group by Hargis (1962).

Lower Claiborne Group

***Annona? foveoreticulata* Zone.** The early middle Eocene *Annona? foveoreticulata* Zone is established for the range of occurrences of *Annona? foveoreticulata* above the *Thomsonipollis magnificus* Top. The *Annona? foveoreticulata* Zone characterizes the lowermost Claiborne Group Reklaw Formation at the surface and shallow subsurface. The zone is used to verify the middle Eocene age of lower Claiborne strata immediately above the Carrizo Formation in most areas other than south Texas.

In south Texas, the section is expanded and the zone apparently spans the early–middle Eocene contact (Text-Figure 1) in the deeper subsurface of south Texas where Hargis (1962) noted the transitional nature of the Carrizo–Reklaw contact. As evidence of the transitional early–middle Eocene contact, *Apectodinium* spp. of early Eocene aspect occur above the *Thomsonipollis magnificus* Top into the lower part of the *Annona? foveoreticulata* Zone.

***Friedrichipollis claibornensis* Base; *Ephedra distachya*-type Base**. Both *Friedrichipollis claibornensis* and *Ephedra distachya*-type occur from the base of the middle Eocene up through the Claiborne Group. The *Annona? foveoreticulata* Zone is effectively split at that datum, e.g. the early–middle Eocene contact.

SYSTEMATIC PALEONTOLOGY

Juglandaceae

Genus *Maceopolipollenites* Leffingwell 1971

Type Species. *Maceopolipollenites triorbicularis* Leffingwell 1971.

Discussion. *Maceopolipollenites* is retained to facilitate partitioning of the great diversity of Paleogene Juglandaceae pollen. The type species, *Maceopolipollenites triorbicularis*, clearly shows a different pattern of polar thinning than *Momipites*.

Maceopolipollenites granulatus (Simpson 1961) comb. nov.
Engelhardtia granulata Simpson, 1961, p. 444, pl. 14, fig. 3-4 (cotypes).
Triporopollenites tectus Newman, 1965, p. 12, pl. 1, fig. 6 (holotype).
Triatriopollenites granulatus (Simpson) Leffingwell, 1971, p. 51, pl. 10, fig. 8-9; Pocknall and Nichols, 1996, p. , pl. 7, fig. 8-9.

Basionym. *Engelhardtia granulata* Simpson 1961, Transactions of the Royal Society of Edinburgh, v. 64, no. 16, p. 444, pl. 14, fig. 3-4.

Discussion. The species is here combined with *Maceopolipollenites* on the basis of the three atriate apertures and on the presence of three thinned areas modifying the exine into a broadly Y-shaped structure at the pole. The ends of the feature are interporal, but this is in keeping with the nature of diversity of the Juglandaceae during the Paleocene and early Eocene.

Leffingwell (1971) placed this species in the form genus *Triatriopollenites*. Pocknall and Nichols (1996) also recognized the atriate nature of the apertures.

Newman's (1965) *Triporopollenites tectus* is apparently a variety of *Maceopolipollenites granulatus*. *Engelhardtioidites minutus* Newman 1965 also appears to be very similar.

Occurrence. Maastrichtian to middle late Paleocene.

ACKNOWLEDGMENTS

We thank the Bureau of Economic Geology, The University of Texas at Austin for samples of lignites and cores; the Texas Water Development Board for access to cores; Ted Delevoryas, The University of Texas at Austin, for a sample from the Marshall Pottery clay pit; and R. E. Carroll, Alabama Geological Survey, for a slide with common specimens of *Bagelopollis verrucatus*. Our appreciation also goes to Joyce Lucas-Clark and Douglas J. Nichols for a thorough review of the paper. This study was supported in part by National Science Foundation grant EAR-9205946 (Earle F. McBride, The University of Texas at Austin), and in part by The MycoStrat Connection.

References Cited

CARROLL, R.E.
1993 Palynofacies in the upper Paleocene Tuscahoma Sand of western Alabama and eastern Mississippi. *Twenty-sixth Annual Meeting of the American Association of Stratigraphic Palynologists*, Baton Rouge, Abstracts, p. 11.
1999 Morphology and distribution of *Bagelopollis verrucatus* (Frederiksen 1988) gen. et comb. nov. from lower Tertiary coastal plain deposits of the southeastern U.S.A. *Palynology*, 23: 55–66.

CARROLL, R.E., and NUNLEY, J.W.
1993 Occurrence of *Corsinipollenites? verrucatus* Frederiksen, a uniquely annulate pollen from upper Sabinian (Paleocene) coastal plain sediments, southeastern United States. *Twenty-sixth Annual Meeting of the American Association of Stratigraphic Palynologists*, Baton Rouge, Abstracts, p. 12.

CRABAUGH, J.P., and ELSIK, W.C.
1996 Biostratigraphy and paleoecology of the late Paleocene and early Eocene Wilcox deltaic systems in the northwestern Gulf of Mexico Basin. *Ninth International Palynological Congress*, Houston, Texas, June 23–28, Program and Abstracts, p. 26.

CRABAUGH, J.P., and MCBRIDE, E.F.
1993 Episodicity in Gulf of Mexico sedimentation, Rocky Mountain tectonism, and early Paleogene climate at the scale of several million years. *Geological Society of America*, Abstracts with Programs, A: 68.

EDWARDS, M.B.
1981 Upper Wilcox Rosita Delta System of south Texas, growth-faulted shelf-edge deltas. *American Association of Petroleum Geologists, Bulletin*, 65: 54–73.

ELSIK, W.C.
1968a Palynology of a Paleocene Rockdale lignite, Milam County, Texas. I. Morphology and taxonomy. *Pollen et Spores*, 10: 263–314.
1968b Palynology of a Paleocene Rockdale lignite, Milam County, Texas. II. Morphology and taxonomy (end). *Pollen et Spores*, 10: 599–664.
1968c Palynology of the Wilcox Group. *In*: Atlee, W. A., Elsik, W. C., Frazier, D. E., and Zingula, R. P., *Environments of Deposition, Wilcox Group, Texas Gulf Coast*, Houston Geological Society, Field Trip Guidebook, p. 22–27.
1968d Morphologic variation in the palynomorphs *Thomsonipollis* and *Pistillipollenites*. *2nd Annual Meeting, South-Central Section, Geological Society of America, Program with Abstracts*, p. 18–19.
1974 Characteristic Eocene palynomorphs in the Gulf Coast, U.S.A. *Palaeontographica*, Abteilung B, 149(1–4): 90–111.
1978 Palynology of Gulf Coast lignites: the stratigraphic framework and depositional environments. *In*: Kaiser, W. R. (ed.), Gulf Coast Lignite Conference: geology, utilization, and environmental aspects, *Bureau of Economic Geology, The University of Texas at Austin, Report of Investigations Number* 90, p. 21–32.

ELSIK, W.C., and CRABAUGH, J.P.
1996 Late Paleocene and early Eocene Wilcox palynomorph assemblages from a south Texas core. *Ninth International Palynological Congress*, Houston, Texas, June 23–28, Program and Abstracts, p. 39–40.

FAIRCHILD, W.W., and ELSIK, W.C.
1969 Characteristic palynomorphs of the lower Tertiary in the Gulf Coast. *Palaeontographica*, Abteilung B, 128(3-6): 81–89.

FISHER, W.L., and MCGOWEN, J.H.
1967 Depositional systems in the Wilcox Group of Texas and their relationship to occurrence of oil and gas. *Transactions—Gulf Coast Association of Geological Societies*, 17: 105–125.

FREDERIKSEN, N.O.
1979 Paleogene sporomorph biostratigraphy, northeastern Virginia. *Palynology*, 3: 129–167.

GLAWE, L.N.
1989 Stratigraphic relationships between *Odontogryphaea thirsae* beds and the Big Shale of the Wilcox (Paleocene–Eocene) in Louisiana. *Transactions—Gulf Coast Association of Geological Societies*, 39: 375–383.

HAMLIN, H.S.
1988 Depositional and ground-water flow systems of the Carrizo–upper Wilcox, south Texas. *Bureau of Economic Geology, The University of Texas at Austin, Report of Investigations Number* 175, 61 pgs.

HARGIS, R.N.
1962 Stratigraphy of the Carrizo–Wilcox of a portion of south Texas and its relationship to production. *In*: *Contributions to Geology of South Texas 1962*, South Texas Geological Society, p. 129–145.
1986 Proposed stratigraphic classification of the Wilcox Group of south Texas. *In*: Stapp, W. L. (ed.), *Contributions to the Geology of South Texas*, South Texas Geological Society, p. 135–159.

LEFFINGWELL, H.A.
1971 Palynology of the Lance (late Cretaceous) and Fort Union (Paleocene) formations of the type Lance area, Wyoming. *In*: Kosanke, R. M., and Cross, A. T. (eds.), Symposium on Palynology of the late Cretaceous and early Tertiary, *Geological Society of America Special Paper* 127, p. 1–64.

NEWMAN, K.R.
1965 Upper Cretaceous–Paleocene guide palynomorphs from northwestern Colorado. *University of Colorado Studies, Series in Earth Sciences*, 2: 1–21.

POCKNALL, D.T.
1987 Palynomorph biozones for the Fort Union and Wasatch formations (upper Paleocene–lower Eocene), Powder River Basin, Wyoming and Montana, U.S.A. *Palynology*, 11: 23–35.

POCKNALL, D.T., and NICHOLS, D.J.
1996 *Palynology of coal zones of the Tongue River Member (upper Paleocene) of the Fort Union Formation, Powder River Basin, Montana and Wyoming*. The American Association of Stratigraphic Palynologists Foundation, Contribution Series, No. 32, 57 pgs.

SIMPSON, J.B.
1961 The Tertiary pollen-flora of Mull and Ardnamurchan. *Transactions of the Royal Society of Edinburgh*, 64: 421–468.

STOVER, L.E., ELSIK, W.C., and FAIRCHILD, W.W.
1966 New genera and species of early Tertiary palynomorphs from Gulf Coast. *University of Kansas Paleontological Contributions, Paper* 5, 11 pgs., *Paper* 6, p. 7.

XUE, L., and GALLOWAY, W.E.
1993 Sequence stratigraphic and depositional framework of the Paleocene lower Wilcox strata, northwest Gulf of Mexico basin. *Transactions—Gulf Coast Association of Geological Societies*, 43: 453–464.
1995 High-resolution depositional framework of the Paleocene middle Wilcox strata, Texas Coastal Plain. *American Association of Petroleum Geologists, Bulletin*, 79: 205–230.

Authors' addresses

WILLIAM C. ELSIK
The MycoStrat Connection
P.O. Box 549
Snook, TX 77878-0549
U.S.A.

JEFF P. CRABAUGH
Department of Geological Sciences
The University of Texas at Austin
Austin, TX 78712-7909
U.S.A.

Present Address:
Department of Geology and Geophysics
The University of Wyoming
Laramie, WY 82070
U.S.A.

PEDIASTRUM AS A GUIDE FOSSIL IN SEQUENCE STRATIGRAPHY

Patrice Brenac and Keith Richards

Abstract

Pediastrum are easily transported freshwater algae which are often found in abundance in marine sediments in some tropical areas. Several species of *Pediastrum* are illustrated and their distribution in a typical marine sedimentary record is shown. In marine sediments, high numbers of *Pediastrum* often coincide with reduced numbers of marine microfossils and are associated with periods of low relative sea level. Reduced numbers of *Pediastrum* are usually recorded during periods of high relative sea level. The record of *Pediastrum* in marine sediments can be useful for sequence stratigraphic interpretation, with maximum abundances usually associated with late highstand and lowstand systems tracts.

PEDIASTRUM AS A GUIDE FOSSIL

Pediastrum are colonial green algae which occur naturally in freshwater environments, in particular, in standing bodies of water such as ponds and lakes. There are many living species of *Pediastrum*, though all are morphologically similar and vary only in shape, wall structure and the number of cells in the colony. The cell walls of *Pediastrum* are highly resistant to decay. Several species occur commonly in the fossil record and are frequently observed in palynological preparations (Plate 1). Palynological studies from offshore wells in the low latitude tropics, for example in west Africa and south east Asia, have shown that *Pediastrum* also occur, often in significant numbers, in sediments deposited on or beyond the marine continental shelf. In the marine sedimentary record, *Pediastrum* occur as part of an allochthonous association brought by rivers and streams from the freshwater catchment areas. There is often a direct relationship between the presence of high numbers of *Pediastrum* and reduced numbers or, in some cases, total absence of planktonic foraminifera and marine microplankton as a result of freshwater rafting into the marine environment. Text-Figure 1 is a generalized illustration showing the total numbers of *Pediastrum*, marine microplankton, mangrove pollen and inland (hinterland) derived palynomorphs found in part of a well section from offshore west Africa, in which the sediments were deposited at water depths ranging from middle shelf to bathyal. Part of each depositional cycle, or "sequence," coincides with generally high proportions of marine and/or brackish water microfossils in response to high relative sea level (transgressive systems tract). The major transgressive events are also marked by increased representation of planktonic and/or benthonic foraminifera (data not shown). The uppermost (and sometimes lowermost) part of each cycle is characterized by increasing numbers of *Pediastrum*, and often with increased representation of hinterland derived palynomorphs, reflecting deposition in response to a lowering of relative sea level. These intervals relate to the late highstand and/or lowstand phases of deposition (i.e. highstand and lowstand systems tracts). Further study will determine if the relationships observed between *Pediastrum* and other palynomorph groups in some tropical areas are also applicable outside tropical latitudes.

Brenac, Patrice, and Richards, Keith
2001 *Pediastrum* as a guide fossil in sequence stratigraphy; *In*: Goodman, D.K., and Clarke, R.T. (eds.), *Proceedings of the IX International Palynological Congress, Houston, Texas, U.S.A., 1996*; American Association of Stratigraphic Palynologists Foundation, p. 239–241.

ISBN 0–931871–06–9

ACKNOWLEDGMENTS

The authors would like to thank the directors of Robertson Research International Limited for permission to present and publish this study. Thanks are also due to Mary Wood for assistance with the poster display and to Bob McDonnell for Text-Figure 1.

Authors' addresses

PATRICE BRENAC
Robertson Research International Limited
Llandudno, North Wales, LL30 1SA
United Kingdom

Present addresses:
Brenac Stratigraphy International
133 Peulwys Lane
Colwyn Bay, North Wales, LL29 8YF
United Kingdom
E-mail: parb@netcomuk.co.uk

KEITH RICHARDS
Robertson Research International Limited
Llandudno, North Wales, LL30 1SA
United Kingdom

Present address:
KrA Stratigraphic
43 Ty Mawr Road
Deganwy, Conwy, LL31 9UD
United Kingdom
E-mail: keith@paly.co.uk
Web: www.paly.co.uk

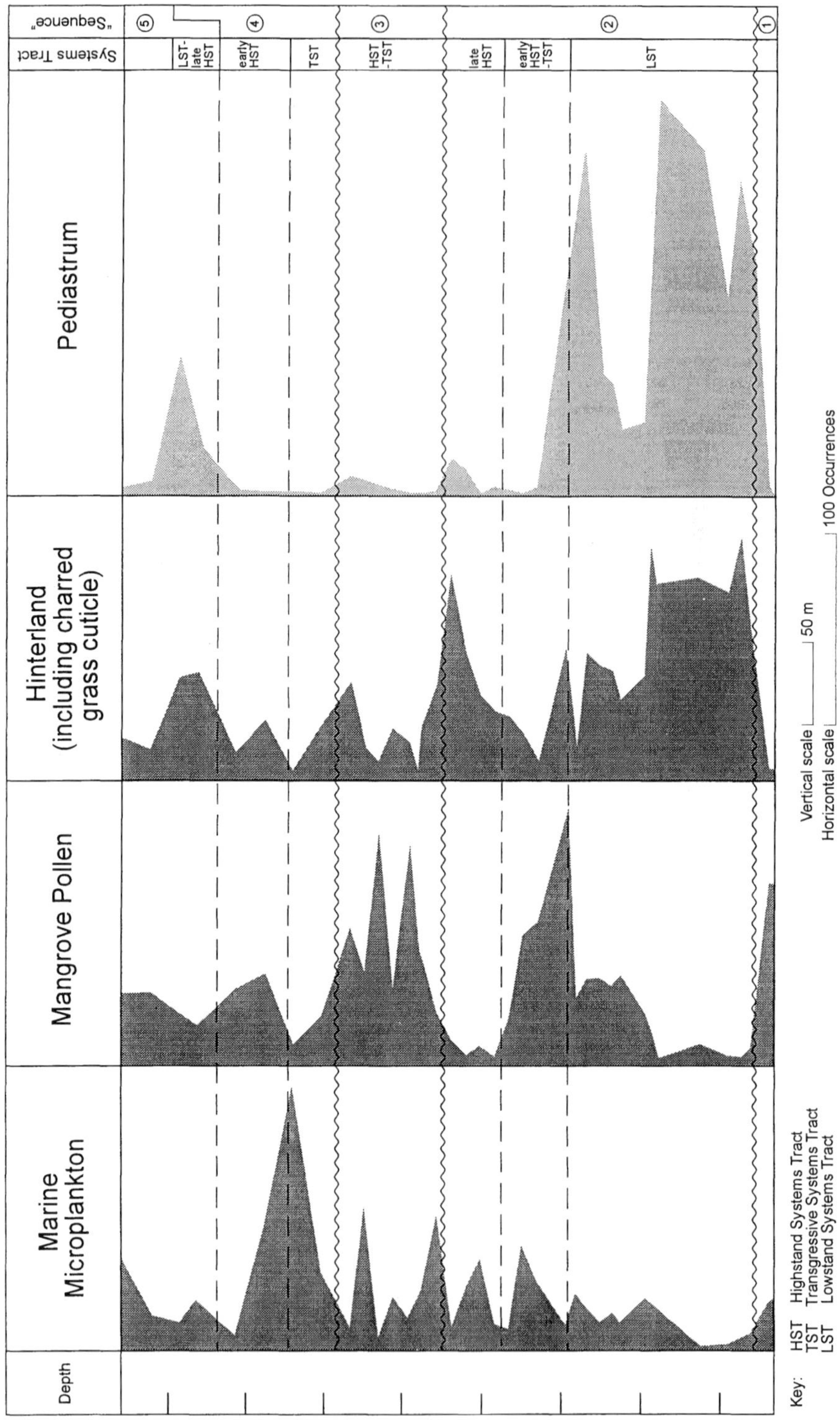

Text-Figure 1. Generalized figure to show the abundances and relative proportions of *Pediastrum*, marine microplankton, mangrove pollen and inland (hinterland) derived palynomorphs (mostly grass pollen and cuticle) in part of a well section from offshore west Africa. The section illustrated represents approximately 400 m vertical thickness of marine sediment.

PLATE 1

Examples of the freshwater algae *Pediastrum* obtained from marine sediments.

1 *Pediastrum delicatites* (x1,000)
2 *Pediastrum kajaites* (x1,000)
3 *Pediastrum* sp. (x1,000)
4 *Pediastrum bifidites* (x1,000)
5 *Pediastrum delicatites* (x400)
6 *Pediastrum kajaites* (x1,000)
7 *Pediastrum kajaites* (x1,000)

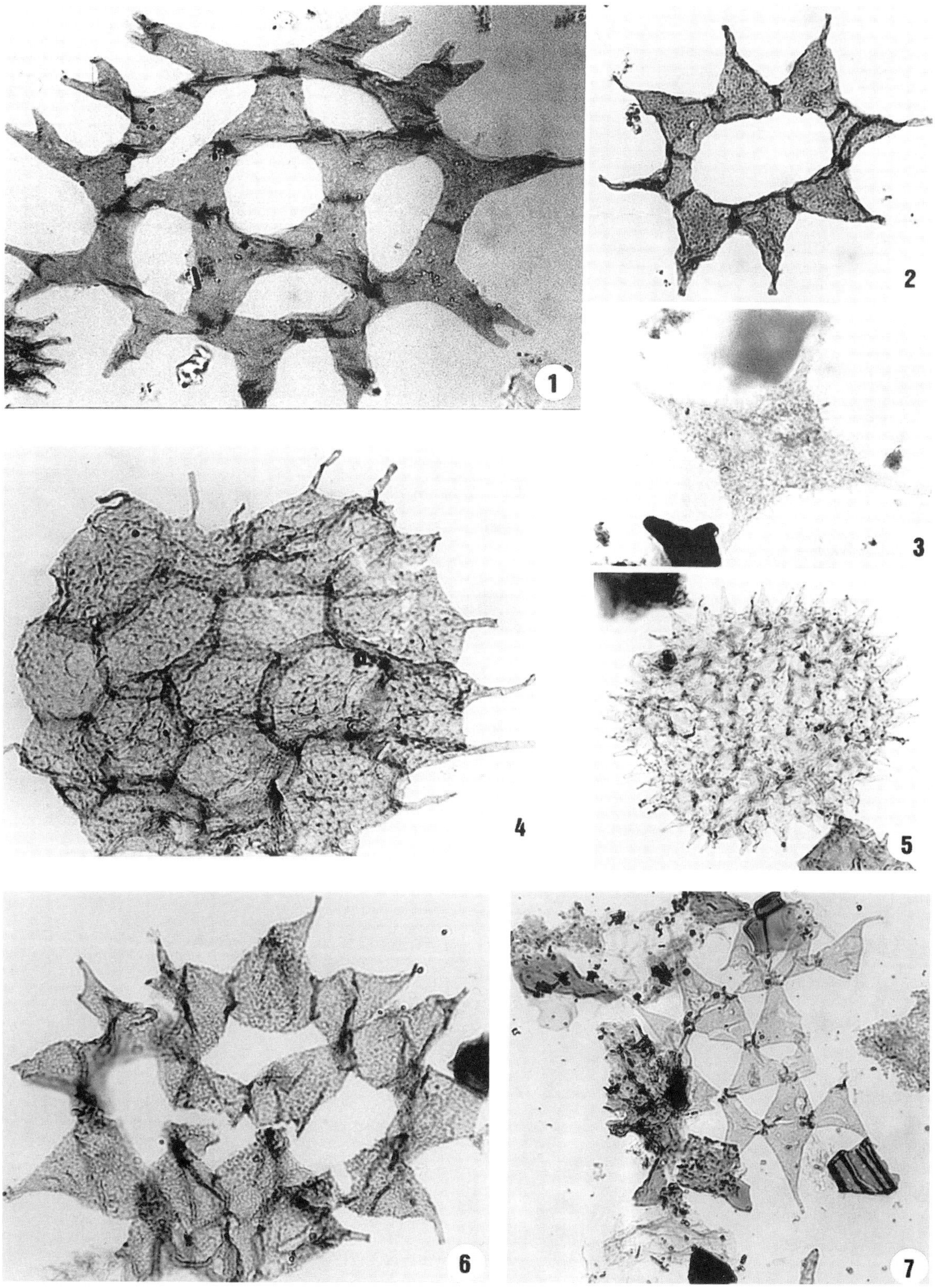
1
2
3
4
5
6
7

INTEGRATION OF SEQUENCE STRATIGRAPHY AND BIOSTRATIGRAPHY AT THE PALEOCENE/EOCENE BOUNDARY IN THE SEDIMENTS OF INTERMEDIATE AND DEEP WATER FROM THE BASQUE BASIN, WESTERN PYRENEES

Koldo Núñez-Betelu, Xabier Orue-Etxebarria, Estibaliz Apellaniz, Aitor Payros, Victoriano Pujalte, Juan Ignacio Baceta, Rodolfo Coccioni, Rita Di Leo, Jaume Dinares-Turell, Josep Maria Pares, Simone Galeotti, Simonetta Monechi, Josep Maria Samso, Josep Serra-Kiel, Josep Tosquella, and Birger Schmitz

Abstract

This study of intermediate-water depth sections in the Basque Basin is of great interest because it allows correlations between deep-sea and shallower sequences, and because it is strategically located between the North Sea and the expanded sections at low latitudes in the Tethys. Two sections, Trabakua Pass and Ermua, have been studied in detail. The Trabakua Pass section comprises marlstones belonging to a basinward location; the Ermua section includes hemipelagic limestones and resedimented deposits which typify a base-of-slope carbonate apron accumulation. A prominent clayey interval can be recognised not only in these two sections but also through the basin and flanking shelves. This interval corresponds to the lowstand wedge deposits of Depositional Sequence 7 and includes in its lower part the carbonate dissolution interval (CDI) which is largely devoid of carbonate microfossils. The CDI may be related to either a large flooding of CO_2 caused by volcanism linked to the early Tertiary opening of the North Atlantic or to a huge release of methane form the sea-floor.

The Trabakua Pass and Ermua sections can be confidently correlated using lithostratigraphy and biostratigraphy. Fossil groups such as marine and terrestrial palynomorphs, planktonic, deep-water and shallow benthic foraminifera, and calcareous nannofossils, as well as the isotopic records have been studied in detail.

Our intercalibration shows that, in these sections, a strong carbon-isotopic shift accompanies the base of calcareous nannofossil zone NP 10. The deep-water benthic extinction occurs below the base of NP 10 whereas the base of the *Wetzeliella astra* (dinocyst) zone occurs above it.

INTRODUCTION

Correlation between distant sections and positioning of stratigraphic boundaries requires integration of many geological tools. Not all sections contain the same type of fossils and, therefore, various zonation schemes and multiple boundary markers have been proposed to the point that intercorrelation and placement of boundaries is often difficult. To help improve the understanding of the Paleocene/Eocene boundary, we present a detailed biostratigraphic and isotopic analyses of two marine sections, Trabakua Pass and Ermua, across the P/E boundary in the Basque Basin. The fossil groups studied include palynomorphs, planktonic foraminifera, deep and shallow water benthic foraminifera, and calcareous nannoplankton. The stable isotope analyses have achieved results similar to other studies (e.g., Kennet and Stott, 1990).

The Trabakua Pass section is hemipelagic and represents continuous (possibly condensed) deposition across the boundary; exhibits very little structural complication; and no evidence of diagenetic alteration or redeposition. In addition, the Trabakua Pass section can be confidently cross-correlated with the Ermua section, which is largely composed of carbonate breccias and turbidites which were resedimented from a coeval carbonate platform that existed to the south of the study area. Since both sections are rich in marine fossils, they offer the possibility to intercalibrate deep and shallow marine fossil groups across the Paleocene–Eocene boundary.

Núñez-Betelu, K., Orue-Etxebarria, X, Apellaniz, E., Payros, A., Pujalte, V., Baceta, J.I., Coccioni, R., Di Leo, R., Dinares-Turell, J., Pares, J.M., Galeotti, S., Monechi, S., Samso, J.M., Serra-Kiel, J., Tosquella, J., and Schmitz, B.
2001 Integration of sequence stratigraphy and biostratigraphy at the Paleocene/Eocene boundary in the sediments of intermediate and deep water from the Basque Basin, western Pyrenees; *In*: Goodman, D.K., and Clarke, R.T. (eds.), *Proceedings of the IX International Palynological Congress, Houston, Texas, U.S.A., 1996*; American Association of Stratigraphic Palynologists Foundation, p. 243–251.
 ISBN 0–931871–06–9

LOCATION, SETTING AND SEQUENCE STRATIGRAPHY

The study area is located on the northern part of the Iberian Peninsula (Text-Figure 1) at the western end of the Pyrenees on what is referred to as the Basque Basin. This basin was an interplate trough of intermediate to deep water depth that evolved during the latest Cretaceous and early Eocene times (Pujalte et al., 1993). It was elongated in the east–west direction, opening westward into the present-day Bay of Biscay, and surrounded by shallow shelf areas to the south, east and north (Text-Figure 1).

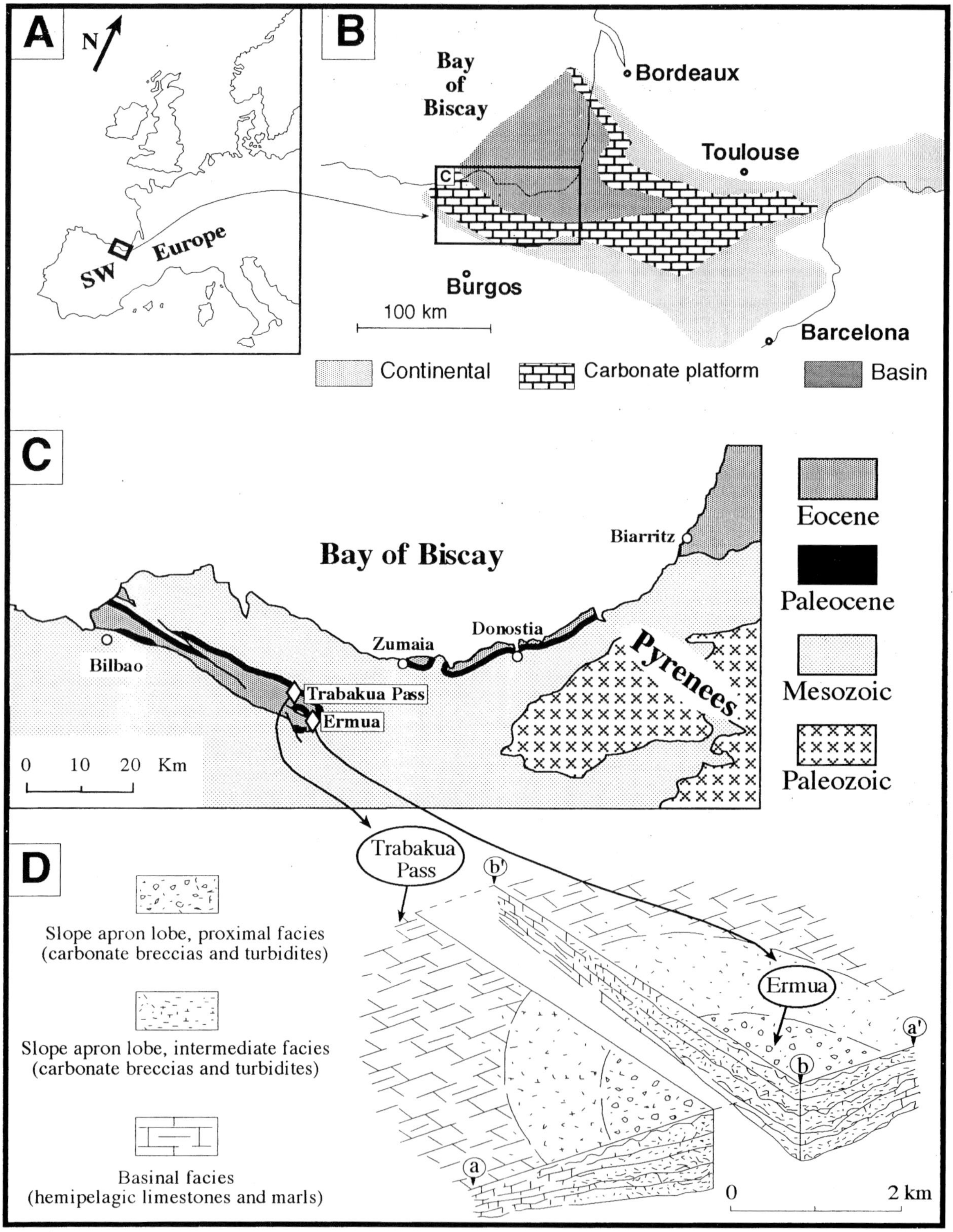

Text-Figure 1. Location and setting of the study area. (A) Location of the study area in southwestern Europe; (B) generalized early Paleogene paleogeography of the northern Iberian Peninsula; (C) simplified geologic map of the Bay of Biscay area with location of Trabakua Pass and Ermua sections; and (D) schematic reconstruction of the depositional setting in the study area.

Throughout most of early Paleogene time, these shallow zones were covered by carbonate platforms, while the central parts of the basin became starved and received little sediments other than hemipelagic rain. Resedimentation from the flanking carbonate platforms, however, created a discontinuous base-of-slope apron of coarse grained breccias and carbonate turbidites, that interfingered basinwards with stacks of hemipelagic limestones.

The Ermua section, comprised of both hemipelagic and resedimented deposits, typifies a base-of-slope carbonate apron accumulation (Text-Figure 1). The Trabakua Pass section occurs in a more basinward position and it is almost exclusively constituted by hemipelagic marls and limestones.

In recent studies (Baceta, 1996, Payros, 1997) of the latest Cretaceous–middle Eocene of the Pamplona and Basque Basins 15 depositional sequences have been recognised. Three of these sequences (DS-6, 7 and 8) were developed during the latest Paleocene–earliest Eocene time span (Text-Figure 2). Of these sequences, lowstand fan system tracts (LSF) are found only in the Ermua section, pinching out basinwards within a few kilometers. In all cases, the LSFs sit on erosional surfaces that irregularly cut the underlying deposits, their features being different in each sequence. Lowstand wedge deposits (LW) account for the bulk of the preserved succession in both the Ermua and Trabakua Pass sections. The LW's of DS-6 and DS-8 are represented in the Ermua section by alternations of globigerinid mudstones/wackstones with fine- and coarse-grained bioclastic grainstones. The globigerinid mudstones/wackstones are hemipelagic and/or very diluted turbiditic deposits of intrabasinal derivation. They form a background sedimentation of this system tract that was interrupted from time to time by the arrival of the bioclastic grainstones. In the Trabakua Pass section, the LW's of DS-6 and DS-8 consist solely of rhythmic alternation of hemipelagic marls and marlstones (Text-Figure 2). The LW of DS-7 of both Trabakua Pass and Ermua sections includes some of the more intriguing sediments of the studied interval. In the Ermua section, its lowermost 4 meters are made up of an alternation of globigerinid mudstones/wackstones and bioclastic grainstones similar to those of the underlying and overlying sequence. The remaining 19 meters consist largely of dark grey claystones with a very low content of $CaCO_3$ (<5%), but with frequent intercalations of bioclastic grainstones. In the Trabakua Pass section, the LW of DS-7 is only 3.5 m thick and is exclusively formed by dark-grey claystones. In both sections, the dark-grey claystone-rich interval (subsequently referred to as "clayey interval") exhibit a rare calcareous micropaleontological content, often reduced to a few dwarfed planktonic foraminifer specimens and/or to agglutinated foraminifera. The paucity of calcareous microfossils is especially prevalent in the lower part of the clayey interval, where samples are mostly barren of carbonate species, both nannofossils and foraminifers. Such absence is tentatively attributed to a deep-water dissolution event and for that reason that part of the clayey interval is thus referred to as the Carbonate Dissolution Interval (CDI). However, the bioclastic turbidites intercalated within the CDI in the Ermua section have a composition similar to the LW's of DS-6 or DS-8. Due to its distinctive character, the clayey interval (and its CDI) is very useful as a lithological marker within the Basque Basin (Text-Figure 2). The CDI could be related to either a large flooding of CO_2 caused by volcanism linked to the early Tertiary opening of the North Atlantic (Eldom and Thomas, 1993) or to a huge release of methane form the sea-floor (Dickens et al., 1997).

Transgressive/Highstand System Tract (TST/HST) deposits are not extensive in the studied succession, and only exist in the Trabakua Pass section. The TST/HST of DS-6 is represented as a thin cap (50 cm) of glauconite-rich pelagic limestones, while that of DS-7 is a 5 meter thick accumulation of pelagic limestones with thin marly intercalations. In both sequences, the TST/HST deposits are separated from the LW by a marly horizon extremely rich in planktonic foraminifera, which probably represents a condensed section.

STABLE ISOTOPE STRATIGRAPHY

Thirty samples from the Trabakua Pass section were studied for bulk-rock isotopic ($\delta^{18}O$ and $\delta^{13}C$) analyses. All results are presented (Text-Figure 2) using the δ notation, which is defined as the per mil deviation from the Peedee Belemnite Standard (PDB). The $\delta^{13}C$ record at the Trabakua Pass section shows a gradual fall from values around 2.5 to 1.3 through the upper Paleocene and the lowest Eocene. Superimposed on this trend is a strong negative shift in the upper part of the clayey interval. The magnitude of the negative $\delta^{13}C$ excursion at Trabakua Pass is more than 4 ‰, as demonstrated by a $\delta^{13}C$ value of -2.9 ‰ at 3 meters above the base of the dissolution interval. This low value probably reflects the global oceanic $\delta^{13}C$ negative excursion in connection with the latest Paleocene benthic foraminiferal extinction event. The $\delta^{18}O$ values at Trabakua Pass show a relatively constant trend around -4.6 to -5 ‰. These values reflect diagenetic reequilibration, however, give no information on the paleoceanic evolution.

BIOSTRATIGRAPHY

The biostratigraphic analyses are based on palynomorphs, planktonic foraminifera, deep and shallow water benthic foraminifera, and calcareous nannoplankton. A zonation scheme has been erected for each one of these groups (Text-Figure 2) to allow comparative analyses. A summary of the results is presented below. For more detailed information see Orue-Etxebarria et al. (1996).

Palynomorphs

The palynological analysis is based on palynomorphs recovered from 13 mudstone–marly samples collected from the Ermua section and two samples from the Trabakua Pass section (Text-Figure 2). The samples from the Ermua section (samples Ep 1 to Ep 13) contain fairly well-preserved, abundant, and rich palynomorph assemblages. Furthermore, a total of seventeen samples from the Trabakua Pass section were analysed but in contrast with the Ermua samples, only two samples (T3 and T5) yielded some poorly preserved and scarce palynomorphs. In the other samples dinocysts were so rare that were not useful for any reliable zonation.

The total palynomorph assemblage includes 47 dinocyst, three acritarch, and 21 spore/pollen taxa. Dinocysts belonging to the *Apectodinium-Wetzeliella* complex are consistently present and relatively well preserved. Marine palynomorphs average between 85 and 90 % of the total palynomorph assemblage and are consistently better pre-

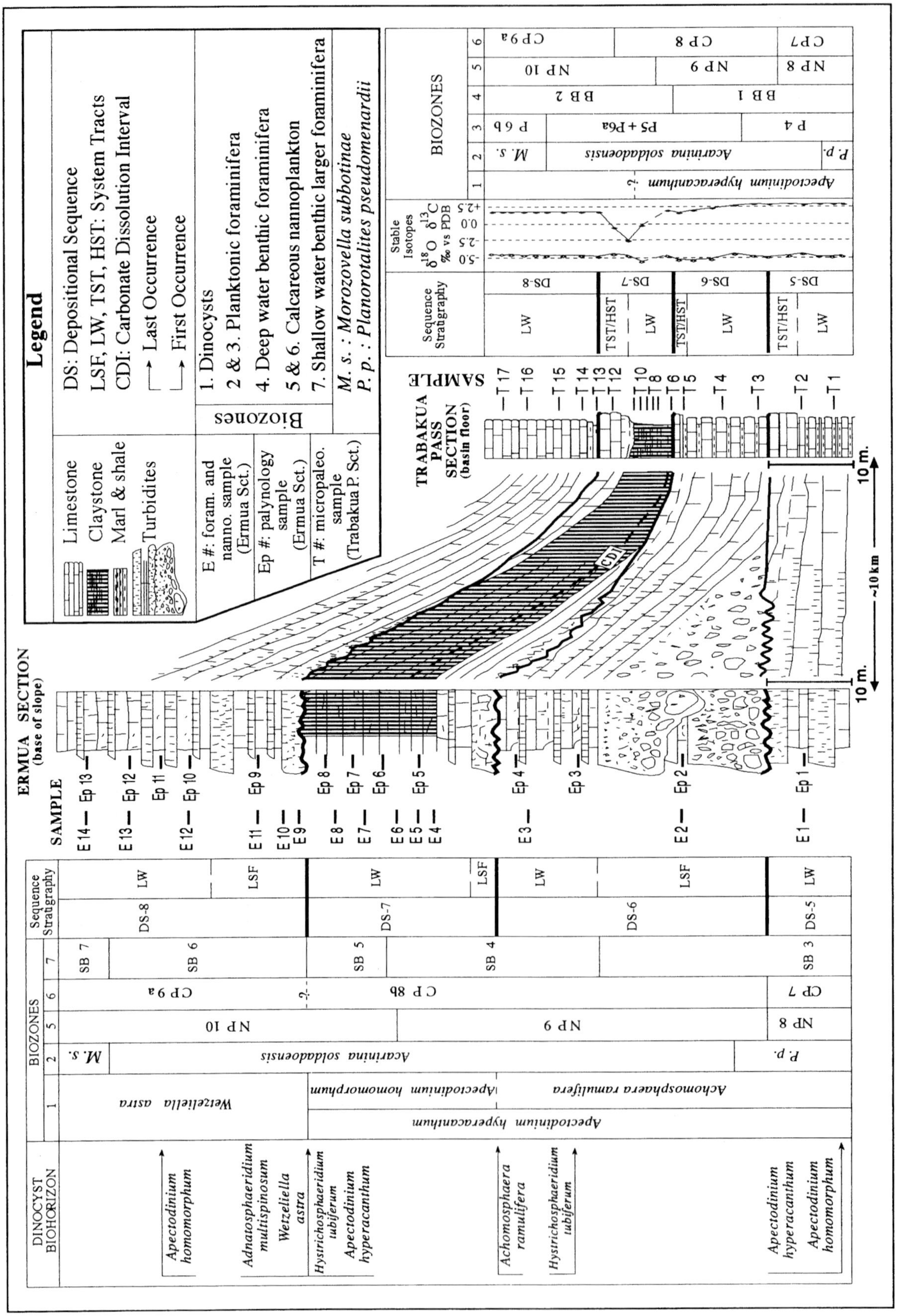

Text-Figure 2. Summary of sequence stratigraphy, stable isotopes and biostratigraphy (dinocysts, planktonic foraminifera, deep water benthic foraminifera, calcareous nannoplankton and shallow water benthic larger foraminifera) across the Paleocene/Eocene boundary at the Trabakua Pass and Ermua sections of the Basque Basin. For explanation on biozonation schemes see main text.

served than terrestrial forms. This palynomorph assemblage is composed of cosmopolitan taxa of which many have been previously reported from Europe (Caro, 1973; Costa et al., 1978; Manum et al., 1989; Powell, 1988) and elsewhere. Low diversity terrestrial palynomorph assemblages are consistently present in the Ermua samples but are absent from the Trabakua Pass section (see Orue-Etxebarria et al., 1996).

In the Ermua section, diversity is high with respect to the marine component. The 13 samples from this section average almost 90 % marine dinocysts. The other 10 % consist of thin-walled, poorly preserved terrestrial spores and pollen. Based on the dinocysts the following biozones are proposed:

***Apectodinium hyperacanthum* Zone.** This zone is recognized following that of Caro (1973). This stratigraphic interval is defined by the first (FO) and last occurrences (LO) of *Apectodinium hyperacanthum*. The zone includes genera such as *Achomosphaera*, *Apectodinium*, *Cometodinium*, *Diphyes*, *Hdespite*Horologinella*, and *Spongodinium*. Two abundance subzones are recognised within this zone:

***Achomosphaera ramulifera* Subzone.** The FO and LO of *Achomosphaera ramulifera ramulifera* define the bottom and top of this subzone, respectively. The genus *Achomosphaera* is remarkably abundant and well preserved throughout this entire abundance subzone of the lower Ermua section. Other common forms present in this subzone include *Achomosphaera ramulifera perforata*, *Achomosphaera improcera*, and *Achomosphaera* sp. Taxa such as *Pyxidinopsis* sp. and *Thalassiphora pelagica* are also noteworthy in this subzone.

***Apectodinium homomorphum* Subzone.** The base of this interval subzone is represented by the LO of *Achomosphaera ramulifera* and the top is marked by the LO of *Apectodinium hyperacanthum*. The genus *Apectodinium* becomes relatively abundant whereas *Achomosphaera* disappears marking the beginning of this subzone. However, *Apectodinium homomorphum* is more abundant than *A. hyperacanthum* and is useful to recognize this subzone. Other forms, such as *Horologinella incurvata* and *Spongodinium reticulatum* are also relatively common in this subzone.

***Wetzeliella astra* Zone.** This zone is herein defined following that of Chateauneuf and Gruas-Cavagnetto (1978). The FO of *Wetzeliella astra* defines the base of this zone. The top of the zone corresponds to the top of the section. *Apectodinium hyperacanthum* is absent and *Apectodinium homomorphum* is rare within this zone. Common taxa include *Adnatosphaeridium multispinosum*, *Cassiculosphaeridia* sp., *Diphyes colligerum*, and *Xenicodinium* sp. Poorly preserved specimens of *Wetzeliella* cf. *articulata* are also present. The palynomorph assemblage is less well preserved than in underlying intervals.

The palynomorph assemblage recovered from the two Trabakua samples (one from the lower part of the section and one from the lower part of the clayey interval) is strictly marine and dominated by *Apectodinium homomorphum*. Other species recovered include *Apectodinium hyperacanthum* and *Spinidinium* cf. *essoi*. Thus, the base of the section to the top of the clayey interval represents the *Apectodinium hyperacanthum* Zone, although neither subzone, 1a or 1b, can reliably be recognized. The samples from the top of the clayey interval to the top of the section are very poor in dinocysts which makes it difficult to correlate with the *Wetzeliella astra* zone of the Ermua section.

Planktonic Foraminifera

The analysis of planktonic foraminifera is based on 31 samples of unweathered marls and marlstones. Of these, 17 are from the Trabakua Pass section (samples T 1 to T17) and 14 are from the Ermua section (samples E 1 to E 14). Samples from the Trabakua Pass section are generally very rich in planktonic foraminifera excluding samples from the clayey interval; two samples (samples T 7 and T 8) from this interval contain only agglutinant foraminifera whereas the other samples, also from this interval, contain mostly dwarfed species. The proportion of planktonic foraminifera versus total foraminifera is generally equal to or greater than 95%. In contrast, most of the samples from the Ermua section are poor to very poor in planktonic foraminifera, especially those from the clayey interval. The preservation of the microfauna also varies greatly. However, the proportion of planktonic versus total foraminifera remains high, as happens in the Trabakua Pass section. The data acquired from the two sections make it possible to differentiate three interval biozones as documented below:

***Planorotalites pseudomenardii* Interval Biozone.** Its lower limit is placed at the FO of *P. pseudomenardii*, and its upper limit at the FO of *A. soldadoensis* which coincides precisely with zone P4 of Blow (1979), and approximately with the biozone P4 of Berggren and Miller (1988) (Text-Figure 2). The most frequent association found within this interval biozone consists of the following taxa: *Eoglobigerina linaperta/triangularis*, *E. velascoensis*, *Morozovella occlusa*, *M. acuta* and *Acarinina nitida*.

***Acarinina soldadoensis* Interval Biozone.** It ranges from the FO of *A. soldadoensis* to the LO of *Morozovella occlusa*, which in the Basque Basin includes species of the *velascoensis* group. The biozone corresponds with biozones P5, P6, the lower part of P7 of Blow (1979), and with the highest part of P4, P5, and P6a of Berggren and Miller (1988) (Text-Figure 2).

The preference of the *A. soldadoensis* biozone is justified mainly by the data obtained in this study, but also finds support elsewhere. In the Trabakua Pass and Ermua sections, the most frequent association within the *A. soldadoensis* biozone is *E. linaperta/ triangularis*, *E. velascoensis*, *M. aequa*, *M. subbotinae* and *M. occlusa*. Although not very abundant, *M. edgari* is found within the lower part of the biozone, *A. wilcoxensis* occurs in the upper part with its FO placed just below that of *A. wilcoxensis berggreni*. It should be mentioned that the latter subspecies was used by Blow (1979) to define the base of his P7 biozone.

***Morozovella subbotinae* Interval Biozone.** The lower limit is placed at the LO of *Morozovella occlusa*, which corresponds to the base of the P6b biozone of Berggren and Miller (1988). The most abundant species are *E. linaperta/triangularis*, *E. velascoensis*, *E. pseudoeocaena*, *M. subbotinae*, *M. marginodentata*, and *A. wilcoxensis*.

Deep Water Benthic Foraminifera

Seventeen samples (samples T 1 to T 17) from the Trabakua Pass section were analysed. The benthic foraminifera occur in low percentages (commonly ≤ 5%) throughout the section. All samples contain a mixed assemblage of agglutinated and calcareous-hyaline forms, except two samples which yield only scattered agglutinants. A total of one hundred and fifty-nine species have been identified, representing seventy-four genera of 40 families among which Gavelinellidae, Parrellodidae, Osangulariidae, Epistomariidae, Heterolepidae, Spiroplectamminidae, Ammosphaeroidinidae, Eggerellidae, and Rhabdamminidae dominate.

All the faunal parameters vary throughout the section with the widest fluctuations occurring across the CDI. Within this interval, the faunal density; the specific diversity; the TAU value (Gibson, 1988); and the infaunal to infaunal and epifaunal forms ratio, I/(I+E) ratio, exhibit the lowest values. Conversely, the ratio agglutinated to agglutinated and calcareous-hyaline ratio, A/(A+H) ratio, peaks within this interval. The faunal density ranges from < 1 to 106 specimens per gram peaking in the upper part of the clayey interval. The specific diversity, which fluctuates more than other parameters, reaches its highest value below the CDI (values range from 52 to 91 peaking just below the clayey interval). On the other hand, the A/(A+H) and I/(I+E) ratios do not change significantly above and below the CDI. As expected the TAU values show the same trend as the specific diversity. The record of the Dissolved Oxygen Index (DOI) expressed as the aerobic to aerobic and dysaerobic forms ratio percent, indicates that throughout the section the benthic foraminiferal assemblages are dominated by aerobic forms, which exhibit the greatest fluctuations above the CDI. Within this interval, the DOI drops to 0% because of the lack of calcareous-hyaline forms.

Based on the distribution of the deep-water benthic foraminifers two biozones of Berggren and Miller (1989), namely BB 1 and BB 2 have been recognised (Text-Figure 2). The boundary between these two zones is located at the base of the clayey interval. Assemblages are mainly characterized by *Anomalinoides welleri, Cibicidoides praemundulus, C.* cf. *pseudoperlucidus, Gavelinella beccariiformis* (which dominates below the CDI, reaching 14.5% of the total fauna), *Karrerulina conversa, Nuttallides truempyi, Oridorsalis umbonatus, Recurvoides walteri, Spiroplectammina lanceolata,* and with a lesser extent *Bulimina bradburyi/tuxpamensis,* and *Rhabdammina* ex gr. *discreta*. Most of these taxa are known as epifaunal species (see Tyszka, 1994). A remarkable crisis of the Textulariacea occurs at, and just below the CDI where all species belonging to this superfamily disappear, and only two species reappear.

In the Trabakua Pass section, a major extinction of deep-water benthic foraminifera occurs at the base of the clayey interval. However, the extinction rate is lower (3–6%) than that documented for the end of the Paleocene at other localities. This major faunal change occurs in the middle part of the *A. soldadoensis* Zone where several taxa locally or temporarily disappear. During that time, stable well-oxygenated conditions were suddenly interrupted in correspondence to a dissolution event which might be linked to the volcanic activity accompanying the early Tertiary opening of the North Atlantic. After the dissolution event the deep-water benthic foraminiferal assemblages demonstrate a rapid pattern of recovery. Small-sized specimens, and infaunal forms thriving under high nutrient level conditions are shown to be favoured in the period of recovery. However, not all the infaunal assemblage shows opportunistic behaviour since all of the genera belonging to the superfamily Textulariacea were relatively unaffected by the dissolution event. Conversely, some taxa characterized by a closely similar morphology and the same inferred trophic strategy of a large group of Textulariacea species show a gradual pattern of recovery. Such a trend points to the necessity of a better understanding of the different trophic strategies of deep-water agglutinated benthic foraminifera within the same morphologic group.

Shallow Water Benthic Larger Foraminifera

The sediments of the Ermua section, while located in relatively deep-waters contain an abundance of larger foraminifera. Larger foraminifera abound, as individual grains, on the bioclastic turbidites of the LW's. They are called "displaced" because most of them were probably derived through the action of storms or similar processes, from contemporaneous or slightly older unconsolidated shallow water sediments.

Fourteen samples (E 1 to E 14) from the Ermua section were studied. An analysis of the samples has resulted in the following biozonation scheme (Text-Figure 2):

***Glomalveolina primaeva* or SB 3 Biozone.** The resedimented larger foraminifera included in the breccia clasts of the first two samples (E 1 and E 2) clearly belong to the middle Thanetian (*sensu* Hottinger, 1960) as they contain the type species of *G. primaeva* Biozone (biozonations of Hottinger, 1960 and Hottinger and Drobne, 1988) or SB 3 (Shallow Benthic Foraminifera Zonations) of Serra-Kiel *et al.* (in press). Other taxa present in this zone include *Ranikothalia sindensis* and *Nummulites heberti, Miscellanea juliettae, Cuvillierina sireli* and *Discocyclina seunesi.*

***Glomalveolina levis* or SB 4 Biozone.** Samples E 3 to E 6 show a mixed association of displaced foraminifera, although based on the presence of *Glomalveolina levis* and *Nummulites* cf. *fraasi*, they clearly belong to SB 4. Due to the mixed character of the assemblages, the boundaries of this zone can not be presently defined.

***Alveolina cucumiformis* or SB 5 Biozone.** The next two samples, E 7 and E 8 come from the upper part of the clayey interval and contain an unmixed derived association, which includes *Nummulites* cf. *gamardensis, Assilina* cf. *ornata, Assilina* cf. *ammonea tectosaga* and *Assilina* cf. *dandotica*. These species are characteristic of the early Ilerdian biozone.

***Alveolina ellipsoidalis* or SB 6 Biozone.** Samples E 9 to E 13 contain mostly derived faunas such as *Alveolina globula, Nummulites* cf. *minervensis, A.* cf. *ornata, A.* cf. *ammonea tectosaga,* and *A.* cf. *arenensis.*

***Nummulites robustiformis* or SB 7 Biozone.** In the uppermost sample, E 14, foraminifera such as *Nummulites* cf. *minervensis, Nummulites* cf. *couiensis, Assilina* cf. *subgranulosa, Alveolina aragonensis* and *Alveolina* cf. *decipiens* indicate that this sample is early middle Ilerdian in age and corresponds to the SB 7 zone.

Calcareous Nannofossils

The study of the calcareous nannofossil assemblages is based on the same samples analyzed for the benthic and planktonic foraminiferal content. The abundant, moderate to poorly preserved calcareous nannofossils recognised in the Paleocene/Eocene transition from Trabakua Pass and Ermua sections have permitted us an identification of a succession of several well-differentiated events which define the interval NP8-NP10 (Text-Figure 2). The calcareous nannofossil assemblage recognised at Trabakua Pass and Ermua have permitted the recognition of the following three zones of Martini (1971) (Text-Figure 2):

***Heliolithus riedelii* NP8 Zone.** Interval characterized by the FO of *H. riedelii* and the FO of *D. multiradiatus*. This is equivalent to CP7 biozone of Okada and Bukry (1980).

***Discoaster multiradiatus* NP9 Zone.** Interval characterized by the FO of *D. multiradiatus* and the FO of *Tribrachiatus bramlettei*. Several bioevents have been observed within this zone: 1) the LO of *Prinsius bisulcus* and *Heliolithus kleinpellii*, at the base of the Trabakua Pass and Ermua sections; 2) the FO of *Campilosphaera eodela*, which allows to recognise in Ermua the Subzone CP8b of Okada and Bukry (1980) but not in Trabakua Pass; and 3) the FO of the genus *Rhomboaster* and *Tribrachiatus spineus*.

***Tribrachiatus contortus* NP10 Zone.** Interval characterized by the FO of *Tribrachiatus nunnii* and the LO of *T. contortus*. This is in part equivalent to the upper part of the CP8 biozone (Okada and Bukry, 1980) and to the lower part of the CP9a biozone. At the Ermua section, the boundary between the CP8b and CP9a biozones is not well defined and tentatively placed at the sequence boundary since this is the case at the Trabakua Pass section.

DISCUSSION

The correlation of the Trabakua Pass and Ermua sections following a sequence stratigraphic approach combined with biostratigraphy offers some advantages over classical methodologies, which use either only sequence stratigraphic criteria or only single biostratigraphic criteria. Furthermore, one of the most important contributions of this work is the potential of integrating multiple fossil groups. For example, shallow benthic larger foraminifera and dinocysts are abundant only in the Ermua section, while calcareous nannofossils, deep-benthic foraminifera and planktonic foraminifera are better preserved in the Trabakua Pass section. Also, the sequence stratigraphic correlation makes clear the existence of minor hiatuses in the section: the LSF is either thin or absent in the Trabakua Pass section, while the record of the TST/HST is missing at Ermua (Text-Figure 2). Note, however, that the extent of both hiatuses is below resolution of biostratigraphy.

Although, it is becoming increasingly clear that the utility of sequence stratigraphy for global correlations should be tackled with caution (Miall, 1992), a recent study (Pujalte et al., 1998) carried out in the Basque Basin has confirmed the synchroneity between the cycles of relative sea-level change that created DS-6 and DS-7 and those shown in the global sea-level chart of Haq et al. (1988).

The correlation of the biostratigraphic scales of the different fossils groups considered in this study (Text-Figure 2) highlights the difficulty to agree on the position of the Paleocene–Eocene boundary. Thus, the data gathered at the Trabakua Pass and Ermua sections demonstrate that these biological events occur at different datum horizons, a conclusion also reached by many previous authors. Most previous attempts to define this boundary have been based on biostratigraphy and the following proposals, among others, having been put forward:

1. top of *Globorotalia velascoensis* biozone (Stainforth et al., 1975); in the studied succession this bioevent coincides with the top of the *A. soldadoensis* biozone (Orue-Etxebarria et al. 1996);
2. base of BB2 deep benthic foraminifera biozone (Berggren and Miller, 1988)
3. base of NP 10 biozone (Martini, 1971);
4. base of CP 9 biozone (Okada and Bukry, 1980);
5. base of *Wetzeliella astra* biozone (Chateauneuf and Gruas-Cavagnetto, 1978)
6. base of the Ilerdian stage (Pomerol, 1975) that is currently placed at the base of the *A. cucumiformis* biozone SB 5 (Serra-Kiel et al., in press).

Furthermore, two dinocyst subzones, the *Achomosphaera ramulifera* Subzone and the *Apectodinium homomorphum* Subzone have been proposed herein. In fact they are based on population acmes for these two species. Recently, Bujak and Brinkhuis (1998) have recognised a world-wide *Apectodinium* Acme characterised by low-diversity dinoflagellate cyst populations dominated by *Apectodinium* . Since this is what happens at the Ermua section it is possible that the *Apectodinium homomorphum* Subzone of this study in fact corresponds to such acme.

CONCLUSIONS

There exits a growing consensus among researchers of the IGCP 308 project that the deep-water benthic extinction and $\delta^{13}C$ isotopic shift events should be used to define the Paleocene/Eocene boundary GSSP. In the Basque Basin these two events occur within the clayey interval, although at slightly different levels: the deep water extinction in the lower part of the interval (e.g., at the CDI) and the isotopic shift in the upper part. Conversely, biozones based on dinocysts, planktonic foraminifers, benthic foraminifers, and nannoplankton occur at different datum horizons but provide enough information on stratigraphic distribution of taxa. In fact, whatever happened at the end of the Paleocene, it influenced very differently the groups included in this study.

ACKNOWLEDGMENTS

The authors are very grateful to Dr. L. Edwards, Dr. D. K. Goodman, and Dr. R. S. Van Pelt for their constructive reviews. This paper is a joint contribution to International Geological Correlation Programme (IGCP) 286 and IGCP 308 projects. Funds for field and laboratory studies were provided by the Spanish Ministry of Science (projects: DGICYT PB 92-0457, DGICYT PB 94-0846, and DGICYT PB 95-0500), the University of the Basque Country (projects:

UPV 121-310-EB 233/93 and UPV 121.310-EB 191/98), and by the Italian Ministry of Science (M.U.R.S.T. to R. Coccioni and S. Monechi).

References Cited

BACETA, J.I.
1996 *El Maastrichtiense superior, Paleoceno e Ilerdiense inferior de la Región Vasco-Cantábrica: secuencias deposicionales, facies y evolución paleogeográfica.* Unpubl. Ph.D. thesis, University of the Basque Country, 372 pgs.

BERGGREN, W.A., and MILLER, K.G.
1988 Paleogene tropical planktonic foraminiferal biostratigraphy and magnetobiochronology. *Micropaleontology,* 34: 362–380.

BLOW, W.H.
1979 *The Cainozoic Globigerinida,* 3 Vols., Leiden: E. J. Brill, 1413 pgs.

BUJAK, J.P., and BRINKHUIS, H.
1998 Global warming and dinocyst changes across the Paleocene/Eocene Epoch boundary. *In:* Aubry, M.-P., Lucas, S., and Berggren, W.A. (eds.), *Late Paleocene–Early Eocene climatic and biotic events in the marine and terrestrial records.* Columbia University Press, New York, p. 277–295.

CHATEAUNEUF, J.J., and GRUAS-CAVAGNETTO, C.
1978 Les zones de Wetzeliellaceae (Dinophyceae) du Bassin de Paris. *Bulletin du Bureau de Recherches Géologiques et Minières,* série 4 (2): 59–93.

COSTA, L.I., and DOWNIE, C.
1976 The distribution of the dinoflagellate *Wetzeliella* in the Paleogene of North-Western Europe. *Paleontology,* 19(4): 591–614.

COSTA, L.I., DENISON, C., and DOWNIE, C.
1978 The Paleocene/Eocene boundary in the Anglo-Paris Basin. *Journal Geological Society London,* 135: 261–264.

GIBSON, T.G.
1988 Assemblage characteristics of modern benthic Foraminifera and application to environmental interpretation of Cenozoic deposits of eastern North America. *Revue de Paléobiologie,* 2: 777–787.

HAQ, B.U., HARDENBOL, J., and VAIL, P.R.
1988 Mesozoic and Cenozoic chronostratigraphy and cycles of sea-level change. *In:* Wilgus, C.K., Hastings, B.S., Kendall, C.G.St.C., Posamentier, H.W., Ross, C.A., and Van Wagoner, J.C. (eds.), *Sea-level change: An integrated approach.* S.E.P.M. Special Publication 42: 71–108.

HOTTINGER, L.
1960 Recherche sur les Alvéolines du Paléocène et de l'Eocène. *Mémoires suisses de Paléontologie,* 75, 76: 1–243.

HOTTINGER, L., and DROBNE, K.
1988 Alvéolines Tertiaires. Quelques problèmes liés a la conception de l'espèce. *Révue de Paléontologie,* Vol. spécial 2 (Benthos'86), p. 665–681.

KENNET, J.P., and STOTT, L.D.
1990 Proteus and Proto-Oceanus: Paleogene oceans as revealed from Antarctic stable isotopic results. *In:* Barker, P., Kennet, J.P., et al. (eds.). *Proceedings ODP, Scientific Results,* Vol. 113, p. 865–880.

MANUM, S.B., BOULTER, M.C., GUNNARSDOTTIR, H., RANGNES, K., and SHOLZE, A.
1989 Eocene to Miocene palynology of the Norwegian Sea (ODP Leg 104). *In:* Eldholm, O., Thiede, J., and Taylor, E. (eds.), *Proceedings ODP, Scientific Results,* Vol. 104, p. 611–662.

MARTINI, E.
1971 Standard Tertiary and Quaternary calcareous nannoplankton zonation. *In:* Farinacci, A. (ed.), *Proceedings 2nd Planktonic Conference Roma, Tecnoscienza,* p. 739–785.

MIALL, A.D.
1992 Exxon global cycle chart: An event for every occasion? *Geology,* 20: 787–790.

OKADA, H., and BUKRY, D.
1980 Supplementary modification and introduction of code numbers to the low-latitude coccolith biostratigraphic zonation (Bukry, 1973; 1975). *Marine Micropaleontolology,* 5: 321–325.

ORUE-ETXEBARRIA, X., APELLANIZ, E., BACETA, J.I., COCCIONI, R., DI LEO, R., DINARES-TURELL, J., GALEOTTI, S., MONECHI, S., NÚÑEZ-BETELU, K., PARES, J.M., PAYROS, A., PUJALTE, V., SAMSO, J.M., SERRA-KIEL, J., SCHMITZ, B., and TOSQUELLA, J.
1996 Physical and biostratigraphic analysis of two prospective Paleocene–Eocene Boundary Stratotypes in the intermediate-deep water Basque Basin, western Pyrenees: The Trabakua Pass and Ermua sections. *Neues Jahrbuch für Geologie und Paläeontologie,* 201: 179–242.

PAYROS, A.
1997 El Eoceno de la Cuenca de Pamplona: Estratigrafía, facies y evolución paleogeográfica. Unpublished Ph.D. Thesis, University of the Basque Country, 403 pgs.

POMEROL, C.
1975 La significaton de l' Ilerdien et l' intéret de cet étage dans la stratigraphie du Paléogène mésogéen. *Bulletin Societé Geologique de France,* 17: 213–217.

POWELL, A.J.
1988 A modified dinoflagellate biozonation for the latest Palaeocene and earliest Eocene sediments from the central North Sea. *Review of Palaeobotany and Palynology,* 56: 327–344.

PUJALTE, V., BACETA, J.I., ORUE-ETXEBARRIA, X., and PAYROS, A.
1998 Paleocene strata of the Basque Country, Western Pyrenees, Northern Spain: Facies and sequence development in a deep-water starved basin. *In: Mesozoic and Cenozoic Sequence Stratigraphy of European Basins.* SEPM Special Publication No. 60, p. 311–325.

PUJALTE, V., ROBLES, S., ROBADOR, A., BACETA, J.I., and ORUE-ETXEBARRIA, X.
1993 Shelf-to-basin Paleocene Palaeogeography and depositional sequences, Western Pyrenees, North Spain. *In:* Posamentier, H.W., Summerhayes, C.P., Haq, B.U., and Allen, G.P. (eds.), *Sequence Stratigraphy and Facies Association.* Special Publication International Association of Sedimentologists, 18, p. 369–395.

SERRA-KIEL, J., HOTTINGER, L., DROBNE, K., FERRANDEZ, C., LESS, G., JAUHRI, A.K., SAMSO, J.M., SCHAUB, H., SIREL, E., TAMBAREAU, Y., TOSQUELLA, J., and ZAKREVSKAYA, E.
in press Benthic foraminifera from Paleocene and Eocene. *In: Mesozoic-Cenozoic sequence stratigraphy of European basins,* Special collection, SEPM.

STAINFORTH, R.M., LAMB, J.L., LUTERBACHER, H., BEARD, J.H., and JEFFORDS, R.M.
1975 Cenozoic planktonic foraminifera zonation and characteristics of index forms. *University of Kansas Paleontological Contributions,* Article 62, 245 pgs.

TYSZKA, J.
1994 Response of Middle Jurassic benthic foraminiferal morphogroups to dysoxic/anoxic conditions in the Pieniny Klippen Basin, Polish Carpathians. *Palaeogeography, Palaeoclimatology, Palaeoecology,* 110: 55–81.

Authors' addresses

KOLDO NÚÑEZ-BETELU
XABIER ORUE-ETXEBARRIA
ESTIBALIZ APELLANIZ
AITOR PAYROS
VICTORIANO PUJALTE
Estratigrafia eta Paleontologia Saila, Zientzi Fakultatea
Euskal Herriko Unibertsitatea
P.O. Box 644,
48080 Bilbao
Basque Country

JUAN IGNACIO BACETA
Depto. Geología, Fac. de Ciencias Experimentales
Universidad de Huelva
La Rábida, 21819 Huelva
Spain

RODOLFO COCCIONI
RITA DI LEO
Istituto di Geologia dell'Università
61029 Urbino
Italy

JAUME DINARES-TURELL
JOSEP MARIA PARES
Instituto de Ciencias de la Tierra "Jaime Almera"
Marti i Franques s/n
08028 Barcelona, Catalonia
Spain

SIMONE GALEOTTI
Istituto di Geologia dell'Università
61029 Urbino
Italy

SIMONETTA MONECHI
Dipt. di Scienze della Terra
Università di Firenze
Via La Pira
50121 Firenze
Italy

JOSEP MARIA SAMSO
JOSEP SERRA-KIEL
JOSEP TOSQUELLA
Dept. Geología Dinámica, Geofísica i Paleontologia
Universitat de Barcelona
Area de Pedralbes
08028 Barcelona, Catalonia
Spain

BIRGER SCHMITZ
Dept. of Marine Geology
University of Göteborg
P.O. Box 7064
40232 Göteborg
Sweden

PALYNOLOGY AND SEQUENCE STRATIGRAPHY OF EARLY MIOCENE–LATE OLIGOCENE DELTAICS IN THE SIHAPAS GROUP, CENTRAL SUMATRA, INDONESIA

Keith Richards

Abstract

This paper presents the results of palynological analyses of 55 samples from four wells drilled in the eastern region of the Central Sumatra Basin, Indonesia. The samples include both ditch cuttings and sidewall cores and are all from the Sihapas Group, which is made up of fluvio-deltaic sediments of Early Miocene–Late Oligocene age. The Sihapas Group is interpreted as an ancient delta system which ceased activity in the Middle Miocene when the marine connection to the Indian Ocean was closed off by mountain uplift.

Most of the samples studied yielded moderate or good recovery of palynomorphs including mixed associations of marine microplankton and pollen and spores from mangrove, backmangrove, freshwater swamp and other terrestrial sources. Samples containing common mangrove pollen also tend to have low to moderate numbers of marine forms present. The same samples often contain frequent pollen from freshwater swamp (probably peatswamp) forests. These probably represent periods of predominantly high relative sea levels and humid climate, which allowed both mangrove and freshwater swamp vegetation to thrive.

Many of the samples which contain low numbers of mangrove pollen have an increased representation of fern spores and pollen from backmangrove vegetation. It is suggested that these represent opportunistic plant communities and may be an indication of lowering relative sea levels.

A synthesis of all stratigraphic data from the study wells shows that the sediments of the Sihapas delta system accumulated predominantly in a coastal plain or marginal marine setting and periodically in a moderately deep marine shelfal setting. At least five major depositional units can be identified within the Sihapas Group which are broadly referred to as "sequences." Each of these probably includes more than one third order cycle of deposition.

INTRODUCTION

Central Sumatra is one of the world's most prolific oil producing regions and includes the giant Duri and Minas fields (Text-Figure 1). The most significant oil production is from Early Miocene–Late Oligocene fluvio-deltaic sediments of the Sihapas Group which extend throughout Central and Southern Sumatra and eastwards into the Malacca Straits.

The Central Sumatra Basin is bounded to the south west by a major fault system related to the uplift of the Barisan Mountains, to the north east by the Asahan Arch and to the south east by the Tigapuluh High (De Coster, 1974). The basin contains sediments of Tertiary age which unconformably overlie pre–Tertiary metamorphic and igneous basement. The oldest sediments are of Paleogene age and consist of non-marine sandstones and claystones/shales of the Pematang Formation. The Sihapas Group (Text-Figure 2) is predominantly of Early Miocene age and is made up of marine, marginal marine and non-marine sandstones and mudstones with localised limestones and volcanics (Mertosono and Nayoan, 1974). Subdivisions of the Sihapas Group can be made based on generalized lithology into the Menggala Formation, Bangko Formation, Bekasap Formation and Duri Formation (Wongsosantiko, 1976). More recent information (e.g. Putrohari, 1992) suggests that the Menggala Formation, at least in the Malacca Straits area, may be of Late Oligocene age.

Richards, Keith
2001 Palynology and sequence stratigraphy of Early Miocene–Late Oligocene deltaics in the Sihapas Group, central Sumatra, Indonesia; *In*: Goodman, D.K., and Clarke, R.T. (eds.), *Proceedings of the IX International Palynological Congress, Houston, Texas, U.S.A., 1996*; American Association of Stratigraphic Palynologists Foundation, p. 253–262.

 ISBN 0–931871–06–9

The onset of deposition within the Sihapas Group proper (i.e., post–Menggala) coincides with a marine transgression during the Early Miocene. The marine transgression probably came from the Indian Ocean (from the south west). There was apparently no significant marine connection to the north east as, at this time, the area was bounded by the ancestral Sunda landmass, which forms the Malay Peninsula at the present time (De Coster, 1974).

The sediments which make up the Sihapas Group are part of an ancient delta system. During the Early Miocene, the area was drained by a series of rivers which brought in significant quantities of coarse grained sediments, sourced mainly from the higher ground situated to the north, east and south east of the catchment. Deltaic sedimentation ceased during the Middle Miocene when the marine connection to the south west was closed off by the uplift of the Barisan Mountains. The modern-day Mahakam Delta of Kalimantan, Indonesia, has been suggested as a possible analog for deposition of at least part of the Sihapas Group (Finger and Drugg, 1992).

In this study, palynological data are presented from four wells located in the eastern region of the Central Sumatra Basin (Text-Figure1) and are used to interpret the paleovegetation of part of the Sihapas delta system.

Palynological analyses were carried out by P.T. Robertson Research Indonesia on 55 samples. These include a mixture of ditch cuttings and sidewall cores, taken at intervals of approximately 20–30 m and represent between 200 m and 500 m thickness of sediment. Where possible, counts of 100 or more pollen, spores, algae and marine microplankton were obtained. In some cases, counts of 100 could not be attained due to the sandy and/or oxidized nature of the sample lithology. Summaries of the data obtained are presented in Text-Figures 3, 4, 5 and 6 where the actual abundances of the most significant taxa are shown. Relative proportions of the principal palynomorph groups and the numbers of specimens re-

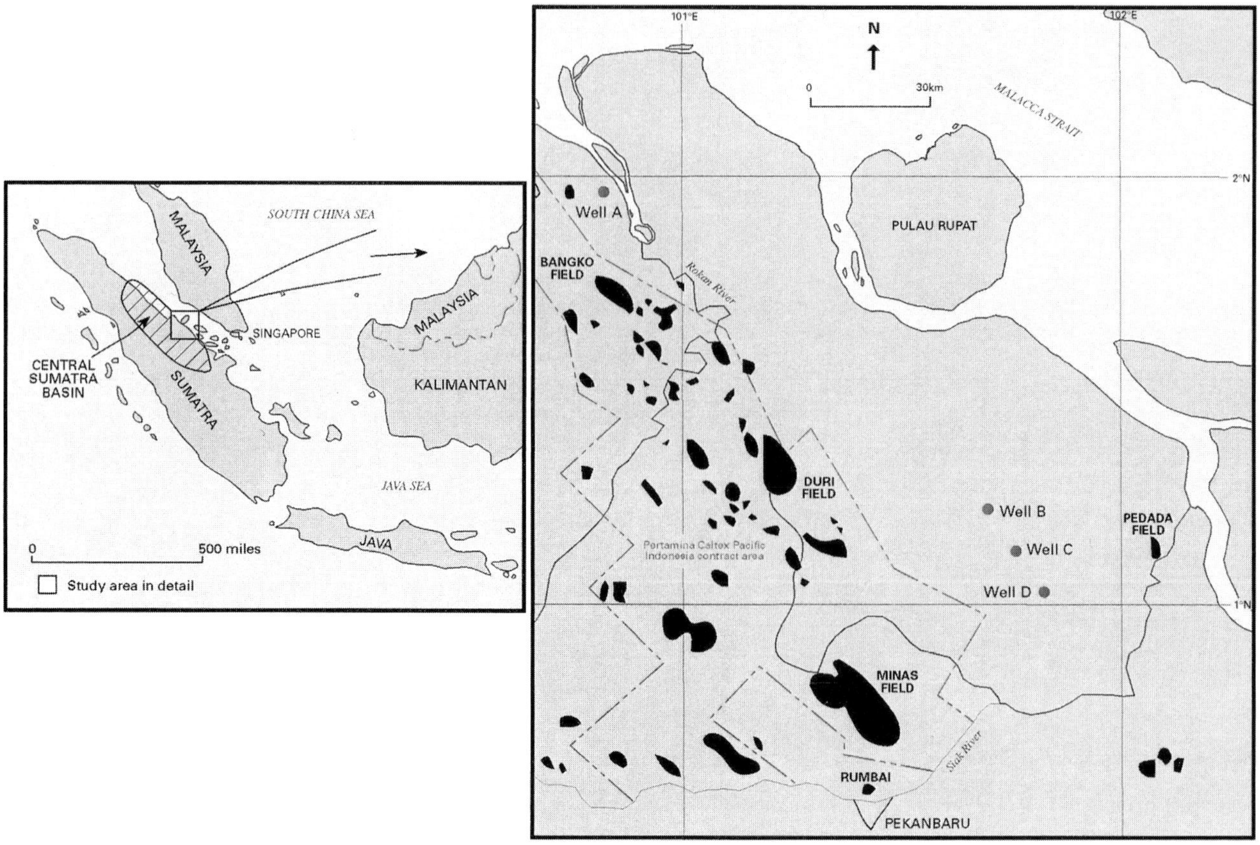

Text-Figure 1. Central Sumatra Basin, location map showing principal oilfields and study wells.

corded in each sample are also shown. Actual depths are not shown but each sample is assigned a number.

THE PALYNOLOGICAL RECORD

Marine microplankton are present in low numbers in most of the studied samples and include dinocysts of genera such as *Spiniferites, Lejeunecysta, Selenopemphix* and *Cribroperidinium*, as well as microforaminferal test linings and leiospheres. These are plotted together in the "Marine Taxa" group (MT). Localised peaks of marine taxa do occur (e.g. towards the base of Well D) which are presumed to represent marine transgressive events.

The "Mangrove Pollen" group (MA) is made up principally of *Zonocostites ramonae* (=*Rhizophora*) but also includes occasional specimens of *Florschuetzia levipoli* (=*Sonneratia caseolaris*) and, very rarely, *Avicennia* pollen. The *Zonocostites ramonae* curves show significant variations in all of the study wells. Increased proportions of this type may be associated with rising sea levels when the mangrove belt reaches it's maximum extent (Poumot, 1989; Morley, 1996).

The "Backmangrove Taxa" group (BM) includes those pollen and spore types most frequently associated with backmangrove and/or coastal settings. These include pollen types such as *Discoidites borneensis* (*Brownlowia*) and *Spinizonocolpites echinatus* (*Nypa*), as well as spores attributed to *Acrostichum aureum* and *A. speciosum* types. The paleoecological significance of this group is difficult to interpret as many modern day plants found in backmangrove habitats are opportunistic species. They are able to tolerate low to medium levels of salinity but may also thrive in non-saline habitats in the absence of competition. This may be significant in a delta plain environment where new areas for plant colonisation can be created, sometimes very rapidly, for example, as a result of river flooding or changes in river course. Equally, the extent of the backmangrove vegetation belt may be influenced by climate and relative sea level. Varying proportions of pollen and spores from the "Backmangrove Group" are present in most of the samples studied, indicating that a near-coastal vegetation belt was more or less continually present in the Sihapas delta system but that it's relative extent changed over time.

The "Freshwater Swamp" group (FS) includes those pollen types likely to be derived from freshwater swamp forest, such as from the Meliaceae and Sapotaceae families and genera such as *Calophyllum, Cephalomappa, Intsia, Calamus, Ilex* and *Alangium*. Pollen derived from freshwater swamp vegetation is present in virtually all of the samples studied. This includes pollen from peat swamp, alluvial swamp and riverine swamp forest and suggests that swamp forest was an important component of the vegetation during the Early Miocene. It is not possible to determine accurately the relative importance of the various types of freshwater swamp forest although several of the pollen types recorded are comparable with tree types found in peat swamp forest in Sumatra at the present time (Whitten et al., 1984).

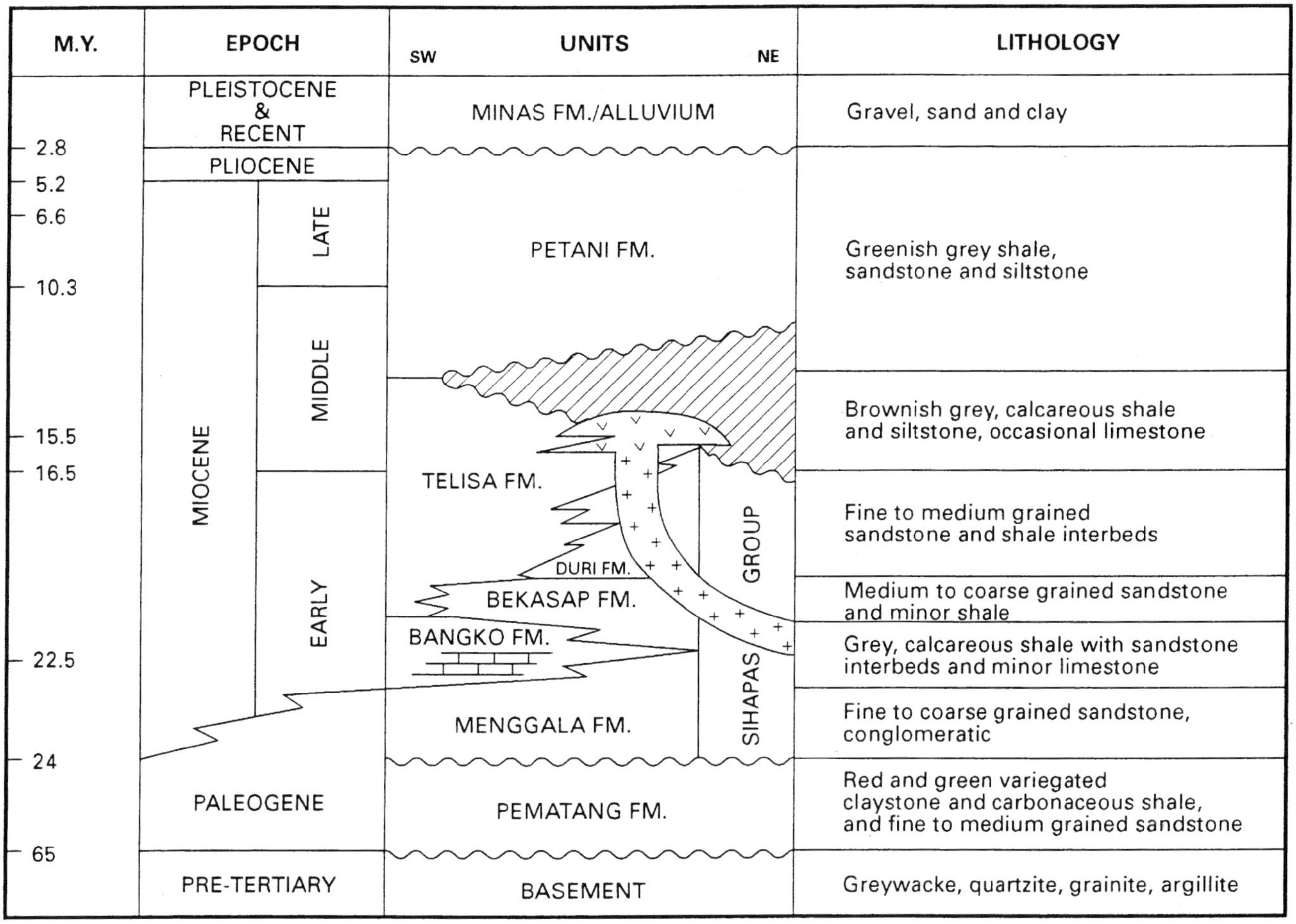

Text-Figure 2. Central Sumatra Basin, generalized lithostratigraphy (after various authors).

Pollen of uncertain ecological affinity is grouped together as "Other Pollen" (OP). This group is made up predominantly of undifferentiated tricolporate pollen types which are assumed to represent mostly arboreal pollen of freshwater origin.

The group "Fern Spores" (FE) includes all pteridophyte spores (e.g. *Verrucatosporites usmensis, Magnastriatites howardi* and *Lycopodium* spp.) but excluding *Acrostichum aureum* type and *Acrostichum speciosum* type. The "Fern Spores" group is well represented in the studied sections. Types such as *Laevigatosporites* spp. and *Deltoidospora* spp. are not restricted to any particular vegetation type but are prevalent in "open" forest habitats, and may be sourced from a coastal or near-coastal locality. The spore type *Verrucatosporites usmensis* (=*Stenochlaena palustris*) is locally common in the palynological record and is probably also largely derived from the coastal plain.

The "Freshwater Algae" group includes isolated occurrences of *Concentricystes circulus, Botryococcus* and *Pediastrum*. The rarity of algae in the studied sections, most significantly of *Pediastrum*, suggests that there were few lakes or ponds present in the Sihapas delta system.

From the data presented in Text-Figures 3–6, it is evident that the palynological records from the four wells are broadly similar in terms of species composition and that most of the taxa recorded are present (though in variable abundances) throughout the greater part of the sampled intervals. There are few stratigraphic tops (extinctions) or bases (evolutionary appearances) that can be reliably used for correlation. This is typical of Miocene palynological records from many tropical areas (Morley, 1990).

Well A is situated in the northern region of the study area. The palynological record (Text-Figure 3) is typified by generally high proportions of the mangrove pollen type *Zonocostites ramonae*. Peak abundances occur in samples 2, 6, 9 and 17. The interval represented by samples 1–6 is characterised by the presence of common freshwater swamp elements and generally low frequencies of backmangrove pollen and spores. The sample interval 7–9 is marked by a downhole increase in backmangrove elements, notably *Acrostichum aureum* type. Samples 10–14 demonstrate a further downhole increase in backmangrove forms, coinciding with the presence of common fern spores. Mangrove pollen is dominant or common in samples 15–18. Sample 19 yielded only few palynomorphs.

Wells B, C and D are situated to the south east of the study region. The palynological record from Well B (Text-Figure 4) shows little overall variation. Samples 1 and 2 are characterised by increased proportions of *Zonocostites ramonae*. Samples 3–8 contain lower levels of mangrove pollen but show increased representation of backmangrove taxa and fern spores. Samples 9–11 are poorly diagnostic.

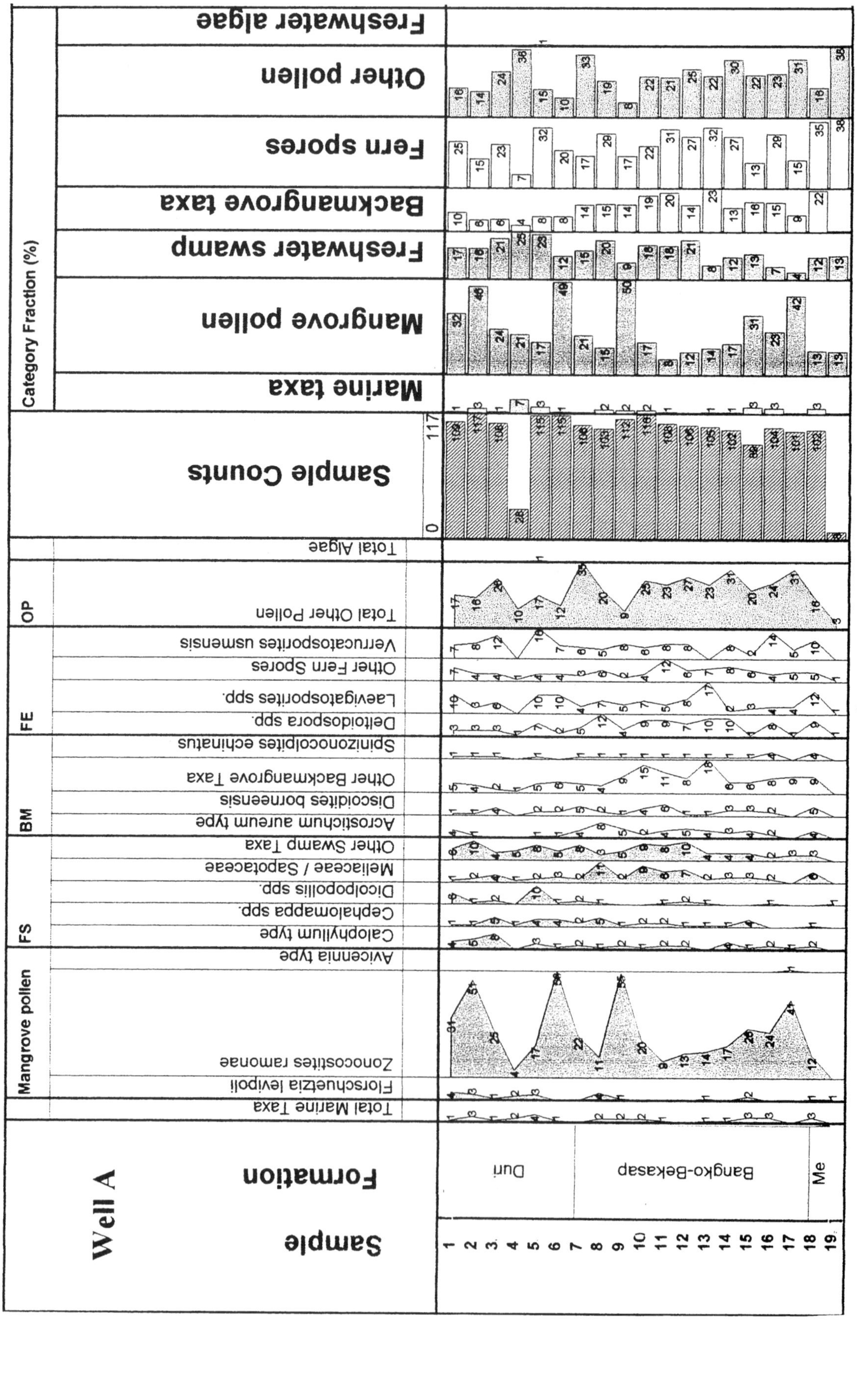

Text-Figure 3. Well A: summary of palynological data. Numbers shown for individual taxa are actual counts recorded. Numbers shown for palynomorph groups are expressed as percentages of the total palynomorph sum.

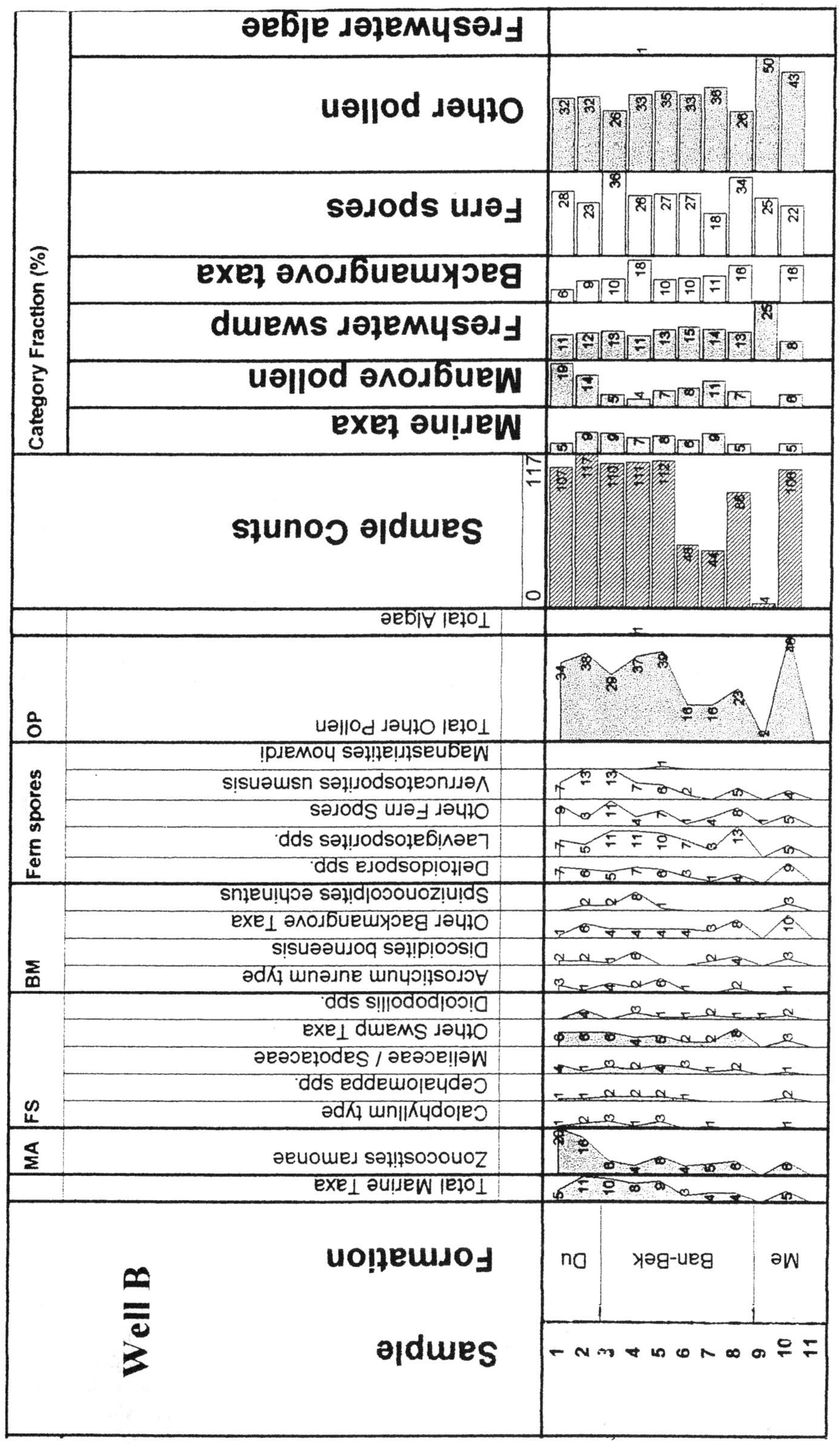

Text-Figure 4. Well B: summary of palynological data. Numbers shown for individual taxa are actual counts recorded. Numbers shown for palynomorph groups are expressed as percentages of the total palynomorph sum.

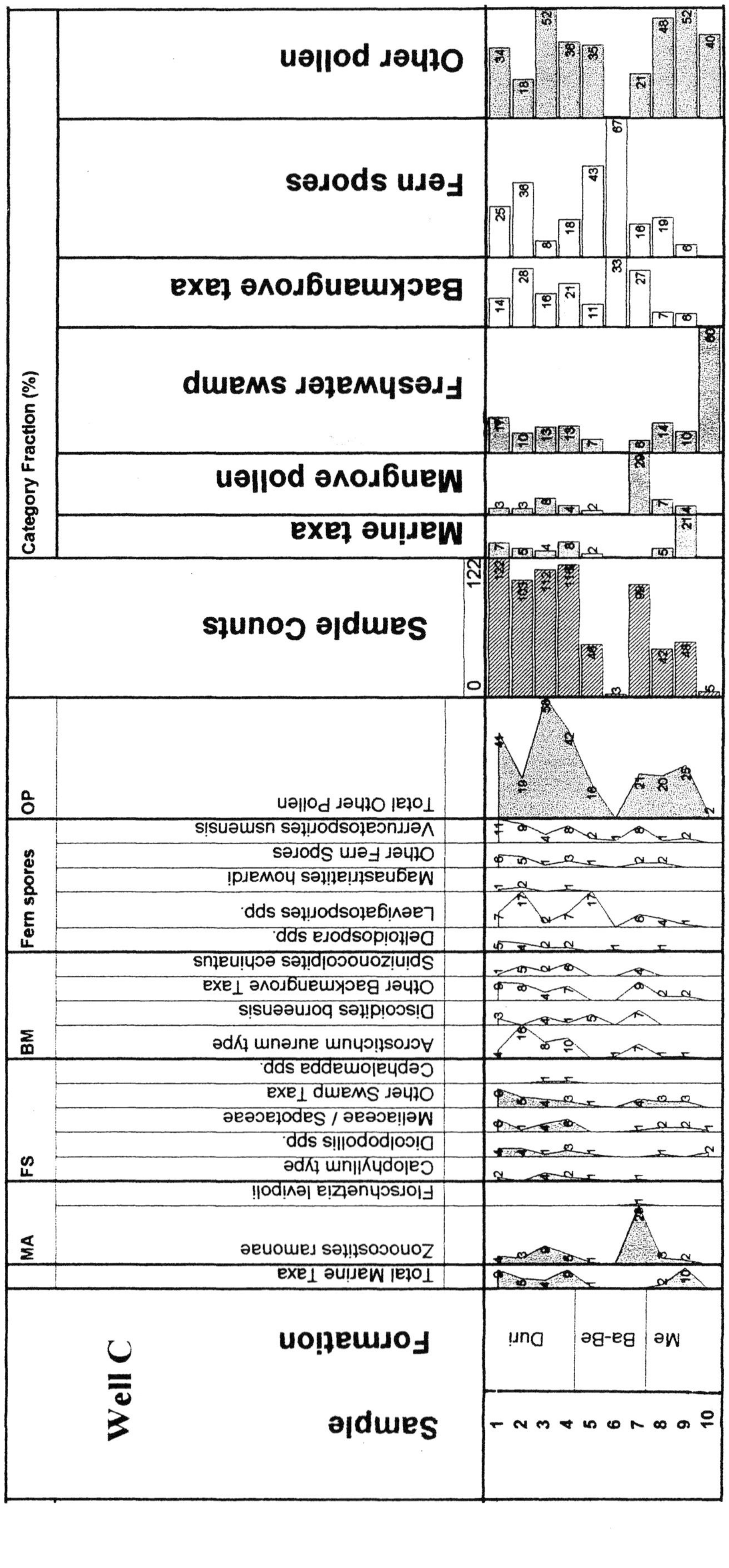

Text-Figure 5. Well C: summary of palynological data. Numbers shown for individual taxa are actual counts recorded. Numbers shown for palynomorph groups are expressed as percentages of the total palynomorph sum.

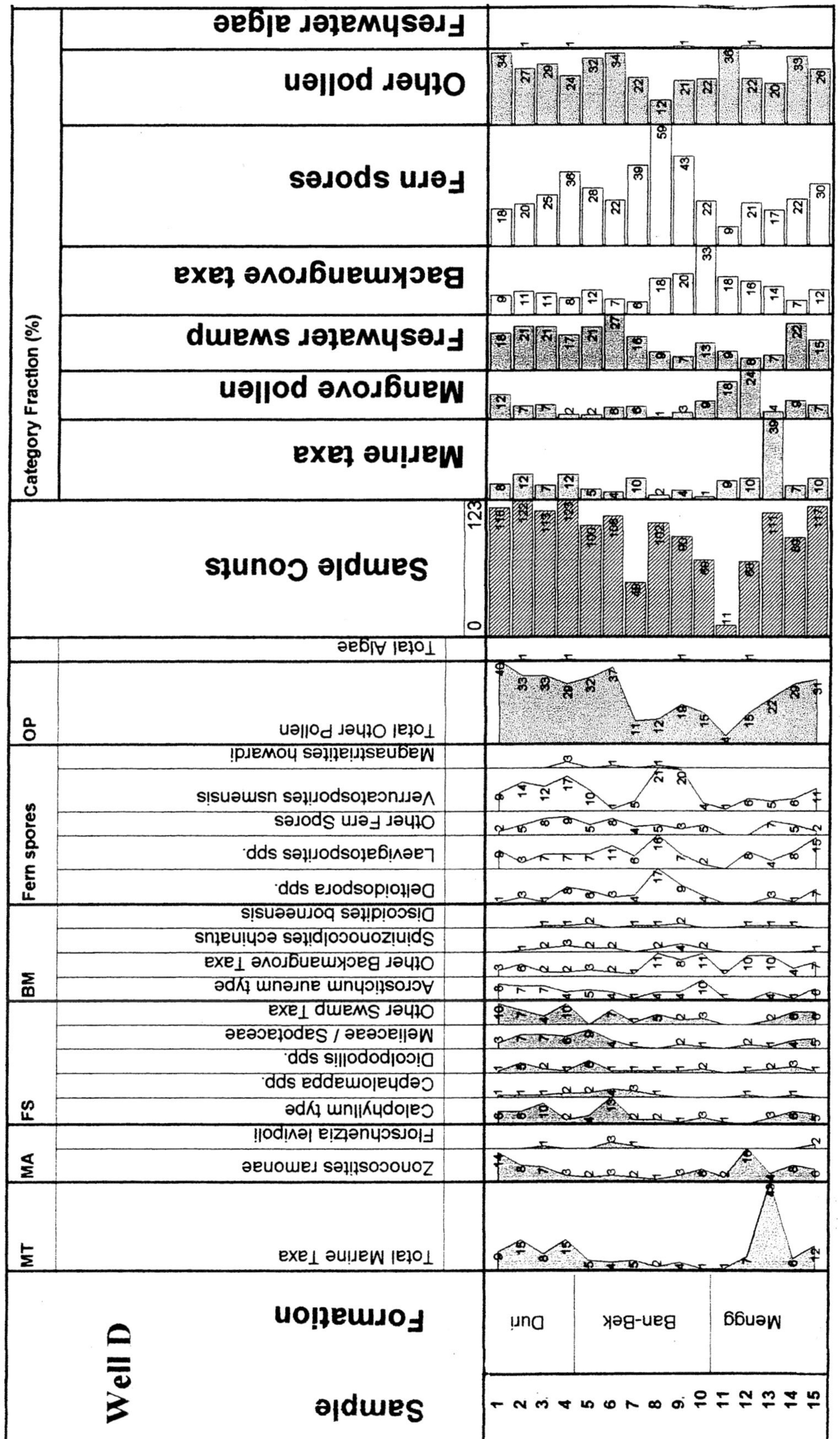

Text-Figure 6. Well D: summary of palynological data. Numbers shown for individual taxa are actual counts recorded. Numbers shown for palynomorph groups are expressed as percentages of the total palynomorph sum.

Data from the upper part of Well C (Text-Figure 5), represented by samples 1–4, include moderately common marine elements, low numbers of mangrove taxa and fairly common backmangrove forms. Recovery from the deeper samples is variable, although peaks of mangrove pollen and marine taxa occur in samples 7 and 9 respectively.

The palynological record from Well D (Text-Figure 6) shows several marked changes. Samples 1–4 are marked by the presence of relatively common marine microplankton and mangrove pollen. Freshwater swamp types are also well represented. Samples 5–10 contain common or abundant fern spores with a significantly lower representation of marine and mangrove indicators. Freshwater swamp pollen is also common but is reduced in abundance below sample 7. A downhole increase in backmangrove elements is evident at this level. The lower part of the well section, represented by samples 11–15, is marked by increased proportions of marine and mangrove elements and variable representation of the other palynomorph groups.

DISCUSSION AND CONCLUSIONS

The palynological associations from the four wells are dominated by pollen and spores derived from a variety of coastal plain, freshwater swamp and other non-marine localities. They are assumed to be broadly representative of the range of vegetation types present during the Early Miocene to Late Oligocene in the Sihapas catchment area. From the palynological data alone, it is suggested that deposition within the Sihapas Group in the studied wells occurred mainly in a coastal plain or nearshore marine setting. There is, however, some indication from foraminiferal and nannofossil associations in some of the studied sections (not presented here) that significant marine influence did occur from time to time, resulting in periodic deposition in an inner to middle neritic realm, with implied water depths of up to ca. 100 m.

From the data presented in Text-Figures 3–6, trends both of individual taxa and groups of taxa are visible. There are several instances (e.g. in Well D) where samples with the highest proportions of the mangrove type *Zonocostites ramonae* coincide, or occur close to, samples with relatively high proportions of marine microplankton. This is consistent with the idea that the mangrove belt may increase in extent during times of marine transgression and that increased representation of mangrove pollen may, in some instances, be a reliable indication of high relative sea levels.

Several of the sections in which common mangrove elements are recorded also contain relatively high proportions of freshwater swamp pollen, which suggests that a predominantly humid tropical climate prevailed in the study area, at least during the Early Miocene. It also suggests that conditions suitable for mangrove development (i.e. high relative sea levels and humid climate) are also suited, in some instances, to the development of freshwater swamp forest, possibly including peatswamps which are fed directly by rainfall (Whitmore, 1984; Whitten et al., 1984). The presence together of mangrove and freshwater swamp vegetation is consistent with the view (e.g. Morley, 1996) that a link exists between high relative sea levels and humid climates (i.e. high rainfall) in the low latitude tropics.

Another possibly significant trend in the palynological record from the study wells is the increased relative abundance of backmangrove pollen and spores in several intervals where proportions of marine, mangrove and sometimes freshwater swamp pollen are reduced. The increases noted in backmangrove elements often coincide with increased proportions of fern spores, suggesting that an association of backmangrove vegetation and ferns could be partly replacing the receding mangrove belt during times of lowering relative sea level.

The Duri, Bekasap-Bangko and Menggala formations, as determined by electric log information (not presented here), are shown on Text-Figures 3–6. In general, the principal changes in the palynological associations (notably the intervals including common marine, mangrove and freshwater swamp elements and those characterised by increased proportions of backmangrove taxa and fern spores) coincide approximately with the principal formation boundaries in these study wells.

Palynological associations from the Duri Formation are characterised by locally common marine, mangrove and freshwater swamp derived elements. These associations indicate that mangrove and freshwater swamp forests were present at this time. Mostly high relative sea levels and a predominantly humid tropical climate are suggested. This unit is best illustrated in Well A.

The Bekasap and Bangko formations were not separated for the purpose of this study. The most extensive record for this combined unit is also in Well A where there are at least two intervals in which mangrove pollen are dominant. The intervening section contains increased relative proportions of backmangrove taxa and fern spores. This is possibly a reflection of at least two transgressive/ regressive depositional cycles over this section.

The Menggala Formation is represented in all four of the study wells. It is often poorly fossiliferous as a result of it's predominantly sandy nature. Red beds are locally present which point to a periodically oxidizing regime. A degree of marine influence is confirmed by the presence of marine and mangrove taxa in Well D.

A synthesis of all available stratigraphic information for the study wells is presented in Text-Figure 7. Paleoenvironmental and age information include data from foraminifera and calcareous nannofossils. At least five major depositional units can be identified within the Sihapas Group which are broadly referred to as "sequences" in Text-Figure 7. Each of these probably includes more than one third order cycle of deposition.

ACKNOWLEDGMENTS

I would like to thank PERTAMINA for permission to use and publish data from the study wells and Robertson Research Utama Indonesia for provision of the data utilized in this study. I would also like to thank the directors of Robertson Research International Limited for permission to present and publish the results of this study and for assistance with drafting of several of the text figures.

References Cited

DE COSTER, G.L.
1974 The geology of the Central and South Sumatra Basins. *Proceedings of the Third Annual Convention Indonesian Petroleum Association*, pp. 77–109.

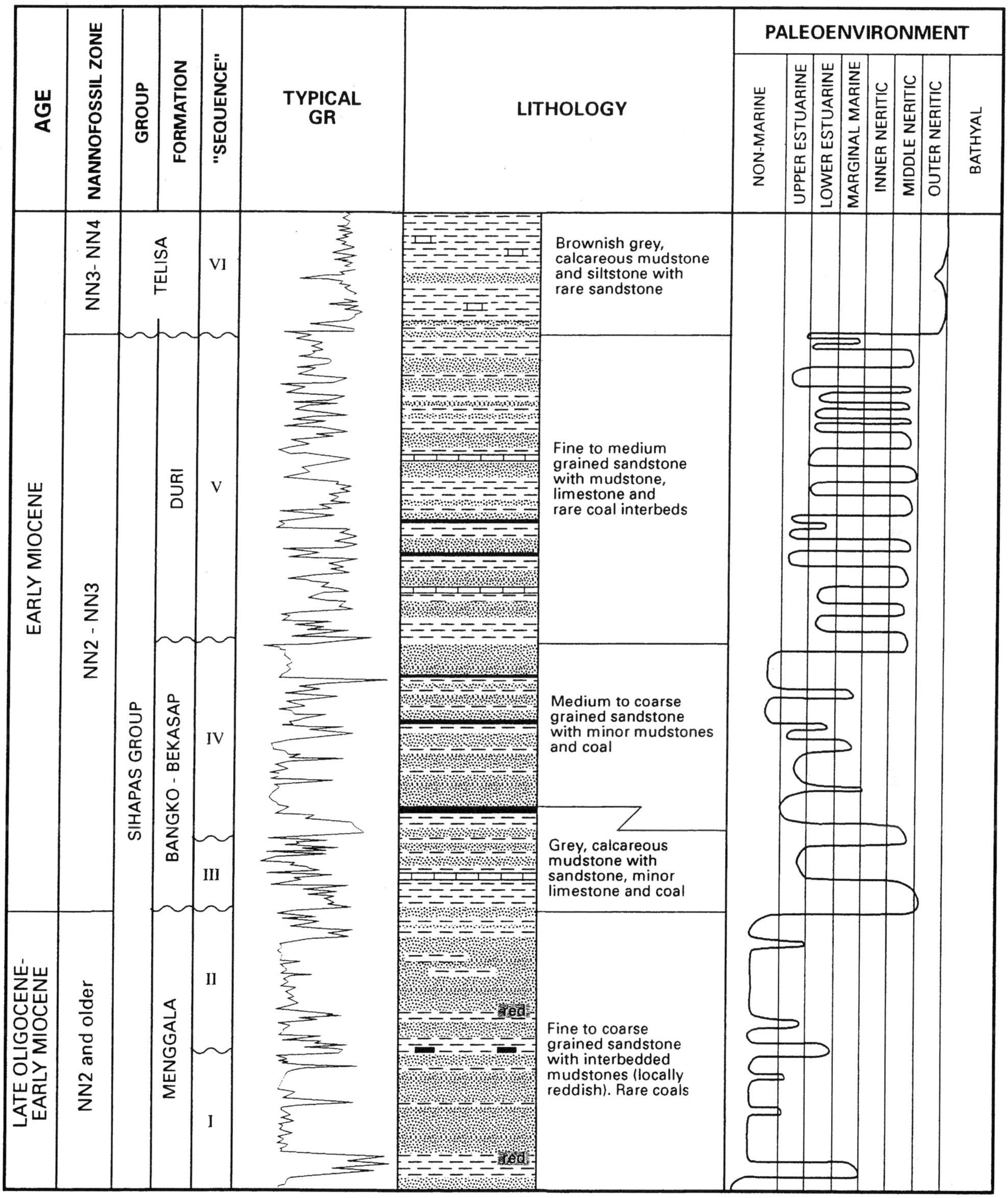

Text-Figure 7. Synthesis of stratigraphic data from the study wells.

FINGER, K.L., and DRUGG, W.S.
1992 Microfossils as indicators of deltaic subenvironments, Minas Field, Central Sumatra. *Proceedings of the Twenty First Annual Convention Indonesian Petroleum Association*, pp. 225–237.

MERTOSONO, S., and NAYOAN, G.A.S.
1974 The Tertiary basinal area of Central Sumatra. *Proceedings of the Third Annual Convention Indonesian Petroleum Association*, pp. 63–76.

MORLEY, R.J.
1990 Tertiary stratigraphic palynology in south east Asia. *Bulletin of the Geological Society of Malaysia*, 10: 1–36.
1996 Biostratigraphic characterization of systems tracts in Tertiary sedimentary basins. *In*: Caughey, C.A., Carter, D.C., Clure, J., Gresko, M.J., Lowry, P., Park, R.K., and Wonders, A. (eds.), *Proceedings of the International Symposium on sequence stratigraphy in south east Asia*. Indonesia Petroleum Association, Jakarta, pp. 49–71.

POUMOT, C.
1989 Palynological evidence for eustatic events in the tropical Neogene. *Bulletin Centres Recherche Exploration-Production Elf-Aquitaine*, 13: 437–453.

PUTROHARI, R.D
1992 MSDC-1: A gas discovery well in the Malacca Straits PSC, Sumatra, Indonesia. *Proceedings of the Twenty First Annual Convention Indonesian Petroleum Association*, pp. 201–223.

WHITMORE, T.C.
1984 *Tropical rain forests of the far east*. Second edition, Oxford University Press, Oxford, 352 pgs.

WHITTEN, A.J., DAMANIK, S.J., ANWAR, J., and HISYAM, N.
1984 *The ecology of Sumatra*. Gadja Mada University Press, Yogyakarta, Indonesia, 583 pgs.

WONGSOSANTIKO, A.
1976 Lower Miocene Duri Formation sands, Central Sumatra Basin. *Proceedings of the Fifth Annual Convention Indonesian Petroleum Association*, pp. 133–150.

Author's address

KEITH RICHARDS
Robertson Research International Limited
Llandudno, North Wales
LL30 1SA, United Kingdom

Present address:
KrA Stratigraphic
43 Ty Mawr Road
Deganwy, North Wales, LL31 9UD
United Kingdom
E-mail: keith@paly.co.uk
Web: www.paly.co.uk

SEQUENCE STRATIGRAPHY OF THE PACHUTA–MARIANNA INTERVAL (UPPER EOCENE–LOWER OLIGOCENE) IN THE U.S. GULF COAST

Carlos A. Jaramillo, and Francisca E. Oboh-Ikuenobe

Abstract

The qualitative and quantitative characteristics of palynofacies and dinocyst assemblages have been used to interpret the sequence stratigraphy of upper Eocene–lower Oligocene strata in southern Mississippi and Alabama (U.S. Gulf Coast). In southern Mississippi, we identified a latest Eocene maximum flooding surface in the middle of the Shubuta Clay, but this surface was identified at the top of Shubuta Clay in southern Alabama, where the Eocene–Oligocene boundary was placed within a condensed section (of approximately 0.19 million years duration). This condensation, which was identified at the Shubuta Clay–Vicksburg Group contact, is equivalent to the accumulation of the upper Shubuta Clay and Red Bluff Clay in southern Mississippi. We also interpreted the Forest Hill–Mint Spring contact as a sequence boundary which coincides with a transgressive surface, but this interpretation differs from that of Pasley and Hazel (1995). In southern Mississippi, the Pachuta Marl and the lower Shubuta Clay comprise a late Eocene transgressive systems tract, whereas the same systems tract consists of the Pachuta and the entire Shubuta Clay in southern Alabama. The overlying early Oligocene highstand systems tract was represented by the upper Shubuta, and the Bumpnose/Red Bluff/Forest Hill Formations. Although a very thin sandstone in the Mint Spring in one southern Mississippi section was interpreted as a possible lowstand deposit, the Mint Spring, along with the Marianna Limestone, constitute the transgressive systems tract in the other four sections.

INTRODUCTION

Relative sea level fluctuations and chronostratigraphic correlations are two important points considered in any sequence stratigraphic analysis of a sedimentary basin. Frequently, in offshore environments and continental deposits, it is difficult to identify key surfaces such as sequence boundaries and their correlative conformities, transgressive surfaces, and maximum flooding surfaces. The best clue to detecting these surfaces are changes in paleobathymetry and locating time condensations or hiatuses in the stratigraphic record. Changes in paleobathymetry can be detected by lithological and paleontological studies. For time analysis, a reliable approach involves an integration of biostratigraphic datums, magnetostratigraphy and radiometric dating through the use of graphic correlation.

We have integrated dinocyst biostratigraphy, dinocyst paleoecology, and palynofacies, in order to obtain chronostratigraphic and paleobathymetric data that were used to interpret the sequence stratigraphy of upper Eocene–lower Oligocene strata in the eastern U.S. Gulf Coast. We studied five localities in southern Mississippi and Alabama: three coreholes and one outcrop in southeastern Mississippi (#1 Young, #1 Ketler, #1 Wayne, and R2089 Type Red Bluff), and one outcrop section in southwestern Alabama (St. Stephens Quarry) (Text-Figure 1). The three coreholes were drilled recently by Mobil exploration (Dockery et al., 1994) and therefore provide fresh information on this largely disputed stratigraphic interval (see discussion below).

Jaramillo, Carlos A., and Oboh-Ikuenobe, Francisca
2001 Sequence stratigraphy of the Pachuta–Marianna interval (Upper Eocene–Lower Oligocene) in the U.S. Gulf Coast; *In*: Goodman, D.K., and Clarke, R.T. (eds.), *Proceedings of the IX International Palynological Congress, Houston, Texas, U.S.A., 1996*; American Association of Stratigraphic Palynologists Foundation, p. 263–276.

 ISBN 0–931871–06–9

The few well exposed Eocene to Oligocene sedimentary rocks in southern Mississippi and Alabama have been extensively studied. These units are more or less continuous and have provided field data for several paleogeographic and biostratigraphic studies (e.g., Hazel et al., 1980; Siesser et al., 1985; Tew and Mancini, 1995). One section in particular, St. Stephens Quarry in southwestern Alabama,

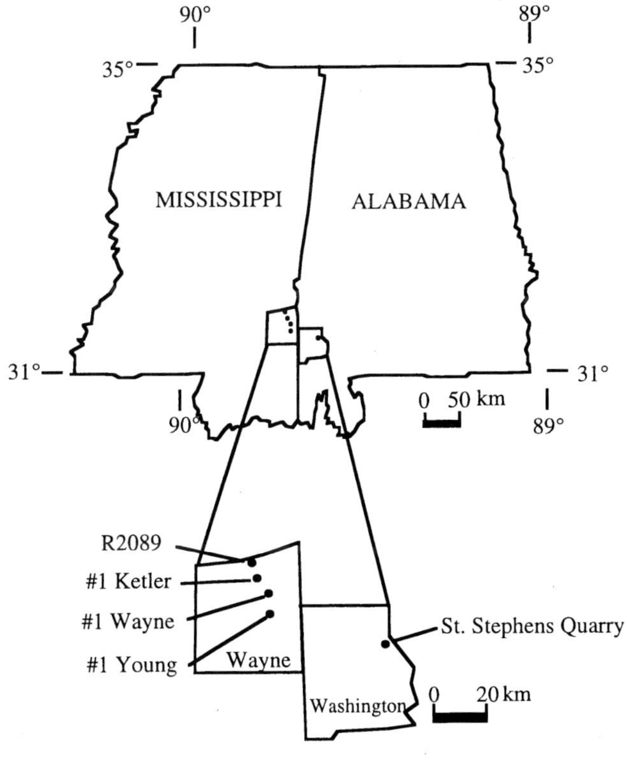

Text-Figure 1. Map of Mississippi and Alabama showing the studied localities in the southern parts of the states.

has been the most studied, providing controversial information on the Eocene–Oligocene boundary (Mancini, 1979; Bybell, 1982; Siesser, 1983; Keller, 1985; Pasley and Hazel, 1990, 1995; and Miller et al., 1993). These biostratigraphic studies have focused mostly on microfossils other than dinoflagellates, in spite of their high abundance and diversity in these units.

Sequence stratigraphic studies for this area, mainly for the St. Stephens Quarry section, have applied biostratigraphy, magnetostratigraphy, stable isotope stratigraphy and lithostratigraphy to their interpretations (Baum and Vail, 1988; Loutit et al., 1983, 1988; Miller et al., 1993; Tew and Mancini, 1995; Pasley and Hazel, 1995). While the various sequence stratigraphic models proposed by these studies largely agree on many issues, the most notable discrepancies occur in the interpretation of the interval represented by the Shubuta Clay, Bumpnose Limestone/Red Bluff Clay/ Forest Hill Sand, Mint Spring Marl and the Eocene–Oligocene boundary (Text-Figure 2A).

Our study provides new lithologic, palynofacies, and dinoflagellate cyst data from three coreholes and two outcrop sections which test the diverse hypotheses that have been formulated for the sequence stratigraphy of this area. Previous interpretations have been based mainly on the St. Stephens Quarry section.

GEOLOGIC SETTING

The Cenozoic history of northwestern Gulf of Mexico is characterized by a rapid sediment input and thick prograding depositional sequences that intertongue with interdeltaic shelf-edge sediments (Galloway, 1989). These units were deposited in a shallow marine to marginal marine environments of a passive margin. Upper Eocene sediments are characterized by fine-grained sediment deposition while the lower Oligocene constitute one of the great progradational wedges in the northwestern Gulf of Mexico (Galloway, 1989).

The latest Eocene time in southern Mississippi and Alabama is recorded in two lithostratigraphic units, the Pachuta Marl and Shubuta Clay members of the Yazoo Formation, Jackson Group (Text-Figure 2A). The lowest Oligocene strata consist of five lithostratigraphic units. In ascending order, the units are the Bumpnose Limestone, Red Bluff Clay, Forest Hill Sand, Mint Spring Marl, and the Marianna Limestone. The Bumpnose Limestone, Red Bluff Clay and Forest Hill Sand are lateral equivalents of each other, with the Bumpnose and Red Bluff intertonguing in some areas in southwestern Alabama (Tew and Mancini, 1995) (Text-Figure 2A).

Four different sequence stratigraphic models have been proposed previously for the St. Stephens Quarry section (Text-Figure 2B). In general, the following facts have been recognized: (a) the Shubuta–Bumpnose contact corresponds to a maximum flooding surface, and (b) a sequence boundary is located at the base of the Mint Spring Marl. Miller et al. (1993) interpreted the Shubuta–Bumpnose contact as a sequence boundary, based on correlation to the Tejas A (TA) 4.4 cycle boundary of Haq et al. (1988) and an oxygen isotope increase at the contact. However, Loutit at al. (1988, text-fig. 21) reported a decrease in oxygen isotope at the same level. Pasley and Hazel (1995) proposed that a surface of maximum starvation was immediately overlain by a sequence boundary at the Shubuta Clay/Bumpnose contact. They also interpreted the Bumpnose Limestone and Red Bluff Clay as constituents of a lowstand systems tract, and noted the absence of a sequence boundary at the base of the Mint Spring Marl.

MATERIALS AND METHODS

Palynological samples for this study were derived from the U.S. Geological Survey (Reston, Virginia) and Mobil Exploration and Production Services (Dallas, Texas). The oxidized samples from all the studied sites were analyzed for dinocyst richness and abundance. The resulting dinocyst ranges for the localities were then correlated using the technique of graphic correlation. The dinocyst distribution was also analyzed using multivariate techniques.

The unoxidized fractions of #1 Young samples provided the main dataset for the palynofacies aspect of this study. At least 400 organic particles per sample were identified with a Zeiss transmitted light microscope (Table 1). We used a classification system adapted from Lorente (1986), Van Vergen et al. (1990), Oboh (1992), and Jaramillo and Yepes (1995). Organic matter results were analyzed using an Euclidean-distance cluster analysis with average linkage, using Systat (1992). This method is useful because it averages all distances between pairs of objects in different clusters and decides how far apart they are (Sokal and Michener, 1958).

PALYNOFACIES ANALYSIS

Palynofacies analysis was only carried out in the #1 Young core. Euclidean-distance cluster analysis identified four groups of samples, based on their organic matter content (Text-Figure 3). In Group A, amorphous organic material dominates the palynofacies assemblage; Group B consists of yellow-brown material, pollen and plant tissue; in Group C, black-brown fragments, black debris and plant tissue dominate the assemblage; while Group D samples are dominated by dinoflagellate cysts, black debris, black-brown fragments and pollen. The differences in organic matter content between the groups are due mainly to changes in their position within the basin in relation to the paleoshoreline (Jaramillo, 1995).

Based on the palynofacies groups identified by cluster analysis, a paleobathymetric curve has been reconstructed for the # 1 Young corehole (right-hand side of Text-Figure 4). This curve shows the relative movements of the shoreline during the latest Eocene and early Oligocene. In order to reconstruct the curve, the sample groups (A to D) were organized in a lateral palynofacies sequence from the deepest to the shallowest depositional environments as follows: Group A, Group D, Group B and Group C. Group A probably accumulated in the deepest environment because the high contents of structureless amorphous material is indicative of relatively deep offshore environments (Lorente 1986; Pasley and Hazel, 1990). Group D has a high diversity of dinoflagellate cysts, combined with terrestrial material. A diverse dinoflagellate cyst assemblage is common in shelf environments (Traverse, 1988), and in this case, we interpret a possible inner to middle shelf which is close enough to land to allow the accumulation of terrestrially derived organic material. Group B is dominated by terrestrial material but still contains dinoflagellate cysts. This assemblage probably accumulated in an inner shelf environment. Finally, Group C represents the shallowest environment because it is comprised almost exclusively of

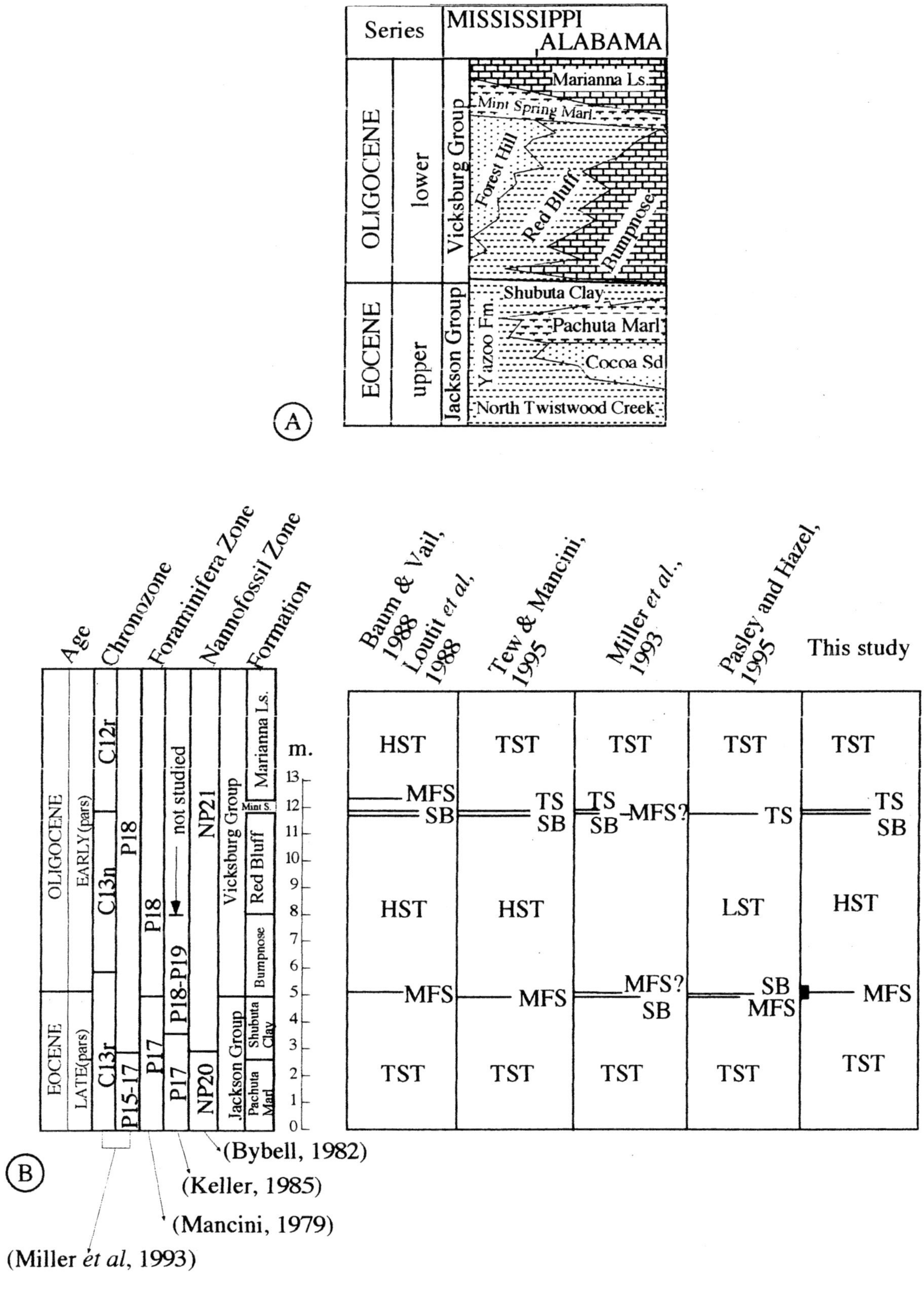

Text-Figure 2. (A) Simplified stratigraphic correlation chart for the late Eocene–early Oligocene of the Mississippi–Alabama area (modified from Galloway et al., 1991 and Pasley and Hazel, 1995). (B) Sequence Stratigraphic, biostratigraphic and magnetostratigraphic interpretations of St. Stephens Quarry section. See discussion in the text for the position of the Eocene–Oligocene boundary. HST = highstand systems tract, TST = transgressive systems tract, LST = lowstand systems tract, TS = transgressive surface, MFS = maximum flooding surface, SB = sequence boundary.

TABLE 1. Palynodebris count data for #1 Young corehole.

SAMPLE		Aquatic		Terrestrial								
		Structureless	Structured	Structureless				Structured				
		amorphous	dinocyst/for.	resins	black debris	yellow brown	black brown	cuticles	plant tissue	woody	sporom.	fungi
meters	ft.	%	%	%	%	%	%	%	%	%	%	%
56.8	186.0–186.5	6.60	31.76	0.00	5.66	5.66	6.92	0.31	13.84	6.29	22.01	0.94
57.8	189.5–190.0	0.29	0.00	0.87	13.70	6.41	11.95	0.29	57.73	5.54	3.21	0.00
59.2	194.0–194.5	1.89	1.35	0.54	13.78	7.03	11.08	1.35	42.16	14.32	6.49	0.00
62.3	204.0–204.5	1.30	0.78	2.08	6.75	15.84	13.77	0.26	36.88	10.65	11.69	0.00
65.3	214.0–214.5	1.31	0.65	1.31	3.59	13.40	14.38	0.00	42.48	15.03	7.84	0.00
66.8	219.0–219.5	0.00	0.56	0.28	19.72	7.04	17.46	0.56	32.39	14.65	7.32	0.00
70.0	229.5–230.0	1.32	0.66	1.66	9.93	5.96	13.91	0.00	43.71	6.62	16.23	0.00
73.1	239.5–240.0	0.30	0.30	0.30	23.08	5.62	22.19	0.59	35.80	6.51	5.33	0.00
76.1	249.5–250.0	0.00	0.99	0.66	20.13	5.94	22.11	0.99	33.00	8.91	7.26	0.00
79.1	259.0–260.0	1.97	0.99	1.32	8.22	11.18	21.05	5.59	28.95	7.89	12.50	0.33
82.2	269.5–270.0	1.33	0.00	2.67	11.00	12.67	13.67	1.33	34.33	7.67	14.67	0.67
85.3	279.5–280.0	0.33	0.33	2.66	5.32	11.96	13.29	5.65	30.23	6.98	22.26	1.00
86.6	284.0–284.5	1.17	0.59	0.59	8.50	20.23	16.42	0.59	34.31	5.87	10.85	0.88
88.3	289.5–290.0	1.62	0.32	1.62	3.56	20.39	11.00	1.62	30.10	7.77	21.04	0.97
91.4	299.5–300.0	2.00	2.33	4.67	5.33	17.33	13.33	1.33	26.00	3.33	23.67	0.67
94.3	309.0–309.5	6.80	17.28	0.00	6.52	9.35	9.92	0.28	17.56	12.75	18.98	0.57
95.8	314.0–314.5	7.44	35.81	1.38	7.16	11.85	3.58	0.00	10.74	4.13	17.08	0.83
97.5	319.5–320.0	33.72	4.61	0.58	8.07	21.90	6.63	0.00	8.93	4.61	9.22	1.73
99.3	325.5–326.0	58.11	7.02	0.00	2.18	12.83	4.12	0.00	1.69	2.66	8.96	2.42
100.5	329.5–330.0	55.52	11.04	0.00	7.36	12.58	1.84	0.00	2.45	1.53	5.83	1.84
103.6	339.5–340.0	73.49	0.30	0.00	5.42	12.95	3.01	0.00	0.60	0.60	2.11	1.51
106.6	349.5–350.0	58.86	6.33	0.00	4.43	13.92	3.16	0.00	1.58	1.58	6.33	3.80
108.0	354.0–354.5	38.59	8.50	0.24	8.01	16.26	5.58	0.00	0.73	1.94	16.26	3.88
109.7	359.5–360.0	23.22	5.01	1.85	7.92	13.19	3.96	0.00	2.11	3.17	29.02	10.55
112.7	369.5–370.0	45.37	6.02	0.00	5.79	16.67	2.55	0.00	1.62	1.62	12.50	7.87
115.3	378.0–378.5	7.94	26.47	0.88	5.59	22.35	11.76	0.00	4.71	8.82	10.29	1.18
118.8	389.5–390.0	3.21	19.87	0.96	30.45	13.46	9.29	0.00	4.17	3.53	14.42	0.64

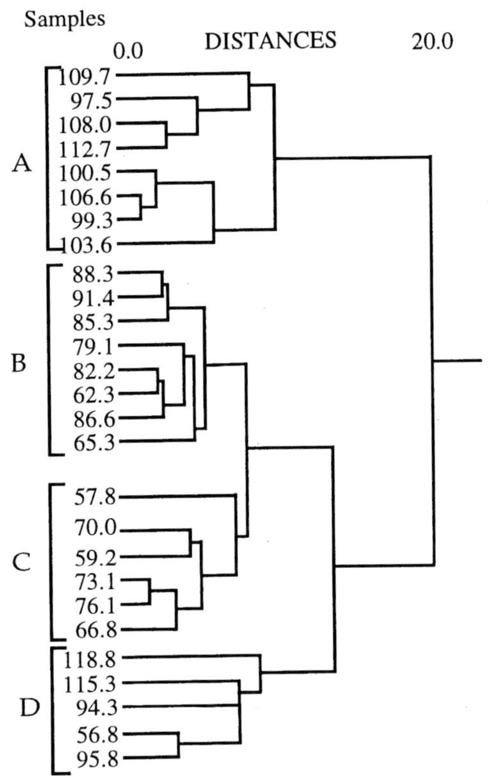

Text-Figure 3. Average linkage cluster analysis (with Euclidean distance) of the organic matter distribution of the #1 Young corehole. Each palynofacies assemblage identified by the cluster analysis represents a group of samples with similar organic matter content which accumulated in a similar position relative to the shoreline. Sample numbers are depths.

terrestrially derived organic material. This group probably accumulated in the coastal plain–innermost shelf transitional environment. Similar palynofacies studies have been successfully carried out by several authors, such as Habib and Miller (1989) and Gregory and Hart (1992).

CHRONOSTRATIGRAPHY

A time-framework based on the stratigraphic distribution of dinoflagellate cysts was developed, using graphic correlation (Shaw, 1964; Edwards, 1989). The dinoflagellate range charts for each section, and illustrations of the most important dinoflagellates and acritarchs used in this study can be found in Jaramillo and Oboh-Ikuenobe (1999). The #1 Young core was chosen as the reference section. Several rounds of correlation were performed on all five stratigraphic sections in order to produce a composite section. The important fossil events in the composite section are listed in Table 2. The composite section was then tied to a worldwide chronostratigraphic framework.

The Eocene–Oligocene boundary is currently defined by the last occurrence of the foraminifera *Hantkenina* spp. in the Massignano section of the Apennine Mountains (Italy), which is located in the youngest part of the magnetostratigraphic Chron C13r (C13r.14) and has an age estimated at 33.7 Ma (Berggren et al., 1995). These markers, last occurrence of *Hantkenina* and Chron C13n, were used to tie our composite section to the world-wide chronostratigraphy of Berggren et al. (1995).

At the St. Stephens Quarry section, Pasley and Hazel (1990, 1995) noted that the last occurrence of *Hantkenina alabamensis* occurred 24 cm above the Shubuta–Bumpnose contact. The location of this boundary is reinforced by the position of the C13n–C13r magnetoboundary in St. Stephens Quarry, located approximately 1.5 m above the Shubuta–Bumpnose contact in a core that was drilled 1.6 km from the St. Stephens Quarry section (Miller et al., 1993). The age of this magnetoboundary is 200ky younger than the Eocene–Oligocene boundary (Berggren et al., 1995). This would indicate that the Eocene–Oligocene boundary is approximately coincident or slightly above the Shubuta–Bumpnose boundary.

The second datum used to tie the composite section to chronostratigraphy was the Chron C13n–C12r magnetoboundary of Miller et al. (1993) at the Red Bluff–Mint Spring contact in the St. Stephens Quarry section. This magnetoboundary is dated as 33.05 Ma (Berggren et al., 1995).

The two datums were projected onto the correlation line drawn by the comparison of the composite section with the St. Stephens Quarry (Text-Figure 5). The line of correlation for St. Stephens Quarry suggests an interval of 80 cm of extreme condensation lasting approximately 13.7 composite units (c.u.). This condensed interval starts 20 cm below the Shubuta–Bumpnose boundary and ends 60 cm above it. Several fossil events that are not coeval appear together in the same sample, thereby indicating condensation. For example, in sample 4.7m (refer to Text-Figures 5 and 7 for position), the following biostratigraphic events occur together even though they occur at different levels in the standard composite section: first occurrence of *Hemiplacophora semilunifera*, peak of *Hemiplacophora semilunifera*, first occurrence of *Wetzeliella lobisca*, last occurrence of *Elytrocysta* sp. A, last occurrence of *Hemiplacophora semilunifera* and first occurrence of taxon *Operculodinium placitum*. A similar case can be made for sample 5.0m. Sample 5.3m has an assemblage which is similar to that in sample 5.0m but it does not show up in the correlation because all of the f.o. datums were already in sample 5.0m. Loutit et al (1988) also interpreted the first 60 cm of the Bumpnose (or unnamed blue clay) as a condensed section, based on the gamma ray and foraminifera fauna. The last occurrence of *H. alabamensis* at 24 cm above the base of the Bumpnose lies in the middle of the condensed section. In order to project this datum on the composite section, we marked the stratigraphic interval containing the condensed section as indicated by the correlation line. Then, we established an average rate of accumulation for this condensed section, and projected the *H. alabamensis* datum into the composite section (Text-Figure 5). A level at 99.5 c.u. in the composite section was found. The second datum, the Bumpnose–Mint Spring boundary, when projected onto the composite section gave a location of 53 c.u. for the C13n–C12r boundary (33.05 Ma) (Text-Figure 5). A duration of 0.014 Ma per composite unit is calculated from the age-data points used.

DINOCYST PALEOBATHYMETRY

Several studies (e.g., Wall et al., 1977; Dale, 1996) have indicated that dinoflagellate cysts are sensitive environmental indicators today, and are therefore of potential use

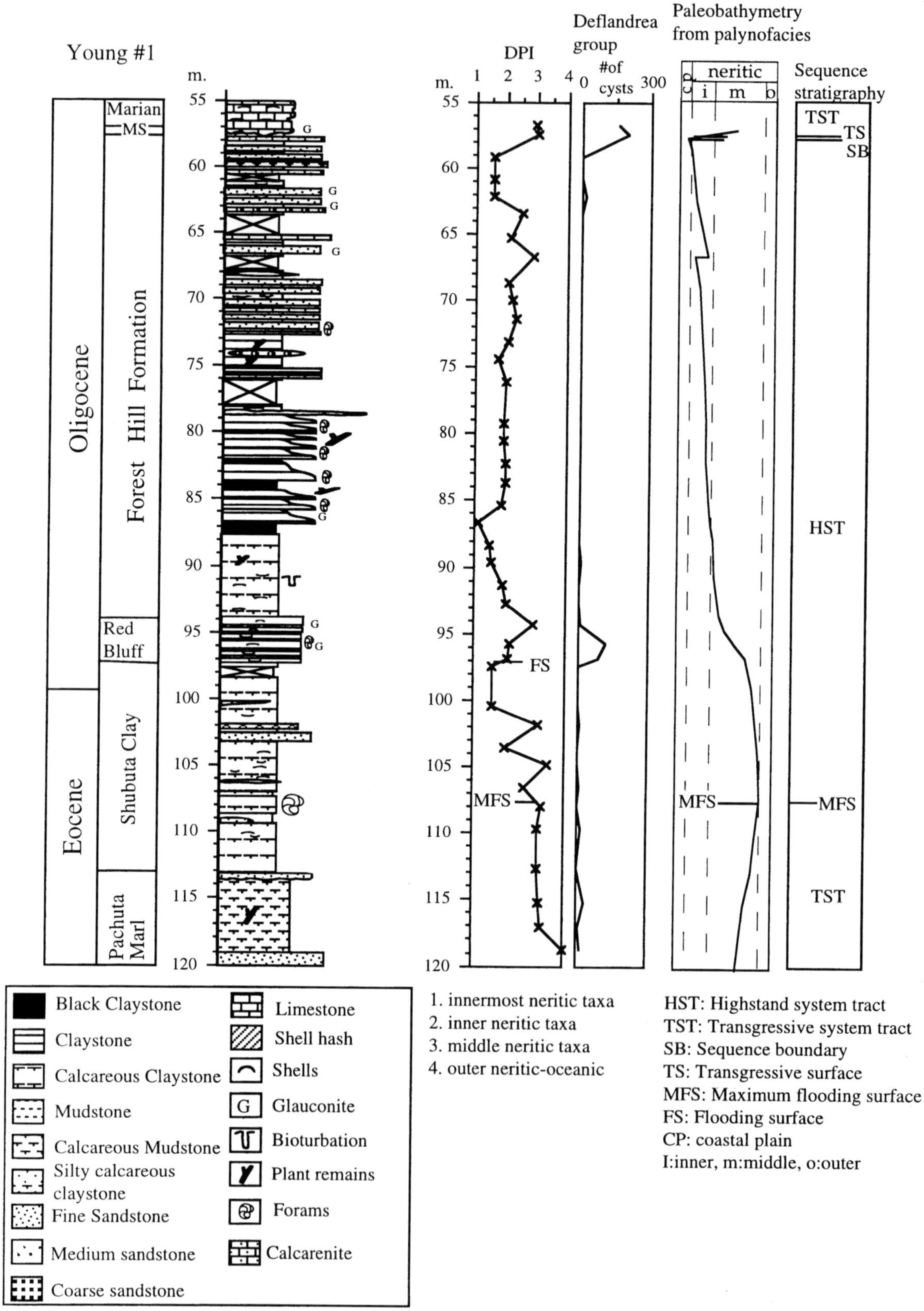

Text-Figure 4. Sequence stratigraphic interpretation for the #1 Young corehole. Refer to Table 3 for taxa used in calculating the dinocyst paleobathymetric index (DPI), and Text-Figure 3 for derivation of paleobathymetric curve from palynofacies. The Eocene–Oligocene boundary is 8 composite units (c.u.) younger (0.112 Ma) than the inferred maximum flooding surface in the middle of the Shubuta Clay.

TABLE 2. Fossil events used in the composite section. C.u. refers to composite unit, f.o. first occurrence, l.o. last occurrence, peak refers to high abundance levels of individual taxa with locally important chronostratigraphic significance. The Eocene–Oligocene boundary (33.7 Ma) occurs at 99.5 c.u., while Chron C13n–C12r magnetoboundary (33.05 Ma) occurs at 53 c.u. Each composite unit has an equivalence of 0.014 Ma.

No.	Species	c.u.	event
73	*Pentadinium* sp. A	52	l.o.
47	*Wetzeliella lobisca*	58	l.o.
82	*Polysphaeridium zoharyi*	59	f.o.
71	*Wetzeliella articulata*	62	l.o.
80	*Systematophora* sp. A	62	l.o.
73	*Pentadinium* sp. A	65	peak
58	*Wetzeliella symmetrica*	65	l.o.
26	*Cribroperidinium tenuitabulatum*	67	l.o.
55	*Heteraulacacysta campanula*	74	l.o.
11	*Cordosphaeridium cantharellum*	73	l.o.
75	*Impagidinium* sp. B	82	f.o.
62	*Cordosphaeridium gracile*	84	l.o.
31	*Dinopterygium cladoides sensu* Morgenroth	85	l.o.
13	*Diphyes colligerum*	86.5	l.o.
71	*Wetzeliella articulata*	93	f.o.
9	*Operculodinium* aff. *centrocarpum*	94	peak
67	*Operculodinium placitum*	94	peak
61	*Tectatodinium pellitum*	94	l.o.
10	*Trigonopyxidia fiscellata*	95	l.o.
65	*Cyclopsiella vieta*	96	f.o.
73	*Pentadinium* sp. A	96	f.o.
80	*Systematophora* sp. A	96	f.o.
19	*Deflandrea phosphoritica*	96	peak
43	*Deflandrea heterophlycta*	97	peak
62	*Cordosphaeridium gracile*	98	f.o.
46	*Areosphaeridium multicornutum*	98	l.o.
68	*Apteodinium australiense*	100	f.o.
77	*Membranophoridium aspinatum*	100	f.o.
60	*Operculodinium divergens*	101	f.o.
67	*Operculodinium placitum*	101	f.o.
56	*Distatodinium* aff. *ellipticum*	102	f.o.
58	*Wetzeliella symmetrica*	102	f.o.
59	*Glaphyrocysta* sp.	102	f.o.
98	*Heteraulacacysta leptalea*	102	f.o.
15	*Muratodinium fimbriatum*	102	l.o.

No.	Species	c.u.	event
29	*Hemiplacophora semilunifera*	102	l.o.
36	*Cannosphaeropsis* sp. A	102	l.o.
38	*Elytrocysta* sp. A of Head and Norris (1989)	103.5	l.o.
55	*Heteraulacacysta campanula*	105	f.o.
14	*Charlesdowniea coleothrypta*	105	peak
26	*Cribroperidinium tenuitabulatum*	105	peak
44	*Systematophora placacantha*	105	peak
39	*Rhombodinium draco*	105	l.o.
42	*Batiacasphaera* aff. *baculata*	104.5	l.o.
46	*Enneadocysta multicornuta*	107	f.o.
47	*Wetzeliella lobisca*	107	f.o.
50	*Cordosphaeridium inodes*	107	f.o.
51	*Ascostomocystis potane*	107	f.o.
52	*Homotryblium vallum*	107	f.o.
29	*Hemiplacophora semilunifera*	107	peak
61	*Tectatodinium pellitum*	106.5	f.o.
1	*Batiacasphaera compta*	107.2	l.o.
24	*Batiacasphaera baculata*	108	l.o.
44	*Systematophora placacantha*	110	f.o.
45	*Homotryblium plectilum*	111	f.o.
43	*Deflandrea heterophlycta*	113	f.o.
49	*Enneadocysta arcuata*	113.5	f.o.
36	*Cannosphaeropsis* sp. A	115	f.o.
37	*Histiocysta* sp. A	115	f.o.
10	*Trigonopyxidia fiscellata*	115	peak
16	*Lingulodinium pugiatum*	115	l.o.
34	*Hystrichokolpoma globulum?*	115	l.o.
41	*Hystrichokolpoma cinctum*	115	l.o.
66	*Nematosphaeropsis pusulosa*	116	f.o.
24	*Batiacasphaera baculata*	117	f.o.
26	*Cribroperidinium tenuitabulatum*	117	f.o.
29	*Hemiplacophora semilunifera*	117	f.o.
31	*Dinopterygium cladoides sensu* Morgenroth	117	f.o.
32	*Rottnestia borussica*	117	f.o.
39	*Rhombodinium draco*	117	f.o.

Eocene–Oligocene boundary 99.5 cu (33.7 Ma)
Chron C13n–C12r boundary 53 cu (33.05 Ma)
Duration of one composite 0.014 Ma

in paleopalynology. Four types of environmental signals can be deduced from recent dinocyst distributions: (1) surface water temperature, (2) a coastal/oceanic trend, (3) salinity, and (4) productivity. The coastal/oceanic trend is a variable which correlates strongly with variations in dinocyst assemblages (Wall et al., 1977, Dale, 1996). The distribution of dinocysts in our study was analyzed using multivariate techniques to identify changes in the composition and abundance of dinocyst assemblages. These changes were interpreted in terms of coastal/oceanic trends which ultimately could be related to changes in paleobathymetry. We used the Spearman rank-order coefficient to perform a non-metric multidimensional scaling (MDS) analysis on the dinocyst and acritarch abundance data. This analysis

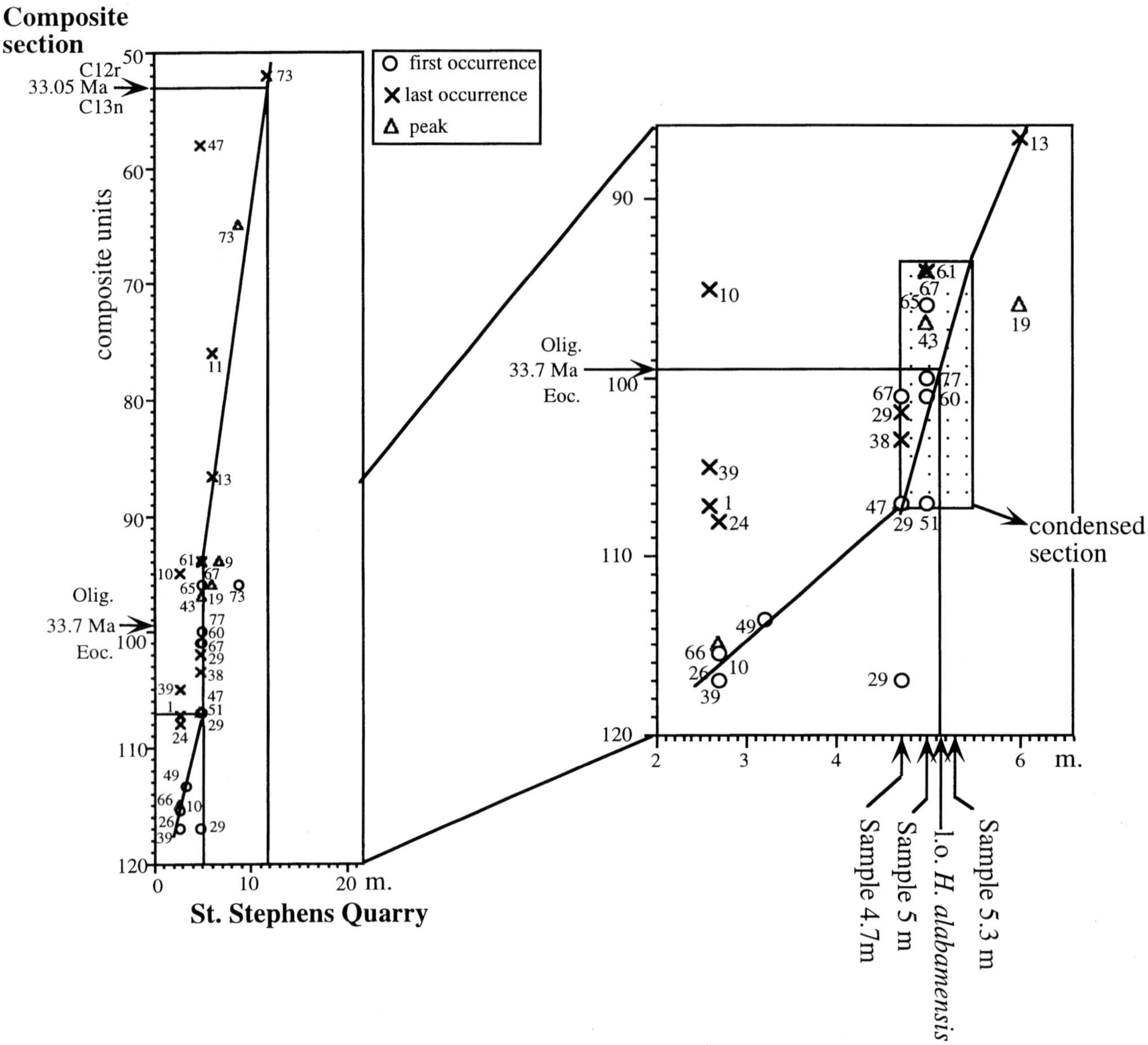

Text-Figure 5. Line of correlation for St. Stephens Quarry section versus the composite section. The position of the Eocene–Oligocene boundary (33.7 Ma) in the composite section is found by projecting the last occurrence of *Hantkenina alabamensis* onto the composite section (i.e., 0.24 m above the Shubuta–Bumpnose contact at the St. Stephens Quarry section, Pasley and Hazel, 1990, 1995). C13n–C12r magnetoboundary (33.05 Ma) occurs at 11.8 m (Miller et al., 1993) and is also projected onto the composite section. Note that the line of correlation shows an 80 cm interval of extreme condensation lasting for 13.7 c.u. Meter 4.9 in this section corresponds to the Shubuta Clay/Red Bluff Clay–Bumpnose boundary (meter 3.95 in the column of Pasley and Hazel, 1990, fig. 4).

identifies major environmental gradients along the first and second axes (Kovach 1989). Based on an extensive literature review of dinocyst paleoecology, some of the groups produced by the cluster analysis could be referred to a relative position of the dinocyst community with respect to the paleoshoreline (Jaramillo and Oboh, 1999 to see literature consulted).

Six dinocyst paleoecological groups were identified in our study, and five of them are interpreted as representing an inshore–offshore trend (Table 3). The environmental interpretations (e.g., inner, middle, outer neritic) are used in a relative sense rather than an absolute one. A dinocyst paleobathymetric index (DPI) was calculated using a weighted mean (Σn_i (i)/n) of the innermost, inner, middle, and outer neritic to oceanic groups, where n represents the total number of specimens in a particular sample, and n_i represents the abundance of each ecological group. The scale of the index goes from 1 to 5,

TABLE 3. Dinocyst ecological groups used in the determination of the paleobathymetric index (Jaramillo, 1995; Jaramillo and Oboh-Ikuenobe, unpubl. results).

Innermost Neritic
Homotryblium plectilum
Pediastrum

Inner Neritic
Ascostomocystis potane
***Glaphyrocysta* group**

Middle Neritic
Charlesdowniea coleothrypta
Hystrichokolpoma rigaudiae

Outer Neritic
***Cannosphaeropsis* sp. A**
Cribroperidinium tenuitabulatum
Hemiplacophora semilunifera
Batiacasphaera compta
***Impagidinium* spp.**
Tectatodinium pellitum

Neritic (inner to outer)
***Cleistosphaeridium* spp.**
***Enneadocysta* group**
Spiniferites pseudofurcatua
Systematophora placantha
***Cordosphaeridium* spp.**
Operculodinium* aff. *centrocarpum
Spiniferites ramosus
***Lingulodinium* spp.**
Pentadinium laticinctum

Deflandrea group
***Deflandrea* spp.**
Samlamdia clamydophora
Corrudinium incomposita

with 1 representing innermost neritic taxa and 5 outer neritic to oceanic neritic taxa. The index value for each sample was plotted for each entire section to obtain a dinocyst paleobathymetric curve for that section. In the reconstruction of the curves, samples with less than 10 dinocyst specimens were excluded from the analysis. A second plotted curve used the abundance of the *Deflandrea* dinocyst group which can be related to, among other factors, run-off changes and the introduction of nutrients in the photic zone associated with flooding events (Brinkhuis, 1994).

SEQUENCE STRATIGRAPHIC INTERPRETATION

#1 Young Corehole

The Eocene–Oligocene boundary was identified at 99.5 m, 2.3 m below the Shubuta Clay–Red Bluff boundary (Eocene/Oligocene boundary is at composite unit 99.5, and #1 Young is the reference section). Two important flooding surfaces were identified in the corehole (see Text-Figure 4). The strongest one occurs at 107.5 m (353 ft), 10.3 meters below the Shubuta Clay–Red Bluff boundary, and other at 57.72 m (189.4 ft) at the Forest Hill–Mint Spring contact. The level at 107.5 m is interpreted as a maximum flooding surface (MFS) because: (a) it is a green calcareous claystone with abundant uvigerinids, which are deep water benthic foraminifera (Dockery et al., 1994), and (b) it occurs between two of the samples with the highest concentration of total marine derived organic matter (amorphous and dinocysts, Table 1). This level also corresponds to a high value (3.2) in the dinocyst paleobathymetric index (Text-Figure 4), and is 0.9 m above a gamma ray peak (on gamma ray log in Dockery et al., 1994). A minor flooding surface is identified at the Shubuta–Red Bluff contact at 97.2 m (319 ft), based on the lithological change (olive gray clay to gray fossiliferous, glauconitic mudstone) and an increase in the *Deflandrea* group.

The paleobathymetry interpreted from palynofacies and dinocyst paleoecology indicates a flooding event at the Forest Hill–Mint Spring contact at 57.72 m (33.1 Ma). We believe that a sequence boundary is merged here with a transgressive surface (Text-Figure 4). The contact is sharp, irregular, burrowed and marked by an abraded shell lag and rounded intraclasts. This hypothesis can be confirmed in up-dip and down-dip sections if forced regressions or incised valleys are found (Posamentier et al., 1992). The proposed lowstand deposit in the #1 Wayne corehole, if proved true, (see below) would confirm this sequence boundary.

#1 Wayne Corehole

The Eocene–Oligocene boundary is projected to be at 55.7 m, 1.9 m above the Shubuta–Red Bluff contact (Text-Figures 6 and 8). A sequence boundary is interpreted at the Forest Hill–Mint Spring contact. This contact is sharp, irregular, and burrowed with a pavement of intraclasts. The facies below the contact is a brown carbonaceous mudstone while above it is a 60 cm-thick, white medium-grained quartz sandstone with calcareous cement. Although samples from this sandstone interval were not analyzed, the unit is interpreted as a forced regression of a possible lowstand system tract based on lithological analysis. This sandstone is overlain by a gray porous limestone that corresponds to the lower Marianna Limestone, and the Mint Spring–Marianna contact is interpreted as a transgressive surface. The paleobathymetric data indicates a deepening across the Mint Spring–Marianna interval.

#1 Ketler Corehole

The Eocene–Oligocene boundary is projected to be at 7.5 m, 3.7 m above the Shubuta–Red Bluff contact (Text-Figures 6 and 8). A maximum flooding surface is identified at 17 m. This maximum flooding surface is characterized by a high index of dinocyst paleobathymetry, a high concentration of marine derived organic matter (amorphous material and dinocysts; Oboh and Yepes, 1995), and a green calcareous claystone with abundant uvigerinids. It is also 0.9 m above a pronounced peak of gamma ray (Dockery et al., 1994). The Pachuta and lower Shubuta constitute a transgressive system tract while the upper Shubuta and Red Bluff represent a highstand system tract.

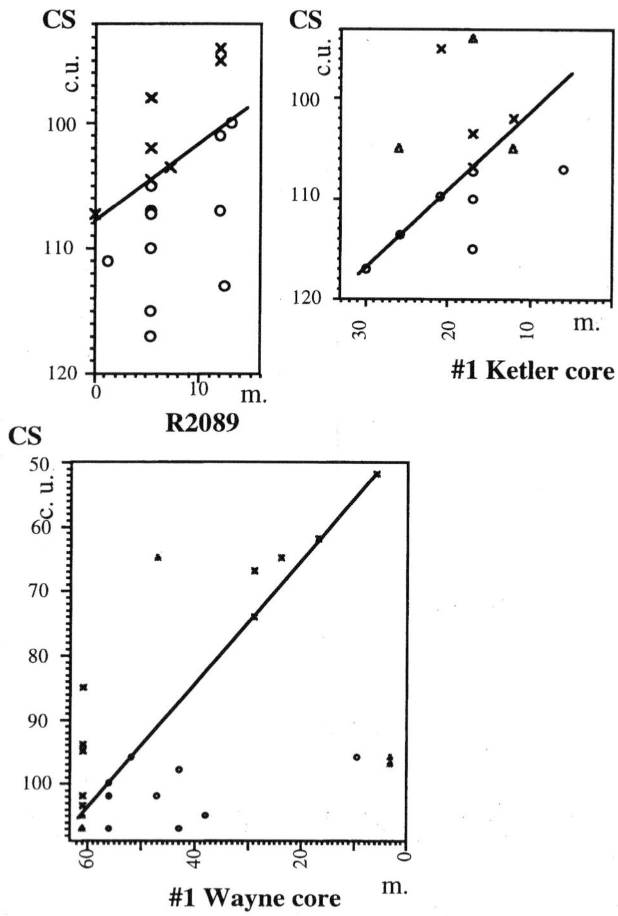

Text-Figure 6. Line of correlation for #1 Wayne corehole, #1 Ketler corehole and R2089 Type Red Bluff section versus the composite section (CS). The Eocene–Oligocene boundary in the #1 Wayne corehole is at 55.7 m, 1.9 m above the Shubuta–Red Bluff contact, while the C13n–C12r magnetoboundary occurs at 7.1 m, 2 m above the Mint Spring–Marianna contact. In the #1 Ketler corehole, the Eocene–Oligocene boundary is located at 7.5 m, 3.7 m. above the Shubuta–Red Bluff contact. At the R2089 section, the Eocene–Oligocene boundary is located at 13.3 m, 0.1 m above the Shubuta–Red Bluff contact. Meter 13.2 in this section corresponds to the Shubuta Clay/Red Bluff Clay contact of Pasley and Hazel (1995, fig. 6).

R2089 Section

The Eocene–Oligocene boundary is identified at 13.3 m measured from the base of the section, which is 0.1 m above the Shubuta–Red Bluff contact (Text-Figures 6 and 8). A tentative marine flooding surface is identified at 0 m, 8.3 c.u. (0.116 Ma) older than the Eocene–Oligocene boundary. Unfortunately, this flooding could not be confirmed because samples below 0m were not observed. A minor flooding surface is identified at the Shubuta–Red Bluff contact based on the *Deflandrea* group.

St. Stephens Quarry Section

A maximum flooding surface, marked by a thin phosphatic shell layer, is identified at the Shubuta–Bumpnose contact (Text-Figure 7). It occurs in an 80 cm-thick condensed section, which begins 20 cm below the Shubuta–Bumpnose contact and ends 60 cm above it. This condensation lasted for 13.7 c.u. (0.19 Ma), and also included the Eocene–Oligocene boundary (see discussion in chronostratigraphy and Text-Figure 5). This interpretation basically agrees with Loutit et al (1988) and Pasley and Hazel (1990). Pasley and Hazel (1995) identified a hiatus lasting 0.288 Ma at the Shubuta–Bumpnose contact. They concluded that this hiatus was the product of starvation during the accumulation of the highstand deposits following the maximum flooding surface. They interpreted this contact as a sequence boundary merged with the surface of maximum starvation. However, the reconstructed paleobathymetric curves (from DPI and *Deflandrea* group) are very similar immediately below and above the contact. Using foraminifera paleoecology, Loutit et al. (1988) also showed that the maximum water depth occurred 60 cm above the contact. If deposits of lowermost Red Bluff/Bumpnose (including unnamed "blue clay") were deposited during the lowstand systems tract, a drastic change in paleobathymetry should be evident. Furthermore, if a significant hiatus were present at the contact, the fossil assemblages immediately above it would show a gap in the fossil record. The dinocyst fossil record, however, does not show a gap. Instead, it registers a condensation of datums in a short stratigraphic interval. Pasley and Hazel's (1995) graphic correlation shows a possible hiatus because of the relative time-coarseness of each composite unit of their composite section in relation with ours. Whereas one composite unit in their scheme corresponds to 0.2 Ma, one unit in our study represents 0.014 Ma. Therefore, even though they observed a similar number of samples than us around the boundary in St. Stephens Quarry, the coarseness of their composite units would create an artificial compression of the correlation line that would suggest a hiatus.

The Red Bluff–Mint Spring contact is suggested here as a sequence boundary that coincides with a transgressive surface. A sequence boundary is inferred based on similar reasons discussed above for the #1 Young corehole. Pasley and Hazel (1990, 1995) indicated that there was no evidence of a sequence boundary at this contact because the contact climbed stratigraphically landward (from SSQ in Alabama toward Jason County in Mississippi, Pasley and Hazel, 1995 their figure 8). Our data shows the opposite: this surface is older in Mississippi and is younger in Alabama (Text-Figure 8). Also, Tew and Mancini (1995) showed several localities in Mississippi and Alabama where this contact was sharp and disconformable with broken and abraded shell material. Furthermore, sandstones in the Mint Spring Marl in the Wayne core could be part of a forced regression that is "the most unequivocal indication of a relative sea level fall in sea level" (Posamentier et al., 1992, p. 1704). The nature of the Red Bluff–Mint Spring remains, however, a problem. The sequence boundary hypothesis could be tested in up-dip and down-dip sections if more forced regressions or incised valleys are found.

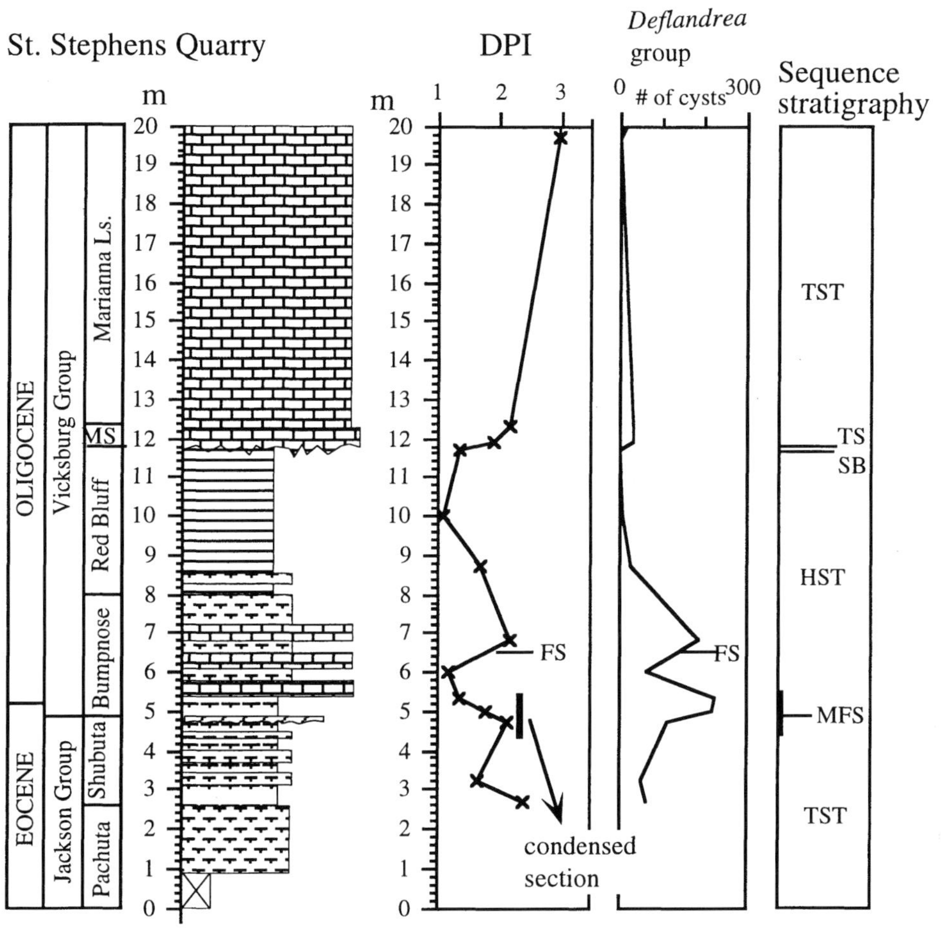

Text-Figure 7. Sequence stratigraphic interpretation for the St. Stephens Quarry section. Refer to Table 3 for the taxa used in calculating the dinocyst paleobathymetric index (DPI), and Text-Figure 4 for the key to lithologic symbols. An 80 cm-condensed section containing the maximum flooding surface and the Eocene–Oligocene boundary begins 20 cm below the Shubuta–Bumpnose contact and ends 60 cm above the contact. Meter 4.9 in this section corresponds to the Shubuta Clay/Red Bluff Clay–Bumpnose boundary (meter 3.95 in the column of Pasley and Hazel, 1990, fig. 4).

GENERAL CORRELATION

A correlation of all the studied sections (Text-Figure 8) shows the maximum flooding surface in all the sections except the St. Stephens Quarry to be coeval and approximately 7.5 c.u. (0.105 Ma) older than the Eocene–Oligocene boundary. At the St. Stephens Quarry section, this surface is 3 c.u. (0.042 Ma) younger than the other sections. This maximum flooding surface occurs in the middle part of the Shubuta Clay in the four southern Mississippi sections and at the Shubuta–Bumpnose contact in the St. Stephens Quarry section. This contact represents a minor flooding surface in southern Mississippi. The Pachuta Marl and the lower Shubuta Clay (the entire Shubuta in Alabama) constitute a transgressive system tract, with the upper Shubuta Clay and Bumpnose/Red Bluff/Forest Hill Formations representing the overlying highstand systems tract (see Text-Figure 2B). The stratigraphic condensation at the St. Stephens Quarry section corresponds to the accumulation of the early Oligocene highstand in southern Mississippi. The continuation of normal sedimentation in the St. Stephens Quarry area appears to correspond to the accumulation of the Forest Hill in southern Mississippi.

Up in the section, the Forest Hill–Mint Spring contact is interpreted as a sequence boundary which coincides with a transgressive surface. This contact is slightly older in the southeastern Mississippi sections relative to southeastern Alabama. The Mint Spring Marl and Marianna Limestone represent a transgressive system tract. However, a possible lowstand deposit (forced regression) is interpreted in the #1 Wayne section where a sandy deposit occur in Mint Spring.

Rates of accumulation increase from southern Alabama (St. Stephens Quarry) to the farthest section (R2089) in southern Mississippi. Areas in southeastern Mississippi experienced more rapid and prograding clastic sedimentation, whereas shelf-edge sediments were deposited much more slowly by longshore transport at St. Stephens Quarry area.

Our sequence stratigraphy interpretation is in general agreement with previous studies (in particular, Tew and Mancini, 1995; see Text-Figure 2B), although there are some distinct differences. First, the maximum flooding surface of the Shubuta–Forest Hill sequence occurs in the middle of the Shubuta Clay in southern Mississippi but at the Shubuta–

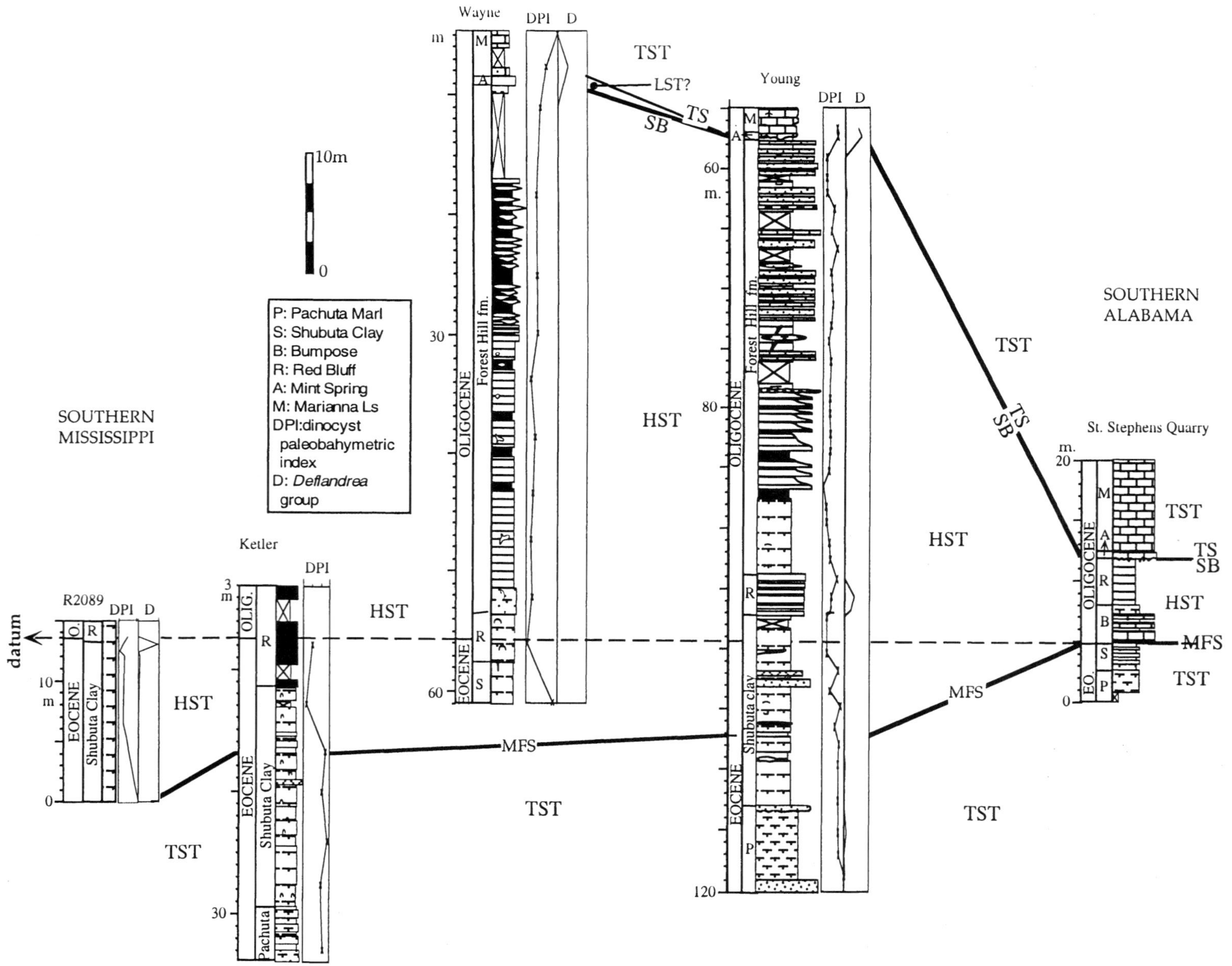

Text-Figure 8. Stratigraphic correlation of all studied sections. The DPI curve represents dinoflagellate paleobathymetric index (scale 1–4), while the D curve shows *Deflandrea* group abundance (scale 0–300 dinocysts). Text-Figures 4 and 7 also show the DPI and D curves for the #1 Young corehole and St. Stephens Quarry section, respectively. The Eocene–Oligocene boundary is chosen as a datum, and is 0.11 Ma younger than a maximum flooding surface identified in the sections. Note the increase in the thicknesses of the stratigraphic sections toward southern Mississippi.

Bumpnose contact in the St. Stephens Quarry section. Second, the age of this maximum flooding surface is latest Eocene. Third, the lower 60 cm of the Bumpnose Limestone in the St. Stephens Quarry section is recognized as a condensed section which accumulated while the upper Shubuta and Red Bluff were being deposited in southern Mississippi.

ACKNOWLEDGMENTS

We would like to thank Norman Frederiksen and Lucy Edwards (U.S. Geological Survey) for loan of the St. Stephens Quarry and R2089 Type Red Bluff samples, Larry Fearn (Mobil Exploration and Producing) for providing the Mississippi corehole samples, and Mobil for permission to publish the results from the Mississippi coreholes. Thanks to Lucy Edwards for the detailed review of this paper which has greatly improved it, and to Martin Farley and David Pocknall for their comments on an earlier version of this manuscript. This research was funded by NSF Grant No. EAR-9405154 to FEO. CJ thanks Colciencias for its support during the research.

References Cited

BAUM, G.R., and VAIL, P.R.
1988 Sequence stratigraphic concepts applied to Paleogene outcrops, Gulf and Atlantic Coastal basins. *In*: Wilgus, C.K., Hastings, B.S., Kendall, C.G.St.C., Posamentier, H.W., Ross, C.A., and Van Wagoner, J.C. (eds.), *Sea Level Changes: An Integrated Approach*. SEPM Special Publication, 42, p. 309–327.

BERGGREN, W.A., KENT, D.V., SWISHER II, C.C., and AUBRY, M.
1995 A revised Cenozoic Geochronology and Chronostratigraphy. *In*: Berggren, W.A., Kent, D.V., Aubry, M.-P., and Hardenbol, J. (eds.), *Geochronology, Time Scales and Global Stratigraphic Correlation*. SEPM Special Publication 54, pp. 129–212.

BRINKHUIS, H.
1994 Late Eocene to early Oligocene dinoflagellate cysts from the Priabonian type-area (Northeast Italy): biostratigraphy and paleoenvironmental interpretation. *Palaeogeography, Palaeoclimatology, Palaeocology*, 107: 121–163.

BYBELL, L.M.
1982 Late Eocene to early Oligocene calcareous nannofossils in Alabama and Mississippi. *Transactions of the Gulf Coast Association of Geological Societies*, 32: 295–302.

DALE, B.
1996 Dinoflagellate cyst ecology: modeling and geological applications. *In*: Jansonius, J., and MacGregor, D.C. (eds.), *Palynology: Principles and Applications*. American Association of Stratigraphic Palynologists Foundation, Dallas, 3, pp. 1249–1275.

DOCKERY, D.T. III., THOMPSON, D.E., and INGRAM, S.L.
1994 The Mobil-Mississippi Office of Geology Core-Hole Project. *Mississippi Geology*, 15: 8–13.

EDWARDS, L.E.
1989 Supplemented graphic correlation: A powerful tool for paleontologists and nonpaleontologists. *Palaios*, 4: 127–143.

GALLOWAY, W.E.
1989 Genetic stratigraphic sequences in basin analysis II: application to Northwest Gulf of Mexico Cenozoic Basin. *AAPG Bulletin*, 73: 143–154.

GALLOWAY, W.E., BEBOUT, D.G., FISHER, W.L., DUNLAP, J.B., CABRERA-CASTRO, R., LUGO-RIVIERA, J.E., and SCOTT, T.M.
1991 Cenozoic. *In*: Salvador, A. (ed.), *The Gulf of Mexico Basin*. Geological Society of America, The Geology of North America Volume J, pp. 245–324.

GREGORY, W.A., and HART, G.F.
1992 Towards a predictive model for the palynologic response to sea-level changes. *Palaios*, 7: 3–33.

HABIB, D., and MILLER, J.A.
1989 Dinoflagellate species and organic facies evidence of marine transgression and regression in the Atlantic coastal plain. *Palaeogeography, Palaeoclimatology, Palaeoecology*, 74: 23–47.

HAQ, B.U., HARDENBOL, J., and VAIL, P.R.
1988 Mesozoic and Cenozoic chronostratigraphy and cycles of sea-level changes. *In*: Wilgus, C.K., Hastings, B.S., Kendall, C.G.St.C., Posamentier, H.W., Ross, C.A., and Van Wagoner, J.C. (eds.), *Sea Level Changes: An Integrated Approach*. SEPM Special Publication, 42, pp. 72–108.

HAZEL, J.E., MUMMA, M.D., and HUFF, W.J.
1980 Ostracode biostratigraphy of the Lower Oligocene (Vicksburgian) of Mississippi and Alabama. *Transactions of the Gulf Coast Association of Geological Societies*, 30: 361–401.

JARAMILLO, C.
1995 *Dinoflagellate cyst biostratigraphy and palynofacies of the Late Eocene–Early Oligocene in southern Mississippi and Alabama, U.S. Gulf Coast*. Unpubl. M.S. thesis., University of Missouri-Rolla, 97 pgs.

JARAMILLO, C.A., and OBOH-IKUENOBE, F.E.
1999 Sequence stratigraphic Interpretations from palynofacies, dinocyst and lithological data of Upper Eocene–Lower Oligocene strata in southern Mississippi and Alabama. *Palaeogeography, Palaeoclimatology, Palaeoecology*, 145: 259–302.

JARAMILLO, C., and YEPES, O.
1995 Palynological recognition of a maximum flooding surface. *AAPG Annual Convention, Official Program*, 4, p. 47A.

KELLER, G.
1985 Eocene and Oligocene stratigraphy and erosional unconformities in the Gulf of Mexico and Gulf Coast. *Journal of Paleontology*, 59: 882–903.

KOVACH, W.L.
1989 Comparisons of multivariate analytical techniques for use in Pre-Quaternary plant paleoecology. *Review of Palaeobotany and Palynology*, 60: 255–282.

LORENTE, M.
1986 Palynology and palynofacies of the upper Tertiary in Venezuela. *Dissertationes Botanicae*, 99: 222 pgs.

LOUTIT, T.S., BAUM, G.R., and WRIGHT, R.C.
1983 Eocene–Oligocene Sea Level Changes as Reflected in Alabama Outcrop sections. AAPG Bulletin, 67: 506.

LOUTIT, T.S, HARDENBOL, J., VAIL, P.R., and BAUM, G.R.
1988 Condensed sections: the Key to age determination and correlation of continental margin sequences. *In*: Wilgus, C.K., Hastings, B.S., Kendall, C.G.St.C., Posamentier, H.W., Ross, C.A., and Van Wagoner, J.C. (eds.), *Sea Level Changes: An Integrated Approach*. SEPM Special Publication, 42, pp. 183–213.

MANCINI, E.A.
1979 Eocene–Oligocene Boundary in Southwest Alabama. *Transactions of the Gulf Coast Association of Geological Societies*, 29: 282–286.

MILLER, K.G., THOMPSON, P.R., and KENT, D.V.
1993 Integrated late Eocene–Oligocene stratigraphy of the Alabama coastal plain: correlation of hiatuses and stratal surfaces to glacioeustatic lowerings. *Paleoceanography*, 8: 313–331.

OBOH, F.E.
1992 Multivariate statistical analyses of palynodebris from the middle Miocene of the Niger Delta and their environmental significance. *Palaios*, 7: 559–573.

OBOH, F.E., and YEPES, O.
1995 Palynofacies analyses of #1 Ketler and #1 Wayne County Lime Company Wells. MOBIL Internal Report, Dallas. 14 pgs.

PASLEY, M.A., and HAZEL, J.E.
1990 Use of organic petrology in graphic correlation of biostratigraphic data in sequence stratigraphic interpretations: example from the Eocene–Oligocene boundary section, St. Stephens Quarry, Alabama. *Gulf Coast Association of Geological Societies Transactions*, 40: 661–683.
1995 Revised sequence stratigraphic interpretation of the Eocene–Oligocene boundary interval, Mississippi and Alabama, Gulf Coast Basin, USA. *Journal of Sedimentary Research*, B65: 160–169.

POSAMENTIER, H.W., ALLEN, G.P., JAMES, D.P., and TESSON, M.
1992 Forced regressions in a sequence stratigraphic framework: concepts, examples and exploration significance: *AAPG Bulletin*, 76: 1687–1709

SHAW, A.B.
1964 *Time in Stratigraphy*. MacGraw-Hill, New York, 365 pgs.

SIESSER, W.G.
1983 Paleogene calcareous nannoplankton biostratigraphy: Mississippi, Alabama and Tennessee. *Mississippi Bureau of Geology Bulletin*, 125, 61 pgs.

SIESSER, W.G., FITZGERALD, B.G., and KRONMAN, D.J.
1985 Correlation of Gulf Coast provincial Paleogene stages with European standard stages. *Geological Society of America Bulletin*, 96: 827–831.

SOKAL, R.R., and MICHENER, C.D.
1958 A statistical method for evaluating systematic relationships. *University of Kansas Sciences Bulletin*, 38: 1409–1438.

SYSTAT
1992 *Statistics, Version 5.2*. SYSTAT, Inc., Evanston, IL, 724 pgs.

TEW, B.H., and MANCINI, E.A.
1995 An integrated lithostratigraphic, biostratigraphic, and sequence stratigraphic approach to paleogeographic reconstruction: examples from the Upper Eocene and Lower Oligocene of Alabama and Mississippi. Palaios, 10: 133–153.

TRAVERSE, A.
1988 *Paleopalynology*. Allen & Unwin, Boston, 600 pgs.

VAN VERGEN, P., N., ALFERINK, J.J., and KERP, J.
1990 Recognition of organic matter types in standard palynological slides. *Mededelingen Rijks Geologische Dienst*, 45: 1–23.

WALL, D., DALE, B., LOHMANN, G.P., and SMITH, W.K.
1977 The environmental and climatic distribution in modern marine sediments from regions in the North and South Atlantic oceans and adjacent seas. *Marine Micropaleontology*, 2: 121–200.

Authors' addresses

CARLOS A. JARAMILLO
Paleobotany Laboratory
Florida Museum of Natural History
University of Florida
Gainesville, Florida 32611
U.S.A.

FRANCISCA E. OBOH-IKUENOBE
Department of Geology and Geophysics
University of Missouri-Rolla
Rolla, Missouri 65409-0410
U.S.A.

PALYNOSTRATIGRAPHIC ZONATION OF THE TERTIARY SEDIMENTS OF THE KERALA BASIN, INDIA

M. R. Rao

Abstract

The Tertiary sedimentary sequence in the Kerala Basin, India, is represented by the Quilon and Warkalli formations. The entire Cenozoic sequence at Kerala has been designated as the Malabar Super Group, which encompasses the Quaternary and Tertiary formations in a recent lithostratigraphic classification based on the overall analysis of surface and subsurface lithofacies and their interrelationships. The palynoflora of the subsurface (Arthungal, Kalarakod boreholes) Tertiary sediments has permitted the recognition of the three cenozones — *Triangulorites bellus* Cenozone, *Crassoretitriletes vanraadshooveni* Cenozone, and *Malvacearumpollis bakonyensis* Cenozone, as well as marker units encompassing Eocene to Lower Miocene strata. Palynological study of the Tertiary sediments of Kerala exposed at Meenkunnu (Cannanore), the Kannanellur–Kundara area, Padappakkara, and Varkala has resulted in recovery of a variety of algal and fungal remains, pteridophytic spores, and angiospermous pollen assemblages indicating an Early to Middle Miocene age.

Arborescent pollen, in general, dominates over herbaceous types in the study area. Composition of the palynological assemblages indicate the existence of tropical rain forests with relatively high rainfall during the time of deposition. Ecological analysis of the assemblage indicates several palaeoassociations of montane, lowland, freshwater swamps and water edge, sandy beach, mangrove, and back-mangrove vegetation. The terrestrial elements of upland relief flora and lowland vegetation tend to merge with the fresh water constituents and dominate the total assemblage. Dinoflagellate cysts (e.g., *Thalassiphora, Achomosphaera*) and pollen (*Brownlowia, Nypa, Rhizophora* and *Calamus)* suggests deposition under brackish-water mangrove swamps all along the coastal line. Luxuriant growth of palms seems to have occurred on sandy beaches. Divergent views on the age of the Quilon and Warkalli formations have been reconsidered on the basis of new palynological evidence.

INTRODUCTION

The Tertiary of the Kerala coast is well known in Indian geology as the "Warkalli" and "Quilon Beds" from the type localities where the sediments were first described by King (1882) and Foote (1883). The most detailed and comprehensive description of the fauna in the Quilon limestone, with special reference to the Miocene mollusca from Quilon and Kerala is by Dey (1962), who compiled an exhaustive collection of fossils from Padappakkara .

STRATIGRAPHY

The Tertiary and recent sediments on the Kerala coast rest directly upon the Archaean crystalline complex along the west coast. The Tertiary formations include (1) the Warkalli beds of variegated sandstones and clays, white plastic clays, carbonaceous clays, sandy clays associated seams of lignite; and (2) the Quilon beds which consist of fossiliferous limestone intercalated with thick beds of variegated sands, carbonaceous clays, sandy clays and sands. These are overlain by sub–recent marine and estuarine formations (Poulose and Narayanaswami, 1968). The recent lithostratigraphic classification (Raha et al., 1983) of the entire Cenozoic sequence of Kerala is based on the analysis of surface and subsurface samples and their relationships, and has been designated as the Malabar Super Group. A laterite horizon marks a distinct unconformity between the Vembanad Formation and the Warkalli Group. The Warkalli Group consists of the Ambalapujha, Quilon and Mayyanad formations. The Ambalapujha and Mayyanad formations are terrigenous whereas the Quilon Formation is marine in nature (Raha et al., 1983, Table 1).

Rao, M. R.
2001 Palynostratigraphic zonation of the Tertiary sediments of the Kerala Basin, India; *In*: Goodman, D.K., and Clarke, R.T. (eds.), *Proceedings of the IX International Palynological Congress, Houston, Texas, U.S.A., 1996*; American Association of Stratigraphic Palynologists Foundation, p. 277–289.
 ISBN 0–931871–06–9

PALYNOLOGY

Palynological studies of the Tertiary sediments of Kerala have been conducted by Rao and Vimal (1953), Potonié and Sah (1960), Ramanujam (1977, 1987), Rao and Ramanujam (1978, 1982), Kar and Jain (1981), Raha et al. (1987), Rajendran et al. (1989), Rao (1990, 1995), Rao and Rajendran (1996), and Rao and Nair (1998).

The palynological assemblages recovered from Arthungal and Kalarakod (boreholes), Meenkunnu (Cannanore), the Kannanellur-Kundara area, Padappakkara, Varkala (surface sections) include 135 genera and 175 species of pteridophytic spores and angiospermous pollen. Dinoflagellate cysts and fungal remains are also present. Detailed descriptions of these fossils are not included in this present paper. However, a list of important taxa recovered is given in Table 2.

PALYNOSTRATIGRAPHIC ZONATION

Based on the first and last appearances of taxa and their maximum development, decline, absence and restricted occurrence, the Arthungal borehole has been divided into three cenozones. In ascending order, these are: (1) *Triangulorites bellus* Cenozone, (2) *Crassoretitriletes vanraadshooveni* Cenozone, and (3) *Malvacearumpollis bakonyensis* Cenozone. Each cenozone has been formally instituted in accordance with the *International Stratigraphic Guide* (Hedberg, 1976). A brief account of each cenozone is discussed below. The palynofloral composition of each of the three cenozones is distinctly different.

TABLE 1. Lithostratigraphic classification of the Cenozoic sedimentary sequence of the Kerala coastal region (modified after Raha et al., 1983).

M A L A B A R S U P E R G R O U P		**Vembanad Formation** (3–60 m)	Soil and alluvium Beach and dune sands, lime-shell deposits, older red 'Teri' sands, marine and estuarine deposits, sands, peat and black steaky and sandy calcareous clays. Gravel beds with laterite gravels at places.
	---	------------------------	----------------- Unconformity (marked by laterite) -----------------
	W A R K A L L I G R O U P	**Ambalapuzha Formation** (3–140 m)	Pebbly and coarse sand with peat and variegated and mottled clay. Grey sand and white plastic steaky clay and lignite. Alternations of sand and steaky grey and carbonaceous clay often with lignite bands, with coarse pebbly sand near the base.
		Quilon Formation (0.5–130 m)	Limestone with bands of calcareous clay and sand in various proportions, grey fossiliferous and calcareous clay near the base.
		Mayyanad Formation (1.0–270m)	Coarse to medium grained sandstones with interbedded grey and white clay and lignite beds. Facies grades laterally to sand clay alternations with lignite bands towards the central part of the basin.
	---	---------------------	------------------------------- Unconformity -------------------------------
			Basement (Gneiss, charnockites, leptynites, etc.)

Triangulorites bellus Cenozone

This cenozone is identified in the Arthungal borehole, Alleppey district, Kerala, between depths of 440.70 m and 272.70 m (Text-Figure 1). The lithology is mainly sand and clayey sand. Characteristic palynofossils include *Ctenolophonidites, Lakiapollis, Cheilanthoidspora, Liliacidites, Neocouperipollis, Polycolpites, Sastriipollenites, Meliapollis, Tricolporopollis, Striacolporites, Polybrevicolporites, Triangulorites, Jacobipollenites,* and *Verrucolporites.* The significant features are the restricted and common occurrence of *Tricolporopollis decoris* (20%), *Jacobipollenites arthungalensis* (18%), *Liliacidites keralaensis* (15%), *Neocouperipollis kutchensis* (11%) and restricted but rare representation of *Triangulorites*

PLATE 1

All photomicrographs are enlarged *ca.* x750. Coordinates of specimens refer to the stage of BH2 Olympus microscope no. 217267.

1 *Palmaepollenites kutchensis* Venkatachala & Kar 1969. Kalarakod borehole. BSIP slide no. 10837, coordinates 12.5x155.0.

2 *Compositoipollenites alleppeyensis* Rao 1995. Kalarakod borehole. BSIP slide no. 10841, coordinates 8.6x155.0.

3 *Ctenolophonidites costatus* (van Hoeken-Klinkenberg) van Hoeken-Klinkenberg 1966. Kalarakod bore hole. BSIP slide no. 10826, coordinates 16.ox132.7.

4 *Ornatetradites kalarakodensis* Rao 1995. Kalarakod borehole. BSIP slide no. 10851, coordinates 8.0x 133.0.

5 *Crassoretitriletes vanraadshooveni* Germeraad et al., 1968, Kalarakod borehole. BSIP slide no. 10828, coordinates 4.5x152.5.

6 *Margocolporites tsukadae* Ramanujam 1966. Kalarakod borehole. BSIP slide no. 10827, coordinates 7.3x137.3.

7 *Sastriipollenites trilobatus* Venkatachala & Kar 1969. Kalarakod borehole. BSIP slide no. 10832, coordinates 21.0x134.0.

8 *Dipterocarpuspollenites retipilatus* (Kar & Jain) Kar 1993. Kalarakod borehole. BSIP slide no. 10825, coordinates 13.0x146.4.

9 *Lakiapollis ovatus* Venkatachala & Kar 1969. Meenkunnu (Cannanore, surface). BSIP slide no. 11434, coordinates 8.1x139.2.

10 *Trisyncolpites ramanujamii* Kar 1979. Kalarakod borehole. BSIP slide no. 10843, coordianates 19.0x134.0.

11 *Periretitricolpites anambraensis* Jan Du Chene et al., 1974. Kalarakod borehole. BSIP slide no. 10846, coordinates 21.0x125.3.

12 *Ericipites congoensis* Sah 1967. Kalarakod borehole. BSIP slide no. 10828, coordinates 6.0x143.6.

13 *Malvacearumpollis grandis* Sah 1967. Kalarakod borehole. BSIP slide no.10853, coordinates 14.0x142.0.

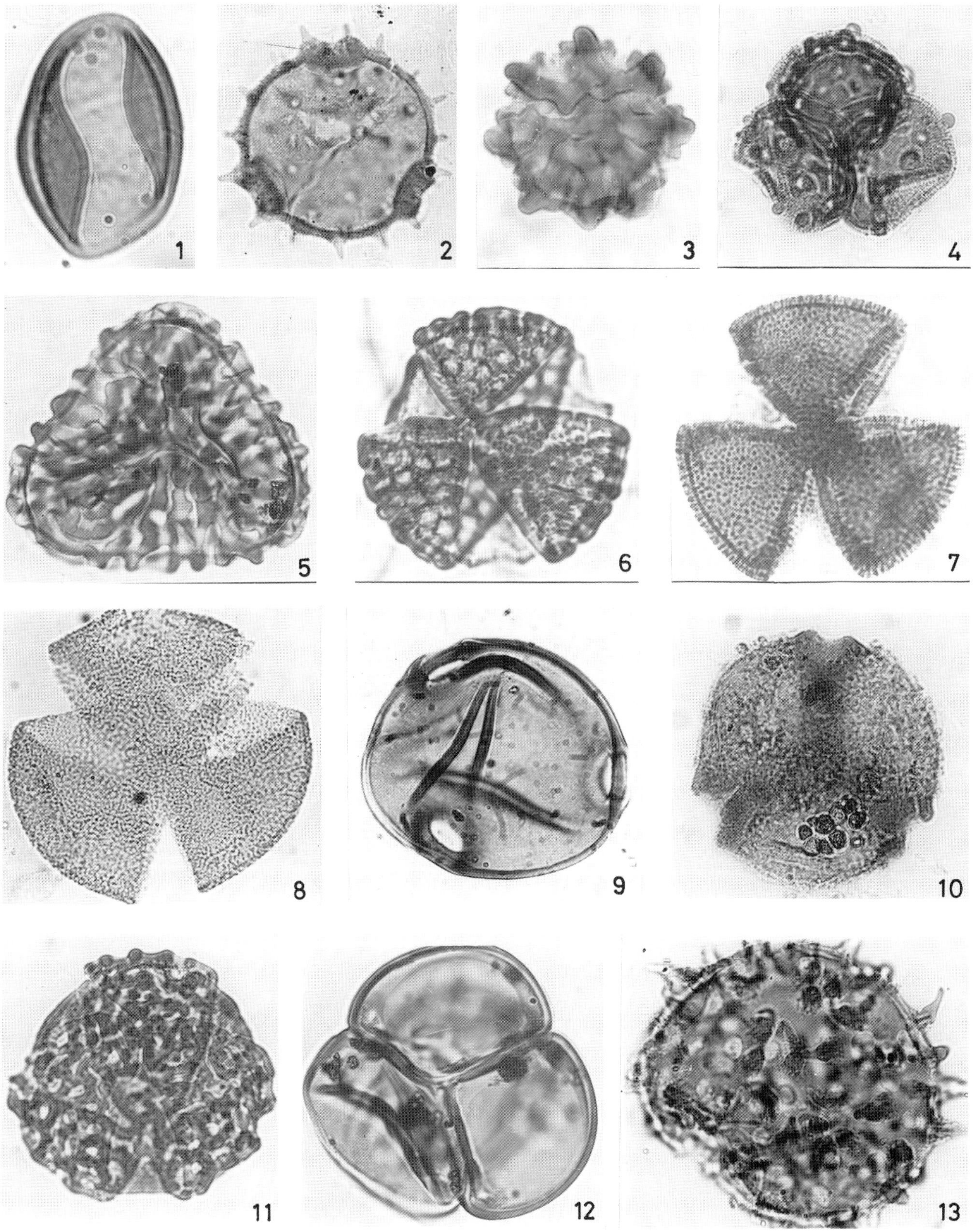
1
2
3
4
5
6
7
8
9
10
11
12
13

TABLE 2. Palynomorph assemblages from Arthungal and Kalarakod (boreholes), Meenkunnu, the Kannanellur-Kundara area, Padappakkara and Varkala (surface) sediments.

Name of taxa	Arthungal	Kala-rakod	Meen-kunnu (Canna-nore)	Kanna-nellur–Kundara Area	Padap-pakkara	Varkala
Pteridophytes						
Alsophilidites kerguelensis	-	-	-	-	+	-
Biretisporites meghalayaensis	-	+	-	-	-	-
Chielanthoidspora monoleta	+	+	+	-	-	-
C. mioceneca	+	-	+	+	+	-
Cibotidites kundavaensis	-	-	-	-	+	-
Cyathidites australis	-	-	+	-	-	-
Crassoretitriletes vanraadshooveni	+	+	+	+	+	+
Dandotiaspora telonata	-	+	-	-	-	-
Dictyophyllidites cannanorensis	-	-	+	-	-	-
Foveosporites miocenicus	-	-	-	-	+	-
Foveotriletes bifurcatus	-	-	-	-	+	-
Gleicheniidites senonicus	-	-	-	-	+	-
Intrapunctisporis intrapunctis	+	-	+	-	+	-
I. harudiensis	-	+	+	+	+	-
Laevigatosporites ovatus	+	-	+	-	-	-
Lycopodiumsporites bellus	-	-	-	-	+	+
L. austraclavatidites	-	-	-	-	+	+
Lygodiumsporites lakiensis	-	+	+	-	+	-
L. eocaenicus	-	+	+	-	-	-
L. padappakkarensis	+	+	+	+	+	+
Neyvelisporites bolkhovitinae	-	+	-	-	+	+
Osmundacidites kutchensis	+	-	-	-	+	-
Pilamonoletes excellensus	-	-	-	+	-	-
Polypodiaceaesporites intrapunctis	+	-	-	-	+	-
P. major	-	+	-	-	-	-
P. chatterjii	-	-	+	-	+	-
Polypodiisporites favus	-	+	+	-	-	-
P. ornatus	-	-	+	-	+	+
P. miocenicus	+	-	-	-	+	-
P. ratnamii	+	-	-	-	+	+
P. speciosus	-	+	-	-	-	-
P. tuberculensis	-	+	+	-	-	-
Pteridacidites vermiverrucatus	-	+	-	-	-	-
P. africanus	-	+	-	-	+	-
Schizaeoisporites grandiformis	-	-	-	-	+	+
Striatriletes susannae	+	+	-	+	+	-
Surmaspora karii	-	-	+	-	-	-
Todisporites major	-	+	-	-	+	-
T. minor	-	-	-	-	+	-
Verrucosisporites sp.	+	-	-	-	-	-

TABLE 2 (continued). Palynomorph assemblages from Arthungal and Kalarakod (boreholes), Meenkunnu, the Kannanellur-Kundara area, Padappakkara and Varkala (surface) sediments.

Name of taxa	Arthungal	Kala-rakod	Meen-kunnu (Canna-nore)	Kanna-nellur–Kundara Area	Padap-pakkara	Varkala
Angiosperm pollen						
Araliaceoipollenites quilonensis	-	-	-	-	+	-
Arecipites keralensis	-	-	-	-	+	-
Assamiapollenites sp.	-	-	+	-	-	-
Anacolosidites luteoides	-	+	-	-	-	-
Bacuspinulopollenites baculatus	-	-	-	+	-	-
Caryophyllidites warkalliensis	-	-	-	-	-	+
Chenopodipollis mioceneca	+	+	-	-	+	-
Clavainaperturites clavatus	+	-	-	-	+	-
Clavaperiporites jacobii	-	+	+	-	+	+
Clavatricolporites brevicolpatus	-	-	-	-	-	+
Compositoipollenites africanus	-	+	-	-	-	-
C. alleppeyensis	-	+	-	-	-	-
C. minimus	-	+	-	-	-	+
Crotonoidaepollenites euphorbiodes	-	-	+	-	-	-
Ctenolophonidites costatus	+	+	+	+	+	+
Cupanidites punctatus	-	-	-	-	+	-
Dermatobrevicolporites dermatus	+	+	+	+	-	-
Dicolpopollis edavensis	-	-	-	-	+	+
Diporites crassiporatus	-	-	-	-	-	+
Dipterocarpuspollenites retipilatus	-	-	+	+	+	-
Droseridites spinosa	-	-	-	-	-	+
Ericipites congoensis	-	+	-	-	+	-
Favitricolporites magnus	+	+	+	+	-	-
Florscheutizia levipoli	-	-	-	-	+	-
Gemmatricolpites saxenii	-	-	+	-	-	-
Gemmastephanocolpites sp.	-	-	+	-	-	-
Grimsdalea sp.	-	-	-	+	-	-
Halorogacidites verrucatus	-	-	-	-	+	-
Heterocolpites combretoides	-	-	-	-	+	+
Hippocrateaceaedites	-	-	-	-	+	-
Iridacidites warkalliensis	-	-	-	+	-	+
Jacobipollenites arthungalensis	+	+	-	-	-	-
Jandufouria searogiformis	-	-	-	-	+	-
Lacrimapollis brownlowioides	-	-	-	-	+	-
Lakiapollis ovatus	+	+	+	+	+	+
Liliacidites keralaensis	+	-	-	-	-	-
Longapertites hammenii	+	-	-	-	+	-
Loranthipites elegans	-	-	-	-	+	-
Maculoporites quilonensis	-	-	-	-	+	-
Malvacearumpollis bakoyensis	+	+	+	+	-	-

TABLE 2 (continued). Palynomorph assemblages from Arthungal and Kalarakod (boreholes), Meenkunnu, the Kannanellur-Kundara area, Padappakkara and Varkala (surface) sediments.

Name of taxa	Arthungal	Kala-rakod	Meen-kunnu (Canna-nore)	Kanna-nellur–Kundara Area	Padap-pakkara	Varkala
M. grandis	-	+	+	-	-	-
Malvacipollis sp.	-	+	-	+	-	+
Marginipollis kutchensis	-	-	-	+	-	-
Margocolporites sahnii	+	-	-	-	-	-
M. sitholeyi	-	+	-	+	-	-
Margocolporites tsukadae	+	+	+	+	+	-
Meliapollis quadrangularis	+	+	+	-	-	-
M. ramanujamii	+	-	+	-	-	+
Meliapollis tamilii	-	-	-	+	-	-
Myricipites singhii	-	+	+	+	-	-
Neocouperipollis kutchensis	+	+	+	-	+	+
Ornatetradites chandae	-	+	-	-	+	-
O. droseroides	-	-	-	-	+	-
O. keralaensis	-	+	-	-	-	-
Palaeosantalaceaepites minutus	-	-	+	-	-	-
P. ovatus	-	-	-	+	-	-
Palmidites maximus	+	+	-	-	-	-
Palmaepollenites eocaenicus	+	-	-	-	-	-
P. keralensis	+	+	-	+	-	-
P. kutchensis	+	+	-	-	+	-
Pelliceroipollenites langenheimii	-	-	+	-	-	-
Periretitricolpites anambraensis	-	+	-	-	-	-
Plumbaginacipites navalii	-	-	-	+	-	-
P. neyvelii	-	-	+	-	-	-
Polybrevicolporites karii	+	-	-	-	-	-
Polyporina globosa	-	-	-	+	-	-
P. multiporosa	+	+	-	+	-	-
Proteacidites triangulus	+	+	+	+	+	-
P. truncatus	+	-	+	+	+	-
Psilodiporites sp.	-	-	+	-	-	-
Quilonipollenites sahnii	+	+	+	+	+	+
Retimonosulcites ramanujamii	-	-	+	-	-	-
R. ovatus	-	-	+	-	-	-
Retipilonapites arcotense	+	-	-	-	-	-
R. delicatissimus	-	-	+	+	-	-
Retistephanocolpites sp.	+	-	-	-	-	-
Retitrescolpites indicus	-	-	+	+	-	-
R. singularis	+	-	-	-	+	-
Retitricolporites crassioratus	+	-	-	-	+	-
Retitriporites quilonensis	-	+	-	-	-	+
Rhoipites anacardioides	-	-	-	+	-	+
Sapotaceoidaepollenites keralensis	-	-	-	-	+	-

TABLE 2 (continued). Palynomorph assemblages from Arthungal and Kalarakod (boreholes), Meenkunnu, the Kannanellur-Kundara area, Padappakkara and Varkala (surface) sediments.

Name of taxa	Arthungal	Kala-rakod	Meen-kunnu (Canna-nore)	Kanna-nellur–Kundara Area	Padap-pakkara	Varkala
Sastriipollenites trilobatus	+	+	-	-	-	-
Sonneratiopollis bellus	-	-	-	-	+	+
Spinizonocolpites quilonensis	-	-	-	-	+	-
Striacolporites cephalus	+	-	+	-	-	-
S. striatus	-	+	+	-	-	-
S. ovatus	+	-	-	-	-	-
Subtriporopollis rotundis	-	+	+	+	-	-
Symplocoipollenites baksii	-	-	-	-	-	+
Talisipites elegans	-	-	-	-	+	-
Thomsonipollis sp.	-	-	-	-	+	-
Tribrevicolporites duttae	-	-	+	-	-	-
T. sarkarii	-	-	+	-	-	-
Triangulorites bellus	+	+	-	-	-	-
Tricolpites crassireticulatus	+	-	+	-	-	-
matanamadhensis	+	+	+	-	-	-
T. retibaculatus	+	+	-	+	-	-
T. reticulatus	-	+	+	-	-	-
Tricolporopilites differentialis	-	-	-	+	-	-
T. pseudoreticulatus	-	+	-	+	-	-
Tricolporopollis alleppeyensis	-	-	+	+	-	-
T. cannanorensis	-	-	+	-	-	-
T. decoris	-	-	+	-	-	-
T. matanamadhensis	+	+	+	-	-	-
T. rubra	+	-	-	-	-	-
Trilatiporites noremii	+	-	-	-	-	-
Triorites quilonensis	-	-	-	+	-	-
Triporopollenites meenkunnuensis	-	-	+	-	-	-
T. minutus	-	-	-	-	-	-
T. robustus	-	+	+	-	-	-
Trisyncolpites ramanujamii	+	+	-	-	-	-
Umbelliferoipollenites sp.	-	-	-	-	+	-
Verrualetes assamicus	-	-	-	+	-	-
Verrutricolporites verrucus	+	-	-	-	-	-
Verrutricolpites sp.	-	-	-	+	-	-
Verrutriporites kundaraensis	-	-	-	+	-	-
V. lundensis	-	-	-	-	+	-
Warkallipollenites erdtmanii	-	-	-	-	+	-
W. ramanujamii	-	-	-	+	-	-
Zonocostites indicus	-	-	-	-	+	-

(after Kar and Jain, 1981; Rao and Ramanujam, 1978, 1982; Ramanujam, 1987; Rao, 1990, 1995; Rao and Rajendran, 1996; Rao and Nair, 1998)

bellus (8%) and *Verrucolporites verrucus* (8%). *Tricolporopollis decoris* is abundant (26%) in the lower part but is rare towards the top (3%).

Crassoretitriletes vanraadshooveni Cenozone

This cenozone occurs between depths of 271.70 m and 151.70 m (Text-Figure 1). The lithology mainly consists of sandy clay with a band of lignite. The upper part of the sediments is consists of sandy clay and it is conformably overlain by the clayey sand of *Malvacearumpollis bakonyensis* Cenozone. Characteristic fossils in this cenozone include *Crassoretitriletes, Polypodiisporites, Lygodiumsporites, Polypodiaceaesporites, Ctenolophonidites, Margocolporites, Lakiapollis,* and *Trisyncolpites.* The appearance of *Crassoretitriletes* and *Trisyncolpites* is significant in these sediments. *Trisyncolpites* is restricted to this cenozone. Germeraad et al. (1968) studied in detail the distribution of *Crassoretitriletes* in pantropical areas and according to them, this genus generally occurs in the Lower Oligocene through the Miocene. *Trisyncolpites* is an important taxon in the Oligocene sediments of Kutch, western India, and a cenozone *Trisyncolpites ramanujamii* has been designated after it (Kar, 1979). *Lygodiumsporites padappakkarensis* (10%) is rare in the lower part of the section and gradually decreases towards the top of the cenozone. *Ctenolophonidites costatus* (49%) and *Lakiapollis ovatus* (25%) are abundant taxa in this cenozone.

Malvacearumpollis bakonyensis Cenozone

This cenozone occurs between depths of 151.70 m and 7.70 m (Text-Figure 1). The lithology mainly consists of clayey sand, however, in the upper part sandy clay alternates with sand. Characteristic taxa include *Quilonipollenites, Striatriletes, Chenopodipollis,* and *Malvacearumpollis.* This cenozone is also characterized by the predominance of *Lakiapollis ovatus* (30%). *Striatriletes susannae, Triporopollenites minutus, Malvacearumpollis bakonyensis* and *Chenopodipollis miocenica* appear for the first time in and are restricted to this cenozone whereas *Lygodiumsporites* and *Palmaepollenites* are absent. The relative increase in *Crassoretitriletes* and *Polypodiacaesporites* has been previously documented.

PALYNOSTRATIGRAPHIC CORRELATION

Palynostratigraphic correlation between two boreholes (Arthungal and Kalarakod) of the Alleppey district and surface sediments (Meenkunnu, K-K Road section, Padappakkara and Varkala) in the Kerala Basin is presented in Text-Figure 2. The Kalarakod borehole has the thickest section of the two boreholes and is represented by three cenozones: *Triangulorites bellus* Cenozone; *Crassoretitriletes vanraadshooveni* Cenozone; and *Malvacearumpollis bakonyensis* Cenozone. These cenozones are correlated with the standard reference section in the Arthungal borehole.

The lowermost *Triangulorites bellus* Cenozone in the Kalarakod borehole is characterized by the presence of *Dandotiaspora telonata, Cheilanthoidspora monoleta, Neocouperipollis kutchensis, Sastriipollenites trilobatus, Triangulorites bellus, Striacolporites striatus, Tricolpites reticulatus, Tricolporopollis matanamadhensis, Periretitricolpites anambraensis* and *Anacolosidites luteoides.* A comparison of the *Triangulorites bellus* Cenozone and the *Triangulorites bellus* Cenozone from the Arthungal borehole reveals that both assemblages are more or less the same except for the presence of *Periretitricolpites, Anacolosidites* and *Dandotiaspora* in the Kalarakod borehole (see Rao, 1995).

The *Crassoretitriletes vanraadshooveni* Cenozone marks the middle part of the Kalarakod borehole; this interval contains *Crassoretitriletes vanraadshooveni, Polypodiisporites speciosus, Trisyncolpites ramanujamii, Margocolporites tsukadae, Lakiapollis ovatus* and *Ctenolophonidites costatus.* A comparison of the *Crassoretitriletes vanraadshooveni* Cenozone with the Arthungal borehole section indicates that both assemblages are homotaxial. *Lakiapollis ovatus* and *Ctenolophonidites costatus* are dominant in both the cenozones (see Rao, 1995).

The uppermost zone in the Kalarakod borehole, the *Malvacearumpollis bakonyensis* Cenozone, contains *Striatriletes susannae, Pteridacidites vermiverrucatus, Quilonipollenites sahnii, Chenopodipollis miocenica* and *Compositoipollenites alleppeyensis.* With the exception of *Pteridacidites* and *Compositoipollenites,* the assemblages of the uppermost cenozone from the Kalarakod borehole and Arthungal borehole are similar in composition (see Rao, 1995).

It is evident from the palynological data that the Arthungal and Kalarakod boreholes penetrate strata ranging from Eocene to Lower Miocene in age. Based on palynological data, an attempt has been made to correlate surface and subsurface Tertiary sediments of the Kerala Basin. It was observed that the *Malvacearumpollis bakonyensis* Cenozone (Lower Miocene) in the borehole are correlative with the lower part of the surface sediments (Text Figure 2).

AGE AND COMPARISON

On the basis of foraminifera, ostracoda, and mollusca occurrences, the Quilon Formation is assigned an Early to Middle Miocene age (Jacob and Sastry, 1952; Dey, 1962). The overlying Warkalli Formation is considered to be Late Miocene to Pliocene in age (Poulose and Narayanaswami, 1968).

Ramanujam (1982) documented a striking similarity between the assemblages from the Quilon and Warkalli formations. He concluded (on the basis of surface and subsurface assemblages) that the entire Tertiary sequence in the Kerala Basin constitutes a single time-transgressive group of Early to Middle Miocene age.

Raha et al. (1987) initially reported Eocene to Early Miocene assemblages from 600 m depth in a borehole near Ambalapuzha in the Alleppey district, Kerala. A recent palynological study (Rao, 1990, 1995) of the Arthungal and Kalarakod boreholes also provides further support for the age interpretation. The palynofloral succession has been divided into three distinct cenozones and each zone contains age definitive palynofossils, the details of which are given in the above mentioned papers.

A rich palynoflora has been recovered from Tertiary sediments from the Quilon and Warkalli formations exposed at Meenkunnu, in the Kannanellur-Kundara area, Padappakkara, and Varkala. The common genera include *Lygodiumsporites, Gleicheniidites, Crassoretitriletes, Pteridacidites, Foveosporites, Foveotriletes, Pteridacidites, Schizaeoisporites, Polypodiisporites, Retipilonapites, Iridacidites, Quilonipollenites, Longapertites, Dicolpopollis, Dipterocarpuspollenites, Plumbaginacipites, Warkallipollenites, Retitrescolpites, Crototricolpites, Marginipollis, Rhoipites,*

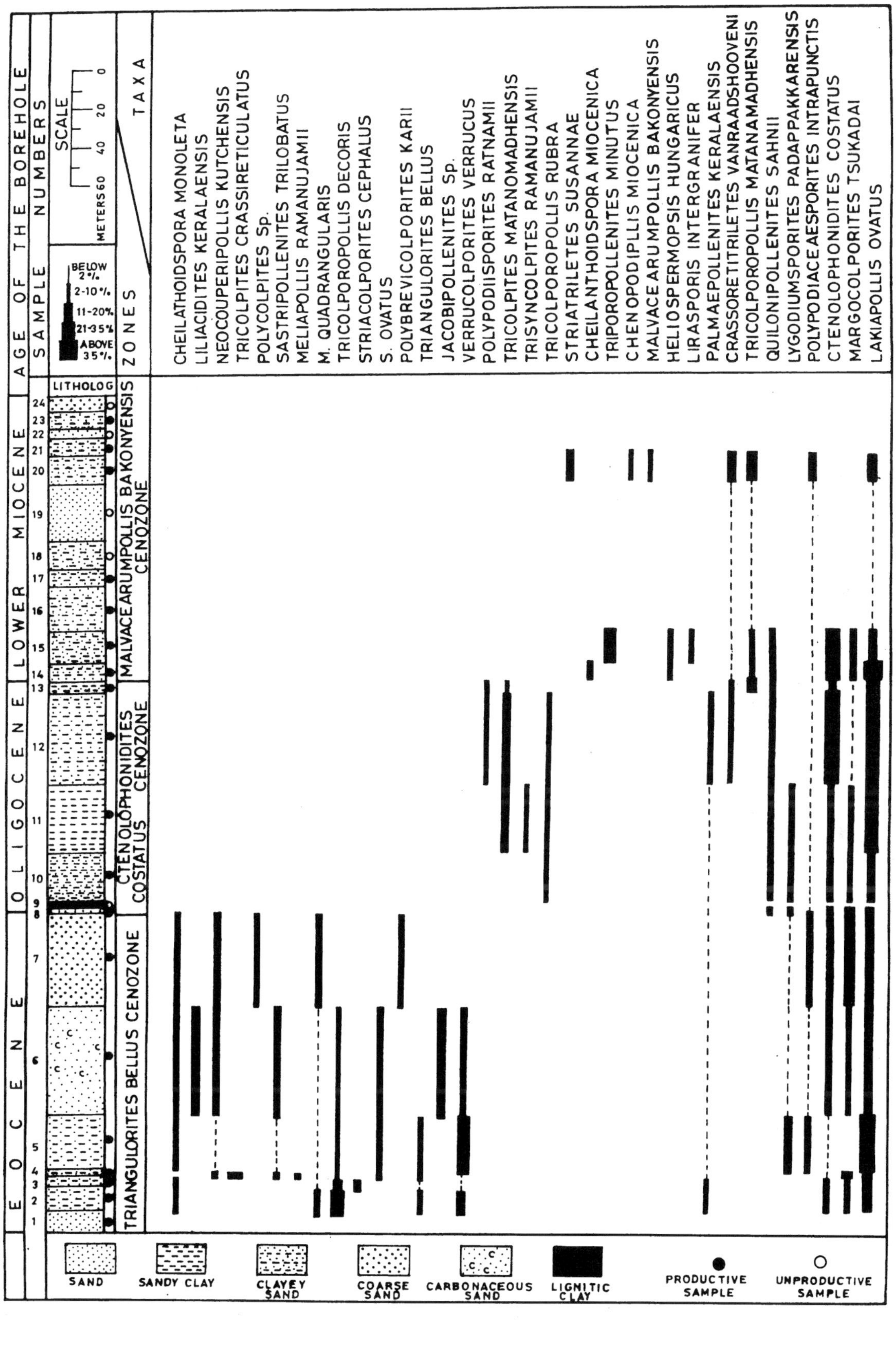

Text-Figure 1. Palynostratigraphic zonation for the Arthungal borehole, Alleppey District, Kerala (after Rao, 1990).

Compositoipollenites, Bombacacidites, Cauveripollis, Margocolporites, Meliapollis, Ctenolophonidites, Heterocolpites, Maculoporites, Verrutriporites, Proteacidites, Clavaperiporites, Chenopodipollis, Malvacearumpollis, Ornatetradites, Ericipites, Florscheutzia, Grimsdalea and *Zonocostites*. Of these, *Crassoretitriletes, Pteridacidites, Polypodiisporites, Quilonipollenites, Dipterocarpuspollenites, Warkallipollenites, Plumbaginacipites, Cauveripollis, Chenopodipollis, Clavaperiporites, Maculoporites, Malvacearumpollis*, and *Zonocostites* are stratigraphically important for Neogene sediments and the association indicates an Early to Middle Miocene. The typical Paleocene to Eocene genera like *Proxapertites, Matanomadhiasulcites, Dandotiaspora, Triangulorites, Sastriipollenites, Striacolporites* and *Neocouperipollis*, which occur in the Paleogene sediments of India, have not been recovered from these localities. Based on the above analyses, the palynoflora recovered from surface sediments in the Kerala Basin confirms an Early to Middle Miocene age.

The Tertiary palynoflora of the Kerala Basin has been compared with the Kutch Basin (Kar, 1985), northeast India (Baksi, 1962; Rao et al., 1985) and South India (the Neyveli lignite; see Ramanujam, 1966, and Singh and Misra, 1991). The common genera recovered include *Bombacacidites, Chenopodipollis, Clavaperiporites, Compositoipollenites, Dipterocarpuspollenites, Malvacearumpollis, Retipilonapites, Malvacipollis* and *Warkallipollenites*. A comparison of these assemblages reveals that they are broadly comparable and confirm a Miocene age for both Quilon and Warkalli sediments.

Thus, on the basis of the subsurface palynofloral assemblage, it is suggested that the entire Tertiary sequence from the Kerala Basin may have to be restudied for finer resolution of age.

DISCUSSION

The Tertiary sediments of the Kerala Basin have yielded a variety of pteridophytic spores and angiospermous pollen. Dinoflagellate cysts and fungal remains are also recorded, however gymnospermous pollen are scarcely represented. The distribution of different plant groups in the Quilon and Warkalli formations, along with common elements between the two units, is presented in Text-Figure 3.

An analysis of the palynoflora indicates that the angiosperm pollen dominate over pteridophytic spores. The pteridophytic spores are represented by 31 genera and 64 species belonging to the Cyatheaceae, Dicksoniaceae, Gleicheniaceae, Lycopodiaceae, Osmundaceae, Ophioglossaceae, Parkeriaceae, Pteridaceae and Schizaeaceae. The Schizaeaceae and Polypodiaceae are the dominant elements of the assemblage. Among angiospermous pollen, 51 families could be identified. Of these, six families belong to the monocots (Arecaceae, Iridaceae, Lemnaceae, Poaceae and Potomogetonaceae) and 45 to the dicots. Arecaceae is the best represented among the monocots, both in Paleogene and Neogene age sediments. Of the 45 families, the Anacardiaceae, Bombacaceae, Dipterocarpaceae, Ebenaceae, Rhizophoraceae, Sonneratiaceae, Euphorbiaceae, Sapotaceae, Caes-alpiniaceae, Clusiaceae, Combretaceae, Asteraceae, Ctenolophonaceae, Oleaceae, Proteaceae, Thymelaeaceae and Malvaceae are most common.

PALAEOCLIMATE

Fossil pollen are reliable indicators of past climates, particularly if they are comparable to modern taxa. Of the 51 angiosperm pollen identified, 35 representing extant families belonging to monocots and dicots (Rao and Ramanujam, 1978, 1982; Ramanujam 1987; Rao, 1990, 1995; Rao and Rajendran, 1996; Rao and Nair, 1998) have been utilized to interpret the palaeoclimate. Of these 35 families, seven are characteristic of tropical climates, 15 families are restricted to tropical to sub-tropical climates, and 13 families pertain to tropical–temperate climatic conditions.

The presence of spores of the Schizaeaceae, Parkeriaceae, Osmundaceae, Lycopodiaceae, Polypodiaceae, and pollen of Anacardiaceae, Dipterocarpaceae, Caesalpiniaceae, Combretaceae, Clusiaceae, Ctenolophonaceae, Euphorbiaceae, Oleaceae and Malvaceae, clearly indicates the prevalence of tropical humid climatic conditions at the time of deposition. The present day climate of Kerala is also of the tropical humid type with heavy precipitation.

The presence of microthyriaceous fruiting bodies (*Phragmothyrites, Notothyrites, Parmathyrites*, etc.) and fungal spores (*Dicellaesporites, Multicellaesporites, Palaeocirrenalia*, etc.) are typical epiphyllous fungi and their occurrence in the present assemblage indicates the existence of a terrestrial plant ecosystem and supports an interpretation of a warm and humid climate with heavy rainfall. Thus the presence of fungal remains, fern spores and angiosperm pollen (tropical rain forest elements) confirms the high degree of rainfall.

Similar paleoclimatic observations have been made by Awasthi and Ahuja (1982) on the basis of the presence of following woods: *Calophyllum, Dryobalanops, Gluta, Swintonia, Terminalia, Diospyros, Listia, Cynometra, Gonostylus* and *Leea*.

ENVIRONMENT OF DEPOSITION

Palynological data were thoroughly scrutinized and ecologically significant taxa were selected and segregated for identifying various habitats. Ecological analysis of the Tertiary fossils from the Kerala Basin identifies several habitats, including montane, lowland, freshwater swamp and water edge, sandy beach, and mangrove and back-mangrove (Table 3).

The ecological analysis at Arthungal, Kalarakod, and Meenkunnu (Cannanore) and the K-K road section, reveals that the freshwater swamp and water edge elements are dominant over the lowland/sandy beach/mangrove, back-mangrove and montane elements. The deposition of the above sediments could have taken place in a cycle of slightly trangressive, regressive and transgressive phases (Rao, 1990, 1995; Rao and Rajendran, 1996; Rao and Nair, 1998).

The presence of dinoflagellate cysts *(Achomosphaera, Spiniferites, Thalassiphora*, etc., see Rao, 1990) and mangrove and back-mangrove elements *Brownlowia (Lacrimapollis), Rhizophora (Zonocostites), Lumnitzera (Heterocolpites), Nypa (Spinizonocolpites), Calamus* (*Dicolpopollis* etc., see Rao and Ramanujam, 1982; Ramanujam; 1987; Rao and Rajendran, 1996) in the palynoflora indicates an influx of brackish water during deposition. The presence of *Palaeocirrenalia*

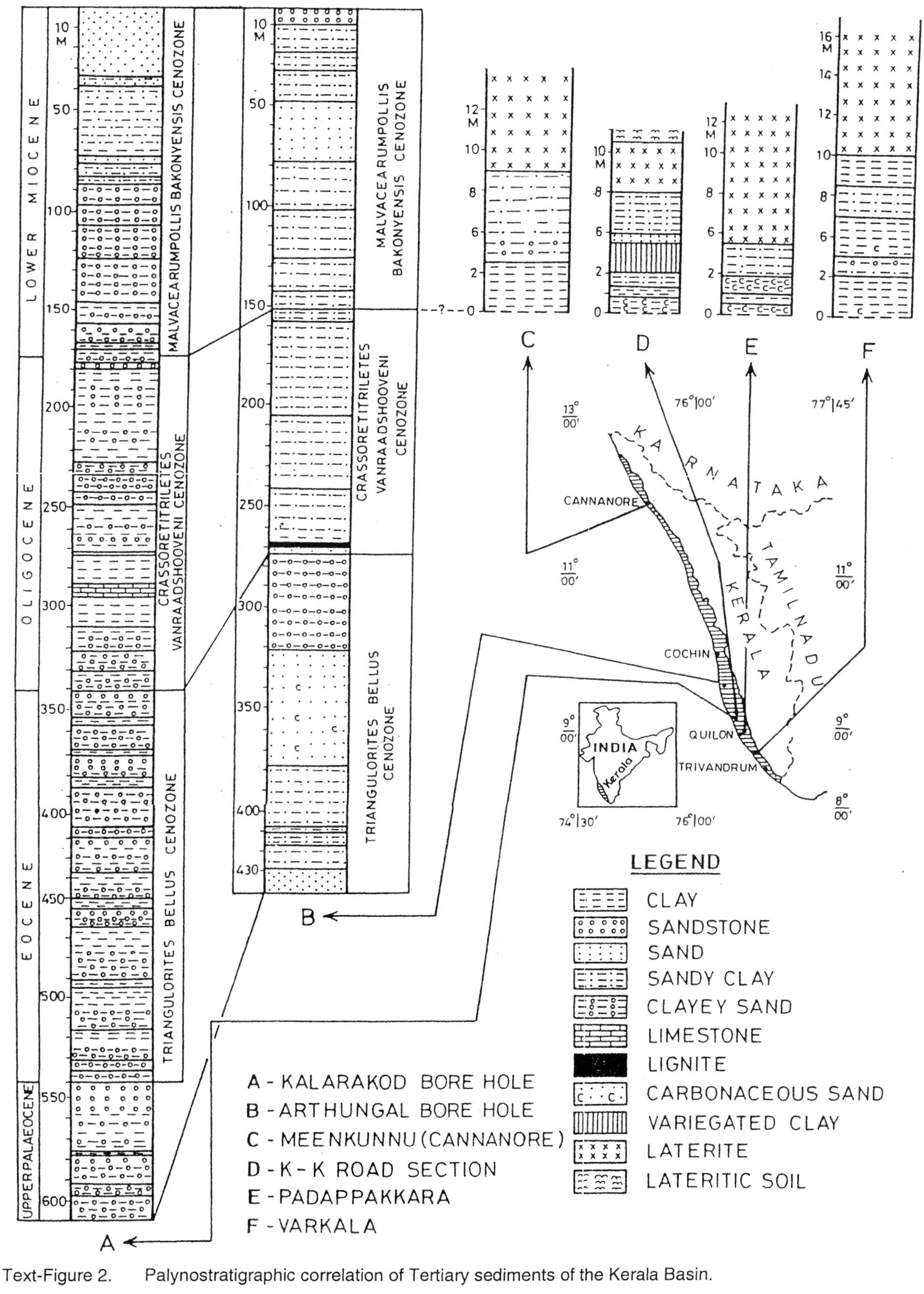

Text-Figure 2. Palynostratigraphic correlation of Tertiary sediments of the Kerala Basin.

TABLE 3. Ecological analysis of the Tertiary palynofossils from the Kerala Basin.

ECOLOGICAL GROUPS	PALYNOTAXA
Montane elements	*Clavaperiporites, Hippocrateaceaedites, Proteacidites, Umbelliferoipollenites*
Lowland elements	*Favitricolporites, Subtriporopollis, Margocolporites, Trisyncolpites, Lakiapollis, Tricolporopollis, Myricipites*
Freshwater swamp and water edge elements	*Lycopodiumsporites, Lygodiumsporites, Crassoretitriletes, Schizaeoisporites, Pteridacidites, Polypodiaceaesporites, Polypodiisporites, Striatriletes, Liliacidites, Ctenolophonidites, Margocolporites, Sapotaceoidaepollenites, Meliapollis, Triporopollenites, Anacolosidites, Ornatetradites, Retipilonapites*
Mangrove and back-mangrove elements	*Heterocolpites, Retitricolporites, Zonocostites, Paleosantalaceaepites, Verrutriporites, Malvacearumpollis, Spinizonocolpites, Warkallipollenites*
Sandy beach elements	*Palmaepollenites, Quilonipollenites, Longapertites, Dicolpopollis, Neocouperipollis*

(which resembles modern *Cirrenalia)* also indicates brackish water to marine conditions (Ellis, 1976). The representation of salt glands (*Heliospermopsis hungaricus* – see Rao, 1990) in the sediments also supports the interpretation above.

The members of the family Palmae are indicative of coastal elements and tend to remain apart from the storm tide. The occurrence of freshwater plants such as Nympheaceae, Potamogetonaceae and Parkeriaceae indicates prevalence of ponds and other freshwater bodies dotting the landscape. Pollen taxa comparable to Chenopodiaceae show a possible halophytic environment. The montane elements of the families Thymelaeaceae, Ericaceae and Proteaceae may have been derived from long distance transport.

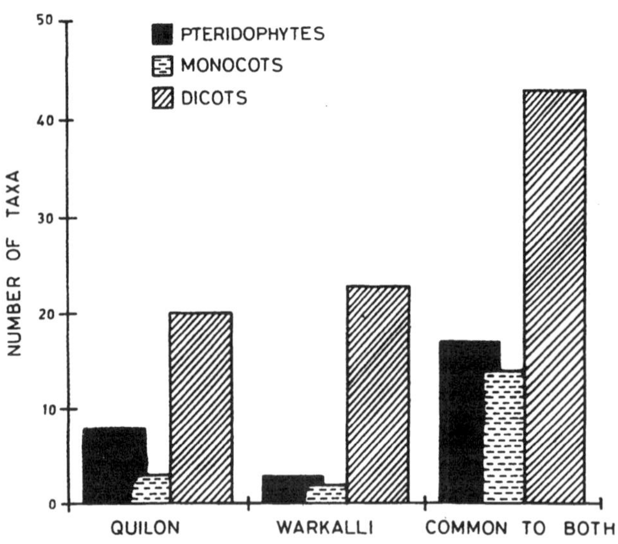

Text-Figure 3. Distribution of different plant groups in the Quilon and Warkalli formations, Kerala Basin.

CONCLUSIONS

1. An analysis of the palynoflora reveals that angiosperm pollen is dominant over pteridophytic spores.
2. The palynosequence in the Arthungal borehole can be divided into three cenozones: *Triangulorites bellus* Cenozone, *Crassoretitriletes vanraadshooveni* Cenozone, and *Malvacearumpollis bakonyensis* Cenozone.
3. Based on palynological data, it was observed that the *Malvacearumpollis bakonyensis* Cenozone (Lower Miocene) in the borehole is correlatable with the lower part of the surface sediments.
4. The palynoflora indicates a humid and tropical climate with high degree of rainfall during the deposition of these sediments.
5. The deposition took place in a nearshore environment with sufficient freshwater or freshwater swamps or both.
6. On the basis of palynological data, the studied sequences have been assigned an Eocene–Lower Miocene age.

ACKNOWLEDGMENTS

The author is thankful to the Director, Birbal Sahni Institute of Paleobotany, Lucknow, India, for providing facilities and partial sponsorship for attending the Ninth International Palynological Congress held in Houston, Texas, USA.

References Cited

AWASTHI, N., and AHUJA, M.
1982 Investigations of some carbonized woods from the Neogene of Varkala in Kerala coast. *Geophytology*, 12: 245–259.

BAKSI, S.K.
1962 Palynological investigation of Simsang river Tertiaries, south Shillong Front, Assam. *Bulletin of the Geological Mining and Metallergical Society of India*, 26: 1–22.

DEY, A.K.
1962 The Miocene mollusca from Quilon, Kerala (India). *Memoirs of the Geological Survey of India. New Series*, 36: 26.

ELLIS, M.B.
1976 More Dematiaceous Hypomycetes. C.M.I: 507.
FOOTE, R.B.
1883 On the geology of south Travancore. *Records of the Geological Survey of India*, 16: 20–35.
GERMERAAD, J.H., HOPPING, C.A., and MULLER, J.
1968 Palynology of Tertiary sediments from tropical areas. *Review of Palaeobotany and Palynology*, 6: 189–348.
HEDBERG, H.D.
1976 *International stratigraphic guide — a guide to stratigraphic classification, terminology and procedure.* John Wiley & Sons, New York, 200 pgs.
JACOB, K., and SASTRI, V.V.
1952 Miocene foraminifera from Chavara near Quilon, Travancore. *Records of the Geological Survey of India*, 82: 342–353.
KAR, R.K.
1979 Palynological fossils from the Oligocene sediments and their biostratigraphy in the districts of Kutch, western India. *Palaeobotanist*, 26: 16–49.
1985 The fossil floras of Kutch-4. Tertiary Palynostratigraphy. *Palaeobotanist*, 34: 1–279.
KAR, R.K., and JAIN, K.P.
1981 Palynology of the Neogene sediments around Quilon and Varakala, Kerala coast, south India-2. Spores and pollen grains. *Palaeobotanist*, 27: 113–131.
KING, W.
1882 General sketch of the geology of Travancore state. The Warkalli beds and associated deposits at Quilon in Travancore. *Records of the Geological Survey of India*, 15: 93–102.
POTONIÉ, R., and SAH, S.C.D.
1960 *Sporae dispersae* of the lignites from Cannanore beach on the Malabar Coast of India. *Palaeobotanist*, 7: 121–135.
POULOSE, K.V., and NARAYANASWAMI, S.
1968 The Tertiaries of Kerala. *Memoirs Geological Society of India*, 2: 300–308.
RAHA, P.K., ROY, S.K., and RAJENDRAN, C.P.
1983 A new approach to the lithostratigraphy of the Cenozoic sequence of Kerala. *Journal of the Geological Society of India*, 24: 325–342.
RAHA, P.K., RAJENDRAN, C.P., and KAR, R.K.
1987 Record of Early Tertiary deposits in Kerala, India and its palaeogeographic significance. *Geophytology*, 17: 209–218.
RAJENDRAN, C.P., RAHA, P.K., and KAR, R.K.
1989 Palynological assemblages from Neogene outcrops of Kerala coast. *Indian Minerals*, 43: 39–46.
RAMANUJAM, C.G.K.
1966 Palynology of the Miocene lignite from south Arcot District, Madras, India. *Pollen et Spores*, 8: 149–204.
1977 A palynological approach to the study of Warkalli deposits of Kerala in south India. *Geophytology*, 7: 160–164.
1982 Tertiary palynology and palynostratigraphy of southern India. *Special Publication, Journal of the Palaeontological Society of India*, 1: 57–64.
1987 Palynology of the Neogene Warkalli beds of Kerala state in south India. *Journal of the Palaeontological of Society of India*, 32: 26–46.
RAO, A.R., and VIMAL, K.P.
1953 Preliminary observation on the microfossil content of some lignites from Warkalli in Travancore. *Current Science*, 21: 210–215.
RAO, K.P., and RAMANUJAM, C.G.K.
1978 Palynology of the Neogene Quilon beds of Kerala state in south India. 1. Spores of pteridophytes and pollen of monocotyledons. *Palaeobotanist*, 25: 397–427.
1982 Palynology of the Quilon beds of Kerala state in south India 2. Pollen of dicotyledons and discussion. *Palaeobotanist*, 30: 68–100.
RAO, M.R.
1990 Palynological investigation of Arthungal bore-hole, Alleppey district, Kerala. *Palaeobotanist*, 38: 243–255.
1995 Palynostratigraphic zonation and correlation of the Eocene–Early Miocene sequences in Alleppey district, Kerala, India. *Review of Palaeobotany and Palynology*, 86: 325–348.
RAO, M.R., and NAIR, K.K.
1998 Palynological investigation of Miocene sediments exposed in Kannanellur-Kundara area, Quilon District, Kerala. *Geophytology*, 27: 49–59.
RAO, M.R., and RAJENDRAN, C.P.
1996 Palynological investigation of Tertiary lignite and associated sediments from Cannanore, Kerala Basin, India. *Palaeobotanist*, 43: 63–82.
RAO, M.R., SAXENA, R.K., and SINGH, H.P.
1985 Palynology of the Barail (Oligocene) and Surma (Lower Miocene) sediments exposed along Sonapur-Badarpur Road Section, Jaintia Hills (Meghalaya) and Cachar (Assam). Part 4. Angiospermous pollen grains. *Geophytology*, 15: 7–23.
SINGH, A., and MISRA, B.K.
1991 New colporate pollen taxa from Neyveli lignite, south India. *Review of Palaeobotany and Palynology*, 67: 205–215.

Author's address

M. R. RAO
Birbal Sahni Institute of Palaeobotany
53 University Road,
Lucknow-226 007
India

LATE CENOZOIC PALYNOLOGY OF THE EASTERN QINGHAI-XIZANG (TIBET) PLATEAU AND DEVELOPMENT OF MONSOON CLIMATE IN CHINA

Gengwu Liu

Abstract

The late Cenozoic pollen flora from the eastern Qinghai-Xizang Plateau can be subdivided into three assemblage zones and two sub-assemblage zones. The Miocene pollen flora contains abundant Fagaceae, Betulaceae and Pinaceae pollen; *Ulmus, Juglans, Carya, "Semitrapa,"* and *Tilia* are common, and *Liquidambar* pollen is rare. This assemblage represents a mixed broad-leaf and conifer forest with a summer wet warm temperate climate. The Pliocene and Early Pleistocene pollen flora in the eastern Qinghai-Xizang (Tibet) Plateau can be separated into two sub-floras: the Ganzi area in north and Panxi area in south. The Pliocene to Pleistocene pollen floras from the high elevation sites in the north contain rich Betulaceae, Compositae, Ulmaceae, Ericaceae and Chenopodiaceae pollen, while those from the low elevation sites in the south contain abundant Fagaceae, Betulaceae, Pinaceae pollen, various tricolpate and tricolporate pollen, and Polypodiaceae spores.

The present pollen data, derived from Miocene pollen and leaf evidences, support the previous conclusion that the uplift of the Qinghai-Xizang Plateau probably reached almost 2000 m in altitude. In the late Pliocene to early Quaternary the pollen data show that the plateau was at least 2500 m, and probably over 3000 m, above sea level.

The modern monsoon climate of China includes summer and winter monsoons. The uplift of the Qinghai-Xizang Plateau was a very important factor in the evolution of the Cenozoic climate. The monsoon climate developed over a long period of geological time. The summer monsoon in China started to develop at least from Miocene time or even earlier. However, the winter monsoon in China developed probably no later than Pliocene time.

INTRODUCTION

The eastern part of the Qinghai-Xizang (Tibetan) Plateau in China consists of two principal areas: the West Sichuan Plateau in the north and the Hengduan Mountains in the south. This is a marginal area of the plateau characterized by extreme topographical differentiation, i.e. very high mountains and very deep valleys. Therefore the climate here is complex; temperature and humidity can vary significantly over a relatively short distance.

The period and amplitude of tectonic uplift in the Qinghai-Xizang Plateau are a very complex scientific problem. Many Chinese scientists have made great contributions in this field in the past decades (Li and Guo, 1976; Li et al., 1979, Song and Liu, G., 1981; Song and Liu, J., 1982; Li, 1991; Wu et al., 1992; Wang et al., 1994). In the early 1980s, pollen and leaf fossil evidence suggested that the elevation of the plateau had reached about 2000 m above sea level during the Miocene or earlier (Li and Guo, 1976; Song and Liu, G., 1981; Song and Liu, J., 1982). In the early 1990s, many scientists drew a similar conclusion and recognized that the elevation of the great plateau had reached approximately 2000 m in the late Miocene (Wu et al., 1992; Wang et al., 1994). The present palynological study will support this conclusion regarding uplift in the Qinghai-Xizang Plateau.

Liu, Gengwu
2001 Late Cenozoic palynology of the eastern Qinghai-Xizang (Tibet) Plateau and development of monsoon climate in China; *In*: Goodman, D.K., and Clarke, R.T. (eds.), *Proceedings of the IX International Palynological Congress, Houston, Texas, U.S.A., 1996*; American Association of Stratigraphic Palynologists Foundation, p. 291–297.

ISBN 0–931871–06–9

The modern monsoon climatic pattern in China was gradually established during the Cenozoic. Uplift of the plateau and a decline in global temperature are especially important in the development of this monsoon climate. In the marginal area of the Qinghai-Xizang Plateau, vegetation and floras are sensitive to the tectonic and climatic changes over geological time. Well-preserved Late Cenozoic pollen and spores from these montane basins in the eastern Qinghai-Xizang Plateau show that the monsoon climate in China was developed over a long period of geological time. The summer monsoon developed from Miocene or earlier time, and the winter monsoon became strong no later than the Pliocene.

MATERIAL AND METHODS

More than one hundred samples analyzed for pollen in this study are from several sections in the eastern Qinghai-Xizang Plateau (Text-Figure 1), where the altitudes vary from 2500 m to over 4000 m above sea level. Most of these samples were collected by the author during a Comprehensive Scientific Expedition to the Eastern Qinghai-Xizang Plateau organized by the Chinese Academy in 1982 and 1983. Only a small number of the core samples of southwestern Sichuan and western Yunnan were collected by local geological teams. All samples were analyzed at the Pollen Laboratory of the Nanjing Institute of Geology and Palaeontology, Academia Sinica, Nanjing. Samples were processed by the traditional method: macerated in 10% HCL to remove carbonates, followed by cold HF treatment to remove silicates. Organic and mineral material were separated by using heavy liquid. Residues were mounted using glycerin jelly.

PRELIMINARY RESULTS

The Late Cenozoic pollen floras from the Ganzi area, which is in the northern part of the eastern Qinghai-Xizang

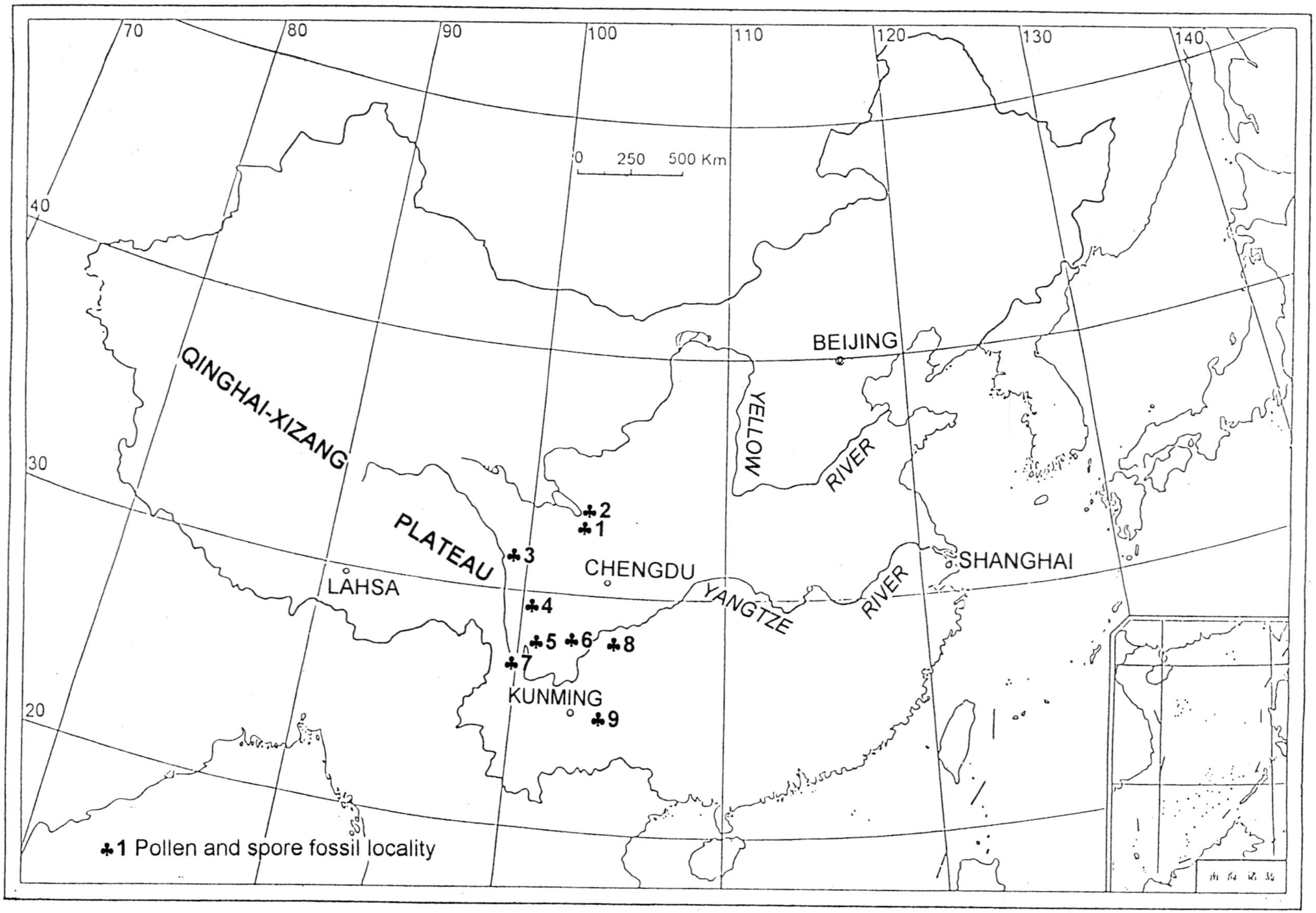

Text-Figure 1. Fossil pollen and spore localities in eastern Qinghai-Xizang Plateau, China. 1. Hongtupo of Songpan County, Sichuan; 2. Zoige area, Sichuan; 3. Changtai Baiyu County, Sichuan; 4. Mula and Jawa of Litang County, Sichuan; 5. Meili of Yanyuan County, Sichuan; 6. Xichang area, Sichuan; 7. Baoshan area, Yunnan; 8. Zhaotong area, Yunnan; 9. Luliang Basin, Yunnan.

Plateau, can be separated into three assemblage zones (labeled zones I, II and III; Text-Figure 2), and two sub-assemblage zones. In the Panxi area, in the southern part of the eastern Qinghai-Xizang Plateau, however, the pollen diagram does not show significant vertical changes (Text-Figure 3) through the section.

The only Miocene assemblage, Assemblage Zone I, is from the Hongtupo Section, Songpan, northwestern Sichuan at an elevation of about 2500 m. It contains abundant Fagaceae, Betulaceae and Pinaceae pollen, together with common *Ulmus, Juglans, Carya, "Semitrapa" and Tilia. Liquidambar* is rare. This pollen assemblage probably represents a mixed broad-leaf and conifer forest or a deciduous broadleaf forest near a conifer forest at a high elevation. The climate is interpreted to be summer wet warm temperate.

The Pliocene and Early Pleistocene pollen flora in the eastern Qinghai-Xizang (Tibet) Plateau can be separated into two sub-floras: the Ganzi area to the north, and the Panxi area to the south.

In the northern Ganzi area, where the mean elevation is about 3000 m, the older assemblage, Assemblage Zone II, (probably middle–late Pliocene in age) is characterized by abundant xerophyte pollen such as *Ephedra, Nitraria*, Chenopodiaceae and Compositae. Woody plant pollen is not common, and consists mainly of Ulmaceae and Pinaceae. Such a pollen flora may represent a steppe or parkland vegetation in a dry climate. The overlying assemblage, Assemblage Zone III (late Pliocene to early Pleistocene) is very different in composition from the underlying assemblage Zone II. Pinaceae pollen, mainly *Picea, Pinus* and *Tsuga*, and Ericaceae and Betulaceae, dominate the assemblage (approximately 60–90% of the assemblage). Other pollen including Fagaceae, Ulmaceae, Juglandaceae, Compositae, etc., is rare in most samples. This younger assemblage probably implies that the earlier vegetation might have developed into a conifer forest with some Ericaceae and Betulaceae shrubs. The climate was likely more humid and much colder that that in the previous period.

Assemblage Zone III can be further divided into two sub-assemblage zones: Sub-assemblage Zones III-1 and III-2. The lower sub-assemblage Zone III-1 contains abundant Betulaceae and *Pinus* pollen, while the upper sub-assemblage Zone III-2 is rich in Ericaceae and *Tsuga*.

In the southern Panxi area, the average elevation is about 2000 m. Pollen floras from the southern area are composed mainly of Fagaceae (mostly *Quercus*), Betulaceae, Compositae, *Ulmus*, Pinaceae and abundant Polypodiaceae spores, which probably suggest a subtropical climatic condition. We temporarily assign a Pliocene age to those pollen assemblages rich in monolete spores of Polypodiaceae. For the Pliocene and lower Pleistocene pollen sequence of the southern area, it is not easy to separate the flora into assemblage zones because the assemblages have a very similar composition. These small quantitative variations in the pollen floras in the southern area probably results from small shifts in vegetation and climate during late Pliocene and early Pleistocene time.

DISCUSSION

Geologic Age of the Palynofloras

The oldest assemblage, Zone I, from the Hongtupo of Songpan, northwest Sichuan, is assigned a mid-Miocene age. It is similar to the mid-Miocene Assemblage III of Liu (1988a). The resemblance between the present assemblage and that from the Lengshuigou Formation of the Weihe Basin suggests a similar geological age for both. The latter is late mid-Miocene in age based on the

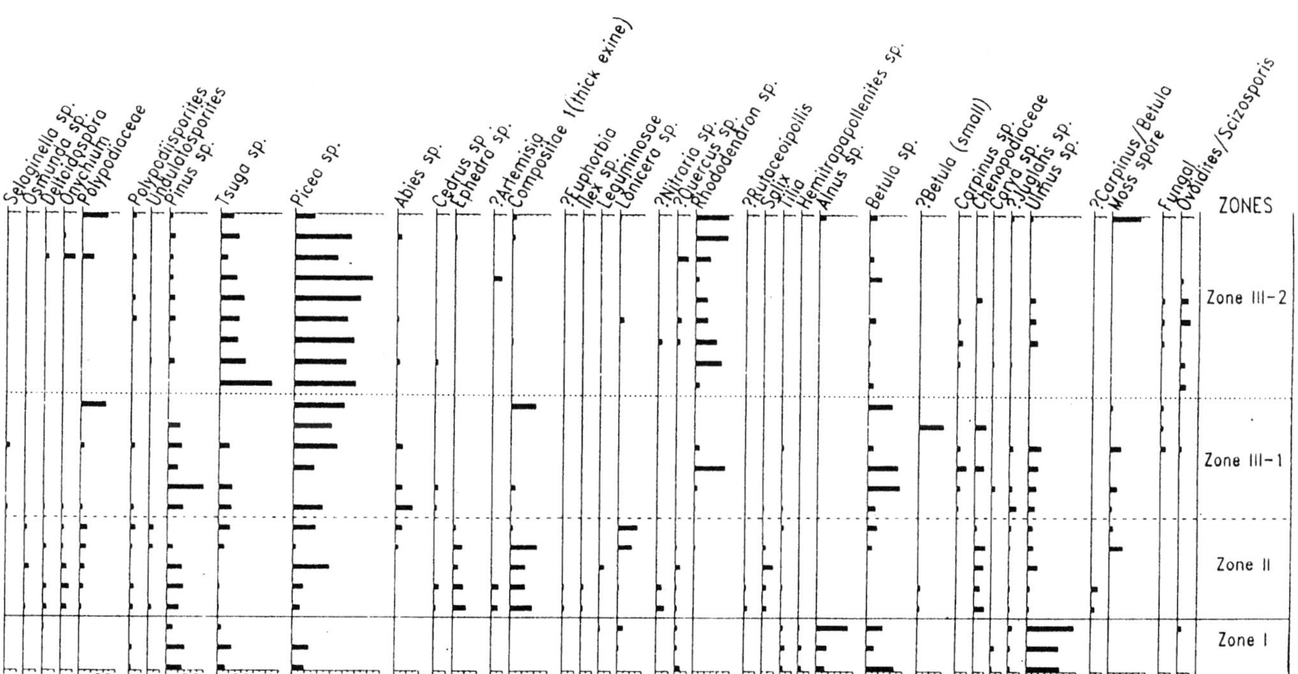

Text-Figure 2. Pollen percentage diagram of selected types from western Sichuan in the northeastern Qinghai-Xizang Plateau (percent total pollen and spores).

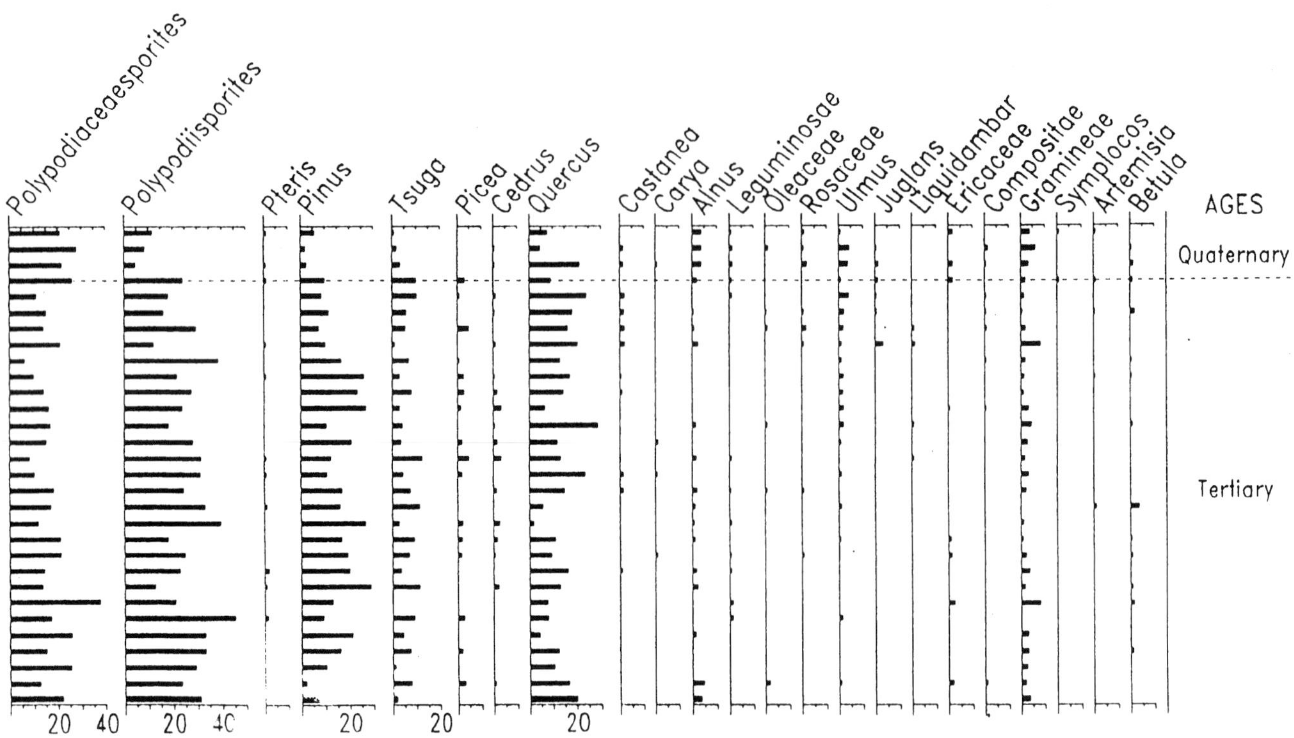

Text-Figure 3. Summary pollen diagram from NW Yunnan and SW Sichuan of the southeastern Qinghai-Xizang Plateau (percent total pollen and spores).

associated vertebrate fossils, which belong to MN 7 (ca. 13 Ma) from the European mammalian fossil zonation (Li et al., 1984; Qiu and Qiu, 1991).

In the northern Ganzi area, a radiometric date of 1.2 Ma is from the upper part of the Relu section of Litang County, some 300 km south to the study area. The pollen assemblage from the upper part of the Relu section is almost identical to Assemblage Zone III in the Changtai Basin. Therefore, Assemblage Zone III of the Ganzi area is approximately 1.2 Ma or slightly older. Herein, Assemblage Zone III is assigned an age of late Pliocene to early Pleistocene. Assemblage Zone II, from the Changtai Basin is possibly mid to late Pliocene in age based on a deduction from the stratigraphic sequence. In addition, Assemblage Zone II is similar to that of the Fifth assemblage of the Yongledian Formation in the Weihe Basin (Sun et al., 1980). Both contain abundant Chenopodiaceae, Ulmaceae, *Ephedra*, and Pinaceae. The Yongledian Formation in the Weihe Basin is regarded as Pliocene based on the vertebrate fossil data (Sun et al., 1980; Qu and Qu, 1991). Assemblage Zone II is also similar to the assemblage recovered from cores in the Zoige region of northwestern Sichuan, about 350 km northeast of the study area. A paleomagnetic date for the Zoige assemblage is 3 Ma (see Text-Figure 2).

In the southern Panxi area, the lower part of the pollen flora which contains abundant Polypodiaceae spores is assigned to the Late Pliocene based on associated vertebrate fossils (Zong, 1987). The upper part, which contains relatively less Polypodiaceae spores, is interpreted as latest Pliocene to Pleistocene in age (Text-Figure 3).

Late Cenozoic Floristic Areas of the Eastern Qinghai-Xizang Plateau

The pollen and spore data show that during and after the Pliocene, there were two different pollen floras in the eastern part of the Qinghai-Xizang Plateau: the Ganzi area in the north, and Panxi area in the south. The main difference between them is that in the north Ganzi area, the Pliocene assemblage contains abundant xerophyte pollen and the early Pleistocene assemblage is dominated by conifer pollen. This pollen and spore flora shows that the climate in the northern Ganzi area experienced an arid phase during mid to late Pliocene time; this arid phase which has not been recogbized in the southern Panxi area. Compared to the Ganzi area in the north, the Pliocene and early Pleistocene pollen flora in the more southern Panxi area is always rich in Fagaceae, Betulaceae, Pinaceae, and various tricolpate and tricolporate pollen, together with abundant monolete Polypodiaceae spores. This flora shows no clear vertical changes, indicating that a relatively stable climate dominated this area during the latest Cenozoic.

Uplift of the Qinghai-Xizang Plateau

The present palynological data reported herein support the conclusion derived from earlier palynological and paleobotanic studies that the altitude of the Qinghai-Xizang Plateau might have been around 2000 m in Miocene time (Li and Guo, 1976; Song and Liu, G., 1981; Song and Liu, J., 1982). The Miocene pollen flora from the Hongtupo of Songpan along the northeastern margin of the eastern Qinghai-Xizang Plateau shows that there a mixed conifer and broadleaf forest consisting mainly of Pinaceae, Betulaceae, Fagaceae (mainly *Quercus*), with some Juglandaceae and few subtropical plants was developed in the area. It most likely represents a lower temperature climate than that of the more cosmopolitan flora from the Weihe Basin (Sun et al., 1980), which implies that the altitude in the Songpan area was probably higher than that of the Weihe Basin (500 km east of the study area). The Pliocene and early Pleistocene pollen data from the Ganzi area indicate that the vegetation during that time was a mixed conifer and broadleaf forest consisting mainly of *Picea, Abies, Tsuga,* and Betulaceae along with shrubs of Ericaceae and *Salix*. In the eastern Qinghai-Xizang Plateau the type of modern vegetation is distributed from about 2500 m to about 4000 m elevation, but mostly above 3000 m (Wu, 1980). Therefore, I conclude that in late Pliocene to early Pleistocene the elevation in the eastern Plateau reached at least 2500 m, and potentially over 3000 m.

Development of a Monsoon Climate in China

The modern monsoon climate in China consists of a summer monsoon, winter monsoon, and plateau monsoon, etc. The discussion in this paper will focus on the development of the summer monsoon dominated by wet south and southeast winds and the winter monsoon dominated by cold and dry northwest winds. Palynological evidence shows that the monsoon climate in China gradually developed over a long period of geological time.

The most important feature of China's topography is the different elevations, i.e. three descending stages from west to east through the country. The highest stage in the west is the magnificent Qinghai-Xizang (Tibet) Plateau with an average elevation of about 4000 m (Wu, 1980; Hou, 1983). This topographic feature has great influence on the characteristics of climate, especially the monsoons. In the geologic past, the uplift of the Qinghai-Xizang Plateau was a very important factor in the development of the Cenozoic climate in China. After the mid-Miocene, the plateau was already over 2000 m in average elevation. Such a height can perturb the western jet stream and split it into southern and northern branches. In addition, the plateau itself became a heat source in the summer and cold source in the winter. Therefore, the summer monsoon developed in the area east to the plateau and strengthened the dry and cold climate in the middle and high latitude area north to the plateau (Ruddiman and Kutzbach, 1991). Thus the Qinghai-Xizang

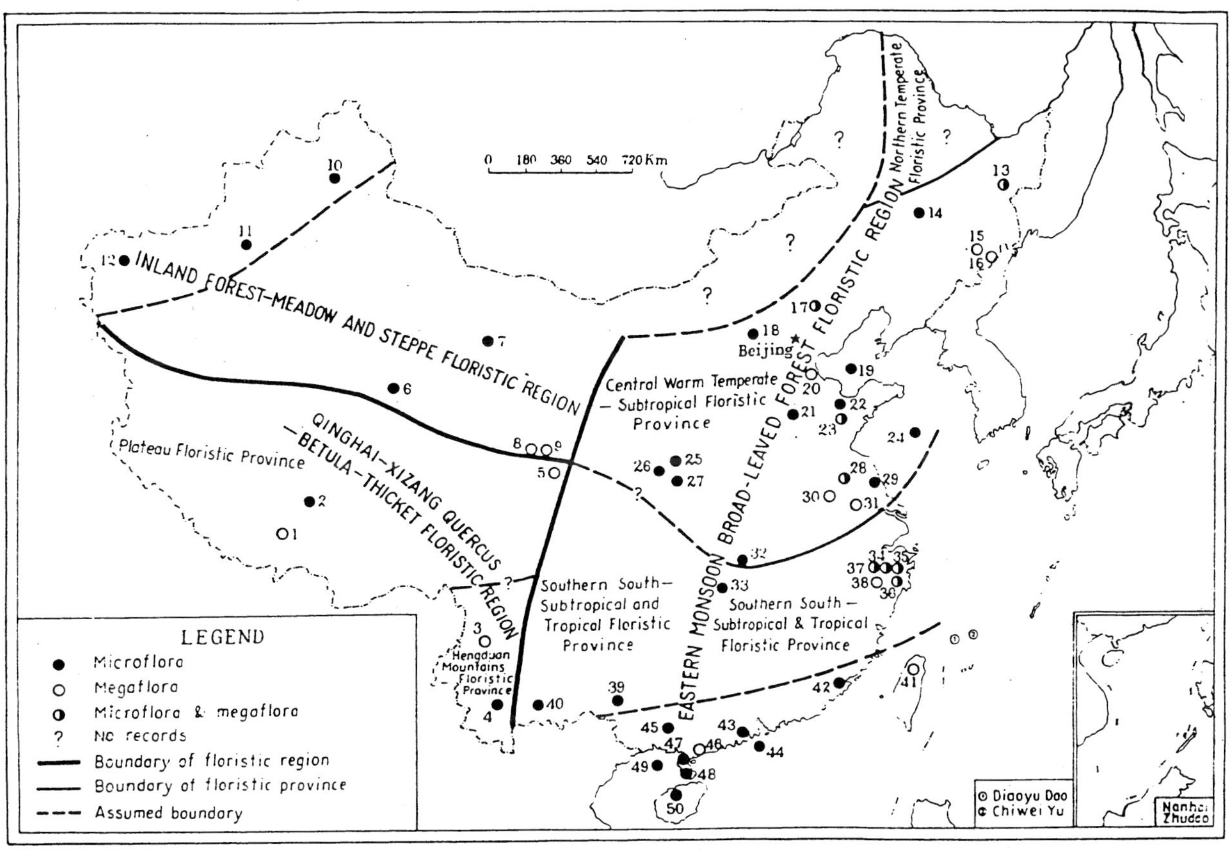

Text-Figure 4. Miocene floristic regions of China (from Song et al., 1981). The three Miocene vegetation regions is similar to the modern ones (see Text-Figure 5). The Eastern Monsoon Broad-leaf Forest Floristic Region represents a summer wet climate in the eastern half of China. The Inland Forest–Meadow and Steppe Floristic Region in the northwestern part of the country shows that the climate there is drier than the other two regions. The Qinghai-Xizang *Quercus–Betula* Thicket Floristic Region developed in the southwestern part of China indicates the elevation there was obviously higher than that in the similar latitudinal area of eastern China.

Plateau served as one of the most important factors driving the evolution of the monsoon climate of China during the Cenozoic.

The summer monsoon in China probably started to develop during the Miocene or perhaps earlier. Pollen and leaf fossil data indicate that the main framework of climate for China was formed as early as the Miocene. During the Miocene, there were three main floristic regions in China: the eastern monsoon forest region, the northwestern xerophyte region, and the Qinghai-Xizang Plateau region (Song et al., 1981). This pattern is very similar to that in present-day China (Text-Figures 4, 5). This suggests that both Miocene and modern China are very similar climatically. The eastern part of the country is dominated by a summer wet climate, while the northwestern part has a dry climate, and the southwestern part displays a unique plateau climate. The author deduces that the climatic difference between the western and eastern parts of the country was caused mainly by the uplift of the plateau which changed the western air jetstream. The wet summer monsoon from the Pacific Ocean (and Indian Ocean) began to dominate the southeastern part of China as the early Tertiary subtropical arid zone retreated from the area. However, the wet summer monsoon rarely affects the inland area in northwestern China because it is too far from the oceans. On the Qinghai-Xizang Plateau, a plateau forest and thicket vegetation developed during the Miocene because of two factors: (1) a relatively short distance to the ancient Indian Ocean and (2) lack of high mountains south to the plateau which would block the wet air from the ocean (Li and Guo, 1976; Song et al. 1981).

The winter monsoon in China probably became strong no later than the Pliocene. Pollen data show that in the mid-Pliocene and early Pleistocene, the eastern part of the Qinghai-Xizang Plateau was at least 2500 m above sea level. Such height of the plateau further enhanced the latitudinal temperature gradient in the mid and high latitudes which in turn would increase energy for air flow. In addition, the global temperature decline after the mid-Pliocene was especially evident in this high latitude area. In the equatorial area this decline was not significant (Kenneth, 1986; Liu, 1988b). Such a differential temperature decline at different latitudes would have strengthened the winter cold air high pressure center in Siberia.

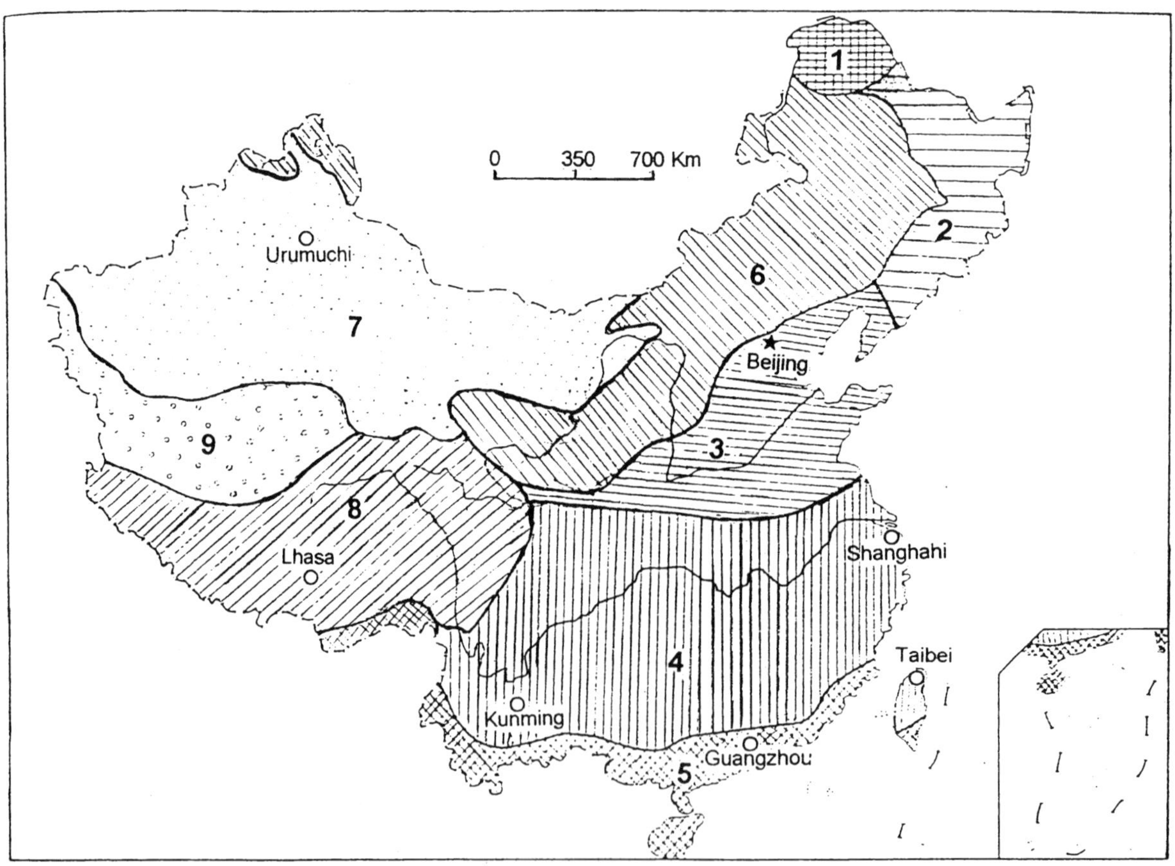

Text-Figure 5. Modern vegetation regions of China (Simplified from Wu, 1980; Hou, 1983, with little modification). 1. Cold-temperate deciduous conifer forest region; 2. Temperate mixed conifer and broadleaf forest region; 3. Warm temperate deciduous broadleaf forest region; 4. Subtropical evergreen broadleaf forest region; 5. Tropical rain forest and monsoon rain forest region; 6. Temperate steppe region; 7. Temperate desert region; 8. Qinghai-Xizang plateau high-cold meadow and steppe region; 9. Qinghai-Xizang plateau high-cold semi-desert and desert region. These vegetation regions may be grouped into three super-regions: the eastern humid forest super-region (regions 1 through 5), northwestern arid and semi-arid super-region (regions 6 and 7), and the southwestern plateau super-region (regions 8 and 9). Framework of the three super regions is quite similar to that of the Miocene vegetation regions (see Text-Figure 4).

When this cold high pressure has enough energy, it develops into a cold air jet moving southward down to the relatively low pressure area in China and southeast Asia, resulting in a strong cold dry winter monsoon, climatic pattern.

GENERAL CONCLUSIONS

The Late Cenozoic pollen flora from the eastern part of the great Qinghai-Xizang Plateau contains three assemblage zones and two sub-assemblage zones: one zone is of Miocene age, and the other two are Pliocene to early Pleistocene in age.

Pollen data support the previous conclusions that in Miocene time the great plateau probably reached 2000 m in altitude. During the late Pliocene to early Quaternary it might already have been above 2500 m.

Palynological evidence shows that the monsoon climate in China was developed over a long geological time. The summer and winter monsoons probably developed at different times: summer monsoon probably began in or before Miocene time, while the winter monsoon became strong no later than Pliocene time. Uplift of the Qinghai-Xizang Plateau acted as a very important factor in strengthening and influencing the modern monsoon climate of China.

REFERENCES

HOU HSIOH-YU (HOU XUE-YU)
1983 Vegetation of China with reference to its geographical distribution. *Annals of the Missouri Botanical Garden*. 70: 509–549.

KENNETH, J.P.
1986 Miocene to early Pliocene oxygen and carbon isotope stratigraphy in the southwest Pacific, Deep Sea Drilling Project Leg 90, *Initial Reports. Deep Sea Drilling Project* 90: 1383–1411.

KONG ZHAOCHEN, LIU LANSOU, and DU NAIQIU
1982 On the uplift of the Qinghai-Xizang Plateau based on the Neogene and Quaternary pollen data from the Kunlun-Tangula Mountain. in *The uplift time, amplitude, and form of the Qinghai-Xizang Plateau, (ed. Comprehensive Scientific Exploration to the Qinghai-Xizang Plateau)*, Science Press, Beijing, pp. 78–88.

LI CHUANKUI, WU WENYU, and QIU ZHUDING
1984 Chinese Neogene: subdivision and correlation. *Vertebrata Palasiatica* 22(3): 163–178.

LI HAOMIN, and GUO SHUANGXIN
1976 Miocene leaf flora from the Namling of Xizang, *Acta Palaeontologica Sinica*, 15(1): 7–18.

LI JIJUN, WEN SHIXUAN, ZHANG QINSONG, WANG FUBAO, ZHENG BENXIN, and LI BINYUAN
1979 A discussion on the period, amplitude, and type of the Uplift of the Qinghai-Xizang Plateau, *Scientia Sinica*, 22(6): 608–616.

LI JIJUN
1991 The Environmental effects of the uplift of the Qinghai-Xizang Plateau. *Quaternary Science Reviews*, 10: 479–483.

LIU GENGWU
1988a Neogene Pollen Sequence of Northern China, *Acta Palaeontologica Sinica*, 27(1): 75–90.
1988b Neogene climatic features and events of northern China. *Neogene biotic evolution and related events, IGCP-246 Symposium special publication, Osaka Museum of Natural History, Osaka*, pp. 21–30.

QU ZHANXIANG, and QIU ZHUDING
1990 The Neogene mammalian fossil sequence and stages of China. *Journal of Stratigraphy*, 14(4): 241–260.

RUDDIMAN, W.F., and KUTZBACH, J.E.
1991 Plateau uplift and climatic change. *Scientific American* 264(3): 66–75.

SONG ZHICHEN, LI HAOMIN, ZHENG YAHUI, and LIU GENGWU
1981 Miocene floristic regions of China. *Geological Society of America, Special Paper* 127: 249–254.

SONG, ZHICHEN, and LIU GENGWU
1981 Tertiary palynological assemblages from Xizang (Tibet) with reference to their palaeogeographical significance. *Proc. Of Symp. On Qinghai-Xizang (Tibet) Plateau (Beijing)*, Science Press, Beijing, Gorden & Breach, vol. I: 207–214.
1982 Early Tertiary palynoflora and its significance on palaeogeography from northern and eastern Xizang (Tibet). in *the series of the Comprehensive Scientific Expedition to the Qinghai-Xizang Plateau. Palaeontology of Xizang* 5, Science Press, Beijing, pp. 165–190.

SONG ZHICHEN, and LIU JINGLING
1982 The Tertiary sporo-pollen assemblages from Namling of Xizang. in *the series of the Comprehensive Scientific Expedition to the Qinghai-Xizang Plateau. Palaeontology of Xizang* 5, Science Press, Beijing, pp. 153–163.

SONG ZHICHEN, and ZHONG BIZHENG
1984 Tertiary sporo-pollen assemblages from Jinggou, Yunnan. *Bulletin of Nanjing Institute of Geology and Palaeontology, Academia Sinica*, No. 8: 1–53.

SUN XIUYU, FAN YONGXIU, DENG CILAN, and YU ZHENGQING
1980 Cenozoic sporo-pollen assemblage of the Weihe Basin, Shaanxi. *Bulletin of the Geological Institute, Academy of Geological Science*, 1(1): 84–109.

WANG FUBAO, SHEN XUHIU, LI SHENGFENG, and ZHANG JIE
1994 Field investigation to the Late Cenozoic sediments from the northern slope of the middle Himalayas. *Abstracts of the Symposium on the National Research Project on the Qinghai-Xizang Plateau*. pp. 40–41.

WU XIHAO, WANG FUBAO, AN ZHISHENG, QIAN FANG, LU YANCHOU, and ZHANG XUANYANG
1992 Episodic uplifts of Qinghai-Xizang Plateau and its corresponding altitude during the late Cenozoic. *In: Loess, Quaternary and Global Change* Part III (chief eds. Liu Dongsheng and An Zhisheng), pp. 1–13.

WU ZHENGYI (CHIEF ED.)
1980 Vegetation of China. *Science Press*, 1382 pgs.

ZONG GUANFU
1987 Mammalian fossils from the Yanyuan Basin of Sichuan and its significance. *Vertebrata Palasiatica* 25(2): 137–145.

Author's address

Gengwu Liu
Nanjing Institute of Geology and Palaeontology
Academia Sinica
Nanjing
People's Republic of China

PALEOGENE POLLEN AND SPORES AS A TOOL FOR QUANTITATIVE PALEOCLIMATE RECONSTRUCTIONS: THE RUPELIAN (OLIGOCENE) OF CENTRAL EUROPE

Jörg Pross, Angela A. Bruch, Volker Mosbrugger, and Zlatko Kvacek

Abstract

This paper presents a quantitative paleoclimate reconstruction for the Rupelian (Oligocene, nannoplankton zone NP 23) of central Europe based on pollen and spore assemblages and megafloras. The calculations were carried out using the coexistence approach proposed by Mosbrugger and Utescher (1997) to climatically evaluate micro- and megafloras from the Mainz Basin (Germany) and the Sava region (Slovenia). The obtained results are compared to data derived from the application of the coexistence approach to other, already published central European micro- and megafloras of the same age.

For all study areas the climatic intervals obtained indicate a mean annual temperature in the range of 16 to 18°C, mean temperatures of the coldest and warmest month of 6 to 12°C and 26 to 28°C, respectively, and a mean annual precipitation of 975 to 1320 mm. This suggests that a very homogenous climate prevailed during the Rupelian in central Europe. A significant climatic gradient between the North and the South had not yet developed.

INTRODUCTION

Determining paleoclimate parameters is an important objective of Tertiary paleobotany, and pollen and spores have proven to be an excellent tool in this field (e.g. Krutzsch, 1967, 1992; Ziembinska-Tworzydlo and Wazynska, 1981; Hubbard and Boulter, 1983; Boulter, 1984; Pocknall, 1990). They are more abundant in the fossil record and can be sampled at closer intervals than megafloral remains, thus allowing a higher stratigraphic and geographic resolution of paleoclimate reconstructions. Whereas quantitative paleoclimate reconstructions on the basis of megafloras are well established (e.g. Wolfe, 1993, 1995), the use of pollen and spores in the field up to now only led to qualitative or semi-quantitative results.

Especially in central Europe, research on Tertiary paleoclimates has a long tradition (e.g. Heer, 1859; Sittler, 1965; Krutzsch and Majewski, 1967; Zagwijn and Hager, 1987; Mai, 1995). Nevertheless, the development of the terrestrial Tertiary paleoclimate in that area is still a matter of debate (e.g. Mai and Walther, 1978, 1983; Krutzsch, 1992; Mai, 1995). In particular, there is need for quantitative palaeoclimate data that allow the reconstruction of the geographic pattern and latitudinal gradient of Tertiary paleoclimate evolution. This will help to address additional questions regarding the climatic impact of the Tertiary Alpine uplift and Paratethys evolution.

Here, we concentrate on a pollen and spore-based quantitative palaeoclimate reconstruction for the Rupelian (Oligocene) of the circumalpine area of central Europe. Following the equable climate of the Eocene (e.g. Sloan & Barron 1990) and the "terminal Eocene cooling event", the Rupelian was thought to have been a time interval with an increased latitudinal temperature gradient and a beginning geographic climate differentiation. To test this hypothesis we applied the coexistence approach proposed by Mosbrugger and Utescher (1997), which can provide quantitative climatic data for the terrestrial realm from all kinds of taphofloras, on Rupelian (Oligocene) pollen and spore assemblages from the North and South of the Alpine area. In addition, megafloras from this region were used to validate the results.

Pross, Jörg, Bruch, Angela A., Mosbrugger, Volker, and Kvacek, Zlatko
2001 Paleogene pollen and spores as a tool for quantitative paleoclimate reconstructions: The Rupelian (Oligocene) of central Europe; *In*: Goodman, D.K., and Clarke, R.T. (eds.), *Proceedings of the IX International Palynological Congress, Houston, Texas, U.S.A., 1996*; American Association of Stratigraphic Palynologists Foundation, p. 299–310.
 ISBN 0–931871–06–9

STUDY AREAS AND MATERIAL

Mainz Basin, Southwest Germany

As a study area north of the Alps, we chose the Mainz Basin in Southwest Germany, a small epicontinental basin at the northern end of the Rhine Graben (Text-Figure 1). During the Rupelian, it was a part of a narrow intracontinental marine strait connecting the Tethys and the North Sea (Rothausen and Sonne, 1984; Ziegler, 1990). Sediments of Rupelian age within the basin form a monotonous succession of marine, locally bituminous clay- and siltstones with occasional intercalations of sandy clays (Text-Figure 2).

The terrigenous palynoflora of the Mainz Basin was studied in 24 core samples from the Bodenheim borehole which was drilled in the central part of the basin (Text-Figure 1). The recovered sediments consist of 59 m of slightly bituminous claystones of the Mittlerer Rupelton which were deposited here at a distance of approximately 15 km from the mid-Rupelian shoreline (Grimm, 1994). Sample positions and lithological details of the Bodenheim core are indicated in Text-Figure 3.

Megafloral data from the Mittlerer Rupelton were obtained from the clay pit of Flörsheim, located close to the northeastern boundary of the Mainz Basin and about 11 km from Bodenheim (Text-Figure 1). The leaf flora of Flörsheim has been extensively studied by early workers (e.g. Geyler, 1883; Engelhardt, 1911). However, more recent investigations that allow a precise determination of the taxa have not been published. The megafloral determinations presented

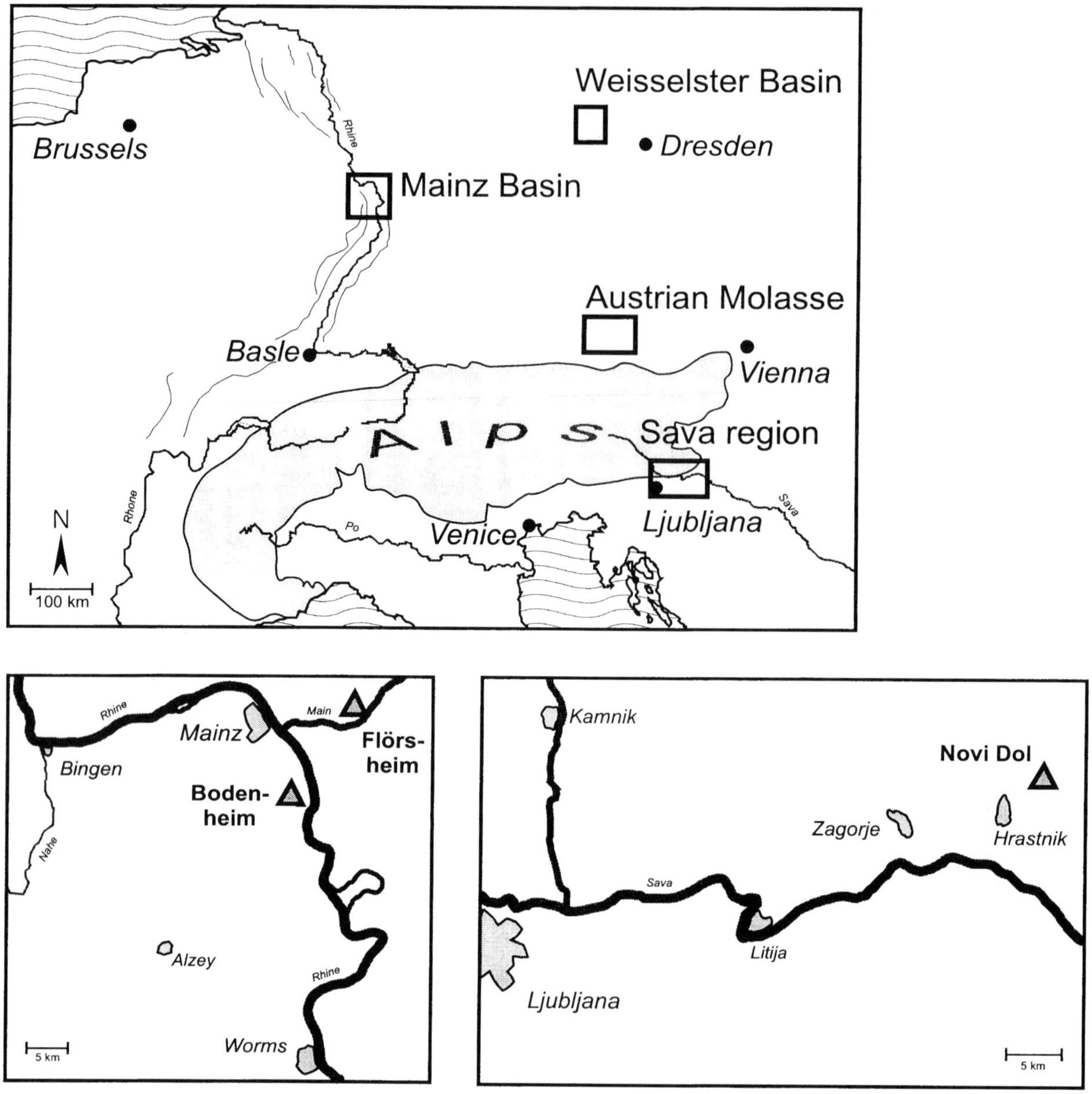

Text-Figure 1. Location of study areas.

in this paper were carried out using the collection of the Senckenberg Museum, Frankfurt/Germany, and are based on cuticular analyses.

The biostratigraphic framework for the Rupelian succession of the Mainz Basin is firmly established due to studies on nannoplankton (Martini, 1971, 1982; Müller, 1971), benthic foraminifera (e.g. Sonne, 1988; Grimm, 1994), and dinoflagellate cyst associations (Pross, 1997). The boundary of nannoplankton zones NP 22 and NP 23 is placed at the base of the Foraminiferenmergel (Müller, 1971; Martini, 1982), and the NP 23/NP 24 boundary correlates with the transition between the Mittlerer Rupelton and Oberer Rupelton (Grimm, written commun., 1996) (Text-Figure 2). Thus, the examined material from the Mittlerer Rupelton of Bodenheim and Flörsheim can be assigned to nannoplankton zone NP 23.

Sava Region, Slovenia

The Sava Region of Slovenia was selected as the study area south of the Alps. During the Rupelian, the Sava region formed the hinterland of the southern coast of the epicontinental Central Paratethys. From the Miocene onward, the sediments of the area were uplifted, dislocated and folded by the Alpine orogenesis (Sztanó, 1994).

The main part of the Rupelian sediments in the Sava region is made up of the Pseudosocka Beds, a succession of coarse clastic sediments, claystones, lignites, and freshwater carbonates. The Pseudosocka Beds were deposited in a low-salinity, probably limnic environment (Cimerman, 1979; Mihajlovic and Jungwirth, 1988) and are overlain by a series of marine clays not yet formally named (Text-Figure 2).

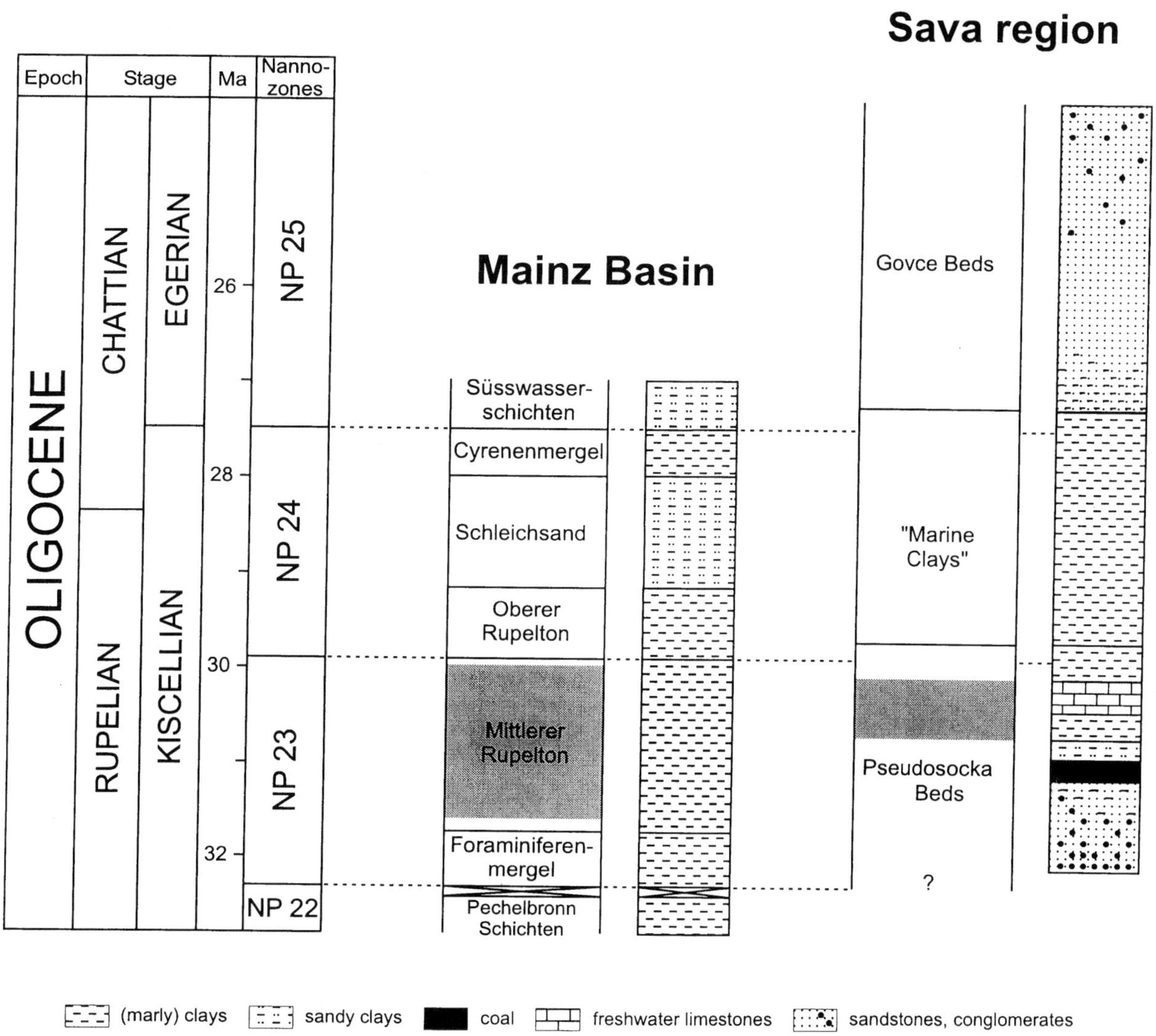

Text-Figure 2. Stratigraphy of the Rupelian in the Mainz Basin and Sava region. Sampled successions (shaded areas) are assigned to nannoplankton zone NP 23. Lithological columns are generalized and not to scale, general Oligocene stratigraphy is compiled after Berggren et al. (1995) and Rögl (1996).

Twenty-seven samples were taken from an outcrop of the Pseudosocka Beds near the village of Novi Dol (Text-Figure 1). Here, the exposed succession consists of marly clays and sandstones containing freshwater molluscs and scattered leafs, and freshwater carbonates discordantly overlain by Middle Miocene units. Sample positions and lithological details of the Novi Dol section are presented in Text-Figure 3.

The Pseudosocka Beds at Novi Dol are assigned to nannoplankton zone NP 23 through a comparison of their mollusc fauna with the Tard Clay Formation in Hungary (Mihajlovic and Jungwirth, 1988; Hably, oral commun., 1996), and thus correspond in age to the Mittlerer Rupelton of the Mainz Basin (Text-Figure 2).

Published Floras

In addition to the floras from the Mainz Basin and the Sava region, we also evaluated two published mid–Rupelian micro- and megafloras from the North of the Alpine area, i.e. from the Austrian Molasse and the Weisselster Basin (Text-Figure 1).

The examined microflora from the Austrian Molasse Basin was taken from the marine Bändermergel ("Banded Marls") Formation which was deposited in the western Paratethys and is dated on the basis of pollen and foraminifera as middle Early Oligocene, thus corresponding to a NP 23 age (Hochuli, 1978).

Furthermore, we calculated the climatic requirements of a megaflora from the Weisselster Basin described by Kunzmann (1995). The fruit and leaf fossils of this flora from the Haselbach-Ton ("Haselbach Clay") were deposited in lakes and meandering rivers draining into the Northwest European Tertiary Basin (Eissmann, 1994) and are assigned to pollen zone 20A/B after the classification of Krutzsch (1992), corresponding to a NP 23 age (Kunzmann, 1995).

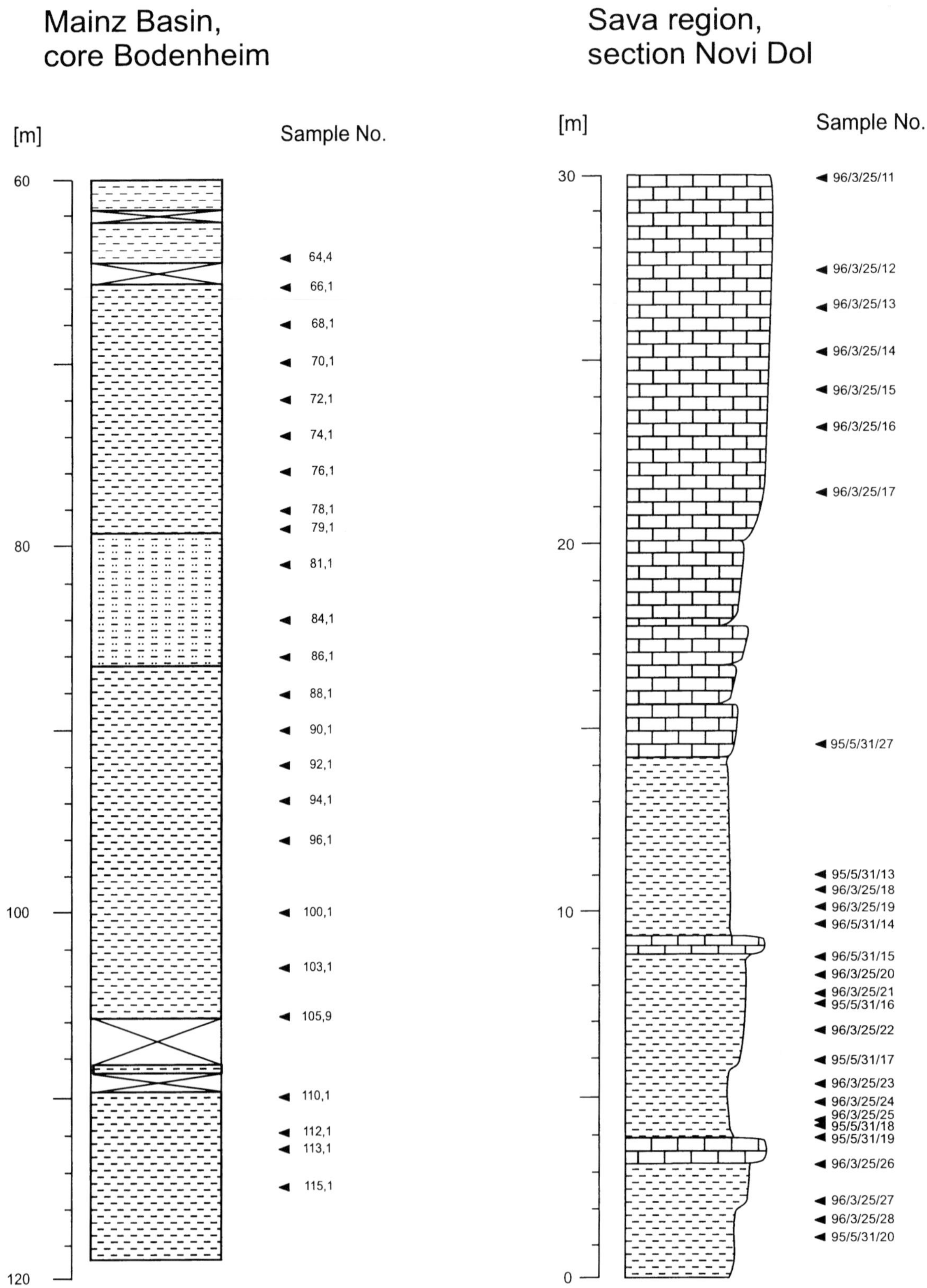

Text-Figure 3. Lithology and sample positions of the core Bodenheim (Mainz Basin) and the Novi Dol section (Sava region). For lithological legend, see Text-Figure 2.

METHODS

Sample Preparation

Palynological preparations were made using standard techniques of rock dissolution with HCl and HF, followed by heavy-liquid centrifugation in $ZnCl_2$ and sieving with a 10 µm nylon mesh. Oxidation, which was necessary in most samples, was achieved with HNO_3. From each residue strew mounts in a mounting medium of Entellan resin were prepared.

The palynological investigation included (1) the counting of up to 200 determinable spores and non-saccate pollen grains per sample, and (2) further examination of the material for additional taxa. Due to the partly poor preservation of palynomorphs, the number of counted individuals in the samples from the Novi Dol section is mostly lower.

Paleoclimatic Reconstructions

Quantitative paleoclimate analyses based on the micro- and megafloral associations were carried out using the coexistence approach proposed by Mosbrugger and Utescher (1997). This technique is based on the "nearest living relative philosophy", i.e. the assumption that the climatic requirements of Tertiary plant taxa are similar to those of their nearest living relatives (NLRs). The aim of the coexistence approach is to find the intervals of various climate parameters for a given fossil flora in which a maximal number of NLRs of this flora can coexist; these coexistence intervals are considered the best description of the paleoclimatic situation under which the fossil flora had lived. The application of the coexistence approach is facilitated by computer programs and a databank which contains the NLRs of more than 400 Tertiary pollen and spore taxa and more than 800 Tertiary plant taxa, together with their climatic requirements for 20 different climatic parameters.

The climatic requirements of the NLRs are derived from meteorological stations located within the distribution areas of the taxa. Therefore the digits after the decimal point for the coexistence intervals do not reflect the real accuracy of the method, but merely the climatic values of the used meteorological stations.

The climatic parameters discussed here are the mean annual temperature (MAT), mean temperature of the coldest month (CMT), mean temperature of the warmest month (WMT), and mean annual precipitation (MAP).

The climatic resolution of the coexistence approach obviously depends on various factors (see Mosbrugger and Utescher, 1997, for a detailed discussion), but doubtless decreases with increasing stratigraphic age as the attribution of NLRs becomes more and more inaccurate. On the other hand, the resolution increases with the number of taxa included in the analysis; generally it can be considered relatively high in floras with about ten or more climatically evaluable taxa. The reliability of the obtained climatic data also varies depending on the climatic parameter considered; the resolution is highest for the temperature-related parameters (MAT, WMT, CMT) where it usually is in the range of 1–2°C. Results for the MAP are less precise and reach an accuracy of 100–200 mm (Mosbrugger, 1995; Mosbrugger and Utescher, 1997).

The coexistence approach is based on the presence/absence of taxa rather than on their relative abundances. This reduces the sensitivity of the method with respect to taphonomy-related changes in the composition of pollen spectra as these types of changes can result from the differential transport of pollen grains.

An initial survey of the sample material revealed that the palynofloras of the samples taken from the Bodenheim profile were not significantly different from each other; the same proved to be true for the samples from the Novi Dol section. Hence, all samples from Bodenheim were put together and considered to represent one single palynoflora; the same was done for the Novi Dol samples. By this means, rare taxa occuring only in a minority of samples could be included into the climatic analysis. The problem of long-distance transport of single pollen grains was addressed by evaluating sporomorphs occuring only in more than one specimen in a given sample.

RESULTS

Mainz Basin

Bodenheim palynoflora

All samples from the Bodenheim core yielded diverse, very homogenous terrigenous palynofloras with good preservation. A total of 95 Oligocene and 14 reworked taxa were identified (Table 1).

Bisaccate pollen of the genera *Piceapollis* and *Pityosporites* is present in high numbers throughout the sequence. The spectrum of non-saccates, which are generally less frequent, comprises 54 taxa. It is dominated by pollen species from anemophilous parent plants and contains elements which can be attributed to different vegetation units, the most abundant being *Caryapollenites simplex* (swamp forest vegetation), *Tricolporopollenites cingulum oviformis*, *T. cingulum pusillus*, *T. edmundi* (mixed mesopytic forest vegetation), *Inaperturopollenites hiatus*, *Sequoiapollenites polyformosus*, *Triatriopollenites rurensis* (peatland vegetation), and *Ephedripites* spp. (vegetation of sandy coastal areas).

Pollen from entomophilous parent plants is less diverse and represented by species such as *Tetracolporopollenites sapotoides* and *Aceripollenites striatus*.

Spores only occur in low numbers in the Bodenheim samples. Among the 37 taxa found, members of the genera *Leiotriletes*, *Polypodiaceoisporites*, *Trilites*, and *Verrucatosporites* are most common.

Reworked taxa from the Permian (*Limitisporites* sp.), Mesozoic (e.g. *Appendicisporites tricornitatus*, *Callialasporites* spp., *Classopollis torosus*, *Foveosporites multifoveolatus*, *Trudopollis succedanus*, *Oculopollis* sp.), and early Paleogene (*Nudopollis endangulatus*, *Vacuopollis concavus*) are present in many samples.

The paleoclimatic reconstruction of the Bodenheim palynoflora relies on a total of 29 taxa with known NLRs (Text-Figure 4). Calculating the MAT with the coexistence approach leads to an interval of 15.7 to 17.1°C. The CMT interval ranges from 6.2 to 7.5°C. The coexistence interval for the WMT is calculated as 25.6 to 26.8°C. No coexistence interval for all 29 NLRs can be found for the mean annual precipitation: Here, the exclusion of *Ephedra*, which forms an outlier with lower precipitation requirements, leads to values between 1096 and 1298 mm.

TABLE 1. Species list of pollen and spores from the Mainz Basin (core Bodenheim). The taxa used for the climatic analysis are marked in boldface type.

In situ Taxa:

***Aceripollenites striatus* (Pflug 1959) Thiele-Pfeiffer 1980**
***Araliaceoipollenites euphorii* (Potonié 1931) Potonié 1951**
Baculatisporites primarius (Wolff 1934) Thomson & Pflug 1953
Boehlensipollis hohli Krutsch 1962
Buxapollis sp.
Camarozonosporites (C.) heskemensis (Pflanzl 1955) Krutzsch 1959
Camarozonosporites (H.) rarus (Doktorowicz-Hrebnicka 1960) Krutzsch 1963
***Carpinuspollenites carpinoides* (Pflug 53) Nagy 1969**
***Caryapollenites simplex* (Potonié 1931) Raatz 1937 emend. Potonié 1960**
Cicatricosisporites chattensis chattensis Krutzsch 1961
Cicatricosisporites dorogensis Potonié & Gelletich 1933
Cupressacites insulipapillatus (Trevisan 1967) Krutzsch 1971
Cyperaceaepollis scholitzensis Krutzsch 1970
Echinatisporites cycloides Krutzsch 1963
***Ephedripites bernheidensis* Krutzsch 1961**
***Ephedripites eocenipites* (Wodehouse 1933) Krutzsch 1961**
***Ephedripites fusiformis* (Shakmundes 1965) Krutzsch 1970**
***Ephedripites lusaticus* Krutzsch 1961**
***Ephedripites tertiarius* Krutzsch 1970**
Ericipites callidus (Potonié 1931) Krutzsch1970
Ericipites ericius (Potonié 1931) Potonié 1960
Ericipites sp.
Foveotriletes crassifovearis Krutzsch 1962
Foveotriletes verrucatoides Krutzsch 1962
Gleicheniidites (L.) elegans Nagy 1963
Gleicheniidites sp.
Graminidites subtiliglobosus (Trevisan 1967) Krutzsch 1970
***Ilexpollenites iliacus* (Potonié 1931) Thomson & Pflug 1953**
***Ilexpollenites margaritatus* (Potonié 1931) Thiergart 1937 ex Potonié 1960**
***Inaperturopollenites concedipites* (Wodehouse 1933) Krutzsch 1971**
***Inaperturopollenites hiatus* (Potonié 1931) Thomson & Pflug 1953**
***Intratriporopollenites insculptus* Mai 1962**
Laevigatosporites hardti hardti Krutzsch 1967
Leiotriletes adriennis (Potonié & Gelletich 1933) Krutzsch 1959
Leiotriletes maxoides maxoides Krutzsch 1962
Leiotriletes maxoides minoris Krutzsch 1962
Leiotriletes triangulatoides Krutzsch 1962
Leiotriletes wolffi wolffi Krutzsch 1962
Lonicerapollis gallwitzi Krutsch 1962
***Magnolipollis magnolioides* Krutzsch 1970**
Microfoveolatisporis neogranuloides Krutzsch 1967
Momipites punctatus (Potonié 1931) Nagy 1969
Momipites quietus (Potonié 1931) Krutzsch 1970
Monoleiotriletes gracilis Krutzsch 1959
Monoleiotriletes minimus Krutzsch 1962
Multiporopollenites polymorphus Schuler 1990
Neogenisporites neogenicus Krutzsch 1962
***Nyssapollenites kruschi accessorius* (Potonié 1931) Nagy 1969**
***Nyssapollenites kruschi analepticus* (Potonié 1931) Nagy 1969**
Pentapollenites cf. *striatus* Krutzsch 1962
Piceapollis spp.
Pityosporites spp.
***Platanoidites* sp.**
Polypodiaceoisporites boerzsoenyensis Nagy 1985
Polypodiaceoisporites marxheimensis (Mürriger & Pflug 1952) Krutzsch 1959
Polypodiaceoisporites semiverrucatus (Krutzsch 1967) Nagy 1985
Polypodiaceoisporites spp.
***Polyporopollenites undulosus* (Wolff 1934) Thomson & Pflug 1953**
***Polyvestibulopollenites verus* (Potonié 1931) Thomson & Pflug 1953**
***Porocolpopollenites vestibulum* (Potonié 1931) Thomson & Pflug 1953**
***Pterocaryapollenites stellatus* (Potonié 1931) Raatz 1937**
Retitricolporites oleoides Roche & Schuler 1976
Retitriletes cf. *pseudoclavatus* Krutzsch 1963
Scabratricolporites cylindricus Roche & Schuler 1976
Sciadopityspollenites serratus (Potonié & Venitz 1934) Thiergart 1937
***Sequoiapollenites polyformosus* Thiergart 1937**
Tetracolporopollenites sapotoides Thomson & Pflug 1953
Toroisporis (T.) neddeni (Potonié 1931) Krutzsch 1959
Toroisporis spp.
***Triatriopollenites coryphaeus* Thomson & Pflug 1953**
Tricolpopollenites liblarensis fallax (Potonié 1934) Thomson & Pflug 1953
Tricolporopollenites cingulum fusus (Potonié 1934) Thomson & Pflug 1953
***Tricolporopollenites cingulum pusillus* (Potonié 1934) Thomson & Pflug 1953**
Tricolporopollenites cingulum oviformis (Potonié 1931) Thomson & Pflug 1953
***Tricolporopollenites edmundi* (Potonié 1931) Thomson & Pflug 1953**
***Tricolporopollenites marcodurensis* Thomson & Pflug 1953**
***Tricolporopollenites megaexactus bruehlensis* (Thomson 1950) Thomson & Pflug 1953**
***Tricolporopollenites megaexactus exactus* (Potonié 1931) Thomson & Pflug 1953**
Tricolporopollenites pseudocingulum (Potonié 1931) Thomson & Pflug 1953
Tricolporopollenites staresedloensis Krutzsch & Pacltova 1970
Tricolporopollenites villensis (Thomson 1950) Thomson & Pflug 1953
Tricolporopollenites spp.
Trilites embryonalis Krutzsch 1967
Trilites multivallatus (Pflug 1953) Krutzsch 1959
Trilites spp.
Triplanosporites microsinuosus Planzl 1955
***Trivestibulopollenites betuloides* Pflug 1953**
Verrucatosporites bockwitzensis Krutzsch 1967
Verrucatosporites megafavus Krutzsch 1967
Verrucatosporites alienus (Potonié 1931) Thomson & Pflug 1953
Verrucatosporites spp.
Verrucingulatisporites treplinensis Krutzsch 1961
Verrucingulatisporites undulatus undulatus Krutzsch 1967
***Verrutricolporites theaceoides* Roche & Schuler 1976**

Reworked Taxa:

Anemiidites cf. *shabashakensis* Ashraf 1977
Appendicisporites tricornitatus Weyland & Greifeld 1953
Callialasporites dampieri (Balme 1957) Sukh Dev 1961
Callialasporites segmentatus (Balme 1957) Srivastava 1963
Cicatricosisporites spp.
Classopollis torosus (Reissinger 1950) Morbey 1975
Concavisporites umbonatus (Bolchovitina 1953) Arjang 1975
Foveosporites multifoveolatus Döring 1965
Limitisporites sp.
Nudopollis endangulatus (Pflug 1953) Pflug 1953
Oculopollis sp.
Trudopollis succedanus Pflug 1953
Trudopollis spp.
Vacuopollis concavus (Pflug 1953) Krutzsch 1960

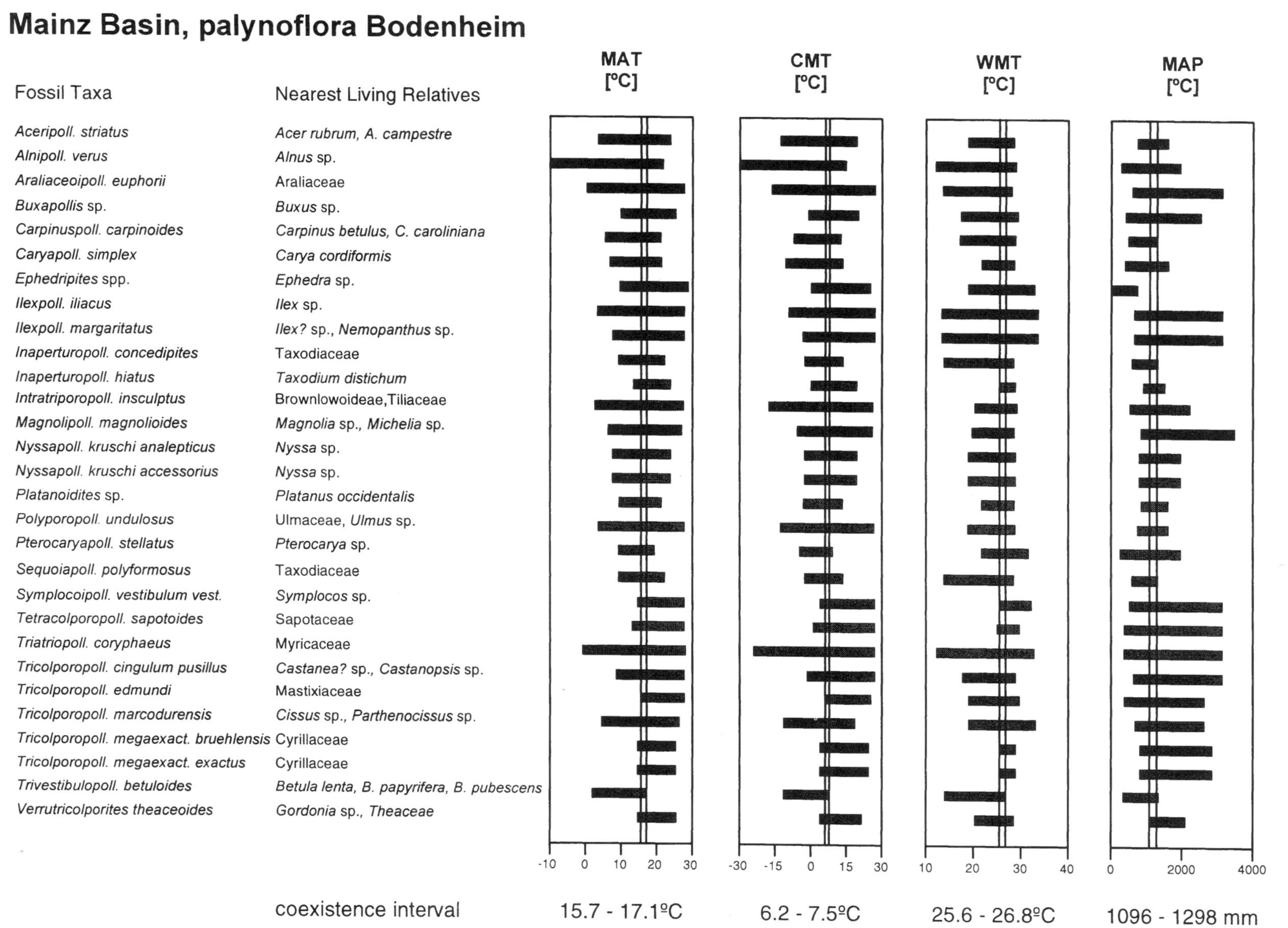

Text-Figure 4. Application of the coexistence approach to the palynoflora from the Bodenheim core (Mainz Basin). The shaded boxes mark the climatic requirements of the taxa, the vertical lines delimit the widths of the coexistence intervals (MAT: mean annual temperature, CMT: mean temperature of the coldest month, WMT: mean temperature of the warmest month, MAP: mean annual precipitation).

Flörsheim megaflora

In the Flörsheim megaflora 24 taxa were identified (Table 2). The assemblage predominantly contains broad-leaved evergreen elements from a mesophytic subtropical forest (e.g. *Daphnogene cinnamomifolia, Platanus neptuni, Ternstroemites floersheimensis*) and from a swampy environment (e.g. *Athrotaxis couttsiae*). Three NLRs of this flora are also found among the NLRs of the Bodenheim palynoflora. However, evidence of the subtropical to tropical palmaceae cf. *Sabal maior* and *Phoenicites* sp. as well as of the climatically significant liana *Smilax weberi* is restricted to megafloral remains.

Nine taxa have NLRs that could be used for calculating the coexistence intervals (Text-Figure 5). For the MAT and the CMT, the coexistence approach yields temperatures from 15.6 to 18.8°C and 7.1 to 10.2°C, respectively. Calculations for the WMT result in values from 25.7 to 28.1°C. In all cases, the intervals cover the temperature requirements of all NLRs. The interval of the mean annual precipitation is determined to be between 1003 and 1250 mm and does not include *Tetraclinis*, which today requires drier climatic conditions.

Sava Region Palynoflora

The samples from the Novi Dol section provided an intermediately- to well-preserved terrigenous palynoflora consisting of 42 taxa (Table 3). Quantitatively, small tricolporate pollen grains are most abundant throughout the succession, whereas bisaccate pollen and spores only occur in subordinate numbers.

Twenty-six taxa of non-saccate pollen were identified (Table 3). Among these, species from anemophilous parent plants assigned to a mixed mesophytic forest environment such as *Tricolporopollenites asper, T. cingulum oviformis, T. cingulum pusillus, T. edmundi,* and *Araliaceoipollenites reticuloides* occur most frequently. Plants from more humid vegetation units are represented by the very abundant pollen *Caryapollenites simplex* (swamp forest vegetation), *Triatriopollenites coryphaeus, T. rurensis,* and *Inaperturopollenites concepidites* (peatland vegetation).

The climatic evaluation of the Novi Dol palynoflora is based on 15 taxa (Text-Figure 5). The calculated coexistence interval for the MAT ranges from 15.7 to 17.2°C. The intervals for the CMT and WMT are determined as 6.2 to 7.0°C and 22.8 to 26.8°C, respectively. For the MAP, the coexistence approach yields values between 975 and 1322 mm.

Austrian Molasse

The palynoflora of Hochuli (1978) from the Austrian Molasse Basin contains a total of 43 taxa, 20 of which were used for calculating the coexistence intervals. The values obtained are 15.7 to 17.1°C for the MAT, 6.2 to 7.5°C for CMT, 25.6 to 26.8°C for the WMT, and 1122 to 1322 mm for the MAP (Table 4).

Weisselster Basin

The megaflora analyzed from the Weisselster Basin (Kunzmann 1995) consists of 33 taxa. Here, the paleoclimatic reconstruction is based on 13 taxa. The resulting coexistence interval for the MAT ranges from 15.9 to 17.4°C. The interval for the CMT is rather broad and ranges from 7.4 to 12.3°C. Calculations of the WMT yield an interval from 25.7 to 26.8°C, and the MAP is suggested to lie between 1096 and 1198 mm (Table 4).

DISCUSSION

General Aspects

The coexistence approach yields very similar values for the calculated MATs, CMTs, WMTs, and MAPs for the different floras and localities (Table 4). The MAT from the Weisselster Basin, the northernmost study area, conforms fully with the data from the Sava region which was situated about 800 km south of the Weisselster Basin during the Rupelian (see Dercourt et al., 1993); a comparable accordance can be observed for the other climatic parameters. These data indicate that the latitudinal climate gradient was very low in the Rupelian of central Europe and, in fact, may have been almost as low as in the Eocene (Greenwood and Wing, 1995). Moreover, the climatic similarity of localities from the north and south of the Alpine area shows that neither the Alps nor the Paratethys played a major role as a climatic divide. Obviously, the mid–Rupelian climate of central Europe was quite homogeneous, and a distinct geographic climatic differentiation did not yet exist.

TABLE 2. Species list of the megaflora from the Mainz Basin (Flörsheim clay pit). The taxa used for the climatic analysis are marked in boldface type.

***Athrotaxis couttsiae* (Heer 1862) Gardner 1884**
Carpinus vel *Ostrya* sp.
***Carya* sp.**
Comptonia cf. *acutiloba* Brongniart 1828
Daphnogene cinnamomifolia (Brongniart 1822) Unger 1850
?*Eotrigonobalanus furcinervis* (Rossmässler 1840) Kvacek & Walther 1989
***Laurophyllum acutimontanum* Mai 1963**
Laurophyllum medimontanum Buzek, Holy & Kvacek 1976
***Laurophyllum* cf. *pseudoprinceps* Weyland & Kilpper 1963**
Leguminosites spp.
Lygodium kaulfussii Heer 1861 emend. Gardner & Ettingshausen 1880
***Myrica longifolia* Unger 1850**
Osmunda lignitum (Giebel 1857) Stur 1870
Phoenicites sp.
Pinus sp.
?*Picea* sp.
Platanus neptuni (Ettingshausen 1866) Buzek et al. 1967
***Pungiphyllum cruciatum* (Al. Braun 1851) Frankenhäuser & Wilde 1995**
cf. *Rosa lignitum* Heer 1896
cf. *Sabal maior* (Unger 1847) Heer 1855
***Smilax weberi* Wessel & Weber 1855**
***Tetraclinis salicornioides* (Unger 1841) Kvacek 1989**
Ternstroemites floersheimensis Kvacek & Walther 1984
Trigonobalanopsis rhamnoides (Rossmässler 1840) Kvacek & Walther 1988

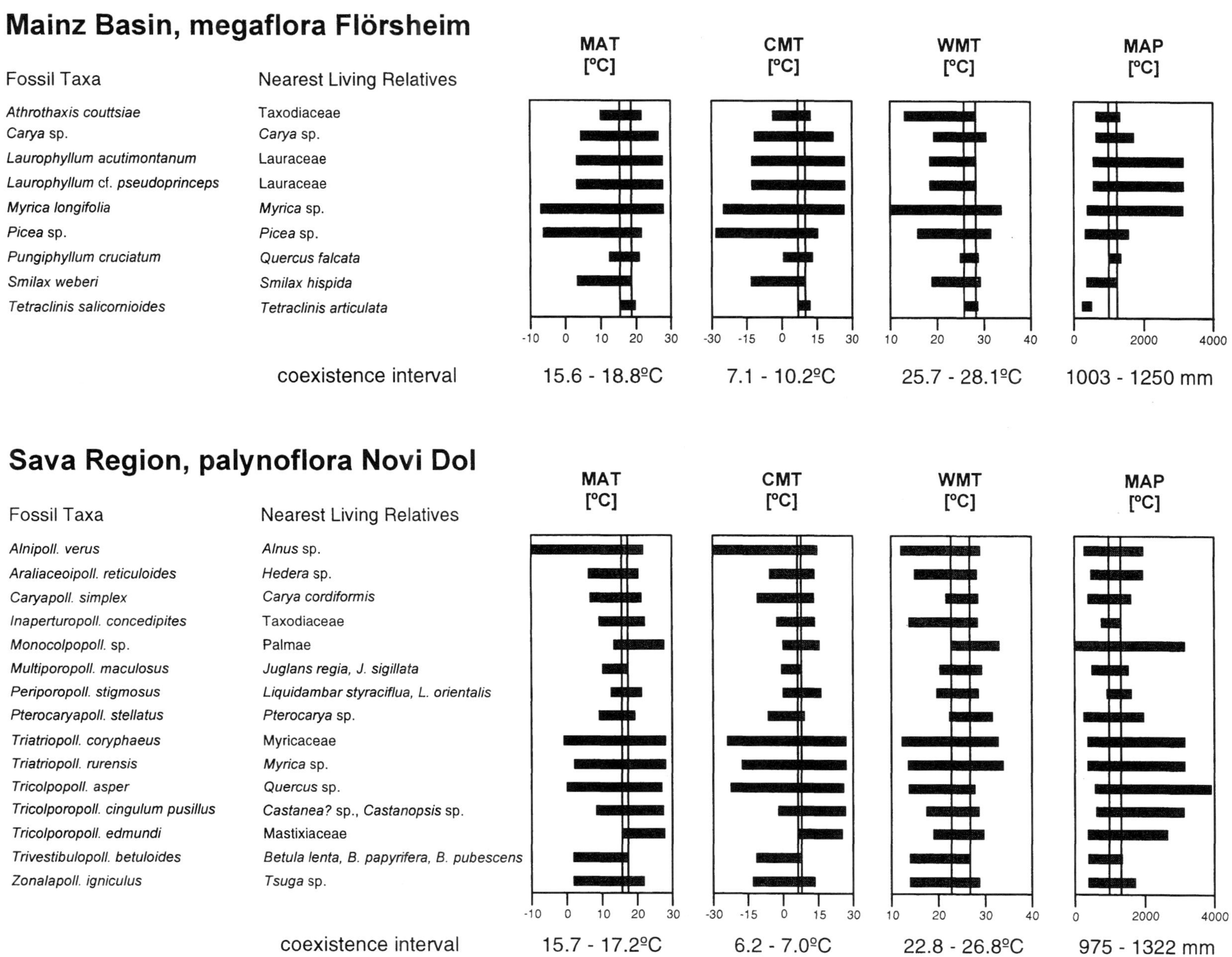

Text-Figure 5. Application of the coexistence approach to the megaflora from Flörsheim (Mainz Basin) and the palynoflora from the Novi Dol section (Sava region). The shaded boxes mark the climatic requirements of the taxa, the vertical lines delimit the widths of the coexistence intervals (MAT: mean annual temperature, CMT: mean temperature of the coldest month, WMT: mean temperature of the warmest month, MAP: mean annual precipitation).

TABLE 3. Species list of pollen and spores from the Sava region (Novi Dol section). The taxa used for the climatic analysis are marked in boldface type.

***Araliaceoipollenites reticuloides* Thiele-Pfeiffer 1980**
***Caryapollenites simplex* (Potonié 1931) Raatz 1937**
Cupressacites insulipapillatus (Trevisan 1967) Krutzsch 1971
Cycadopites microfollicularis Krutzsch 1970
Echigraminidites sp.
Echinatosporites sp.
Engelhardtioidites microcoryphaeus (Potonié 1931) Thomson & Pflug 1953
***Inaperturopollenites concedipites* (Wodehouse 1933) Krutzsch 1971**
Ischyosporites sp.
Laevigatosporites haardti haardti Krutzsch 1967
Leiotriletes maxoides minoris Krutzsch 1962
Lusatisporis perinatus Krutzsch 1963
Lusatisporis punctatus Krutzsch 1963
Momipites quietus (Potonié 1931) Krutzsch 1970
***Monocolpopollenites* sp.**
***Multiporopollenites maculosus* (Potonié 1931) Thomson & Pflug 1953**
Nupharipollis sp.
***Periporopollenites stigmosus* (Potonié 1950) Thomson & Pflug 1953**
Piceapollis spp.
Pityosporites spp.
Plicatopollenites plicatus (Potonié 1934) Krutzsch 1962
Polypodiaceoisporites triangulus trianguloides Krutzsch 1967
Polypodiaceoisporites spp.
***Polyvestibulopollenites verus* (Potonié 1931) Thomson & Pflug 1953**
***Pterocaryapollenites stellatus* (Potonié 1931) Raatz 1937**
Sciadopitys sp.
Tetracolporopollenites sp.
Toroisporites sp.
***Triatriopollenites coryphaeus* Thomson & Pflug 1953**
***Triatriopollenites rurensis* Thomson & Pflug 1953**
***Tricolpopollenites asper* Thomson & Pflug 1953**
Tricolpopollenites henrici (Potonié 1931) Thomson & Pflug 1953
Tricolporopollenites cingulum fusus (Potonié 1934) Thomson & Pflug 1953
Tricolporopollenites cingulum oviformis (Potonié 1931) Thomson & Pflug 1953
***Tricolporopollenites cingulum pusillus* (Potonié 1934) Thomson & Pflug 1953**
***Tricolporopollenites edmundi* (Potonié 1931) Thomson & Pflug 1953**
Tricolporopollenites reticingulum Krutzsch & Vanhoorne 1977
Tricolporopollenites spp.
Triplanosporites microsinuosus Pflanzl 1955
***Trivestibulopollenites betuloides* Pflug 1953**
Verrucatosporites favus favus Krutzsch 1967
Verrucatosporites histiopteroides Krutzsch 1967
Verrucatosporites alienus (Potonié 1931) Thomson & Pflug 1953
Verrucatosporites spp.
***Zonalapollenites igniculus* (Potonié 1931) Thomson & Pflug 1953**

Within the Mainz Basin, however, the temperature data from the Flörsheim megaflora indicate slightly higher values than for the Bodenheim palynoflora, although megaflora and palynoflora yield mutually consistent palaeotemperature results (see Table 4). This may represent an artifact of the methodology, but can also be explained by the different provenances of the micro- and megafloras. Due to their excellent stage of preservation, the megafloral remains from Flörsheim have probably been transported only a short distance and, therefore, represent the vegetation of the warmer coastal area. On the other hand, the palynoflora from Bodenheim can contain pollen from a wider range of sources, reflecting vegetational elements from the coast as well as from its possibly elevated and therefore cooler hinterland. Since elements from both areas are included in the climatic analysis, lower temperature values are obtained.

COMPARISON WITH OTHER PALEOCLIMATIC RECONSTRUCTIONS

Generally speaking, our quantitative palaeoclimatic reconstruction for the mid-Rupelian of central Europe is well in accordance with the view that a Cfa climate (classification of Koeppen, 1931) prevailed in the European Oligocene (e.g. Mai, 1995). Our calculations of the WMT and the MAP correspond well to estimates around 25°C and 700–1300

TABLE 4. Coexistence intervals of the calculated climatic parameters for the evaluated floras (MAT: mean annual temperature, CMT: temperature of the coldest month, WMT: temperature of the warmest month, MAP: mean annual precipitation).

	No. of taxa	MAT [°C]	CMT [°C]	WMT [°C]	MAP [mm]
Mainz Basin, palynoflora Bodenheim	30	15.7–17.1	6.2–7.5	25.6–26.8	1096–1298
Mainz Basin, megaflora Flörsheim	8	15.6–18.8	7.1–10.2	25.7–28.1	1003–1250
Sava Region, palynoflora Novi Dol	16	15.7–17.2	6.2–7.0	22.8–26.8	975–1322
Austrian Molasse, palynoflora (Hochuli, 1978)	20	15.7–17.1	6.2–7.5	25.6–26.8	1122–1322
Weisselster Basin, megaflora (Kunzmann, 1995)	13	15.9–17.4	7.4–12.3	25.7–26.8	1096–1198

mm, respectively, as they were derived from mid–Oligocene floras by most previous authors (e.g. Andreanszky, 1966; Mai and Walther, 1978, 1983, 1991; Mai, 1995). For the MAT, our calculations are consistent with the estimates of Krutzsch (1992) for the late Rupelian of central Germany, which are as high as 17.5°C, and also with the values of Berger (1989) given for the Oligocene of Switzerland (13–17°C, >17°C). They are, however, generally warmer than the temperatures proposed by Mai (1995) which range from 13 to 15°C.

If we consider specific floras, our results from the Flörsheim and Weisselster Basin megafloras for the MAT and the CMT differ from estimates of earlier researchers. For these floras, the intervals for the MAT calculated with the coexistence approach are distinctly warmer than the estimates by Mai and Walther (1983) of 13 to 16.5°C for Flörsheim and 10 to 15.5°C for Weisselster Basin. Similarly, our calculated intervals for the CMT are also warmer than the values proposed by these authors (< 6°C for Flörsheim and -2 to 4°C for Weisselster Basin). They make minimum temperatures as low as -15°C, as proposed by Mai and Walther (1978), seem very improbable and are more consistent with the estimations of Krutzsch (1992) of minima slightly below 0°C.

Generally, our temperature results are supported by the view of Krutzsch (1992) who postulates a relatively warm, subtropical Rupelian climate in central Europe. They are consistently higher than the estimates of Mai and Walter (1978, 1983, 1991) and Mai (1995).

CONCLUSIONS

The application of the coexistence approach to palyno- and megafloras from the Early Oligocene of central Europe leads to paleoclimate data that are internally consistent and supported by results of other authors. This indicates that with the help of the coexistence approach palynofloras as old as the Paleogene can be used routinely for quantitative paleoclimate reconstructions. The climatic ranges obtained for the MAT, CMT, WMT, and MAP are generally narrower than the estimates of previous researchers.

Our quantitative paleoclimate reconstruction for the mid-Rupelian of central Europe is well within the limits of a Cfa climate after the classification of Koeppen (1931). It indicates a fairly homogeneous warm-temperate and humid climate with a very low latitudinal climate gradient and no distinct geographic differentiation. Obviously, neither the Alps nor the Western Paratethys played a role as a climatic divide at that time.

ACKNOWLEDGMENTS

The help of B. Jelen (Ljubljana) for his support in the field work in Slovenia and H. Weiler (Mainz) for providing sample material from the Bodenheim core is gratefully acknowledged. We would also like to thank W. Krutzsch (Berlin) for fruitful discussions of the pollen floras, and E. Kempf (Köln) and F. Schaarschmidt (Frankfurt) for the access to the collections and facilities of the Senckenberg Museum. S. Fowell, F. McCarthy, and P.A. Zippi are thanked for critical reviews of an earlier version of the manuscript. J.P. acknowledges financial support from the Studienstiftung des deutschen Volkes. The work of A.A.B. and V.M. was supported by the Collaborative Research Center 275, University of Tübingen.

References Cited

ANDREANSZKY, G.
1966 On the upper Oligocene flora of Hungary. Analysis of the site at the Wind brickyard, Eger. *Studia Biologica Hungarica*, 5/5: 5–151.

BERGER, J.P.
1989 Paläoökologie und Paläoklimatologie einiger oligozäner Floren der westschweizerischen subalpinen Molasse. *Courier Forschungsinstitut Senckenberg*, 109: 207–228.

BERGGREN, W.A., KENT, D.V., SWISHER, C.C., and AUBRY, M.-P.
1995 A revised Cenozoic geochronology and chronostratigraphy. *In*: Berggren, W.A., Kent, D.V., Aubry, M.-P., Hardenbol, J. (eds.), Geochronology, time scales and global stratigraphic correlation. *Society for Sedimentary Geology Series Publication*, 54: 129–212.

BOULTER, M.C.
1984 Palaeobotanical evidence for land-surface temperature in the European Palaeogene. *In*: Brenchley, P.J. (ed.), *Fossils and Climate*. Wiley, Chichester, pp. 35–48.

CIMERMAN, F.
1979 Oligocene Beds in Slovenia. *In*: Drobne, K. (ed.), Geological Development in Slovenia and Croatia. *Proceedings of the 16th European Microplaeontological Colloquium*, pp. 65–70.

DERCOURT, J., RICOU, L.E., and VRIELYNCK,B.
1993 *Atlas Tethys Palaeoenvironmental Maps*. Gauthier-Villard, Paris, 307 pgs.

EISSMANN, L.
1994 Leitfaden der Geologie des Präquartärs im Saale-Elbe-Gebiet. *In*: Eissmann, L., and Litt, T., Das Quartär Mitteldeutschlands. *Altenburger Naturwissenschaftliche Forschungen*, 7: 11–53.

ENGELHARDT, H.
1911 Über tertiäre Pflanzenreste von Flörsheim a. M. *Abhandlungen der senckenbergischen naturforschenden Gesellschaft*, 29: 307–406.

GEYLER, H.T.
1883 Verzeichnis der Tertiärflora von Flörsheim a. M. *Berichte der senckenbergischen naturforschenden Gesellschaft*, 285–287.

GREENWOOD, D.R., and WING, S.L.
1995 Eocene continental climates and latitudinal temperature gradients. *Geology*, 23/11: 1044–1048.

GRIMM, K.I.
1994 Paläoökologie, Paläogeographie und Stratigraphie im Mainzer Becken, im Oberrheingraben, in der Hessischen Senke und in der Leipziger Bucht während des Mittleren Rupeltons (Fischschiefer/Rupelium/Unteroligozän). *Mitteilungen Pollichia*, 81: 7–193.

HEER, O.
1859 *Flora tertiaria Helvetiae, Band 3*. Wurster and Co., Winterthur, 378 pgs.

HOCHULI, P.A.
1978 Palynologische Untersuchungen im Oligozän und Untermiozän der Zentralen und Westlichen Paratethys. *Beiträge zur Paläontologie von Österreich*, 4: 1–132.

HUBBARD, R.N.L.B., and BOULTER, M.C.
1983 Reconstruction of Palaeogene climate from palynological evidence. *Nature*, 301: 147–150.

KOEPPEN, W.
1931 *Grundriß der Klimakunde*. De Gruyter, Berlin, 388 pgs.

KRUTZSCH, W.
1967 Der Florenwechsel im Alttertiär Mitteleuropas auf Grund von sporenpaläontologischen Untersuchungen. *Abhandlungen des Zentralen Geologischen Instituts*, 10: 17–38.

1992 Paläobotanische Klimagliederung des Alttertiärs (Mitteleozän bis Oberoligozän) in Mitteldeutschland und das Problem der Verknüpfung mariner und kontinentaler Gliederungen (klassische Biostratigraphien—paläobotanisch-ökologische Klimastratigraphie—Evolutions-Stratigraphie der Vertebraten). *Neues Jahrbuch für Geologie und Paläontologie, Abhandlungen*, 186: 137–253.

KRUTZSCH, W., and MAJEWSKI, J.
1967 Zur Methodik der pollenstratigraphischen Zonengliederung im Jungtertiär Mitteleuropas. *Abhandlungen des Zentralen Geologischen Instituts*, 10: 83–98.

KUNZMANN, L.
1995 Rekonstruktion einer tertiären Flußlandschaft. Geologische und makropaläobotanische Untersuchungen am paläogenen Zwickau-Altenburger Flußsystem im Weißelsterbecken, NW–Sachsen. Methoden, Möglichkeiten, Grenzen. *Abhandlungen des Staatlichen Museums für Mineralogie und Geologie zu Dresden*, 41: 73–95.

MAI, D.H.
1995 *Tertiäre Vegetationsgeschichte Europas*. Fischer, Jena, 691 pgs.

MAI, D.H., and WALTHER, H.
1978 Die Floren der Haselbacher Serie im Weißelster-Becken (Bezirk Leipzig, DDR). *Abhandlungen des Staatlichen Museums für Mineralogie und Geologie zu Dresden*, 28: 1–200.
1983 Die fossilen Floren des Weißelster-Beckens und seiner Randgebiete. *Hallesches Jahrbuch für Geowissenschaften*, 8: 59–74.
1991 Die oligozänen und untermiozänen Floren NW–Sachsens und des Bitterfelder Raumes. *Abhandlungen des Staatlichen Museums für Mineralogie und Geologie zu Dresden*, 38: 16–230.

MARTINI, E.
1971 Standard Tertiary and Quaternary calcareous nannoplankton zonation. *In*: Farinacci, A. (ed.): *Proceedings of the II Planktonic Conference, Roma 1970*, 2: 739–785.
1982 Bestandsaufnahme des Nannoplankton im "präaquitanen" Tertiär des Mainzer Beckens. *Mainzer geowissenschaftliche Mitteilungen*, 10: 29–36.

MIHAJLOVIC, D., and JUNGWIRTH, E.
1988 Oligocene aged flora from Novi Dol (Slovenia, Yugoslavia). *Revue de Paléobiologie*, 7/2: 435–447.

MOSBRUGGER, V.
1995 New Methods and Approaches in Tertiary Palaeoenvironmental Research. *Abhandlungen des Staatlichen Museums für Mineralogie und Geologie zu Dresden*, 41: 41–52.

MOSBRUGGER, V., and UTESCHER, T.
1997 The coexistence approach — a method for quantitative reconstructions of Tertiary terrestrial palaeoclimate data using plant fossils. *Palaeogeography, Palaeoclimatology, Palaeoecology*, 134: 61–86.

MÜLLER, C.
1971 Nannoplankton aus dem westdeutschen Mitteloligozän. *Notizblätter des hessischen Landesamtes für Bodenforschung*, 99: 43–53.

POCKNALL, D.T.
1990 Palynological evidence for the early to middle Eocene vegetation and climate history of New Zealand. *Review of Palaeobotany and Palynology*, 65: 57–69.

PROSS, J.
1997 Aquatische Palynomorphe im Rupel des Mainzer Beckens (Oligozän, Süddeutschland): Paläoökologie, Biostratigraphie und Taxonomie. *Tübinger Mikropaläontologische Mitteilungen*, 15: 1–208.

RÖGL, F.
1996 Stratigraphic correlation of the Paratethys Oligocene and Miocene. *Mitteilungen der Gesellschaft der Geologie- und Bergbaustudenten in Österreich*, 41: 65–73.

ROTHAUSEN, K.H., AND SONNE, V.
1984 *Mainzer Becken*. Sammlung geologischer Führer, 79. Borntraeger, Berlin, 203 pgs.

SITTLER, C.
1965 Le paléogène des fossés Rhénan et Rhodanien. Etudes sédimentologiques et paléoclimatiques. *Mémoires du Service de la Carte Geologique d'Alsace et de Lorraine*, 24: 1–392.

SLOAN, L.C., and BARRON, E.J.
1990 "Equable" climates during Earth's history? *Geology*, 18: 489–492.

SONNE, V.
1988 Oberer Rupelton, Schleichsand (Rupel) und Cyrenenmergel (tiefes Chatt) im Mainzer Becken: Können sie mikropaläontologisch definiert werden? *Mainzer Geowissenschaftliche Mitteilungen*, 17: 19–30.

SZTANÓ, O.
1994 The tide-influenced Pétervására sandstone, Early Miocene, northern Hungary: sedimentology, palaeogeography and basin development. *Geologica Ultraiectina*, 120: 1–155.

WOLFE, J.A.
1993 A method of obtaining climatic parameters from leaf assemblages. *United States Geological Survey Bulletin*, 2040: 1–71.
1995 Paleoclimatic estimates from Tertiary leaf assemblages. *Annual Review of Earth and Planetary Sciences*, 23: 119–142.

ZAGWIJN, W.H., and HAGER, H.
1987 Correlation of continental and marine Neogene deposits in the South–Eastern Netherlands and the Lower Rhine district. *Mededelingen van de Werkgroep voor Tertiaire en Kwartaire Geologie*, 24: 59–78.

ZIEGLER, P.A.
1990 *Geological atlas of western and central Europe*. Shell Internationale Petroleum Maatschappij B.V., The Hague, 239 pgs.

ZIEMBINSKA-TWORZYDLO, M., and WAZYNSKA, H.
A palynological subdivision of the Neogene in western Poland. *Bulletin de l'Académie Polonaise des Sciences*, 29: 29–43.

Authors' addresses

Jörg Pross*
Angela A. Bruch
Volker Mosbrugger
Institut und Museum für Geologie und Paläontologie
Universität Tübingen
Sigwartstr. 10
D-72076 Tübingen
Germany
e-mail: joerg.pross, angela.bruch, volker.mosbrugger@uni-tuebingen.de

*Present address:
Department of Geology
Colby College
5820 Mayflower Hill
Waterville, ME 04901-8858
U.S.A.

Zlatko Kvacek
Department of Paleontology
Charles University
Albertov 6
CZ Praha 2
Czech Republic
e-mail: kvacek@mail.natur.cuni.cz

FOSSIL POLLEN OF THE MYRTACEAE IN THE LATE TERTIARY OF NORTHEASTERN ARGENTINA: TAXONOMIC CONSIDERATIONS

Liliana T. Acevedo and Luisa M. Anzotegui

Abstract

Fossil pollen of the Family Myrtaceae from the Pliocene Ituzaingó Formation and the Late Miocene Paraná Formation in northeastern Argentina have been studied. The comparative analysis with extant species reveals that species, groups of species, as well as genera can be identified using a combination of morphological features.

Myrtaceidites Cookson & Pike 1954 is the genus usually used to name fossils attributed to the Myrtaceae. However, it does not include all the variability found in specimens from northern Argentinian formations. One possibility would be to emend only *Myrtaceidites* which is not advisable because it would become a "macrogenus" with minimal biological integrity. Taking these considerations into account we have employed several established morphogenera with the required emendations to host the fossil grains with Myrtaceaus affinities. They are: *Myrtipites* Norton & Hall 1969, emend. for tricolporate grains; *Syncolporites* Van der Hammen 1954, emend for syncolporate grains; *Myrtaceidites* Cookson & Pike 1954, emend for parasyncolporate grains.

In the Ituzaingó Formation, several new species have been identified, including *Myrtipites annulatus, Syncolporites fastigiosus, S. rugulosus* and *Myrtaceidites triangularis.* Two other new species, *Myrtipites miocenica* and *Myrtaceidites triangularis,* are described from the Paraná Formation.

INTRODUCTION

Palynological investigation of the Ituzaingó Formation (Anzótegui, 1974; Anzótegui and Lutz, 1987) and Paraná Formation (Anzótegui and Garralla, 1986) has indicated the presence of various well-preserved tricolporate, parasyncolporate and syncolporate pollen grains, related to the family Myrtaceae. Cookson and Pike (1954) created the genus *Myrtaceidites* and proposed that it should be used for all the fossil grains related to Myrtaceae, because they considered it impossible to differentiate them generically due to the minimal variability that these species presented. This proposal became generalized and hence most of grains with these characteristics are now included in that genus, regardless their age, origin, and aperture types. However, other morphogenera have been proposed subsequently for these grains. These include *Myrtaceiopollenites* Potonié 1960, *Myrtipites* Norton & Hall 1969, *Myrtacidites* Belsky et al. 1965, but none of these genera has been used with the same frequency as *Myrtaceidites.*

Papers on recent Myrtaceae pollen such as those of Pike (1956), Gadek and Martin (1981), Barth and Barbosa (1972), Patel et al. (1984), and Chalson and Martin (1995), as well as our own observations on grains of Myrtaceae present in northeastern Argentina (Acevedo and Anzótegui, 1998) show that, using combinations of different morphological features, it is possible to characterize and identify species, groups of species, or genera. The following features of this difficult family can be employed for its identification the following order of importance:

Acevedo, Liliana T. and Anzotegui, Luisa M.
2001 Fossil pollen of the Myrtaceae in the Late Tertiary of northeastern Argentina: Taxonomic considerations; *In*: Goodman, D.K., and Clarke, R.T. (eds.), *Proceedings of the IX International Palynological Congress, Houston, Texas, U.S.A., 1996*; American Association of Stratigraphic Palynologists Foundation, p. 311–318.
 ISBN 0–931871–06–9

a. aperture type (tricolporate, syncolporate, parasyncolporate, heterosyncolporate);
b. presence of fastigium (Reistma, 1966, in Punt et al., 1994) or thickening in the pores;
c. colpi form (brevicolpate, longicolpate, linear, wide, striaght or curved;
d. amb outline (straight, concave or convex sides) rounded or truncated angles; and
e. sculpture (psilate or rugulate).

In order to clearly establish the features related to the apertures, and with the works of Erdtman (1952), Pike (1956), Krutzsch (1969b), Barth and Barbosa (1972), Patel et al. (1984) and Acevedo and Anzótegui (1998) as a basis, the following definitions are proposed.

Parasyncolporate. The colpi or their extensions branch and meet at the poles leaving an area with sexine known as polar islands in its center. This type of aperture defines the genus *Myrtaceidites* (Plate 1, figs. 3, 4; Plate 2, figs. 8, 9).

Syncolporate. The colpi join at the poles without forming a polar island. This type of aperture defines the genus *Syncolporites* (Plate 1, fig. 2; Plate 2, figs. 4–7).

Heterosyncolporate. Colpi meet at a pole forming a polar island and the opposite one without a polar island; it may happen that colpi do not join, having been named as *Bohlensipollis* Krutzsch 1962 (Plate 1, fig. 5).

Tricolporate. The colpi do not join at the poles being brevicolpate or longicolpate (Plate 1, fig. 1).

Geological Setting

Palynomorphs were collected from the Paraná and Ituzaingó formations. The Paraná Formation is composed of shales, sandstones, and organogenic limestones which are generally yellowish and green to gray. It is up to 300 m thick and has been recorded in the subsurface of great Chaco–Pampean Plain (Text-Figures 1a and 1c) in the provinces of Formosa, Chaco, Santa Fé, east Santiago del Estero and Córdoba, and western Entre Rios and

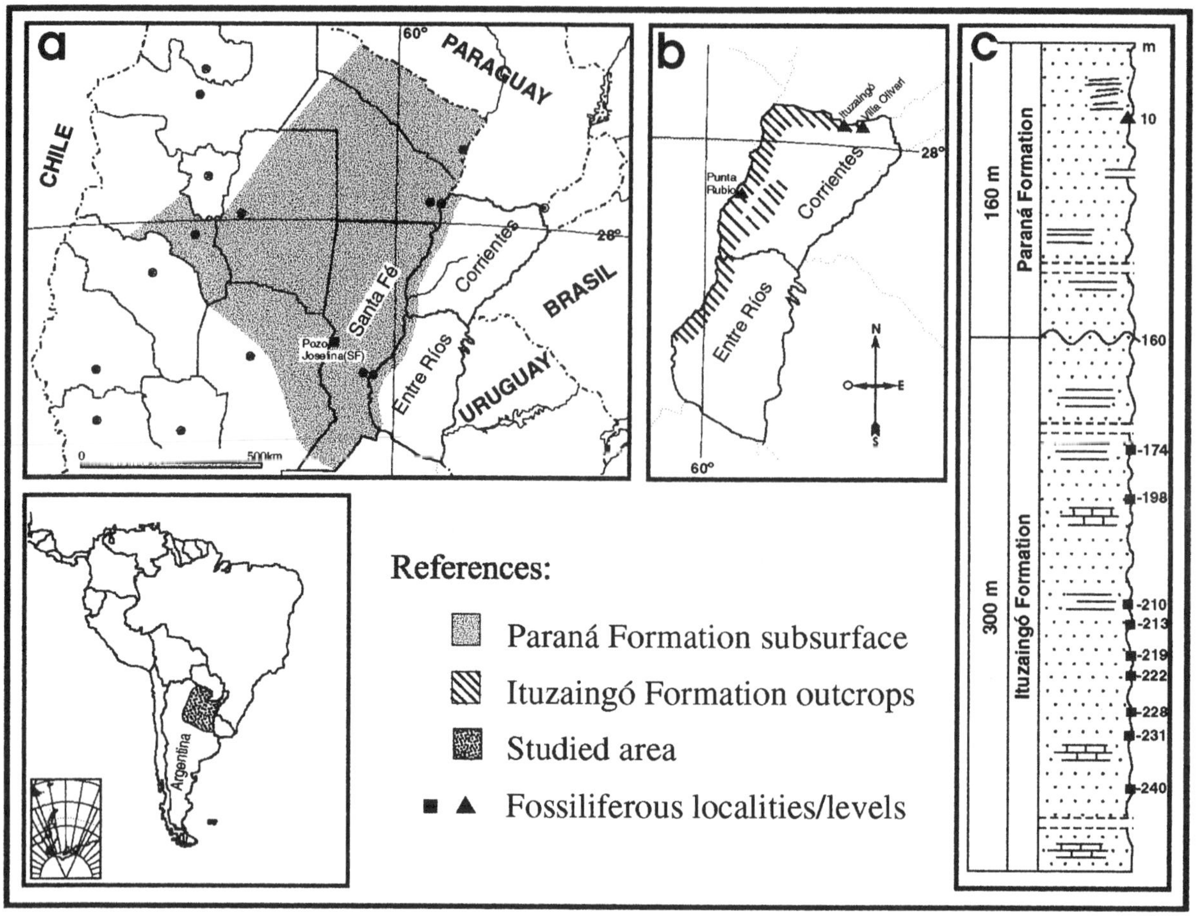

Text-Figure 1a, b, c. Geological sketch maps and stratigraphic column with positions of fossiliferous localities.

Corrientes (Herbst and Zabert, 1987). Its outcrops are limited to the southern half of Entre Rios, mainly along the Paraná River. The samples studied here are from the Josefina (SF) well in Santa Fé Province. The age assigned to this unit is Late Miocene (Herbst and Santa Cruz, 1985) but a somewhat older age is also possible. The Paraná Formation represent the sediments of the so called "Paraná Ingression," but in part they are of fluvial origen.

The Ituzaingó Formation is composed of fluviatile sands and sanstones which are yellow or whitish to dark red; only few shales are present as is characteristic of a Platte-type river (Jalfin, 1988). It is up to 160 m thick in the subsurface near the city of Corrientes, and it outcrops with thicknesses of up to 15–20 m along the Paraná River from Ituzaingó to Paraná City, and alongside some interior rivers (Text-Figures 1b and 1c). Palynomorphs have been recovered from thin grey to black sahles at different localities. Its age has been established as Late Pliocene (Herbst and Santa Cruz, 1985).

MATERIALS AND METHODS

The fossil material has been treated with the standard technique of dispersion with sodium hexametaphosphate, separation of organic matter with heavy liquids, removal of silicates with hydrofluoric acid, and oxidation with nitric acid. The recent material was stained using the acetolysis technique (Erdtman, 1966). Specimens were been observed using a LEITZ–WETZLAR microscope (Diaplan) No. 61184 of the PRINGEPA. The photographs with optical microscope were taken with a low speed film (25 ASA). Scanning electronic microscopy used a JEOL instrument at the Facultad de Ciencias Naturales y Museo in La Plata, Argentina.

PLATE 1

Extant species; polar views. The bar = 6 µm in fig. 1; 5 µm in fig. 2; 4 µm in figs. 3 and 4.

1 *Psidium cuneatum* Camb., PAL–CTES 3582.
2 *Campomanesia guaviroba* (D.C.) Kiaersk., PAL–CTES 5039.
3 *Myrcianthes pungens* (Berg.) Legrand, PAL–CTES 3580.
4 *Myrceugenia glaucescens* (Camb.) Legrand et Kausel, PAL–CTES 3557.
5 *Myrceugenia euosma* (Berg.) Legrand, PAL–CTES 3556.

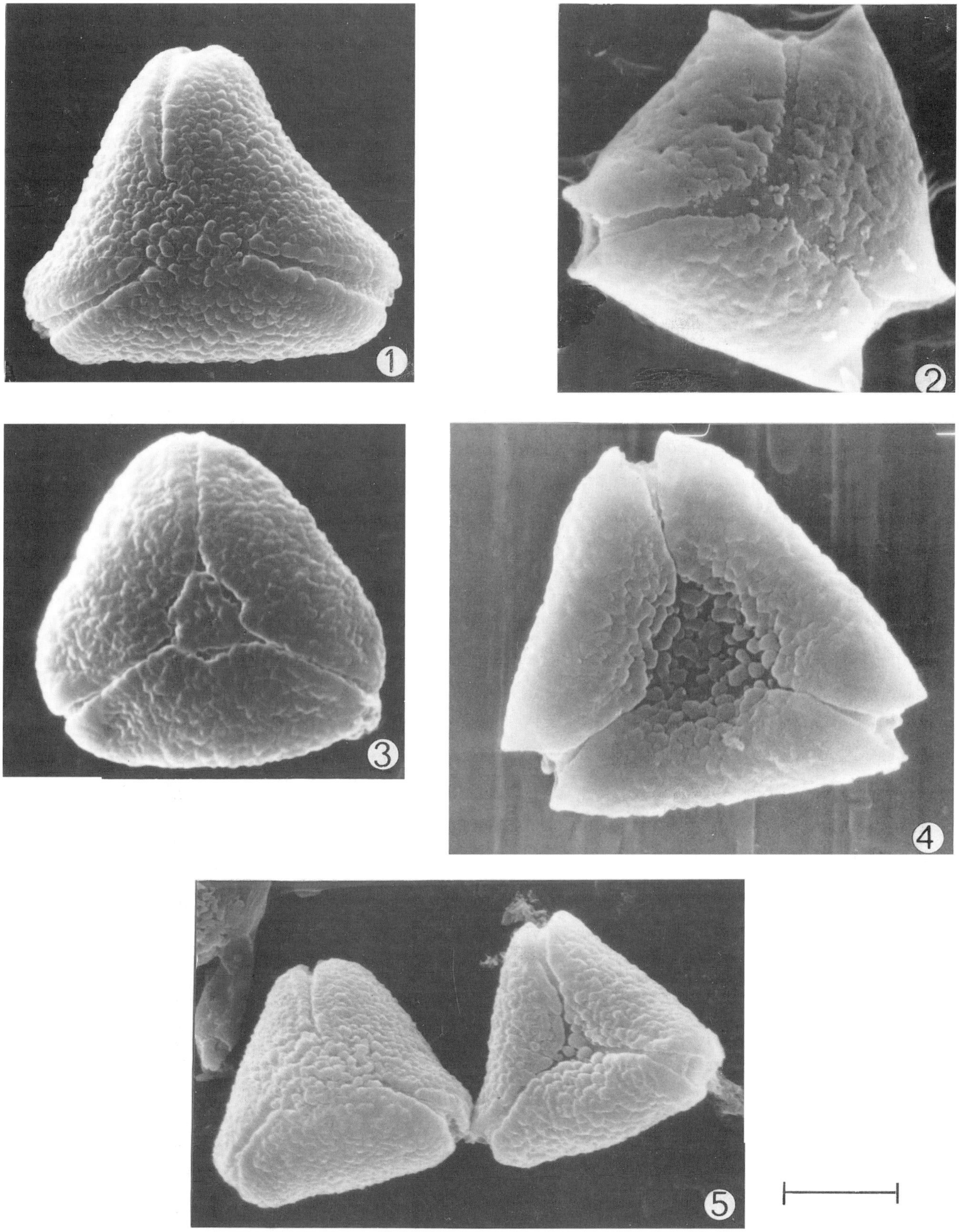
1
2
3
4
5

Repository

Both sets of slides, fossil and recent, are deposited in the CTES–PMP and CTES–PAL collections, respectively, at the Facultad de Ciencias Exactas y Naturales y Agrimensura of the Universidad Nacional del Nordeste, Argentina.

SYSTEMATIC PALEONTOLOGY

Genus *Myrtipites* Norton & Hall 1969 emend.

Type. *Myrtipites granulatus* Norton & Hall 1969.

Emended Diagnosis. Grains oblate, small to medium, isopolar to subisopolar, amb triangular to subtriangular, straight, convex or concave sides. Tricolporate; colpi never joined at poles, short or long. Pores with or without fastigium (vestibulum), with or without sexine thickening. Tectate, psilate, rugulate or granulate exine.

Discussion. This emendation is made for tricolporate species related to the family Myrtaceae. The following characters are added to the original diagnosis: oblate, subisopolar or isopolar, because, at equatorial view they can be either double-convex, or one of the poles may be straight and the other one convex. As regards the apertures, long colpi and pores with or without fastigium, with or without thickenings, psilate or rugulate sculpture are included. All these variations have been observed in recent species of Myrtaceae. We think it is more appropiate to use this emended genus because the original diagnosis was restrictive. These features allow us to incorporate the variability of forms that these grains present. On the other hand, *Myrtacidites* Belsky et al. 1965 is restricted for the presence of large camerae below the colpi and *Myrtaceiopollenites* Potonié 1960, because their apertures are surrounded by a short tubular structure which makes it more similar to grains of the Sapindaceae family.

Myrtipites annulatus sp. nov.
Plate 2, fig. 1

Diagnosis. Grains oblate, isopolar, amb triangular, straight sides, rounded angles. Tricolporate, angulaperturate. Straight and narrow colpi, not joined at the poles. Pores with sexine thickenings up to 2 µm. Tectate, psilate–rugulate exine. Sometimes stronger rugula at the poles.

Dimensions. (7 specimens). Equatorial diameter: 14 µm.

Holotype. Polar view PMP–CTES 1281: 45.7/94. Plate 2, fig.1.

Type Locality. Villa Olivari, Corrientes province.

Stratigraphy. Ituzaingó Formation.

Age. Late Pliocene.

Other Locality. Ituzaingó, Corrientes province.

Derivation of Name. It refers to the annular thickening of the pores.

Comparison. *Myrtipites granulatus* Norton & Hall, from the Cretaceous and Tertiary of the U.S.A. and *Eugenia/Myrcia* from the Miocene in Costa Rica, in Graham (1987, fig. 50) are different from the species here described because they are of larger size 24–26 µm and 32 µm respectively and do not present pore thickening. *Myrtaceidites parvus* Cookson & Pike, in Frederiksen et al. (1983, pl. 17, figs. 16–17) and *Myrtaceidites verrucosus* Stover & Partridge, in Baldoni (1987, pl. VI, figs. 16–17) from the Tertiary of Argentina, are different because they do not have these thickenings at the pores.

Myrtipites miocenica sp. nov.
Plate 2, figs. 2, 3

Diagnosis. Grains oblate, isopolar, amb subtriangular to subcircular, convex sides, rounded angles. Tricolporate, angulaperturate. Straight and narrow colpi, not joined at the poles. Pores with fastigium about 2–3 µm height. Tectate and rugulate exine, 1.5–2 µm thick.

Dimensions. (15 specimens). Equatorial diameter: 15–18 µm.

Holotype. Polar view PMP–CTES 1166 (-228 m): 39.8/104.3. Plate 2, fig. 2.

Type Locality. Josefina, Santa Fe province.

Stratigraphy. Paraná Formation.

Age. Late Miocene.

Other Levels. -174 m, -198 m, -210 m, -213 m, -219 m, -222 m, -240 m.

Derivation of Name. Refers to the age of the sediment from which the species was recovered.

Comparisons. This species differs from *Myrtipites annulatus* sp. nov. because it has a fastigium and lacks a sexine thickening in the pores; it differs from *Eugenia/Myrcia* in Graham (1987), and from *Myrtipites granulatus* because they are larger, 32 µm and 24–26 µm, respectively. From the illustration in Frederiksen et al. (1983, pl. 17, figs. 16–17) it is deduced that *Myrtaceidites parvus* also differs because it does not have a fastigium. Even though the description points out that *Myrtaceidites verrucosus* reported in Baldoni (1987, pl. VI, figs. 16–17) has a fastigium, it differs because it is verrucate and the species here described is rugulate.

Genus *Syncolporites* van der Hammen 1954 emend.

Type. *Syncolporites lisamae* van der Hammen 1954.

Emended Diagnosis. Grains oblate, isopolar to subisopolar, syncolporate, amb triangular to subtriangular, straight to convex or concave sides, angulaperturate. Long, straight or curved, narrow or wide colpi, joined at the poles. Polar island absent. Pores with or without fastigium, with or without sexine thickening. Tectate, psilate, rugulate scabrate or microverrucate exine.

Discussion. This emendation is made to include syncolporate species related to the Myrtaceae. The following features are added to the original diagnosis: isopolar or subisopolar, form of the amb, form of the colpi (straight, curved, wide, or linear), absence of polar islands, whit or whitout fastigium at the pores, with or without sexine thickenings, and psilate or rugulate sculpture. All these features have been found in syncolporate grains either fossil or recent in the Myrtaceae.

Syncolporites fastigiosus sp. nov.
Plate 2, figs. 4, 5

Diagnosis. Grains oblate, isopolar, amb triangular, straight side, rounded angles. Syncolporate, wide colpi, joined at the poles developing a syncolpium without polar island. Apertures with fastigium of 2 μm height. Tectate, psilate–fosulate exine, 1.5 μm thick.

Dimensions. (6 specimens) Equatorial diameter: 13 μm.

Holotype. Polar view PMP–CTES 1286: 43.9/102. Plate 2, fig. 4.

Type Locality. Villa Olivari, Corrientes province.

Stratigraphy. Ituzaingó Formation.

Age. Late Pliocene.

Other Localities. Ituzaingó and Punta Rubio, Corrientes province.

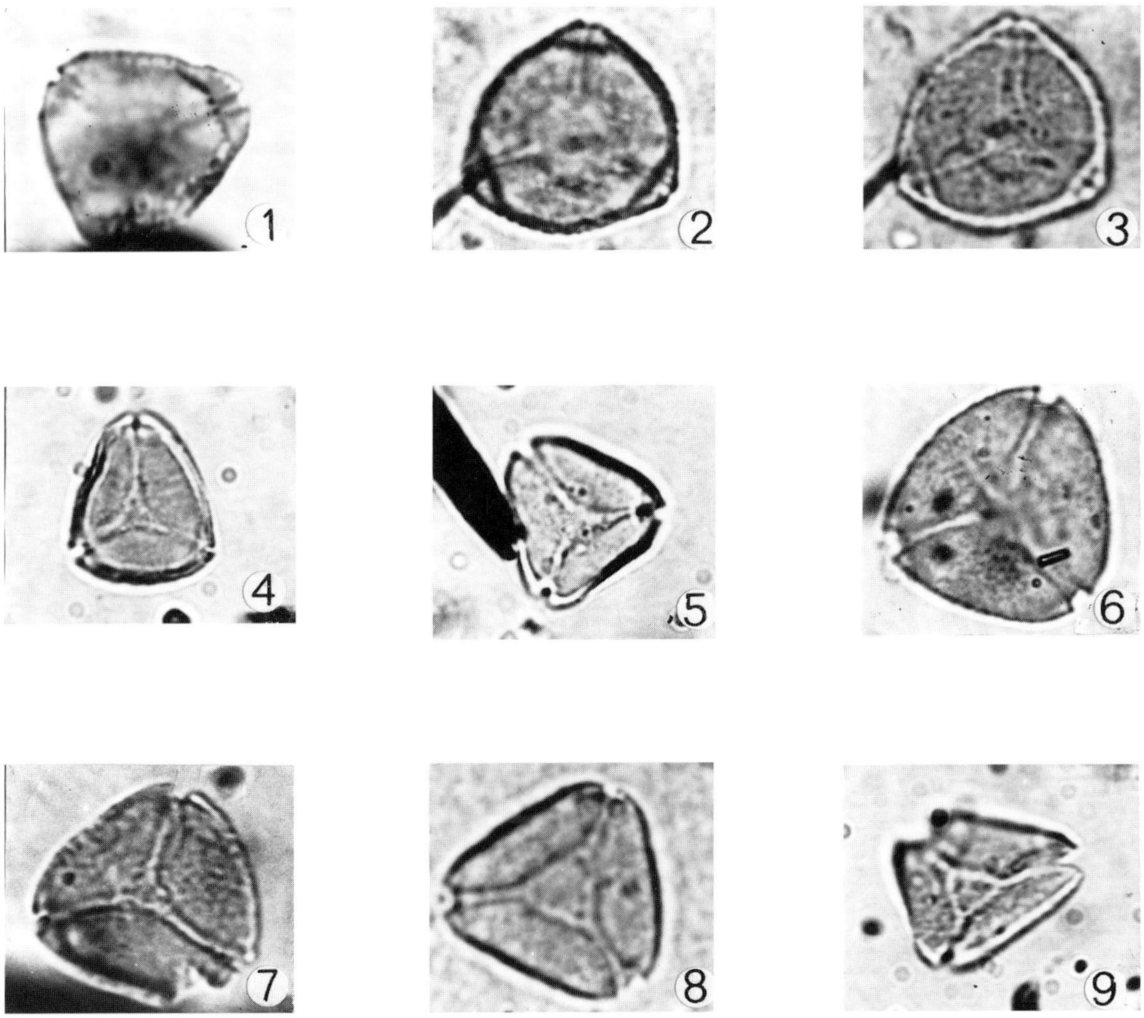

PLATE 2

Fossil species; polar views. The bar = 7 μm in figs. 2–4; 6 μm in figs. 1, 6, 7 and 9; 5 μm in figs. 5 and 8.

1 *Myrtipites annulatus* sp. nov. Holotype: PMP–CTES 1281: 45,6/99,5.

2, 3 *Myrtipites miocenica* sp. nov. Holotype: PMP–CTES 1166: (-228 m). 39,8/104,3. Fig. 2 optical section; fig. 3 surface view at high focus.

4 , 5 *Syncolporites fastigiosus* sp. nov. Fig. 4, Holotype: PMP–CTES 1286: 43,9/102; fig. 5, Paratype: PMP–CTES 1281: 34,8/98,4.

6, 7 *Syncolporites rugulosus* sp. nov. Fig. 6, Holotype: PMP–CTES 305: 39,6/98; fig. 7, Paratype: PMP–CTES 305: 38,9/99.

8, 9 *Myrtaceidites triangularis* sp. nov. Fig. 8, Holotype: PMP–CTES 1166: (-231 m): 43,1/97,4; Fig. 9, Paratype: 1165 (-222 m): 41,2/95,7.

Botanical Affinity. *Plinia rivularis* (Camb) Rotman.

Derivation of Name. Refers at fastigium that is present in the apertures.

Comparison. *Syncolporites fastigiosus* differs from the following species which lack a fastigium: *Myrtaceidites eugenioides* Cookson & Pike, from the Cainozoic of Queensland (Hill et al., 1970, pl. XV, fig. 10); *Myrtaceidites parvus* forma *anesus* Cookson & Pike from the Cainozoic of Australia (Cookson and Pike, 1954); the Eocene of Chile (Troncoso and Barrera, 1980, pl. I, fig. 28), and Takahashi (1977, pl. 12, figs. 7–11); *Myrtaceidites* sp. B from the Paleocene of Canada (Jarzen, 1982, pl. I, fig. 15); *Myrtaceidites* sp. from the Cretaceous–Eocene of Malaysia (Müller, 1968, pl. 5, fig. 4) and *Eugenia/Myrcia* from the Late Miocene of Panamá (Graham, 1988, figs. 65–66).

Syncolporites rugulosus sp. nov.
Plate 2, figs. 6, 7

Diagnosis. Grains oblate, isopolar, amb triangular, straight or slightly convex sides. Syncolporate, wide colpi, joined at the poles developing a syncolpium, without polar island. Aperturas without fastigium nor sexine thickening. Rugulate sculpture. Exine about 1.5 µm thick.

Dimensions. (11 specimes). Equatorial diameter: 13–17 µm.

Holotype. Polar view PMP–CTES 305: 39.6/98. Plate 2, fig. 6.

Type Locality. Punta Rubio, Corrientes province.

Stratigraphy. Ituzaingó Formation.

Age. Late Pliocene.

Other Locality. Villa Olivari, Corrientes province.

Derivation of Name. In reference to the sculpture of the grains.

Comparisons. This form differs from the following species: *Myrtaceidites eugeniioides* in Hill et al. (1970, pl. XV, fig. 10); *Myrtaceidites parvus* forma *anesus* in Takahashi (1977, pl. 12, figs. 7–11) and Troncoso and Barrera (1980, pl. I, fig. 28), *Myrtaceidites* sp. in Müller (1968, pl. 5, fig. 5), and *Myrtaceidites* sp. B in Jarzen (1982, pl. I, fig. 15), because the sculptures are psilate, escabrate, chagrenate, slightly striated or granular. In general, *Myrtaceidites* sp. in Elsik and Dilcher (1974, pl. 31, figs. 179–180), is apparently similar, but no descriptive details were given.

Genus *Myrtaceidites* Cookson & Pike 1954 emend.

Lectotype. *Myrtaceidites mesonesus* Cookson & Pike 1954.

Emended Diagnosis. Grains small to medium, oblate to subspherical, isopolar to subisopolar, amb triangular to subtriangular, straight or slightly convex or concave sides. Parasyncolporate. Straight or curved, narrow or wide colpi, with or without arci, enclosing a more or less developed polar area, called polar island, clearly or poorly definite. Pores with or without fastigium, with or without sexine thickening. Tectate, psilate, granulate, rugulate, never reticulate exine.

Discussion. Potonié (1960) emended the original diagnosis, adding features that refer to the form of the grain from equatorial (oblate) and polar views (angle-aperturated, slightly convex or concave sides); he also pointed out that the arci are distinguishable and clearly delineated, and that it does not always leave a free polar area. This observation creates more confusion for the interpretation of these grains. Krutzsch (1969) added the that pores might lack thickenings. Our emendation attempts to incorporate the term "parasyncolporate" to the original diagnosis with the meaning proposed here; it is clear that the arci are not alway present. The following features are also added: isopolar, subisopolar, straight or curved, narrow or wide colpi, pores with or without fastigium, with or without sexine thickening.

Myrtaceidites triangularis sp. nov.
Plate 2, figs. 8–9

Diagnosis. Grains oblate, isopolar, amb triangular, straight sides, truncate angles. Parasyncolporate, straight colpi with branches bifurcated at the poles, enclosing a well definite polar island, of triangular shape, about 3.5–5 µm diameter. Without fastigium nor sexine thickening at the pores. Tectate, psilate to slightly rugulate.

Dimensions. (12 specimens). Equatorial diameter: 14–15 µm.

Holotype. Polar view PMP–CTES 1166 (-231 m): 43.2/97.4. Plate 2, fig. 8.

Type Locality. Josefina, Santa Fe province.

Stratigraphy. Paraná Formation.

Age. Late Miocene.

Other Levels. -210 m, -213 m, -222 m, -240 m.

Other Localities. Villa Olivari and Ituzaingó, Corrientes, province (Ituzaingó Formation).

Botanical Affinity. *Eugenia involucrata* DC.

Derivation of Name. in reference to form of the polar islands

Comparison. *Myrtaceidites parvus* forma *nesus*, Cookson & Pike, from Cainozoic of Australia (Cookson and Pike, 1954) differs from *Myrtaceidites triangularis* because it has convex sides, rounded angles, and the exine is granular. *Myrtaceidites triangularis* differs from *M. verrucosus* in Stover and Partridge (1973, pl. 17, figs. 17–19), Romero and Zamaloa (1985, figs 15–18), Baldoni (1987, pl. VI, fig. 13), and Barreda (1997), and from *Myrtaceidites* sp. in Guerstein (1990, pl. VI, fig. 2) because it has a fastigium. *Myrtaceidites* sp. in Fasola (1969), which is not have fastigium, differs because it has a granular exine and the sculture is more marked over the polar islands. *Myrtaceidites* sp A in Jarzen (1982, pl. I, figs. 8–9) differs because it is heteroparasyncolporate.

RESULTS

From the above revision and observations we conclude:

a. To distinguish the different fossil pollen types of the Myrtaceae, the following emended morphogenera are proposed: *Myrtipites* (Norton & Hall) emend., for tricolporate grains; *Syncolporites* (Van der Hammen) emend., for syncolporate grains; *Myrtaceidites* (Cookson & Pike) emend., for parasyncolporate grains.
b. The following taxa should be included in *Myrtipites* as new combinations, because they are tricolporate:

Myrtaceidites verrucosus Stover & Partridge, in Baldoni (1987, figs. 16 and 17); *Myrtaceidites parvus* Cookson & Pike, in Frederiksen et al. (1983).

c. The following taxa should be included in *Syncolporites* as new combinations, because they lack polar islands: *Myrtaceidites parvus* forma *anesus* Cookson & Pike, in Cookson and Pike (1954), Troncoso and Barrera (1980), and Takahashi (1970); *Myrtaceidites eugenioides* Cookson & Pike, in Hill et al. (1970), *Myrtaceidites* sp. B in Jarzen (1982), and *Myrtaceidites* sp. in Elsik and Dilcher (1974).
d. In the Ituzaingó Formation (late Pliocene) the species formerly named *Eugenia* aff. *burkartiana* (Anzótegui and Lutz, 1987), is the new species *Syncolporites fastigiosus*. There are also three other new species in this formation: *Myrtipites annulatus*, *Syncolporites rugulosus*, and *Myrtaceidites triangularis*.
e. In the Paraná Formation the taxa named as *Myrtaceidites* sp. 3 and *Myrtaceidites* sp. 4 by Anzótegui and Garralla (1986), resulted in two new species, *Myrtipites miocenica* and *Myrtaceidites triangularis*. The latter is present in both formations.
f. Comparisons with extant species suggest the following botanical affinities: *Syncolporites fastigiosus* sp. nov. is very similar to *Plinia rivularis* (Camb.) Rotman, and *Myrtaceidites triangularis* is very close to *Eugenia involucrata* D.C.

ACKNOWLEDGMENTS

We thank Drs. R. Herbst and J.C. Gamerro for their reading of the paper. Thanks also to the authorities of the IBONE (Corrientes) for access to the Herbarium and to those of the Museum of La Plata for the use of the scanning electronic microscope. The Secretary of Science and Technology of the University of Northeast (Corrientes) financed the Project. Dr. U. Eskuche kindly translated the paper by Krutzch (1969), and Mr. Ramón Z. Rodriguez produced the photographs.

References Cited

ACEVEDO, T.L., and ANZÓTEGUI, L.M.
1998 Myrtaceae. In. S.M. Pire, L.M. Anzótegui y A.G. Cuadrado (eds). *Flora Polínica del Nordeste Argentino*, I: 67–80. EUDENE.

ANDERSON, R.I.
1960 Cretaceus–Tertiary Palynology, Eastern side of San Juan Basin, Nuevo Mexico. *New Mexico Bureau Mines and Mineral. Resources Memory*, 6: 1–59.

ANZOTEGUI, L.M.
1974 Esporomorfos del Terciario Superior de la Provincia de Corrientes, Argentina. *Primer Congreso Argentino de Paleontología y Bioestratigrafía*, Actas II: 329–348.

ANZOTEGUI, L.M., and GARRALLA, S.S.
1986 Estudio Palinológico de la Formación Paraná (Mioceno superior), Pozo Josefina, Provincia de Santa Fe, Argentina. Parte I. Descripciones sistemáticas. *FACENA*, 6: 101–176.

ANZOTEGUI, L.M., and LUTZ, A.
1987 Paleocomunidades Vegetales del Terciario Superior (Formación Ituzaingó) de la Mesopotamia Argentina. *Revista de la Asociación de Ciencias Naturales del Litoral*, 18: 131–144.

BALDONI, A.M.
1987 Estudios Palinológicos de la zona de Collon Cura, Provincia de Neuquén, sobre elementos del Terciario Inferior y redepositados del Cretácico Inferior. *Revista Española de Micropaleontología*, 19: 367–411.

BARREDA, V.
1997 Palynomorph assemblage of the Chenque Foramtion, Late Oligocene? — Miocene from Golfo San Jorge Basin, Patagonia, Argentina. Part 3. Polycolpate and tricolporate pollen. *Ameghiniana*, 34: 131–144.

BARTH, O.M., and BARBOSA, A.F.
1972 Catálago sistemático dos pólen das plantas arbóreas do Brasil Meridional. XV—Myrtaceae. *Memorias Instituto Oswaldo Cruz*, 70: 467– 496.

BELSKY, C.I., BOLTENHAGEN, E., and POTONIE, R.
1965 Sporae dispersae der Oberen Kreide von Gabun, Äquatoriales Africa. *Paläeontologische Zeitschrift*, 39: 72–83.

COOKSON, I.C., and PIKE, K.M.
1954 Some dicotyledonous pollen types from Cainozoic deposits in the Australian region. *Australian Journal of Botany*, 2: 197–219.

COOKSON, I.C.
1956 On some Australian Tertiary spores and pollen grains that extends the Geological and Geographical distribution of living genera. *Proceedings Royal Society of Victoria*, 69: 40–51.

CHALSON, J.M., and MARTIN, H.A.
1995 The pollen morphology of some co-occurring species of the Family Myrtaceae from the Sydney region. *Proceedings Linnean Society New South Wales*, 115: 164–191.

ELSIK, W.C., and DILCHER, D.L.
1974 Palynology and age of clays exposed in Lawrence clay pit, Henry County, Tennessee. *Palaeontographica*, Abteilung B, 146: 65–87.

ERDTMAN, G.
1966 Pollen morphology and plant Taxonomy. Angiosperm (An Introduction to Palynology, I *Almqvist and Wiksell, Stockholm.* 553 pgs.

FASOLA, A.
1969 Estudio palinológico de la Formación Loreto (Terciario Medio). Provincia de Magallanes (Chile). *Ameghiniana*, 6: 3–49.

FREDERIKSEN, N., CAR D.R., and GARY, D.
1983 Middle Eocene palynomorphs from San Diego, California. Part II, Angiosperm pollen and Miscellanea. *American Association of Stratigraphic Palynologists Contributions Series*, 12: 32–155.

GADEK, P.A., and MARTIN, H.A.
1982 Pollen morphology in the subtribe *Metrosiderinae* of the *Leptospermoideae* (Myrtaceae) and its taxonomic significance. *Australian Journal of Botany*, 29: 155–184.

GRAHAM, A.
1987 Miocene communities and paleoenvironments of southern Costa Rica. *American Journal of Botany*, 74: 1512–1513.
1988 Studies in neotropical paleobotany V. The lower miocene commnities of Panamá. The Culebra Formation. *Annals Missouri Botanical Garden*, 75: 1440–1446.

GUERSTEIN, R.
1990 Palinología del Terciario de la Cuenca del Colorado, República Argentina. Parte I. Especies terrestres de la perforación Nadir N°1. *Revista Española de Micropaleontología*, 22: 33–61.

HERBST, R., and SANTA CRUZ, J.N.
1985 Mapa litoestratigráfico de la Provincia de Corrientes. *D'Orbignyana*, 2: 1–51

HERBST, R., and ZABERT, L.
1987 Microfauna de la Formación Paraná (Mioceno superior) de la Cuenca Chaco-Paranense (Argentina). *FACENA*, 7: 165–206

HILLS, L.V., PLAYFORD, G., and WOODS, J.T.
1970 Cainozoic Fossils of Queensland. *Queensland Palaeontographical Society Brisbane*, 36 pgs.

JALFIN, G.A.
1988 Formación Ituzaingó (Plio-Pleistoceno) en Empedrado, provincia de Corrientes: un paleorio arenoso entrelazado tipo Plate. *Segunda reunión Argentina de Sedimentología*, Actas: 130–134
JANSONIUS, J., and HILLS, L.V.
1976 Genera file of Fossil Spores. *Special Publication. Department Geology University of Calgary*. Canada.
JARZEN, D.M.
1982 Paleocene Angiosperm pollen from the Ravenscrag Formation (Paleocene), Southern Saskatchewan, Canada. *Pollen et Spores*, 24: 119–155.
KRUTZSCH, W.
1969 Taxonomie syncolp(or)ate und morphologisch benachbarter Pollengattungen und Arten (Sporae dispersae) aus der Oberkreide und dem Tertiär. I. Syncolp (or) ate und Syncolp (or) atoide pollenformen. *Pollen et Spores*, 11: 397–424.
MÜLLER, J.
1968 Palynology of the Pedawan and Plateau Sandstone Formation (Cretaceous–Eocene) in Sarawak, Malaysia. *Micropaleontology*, 14: 1–37.
1975 Palynolgical study of a Holoceno peat and a Miocene coal deposits from NW Borneo. *Rewiew of Palaeobotany and Palynology*, 19: 304–306.
NORTON, N.J., and HALL, J.W.
1969 Palynology of the Late Cretaceous and Lower Tertiary in the type locality of the Hell Creek Formation, Montana (U.S.A.) *Paleontographica*, Abteilung B, 125: 1–64.
PATEL, V.C., SKVARLA, J.J., and RAVEN, P.H.
1984 Pollen characters in relations to delimitation of Myrtales. *Annals Missouri Botanical Garden*, 71: 858–969.
PIKE, K.M.
1956 Pollen morphology of Myrtaceae from south–east Pacific area.*Australian Journal of Botany*, 4: 13–53.
POTONIÉ, R.
1960 Sporologie der Eozänen Kohle von Kalewa in Burma. *Senckenbergiana Lethaea*, 41, S: 451–481.
PUNT, W., BLACKMORE, S., NILSSON, S., and LE THOMAS, A.
1994 *Glossary of Pollen and Spores Terminology*. LPP Foundation, LPP Contributions Serie N° 1. University of Utrecht, The Netherlands. 71 pgs.
ROMERO, E., and ZAMALOA, M.C.
1985 Polen de Angiospermas de la Formación Río Turbio (Eoceno), Provincia de Santa Cruz, República Argentina. *Ameghiniana*, 22: 43–51.
SHALKE, H.J.N.G.
1973 The Late Quaternary of cape flats area (Cape Province, South Africa). *Scripta Geologica*, 15: 22–49
STOVER, L.E., and PARTRIDGE, A.D.
1973 Tertiary and Late Cretaceous spores and pollen from the Gippsland Basin, southestern Australia. *Proceedings Royal Society Victoria*. 85: 237–286.
TAKAHASHI, K.
1977 Palynology of the lower Tertiary Concepcion Formation Central Chile. *Palaeontologycal Society of Japan*, 106: 71–88.
TRONCOSO, A., and BARRERA, E.
1980 Polen del Eoceno de Osorno (Chile). *Boletín del Museo Nacional de Historia Natural*, 37: 179–203.
VAN DER HAMMEN, T.
1954 El desarrollo de la flora colombiana en los períodos geológicos I. Maestrichtiano hasta Terciario más inferior. *Boletín Geologico de Bogotá*, 2: 4–106.
VAN DER HAMMEN, T., and GONZALEZ, E.
1960 Late Pleistocene and Holocene climate and vegetation of the sabana de Bogotá (Colombia South America). *Leidse Geologische Mededelingen*, 25: 289–315.

Authors' addresses

LILIANA T. ACEVEDO
Facultad de Ciencias Exactas y Naturales y Agrimensura
UNNE and PRINGEPA–CONICET
Casilla de Correo 128. 3400 Corrientes
Argentina
E-mail: cecoal@arnet.com.ar

LUISA M. ANZOTEGUI
Facultad de Ciencias Exactas y Naturales y Agrimensura
UNNE and PRINGEPA–CONICET
Casilla de Correo 128. 3400 Corrientes
Argentina

INTEGRATED PALEONTOLOGICAL STUDIES OF PLIOCENE TO PLEISTOCENE DEPOSITS OF THE ORINOCO DELTA, EASTERN VENEZUELA AND TRINIDAD

David T. Pocknall, Lesli J. Wood, Alfred F. Geen, Brian E. Harry and Richard W. Hedlund

Abstract

Paleontological studies are an integral component of investigations into the age relationships and depositional environments of Pliocene and Pleistocene-aged deltaic sediments of the Eastern Venezuela basin. A chronostratigraphic framework for the basin has been developed through studies of well cuttings, sidewall cores, and onshore Trinidad surface exposures. Rapid rates of sedimentation in response to progradation of the Orinoco River Delta since the Late Miocene produced highly turbid water conditions that were unfavorable for planktonic marine organisms (foraminifera, nannofossils, dinoflagellates). Limited planktonic fauna make it difficult to use the conventional and internationally known faunal biochronologies, and the paucity of Pliocene–Pleistocene extinctions in the plant realm make it difficult to employ traditional stratigraphic extinction (tops) methodology. Several palynological datums including the first appearance of Compositae (*Echitricolporites mcneillyi, E. spinosus, Fenestrites* species [Liguliflorae]) and *Alnus* (*Alnipollenites verus*), and an acme of *Grimsdalea magnaclavata* have been used to create a chronostratigraphic framework for the Eastern Venezuela basin. This framework has been tested and proven accurate by recent exploration wells in the basin. Within the enormous datasets of palynological information from outcrop and wells in the basin there remains the likelihood of identifying new datums, perhaps through the application of improved data management and assessment techniques.

Knowledge of the depositional conditions and facies associations in ancient deltaic settings is crucial to understanding seal and reservoir relationships, and the development of exploration play concepts. The similarity between the vegetation of the past with that of the present and the absence of significant species turnover since the onset of the Pliocene, makes it possible to identify the importance of several plant associations in these studies. Some of the key palynomorphs and their associations with other fossils are discussed in the light of integrated studies of well logs, foraminifera, and palynomorphs.

INTRODUCTION

The Eastern Venezuela basin is located along the southern margin of the obliquely converging Caribbean–South American plate boundary, a zone of intense structural deformation, and includes the Maturin and Columbus sub-basins (Text-Figure 1). The entire basin is a prolific hydrocarbon province, having produced more then 7 billion barrels of oil in the Venezuelan part of the basin alone (Erlich and Barrett, 1994). Exploration in the eastern extensions of the basin (Columbus basin of Leonard, 1983) began in the late 1960s resulting in the discovery of over 3.27 billion barrels of oil in place and 20 TCF of gas in place. The Pliocene to Pleistocene strata of the Eastern Venezuela basin consist of marine and terrigineous clastic sediments (Wood, 1995; Wood et al., 1994). The Orinoco and proto-Orinoco rivers have drained the Andean highlands since Miocene times (Hoorn, 1995), and their associated marginal and deep marine depositional systems have fed >14,000 meters (45,000 feet) of post–Miocene sediment into the basin (Erlich and Barrett, 1990). Early work in Trinidad for the most part focused on understanding the geological relationships of small areas around discoveries and producing fields (e.g., Teak, Samaan), however, recently, following the work of Leonard (1983), regional studies have been undertaken, applying an integrated approach to understanding the age relationships and geological development of this complex basin. These studies have resulted in accurate interpretations of facies and their stratigraphic relationships, and improved our ability to predict reservoir character and presence.

This paper reviews aspects of ongoing regional geological studies, aimed at understanding the age relationships of the stratigraphic section and the environments of deposition of thick stratigraphic sequences encountered in the subsurface of the Eastern Venezuela basin. Interpretations are based on palynological and foraminiferal data obtained from well samples (cuttings and core) and, to a lesser degree, from analysis of surface outcrops exposed along the eastern and southern coasts of Trinidad. The majority of the data are proprietary and we are not able to present detailed checklists of data for confidentiality reasons; we have, however, been granted permission to provide this general overview.

Pocknall, David T., Wood, Lesli J., Geen, Alfred F., Harry, Brian E., and Hedlund, Richard W.
2001 Integrated Paleontological Studies of Pliocene to Pleistocene Deposits of the Orinoco Delta, Eastern Venezuela and Trinidad; *In*: Goodman, D.K., and Clarke, R.T. (eds.), *Proceedings of the IX International Palynological Congress, Houston, Texas, U.S.A., 1996*; American Association of Stratigraphic Palynologists Foundation, p. 319–326.
 ISBN 0–931871–06–9

METHODS

Outcrop or core (conventional or sidewall) are the preferred sample types for paleontological examinations but, typically in this study, only cuttings samples were available. High volumes of sediment being deposited in the basin over very short time periods have resulted in an extremely young and unconsolidated section in the subsurface. In an unconsolidated section of this type caving during drilling and, thus, contamination of cuttings by

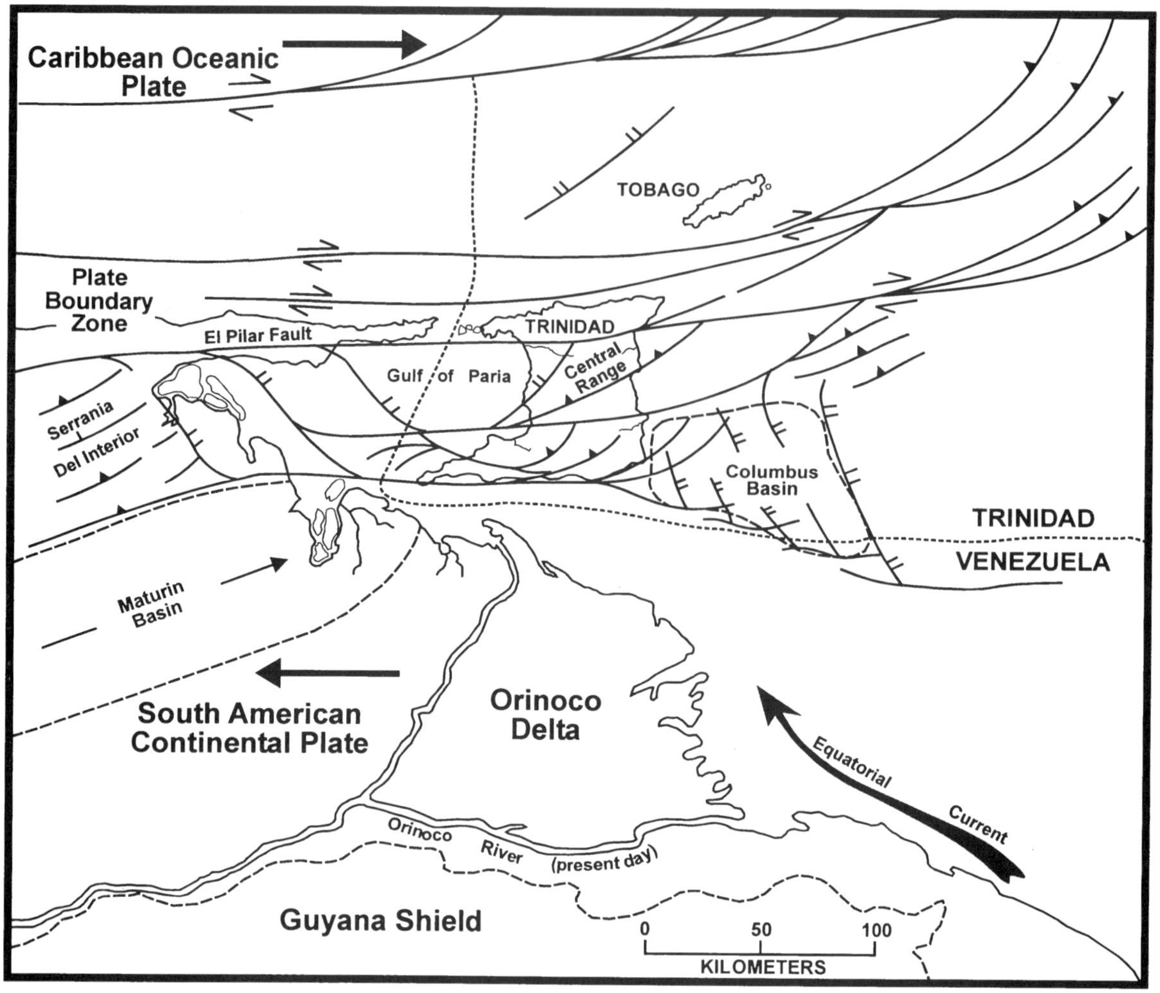

Text-Figure 1. Location of the Maturin and Columbus sub-basins of the Eastern Venezuela basin, showing the major structural features associated with the Caribbean and South American plate boundary zone.

overlying strata is always a problem. Casing point locations were used by interpreters in evaluating the *in situ* nature of the floral and faunal occurrences. This assisted in reducing the influence of contamination in interpretations. Palynological sample preparations followed standard techniques as outlined by Wood et al. (1996).

DISCUSSION

Significant information regarding the palynology of this region is contained in unpublished oil company reports, and there are very few publications on the subject. Muller's (1959) study of the recent sediments of the Orinoco was the first palynological publication on this region. His observations have improved our understanding of palynomorph dispersal in the delta region and therefore assisted our use of palynomorphs, in this region and worldwide, as indicators of depositional environment. Palynological zonations have been reported for the Trinidad and eastern Venezuela Miocene through Pleistocene sediments by Germeraad et al. (1968), Lamy and Scott (1980), Lamy (1985), Lorente (1986), and Muller et al. (1987) (Text-Figure 2). In addition, Batjes (1965) discussed palynology with respect to the Cruse and Forest formations in southern Trinidad, and Payne (1991) utilized an unpublished Texaco palynological zonation for Trinidad developed by E. Gonzalez (oral commun., 1995). A limited number of papers have been published on the palynomorph taxonomy for this region, although it is possible to use Germeraad et al. (1968) as a guide to taxonomy and assemblages.

The terrestrial palynomorph assemblages of the Miocene to Pleistocene stratigraphic section in northern South America are known from the above cited publications but their precise age is often difficult to determine. This is largely because in nonmarine and marginal marine deposits there is a paucity of marine fauna and, therefore, no consistent means of correlating to the global marine fossil-record. In addition, very few plant extinctions occurred during the Neogene. It is estimated that by Late Miocene time 95% of modern plant families existed and by the Late Pliocene all of the modern families were present (Muller, 1981). Germeraad et al. (1968) endeavored to relate the age of their palynofloral zones to the foraminiferal record, thereby achieving some degree of consistency in their age assignments. However, they encountered problems in the Late Miocene to Pleistocene section of Trinidad and eastern Venezuela because of the absence, or lack of abundance, of planktonic organisms (foraminifera, nannofossils,

AGE	TROPICAL ZONE	VENEZUELA	NORTHERN SOUTH AMERICA	TRINIDAD	
	Germeraad, Hopping & Muller (1968)	Lorente (1986)			
	Caribbean	Eastern Venezuela Basin	Muller, de Giacomo & Van Erve (1987)	Lamy & Scott (1980)	Gonzalez in Payne (1991)
PLEISTOCENE	*Alnipollenites verus*		*Alnipollenites verus*	△IVE	PL
PLIOCENE	*Echitricolporites mcneillyi*	*Psilatricolporites caribbiensis*	*Echitricolporites mcneillyi*	△IVD	PIII
					PII
				△IVC	PI
LATE MIOCENE	*Pachydermites diederixi*	*Stephanocolpites evansii*	*Echitricolporites spinosus*	△IVB4	UMII
		Asteraceae: *Fenestrites*		△IVB3	
		Asteraceae: *Bombacacidites ciriloensis*		△IVB2	UMI

Text-Figure 2. Summary chart of palynological zones used in the Late Miocene to Pleistocene of Eastern Venezuela and Trinidad.

dinoflagellates). High sedimentation rates and associated increases in the suspended sediment load of the water column provided unfavorable living conditions and account for the scarcity of planktonic organisms. When present, planktonic tops are often suppressed as a result of the regions consistently shallow nearshore or estuarine depositional environments.

Previously, benthic foraminifera (e.g., *Saccammina, Eggerella, Textularia*) have been used in the basin as chronostratigraphic datums; however, these extinctions are diachronous across the basin. Benthic fauna live on the ocean bottom, hence their occurrences (first and last appearances) are strongly facies dependent and diachronous across the basin, as the shoreline has shifted through time. The first downhole occurrence of some benthic assemblages is often associated with unconformities developed through rapid changes in water depth throughout the basin (basinward or landward shifts in facies). These occurrences are often valid datums between wells that are along depositional strike from one another. In a depositional dip direction these datums are difficult to apply.

Active thrusting on the island of Trinidad during the Pliocene and Pleistocene, and high relief of previously thrusted terrains, has resulted in frequent reworking of faunal and floral assemblages. Additionally, in high energy deltaic settings there is a strong tendency to rework tops. Reworking and downhole caving of fauna and flora complicates not only the age relationships but also contaminates the *in situ* environmentally diagnostic assemblages. In the light of these problems a multitude of factors, including the regional context of each well, engineering considerations (mud additives, casing points), and potential surfaces of reworking (i.e., transgressive surfaces of erosion, lowstand unconformities, slump scars), were carefully considered when examining *in situ* fossil assemblages.

Biostratigraphic and Chronostratigraphic Events

Evolutionary change in the world's land-plant biota has been slow in the Neogene, relative to other periods of geologic time; consequently there are low numbers of stratigraphic tops and bases. In the Eastern Venezuela basin, two datums — the extinction of *Grimsdalea magnaclavata* in the Early Pleistocene and the first appearance of *Alnipollenites verus* near the Pliocene/Pleistocene boundary (Text-Figure 3) — are proving to be of value as correlatable biostratigraphic datums.

Grimsdalea magnaclavata was common during much of the Late Miocene and Pliocene in the northern part of South America, but with the onset of much drier conditions in the region during the Late Pliocene it became less widely distributed, finally becoming extinct in the Pleistocene. An acme of *G. magnaclavata* occurs in the regions' stratigraphic section between 2.5 and 2.1 Ma (Late Pliocene). This dominance and subsequent decline of *G. magnaclavata* is possibly related to climatic warming and cooling, respectively. A major regional lowstand shown on the Haq et al. (1987) sea level curve at around 2.1 to 2.0 Ma may have been coeval with this climatic change.

Germeraad et al. (1968) noted that *G. magnaclavata* became extinct in the Early Pleistocene. Our recent work in the Eastern Venezuela basin shows that the extinction occurred around 1.3 Ma calibrated with the non-overlapping ranges of *G. magnaclavata* and the planktonic foraminifera *Globorotalia flexuosa.* This extinction correlates with the emplacement of a large, prograding clastic wedge of Pleistocene age along the eastern margins of the Columbus basin. This extensive and long-lived basinward shift in shoreline was likely associated with a global cooling event that influenced climate in the southern Caribbean region, and which is documented on the Haq et al. (1987) sea level curve and by many other workers (e.g., Hoorn, 1995).

The genus *Alnus* (which produces the palynomorph *Alnipollenites verus*) does not grow in the Orinoco Delta today and probably never did. But, it has been much talked about in Trinidad palynology. It is usually associated with the montane Andean vegetation forest belt (van der Hammen, 1974). Muller (1959) recorded low numbers of specimens of *Alnus* pollen in modern samples from the Orinoco Delta and its offshore areas, and its presence is recorded in most wells examined in the Columbus basin. These pollen are believed to have been carried from the Andean region via the Orinoco, a distance of some 850 to 900 miles. The appearance of *Alnus* pollen in the basin has

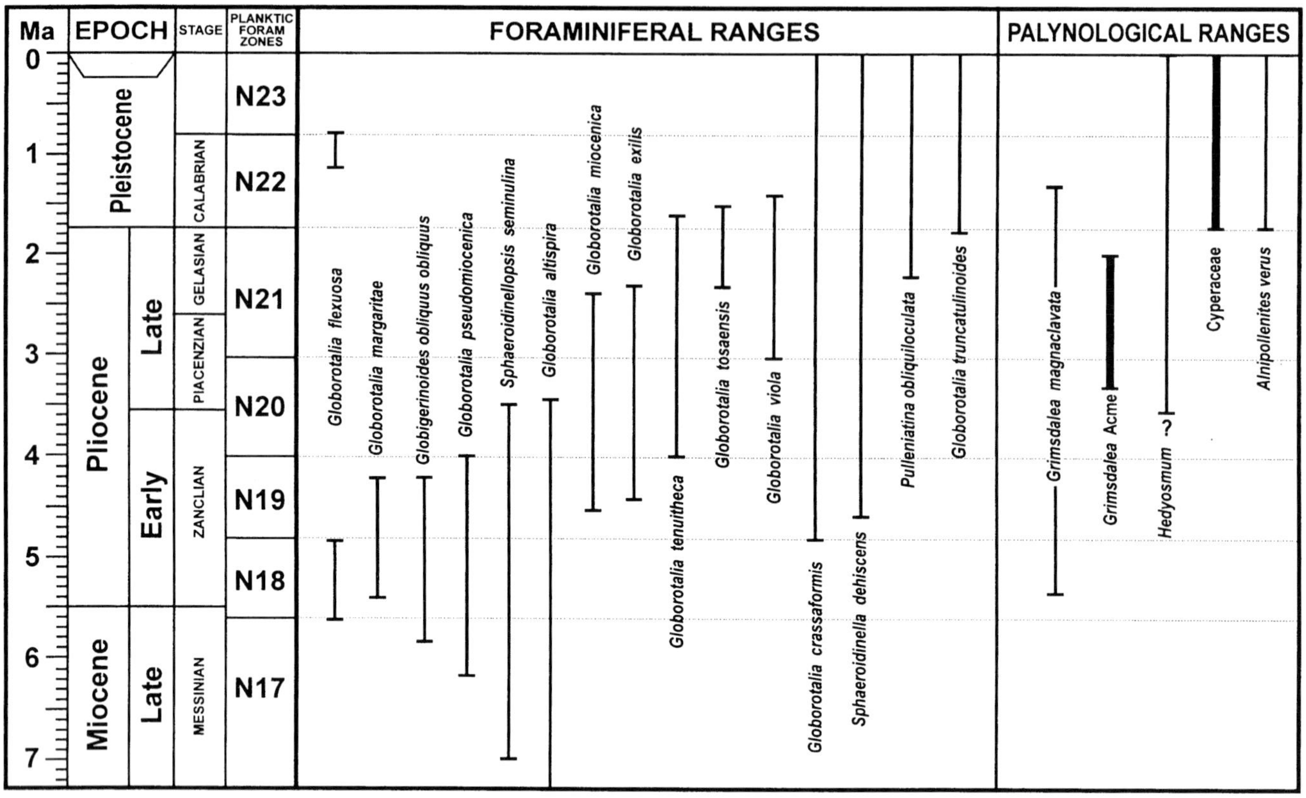

Text-Figure 3. Range chart showing age ranges for key palynomorphs and planktonic foraminifera in the Eastern Venezuela basin.

long been considered by hydrocarbon industry operators in the region to coincide with the base of the Pleistocene.

Lamy and Scott (1980) and Lamy (1985) used the first evolutionary occurrence (or stratigraphic base) of *Alnipollenites verus* to mark the base of Zone E which they considered to be of Pleistocene age (see Text-Figure 2). Germeraad et al. (1968) indicated that there is no means of independently dating the first appearance of A. *verus*, or the boundary of the *A. verus* Zone from the underlying *Echitricolporites mcneillyi* Zone, and placed the base of the *A. verus* Zone at the first regular occurrence of *A. verus*. The different application of the same fossil marker horizon has caused confusion and is, in part, influenced by evolving hypotheses about the time of appearance of A. *verus* in the Caribbean.

Van der Hammen (1974) suggested that *Alnus* arrived in South America in the earliest Pleistocene. In DSDP 502, situated off the northern coast of Colombia, *Alnus* first appears at the Pliocene-Pleistocene boundary, as dated by nannofossils and paleomagnetics (Pocknall and Hedlund, in prep.). Integrated chronostratigraphic analysis in the Eastern Venezuela basin (Columbus sub-basin) has verified this marker as chronologically significant, appearing at or near the beginning of the Pleistocene. Lorente (1986) did not record *Alnipollenites verus* in her studies, although recent work has confirmed its existence in El Furrial Trend and Gulf of Paria wells (Amoco unpublished reports). In view of climatic considerations (cooling) and the uplift of the Andes (Hoogheimstra, 1989), it is estimated that the appearance of *Alnus* closely coincides with the Pliocene–Pleistocene boundary.

Coincident with the first appearance of *Alnus* is the appearance of *Avicennia* and an increase in the abundance of Cyperaceae pollen. Although present in Miocene and Pliocene sediments, the increase in abundance of Cyperaceae in the Pleistocene appears to be in response to the same climatic events that led to the introduction of *Alnus* into the Andes. In wells where *Alnus* appears to be absent, the sudden increase in abundance of *Avicennia* or Cyperaceae can be used to constrain the base of the Pleistocene.

A further chronostratigraphically significant palynomorph appearance in this region is that of the Compositae pollen (*Echitricolporites mcneillyi, E. spinosus, Fenestrites* species [Liguliflorae]). Our studies corroborate the observations made by Germeraad et al. (1968) and Lorente (1986) who indicated a Late Miocene appearance for Compositae and an age range through to the present. The individual components of the Compositae assemblage probably make their first appearances at different times in the stratigraphic record. *Echitricolporites mcneillyi* is thought to consistently appear near the Miocene/Pliocene boundary (Germeraad et al., 1968; Lorente, 1986; Muller et al., 1987), although calibration to the marine fossil record is needed to substantiate this observation.

Additional palynomorph taxa that hold promise as datums in the region are: the extinctions of *Psilatricolporites crassus* and *Proxapertites operculatus*, the first appearances of *Hedyosmum, Psilatricolporites carribiensis*, and *Bombacacidites ciriloensis*; and the acme events of *Magnastriatites howardi* and *Pachydermites diederixi* (in the Pliocene section). At present, these events are largely undocumented and further calibration work on more sections is warranted.

Planktonic foraminifera, although less abundant in the stratigraphic sections of the region, still provide an important means of age dating in the basin. Two specific planktonic occurrences are used most extensively across the basin to define the Pleistocene. These are the second evolution of *Globorotalia flexuosa* and the evolution of *Globorotalia truncatulinoides* (Text-Figure 3). As noted, the second evolution of *G. flexuosa* occurs in the Pleistocene, with its first apparent disappearance documented regionally and worldwide at approximately 4.8 Ma (Early Pliocene) (Bolli et al., 1985). The age ranges of these species have been integrated with specific palynomorph occurrences (discussed above) to define the base of the Pleistocene, and a younger Pleistocene, 1.2 Ma flooding event that culminates a major period of clastic wedge progradation and reservoir development across the basin.

Columbus basin Pliocene strata generally contain more planktonic foraminifera than the overlying Pleistocene section. Faunal occurrences found in the basin and in the Caribbean region that are used in conjunction with palynomorph data to constrain correlations, and the ages in the Pliocene section, include *Globorotalia viola, G. exilis, G. miocenica, G. tosaensis, G. tenuitheca, G. altispira, G. margaritae, Sphaeroidinellopsis seminulina, Globigerinoides obliquus*, and *Pulleniatina obliquiloculata* (Text-Figure 3). The worldwide age ranges of all of these occurrences are well documented, most notably by Bolli et al. (1985).

Paleoenvironments

Well-log motif analysis has been used for many years to identify facies types (Galloway and Hobday, 1983; Van Wagoner et al., 1990; Mitchum et al., 1991). However, this approach can be inaccurate because similar log motifs can be obtained from sediments deposited in different environments, or complex geologic processes may result in log motifs that are uncharacteristic of certain environments. Accurate interpretation of environments of deposition in marine sequences is aided by a thorough understanding of the environmental preferences of sea-bottom dwelling benthic foraminifera, which are critical to the interpretation of water depth. A thorough understanding of facies relationships leads to important insights regarding reservoir and seal distribution, as well as the influence of structure on shoreline and offshore facies development.

Studies of the Pliocene to Pleistocene section of the Eastern Venezuela basin offer an opportunity to integrate log motif with faunal and palynofloral assemblages to create accurate facies interpretations. Pocknall et al. (1998) identified six depositional facies in the Columbus basin. These six include fluvial/estuarine, transitional barrier, prograding shoreface, slope fan, basin floor fan, and condensed section. These are shown in Text-Figure 4, a cartoon of faunal and floral distribution along a coastal to deep marine profile. The environmental and depth interpretations are based on the published works of Drooger and Kaasschieter (1958) for foraminifera, and Muller (1959) for palynomorphs. Only those fossils that are considered key indicators of continental, shelf, or slope deposition, as well as water depth, are shown on Text-Figure 4, although many taxa provide corroborative information. These are not discussed herein.

Palynomorph occurrence and abundance data provide important environmental information, especially in nearshore and continental environments. Some of the more important palynomorph groups in the Late Miocene to Pleistocene section in the Eastern Venezuela basin include the mangrove genera *Rhizophora, Avicennia, Pelliciera*, and *Acrostichum*, savannah grasses, *Symphonia globulifera*, the water fern *Ceratopteris*, and dinoflagellates.

Fluvial channel and valley fill sequences of the Columbus basin contain an abundance of palynomorphs derived from the vegetation growing along the coastal margin and in the immediate hinterland. Mangrove pollen (derived from *Rhizophora, Avicennia*, and *Pelliciera*) are common in estuarine valley-fill sediments. Other significant components of this complex are: the benthic foraminifera *Miliammina telemaquensis, Arenoparrella mexicana, Ammonia beccarii*, and *Ammobaculites dilitatus*; pollen derived from swamp plants such as *Symphonia (Pachydermites diederixi)* and *Ceratopteris (Magnastriatites howardi)*; and Gramineae pollen derived from the swamp marsh grasses. *Miliammina telemaquensis* is an extinct benthic foraminifera closely related to *M. fusca*, a modern species that favors the brackish and hypersaline environments of mangrove swamps and marginal marine environments (Saunders, 1958). *Arenoparrella mexicana* living offshore Trinidad and in the Gulf of Paria, also favors hypersaline environments (Drooger and Kaasschieter, 1958). Modern beach, tidal flat, and near coastal sediments often contain abundant pollen derived from plants in the Palmae (palms) and Chenopodiaceae (chenopods). These pollen types are noted to occur in the same facies in ancient sections.

Palynomorphs are not common in deep-marine facies, with the exception of *Rhizophora* pollen. Slope fan facies deposited in water depths ranging from 600 to 3000 ft as well as basin floor fan and deep water condensed deposits, all contain high abundances of *Rhizophora* pollen. Observations made by Muller (1959) help to explain this apparent anomaly. *Rhizophora* produces copious quantities of small pollen which are easily dispersed by wind and water. Muller (1959) studied the modern pollen distribution in the Orinoco Delta and found that seaward from the eastern part of the delta there was a distinct, but gradual increase in the percentage of *Rhizophora*, although the total pollen content decreased. This clearly demonstrates that the *Rhizophora* percentage is inversely proportional to total pollen content. It probably represents an example of selective sorting resulting in a relative increase in the smallest pollen grains eastward. An alternative hypothesis to preferential transport is that the high abundances of *Rhizophora* indicate an expansion of the pollen-producing mangrove belt as a result of marine transgression time equivalent with deepwater deposition onto a broad, tidally influenced delta plain.

Marine and estuarine dinoflagellates provide a means of discriminating periods of marine and brackish conditions. In the case of the Orinoco Delta deposits, they indicate periods of marine sedimentation when the water column was less turbid. Clastic rich deltaic systems, such as the Orinoco, in which the water column is almost constantly turbid, do not provide the best environmental conditions

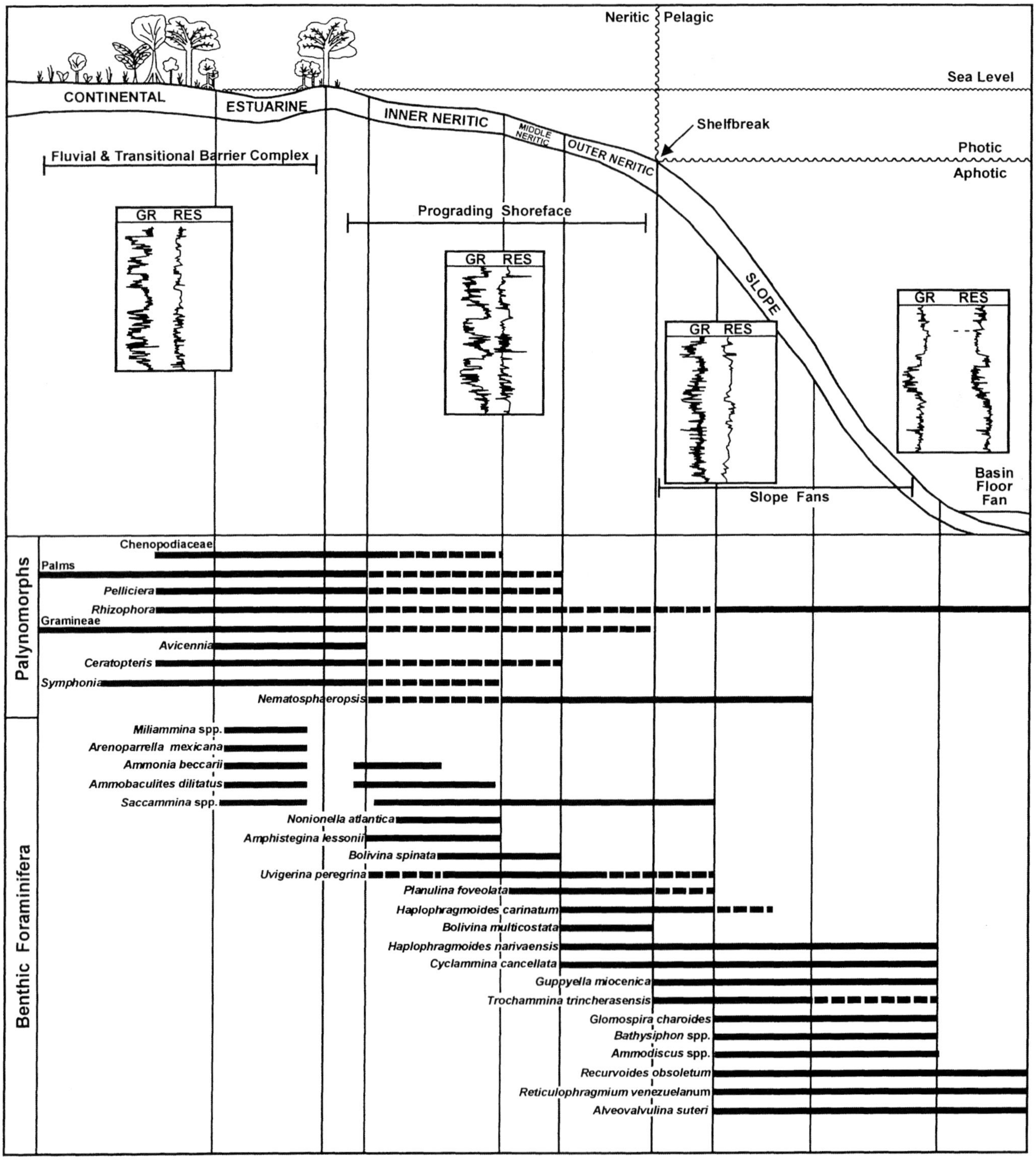

Text-Figure 4. Chart showing the distribution of flora and fauna and the typical well-log motif pattern which characterize different depositional environments in the Pliocene to Pleistocene section in Eastern Venezuela and Trinidad (after Pocknall et al., 1998).

for dinoflagellates, or other planktonic organisms for that matter, and their occurrence is usually limited. When there is an influx (or bloom) of dinoflagellates and planktonic foraminifera we interpret the conditions to represent a marine flooding event and possibly a period of quiet sedimentation. These events are often observed in deep water condensed sections (Pocknall et al., 1998).

CONCLUSIONS

The use of palynomorph assemblages, occurrence, and extinction data is not straight forward in ancient deltaic deposits such as those of the Orinoco River, and caution must be exercised when making interpretations of the geological record. The Eastern Venezuela basin contains thick sections of Pliocene–Pleistocene strata deposited in environments ranging from nonmarine to basin floor. Palynomorphs are found in abundance throughout and detailed sampling of well and outcrop sections has provided a useful framework from which to build a chronostratigraphic and depositional history for the basin. Several chronostratigraphically significant events calibrated to a fossil record of marine fauna occur in the palynological datasets although, in general, extinctions are rare. This is because the flora has not changed significantly in composition since the Late Miocene and climate has remained relatively stable. It is our opinion that a more detailed examination of the assemblages could reveal a palynofloral record that reflects climatic cyclicity and which could provide additional correlation potential in the basin. The design of a facies-based assemblage framework integrating both faunal and palynofloral data, with well-log motif, stacking patterns analysis, and available seismic data will provide the means of deciphering environments of deposition, therefore permitting us to better predict resorvoir and seal presence.

ACKNOWLEDGMENTS

We wish to thank Amoco Trinidad Oil Company for permission to publish this work. Over the past few years many people have contributed to our regional studies in Trinidad, including Maria Henry, Nancy Engelhardt-Moore, and Paul Belanger. Their geological and paleontological contributions are greatly appreciated. Teresa Holmes and Mark Schendel drafted the figures. Robert Gastaldo provided constructive review of the manuscript.

References Cited

BATJES, D.A.J.
1965 Palaeoecology of foraminiferal assemblages in the Late Miocene Cruse and Forest Formations of Trinidad, Antilles. *Transactions of the Fourth Caribbean Geological Conference*, Port-of-Spain, Trinidad and Tobago, pp. 141–156.

BOLLI, H. M., SAUNDERS, J. B., and PERCH-NIELSEN, K.
1985 Comparison of zonal schemes for different fossil groups. *In*: Bolli, H. M., Saunders, J.B., and Perch-Nielsen, K. (eds.), 1985, *Plankton Stratigraphy*, Cambridge University Press, pp. 11-16.

DROOGER, C.W., and KAASCHIETER, J.P.H.
1958 Foraminifera of the Orinoco-Trinidad-Paria Shelf. Reports of the Orinoco Shelf Expedition, 4. N.V. Noord-Holtandsche Uitgevers Maatschappij, Amsterdam, 108 pgs.

ERLICH, R.N., and BARRETT, S.F.
1990 Cenozoic plate tectonic history of the northern Venezuelan–Trinidad area. *Tectonics*, 9: 161–184.
1994 Petroleum geology of the eastern Venezuela Foreland Basin: Foreland Basins and Fold Belts. *American Association of Petroleum Geologists Memoir*, 55: 341-362.

GALLOWAY, W.E., and HOBDAY, D.K.
1983 *Terrigenous Clastic Depositional Systems: Applications to petroleum, coal and uranium exploration*. Springer-Verlag, New York, 423 pgs.

GERMERAAD, J.H., HOPPING, C.A., and MULLER, J.
1968 Palynology of Tertiary sediments from tropical areas. *Review of Palaeobotany and Palynology*, 6: 189-348.

HAQ, B.U., HARDENBOL, J., and VAIL, P.R.
1987 Chronology of fluctuating sea levels since the Triassic. *Science*, 235: 1156–1167.

HOOGHEIMSTRA, H.
1989 Quaternary and Upper Pliocene glaciations and forest development in the tropical Andes: evidence from a long high resolution pollen record from the sedimentary basin of Bogota, Colombia. *Palaeogeography, Palaeoclimatology, Palaeoecology*, 72: 11–26.

HOORN, C.
1995 *Miocene palynostratigraphy and paleoenvironments of northwestern Amazonia: evidence for marine incursions and the influence of Andean tectonics*. Ph.D. dissertation, University of Amsterdam, 156 pgs.

LAMY, A.
1985 Plio-Pleistocene palynology and visual kerogen studies, Trinidad, W.I., with emphasis on the Columbus Basin. *Transactions of the First Geological Conference of the Geological Society of Trinidad and Tobago*, Port-of-Spain, Trinidad and Tobago, pp. 115- 125.

LAMY, A., and SCOTT, R.W.
1980 Summary of palynological biostratigraphy of Neogene Quaternary strata, Offshore Trinidad. Amoco Production Company Research Report P80-G-12.

LEONARD, R.
1983 Geology and hydrocarbon accumulations, Columbus Basin, offshore Trinidad. *American Association of Petroleum Geologists Bulletin*, 67: 1081-1093.

LORENTE, M.A.
1986 Palynology and palynofacies of the Upper Tertiary in Venezuela. *Dissertationes Botanicae*, Band 99, 222 pgs.

MITCHUM, R.M., SANGREE, J.B., VAIL, P.R., and WORNARDT, W.W.
1991 Sequence stratigraphy in late Cenozoic expanded sections, Gulf of Mexico. *In*: Armentrout, J. and Perkins, B. (eds), *Sequence Stratigraphy as an Exploration Tool, Concepts and Practices in the Gulf Coast*. Gulf Coast Section Society of Economic Paleontologists and Mineralolgists, pp. 237–256.

MULLER, J.
1959 Palynology of the Recent Orinoco Delta and shelf sediments: reports of the Orinoco Shelf expedition. *Micropaleontology*, 5: 1-32.

MULLER, J.
1981 Fossil pollen records of extant angiosperms. *The Botanical Review*, 47: 1–142.

MULLER, J., DE DI GIACOMO, E., and VAN ERVE, A.W.
1987 A palynological zonation for the Cretaceous, Tertiary, and Quaternary of Northern South America. *American Association of Stratigraphic Palynologists, Contributions Series*, 19: 7-76.

PAYNE, N.
1991 An evaluation of post Middle Miocene geological sequences, Offshore Trinidad. *In*: Gillezeau, K.A. (ed.), *Transactions of the Second Geological Conference of the Geological Society of Trinidad and Tobago*, Port-of-Spain, Trinidad and Tobago, pp. 70-87.

POCKNALL, D.T., WOOD, L.J., and GEEN, A.F.
1998 Depositional facies in the Pliocene–Pleistocene section, Offshore Eastern Trinidad. *Transactions of the Third Geological Conference of the Geological Society of Trinidad and Tobago and the Fourteenth Caribbean Geological Conference*, Port-of-Spain, Trinidad and Tobago, pp. 713–732.

SAUNDERS, J.B.
1958 Recent Foraminifera of mangrove swamps and river estuaries and their fossil counterparts in Trinidad. *Micropaleontology*, 4: 79-92.

VAN DER HAMMEN, T.
1974 The Pleistocene changes of vegetation and climate in tropical South America. *Journal of Biogeography*, 1: 3-26.

VAN WAGONER, J.C., MITCHUM, R.M., CAMPION K.M., and RAHMANIAN, V.D.
1990 Siliciclastic Sequence Stratigraphy in Well Logs, Cores and Outcrops. *American Association of Petroleum Geologists Methods in Exploration Series*, 7, 53 pgs.

WOOD, G.D., GABRIEL, A.M., and LAWSON, J.C.
1996 Chapter 3. Palynological techniques — processing and microscopy. *In*: Jansonius, J. and McGregor, D.C. (eds.), *Palynology: Principles and Applications*. American Association of Stratigraphic Palynologists Foundation, Vol. 1, p. 29–50.

WOOD, L.J.
1995 Stratigraphic sequence development in high sediment supply, high accommodation space basins. *Program and Abstracts Third Geological Conference of the Geological Society of Trinidad and Tobago and the Fourteenth Caribbean Geological Conference*, Port-of-Spain, Trinidad and Tobago, July 1995, p. 87.

WOOD, L.J., HENRY, M.W., LANAN, B., SPENCER, L., BELANGER, P., ORTMANN, K., ENGELHARDT-MOORE, N., POCKNALL, D.T., and ROMERO, T.
1994 Sequence stratigraphy in a structurally complex area: Integration of multidisciplinary data sets, Columbus basin, Trinidad. *Abstracts and Program American Association of Petroleum Geologists Meeting*, Denver, p. 286.

Authors addresses

DAVID T. POCKNALL
LESLI J. WOOD
ALFRED F. GEEN
BRIAN E. HARRY
Amoco Corporation
P.O. Box 3092
Houston, Texas 77253
U.S.A.

RICHARD W. HEDLUND
6923 S. Knoxville
Tulsa, Oklahoma 74136
U.S.A.

Part 4: Quaternary/Pleistocene Palynology

PLEISTOCENE DEPOSITS OF THE AMAZON FAN: A PALYNOLOGICAL ANALYSIS OF HOLES 940A, 944A, AND 946A (ODP LEG 155)

Carina Hoorn

Abstract

This paper presents a summary of a palynological study carried out for the Ocean Drilling Program, Leg 155. The study focused on the composition and distribution of organic matter in sediments of the Amazon Fan during the Pleistocene glacial/interglacial sequences. Samples from holes 940A, 944A, and 946A showed that glacial lowstand deposits are characterized by large concentrations of a palynological assemblage composed of Andean and tropical lowland taxa, minor amounts of reworked Paleozoic, Cretaceous, and Tertiary taxa, and large volumes of organic debris (e.g. wood, vessels, epidermis, cuticles). In general, these intervals do not contain many taxa diagnostic for the lowland savannah's. In all the studied sites, there is an increase of ferns, fern allies, reworked and corroded palynomorphs, and fungal spores at the time of the last glacial episode.

Interglacial highstand deposits are characterized by low concentrations of the terrigenous palynological assemblage and a predominance of marine microfossils. Although no marked differences exist in the composition of the organic matter along the fan, there is a decrease in the concentration of palynomorphs from proximal to distal sites. No compositional difference between the levees and the debris flow was observed. The palynological sequences in the Amazon deep sea fan give no indication of major vegetational changes in the lowlands. If extensive savannah vegetation existed in Amazonia during the Pleistocene glacial intervals, as claimed in the refugia theory, they were either obliterated or not recorded in the sediments of the Amazon Fan.

INTRODUCTION

The Amazon Fan is a large muddy fan that has been active since Miocene times (Damuth and Kumar 1975; Hoorn et al., 1995). The fan receives sediments from the entire Amazon drainage basin and has a large riverine input of terrigenous organic matter from the Andean region, the lowland savannahs, and the Amazon rainforest. Fan growth has occurred primarily during phases of global sea level lowstand when the clastic sediments from the drainage basin bypassed the shelf and were transported directly into the submarine fan (Damuth and Kumar, 1975; Manley and Flood, 1988). During these periods of global sea level lowstand, there was little or no deposition in the Amazon fluvial system and deep incision in the fluvial valleys occurred (Muller et al., 1995). However, during the periods of global sea level highstand, hardly any sediment reached the fan and deposition took place in the terrestrial riverine and lacustrine environments, and on the continental shelf.

The process of sediment deposition in the Amazon Fan is comparable with other fan systems in the world. For instance, the Indus and Bengal fans are fed by sediments from the Himalayan hinterland. Recent palynological studies have shown that Pleistocene deposits in these fan systems reflect climatic change and composition of the vegetation in the drainage basins (Yoshinori et al., 1990; Yoshinori et al., 1991).

The Amazon Fan, a sink for Pleistocene glacial sediments, provides an opportunity to examine the nature of glacial vegetation cover in the basin of which only rare relicts are left in the terrestrial environment. The debate on the origins of diversity of the Amazon rainforest and its relation to Pleistocene climatic fluctuations (Haffer, 1969; Absy and Van der Hammen, 1976; Colinvaux, 1987; Salo et al., 1986; Bush, 1994; Van der Hammen and Absy, 1994) has searched for explanations in the recent geological past. However, Pleistocene (glacial) deposits are rare on the continent and do not provide complete answers.

The palynological composition of the Pleistocene glacial/interglacial sediments within the Amazon Fan was documented by Hoorn (1997) on the base of three selected holes (940A, 944A, and 946A). Moreover, Haberle (1997) provided a marine/terrestrial correlation based on a selection of core intervals and samples from the Amazon River. This paper presents an overview of the results of the palynological study carried out for ODP Leg 155 and provides an insight into palynological composition of the Pleistocene deposits of the Amazon Fan. Furthermore, the palynological composition between sediments of the middle and lower fan is compared. Finally, this paper discusses if Pleistocene climatic change can unambiguously be recognized in the Amazon Fan.

Hoorn, Carina
2001 Pleistocene deposits of the Amazon Fan: A palynological analysis of holes 940A, 944A, and 946A (ODP Leg 155); *In*: Goodman, D.K., and Clarke, R.T. (eds.), *Proceedings of the IX International Palynological Congress, Houston, Texas, U.S.A., 1996*; American Association of Stratigraphic Palynologists Foundation, p. 329–338.
 ISBN 0–931871–06–9

SELECTED SITES AND LITHOLOGIES

Three sites were chosen to provide an overview of the palynological signal throughout the fan in time and space. Hole 940A (Text-Figure 1), in the uppermost middle fan, consists of sediments composed of late Pleistocene levee deposits and is covered by a thin interval of Holocene sediments. The Pleistocene sequence consists of mud with laminae and beds of silt and sand whereas the Holocene cover consists of calcareous mud.

Hole 944A (Text-Figure 1), in the lower part of the middle fan, consists of middle and late Pleistocene levee deposits and a debris flow deposit. In the lower part of the hole, an interglacial interval (isotopic stage 9, Maslin and Mikkelsen, 1997) was recovered. The Pleistocene sequence

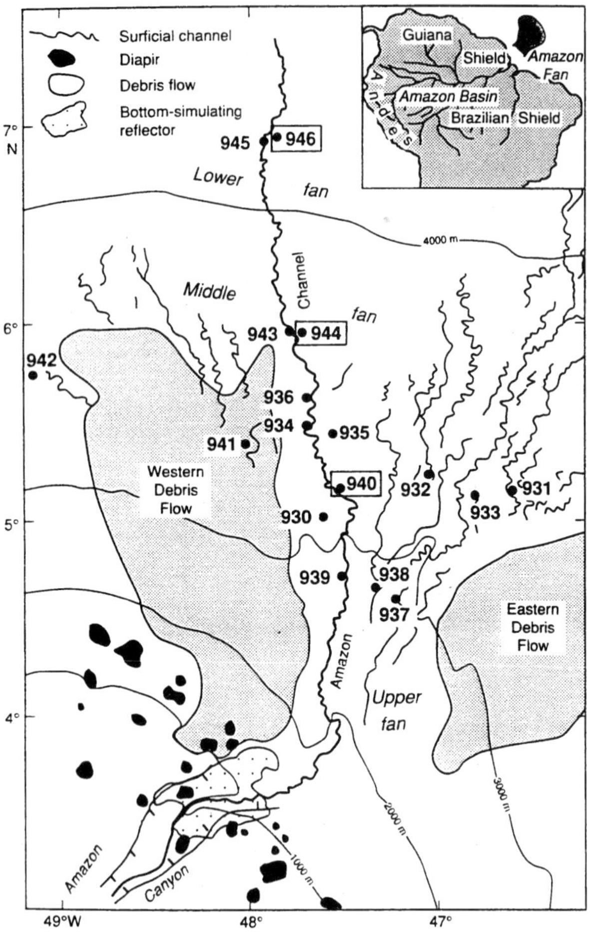

Text-Figure 1. The Amazon Fan with the locations of the sites where the holes were drilled (from Flood et al., 1997).

consists of mud with laminae and beds silt and sand whereas the debris flow deposit consists of mud blocks. The Holocene cover and the interglacial sequences consist of calcareous mud.

Hole 946A (Text-Figure 1), near the transition from the middle to the lower Amazon Fan, consists of middle and late Pleistocene muddy levee deposits with a predominance of sands. These sands were formed in a sandy lobe beyond the contemporary channel and were afterwards covered by the Amazon levee. Two interglacial intervals (isotopic stages 5 and 11, Maslin and Mikkelsen, 1997) are separated from each other by a sandy levee. The interglacial intervals and the Holocene cover consist of calcareous mud.

MATERIALS AND METHODS

For the palynological study a total of 102 samples was analyzed. The samples were taken in clusters of two or three and at intervals ranging between 10 and 30 m. The samples were processed following the standard procedures of the Hugo de Vries Laboratory (Hoorn, 1997). The taxa were identified based on the work of Van der Hammen (1963), Germeraad et al. (1968), Absy (1979), Hooghiemstra (1984), Roubik and Moreno (1991), and Van Waveren (1993). Pollen, spores, marine microfossils (microforaminifers, dinoflagellates, copepod fragments, tintinnids), reworked and corroded pollen and spores, and fungal spores were counted and absolute concentrations were calculated for each group. A pollen sum of about 300–400 Quaternary angiosperm pollen grains was used although in some samples a minimum sum of 100 pollen grains was accepted. Not included in the pollen sum were pteridophytic spores, bryophytic spores, fungal spores, reworked pre-Quaternary and/or corroded pollen and spores, marine microfossils, and algae.

The palynological summary diagrams (Text-Figure 2, 3, and 4) were made with computer applications Tilia and Tilia Graph 2.00 (Grimm and Troostheide, 1994). Full diagrams representing the percentages of the individual taxa, concentrations and granulometric distributions can be found in Hoorn (1997). The groups in the summary diagrams are chosen based on information of the present geographical occurrence and ecological affinity (e.g. Van der Hammen and Gonzalez, 1960; Van der Hammen, 1963; Van der Hammen et al., 1973; Absy, 1979; Hooghiemstra, 1984; Duivenvoorden and Lips, 1995; A.M. Cleef, pers. commun., 1995). The composition of the palynological groups (Andean–Paramo, Subandean–Tropical Lowland Forest, Lowland Savannah, Coastal, and Marine organisms) is indicated in Table 1. In addition, a correlation of the sites, based on the palynological data, is shown in Text-Figure 5.

OVERVIEW OF PALYNOLOGICAL RESULTS

The palynological assemblages of the Pleistocene sediments in the Amazon Fan are composed of a large number of pollen and spores representative of the tropical vegetation in the Amazon drainage. Common taxa include the Andean related elements *Alnus, Hedyosmum,* and *Podocarpus* and the Subandean and Tropical Lowland Forest related taxa such as *Alchornea, Mauritia,* Malpighiaceae, Palmae, Combretaceae/Melastomataceae, and Sapotaceae. In addition, there are many taxa common in a wide range of vegetation types such as Gramineae, Compositae, Myrtaceae, and *Ilex*. The occurrence of the coastal genus *Rhizophora* and the savannah genera *Byrsonima* and *Curatella* is noticeably low.

The palynological assemblages are largely dominated by fern, fern allies, and fungal spores, which form the most important indication of any change in the hinterland. The most abundant spores are Polypodiaceae and Cyatheaceae. Andean related fern and moss spores such as *Cyathea cuspidata, C. horrida, Lophosoria, Anthoceros,* Adiantaceae, Aspleniaceae, and Thelypteridaceae were also observed. The large range of fungal spores was not differentiated and is represented as a single group in the palynological diagrams.

In addition to the Quaternary assemblage, a considerable number of reworked pollen and spores from Paleozoic, Cretaceous, and Tertiary formations were found which were grouped as 'reworked pre-Quaternary'. Among these were Paleozoic megaspores, gymnosperms, and acritarchs; Cretaceous dinoflagellate cysts and pollen such as *Spinozonocolpites* and *Buttinia andreevi,* and Tertiary pollen and spores such as *Echitriporites trianguliformis, Verrutricolporites rotundiporus, Grimsdalea magnaclavata,* and

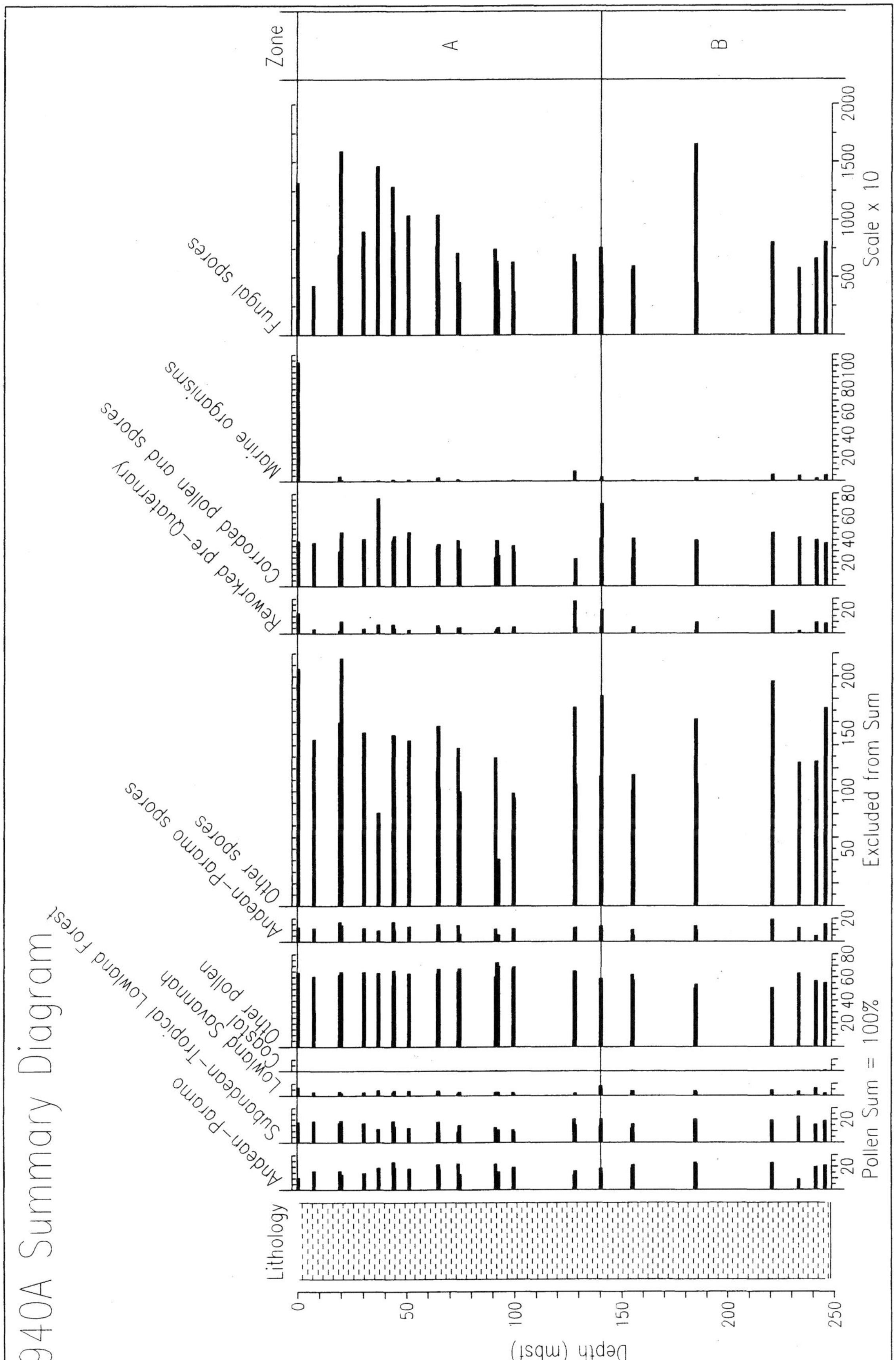

Text-Figure 2. Pollen diagram summarizing the data from hole 940A. The legend for Text-Figures 2, 3, and 4 is given in Text-Figure 5.

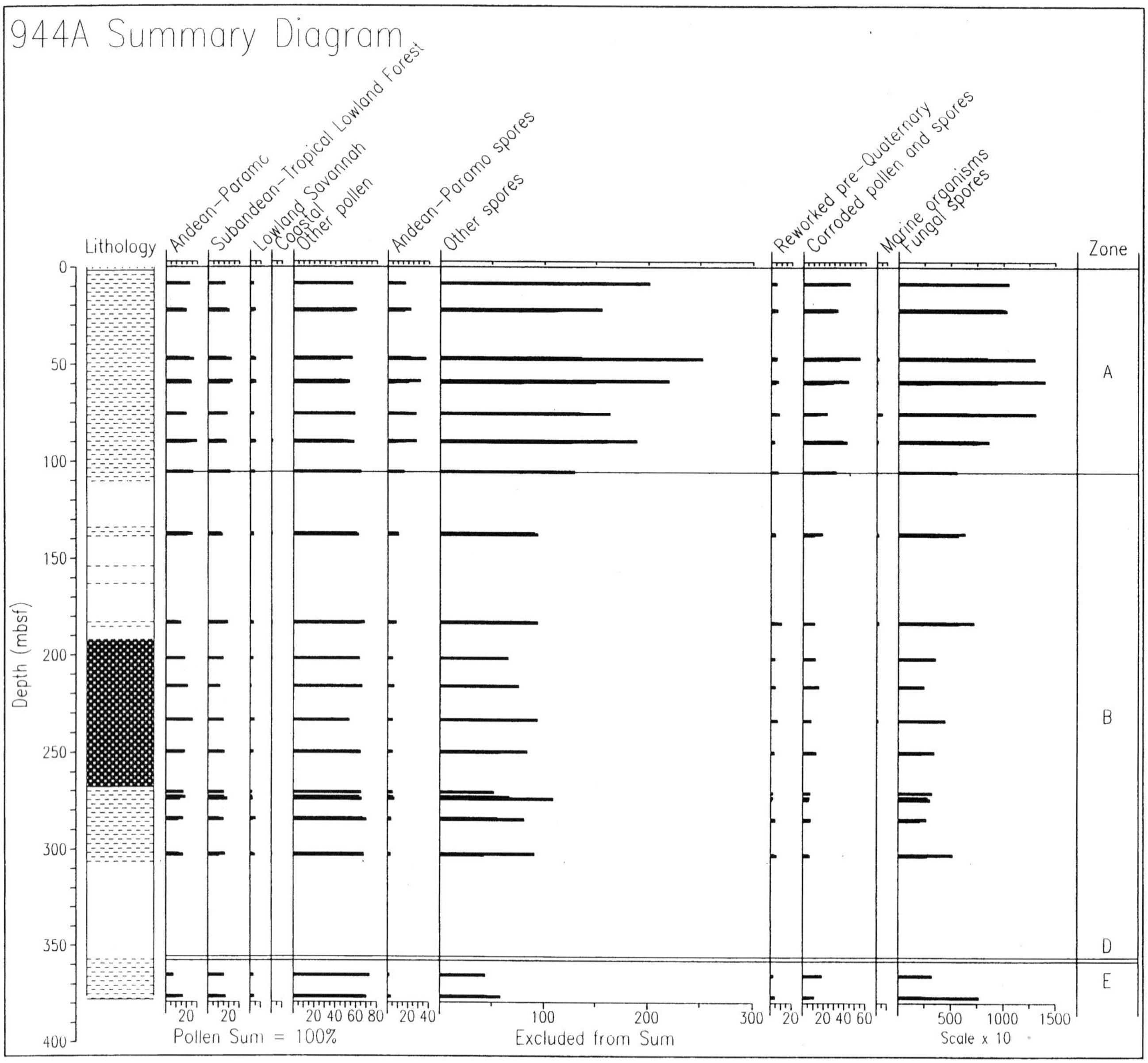

Text-Figure 3. Pollen diagram summarizing the data from hole 944A.

Crassoretitriletes vanraadshoovenii. A separate curve represents the large percentages of corroded pollen and spores of Quaternary age and gives an indication of the strong effect of weathering due to transport from the continental into the deep sea environment. In Hole 946A the occurrence of *Nothofagus* was observed. This southern South American taxon is possibly related to another path of long distance transport, and suggests a combination of wind transport and the Guyana Coastal Current bringing pollen from more southerly regions into the deep sea fan.

The palynological assemblages have an extremely homogeneous character and do not allow us to distinguish zones based on the changing composition of the pollen assemblage. However, the palynological diagrams (Text-Figures 2, 3 and, 4) were divided into seven zones (A–G) based on the frequencies of fern, fern allies, fungal spores, and marine microfossils (Hoorn, 1997). These zones combined with the ages provided by Maslin and Mikkelsen (1997), and Cisowski and Hall (1997) allow us to correlate the sites and demonstrate the consistency of the palynological signal within the fan (Text-Figure 5).

Zone A is characterized by abundances of fungal spores (all sites), and fern spores (holes 944A and 946A). Additionally, a predominance of corroded pollen and spores was observed in Hole 944A. Zone B has, in comparison to zone A, relatively lower values of fern and fungal spores (all sites). Paleomagnetic and micropaleontological dating of Hole 946A (Cisowski and Hall, 1997; Maslin and Mikkelsen, 1997) suggest these zones belong to the last glacial interval. Based on the date and stratigraphic position of the Lake

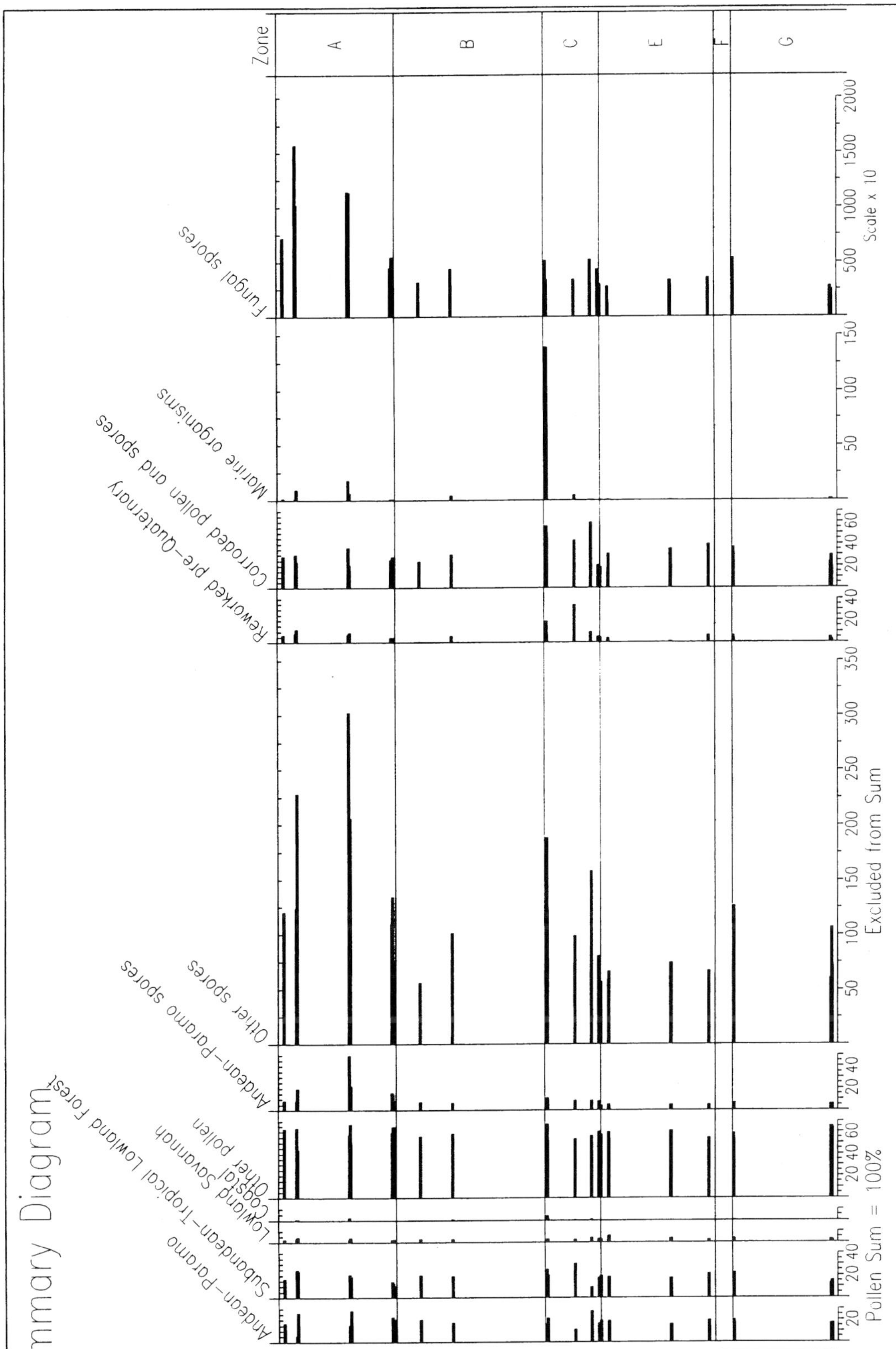

Text-Figure 4. Pollen diagram summarizing the data from hole 946A.

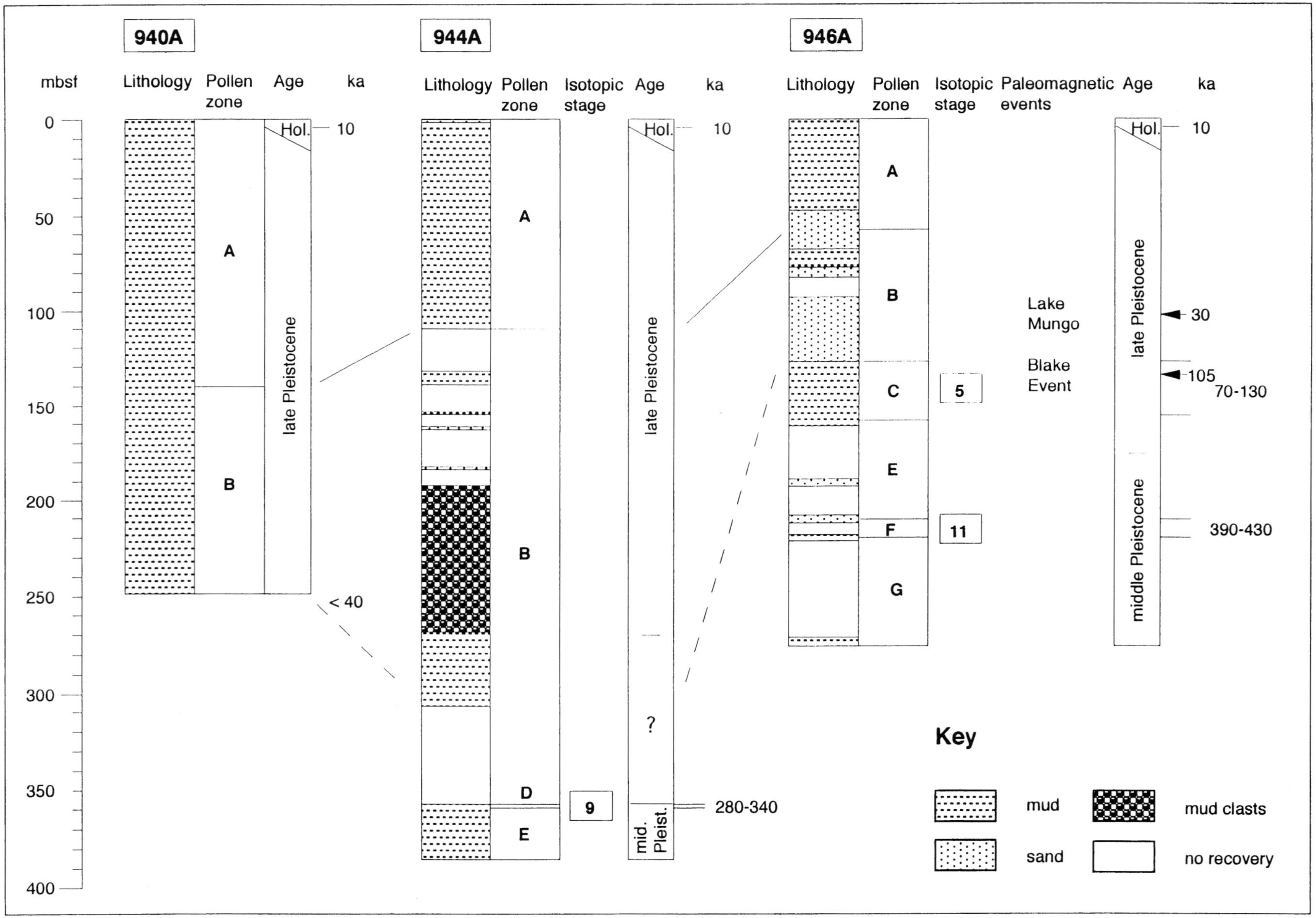

Text-Figure 5. Correlation figure (from Hoorn, 1997) showing the three holes with the palynological zonation and the absolute and relative ages based on Maslin and Mikkelsen (1997) and Cisowski and Hall (1997) .

Mungo paleomagnetic excursion (ca. 30 ka) zone A is probably younger than 30 ka.

Zones C, D, and F represent the interglacial intervals (isotopic stages 5, 9, and 11, Holes 944A and 946A; Maslin and Mikkelsen, 1997). These zones are characterized by spikes of marine microfossils (zone C in 946A) and low concentrations of terrigenous palynomorphs in zones D and F. Relatively higher concentrations of terrigenous palynomorphs were observed in zone C, particularly in the group of "reworked pre-Quaternary" and corroded pollen and spores. Concentration diagrams can be observed in Hoorn (1997).

The middle Pleistocene zones E and G are situated below the isotopic stages 9 and 11 respectively. Zones E and G are similar to zone B though concentrations of terrigenous palynomorphs are higher in G. No zone was defined for the Holocene section which in terms of palynological composition resembles the earlier interglacial intervals. Moreover, pollen sums of those samples which fall in the Holocene are too small to allow representation in the main and summary diagrams.

DISTRIBUTION AND CONCENTRATION OF PALYNOMORPHS THROUGHOUT THE FAN

As expected, there is a noticeable difference between Holocene and interglacial samples versus glacial samples. The three Holocene samples and the samples of the three earlier interglacial intervals have very low concentrations of terrigenous palynomorphs (less than 200 grains/cm^3) and relatively high concentrations of marine microfossils such as copepod fragments, microforaminifers, dinoflagellate cysts, and tintinnids. Whereas, the glacial intervals have large concentrations of terrigenous palynomorphs ranging from pollen and spores to large fragments of wood and seeds. The finer fractions of plant debris include woody fragments, vessels, cuticles, epidermis, fungal remains, and resins.

In Hoorn (1997) it is shown that the concentration of terrigenous palynomorphs decreases seawards from middle fan (holes 940A and 944A) to lower fan (hole 946A). This is particularly noticeable in zones A and B (last glacial), which are represented in all of the three sites. Pollen concentrations in the glacial intervals in the proximal sites range from ca. 300 to 3500 grains/cm^3, whereas at the distal site 300 to 2500 grains/cm^3 were counted . Pteridophytic spore concentrations fluctuate between ca. 250 and 8500 grains/cm^3 at the proximal sites versus a maximum of ca. 2500 grains/cm^3 at the distal site. Striking are the concentration values of the fungal spores which are 1000 to 50,000 grains/cm^3 at holes 940A and 944A, and 500 to 10,000 grains/cm^3 at hole 946A. This clearly indicates a relative decrease of spores from proximal to distal sites suggesting selective transport of the organic matter.

No pronounced difference in palynomorph concentrations were observed according to the environment of deposition (e.g. levee, debris flow). However, concentrations seem to be slightly higher in mud samples and lower in sandy mud samples. A pilot study on granulometry of organic particles in hole 944A (Hoorn, 1997) showed that in sandy mud, organic particles range in size from fine to coarse sand size whereas in mud they vary between very fine silt to fine sand size. This suggests a size sorting of the organic matter particles according to the lithology and the sedimentary environment within the fan.

DISCUSSION AND CONCLUSIONS

The results of this study on Pleistocene Amazon Fan deposits give new information concerning the organic matter composition and distribution within the fan. Furthermore, the results contribute to the discussion on the vegetational history in the Amazon Basin during the Pleistocene glacial periods.

From the observations of the organic matter composition within the fan it can be concluded that the glacial lowstand deposits of the Amazon Fan, as observed in holes 940A, 944A, and 946A, contain high concentrations of terrigenous palynomorphs. The terrigenous material is composed of a mixture of Andean montane and tropical lowland pollen and spore taxa, with minor amounts of reworked Paleozoic, Cretaceous, and Tertiary taxa, and large volumes of organic debris (e.g. wood, vessels, epidermis, cuticles). Interglacial highstand deposits are characterized by low concentrations of terrigenous palynomorphs and high concentrations of marine microfossils.

Within the fan sediments seven palynological zones were distinguished based on the frequencies of fern and fungal spores, pollen concentrations and, abundance of marine organisms. From these zones, A and B correlate with the last glacial period. Zone A is younger than 30 ka (hole 946) and is characterized by relatively high values of fern and fungal spores, and corroded pollen and spores. This palynological signal is comparable to observations made by Poumot (1989) who reports on increases of fern and fungal spores during periods of sea level lowstand in Neogene tropical marine settings.

The reason for the increase in spores in the Amazon Fan is not clear. One explanation could be stronger erosion in the hinterland due to climatic cooling. Destruction of peat layers and proliferation of ferns and fern allies over barren areas could also cause an increase of pteridophytic and fungal spores in the fan. This would agree with Haberle (1997) who observed in his study a general decrease of Andean cold-adapted taxa (*Podocarpus*, *Hedyosmum* and *Alnus*) between the last glacial and the Holocene. Haberle's land–sea correlation suggests that the extreme climatic conditions of the glacial periods caused an expansion of the Andean cold adapted taxa.

In spite of this, no increase in savannah woodland taxa and herbs such as Gramineae and Compositae were observed in the fan. However, pollen of herbs such as Gramineae and Compositae, which predominate in the fan sediments, occur in a wide range of environments and thus are no indication of large scale vegetation change in the Amazon lowlands. An increase of Gramineae in any particular area of the drainage basin is unlikely to be noticeable in the organic matter record of the Amazon Fan.

The organic matter distribution in zones A and B show a seawards decrease in the concentration of terrigenous palynomorphs from middle fan to lower fan. Furthermore, concentrations and granulometry of the organic particles vary according to the lithology and sedimentary environment within the fan. Sand-rich lithologies tend to have low concentrations and large sized organic particles whereas mud-dominated lithologies have high concentrations and small sized organic particles. Additionally, the occurrence of *Nothofagus* in Hole 946A suggests the influence of the

TABLE 1. Composition of the palynological groups represented in the summary diagrams.

Andean/Paramo Pollen
Alnus
Caryophyllaceae
Cordia
Dicliptera
Dodonea
Fuschia
Hedyosmum
Juglans
Lepidium
Merremia
Myrica
Myriophyllum
Pinus
Podocarpus
Polygonaceae
Quercus
Rapanea
Spatiphyllum
Symplocos
Tristerix
Valerianaceae
Weinmannia

Subandean/Tropical Lowland Forest
Aeschynomene
Alchornea
Amanoa
Anacardiaceae
Annonaceae
Antidaphne
Apocynaceae
Bombacaceae
Byttneria
Croton
Dalechampia
Eriosema
cf. *Eschweilera*
Ficus
Humiria
Hybiscus
cf. *Hieronima*
Macrolobium
Macrophyllum
Malpighiaceae
Mauritia
Palmae
Rhus
Sapium
Sapotaceae
Tournefourtia
Trema
Urticariaceae–Moraceae
Virola

Lowland Savannah
Borreria
Byrsonima
Curatella
Hippocratea
Ilex

Coastal
Rhizophora

Other Pollen
Acanthaceae
Althernanthera/Gomphrena
Amarantaceae–Chenopodiaceae
Anacardiaceae
Combretaceae–Melastomataceae
Compositae
Cuphea
Didimopanax
Duroia
Ericaceae
Euphorbiaceae
Gramineae
cf. Leguminosae
Loranthaceae
Ludwigia
Mabea
Malvaceae
Mimosoide
Myrtaceae
Onagraceae
Passifloraceae
Polygalaceae
Proteaceae–Sapindaceae
Rubiaceae
Salix
Symphonia
cf. *Solanaceae*
diverse indeterminate

Andean/Paramo Spores
Adiantaceae (*Jamesonia, Pteris, Pityrogramma*)
Aspleniaceae (*Asplenium, Ctenitis, Cystopteris*)
Anthoceros
Cyathea cuspidata
Cyathea horrida
Lophosoria
Lycopodium
Thelypteridaceae

Other Spores
Acrostichum
Ceratopteris
cf. Polypodiaceae
cf. Cyatheae
bryophytic spores
diverse indeterminate

Marine Organisms
dinoflagellate cysts
copepods (fragments)
microforaminifers
tintinnids

wind and water transport, bringing pollen from more southerly regions into the Amazon Fan.

Finally, it can be concluded that Pleistocene sediments of the Amazon Fan provide mostly circumstantial evidence about the history of the drainage basin. This is due to that the palynological assemblage in the fan deposits form a homogeneous mixture of pollen and spore taxa of the many vegetation types that at present occur in the drainage basin. This homogenized pollen signal does not register large scale changes of the vegetation in the Amazon lowlands and thus no conclusive answers can be given to the question of what happened during the late Pleistocene to the Amazon rain forest. If changes in the vegetation of the Amazon lowlands occurred during the Pleistocene glacial intervals they were either obliterated or not recorded in the fan. To, unambiguously, test the current hypothesis on the vegetation cover of the Amazon lowlands during the Pleistocene a wide range of data from land, shelf, and fan should be compared.

ACKNOWLEDGMENTS

I am grateful to Alastair Milne for his numerous contributions to this work. Henry Hooghiemstra, Bas van Geel, and Simon Haberle are thanked for valuable discussions on tropical Quaternary palynology. Elly Beglinger and Annemarie Phillip are acknowledged for processing the palynological samples, and Jody dos Santos is thanked for her support in all sorts of practical and technical matters.

References Cited

ABSY, M.L.
1979 A palynological study of Holocene sediments in the Amazon Basin. Unpubl. Ph.D. thesis, University of Amsterdam, 86 pgs.

ABSY, M.L., and VAN DER HAMMEN, T.
1976 Some palaeoecological data from Rondonia, southern part of the Amazon Basin. *Acta Amazonica*, 6: 293–299.

BUSH, M.B.
1994 Amazonian speciation: a necessarily complex model. *Journal of Biogeography*, 21: 5–17.

CISOWSKI, S.M., and HALL, F.R.
1997 An examination of the paleointensity record and geomagnetic excursions recorded in Leg 155 cores. *In*: Flood, R.D., Piper, D.J.W., Klaus, A., and Peterson, L.C. (eds.), *Amazon Fan, Proceedings of the ODP, Scientific Results*, 155: 231–244.

COLINVAUX, P.
1987 Amazon diversity in light of the paleoecological record. *Quaternary Science Reviews*, 6: 93–114.

DAMUTH, J.E., and KUMAR, N.
1975 Amazon Cone: Morphology, sediments, age, and growth pattern. *Geological Society of America Bulletin*, 86: 863–878.

DUIVENVOORDEN, J.F., AND LIPS, J.M.
1995 A land-ecological study of soils, vegetation, and plant diversity in Colombian Amazonia. Ph.D. thesis, University of Amsterdam, *Tropenbos Series*, 12: 438.

FLOOD, R.D., PIPER, D.J.W., and SHIPBOARD SCIENTIFIC PARTY.
1997 Introduction. *In*: Flood, R.D., Piper, D.J.W., Klaus, A., and Peterson, L.C. (eds.), *Amazon Fan, Proceedings of the ODP, Scientific Results*, 155: 5–16.

GERMERAAD, J.H., HOPPING, C.A., and MULLER, J.
1968 Palynology of Tertiary sediments from tropical areas. *Review of Palaeobotany and Palynology*, 6: 189–348.

GRIMM, E.C., and TROOSTHEIDE, C.D.
1994 Tilia 2.00, program for plotting palynological diagrams. Illinois State Museum, Springfield, IL, USA.

HABERLE, S.
1997 Late Quaternary vegetation and climate history of the Amazon Basin: Correlating marine and terrestrial pollen records. *In*: Flood, R.D., Piper, D.J.W., Klaus, A., and Peterson, L.C. (eds.), *Amazon Fan, Proceedings of the ODP, Scientific Results*, 155: 381–396.

HAFFER, J.
1969 Speciation in Amazonian forest birds. *Science* 165: 131–137.

HOOGHIEMSTRA, H.
1984 Vegetational and climatic history of the high plain of Bogota, Colombia: a continuous record of the last 3.5 million years. Ph.D. thesis, University of Amsterdam, *Dissertationes Botanicae*, 79: 368.

HOORN, C.
1997 Palynology of the Pleistocene glacial/interglacial cycles of the Amazon Fan (Holes 940A, 944A, and 946A). *In*: Flood, R.D., Piper, D.J.W., Klaus, A., and Peterson, L.C. (eds.), *Amazon Fan, Proceedings of the ODP, Scientific Results*, 155: 397–410.

HOORN, C., GUERRERO, J., SARMIENTO, G.A., and LORENTE, M.A.
1995 Andean tectonics as a cause for changing drainage patterns in Miocene northern South America. *Geology*, 23: 237–240.

MANLEY, P.L., and FLOOD, R.D.
1988 Cyclic sediment deposition within Amazon deep-sea fan. *American Association of Petroleum Geologists Bulletin*, 72: 912–925.

MASLIN, M., and MIKKELSEN, N.
1997 Amazon Fan mass-transport deposits and underlying interglacial deposits: Age estimates and fan dynamics. *In*: Flood, R.D., Piper, D.J.W., Klaus, A., and Peterson, L.C. (eds.), *Amazon Fan, Proceedings of the ODP, Scientific Results*, 155: 353–366.

MULLER, J., IRION, G., NUNES DE MELHO, J., and JUNK, W.
1995 Hydrological changes of the Amazon during the Last Glacial/Interglacial cycle in Central Amazonia (Brazil). *Naturwissenschaften*, 82: 232–235.

POUMOT, C.
1989 Palynological evidence for eustatic events in the tropical Neogene. *Bulletin du Centre de Recherche d'Exploration et de Production Elf-Aquitaine*, 13: 437–453.

ROUBIK, D.W., and MORENO, J.E.
1991 Pollen and Spores of Barro Colorado Island. *Monography in Systematic Botany of the Missouri Botanical Garden*, 36: 269pgs.

SALO, J.S., KALLIOLA, R., HÄKKINEN, I., MÄKINEN, Y., NIEMELÄ, P., PUHAKKA, M., and COLEY, P.D.
1986 River dynamics and the diversity of Amazon lowland forest. *Nature*, 322: 254–258.

VAN DER HAMMEN, T.
1963 A palynological study on the Quaternary of British Guiana. *Leidse Geologische Mededelingen*, 29: 125–180.

VAN DER HAMMEN, T., and ABSY, M.L.,
1994 Amazonia during the last glacial. *Palaeogeography, Palaeoclimatology, Palaeoecology*, 109: 247–261.

VAN DER HAMMEN, T., and GONZALEZ, E.
1960 Holocene and Late Glacial climate and vegetation of Paramo de Palacio (Eastern Cordillera, Colombia, South America). *Geologie en Mijnbouw*, 39: 737–746.

VAN DER HAMMEN, T., WERNER, J.H., and VAN DOMMELEN, H.
1973 Palynological record of the upheaval of the Northern Andes: a study of the Pliocene and Lower Quaternary of the Colombian Eastern Cordillera and the early evolution of its High-Andean biota. *Review of Palaeobotany and Palynology*, 16: 1–122.

VAN WAVEREN, I.M.
1993 Planktonic organic matter in surficial sediments of the Banda sea (Indonesia) — A palynological approach. Ph.D. thesis, University of Utrecht, *Geologica Ultraiectina*, 104: 237 pgs.

YOSHINORI, Y., AMANO, K, and YAMANOI, T.
1990 Pleistocene climatic changes as deduced from a pollen analysis of site 717 cores. *In*: Cochran, J.R., Stow, D.A.V., et al. (eds.), *Bengal Fan, Proceedings of the ODP, Scientific Results*, 116: 249–414.

YOSHINORI, Y., NIITSUMA, N, and HAYASHIDA, A.
1991 A pollen analysis of the Indus deep sea fan from site 720 cores. *In*: Prell, W.J., Niitsuma, N., et al. (eds.), *Indus Fan, Proceedings of the ODP, Scientific Results*, 117: 283–290.

Author's address

CARINA HOORN
Hugo de Vries Laboratory / Centre for Geo-ecological Research (ICG)
University of Amsterdam
Kruislaan 318
1098 SM Amsterdam
The Netherlands

Present address:
The Herbarium
Royal Botanic Gardens
Kew Richmond
Surrey TW9 3AB
U.K.
E-mail: carina.hoorn@milne.cc

WHAT IS THE MEANING OF THE HIGH PERCENTAGES OF FERN SPORES IN ARCHEOLOGICAL SEDIMENT PALYNOLOGICAL ANALYSES

Jacqueline Argant

Abstract

Among different European sites (mainly caves), pollen analysis of archaeological levels sometimes has low diversity and very high percentages of Pteridophyte spores (smooth monolete spores or *Polypodium*, mainly). The interpretation, in such cases is difficult and varies according to authors. Suggested reasons are:

1. The effects of the differential pollen and spore preservation.
2. Environmental causes either climatic or anthropic (forest-clearing).
3. Human introduction of ferns for forage or bedding.

The study of cave sediment as well as the study of the today's fern spore dispersal make it possible to eliminate hypotheses one and two.

To investigate the third proposal, in the precise case of *Polypodium*, an experiment was conducted involving sheep. It shows that *Polypodium* was not eaten by sheep, which means that food or sheep dung are not responsible for the high spore rate recorded. However *Polypodium* fronds may be used as beddings which may account for the high spore percentages in the sediments.

INTRODUCTION

In numerous archaeological sites in Europe (Table 1) belonging to different times (ranging from the Aurignacian period to present), in different environments such as caves, lakes and megaliths, we have found very high fern spores percentages. These are mostly smooth, monolete spores or *Polypodium*. Such occurrences are rare in the current natural environment.

When not known, those results are explained as: (1) corresponding to a forest clearing or as an increase in humidity, (2) as a human introduction, (3) or as poor preservation.

Often, the high fern percentages are associated with cave-sheepcot. In the Southern France Provence area, many caves, inhabited from the Neolithic onwards have been used as shepherds' shelters. These sites often show a series of burning levels with in between layers of anthropic refuse (Text-Figure 1). For instance, in the Vieux-Mounoï cave (Signes, Var), pollen analysis of these different levels (Text-Figure 2) shows that pollen is present but in low amounts, whereas fern spores, *Polypodium* particularly, are abundant in certain layers. This leads me to ask the following questions:

1. Can *Polypodium* spores' natural dispersion alone account for such a heavy concentration?
2. Does differential preservation explain the high fern spore percentages ?

Argant, Jacqueline
2001 What is the meaning of the high percentages of fern spores in archeological sediment palynological analyses; *In*: Goodman, D.K., and Clarke, R.T. (eds.), *Proceedings of the IX International Paly-nological Congress, Houston, Texas, U.S.A., 1996*; American Association of Stratigraphic Palynologists Foundation, p. 339–345.
 ISBN 0–931871–06–9

3. Can ferns be used as food or litter? If so, how do they occur in pollen spectra ?
4. What is the effect of fire on plant litter or on *Polypodium* in particular ?

To answer these questions, a series of experiments using *Polypodium vulgare* were conducted in the winter of 1995–1996.

EXPERIMENT 1

Study of *Polypodium* spore Dispersion

Contemporary studies have shown that Pteridophyte spores do not contribute much influence to the pollen rain of either open areas or caves (Bui-Thi-Mai 1974; Diot 1991). It was thus necessary to further evaluate *Polypodium* spore dispersal.

To do so, I chose a wall on which a large *Polypodium* stem was growing. Nineteen moss samples were taken from the wall at one-meter intervals to the north and at half m intervals to the south. One sample was taken below the stem and three others were taken at the same vertical position on the wall surface (Text-Figure 3).

The results are shown in Text-Figure 3. The pollen preservation was good in the cushions and the spectra that were obtained give a good view of the distribution within a 50 meter range. *Polypodium* spores are poorly dispersed because spores were rare farther than one meter from the plant. The part played by gravity and water runoff is important but the rate of deposition is moderate. If we exclude the base of the plant where spore counts go up to 40%, the highest other average elsewhere never goes above 15%.

To account for the very high proportion of *Polypodium* spores (90%) observed in a cave, such as Vieux-Mounoï's, one would have to suspect the presence of a very large *Polypodium* stand growing on the wall precisely above where the 90% sample was collected as well as a very heavy and steady water runoff. Neither have been documented in this

TABLE 1. European sites belonging to different periods. Diverse interpretations of high fern rates. F = France, B = Belgium, E = Spain.

	Site	Author, Year	Site datation	Site type	Monolete spores M (%) *Polypodium* P (%)	Taxon number	Occupation (O) Breeding (B)	Proposed interpretation
1	Lann Gouh, Melrand (Morbihan, F.)	Marguerie D., 1992	Middle-Age	pit below human dwelling	P (> 70%)			differential conservation
2	Talhouët, Pluvigner (Morbihan, F.)	Marguerie D., 1992	Iron Age	village	P (up to 80%)			differential conservation
3	Grotte des Sarrasins Seyssinet-Pariset (Isère, F)	Borel J-L.	Bronze Iron Age	cave	M (->90%)	33		clearings, connection with agricultural human activity and fallows.
4	Abri du Mollard, Lavours (Ain, F.)	Argant J., (inédit)	Bronze Iron Age	cave	M (100%)	1	O	litters ?
5	Kerfandol, Ploërdut (Morbihan)	Marguerie D., 1992	Bronze	tumulus	M, P (up to 95%)	15		differential conservation
6	Lanner Scot, Pluvigner (Morbihan, F.)	Marguerie D., 1992	Bronze	grave	M (and trilete) (90%)			differential conservation
7	Abauntz, Arraiz (Navarre, E.)	Lopez P., 1982	Chalcolthic	grotte	P (up to 341%)	<10		moisture, warming up
8	Gué du Plantain (B)	Bastin B., 1977	Late Atlantic Subatlantic	open site, way	P M			*Alnus* epiphyte, open site
9	Saldopro (Biscaye, E)	Penalba M-C., 1989	7000 BP	peat bog	M (80%)			differential conservation , water runoff
10	Château Bû de la Grée de Cojoux, St-Just (Ile-et-Vilaine, F.)	Marguerie D., 1992	Neolithic	dolmen	P (15 %)			differential conservation
11	Table des Marchands, Locqmariaquer (Morbihan, F.)	Marguerie D., 1992	Neolithic	dolmen	P (40%)	>15		differential conservation
12	Petit Mont, Arzon (Morbihan, F.)	Marguerie D., 1992	Neolithic	megalith	P	weak		differential conservation
13	Cairn de Vierville (Manche, F.)	Clet -Pellerin M., 1986	Neolithic	tumulus	M et P (up to. 70%)	>13		intentional bringing in burial
14	La Cova de l'Or, Beniarrès (Alicante, E.)	Dupré-Ollivier M., 1988	Neolithic	cave	P (jq. 63%)	22	B(Ovicapridae)	?
15	Le Trou-qui-Fume, St-Romain (Côte-d'Or, F.)	Argant J., 1990	Neolithic	gap	M	4 à 11		anthropic refuse (litters?) water runoff ?
16	Vieux Mounoï, Signes (Var, F.)	Argant J., (inédit)	Neolithic	cave sheepcot	P, M (up to 93 %)	10 à 24	B (Ovicapridae)	litters
17	Baume de Ronze, Orgnac (Ardèche, F.)	Argant J., (inédit)	Neolithic	cave sheepcot	P	?	B(Ovicapridae)	litters
18	Charavines (Isère, F.)	Emery-Barbier A., 1989	Neolithic	lacustrine site	P	?	B (coproliths)	litters
19	Le Pirou, Polignac (Hte-Loire, F.)	Paquereau M-M., 1979	Chassean (c. 1, 2 et 2/3)	open site	M, P	?	B(Bos dominant)	?
20	La Balma Margineda (Pté d'Andorre)	Leroyer C., 1985	c.5 c.4=Mésolith. c.3-3b=Néol. c.1-2=Hist.	cave	M 30 to 40 %	15 16 15 15	B (Ovicapridae, pig, ox)	clearing increase in humidity dense occupation
21	La Pente des Brosses Montigny sur Loing (Seine-et-Marne, F)	Boyer-Klein A., 1983	Magdalenian	plein air	P (->97%)	10		moisture
22	El Juyo, Santander (Cantabria, E.)	Leroi-Gourhan A., 1994	Magdalenian Bronze	cave	M (>50%) M and P (450%)	>14 >12	O B	litter clearing
23	Traweye Rotche (caverne de), Sprimont (Prov. de Liège, B.)	Heim J., 1986	Aurignacian	cave	M (81%)			human clearings, increase in humidity

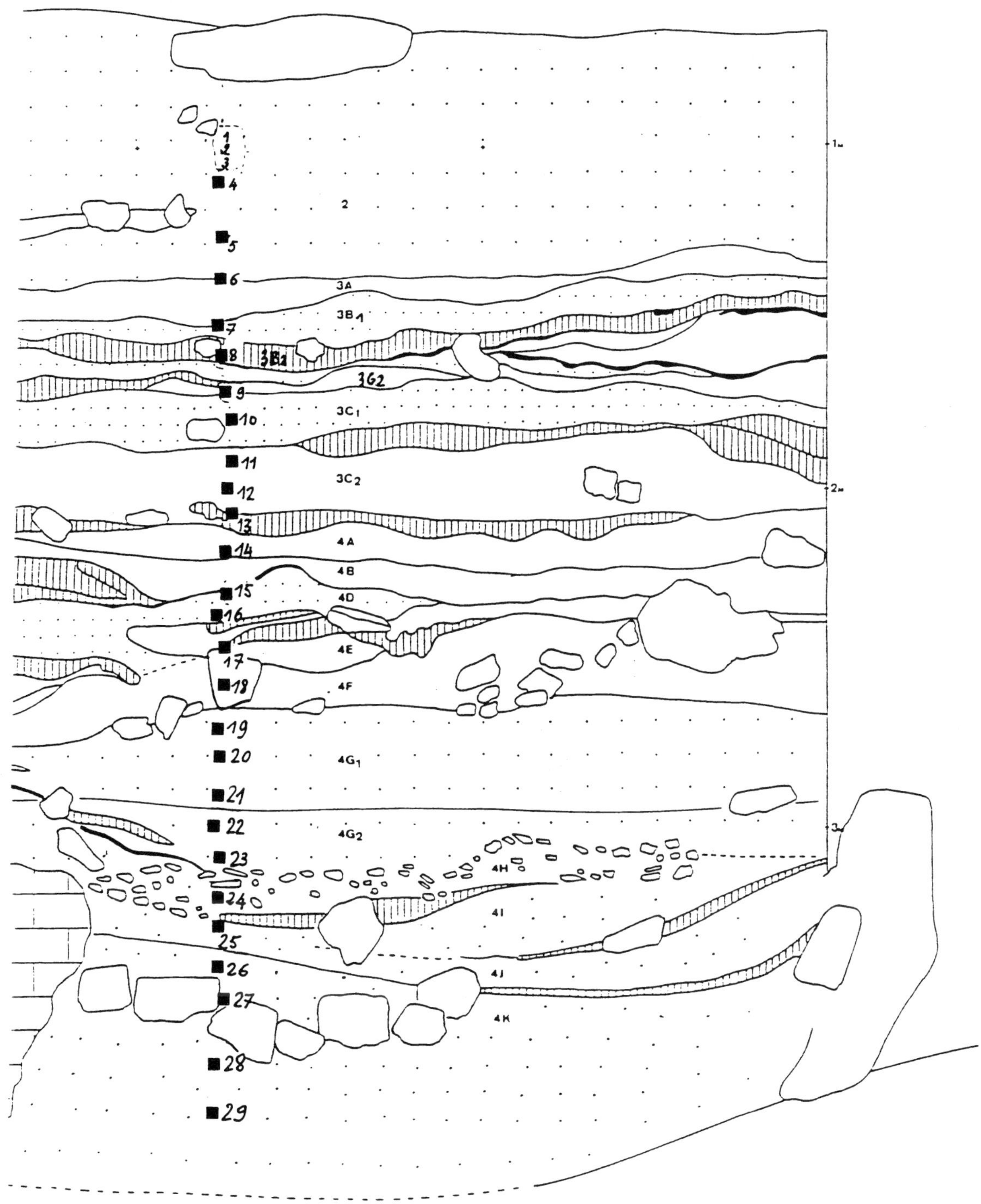

Text-Figure 1. Le Vieux Mounoï (Signes, Var, France). Sondage 4. Coupe 8. Samples for pollen analysis (after Hameau, 1993) + legend. Very dark brown layers (ashes + charcoal). Gray to dark grayish brown layers. White layers (ashes only).

cave. One would also have to imagine that such *Polypodium* stands were abruptly replaced by another fern species at certain periods since *Polypodium* spores are not always dominant (see for example samples 20 and 22 s in Text-Figure 2). I conclude that natural spore dispersion is not the reason for high fern spore percentages in archaeological sites.

Role of Differential Pollen Destruction

Many articles have already studied the exine destruction of pollen grains and spores (Havinga, 1963, 1964, 1971; Cushing, 1967; Elsik, 1971; Stanley and Linskens, 1974; Lowe, 1982; Gröner, 1985). They have shown how sporopol-

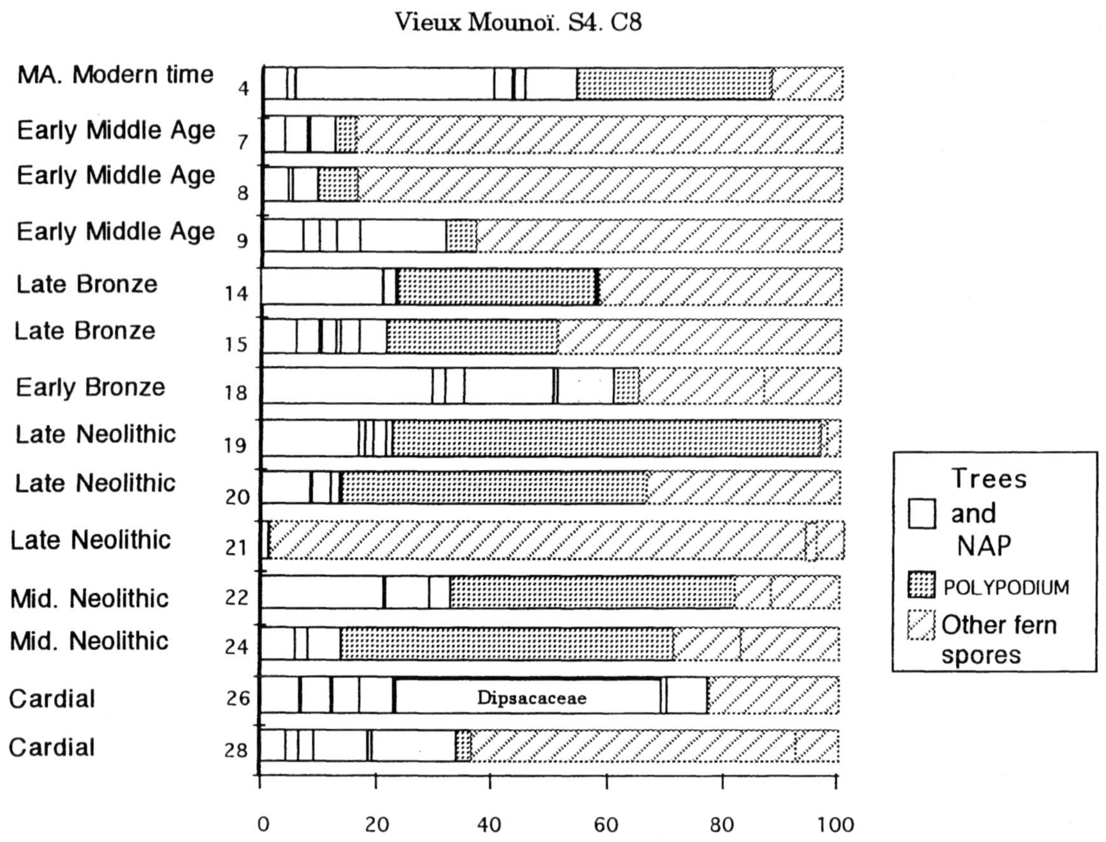

Text-Figure 2. Le vieux Mounoï (Signes, var). Palynological diagram (simplified).

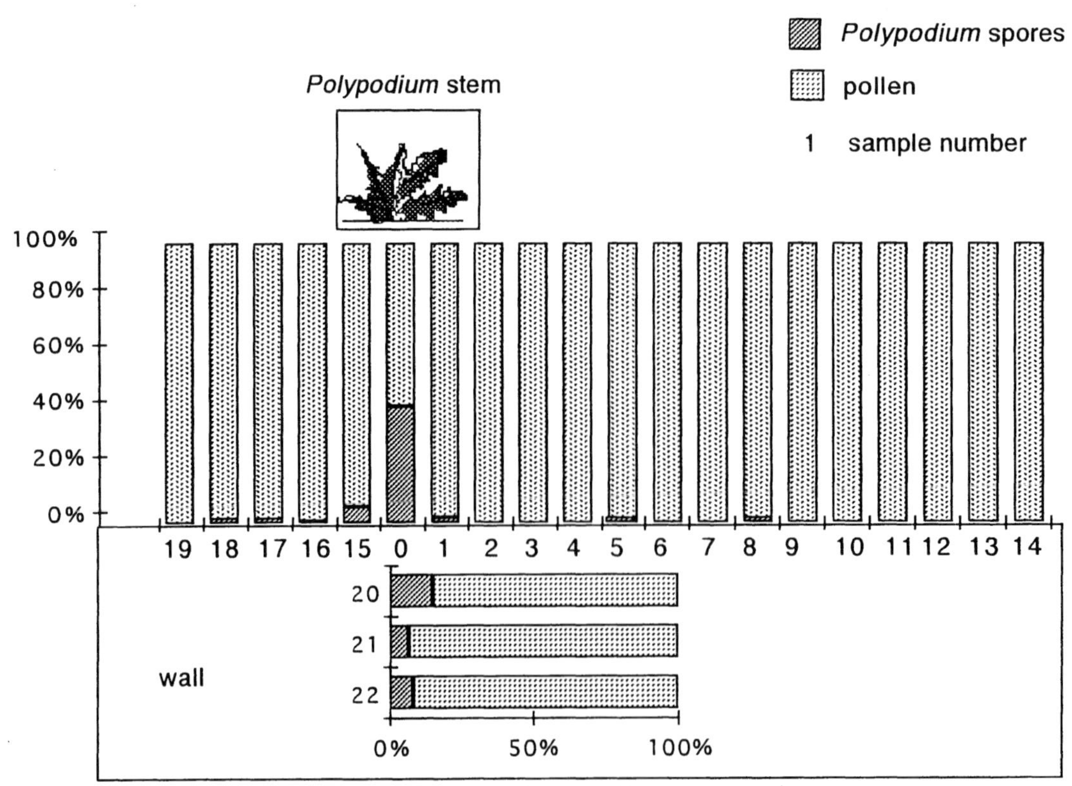

Text-Figure 3. The study of *Polypodium* spore dispersion in space. Samples position and simplified spectrum obtained.

lenin composition, ground pH, weather conditions and oxydation influence microorganism activity, and they produced a scale of pollen grain resistance to "corrosion." Such differential destruction phenomena reduces taxon diversity. Fern spores are amongst those with the highest rate of resistance. So it seems logical to consider that differential preservation might have taken place at Vieux-Mounoï, a place where the taxon number ranges from 9 to 24 (Table 2). However, the fern spore percentage is higher (88%) in samples with 19 taxa than it is in samples with 23 taxa (79%).

The lack of correlation between diversity and spore percentages suggest that other factors must play a part in the pollen assemblage of archaeological sediments. Former studies of cave-sheepcots have shown that human activities may be responsible for anomalies (Argant, 1990, Argant et al., 1991).

EXPERIMENT 2

Effects of *Polypodium* Introduction in a Sheepcot (Food and Litter)

Polypodium fronds were gathered in woods where this fern spreads over sandstone rocks in large carpets. Ten kg of leaves were cut with a knife and a sickle.

Dry *Polypodium* was given as the only food to a starving ewe. It refused to eat the fern which obviously cannot be eaten by sheep today. The same experiment was performed on lambs.

Dry *Polypodium* was used mixed with hay to make a comfortable litter for the ewe. After a week, the manure was 8 cm thick. Five samples were then gathered every two centimeters. Analysis (Text-Figure 4) showed abundant pollen. Poaceae largely dominate the spectra because of the hay. Spores are present in limited numbers. They are most numerous (10%) at the bottom where the hay has decayed. Their color, green at the beginning, has turned to brown.

Sheep pellets were also analyzed (Text-Figure 5). They show very few spores, which is logical because the animal did not eat *Polypodium* leaves and the few spores that were found could easily have been eaten with the hay.

EXPERIMENT 3

Heat Effect

Dung was burned, which was rather difficult because it was wet. The analysis (Text-Figure 4) shows that the spore percentage is twice as high in burned litter than it is in the fresh litter. All the spores became brown.

Because of the heat: 1) a part of the sporopollinic material was destroyed. Fern spores have a relative better resistance to fire than the other taxa, perhaps thanks to the protection of the sporangium hull. 2) sporangia burst and thus set free what they hold; hence a better dispersion. Resistance to heat was verified by submitting a piece of *Polypodium* leaf to the heat of an oven (220°C, for one hour and a half).

EXPERIMENT 4

After complete combustion of dry *Polypodium* leaves, some spores were preserved in the ashes but a litter sample of straw and hay taken under the same conditions did not have any pollen left.

Most archaeological layers in cave-sheepcots show fire traces sometimes ascribed to dung burning. So it is possible that the larger number of *Polypodium* spores could be explained by the heat effect.

Other possible uses of ferns

The ferns may have also been used as beds for the shepherds or as shelters against draughts. They may have been used as corral walls. Also, these plants are said to repel some parasites, particularly fleas.

CONCLUSIONS

Natural dispersion alone cannot account for the large amount of fern spores in certain sediments, more particularly in cave-sheepcots. *Polypodium* and other fern were introduced by humans into cave-sheepcots to make litters for animals or for men. Because *Polypodium* spores resist heat better than plant pollen, its presence is increased by dung combustion, either through accidental or intentional fires.

Generalizing the observations made from *Polypodium* will not be possible before experiments are repeated with other fern species, particularly concerning how spores react to heat.

ACKNOWLEDGMENTS

Pascale and Jacques Nuttall were willing to make the experiment concerning the behaviour of sheep in their sheepcot in Brittany. Claude Allibert translated the French version into English. Owen Davis helped proof the article, as well as V. Bryant, G. Jones and D. Goodman. I thank them all for their help.

TABLE 2. Vieux Mounoï (Signes, Var). taxon, spore and pollen number in 14 layers documented. Classification according to a range showing an increase in taxon amounts.

Sample Number	20	21	14	24	8	22	9	7	18	19	26	28	14	15
Taxon amount	9	9	12	15	13	15	16	19	20	20	20	23	23	24
Pollen + spores total	235	1010	254	500	457	563	129	638	129	1065	570	303	662	302
Spore total	202	992	195	456	397	380	89	565	51	830	125	140	526	200
Polypodium	123	942	88	34	266	277	6	24	7	793	112	102	321	7
Spore/total %	86	98	77	87	91	67	69	88,6	39,5	78	22	46	79	66

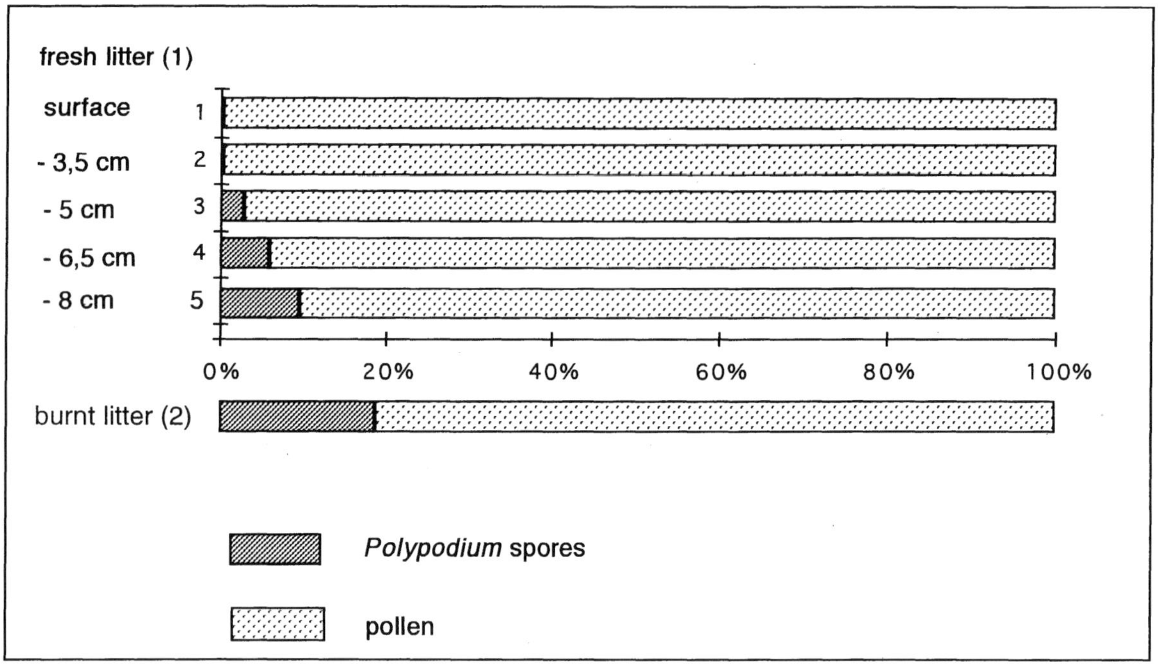

Text-Figure 4. Litter made of *Poypodium* + hay: 1– Pollen analysis of fresh litter. 2– Pollen analysis of burnt litter

References Cited

ARGANT, J.
1990 Climat et environnement au Quaternaire dans le Bassin du Rhône d'après les données palynologiques. *Documents du Laboratore de Géologie Lyon*, n°111, 199 pgs.

ARGANT, J., HEINZ, C., and BROCHIER, J-L.
1991 Pollens, charbons de bois et sédiments: l'action humaine et la végétation, le cas de la grotte d'Antonnaire (Montmaur-en-Diois, Drôme). *Revue d'Archéométrie* n°15: 29–40.

BASTIN, B.
1977 Le gué du Plantain (Neufvilles, Hainaut) site néolithique et romain. IV. Palynologie. *Dissertationes Archaeologicae*, Gand, 27: 31–43.

BOYER-KLEIN, A., DAVID, F., JOUVE, A., and LALOY, J.
1983 Le gisement magdalénien de la Pente-des-Brosses à Montigny-sur-Loing (Seine-et-Marne). *Gallia Préhistoire*, 26: 129–138.

BUI THI MAI
1974 Contribution à l'étude du transport et de la sédimentation des pluies polliniques dans un abri sous roche: l'abri Vaufrey (Dordogne); Thèse 3ème cycle, n°1155, Université de Bordeaux, 167 pgs.

CLET-PELLERIN, M.
1986 Analyses polliniques dans quelques sites néolithiques de Normandie. *Revue Archéol.ogique de l'Ouest*, suppl. n°1: 279–284.

CUSHING, E.J.
1967 Evidence for differential pollen preservation in late Quaternary sediments in Minnesota. *Revue of Palaeobotany and Palynology*, 4: 87–101.

DIOT, M-F.
1991 Apport et conservation sporo-pollinique dans les grottes: relation avec la fréquentation humaine et animale. *In:* Archéologie expérimentale, T.2: La Terre; Actes colloque *Expérimentation en archéologie: bilan et perspectives*, Beaune, 6–9 avril 1988. Paris, Errance: 236–244.

DUPRE-OLLIVIER, M.
1988 Palinologia y paleoambiente. Nuevos datos españoles. Referencias. *Servicio de Investigacion Prehistorica*, Ser. de Trabajos Varios, 84, 160 pgs.

EMERY-BARBIER, A.
1989 Variation du milieu et occupation du sol à Charavines (Isère). *Cahiers du Quaternaire*, 13, CNRS édit.: 123–128.

ELSIK, W.C.
1971 Microbiological degradation of sporopollenin. *In* Brooks, J., Grant, P.R., Muir, M.D., Gijzel, P van, and Shaw, G. (eds.), *Sporopollenin*. Academic Press, London, New York, pp. 480–511.

GRONER, U.
1985 Palynologie der Karsthöhlensedimente im Hölloch, Zentralschweiz. *Dissertation Universität Zurich*, 172 pgs.

HAMEAU, P.
1993 La grotte du Vieux Mounoï (Signes, Var). Rapport de fouilles, 18 pgs., annexes.

HAVINGA, A.J.
1963 A palynological investigation of soil profiles developed in cover sand. *Meded. Land bouwhogeschool*, Wageningen, Nederland, 63. (Abstr. of Diss.).
1964 Investigation into the differential corrosion susceptibility of pollen and spores. *Pollen et Spores*, 6: 621–635.
1971 An experimental investigation into the decay of pollen and spores in various soil types. *In* Brooks, J., Grant, P.R., Muir, M.D., Gijzel, P van, and Shaw, G. (Eds), Sporopollenin. Acad. Press, London, New York, pp. 446–479.
1984 A 20 year experimental investigation into the differential corrosion susceptibility of pollen and spores in various soil types. *Pollen et Spores*, Paris, vol.XXVI, n° 3–4: 541–557.

HEIM, J.
1986 Analyse palynologique. Paléoenvironnement du paléolithique supérieur ancien de la caverne de la Traweye Rotche à Sprimont (Province de Liège, Belgique). *Bulletin Société Royale belge d'Anthropologie et Préhistoire* 97: 113–120.

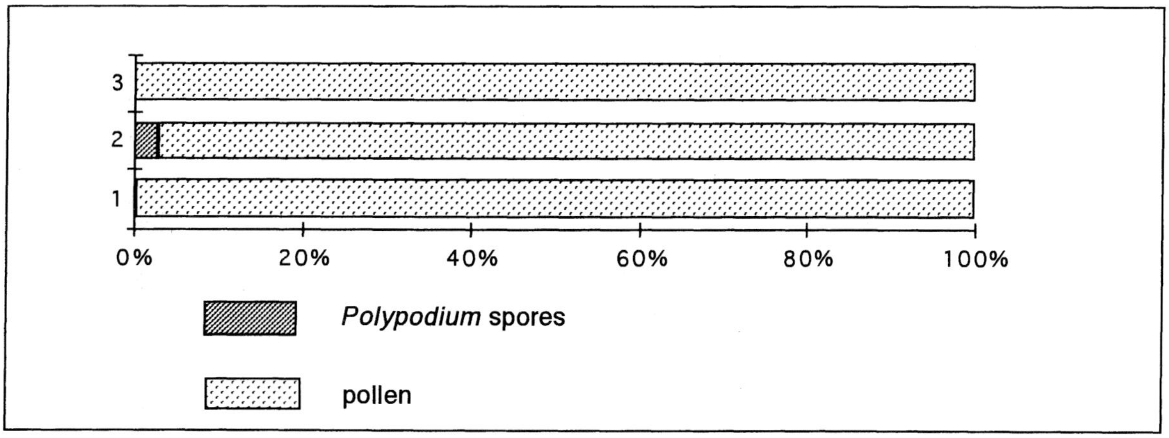

Text-Figure 5. Three sheep drops: pollen analysis. February 1996. Litter: *Polypodium* + hay; food: hay.

LEROI-GOURHAN, A.
1994 Le Temps d'el Juyo. *Museo y Centro de Investigation de Altamira. Monografias n°17:* 123–129.

LEROYER, C.
1985 Evolution du milieu pendant le post-glaciaire dans les Pyrénées andorranes: premiers résultats des recherches à La Balma Margineda. *Cahiers Ligures de Préhistoire et de Protohistoire,* N.S., n° 2: 192–214.

LOPEZ, P.
1982 Abauntz: analisis polinico. *Trabajos de Arqueologia Navarra,* 3: 355–358.

LOWE, J-J.
1982 Three Flandrian pollen profiles from the Teith Valley, Perthshire, Scotland. II. Analysis of deteriorated pollen. *New Phytologist,* 90: 371–385.

MARGUERIE, D.
1992 Evolution de la végétation sous l'impact humain en Armorique du Néolithique aux périodes historiques. *Travaux du Laboratoire d'Anthropol.ogie de Rennes,* n°40, 313 pgs.

PAQUEREAU, M-M. *in* HOUDRE, J-J., and VITAL J.
1979 Le gisement chasséen ancien du Pirou (commune de Polignac). Analyse palynologique. *Bulletin Société Préhistorique Française,* 76, 10–12: 371.

PENALBA GARMENDIA, M.C.
1989 Dynamique de végétation Tardiglaciaire et Holocène du Centre-Nord de l'Espagne d'après l'analyse pollinique. *Thèse de Doctorat, Univ. Aix-Marseille III,* 165 pgs.

STANLEY, R.G., and LINSKENS, H.F.
1974 *Pollen - Biology, Biochemistry, Management.* Springer-Verlag, 307 pgs.

Author's address

JACQUELINE ARGANT
A.R.P.A., U.F.R. des Sciences de la Terre
Université Claude Bernard
27–43 Bd du 11 novembre 1918
69622 Villeurbanne Cedex
France

PALYNOLOGICAL ANALYSIS AT THE GORNY SITE (KARGALY REGION): THE EARLIEST METALLURGICAL CENTRE IN NORTHERN EURASIA (RUSSIA)

Pilar López, Eugène N. Chernykh and José Antonio López-Sáez

Abstract

The Kargaly site, situated in northern Eurasia (Russia), offers an interesting cultural sequence. Palynological and charcoal analysis reveal various paleoecological phases from the Bronze Age (Subboreal period) to the 18th–19th centuries. Metallurgical activities were the cause of a slight deforestation during the Bronze age, mainly of the *Betula* forests.

INTRODUCTION

The purpose of this article is to consider the human factor in the interpretation of changes in vegetation for an archaeological site in northern Eurasia. The area under study is the Kargaly metallurgic-mining complex, which is known for the mining of extensive copper deposits at two different historical periods: the second millennium BC (Bronze Age), and the 18th–19th centuries AD. The two mining periods ended abruptly, possibly due to ecological catastrophes set off by the overexploitation of the local energy resources, mainly the gallery forest around the archaeological settlement, which provided the only fuel available on the steppes (Chernykh, 1994a, 1995).

PREHISTORICAL AND HISTORICAL EVIDENCE

Kargaly is one of the mining-metallurgical areas of eastern Europe and it played an exceptional role in the history of the peoples in the region (Chernykh, 1992, 1993, 1994a, 1994b, 1997a, 1997b). Located at the extreme eastern limit of the European continent (only 50 km from the Ural river, the conventional border with Asia), Kargaly is a mining locality that covers about 500 km^2 (Text-Figure 1). The ores (malachite and azurite) are found in abundance at the surface in layers of reddish-brown sandstone. The outcrops are concentrated in eleven areas with mineral veins and lenses that reach a depth of up to 90 m (Chernykh, 1994b, 1997a, 1997b).

One of the main goals of the Kargaly project is the evaluation of the impact of mining and metallurgy on the environment (Chernykh, 1994a, 1995). Therefore the most important factor to consider is the energy source needed for smelting. The most important fuel available on the steppes is the relatively scarce birch wood, available in the gallery forest found in the humid bottoms of ravines. About 400 m^3 of birch wood (1.5 to 2 ha. of forest) are needed to smelt a ton of copper (Chernykh, 1995). Birch was also needed as a resource for the human inhabitants of the region, who used the wood for construction, heating, and many other needs. The impact of the intensive metallurgical production on the fragile steppe ecosystem must, therefore, have been enormous. In the 18th and 19th centuries AD the metal from Kargaly was smelted in the Ural Mountains, more than 500 km away, in order to obtain wood from forests located there. More than 30,000 km^2 of forest were consumed in that phase of metallurgical activity. Prehistoric miners at Kargaly must have used locally available fuel, which would have soon become a limiting factor for production.

These facts suggest that the sudden abandonment of the Kargaly region at the end of the Srubnaya culture (Bronze Age) may have been caused by the degradation of forest resources due to metallurgical activity (Chernykh, 1995).

Archaeobotanical studies try to test this hypothesis although much still remains to be done to clarify the role of human activity in determining the course of steppe vegetation history.

López, Pilar, Chernykh, Eugène N., and López-Sáez, José Antonio
2001 Palynological analysis at the Gorny Site (Kargaly region): The earliest metallurgical centre in northern Eurasia (Russia); *In*: Goodman, D.K., and Clarke, R.T. (eds.), *Proceedings of the IX International Palynological Congress, Houston, Texas, U.S.A., 1996*; American Association of Stratigraphic Palynologists Foundation, p. 347–355.
 ISBN 0–931871–06–9

BACKGROUND

The Environnment

The region of Oremburg — which includes the Kargaly mining region — includes a notable variety of landscapes and plant formations. Areas vary from the forests–steppe which is very rich in birches, oaks and lindens, to the typical hill vegetation containing scarce forest.

The region of Oremburg is biogeographically framed with a landscape attributable to the "forest–steppe." However, this zone constitutes a very peculiar form of steppic vegetation, strongly influenced by the terminal mountainous areas of the South Ural. This notable influence clearly affects the bioclimatology and regional geography and directly impacts on the survival of some floristic elements with a residual character in the area of Kargaly, including *Betula pendula*, *Fraxinus excelsior*, *Hypericum perforatum*, *Convallaria majalis*, *Delphinium consolida*, *Viburnum opulus*, *Paris quadrifolia*, *Quercus robur* and *Ulmus laevis*, which represent the most characteristic examples. The region cannot be defined as a "typical steppe," which is usual in having hills and dug ravines, as well as small ripparian

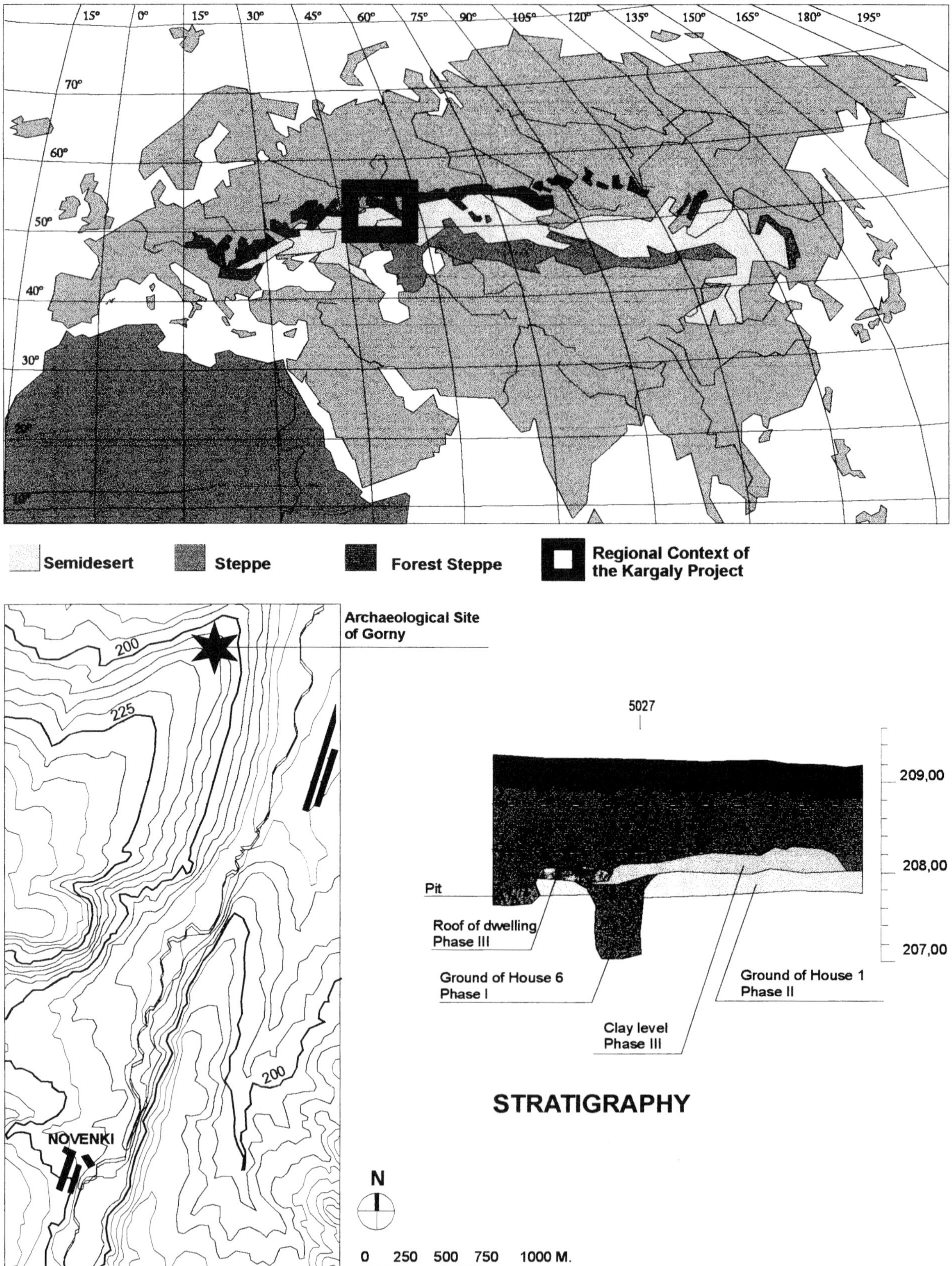

Text-Figure 1. Location map of the Kargaly mining area, including stratigraphy and cultural phases at the Gorny site.

forest environments which are always protected in moderately incised valleys. These characteristics permit the region of Kargaly to be considered within context of the "forest–steppe" (Novikov and Gubanov, 1985).

Three districts in the mining region of Kargaly — Panika, Miasnikovski and Staro-Ordinski — are defined with quite characteristic mining and anthropic landscapes (Chibilov, 1996) as well as the forests model of Sirtovo-Kargalynski.

In the district of Panika, it is possible to find some very well-preserved steppe associations with a predominance of the high steppe vegetation. Among herbaceous plants the most important are *Filipendula hexcapetea*, *Pulsatilla patens*, *Polygala comosa*, *Scorzonera humilis*, *Lynosyris villosa*, *Galatella villosa* and *Stipa* sp. The entire district is occupied by slag heaps. Abundant steppe shrubs (*Thymus* sp. and *Ephedra distachya*) grow on these slag heaps, whereas in the ravines isolated trees (*Populus tremula*, *Betula pendula*) are more typical.

The district of Miasnikovski includes five areas; the central area corresponds to a mining knoll with the same name. It is a hillside with outcrops of reddish sandstone where numerous mining wells are located. The slope is quite pronounced, which causes for a very peculiar vegetation covered by gramineae and thymes as *Thymus mugodzharicus* and *Ephedra distachya*. Conversely, the mining wells are colonized by Rosaceae including *Rosa cinnamomea*, *Spiraea crenata* and *Cotoneaster melanocarpus*, among others.

The mining field of Miasnikovski is located close to the knoll; it is one of the most ancient areas and where the most common grass is *Hedisarum grandiflorum*. In the bottom of the ravines grows a very picturesque grove known as *miasnikovskaya*. It is considered a birch forest of *Betula pendula* with shaking poplars (*Populus tremula*) that has been preserved perfectly by the low anthropic influence. The shrub horizon is composed by *Frangula alnus*, *Rhamnus cathartica*, *Viburnum opulus*, *Prunus avium*, *Sorbus aucuparia* and *Padus racemosa*. Between the grasses gtow *Aegopodium podagraria*, *Polygonatum officinalis*, *Campanula rotundifolia* and *C. persicifolia*, *Melandrium album*, *Veronica teucrium*, *Convallaria majalis*, *Trifolium pratense*, *Sanguisorba officinalis* and *Chelidonium majus*. In the forest are preserved isolated examples of oaks, which represent remnants of the groves destroyed by the miners during the 18th and 19th centuries (Chibilov, 1996); these relict oaks represent a very important factor in the palaeoecological understanding of the territory, above all in reference to the possible selective performance of man with regard to certain elements of the riparian forest (in particular birches, oaks and alders). One of the peculiarities of this district lies in the isolated fragments with a meadow steppe vegetation, that grows in the chernozems of the edges of the groves and of the ravines. It is one of the scarce zones of the region of Oremburg where the former associations (high steppe) are preserved. The most characteristic plants of these areas belong to genus *Stipa*, such as *S. zalesski-wilensky*, *S. dasyphylla* and *S. pennata*, near *Campanula rotundifolia*, *Oxytropis spicata*, *Centaurea ruthenika*, *Trifolium pratense* and *Pedicularis palustris* (Novikov and Gubanov, 1985; Riabinina, 1995).

The third district is called Staro-Ordinski. It occupies the mining field which extended throughout the ravines of the Usolka River. Its relief is damaged by the slag heaps, destroyed galleries and other remains resulting from mining work. As a result, some blind ravines up to 300 m long were developed, occupied by a vegetation of wet environments, including *Betula pendula*, *Padus racemosa*, *Rosa cinnamomea*, *Frangula alnus*, etc. Some of these damaged galleries became ice natural refuges where snow now remains until July or August; the areas are very favorable habitats for the survival of various kinds of ferns and bryophytes.

The forest model Sirtovo-Kargalynski is included in the mining-metallurgical complex of Kargaly as a landscape without apparent disturbance, corresponding to high steppe meadows. Birches and lindens are characteristic, as are steppe cherries (*Prunus avium*), dwarf almond-tree (*Amygdalus nana*) and rose-colored between shrubs. *Aegopodium podagraria*, *Convollaria majalis*, *Chelidonium majus*, *Scrophularia umbrosa*, *S. nodosa*, *S. vernalis*, *Lychnis chalcedonica* and *Origanum vulgare* are the most common grasses.

The Site

Gorny is one of the numerous habitation sites in the region and was used by the miners and metallurgist of the Srubnaya culture. The Srubnaya complex is one of the main historic-cultural communities of the Eurasian Metallurgical Province during the Late Bronze Age. It has three phases of chronological development (Text-Figure 1). In the first phase (18th to 16th centuries BC), its features are rather unstable. During the second phase (16th to 15th centuries BC) these features become stable. The third phase (15th to 14th centuries BC) is characterized by contact with the outside world and the collapse of the eastern periphery of the culture between the Volga and Ural rivers.

The miners exploited the copper resources during the 2nd millennium BC, and later during the 18th and 19th centuries AD. The site covers at least 3.5 to 4 ha. of the upper part of a flat-topped hill. The entire hill is covered with remains of the mining and smelting work undertaken in prehistoric and modern times, including the opening of mine shafts and wells, cave-ins, smelting locales, and spoil heaps.

Five separate excavations have been carried out in the area since 1992. The rich archaeological deposits have a maximum thickness of 2 to 2.5 m. The layers are so full of ore that it appears that malachite and azurite were deposited intentionally in certain areas. Slag was usually crushed to extract the copper droplets trapped in its interior. More than 100 casting molds have been found including some used to cast the picks and shovels that were used locally in the pits and galleries. Most of the cast copper was used to make these mining tools, but both these casts and examples of daggers, blades, chisels, and other tools, have also been found (Chernykh, 1994a, 1994b, 1995, 1997a, 1997b).

The archaeological deposit lacks any indication of farming, but it contains more than 700,000 fragments of bones of domesticated animals. Many types of seeds, stones and fruits have been found in the sediments. The bones were intentionally broken into pieces, and were often used to make tools and other objects such as the large collection of dice made from the radii of adult cattle. Almost 99% of the bones belong to six domesticated species: cattle, sheep, goats, horses, pigs and dogs. Except for dogs, all the bones were butchered to remove meat, and beef seems to have been the main meat consumed by the inhabitants of the site. The bones might suggest that the economy was principally devoted to livestock raising, but the archaeological context

TABLE 1. Radiocarbon ages.

British Museum Number	Age BP	Sample	Possible calibrated age ranges (calendar years BC), 95% probabl.
BM-2945	3270 ± 40	*Alnus* charcoal	1680–1450
BM-2962	3340 ± 35	*Bos* femur and tibia	1740–1520
BM-2963	3380 ± 35	*Bos* radius	1870–1840 or 1770–1600
BM-2964	3170 ± 50	*Bos* radius and metatarsus	1660–1560 or 1530–1380 or 1350–1310
BM-3016	3120 ± 45	*Bos* radius	1520–1300

also shows the importance of mining and metallurgical activity (Chernykhj, 1992, 1993, 1994, 1995).

Methods

Twenty samples of 40 g each were analyzed for pollen. Sample intervals of 5–15 cm were analysed in 1996 from the West profile, which is 210 cm thick. Five samples of bone were radiocarbon dated (Table 1).

Chemical treatment for pollen extraction follows López-García (1984). Pollen concentrations were estimated by adding a Thoulet solution to sediment (Goeury and Beaulieu, 1979). Particles < 10 mm were removed by ultrasonic sieving.

All pollen grains and spores encountered were identified and counted. For pollen and spore identification, we used the pollen keys of Moore and Webb (1978), Moore et al. (1991), Reille (1992, 1995), as well as the pollen reference collection in the Laboratory of Archaeobotany of the Department of Prehistory (Madrid, Spain).

The results of the pollen counts are presented as a percentage diagram based on the sum of terrestrial pollen (Text-Figures 2 and 3). Pollen from aquatic plants and spores were excluded. Samples with less than 100 pollen grains were not used in the calculations. For interpretation and reconstruction of the palaeoenvironment, 14 samples were ultimately used with pollen sums varying between 140 and 381 grains.

Local pollen assemblage zones have been adopted throughout and are based on the observed pollen stratigraphy, placing special emphasis on important changes in frequencies of arboreal and anthropogenic taxa.

Palaeobotanical Evidence

Five pollen zones (KG-1 to KG-5) could be recognized in the pollen record. They are believed to represent the dominant vegetation development (Text-Figures 2 and 3). Fluctuations in the curves of *Pinus sylvestris* t., *Betula* and Asteraceae t. Cichorioideae are critical for the zonation.

Zone KG-1

This interval is considered as a zone because the upper layers belong to a sterile clay without any pollen or spores. The relation AP/NAP is 60/40. It contains 15% birch pollen. The percentages of *Quercus robur* t. are up to 15%, *Prunus* t. to 10% and *Salix* to 20% of the pollen sum. *Picea* is also present with values near 5% (Text-Figure 2).

Of the upland herbs, the most outstanding are the Asteraceae, mainly the Cichorioideae with 10% of the pollen sum and *Anthemis* t., also with 10%. *Polygonum aviculare* t. is also present (Text-Figure 3).

Zone KG-2

This zone is practically sterile. The lithology of the sediment is clayey and preserved pollen is absent. Two 14C data are available (3380 ± 35 BP, and 3340 ± 35 BP).

Zone KG-3

Three ^{14}C dates have been obtained in this zone: 3270 ± 40 BP, 3170 ± 50 BP, and 3120 ± 45 BP.

This zone contains high *Pinus sylvestris* t. values (up to 60%) and moderate amounts of *Quercus robur t.* (about 15%). The AP is about 50% at the archaeological height 20,775, increasing up to 80% at 20,787. *Ephedra distachya t.* is as high as 10%. *Abies* is almost always present except at 20810 and 20840 heights. *Alnus* is only up to 5% of the pollen sums but is clearly present in all the sequence, with *Acer* and *Corylus*. *Betula* showing values down to 5% except at 20,850 where it increases up to 10%. *Picea, Juniperus* t., *Populus, Salix, Tamarix, Tilia* and *Ulmus* are well-represented (Text-Figure 2).

Among the NAP, the Asteraceae are well represented, mainly the Cichorioideae which represent up to 15% of the pollen sum at 20,810. *Anthemis* t., *Aster* t., Boraginaceae, *Convolvulus arvensis* t., *Malva sylvestris* t., *Plantago lanceolata* t., *Polygonum aviculare* t. and *Rumex acesota* t. are well-presented among the taxa with anthropogenic origin (Text-Figure 3).

Taxa that grow on moist soils or aquatic plants are also represented in this zone (Cyperaceae, *Nymphaea alba t.* and *Typha),* together with some spores (Text-Figure 3).

Zone KG-4

This zone corresponds to a post-Srubnaya phase. Between KG-4 and KG-5 there is a pollen hiatus where preserved pollen is absent. *Pinus sylvestris* t. has values up to 50% and *Quercus robur* t. is less than 5%. *Abies, Alnus, Populus, Tilia* and *Betula* are also present (Text-Figure 2). The AP shows values vary up to 60%.

Zone KG-5

Archaeologically this zone belongs to 18th–19th centuries AD. The pollen diagram shows a rise in *Pinus* pollen and an increase in deciduous tree pollen such as *Quercus* and *Betula*. The upper part of the zone has an accumulation of organic deposits following the abandonment of the site. Finally, in this zone, *Tilia* appears and both *Corylus* and

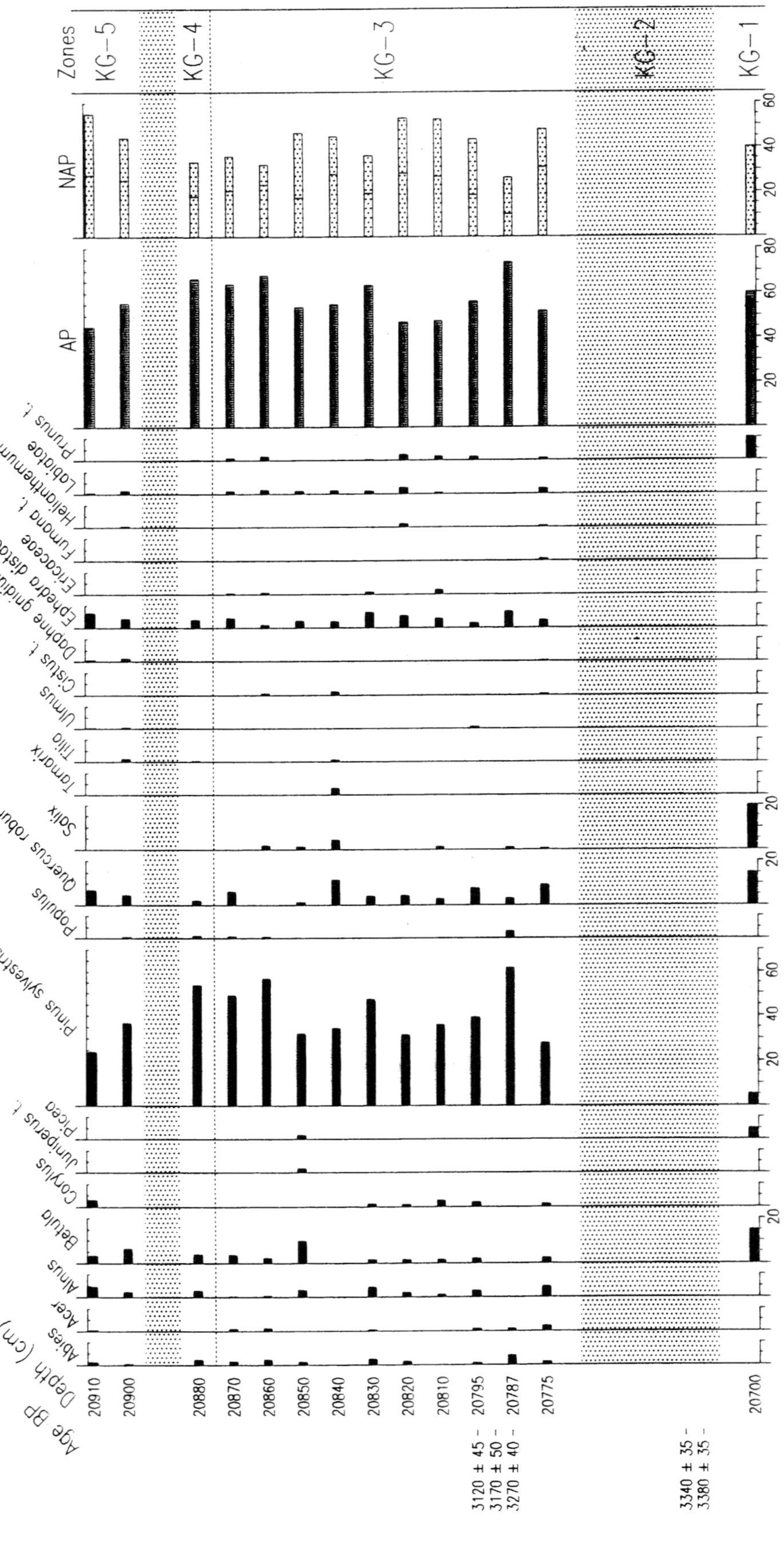

Text-Figure 2. Percentage pollen diagram: trees and shrubs.

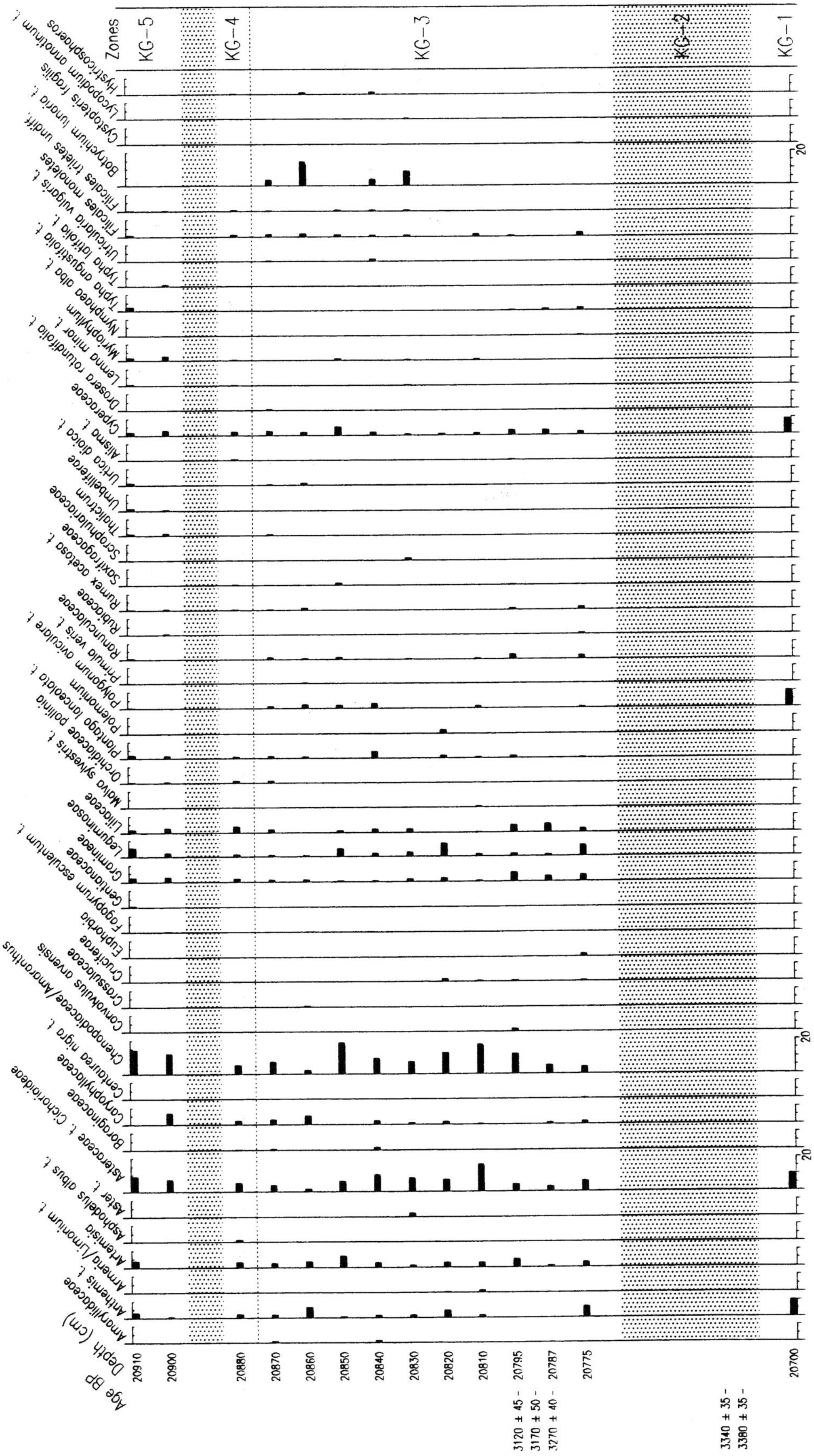

Text-Figure 3. Percentage pollen diagram: herbs, aquatics and cryptogams.

Ulmus pollen are again present (Text-Figure 2).This zone shows a new increase of the percentages of nitrophilous plants such as Cichorioideae, and both *Plantago lanceolata* t. and *Fagopyrum esculentum* t. are present (Text-Figure 3).

PALAEOENVIRONMENTAL INTERPRETATION

KG-1 and KG-2 belong to a period before 3200 BP, denoted by the two C^{14} dates at KG-2. The high percentages of birch in KG-1, along with the absence of *Abies* and *Ephedra distachya*, as well as 15% abundance of *Quercus robur* t., could suggest that it was a cool phase, prior to the Srubnaya culture. We cannot obtain any information for zone KG-2 using palynological information. Low percentages of pollen from the clay layer at the bottom of the profile is expected, but we also have other botanical data from charcoal remains (Table 2); the most important charcoal samples represent *Salix* sp., *Alnus incana*, and *Alnus viridis*. Some remains of bark have been found, possibly belonging to *Betula pendula*, suggesting its use by the inhabitants to cover the ground in their houses which were partially dug into the clay subsoil.

These archaeological deposits are full of bones of domesticated animals. The predominance (82%) of cattle indicates that beef was the main meat consumed by the site's inhabitants and might suggest that the economy was principally devoted to livestock raising.

Zone KG-3 corresponds to the Srubnaya occupation of the site between 3300 and 3100 BP (Chernykh, 1994a, 1997a). Taxa with an anthropogenic origin are evident, mainly Cichorioideae, *Polygonum aviculare* t. and *Plantago lanceolata* t. At the same time, Gramineae pollen shows an increase in abundance. This phenomenon is well-represented in zone KG-3. The KG-3 zone is also defined by an increase in *Pinus sylvestris* t. pollen showing both regional and extra-regional vegetation factors, and the development of pine forests in the South Urals. There is also an increase of *Abies* pollen at this time. The north winds were extremely strong, explaining the high percentage of *Pinus* and the presence of *Abies*. *Acer* is present in six pollen spectra with a percentage averaging about 2% which suggests the presence as a local forest near the site (Vuorela, 1993). Expansion of grazing is suggested by increased abundance of herbaceous and shrub taxa such as Chenopodiaceae/*Amaranthus*, *Artemisia*, *Centaurea* and *Ephedra distachya t.* Zone KG-3 could be placed into a cool phase, but warmer than the pre-Srubnaya period, dated between 3300 and 2500 BP, very well-defined on the Russian Holocene (Khotinsky, 1984; Hahne and Melles, 1997). *Betula* pollen is poorly represented (in comparison with the KG-1 zone) coinciding with a strong increase of human activity, suggesting that the most important fuel available on the steppes was the wood of birches located on the forest galleries in the humid bottoms of ravines near the site. Charcoal remains indicates the use of *Betula*, *Salix–Populus* and *Alnus* as main wood resources (Table 2). *Betula* pollen disappears at 20787 and 20840, suggesting an intensified use by humans. Birch was also apparently needed as a resource for human inhabitants of the region, who could use the wood for construction (some charcoal remains could be related with this activity), heating, and other needs. *Quercus robur* t. and *Betula* present opposite curves which would permit the supposition of a selective use of both trees at different periods during KG-3, time, even though *Quercus* is very rare in charcoal remains. The landscape during KG-3 time would be very similar to the present-day type with a forest of *Populus*, *Alnus* and *Betula* near the rivers. The charcoal remains corroborate this theory. The two riparian forests situated near the site would represent the only wood supply during the occupation of the site. *Salix* grains reflect the introduction of willow as a pioneer species in deforested areas. *Juniperus* appears in this zone suggesting the presence of open areas on well-drained soils (Asplund and Vuorela, 1989). High percentages of *Artemisia*, *Ephedra distachya* t. and Chenopodiaceae/*Amaranthus* confirm cool and dry conditions during KG-3 time. The most remarkable feature during this time is the strong human impact on vegetation near the Gorny site. The presence of Ericaceae pollen suggests increased grazing. The increasing frequencies of pollen from cultural indicators such as *Rumex acetosa* t., *Plantago lanceolata* t., Caryophyllaceae and Cruciferae suggest increased human activities not far from the site (Behre, 1981; Asplund and Vuorela, 1989; Vuorela, 1993), as well as no substantial evidence of arable lands.

The pollen diagram indicates that the local landscape became open during the course of the Bronze Age with large parts of the area becoming covered by grass. The vegetation was probably a mosaic of grassland and forest

TABLE 2. Charcoal remains. ^{14}C dates: 3380 ± 35; 3270 ± 40; 3120 ± 45; 3170 ± 50; 3340 ± 35.

TAXA	Phase I N %	Phase II N %	Phase III N %	Phase IV N %
Betula	2	87	51	35
Alnus	13	54	7	
Quercus f.c.		6		
Salix-Populus	9	181	48	3
Acer		5		
c.f. *Prunus*		1		
Rhamnus		3		
Leguminoseae		1		
Inde..........		11		
	24	349	106	38

on wet areas, and conifers (*Pinus, Abies* and *Picea)* in the mountains. The grazing of livestock was probably the most important land-use in the area.

Zone KG-4 is very similar to zone KG-3 with a predominance of *Pinus sylvestris* t. and *Abies* around the mountains and a forest of *Alnus, Betula* and *Populus* near the rivers of Kargaly. A marked increase of AP coincides with the abandonment of the site, suggesting a marked forest regeneration. The changes in vegetation suggest a significant reduction of human activity.

Following the hiatus between zones KG-4 and KG-5, noted by an abrupt lithological change from a compact sediment to a soft humus, it is possible to see a change in the landscape occurring around the 18th to 19th centuries even though this period (KG-5) was not characterized by any major changes in the cultivation of crops. There is a gradual increase in the pollen of Asteraceae, Chenopodiaceae/*Amaranthus*, and pastoral pollen indicators such as *Urtica dioica* t., Boraginaceae and Rubiaceae.

An increase in *Quercus, Alnus* and *Corylus* pollen in zone KG-5, as well as the presence of *Tilia* and *Ulmus* pollen and a decline of *Pinus*, suggest a progressive climatic change to more temperate conditions. The decline of *Pinus sylvestris* t. could be also explained by the use of pine wood in the mining industries of the South Urals during the 18th and 19th centuries where copper of Kargaly was transferred for this metallurgical activity (Chernykh, 1994a, 1995, 1997a). The presence of *Ephedra distachya* t. and *Artemisia* confirms the steppic conditions of this area.

CONCLUSIONS

The predominant groups of NAP from the period between the Bronze Age and the 18th–19th centuries were anthropochores and apophytes. Much of these open areas might have been occupied by vegetation types characteristic of settlement areas with a rapid recolonization by trees and shrubs when the sites abandoned. There are ecological factors which play an important role, yet we are not able to evaluate their importance on society during different times. The natural trend in the woodland development has been a gradual replacement of gallery forest and shrubs on wet soils. This may be regarded as being caused partly by climatic conditions that caused a change among the succession of tree taxa. Human exploitation could cause a slight deforestation by the Bronze age, and the remaining woodlands were transformed by human activities into woodlands.

Damp soils with birch, elm and alder forests were exploited for fodder production and fuel beginning in the Bronze Age and continuing into the 19th century. These areas were transformed into open lands. The impact of the intensive metallurgical production on the fragile steppe ecosystem must have been enormous. More than 30,000 km^2 of forest were consumed during that phase of metallurgical activity, with catastrophic results on the environment. We cannot infer by pollen analysis what drastic changes on the vegetation were caused by this land-use strategy. The region is large enough to include areas for mining activity as well as ample pasture land. At least during prehistoric times the inhabitants of the steppes must have adapted their way of life to changes in the very harsh climate. This emphasizes the close relationship between environmental conditions and human adaptation.

References Cited

ASPLUND, H., and VUORELA, I.

1989 Settlement studies in Kemio. Archaeological problems and palynological. *Fennoscandia Archaeologica*, 6: 67–79.

BEHRE, K.E.

1981 The interpretation of anthropogenic indicators in pollen diagrams. *Pollen et Spores*, 23: 225–245.

CHERNYKH, E.N.

1992 *Ancient Metallurgy in the USSR: The Early Metal Age.* Cambridge University Press, Cambridge, 416 pgs.

1993 Ritmo y modelos de las destrucciones tecnológicas y culturales fundamentales a partir del descubrimiento del metal. *In*: Martínez Navarrete, I. (ed.), *Teoría y práctica de la Prehistoria: Perspectivas desde los Extremos de Europa.* Santander: Universidad de Cantabria — CSIC, pp. 275–300.

1994a L'ancienne production minière et métallurgique et les catastrophes écologiques anthropogènes: introduction au problème. *Trabajos de Prehistoria,* 51: 55–68.

1994b Kargaly. Orígenes de la metalurgia en Eurasia Central. *Revista de Arqueología*, 153: 12–19.

1995 Kargaly. La energía de producción y las catástrofes ecológicas. *Revista de Arqueología*, 168: 30–35.

1997a Kargaly: the most oldest mining and metallurgical huge complex in Northern Eurasia. *Russian Archaeology*, 1: 21–36.

1997b *Kargaly, the Forgotten World.* Nox, Moscow (in Russian), 175 pgs.

CHIBILOV, A.A.

1996 *The Natural Heritage of the Oremburg Region.* Oremburg Books, Oremburg (in Russian), 228 pgs.

GOEURY, C.L., and BEAULIEU, J.L.

1979 A propos de la concentration du pollen a l'aide de la liqueur de Thoulet dans les sédiments minéraux. *Pollen et Spores*, 23: 239–251.

HAHNE, J., and MELLES, M.

1997 Late- and post-glacial vegetation and climate history of the south-western Taymyr Peninsula, central Siberia, as revealed by pollen analysis of a core from Lake Lama. *Vegetation History and Archaeobotany*, 6: 1–8.

KHOTINSKY, N.A.

1984 Holocene vegetation history. *In*: Velichko, A.A. (ed.), *Late Quaternary Environments of the Soviet Union.* Longman, London, pp. 179, 200.

LÓPEZ GARCÍA, P.

1984 Aplicaciones de la palinología a la prehistoria: métodos utilizados y resultados. *Actas Primeras Jornadas de Metodología de Investigación Prehistórica*, 1: 309–317.

MOORE, P., and WEBB, J.S.

1978 *An Illustrated Guide to Pollen Analysis.* Hodder & Stroughton, London, 133 pgs.

MOORE, P., WEBB, J.S., and COLLINSON, M.E.

1991 *Pollen Analysis.* 2nd edition. Blackwell Scientific Publications, London, 216 pgs.

NOVIKOV, V.S., and GUBANOV, I.A.

1985 *Scholar Atlas: Key of the Higher Plants*, Prosveschenie, Moscow (in Russian), 239 pgs.

REILLE, M.

1992 *Pollen et Spores d' Europe et d' Afrique du Nord.* Laboratorie de Botanique Historique et Palynologie, Marseille, 520 pgs.

1995 *Pollen et Spores d' Europe et d' Afrique du Nord. Supplement 1.* Laboratorie de Botanique Historique et Palynologie, Marseille, 327 pgs.

RIABININA, Z.N.

1995 *The rare plants of the Oremburg region and its protection* (in russian). UIF Nauka, Ekaterinburg, 166 pgs.

VUORELA, I.
1993 Pollen analysis of a subrecent fossil soil in Haltiala, Northern Helsinki, Finland. *In*: Autio, S. (ed.), *Geological Survey of Finland (Current Research 1991–1992)*, Special Paper 18, pp. 79–85.

Authors' addresses

Pilar López
Laboratorio de Arqueobotánica
Centro de Estudios Históricos
CSIC
Duque de Medinaceli 8
28014 Madrid
Spain

Eugène N. Chernykh
Laboratory of Natural Sciences in Archaeology
Institute of Archaeology
Russian Academy of Sciences
Dm. Ulyanova 19
117036 Moscow
Russia

José Antonio López-Sáez
Paloma Uzguiano
Laboratorio de Arqueobotánica
Centro de Estudios Históricos
CSIC
Duque de Medinaceli 8
28014 Madrid
Spain

A LATE QUATERNARY POLLEN RECORD OF VEGETATION AND CLIMATE CHANGE FROM KAITAIA BOG, FAR NORTHERN NEW ZEALAND

Mike Elliot

Abstract

Pollen analysis from Kaitaia Bog, far northern New Zealand, provides evidence of vegetational history and climatic change over the last ca. 25 ka. Throughout this period the region was occupied by mature forest. Regional forest of the Last Glacial Maximum comprised a tall, complex, conifer–beech–hardwood association. The most common trees were *Dacrydium cupressinum* and *Fuscospora*. Warmth-loving, moist elements such as *Ascarina lucida* and *Agathis australis* were scarce. High concentrations of microscopic charcoal fragments indicate frequent fires. The presence of continuous forest cover in far northern New Zealand during the Last Glacial Maximum (LGM) contrasts sharply with grass and shrub-dominant landscapes south of 38°S. Cool climate species such as *Fuscospora* declined in the Lateglacial, and by ca. 11.3 ka *Ascarina lucida* began to increase rapidly. From the early Postglacial, conifer–beech–hardwood forest was replaced by a *Dacrydium cupressinum*-dominant conifer–hardwood association. At the same time *Agathis australis* and other frost-tender and drought-intolerant species expanded throughout the region.

INTRODUCTION

Palaeoenvironmental data from sites throughout New Zealand indicate that during the Last Glacial Maximum (LGM) most of the country south of 38°S was deforested. Grass and shrub-dominant communities were widespread, and forest species survived in small sheltered areas where there was protection from heavy frosts and strong, cold winds. However, the lack of information available for Northland does not allow a reconstruction of LGM vegetation cover with any certainty. Only four published terrestrial pollen diagrams cover this period (Dodson et al., 1988; Newnham, 1992; Newnham et al., 1993). One pollen record from near Cape Reinga indicates persistent podocarp–hardwood forest for the past 17 ka (Dodson et al., 1988). There is no evidence for the period immediately prior to this. Two pollen profiles from Aupouri Peninsula have radiocarbon dates in the range 30 to 4 ka (Newnham et al., 1993), but provide only a tentative reconstruction of far northern vegetation during the Last Glaciation and Holocene because of uncertainty over dating. There is evidence of mixed podocarp–hardwood forest prior to the LGM, but the LGM record itself may be missing from these sequences owing to possible hiatuses in sedimentation (Newnham et al., 1993). A 30 ka profile from near Whangarei provides a rather better record indicating persistence of diverse podocarp–hardwood forest throughout the sequence, at times dominated by *Nothofagus* (Newnham, 1992). However, like the Aupouri Peninsula sites, this record also suffers from an uncertain chronology. At a time when much of the rest of New Zealand was dominated by grass and shrubland, Northland's vegetation remains inadequately described. This paper addresses this issue and evaluates the manner in which northern forests responded to ameliorating conditions in the Postglacial transition.

Elliot, Mike
2001 A late Quaternary pollen record of vegetation and climate change from Kaitaia Bog, far northern New Zealand; *In*: Goodman, D.K., and Clarke, R.T. (eds.), *Proceedings of the IX International Palynological Congress, Houston, Texas, U.S.A., 1996*; American Association of Stratigraphic Palynologists Foundation, p. 357–365.

ISBN 0–931871–06–9

STUDY AREA

Kaitaia Bog is situated at the north-western margins of Kaitaia township (35°06'52"S, 173°12'34"E) on the southern end of the Aupouri Peninsula in far northern New Zealand (Text-Figure 1). Though substantially modified now, the Kaitaia Bog was formerly an ombrogenous, raised bog, dominated by the restiad *Sporadanthus traversii* (Cranwell, 1953). It has a low-lying surface, mostly below 10 m asl, and developed as a consequence of progressive sand dune encroachment on drainage from the Herekino Range in the south and lower eastern hills. The Herekino Range reaches 300 m and comprises Tertiary marine volcanics. The northern and western margins are all bounded by late Quaternary sand dunes. The low hills to the east rise up to 100 m, comprising Oligocene calcareous siltstones and sandstones (Kear and Hay, 1961; Hay, 1975). The main natural drainage of the bog now is via the Awanui River, north–eastward into the Rangaunu Harbour. Most of the bog is now artificially drained and the dominant soils are loamy peats. Only traces of the original swamp forest persist on isolated patches of poorly drained soils, including *Cordyline australis* and *Dacrycarpus dacrydioides*. *Polygonum aviculare* has invaded partially drained areas, while *Typha orientalis* dominates margins of the large drains. Better drained, elevated tracts are often covered by a *Leptospermum scoparium* shrubland (Dick, 1950). Prior to European arrival large areas were gardened by Polynesians (Wilson, 1921).

The climate of Northland is typically warm and humid. The rainfall varies from ~1100 mm yr^{-1} in coastal areas to 2500 mm yr^{-1} in upland areas. Most of this falls during the winter half year, while summers are usually dry. Mean monthly air temperatures range from 19.5°C to 10.5°C, and mean annual temperatures range from 14–15.5°C in the west and south, to 15.5–16°C in the north and east. Winters are mild with many parts experiencing only a few shallow ground frosts each year, generally in sheltered inland areas (Moir et al., 1986).

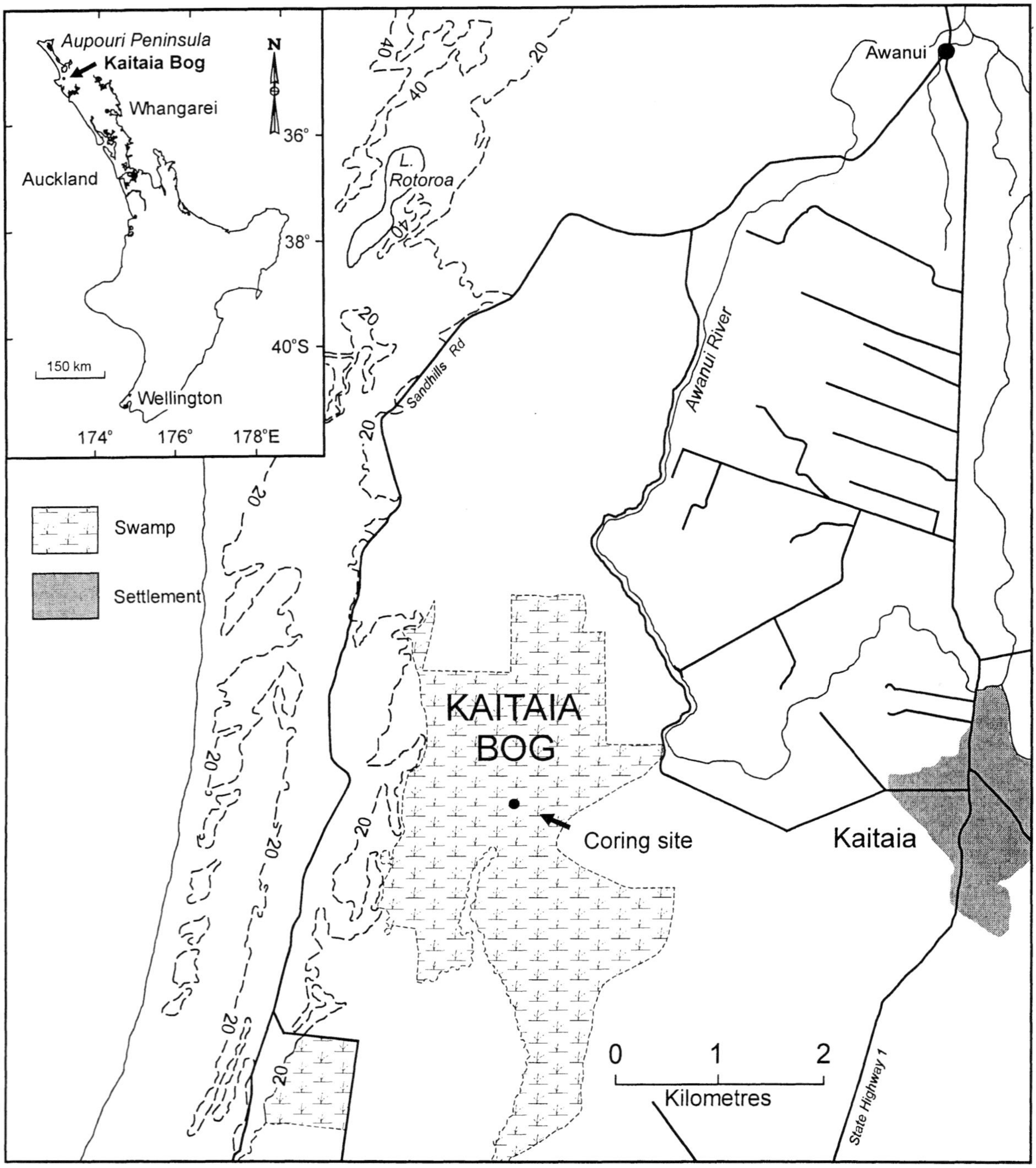

Text-Figure 1. Map of North Island, New Zealand showing the study area and coring site.

METHODS

Sediments from a 6 m core were analysed for pollen. Nine samples of 5 cm length were submitted for AMS radiocarbon dating to the Rafter Radiocarbon Laboratory, Lower Hutt, New Zealand. Pollen samples were taken at 10 cm intervals except between 4.55 m and 5.45 m where samples were taken at 5 cm intervals. Pollen preparation followed standard procedures (Moore et al., 1991). Minimum counts of 200 dryland pollen types were made for each sample. Results are presented as percentage data (Text-Figures 3A-C), and include pollen and charcoal concentration data in grains cm^{-3}. The pollen sum consists of all dryland taxa. Four pollen zones are defined based on cluster analysis (Grimm, 1987). Counting of microscopic charcoal fragments follows the method used by Bush et al. (1992). Taxonomic nomenclature follows that of Allen (1961), Connor and Edgar (1987) and Molloy (1995). *Nothofagus fusca*-type pollen species are designated *Fuscospora* after McGlone et al. (1996).

SEDIMENT STRATIGRAPHY

Sediment stratigraphy (Text-Figure 2) consists of an upper sequence of dark brown peat from the surface to 5.42 m. A narrow layer of sandy organic mud follows between 5.42 and 6.0 m. The underlying material consists of sands grading into a stiff basal clay. A suite of nine AMS radiocarbon dates (Table 1) provides a chronology including the LGM (Text-Figure 2). The lowermost date at 5.5m is 22, 410 ±450 yr B. P. Extending the age-depth line suggests that the base is at least 25, 000 yr. B. P. The steepness of the age-depth curve between ca. 22.4 and 11.6 ka, based on linear interpolation, suggests a period of very slow sedimentation, although it is possible there may be a hiatus in sedimentation. The remainder of the curve shows a consistent rate of sedimentation indicating the age at the surface is ca. 2.4 ka.

RESULTS

Pollen Zone K4 is characterised by high levels of *Dacrydium cupressinum* and *Fuscospora*. Other significant pollen types include *Podocarpus, Prumnopitys ferruginea, P. taxifolia, Libocedrus, Dacrycarpus dacrydioides* and *Metrosideros*. Also present are *Halocarpus, Manoao colensoi, Phyllocladus, Nestegis, Coprosma, Myrsine* and various Myrtaceae species. Tree ferns are common, especially *Cyathea smithii*-type. *Ascarina lucida* and *Agathis australis* values are low. *Laurelia novae-zelandiae* is present in two samples and *Elaeocarpus* is consistently recorded. Bog species are dominated by Restionaceae pollen and *Gleichenia* spores. The charcoal concentration was high during this zone.

At the beginning of Zone K3 sharp changes occur. *Fuscospora* increases as *Dacrydium cupressinum, Dacrycarpus dacrydioides, Libocedrus,* and tree ferns decline. *Fuscospora* values (30%+) peak between 5.15 and 5.30 m. Tall conifers *Podocarpus, Prumnopitys taxifolia* and *Manoao colensoi* become more common. *Agathis australis* is present in low percentages throughout the zone. *Metrosideros* increases upwards. *Ascarina lucida,* from initially low abundance, becomes common toward the top of the zone. Significant concentrations of charcoal fragments are recorded with a peak at 5.20 m, but decline towards the top of the zone. The first pollen of *Dodonaea viscosa* is recorded at 5.10 m and persists thereafter. *Nestegis* pollen is common. Bog taxa are dominated by Cyperaceae, but after the charcoal peak *Leptospermum* and Restionaceae rise sharply.

During Zone K2 time, *Fuscospora* abundance declines rapidly from ca. 9.5 ka, and *Dacrydium cupressinum,* rises to 40 percent. *Ascarina lucida* is abundant, particularly in the lower half of this zone. *Metrosideros* species are common. From ca. 9 ka *Agathis* and *Phyllocladus* pollen increase markedly. *Dacrycarpus* increases towards the top of the zone. *Elaeocarpus, Knightia excelsa* and *Nestegis* are consistently present, and *Cyathea* is abundant. *Prumnopitys ferruginea, Halocarpus* and *Manoao* are less common than previously, but *Libocedrus* is consistently recorded. Restionaceae and *Leptospermum* dominate the bog flora initially, but from mid-zone *Gleichenia* becomes increasingly important. Charcoal concentration is low throughout.

Zone K1 is characterised by very high levels of *Dacrydium* (45–60 percent). *Prumnopitys taxifolia* pollen is abundant. Pollen grains of *Phyllocladus, Podocarpus, Prumnopitys ferruginea* and *Agathis australis* are significant. *Ascarina lucida* pollen becomes slightly less common than previously, and *Metrosideros* values also decline. *Halocarpus, Libocedrus* and *Manoao,* initially common, are scarce in upper samples. *Gleichenia* and Restionaceae dominate the bog flora. Microscopic charcoal fragments are scarce. In the uppermost sample, *Pteridium esculentum* spores and Poaceae pollen are abundant along with reduced tree pollen generally.

DISCUSSION

Prior to the LGM, a tall, complex, conifer–beech forest covered the region dominated by podocarp trees. The most abundant of these was *Dacrydium cupressinum. Fuscospora* (probably *Nothofagus truncata,* hard beech; see Newnham, 1992) was also an important element. The diversity of regional forest is apparent in the variety of conifer types recorded. All conifers which are present in the region today were represented. Compared to more recent spectra, warmth-loving, drought-intolerant elements such as *Agathis australis* and *Ascarina lucida* were less common. The high concentrations of charcoal fragments suggest fires were frequent. *Cyathea smithii*-type tree ferns were common. They are typical of present-day cool, moist upland areas in Northland. The occurrence of *Laurelia novae-zelandiae* and *Syzygium maire* is significant because they are often extremely underrepresented in pollen records. Both trees are typical of swamp forests and their presence here suggests the water table was high. Regional climate during this period is interpreted as being cool, moist and probably windy. Records from south of Auckland indicate similar deteriorating climates from ca. 28 ka (McGlone and Topping 1983; McGlone et al., 1984).

TABLE 1. Radiocarbon dating results.

Lab. No.	Depth (m)	Material dated	Age ^{14}C yr BP	$\delta^{13}C$‰	Cal. Age BP
NZA-4713	0.20–0.25	sieved peat	2639 ± 68	-25.51	2849–2401
NZA-4714	0.75–0.80	sieved peat	3505 ± 61	-23.90	3874–3568
NZA-6132	1.95–2.00	treated peat	5521 ± 75	-26.20	6462–6176
NZA-3817	2.95–3.00	treated peat	7660 ± 100	-23.78	8551–8163
NZA-5801	3.49–3.54	sieved peat	9357 ± 78	-27.50	10490–10055
NZA-5735	3.95–4.00	treated peat	10011 ± 94	-27.20	11867–10990
NZA-5734	4.49–4.54	treated peat	11030 ± 110	-26.10	13137–12674
NZA-5736	4.95–5.00	treated peat	11640 ± 130	-26.90	13878–13235
NZA-3818	5.49–5.54	treated peat	22410 ± 450	-27.68	out of range

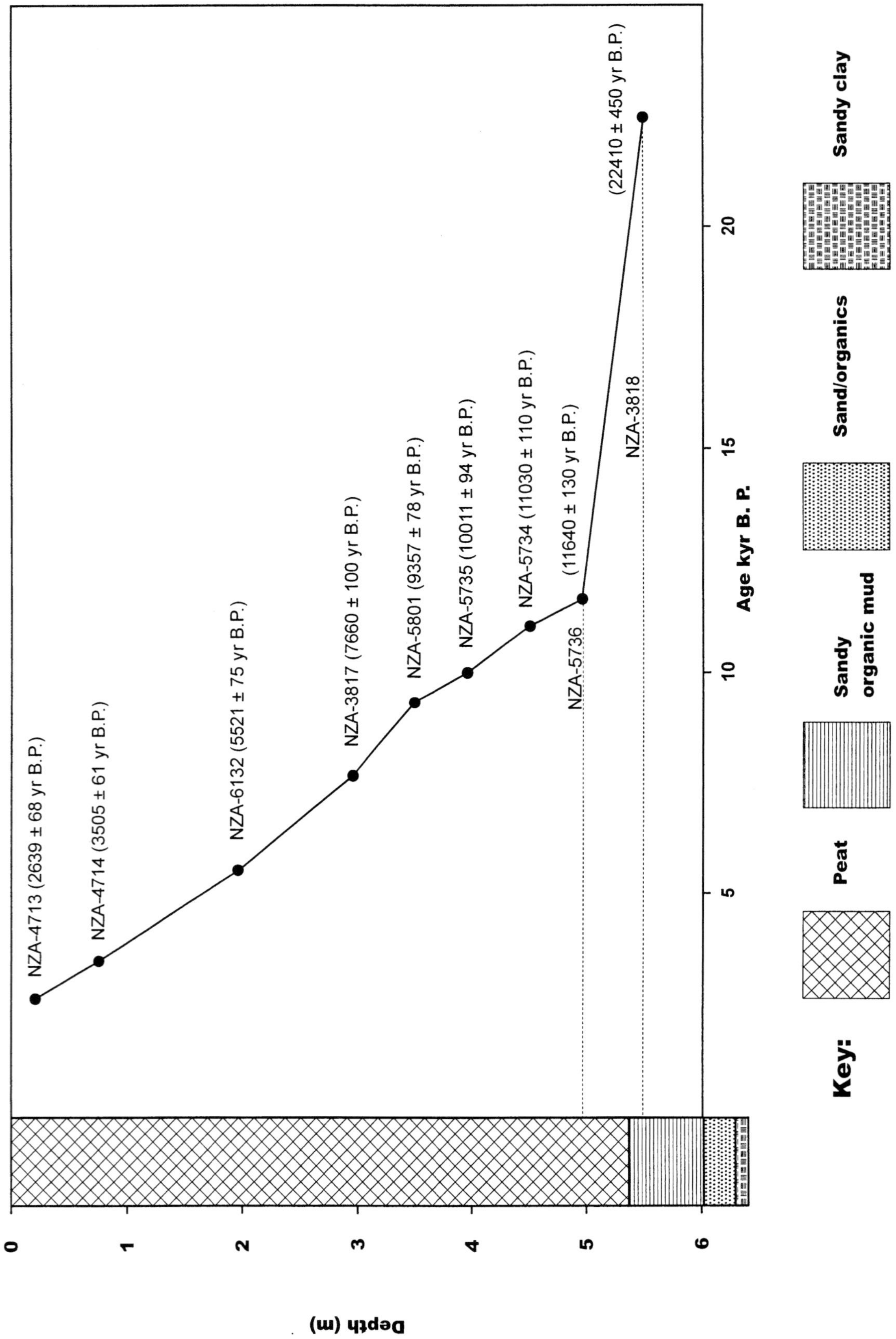

Text-Figure 2. Stratigraphy and age-depth curve of the pollen core based on linear interpolation of uncorrected radiocarbon dates.

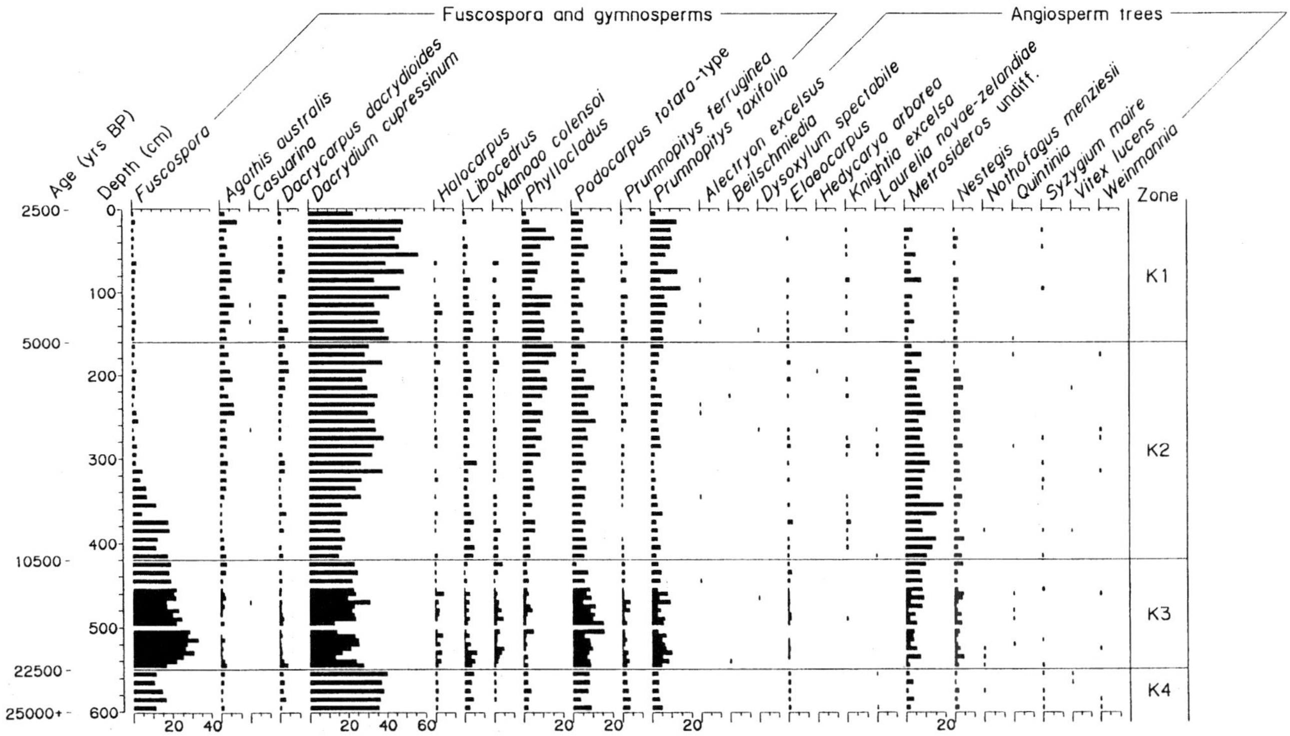

Text-Figure 3A. Percentage pollen diagram showing *Fuscospora*, gymnosperms and angiosperm trees. Pollen sum includes all dryland taxa.

The sharp changes which followed in regional vegetation characterise the LGM. *Fuscospora* dominated the forest flora. *Dacrydium cupressinum, Dacrycarpus dacrydioides* and *Libocedrus* became less common as did tree ferns. At the same time hardy podocarps—*Podocarpus, Prumnopitys taxifolia* and *Manoao colensoi*—increased their abundance. These trees are associated with cooler/lower insolation environments (Leathwick, 1995). *Manoao colensoi* is frost tolerant (Sakai and Wardle, 1978), and its increased presence here suggests cooler temperatures. In northern forests today it is found at frosty sites or on infertile swamp soils (Newnham, 1992). The scarcity of *Ascarina lucida* and low frequencies of *Agathis australis* provide further evidence that conditions were not only cooler, but also probably much drier and windier than present, with frequent fires. *Ascarina* is a small understorey tree intolerant of frost and drought (McGlone and Moar, 1977), and *Agathis* requires a warm humid climate with annual rainfall between 1000 and 2500 mm (Ecroyd, 1982). Climatic conditions appear to have been coolest and driest at about 5.25 m, which read from the age-depth curve (Text-Figure 2) correlates to ca. 18 ka. Dodson et al. (1988) do not record a similarly *Fuscospora*-dominant flora from their Cape Reinga site. Comparison with Newnham's (1992) site at Otakairangi is difficult owing to an uncertain chronology. *Fuscospora* forest is scarce in present-day Northland. Small stands of *Nothofagus truncata*, of at most 2–3 acres, exist in scattered locations in Northland (MacDonald, 1984; Wardle, 1984). The fossil record (Newnham, 1992; Kershaw and Strickland, 1988) suggests that *Nothofagus* in Northland may have been much more widespread in the past, but its occurrence and distribution are poorly defined.

The beginning of the Lateglacial is not easily defined owing to the slow sedimentation rate between 5.49–5.00m. However, the first occurrence of *Dodonaea viscosa* pollen at 5.10 m correlates with an interpolated age of ca. 14 ka. *Dodonaea* prefers warm sunny sites, and is confined largely to coastal locations. Its presence here implies a transition to more favourable climate. From ca. 11.3 ka this transition is more pronounced as *Ascarina lucida* becomes more common. The early Postglacial rise of *Ascarina* has been well documented in more southern records (McGlone and Moar, 1977; Newnham et al., 1989).

There are a number of features in the Postglacial section of the Kaitaia Bog record which compare with other Northland pollen sequences (MacDonald, 1984; Kershaw and Strickland, 1988; Elliot et al., 1995), as well as some more southern sites in the North Island (McGlone, 1988). All share a regional forest cover of diverse conifer–hardwood trees of which *Dacrydium cupressinum, Phyllocladus* and *Prumnopitys taxifolia* were the most common species. By 10 ka climate was warm and moist. Fires were still common but burning was much reduced. At Kaitaia, *Fuscospora* declined sharply after 9.5 ka and by ca. 8 ka was only a minor element in northern forests. *Dacrydium cupressinum* and *Metrosideros* were the most common emergent trees, and *Agathis australis* became more abundant from ca. 9 ka. *Phyllocladus* increased steadily throughout the Postglacial. This successional species is often associated with *A. australis* communities (Ecroyd, 1982). Reduced frequencies for *Prumnopitys* sp., *Manoao colensoi* and *Halocarpus* together with increased abundance of *Libocedrus* lend support for a moister, milder early to mid–Holocene climate. The period from ca. 10.5–7.6 ka is identified with greatest abundance of *Ascarina lucida* suggesting the early Holocene experienced the warmest, most equable climate over the last 25 ka. Speleothem records from the South Island suggest early Holocene temperatures were 1–2°C warmer than present (Hendy and Wilson, 1968). Pollen

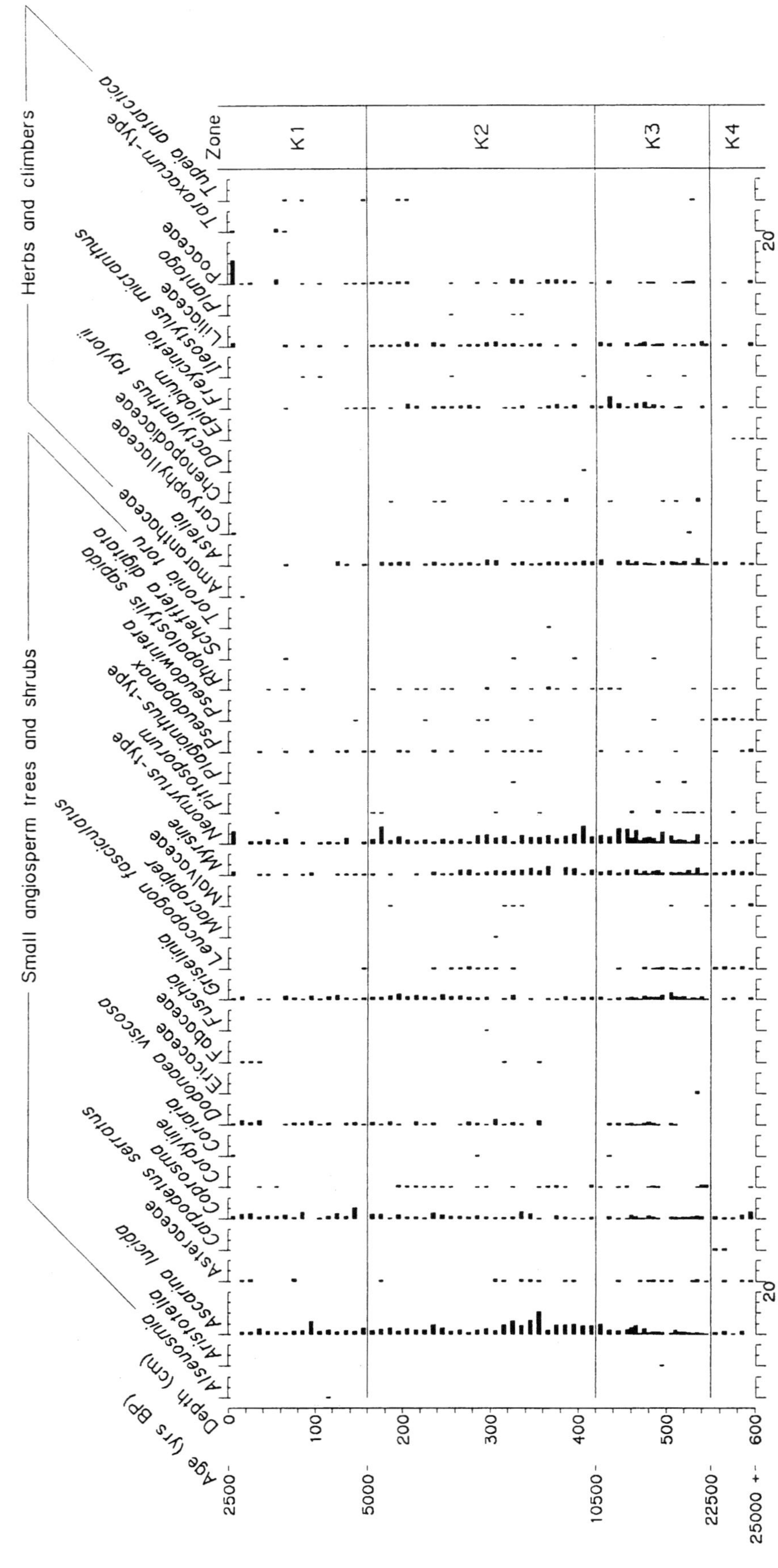

Text-Figure 3B. Percentage pollen diagram showing small angiosperm trees, shrubs, herbs and climbers.

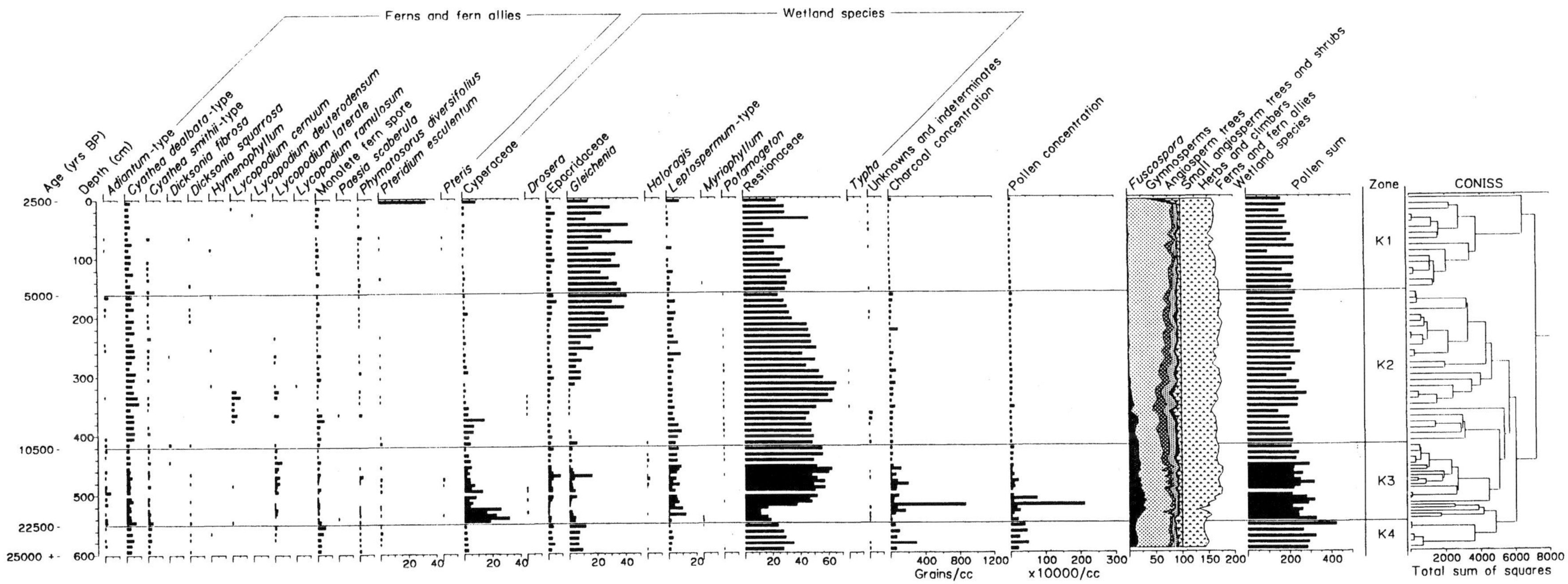

Text-Figure 3C. Percentage pollen diagram showing ferns, fern allies and wetland species. Pollen concentration (x10,000) and charcoal concentration are shown in grains cm^{-3}.

and plant macrofossil evidence from southern North Island records also indicates milder, less frost-prone climates at this time (McGlone et al., 1993). Postglacial expansion of *Agathis australis* is reported from elsewhere in Northland from ca. 8 ka (MacDonald, 1984; Newnham, 1992). Warm successional elements such as *Knightia excelsa* and *Phyllocladus* also became more common during the early Postglacial from ca. 7 ka at other northern North Island sites (McGlone, 1988; Newnham et al., 1989).

Hardy podocarps, especially *Prumnopitys taxifolia* and *Manoao colensoi* increased in the mid– to late Holocene, while *Metrosideros* sp., *Ascarina lucida* and *Libocedrus* were less abundant. This suggests a return to cooler/drier climate, possibly indicating a more frequent occurrence of summer drought. McGlone et al. (1993) have argued that drought was more common from ca. 6 ka, and more frequent invasion of cold polar air masses caused average temperatures to decline. Elliot et al. (1995) report cooler, drier climate from Lake Taumatawhana (Aupouri Peninsula) between ca. 5–3.4 ka.

The occurrence of high levels of *Pteridium esculentum* spores, grass pollen and charcoal fragments in the uppermost samples is typical of spectra associated with human disturbance (McGlone, 1983). These features are common in many late Holocene pollen diagrams (e.g. Elliot et al., 1995; Newnham et al., 1995) and have been attributed to Polynesian deforestation. The onset of major Polynesian land clearance in New Zealand is generally accepted as having occurred at ca. 700 yr. B. P. (McGlone et al., 1994). However, these samples are so close to the surface no conclusions can be made about their age.

CONCLUSIONS

The pollen zones and their associated palaeoenvironmental and palaeoclimatic inferences are summarised in Table 2. It is likely that forest cover has been continuously present at Kaitaia throughout the past 25 ka. *Fuscospora* was a dominant element of regional vegetation at the LGM in sharp contrast to the typical podocarp–hardwood forests of the Postglacial prior to human arrival. This also contrasts sharply with LGM vegetation cover in more southern regions where forest was very restricted in extent, and grassland–shrub communities were widespread (McGlone et al., 1993). Climate sensitive species were restricted during the LGM. Typical northern elements were always present. The availability of moisture may have been the most limiting factor for forest species in the far north. The LGM forests of the north share many similarities with those of the present day Nelson-Marlborough Sounds area (South Island). This commonality is expressed by a mixed podocarp–hardwood assemblage dominated by *Nothofagus truncata* and *Dacrydium cupressinum*, and suggests that LGM climate in far northern New Zealand may have been similar to that of the Nelson area today (i.e. annual rainfall ~1000 mm, mean annual temperature ~12.5°C). Kaitaia currently experiences an annual rainfall of ~1450 mm, and average mean annual temperature of ~15.9°C, which suggests LGM climates at Kaitaia may have been 3–3.5°C cooler with only 2/3 the annual rainfall.

Lateglacial climatic conditions underwent rapid transition from cool/dry to warm/moist environments expressed most clearly by the sharp rise in *Ascarina lucida*. *Fuscospora* was only a minor component of Postglacial northern forests. The expansion of *Agathis australis* from ca. 9 ka is consistent with other Northland records (Newnham, 1992). A warm, moist early Postglacial climate is in agreement with other parts of New Zealand (McGlone, 1988). From the mid–Holocene climate became slightly drier and cooler. McGlone et al. (1993) have argued for similar environments in more southern regions.

ACKNOWLEDGMENTS

I thank M. McGlone, V. Neall, P. Kershaw and G. Hope for their helpful advice on an earlier version of this paper. Special thanks to O. Davis and J. Ward for their comments in reviewing the submitted manuscript. Field research and five radiocarbon dates were funded by FRST grant FLE102. Other dates were funded by a MUGRF grant.

TABLE 2. Pollen zones showing palaeoenvironment and palaeoclimate interpretation.

Pollen Zone	Timespan	Dominant/indicator species	Palaeoenvironment	Palaeoclimate
K1	*ca* 2,500 to 5,000	*Dacrydium, Agathis, Prumnopitys taxifolia*	Kauri–podocarp–hardwood forest	Mild, summer drought
K2	*ca* 5,000 to 10,500	*Dacrydium, Agathis, Metrosideros, Ascarina*	Kauri–podocarp–hardwood forest	Warm, moist
	ca 10,500 to 14,000	*Ascarina, Dacrydium, Metrosideros, Podocarpus, Prumnopitys taxifolia, Nothofagus* cf. *truncata*	Podocarp–beech–hardwood forest	Cool-mild, moist, windy
K3	*ca* 14,000 to 22,500	*Nothofagus* cf. *truncata, Dacrydium, Podocarpus, Prumnopitys taxifolia*	Beech–podocarp–hardwood forest	Cold, dry, windy
K4	*ca* 22,500 to > 25,000	*Dacrydium, Nothofagus* cf. *truncata, Podocarpus, Prumnopitys taxifolia, P. ferruginea*	Conifer–beech–hardwood forest	Cool, moist, windy

References Cited

ALLEN, H.H.
1961 *Flora of New Zealand* Vol. I. Government Printer, Wellington, 1085 pgs.

BUSH, M.B., PIPERNO, D.R., COLINVAUX, P.A., DE OLIVEIRA, P.E., KRISSEK, L.A., MILLER, M.C., AND ROWE, W.L.
1992 A 14,300-year palaeoecological profile of a lowland tropical lake in Panama. *Ecological Monographs*, 62: 251–276.

CONNOR, H.E., AND EDGAR, E.
1987 Name changes in the indigenous New Zealand flora, 1960–1986 and Nomina Nova IV, 1983–1986. *New Zealand Journal of Botany*, 25: 115–170.

CRANWELL, L.M.
1953 An outline of New Zealand peat deposits. *Proceedings of the Seventh Pacific Science Congress of the Pacific Science Association*, held at Auckland and Christchurch 1949. Volume V, Botany, Government Printer, Wellington, pp. 186–208.

DICK, R.S.
1950 *The Plant Geography of the "Far North", New Zealand.* Unpubl. M.Sc. thesis, University of New Zealand, Auckland, 157 pgs.

DODSON, J.R., ENRIGHT, N.J., AND MCLEAN, R.F.
1988 A late Quaternary vegetation history for far northern New Zealand. *Journal of Biogeography*, 15: 647–656.

ECROYD, C.E.
1982 Biological flora of New Zealand 8. *Agathis australis* (D. Don) Lindl. (Araucariaceae) Kauri. *New Zealand Journal of Botany*, 20: 17–36.

ELLIOT, M.B., STRIEWSKI, B., FLENLEY, J.R., AND SUTTON, D.G.
1995 Palynological and sedimentological evidence for a radiocarbon chronology of environmental change and Polynesian deforestation from Lake Taumatawhana, Northland, New Zealand. *Radiocarbon*, 37: 899–916.

GRIMM, E.C.
1987 CONNISS: A FORTRAN 77 programme for stratigraphically constrained cluster analysis by the method of incremental sum of squares. *Computers and Geosciences*, 13: 13–35.

HAY, R.F.
1975 Sheet N7, Doubtless Bay (1st edition). *Geological Map of New Zealand 1:63 360*. DSIR, Wellington.

HENDY, C.H., AND WILSON, A.T.
1968 Palaeoclimatic data from speleothems. *Nature*, 219: 48–51.

KEAR, D., AND HAY, R.F.
1961 Sheet 1, North Cape (1st edition). *Geological Map of New Zealand 1:250, 000*. DSIR, Wellington.

KERSHAW, A.P., AND STRICKLAND, K.M.
1988 A Holocene pollen diagram from Northland, New Zealand. *New Zealand Journal of Botany*, 26: 145–152.

LEATHWICK, J.R.
1995 Climatic relationships of some New Zealand forest tree species. *Journal of Vegetation Science*, 6: 237–248.

MACDONALD, C.
1984 The ecology of *Nothofagus truncata* in northern New Zealand. Unpubl. MSc thesis, University of Auckland, Auckland, 128 pgs.

MCGLONE, M.S.
1983 Polynesian deforestation of New Zealand: a preliminary synthesis. *Archaeology in Oceania*, 18: 11–25.
1988 History of the New Zealand vegetation. *In*: Huntley, B., and Webb III, T. (eds.), *Handbook of Vegetation Science Volume 7: Vegetation History*. Kluwer Academic Publishers, Amsterdam, pp. 558–599.

MCGLONE, M.S., ANDERSON, A.J., AND HOLDAWAY, R.N.
1994 An ecological approach to the Polynesian settlement of New Zealand. *In*: Sutton, D.G. (ed.), *The Origins of the First New Zealanders*. Auckland University Press, Auckland, pp. 36–163.

MCGLONE, M.S., HOWORTH, R., AND PULLAR, W.A.
1984 Late Pleistocene stratigraphy, vegetation and climate of the Bay of Plenty and Gisborne regions, New Zealand. *New Zealand Journal of Geology and Geophysics*, 27: 327–350.

MCGLONE, M.S., MILDENHALL, D.C., AND POLE, M.S.
1996 History and paleoecology of New Zealand *Nothofagus* forests. *In*: Veblen, T.T., Hill, R.S., and Read, J. (eds.), *The Ecology and Biogeography of Nothofagus Forests*. Yale University Press, New Haven, pp. 83–130.

MCGLONE, M.S., AND MOAR, N.T.
1977 The *Ascarina* decline and post-glacial climatic change in New Zealand. *New Zealand Journal of Botany*, 15: 485–489.

MCGLONE, M.S., SALINGER, M.J., AND MOAR, N.T.
1993 Paleovegetation studies of New Zealand's climate since the Last Glacial Maximum. *In*: Wright, J.H., Kutzbach, E.E., Webb III, T., Ruddiman, W.F., Street-Perrott, F.A., and Bartlein, P. J. (eds.), *Global Climates Since the Last Glacial Maximum*. University of Minnesota Press, Minneapolis, pp. 294–317.

MCGLONE, M.S., AND TOPPING, W.W.
1983 Late Quaternary vegetation, Tongariro region, central North Island, New Zealand. *New Zealand Journal of Botany*, 21: 53–76.

MOIR, R.W., COLLEN, B., AND THOMPSON, C.S.
1986 The climate and weather of Northland. *New Zealand Meteorological Service Miscellaneous Publication*, 115, 37 pgs.

MOLLOY, B.P.J.
1995 *Manoao* (Podocarpaceae), a new monotypic conifer genus endemic to New Zealand. *New Zealand Journal of Botany*, 33: 183–201.

MOORE, P.D., WEBB, J.A., AND COLLINSON, M.E.
1991 *Pollen Analysis* (2nd edition). Blackwell Scientific, Oxford, 216 pgs.

NEWNHAM, R.M.
1992 A 30,000 year pollen, vegetation and climatic record from Otakairangi (Hikurangi), Northland, New Zealand. *Journal of Biogeography* 19: 541–554.

NEWNHAM, R.M., DE LANGE, P.J., AND LOWE, D.J.
1995 Holocene vegetation, climate and history of a raised bog complex, northern New Zealand based on palynology, plant macrofossils and tephrochronology. *The Holocene*, 5: 267–282.

NEWNHAM, R.M., LOWE, D.J., AND GREEN, J.D.
1989 Palynology, vegetation and climate of the Waikato lowlands, North Island, New Zealand, since c. 18,000 years ago. *Journal of the Royal Society of New Zealand*, 19: 127–150.

NEWNHAM, R.M., OGDEN, J., AND MILDENHALL, D.
1993 A vegetation history of the far north of New Zealand during the late Otira (last) glaciation. *Quaternary Research*, 39: 361–372.

SAKAI, A., AND WARDLE, P.
1978 Freezing resistance of New Zealand trees and shrubs. *New Zealand Journal of Ecology*, 1: 51–61.

WARDLE, J.A.
1984 *The New Zealand Beeches: Ecology, Utilisation and Management*. New Zealand Forest Service, Christchurch, 447 pgs.

WILSON, D.M.
1921 Ancient drains Kaitaia Swamp. *Journal of the Polynesian Society*, 30: 184–188.

Author's address

MIKE ELLIOT
National Institute of Water
and Atmospheric Research
P.O. Box 14901
Kilbirnie, Wellington
New Zealand

POLLEN ANALYSIS OF AN 800M CORE SAMPLE FROM LAKE BIWA, JAPAN

T. Fujiki, N. Miyoshi, Y. Morita, S. Horie, and K. Takemura

Abstract

Sediment cores were collected in Lake Biwa in 1982 and 1983. The upper layer from a depth of approximately 900 m consisted of unconsolidated sediments. These were divided into five layers (T, S, R, Q, P). In the present investigation, pollen analysis of layers S, R and Q (250–800 m) was performed.

On the basis of the appearance of fossil pollen grains, interglacial periods were considered to be represented by layers containing fossil pollen from the evergreen broadleaf trees, *Cyclobalanopsis* and *Castanopsis*, and a few fossil pollen grains of *Lagerstroemia*. Glacial periods were considered to be represented by layers in which fossil pollen grains of Pinaceae conifers, *Abies, Tsuga, Picea,* and *Pinus,* were dominant. Consequently, these sediments may include up to eight glacial and eight interglacial periods. Therefore, the Lake Biwa 800-m core sample may contain a record of thirteen repeated warm and cold periods. Short-term interglacial periods separated by long glacial periods occurred in cycles during the Pleistocene. The present pollen analysis involved the use of only single samples at intervals of 5 m. If more detailed analysis could be done, more pollen zones might be delineated.

INTRODUCTION

Lake Biwa was formed in the Iga Basin about 5 million years ago. It then extended north to its present position about 3 million years ago. Boring to recover sediment cores was carried out in Lake Biwa (35°13'06" N, 136°00'49" E) between 1982 and 1983 by Horie (1984) (Text-Figure 1). The upper layer from a depth of approximately 900 m consisted of unconsolidated sediments, and these were divided into five layers (T, S, R, Q, P). General pollen analysis had previously been carried out on the same five layers by Fuji (1988). Among the pollen types, five elements were distinguished and 37 pollen zones identified. Detailed pollen analysis of layer T and an upper section of layer S (17–250 m) was done by Miyoshi (1994). The layers containing *Cyclobalanopsis, Castanopsis* and *Lagerstroemia* are regarded as interglacial periods. Miyoshi (1994) showed that five glacial and five interglacial periods occurred over a period of about 500,000 years. The interglacial periods consist of two types, one in which evergreen broadleaf trees are dominant and *Lagerstroemia* never appears, such as the present time and BW-9, and one in which a few *Lagerstroemia* appear and evergreen broadleaf trees are not dominant, as in other pollen zones. In the present investigation, pollen analysis of the remainder of layer S as well as layers R and Q (250–800 m) was performed.

MATERIALS AND METHODS

In the core, from the surface to a depth of 250 m (layer T), a layer of clay developed. Many layers include sand or gravel below 250 m. There is a layer of alternating sand and silt with gravel at a depth of 250–580 m (layer S), and beneath there is a layer of clay and sand at 580–730 m (layer R). At 730–800 m, there is a layer of gravel, sand and silt. The samples for pollen analysis were picked up at intervals of 3–5 m. Fission track dating was performed at eight points in all layers (Takemura, 1990).

Fossil pollen and spores were extracted from samples by boiling in 10% KOH solution and applying mineral separation with $ZnCl_2$ (sp gr = 1.78). The samples were subsequently treated with 40% HF and then subjected to Erdtman's acetolysis method. The treated materials were mounted on slides in glycerin jelly. Samples for scanning electron microscope were fixed with 2% osmium tetroxide after acetolysis. They were then dehydrated with 99.5% ethanol, and substituted with xylene. The samples were placed on sample stubs, sputter-coated with gold-palladium for 5 minutes, and observed using a high-resolution scanning electron microscope (JSM-890).

Fujiki, T., Miyoshi N., Morita, Y., Horie, S., and Takemura, K.
2001 Pollen analysis of an 800 m core sample from Lake Biwa, Japan; *In*: Goodman, D.K., and Clarke, R.T. (eds.), *Proceedings of the IX International Palynological Congress, Houston, Texas, U.S.A., 1996*; American Association of Stratigraphic Palynologists Foundation, p. 367–373.
 ISBN 0–931871–06–9

RESULTS

In this analysis, 76 kinds of fossil pollen grains and spores were identified. These were divided into AP (arboreal pollen), NAP (non-arboreal pollen) and FS (fern spores), and are listed as follows:

AP: *Podocarpus, Abies, Tsuga, Picea, Pinus diploxylon* type, *Pinus haploxylon* type, *Sciadopitys, Cryptomeria,* Cupressaceae, *Ephedra, Platycarya, Juglans,* and *Pterocarya, Carya, Carpinus, Betula, Fagus, Lepidobalanus, Castanea, Ulmus, Zelkova, Celtis* and *Aphananthe, Euptelea, Acer, Aesculus, Tilia, Corunus, Fraxinus, Styrax, Sapium, Lagerstroemia, Liquidamber, Myrica, Cyclobalanopsis, Castanopsis, Symplocos.*

NAP: *Salix, Corylus, Alnus, Buxus, Ilex, Elaeagnus, Ligustrum, Clethra,* Ericaceae, *Weigera, Typha,* Gramineae, Cyperaceae, *Eriocaulon,* Liliaceae, *Persicaria,* and other Polygonaceae, Chenopodiaceae, Caryophylaceae, *Thalictrum* and other Ranunculaceae, *Sanguisorba, Geranium, Impatiens, Viola, Haloragis,* Umbelliferae, *Menyanthes* and other Gentianaceae, *Lonicera,* Labiatae, *Scabiosa, Artemisia* and other Compositae.

FS: monolete type, trilete type, Polypodiaceae, Lycopodiaceae, Osmundaceae.

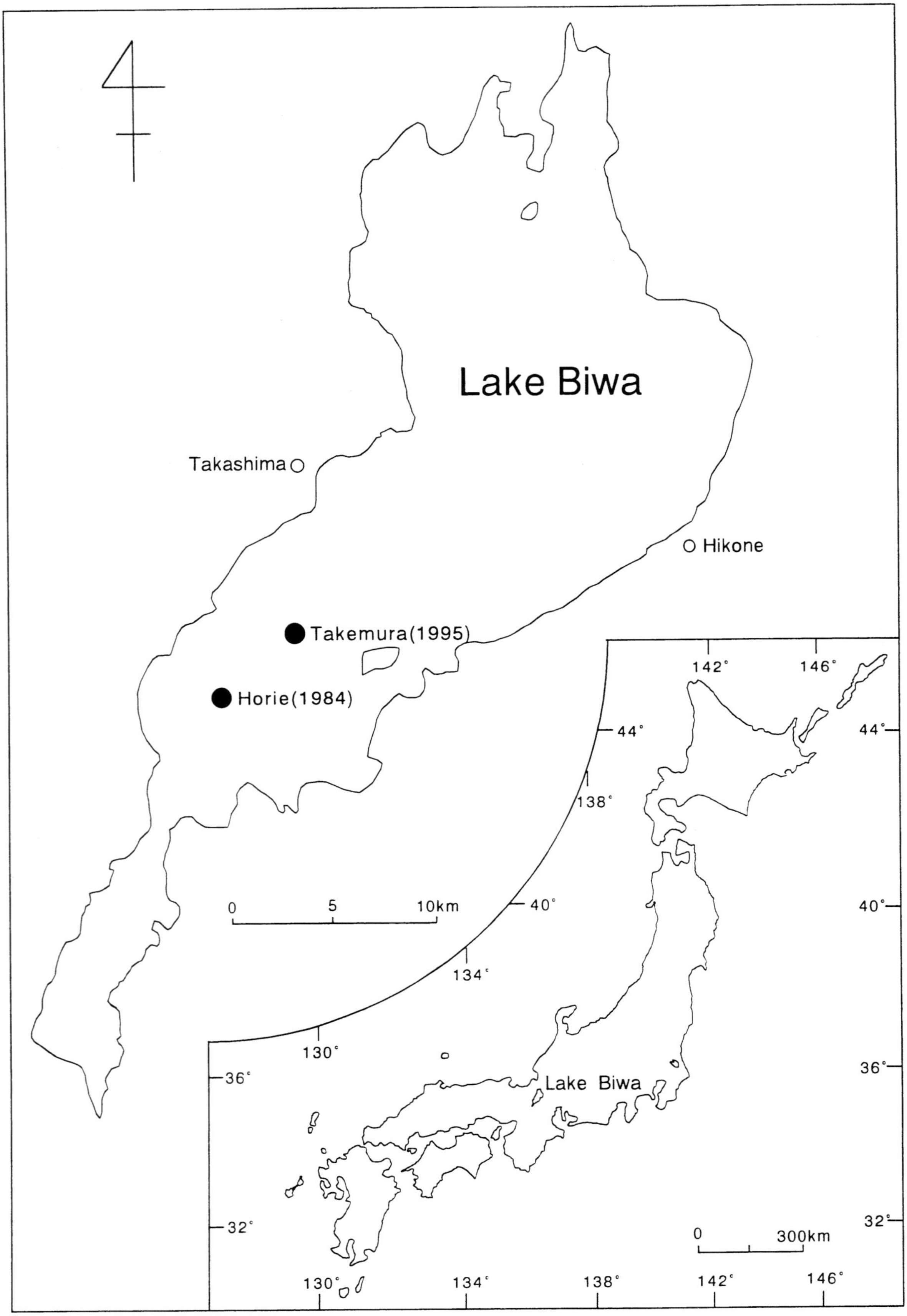

Text-Figure 1. Map showing the position of the core.

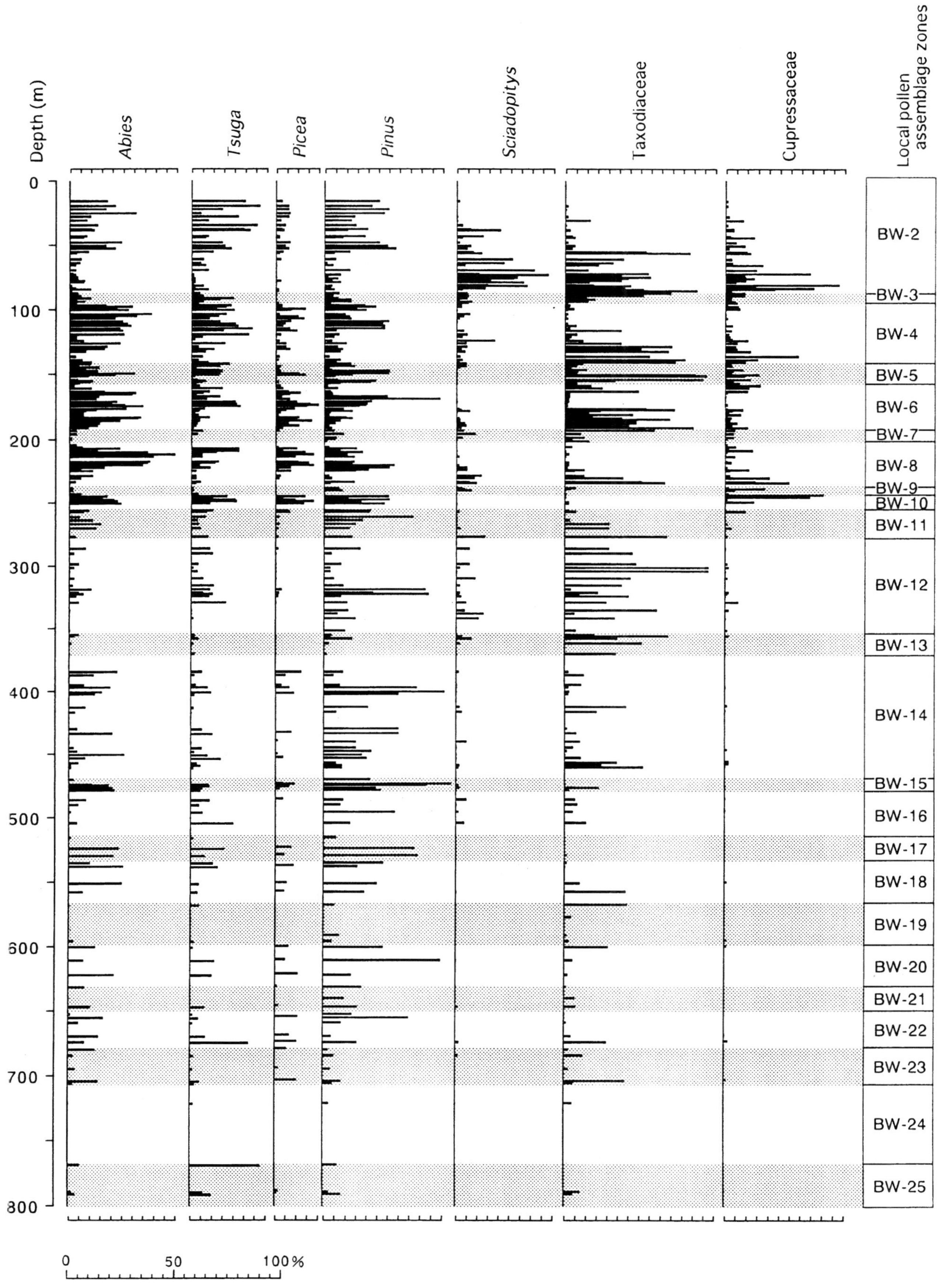

Text-Figure 2. Stratigraphic sequence and pollen diagram of gymnosperms from Lake Biwa.

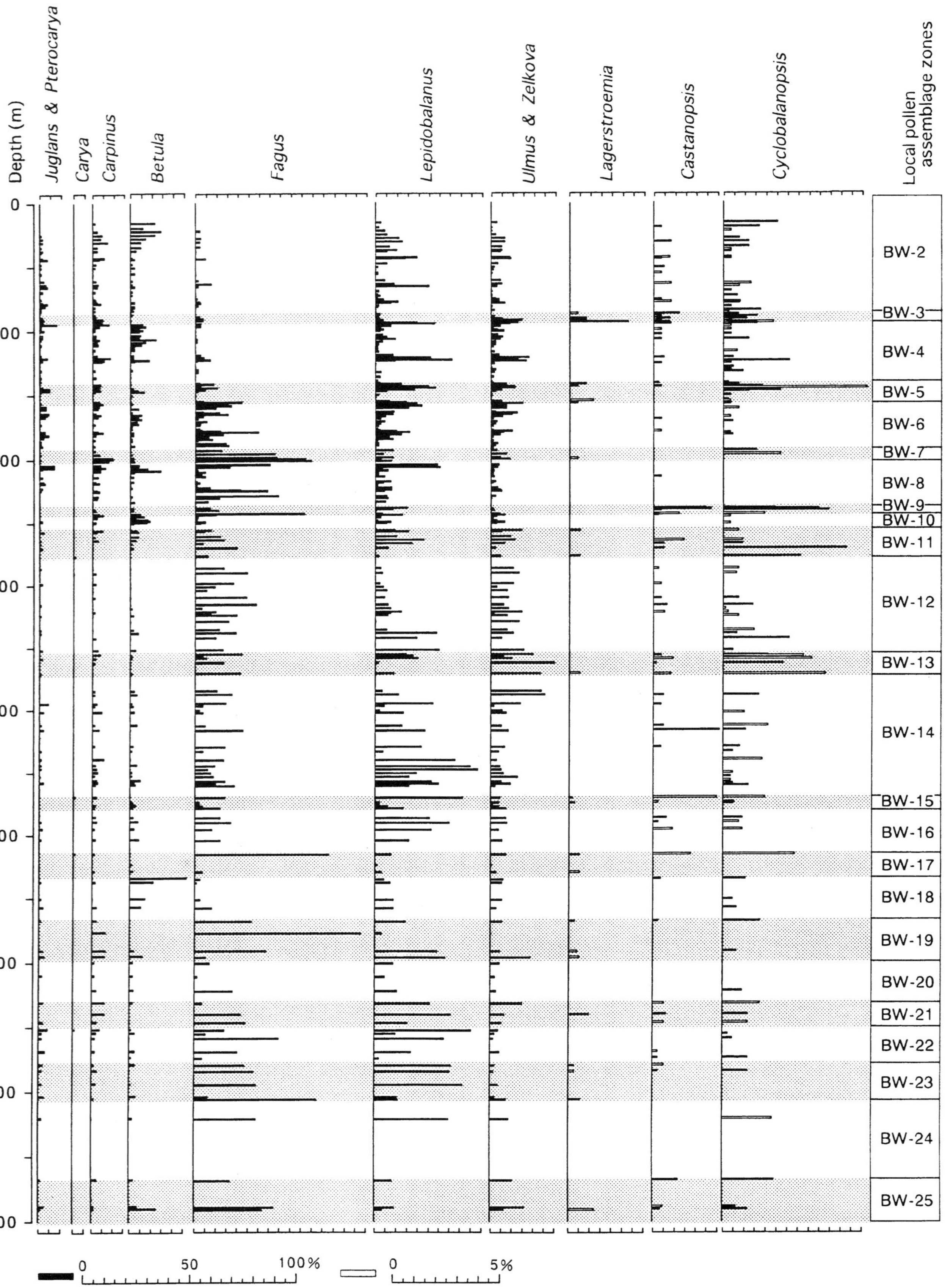

Text-Figure 3. Stratigraphic sequence and pollen diagram of deciduous and evergreen broadleaf trees from Lake Biwa.

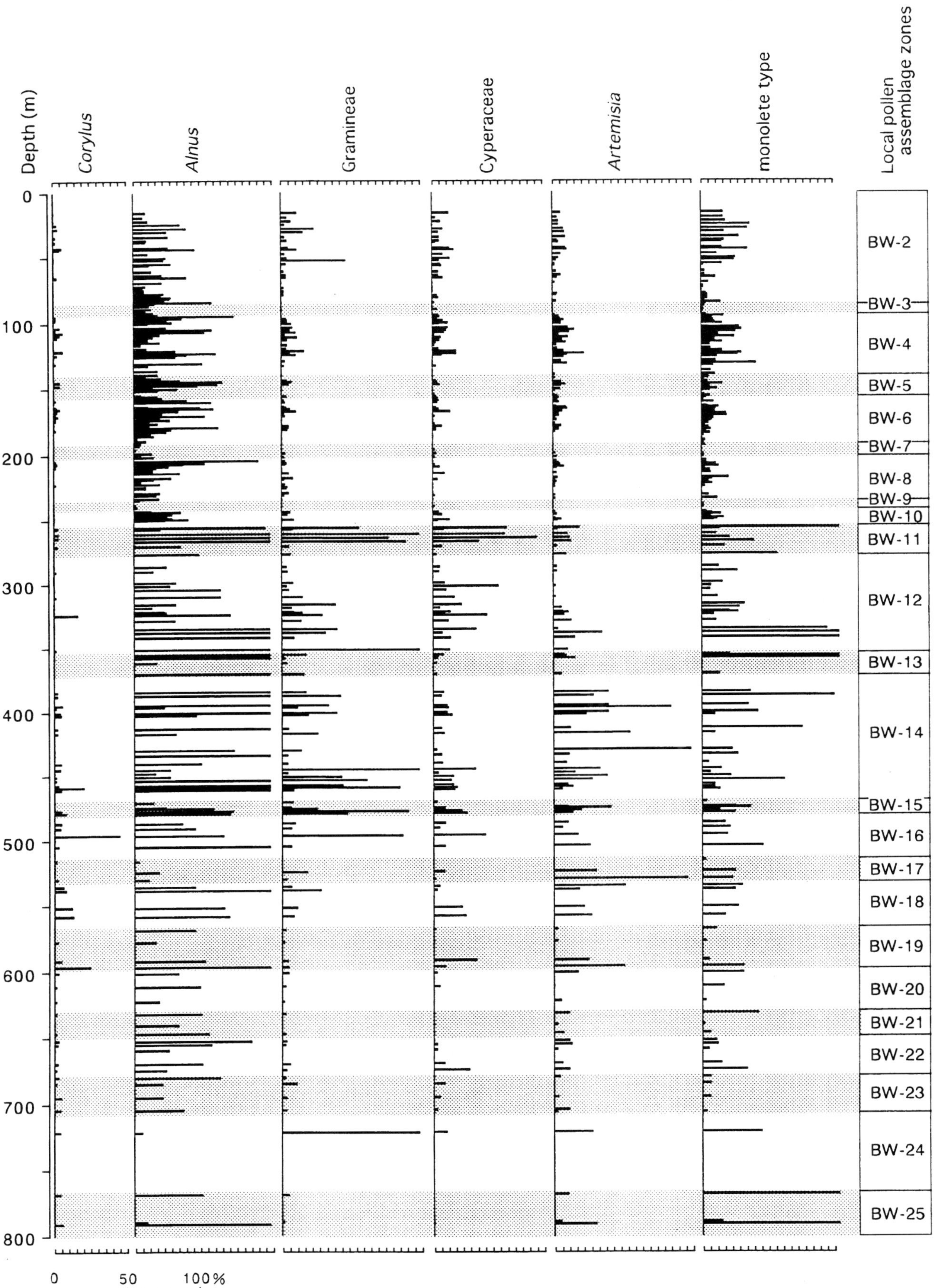

Text-Figure 4. Stratigraphic sequence and pollen diagram of shrubs and NAP from Lake Biwa.

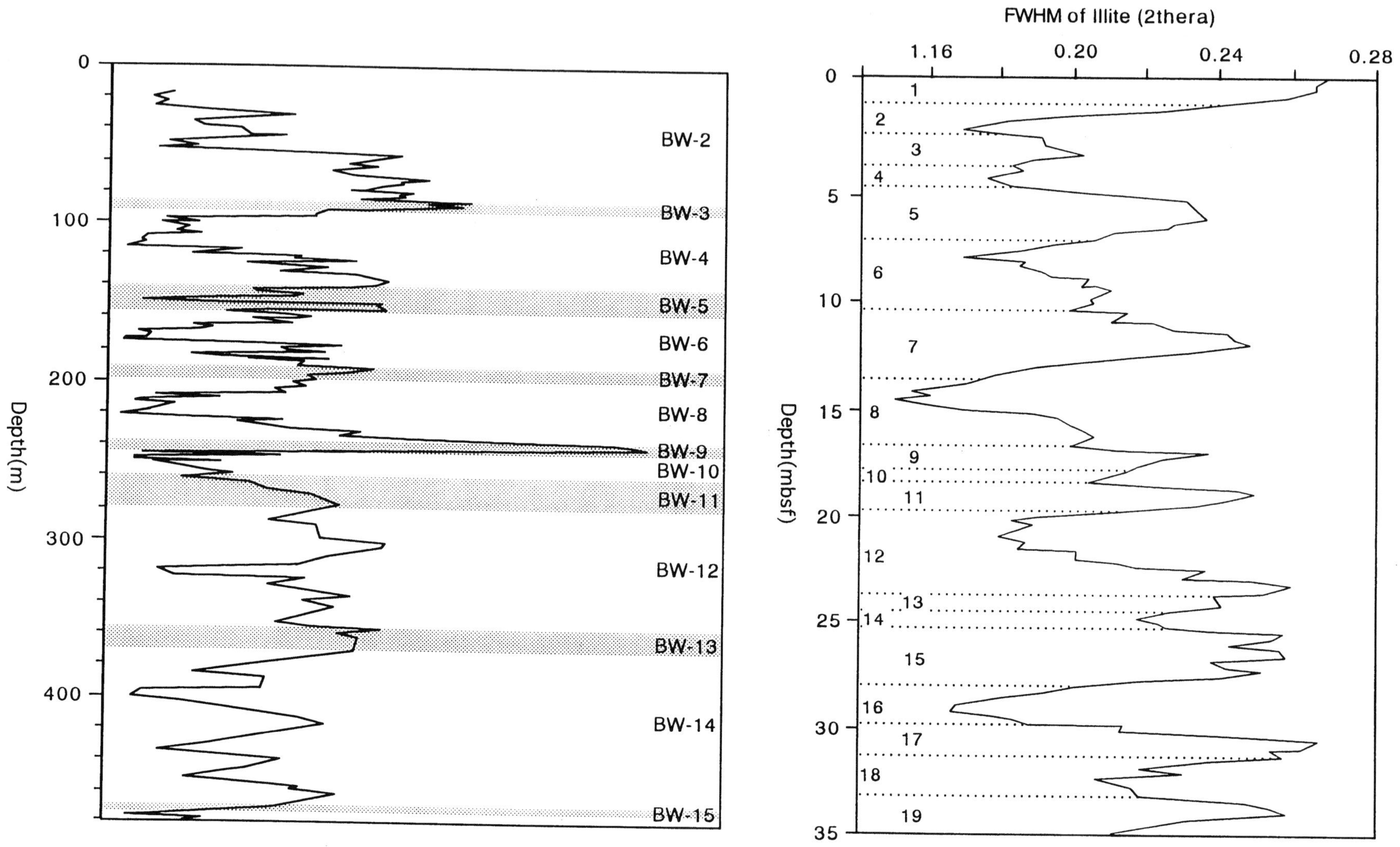

Text-Figure 5. The illite crystal degree curve derived from marine sediments taken from Japan Sea (Fukusawa and Koizumi, 1994) and the paleotemperature curve constructed from the percentages of arboreal fossil pollen grains.

Increases and decreases of principal pollen types were as follows. Among gymnosperms, Pinaceae conifers were dominant in all layers. Taxodiaceae were dominant in the upper layer (Text-Figure 2). Among deciduous broad-leaved trees, *Lepidobalanus* and *Fagus* were dominant in all layers. *Lagerstroemia* was identified in 15 samples, but at a very low percentage abundance. Among evergreen broadleaf trees, *Cyclobalanopsis* and *Castanopsis* appeared in very low percentages, but increased slightly during interglacial periods (Text-Figure 3).

Among shrubs and NAP, *Alnus* was dominant in all layers. Herbs, Gramineae, Cyperaceae and *Artemisia* were dominant in the middle layer. Pteridophyta spores were dominant in all layers (Text-Figure 4).

DISCUSSION

According to studies conducted in Lake Biwa (Miyoshi, 1994) and the Tokusa Basin (Miyoshi, 1991), *Lagerstroemia* appears in interglacial periods. In this study also, layers containing *Cyclobalanopsis, Castanopsis* and *Lagerstroemia* were considered to represent interglacial periods. Thus BW-11, 13, 15, 17, 19 21, 23 and 25 (odd numbers) are interglacial periods and BW-10, 12, 14, 16, 18, 20, 22 and 24 (even numbers) are glacial periods. Therefore, this sediment may contain eight interglacial periods and eight glacial periods. Below at depth of 500 m, glacial periods become short and interglacial periods become long. In this study, interglacial periods, such as BW-9 were not identified, and only interglacial periods such as BW-3, 5 and 7, were evident. This type of interglacial period may have had a lower temperature than the Jomon transgression in the Holocene, since the percentage of evergreen broadleaf trees was very low. However, the climate may have been mild, since *Lagerstroemia* was able to survive the winter. It is presumed that climate during this interglacial may have been different from that of the present because evergreen broadleaf trees were not dominant. It has been reported that fossil pollen grains of *Lagerstroemia* appear with those of *Fagus*. Therefore, *Fagus* must have flourished in mountainous districts and the climate may have been wet and mild. In the glacial periods, Taxodiaceae, *Lepidobalanus* and *Fagus* were dominant at the beginning, and Pinaceae conifers were dominant at the end. Taxodiaceae and *Fagus* grow in a cool, wet climate, whereas Pinaceae conifer grow in a cold, dry climate. Therefore, the beginning of the glacial periods may have been rainy and cool, and then became rather drier and colder. From Miyoshi (1994) and this study, the 800 m core appears to include up to twelve glacial periods and thirteen interglacials .

The paleotemperature curve constructed from the percentages of arboreal fossil pollen grains is high on the right and low on left. In the layers considered to represent interglacials, the temperature is high and these mostly correspond with each other (Text-Figure 5). The illite crystal degree curve derived from marine sediments taken from the Japan Sea (Fukusawa and Koizumi, 1994) shows that ten interglacial periods have occurred over about 700,000 years. This figure agrees with the oxygen isotope ratio curve from Emiliani (1978). However, in this pollen analysis, only eight interglacial periods were recognized. It is possible that new interglacial periods may be found in BW-12 and BW-14.

References Cited

EMILIANI, C.
1978 The cause of the ice ages. Earth and Planetary Science Letters, 37: 349–352.

FUJI, N.
1988 Palaeovegetation and palaeoclimate changes around Lake Biwa, Japan during the ca. 3 million years. *Quaternary Science Review,* 7: 21–28.

FUKUSAWA, H., and KOIZUMI, I.
1994 The climate change from the sediment of the depths of Japan Sea on East Asia. *Earth,* 16: 678–684.

HORIE, S.
1984 Lake Biwa. *Monographiae biologiae,* 54. W. Junk Publishers, Dordrecht, 654 pgs.

MIYOSHI, N.
1991 Polenanalytische Untersuchungen an einem 85-Meter-Bohrkern aus dem Tokusa-Becken, Prafektur Yamaguchi, West-Japan. *In:* Horie, S. (ed.), *Die Geschichte des Biwa ,* pp. 243–256.
1994 Vegetation and climate from pollen analysis of Lake Biwa. *Kagaku Asaahi,* 11: 24–26.

TAKEMURA, K.
1989 Tectonic and climatic record of the Lake Biwa, Japan, region provided by the sediments deposited since Pliocene times. *Palaeogeography, Palaeoclimatology, Palaeoecology,* 78: 185–193.

Authors' addresses

T. FUJIKI
International Research Center for Japanese Studies
Oeyama-cho 3-2, Goryo, Nishikyo-ku
Kyoto 610-1192
Japan

N. MIYOSHI
Y. MORITA
Biological Laboratory
Okayama University of Science
Ridai-cho 1-1
Okayama 700-0005
Japan

S. HORIE
K. TAKEMURA
Kyoto University
Kitashirakawa Oiwake-cho, Sakyou-ku
Kyoto 606-8224
Japan

COMPOSITION OF DINOFLAGELLATE CYST ASSEMBLAGES IN SEDIMENTS DEPOSITED OFF THE RHÔNE DELTA DURING THE LAST GLACIAL- INTERGLACIAL CYCLE

Marie-Therese Morzadec-Kerfourn

Abstract

On the continental slope and the basin off the Rhône Delta, the sedimentary succession is composed of grey muds, dated as belonging to the last glacial phase, which are separated by a hiatus from the overlying beige calcareous mud of Holocene age. As with the sediments, the estuarine dinoflagellate cysts and the reworked forms were for the most part stored on the continental shelf during the Holocene. In sediments of the deep sea fan, the maximum abundance of reworked forms is coeval with the climatic warming that occured at the end of the isotope stage 3. The input of estuarine cysts ceased at the end of this warm phase. This change coincides with the accentuation of the cold climatic conditions that took place during the isotope stage 2. The time of maximum cooling does not correspond to the maximum transport rate for reworked and redeposed microfossil material because the cold phase may have been accompanied by a decrease in humidity.

INTRODUCTION

Dinoflagellate cyst assemblages in Quaternary pelagic sediments are made up of cysts derived from plankton floating in the surface water layer, neritic to estuarine cysts displaced by bottom currents, and fossil cysts reworked following the erosion of older terrains (Wall et al., 1977; Morzadec-Kerfourn, 1992b). The relative distribution of these three groups of cysts in sediments is a function of the depositional environment (i.e. the morphology of both the continental shelf and the slope), the climatic variations associated with sea-level changes, and the intensity of erosion.

Data concerning sedimentation off the Rhône Delta are sufficiently numerous (Bellaiche et al., 1984; Mear, 1984; Droz, 1983, 1991; Monaco et al., 1984) to allow an interpretation of the whole set of dinoflagellate cysts encountered in pelagic sediments deposited during the last interglacial to glacial cycle at the foot of the continental slope (Morzadec-Kerfourn, 1984).

SEDIMENTARY DOMAINS OFF THE RHONE DELTA

Offshore from the Rhône Delta, the continental shelf extends outwards beyond the 100 m isobath. This shelf constitutes a zone for the accumulation of Holocene terrigenous sediments (Monaco et al., 1984); most of the fluvial input from the Rhône is currently deposited in the subaerial delta.

The shelf edge is situated at the 120 m isobath, a contour which corresponds to the maximum regression of the last glacial epoch. Beneath 200 m, the slope becomes steeper and leads down towards the ocean basin. An extensive network of submarine canyons (or channels) cut into the continental rise and slope. From west to east, these include the Petit-Rhône (PR), Grand-Rhône (GR), Marseilles (M) and Planier (P) canyons (Text-Figure 1). This domain is strongly influenced by erosion, and the thickness of Holocene deposits is only of the order of tens of cm (Monaco et al., 1984; Droz, 1991). Farther east, towards the Planier canyon, Pliocene sediments occur near the surface beneath a few tens of cm of Holocene sediments (Fernex et al., 1984).

The submarine delta fan of the Rhône is built up on the Balearic–Provencal basin, between the 1800 and 2500 m isobaths (Bellaiche et al., 1984), in relation to sedimentation during glacial phases. This fan is located at the outlet of the Petit-Rhône canyon, and to the east, a ridge is built at the outlet of the Grand-Rhône canyon (Text-Figure 1) (Droz, 1991). The sedimentary succession is composed of grey muds and silts, dated as belonging to the last glacial phase, separated by a hiatus or by an oxidized horizon from the overlying beige calcareous muds of Holocene age—which are of variable thickness according to topographic location. This succession corresponds to the last glacial-interglacial cycle, i.e. from the end of isotope stage 3 up to isotope stage 1. The numerous available radiocarbon dating results make it possible to bracket the age of the hiatus between 20,000 and 11,000 years BP. Such a hiatus could be related to the eastward migration of the mouth of the Rhône (Mear, 1984, Monaco et al., 1984; Droz, 1991).

DINOFLAGELLATE CYSTS IN SURFACE SEDIMENTS

The spatial distribution of dinoflagellate cysts in recent sediments was studied using samples collected by grab bucket (BF) or taken from the tops of cores (MKS = ECOMED program; KSPF = Profans program) (Text-Figure 1). Although all these samples are Holocene, they are not strictly coeval among themselves. Sample treatment was carried out according to the conventional method (50% HCl, 70% HF) at the Palynology Laboratory of the Compagnie Française de Pétroles Total. Cysts of quaternary het-

Morzadec-Kerfourn, Marie-Therese
2001 Composition of dinoflagellate cyst assemblages in sediments deposited off the Rhône Delta during the last glacial–interglacial cycle; *In*: Goodman, D.K., and Clarke, R.T. (eds.), *Proceedings of the IX International Palynological Congress, Houston, Texas, U.S.A., 1996*; American Association of Stratigraphic Palynologists Foundation, p. 375–383.
 ISBN 0–931871–06–9

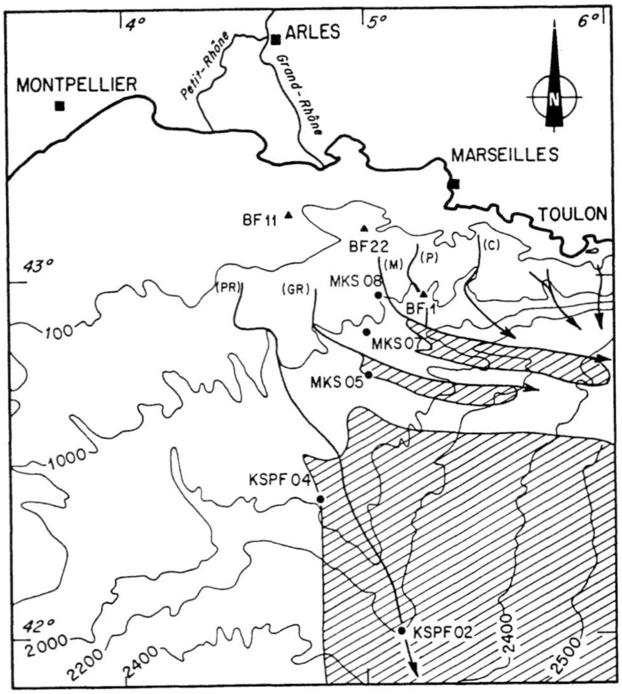

Text-Figure 1. The continental slope and the basin off the Rhône Delta (France). Bathymetry and location of studied samples and cores. (PR), Petit-Rhône canyon; (GR), Grand-Rhône canyon; (M), Marseilles canyon; (P), Pannier canyon; (C), Cassis canyon. Diagonal shading, submarine delta fan and ridges. Samples from grab buckets, BF ; samples from top of cores, MKS = ECOMED program; KSPF= Profans program.

erotrophic Protoperidinoids are rare and not taken into account for the interpretation of the paleoenvironment.

On the continental shelf, the gonyaulacoid cyst assemblages are dominated by *Lingulodinium machaerophorum* (Deflandre & Cookson 1955) Wall 1967 and *Spiniferites bentorii* (Rossignol 1964) Wall & Dale 1970. These forms are considered as typical of cysts from estuarine to neritic environments (Wall et al., 1977), being derived from dinoflagellates that are able to adapt to variations in salinity. The most abundant species, *L. machaerophorum*, makes up a total of 72% of all cysts in the shelf sediments from -112 m (BF 22). However, its abundance falls off rapidly at greater depth on the continental slope (Text-Figure 2a). A similar trend is followed by the reworked forms, whose relative abundance decreases from 13% at -92.5 m (BF 11) to 1.3% at -2,300 m (KSPF 02). These cysts are mostly Pliocene in age, being derived from Plaisancian deposits present around the delta of the Rhône near Arles and Montpellier. As with the sediments, the estuarine dinoflagellate cysts and the reworked forms were for the most part stored on the continental shelf during the Holocene.

On the continental slope and within the basin, the gonyaulacoid cyst assemblages are formed of neritic to oceanic species, containing predominant *Impagidinium aculeatum* (Wall 1967) Lentin & Williams 1981 as well as abundant *Impagidinium patulum* (Wall 1967) Stover & Evitt 1978 and *Nematosphaeropsis labyrinthea* (Ostenfeld 1903) Reid 1974 (Text-Figure 2b). Estuarine dinoflagellate cysts (*L. machaerophorum* and *S. bentorii*) are absent, while reworked forms are present only in core KSPF 04 at the foot of the continental slope.

DATING OF SEDIMENT CORES AND DINOFLAGELLATE CYSTS IN RELATION TO CLIMATIC VARIATIONS

The five cores studied here come from the continental slope and the basin. Two cores were taken from the Rhône fan (Profans program), at the outlet of the Petit-Rhône canyon: KSPF 04 at -1850 m in its upstream part and KSPF 02 at -2300 m in its downstream part. Three cores come from the continental slope (ECOMED program): MKS 05 at -1590 m from a ridge situated on the SSW side of the Grand-Rhône canyon, MKS 07 at -1195 m from the axis of a secondary canyon situated between the Grand-Rhône and Marseilles canyons, and MKS 08 at -1155 m from the axis of the Marseilles canyon (Text-Figure 3).

C^{14} radiometric dating was carried out at the Radiocarbon Laboratory of the University of Lyons (Ly) on samples from different core intervals, providing results that establish a chronological framework. However, a slight overestimation of the depositional ages may be envisaged due to the reworking of pre-existing sediments (J. Evin, communication). The dating results indicate that the sedimentary hiatus recognized throughout this area between the grey muds of the last glaciation and the Holocene beige muds (Mear, 1984) is represented in the Rhône fan by a gap from 23,000 to 11,000 years BP in core KSPF 02, and from 21,000 to 11,000 years BP in core KSPF 04 (cf. Text-Figures 4, 5). In core MKS 07 from the continental slope, the grey muds correspond to the end of the last glacial epoch between ca. 30,000 and 14,000 years BP (Text-Figure 7). No dating work has been carried out on the sediments of cores MKS 05 and MKS 08. Indications of climatic trends are provided by the composition of gonyaulacoid dinoflagellate cyst assemblages of oceanic origin, in particular the relative abundance of temperate–cold species (*Operculodinium centrocarpum* (Deflandre & Cookson 1955) Wall 1967, *N. labyrinthea)* as compared with temperate–warm species (*Impagidinium strialatum* (Wall 1967) Stover & Evitt 1978, *I. aculeatum, I. patulum*). These data also make it possible to establish a relative chronology (Morzadec-Kerfourn, 1984, 1992a). The climatic pattern of the last glacial–interglacial cycle is clearly reflected in core KSPF 04 (Text-Figure 5). Here, the cold climatic fluctuation which corresponds to isotope stage 2, is characterized, at around 21,000 years BP, by an assemblage with *O. centrocarpum* and *N. labyrinthea* accounting for more than 60% of the total Quaternary gonyaulacoid cysts. The Holocene warming of isotope stage 1 is associated with an assemblage where *I. aculeatum* exceeds 60% of the total gonyaulacoids cysts, reaching abundances comparable to those in surface sediments.

DINOFLAGELLATE CYSTS AND REWORKING

Each of the areas located offshore from the Rhône Delta—including the Rhône deep-sea fan and canyons on the continental slope—functioned as an independent system between 30,000 BP and the present day (Mear, 1984; Droz,

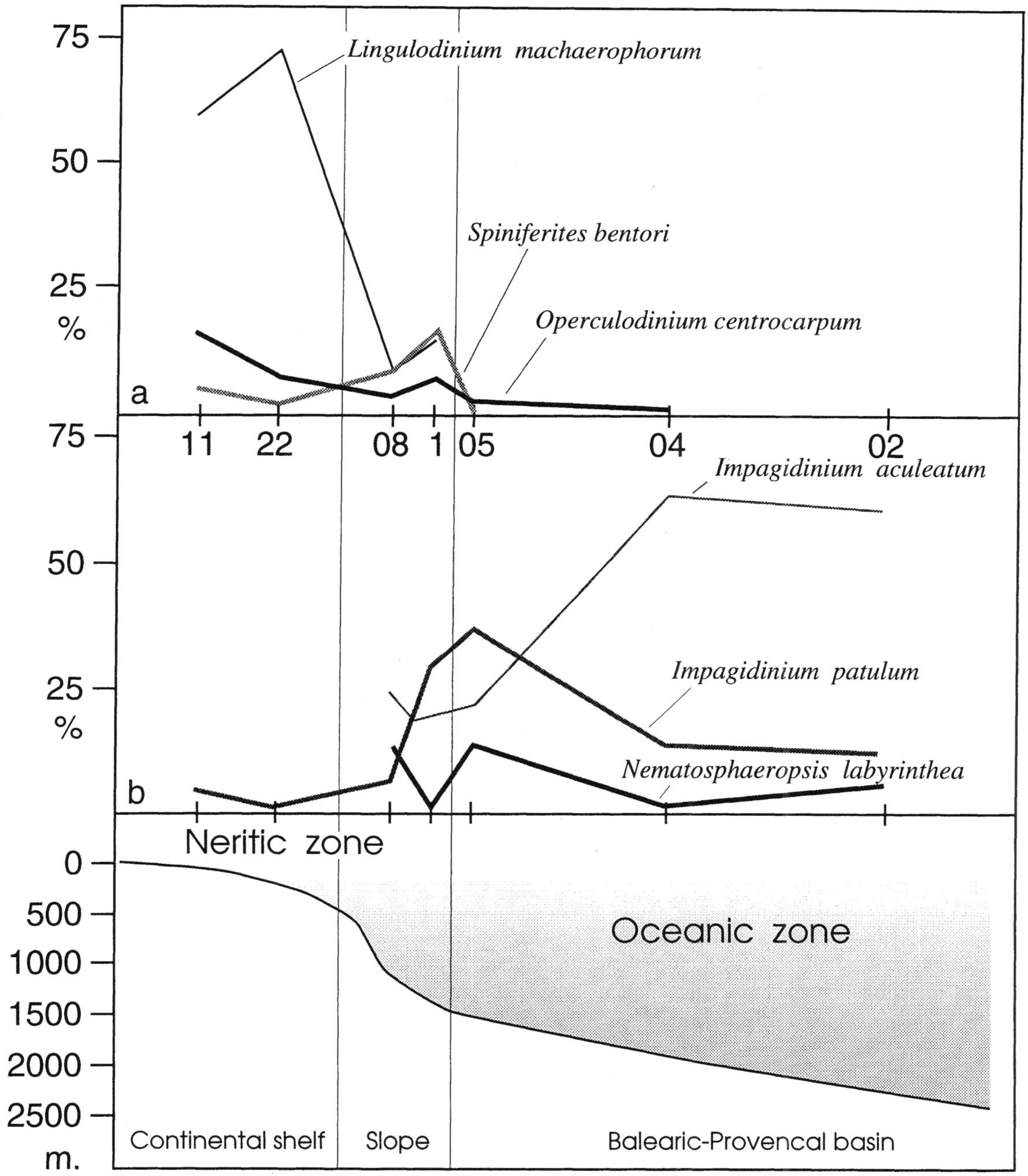

Text-Figure 2. Distribution of Quaternary dinoflagellate cysts in recent sediments off the Rhône Delta. 2a– estuarine to neritic species; 2b– oceanic species.

1991), so the distribution of dinoflagellate cysts in these different domains should be examined and interpreted separately.

Rhône Deep-Sea Fan, Outlet of the Petit-Rhône Canyon

KSPF 02, -2300 m, downstream end of the Petit-Rhône canyon (Text-Figure 4). The onset of cold conditions, as indicated by the increased abundance of *O. centrocarpum* and *N. labyrinthea* in Pleistocene grey muds, corresponds to a reduced supply of redeposited and reworked cysts. This overall decrease in externally derived cysts is accompanied by qualitative changes in the assemblages, *viz:*

- disappearance of reworked dinoflagellate cysts derived from Cretaceous sediments (*Gonyaulacysta* sp., *Muderongia staurota* ...) at a depth of 250 cm in the core;
- disappearance of redeposited estuarine forms (mainly *L. machaerophorum*) at 200 cm;
- change of abundance of Pliocene reworked forms from 70% at the base of the core to 30% at 200 cm.

A modification in the magnitude of the input (decreased abundance) as well as provenance direction (disappear-

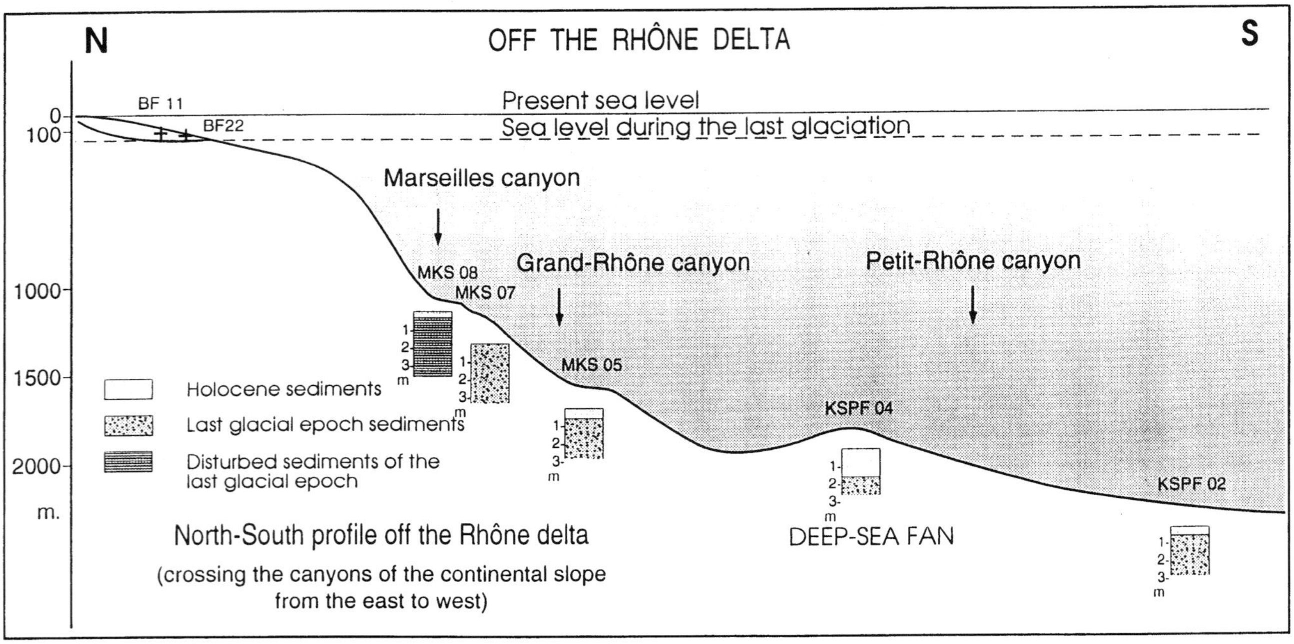

Text-Figure 3. North–South profile off the Rhône Delta. Location and lithology of studied cores and grab bucket samples.

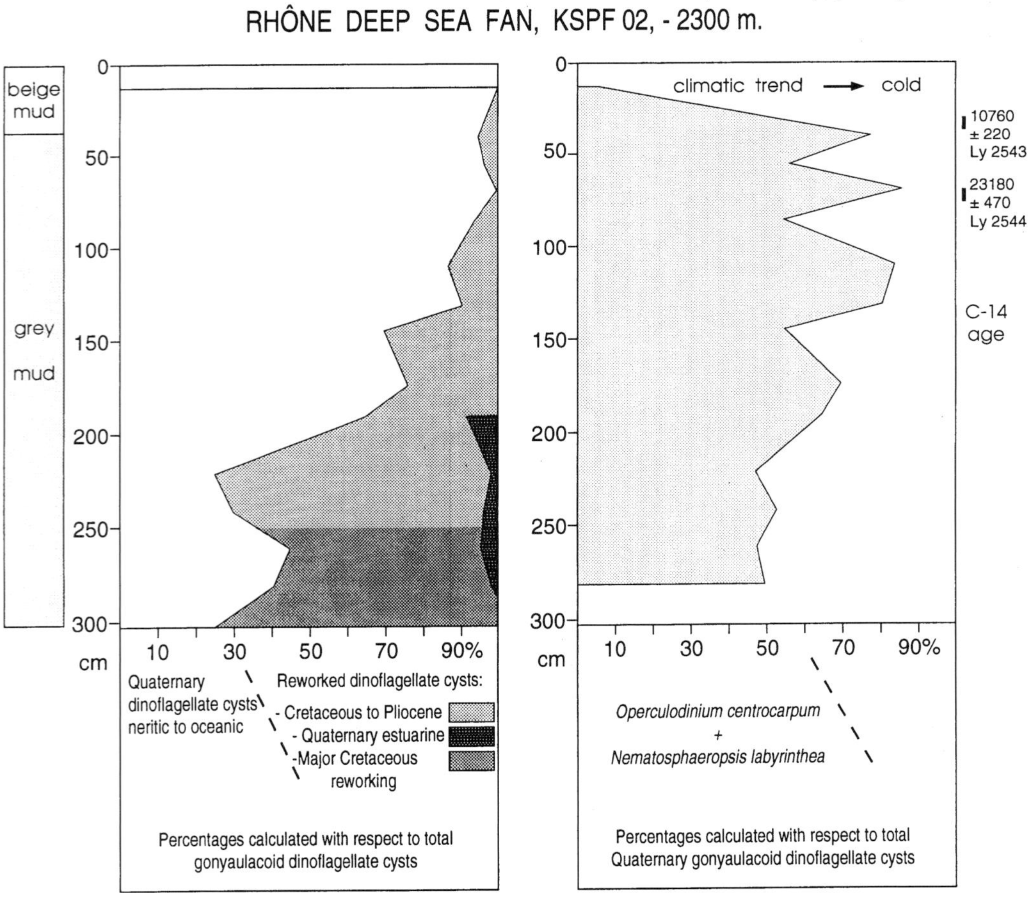

Text-Figure 4. The Rhône deep-sea fan, core KSPF 02, downstream end of the Petit-Rhône canyon.

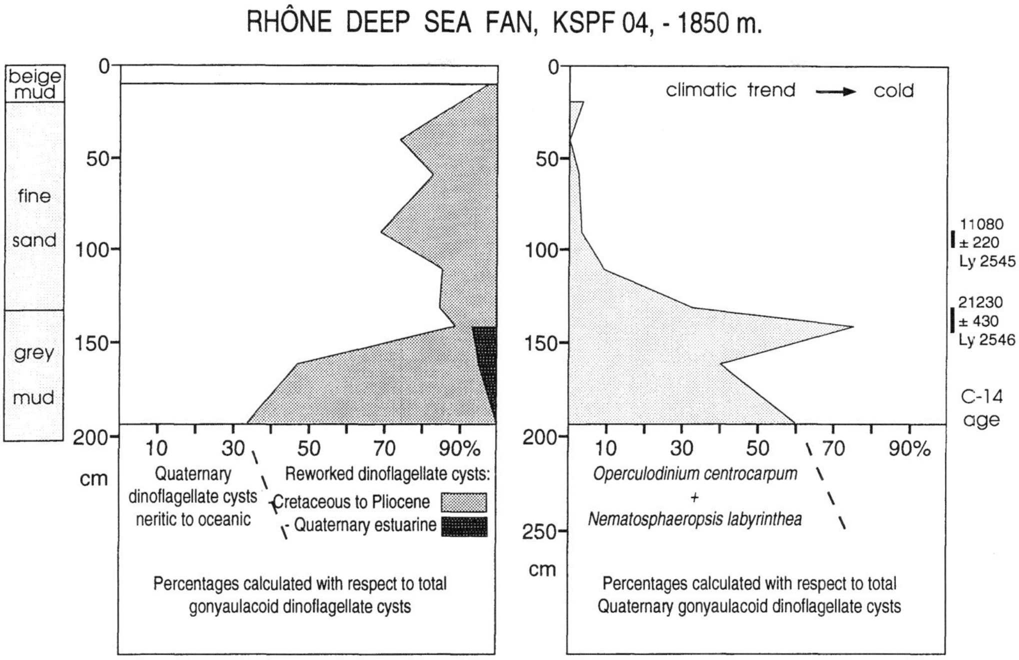

Text-Figure 5. The Rhône deep-sea fan, core KSPF 04.

ance of Cretaceous inputs) took place at the end of isotope stage 3, before 23,000 years BP, as indicated by the radiocarbon age obtained from the sample taken at 60 cm. Between 23,000 and 11,000 years BP, i.e. during stage 2, sedimentation on the Rhône fan at the outlet of the Petit-Rhône canyon is limited to about 10 cm of grey mud.

The Holocene sedimentation, beginning after 11,000 years BP, is represented by 40 cm of beige muds containing an assemblage of oceanic cysts with predominant *I. aculeatum*. The displacement or reworking of dinoflagellate cysts is of minor importance in these sediments.

KSPF 04, -1850 m, upslope part of the fan (Text-Figure 5). The cored sequence contains younger material than that observed in KSPF 02, thus providing a more representative view of Holocene sedimentation.

The grey muds deposited before 21,000 years BP contain an oceanic cyst assemblage with temperate–cold affinities (dominant species: *O. centrocarpum* and *N. labyrinthea*). Reworked cysts make up more than 60% of the total cysts present, but no significant supply from Cretaceous formations can be detected. The supply of estuarine cysts is abruptly halted at a depth of 150 cm at the top of the grey mud interval.

The base of the post-glacial period, at around 11,000 years BP, corresponds to a depth of 140 cm in the core, being represented by fine sands with an oceanic cyst assemblage of temperate–warm affinity containing predominant *I. aculeatum*. Nevertheless, a slight drop in the abundance of this species between 60 cm and 40 cm core depth possibly reflects a cooling that could be attributed to the Younger Dryas phase (Morzadec-Kerfourn, 1984). The Holocene sands contain a relatively high abundance (30% maximum) of reworked Pliocene forms. Between 60 cm and 40 cm, the presence of some Cretaceous forms coincides with the slight cooling inferred from the distribution of oceanic forms. In the overlying beige muds, the reworked forms display a strong reduction in abundance.

Submarine Canyons on the Continental Slope

MKS 05, -1590 m, on the ridge flanking the southern side of the Grand-Rhône canyon (Text-Figure 6). No radiocarbon ages are available for this core. The shift in the proportion of temperate–cold in favour of temperate–warm cysts indicates the end of isotope stage 3, the transition into stage 2 and the beginning of stage 1. It appears that the sedimentary hiatus identified in the Rhône fan during stage 2 is less well marked in the MKS 05 core. The reworked forms—mostly of Pliocene age—are relatively less abundant in these sediments. The maximum abundance of 50% (samples between 170 and 130 cm in the core) is coeval with the climatic warming that occurred at the end of stage 3, around 29,000–30,000 years BP. At the end of this warm phase, the input of estuarine cysts ceased as also observed in the Rhône deep-sea fan. The proportion of reworked forms decreases above a core depth of 130 cm, at the same point as the disappearance of estuarine forms. This change coincides with the accentuation of cold climatic conditions that took place during stage 2.

At the top of the core (sample at 20 cm), the Holocene mud contains 60% temperate–warm species (*I. aculeatum* + *I. patulum*) with respect to total cysts and lacks any reworked or estuarine forms.

MKS 07, -1195 m, axis of a secondary canyon (Text-Figure 7). This core is composed only of end-glacial grey muds. Radiometric dating indicates an age between 30,000 and 14,000 years BP, with relatively high sedimentation rates. The temperate–cold species *O. centrocarpum* and *N. labyrinthea* show mean abundances of 60%. Reworking of

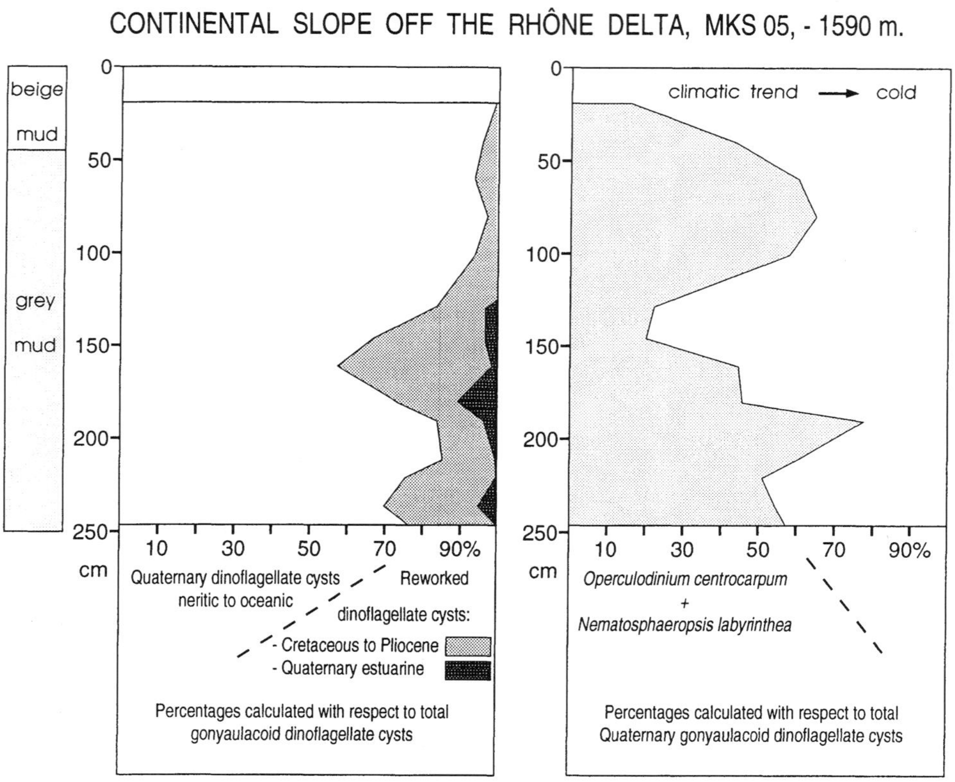

Text-Figure 6. The continental slope off the Rhône delta, core MKS 05, on the ridge flanking the southern side of the Grand-Rhône canyon.

the sediment is important at the base of the core, but diminishes above 200 cm. The reduction in the degree of reworking corresponds to the onset of climatic cooling after 29,000 years BP, at the beginning of isotope stage 2, as observed in the Rhône fan.

MKS 08, -1155 m, axis of the Marseilles canyon (Text-Figure 8). This core is mainly composed of black muds deposited during the last glacial epoch. The sediments are poor in planktonic foraminifera (Blanc-Vernet, 1984) and dinoflagellate cysts, probably because they are diluted by a considerable input of detrital material. At a depth of 40 cm in the core, the black muds are separated from the overlying Holocene beige muds by a layer containing poorly preserved and highly degraded organic matter made up of pollen and dinoflagellate cysts of temperate–cold affinity. This suggests, as in the Rhône fan, the existence of a discontinuous transition between the black mud and the beige mud facies.

At the top of the sequence, the Holocene beige muds contain 60% temperate–warm species (*I. aculeatum* + *I. patulum*) with respect to total cysts, but has few reworked (1.7%) or displaced forms (*L. machaerophorum*: 1.7%).

CONCLUSION: EROSION AND TRANSPORT OF SEDIMENTS OFF THE RHONE DELTA DURING THE LAST INTERGLACIAL–GLACIAL CYCLE

By considering the sedimentary criteria, a reduction in sediment supply—particularly from the Rhône river—is shown to have occurred in the Rhône deep-sea fan after 28,000 years BP (Mear, 1984, Monaco et al.,1984; Droz, 1991). This is interpreted by Mear (1984) as being due to the collapse, between 28,000 and 20,000 years BP, of a levée which led to the blocking of the main valley of the Petit-Rhône, thus causing the trapping of sediments upstream and a lateral shift of the depocentre. Such a process is responsible for the sedimentary hiatus between 23,000 and 11,000 years BP that is identified in the submarine fan of the Rhône at the outlet of the Petit-Rhône canyon. The terrigenous input from the Rhône thus supplies the ridge flanking the southern side of the Grand-Rhône canyon (Droz, 1991; Mear, 1984; Aloisi et al., 1975). Farther east, the Marseilles and Planier submarine canyons received their supply from the Durance river during the last glacial epoch (Montjuvent et al., 1984; Droz, 1991).

The results provided by the analysis of dinoflagellate cysts in the deep-sea fan of the Rhône are consistent with the above interpretation. The decrease in the supply of reworked dinoflagellate cysts in cores KSPF 02 and KSPF 04 from the Rhône fan is coeval with cooling that took place during isotope stage 2. This would be contemporaneous with the blocking of the Petit-Rhône canyon, noted by Mear (1984). The reduction in supply of Cretaceous material observed at the outlet of the Petit-Rhône canyon, in core KSPF 02, could be directly linked to a change in the flow of fluvial supply from the Rhône at the end of the last glacial epoch. Subsequently, the Pleistocene course would have been abandoned since, as observed in core KSPF 04, the fine sands at the base of the Holocene reflect the renewal of sedimentation in the Petit-Rhône canyon, with sediments

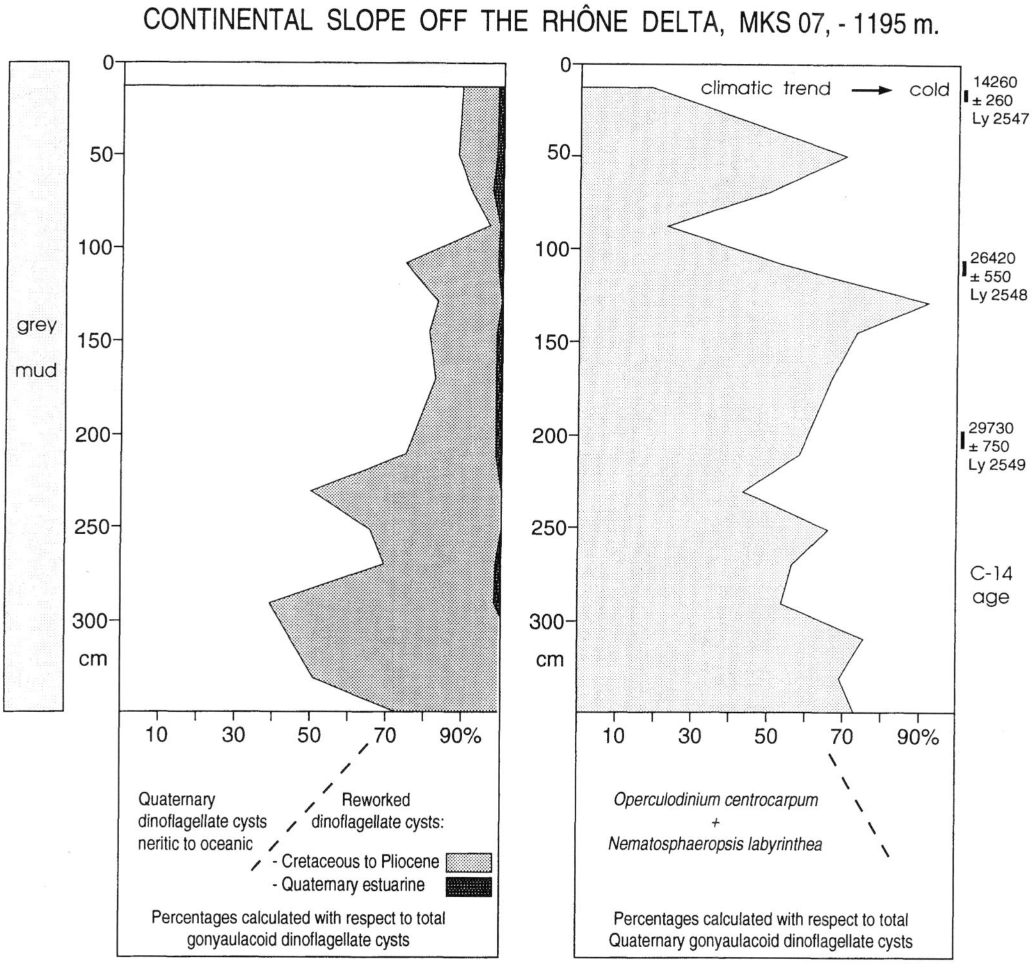

Text-Figure 7. The continental slope off the Rhône delta, core MKS 07. axis of a secondary canyon between the Grand-Rhône and the Marseilles canyons.

containing relatively high amounts of forms derived solely from Pliocene formations. Moreover, the Holocene sands are partly derived from reworked Pliocene sands and sediments produced by overdeepening of the channel (Droz, 1991; Mear, 1984).

According to the interpretation of Mear (1984) and Droz (1991), sediment supply to the Grand-Rhône canyon and on the axis of a secondary canyon situated farther east would have received inputs from the Rhône river after blocking of the Petit-Rhône. The reworked and redeposited dinoflagellate cysts in grey mud at these sites (MKS 05) and on the interfluve between the Grand-Rhône et the Marseilles canyons (MKS 07) show a fall in their abundance during stage 2, in a similar way as observed in the deep-sea fan of the Rhône. This lowering in the abundance of reworked and redeposited forms is well calibrated by radiocarbon dating at MKS 07 and by bioclimatic indicators at MKS 05, in particular the ratio of temperate–cold to temperate–warm species which leads to correlations with the Rhône deep-sea fan.

It is no longer possible to invoke only the blocking of distributary channels in order to explain the decrease in the supply of reworked and redeposited dinoflagellate cysts in the canyons to the east of the Petit-Rhône. Such a decrease in the supply of externally derived material could be indicative of a lower rate of erosion linked to a lowering of humidity in the Mediterranean domain during the glacial maximum. High levels of aridity were attained during the late Pleistocene, between 20,000 and 12,500 years BP, in the south of the Mediterranean Basin, in the Nile valley, (Adamson et al., 1980), in the Gulf of Gabes, offshore Tunisia (Morzadec-Kerfourn, 1979), and could have developed to a lesser extent in the northern Mediterranean region. In the lower Rhône valley, an arid and cold vegetation with *Artemisia* and Chenopodiaceae, is present during the Older Dryas period, before 13,300 B.P (Triat-Laval, 1979)

Droz (1991) pointed out that "eustatic variations are not the only factors influencing the rate of continental and submarine erosion, which itself affects the sediment supply to the basin and thus the rate of sedimentation". This author also refers to the possible blocking of distributaries, as in the example of the Petit Rhône canyon, described by Mear (1984). In addition to this process, the climatic conditions on the continent control both the volume of supply as well as the transport of sediment along the bottom. Such a phenomenon is well shown offshore from the Rhône delta, where the cold climatic maximum does not correspond to the maximum transport rate for reworked and redeposited microfossil material. This is

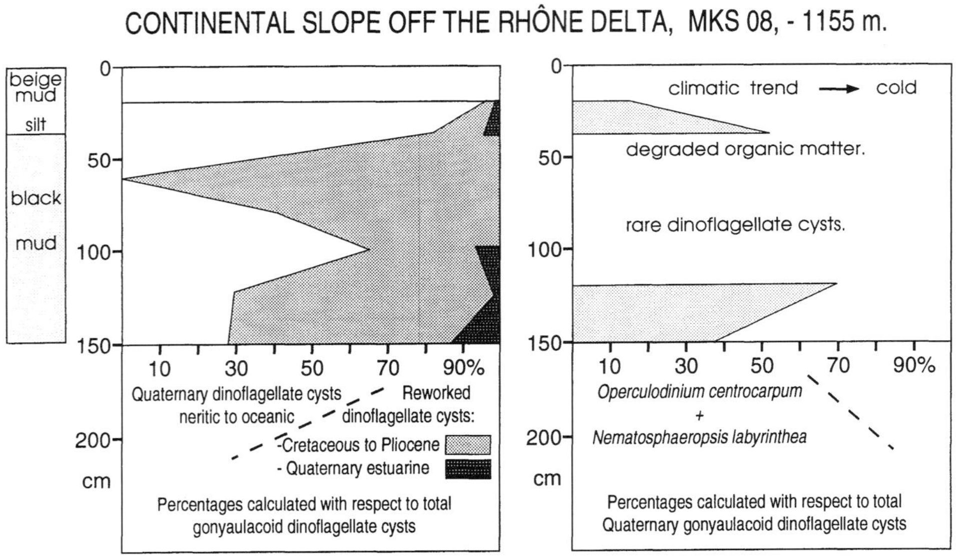

Text-Figure 8. The continental slope off the Rhône delta, core MKS 08, axis of the Marseilles canyon.

because the cold phase may have been accompanied by a decrease in humidity in the northwestern part of the Mediterranean Basin.

ACKNOWLEDGMENTS

This research has benefited from support of the French Petroleum Company Total within the context of the ECOMED program: ecology of microorganisms in the Western Mediterranean Sea. I am especially grateful to P.F. Burollet and Y. Caro for their scientific assistance, to M.S.N. Carpenter for translating this paper into English, and to M. Lautram for the drawing of the figures. I thanks the two reviewers, Lucy E. Edwards and Joyce Lucas-Clark, for the constructive comments.

References Cited

ADAMSON, D.A., GASSE, F., STREET, F.A., and WILLIAMSON, M.J.A.

1980 Late Quaternary history of the Nile. *Nature*, 288: 50–55.

ALOISI, J.C., MONACO, A., THOMMERET, J., and THOMMERET, Y.

1975 Evolution paléogéographique du plateau continental languedocien dans le cadre du golfe du Lion. Analyses comparées des données sismiques, sédimentologiques et radiométriques concernant le Quaternaire récent. *Revue de Géographie Physique et de Géologie Dynamique, Paris*, 2, XVII: 13–22.

BELLAICHE, G., DROZ, L., COUTELLIER, V., ORSOLINI, P., BERTHON, J.L., RAVENNE, C., ALOISI, J.C., GOT, H., and MONACO, A.

1984 Morphologie détaillée, structure superficielle et évolution plio–quaternaire du golfe du Lion et de l'éventail sous-marin profond du Rhône (Rhone deep sea Fan) *In*: Bizon, J.J., and Burrollet P.F. (eds.). Ecologie des micro-organismes en Méditerranée occidentale "ECOMED." Etude morphologique, hydrologique, sédimentologique, géochimique et écologique du bassin et de ses marges entre Languedoc et Corse. *Association Française des Techniciens du Pétrole, Paris*: 8–13.

BLANC-VERNET, L.

1984 Etude des carottes, évolution du climat d'après les Foraminifères planctoniques. *In*: Bizon, J.J., and Burrollet, P.F. (eds.). Ecologie des microorganismes en Méditerranée occidentale "ECOMED." Etude morphologique, hydrologique, sédimentologique, géochimique et écologique du bassin et de ses marges entre Languedoc et Corse. *Association Française des Techniciens du Pétrole, Paris*: 140–146.

DROZ, L.

1983 *L'éventail sous-marin profond du Rhône (Golfe du Lion): grands traits morphologiques et structure semi-profonde.* Thèse de 3ème cycle, Université Pierre et Marie Curie, Paris VI: 195 pgs.

1991 *Les éventails sous-marins profonds: structure et évolution sédimentaire. A partir de l'étude comparative de trois édifices: l'éventail du Rhône, la ride du Var, le cône de l'Indus.* Mémoire d'Habilitation, Université Pierre et Marie Curie, Paris VI, Tome 1: 254 pgs.

FERNEX, F., ADDED, A., and SPAN, D.

1984 Les sédiments de la pente au Sud et au Sud-Est du Delta du Rhône (canyons du Grand Rhône, de Marseille, du Planier et de Cassis). *In*: Bizon, J.J., and Burrollet P.F. (eds.). Ecologie des microorganismes en Méditerranée occidentale "ECOMED". Etude morphologique, hydrologique, sédimentologique, géochimique et écologique du bassin et de ses marges entre Languedoc et Corse. *Association Française des Techniciens du Pétrole, Paris*: 42–44.

MEAR, Y.

1984 *Séquences et unités sédimentaires du glacis rhodanien (Méditerranée occidentale).* Thèse 3ème Cycle, Université de Perpignan: 214 pgs.

MONACO, A., ALOISI, J.C., GOT, H., MEAR, Y., and BELLAICHE, G.

1984 Sédimentogenèse en Méditerranée nord-occidentale. *In*: Bizon, J.J., and Burrollet P.F. (eds.). Ecologie des microorganismes en Méditerranée occidentale "ECOMED". Etude morphologique, hydrologique, sédimentologique, géochimique et écologique du bassin et de ses marges entre Languedoc et Corse. *Association Française des Techniciens du Pétrole, Paris*: 24–32.

MONJUVENT, G., et al. (eds.)
1984 Quaternaire. *In*: Synthèse géologique du Sud-Est de la France. *Mémoire du Bureau de Recherche Géologiques et Minières, Orléans,* 125: 521–580.

MORZADEC-KERFOURN, M.T.
1979 Les kystes de Dinoflagellés. *In*: La mer pélagienne. Etude sédimentologique et écologique du plateau tunisien et de ses abords. *Géologie Méditerranéenne, Marseille,* VI, 1: 221–246.
1984 Les kystes de dinoflagellés dans les sédiments pléistocènes supérieurs et holocènes au large du delta du Rhône et de la Corse. *In*: Bizon, J.J.,and Burrollet P.F. (eds.). Ecologie des microorganismes en Méditerranée occidentale "ECOMED." Etude morphologique, hydrologique, sédimentologique, géochimique et écologique du bassin et de ses marges entre Languedoc et Corse. *Association Française des Techniciens du Pétrole, Paris*: 170–183.
1992a Upper pleistocene and holocene dinoflagellate cyst assemblages in marine environments of the Mediterranean Sea and the northwest Atlantic coast of France. *In*: Head, M.J., and Wrenn, J.H. (eds.), *Neogene and Quaternary Dinoflagellate Cysts and Acritarchs.* American Association of Stratigraphic Palynologists Foundation, Dallas: 121–132.
1992b Estuarine dinoflagellate cysts among oceanic assemblages of Pleistocene deep-sea sediments from the west african margin and their paleoenvironmental significance. *In*: Head, M.J., and Wrenn, J.H. (eds.), *Neogene and Quaternary Dinoflagellate Cysts and Acritarchs.* American Association of Stratigraphic Palynologists Foundation, Dallas: 133–146.

TRIAT-LAVAL, H
1979 *Contribution pollenanalytique à l'histoire Tardi et Postglaciaire de la végétation de la basse vallée du Rhône.* Thèse de Doctorat d'Etat, Université d'Aix-Marselle III, 380p.

WALL, D., DALE, B., LOHMAN, G.P., and SMITH, W.K.
1977 The environnemental and climatic distribution of dinoflagellate cysts in modern marine sediments from regions in the North and South Atlantic Oceans and adjacent seas. *Marine Micropaleontology,* 2: 120–200.

Author's address

MARIE-THERESE MORZADEC-KERFOURN
Laboratoire de Micropaléontologie et Paléontologie Marines
Institut de Géologie
Université de Rennes 1
35042 Rennes Cedex
France

PALEOECOLOGY AND PALEOLIMNOLOGY OF THE LAST PLENIGLACIAL IN THE VENEZUELAN ANDES

Valentí Rull, and Teresa Vegas-Vilarrúbia

Abstract

The present paper reviews the results of paleoecological and paleolimnological analyses of a sedimentary sequence deposited during the Last Glacial Maximum, in the Venezuelan Andes. The sediment accumulated in a pre-glacial lake which existed until after 16 ka, when it dried. Pollen and algal remains of the sediments show three intervals of high Andean vegetation (stadials), interrupted by two intervals of superpáramo (interstadials). One of these stadials was called *El Caballo* and dated to 16.5 ^{14}C ka BP. Estimated temperatures were around 7°C lower than today. After that, the glacier retreat proceeded at an average of about 7.3 cm altitude/y, but with oscillations, due to Postglacial and Holocene climate shifts. A pre-glacial lake existed in the site prior to 16.5 ka BP, with well-developed pelagic and littoral communities. The latter disappeared during the colder interval, and only pelagic communities dominated by *Botryococcus* survived. The lake dried out just after the *El Caballo* stadial.

A total of 28 unidentified fungi spores and 30 other unknown palynomorphs were documented and related to pollen and algae assemblages, in order to test their palaeoecological usefulness. Several fungal spores resulted to be associated to the superpáramo vegetation, while others show an inverse trend, suggesting different temperature controls. Among unknown remains, a substantial number seems to be originated from aquatic organisms. The affinities of several of them with either littoral, pelagic or bog assemblages could be established. Continued taxonomic work and intensive studies on modern sedimentation patterns of unidentified forms is proposed to improve their ecological characterization.

INTRODUCTION

Most of earlier palynological work in the Venezuelan Andes has been developed for the *páramo* region, focusing on Late Glacial and Holocene sediments and their paleoclimatic inferences (reviews in Salgado-Labouriau, 1989, 1991; Rull, 1996, 1999). Pollen analysis of older sediments (middle Wisconsin) has been carried out only in two sites (Salgado-Labouriau, 1984; Rull, 1999). Pleniglacial sediments were found during the last decade in the site called *Mesa del Caballo*, situated at about 8°47' N to 71°50' W (Schubert and Rinaldi, 1987), but their palaeoecological study has been completed only recently (Rull, 1998; Rull and Vegas-Vilarrúbia, 1996, 1997, 1998). These sediments were deposited during the worldwide Last Glacial Maximum, and allowed determination of the main vegetation, climate and limnological conditions of the *páramos*, during this time. The present paper reviews the results of the different studies conducted so far on sediments from the *Mesa del Caballo* site, which include analysis of pollen and fern spores, fungi, algae and animal remains, as well as a large amount of unknown fossils.

THE *PÁRAMOS* AND THE STUDY SITE

The studied area is located in the Andean *páramo* belt, with its unique geomorphologic, climatic and vegetational features. In the Northern Andes, *páramo* (*s.l.*) refers to the altitudinal belt between tree line and permanent snow. In Venezuela it extends from about 3000 to 4700 m above the sea level. The bedrock is composed of Precambrian and Paleozoic rocks, which have been modeled by glacial erosion. The *páramos* can be subdivided into three altitudinal belts: (1) the *subpáramo*, or transition zone between the Andean forest and (2) the *páramo* proper, which is up to the line of nightly freezing, and (3) the *superpáramo*, between the *páramo* proper and the permanent snowline, around 4700 m altitude. The tree line (upper limit of the tropical montane forest) is around 3000 m altitude, with variations depending on local climates and slope orientation. The vegetation is open, with two well-differentiated strata: the lower consists of gramineae and other herbs, while the upper one is formed of shrubs and the typical 'caulirrossuae' (rosettes above a columnar stem) of the compositae *Espeletia*, which are the most conspicuous plant elements of these communities. The climate of the *páramos* is cold and humid. Annual average temperatures vary from 0°C to 10°C, and are dependent of the altitude, decreasing at a rate of approximately 0.6°C/100 m altitude. Thermal seasonality is not significant, but daily variations are high. Mean annual precipitation ranges from 600 to 1800 mm, with a dry season of four months. This parameter does not show altitudinal dependence, but it varies with slope orientation (Salgado-Labouriau, 1979).

The section studied (*Mesa del Caballo*) lies in the Mucubají region, where morainic complexes of the Late Glacial were originally described by Schubert (1974). Text-Figure 1 shows the location of the sampling site and its glacio-geomorphological context. The site is located at about 3500 m a.s.l., within the *páramo* proper vegetation belt. Cattle raising and cultivation of potatoes are the main human activities in the region, but the disturbance is small, because this is close to the upper limit of human activities (Salgado-Labouriau, 1979). The *Mesa del Caballo* is a triangular terrace with an area of about 1 km^2 and 30 m thickness, developed by in-filling of a small intramontane basin between the lateral and terminal moraines *El Caballo* and *La Mucuchache*, which was subsequently cuttdown by *La Cañada* creek. Schubert and Rinaldi (1987) accurately described its sediments, and radiometrically dated its base and top as 19.1 and 16.5 ka BP, respectively (Table 1).

Rull, Valentí, and Vegas-Vilarrúbia, Teresa
2001 Paleoecology and paleolimnology of the last pleniglacial in the Venezuelan Andes; *In*: Goodman, D.K., and Clarke, R.T. (eds.), *Proceedings of the IX International Palynological Congress, Houston, Texas, U.S.A., 1996*; American Association of Stratigraphic Palynologists Foundation, p. 385–401.
 ISBN 0–931871–06–9

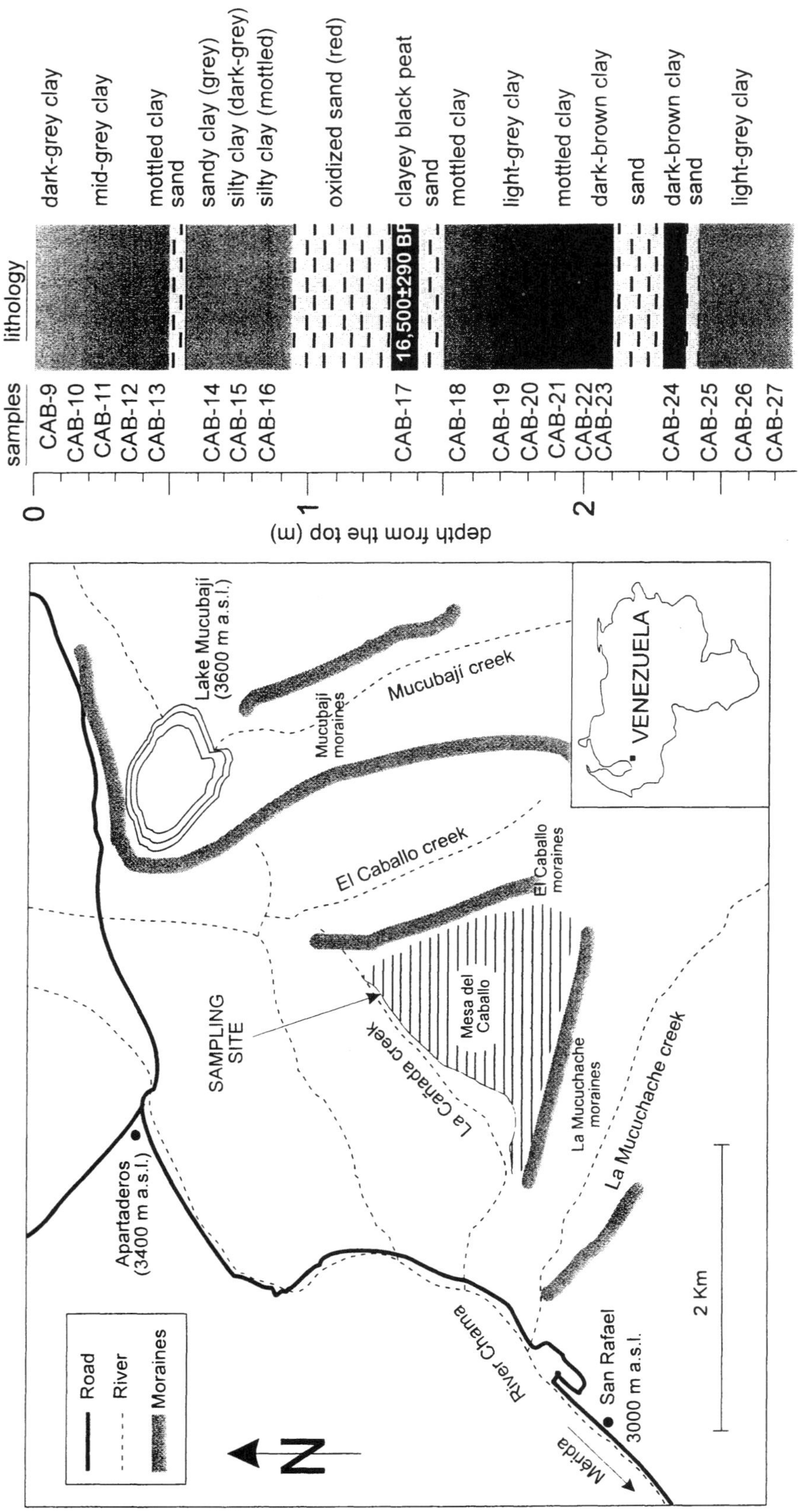

Text-Figure 1. Location map of the sampling site, with the lithostratigraphy of the section studied and the samples taken (modified from Rull, 1998).

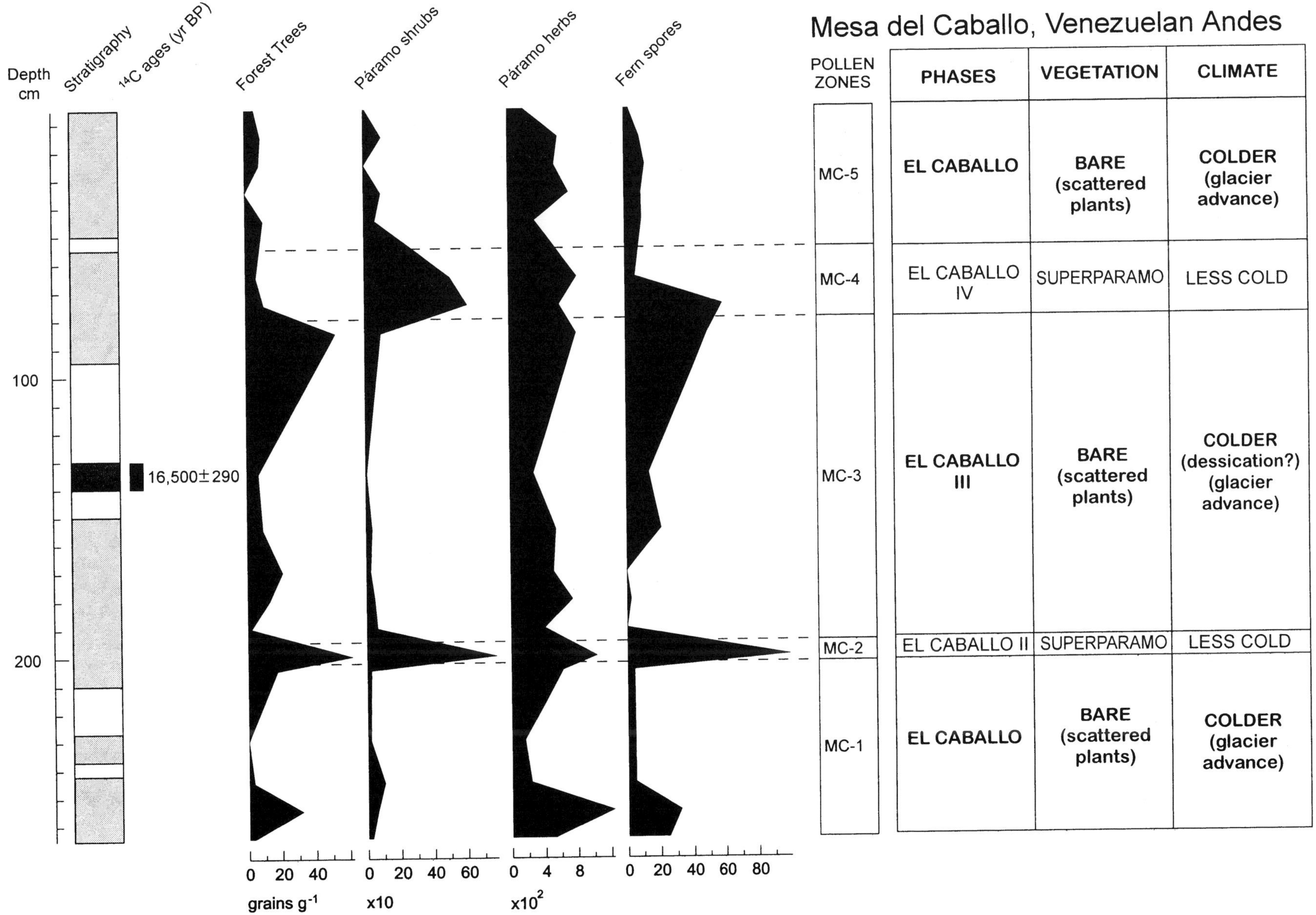

Text-Figure 2. Summary of pleniglacial vegetation and climatic trends in the *Mesa del Caballo* site, as deduced from pollen analysis (after Rull and Vegas-Vilarrúbia, 1998). Forest trees: *Alnus, Hedyosmum, Podocarpus, Cecropia, Vallea, Juglans*; Páramo shrubs: Compositae; Páramo herbs: Gramineae, Cyperaceae, *Montia, Gentiana, Arenaria*-type, Portulacaceae; Ferns and allied spores: *Isoëtes, Polypodium* (verrucate), *Alsophila*-type, *Jamesonia, Lycopodium.*

TABLE 1. Radiocarbon dates of Schubert and Rinaldi (1987) for the Mesa del Caballo section studied. Anaysis by Teledyne Isotopes, Westwood, NJ, U.S.A.

SAMPLE	LAB NUMBER	DEPTH (m)	MATERIAL	^{14}C AGE
I	11,021	2	black peat	16,500 ± 290
II	11,022	28.5	dark-grey clay	19,080 ± 820

Sediments consist of an alternating accumulation of clay, silt, sand and gravel layers of variable thickness, was interpreted to represent a fluvio-glacial (outwash) plain, progressively accumulated when the terminal and lateral moraines were developing under the action of the *Mucuchache* and *Mucubají* glaciers.

The results discussed in the present report are from a 3 m thick section, situated at the top of the *Mesa del Caballo* sequence, where the black peat top layer dated by Schubert and Rinaldi (1987) could be easily recognized (Text-Figure 1).

PLENIGLACIAL CLIMATES AND VEGETATION

Changes in pollen and fern spores assemblages (Text-Figure 2; Plate 1) provided the primary climatic trends, reconstructed by comparison with present-day assemblages (Rull, 1998; Rull and Vegas-Vilarrúbia, 1996). The diagrams show two alternating pollen assemblages. Assemblage A is represented in pollen zones MC-1, MC-3 and MC-5, and is interpreted as indicative of bare terrain with scattered vegetation (mainly gramineae), characteristic of the uppermost altitudinal levels near the snowline (~ 4700 m). Assemblage B, found in pollen zones MC-2 and MC-4, is interpreted to represent the beginning of the establishment of incipient *superpáramo* plant communities from lower elevations. From a paleoclimatic perspective, assemblage A indicates colder phases, related probably to glacier readvances, whereas assemblage B represents less cold times, associated to times of glacier retreats. The *Mesa del Caballo* sequence thus shows an alternating movement of vegetation belts closer to snowline upwards and downwards. The radiocarbon-dated *El Caballo* III phase at 16.5 ka BP (pollen zone MC-3) is considered a stadial, with estimated average temperatures around 7°C lower than today. This is deduced from an estimated altitudinal shift of about 1200 m in the climatic and vegetation belts, during *El Caballo* III phase, and considering the same lapse rate than today (-0.6°C/100m altitude).

SEDIMENTATION RATES

Based on radiocarbon dates of Schubert and Rinaldi (Table 1), sedimentation rates at *Mesa del Caballo* are of 1.07 cm/y, up to 50 times higher than in other sites from Venezuela (Rull, 1998). A negative exponential relationship was documented between sedimentation rates and altitude, for fluvioglacial terraces (Text-Figure 3). This is probably due to the fact that the amount of sediments carried and deposited increases downstream as a consequence of the increasing amount of glacial and fluvial erosion and transport, due to the increase in size of catchment area with decreasing altitude. The departure of Piedras Blancas from this trend is probably due to its different sedimentological conditions. It is a peat bog within a high-altitude (> 4000 m) glacier cirque and consists mainly of plant remains with few siliciclastic sediments (Rull et al., 1987). The same tendency has been

PLATE 1

1 *Hedyosmum*. CAB-26. Film VR50, 29-29A. 40x2.5. Bar = 10 µm. CAB-16. Film VR65, 17-17A. Bar = 8 µm.
2 *Podocarpus*. CAB-26. Film VR50, 10-10A. 40x2.5. Bar = 10 µm.
3 Gramineae. CAB-9. Film VR67, 18. 100x1.25. Bar = 8 µm.
4 Reticulate, monocolpate pollen. CAB-16. Film VR65, 22-22A. 100x1.25. Bar = 8 µm.
5 Urticales. CAB-26. Film VR50, 37-37A. 40x2.5. Bar = 10 µm.
6 *Draba*. CAB-15. Film VR65, 30-30A. 100x1.25. Bar = 8 µm. CAB-25. Film VR61, 5. 100x1.25. Bar = 8 µm.
7 *Acaena-Polylepis*. CAB-16. Film VR65, 24-24A. 100x1.25. Bar = 8 µm.
8 Myrtaceae. CAB-20. Film VR64, 1A-2. 100x1.25. Bar = 8 µm.
9 Compositae. CAB-13. Film VR66, 14. 100x1.25. Bar = 8 µm. CAB-13. Film VR66, 13. 100x1.25. Bar = 8 µm.
10 Compositae. CAB-25. Film VR61, 3. 100x1.25. Bar = 8 µm. CAB-25. Film VR61, 2. 100x1.25. Bar = 8 µm.
11 Compositae. CAB-14. Film VR66, 5. 100x1.25. Bar = 8 µm. CAB-14. Film VR66, 6. 100x1.25. Bar = 8 µm.
12 *Alnus*. CAB-26. Film VR50, 35-35A. 40x2.5. Bar = 10 µm.
13 *Arenaria*-type. CAB-10. Film VR67, 2. 100x1.25. Bar = 8 µm. CAB-10. Film VR67, 3. 100x1.25. Bar = 8 µm.
14 *Montia*. CAB-24. Film VR62, 1A-2. 100x1.25. Bar = 8 µm. CAB-24. Film VR62, 0. 100x1.25. Bar = 8 µm.
15 Psilate monoletes. CAB-10. Film VR67, 1. 100x1.25. Bar = 8 µm. CAB-10. Film VR67, 0. 100x1.25. Bar = 8 µm.
16 *Isoëtes*. CAB-25. Film VR61, 8. 100x1.25. Bar = 8 µm.
17 *Lycopodium*. CAB-26. Film VR50, 33-33A. 40x2.5. Bar = 10 µm.
18 *Polypodium*. CAB-26. Film VR50, 31-31A. 40x2.5. Bar = 10 µm.
19 *Jamesonia*. CAB-16. Film VR65, 27-27A. 100x1.25. Bar = 8 µm.

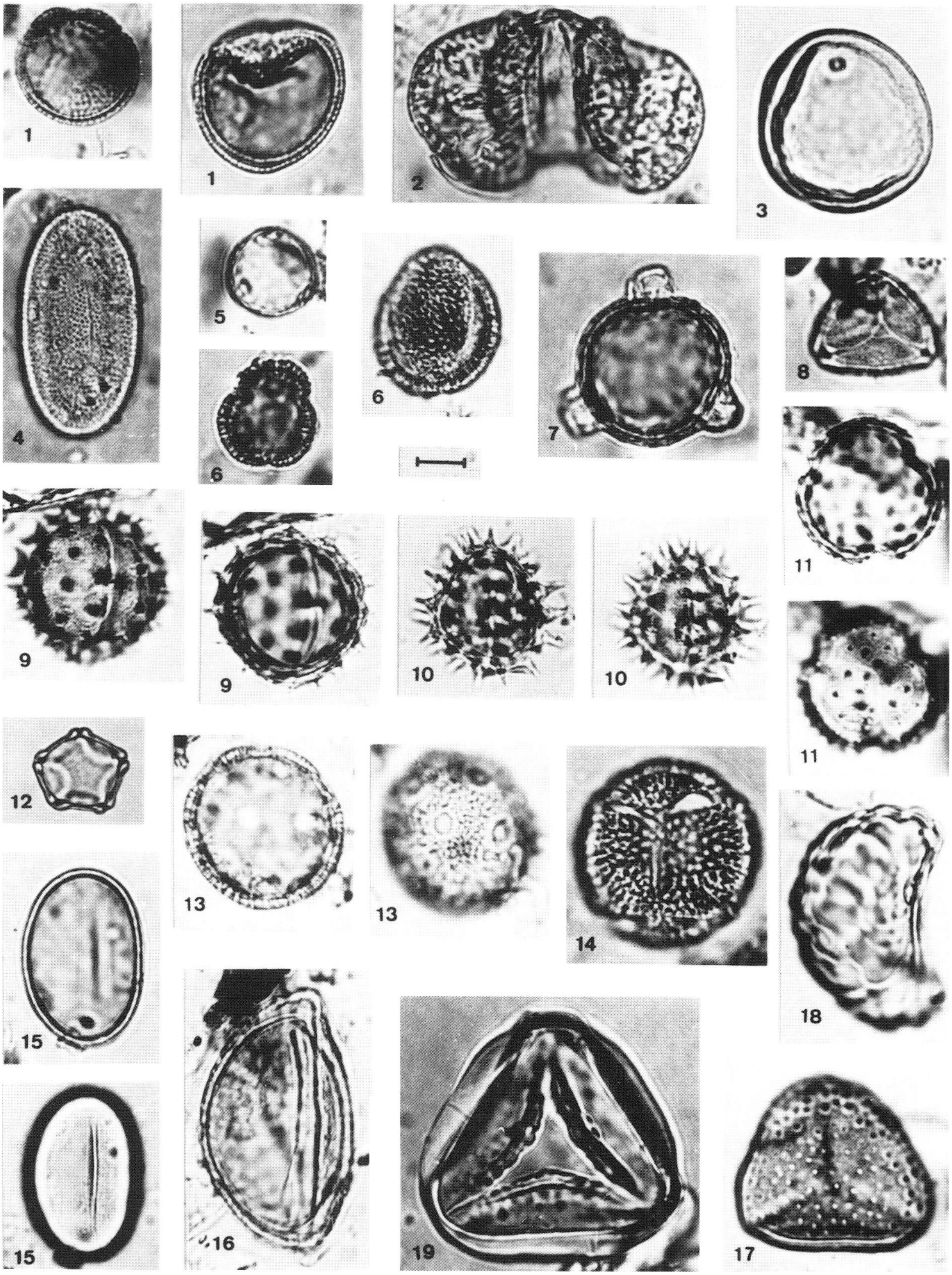
1
1
2
3
4
5
6
6
7
8
9
9
10
10
11
11
12
13
13
14
15
15
16
18
19
17

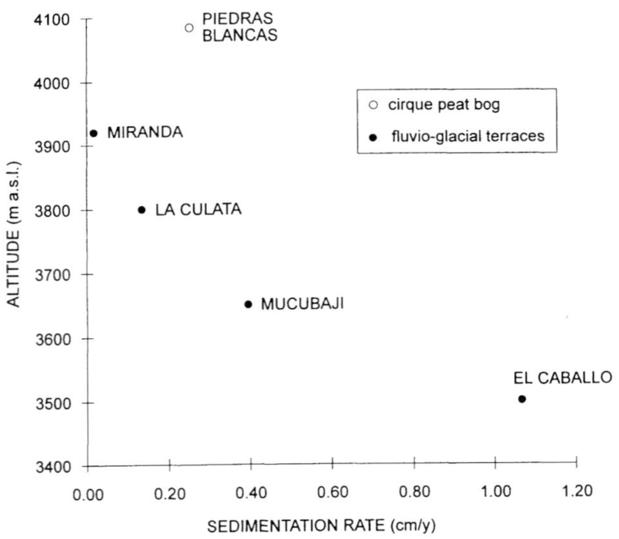

Text-Figure 3. Average sedimentation rates in relation with altitude in the Venezuelan Andes. Data from Rull (1998), Rull et al. (1987), Salgado-Labouriau and Schubert (1976), and Salgado-Labouriau et al. (1977, 1988).

observed in lake sediments but with lower absolute values, which is probably due to sorting, because the different transport energies involved (Rull, 1998). Contrarily to their inverse relationship with altitude, sedimentation rates showed an increasing tendency with age, due to the inverse relationship between age and altitude of sediments (see below).

GLACIER RETREAT

The oldest sediments found in the Venezuelan Andes are at the lowest elevations reached by glaciers at the Last Glacial Maximum (Text-Figure 4). The sediments formerly deposited at higher elevations had been removed by the descending glaciers, and only new, younger sediments could be deposited after the glacier retreat. Since the altitude of the glaciers is an inverse function of time, the age of its sediments should be related to it in the same way. With the available data (Text-Figure 4), it is possible to roughly estimate that the velocity of the glacier retreat between the *El Caballo* stadial and the present averages 7.3 cm(altitude)/y. However, retreat was probably not been constant. Average retreat velocities were less than 3 cm/y during the Late Glacial, and a 'negative velocity' (glacier readvance) was also recorded around 6 ka BP. The retreat proceeded later at low velocities until the Little Ice Age, after which the speed dramatically increased, until present-day. The relationship between the altitude and the age of sediments is an interesting factor to take into account for developing a sampling planning.

PALEOLIMNOLOGY

Statistical association among remains of aquatic organisms provided three assemblages representing respectively littoral, pelagic and bog conditions (Rull & Vegas-Vilarrúbia, 1996, 1997). The littoral assemblage is composed by two filamentous algae of the genus *Spirogyra* and the oribatid mite genus *Hydrozetes* (Plates 3 and 6). The species of the genus *Hydrozetes* are cosmopolitan and fully aquatic, yielding abundant remains in lake sediments (Erickson, in press). According to Erickson (1988), they live in the littoral zone, and are intimately associated to the aquatic vegetation, being opportunistic in their feeding requirements, that include filamentous algae, wood fibers, detritus, bacterial slime, and decaying plant and animal material. The pelagic assemblage is entirely dominated by *Botryococcus*, with the occurrence of *Bulbochaete* and dinoflagellate cysts (Plates 3 and 4). The bog assemblage is represented by *Isoëtes* and *Zygnema*-type (Plates 1 and 3).

From the stratigraphic sequence of these three assemblages (Text-Figure 5), a lake with well-developed nearshore communities was inferred for pollen zone MC-1. During MC-2, a bog expansion began and near-shore communities declined. A maximum of pelagic environments in the middle of MC-3, together with a drastic reduction of littoral assemblage, coincides with minimum estimated

PLATE 2

20 Film VR50, 30-30A. 40x2.5. Bar = 10 µm.
21 Film VR66, 34. 100x1.25. Bar = 8 µm.
22 Film VR61, 29. 100x1.25. Bar = 8 µm.
23 Film VR61, 26. 100x1.25. Bar = 8 µm.
24 Film VR67, 10. 100x1.25. Bar = 8 µm.
25 Film VR49, 2A-3. 40x2.5. Bar = 10 µm.
26 Film VR66, 30. 100x1.25. Bar = 8 µm.
27 Film VR62, 29A-30. 100x1.25. Bar = 8 µm.
28 Film VR61, 25. 100x1.25. Bar = 8 µm.
29 Film VR63, 27A-28. 100x1.25. Bar = 8 µm.
30 Film VR62, 27. 100x1.25. Bar = 8 µm.
31 Film VR67, 9. 100x1.25. Bar = 8 µm.
32 Film VR67, 4. 100x1.25. Bar = 8 µm.
33 Film VR67, 12. 100x1.25. Bar = 8 µm.
34 Film VR66, 35. 100x1.25. Bar = 8 µm.
35 Film VR63, 26-27A. 100x1.25. Bar = 8 µm.
36 Film VR66, 11. 100x1.25. Bar = 8 µm.
37 Film VR66, 4. 100x1.25. Bar = 8 µm.
38 Film VR66, 1. 100x1.25. Bar = 8 µm.
39 Film VR67, 8. 100x1.25. Bar = 8 µm. Film VR67, 6. 100x1.25. Bar = 8 µm.
40 Film VR67, 13. 100x1.25. Bar = 8 µm.
41 Film VR67, 24. 100x1.25. Bar = 8 µm.
42 Film VR64, 26A-27. 100x1.25. Bar = 8 µm.
43 Film VR66, 32. 100x1.25. Bar = 8 µm. Film VR66, 33. 100x1.25. Bar = 8 µm.
44 Film VR67, 14. 25x2.5. Bar = 16 µm.
45 Film VR66, 7. 100x1.25. Bar = 8 µm.
46 Film VR66, 16. 100x1.25. Bar = 8 µm.
47 Film VR61, 24. 100x1.25. Bar = 8 µm.

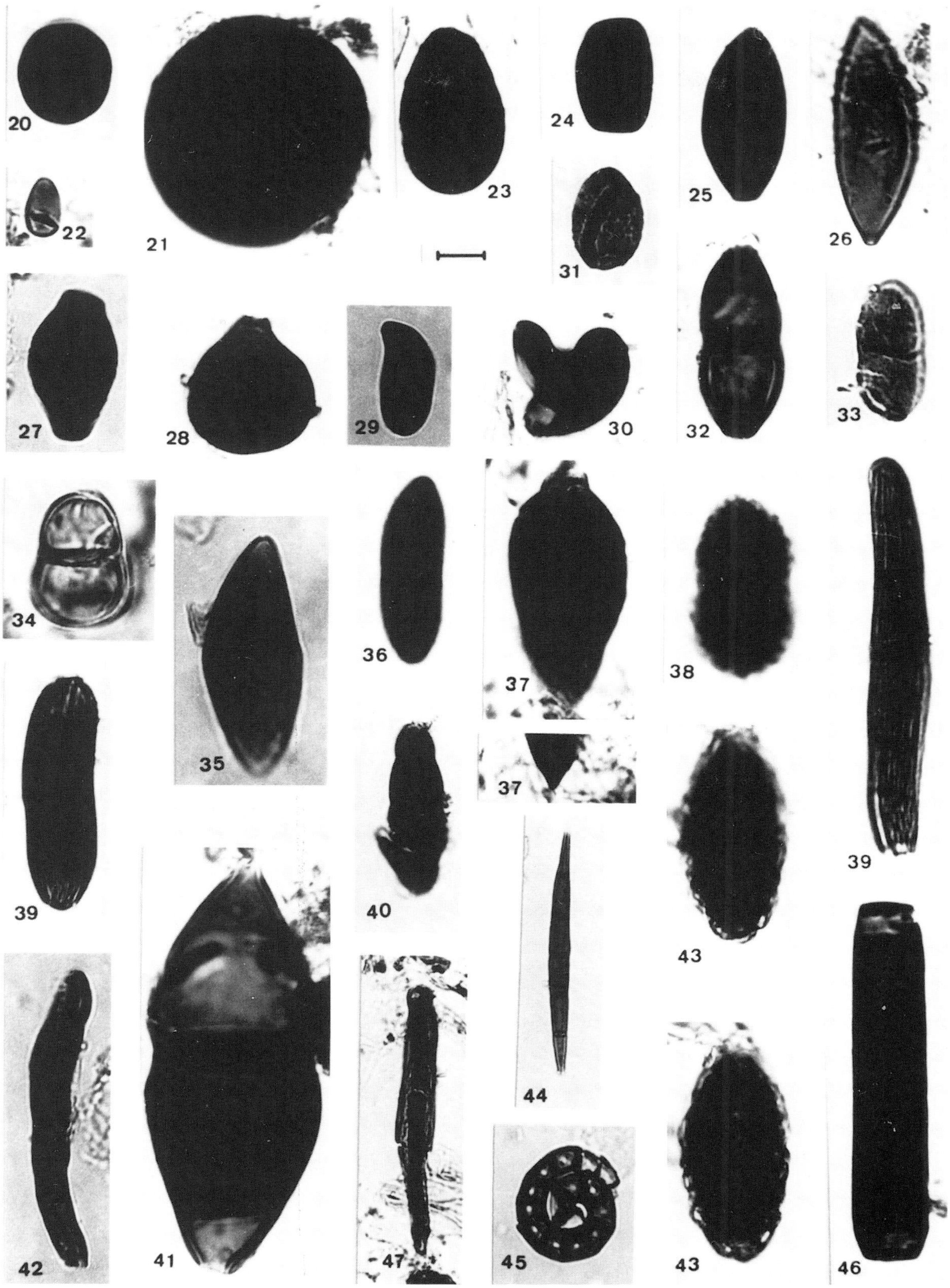
20
21
22
23
24
25
26
27
28
29
30
31
32
33
34
35
36
37
37
38
39
39
40
41
42
43
43
44
45
46
47

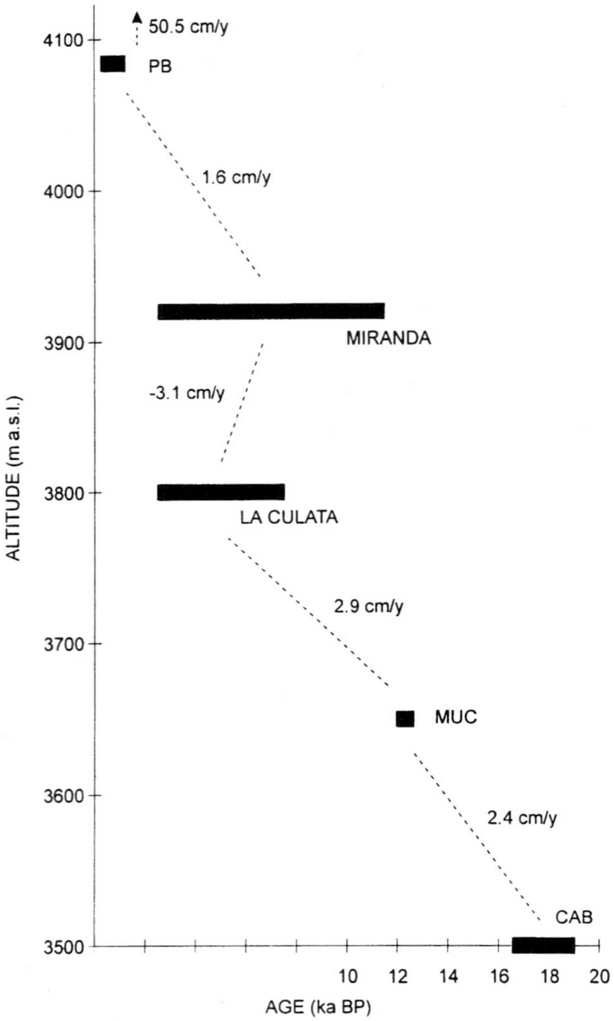

Text-Figure 4. Age-altitude relationships for the sites studied so far in the Venezuelan Andes. PB = Piedras Blancas, MUC = Mucubají, CAB = El Caballo. Source data as in Text-Figure 3.

temperatures (*El Caballo* stadial). This suggests that the lake was probably frozen, thus preventing the littoral communities to develop, but allowing the plankton to persist below the ice cover. An alternative explanation is that *Botryococcus* was the only preserved remains from the few organisms that could tolerate glacial temperatures, while zygospore producers could not grow under these environmental conditions. This is supported by the assessment of Salgado-Labouriau (1979), who suggested that zygospores preserved in Andean sediments have a lower altitudinal — and, therefore, thermal- limit than *Botryococcus.* Furthermore, Vareschi (1970) observed that Andean lakes above 4000 m altitude have very well developed plankton communities, but filamentous algae are scarce or absent. Whatever the case, the existence of a glacial lake without or with poorly developed littoral communities seems the reasonable interpretation for the *El Caballo* stadial. This lake disappeared immediately after.

FUNGI SPORES AND UNKNOWN FORMS

An important part of the present study is the documentation of unknown fossils, and their ecological characterization, for further use as potential palaeoecological indicators. Two main approaches are used for this study: the taxonomic one, based on the identification of the remains by comparison with living materials, and the statistical one, which attempts to objectively associate the stratigraphic distribution of unknown forms with others (or their associations) having well-known indicator properties (Rull, 1991).

A total of 28 fungal types classes and 30 unknown forms have been reported thus far (Rull and Vegas-Vilarrúbia, 1998). Concerning fungal spores, six forms (20, 28, 29, 30, 35 and 37; Plate 2) had significant correlations with both pollen and fern spores, suggesting some degree of ecological association between them. Quantitative association within fungal spores was also tested, and six groups could be found, although only two of them seem to be paleoecologically significant (Text-Figure 5). Assemblage I is represented by forms significantly correlated with *páramo* pollen and spores (20, 23, 28, 29, 30, 35, 37), while assemblage B is constituted by forms 24, 31, 38, 39 and 44, all of them having significant negative correlations with *páramo* plants. The highest values of assemblage II correspond to the *El Caballo* stadial, suggesting some relation to low temperatures. It is proposed tentatively that these two fungal associations have different temperature tolerances, and could be used as paleoclimatic indicators. This possibility is worth to be evaluated, via the taxonomic identification of spores, and the study of their distribution in modern sediments from different altitudes.

Several of the unidentified forms show the same trends as aquatic organisms, which suggests similar habitat requirements (Rull & Vegas-Vilarrúbia, 1998). Forms 58, 69, 71, 72, 76, 79 and 84 (Plates 3, 4 and 5) are significantly correlated with the littoral assemblage, therefore, they are probably remains of elements from shore-lake communities. Two forms (54 and 68) show significant correlations with pelagic assemblage. Finally, forms 70 and 77 show good statistical agreement with the bog assemblage.

In general, the statistical approach used in the present work proved its potential usefulness for the ecological characterization of unknown fossil remains. However, it is still necessary to develop more of these studies to confirm or reject particular conclusions. A survey oriented towards the characterization of modern sediments from different environments and (aquatic) sub-environments by their total fossil content would be especially informative. Moreover, the effort to appropriately document and identify as much of these forms as possible should continue.

ACKNOWLEDGMENTS

The authors are grateful to B. van Geel and J. M. Erikson for the identification of some unknown forms, and to Elisa and Montserrat Rull for their help in the field. The critical review of two unknown referees contributed to the improvement of the original manuscript. Editorial facilities were provided by PDVSA Exploration & Production.

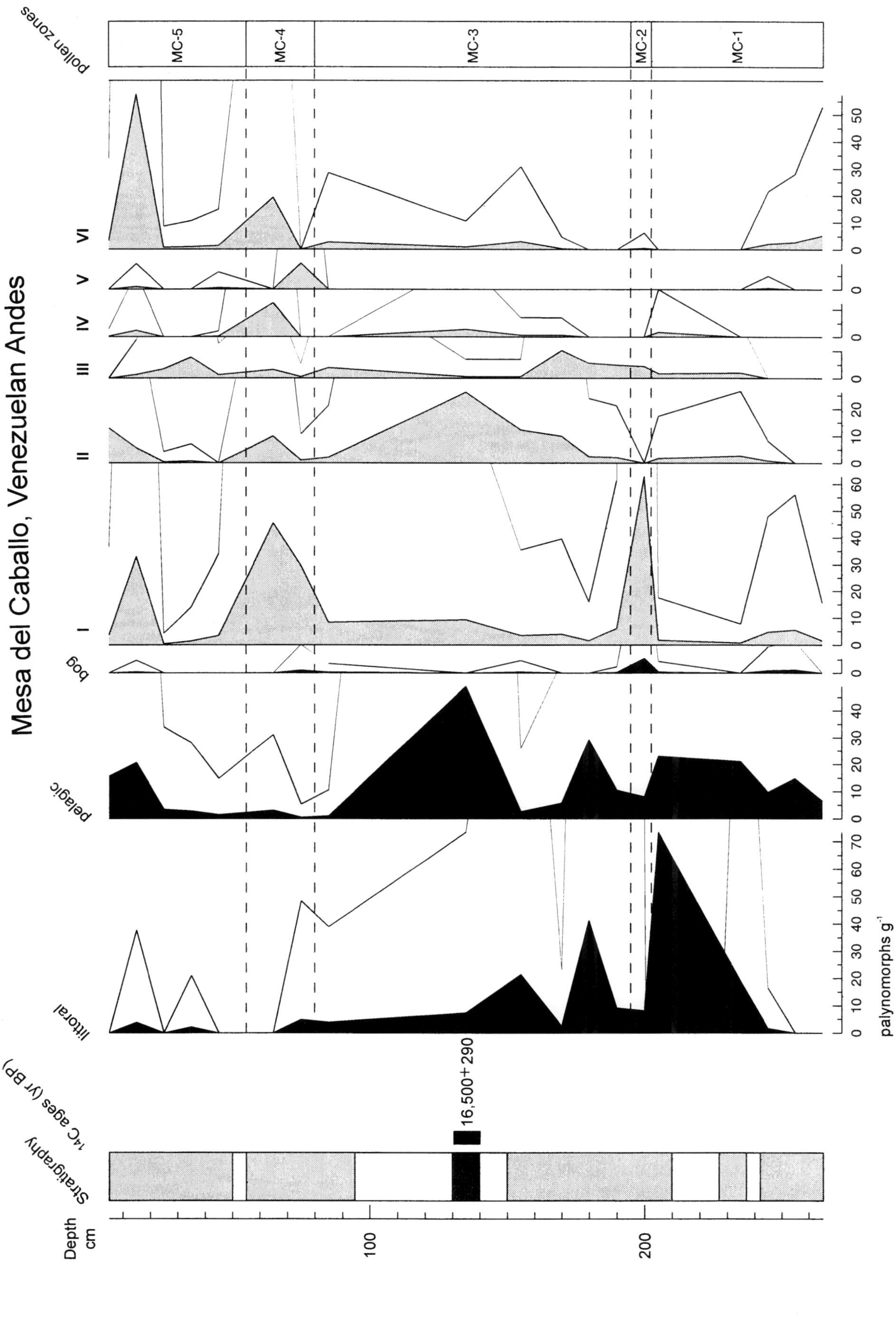

Text-Figure 5. Stratigraphic distribution of aquatic remains (black) and fungal assemblages (gray), in concentration values (modified from Rull and Vegas-Vilarrúbia, 1998). Solid lines represent x10 exaggeration.

References Cited

ERICKSON, J. M.

1988 Fossil oribatid mites as tools for Quaternary paleoecologists: preservation quality, quantities, and taphonomy. *Bulletin of Buffalo Society of Natural Sciences*, 33: 207–226.

in press Can paleoacarology contribute to global change research? *Proceedings of the International Congress Acarology*.

RULL, V.

1991 Palaeoecological significance of Chryosphycean stomatocysts: a statistical approach. *Hydrobiologia*, 220: 161–165.

1996 Late Pleistocene and Holocene climates of Venezuela. *Quaternary International*, 31: 85–94.

1998 Palaeoecology of Pleniglacial sediments from the Venezuelan Andes. Palynological record of El Caballo Stadial. *Review of Palaobotany and Palynology*, 99: 95–114.

1999 Palaeoclimatology and sea-level history in Venezuela. New data, land-sea correlations, and proposals for future studies in the framework of the IGBP-PAGES Project. *Interciencia*, 24: 92–101.

1999 The middle Wisconsin El Pedregal interstadial in the Venezuelan Andes: palynological record. *Current Research in the Pleistocene*, 16: 111–113.

RULL, V., and VEGAS-VILARRÚBIA, T.

1996 Palynological evidence for the Last Glacial Maximum in the Venezuelan Andes. *Current Research in the Pleistocene*, 13: 89–91.

1997 Acari remains from pleniglacial sediments of the Venezuelan Andes and their paleolimnological significance. *Current Research in the Pleistocene*, 14: 153–155.

1998 Palaeoecology and palaeolimnology of pleniglacial sediments from the Venezuelan Andes. Palynomorphs other than pollen and fern spores. *Boletín de laSociedad Venezolana de Geólogos*, 23: 5–27.

RULL, V., SALGADO-LABOURIAU, M. L., SCHUBERT, C., and VALASTRO, S.

1987 Late Holocene temperature depression in the Venezuelan Andes: palynological evidence. *Palaeogeography, Palaeoclimatology, Palaeoecology*, 60: 109–121.

SALGADO-LABOURIAU, M. L.

1979 Modern pollen deposition in the Venezuelan Andes. *Grana*, 18: 53–68.

1984 Late Quaternary palynological studies in the Venezuelan Andes. *Erdwissenschaftliche Forschung*, 18: 279–293.

1989 Late Quaternary climatic oscillations in the Venezuelan Andes. *Biology International*, 18: 12–14.

1991 Palynology of the Venezuelan Andes. *Grana*, 30: 342–349.

SALGADO-LABOURIAU, M. L., and SCHUBERT, C.

1976 Palynology of Holocene peat bogs from central Venezuelan Andes. *Palaeogeography, Palaeoclimatology, Palaeoecology*, 19: 147–156.

SALGADO-LABOURIAU, M. L., SCHUBERT, C., and VALASTRO, S.

1977 Paleoecologic analysis of a Late Quaternary terrace from Mucubají, Venezuelan Andes. *Journal of Biogeography*, 4: 313–325.

SALGADO-LABOURIAU, M. L., RULL, V., SCHUBERT, C., and VALASTRO, S.

1988 The establishment of vegetation after late Pleistocene deglaciation in the Páramo Miranda, Venezuelan Andes. *Review of Palaeobotany and Palynology*, 55: 5–17.

SCHUBERT, C.

1974 Late Pleistocene Mérida Glaciation, Venezuelan Andes. *Boreas*, 3: 147–152.

SCHUBERT, C., and RINALDI, M.

1987 Nuevos datos sobre la cronología del estadio tardío de la Glaciación Mérida, Andes Venezolanos. *Acta Científica Venezolana*, 38: 135–136.

VARESCHI, V.

1970 *Flora de los páramos de Venezuela*. Universidad de los Andes. Ediciones del Rectorado, Mérida. 429 pgs.

Authors' addresses

VALENTÍ RULL
PDVSA Exploration & Production
PO Box 829
Caracas 1010-A, Venezuela
rullv@pdvsa.com

TERESA VEGAS-VILARRÚBIA
GEOHIDRA
División Ambiente
PO Box 47851
Caracas, Venezuela
terveg@telcel.net.ve

PLATE 3

48 *Botryococcus*. Film VR66, 20. 100x1.25. Bar = 8 µm. Film VR62, 5A-6. 100x1.25. Bar = 8 µm.

49 *Zygnema*-type. Film VR61, 12. 100x1.25. Bar = 8 µm. Film VR61, 11. 100x1.25. Bar = 8 µm.

50 *Spirogyra-1*. Film VR63, 5A-6. 100x1.25. Bar = 8 µm. Film VR63, 6A-7. 100x1.25. Bar = 8 µm.

51 *Spirogyra-2*. Film VR49, 5A-6. 40x2.5. Bar = 10 µm.

52 Oedogoniaceae. Film VR49, 27A-28. 40x2.5. Bar = 10 µm. Film VR49, 28A-29. 40x2.5. Bar = 10 µm. Film VR49, 29A-30. 40x2.5. Bar = 10 µm.

53 Unknown. Film VR61, 32. 100x1.25. Bar = 8 µm. Film VR61, 33. 100x1.25. Bar = 8 µm.

54 Desmidiaceae?. Film VR62, 23A-24. 25x2.5. Bar = 16 µm. Film VR62, 24A-25. 25x2.5. Bar = 16 µm.

55 Unknown. Film VR66, 9. 100x1.25. Bar = 8 µm.

56 Unknown. Film VR66, 10. 100x1.25. Bar = 8 µm.

57 *Mougeotia?*. Film VR64, 25A-26. 100x1.25. Bar = 8 µm.

58 Unknown. Film VR63, 13A-14. 25x2.5. Bar = 16 µm. Film VR63, 12A-13. 25x2.5. Bar = 16 µm.

59 Type 179. Film VR65, 35-35A. 100x1.25. Bar = 8 µm.

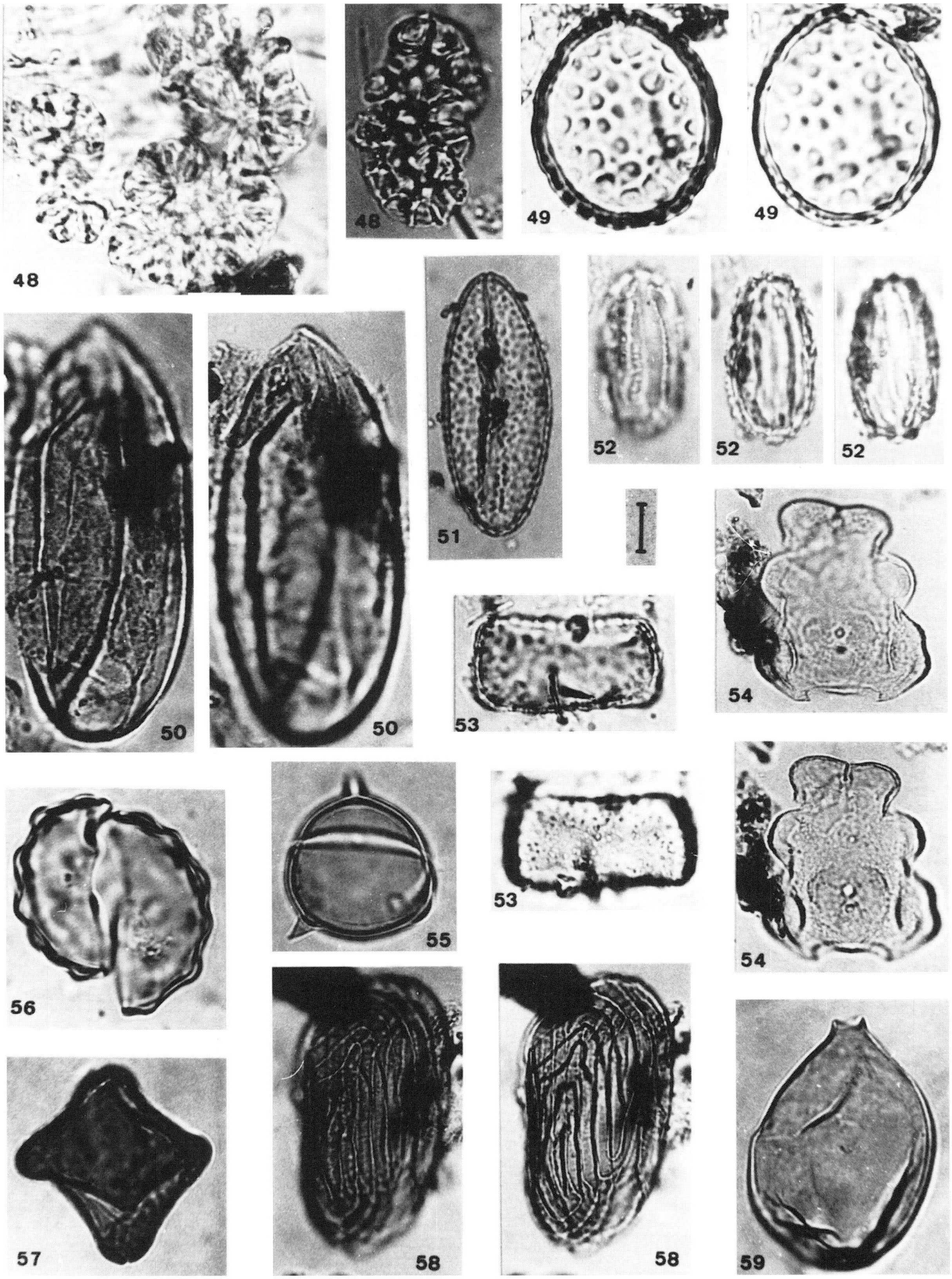
48
48
49
49
52
52
52
51
50
50
53
54
53
54
55
56
57
58
58
59

PLATE 4

60 Unknown. Film VR62, 7A-8. 100x1.25. Bar = 8 µm. Film VR62, 8A-9. 100x1.25. Bar = 8 µm.

61 Unknown. Film VR63, 7A-8. 100x1.25. Bar = 8 µm.

62 Unknown. Film VR49, 24A-25. 100x1.25. Bar = 8 µm.

63 Unknown. Film VR49, 13A-14. 40x2.5. Bar = 10 µm.

64 Unknown. Film VR67, 23. 100x1.25. Bar = 8 µm.

65 Unknown. Film VR49, 23A-24. 40x2.5. Bar = 10 µm.

66 Unknown. Film VR66, 0. 100x1.25. Bar = 8 µm. Film VR65, 36-36A. 100x1.25. Bar = 8 µm.

67 Unknown. Film VR66, 18. 100x1.25. Bar = 8 µm. Film VR66, 17. 100x1.25. Bar = 8 µm.

68 Unknown. Film VR67, 26. 25x2.5. Bar = 16 µm.

69 Unknown. Film VR65, 4-4A. 100x1.25. Bar = 8 µm.

70 Unknown. Film VR63, 23A-24. 100x1.25. Bar = 8 µm. Film VR63, 24A-25. 100x1.25. Bar = 8 µm.

71 Unknown. Film VR63, 32A-33. 100x1.25. Bar = 8 µm.

72 Unknown. Film VR62, 9A-10. 100x1.25. Bar = 8 µm. Film VR62, 10A-11. 100x1.25. Bar = 8 µm.

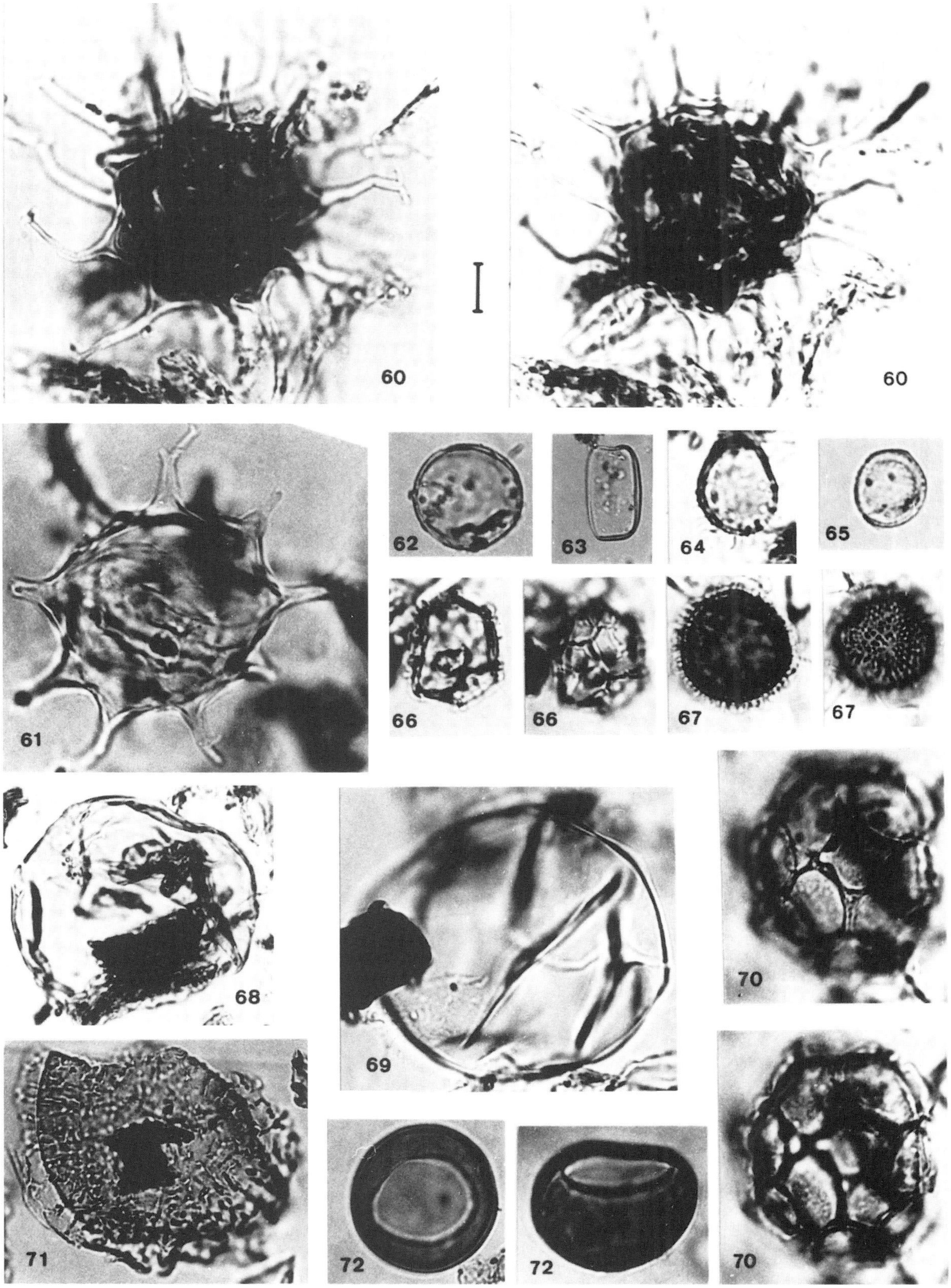
60
60
61
62
63
64
65
66
66
67
67
68
69
70
70
71
72
72

PLATE 5

73 Unknown. Film VR50, 15-15A. 40x2.5. Bar = 10 µm. Film VR49, 16-16A. 40x2.5. Bar = 10 µm.

74 Unknown. Film VR50, 14-14A. 40x2.5. Bar = 10 µm.

75 Unknown. Film VR63, 31A-32. 25x2.0. Bar = 20 µm.

76 Type 353B. Film VR63, 9A-10. 25x2.0. Bar = 20 µm. Film VR64, 3A-4. 25x1.6. Bar = 25 µm.

77 Unknown. Film VR63, 30A-31. 100x1.25. Bar = 8 µm. Film VR63, 29A-30. 100x1.25. Bar = 8 µm.

78 Unknown. Film VR65, 33-33A. 25x2.5. Bar = 16 µm. Film VR65, 34-34A. 25x2.5. Bar = 16 µm. Film VR65, 13-13A. 25x2.5. Bar = 16 µm. Film VR65, 10-10A. 25x2.5. Bar = 16 µm. Film VR65, 11-11A. 25x2.5. Bar = 16 µm.

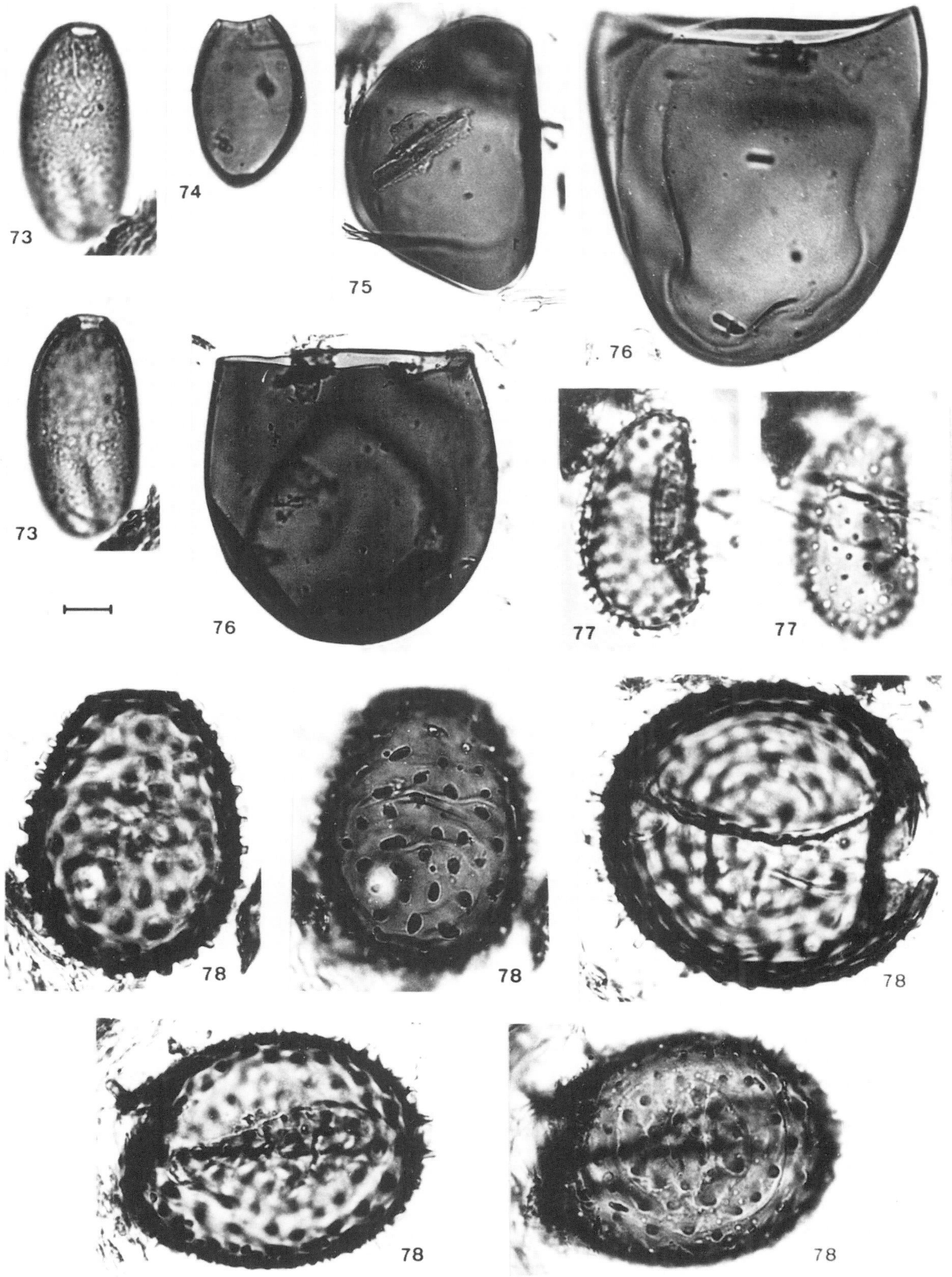
73
74
75
76
73
76
77
77
78
78
78
78
78

PLATE 6

79 Unknown animal remains. Film VR6, 16-16A. 25x2.0. Bar = 20 µm. Film VR49, 8A-9A. 25x2.5. Bar = 16 µm. Film VR61, 19. 25x2.5. Bar = 20 µm. Film VR61, 20. 100x1.25. Bar = 8 µm.

80 Unknown animal remains. Film VR61, 18. 100x1.25. Bar = 8 µm. Film VR67, 17. 25x1.6. Bar = 25 µm.

81 Unknown animal remains. Film VR64, 6A. 100x1.6. Bar = 6.25 µm.

82 Unknown animal remains. Film VR62, 13A-14. 25x2.5. Bar = 16 µm.

83 Oribatei, Acari (chela). Film VR6, 17A-18. 100x2. Bar = 5 µm.

84 Cladocera (postabdomen). Film VR62, 22A-23. 100x1.25. Bar = 8 µm.

85 *Hydrozetes* (Oribatei, Acari). Film VR64, 7A-8. 100x1.25. Bar = 8 µm. Film VR64, 8A-9. 100x1.25. Bar = 8 µm. Film VR65, 9-9A. 25x2.5. Bar = 16 µm. Film VR65, 8-8A. 25x2.5. Bar = 16 µm. Film VR64, 6A-7. 100x1.25. Bar = 8 µm. Film VR62, 16A-17. 100x1.25. Bar = 8 µm. Film VR64, 24A-25. 25x2.5. Bar = 16 µm.

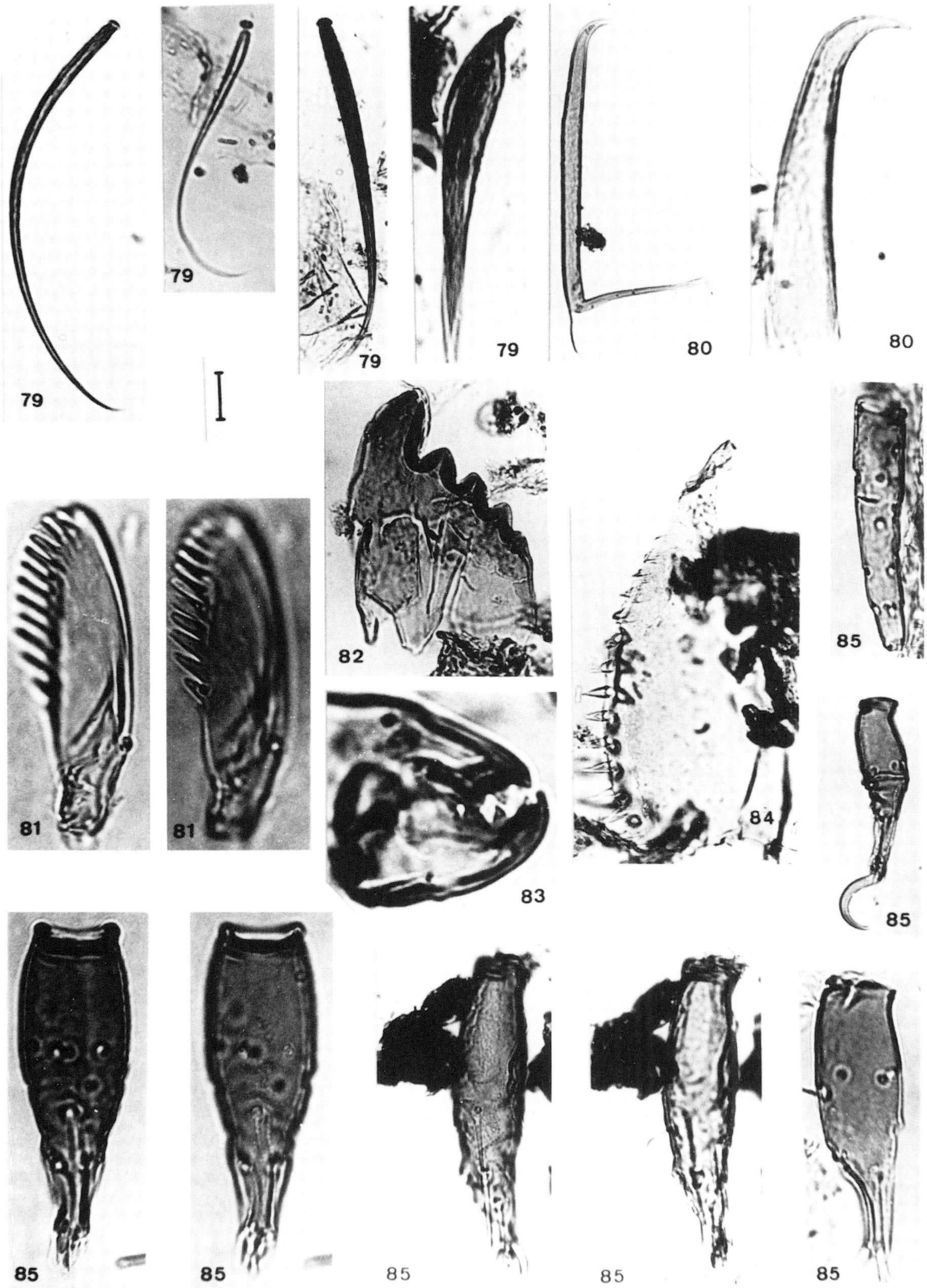
79
79
79
79
80
80
79
82
85
81
81
84
83
85
85
85
85
85
85

POLLEN RECORD REFLECTING GLACIAL–INTERGLACIAL CYCLES AROUND 1.1 MA: AN EXAMPLE FROM THE YACHIHO GROUP, CENTRAL JAPAN

Takeshi Saito

Abstract

Sediments, approximately 1.1 Ma, collected from the Yachiho Group in central Japan were palynologically studied in order to reconstruct the changes in paleovegetation and paleoclimate. The pollen assemblages from 39 samples consist mainly of Pinaceae including *Picea, Tsuga, Abies,* and *Pinus,* associated with the broad-leaved deciduous trees such as *Fagus.* These assemblages reflect the Pinaceae-dominated coniferous vegetation, associated with broad-leaved deciduous forests. The pollen assemblages of lower samples contain a higher abundance of *Larix*-type pollen than those from upper samples. This suggests the vegetation recovery by a pioneer species such as *Larix kaempferi* after destruction by mudflow.

Evaluation of variability among pollen assemblages allows to detect that small changes are significant level in the percentages of the temperate elements. The pattern of these small changes shows cyclicity, and at least three cycles are recognizable. These cyclic changes probably resulted from fluctuations in Early Pleistocene climate, representing glacial–interglacial cycles. The paleoclimate changes seem to be consistent with low-amplitude fluctuations in the oxygen isotope records around 1.1 Ma. The cycles might be correlated with oxygen isotope stages 38 to 34.

INTRODUCTION

Glacial–interglacial cycles, represented by changes in continental ice volume in Quaternary, have been recognized in the oxygen isotope records of deep-sea cores (e.g. Shackleton and Opdyke, 1976; Williams et al., 1988; Ruddiman et al., 1989; Shackleton et al., 1990; Tiedemann et al. 1994). The Milankovitch theory and other theoretical studies (e.g. Berger and Loutre, 1991) indicate that astronomic forcing has caused glacial–interglacial cycles. Palynology has also made significant contributions to understand the glacial–interglacial cycles in latest Quaternary. In contrast, it seems that the number of palynological studies on glacial–interglacial cycles in Early Pleistocene is limited. This is most likely because of generally small changes in climate between glacial and interglacial conditions, and the comparatively short period of the cycles in Early Pleistocene as seen in the oxygen isotope records characterized by low-amplitude (Williams et al., 1988; Shackleton, 1995) and high-frequency (Williams et al., 1988; Ruddiman et al. 1989; Tiedemann et al. 1994). Some high-resolution palynological studies (Combourieu-Nebout, 1993; Hooghiemstra and Ran, 1994; Leroy and Dupont, 1994; Ravazzi and Strick, 1995) can show pollen profiles indicating glacial–interglacial cycles in Early Pleistocene and/or Late Pliocene. Temperature changes in the cycles, with some exceptions, are generally smaller than those in latest Quaternary.

The objective of this paper is to try to present a pollen record reflecting Early Pleistocene glacial–interglacial cycles by using a new sampling procedure. Then, the pollen record is compared to global climatic changes expressed by the oxygen isotope records. Rapid vegetation recovery after destruction by mudflow is also discussed.

Saito, Takeshi
2001 Pollen record reflecting glacial–interglacial cycles around 1.1 Ma: An example from the Yachiho Group, central Japan; *In*: Goodman, D.K., and Clarke, R.T. (eds.), *Proceedings of the IX International Palynological Congress, Houston, Texas, U.S.A., 1996;* American Association of Stratigraphic Palynologists Foundation, p. 403–409.

 ISBN 0–931871–06–9

GEOLOGY

The samples for this study come from Unit II of the Uppermost Formation of the Yachiho Group (Research Group for Mt. North Yatsugatake, 1988). The formation is distributed along the north and east sides of Mt. Yatsugatake in central Japan. It consists of lavas, volcaniclastic rocks, and intercalated non-volcanic sediments. The non-volcanic sediments were examined.

Unit II corresponds to the Hikage Formation of Kawachi (1974), consisting of organic-rich sandy mudstones, organic-rich mudstones, tuffaceous sandstones, and tuffs. The marker tuff layers, "Sakura" and "Kirapi" (Research Group for Mt. North Yatsugatake, 1988), are intercalated in the upper part of the unit. The organic-rich sediments yield non-marine diatom fossils. The sedimentary basin was formed by a mudflow creating a dam (Kawachi, 1974).

K-Ar dating by Kaneoka and Kawachi (1983) gave an age of 1.17 ± 0.05 Ma for the lowermost part of the Uppermost Formation, and the normal magnetic polarity in Unit IV reported by Kondo (1988) is correlated with the Jaramillo subchron (Research Group for Mt. North Yatsugatake, 1988). Therefore, the age of Unit II is around 1.1 Ma in Early Pleistocene (Text-Figure 1).

MODERN VEGETATION

The present-day altitudinal zonation of vegetation in central Japan is summarized as follows to understand paleovegetational and paleoclimatic interpretation. For the basal zone (0 to ca. 700 m): the main taxa are *Castanopsis sieboldii, C. cuspidata,* evergreen *Quercus* species (*Q. glauca, Q. myrsinaefolia,* etc.), Lauraceae (*Cinnamomum camphora, Machilus thunbergii,* etc.), and *Camellia japonica.* For the

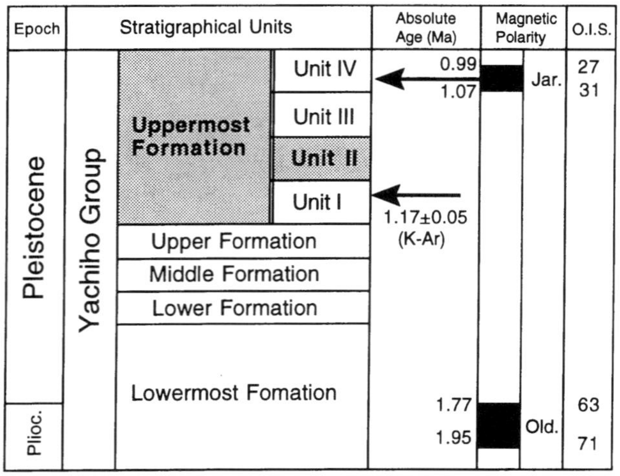

Text-Figure 1. Stratigraphical position for Unit II of the Uppermost Formation, Yachiho Group. Magnetic polarity time scale and oxygen isotope stages (O. I. S.) for magnetic boundaries are taken from Shackleton et al. (1990). Jar.=Jaramillo subchron, Old.=Olduvai subchron.

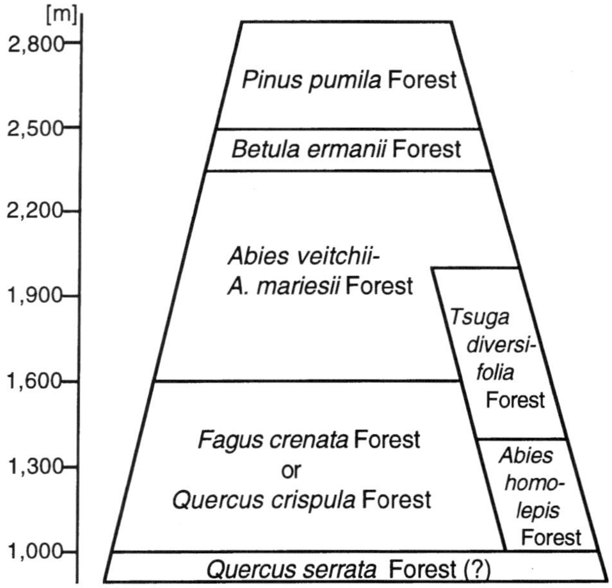

Text-Figure 2. Potential natural vegetation of Mt. Yatsugatake (adapted from Tsuchida, 1991).

montane zone (ca. 700 to 1700 m): the main taxa of the lower part are *Castanea crenata, Zelkova serrata, Quercus acutissima* (deciduous), and *Q. serrata* (deciduous). The main taxa of the upper part are *Fagus crenata, Quercus crispula* (deciduous), and *Betula platyphylla* var. *japonica*. For the subalpine zone (ca. 1700 to 2500 m): the main taxa are *Abies veitchii, A. mariesii, Picea jezoensis* var. *hondoensis*, and *Tsuga diversifolia*. For the alpine zone (over ca. 2500 m): main taxa are *Pinus pumila, Betula ermanii, Alnus maximowiczii, Sorbus commixta*, and *Rhododendron aureum*. More complete accounts of the modern vegetation in central Japan are given in Miyawaki (1985). Text-Figure 2 shows a simplified diagram of the potential natural vegetation of the Mt. Yatsugatake (Tsuchida, 1991).

SAMPLES AND METHODS

The sampling locality is Minamiaiki, Nagano Prefecture, at an altitude of about 1000 m (Text-Figure 3). The columnar sections in Text-Figure 4 show the horizons of the samples used for pollen analysis. Two series of samples were collected, each having different resolutions. One series of samples (Series I) was successively taken from the lowest organic-rich sediment of the column, which was 20 cm thick. Series I was analyzed to evaluate the variability of pollen percentages and to reveal the vegetation recovery after destruction by mudflow. The other series of samples (Series II) was collected at the interval of about 75 cm in order to investigate the vegetation response to climate changes.

The samples were prepared palynologically as follows. The crushed samples were treated with KOH (10%), HF (46%) and the acetolysis. A solution of $ZnCl_2$ (s.g. 1.9) was added to concentrate pollen grains. The treated residue was mounted in glycerine jelly on glass slides. More than 200 pollen grains and *Lycopodium* spores were identified in every sample. The sum of the temperate elements and that of non-arboreal pollen (NAP) were used for percentages. The temperate elements are as follows: *Juglans–Pterocarya, Corylus, Carpinus, Fagus*, deciduous *Quercus, Castanea*-type, *Ulmus–Zelkova*, and *Tilia*. Deciduous *Quercus* is included as a temperate element because the genus is one of the most important temperate elements in Japan, although a species of this genus (*Q. crispula* var. *horikawae*) now grows in the alpine zone in central Japan.

RESULTS

Text-Figure 5 shows the pollen and spore assemblages (simply called pollen assemblages below) of Series I and Series II. As a whole, the pollen assemblages consist mainly of Pinaceae including *Picea, Tsuga, Pinus*, and *Abies*, associated with broad-leaved deciduous trees including *Fagus*, deciduous *Quercus, Corylus*, and *Carpinus*. No major palynofloral change can be recognized.

In Series I, percentages of each taxon, temperate elements, and NAP do change. These changes provide a basis to evaluate pollen assemblages of Series II. For example, percentages of the temperate elements fluctuate, but they are relatively stable within 7.0 %. No significant changes for vegetation recovery after destruction by mudflow can be recognized in pollen assemblages in Series I. However, Series I and the lower part of Series II do contain a higher abundance of *Larix*-type pollen than the upper part of Series II.

In Series II pollen assemblages, percentages of the temperate elements show significant small changes (0–23.7 %), and the changes exhibit cyclicity. *Fagus* is the major contributor to the pollen assemblages rich in the temperate elements.

The pollen assemblages of the top four samples of Series II are characteristic. *Pinus* pollen is dominant except for the top third sample, whereas the pollen assemblages of other samples are dominated by *Picea* or *Tsuga*. The percentages of the temperate elements are very low. The *Tsuga* percentages decrease abruptly, and in contrast, the percentages of NAP and *Lycopodium* spores increase.

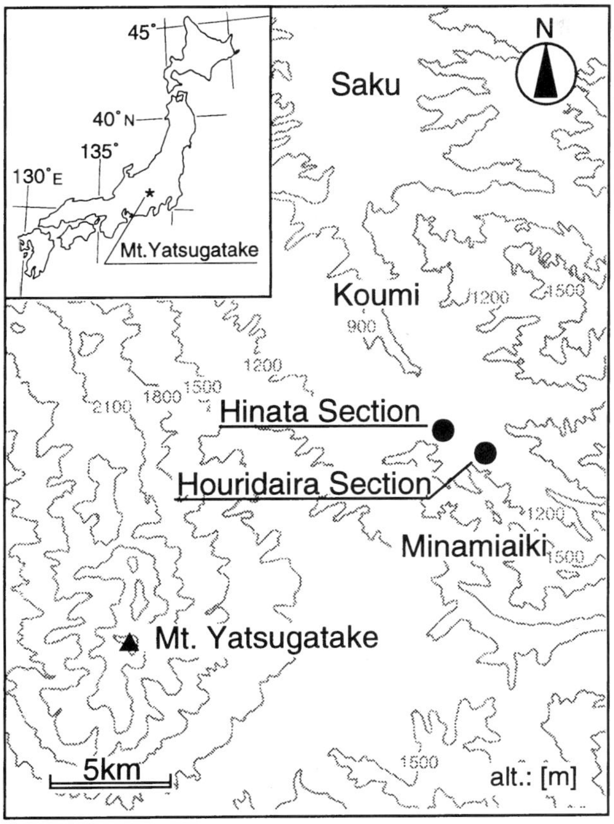

Text-Figure 3. Map showing localities of the sections in Text-Figure 4.

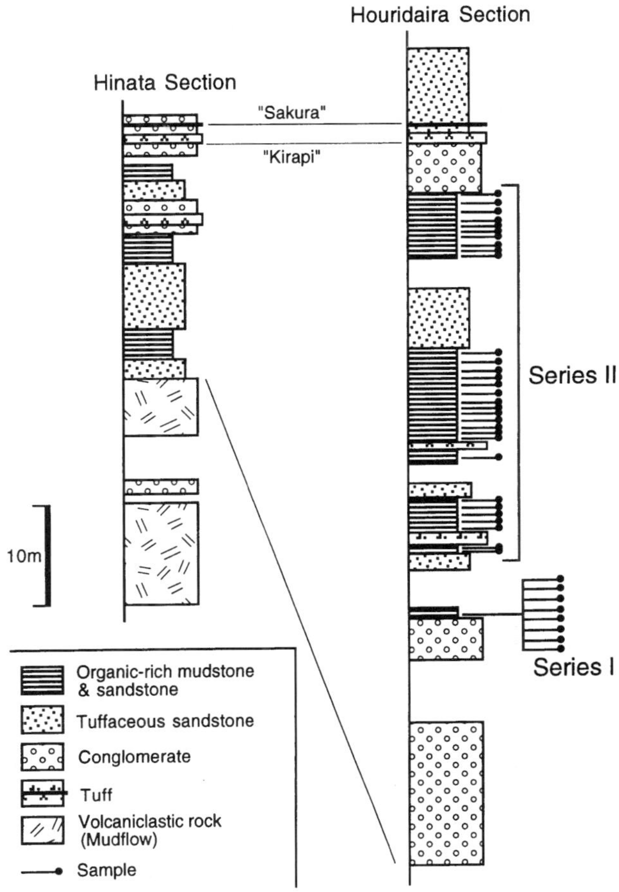

Text-Figure 4. Geologic columns showing horizons of samples.

DISCUSSION

Paleovegetation

The pollen assemblages in Text-Figure 5 reflect Pinaceae-dominated coniferous vegetation with broad-leaved deciduous forests. With the exception of the assemblages of the top four samples, the paleovegetation is comparable with the lower part of the subalpine zone or the uppermost part of the montane zone of present-day vegetation. The pollen assemblages of the top four samples probably indicate the alpine zone vegetation such as the dwarf pine zone (*Pinus pumila* forest). According to Tsukada (1967), *Pinus* pollen dominates the pollen assemblages from the surface sediments of the dwarf pine zone in central Japan, while *Tsuga* pollen and the other boreal elements such as *Abies, Larix,* and *Picea* are present in decreasing amounts, with *Fagus* and the other temperate elements being scarce. Furthermore, *Lycopodium* spores commonly appear only in the middle part of the dwarf pine zone, whereas they are scarce in other vegetation zones. These data are in good agreement with the pollen assemblages of the top four samples.

The pollen assemblages of Series I do not clearly reflect a vegetation recovery after destruction by mudflow, owing to pollen dispersal from the areas unaffected by the mudflow, and the passage of time after the occurrence of the mudflow. However, *Larix*-type pollen in the lower part of the studied sediments suggest that a pioneer species such as *Larix kaempferi* recovered the area damaged by the mudflow.

Paleoclimate

The relative paleoclimate change is inferred mainly on the basis of the total percentage of the temperate elements. In the above discussion, paleovegetation is interpreted by comparison with present-day vegetation zonation. However, paleovegetation zonation appears to have been neither quite the same as modern zonation, nor the same as one vertically shifted. This must be taken into consideration if we are to infer paleoclimate. For example, temperate coniferous forests, which now exist in restricted areas of the basal or montane zones, were often dominant in areas of Japan with a temperate climate during Pleistocene (Minaki, 1989). One of the extinct *Picea* species, possibly one of the dominant species in this study, seems to have grown in warm-temperate or subtropical areas; warmer conditions than for modern *Picea* (Momohara et al., 1990). It is because of this suspicion regarding conifers that the relative paleoclimate change is inferred almost entirely on the basis of the total percentage of the temperate elements, which presently grow and previously grew in temperate climate areas.

The cyclic nature of the changes in temperate element percentages, and the geological age of the studied sediments probably indicate that such cyclic changes resulted from climate fluctuations caused by glacial–interglacial cycles. The temperate elements were more abundant in

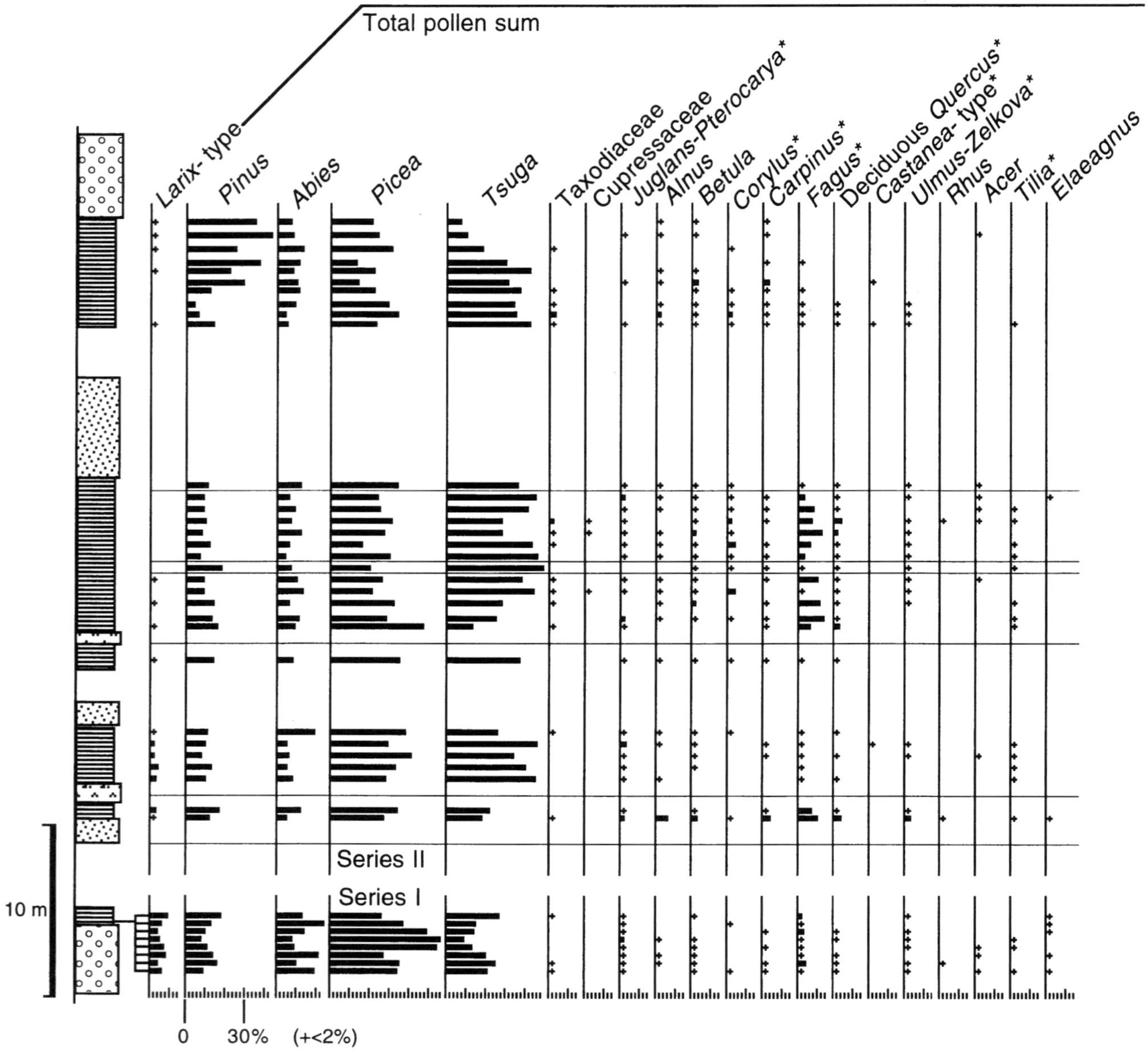

Text-Figure 5. Percentage diagram of the pollen and spore assemblages from Unit II of the Uppermost Formation, Yachiho Group. Pollen zones, climatic interpretation, and tentative correlation with oxygen isotope stages are shown on the right. Percentages of *Lycopodium* spores are calculated on the basis of total pollen sum. Asterisk represents the temperate element. GL=glacial, IG=interglacial, NAP=non-arboreal pollen.

the interglacial stages than during the glacial stages. The small changes observed in the percentages of the temperate elements are not due to inevitable variability in the pollen assemblages as pointed out above; percentages of the temperate elements of Series I are relatively stable. The amplitude of the climate fluctuations seems to be relatively small, since the palynoflora does not show major changes. Changes in pollen assemblages from the last glacial to postglacial periods near the studied locality (Tsuda, 1990) were bigger than those of this study. The paleoclimate indicated by the pollen assemblage from the top four samples was probably the coldest.

Cold zones (C1 to C4) and warm zones (W1 to W3) are distinguished by the percentage of the temperate elements (Text-Figure 5). The boundary value was conveniently set at 7.0 % on the basis of the percentages of the temperate elements of Series I. At least three cycles can be recognized from the pollen assemblages of Series I and II: C1-W1, C2-W2, C3-W3, and C4. It should be noted that the pollen assemblages of Series I correspond to the pollen assemblage of one sample of Series II. These cycles probably show the glacial–interglacial cycles and my interpretation is in Text-Figure 5. C3 is considered to be a cooling within an interglacial, because the term is rather shorter than other cold zones except C1.

The fluctuations in the temperate elements are broadly consistent with those of oxygen isotope fluctuations around 1.1 Ma. It is recognized that the oxygen isotope fluctuations

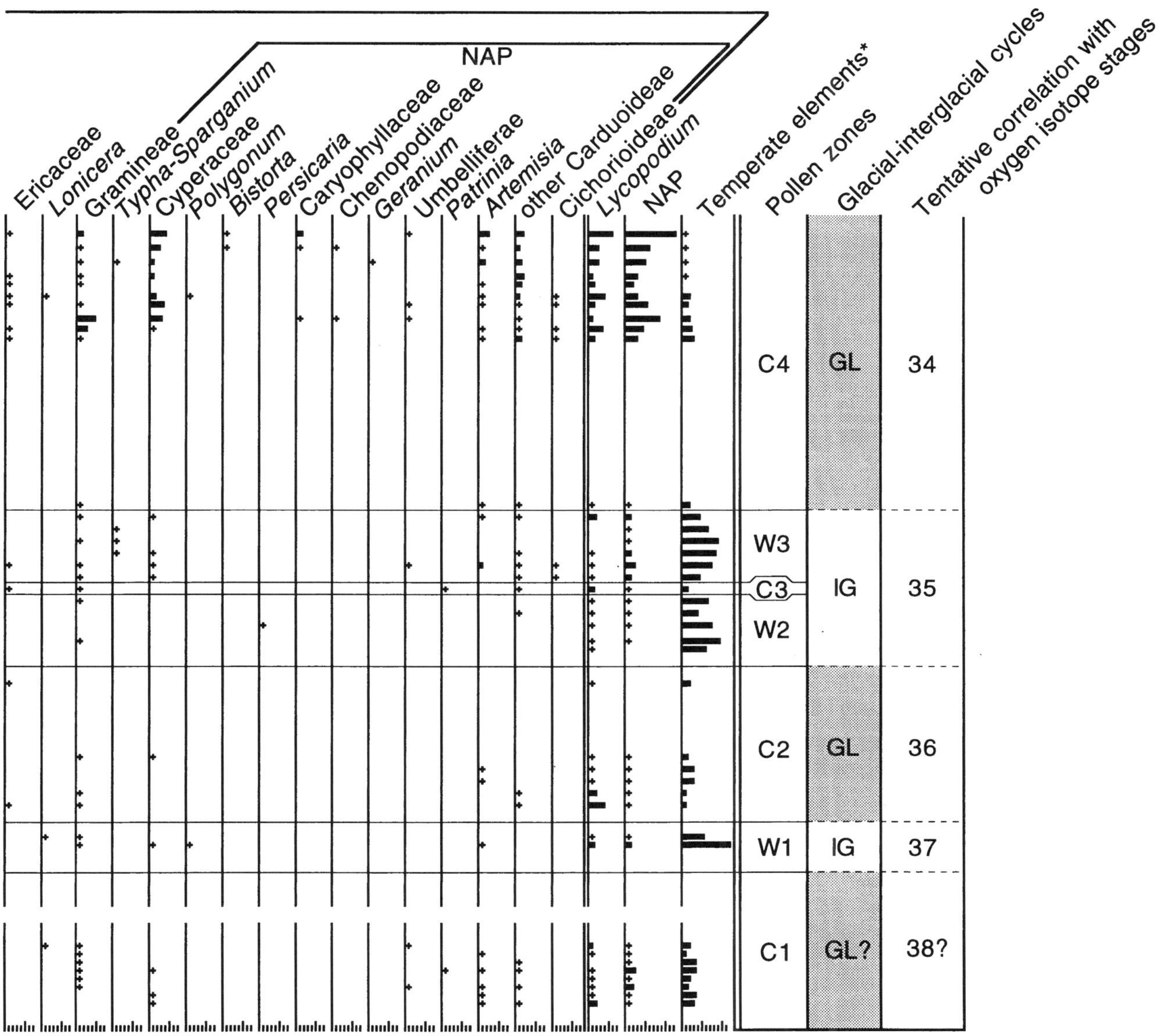

Text-Figure 5 (continued). Percentage diagram of the pollen and spore assemblages from Unit II of the Uppermost Formation, Yachiho Group. Pollen zones, climatic interpretation, and tentative correlation with oxygen isotope stages are shown on the right. Percentages of *Lycopodium* spores are calculated on the basis of total pollen sum. Asterisk represents the temperate element. GL=glacial, IG=interglacial, NAP=non-arboreal pollen.

in Early Pleistocene are generally characterized by low-amplitude (Williams et al., 1988; Shackleton, 1995) and high-frequency (Williams et al., 1988; Ruddiman et al., 1989; Tiedemann et al. 1994), in contrast with those in late Pleistocene. The low-amplitude fluctuations indirectly indicate that the climate changes occurring around 1.1 Ma were smaller than those in late Pleistocene. This is consistent with the small changes in climate observed in this study. In contrast, pollen floral changes reflecting glacial–interglacial cycles in middle and late Pleistocene are drastic: dominant taxa clearly change (e.g. Kishi, 1992). The period of the cycles in the temperate elements can not be estimated because the sedimentation rate is unknown. It is likely, however, that the period of the cycles is several tens of thousands of years, which relates to the dominant period (41,000-year obliquity cycle) of the isotope records between 1 to 1.5 Ma (Tiedemann et. al, 1994).

I attempted a tentative correlation of the pollen record of this study with the oxygen isotope stages of ODP Site 677 (Shackleton et al. 1990) (Text-Figure 5). Stage 34 is assigned to the C4 pollen zone on the basis of the age data and inferred coldest paleoclimate. Then, stage 35 could follow for W3, C3, and W2. C3 may be a cooling within stage 35. This cooling might be related to the two obliquity cycles in stage 35 (Shackleton et al., 1990). And then, stages 36, 37, and 38 could be selected for the pollen zones C2, W1, and C1, respectively. However, it is not evident that C1 corresponds to one oxygen isotope stage. This tentative correlation can be ascertained by the further paleomagnetic study of the Houridaira section to find the Cobb

Mountain normal event at the base of stage 35 (Shackleton et. al. 1990).

I can not exclude the possibility that the cycles of the percentages of the temperate elements show small fluctuations within one glacial–interglacial cycle and one or two oxygen isotope stages, because sedimentation rate is unknown.

In the above discussion, the possible effect of the uplift of the mountains in central Japan is not taken into consideration. The mountains area in central Japan including Mt. Yatsugatake has been abruptly uplifted since Pleistocene (Sakaguchi et al., 1976), but the details are not known. The pollen assemblage of this study may have been affected by this uplift of the mountains. However, since the pollen assemblages show cyclic change, the effect is probably small.

One of the reason this study's pollen record reflects the glacial–interglacial cyclicity is that the sampling locality seems to have been surrounded by relatively high mountains, and was a high-plain during the deposition of the studied sediments, as it is today. This may have prevented pollen dispersal from the lower elevations to the locality.

CONCLUSIONS

The pollen assemblages from Unit II of the Uppermost Formation of the Yachiho Group consist mainly of Pinaceae including *Picea, Tsuga, Abies,* and *Pinus,* associated with the broad-leaved deciduous trees such as *Fagus.* These assemblages reflect the Pinaceae-dominated coniferous vegetation associated with broad-leaved deciduous forests, which is comparable with the lower part of the subalpine zone or the uppermost part of the montane zone of present-day vegetation except for top four samples. The pollen assemblages of the top four samples are characterized by dominant *Pinus* and low percentages of the temperate elements, suggesting the modern alpine zone vegetation. The pollen assemblages of lower samples contain a higher abundance of *Larix*-type pollen than those of upper samples. This suggests the rapid vegetation recovery by a pioneer species such as *Larix kaempferi* after destruction by mudflow.

The percentages of the temperate elements show significant small changes exhibiting cyclicity, while no major palynofloral change can be recognized. Two series of samples having different resolutions demonstrate that these small changes are not due to inevitable variability in the pollen assemblages but due to climatic changes such as glacial–interglacial cycles. At least three cycles are recognizable and the cycles are interpreted as glacial–interglacial cycles. The nature of the small changes is broadly consistent with that of the oxygen isotope records around 1.1 Ma generally characterized by low-amplitude fluctuations. The cycles of the temperate elements are tentatively correlated with stages 38 to 34 mainly on the basis of the age data and inferred paleoclimate.

The evaluation of the variability of pollen assemblages is necessary in order to discern small changes in pollen assemblages, such as in this study. This is important for Early Pleistocene palynology.

ACKNOWLEDGMENTS

I am indebted to Professor Junji Itoigawa (Sugiyama Jogakuen University) and Professor Hiroshi Shibata (Nagoya University) for critical reading of the manuscript. I also thank Professor Tohru Yamanoi (Yamagata University) for suggestions about pollen identification. The comments by Dr. Owen Davis and an anonymous reviewer greatly improved the manuscript. This work was supported in part by a grant from Meijo University.

References Cited

BERGER, A., and LOUTRE, M.F.
1991 Insolation values for the climate of the last 10 million years. *Quaternary Science Reviews*, 10: 297–317.

COMBOURIEU-NEBOUT, N.
1993 Vegetation response to Upper Pliocene glacial/interglacial cyclicity in the central Mediterranean. *The Quaternary Research*, 40: 228–236.

HOOGHIEMSTRA, H., and RAN, E.T.H.
1994 Late Pliocene–Pleistocene high resolution pollen sequence of Colombia: an overview of climatic change. *Quaternary International*, 21: 63–80.

KANEOKA, I., and KAWACHI, S.
1983 K-Ar ages of volcanic rocks from the northern area of the Yatsugatake volcanic chain, central Japan. *The Journal of the Geological Society of Japan*, 89: 359–361.

KAWACHI, S.
1974 Geology of the Tateshinayama district, with geological sheet map at 1:50,000. *Geological Survey of Japan*, 117 pgs.

KISHI, H.
1992 Pleistocene climatic history around Lake Biwa, based on the pollen analysis of the Katata Formation, Shiga Prefecture, Japan. *Bulletin of the Mizunami Fossil Museum*, No. 19: 495–507.

KONDO, Y.
1988 Palaeomagnetism of volcanic rocks in the northeast and south part of the piedmont area of Mts. Yatsugatake, central Japan. *In*: Collaborative Research Group for Yatsugatake (ed.), Quaternary system around Mts. Yatsugatake. *Monograph of the Association for the Geological Collaboration in Japan*, No. 34: 179–190.

LEROY, S., and DUPONT, L.
1994 Development of vegetation and continental aridity in northwestern Africa during the Late Pliocene: the pollen record of ODP Site 658. *Palaeogeography, Palaeoclimatology, Palaeoecology*, 109: 295–316.

MINAKI, M.
1989 Fossil conifers from the Middle to Late Pleistocene of Japan and three types of plant macrofossil assemblages. *Japanese Journal of Historical Botany*, 4: 19–31.

MIYAWAKI, A.
1985 *Vegetation of Japan*, vol. 6, Chubu. Shibundo Co. Ltd. Publishers, Tokyo, 604 pgs.

MOMOHARA, A., MIZUNO, K., TSUJI, S., and KOKAWA, S.
1990 Early Pleistocene plant biostratigraphy of the Shobudani Formation, southwest Japan, with reference to extinction of plants. *The Quaternary Research*, 29: 1–15.

RAVAZZI, C., and STRICK, M. R.
1995 Vegetation change in a climatic cycle of Early Pleistocene age in the Leffe Basin (Northern Italy). *Palaeogeography, Palaeoclimatology, Palaeoecology*, 117: 105–122.

RESEARCH GROUP FOR MT. NORTH YATSUGATAKE
1988 Pliocene–lower Pleistocene around Mts. Yatsugatake—especially on the Yachiho Group. *In*: Collaborative Research Group for Yatsugatake (ed.), Quaternary system around Mts. Yatsugatake. *Monograph of the Association for the Geological Collaboration in Japan*, No. 34: 15–52.

RUDDIMAN, W.F., RAYMO, M.E., MARTINSON, D.G., CLEMENT, B.M., and BACKMAN, J.
1989 Pleistocene evolution: northern hemisphere ice sheets and North Atlantic Ocean. *Paleoceanography*, 4: 353–412.

SAKAGUCHI, Y., TAKAHASHI, Y, and CHINZEI, K.
1976 "Topography of Japan". *Kagaku*, 46: 223–234 (title is written only in Japanese).

SHACKLETON, N.J.
1995 New data on the evolution of Pliocene climatic variability. *In*: Vrba, E.S., Denton, G.H., Partridge, T.C., and Burckle, L.H. (eds.), *Paleoclimate and evolution, with emphasis on human origins.* Yale University Press, New Haven and London, pp. 242–248.

SHACKLETON, N.J., and OPDYKE, N.D.
1976 Oxygen-isotope and paleomagnetic stratigraphy of Pacific core V28-239, late Pliocene to latest Pleistocene. *Geological Society of America Memoir 145*, 449–464.

SHACKLETON, N.J., BERGER, A., and PELTIER, W.R.
1990 An alternative astronomical calibration of the lower Pleistocene timescale based on ODP Site 677. Transactions of the Royal Society of Edinburgh: *Earth Sciences*, 81: 251–261.

TIEDEMANN, R., SARNTHEIN, M., and SHACKLETON, N.J.
1994 Astronomic timescale for the Pliocene Atlantic $\delta^{18}O$ and dust flux records of Ocean Drilling Program site 659. *Paleoceanography*, 9: 619–638.

TSUCHIDA, K.
1991 *Nature in Mt. Yatsugatake.* The Shinano-Mainichi-Shimbun, Nagano, 224 pgs.

TSUDA, M.
1990 Pollen analysis of core from the Ooahara Bog at Nyugasa-yama in Nagano Prefecture. *The Quaternary Research*, 29: 439–446.

TSUKADA, M.
1967 Vegetation and climate around 10,000 B.P. in central Japan. *American Journal of Science*, 265: 562–585.

WILLIAMS, D.F., THUNELL, R.C., TAPPA, E., RIO, D., and RAFFI, I.
1988 Chronology of the Pleistocene oxygen isotope record: 0–1.88 m.y. B.P. *Palaeogeography, Palaeoclimatology, Palaeoecology*, 64: 221–240.

Author's address

TAKESHI SAITO
Department of Environmental Science and Technology
Faculty of Science and Technology
Meijo University
Nagoya 468-8502
Japan

THE LAST INTERGLACIAL/GLACIAL CYCLE IN TARANAKI, WESTERN NORTH ISLAND, NEW ZEALAND: A PALYNOSTRATIGRAPHIC MODEL

Rewi M. Newnham and Brent V. Alloway

Abstract

Previous late Quaternary palynological investigations from Taranaki, western North Island, New Zealand, are reviewed and a climatostratigraphic model based on key pollen taxa is developed for future work in this important region. Full interglacial environments are indicated by podocarp–angiosperm forests with *Ascarina* and high levels of *Dacrydium* pollen. Interstadial and stadial assemblages represent a continuum from podocarp–angiosperm forest–shrubland communities with *Prumnopitys* and *Dacrydium* dominant, to beech forest–shrubland–grassland communities, including *Halocarpus* and *Libocedrus*. Full glacial conditions are likely to be represented by assemblages indicating grassland–shrubland communities with scattered pockets of beech forest. This model is corroborated using a new high resolution late glacial/early Holocene pollen record at Umutekai Bog, then applied to a tephropalynological record of environmental change from Airedale Reef, spanning the interval from oxygen isotope Substage 5e to Stage 4 inclusive. This record indicates that the 5-substage subdivision of Stage 5 is applicable in New Zealand. At Airedale Reef. Substage 5d appears to have been more humid than Substage 5b while the transition from Substage 5a to Stage 4 was marked by a comparatively dry climate and is characterised by evidence of natural fire, perhaps associated with volcanism. Several examples of volcanic impact on vegetation and landscapes are evident in these records and the rich suite of volcaniclastic deposits at these Taranaki sites demonstrates their strong potential for further age refinement and correlation.

INTRODUCTION

The Taranaki landscape of western North Island, New Zealand, is dominated by Egmont Volcano (Mt. Egmont)—a 2518 m high late Quaternary andesitic stratovolcano (Text-Figure 1). To the northwest are two progressively older and eroded volcano remnants—the ca. 250 ka Pouakai Volcano (1399 m) and the ca. 500 ka Kaitake Volcano (684 m).

The latest definitive eruption of Mt. Egmont occurred at approximately 1755 A.D., just before European settlement (Neall and Alloway, 1986) but stratigraphic studies (Neall,1972; Alloway, 1989; Alloway et al., 1995) indicate that tephra emission and lahar inundation are typical and frequently recurring features of this Volcano. A more complete record of these eruptive events is found on the lower flanks of the volcano rather than on the volcano itself, where surficial deposits are readily removed by erosion or deeply buried beneath the products of more recent eruptions. The lower flanks consist of an extensive, gently sloping apron of coalesced fans of laharic, pyroclastic and alluvial volcaniclastic detritus and are referred to as the Egmont ring plain (Neall, et al., 1986). On the depositional surfaces that comprise the ring plain, allophane-dominated soils (Andisols) are extensively distributed and have accumulated from episodic accretion and subsequent weathering of aerially transported fine-grained sediment of dominantly andesitic provenance. Andisols in central Taranaki are often interbedded by numerous millimeter and centimeter-thick Egmont-sourced tephra and occasional silicic tephra from the Taupo Volcanic Zone (TVZ; Text-Figure 1) that, where dated, permit correlation to diverse depositional environments such as lakes and peats.

Peats are found in areas of Taranaki that lie beyond the overwhelming influence of significant quantities of accreting tephra that would otherwise engulf and transform the peats into deep, imperfectly or poorly drained andisols. The genesis of peat deposits is influenced more by the deposition of volcaniclastic detritus transported by lahars than by any other eruptive event. For instance, in areas of east-central Taranaki that border the ring plain, the formation of extensive peat deposits has resulted from blockage of Tertiary hill country drainage outlets by large volumes of detritus deposited by laharic events (Alloway, 1989). On the ring plain, however, localised peat deposits have formed in enclosed depressions developed at the margins of coalesced fans as well as upon the initial depositional surfaces. Many of these peat deposits have been buried and preserved beneath subsequent laharic detritus.

These buried peat deposits containing tephra interbeds therefore present opportunities to assess the vegetation and climate change in Taranaki that can be compared with records obtained elsewhere in central and southern North Island over a similar time interval. In this paper, previous late Quaternary palynological investigations in the Taranaki region are briefly reviewed and from these a model of key palynological criteria for determining climatostratigraphic intervals is developed. This model is demonstrated at Umutekai Bog, a late glacial–early Holocene pollen site from northern Taranaki, and then applied to a peat deposit at Airedale Reef on the north Taranaki coast where a near complete record of Last Interglacial environmental change is preserved.

Newnham, Rewi M., and Alloway, Brent V.
2001 The last interglacial/glacial cycle in Taranaki, western North Island, New Zealand: A Palynostratigraphic Model; *In*: Goodman, D.K., and Clarke, R.T. (eds.), *Proceedings of the IX International Palynological Congress, Houston, Texas, U.S.A., 1996*; American Association of Stratigraphic Palynologists Foundation, p. 411–422.
 ISBN 0–931871–06–9

METHODS

All pollen slides were prepared using standard procedures (Moore et al., 1991) including the addition of exotic *Lycopodium* spores to permit calculation of pollen concentrations. Pollen percentages were calculated using a pollen sum of trees, shrubs and dryland herbs. Stratigraphically constrained cluster analysis using least sums of squares (CONISS) and detrended correspondence analysis

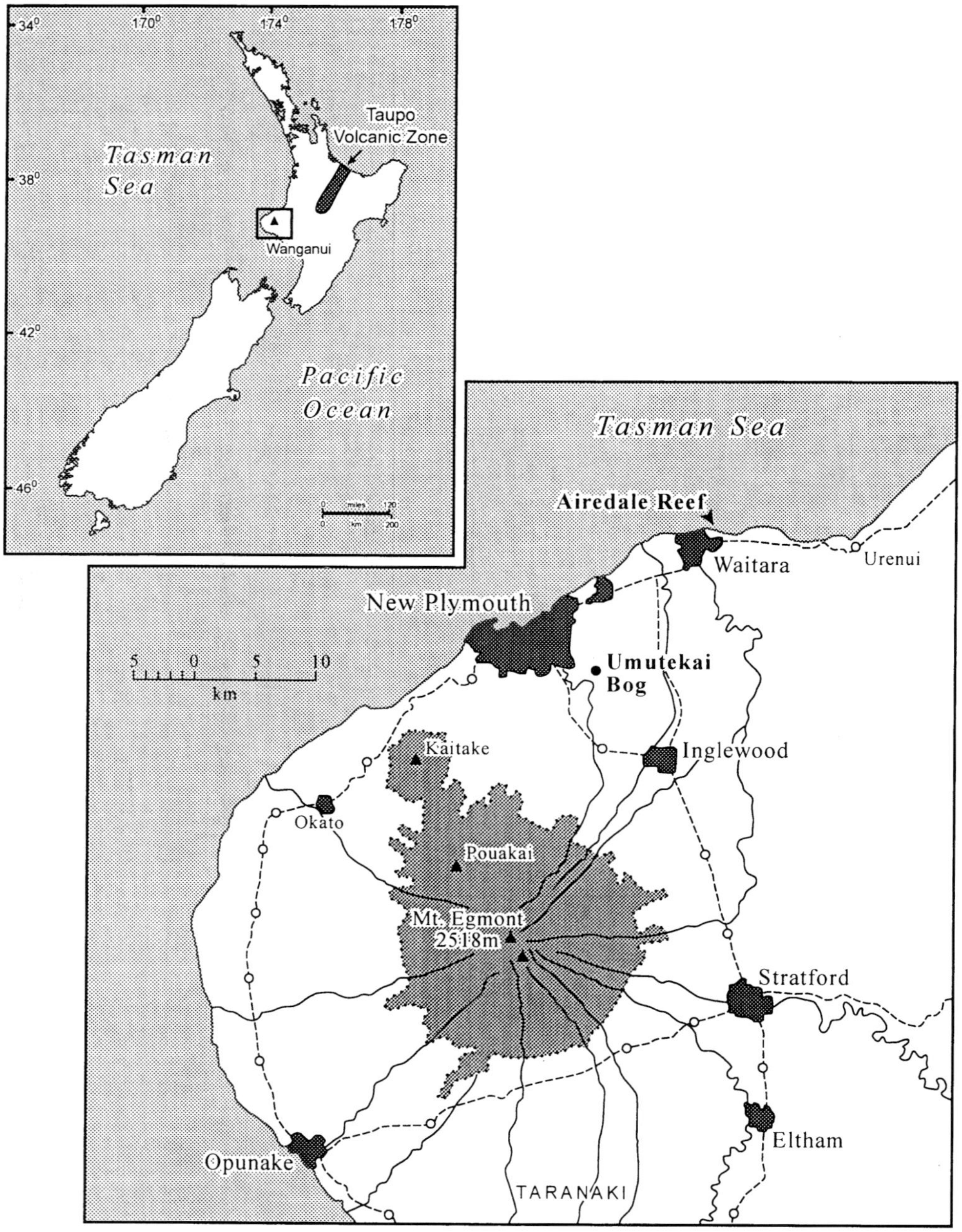

Text-Figure 1. Taranaki localities including pollen sites at Airedale Reef and Umutekai Bog. Lightly-shaded area indicates Egmont National Park.

(DECORANA) were performed on the pollen data using the software package TILIA. Volcanogenic sediment (tephras and the debris avalanche deposit) were not sampled for pollen. All radiocarbon dates are reported as conventional radiocarbon years and were obtained on bulk peat samples assayed at the Rafter Radiocarbon Laboratory, Lower Hutt, New Zealand.

CLIMATE AND VEGETATION OF WESTERN TARANAKI

Coastal areas have a mild warm-temperate climate while inland areas have cooler annual temperatures and higher average rainfall with frosts and fogs common (New Zealand Meteorological Service, 1963). Mean annual temperatures decline from about 13°C at the coast at New Plymouth to 9.5°C at 850 meters above sea level. Radiosonde plot data show the temperature lapse rate to be about 0.6°C per 100 m altitude. Annual rainfall varies between 900–1600 mm, increasing with altitude, and is evenly distributed throughout the year. The climate is windy and gales are relatively frequent.

Strong temperature and rainfall gradients from the coast to summit of Mt. Egmont are reflected in a distinctive zonal vegetation pattern (Clarkson, 1986; Text-Figure 2). Little remains of the coastal, semi-coastal and lowland forests

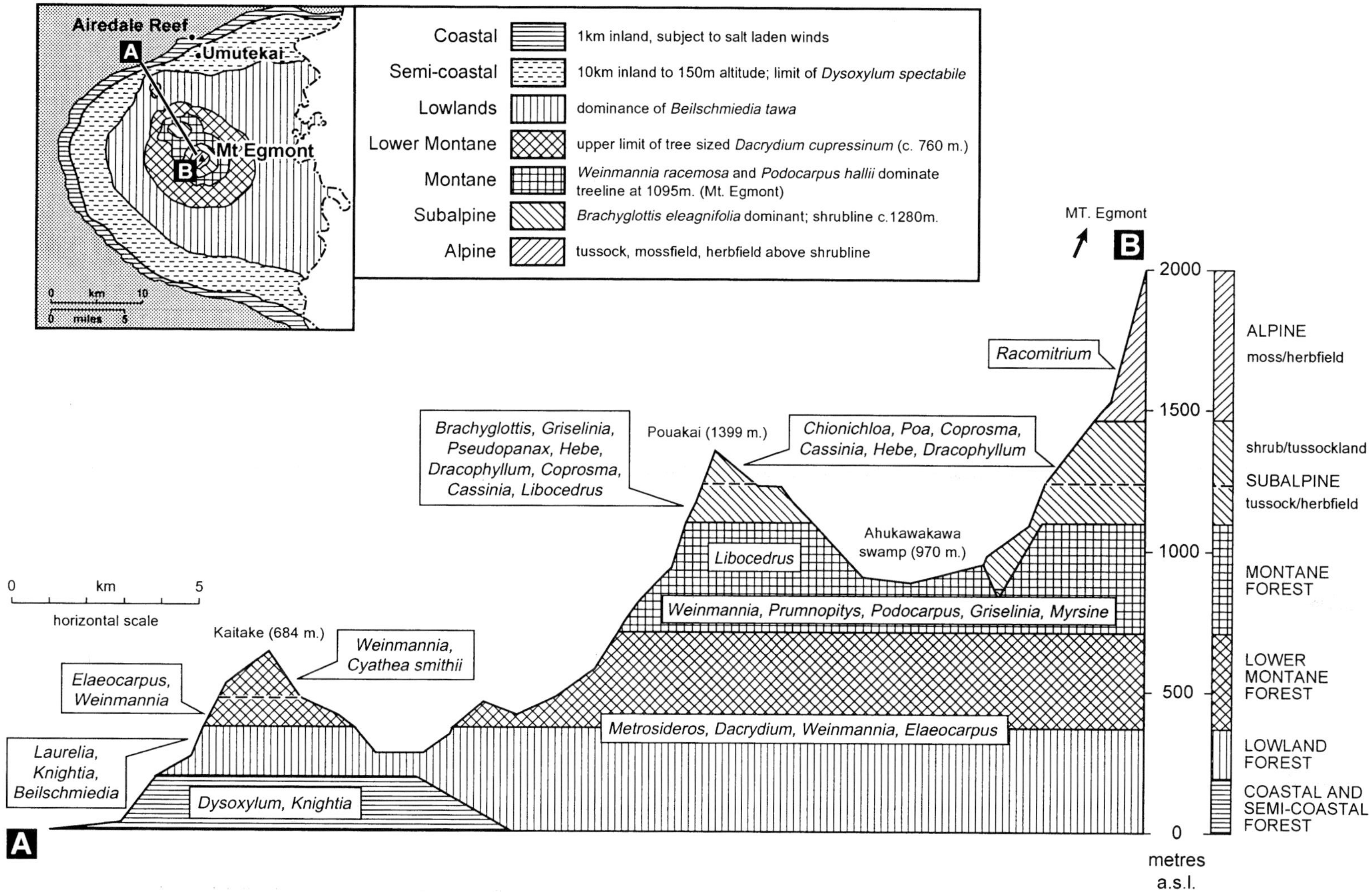

Text-Figure 2. The vegetation of Egmont National Park along a transect from the north-west coast to near the summit of Mt. Egmont. INSET: Bioclimatic zonation of western Taranaki (after Bayfield and Benson, 1986).

due to human settlement and deforestation for agriculture, but remnants occur which have provided a basis for characterising the likely original vegetation. Semi-coastal forests are characterised by broadleaved angiosperms *Dysoxylum spectabile, Alectryon excelsus, Beilschmiedia tawa, Corynocarpus laevigatus, Laurelia novae-zelandiae, Vitex lucens,* and *Knightia excelsa,* together with tree ferns (*Cyathea* spp.) and the palm *Rhopalostylis sapida* (Clarkson and Boase, 1982; Clarkson, 1985). Further inland, dense, warm-temperate podocarp–angiosperm forest probably covered the lowlands and lower slopes of Mt. Egmont (McGlone and Neall, 1994). The dominant canopy trees were likely to have included *B. tawa, Dacrydium cupressinum, Metrosideros robusta,* and *Weinmannia racemosa.* Above 760 m, lowland forest is gradually replaced by montane podocarp–angiosperm forest dominated by *W. racemosa* and *Podocarpus hallii* with *Libocedrus bidwillii* important above 900 m (Clarkson, 1986).

Above the treeline (1050 m), the subalpine shrubland is characterised by *Brachyglottis eleagnifolia* (Asteraceae), with common associates *Cassinia vauvilliersii* (Asteraceae), *Coprosma* spp., *Dracophyllum filifolium, Hebe* spp., and *Pseudopanax* spp. Scattered emergent *Libocedrus bidwillii* occur at lower elevations within this scrub belt, along with two other montane forest species, *Griselinia littoralis* and *Podocarpus hallii.* Shrubland gives way to shrub–tussockland and ultimately, above 1400 m, to tussockland. Above 1200 m, *Chionochloa rubra* becomes dominant, while *Schoenus pauciflorus* (on Pouakai Range only) and *Poa* spp. are common. Above 1500 m, herbs predominate over tussock, and mosses, especially *Racomitrium* spp. form dense patches, while above 1600 m the plant cover becomes patchy, and above 1750 m, inconsequential (McGlone et al., 1988).

Although this bioclimatic zonation (Text-Figure 2) provides a basic framework for interpreting pollen records from this region, modern vegetation patterns also reflect the effects of volcanic disturbance and of local soil and drainage characteristics. Much of the vegetation on Mt. Egmont is still in the process of readjustment following disturbance by volcanic activity. Primary succession has occurred where the former vegetation has been completely destroyed by lavas, lahars or floods. Thus on the north–west slopes of Mt. Egmont, the treeline is depressed to about 790 m a.s.l (Text-Figure 2). Approximately 460 species found at comparable North Island localities do not occur on Mt. Egmont (Clarkson, 1986). This deficiency, most noticeable among the subalpine and alpine species, is thought to be a consequence of both geographic isolation and continuing volcanic activity. Palynologically prominent taxa missing from Mt. Egmont today include *Nothofagus, Phyllocladus, Halocarpus* and *Empodisma,* species of which are common in montane and subalpine forests or bogs elsewhere in North Island.

PREVIOUS LATE QUATERNARY PALYNOLOGY IN TARANAKI

The stratigraphic record, constant Quaternary uplift, and volcanic activity in the Taranaki region have provided an excellent framework for major palynological investigations (e.g., McGlone, 1980; Lees, 1987; Bussell, 1988a). Interpretation of the fossil pollen record is supported by modern pollen rain studies in the region (McGlone, 1982; Bussell, 1988b). Key pollen sites and their inferred oxygen isotope stage correlations are listed in Table 1 together with those for the Wanganui region adjacent to the south (Text-Figure 1). Correlations and chronologies are based on stratigraphic position in relation to wave cut terraces, tephrochronology, and independent radiometric dating, supported in some cases by palynostratigraphy. The record is fragmentary. For example, no single site covers all of oxygen isotope Stage 5. However, each of the major warm intervals for the late Quaternary back to Stage 9 are represented at one site at least (Table 1). Not surprisingly, coverage is best for the late glacial and Holocene and weakest generally for cold stages. On the basis of the key palynological criteria used to establish these stratigraphic relationships (Table 1), a tentative model is proposed here to assist with interpretations and correlations at new sites in the Taranaki region (Table 2).

Full Interglacial conditions are indicated by podocarp–angiosperm forest, analogous to pollen assemblages of the early Holocene in the region. A key taxon for interglacial peaks appears to be *Ascarina lucida,* a frost sensitive angiosperm shrub, comparatively rare in New Zealand today, but abundant in the early–mid Holocene at many western

TABLE 1. Key late Quaternary Taranaki and Wanganui pollen sites and their inferred correlation with marine oxygen isotope stratigraphy

SUBSTAGES	SITE	REFERENCE
Taranaki		
2, 1	Eltham Swamp, etc.	McGlone and Neall (1994)
2, 1	Durham Rd	Alloway et al. (1992a)
5c, 5b, 5a, 4, 3	Inaha Terrace	McGlone et al. (1984)
5d, 5c, 5b	Ararata Rd	Bussell (1993)
5e, 5d	Rapanui	Bussell (1992)
6, 5e	Ohawe Beach	Bussell (1990)
Wanganui		
7b, 7a	Landguard Bluff	Pillans et al. (1988)
9b,9a	Fordell	Bussell and Pillans (1992)

TABLE 2. Pollen-climatostratigraphic model for Late Quaternary lowland Taranaki.

Climate	Vegetation	Key taxa
Full Inter-glacial	Podocarp–angiosperm forest	*Ascarina**, *Dacrydium* (high)
Inter-stadial	Podocarp–angiosperm forest–shrubland	*Prumnopitys, Dacrydium* (mod.)
Stadial/ Glacial	shrub/ grassland/ beech forest mosaic (raised bog?)	*Fuscospora*, Nothofagus menziesii** *Halocarpus* Libocedrus* *Coprosma, Myrsine, Leptospermum* Asteraceae, Poaceae (*Empodisma,* Gleichenia**)

* rare or not found in Taranaki today

sites, including lowland Taranaki. *Ascarina* is abundant in all inferred full interglacial spectra in these Taranaki records, always in association with other taxa characteristic of podocarp–angiosperm forests including high levels of *Dacrydium cupressinum* pollen and usually high levels of *Cyathea* tree fern spores. Interstadial pollen floras may be similar but lack *Ascarina* and other thermophilous angiosperm taxa, and have lower levels of the tall podocarp *Dacrydium*. A key interstadial taxon appears to be *Prumnopitys taxifolia*, another tall podocarp tree, which is more tolerant of drier and cooler conditions. Interstadial pollen floras may also have higher levels of shrub taxa broadly analogous to the cooler montane forests on the flanks of Mt. Egmont.

Stadial pollen floras tend to have low levels of podocarp pollen, except for the shrub *Halocarpus* which is absent from Mt.. Egmont today, but elsewhere in North Island occurs in wet areas, often towards or at the upper limits of forest. Stadial and interstadial assemblages may also have higher levels of *Nothofagus* and *Libocedrus bidwillii*. The former is absent from Mt.. Egmont today, but from modern pollen rain studies, both may be indicative of the shrubland and grassland communities found close to both sides of the treeline. Full glacial floras are harder to characterise as this environment is poorly represented in these records. They may contain high levels of grass pollen together with more shrub and less tree pollen taxa of the interstadial floras. At times during full glacial conditions, high levels of *Nothofagus* pollen are recorded, although this is usually interpreted as over-representation of scattered pockets of beech trees, the pollen of which may mask other taxa in the absence of other big pollen producers (Bussell, 1988b).

UMUTEKAI BOG: TESTING THE POLLEN–CLIMATOSTRATIGRAPHIC MODEL

The model based on previous palynological work in Taranaki and Wanganui (Table 2), provides a basic framework for interpreting the pollen assemblages in this investigation. A new radiocarbon-dated high resolution pollen diagram from Umutekai Bog, northern Taranaki (Text-Figure 3), spanning the late glacial to early Holocene demonstrates how this model can be applied to determine broad climatic intervals in pollen records from this region. The interpretation of the pollen diagram is as follows.

The lowermost samples in zones UM1-3, dated ca. 11–10,000 BP, indicate different types of interstadial assemblages. The earlier assemblages (UM1and 2) have high levels of *Libocedrus, Nothofagus, Halocarpus*, and *Phyllocladus* but also include grasses and a substantial shrub component (not shown). Zone Um3 indicates a different type of interstadial assemblage with high levels of *Prumnopitys*, declining levels of those taxa previously dominant and increasing levels of *Dacrydium* pollen. These assemblages are transitional to the full Interglacial spectra (Um4) which are marked by relatively high levels of *Ascarina* and *Dacrydium*. Mid–to–late Holocene pollen spectra at this site and others in northern Taranaki show a decline in *Ascarina* coincident with generally lower *Dacrydium* pollen levels relative to other tall podocarps (Newnham, 1990) and increases in *Lagarostrobos* (McGlone and Neall, 1994) and *Libocedrus* (McGlone et al., 1988). These palynological changes within conifer–angiosperm spectra may be indicative of regional pollen assemblages for the latter stages of interglacials in Taranaki.

The Umutekai record also illustrates the role of volcanism in Taranaki pollen sequences. The profile includes several important tephra marker beds, deposited ca. 23 km north of their Mt. Egmont source, that support the radiocarbon chronology at this site and permit correlation with other sites in the region. There are also examples where tephra layers precede significant change points in the vegetation succession. For example, the 5 cm thick Konini Tephra immediately precedes an abrupt rise in *Prumnopitys taxifolia*; whilst the sharp rise in *Ascarina* follows the deposition of a thinner unnamed tephra layer (Text-Figure 3). It may be hypothesised that the deposition of these tephra layers, although not directly triggering vegetation change, nevertheless may act as a catalyst to vegetation developments already underway. Canopy trees in particular may have been more susceptible to physical or toxic effects of tephra fall, resulting in the creation of canopy gaps that encourage the development of saplings or subcanopy shrubs. This hypothesis, based on a single site, can be tested at other sites from the region and provides an additional focus for investigation at another new pollen profile, at Airedale Reef.

AIREDALE REEF

At Airedale Reef on the north Taranaki coast (Text-Figure 1), a cliff section exposes a 9m sequence of coverbeds above an uplifted marine terrace (NT2) which has been correlated to the Last Interglacial, oxygen isotope stage 5e (Dickson et al., 1974; Chappell, 1975; Alloway, 1989). The coverbeds consist of carbonaceous brown and grey muds, interbedded with numerous ash to lapilli sized tephras of

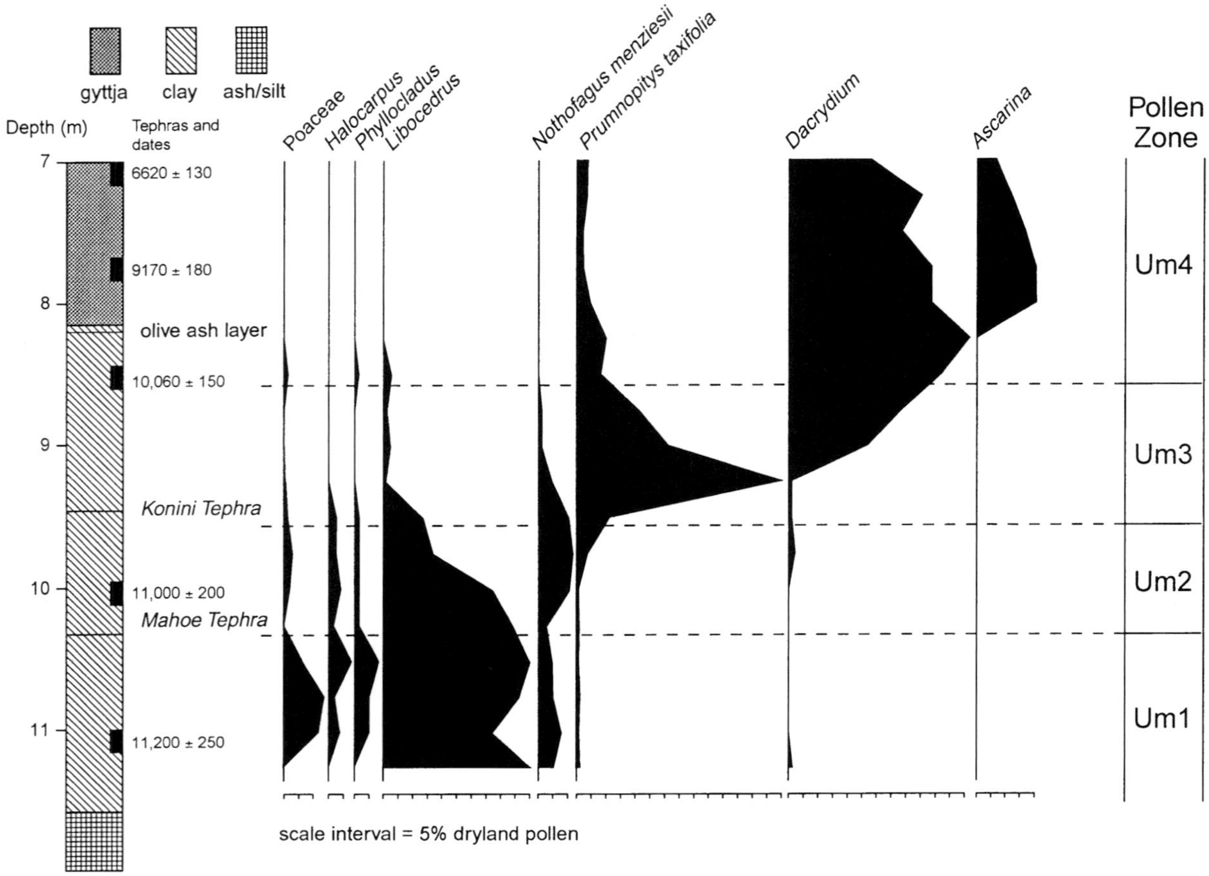

Text-Figure 3. Umutekai Bog pollen percentages for selected taxa, late glacial to early Holocene. Dates are radiocarbon years BP.

both andesitic and rhyolitic origin and a 3–4 m thick diamicton interpreted as a debris avalanche deposit (Text-Figure 4). The andesitic tephra units include the Epiha a - e series and the Te Arei tephra (Alloway, 1989). Higher up in the sequence, above the uppermost organics are a suite of palaeosols and several andesitic and rhyolitic tephras including three key well known and widely-dispersed units from the Taupo Volcanic Zone: the Rotoehu Tephra with a K/Ar age of 64 ± 4 ka (Wilson et al., 1992); the Kawakawa Tephra, radiocarbon dated to around 22.6 ka (Pillans et al., 1993) and the ca. 4 ka Stent Tephra (Alloway et al., 1994). The organic sediments at Airedale Reef, therefore, are well constrained chronologically between oxygen isotope Substage 5e and around the Stage 4/3 boundary (a complete description of the sedimentary sequence, stratigraphic relationships and palynology at Airedale Reef will be published elsewhere).

The pollen diagram is subdivided into 6 local pollen assemblage zones with two obvious breaks in the sequence (Text-Figure 4). The cluster analysis program CONISS assisted with pollen zonation.

Zone AR1: Below 900 cm

The lowermost zone, consisting of carbonaceous muds, is dominated by *Dacrydium* pollen (47–60%), with *Prumnopitys, Dacrycarpus, Podocarpus,* and *Fuscospora* comprising most of the remaining tree pollen. These assemblages record the highest percentages of tree pollen in the entire profile. Shrub (mostly *Myrsine*) and herb pollen levels are low but rise towards the top of the zone. Mire and aquatic pollen is rare, while *Cyathea* tree fern spores and monolete fern spores are common.

These pollen assemblages indicate that a dense and probably extensive podocarp forest occupied the area around the site, with beech (*Nothofagus*) forest probably further inland at higher elevations. Tree ferns and forest floor or epiphytic ferns were abundant in the podocarp forest. The depositional site may have been a *Dacrydium–Dacrycarpus* (podocarp) swamp forest. These assemblages are similar to those recorded at Late Holocene sites in lowland Taranaki (McGlone and Neall, 1994) except that they include significant levels of *Nothofagus* (*Fuscospora*) pollen. This, together with the overall dominance of lowland tree pollen, but lack of coastal elements, suggest a time of deposition towards the latter stages of an Interglacial, with the coastline beyond its present position at Airedale Reef.

Zone AR2: 900 - 849 cm

The lower boundary of this zone coincides with a lithological change from slightly carbonaceous olive-brown

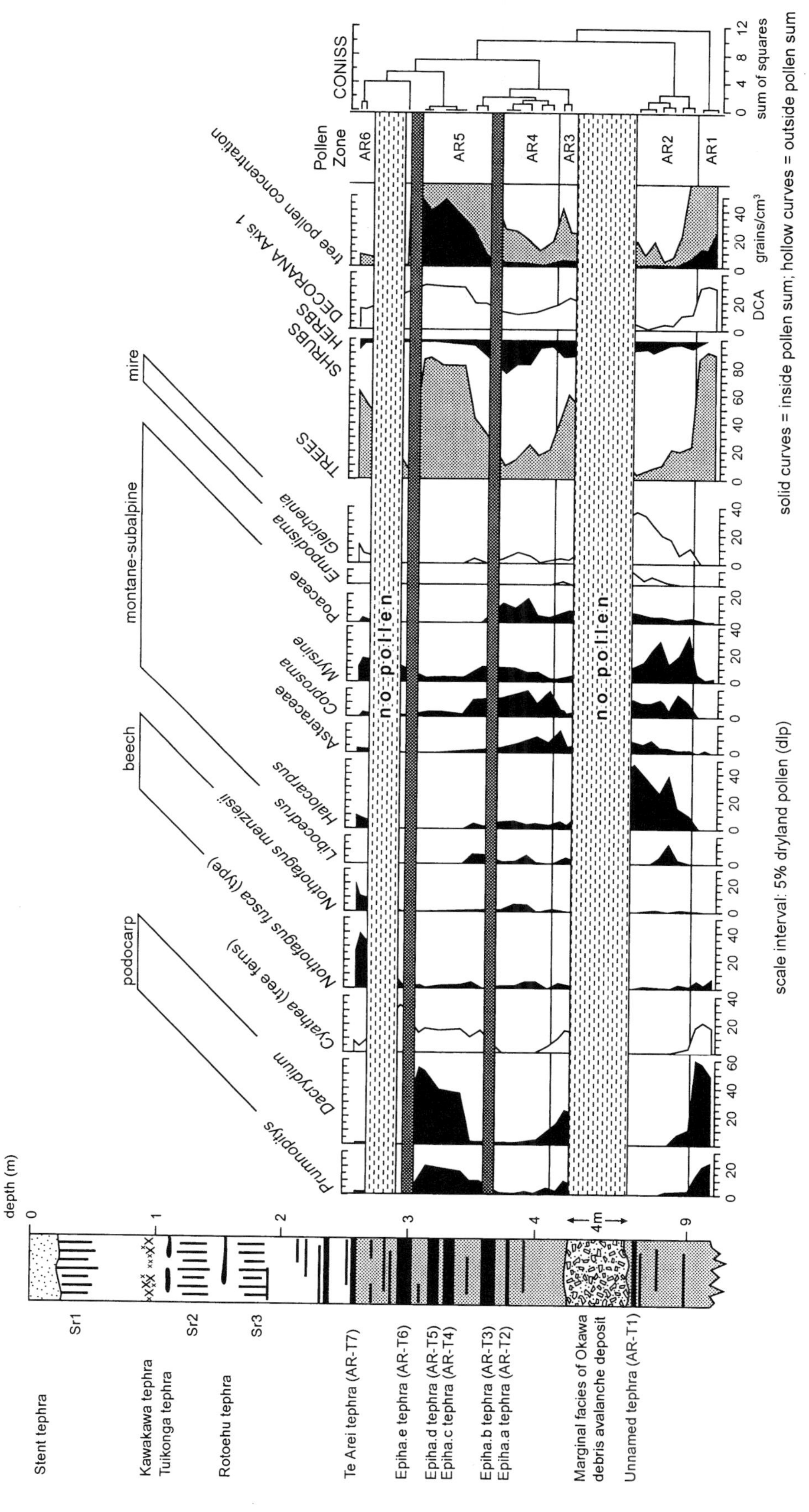

Text-Figure 4. Airedale Reef pollen percentage diagram for selected key taxa.

muds (below) to highly carbonaceous muds and peat. The upper boundary is formed by the base of the Okawa debris avalanche deposit.

Tree pollen percentages fall dramatically at the beginning of this zone to ca. 20% and then decline further to ca. 10% at the top of the zone. Shrub taxa rise to a dominance of ca. 80% dryland pollen; while herb pollen, mostly Poaceae, increases to between 5 and 10%. Successive peaks in the shrub taxa *Myrsine, Coprosma* and *Leptospermum, Halocarpus* (coinciding with a *Libocedrus* peak) then *Myrsine* again characterise the lower half of the zone, while *Halocarpus* remains dominant throughout the upper half. Wetland taxa also increase dramatically, in particular Cyperaceae, *Gleichenia* and *Empodisma,* with the latter two taxa expanding towards the top of the zone.

These assemblages indicate that podocarp forest declined rapidly at the base of the zone, to be replaced by extensive shrubland vegetation, probably similar to that occuring at or near the treeline in the region today. A range of shrub taxa became common, including species of Asteraceae, *Coprosma,* Epacridaceae, *Griselinia, Hebe, Leptospermum* and *Myrsine* (not all of which are shown in Text-Figure 5). At the same time, the depositional site changed from podocarp swamp forest to a peat swamp that became increasingly oligotrophic.

The occurrence of subalpine vegetation at a site currently at sea level attests to a climate considerably cooler than the present, indicating stadial conditions. The development of peatland at the site indicates moist conditions although, initially, the climate need not have been wetter than in the previous zone when swamp forest prevailed under warmer conditions. However, parallel increases in taxa common in ombrogenous peats—*Gleichenia, Empodisma* and *Halocarpus*—may point to rising precipitation through the upper half of the zone.

Zone AR3: 424 - 410 cm

This zone, which consists of olive-grey muds, is separated from the previous zone by the ca. 425 cm-thick Okawa debris avalanche deposit.

The two pollen assemblages in this zone are markedly different from those in Zone Ar2. The shrub-dominant assemblages give way to higher levels of tree pollen, in particular *Dacrydium, Metrosideros, Prumnopitys, Podocarpus, Dacrycarpus, Fuscospora, Libocedrus* and *Lagarostrobos. Cyathea* tree fern spores and monolete fern spores remain common. *Halocarpus* pollen levels are dramatically reduced although this taxon persists together with Asteraceae, *Coprosma, Leptospermum, Myrsine* and Poaceae. Although less prominent than in the previous zone, shrub pollen levels increase during this zone, at the expense of tree pollen. *Empodisma* pollen and *Gleichenia* spore levels decline, while Cyperaceae levels increase along with *Gunnera, Myriophyllum, Potamogeton,* and the colonial alga *Pediastrum.*

This zone signals the return of podocarp forest, with some angiosperm trees and tree ferns prominent, although this vegetation type appears to be declining as the shrub vegetation expands. In the upper part of the zone the occurrence of macrophytic pollen and spores together with the decline in terrestrial peatland taxa reflects the development of a lacustrine environment at the site.

The assemblages in this zone may be seen as being intermediate between those of the previous two zones and are interpreted as indicating interstadial conditions. However, the decline in podocarp forest suggests a cooling cli-

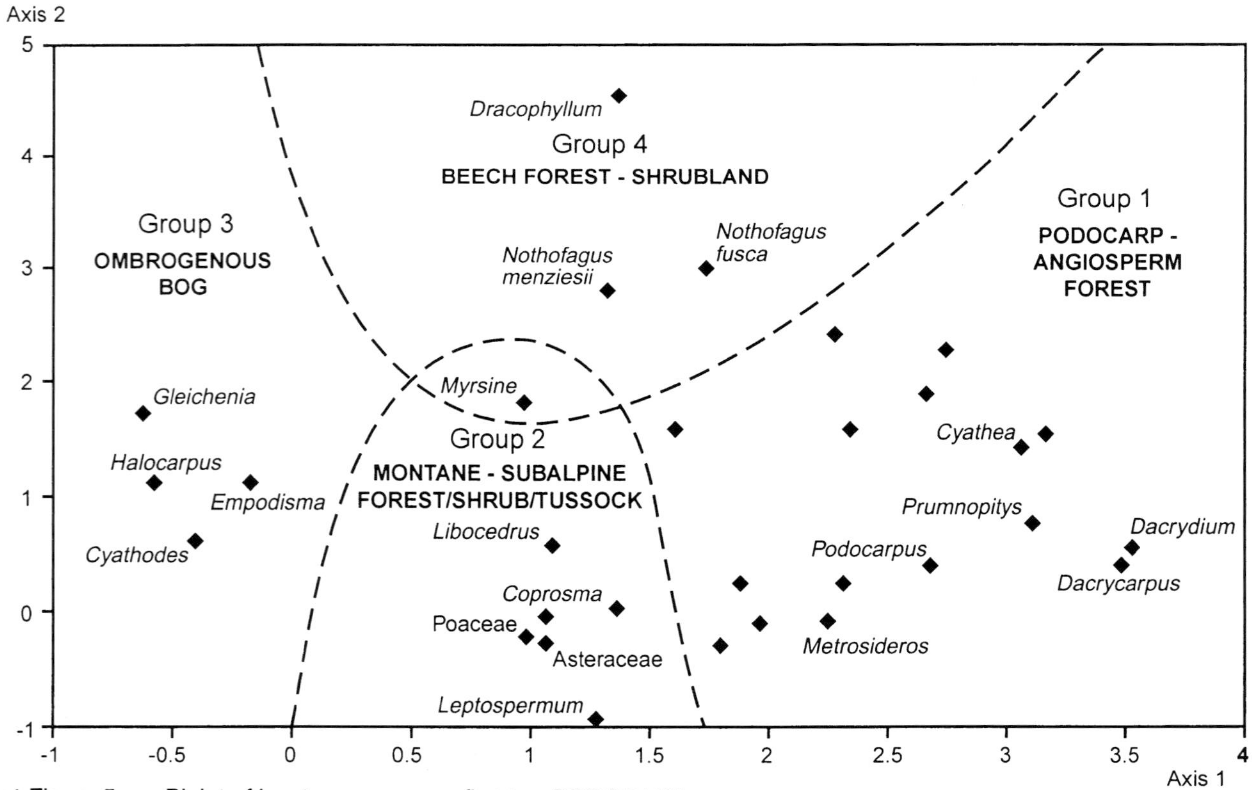

Text-Figure 5. Biplot of key taxa scores on first two DECORANA axes.

mate during this zone. Although the deposition of the debris avalanche deposit is likely to have been geologically instantaneous, it is apparent from the marked differences with the top of the previous zone that some of the organic record has been lost. The occurrence of peat rip-clasts within the matrix of the debris avalanche, and along section evidence of basal scouring and partial detachment of the carbonaceous muds immediately underlying the debris avalanche deposit, supports the conclusion that sediments at the top of Zone Ar2 were sheared off when the debris deposit was emplaced. The development of aquatic conditions following debris avalanche emplacement resulted from drainage impedance and ponding of water within small depressions formed on the debris avalanche upper surface.

Zone AR4: 410 - 370 cm

The lower boundary of this zone coincides with a lithological change from pale (olive-grey) muds to highly carbonaceous muds and peats. The upper boundary coincides with the 10 cm thick Epiha.d tephra layer.

Tree pollen levels continue to decline from the previous zone, while the shrub taxa *Leptospermum, Coprosma, Myrsine* and Asteraceae return to dominance. Pollen of *Nothofagus menziesii*, Poaceae and *Astelia* is more prominent than previously recorded. Cyperaceae pollen persists albeit at lower levels than in Zone Ar3 whilst *Phormium, Gleichenia*, and monolete fern spores are prominent in some assemblages.

This zone records the continuing decline of podocarp–angiosperm forest and the expansion of shrubland–grassland vegetation. The inferred vegetation is similar to that of Zone AR2 and, therefore, is interpreted as indicating a return to stadial conditions. However this was a much drier climate than the previous stadial, judging by the lower levels of *Halocarpus* and associated peatland plants, in particular *Empodisma* and *Gleichenia*.

Zone AR5: 370 - 304 cm

The lower and upper boundaries of this zone coincide with the Epiha.d and Epiha.a tephras respectively.

The highly carbonaceous muds in this zone are rich in tree pollen, in particular *Dacrydium, Prumnopitys, Dacrycarpus, Podocarpus, Lagarostrobos*, and *Nestegis. Prumnopitys taxifolia* is initially prominent, along with *Libocedrus* and *Elaeocarpus*, before giving way to *Dacrydium* as the dominant taxon. Shrub pollen levels, in particular, *Leptospermum, Coprosma, Myrsine*, Asteraceae, *Pseudopanax* and *Griselinia*, are initially high but decline to ca. 15% at the top of the zone. *Cyathea smithii* and several other fern spores are prominent while wetland taxa are much reduced.

These assemblages indicate the return of lowland podocarp forest in the region and possibly swamp forest at the site. The presence of a significant shrub element together with the lack of key interglacial angiosperm taxa, in particular *Ascarina*, favours a classification of interstadial rather than interglacial. The lower part of the zone, immediately above the 10 cm thick Epiha.d tephra layer, is characterised by seral shrubs and trees such as *Prumnopitys taxifolia* and *Libocedrus bidwillii* that may rapidly colonise fresh volcanic or alluvial surfaces. These assemblages indicate that tephra fall had an impact on the vegetation.

Hiatus 304 - 265 cm

These sediments, immediately overlying Epiha.a tephra layer, consist of carbonaceous to slightly carbonaceous muds containing several unnamed tephra layers as well as two prominent charcoal layers. The pollen content in this section of the sequence is impoverished and not quantifiable. Some pollen is preserved, but in low concentrations and mostly in unidentifiable condition. It is possible that these characteristics are related as there appears to have been significant environmental change in this region during the period following deposition of the 10-cm thick Epiha.a tephra layer. The lack of well preserved pollen together with charcoal layers and deposition of additional tephras indicate a period of environmental instability, during which fires, possibly triggered by volcanism, damaged the vegetation or burnt the surface at the site.

Zone AR6: 265 - 254 cm

The uppermost two assemblages, straddling the Te Arei tephra layer, reveal a different pollen spectrum to any others in the sequence. Pollen concentrations are comparatively low and the spectra are dominated by *Fuscospora* and *Nothofagus menziesii*. Associated taxa include *Myrsine, Dracophyllum, Coprosma, Halocarpus, Nestegis, Lagarostrobos*, and *Astelia*. *Gleichenia* spores are common.

Interpretation of these assemblages is problematic. The pollen percentage data suggest the expansion of beech forest in the region, yet the relatively low pollen concentrations seem contradictory. Alternatively, the higher *Nothofagus* levels may reflect long distance pollen transport to the site accentuated in these pollen spectra by a decline in local vegetation. The low pollen concentrations together with the evidence for environmental stability prior to this period lend weight to this contention. These difficulties in interpretation together with the lack of a modern analogue in the Taranaki vegetation—where *Nothofagus* is largely absent—make it difficult to decipher the climatic record from these assemblages. However the scarcity of pollen indicative of Holocene vegetation in the region, in particular lowland podocarp taxa, rules out an interglacial or interstadial climate.

DECORANA

Detrended correspondence analysis (DECORANA) of these data reveals some interesting patterns. A biplot of taxa scores on the first two axes (Text-Figure 5) indicates four distinct groups.

1. Podocarp–angiosperm forest taxa, analogous to the lowland to montane forests in the modern (Holocene) vegetation, have high positive scores on the first axis.
2. Montane–subalpine forest / shrubland / tussock taxa, reflecting vegetation of cooler climate in the region today, have low positive scores on the first axis.

The other two groups effectively have no modern analogues in Taranaki.

3. A group of taxa characteristic of peat bogs elsewhere in New Zealand—*Halocarpus, Empodisma, Gleichenia, and Cyathodes cf. juniperina*—but which are either absent or rare in Taranaki today, have negative scores on the first axis. Stratigraphically these taxa are most closely associated with the previous group and ecologically they are also consistent with cooler (but moister) climate.
4. A group indicative of beech forest and shrubland, also without a modern analogue, has high scores on the second DECORANA axis. This group is only prominent during Zone Ar6 and is also interpreted as reflecting cooler climate than exists at this locality today.

This analysis reveals some interesting climatic patterns. Axis 1 clearly separates the warm-temperate interglacial or interstadial taxa from two types of cool stage assemblage—one indicating a comparatively wet climate (group 3), the other comparatively dry (group 2). A stratigraphic plot of the first axis, therefore, may be used to illustrate the extent of warm-temperate forest development in the region (Text-Figure 4). Axis 2 separates the *Nothofagus*-dominated assemblages from all others in the sequence.

These results provide some insight into why taxa in groups 3 and 4 are rare or absent from the Taranaki region today. Taxa from both groups are abundant during cooler episodes of the Pleistocene, but have been replaced by more thermophilous elements during the Holocene. As the Umutekai Bog pollen profile (Text-Figure 3) illustrates, late glacial and Holocene pollen diagrams from Taranaki record the gradual diminution of habitats suitable for these cool stage elements (McGlone et al., 1984; Newnham, 1990; McGlone and Neall, 1994).

CORRELATION WITH THE MARINE OXYGEN ISOTOPE RECORD (TABLE 3)

As discussed earlier, the Airedale Reef pollen record is constrained to the interval between Subtage 5e and early Stage 3.

Zonation of the Airedale Reef pollen record, supported by cluster analysis (CONISS) and DECORANA, reveals six distinct intervals alternating between cool and mild or warm climate. These variations correspond strongly with fluctuations in marine oxygen isotope records during the interval represented by late substage 5e to early Stage 4 (Table 3). The lowermost olive-brown muds and underlying andesitic sands, therefore, were deposited late in Substage 5e, while the overlying carbonaceous muds and peats beneath the Okawa Formation accumulated during Substage 5d. The Okawa debris avalanche probably occurred during Substage 5c and removed organic sediments corresponding to 5c and late 5d. Alternatively, this event may have occurred late in Substage 5d with no subsequent deposition until late in Substage 5c, when the olive-grey lake muds began to accumulate. The overlying carbonaceous muds and peats up to the base of Epiha.b tephra, and including Epiha.a tephra correspond to Substage 5b. Deposition of the remainder of the Epiha tephra series, during Subtage 5a, was followed by a period of environmental instability characterised by continuing volcanic activity and fire. The uppermost sediments that accumulated around

TABLE 3. Summary of Airedale Reef palynology, volcanic impacts and inferred correlation with marine oxygen isotope stratigraphy.

Pollen Zone / volcanic unit	Key Taxa	Inferred Climate	MOI (Sub) stage	Volcanic Impacts
AR6	*Fuscospora, Nothofagus menziesii*	cool, wet	4	
	no pollen	hiatus		volcanic induced fires; vegetation supression and/or removal of peat
Epiha.e tephra			?late 5a	widespread destruction of Interstadial forest
AR5	*Dacrydium, Prumnopitys, Cyathea*	mild, moist	5a	
AR4	*Leptospermum, Coprosma, Myrsine.* Poaceae	cool, dry	5b	Epiha.b tephra (late 5b): stadial vegetation stressed by tephra fall; soil rejuvenation - expansion of podocarps
AR3	*Dacrydium, Cyathea*	mild (cooling)	late 5c	
Okawa Debris Avalanche deposit			?late 5c	vegetation destruction and removal of unknown thickness of sediment
AR2	*Halocarpus ,Libocedrus, Coprosma ,Myrsine, Gleichenia*	cool, moist	5d	
AR1	*Dacrydium, Prumnopitys, Cyathea*	mild, moist	late 5e	

the time of the Te Arei tephra are likely to correspond with early Stage 4.

The significance of the Airedale Reef deposit is threefold. Firstly, several newly defined Egmont-sourced tephra (Alloway, 1989) occur at the site interbedded with the peat. The approximate ages for these tephra are broadly constrained by overlying tephra marker beds and the underlying NT2 wave-cut surface, and further constrained by correlation with the marine oxygen isotope record using the pollen zonation. Secondly, the site provides a near complete terrestrial record of Last Interglacial vegetation and climate change in New Zealand that can be compared with fragmentary records obtained from elsewhere in central and southern North Island over a similar time interval. Thirdly, the timing of fluctuating climatic conditions identified from pollen changes in the peat in relation to tephra marker inter-beds can be approximately determined in the Andisol-forming environment using the same tephra markers, and confirms the morpho- and mineralogical variations recorded in Andisol sequences of Last Interglacial age (Alloway et al., 1988, 1992b).

The Airedale Reef profile also has important implications for models of inter-hemispherical climate change. This record clearly supports the 5 stage subdivision of marine oxygen isotope Stage 5 and indicates that these fluctuations should be detected in other terrestrial sequences in New Zealand. The vegetation patterns revealed at this site provide insights into the type of climate prevailing during substages 5a–d and Stage 4. In particular, the climate of western Taranaki during Substage 5b appears to have been considerably drier than during Substage 5d, and the transition from Substage 5e to Stage 4 may also have been comparatively dry.

VOLCANIC IMPACTS

As well as these broad climatic signals, the Airedale Reef sequence provides some clear examples of significant regional volcanic impact on the vegetation and landscape (Table 3). The 4.25 m debris avalanche deposit (diamicton) is supposed to have had a devastating effect on the environment. However, as the pollen record suggests a considerable time gap exists between the two organic units separated by the diamicton, it is not possible to determine either any destruction of existing vegetation or the successional vegetation that may have followed.

There are also signs of significant vegetation change following deposition of the thickest and coarsest tephra units. Epiha.d Tephra (10 cm thick) occurs at the end of a stadial during which subalpine-like shrub and herb vegetation occupied the area around the site, with forest trees rare. Immediately above the tephra unit the pollen of tall tree species increases substantially, initially *Prumnopitys taxifolia* and *Libocedrus bidwillii*, and then *Dacrydium cupressinum*. It is possible that deposition of Epiha.d Tephra accelerated the reafforestation process that was otherwise occurring gradually as climate shifted from stadial to interstadial mode. *P. taxifolia* is known to be an early coloniser of fresh volcanic surfaces in central North Island (Beveridge, 1983) and a similar possibly accelerated invasion by this species occurred in the Waikato lowlands following the deposition of the 14.5 ka Rerewhakaaiiutu Tephra as the climate switched from stadial to interstadial mode (Newnham et al., 1989). Similarly *L. bidwillii* is known to be able to colonise freshly disturbed surfaces in Taranaki and elsewhere (Veblen and Stewart, 1982; Norton, 1983; McGlone et al., 1988).

A much more dramatic volcanic impact followed deposition of the coarse-grained 10 cm Epiha.a Tephra at the top of this zone. Although the immediately overlying sediments are devoid of pollen, two macroscopic charcoal layers attest to a period of burning perhaps associated with continuing volcanic activity indicated by several thin unnamed tephra layers. The pollen record resumes with distinctly different assemblages to those immediately beneath the Epiha.a Tephra. Our interpretation of this evidence is that this eruption caused widespread damage to the vegetation, probably through a variety of mechanisms that most likely include inundation by ash falls, pyroclastic flows and lahars. This activity was then followed by a phase of sustained volcanism of unknown duration, during which fires appeared to be frequent.

ACKNOWLEDGMENTS

The fieldwork of BVA was funded by a DSIR-funded Ph.D. scholarship at Massey University as well as internal research grants from Auckland University. We thank Vince Neall and Matt McGlone for assisting with fieldwork at Umutekai. We are grateful to Mairie Fromont for reviewing the manuscript; to Cristina Penalba and David Pocknall for comprehensive and helpful referee reports; and to Javier Helenes and David Goodman for further comment and editorial assistance. Brian Rogers drafted the figures.

References Cited

ALLOWAY, B.V.

1989 Late Quaternary cover bed stratigraphy and tephrochronology of north-eastern and central Taranaki, New Zealand. Unpubl. Ph.D. thesis, lodged in the library, Massey University, Palmerston North.

ALLOWAY, B.V., LOWE, D.J., CHAN, R.P.K., EDEN, D.N., and FROGGATT, P.C.

1994 Stratigraphy and chronology of the Stent Tephra, a c. 4000 year old distal silicic tephra from Taupo Volcanic Centre, New Zealand. *New Zealand Journal of Geology and Geophysics*, 37: 37–47.

ALLOWAY, B.V., McGLONE, M.S., NEALL, V.E., and VUCETICH, C.G.

1992a The role of Egmont-sourced tephra in evaluating the paleoclimatic correspondence between the bio- and soil-stratigraphic records of central Taranaki. *Quaternary International*, 13/14: 187–194.

ALLOWAY, B.V., NEALL, V.E., and VUCETICH, C.G.

1988 Localised volcanic loess deposits in North Taranaki, New Zealand. In D.N. Eden and R.J. Furkert (Eds.) *Loess - its distribution, geology, and soils*. A. A. Balkema Publishers, Rotterdam. pp. 1–8.

1995 Late Quaternary (post c. 28,000 year B.P.) tephro-stratigraphy of northeast and central Taranaki, New Zealand. *Journal of the Royal Society of New Zealand*, 25: 385–458.

ALLOWAY, B.V., STEWART, R.B., NEALL, V.E., and VUCETICH, C.G.

1992b Climate of the Last Glaciation in New Zealand, based on aerosolic quartz influx in an andesitic terrain. *Quaternary Research*, 38: 170–179.

BAYFIELD, M.A., and BENSON, M.A.

1986 Egmont Ecological Region surver report for the New Zealand Protected Natural Areas Programme. Department of Lands and Survey, Wellington.

BEVERIDGE, A.S.
1983 Regeneration of podocarp forests and maintenance of productivity. *In*: Thompson, K., Hodder, A.P.W., and Edmonds, A.S. (eds.), *Lowland Forests of New Zealand.* University of Waikato, Hamilton, p. 93–112.

BUSSELL, M.R.
1988a Quaternary vegetational and climatic changes recorded in cover beds of the South Wanganui Nasin marine terraces, North Island, New Zealand. Unpubl. Ph.D. thesis, Australian National University.
1988b Modern pollen rain, central–western North Island, New Zealand. *New Zealand Journal of Botany*, 26: 431–451.
1990 Palynology of oxygen isotope stages 6 and 5e from the coverbeds of a marine terrace, Taranaki, New Zealand. *Quaternary Research*, 34: 86–100.
1992 Late Pleistocene palynology of terrestrial coverbeds at the type section of the Rapanui Terrace, Wanganui, New Zealand. *Journal of the Royal Society of New Zealand*, 22: 77–90.
1993 A late Pleistocene vegetation and climatic history of part oxygen isotope stage 5, Ararata, south Taranaki, New Zealand. *Journal of the Royal Society of New Zealand*, 23: 129–145

BUSSELL, M.R ., and PILLANS, B.J.
1992 Vegetational and climatic history during oxygen isotope stage 9, Wanganui District, New Zealand, and correlation of the Fordell Ash. *Journal of the Royal Society of New Zealand*, 22: 41–60.

CHAPPELL, J.
1975 Upper Quaternary warping and uplift rates in the Bay of Plenty and West Coast, North Island, New Zealand. *New Zealand Journal of Geology and Geophysics*, 18: 129–155.

CLARKSON, B.D.
1985 The vegetation of the Kaitake Range, Egmont National Park, New Zealand. *New Zealand Journal of Botany*, 23: 15–31.
1986 *Vegetation of Egmont National Park, New Zealand.* Science Information Publishing Centre DSIR, Wellington.

CLARKSON, B.R., and BOASE, M.R.
1982 Scenic reserves of West Taranaki. *Biological Survey of Reserves Series 10.* Department of Lands and Survey, Wellington.

DICKSON, M., FLEMING, C.A., and GRANT-TAYLOR, T.L.
1974 Ngarino Terrace: an addition to the Late Pleistocene standard sequence in the Wanganui–Taranaki District. *New Zealand Journal of Geology and Geophysics*, 17: 789–798.

LEES, C.M.
1987 The late Holocene changes in the vegetation of western Taranaki by soil palynology. Unpubl. Ph.D. thesis, Massey University, Palmerston North.

McGLONE, M.S.
1980 Late Quaternary vegetation history of central North Island, New Zealand. Unpubl. Ph.D. thesis, University of Canterbury, Christchurch.
1982 Modern pollen rain, Egmont National Park, New Zealand. *New Zealand Journal of Botany*, 20: 253–262.

McGLONE, M.S., and NEALL, V.E.
1994 The late Pleistocene and Holocene vegetation history of Taranaki, North Island, New Zealand. *New Zealand Journal of Botany*, 32: 251–269.

McGLONE, M.S., NEALL, V.E., and CLARKSON, B.D.
1988 The effect of recent volcanic events and climate changes on the vegetation of Mt. Taranaki (Egmont), New Zealand. *New Zealand Journal of Botany*, 26: 123–144.

McGLONE, M.S., NEALL, V.E., and PILLANS, B.J.
1984 Inaha Terrace deposits: a late Quaternary terrestrial record in south Taranaki, New Zealand. *New Zealand Journal of Geology and Geophysics*, 27: 35–49.

MOORE, P.D., WEBB, J. and ALLINSON
1991 Pollen Analysis. Blackwell Scientific Publications, London.

NEALL, V.E.
1972 Tephrochronology and tephrostratigraphy of western Taranaki (N108-109), New Zealand. *New Zealand Journal of Geology and Geophysics*, 15: 507–557.

NEALL, V.E., and ALLOWAY, B.V.
1986 International Volcanological Congress, New Zealand. Tour Guide C3: Quaternary volcaniclastics and volcanic hazards of Taranaki. *New Zealand Geological Survey Record 12.*

NEALL, V.E., STEWART, R.B., and SMITH, I.E.M.
1986 History and petrology of the Taranaki Volcanoes. *Royal Society of New Zealand Bulletin*, 23: 251–263.

NEWNHAM, R.M.
1990 Late Quaternary palynological investigations into the history of vegetation and climate in northern New Zealand. Unpubl. Ph.D. thesis, lodged in the library, University of Auckland.

NEWNHAM, R.M., LOWE, D.J., AND GREEN, J.D.
1989 Palynology, vegetation and climate of the Waikato lowlands, North Island, New Zealand, since *c.* 18,000 years ago. *Journal of the Royal Society of New Zealand*, 19: 127–150.

NEW ZEALAND METEOROLOGICAL SERVICE
1963 Summaries of radiosonde data. *New Zealand Meteorological Service Miscellaneous Publication*, 119.

NORTON, D.A.
1983 Population dynamics of subalpine *Libocedrus bidwillii* forests in the Cropp River Valley, Westland, New Zealand. *New Zealand Journal of Botany*, 21: 127–140.

PILLANS, B.J., HOLGATE, G., and McGLONE, M.S.
1988 Climate and sea level during oxygen isotope stage 7b: On land evidence from New Zealand. *Quaternary Research*, 29: 176–185.

PILLANS, B.J., McGLONE, M.S., PALMER, A.S., MILDENHALL, D.C., ALLOWAY, B.V., and BERGER, G.
1993 The Last Glaciual Maximum in central and southern North Island, New Zealand: a palaeoenvironmental reconstruction using the Kawakawa Tephra Formation as a chronostratigraphic marker. *Palaeogeography, Palaeoclimatology, Palaeoecology*, 101: 283–304.

VEBLEN, T.T., and STEWART, G.H.
1982 On the conifer regeneration gap in New Zealand: the dynamics of *Libocedrus bidwillii* stands on South Island. *Journal of Ecology*, 70: 413–436.

WILSON, C.J.N., HOUGHTON, B.F., LANPHERE, M.A., and WEAVER, S.D.
1992 A new radiometric age estimate for the Rotoehu Ash from Mayor Island volcano, New Zealand. *New Zealand Journal of Geology and Geophysics*, 35: 371–374.

Authors' addresses

REWI M. NEWNHAM
Department of Geographical Sciences
University of Plymouth
Plymouth PL4 8AA
England

BRENT V. ALLOWAY
Institute of Geological and Nuclear and Sciences
Private Bag
Lower Hutt
New Zealand

THE HISTORY OF THE HOLOCENE VEGETATION IN NW SPAIN FROM POLLEN ANALYSIS

M. P. Saa Otero, E. Diaz Losada, and A. V. Gonzalez Porto

Abstract

Five pollen diagrams from four peat bogs located in the Atlantic zone of NW Spain form the basis to define a Holocene pollen reference sequence for the region. The main conclusion concerns the abundance of deciduous tree pollen. Development of the vegetation is defined mainly by the AP/NAP ratio. AP for most levels is higher than NAP, with the exception of the surface levels in which the decrease of tree vegetation may be attributed to human activity, but not to climatic change. The importance of individual tree taxa differs between pollen zones; especially the proportions of *Quercus*, *Corylus* and *Betula* which are important to characterize the past vegetation.

This study is a contribution to the knowledge of the development of vegetation during the Preboreal, Boreal and Atlantic in the NW Iberian Peninsula. The study shows the importance of the deciduous tree species *Quercus*, *Betula*, *Corylus* and *Alnus* in the arboreal vegetation during the Holocene and their progressive decline through the climatic period after the Atlantic. Cultivated plants, *Pinus pinaster*, *Castanea*, uncultivated herbs, and the increase of heath lands are present in the youngest periods. *Fagus* is present in low percentages only, just as in other pollen diagrams of the NW Iberian Peninsula.

INTRODUCTION

This study of pollen was carried out in medium-altitude (up to 700 m) peat bogs in northwestern Iberian Peninsula (Text-Figure 1). The present climate, influenced by the North Atlantic, is characterized by high annual rainfall (snow on the coldest winter days) and a thermal pattern which, because of moderate summer temperatures, can be defined as mild with moderate temperature oscillations. The average temperatures in the coldest months vary from 6–8°C (38–40°F) and in the warmest months from 16–20°C (48–52°F). All study sites are located in the Eurosiberian biogeographic region (Rivas Martínez et al., 1987).

The modern vegetation of the northwestern Iberian Peninsula has very interesting peculiarities, including the coexistence of elements characteristic of both Atlantic and Mediterranean vegetation. Despite modern pollen data (Menéndez Amor, 1971; Van Mourik, 1986; Saa Otero and Díaz Fierros, 1986), it has not been possible to establish definitive models of vegetation development during the Pleistocene and Holocene periods. Such information would allow the proposal of a vegetation development model for the Quaternary period following the Würm glaciation.

For the northern Iberian Peninsula, Peñalba (1989) proposed, on the basis of palynological data, vegetation dynamics having Mediterranean characteristics and a Central European influence. The Cantabrian coastal belt (including the area of the present study) and the northeastern region were specifically mentioned in this study. However, Peñalba also mentioned the drier character of southern Galicia (Spain) as being a slight variant, a point which was also noted by Saa Otero (1985).

In this context, one of the most debated questions is the relative importance that each of the arboreal taxa (such as *Quercus*, *Corylus* and *Betula*) have in the development of the post-Würm vegetation, as well as the date of the appearance and expansion of *Corylus*. The presence of *Ulmus* and *Tilia*, and their absence from modern floras in Galicia, is also an object of debate (Van Mourik, 1986; Díaz Fierros et al., 1979; Saa Otero, 1985).

The purpose of this study is to contribute to a better understanding of the vegetation dynamics during the Holocene in northern Spain, through pollen analysis and radio carbon dating of the peat bogs in Lobeiras, Chao de Lagozas and Cuadramón, and of coastal sediment in Xove (Text-Figure 1).

Saa Otero, M.P., Diaz Losada, E. and Gonzalez Porto, A.V.
2001 The history of the Holocene vegetation in NW Spain from pollen analysis; *In*: Goodman, D.K., and Clarke, R.T. (eds.), *Proceedings of the IX International Palynological Congress, Houston, Texas, U.S.A., 1996*; American Association of Stratigraphic Palynologists Foundation, p. 423-433.
 ISBN 0-931871-06-9

MATERIAL AND METHODS

Three peat sections were analyzed (Text-Figures 1a and 1b): the Chao de Lagozas peat-bog (520 m a.s.l.), the Lobeiras peat-bog (600 m a.s.l.), and the Cuadramón peat-bog (700 m a.s.l.). An organic sediment section labeled as Xove was also analyzed; samples were collected near the town of the same name, at an altitude of 110 m.

The samples for pollen analysis were processed according to the method of extraction of pollen from mineral sediment presented by Saa Otero (1985), including differential flotation in a dense solution of ZnBr, or, for peat samples, by the method of Beaulieu (1977), including the use of acids, acetolysis and filtration. Pollen concentrations were established using the volumetric method.

The following ^{14}C data were obtained: for 65–85 cm of Lobeiras I: 2,080 ± 60 yr B.P. UGRA-347, and for 100–117 cm: 7,930 ± 70 yr B.P. ICEN-554; for 107–120 cm of Lobeiras II: 9,800 ± 120 yr B.P. UGRA-348; for 61–70 cm of Cuadramón: 8,770 ± 170 yr B.P. UGRA-491; for 100–110 cm of Chao de Lagozas: 4,100 ± 60 yr B.P. UBAR-515; for 185–200 cm of Xove: 770 ± 60 yr B.P. UBAR-517.

RESULTS AND DISCUSSION

Two sections were analyzed in the Lobeiras peat-bog: in Lobeiras-I (total a depth 162 cm), 26,051 pollen grains and spores were identified in 32 samples, corresponding to 77 different palynological types; 21 of these are arboreal or shrub-like. Lobeiras-II, collected a few cm from the first core, which is 14 cm long (between 107 and 121 cm below

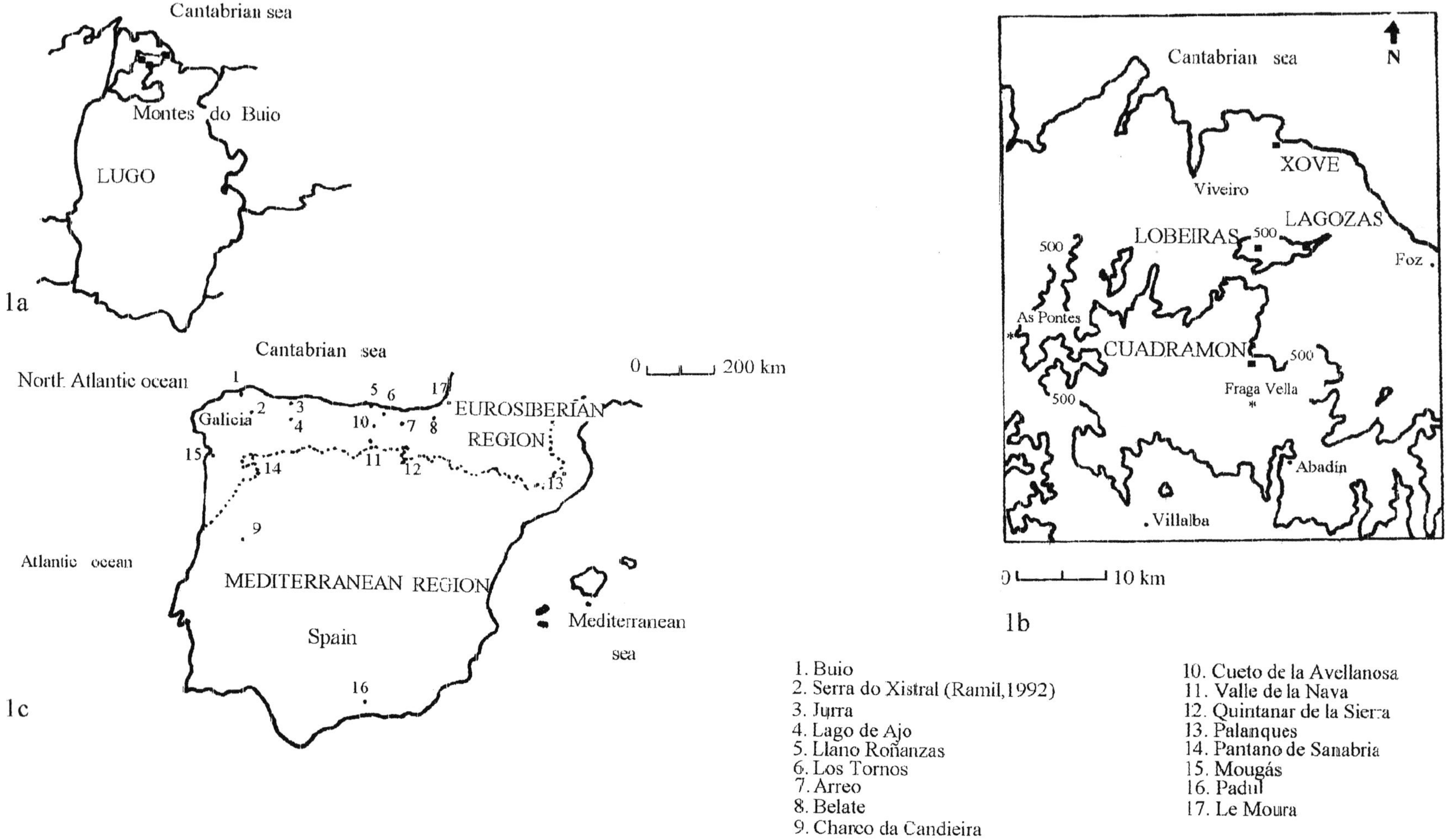

Text-Figure 1. The geographical and topographical situation (1a and 1b). Location of other available data of interest (1c).

the surface), is situated between two sandy beds; three samples were analyzed and 861 pollen and spores were recovered, corresponding to 54 different palynological types, 12 of which are arboreal or shrub-like taxa. At Cuadramón, 93 cm of peat was analyzed in 20 samples: 10,947 grains were identified, corresponding to 64 different palynological types, 17 of which are arboreal or shrub-like. At Chao de Lagozas, 125 cm of peat was analyzed in 19 samples; 7,464 grains were identified, corresponding to 57 palynological types, 14 of which are arboreal or shrub-like. The organic sediment core at Xove was 215 cm deep; 22 samples were collected, with 20,757 pollen grains and spores counted; these represent 71 palynological types identified, 17 of which are arboreal or shrub-like. The results are presented in the form of percentage pollen diagrams (Text-Figures 2–6).

Fourteen pollen zones were established (Table 1). The zones were defined according to the degree of oscillation shown by the most variable deciduous arboreal taxa (such as *Corylus*, *Quercus* and *Betula*) and by relatively abundant non-arboreal taxa, especially the Poaceae and Ericaceae (*Erica* and *Calluna*).

A high diversity of non-arboreal plants, was found periods of human activity as much as in the first part of the Holocene and the end of the Late–glacial. This is almost certainly due to the dominance influence of *Corylus* during the entire Holocene period which, due to the configuration of their crowns, allow the passage of light to the herbaceous stratum of the ecosystem.

The taxa that dominated the arboreal vegetation were, in order of importance, *Quercus*, *Corylus* and *Betula*, which never formed a closed canopy. *Alnus* has had a discrete and constant presence dating from the Atlantic period.

Arboreal taxa typical of nearby mountain areas (for example, *Fagus*, *Abies* and *Pinus*) have never been important in the regional lowland vegetation. It is possible to speculate about a "*Pinus* episode" only in reference to recent periods (due to reforestation campaigns), even though its presence is constant (but sporadic) in all the Galician pollen diagrams. The occurrence of *Pinus* pollen is due to either long-distance transport or the possible existence of small local populations of pine trees.

The early and possibly synchronous expansion of *Quercus* and *Corylus* (González et al., 1996), is an other event that should be reconsidered also in the context of the recently published results for the Serra da Estrela, in Portugal (Van der Knaap and Van Leeuwen, 1995).

The present data suggest a new interpretation of the dynamics of Holocene vegetation in the Cantabrian coastal sector in northwestern Spain. However, the most difficult obstacle to overcome is the scarcity of sediments with an adequate independent chronology.

CONCLUSIONS

The vegetation dynamics of the Holocene are characterized by variations in abundance of *Quercus*, *Corylus* and *Betula* (arboreal), and Poaceae, *Calluna* and other Ericaceae (non-arboreal). Non-arboreal vegetation from the beginning of the Holocene period is characterized by *Thalictrum*, *Armeria* and *Artemisia*.

The oldest pollen zones found in the Lobeiras and Cuadramón peat-bogs offer a reliable record which, together with radiocarbon dating, allows us to reconstruct the development of post–Würm Glaciation Period vegetation in this region. Data published to date for the northern Iberian Peninsula points to the appearance of *Quercus* before *Corylus*, at 9,800 yr B.P. Data shown in this paper indicate these events to be synchronous and earlier than 9,800 ± 120 yr B.P. Therefore, an earlier date for the appearance of both taxa seems plausible.

The expansion of *Corylus* in northwestern Spain has been placed between 8,000 and 7,930 ± 70 yr B.P. From the results indicated here, an earlier expansion prior to 8,770 ± 170 yr B.P., in accordance with parts of Europe with a strong Atlantic influence, could be considered.

An early appearance of *Alnus* around 9,800 ± 120 yr B.P., together with the beginning of the expansion of deciduous tree such as *Quercus* and *Corylus* around 8,770 ± 170 yr B.P., is proposed.

Even though a spontaneous presence of *Ulmus* in this region has been discussed (Van Mourik, 1986), it is worth pointing out a definitive appearance of this taxon at 9,800 ± 120 yr B.P.

The low abundances of *Pinus sylvestris* in the diagrams presented here lead to the proposal of an earlier retreat of this taxon, before 9,800 yr B.P., compared to the known data from other localities in northern Spain (Basque Country, Asturias).

References Cited

BEAULIEU, J.L. DE
1977 Contribution pollenanalytique à la histoire tardiglaciaireet holocène de la végétation des Alpes méridionales francaises. Thèse ès Sciences, Université d'Aix-Marseille III, 358 pgs.

DÍAZ FIERROS, F., TORRAS TRONCOSO, L., and VÁZQUEZ VARELA, J.M.
1979 La evolución de la flora gallega durante el Holoceno. *El Museo de Pontevedra, Pontevedra*, 33: 3–8.

GONZÁLEZ, A.V., and SAA OTERO, M.P.
in press Analyse pollinique d'une tourbière dans la Sierra del Buyo: Cuadramón (Galice, Nord–Ouest de l'Espagne). *Ecologia Mediterranea.*

MENÉNDEZ AMOR, J.
1971 Estudio esporo-polínico de dos turberas en la Sierra de Queija (Orense). *Boletin de la Real Sociedad Española de Historia Natural (sección Geología)*, 69: 85–92.

PEÑALBA, M.C.
1989 Dynamique de végétation tardiglaciaire et holocène du centre-nord de l'Espagne d'après l'analyse pollinique. Ph.D. thesis, Université d'Aix-Marseille III, 165 p.

RIVAS MARTÍNEZ, S., GANDULLO GUTIERREZ, J.M., SERRADA, R., ALLUÉ ANDRADE, J.L., MONTERO DE BUROS, J.L., and GONZÁLEZ REBOLLAR, J.L.
1987 Memoria del mapa de series de vegetación de España. *Ministerio de Agricultura, Pesca y Alimentación, ICONA, Serie Técnica, Madrid*, 268 pgs.

SAA OTERO, M.P.
1985 Contribución a la cronología de sedimentos costeros por análisis polínico. Tesis doctoral, Facultad de Biología, Univerisdad de Santiago, 337 pgs.

SAA OTERO, M.P., and DÍAZ-FIERROS, F.
1986 Contribution à la connaissance, à partir de l'analyse pollinique, des différences existant entre la végétation du N. et du S. de la Galice (Espagne) pendant les périodes pré-boréale, atlantique et subatlantique. *Revue de Paléobiologie*, 5: 325–336.

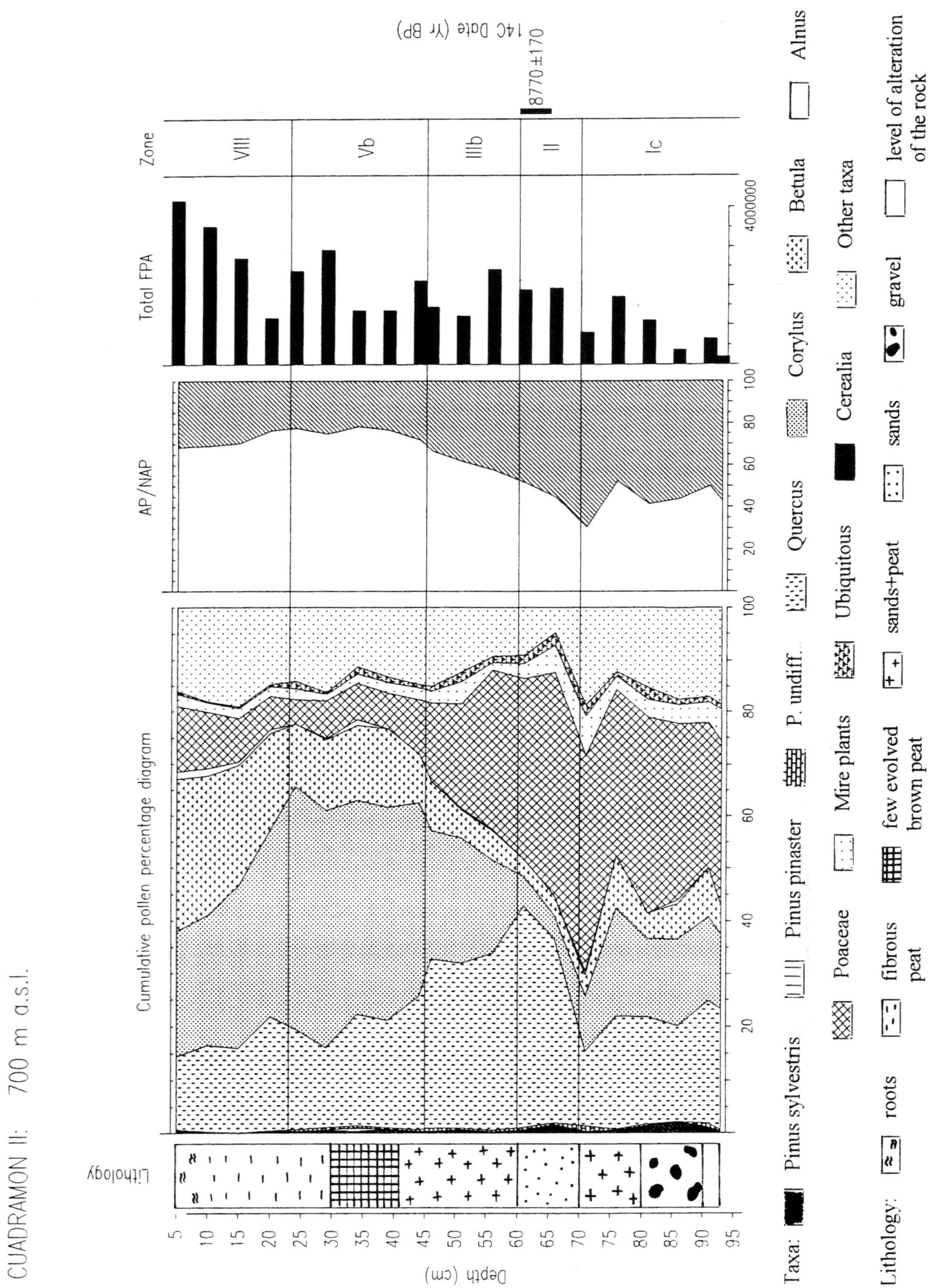

Text-Figure 2. Synthetic pollen percentage diagram and total pollen concentration of Cuadramón II peat bog.

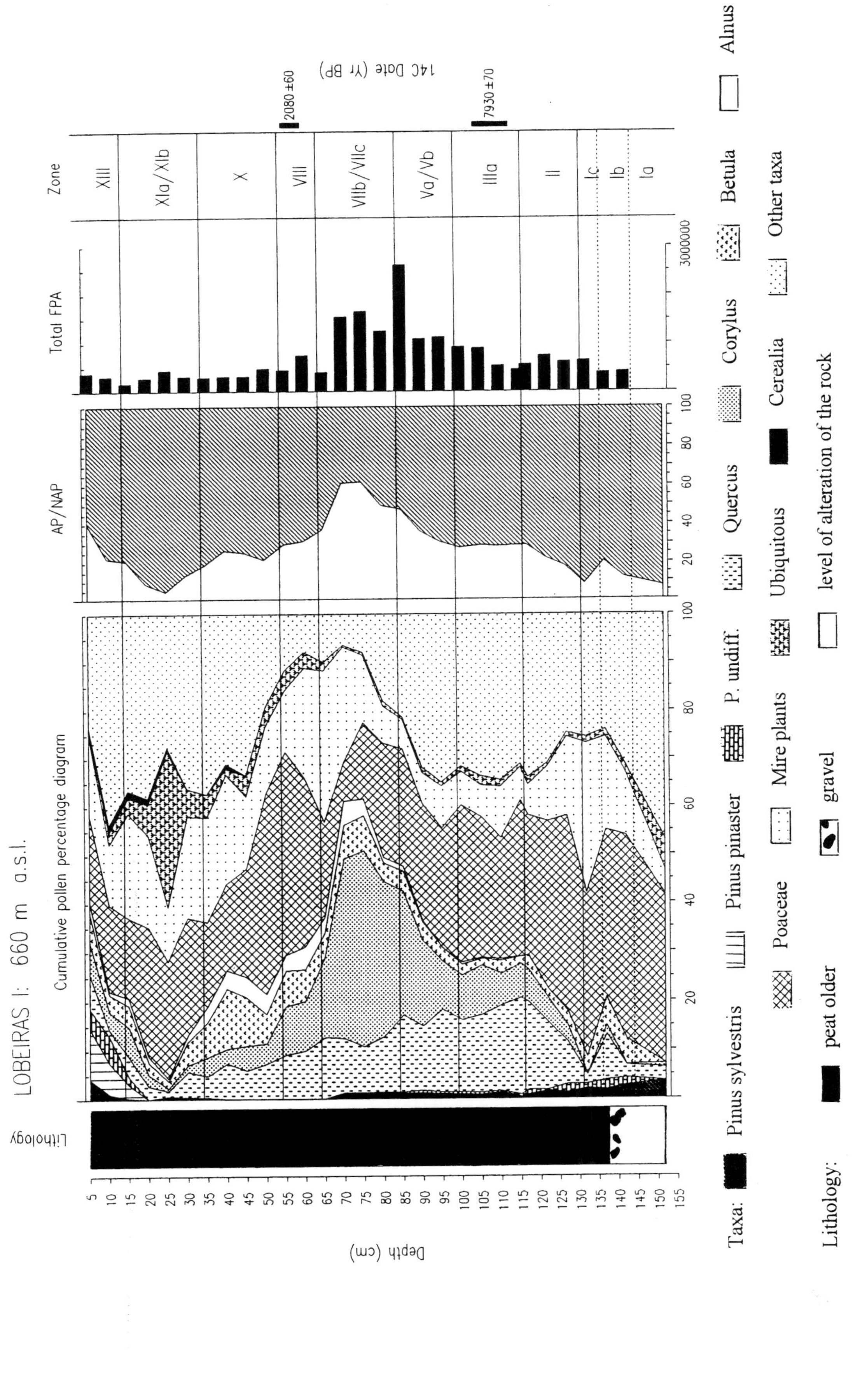

Text-Figure 3. Synthetic pollen percentage diagram and total pollen concentration of Lobeiras I peat bog.

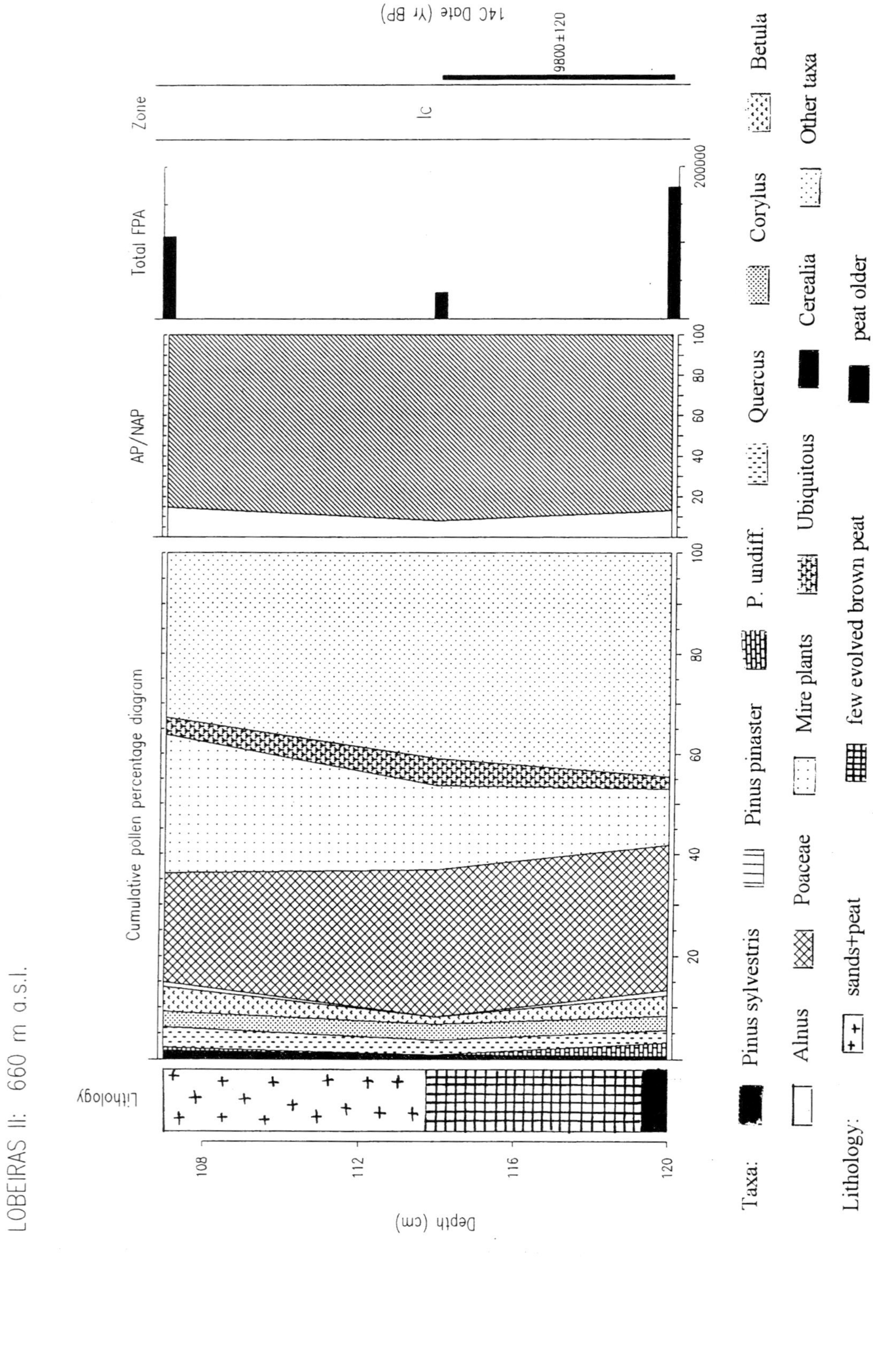

Text-Figure 4. Synthetic pollen percentage diagram and total pollen concentration of Lobeiras II peat bog.

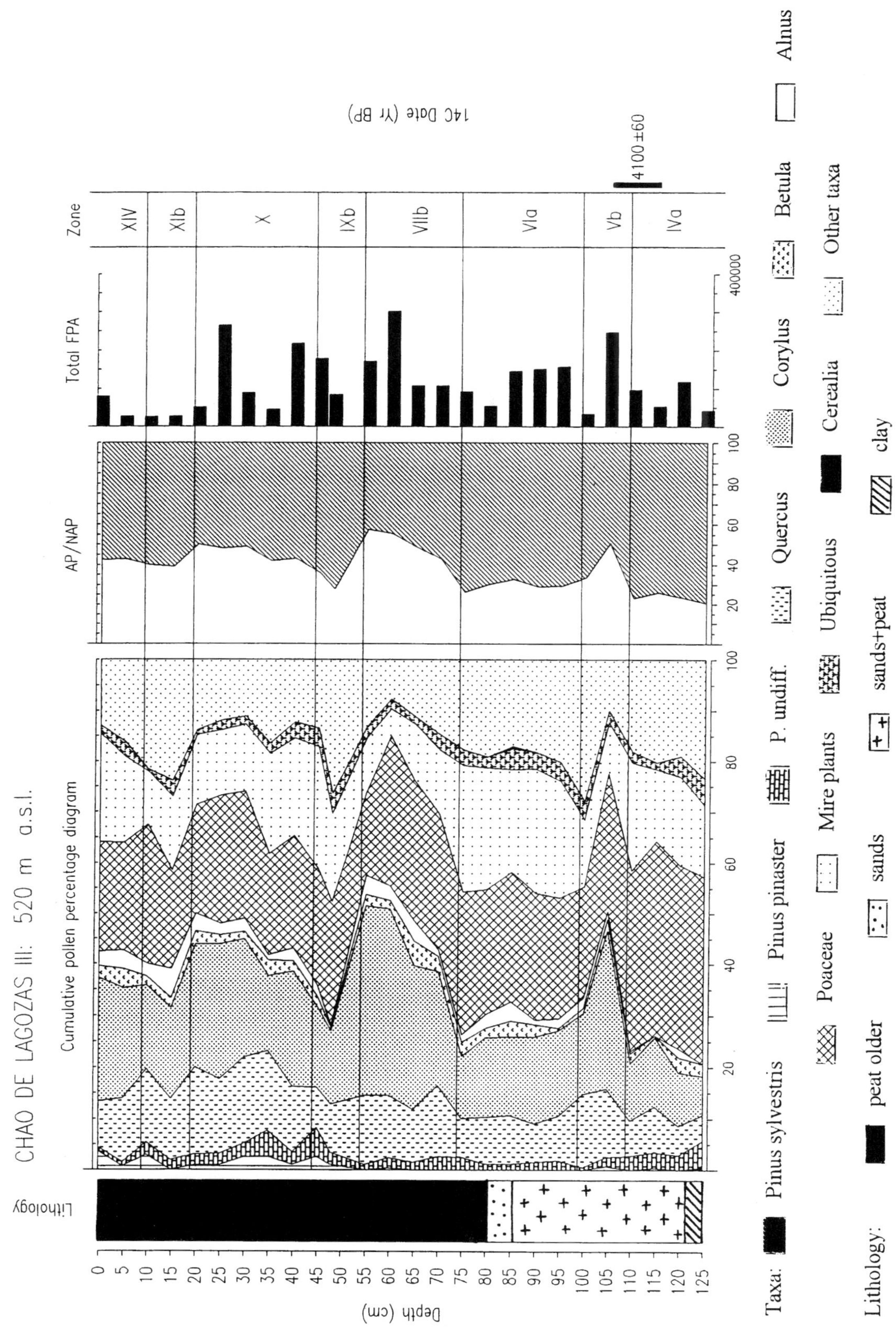

Text-Figure 5. Synthetic pollen percentage diagram and total pollen concentration of Chao de Lagozas peat bog.

TABLE 1. Pollen zonation of the study area.

Zone	Characterization of pollen zones		Characteristic Taxa	700m u.s.l. Cua II	660m u.s.l. Lob. I II	520m u.s.l. Lag III	110m u.s.l. Xov	^{14}C data (Yr. BP)
XIV	Recovery of arboreal vegetation: *Corylus, Quercus*. Decrease of *Castanea* and absence of *Cerealia*. Increase of Ericaceae (4–8%) and *Calluna* (7–15%). Poaceae 20–22%		First presence of Oleaceae			*	*	
XIII	*Eucalyptus* is present. Increase of *Pinus. pinaster* T. Increase of *Castanea*. Ericaceae (0–9%) and *Calluna* (0–18%). *Cerealia* curve constant. Poaceae 3–34%				*			
XII	Increase of the tree and shrub pollen: *Corylus*. Decrease of *Castanea* and Ericaceae. Increase of *Calluna*. Absence of *Cerealia*. Poaceae 26%						*	
XI	Herbs and shrub dominance. Poaceae 10–29%. Increase of Ericaceae (25%). *Calluna*: 0–32%. *Castanea* curve constant	c	*Cerealia*, *Tilia* and *Fraxinus* are present. *Calluna* minimum (1%)				*	770 ± 60
		b	Increase of *Castanea* (máximum). Increase of *Calluna* (5–32%). Presence of *Cerealia*		*	*		
		a	Mínimum of *Castanea*. Increase of marsh–peat herbs. *Calluna* (0–13%). *Cerealia* is present		*			
X	Recovery of arboreal vegetation: *Betula, Corylus, Quercus*. Increase of *Pinus pinaster* T. Minimum of Poaceae (2–41%), Ericaceae (0–7%) and *Calluna* (0–16%). Presence of *Cerealia*. Decline of *Castanea*.		Presence of *Juglans* and *Ulmus*		*	*		
IX	Minimum of arboreal vegetation: Poaceae 18–35%. Increase of Ericaceae and of *Calluna*. Increase of *Castanea*.	b	Increase of *Pinus pinaster* T.			*		
VIII	Decrease of arboreal vegetation. *Pinus pinaster* curve constant. Increase of antrophic taxa. Minimum of *Castanea*. Minimum of Ericaceae (0–9%). Decrease of *Calluna* (0–17%). Poaceae 3–42%			*	* *Taxus*			2080 ± 60

TABLE 1 (continued). Pollen zonation of the study area.

Zone	Characterization of pollen zones		Characteristic Taxa	700m u.s.l. Cua II	660m u.s.l. Lob. I II	520m u.s.l. Lag III	110m u.s.l. Xov	^{14}C data (Yr. BP)
VII	Dominance of arboreal vegetation : *Corylus*, *Quercus*. Poaceae 6–29%. *Betula* 2–16%. Decrease of Ericaceae and *Calluna*. *Pinus pinaster* T. and *Castanea* are present	c	*Betula* is abundant. First presence of *Taxus*		*			
		b	Presence of *Taxus* and *Juglans*.		*	*		
VI	Minimum of arboreal vegetation: Poaceae 6–35%. Ericaceae 1–21%. *Corylus* 1–26%. Increase of *Calluna* and Asteroideae. Expansion of *Pinus pinaster* T. *Cerealia* is present	a	Presence of Oleaceae			*		
V	Increase of tree and shrub pollen: *Corylus*, *Quercus*. Presence of *P.pinaster* T. Decrease of *Castanea*. Decrease of *Calluna* (0–10%). Poaceae 4–27%	b	Smallest frequency of Ericaceae (0–4%). Prompt presence of *Fraxinus* and *Juglans*.	*	*	*		4100? 60
		a	Increase of Ericaceae (2–13%). Presence of *Juglans*		*			
IV	Decline of arboreal vegetation: Poaceae 17–48%. Increase of Ericaceae (4–11%). *Calluna* (0–16%). *Castanea's* curve constant.	a	*Pinus pinaster* T present. Decline of *Calluna*.			*		
III	Increase of arboreal vegetation: *Quercus* and *Corylus*. Poaceae 15–32%. Decline of *Calluna* (1–4%).	b	Absence of *Castanea*. Decline of Ericaceae (1%).	*				
		a	*Castanea* is present. Increase of Ericaceae (2–7%). Presence of *Ulmus*.		*			7930 ± 70
II	Increase of tree and shrub pollen: *Quercus*. Presence of *Castanea* and *Alnus*. Decrease of Ericaceae and *Calluna*. Poaceae 19–43%			*	*			8770 ± 170
I	Herbs dominance. Poaceae 21–42%. *Calluna* 2–28%. Ericaceae 1–7%	c	Presence of *Fagus* and *Ulmus*.	*	*	*		9800 ± 120
		b	Increase of tree pollen. Appearance of *Alnus*, *Castanea* is present, Increase of Ericaceae and *Calluna*.		*			
		a	*Armeria* and *Thalictrum*.		*			

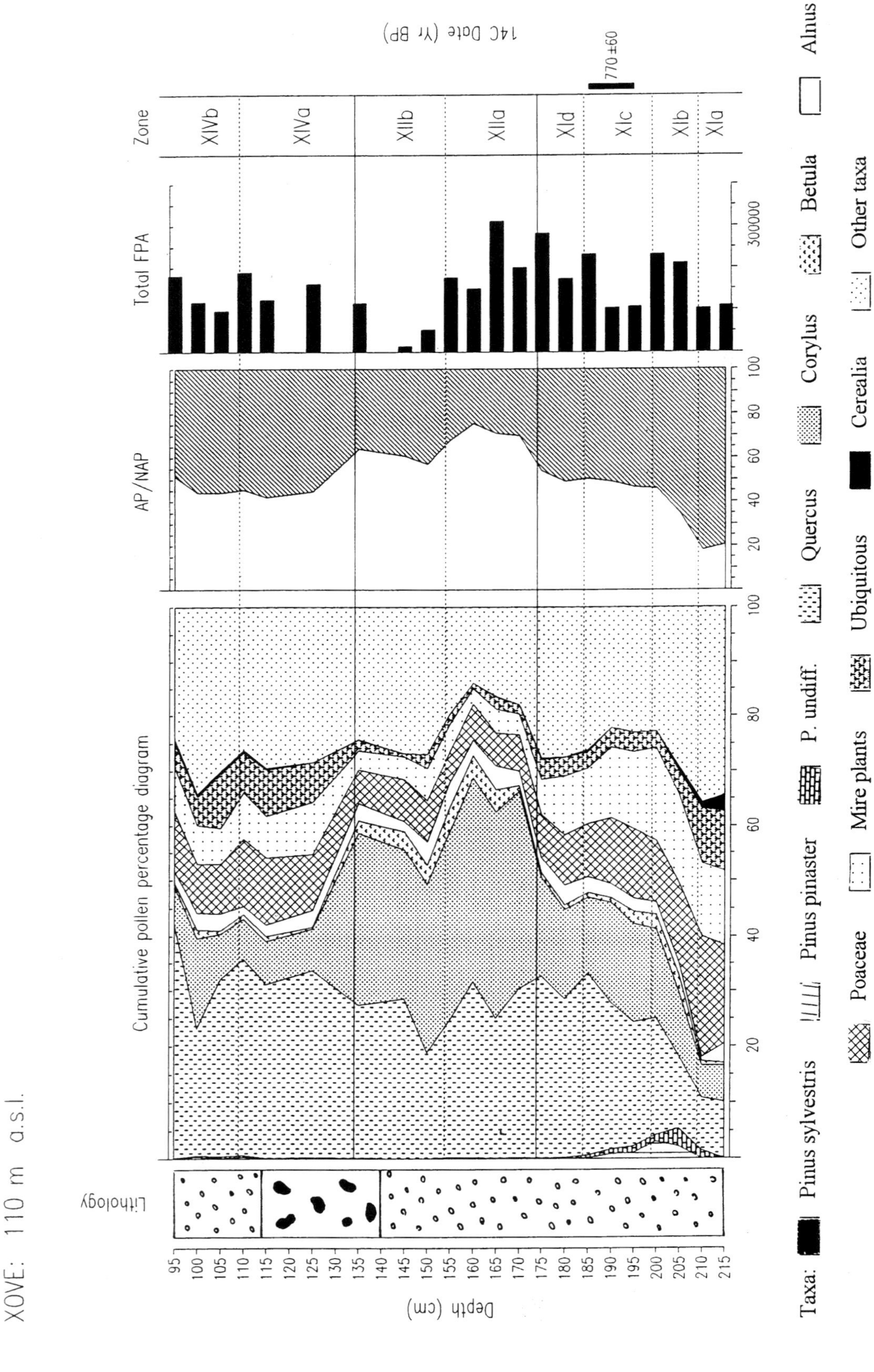

Text-Figure 6. Sinthetic pollen percentage diagram and total pollen concentration organic sediment in Xove.

TABLE 2. Data in Carbon dating of peat-bogs and coastal sediment.

peat bog	zone	level (cm)	absolute data	ref. lab.
Lobeiras II	a3	127–120	9,800 ± 120 BP	UGRA–348
Lobeiras I	c1	100–117	7,930 ± 70 BP	ICEN–554
Lobeiras I	h	65–85	2,080 ± 60 BP	UGRA–347
Cuadramon II	b	61–70	8,770 ± 170 BP	UGRA–491
Chao de Lagozas III	e2	100–110	4,100 ± 60 BP	UBAR–515
Xove	K3	185–200	770 ± 60 BP	UBAR–517

VAN DER KNAAP, W.O., and VAN LEEUWEN, J.F.N.
1995 Holocene vegetation succession and degradation as response to climatic change and human activity in the Serra da Estrela, Portugal. *Review of Palaeobotany and Palynology*, 89: 153–211.

VAN MOURIK, J.M.
1986 Pollen profiles of slope deposits in the Galician area (NW Spain). *Nederlandse Geografische Studies*, 12, 171 pgs.

Authors' addresses

M. P. SAA OTERO
E. DIAZ LOSADA
A. V. GONZALEZ PORTO
Departamento de Bioloxía Vexetal e Ciencias do Solo
Facultad de Ciencias de Orense
Campus Universitario As Lagoas s/m
32004 Orense
Spain

ENVIRONMENTAL CHANGES SINCE THE LATE PLEISTOCENE AT LOP NOR LAKE IN XINJIANG, CHINA

Yushu Wu

Abstract

Pollen, ostracoda, sediment analyses, and radiocarbon dating of samples from the F4 pit suggest that the environmental changes in Lop Nor Lake, since the late Pleistocene can be divided into four stages. Stage I: the late Pleistocene (450–375 cm) is represented by a gypsum-bearing clay deposited in a rather deep and saline lake, with dry climate; a vegetational assemblage of *Ephedra* desert shrubs was present in the area. Stage II: during the late Pleistocene to early Holocene time (375–187 cm) silty and gypsum-bearing clay was deposited in a turbulent, saline lake environment, with dry climate: few pollen taxa were recovered from this interval, including *Ephedra*, *Artemisia*, Chenopodiaceae and Gramineae. Stage III: the middle Holocene (187–78 cm) is represented by a gypsum-bearing clay and mud deposited in an alkaline-saline lake that apparently expanded in the coolest, most humid climate of the entire section; *Artemisia* desert steppe was the main vegetation. Stage IV: during the late Holocene (78–0 cm), a gypsum-bearing carbonate clay was deposited in a saline lake which gradually disappeared; the climate was very dry and a desert vegetation with Chenopodiaceae was present in the area. Since the late Pleistocene, the climate has been dry in the area, with only a slight climatic change in the middle Holocene. Lop Nor Lake was saline until the early 1970s, when it became a playa due to human influence.

INTRODUCTION

In the past century, several explorers and geographers have visited the Lop Nor area, but the information they obtained was fragmentary. The biggest and the most comprehensive scientific expedition was organized by the Xinjiang Branch of Academia Sinica (1987) in 1980–1981. This expedition team was composed of experts in different fields, including geology, geomorphology, soil, animals, plants, chemistry and historic geography. For the first time, the study of palynology was done only in the southern area of Lop Nor Lake (Yan et al., 1983).

In October 1989, the Saline Lake Team of Chinese Academy of Geological Sciences reached the center of the Lop Nor area. Seventeen pits were dug in the salt crust of Lop Nor (Zheng et al., 1992). Pit F4 was the deepest of these 17 pits. Samples were collected for various types of analysis. Wu (1994) studied the pollen of pit F4 and published the results as the second paper about pollen studies in the Lop Nor area. Until now, the information about the environmental changes in this area since the late Pleistocene was still very limited.

This paper reports the environmental studies in the Lop Nor area, based mainly on the pollen analysis, even though it is still preliminary.

GEOGRAPHY AND PRESENT VEGETATION

Lop Nor was one of the most famous lakes in the arid land of China. Now it has developed into a playa. The Lop Nor area (88°–94°E, 39°30'–41°N) is located in the eastern Taklimakan desert, NW China (Text-Figure 1). The altitude is 780–795 m above sea level at the center, which is the lowest region in the Taklimakan desert. Lop Nor Lake was formed by faulting during the late Tertiary to early Quaternary. Based on the information obtained by the Team of the Scientific Expedition of Academia Sinica to Lop Nor (1987), the annual average temperature is 10°C, the annual average precipitation is 20 mm, and the annual average evaporation is 2899–3000 mm in this area. The climate is extremely dry. The atmospheric circulation in the area is affected only by strong winds from the northeast. The desert vegetation has the fewest varieties of any area in China. The modern flora consists of 13 families, 26 genera and 36 species, mainly Chenopodiaceae, Compositae, Tamaricaceae, Lequminosae, Zygophyllaceae and Apocynaceae (Table 1); bryophyta and pteridophyta are absent. A map of the present distribution of desert vegetation in the region is shown in Text-Figure 2.

METHODS

Treatment of the 17 samples follows methods similar to those described by Faegri and Iverson (1964). The total pollen from the 17 samples was 1638 grains. Slides were traversed under a magnification of x400. The pollen of *Pinus*, *Picea*, and *Betula* trees is present throughout in the diagram, but is less than 5% of the total count per taxon. Based on surface analysis, the tree pollen may be mostly transported. Only a select number of taxa are included in the pollen diagram (Text-Figure 3). Based on the information obtained by the Team of the Scientific Expedition of Academia Sinica to Xinjiang (1978), the taxa are grouped into three distinct, ecologically significant categories (Text-Figure 2), as follows:

(a) Shrub-Desert Taxa
- *Ephedra*
- *Nitraria*
- Chenopodiaceae
- Gramineae
- *Artemisia*

(b) Succulent-woody and Salt-bushland Desert Taxa
- Chenopodiaceae
- *Artemisia*
- Gramineae
- *Tamarix*

Wu, Yushu
2001 Environmental changes since the Late Pleistocene at Lop Nor Lake in Xinjiang, China; *In*: Goodman, D.K., and Clarke, R.T. (eds.), *Proceedings of the IX International Palynological Congress, Houston, Texas, U.S.A., 1996*; American Association of Stratigraphic Palynologists Foundation, p. 435–442.

ISBN 0–931871–06–9

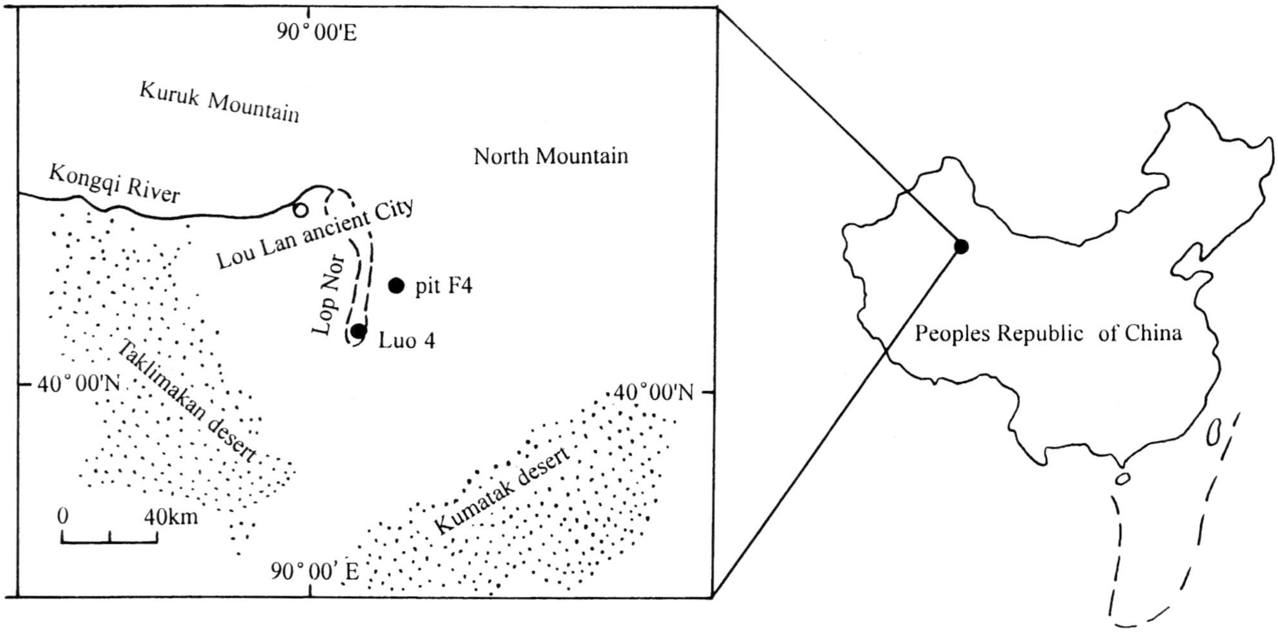

Text-Figure 1. Map of Lop Nor showing the locations of pit F4, core Luo4 and Lou Lan ancient city.

(c) **Desert-Steppe Taxa**
Artemisia
Gramineae
Chenopodiaceae
Lequminosae

STRATIGRAPHY AND CHRONOLOGY

Pit F4 lies at the eastern playa of Lop Nor Lake (Text-Figure 1). The stratigraphy is described in Zheng et al. (1992). The lithology is presented as a stratigraphic column in Text-Figure 6, and can be divided into five sections:

1. 450–375 cm depth: grayish-red and brownish-yellow carbonate-silty clay.
2. 375–213 cm depth: brown clay with silt and gypsum accumulates.
3. 213–187 cm depth: grayish-green gypsum-bearing clay.
4. 187–78 cm depth: black gray and grayish-brown gypsum-bearing clay and mud.
5. 78–0 cm depth: gypsum-bearing carbonate clay and halite. The top salt crust is 23 cm thick.

The stratigraphic profile was dated by means of four radiocarbon determinations given in Text-Figure 3 and Table 2 (Zheng et al., 1992).

TABLE 1. Present flora in the Lop Nor Area. (From "The Team of the Scientific Expedition of Academia Sinica to Lop Nor," 1987).

Families	Genera	Species
Chenopodiaceae	5	5
Compositae	4	5
Tamaricaceae	1	5
Lequminosae	4	4
Zygophyllaceae	2	4
Apocynaceae	2	3
Polygonaceae	2	3
Gramineae	1	2
Salicaceae	1	1
Eleagnaceae	1	1
Cynomoriaceae	1	1
Solanaceae	1	1
Ephedraceae	1	1
Total 13	26	36

THE NEW POLLEN RECORD AT PIT F4

The pollen data from the dated profile are presented in Text-Figure 3. Frequencies of a select group of pollen taxa are expressed as percentages of the total plant pollen.

The pollen diagram is divided into four pollen zones.

Zone I

In the diagram from pit F4, *Ephedra* is the most important pollen type during the late Pleistocene, values of up to 48 %. Further upwards *Ephedra* continues to be present in lower frequencies. The pollen of *Artemisia*, Chenopodiaceae and Gramineae was less important. Few pollen grains were recovered, including *Nitraria*, *Euphorbia*, Leguminosae, Caryophyllaceae, Compositae and *Calligonum*. The pollen of the aquatic plant *Typha* is present.

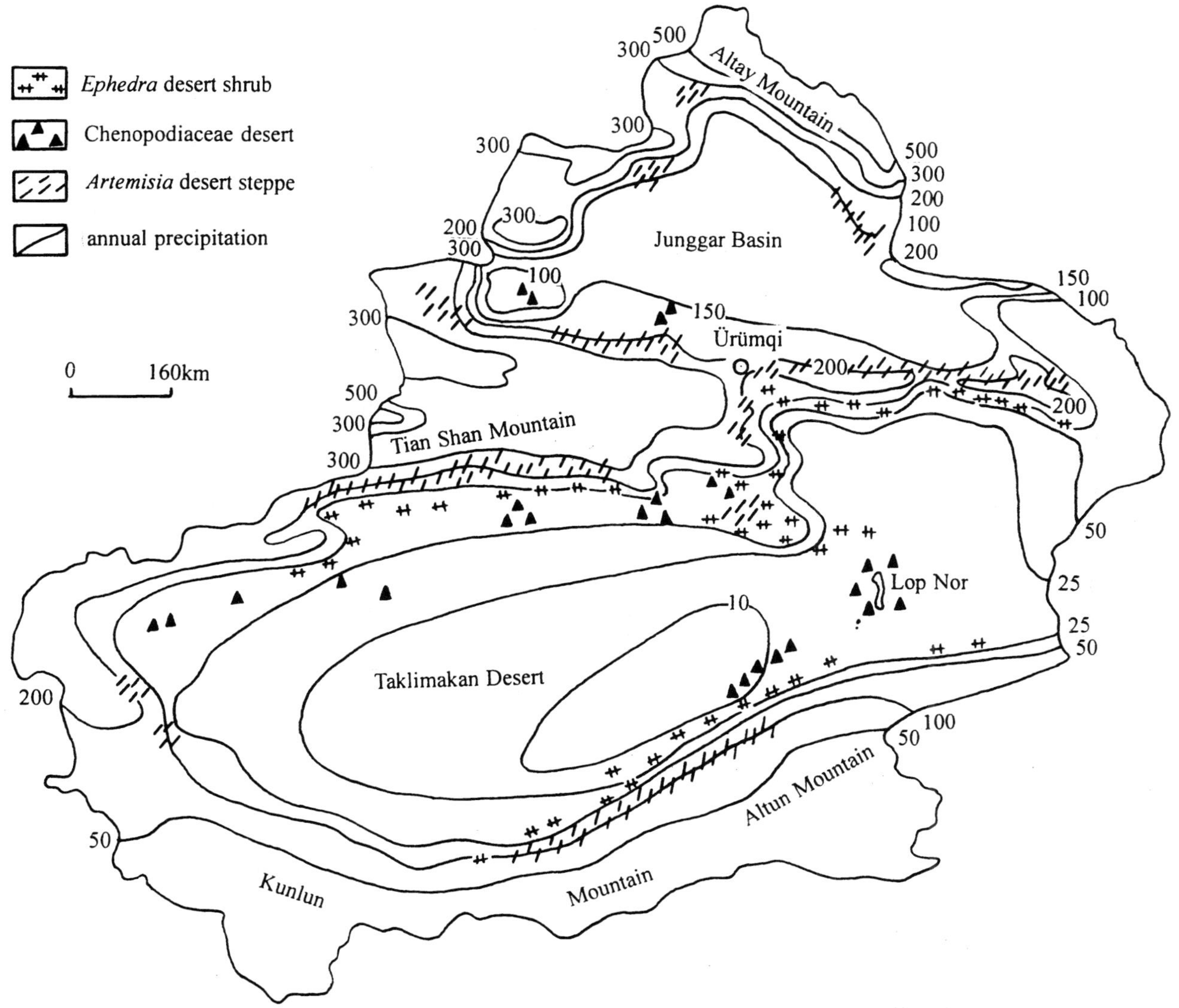

Text-Figure 2. Sketch map of the present distribution of desert vegetation in Xinjiang, China.

Zone II

Pollen assemblages were poor and difficult to count. Pollen of *Artemisia*, Chenopodiaceae, *Ephedra*, Gramineae, Caryophyllaceae, *Euphorbia* and *Typha* are present, but only a few pollen grains were found (Text-Figure 3).

Zone III

This is the most diverse pollen assemblage in the profile. Pollen of *Artemisia* is present in higher frequencies in this zone, with values of up to 41.4%. Pollen of *Ephedra*, Chenopodiaceae and Gramineae are less important. A few pollen grains of *Nitraria*, Compositae, Caryophyllaceae, *Polygonum*, Cyperaceae, *Colligonum* and Umbelliferae appear at about the same time in the interval. The occurrence at a higher level (12%) of *Typha* and a few *Potamogeton* grains possibly indicate freshwater environments.

Zone IV

In the lower part of the zone, the major feature is dominant Chenopodiaceae, with values of up to 70%; the frequencies of other pollen taxa in zone III decreased or disappeared in zone IV. Further upwards all pollen frequencies continue to decrease or disappear.

VEGETATIONAL AND CLIMATIC CHANGES

Pollen analysis of a few surface samples has been conducted in the area. The pollen of *Pinus*, *Picea* and *Betula* is commonly found in surface samples, but always represents less than 5% per taxon. Yan Fuhua et al. (1983) completed the pollen analysis of core Luo 4, located in the southern part of Lop Nor Lake (Text-Figure 1). This figure shows that the desert vegetation was dominated

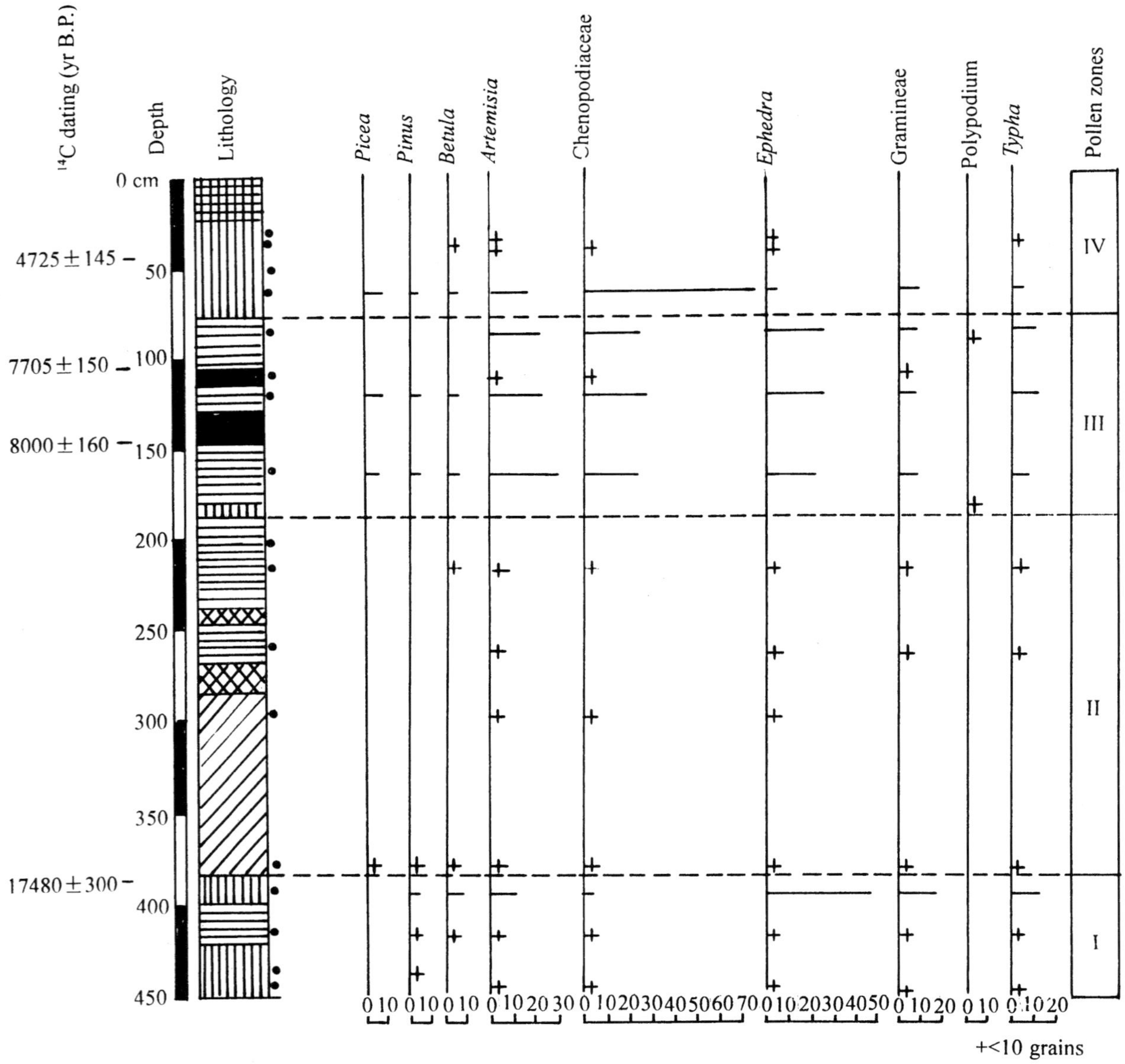

Text-Figure 3. Pollen percentage diagram for pit F4.

mainly by Chenopodiaceae, *Artemisia* and *Ephedra* since 20, 000 yr B.P. (Text-Figure 4). Based on the results of pollen analysis from pit F4 (Text-Figure 3), the vegetational changes are divided into four periods since the late Pleistocene.

Late Pleistocene

In the pollen diagrams of both pit F4 and core Luo 4, there is a predominance of *Ephedra* pollen (Text-Figure 5). The highest percentages were 80% in core Luo 4 and 48%

TABLE 2. Radiocarbon dates from Pit F4.

Lab. No.	Depth/cm	Radiocarbon yr B.P.
CG-2433	45-55	4725±145
CG-2526	95-105	7700±150
CG-2434	125-135	8000±160
CG-2527	385-395	17,480±300

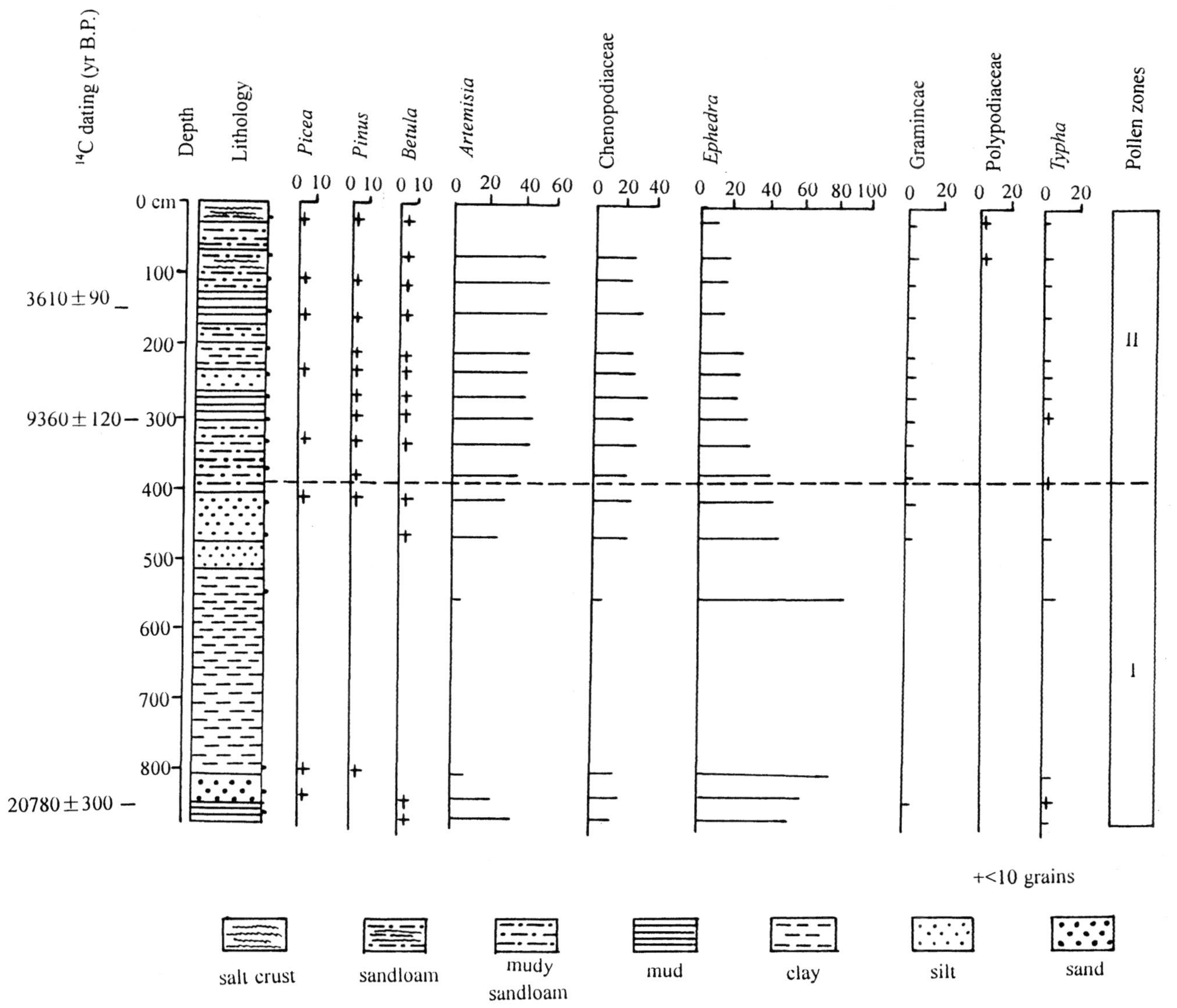

Text-Figure 4. Pollen percentage diagram of the core Luo 4 (after Yan Fuhua et al. 1983).

in pit F4. The pollen of *Artemisia* and Chenopodiaceae was less important. The vegetation of *Ephedra* desert shrub might have been widely distributed in the Lop Nor area at this period. At present, the distribution of *Ephedra* desert shrub is limited to the piedmont alluvium of Kunlun, Tian Shan, and Altay mountains (Text-Figure 2). The annual average precipitation in the area is about 50–150 mm. This suggests that during the late Pleistocene the annual average precipitation was more than the current one, but the climate was still dry.

Late Pleistocene–Early Holocene

The late Pleistocene and early Holocene cannot be divided at present, because of rare pollen in pit F4. In contrast, in core Luo 4, the vegetation during this period could have been desert shrub. Later, the pollen of *Artemisia* and Chenopodiaceae increased gradually, perhaps by the beginning of the Holocene. The vegetation and climate were probably similar to that of the late Pleistocene period.

Middle Holocene

Pollen is abundant in sediments of this period. *Artemisia* pollen increased in both pit F4 and core Luo 4. Chenopodiaceae pollen was less important, and *Ephedra* pollen frequencies decreased. These conditions suggest that the vegetation of the *Artemisia* desert steppe occupied a large area during the middle Holocene. At present, the *Artemisia* desert steppe is distributed on the mountains of Xinjiang, where the altitude is over 1, 000 m and the annual average precipitation approximately 150 mm (Text-Figure 2). The results of the pollen analysis show that there was an expansion of the vegetation

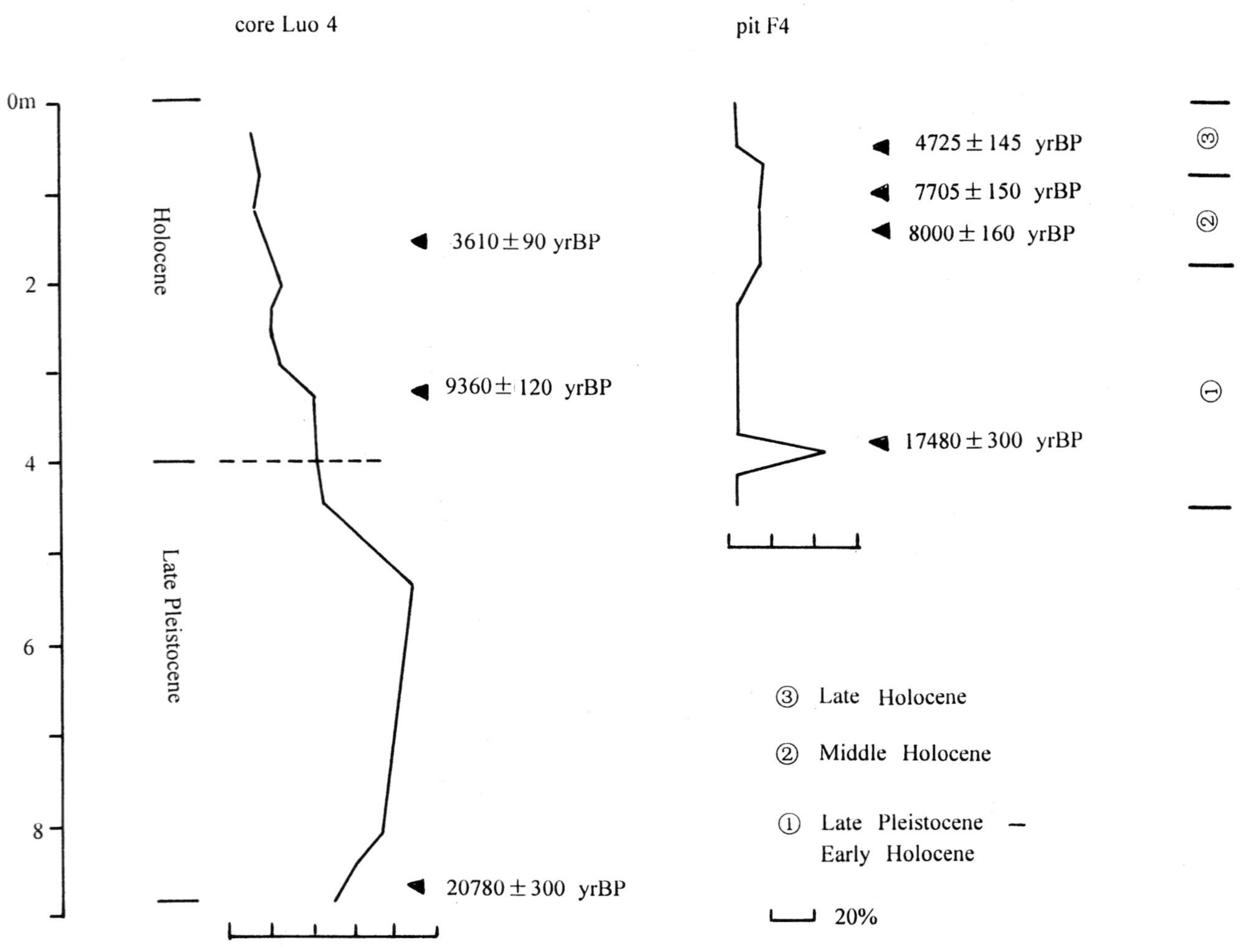

Text-Figure 5. Stratigraphic distribution of *Ephedra* pollen.

of *Artemisia* desert steppe from the mountains to the Lop Nor area. The climate was cooler and more humid than at any other time represented in the profile.

Late Holocene

The pollen frequencies for the Chenopodiaceae increased rapidly at the beginning of this period at pit F4. Pollen of *Artemisia* and *Ephedra* decreased gradually. Now the desert vegetation, which consists primarily of fleshy Chenopodiaceae plants, is widely distributed on the saline alkali soil of the hollow zone in Xinjiang, where the annual average precipitation is 10–50 mm. The area is very dry, and has high average temperatures. Most probably, these conditions have not changed since late Holocene times.

Another major event during this period was the destruction of the Lou Lan ancient city (89°55'22"E, 40°29'55"N) near Lop Nor Lake (Text-Figure 1). On the basis of the results from the Team of the Scientific Expedition of Academia Sinica to Lop Nor (1987), Lou Lan city was a flourishing city approximately 2, 000 years ago. At that time the rivers formed a network, and a dense forest of *Populus diversifolia* was distributed in delta zones of the Kongqi River. Lou Lan ancient city was destroyed after approximately 330 years A.D. The main reason probably resulted from the introduction of irrigation and hydrological engineering works in mid-upper reaches of the Kongqi and Tarim rivers, which brought the change of water course. This resulted in a reduction of water discharge in the downstream river, and the eventual cut-off of the water discharge. Therefore, the recent environmental changes in this area were not due to the natural climatic development, but can be attributed to human activities.

HYDROSERE DEVELOPMENT

Analyses of the sediments, pollen and ostracoda show changes in lake water condition since late Pleistocene. These changes can be divided into four stages (Text-Figure 6):

1. **Stage I (450–375 cm depth):** grayish-red and brownish-yellow color section, consisting of carbonate-silty clay, partly laminated, bearing a little wedge-shaped gypsum and porespace brine (Zheng et al., 1992). The saline water fossil ostracoda *Cyprideis littoralis* (Brandy) found at the lake bottom indicates that the lake was

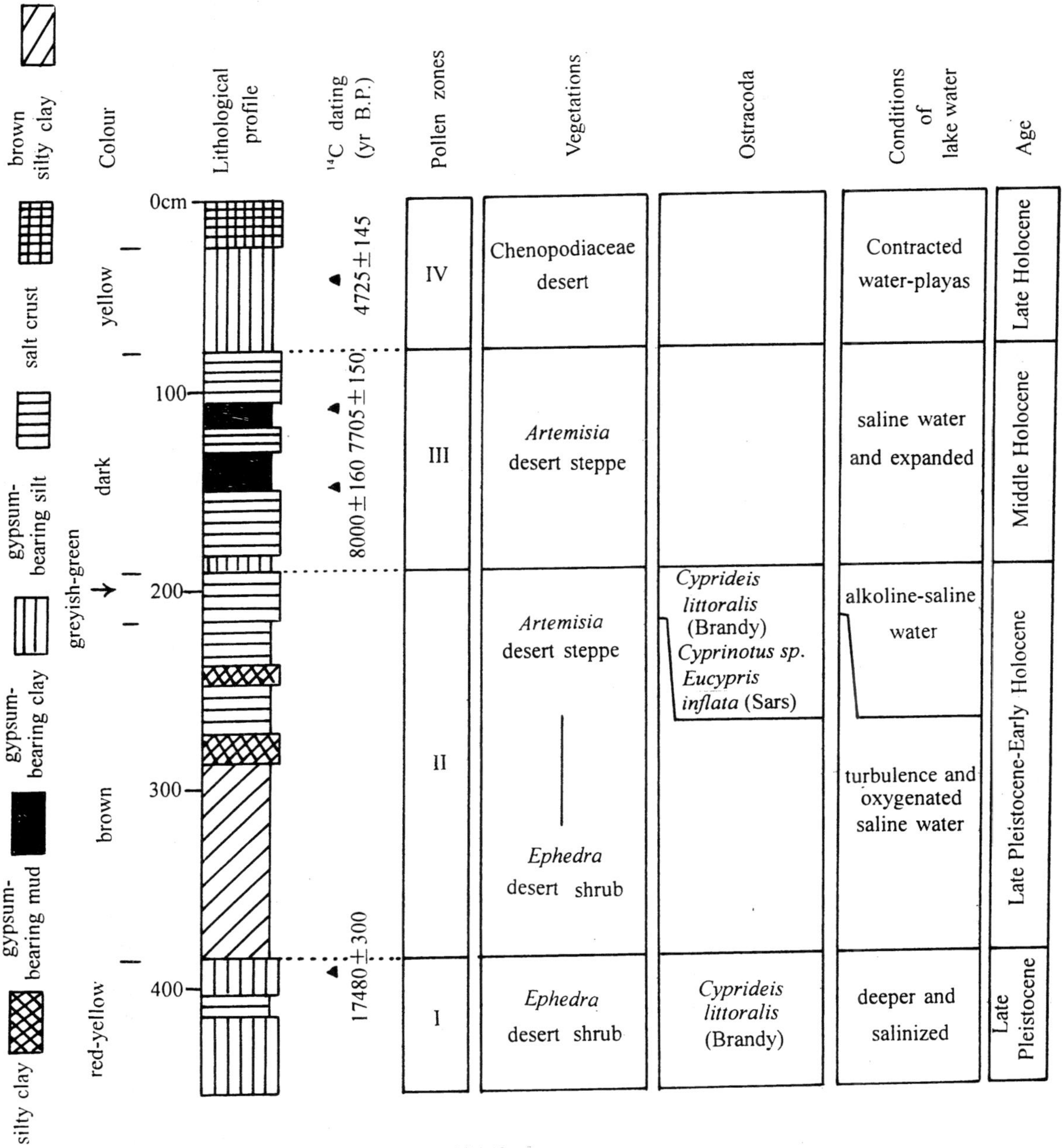

Text-Figure 6. A synthetic study of pit F4 in Lop Nor.

rather deep and saline. A few pollen grains of *Typha* (Text-Figure 3) show that fresh water environment was present along part of the lake margin.

2. **Stage II (375–187 cm depth):** brown color clay section. The silt content increased and the gypsum was dispersed. The sediments reflect the turbulence of the lake water and an oxygenated environment. The situation was unfavorable for pollen preservation, which is why few fossil pollen grains are found in this period (Text-Figure 3). The ostracoda *Cyprideis littoralis* (Brandy), *Cyprinotus* sp. and *Eucypris inflata* (Sars) were present in the upper section, indicating alkaline, saline lake water.

3. **Stage III (187–78 cm depth):** black, dark gray and grayish-brown gypsum-bearing clay and mud, indicating a saline lake with stable water. Pollen of the aquatic plant *Typha* is continuously present in the sediments, with the highest frequencies since the late Pleistocene. A possible explanation is that the Kongqi River entered the lake, which increased the water discharge and precipitation slightly. Aquatic and swamp plants (e.g. Cyperaceae, Umbelliferae, *Potamogeton* and *Typha*) grew in the marginal zones of the salt lake and in the delta zones of the rivers. Afterwards, with the drying of Lop Nor Lake, the aquatic and swamp plants disappeared in the area.

4. **Stage IV (78–0 cm depth):** gypsum-bearing carbonate clay. The lake water was saline. Later, the water disappeared gradually. The salt crust is 23 cm thick at the top of the profile, indicating that Lop Nor Lake recently became a playa. Because of the human activities, the entire Lop Nor Lake dried up very rapidly during the early 1970s.

CONCLUSIONS

Since the late Pleistocene, the desert vegetation consisted of *Ephedra, Artemisia* and Chenopodiaceae over the entire Lop Nor area. The dry climate has not changed much until the present day. There was only a slight fluctuation to a cooler and more humid climate during the middle Holocene, and drier conditions were established again at the beginning of the late Holocene. The desert vegetation was then dominated, as it is today, by Chenopodiaceae, *Artemisia* and *Ephedra.*

Lop Nor Lake has been a salt lake since the late Pleistocene. It became a playa in the early 1970s, when human activities were responsible for the environmental changes.

References Cited

FAEGRI, K., and IVERSON, J.
1964 *Textbook of Pollen Analysis*, Hafner Press. New York.

THE TEAM OF THE SCIENTIFIC EXPEDITION OF ACADEMIA SINICA TO XINJING
1978 *The Vegetation of Xinjiang and its Utilization*. Science Press, China. (in Chinese).

THE TEAM OF THE SCIENTIFIC EXPEDITION OF ACADEMIA SINICA TO LOP NOR
1987 *The Research and Scientific Expedition to Lop Nor.* Science Press, China. (in Chinese).

WU, YUSHU
1994 The sporo-pollen assemblage and its significance of pit F4 from Lop Nor area in Xinjiang. *Arid Land Geography*, 17: 24–29. (in Chinese).

YAN FUHUA, YE UENGYING, and MAI XUISHUN
1983 The sporo-pollen assemblage in the Luo 4 drilling of Lop Nor Lake in Uygur Autonomous Region of Xinjiang and its significance. *Seismology and Geology*, 5(4): 75–80. (in Chinese).

ZHENG MIANPING, WU YUSHU, LIU JUNYING, and QI WEN
1992 Preliminary study on sedimentary environment of the Lop Nor Salt Lake and its prospect for potassium. *Chinese Science Bulletin*, 37: 935–939.

Author's address

WU YUSHU
Department of Palaeobotany
Institute of Botany
Chinese Academy of Sciences
Xiangshan, Beijing, 100093
China

PALYNOLOGICAL DATA FROM QUATERNARY DEPOSITS OF TWO LAKES IN THE NORTHERN REGION OF THE STATE OF RIO DE JANEIRO

Ortrud Monika Barth, Cynthia F. P. da Luz, Mauro Bevilacqua de Toledo, Marcia Aguiar de Barros, and Cleverson Guizan Silva

Abstract

The interpretation of vegetation and climatic changes, during the last 10,000 years of the Quaternary period, are the goals to be achieved using palynological data. We have analyzed the sediments of two lakes in reference to pollen and spore spectra fluctuation during the last 6,000 years. The Lagoa Salgada is situated next to the ocean coast, while the Lagoa de Cima is located at a higher region; both are situated in the northern region of the state of Rio de Janeiro. The core from the Lagoa Salgada is younger than 3,000 years, while the core from the Lagoa de Cima represents a period starting at more or less 6,100 years B.P. (uncalibrated years). The pollen analysis of the sediments from the Lagoa Salgada indicates a humid climate at the bottom, a drier period in the middle, and a return to a more humid climate at its top layer. Pollen analysis of the older sediments of the Lagoa de Cima indicate in its lower part the existence of the lake that dries up during the following periods and is partially occupied by trees and shrubs, sometimes drying up completely.

INTRODUCTION

The cold fronts coming from the south are blocked in the northern part of the state of Rio de Janeiro during "El Niño" occurrences. No significant data with respect to the paleoclimate and the succession of vegetation during the Holocene in this region were previously obtained.

The northeast Precambrian area of the state of Rio de Janeiro comprises the Serra do Mar that finishes in the delta of the river Paraíba do Sul. Its east side is constituted by Pleistocene relief of the Formação Barreiras ("tabuleiros costeiros") and by the Quaternary coastal plain of the delta. One lake and one lagoon located in the municipality of Campos, State of Rio de Janeiro, were chosen for this study: the Lagoa Salgada (without any tributary) and the Lagoa de Cima (an open lake that receives the tributaries Imbé and Urubu). The first is located near to the seashore and its depth is dependent on seasons and pluviosity, sometimes remaining dry (Lemos, 1995). The second is situated at the limit of the pre–Cambrian area and the "tabuleiros costeiros," at 50 km west from the seashore (Text-Figure 1). The Lagoa Salgada is 8.5 km in length, 1.9 km wide and 1 m deep. The Lagoa de Cima has a water surface of ca 12.5 km^2 and is 2 m deep.

The northern region of the state of Rio de Janeiro has a hot and humid climate near to the ocean coast, with a dry season during the winter months from July to September. The higher regions have a better distribution of rainfall, so that a totally dry season does not occur.

The vegetation of the lowlands consists of pioneer plant associations of the "restinga" formation, regions covered by grass savannas and hydrophylic and tropical semidecidual seasonal forests. At higher altitudes (600 m) a dense tropical ombrophilous forest is present (Radambrasil Project, 1983) (Text-Figure 2).

Barth, Ortrud Monika, Luz, Cynthia F. P. da, de Toledo, , Mauro Bevilacqua, Barros, Marcia, Aguiar de, and Silva, Cleverson Guizan

2001 Palynological data from Quaternary deposits of two lakes in the northern region of the swtate of Rio de Janeiro; *In*: Goodman, D.K., and Clarke, R.T. (eds.), *Proceedings of the IX International Palynological Congress, Houston, Texas, U.S.A., 1996*; American Association of Stratigraphic Palynologists Foundation, p. 443–450.

 ISBN 0–931871–06–9

MATERIAL AND METHODS

Sediment Core of the Lagoa Salgada

The 120 cm long core of organic mud and carbonate sediments was collected by the percussion method by a team from the LAGEMAR department (Laboratório de Geologia Marinha) of the Universidade Federal Fluminense, Niterói, state of Rio de Janeiro. Ten samples, each 2 cm thick, collected at different levels of the core were analyzed. Radiocarbon dates obtained from shells of the marine sequence below the lagoonal sediments dated around 3,000 years B.P. (uncalibrated years) (Martin et al., 1993).

Sediment Core of the Lagoa de Cima

The 417 cm long core of organic clay and sands, was obtained using a vibro-coring sampler (Martin and Flexor, 1989) by the team from ORSTOM, France (number RJ-92/6). Twenty samples, each 1 cm thick, collected at different levels of the core, were analyzed. The basal sample of the core was dated by the radiocarbon method at the laboratories in Bondy, France, and dated circa 6,100 years B.P. (uncalibrated years).

Preparation of Palynological Slides

Samples were processed by the methodology previously proposed (Ybert et al., 1992) and the concentration of palynomorphs was obtained by the introduction of tablets of a exotic plant species (*Lycopodium*). The slides are kept at the Laboratório de Palinologia (UFRJ).

RESULTS AND DISCUSSION

Lagoa Salgada

The results of the palynological analysis of the core define two zones that can be divided into several levels each (Text-Figure 3). The main Zone I, begins at the bottom of the core and comprises levels 1–5. It is characterized by a low frequency of pollen types (Poaceae, Cyperaceae, *Amaranthus*/Chenopodiaceae, Mimosoideae and Asteraceae) but by a higher concentration of fern spores (Cyatheaceae/Polypodiaceae type; referred to as "Trilete–Psilate" and "Monolete" on the diagram). In this regard a humid environment similar to the actual climate may be assigned to the bottom of Zone I. Levels 4 and 5 indicate a drier period and are characterized by an increasing concentration of *Amaranthus*/Chenopodiaceae pollen type and a reduction of Cyperaceae pollen and fern spores. The sudden increase of Cyperaceae pollen grains at level 6 confirms that a more humid phase is starting and that the water level rises. A lens of limestone (Lemos, 1995) is responsible for the interruption observed at level 4 and is thicker at level 5, where a higher concentration of grass pollen and a lower of fern spores occur.

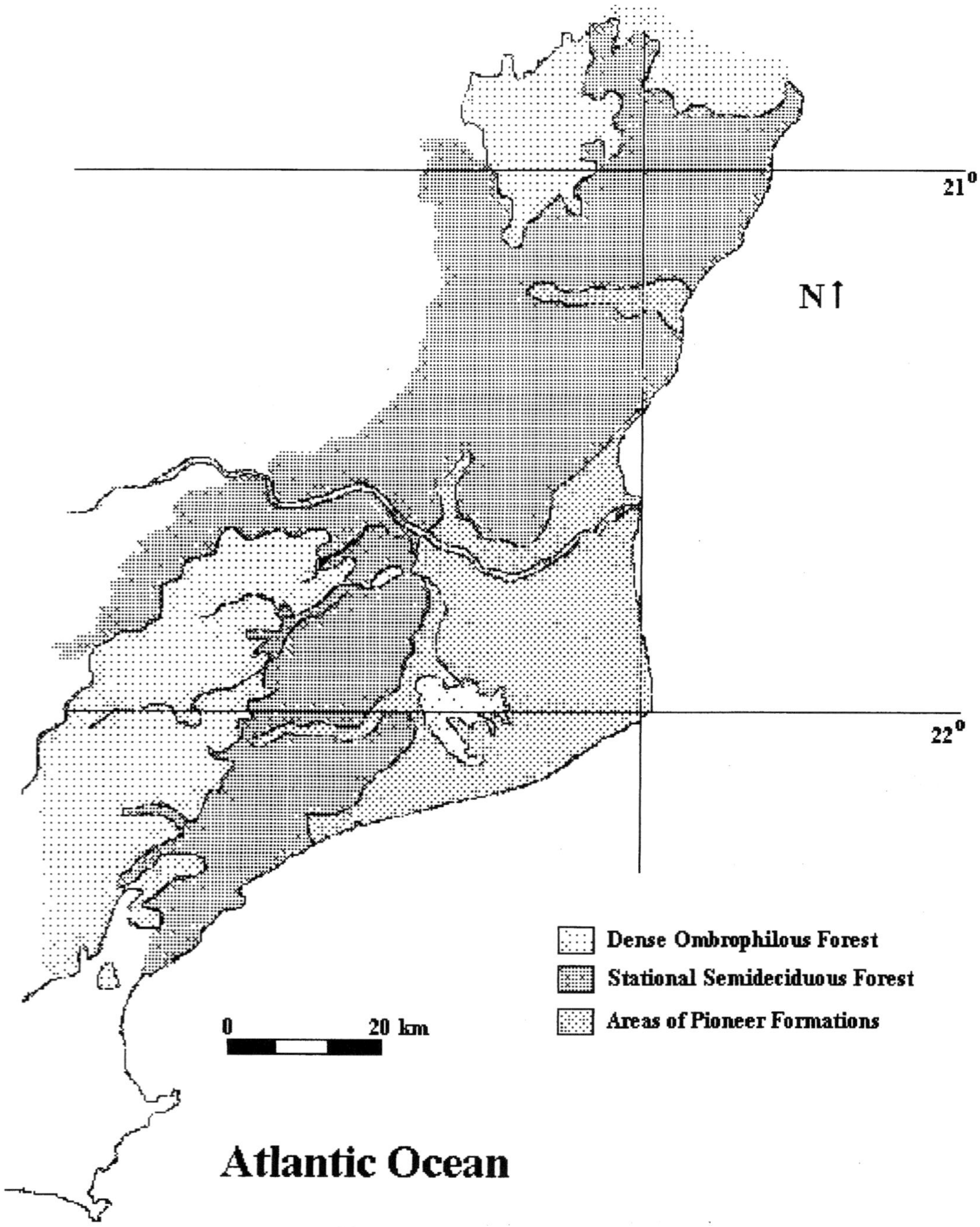

Text-Figure 1. Location of the study sites (Martin et al., 1993).

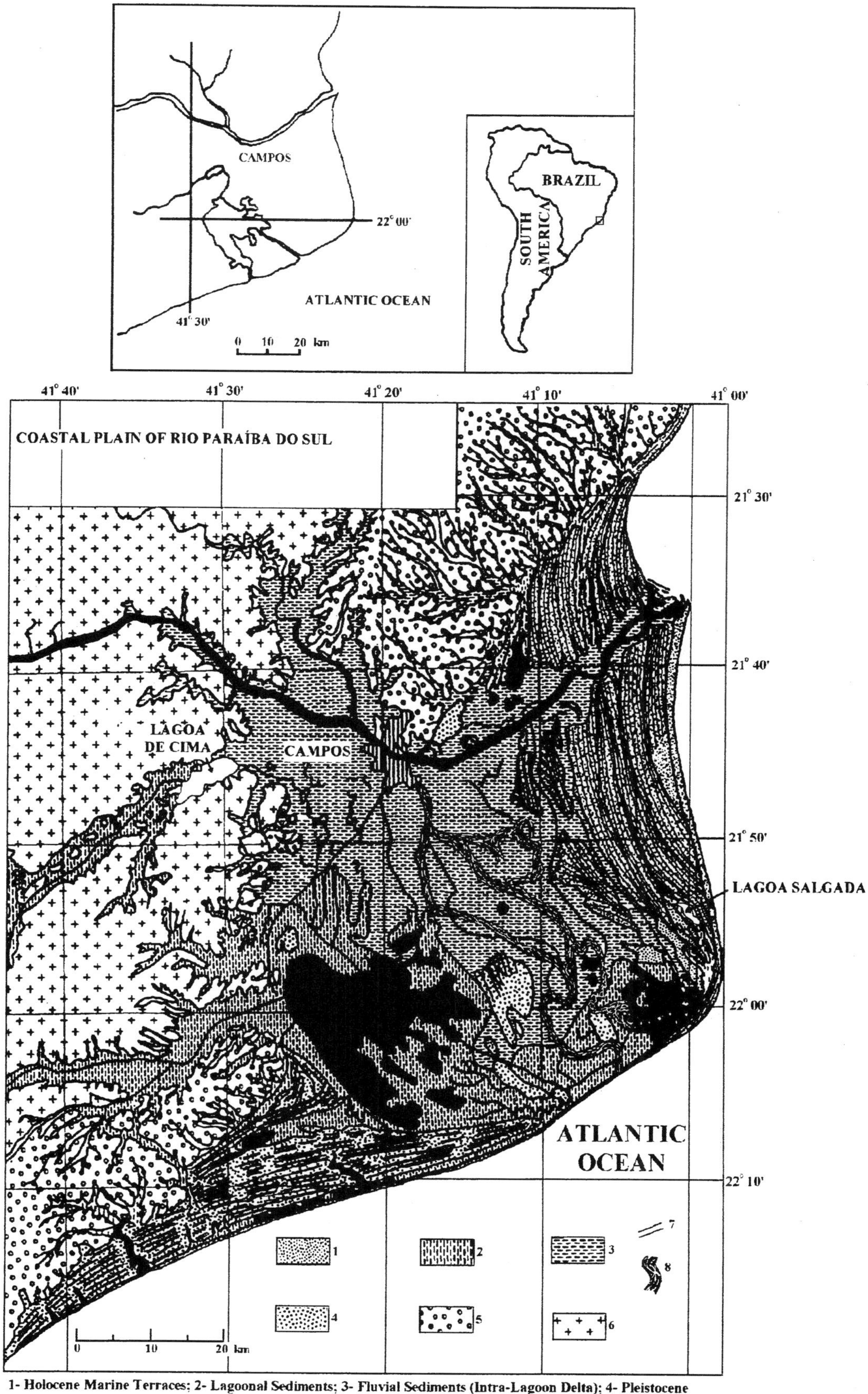

1- Holocene Marine Terraces; 2- Lagoonal Sediments; 3- Fluvial Sediments (Intra-Lagoon Delta); 4- Pleistocene Marine Terraces; 5- Continental Sediments of the Barreiras Group (Upper Tertiary); 6- Pre-Cambrian Crystalline Basement Rocks; 7- Direction of the Beach Ridges; 8- Fluvial Palaeo-Channels.

Text-Figure 2. Phytogeographical formation of the study area (Radambrasil Project, 1983).

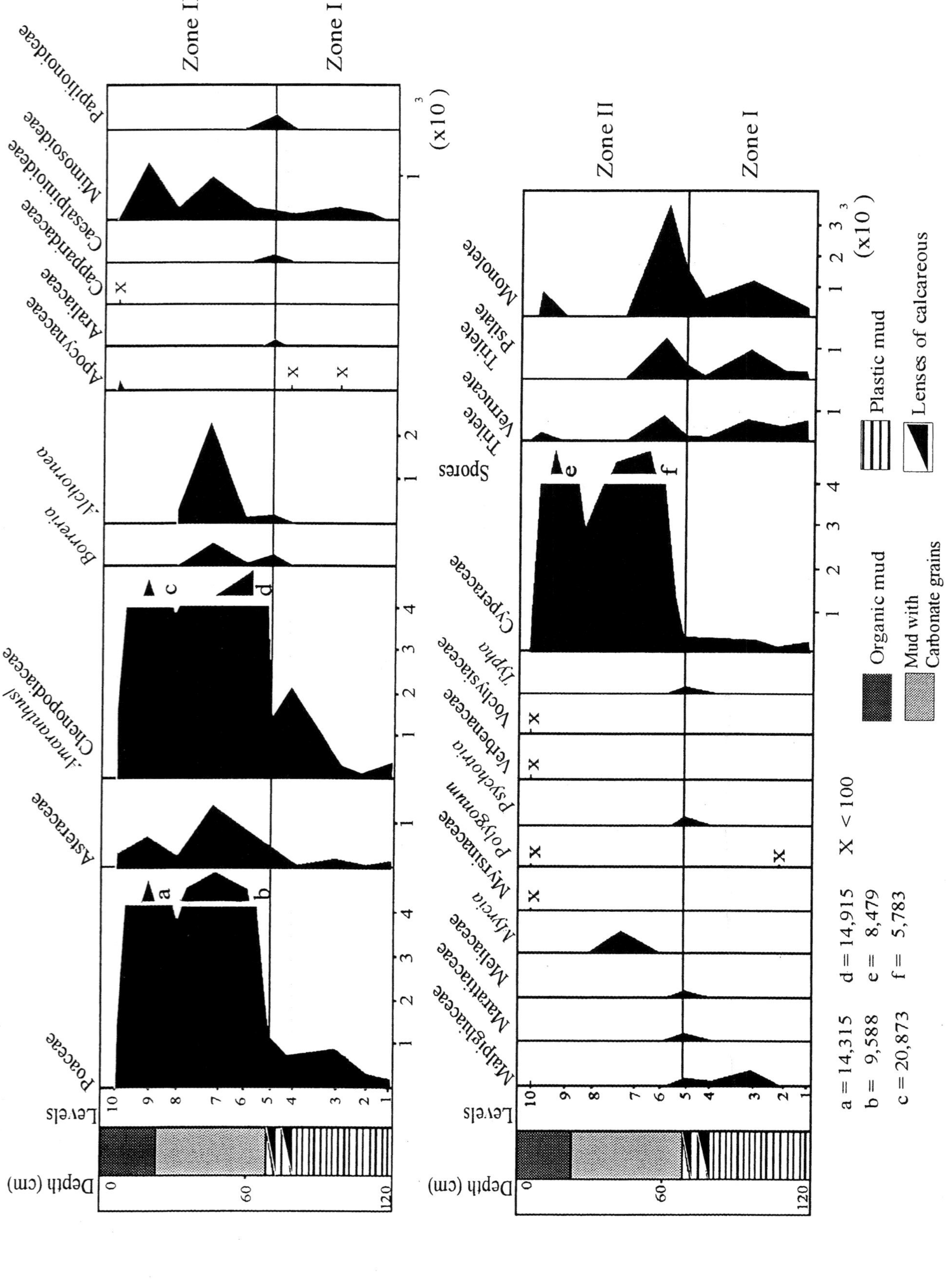

Text-Figure 3. Concentration (palynomorphs/g of sediment), Lagoa Salgada.

Zone II comprises levels 6–10. The climatic conditions changed again and a drier environment can be attributed to this zone. It is characterized by a higher frequency of Poaceae, Cyperaceae, *Amaranthus/* Chenopodiaceae and Asteraceae pollen and a reduction of fern spores at levels 7–9. The levels 6–8 present a mudy carbonate sediment (Lemos, 1995) that result from water evaporation caused by higher temperatures and a lowering of the water level. The higher concentration of pollen of Cyperaceae (levels 6–10) and the absence of fern spores at levels 7–9 suggest a reduction of the lagoon depth. A special situation is found at level 8 where the pollen concentration is reduced; no additional pollen types of forest species are recognized and it seems that this driest phase results in poor conditions for pollen grain preservation. A more humid phase begins when fern spores reappear at level 10.

Lagoa de Cima

Four main zones can be established in the palynological diagrams (Text-Figure 4), based in the obtained data. Starting at the bottom, Zone I (levels 1–5) starts with a warm phase and low humidity, presenting well developed swampy areas around the lake. The forest is of the semideciduous type and pioneer species. In sequence, possibly one of the driest phases resulted in fluvial reworking when the sea level was low. The semideciduous forest remains during this dry period.

Zone II (levels 6–11) the humidity increase and the lake fills again. The semideciduous forest presents a more advanced stage of ecological succession with dry grasslands from the coastal plain and evergreen rainfall forests in the valleys. At the end of this period the climate dries again and all the vegetation is affected.

Zone III (levels 12–17) is characterized by a return of more humid conditions. During this period coastal grasslands coexists with the semideciduous forest and swampy vegetation. The evergreen rainfall forest is more developed inside the valleys and is expanded. A mountain rainfall forest also develops at the west side of the lake. At the end of this zone a new dry phase (about 4,000 years B.P.) occurs and the size of the lake decrease.

Zone IV (levels 18–20) (after 4,000 years B.P.) remains dry, reflecting high temperatures in the plain. Dry grassland plant species and some isolated trees from the semideciduous forest appear around the lake. The evergreen rainfall forest remains only in the mountains.

CONCLUSIONS

Changes of climatic conditions during the Holocene period have influenciated the environment of the Lagoa Salgada and the Lagoa de Cima. Drier and more humid phases are detected. The variation of pollen and spore concentration and assembly suggest a succession of plant species dependent on changing of the climatic conditions during this period.

Before 6,000 years B.P. a period of a relative humidity existed. After this time drier and humid phases alternated. A drier period began around 4,000 years B.P., while humid conditions were maintained in the mountains. After 3,000 years B.P. a more humid phase was temporarily interrupted by a very dry period.

A possible drying of the Lagoa Salgada can be defined by the absence of palynomorphs (level 8). The humid phases at the bottom and at the top of this core correlate with a higher water level when the lagoon was deeper and palynomorphs were well preserved; pollen grains of Atlantic forest plant species are present, as well as a great amount of fern spores. The drier phase in the upper half of the core may be related to a lower sea level in correlation with warmer climatic conditions.

Starting during the warm phase, the Lagoa de Cima developed into a swampy area and finally into a possible fluvial process. With increasing humidity the lake became surrounded by a typical semideciduous forest that coexisted with perennial grasslands. In our view, a continuous forest vegetation never was installed, so that the water level of the lagoon changed frequently, reflecting the effect of the climate from the lower lands and being next to the ocean.

The results confirm the occurrence of short and frequent dry periods during the middle Holocene, possibly attributed to the "El Niño" phenomenon. A well defined phase occurs at about 4,000 years B.P. After 3,000 years B.P. occurred another variability of climate conditions. The last period is more humid, similar to modern conditions, and may be related to a marine transgression.

ACKNOWLEDGMENTS

We are gratefully indebt to Dr. L. Martin for long time and stimulating discussions. Financial support: Conselho Nacional de Desenvolvimento Científico e Tecnológico (CNPq), Fundação de Amparo à Pesquisa do Estado do Rio de Janeiro (FAPERJ) and ORSTOM, França.

References Cited

LEMOS, R.M.
1995 História evolutiva dos sedimentos Quaternários e das estruturas estromatolíticas da Lagoa Salgada, Rio de Janeiro. M.Sc. dissertation, Universidade Federal Fluminense, Departamento de Geologia, Niterói, 122 pgs.

MARTIN, L., and FLEXOR, J.-M.
1989 Vibro-testemunhador leve: construção, utilização e possibilidades. *Proceedings 2nd Congress ABEQUA, Rio de Janeiro, Special Issue* n, 1, 14 pgs.

MARTIN, L., SUGUIO, K., and FLEXOR, J.-M.
1993 As flutuações do nível do mar durante o Quaternário Superior e a evolução geológica de "deltas" Brasileiros. *Boletim IG-USP*, Publicação Especial 15, São Paulo, SP, 186 pgs.

RADAMBRASIL PROJECT
1983 *Levantamento de recursos naturais. Geologia, Geomorfologia, Pedologia, Vegetação, Uso Polencial da Terra.* Folhas SF 23/24 – Rio de Janeiro – Vitória, *Publicação* MME, Vol. 32, 78 pgs.

YBERT, J.-P., SALGADO-LABOURIAU, M.L., BARTH, O.M., LORSCHEITTER, M.L., BARROS, M.A., CHAVES, S.A.M., RIBEIRO, M., SCHEEL, R., and VICENTINI, K.R.F.
1992 Sugestões para padronização da metodologia empregada em estudos palinológicos do Quaternário. *Revista do Instituto Geológico XIII*, pp.47–49.

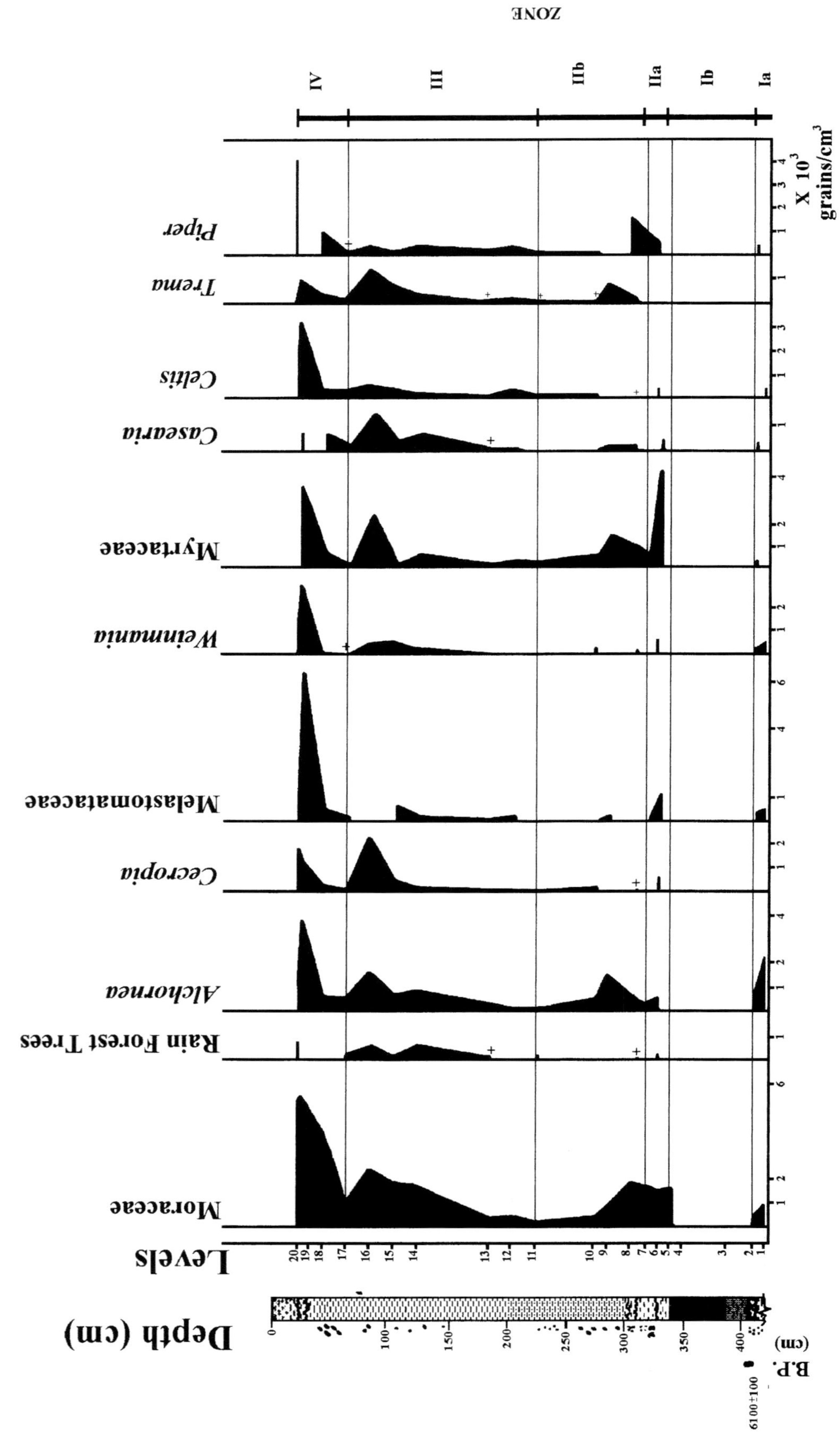

Text-Figure 4. Concentration (palynomorphs/cm³ of sediment), Lagoa de Cima.

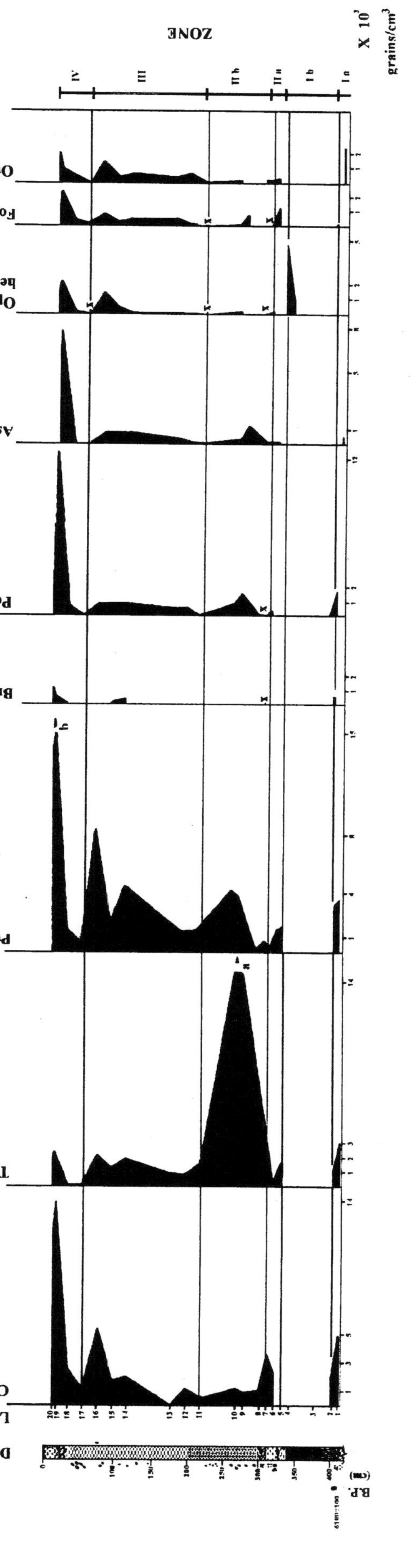

Text-Figure 4 (continued). Concentration (palynomorphs/cm³ of sediment), Lagoa de Cima.

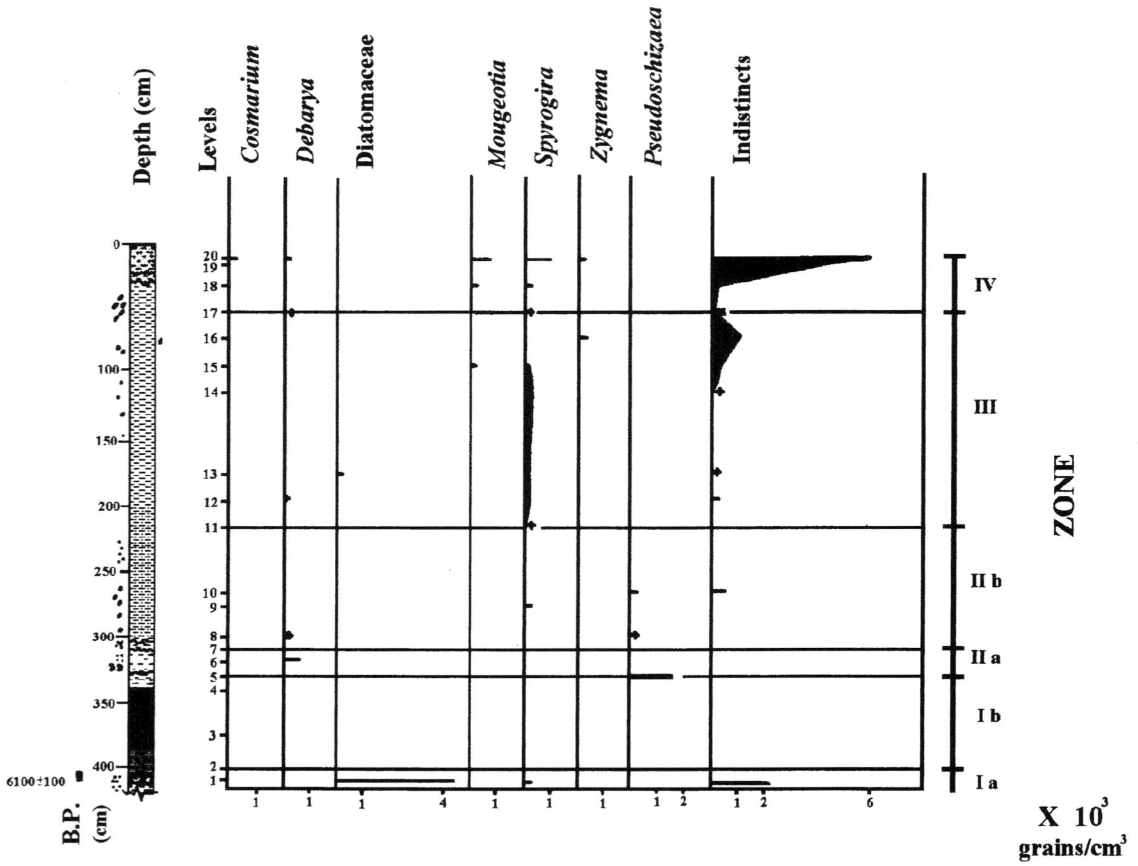

Text-Figure 4 (continued). Concentration (palynomorphs/cm³ of sediment), Lagoa de Cima.

Authors' addresses

ORTRUD MONIKA BARTH
Departamento de Botânica
Instituto de Biologia
Universidade Federal do Rio de Janeiro
CCS, Bloco A
22941–900 Rio de Janeiro
Brazil
and
Departamento de Virologia
Instituto Oswaldo Cruz
Avenida Brasil 4365
21045–900 Rio de Janeiro
Brazil

CYNTHIA F.P. da LUZ
Departamento de Botânica
Instituto de Biologia
Universidade Federal do Rio de Janeiro
CCS, Bloco A
22941–900 Rio de Janeiro
Brazil

MAURO BEVILACQUA DE TOLEDO
Departamento de Botânica
Instituto de Biologia
Universidade Federal do Rio de Janeiro
CCS, Bloco A
22941–900 Rio de Janeiro
Brazil
and
Departamento de Geologia (LAGEMAR)
Universidade Federal Fluminense
Av. Litorânea s/n
24210–340 Niterói, Rio de Janeiro
Brazil

MARCIA AGUIAR de BARROS
Departamento de Botânica
Instituto de Biologia
Universidade Federal do Rio de Janeiro
CCS, Bloco A
22941–900 Rio de Janeiro
Brazil

CLEVERSON GUIZAN SILVA
Departamento de Geologia (LAGEMAR)
Universidade Federal Fluminense
Av. Litorânea s/n
24210–340 Niterói, Rio de Janeiro
Brazil

Part 5: Recent Palynology

(Melissopalynology, Vegetational Palynology, Archaeopalynology, Pollen Spectra, Ecological Palynology, Aeropalynology, Pollen Morphology, and Forensic Palynology)

ALCOHOL DILUTION OF HONEY[1]

Gretchen D. Jones and Vaughn M. Bryant, Jr.

Abstract

Plants used by honeybees can be determined by identifying and quantifying pollen in honey. Techniques for pollen recovery from honey vary. Water is always used to dilute honey prior to pollen extraction, but the amount of water used varies, as does speed and length of centrifugation. A liquid with a specific gravity less than water was examined for diluting honey. Ten samples were examined to compare two centrifugation times of water-diluted honey and alcohol-diluted honey. Water-diluted samples were centrifuged for one (short spin) and 10 (long spin) minutes. Alcohol-diluted samples were centrifuged for three minutes. In addition, alcohol-diluted samples were compared to the Lutier-Vaissiere filtration technique.

Pollen concentration values of alcohol-diluted samples were significantly higher than those of water-diluted samples. There was no significant difference in pollen diversity between alcohol-diluted and long spin water-diluted samples. However, alcohol-diluted samples had significantly greater pollen diversity than short spin water-diluted samples.

Alcohol-diluted samples also had significantly higher pollen concentration values than filtered samples. Higher pollen counts yielded significantly more taxa than lower counts. There was little difference in frequency class between alcohol-diluted and filtered samples.

INTRODUCTION

To date, except for the United States, many major honey-producing regions require certification of their honey. Data collected by these nations enables them to impose laws governing importation and exportation of honey products (Johansson and Johansson, 1968). These laws require certification that includes verification of a honey's floral type, quality, and precise place of origin. Without similar data, United States' domestic honey is limited to exportation to countries not requiring certification

With the advent of free trade between the United States, Canada, and Mexico, expensive domestic honey can be replaced by inexpensive foreign honey. To protect domestic honey producers and increase exportation of domestic honey, the National Honey Board and United States Department of Agriculture are interested in establishing quality standards for honey.

Pollen analyses are an essential tool for honey standards and certification. Data from pollen analyses assess the botanical and geographical origin of honey and foraging ecology of honeybees (Agwu and Akanbi, 1985; Barth, 1990; Crane, 1980; Feller-Demalsy et al., 1987; Moar, 1985). Although there are standard techniques for pollen analysis of honey (Louveaux et al., 1978), extraction techniques vary (Lieux 1980, 1981; Louveaux, 1966; Louveaux and Maurizio, 1963; Maurizio, 1951, 1975; White et al., 1991).

The standard technique (Louveaux et al., 1970, 1978) is diluting a honey sample 1:2 with water, then centrifugation at a low speed (2,500 r.p.m.) for 10 minutes. However, in melissopalynological techniques centrifugation times range from five minutes at 2,500 r.p.m. Gadbin (1979), to 12 minutes at 2,000 r.p.m. (Lieux, 1980) to one minute at 4,000 r.p.m. (White et al., 1991). Low et al. (1989) found that by diluting honey with more water, and centrifuging for 10 minutes at 3,500–4,000 r.p.m., pollen recovery was increased. In addition, a new filter technique was devised for pollen recovery from honey (Lutier and Vaissiere, 1993).

The problem of pollen recovery from honey is that honey contains fresh pollen that can have a specific gravity < 1, despite the time it remains in honey. This low specific gravity hampers pollen recovery from aqueous solutions when using centrifugation (Low et al., 1989; Jemmett and Owen, 1990). If a liquid with a specific gravity < 1 is used to dilute the honey, the resulting solution will be < 1, thus increasing pollen recovery. There is no published information on the use of liquids with a specific gravity lower than water for honey dilution.

The goals of this research were threefold. First, which previously described technique had the best pollen recovery? Second, could a liquid medium with a specific gravity less than water be used for diluting honey? Third, how would this low specific gravity liquid dilution compare to the new filtration technique?

METHODS AND PROCEDURES

Short Spin, Long Spin, and Alcohol Techniques

Samples were donated by beekeepers in East Texas and were said to be from mixed wildflower sources. Ten honey samples were selected at random to compare water-diluted honey at two different centrifugation times [short spin (**SH**) and long spin (**LG**)] to alcohol-diluted honey (**AL**).

Each sample was heated in a microwave oven to 38° C and thoroughly stirred to ensure uniform mixing of pollen. Ten grams of honey [the internationally preferred standard (Louveaux et al., 1978)] were extracted from each sample and poured into separate, marked 150 ml beakers.

For **SH** and **LG**, 100 ml of distilled water were added (White et al. 1991). This honey to water ratio was used because it represented an average ratio used by other melissopalynologists.

Jones, Gretchen D., and Bryant, Vaughn M., Jr.
2001 Alcohol dilution of honey; *In*: Goodman, D.K., and Clarke, R.T. (eds.), *Proceedings of the IX International Palynological Congress, Houston, Texas, U.S.A., 1996*; American Association of Stratigraphic Palynologists Foundation, p. 453–458.
 ISBN 0–931871–06–9

[1]Mention of a commercial or proprietary product in this paper does not constitute an endorsement by the USDA, nor does it imply the recommendation of the product by the USDA to the exclusion of similar products.

Five *Lycopodium* tablets (each containing 11,300 ± 300 spores, batch #201890) dissolved in five ml of 5% hydrochloric acid (**HCL**) were added to each sample. A 1:1 spore to pollen ratio was desired for a more accurate comparison of concentration values.

Honey did not dissolve well in 95% ethyl alcohol (**ETOH**). Therefore, for **AL** samples, 10 g of honey were dissolved first in 10 ml of water. Next, five dissolved *Lycopodium* tablets and 100 ml of 95% **ETOH** were added to the **AL** samples. Adding 100 ml of 95% **ETOH** reduced the samples' specific gravity to 0.83–0.88. This variation is probably due to variation in specific gravity of the individual honeys.

Samples were kept on a warm hot plate and stirred occasionally to ensure mixing of pollen and spores. To reduce the liquid and concentrate the pollen, each honey solution was poured into labeled 12 ml centrifuge tubes. Samples were centrifuged at 4,000 r.p.m. **SH** samples were centrifuged for one minute (White et al., 1991), **LG** samples for 10 (Low et al., 1989), and **AL** samples for three. Three minutes was chosen for the **AL** technique because it represented a more or less middle time between the **SH** and **LG** techniques.

After decanting the supernatant, the residue was thoroughly mixed. Additional honey solution was added to the residue. This process of centrifuging, decanting, mixing, and adding more solution was repeated until the volume was reduced to only a residue. **SH** and **LG** beakers were rinsed twice with distilled water and once with 95% **ETOH**; **AL** beakers were rinsed three times with 95% **ETOH**.

Samples were acetolyzed (Erdtman, 1960; Lieux, 1980) for 12 minutes, rinsed twice with distilled water, and once with 95% **ETOH**; centrifuging, decanting, and mixing each time. Samples were stained with Safranin O and poured into marked vials. Five drops of glycerin were added to each vial. Vials were left on a hot plate until the **ETOH** evaporated. Samples were mixed for 30 seconds prior to making slides.

Pollen from each sample was identified and counted. Counts of 200 pollen grains were made for all samples unless 1000 spores were counted first. Pollen concentration values per 10 g of sample were calculated by using the following formula:

$$\frac{\text{(number pollen grains counted)(number of spores added)}}{\text{(number spores counted)}}$$

Number of taxa per technique was totaled and floral frequency classes were calculated (Louveaux et al., 1970). "Predominant" pollen occur in excess of 45%, "secondary" between 16–45%, "important minor" between 3–15%, and "minor" below 3% (Louveaux et al. 1970, 1978).

Filter and Alcohol Techniques

Honey samples used for water- and alcohol-dilution comparisons were unfilterable because they would not go through the filter. Apparently microwaving the honey, lengthened the honey's carbon chains and prevented filtration. Therefore, five new samples were used to compare alcohol-diluted honey (**AL**) to the new filtration technique (**FIL**) (Lutier and Vaissiere, 1993).

Two *Lycopodium* tablets were used in comparing **AL** and **FIL**, because five tablets effervesced over the centrifuge tube top. Alcohol-diluted samples were prepared as previously described. A modified filtration technique was used for **FIL** samples (Lutier and Vaissiere, 1993). The filtration technique was followed (Lutier and Vaissiere, 1993) until the filter was removed from the filtration unit. At this time instead of putting the filter into a microcentrifuge tube, it was gently folded and placed into a 12 ml centrifuge tube. *Lycopodium* tablets were placed on top of the filter to prevent spore loss and ensure dissolving of the tablets by the acetolyzation mixture. The filter and tablets were acetolyzed for 12 minutes. Filtered samples were finished as described for water-diluted samples.

For **AL** and **FIL** comparisons, 500 grain counts were made by first counting 200 grains, then counting an additional 300 grains in sets of 100. Pollen concentration values, percentages, and floral frequency classes were calculated, and number of taxa (diversity) per count were totaled.

RESULTS

Short Spin, Long Spin, and Alcohol Techniques

In some **SH** and **LG** samples, 200 grains were not counted because 1000 spores were counted first (Table 1). Alcohol-diluted samples yielded the highest pollen concentration values, and **SH** the lowest (Table 1). Half of the **AL** samples had over 100,000 pollen grains per 10 g sample.

Alcohol-diluted samples had significantly higher pollen concentration values than **SH** or **LG** (**ANOVA**, $P = 0.0001$, Tukey's studentized range test < 0.05). Short spin and **LG** samples were not significantly different from each other (**ANOVA**, $P > 0.05$). On average, **AL** samples had double the concentration values of **LG** samples (226%) and quadruple **SH** (426%).

Pollen diversity varied from 18 (**LG** sample 22) to 40 taxa (**LG** sample 1) (Table 2). No technique consistently had more or fewer taxa than the others (Table 2). Mean taxa diversity of **AL** samples was slightly more than that of **SH** or **LG** samples. **AL** samples had significantly more taxa than **SH** samples (t-test for paired comparisons, $P < 0.05$). On average, **AL** samples had 10% more taxa than **LG** samples, and 16% more than **SH** (calculated from Table 2).

Frequency classes were not consistent across the three techniques. Some species had the same frequency in all techniques, but others were found in only one technique of one sample. Thus, frequency classes could not be used to test the different techniques.

Filter and Alcohol Techniques

There was no evidence that pollen in **AL** honey was broken more than that in **FIL** honey. Therefore, we concluded that short period microwaving of honey does not destroy the pollen found inside.

Pollen concentration values ranged from 53,176 (**FIL** sample 5, the 200 grain count) to 382,317 (**AL** sample 11, the 200 grain count) (Table 3). There were no significant differences in concentration values among the 200, 300, 400, and 500 grain counts (**ANOVA**, $P = 0.41$). However, **AL** samples had significantly higher concentrations values than **FIL** samples (t-test for paired comparisons, $P < 0.05$).

TABLE 1. Number of pollen grains and *Lycopodium* spores counted, and pollen concentration values per 10 g of honey from samples used in the water and alcohol comparison techniques.

Sample Number	Treatment	Pollen Counted	Lycopodium Counted	Pollen Concentration
1	short	201	826	13,748.79
	long	201	529	21,467.86
	alcohol	202	220	51,877.27
6	short	203	410	27,974.39
	long	200	333	33,933.93
	alcohol	200	291	38,831.62
19	short	203	223	51,432.74
	long	235	147	90,323.13
	alcohol	211	82	145,384.15
22	short	200	887	12,739.57
	long	207	338	34,602.07
	alcohol	205	151	76,705.30
23	short	207	479	24,416.49
	long	205	265	43,707.55
	alcohol	208	92	127,739.13
27	short	115	1089	5,966.48
	long	208	1142	10,290.72
	alcohol	200	169	66,863.91
29	short	204	642	17,953.27
	long	207	371	31,524.26
	alcohol	202	94	121,414.89
30	short	217	419	29,261.34
	long	203	339	33,833.33
	alcohol	203	74	154,993.24
31	short	119	1022	6,578.77
	long	126	1024	6,952.15
	alcohol	204	354	32,559.32
36	short	100	1000	5,650.00
	long	154	1000	8,701.00
	alcohol	200	53	213,207.55
Mean	**short**	**177**	**700**	**19,572.18**
	long	**195**	**549**	**31,533.60**
	alcohol	**204**	**158**	**102,957.64**

Pollen diversity ranged from 17 (**AL** sample 5, the 200 grain count) to 65 (**AL** sample 16, the 500 grain count) (Table 4). **AL** samples had greater diversity in 12 of the 20 counts. Pollen diversity significantly increased as counts increased (**ANOVA**, P = 0.0029). Pollen diversity in **FIL** 500 grain count was not significantly different from **AL** 400 or 500 grain count (**ANOVA**, P = >0.05). However, the 400 and 500 **AL** counts and 500 **FIL** count had a significantly greater diversity than other counts (**ANOVA**, P = 0.0029, Table 4).

Frequency classes between the two techniques were relatively consistent. However, in sample 5, *Berchemia scandens* was a predominant type (over 45%) in **AL**, but a secondary type (16–45%) in **FIL**.

DISCUSSION

Short Spin, Long Spin, and Alcohol Techniques

Historically, water is used to dilute honey. Although 10 g of honey diluted with 20 ml of water is recommended (Louveaux et al., 1970; 1978), ratios vary from 1 to 20 g of honey diluted with 20 to 100 ml of water. The reason for increased dilution is better recovery of pollen. Since the specific gravity of honey is about 1.44, adding water reduces the specific gravity to between 1.44 and 1.0. Any pollen grain with a specific gravity greater than the solution should sink, especially when centrifuged. Most acetolyzed pollen grains have a specific gravity around 1.4, and can be

TABLE 2. Number of taxa found in honey samples processed by the short spin and long spin water-diluted technique and the alcohol-dilution technique.

Sample Number	Treatments Short	Long	Alcohol
1	36	40	35
6	19	28	32
19	36	26	37
22	34	18	32
23	25	35	31
27	22	26	25
31	19	27	33
36	27	34	28
29	27	22	27
30	29	34	39
Mean	27.4	29.0	31.9

recovered easily from aqueous suspensions by centrifugation (Flenley, 1971; Juvigne, 1973; Traverse, 1988; Lutier and Vaissiere, 1993).

Low et al. (1989) found a significant increase in pollen recovery by increasing both the amount of water and centrifugation time. Our results agree with Low et al. (1989) that lengthening centrifugation time increases pollen recovery. Long spin samples, on average, had higher pollen concentration values than short spin. Although, there was no significant difference in pollen diversity between short and long spin samples, long spin samples more often had greater pollen diversity.

Increasing water and centrifugation times attempt to resolve the problem of the specific gravity of the solution and of the pollen in the solution prior to acetolyzation. Pollen in honey reacts like fresh pollen. Specific gravity of fresh pollen can be < 1.0 and may vary considerably, depending on the taxon and the type of pollen grain (Brush and Brush, 1972; Juvigne, 1973; Lutier and Vaissiere, 1993). For example, pine pollen can float on the surface of water for over a year (Traverse, 1988).

TABLE 3. Pollen concentration values of the alcohol-diluted and filtered honey samples per 10 g of honey at 200, 300, 400, and 500 pollen grain counts. Pollen counted = pollen, *Lycopodium* spores counted = Lyco, pollen concentration values = PC.

	Alcohol-diluted			Filtered		
Sample#	Pollen	Lyco	PC	Pollen	Lyco	PC
5	200	64	70,625.00	200	85	53,176.47
	300	95	71,368.42	300	120	56,500.00
	400	125	72,320.00	400	159	56,855.35
	500	150	75,333.33	500	193	58,549.22
11	200	43	105,116.28	200	39	115,897.44
	301	62	109,719.35	300	66	102,727.27
	402	81	112,162.96	400	93	96,961.29
	502	106	113,452.00	500	114	99,122.81
16	203	12	382,316.67	200	33	136,969.70
	308	25	278,432.00	300	49	138,367.35
	410	31	298,903.23	400	65	139,076.92
	510	33	349,272.73	500	83	136,144.58
18	201	40	113,565.00	200	35	129,142.86
	301	59	115,598.31	300	55	123,272.73
	401	73	124,145.21	400	79	114,430.38
	501	99	114,369.70	501	107	105,818.69
35	200	49	92,244.90	200	58	77,931.03
	301	68	100,038.24	301	84	80,983.33
	401	94	96,410.64	401	114	79,496.49
	501	118	95,954.24	501	141	80,302.13

TABLE 4. Pollen diversity in alcohol-diluted (**AL**) and filtered (**FIL**) samples per 200, 300, 400, and 500 pollen grain counts.

Sample #	Count	Number of Taxa AL	FIL
5	200	17	19
	300	21	22
	400	27	26
	500	32	30
11	200	31	31
	300	36	34
	400	40	38
	500	45	41
16	200	35	18
	300	45	32
	400	57	42
	500	65	48
18	200	32	20
	300	35	30
	400	44	43
	500	47	50
35	200	25	26
	300	31	32
	400	35	35
	500	37	40

Furthermore, pollen grains with the same specific gravity can have different sinking speeds due to variations in diameter and ornamentation (Low et al., 1989). Pollen's low specific gravity and different sinking rates hamper its recovery from aqueous suspensions (Low et al., 1989; Jemmett and Owen, 1990; Lutier and Vaissiere, 1993).

Pollen grains with a specific gravity of 1.0 or less float in a honey and water solution. Even after centrifugation, many of these types can be decanted with the supernatant and accidentally lost. When a liquid with a specific gravity lower than water is used to dilute honey, the resulting solution will have a specific gravity less than 1.0, thus allowing sedimentation of all pollen, especially when centrifuged. Random checks were made on the supernatant and no pollen was found in the supernatant indicating that little or no pollen was accidentally decanted.

At 20°C, a solution of 10 g honey and 20 ml water has a specific gravity of about 1.3 (Low et al., 1989). The specific gravity of the samples was not significantly reduced by heating. Absolute ethyl alcohol (**ETOH**) (100%) has a specific gravity of 0.78945 at 20°C (Wheast et al., 1964). When absolute **ETOH** is heated from 0 to 30°C, its specific gravity is reduced from 0.80625 to 0.78097 (op. cit.). Ethyl alcohol is economical, easily available, relatively inexpensive, non-toxic, and can be purchased in various grades at various prices. The specific gravity of 95% **ETOH** is only 0.02 greater than absolute. Since 95% **ETOH** was cheaper than 100%, and its specific gravity was low, 95% **ETOH** was ideal to test as a low specific gravity liquid for honey dilution.

The best technique for recovering pollen is alcohol-dilution. Alcohol-diluted samples significantly recovered more pollen and yielded higher pollen concentration values than water-diluted samples. Alcohol-diluted samples had significant greater diversity of taxa than short spin samples.

During preliminary observations, alcohol-dilution seemed to increase sedimentation of smaller grains, which can be discarded with the supernatant. However, larger grains increased in some samples. The same was true for ornamentation. In some samples, highly ornamented grains increased in number, and decreased in others. Thus, there was no uniform correlation between a pollen grain's size or ornamentation in regard to its recovery.

Filter and Alcohol Techniques

Filtration attempts to solve the specific gravity of pollen problem in a different way. By filtering the honey, pollen becomes trapped in the filter (Lutier and Vaissiere, 1993). The filtration technique is fast (about one hour), and reduces amount and cost of chemicals and lab equipment. Using plastic microcentrifuge tubes with lids not only reduces broken glassware, but also reduces sample loss due to breakage.

However, when trying to duplicate the filter technique, several problems occurred. First, equipment (filters, filtering units, and pumps) had to be purchased. Recommended filters (Lutier and Vaissiere, 1993) were 100% cellulose acetate, and sold by only one company (Sartorius, Inc.) in the United States. Therefore, this technique totally depends on availability of these filters.

Second, filtration has a narrow temperature and time range at which honey can be filtered. Samples below 78°C or above 83°C did not filter. Also, samples did not filter if heated longer than 15 minutes. Using Pyrex glassware lengthened heating time. This additional heating time increases the likelihood of polymerization. Lutier and Vaissiere (1993) used pressed glass "yogurt jars" that heat up quickly, but are not easily obtainable in the United States.

Third, dissolved *Lycopodium* tablets clogged the filter. Lutier and Vaissiere (1993) did not use marker spores to determine concentration values, thus they did not encounter this problem.

Alcohol-dilution technique equaled and recovered more pollen than filtration. Alcohol-dilution had significantly higher pollen concentrations. The 400 and 500 alcohol-dilution and 500 filtration grain counts produced significantly more taxa than other counts. When pollen diversity is important, either a 400 or 500 grain count (alcohol-diluted) or a 500 grain count (filtered) should be made. Obviously the greater the count, the better the definition of honeybee foraging resources and floral sources of the honey.

Frequency class types between the two techniques were relatively consistent. In sample 5, when filtered, *Berchemia scandens* was a secondary type (16–45%), but was a predominant type (over 45%) when diluted with alcohol. However, the difference between under or over 45% can be only a single grain or two.

Recommendations

Although filtering is fast and uses fewer chemicals, there are several disadvantages to filtering. First, the 1.5 microcentrifuge tubes are small and can be awkward to use causing chemical spillage and pollen loss. Second, if scanning electron microscopy is used in analyzing the pollen residue, filter remnants will obscure some pollen grains. Third, the equipment itself presents a host of problems: microcentrifuge tubes are costly; centrifuges must be modified, adjusted, or purchased to accommodate microcentrifuge tubes; and, labeling is difficult because most labels are too large or do not stick. Finally, the whole filter technique depends on availability and purchase of 100% cellulose acetate filters.

The alcohol-dilution technique uses equipment found in most pollen laboratories and techniques familiar to palynologists. Any previous treatment of the honey is often unknown prior to its arrival in the lab. This creates a question as to filterability of new samples.

Because pollen analyses of honey are concerned with pollen concentration values, pollen diversity and percentages, and floral resources used by honeybees, the alcohol-dilution technique should become the standard pollen extraction technique for honey. Whether a shortened centrifugation time of alcohol-diluted samples would yield the same results is a subject for further research.

References Cited

AGWU, C.O.C., and AKANBI, T.O.
1985 A palynological study of honey from four vegetation zones in Nigeria. *Pollen et Spores*, 27: 335–348.

BARTH, O.M.
1990 Pollen in monofloral honeys from Brazil. *Journal of Apicultural Research*, 29: 89–94.

BRUSH, G.S., and BRUSH, L.M., Jr.
1972 Transport of pollen in a sediment-laden channel: a laboratory study. *American Journal of Science*, 272: 359–381.

CRANE, E.
1980 *A Book of Honey*. Oxford University Press, Oxford. 193 pgs.

ERDTMAN, G.
1960 The acetolysis method. A revised description. *Svensk Botanisk Tidskrift*, 54: 561–564.

FELLER-DEMALSY, M.-J., PARENT, J., and STRACHAN, A.
1987 Microscopic analysis of honeys from Alberta, Canada. *Journal of Apicultural Research*, 26: 123–132.

FLENLEY, J.R.
1971 Measurements of the specific gravity of the pollen exines. *Pollen et Spores*, 13: 170–186.

GADBIN, C.
1979 L'interel de l'acetolyse en melissopalynologie. *Apidologie*, 10: 23–28.

JEMMETT, G., and OWEN, J.A.K.
1990 Where has all the pollen gone? *Review of Palaeobotany and Palynology*, 64: 205–211.

JOHANSSON, T.S.K., and JOHANSSON, M.P.
1968 Pollen analysis of honey — an appraisal. *American Bee Journal*, 109: 57–58.

JUVIGNE, E.
1973 Desite des exines de quelques especes de pollen et spores fossiles. *Annales. Societe Geolgique de' Belgique*, 96: 363–372.

LIEUX, M.H.
1980 Acetolysis applied to microscopical honey analysis. *Grana*, 19: 95–61.
1981 An analysis of Mississippi USA honey: pollen, color and moisture. *Apidologie*, 12: 137–158.

LOUVEAUX, J.
1966 Pollenanalyse einiger kanadischer Honige. *Zeitschrift für Bienenforschung*, 8: 195–202.

LOUVEAUX, J., and MAURIZIO, A.
1963 Methodes d'analyse pollinique des miels. *Annales de l'Abielle*, 6: 75–76.

LOUVEAUX, J., MAURIZIO, A., and VORWOHL, G.
1970 Methods of melissopalynology. *Bee World*, 51: 125–131.
1978 Methods of melissopalynology. *Bee World*, 59: 139–157.

LOW, N.H., SCHWEGER, C., and SPORNS, P.
1989 Precautions in the use of melissopalynology. *Journal of Apicultural Research*, 28: 50–54.

LUTIER, P.M., and VAISSIERE, B.E.
1993 An improved method for pollen analysis of honey. *Review of Palaeobotany and Palynology*, 78: 129–144.

MAURIZIO, A.
1951 Pollen analysis of honey. *Bee World*, 32: 1–5.
1975 Microscopy of honey. *In*: Crane, E. (ed.), *Honey, a Comprehensive Survey*. Crane, Russak & Co., Inc., New York, p. 240–257.

MOAR, N.T.
1985 Pollen analysis of New Zealand honey. *New Zealand Journal of Agricultural Research*, 28: 39–70.

TRAVERSE, A.
1988 *Paleopalynology*. Unwin Hyman, London. 600 pgs.

WHEAST, R.C., SELBY, S.M., and HODGMAN, C.D. (eds.).
1964 *Handbook of Chemistry and Physics*. 45th ed. The Chemical Rubber Co., Cleveland, Ohio.

WHITE, J.W., Jr., BRYANT, V.M., Jr., and JONES, J.G.
1991 Adulteration testing of southwestern desert honeys. *American Bee Journal*, 131: 123–126.

Authors' addresses

GRETCHEN D. JONES
USDA, ARS, Southern Plains Agricultural Research Center
Areawide Pest Management Research Unit
2771 F&B Rd.
College Station, TX 77845
U.S.A.
E-mail: g-jones@tamu.edu

VAUGHN M. BRYANT, JR.
Palynology Laboratory
Department of Anthropology
Texas A&M University
College Station, TX, 77845-4352
U.S.A.
E-mail: vbryant@neo.tamu.edu

PLANTS OF APICULTURAL INTEREST IN THE PLUMA HIDALGO ZONE, OAXACA, MEXICO

Salvador Acosta-Castellanos and Rodolfo Palacios-Chavez

Abstract

The Pluma Hidalgo zone is located on the Sierra Madre del Sur in Oaxaca State. The main vegetation types in this area are cloud forest and semi-evergreen montane tropical forest. Coffee plantations are common in the area. The climate is semi-warm, humid, with summer rainy season.

Palynological analysis was conducted on two honey samples produced by *Apis mellifera ligustica*, and pollen loads of these bees were collected periodically during an annual cycle in two separate hives.

Pollen diagrams of the honey show 31 pollen types. Although *Coffea arabica* L. occurred in a sample, its abundance was not significant. A large proportion of pollen belongs to native tree species, including Actinidiaceae, Anacardiaceae, Asteraceae, Burseraceae, Fabaceae and Sapindaceae.

Pollen diagrams of bee loads show 75 pollen types, with a predominance of woody species throughout the year and an increasing importance of herbaceous species during the dry season. Important polliniferous species include taxa from the Anacardiaceae, Asteraceae, Caprifoliaceae, Fabaceae, Fagaceae, Myrtaceae, Rubiaceae and Ulmaceae.

INTRODUCTION

Mexico is an important honey producer and one of the principal honey exporting countries. From the standpoint of economic and scientific interest, it is necessary to know the major plants used as food sources by honeybees and the flowering phenology of vegetation to improve apicultural management.

Since the early nineteenth century, many authors have registered the value of bee foraging plants, including citrus and other fruit trees as important food sources for honeybees. Other authors have mentioned the role that honeybees play on the yield of different crops and their importance as pollinator agents (Lindberg, 1989).

Several researchers have investigated on plants of apicultural and melissopalynological value in different places of Mexico (Cabrera-Pech, 1966; Ordetx et al., 1972; Carmona, 1980; Chazaro, 1982; Roldan-Ramos, 1984; Villanueva, 1984; Alvarado and Delgado-Rueda, 1985, 1988). In Oaxaca State, only Piña-Lujan (1975) reported apicultural-ecological studies, in which he mentioned the environmental conditions and the principal plant resources of nectar and pollen and their phenology.

The Pluma Hidalgo region has a high biodiversity. This region is an important location for coffee plantations, but the people need to diversify production activities to improve their life conditions and to preserve natural resources. The present work aims to provide knowledge of the melliferous flora of this zone, as a base for better apicultural management.

Acosta-Castellanos, Salvador and Palacio-Chavez, Rodolfo
2001 Plants of apicultural interest in the Pluma Hidalgo Zone, Oaxaca, Mexico; *In*: Goodman, D.K., and Clarke, R.T. (eds.), *Proceedings of the IX International Palynological Congress, Houston, Texas, U.S.A., 1996*; American Association of Stratigraphic Palynologists Foundation, p. 459–469.
 ISBN 0–931871–06–9

Environmental Setting

Pluma Hidalgo is located on the southwestern slope of Sierra Madre del Sur in the Pochutla District of Oaxaca State (Text-Figure 1). The topography is very steep. The geological substrate is composed of Mesozoic acid intrusive igneous rocks. The soil types are classified as haplic feozem, eutric regosol, lithosol, and eutric cambisol (INEGI, 1980). The climate is semi-warm and humid with summer rainy season. According to the Köppen classification modified by Garcia (1981), the climatic formula is: (A) Cm (w) b i g (Text-Figure 2).

The main vegetation type in the zone is cloud forest, distributed within an altitudinal range of 900–1700 m. Below an altitude of 900 m the vegetation type becomes semi-evergreen tropical montane forest.

The cloud forest is a dense community, with subperennial leaves, composed of three tree strata. The highest tree stratum, 20–35 m tall, is made up of *Alchornea, Clethra, Persea, Phoebe, Nectandra, Quercus* and *Ulmus*. The middle tree strata, 12–20 m tall, has trees of *Coccoloba, Cupania, Ficus, Guarea, Hasseltia, Inga, Styrax, Zanthoxylum*. The lowest tree stratum, 4–10 m tall, contains plants of the genera *Cyathea, Leucaena, Miconia, Myrcia, Oreopanax, Rapanea, Saurauia, Sommera, Swartzia, Viburnum*. There is also a shrubby stratum, 1–4 m tall, composed of the following genera: *Acacia, Ardisia, Brugmansia, Calliandra, Conostegia, Hamelia, Malvaviscus, Sambucus, Senna, Tibouchina, Vernonia*. The herbaceous stratum is composed of plants of the Acanthaceae, Asteraceae, Fabaceae, Melastomataceae, Poaceae, and a great variety of ferns. Among the vines and scandent plants are *Serjania, Dioscorea, Smilax, Vitis, Dalechampia, Senecio, Ipomoea, Monstera, Anthurium* and *Schrankia*. Finally, there are numerous epiphytic ferns and plants of Orchidaceae and Bromeliaceae families.

MATERIALS AND METHODS

Two representative honeybee hives of the zone were selected: one at 1 km W of the town of Pluma Hidalgo and

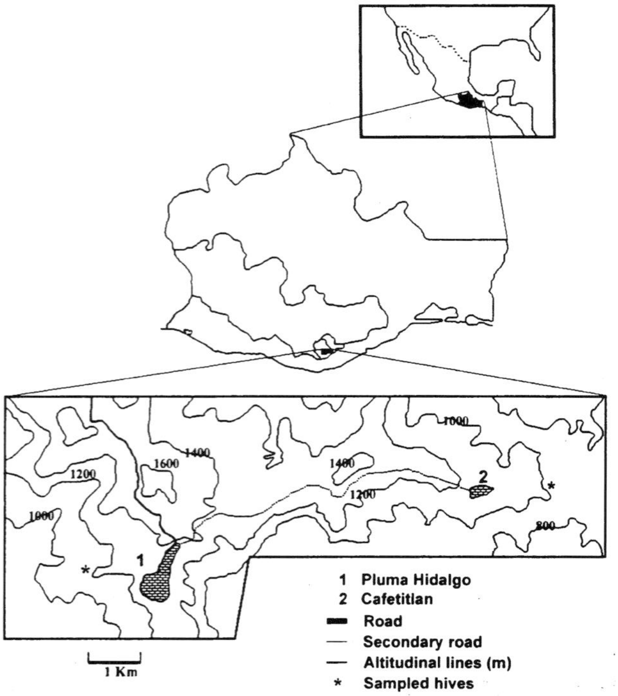

Text-Figure 1. Location of the area of study.

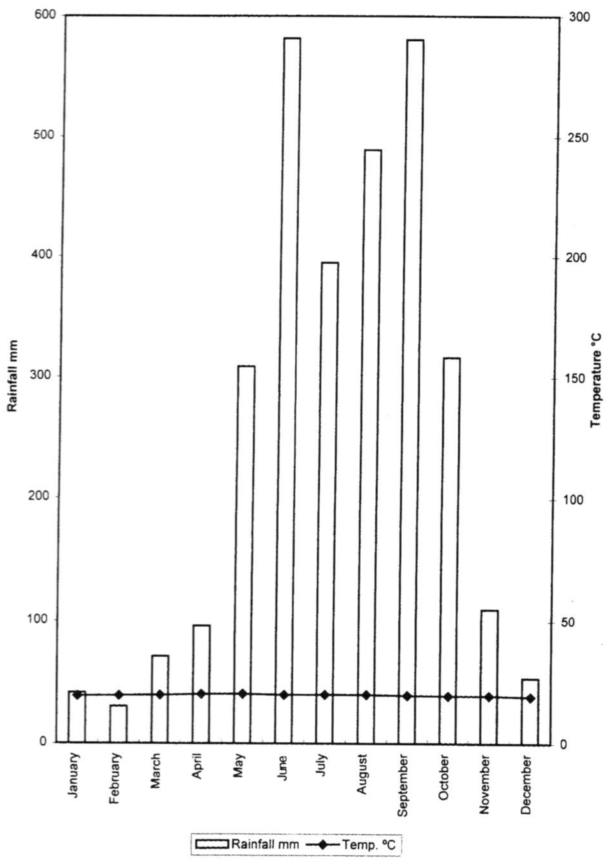

Text-Figure 2. Diagram of the mean monthly temperatures and rainfall registered in the climatological station of Pluma Hidalgo, Oaxaca.

the other at 1.5 km east of Cafetitlan Village (Text-Figure 1). We collected plant specimens every month during an annual cycle (March, 1989, to February, 1990). Plants were taken from the area around both honeybee hives and the flowering plants were vouchered. An elaborate palynological reference collection was used to help identify the pollen present in honey and pollen loads. We determined the plants frequently visited by honeybees (*Apis mellifera ligustica*) through both direct observation and information from the beekeepers.

Pollen loads of both hives were sampled monthly, using entrance pollen traps. These loads were dried, homogenized, and then acetolyzed (Erdtman, 1943). A honey sample of each hive was taken and processed according to Lieux (1972) to identify and quantify the pollen present. At least 500 pollen grains were counted in each sample (pollen loads and honey). The ENCB (Escuela Nacional de Ciencias Biologicas) palynological reference collection also was used to identify some species. Respective pollen diagrams were drawn.

RESULTS

The main flowering season seems to be associated with low rainfall, although flowering increased in September, a month with high rainfall (Text-Figure 3). Sixty-six taxa of plants are potentially important sources of pollen and nectar, but only twenty-five were corroborated in the pollen analysis.

In the two honey samples we found 31 pollen types, 24 were identified at least to the generic level. The pollen diagram of the Pluma Hidalgo honey shows taxa of natural vegetation. In the honey from Cafetitlan the presence of two cultivated plant species was detected (*Coffea arabica* and *Spondias purpurea* L.); only *Coffea arabica* occurs in a significant amount. Both honey samples are classified as monoflorous (Text-Figures 4 and 5). Almost all the taxa found in the honey appeared in the pollen loads gathered by *Apis mellifera ligustica*. This indicates that taxa important as nectar sources are pollen suppliers as well. We found that the following species in order of importance are nectar and pollen sources for honeybees: *Bursera simaruba* (L.) Sarg., *Saurauia scabrida* Hemsl., *Coffea arabica*, *Cupania glabra* Swartz, *Viburnum hartwegii* Benth., *Rhus striata* Ruiz & Pavon, *Trema micrantha* (L.) Blume, and Asteraceae. It should be noted the presence of *Quercus* pollen as well as another polliniferous species such as *Mimosa albida* H. & B. and *Acacia* spp. in both honey samples.

Pollen loads from both sites contained 75 pollen types (Tables 1 and 2). Pluma Hidalgo pollen samples had 53 pollen types, with a monthly average of nine types. Thirteen types occurred in February (low rainfall season) and only two types in July (rainy season) (Text-Figure 6). During the annual cycle the number of woody species was always greater than herbaceous species. Herbaceous taxa increased toward the low rainfall season (from November to March). During this period, the pollen proportion of herbaceous species was more abundant, as opposed to the rest of year, in which the pollen of woody species prevailed (Text-Figure 7).

Cafetitlan pollen loads contained 46 pollen types, with a monthly average of 8 types. Eleven types appeared in October and five types in May and in July (Text-Figure 8). From February to October, pollen from woody species was

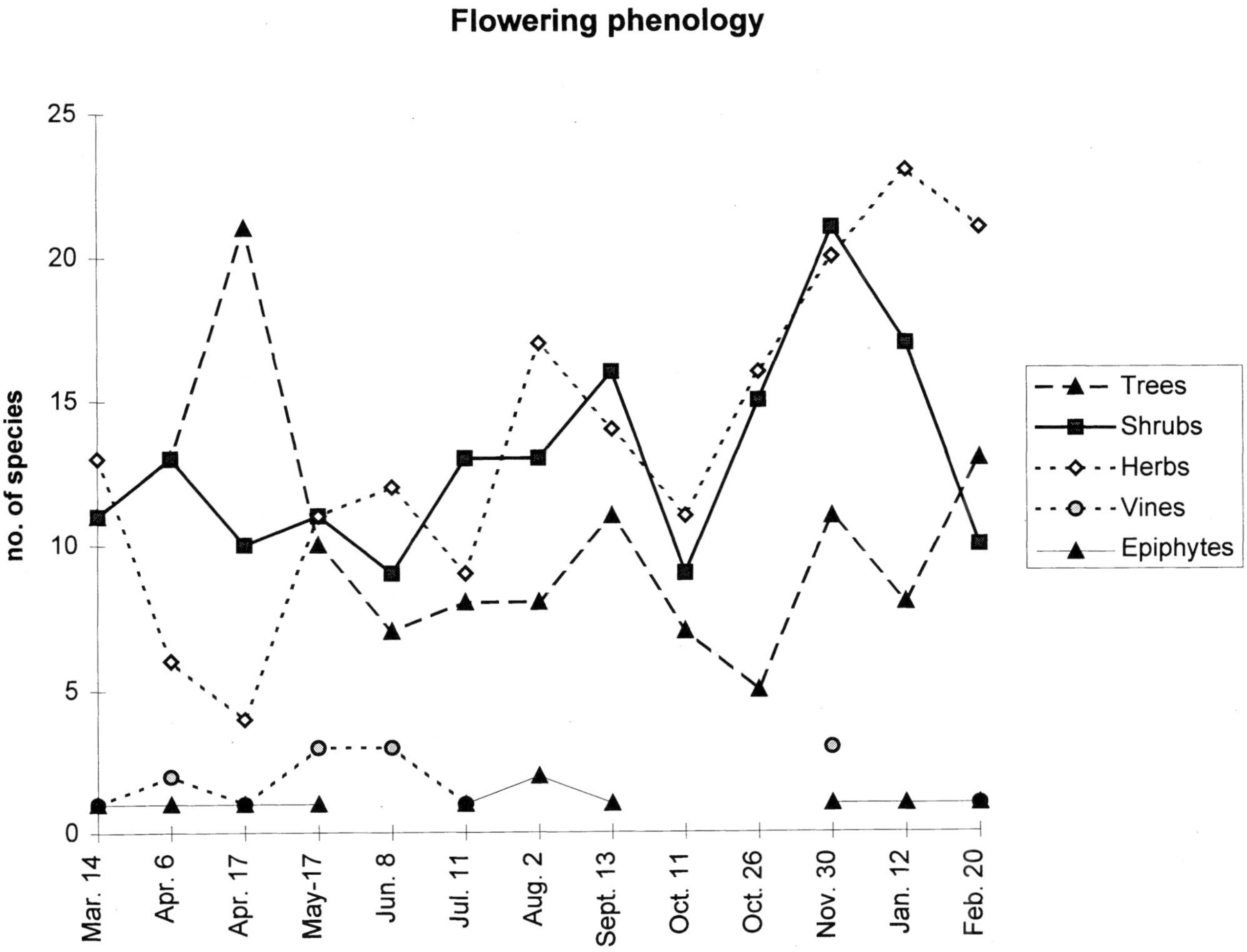

Text-Figure 3. Diagram of the flowering plant species registered during the annual cycle (March 1989 to February 1990) in the Pluma Hidalgo zone, Oaxaca.

predominant. The herbaceous species types begin to increase in September, with a predominance in November and January. In this period it could be noted that the pollen proportion of herbaceous species increases, being most abundant in November. Pollen of woody species predominated from February to June (Text-Figure 9).

Among the outstanding taxa of plants that are pollen sources for honeybees, are in order of importance: Asteraceae (including *Vernonia* sp. as a noteworthy polliniferous species), *Quercus* spp., *Myrcia oerstediana* Berg., *Rhus striata, Coffea arabica, Viburnum hartwegii, Bursera simaruba, Acacia cornigera* (L.) Willd., *Gliricidia sepium* (Jacq.) Steud., *Mimosa albida, Vitis tiliifolia* Humb. & Bonpl. ex Roem. & Schult., *Ulmus mexicana* (Liebm.) Planch., *Piper* spp., two cultivated plants *Citrus sinensis* Osbeck and *Zea mays* L. and an ornamental *Brugmansia* x *candida* Pers.

Plant species that appear in pollen loads during three or more months are important because they provide food for honeybees through a long period. They include: Asteraceae, Caprifoliaceae, Fabaceae, Fagaceae, and Myrtaceae. *Acacia cornigera, Alchornea latifolia* Swartz, *Bursera simaruba, Gliricidia sepium, Iresine* sp., *Quercus* spp., *Ulmus mexicana, Vernonia* sp. are important polliniferous species only during the dry season. Important food sources in the rainy season are: *Acacia* sp., *Citrus sinensis, Colubrina* sp., *Cupania glabra, Myrcia oerstediana, Viburnum hartwegii, Vitis tillifolia, Zea mays.*

DISCUSSION

Villanueva (1984) mentioned the great importance of woody species as food sources for *Apis mellifera* throughout the year. In accordance with his results, we also found that most taxa represented in the honey and pollen loads samples are native plants. Coffee is the most extensive crop in the zone, and it is planted along with a variety of fruit trees. Both coffee and its associated fruit trees are considered as important polliniferous or nectariferous plants. However, few cultivated plants in addition to *Coffea arabica* had a significant role as food sources for honeybees. This could be due to the low density and scattered distribution of these plants, which do not reward the pollen and nectar gathering efforts of honeybees. *Bixa orellana* L., *Byrsonima crassifolia* (L.) HBK, *Mangifera indica* L., *Musa* spp., *Syzygium jambos* (L.) Alston did not appear in the pollen record. In the Cafetitlan pollen load samples, *Zea mays* occurred for three

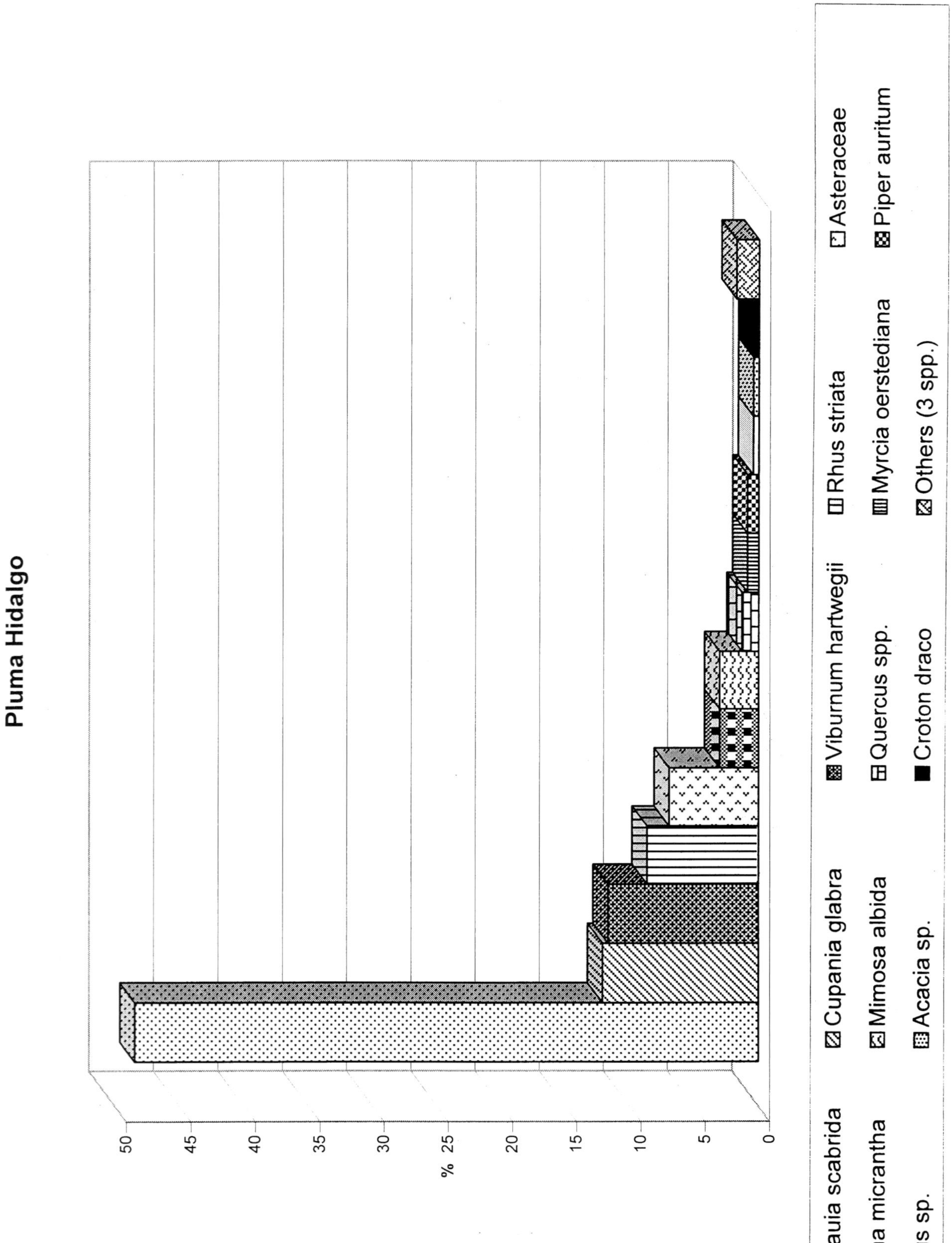

Text-Figure 4. Pollen spectrum of the Pluma Hidalgo honey sample.

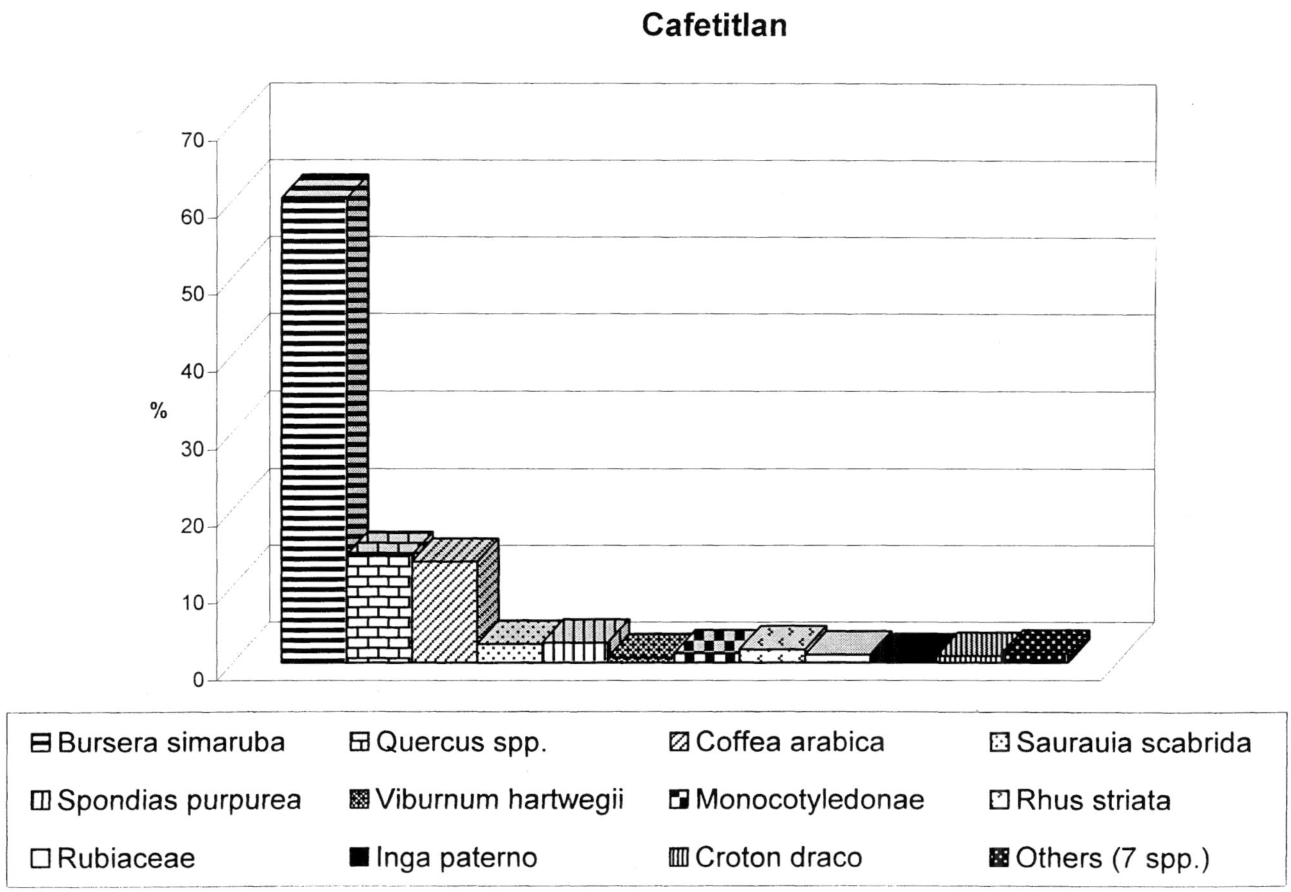

Text-Figure 5. Pollen spectrum of the Cafetitlan honey sample.

months (similar to that of *Coffea arabica*), but in larger amounts, perhaps because corn is more widely planted and is grown in greater density. However, in the Pluma Hidalgo samples corn pollen was absent and *Coffea arabica* reached 87 % of the pollen in May.

We agree with Proctor and Yeo (1973) and Alvarado and Delgado-Rueda (1985), that honeybees more frequently visit plants that provide both nectar and pollen foods. On other hand, Villanueva (1984) mentioned that due to their short flowering periods, no tropical deciduous forest plants were found in both dry and rainy season samples. In the Pluma Hidalgo samples, some taxa that appeared abundantly by the end of the dry season, also were present during a part of the rainy season (*Quercus* spp., *Rhus striata, Sambucus mexicana* and *Brugmansia* x *candida*). Other taxa such as *Trema micrantha, Mimosa albida,* and several Asteraceae, appeared in the rainy season and extended their presence through the dry season. This fact can be explained because seasonality is not well defined in the cloud forest and the plants respond to other factors such as temperature and photoperiod.

According to Villanueva (1984), taxa that blossom at the end of the rainy season are of great value for beekeepers, because they produce sufficient pollen and nectar that honeybees can store in great amounts. Plants of this kind found in the studied area include Asteraceae, Fabaceae (including *Mimosa albida*), *Alchornea latifolia, Zea mays, Ulmus mexicana*. *Coffea arabica* is important only for the honey yield of May–June.

It should be noted that the Lauraceae (*Persea americana, Phoebe mexicana, Nectandra sinuata*) are very common in the cloud forest and they have been registered as important nectar-polliniferous species; however, they did not appear in the pollen record.

Dicliptera unguiculata Nees, *Rhus striata, Viburnum hartwegii, Saurauia scabrida, Myrcia oerstediana, Colubrina* sp., *Vitis tiliifolia, Eleocharis* sp., *Tabebuia rosea* (Bertol) DC., *Belotia mexicana* (DC.) K. Schum., *Triumfetta copiacea* Hochr. are new records of plants of apicultural interest.

ACKNOWLEDGMENTS

This work was initiated through a Perfectioning Fellowship of the Red Latinoamericana de Botanica (RLB) granted to the first author in the Palaeobotany and Palynology Laboratory of the Biological Sciences Department, University of Buenos Aires, under the direction of Dr. Edgardo Romero. Dr. Alejandro Camacho kindly revised the manuscript. Drs. Gretchen Jones and Vaughn Bryant made valuable suggestions to improve this paper. The authors also appreciate the valuable help of M. Sc. Leonor Quiroz to identify some pollen grains.

TABLE 1. Pollen percentages in load samples of the Pluma Hidalgo hive.

Species	3/14/89	4/6/89	5/18/89	5/28/89	6/9/89	7/12/89	8/2/89	3/12/89	11/30/89	1/12/90	2/21/90
Pinus spp.		0.15	0.16	0.78							
Dicliptera sp.											0.16
Justicia sp.	0.6										0.96
Saurauia sp.								0.33			
Alternanthera sp.										0.17	
Iresine sp.	31.15										
Anacardiaceae										0.17	
Rhus striata				0.31		99.53	23.54				1.76
Spondias purpurea			1.48								0.16
Oreopanax langlassei				0.62							
Arecaceae					0.16						
Asteraceae	43.37	19.79							56.94	92.01	86.08
Tabebuia rosea										1.74	
Ananas comosus							0.54				
Sambucus mexicana		22.9		3.9							
Viburnum hartwegii								70.81			
Celastraceae							73.6				
Clethra mexicana	4.17	0.15									
Commelinaceae		0.15							2.38		
Eleocharis sp.								10.78			
Alchornea latifolia									3.37		
Croton sp.	0.89	2.36		0.47		0.47			0.28		
Fabaceae(Papilionoideae)							0.81		0.7		0.96
Acacia cornigera			0.32								
Acacia sp.								7.46			
Cassia sp.										0.87	
Gliricidia sepium		38.7									
Quercus spp.	16.39	1.03	0.83	90.48	75.41		0.27		1.82		0.48
Casearia spp.				1.09	6.06						
Malpighiaceae	0.45										2.08
Melastomataceae											0.96
Trichilia sp.							0.41				
Myrcia oerstediana			0.49								
Onagraceae	0.3										
Oxalis sp.	1.34										1.12
Piper auritum				0.31	0.16		0.14				
Piperaceae										0.17	
Poaceae		1.18									
Coffea arabica			87.83	0.94	0.33						2.08

TABLE 1 (continued). Pollen percentages in load samples of the Pluma Hidalgo hive.

Species	3/14/89	4/6/89	5/18/89	5/28/89	6/9/89	7/12/89	8/2/89	3/12/89	11/30/89	1/12/90	2/21/90
Rubiaceae sp.1			0.16								
Rubiaceae sp. 2					0.16						
Citrus sinensis					14.92		0.14				
Cupania glabra							0.14	6.8			
Sapindaceae			3.78		2.62		0.14		0.42		
Brugmansia X candida		13.44		1.09							
Solanaceae					0.16						
Belotia mexicana								0.66			
Triumfetta sp.											0.16
Trema micrantha	1.34	0.15						3.15		4.17	3.04
Ulmus sp.									34.08	0.69	
Vitis tiliifolia							0.27				
undetermined			4.94								
Total	**32681**	**32704**	**32745.99**	**32755.99**	**32767.98**	**32801**	**32822**	**32678.99**	**32941.99**	**32984.99**	**33025**

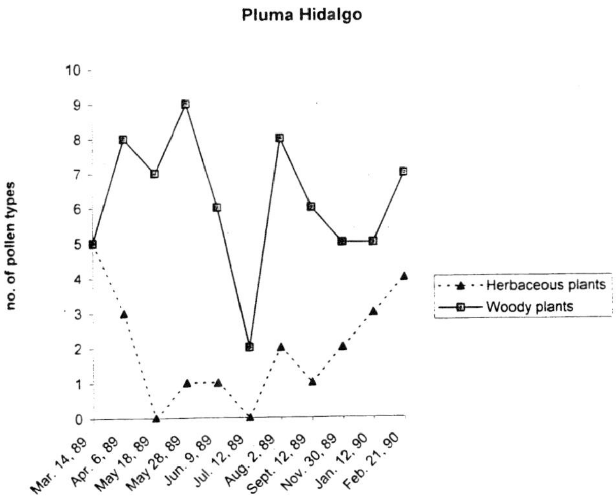

Text-Figure 6. Diagram of pollen types found in the load samples of the Pluma Hidalgo hive.

References Cited

ALVARADO, J. L., and DELGADO-RUEDA, M.
1985 Flora apicola en Uxpanapa, Veracruz, Mexico. *Biotica*, 10: 257–275.
1988 Flora melifera en una region calido-humeda de Veracruz (Uxpanapa), Mexico. II. *Biotica*,13: 69–79.

CABRERA-PECH, J. V.
1966 *Apicultura y flora apicola en el mpio. Villa de Arriaga, S.L.P., Mexico.* Professional Thesis. Escuela Nacional de Ciencias Biologicas, Instituto Politecnico Nacional. Mexico, D.F., 50 pgs.

CARMONA, M.L.
1980 *Contribucion al conocimiento de la flora melifera del Estado de Morelos.* Professional Thesis. Universidad Autonoma del Estado de Morelos. Cuernavaca, 86 pgs.

CHAZARO, M.
1982 Flora apicola de la zona cafetalera de Coatepec, Veracruz. In: Jimenez-Avila, E., and Gomez-Pompa, A. (eds.). *Estudios ecologicos en el agroecosistema cafetalero.* INIREB–CECSA. Mexico, D.F., p. 95–102.

ERDTMAN, G.
1943 *An introduction to Pollen Analysis.* The Ronald Press Company. New York, 239 pgs.

GARCIA, E.
1981 *Modificacion al sistema climatico de Köppen (para adaptarlo a las condiciones de la Republica Mexicana).* Third Ed. Instituto de Geografia, Universidad Nacional Autonoma de Mexico. Mexico, D.F.

INEGI
1980 Cartografia Edafologica. Hoja Mexico. scale 1: 1,000,000. Mexico, D.F.

LIEUX, M.
1972 A melissopalynological study of 54 Louisiana (U.S.A.) honeys. *Review of Palaeobotany and Palynology*, 13: 95–124.

LINDBERG, C.
1989 The economic value of insects. *Traffic Bulletin*, 10: 32–36.

ORDETX, G. S., ZOZAYA-RUBIO, J. A., and MILLAN, W. F.
1972 Estudio de la flora apicola nacional. Direccion General de Extension Agricola. Chapingo, 95 pgs.

TABLE 2. Pollen percentages in load samples of the Cafetitlan hive.

Species	**3/14/89**	**4/6/89**	**4/15/89**	**5/20/89**	**6/2/89**	**7/12/89**	**8/2/89**	**9/13/89**	**10/18/89**	**11/29/89**	**1/12/90**	**2/6/90**
Pinus sp.	0.33		0.3									0.15
Dicliptera sp.										0.29		
Alternanthera			0.3									
Rhus striata			9.88		0.15	19.93					0.17	
Arecaceae							0.14					
Asteraceae	0.5	0.13				0.33	20.56	9.69	45.18	61.29	23.26	10.3
Vernonia sp.	3.47										72.66	2.47
Bursera simaruba			50.6		0.15							
Viburnum hartwegii								65.58	1.36			
Clethra mexicana		0.66										
Cyperaceae										0.14		
Saurauia sp.									0.41			
Alchornea latifolia										7.57		
Croton draco							2.36					
Acacia cornigera	23.76	5.45										
Inga paterno			0.6									
Mimosa albida							47.78		48.03	20.71		
Schrankia distachya									0.41			
Senna sp.											0.85	
Rhynchosia sp.										4.29		
Lonchocarpus sp.											0.85	
Quercus sp.	55.27	86.57	36.82	19.28	13.67		0.69			5.14		85.63
Hyptis sp.										0.57		
Sida sp.									0.54			
Malvaceae									0.27			
Melastomataceae											0.34	
Cecropia obtusifolia		0.27									0.34	
Moraceae-Urticaceae	1.98		0.3					1.12				
Myrcia oerstediana				76.76	85.14	5.72	11.53	2.98				
Piper auritum		0.53										0.58
Piper sp.	3.8					4.9		14.29			0.51	
Poaceae											0.17	
Zea mays							3.33	4.35	1.9			
Colubrina sp.							4.86					
Coffea arabica	0.33			2.97	0.59							
Hamelia patens	10.56	6.12	1.2									0.58
Psychotria sp.									0.41			
Rubiaceae				0.66	0.3							
Cupania glabra								0.25	0.27			
Heliocarpus sp.									1.22			
Trema micrantha							0.14	1.74				
Urera sp.											0.85	
Vitis tillifolia						69.12	8.61					
undetermined		0.27		0.33								0.29
Total	**100**	**100**	**100**	**100**	**100**	**100**	**100**	**100**	**100**	**100**	**100**	**100**

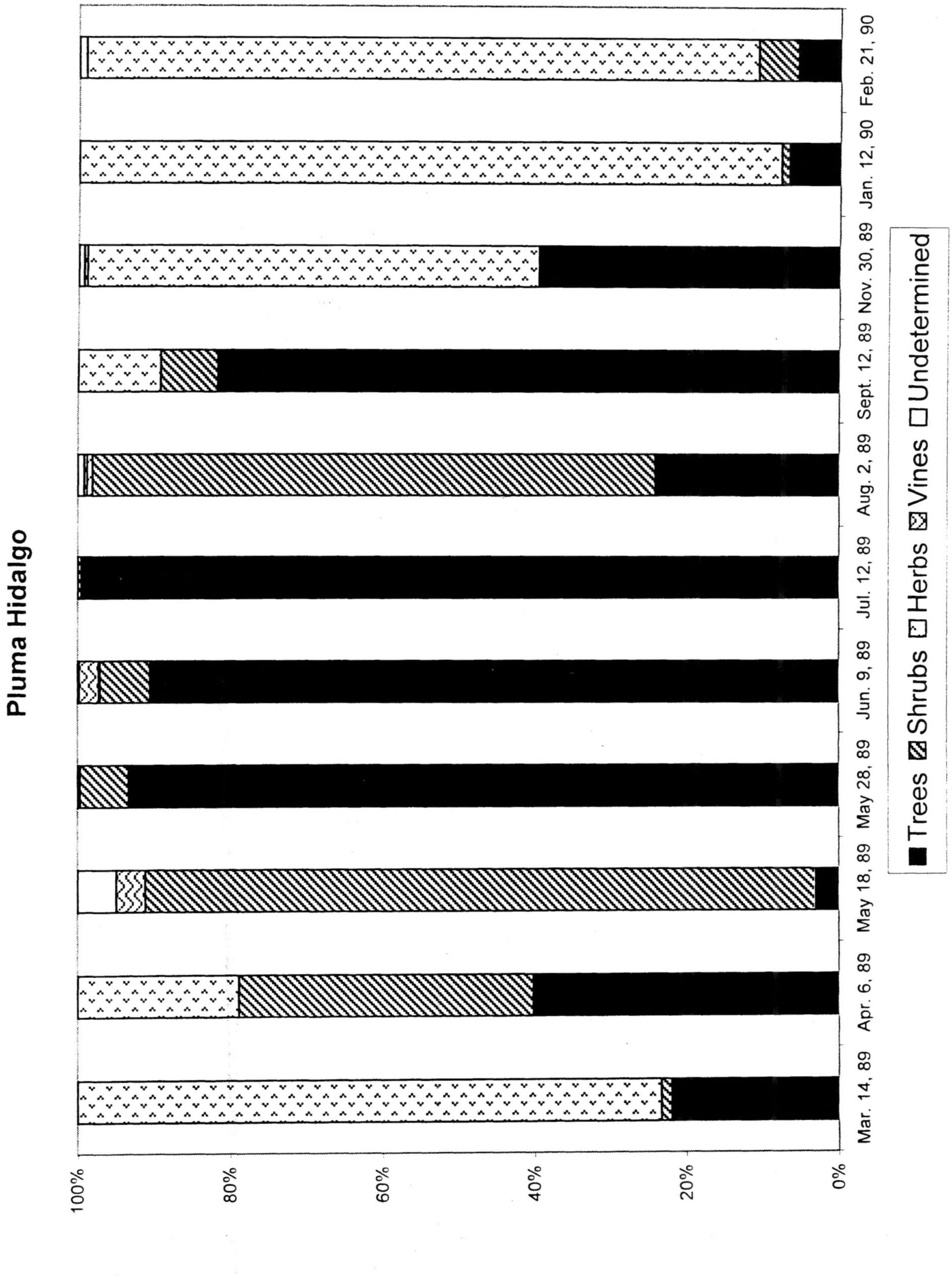

Text-Figure 7. Proportions of pollen of the different life forms found in samples of the Pluma Hidalgo hive.

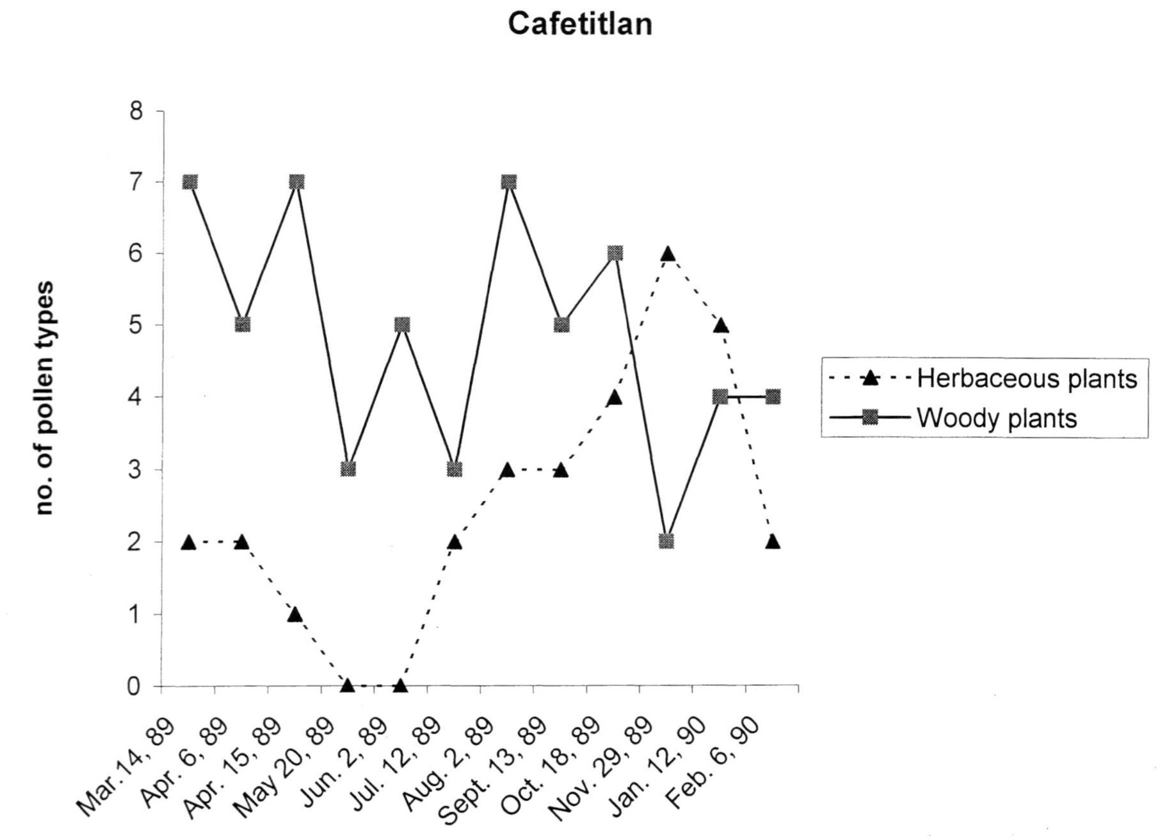

Text-Figure 8. Diagram of pollen types found in the samples of the Cafetitlan hive.

PIÑA-LUJAN, I.
1975 *Estudios ecologico-apicolas del estado de Oaxaca, 1971–1973.* Laboratorios Nacionales de Fomento Industrial. Mexico, D.F. (unpublished), 157 pgs.

PROCTOR, M., and YEO, P.
1973 *The Pollination ofFlowers.* W. Collins Sons & Co. Ltd., Glasgow, 418 pgs.

ROLDAN-RAMOS, L. A.
1984 *Flora melifera de la zona de Tixcacaltuyub, Yucatan.* Professional Thesis. Facultad de Ciencias, Universidad Nacional Autonoma de Mexico. Mexico, D.F., 95 pgs.

VILLANUEVA, R.
1984 Plantas de importancia apicola en el ejido de Plan del Rio, Veracruz, Mexico. *Biotica*, 9: 279–340.

Authors' addresses

SALVADOR ACOSTA-CASTELLANOS
Centro Interdisciplinario de Investigacion para el Desarrollo Integral Regional-Unidad Oaxaca
Instituto Politecnico Nacional
Apartado Postal 24 Administracion 3
68101 Oaxaca, Oaxaca
Mexico

RODOLFO PALACIOS-CHAVEZ
Escuela Nacional de Ciencias Biologicas
Instituto Politecnico Nacional
Apartado Postal 17–524
Mexico 11410, D.F.
Mexico

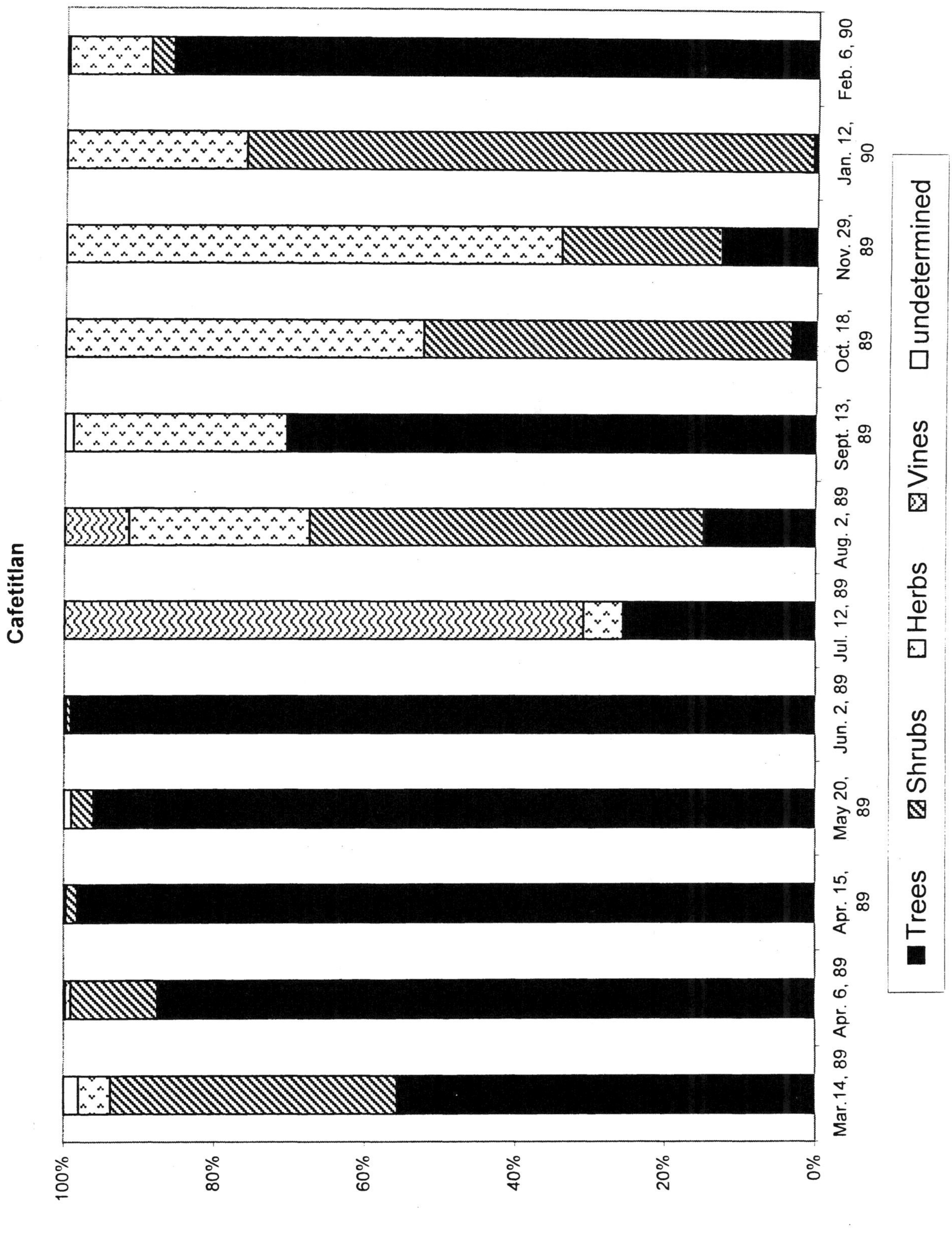

Text-Figure 9. Pollen proportions of the different plant life forms found in the pollen load samples of the Cafetitlan hive.

THE IDENTIFICATION OF NECTAR SOURCES FOR THE AFRICAN HONEYBEE, *APIS MELLIFERA ADANSONII*, BY THE EXAMINATION OF GHANAIAN HONEY

Joyce Amoako and Robert S. Pickard

Abstract

Nectar sources for the African honeybee, *Apis mellifera adansonii*, in seven geographical regions of Ghana were investigated. This study involved the pollen analysis of 47 samples of honey harvested after the major and minor honey flow seasons in 1994. The pollen sediment of the honeys was acetolyzed, mounted on slides and pollen types were identified.

A total of 244 taxa were identified from plants acting as nectar sources. Qualitatively, the most common nectar sources in order of decreasing rate of occurrence were: *Elaeis guineense, Tridax procumbens, Panicum maximum, Ceiba pentandra, Mimosa pudica, Zea mays, Leucena glauca, Securinega virosa, Morus mesozygia* and *Cocos nucifera*.

Quantitatively, the most abundant nectar sources in order of decreasing abundance were: *Tridax procumbens, Elaeis guineense, Morus mesozygia, Securinega virosa, Ceiba pentandra, Tectona grandis, Borreria* sp.#1, *Cocos nucifera, Ximenia americana, Chromolaena odorata, Dialium guineensis, Leucena glauca, Blighia sapida, Zea mays, Eucalyptus* sp., *Melanthera scandens, Erythrophleum guineensis, Cassia* sp. #2, *Panicum maximum* and *Mallotus oppositiifolius*. There was no indication of honeydew elements in the honey samples.

INTRODUCTION

The Republic of Ghana lies on the West Coast of Africa and covers an area of 239,460 sq. km. It has a population of 16 million. The southernmost tip reaches latitude 4.5° North while the northernmost tip extends to 11° North. It is divided into ten regions for administrative purposes. The coast is covered with grass and in some places scrubby mangroves. The Volta Basin has savanna woodland and swamps, a tropical rain forest occurs in the southwestern part, the semi-deciduous forest is in the southern part, and there is mostly grassland in the northern part. Annually, two wet seasons (March–June, September–October) and one dry season (November–February) are distinguishable.

Ghana, like many other tropical countries, has a rich and varied flora and ideal climatic conditions favorable for beekeeping. There is considerable scope for harnessing the natural renewable resources for the production of large quantities of honey through beekeeping. But beekeeping is not well organized and utilized to its fullest capacity due to a lack of knowledge of plants that are important to honeybees and to the severe deforestation that is occurring in most parts of the country. The only information that exists on plants important to honeybees comes from D'Albore and Owusu (1994) which determined the honeybee flora through pollen analysis. Therefore there is little economic benefit of the very numerous nectar and pollen producing plants in the country.

This research was carried out in 1994 to identify nectar sources of the African honeybee, *Apis mellifera adansonii*, through pollen analyses of 47 honey samples collected after the major and minor honey flow seasons in seven regions located within four vegetational zones in Ghana.

Amoako, Joyce, and Pickard, Robert S.
2001 The identification of nectar sources for the African honeybee, *Apis mellifere adansonii*, by the examination of Ghanaian honey; *In*: Goodman, D.K., and Clarke, R.T. (eds.), *Proceedings of the IX International Palynological Congress, Houston, Texas, U.S.A., 1996*; American Association of Stratigraphic Palynologists Foundation, p. 471–482.
 ISBN 0–931871–06–9

MATERIALS AND METHODS

Pieces of fully capped combs containing only honey were cut from selected hives in each apiary. Forty-seven honey samples, of which 34 were collected after the September–December nectar flow and 13 after the December–February flow, were collected from 19 sites in the 7 regions.

Slide preparation of honey followed the methods published by the International Commission for Bee Botany (Louveaux et al., 1978) with minor modifications (Low et al., 1989). In these modifications, 10 g of honey was thoroughly dissolved in 200 ml of warm water and centrifuged at 3500–4000 r.p.m. for 10 min instead of the usual 10 g of honey in 20 ml of water and centrifuged at 2000 r.p.m. This was necessary to get maximum sedimentation of the pollen, especially those with low densities. *Lycopodium clavatum* spore tablets were added (12,542 spores per tablet) to serve as an internal calibration standard. Pollen sediment of the honeys was acetolyzed and mounted on slides with glycerine jelly. The pollen grains were counted, 1200 per 10 g of honey sample, to determine the relative frequency of the different pollen types in the honey samples. In addition to using relevant literature, pollen grain identification was based on 399 reference pollen samples collected in the field and from the herbarium in Ghana.

RESULTS

In the present melissopalynological studies, 244 taxa were identified as nectar sources for the African honeybee. Mean number of pollen types in the honey samples ranged between 31.78 ± 1.93 and 17.75 ± 6.21 depending on the region.

The most abundant and common taxa identified from the regions and their frequencies, arranged in decreasing order, were determined. In the Ashanti region, the most abundant taxa were *Tridax procumbens, Elaeis guineense,*

Bidens pilosa, Cocos nucifera, Morus mesozygia, Schrankia leptocarpa, Commelina umbellata, Securinega virosa, Mallotus oppositiifolius, Eucalyptus sp., *Ceiba pentandra* and *Cassia* sp.#1. The most common taxa identified from the honey samples collected after the September–December flow were *Tridax procumbens, Commelina. umbellata, Elaeis guineense, Panicum maximum, Mallotus oppositiifolius, Euphorbia heterophylla, Leucena glauca, Boerhaevia* sp., Asteraceae # 1, 2, *Bidens pilosa, Sida acuta* and *Ceiba pentandra*. The most common taxa in the December–February flow were *Elaeis guineense, Ceiba pentandra, Bombax buonopozense, Terminalia catappa, Malotus opposittifolius, Tridax procumbens, Morus mesozygia, Mimosa pudica, Cocos nucifera, Leucena glauca, Celtis iguanea, Amaranthus hybridus, Panicum maximum* and *Melanthera scandens*.

In the Brong Ahafo region, the most abundant taxa identified were *Ceiba pentandra, Elaeis guineense, Morus mesozygia, Chromolaena odorata, Cassia ligustrina, Synedrella* sp., *Zea mays, Melanthera scandens, Leucena glauca, Securinega virosa, Emilia* sp., *Tridax procumbens, Lannea welwitschii, Terminalia catappa* and *Mimosa pudica*. The most common taxa in the honeys collected after the September–December flow were *Ceiba pentandra, Elaeis guineense, Panicum maximum, Zea mays, Tridax procumbens, Bombax buonopozense, Sida acuta, Morus mesozygia, Cynodon dactylon, Leucena glauca, Euphorbia heterophylla, Lannea welwitschii, Mimosa pudica, Mallotus oppositiifolius,* Asteraceae #2, *Elaeis guineensis, Terminalia catappa* and *Synedrella* sp. The most common taxa identified in the honeys collected after the December–February flow were *Elaeis guineense, Morus mesozygia, Securinega virosa, Chromolaena odorata, Terminalia catappa, Borassus* sp., *Chromolaena* sp. #1, *Khaya senegalense, Parkia clappertoniana., Panicum maximum, Mallotus oppositiifolius, Bombax buonopozense, Tridax procumbens,* and *Celtis* sp.

In the central region, the most abundant taxa identified were *Tectona grandis, Dialium guineensis, Elaeis guineense, Cocos nucifera, Ximenia americana, Securinega virosa, Cyperus esculentus, Erythrococca anomala* and *Manilkara obovata*. The most common taxa identified from the honey samples collected after the September–December flow were *Mimosa pudica, Elaeis guineense, Cyperus esculentus, Dialium guineensis, Cocos nucifera, Tectona grandis, Psidium guajava, Tamarindus indica, Azadiracta indica, Schrankia leptocarpa, Emilia* sp., *Fagara xanthoxyloides, Ximenia americana, Talinum triangulare, Securinega virosa, Manilkara obovata, Erythrococca anomala,* Fabaceae # 2, *Erythroxyllum emarginatum, Mitragyna inermis,* and *Lantana camara*.

The most abundant taxa identified from the honeys from the Eastern region collected after the December–February flow were *Chromolaena odorata, Lannea nigritana, Morus mesozygia, Chromolaena* sp.#1 and *Elaeis guineense*.

Honey samples from the Greater Accra region showed *Securinega virosa, Morus mesozygia, Blighia sapida, Tectona grandis, Cleome viscosa,* Melastomataceae, *Dialium guineensis, Eucalyptus* sp., *Gynandropsis gynandra,* and *Elaeis guineense* were the most abundant taxa. The most common taxa from the honey samples collected after the September–December flow were *Elaeis guineense, Securinega virosa, Cocos nucifera, Delonix regia, Leucena glauca, Mimosa pudica, Tectona grandis, Cleome viscosa, Eucalyptus* sp., *Schrankia leptocarpa, Blighia sapida, Peltophorum pterocarpum, Panicum maximum, Morus mesozygia, Polygonon pulchrum, Cyperus esculentus* and *Psidium guajava*.

In the Northern region, the most abundant taxa identified were *Borreria* sp.1, *Capsicum frutescens, Panicum maximum, Jussiaea* sp., *Securinega virosa, Cassia siamea,* and *Bidens pilosa*. The most common taxa were *Borreria* sp.#1, *Hyptis* sp., *Panicum maximum, Capsicum frutescens, Justicia* sp.#1, *Borreria verticillata* and *Cassia* sp. #3.

In the Western region, the honey samples collected after the December–February flow presented *Ximenia americana, Mimosa pudica, Dialium guineensis* and *Chromolaena odorata* as the most abundant taxa.

The most abundant taxa identified from the honey samples collected after the September–December flow were *Tridax procumbens, Elaeis guineense, Ceiba pentandra, Tectona grandis, Securinega virosa, Bidens pilosa, Borreria* sp.#1, *Cocos nucifera, Schrankia leptocarpa, Dialium guineensis, Gynandropsis gynandra, Commelina umbellata, Blighia sapida, Zea mays, Leucena glauca, Melanthera scandens, Cassia* sp.#2, *Panicum maximum, Chromolaena odorata* and *Emilia* sp.

The most abundant taxa identified from the honey samples collected after the December–February flow were *Morus mesozygia, Securinega virosa, Lannea nigritana, Ximenia americana, Elaeis guineense, Cocos nucifera, Chromolaena odorata, Erythrophleum guineensis, Tridax procumbens, Ceiba pentandra, Mallotus oppositiifolius,* Melastomataceae, *Cassia* sp.#1, *Celtis* sp., *Terminalia catappa, Cassia siamea, Bidens pilosa,* Caesalpinoideae #1 and *Syzygium guineensis*.

The percentage occurrence of the pollen types found in more than 20% of the honey samples collected after the December–February and September–December flow and expressed as a percentage according to frequency classes are summarized (Text-Figures 1, 2). The taxa identified from the honey samples is presented (Table 1).

Estimated Absolute Number of Pollen in 10 g of Honey Samples

The estimated absolute pollen content per 10 g of honey sample ranged between 5,850 and 1,309,200 pollen grains. The mean taxa identified and the estimated absolute number of pollen grains per 10 g of honey samples collected during the December–February nectar flow were 32.92 ± 2.89 and 360,846 ± 79,103.47 respectively. The mean taxa and the estimated absolute number of pollen grains in 10 g of honey samples collected during the September–December nectar flow were 27.38 ± 1.54 and 249,106.38 ± 50,500.85.

Families with Larger Number of Plant Species Represented

Families that contributed the largest number of taxa in order of decreasing abundance were Fabaceae, Asteraceae, Euphorbiaceae and Rubiaceae. The number of taxa contributed by Fabaceae was 54, Asteraceae 33, Euphorbiaceae 12 and Rubiaceae 12.

DISCUSSION

The results of the present study demonstrate that melissopalynology can be successfully applied to the study of honeys from Ghana. The acetolyzation procedure adopted permitted detailed observation of the exine structure making identification possible.

Ghanaian honeys are multifloral, reflecting the rich flora and high species diversity. In this study, 244 plant species were identified in 47 honey samples. This sharply contrasts

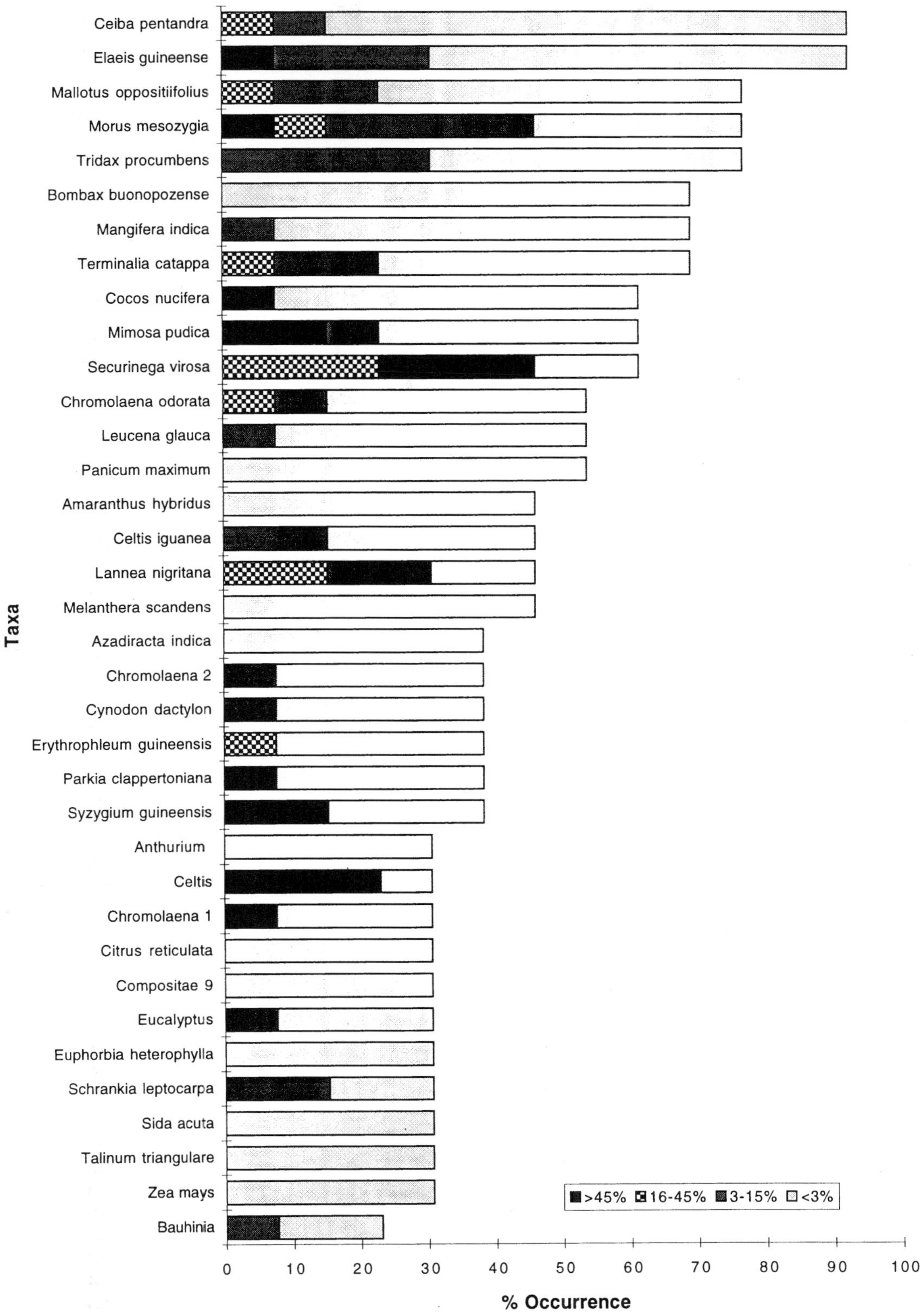

Text-Figure 1. Taxa occurring in more than 20% of the post December–February nectar flow honey samples expressed as a percentage according to frequency classes.

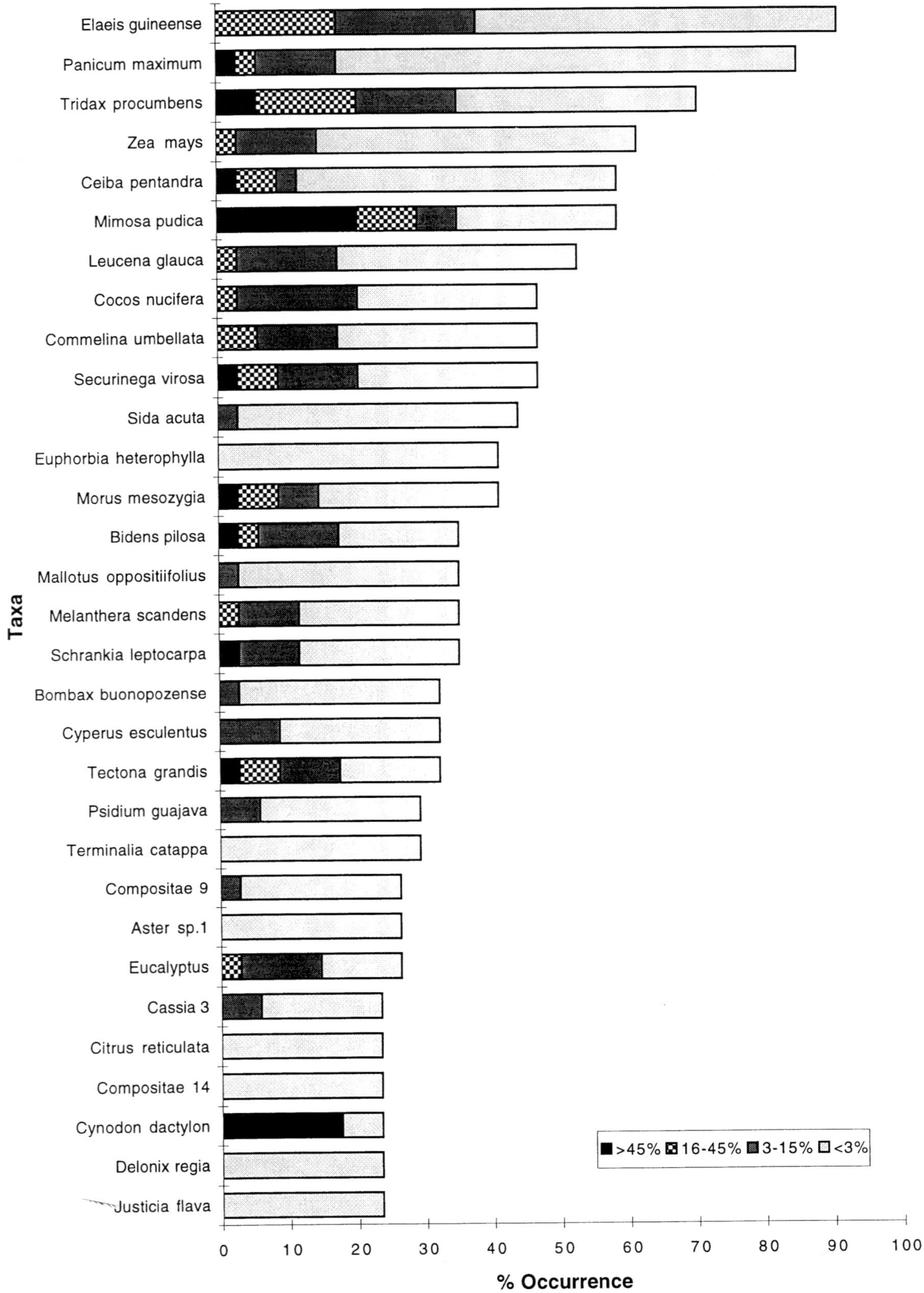

Text-Figure 2. Taxa occurring in more than 20% of the post September–December nectar flow honey samples expressed as a percentage according to frequency classes.

the data obtained for some temperate countries; 67 pollen types in 130 honey samples from Luxemburg (Maurizio, 1971); 58 pollen types in 54 honey samples from Louisiana (Lieux, 1972); and 61 pollen types in 115 honey samples from the Basque country of Spain (Sancho et al., 1991). Of greater importance is the number of identified pollen types per honey sample. The highest number obtained for the Ghanaian samples was 48 in comparison to 25 and 20 in honeys from Louisiana and Luxemburg respectively.

Estimated absolute pollen grain content per 10 g of honey samples indicates that Ghanaian honeys are rich in pollen varying within the normal floral honey, most of them from over-represented sources (Sawyer, 1988). Varying amounts of pollen grains come into contact with nectar which is subsequently recovered in the honey (Demianowicz and Jablonski, 1959). This might be due to the differences in blossom structure, arrangement of the floral parts and the fact that different plants produce pollen in greater or lesser amounts. Honeybees also filter pollen in suspension in nectar through the action of the proventriculus (Bailey, 1951). Thus a number of factors might influence the final pollen content of the honey. The absolute number of the various pollens in the honeys therefore does not exactly reflect the floral nectar contributions of the individual plants.

There is a decreasing number of pollen grains in the honey samples as grain size increased. Plants in the Cucurbitaceae, Malvaceae, Mimosaceae, Euphorbiaceae and Rubiaceae are under-represented by pollen although their nectar contribution to honey is high. Some plants are over-represented such as *Mimosa pudica, Morus mesozygia* and *Ximenia americana.*

Pollen types from plants that do not produce nectar also were found in the honey samples. Among the nectarless plants, *Cynodon dactylon, Cyperus esculentus Panicum maximum,* and *Zea mays* were the most important. *Panicum maximum* appeared in about 70 % of the honey samples. Many activities within the colony contribute to the introduction of pollen from nectarless plants into the honey. Pollen enters into unripe honey in the hive as bees brush against other foragers, through grooming of the body, food transmission and cleaning. Moreover, many bees that collect pollen digress later to nectar collection, or gather nectar and pollen at the same visit.

The most reliable nectar sources for the African honeybee were: *Elaesis guineense, Tridax procumbens, Ceiba pentandra, Mimosa pudica, Leucena glauca, Morus mesozygia,* and *Securinega virosa.* The most abundant taxa which occurred in order of decreasing abundance, were *Tridax procumbens, Elaesis guineense, Morus mesozygia, Securinega virosa, Ceiba pentandra, Tectona grandis, Borreria* sp.#1, *Cocos nucifera, Ximenia americana, Chromolaena odorata, Dialium guineensis, Leucena glauca, Blighia sapida, Eucalyptus* sp., *Melanthera scandens, Erythrophleum guineensis, Cassia* sp. #2 and *Mallotus oppositiifolius.* These important taxa visited represented a small number of the total plant population.

This study confirms some of the plants identified in Ghanaian honeys by D'Albore and Owusu (1994). Pollen grains which are not found or are rare in the honey do not mean that the plants are not visited. Only pieces of honeycombs were used for the analysis. Detailed results could be obtained by taking samples of honey more regularly throughout the year.

In Ghana, hives are left unattended until harvesting. Ideal climatic conditions allow the honeybees to collect nectar and pollen throughout the year which remain in the combs. This, and the fact that some of the plants have long flowering periods and also flower more than once a year, probably explains why the common plants identified in the honeys can occur in both harvesting seasons.

The period of December to February is when most of the plant species visited were flowering. This coincides with the months in which the bulk of honey is harvested. Families with the highest number of pollen species identified are Fabaceae and Asteraceae.

Ghanaian honeys do not contain honeydew elements. Honeydew may not be exposed for honeybee use due to the high density of ants in the tropics. The ants feed on the honeydew and probably do not allow the bees to collect it. Moreover, the bees have access to abundant nectar and pollen sources throughout the year and probably do not collect honeydew.

Knowledge of honeybee flora provides information to beekeepers for apiary sites where floral resources are plentiful. This will boost the beekeeping industry, which will help communities be economically productive. In turn, this will lead to an increase in honey production and other important products like beeswax and propolis for local consumption and export, thus earning foreign exchange for national development. The beeswax would be of great importance to the "lost wax" casting industry which is a tourist attraction. Recommendations can be made for planting and conservation as to which renewable natural resources should be exploited by the country. In addition, pollination of plants by the honeybees would increase the yield of crops.

References Cited

BAILEY, L.
1951 The action of the proventriculus of the honeybee *Apis mellifera* L. *Bee World,* 32: 92.

DEMIANOWICZ, Z., and JABLONSKI, B.
1959 The pollen coefficient of three unifloral honeys, from the point of view of statistical analysis. *Pszczelnicze Zeszty Naukowe,* 3: 25–34.

LIEUX, M.H.
1972 A melissopalynological study of 54 Louisiana (USA) honeys. *Review of Palaeobotany and Palynology,* 13: 95–124.

LOUVEAUX J., MAURIZIO, A., and VORWOHL, G.
1978 Methods of melissopalynology. *Bee World,* 59: 139–157.

LOW, N.H., SCHWEGER, C., and SPORNS, P.
1989 Precautions in the use of melissopalynology. *Journal of Apicultural Research* 28: 50–54.

MAURIZIO, A.
1971 Le spectre pollinique des miels Luxembourgeois. *Apidologie,* 2: 221–238.

RICCIARDELLI D'ALBORE, G.C., and OWUSU, M.
1994 Beekeeping and pollen spectra of some honeys from Ghana. *Proceedings of the fifth international conference on apiculture in tropical climates,* pp. 106–109.

SANCHO, M.T., MUNIATEGUI, S., HUIDOBRO, J.F., and SIMAL-LOZANO, J.
1991 Discriminant analysis of pollen spectra of Basque country (Northern Spain) honeys. *Journal of Apicultural Research,* 30: 162–167.

SAWYER, R.W.
1988 *Honey Identification.* Cardiff Academic Press, 115 pgs.

TABLE 1. Percentage of occurrence (% Occur), frequency classes of the taxa found (D-Predominant pollens, >45 %; S-Secondary pollen, 16-45 %; s-Important minor, 3-15 %; r-Minor, <3 %), number of samples of each taxon in each frequency class and total number of samples (Total) of the Ghanaian honey examined.

Taxon	% Occur	Frequency Classes				Total
		D	S	s	r	
CRYPTOGAMAE						
Polypodiaceae	2.13				1	1
MONOCOTYLEDONAE						
ARACEAE						
Anthurium sp.	19.17			1	8	9
COMMELINACEAE						
Commelina umbellata	38.34		2	4	12	18
Commelina erecta	2.13				1	1
CUPRESSACEAE						
Cupressus sp.					1	1
CYPERACEAE						
Cyperus esculentus	29.82			3	11	14
Cyperus rotundus	6.39				3	3
POACEAE						
Cynodon dactylon	25.56					
Panicum maximum	70.29	1	1	4	27	33
Panicum sp.	17.04			1	7	8
Zea mays	53.25		1	4	20	25
MUSACEAE						
Musa sp.	6.39				3	3
ARECACEAE						
Cocos nucifera	51.12	1	1	6	16	24
Elaeis guineense	91.59	1	5	11	26	43
Borassus sp.	17.04			1	7	8
DICOTYLEDONAE						
ACANTHACEAE						
Hypoestes verticillaris	4.26				2	2
Justicia flava	21.30				10	10
Justicia sp.	8.52				4	4
Ruellia sp.	2.13				1	1
Acanthaceae	4.26				2	2
AMARANTHACEAE						
Amaranthus hybridus	21.30				10	10
Cyathula sp.	6.39				3	3
ANACARDIACEAE						
Lannea nigritana	17.04		2	2	4	8
Lannea welwitschii	17.04		1	1	6	8
Mangifera indica	31.95			1	14	15
ANNONACEAE						
Annona sp.	2.13				1	1

TABLE 1 (continued). Percentage of occurrence (% Occur), frequency classes of the taxa found (D-Predominant pollens, >45 %; S-Secondary pollen, 16-45 %; s-Important minor, 3-15 %; r-Minor, <3 %), number of samples of each taxon in each frequency class and total number of samples (Total) of the Ghanaian honey examined.

Taxon	% Occur	Frequency Classes				Total
		D	S	s	r	
BIGNONIACEAE						
Tecoma stans	4.26				2	2
Tabebuia sp.	2.13				1	1
Arrabidaea	2.13				1	1
Bignoniaceae	8.52				4	4
BOMBACACEAE						
Adansonia digitata	6.39				3	3
Bombax buonopozense	42.60			1	19	20
Ceiba pentandra	68.16	1	3	2	26	32
Bombacaceae	2.13				1	1
BORAGINACEAE						
Cordia sp.	4.26				2	2
Heliotropium sp.	2.13				1	1
CACTACEAE						
Opuntia	6.39				3	3
CAPPARIDACEAE						
Cleome viscosa	8.52		1	2	1	4
Gynandropsis gynandropsis	19.17		2	1	6	9
CARICACEAE						
Carica papaya	8.52				4	4
COMBRETACEAE						
Anogeissus schimperi	10.65		1		4	5
Terminalia catappa	40.47		1	2	16	19
Combretum	8.52			1	3	4
Combretaceae	10.65				5	5
ASTERACEAE						
Ageratum conyzoides	8.52				4	4
Aster	34.08				16	16
Asteraceae # 1	19.17			2	7	9
Asteraceae # 2	27.69			1	12	13
Asteraceae # 3	76.07			5	34	39
Bidens pilosa	29.82	1	2	4	7	14
Chromolaena odorata	27.69		3	3	7	13
Chromolaena	29.82			4	10	14
Emilia sp.	17.04		1	1	6	8
Inula sp.	6.39				3	3
Melanthera scandens	36.21		1	3	13	17
Mikania sp.	12.78			2	4	6
Parthenium sp	21.30				10	10
Pyrethrum sp.	10.65		1	2	2	5
Senecio sp.	10.65			2	3	5

TABLE 1 (continued). Percentage of occurrence (% Occur), frequency classes of the taxa found (D-Predominant pollens, >45 %; S-Secondary pollen, 16-45 %; s-Important minor, 3-15 %; r-Minor, <3 %), number of samples of each taxon in each frequency class and total number of samples (Total) of the Ghanaian honey examined.

Taxon	% Occur	Frequency Classes				Total
		D	S	s	r	
ASTERACEAE (continued)						
Synedrela sp.	21.30			1	9	10
Tagetes sp.	10.65			1	4	5
Tithonia sp.	6.39			1	2	3
Tridax procumbens	70.29	2	5	8	18	33
Vernonia djalonensis	6.39				3	3
Vernonia guineensis	2.13				1	1
Vernonia	17.04				8	8
CONVOLVULACEAE						
Merremia sp.	2.13				1	1
CUCURBITACEAE						
Citrullus vulgaris	4.26				2	2
Cucurbita pepo	8.52				3	3
Luffa cylindrica	6.39				3	3
DICHAPETALACEAE						
Dichapetalum sp.	2.13				1	1
ERYTHROXYLACEAE						
Erythroxyllum emarginatum	6.39			1	2	3
EUPHORBIACEAE						
Croton sp.	6.39				3	3
Erythrococca anomala	6.39			1	2	3
Euphorbia heterophylla	38.34				18	18
Euphorbia sp.	8.52				4	4
Euphorbiaceae	2.13				1	1
Hura sp.	8.52				4	4
Jatropha curcas	4.26				2	2
Mallotus oppositiifolius	46.86		1	3	18	22
Mallotus sp.	14.91			3	4	7
Manihot esculenta	12.78				6	6
Ricinus communis	10.65			1	4	5
Securinega virosa	51.12	1	5	7	11	24
FLACOURTIACEAE						
Casearia sp.	8.52			1	3	4
GUTTIFERAE						
Guttiferae	2.13				1	1
LAMIACEAE						
Hyptis sp.	21.30			1	9	10
FABACEAE						
CAESALPINIOIDEAE						
Afzelia africana	4.26				2	2
Afzelia sp.	2.13				1	1

TABLE 1 (continued). Percentage of occurrence (% Occur), frequency classes of the taxa found (D-Predominant pollens, >45 %; S-Secondary pollen, 16-45 %; s-Important minor, 3-15 %; r-Minor, <3 %), number of samples of each taxon in each frequency class and total number of samples (Total) of the Ghanaian honey examined.

Taxon	% Occur	Frequency Classes				Total
		D	S	s	r	
CAESALPINIOIDEAE (continued)						
Anthonatha macrophylla					1	1
Bauhinia tomentosa	2.13				1	1
Bauhinia variegata	2.13				1	1
Bauhinia sp.	14.91		1	1	5	7
Cassia rotundifolia	14.91			2	5	7
Cassia siame	12.78		1	1	4	6
Cassia	46.86		3	6	13	22
Caesalpinioideae	10.65				5	5
Delonix regia	19.17				9	9
Detarium senegalense	6.39				3	3
Dialium guineensis	17.04		3	3	2	8
Erythrophleum guineensis	23.43		1	2	8	11
Griffonia simplicifolia	12.78			2	4	6
Haematoxylon campichianum	2.13				1	1
Hymenostegia sp.	12.78				6	6
Julbernardia	6.39				3	3
Macrolobium sp.	2.13				1	1
Peltophorum pterocarpum	19.17				9	9
Tamarindus indica	10.65			1	4	5
Tephrosia	14.91			2	5	7
MIMOSOIDEAE						
Acacia sp.	2.13				1	1
Albizia lebbeck	8.52				4	4
Entada	8.52				4	4
Leucena glauca	53.25		1	6	18	25
Mimosa pudica	61.77	9	2	*5*	13	29
Mimosa sp.	17.04				9	9
Mimosoideae	4.26				2	2
Parkia clappertoniana	14.91			1	4	5
Pithecollobium dulce	8.52				4	4
Schrankia leptocarpa	34.08	1		5	10	16
PAPILIONOIDEAE						
Arachis hypogaea	6.39				3	3
Cajanus cajan	4.26				2	2
Centrosema pubescens	2.13				1	1
Fabaceae	8.52				4	4
Machaerium	10.65				5	5
Papilionoideae	12.78				6	6
Sesbania pachycarpa	2.13				1	1
Vigna sinensis	2.13				1	1

TABLE 1 (continued). Percentage of occurrence (% Occur), frequency classes of the taxa found (D-Predominant pollens, >45 %; S-Secondary pollen, 16-45 %; s-Important minor, 3-15 %; r-Minor, <3 %), number of samples of each taxon in each frequency class and total number of samples (Total) of the Ghanaian honey examined.

Taxon	% Occur	Frequency Classes				Total
		D	S	s	r	
LYTHRACEAE						
Lagerstroemia indica	12.78				6	6
Lawsonia inermis	8.52				4	4
MALPIGHIACEAE						
Malpighiaceae	12.78				6	6
MALVACEAE						
Abutilon mauritianum	2.13				1	1
Gossypium hursutum	2.13				1	1
Gossypium	2.13				1	1
Hibiscus esculentus	4.26				2	2
Malvaceae	19.17			1	8	9
Sida acuta	40.47			1	18	19
Sida rhombifolia	4.26				2	2
Urena lobata	4.26			1	1	2
MELASTOMATACEAE						
Melastomataceae	4.26		1		1	2
MELIACEAE						
Azadiracta indica	25.56				12	12
Cedrella sp.	2.13				1	1
Khaya senegalense	4.26				2	2
Meliaceae	4.26				2	2
Trichilia acutifoliola	6.39				3	3
Trichilia sp.	6.39			2	1	3
MORACEAE						
Morus mesozygia	51.12	2	3	6	13	24
MYRTACEAE						
Eucalyptus sp.	27.69		1	6	6	13
Psidium guajava	25.56			2	10	12
Syzygium guineense	14.91			3	4	7
NYCTAGINACEAE						
Boerhaevia sp.	14.91				7	7
OLACACEAE						
Ximenia americana	10.65	1	1		3	5
ONAGRACEAE						
Jussiaea sp.	2.13		1			1
PEDALIACEAE						
Sesamum sp.	4.26				2	2
PHYTOLACCACEAE						
Phytolacca dodecandra	4.26				2	2
POLYGONACEAE						
Antigonon leptopus	6.39				3	3
Polygonon pulchrum	6.39				3	3

TABLE 1 (continued). Percentage of occurrence (% Occur), frequency classes of the taxa found (D-Predominant pollens, >45 %; S-Secondary pollen, 16-45 %; s-Important minor, 3-15 %; r-Minor, <3 %), number of samples of each taxon in each frequency class and total number of samples (Total) of the Ghanaian honey examined.

Taxon	% Occur	Frequency Classes				Total
		D	S	s	r	
PORTULACACEAE						
Portulaca grandifolia	10.65				5	5
Talinum triangulare	17.04				8	8
RHAMNACEAE						
Rhamnaceae 1	8.52			1	3	4
RUBIACEAE						
Borreria compressa	6.39				3	3
Borreria scabra	6.39				3	3
Borreria verticillata	19.17				9	9
Borreria sp.	12.78	2			4	6
Crossopteryx febrifuga	6.39				3	3
Mitragyna inermis	6.39				3	3
Morinda morindoides	6.39				3	3
Rubiaceae	17.04				8	8
RUTACEAE						
Citrus aurantifolia	6.39				3	3
Citrus reticulata	25.56				12	12
Fagara xanthoxyloides	14.91			2	5	7
Murraya paniculata	2.13				1	1
SAPINDACEAE						
Allophyllus sp.	2.13				1	1
Blighia sapida	12.78	1			5	6
Cardiospermum halicacabum	12.78				6	6
Majidea fosteri	2.13				1	1
Paullinia pinnata	4.26				2	2
Paullinia sp.	6.39				3	3
SAPOTACEAE						
Manilkara obovata	6.39			1	2	3
SOLANACEAE						
Capsicum frutescens	10.65		1		4	5
Solanum melongena	6.39				3	3
Solanaceae	6.39				3	3
STERCULIACEAE						
Theobroma cacao	4.26				2	2
TILIACEAE						
Corchorus aestuans	6.39				3	3
Grewia carpinifolia	8.52				4	4
Tiliaceae	10.65				5	5
Triumfetta cordifolia	4.26			1	1	2
Triumfetta	21.30			1	11	12

TABLE 1 (continued). Percentage of occurrence (% Occur), frequency classes of the taxa found (D-Predominant pollens, >45 %; S-Secondary pollen, 16-45 %; s-Important minor, 3-15 %; r-Minor, <3 %), number of samples of each taxon in each frequency class and total number of samples (Total) of the Ghanaian honey examined.

Taxon	% Occur	Frequency Classes				Total
		D	S	s	r	
ULMACEAE						
Celtis iguanea	19.17			3	6	9
Celtis sp.	21.30			3	7	10
APIACEAE						
Umbelliferae	2.13				1	1
VERBENACEAE						
Lantana camara	8.52				4	4
Stachytarpheta	2.13				1	1
Tectona grandis	27.69	1	2	3	7	13
Vitex agnus-castus	2.13				1	1
Vitex grandifolia	4.26			1	1	2
VITACEAE						
Vitaceae	2.13				1	1
ZYGOPHYLLACEAE						
Tribulus sp.	4.26				2	2
UNKNOWN	6.39			1	2	3

Authors' addresses

JOYCE AMOAKO*
ROBERT S. PICKARD
School of Pure and Applied Biology
University of Wales
College of Cardiff
P.O. Box 915
Cardiff CF1 3TL
Wales
U.K.

*Present Address:
113 Persimmon Drive
Collegeville, PA 19426
U.S.A.

IS ONE DROP ENOUGH?[1]

Gretchen D. Jones and Vaughn M. Bryant, Jr.

Abstract

Approximately 250 million pounds of honey are produced annually in the United States with a value of $200 million. Pollen taxa found in honey indicates its floral sources. Honey made from a nectar of a specific plant (citrus, clover, or mesquite) can be sold for a higher price than honey from mixed floral sources. To determine a honey's floral sources, only one drop of honey is analyzed. It is assumed that all drops from a honey's pollen residue are equal in pollen concentration values and diversity (number of taxa). To test this assumption, five drops of one honey sample were examined. A 500 grain count for each drop was made by first counting 200 grains and then three consecutive sets of 100 grains. For each count, pollen concentration values, percentages, and floral frequency classes were calculated, and total number of taxa per count summed.

Mean concentration values ranged from 69,496 grains per 1 g of honey to 131,657 grains. There was no significant difference in concentration values of counts among the drops, nor among the counts of the same drop.

Pollen diversity varied from 34 to 75 taxa. No single drop consistently had more or fewer taxa. Each successively higher count had significantly more taxa. Differences in pollen diversity among the different drops were greater than expected. No drop accounted for more than 60% of the 130 taxa encountered. Four taxa occurred in only one drop: *Cardiospermum corindum*, *Polygonum* sp., *Liquidambar styraciflua*, and *Acer* sp.

INTRODUCTION

Social honeybees (*Apis* spp.) evolved in the Old World tropics and have been producing and storing honey for probably more than 20 million years (Crane, 1980). Today, honey is an important food and carbohydrate source throughout the world. Approximately 250 million pounds of honey are produced annually in the United States and have a value of $200 million. In addition, four million pounds of beeswax are produced. Lesser known substances, like bee pollen and royal jelly, are important ingredients in a variety of products. It has been estimated that honeybees (*Apis mellifera* Linnaeus) pollinate more than 90 cultivated crops (apples, almonds, blueberries, melons, cucumbers, etc.,) in the United States with a combined annual value of $20 billion (Tew, 1989).

Taxa of pollen found in honey indicate a honey's floral sources. Thus, pollen frequency indicates major and minor plants utilized by honeybees. This information is important because honey of a dominant floral type (e.g. citrus, clover, or mesquite honey) can bring a higher price than honey of mixed or unknown floral sources.

When a honey sample is examined for pollen, a small subsample is removed for pollen analysis. From this subsample, one drop is removed and examined. The entire honey will be classified from this one drop. It has always been assumed that all drops from the same sample were equal with respect to concentration values and number of taxa. The purpose of this study was to determine if drops from a single honey sample had similar pollen concentration values and number of taxa.

METHODS AND PROCEDURES

A mixed floral honey rich in pollen was used to test differences in pollen concentration values and number of taxa. In a sample with only 20 to 30 pollen types, all taxa should be counted within the first 200–300 grains. But in a sample containing over 50 taxa, some taxa should not be encountered within the first 200–300 grain count.

The honey sample was heated in a microwave oven to 38°C and thoroughly stirred to ensure uniform mixing of the pollen. Ten grams of honey [the internationally preferred standard for testing pollen content in honey samples (Louveaux et al., 1978)], were extracted and poured into a 150 ml beaker. Ten ml of water were added and the resulting solution was stirred until the honey dissolved.

Five *Lycopodium* tablets (containing 11,300 ± 300 spores, batch #201890) were dissolved in 5 ml of 5% hydrochloric acid (HCL), and added to 10 ml of distilled water. Five tablets were used as a marker to ensure that the pollen to spore ratio would be closer to 1:1. Addition of one or two tablets probably would not have provided an accurate comparison of concentration values nor an accurate estimate of pollen concentration values.

The dissolved *Lycopodium* tablets were added to the honey, and the solution was stirred. One hundred mls of 95% ethyl alcohol (ETOH) were added to the honey-water solution and stirred well. The beaker was then placed on a warm hot plate and periodically stirred.

The honey solution was poured into a 12 ml centrifuge tube and centrifuged at 4,000 r.p.m for three minutes. After decanting the supernatant, the residue was mixed for 15 seconds. Additional honey solution from the beaker was added to the residue. This process of centrifuging, decanting, mixing, and adding more honey solution was repeated until the beaker was empty. Once empty, the beaker sides were rinsed with 95% ETOH. The rinse liquid was poured into the centrifuge tube, then centrifuged, decanted, and mixed.

The sample was acetolyzed (Erdtman, 1963; Lieux, 1980; Low et al., 1989) for 12 minutes. After acetolyzation, it was rinsed twice with distilled water and once with 95% ETOH; centrifuging, decanting, and mixing each time. Safranin O

Jones, Gretchen D., and Bryant, Vaughn M., Jr.
2001 Is one drop esnough?; *In*: Goodman, D.K., and Clarke, R.T. (eds.), *Proceedings of the IX International Palynological Congress, Houston, Texas, U.S.A., 1996*; American Association of Stratigraphic Palynologists Foundation, p. 483–487.
 ISBN 0–931871–06–9

[1] Mention of a commercial or proprietary product in this paper does not constitute an endorsement by the USDA, nor does it imply the recommendation of the product by the USDA to the exclusion of similar products.

stain was added and the pollen residue was poured into marked 2 dram vial. Ten drops of glycerin were added, and the vial was placed on a hot plate and left over night to evaporate the ETOH.

The vial of pollen residue was mixed for 30 seconds before slides were made. Five slides were made by placing one drop onto five separate glass slides. The drop was allowed to spread for 30 seconds.

A 500 pollen grain count was made of each drop by first counting 200 grains and then three consecutive sets of 100 grains. *Lycopodium* spores were counted but their totals were kept separate from the pollen counts. After finishing the first drop's 500 grain count, a cursory examination of the rest of the slide was made.

For each count, the number of taxa was summed and pollen concentration values, percentages, and floral frequency classes were calculated. Pollen concentration values per 1 g of honey were calculated by using the following formula:

$$\frac{\text{(number pollen grains counted)(number spores added)}}{\text{10 g (number of spores counted).}}$$

Frequency classes followed Louveaux et al. (1970, 1978): "predominant" pollen occur in excess of 45%, "secondary" between 16–45%, "important minor" between 3–15%, and "minor" below 3%. One-way analysis of variance (ANOVA) and Tukey's studentized range test were used to test differences in pollen concentration values and number of taxa.

RESULTS

Calculated mean concentration values ranged from 66,496 (Drop 4, 300 grain count) to 131,687 (Drop 3, 300 grain count) (Table 1). Differences in pollen concentration values among drops were not significant, nor were pollen concentration values among different counts within the same drop (ANOVA, $P > 0.05$).

From the 2,500 pollen grains counted, 130 taxa were encountered. Four taxa occurred in only one drop: *Cardiospermum corindum* C. Linnaeus, *Polygonum* sp., *Liquidambar styraciflua* C. Linnaeus, and *Acer* sp. Taxa diversity ranged from 34 (Drop 2, 200 grain count) to 75 (Drop 5, 500 grain count) (Text-Figure 1). No single drop consistently had more or fewer taxa. However, as each pollen count increased, pollen diversity per drop significantly increased (ANOVA $n = 20$, $F = 41.79$, $P = 0.0001$, Tukey's studentized

TABLE 1. Concentration value ranges and means, and number of pollen grains and *Lycopodium* spores counted in 1 g of honey per individual pollen count.

Sample	Pollen counted	Spores counted	Pollen Concentration Values Low	High	Mean
Drop 1	203	12	93,042	98,117	95,579
	308	25	67,760	71,456	69,608
	409	31	72,565	76,523	74,544
	510	33	85,000	89,636	87,318
Drop 1	204	11	102,000	107,564	104,782
	305	16	104,844	110,563	107,703
	406	23	97,087	102,383	99,735
	507	27	103,278	108,911	106,094
Drop 3	203	9	124,056	130,822	127,439
	303	13	128,192	135,185	131,687
	403	20	110,825	116,870	113,848
	511	26	108,096	113,992	111,044
Drop 4	201	14	78,964	83,271	81,118
	306	26	64,731	68,262	66,496
	406	28	79,750	84,100	81,925
	509	32	87,484	92,256	89,870
Drop 5	200	9	122,222	128,889	125,556
	303	16	104,156	109,838	106,997
	408	23	97,565	102,887	100,226
	509	26	107,673	113,546	110,610

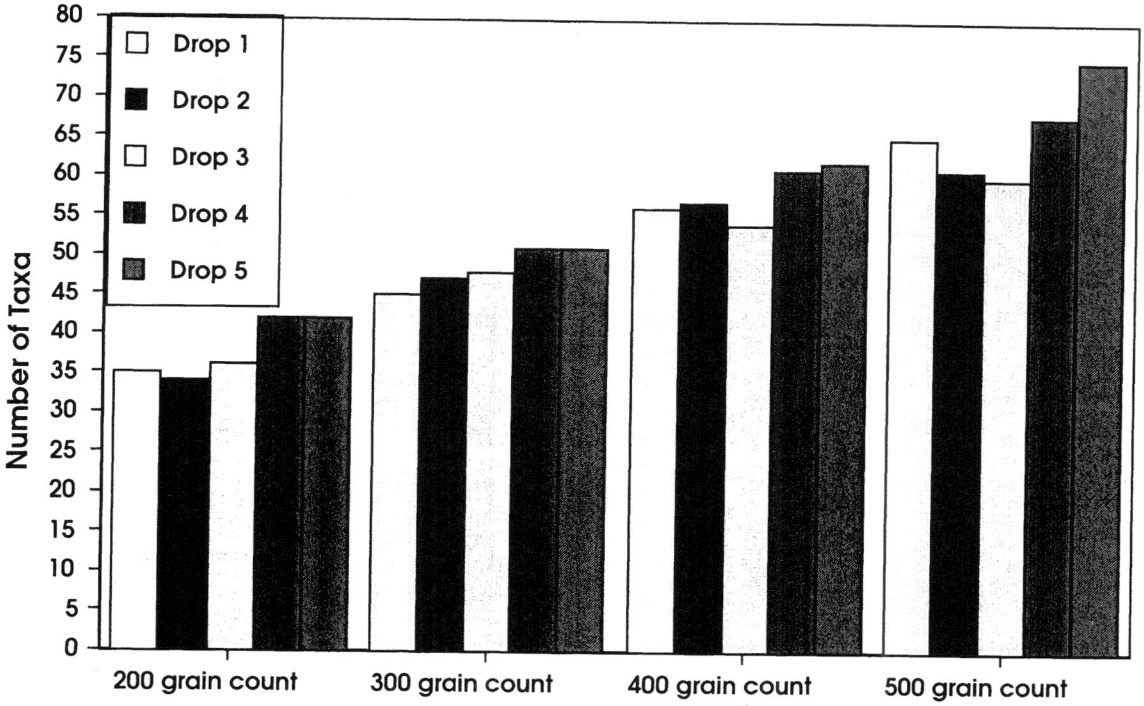

Text-Figure 1. Cumulative total number of taxa per 200, 300, 400 and 500 pollen grain counts within each of the five drops of a honey sample.

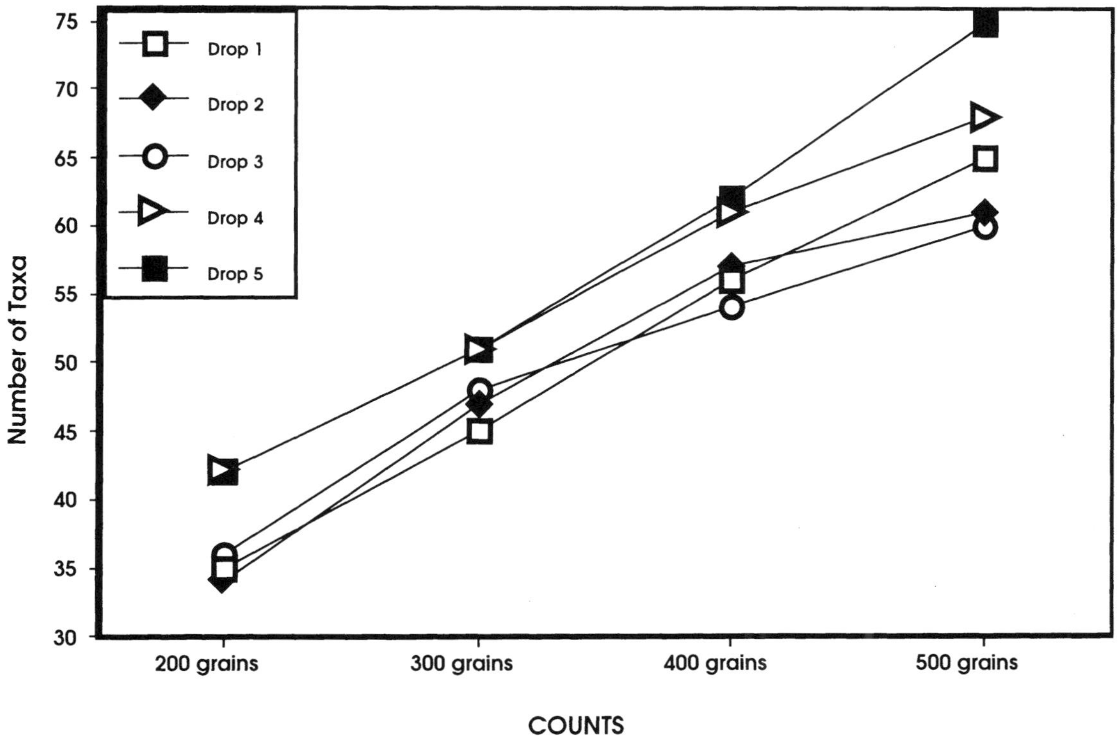

Text-Figure 2. Taxa diversity (number of taxa) per 200, 300, 400 and 500 pollen grain counts for each of the five drops of a honey sample.

range test $P < 0.05$). Drop 5 had significantly more taxa than the others (ANOVA $P = 0.0001$, Tukey's studentized range test). When plotted together, Drops 1 and 5 had the most vertical assent and do not approach an asymptotic number of taxa (Text-Figure 2), indicating that if counts were continued more taxa would be encountered. In the 200 grain counts, pollen diversity ranged from 26 to 33% of the 130 taxa (extrapolated from Text-Figure 2). For drop 5, the 500 grain count accounted for greatest pollen diversity, 60%. No counts accounted for more than 60% of the total taxa encountered in all the drops.

After the counts were made, 19 pollen types not previously encountered cursorial were found in a cursorial examination of Drop 1 These 19 types included large grains such as *Nyssa sylvatica* H. Marshall and relatively small grains like *Phytolacca americana* C. Linnaeus. Of those 19 types, all but two were encountered in later counts.

Within the five different drops, there was no predominant pollen that occurred in excess of 45% of the total pollen count. *Berchemia scandens* (J. Hill) K. Koch was a secondary pollen type in all samples (16–45%), and in all counts except the 300, 400, and 500 grain counts in drop 5. In those counts, it was a important minor type occurring between 3–15% of the total pollen count. *Brunnichia ovata* (T. Walter) L. Shinners was a secondary type (16–45%) in all counts of Drop 3 but an important minor type (3–15%) in all counts of Drops 1 and 5. *Brunnichia ovata* was not encountered in any count of Drops 2 and 4. *Rhamnus caroliniana* Walt. and *Toxicodendron radicans* (C. Linnaeus) K. E. O. Kuntze (sy = *Rhus toxicodendron* L.) were important minor pollen types in all five drops and in all counts.

DISCUSSION

All five drops were similar in terms of pollen concentration values. There were no significant differences in pollen concentration values among the five different drops nor among the different counts of an individual drop (ANOVA $P = > 0.05$, Table 1).

However, different drops from the same sample were not the same in pollen diversity. No pollen count accounted for more than 60% of the total number of taxa. Only in the 500 grain counts was more than 50% of the total number of taxa encountered.

In each drop, as the pollen count increased, pollen diversity significantly increased. Drop 5 had significantly more taxa than the others (ANOVA $P = 0.0001$, Tukey's studentized range test). This difference may be due to the amount of *Berchemia scandens* pollen in the honey sample. In all drops and counts, except Drop 5, *B. scandens* accounted for between 16% and 45% of the pollen types. In Drop 5, it fell below 16% after the first 200 grain count. *Berchemia scandens* pollen is small, about 10 µm in length. As such, it might easily be stirred to the top of the pollen residue and withdrawn in the first couple of drops. Once the majority of it is removed, other taxa would take its place.

Most floral sources of honey are usually determined by the dominance of one pollen grain over another and not the number of taxa within the honey sample. However, the frequency classes of the five drops indicates a possible problem with the using frequency classes to determine floral sources. *Berchemia scandens* and *Brunnichia ovata* were secondary pollen types (16–45%) is some drops and important minors (3–15%) in others. *Brunnichia ovata* was not encountered in any count of Drops 2 and 4. This inconsistency indicates that a 500 grain count may be too low a count to determine a honey's floral sources.

It was expected that an asymptotical number of taxa would be approached by each 500 grain counts; however, this did not occur. Drops 1 and 5 continued increasing in pollen diversity. At what point the addition of new taxa stopped in Drops 1 and 5 was beyond the scope of this project. It is possible that an asymptotical number would be reached in all five drops with a 1,000 or a 1,200 grain count. Vergeron (1964) reached an asymptotical number of taxa by the 500 grain count. However, there were only 30 pollen types in his study. Obviously, when taxa are lumped together into family or type categories, taxa diversity is less than when they are segregated.

Of the 19 grains found in the cursory exam of Drop 1, all but two were encountered in later counts. In other words, in the 2,500 grains counted from the five drops, at least two taxa were missed. This indicates that if a sample is diverse and taxa-rich, a 500 grain count might not be sufficient to sample the diversity in taxa, nor will it be close to the asymptotical number of taxa.

Although the total number of taxa may not be important, when a taxon is restricted geographically it becomes an important indicator of the geographic origin of the honey. If it is not encountered within the sample, determination of the geographic origin of the honey becomes more difficult.

CONCLUSIONS

Pollen concentration values within drops from the same sample are not significantly different. If the main objective is to determine pollen concentration values, one drop of pollen residue is ample. However, pollen analyses of honey is usually used to determine the floral sources and geographic origins of the honey.

If the object is to determine honeybee foraging resources, pollen diversity, a honey's floral sources, or the geographic origin of the honey counting less that 500 grains from one drop of pollen residue is inadequate. Over half the taxa in the sample were missed by counting less than 500 grains. Even after 2,500 pollen grains were counted, two taxa were never included in any of the counts. In addition, a 500 grain count is not adequate for obtaining a good representation of the taxa within a pollen residue that is taxa-rich. It is not known if a 200 grain count would be sufficient for samples that are not taxa rich. Whether or not a 1000 + grain count would account for those taxa missed was beyond the scope of this study. However, in order to obtain a relatively comprehensive list of the taxa and determine the frequency classes within a honey sample, a 500 pollen grain count is the minimum one should use.

References Cited

CRANE, E.
1980 *A book of honey.* Oxford University Press, Oxford, 193 pgs.
ERDTMAN, G.
1963 Palynology. *Advances in Botanical Research,* 1: 149–208.
LIEUX, M.H.
1980 Acetolysis applied to microscopical honey analysis. *Grana,* 19: 95–61.

LOUVEAUX, J., MAURIZIO, A., and VORWOHL, G.
1978 Methods of melissopalynology. *Bee World*, 59: 139–157.
LOW, N.H., SCHWEGER, C., and SPORNS, P.
1989 Precautions in the use of melissopalynology. *Journal of Apicultural Research*, 28: 50–54.
TEW, J.
1989 Honey bees and beekeepers: working together for a bountiful food supply. *Apiculture Extension Service Circular*, U.S. Department of Agriculture.
VERGERON, P.
1964 Interprétation os statistique des résultats en matière d'analyse pollinique des miels. *Annales de l'Abeille*, 7: 349–364.

Authors' addresses

GRETCHEN D. JONES
USDA, ARS, Southern Plains Agricultural Research Center
Areawide Pest Management Research Unit
2771 F&B Road
College Station, TX 77845
U.S.A.
E-mail: g-jones@tamu.edu

VAUGHN M. BRYANT, JR.
Department of Anthropology
Texas A&M University
College Station, TX 77843-4352
U.S.A.
E-mail: vbryant@neo.tamu.edu

MELISSOPALYNOLOGICAL OBSERVATIONS IN A MANGROVE AREA NEXT TO RIO DE JANEIRO, BRAZIL

Cynthia Fernandes P. da Luz and Ortrud Monika Barth

Abstract

Palynological analyses of honeys and pollen loads collected monthly of *Apis mellifera* L., and stored during one year in the region of Barão do Iriri, municipality of Magé, state of Rio de Janeiro, are presented. Beehives were placed on a narrow land belt in front of a mangrove and parallel to the coast side. *Laguncularia racemosa* is the most frequent species in this area. Some individual plants of *Avicennia tomentosa* also occur. Pollen analyses shows that the bees in search of pollen grains visited wild plant species, grasses and crops. The most common polliniferous species belong to the Asteraceae, Euphorbiaceae, Fabaceae, Myrtaceae and Arecaceae. The most important nectariferous species are *Croton sp., Eucalyptus sp., Eupatorium maximillianii, Gochnatia polymorpha, Mimosa bimucronata, Mimosa pudica, Spondias* sp. and species of Sapindaceae. Typical species of the mangrove vegetation, except *Laguncularia racemosa*, were not commonly detected. This fact reflects the actual environment of this region, as well as the value of these species for beekeeping activities.

INTRODUCTION

Mangrove areas in Brazil occupy approximately 25,000 km^2 at the Atlantic Coast, from Cape Orange in the state of Amapá, to Araranguá, state of Santa Catarina (Lacerda, 1984; Carmo et al., 1995). This coastal area has not been previously evaluated for beekeeping potential, especially with regard to *Laguncularia racemosa* (L.) Gaertn. and *Avicennia tomentosa* Jacq., which occur along the Brazilian Coast, as well as in other countries of the world (Silveira, 1937; Thanikaimoni, 1987). The definition of apicultural values and the periods of flowering for native, wild and traditionally cultivated crop species on the Atlantic Coast, including mangrove areas, are important for apicultural activities in the southeast region of Brazil. Evaluation of the apicultural importance of mangrove plant species that furnish nectar and pollen grains for the bees, may be a valuable option for the rational management of this ecosystem, which is currently under a permanent negative impact by man. Most of the mangrove areas around the Guanabara Bay of Rio de Janeiro have been covered with soil. Today only a mangrove belt 60 km^2 wide, extending from the estuaries of the Meriti River to the Itaoca River (FEEMA, 1979), remains (Text-Figure 1).

Honey samples collected in several areas of the state of Rio de Janeiro were previously examined with regard to their pollen spectra by Barth (1970) and Luz and Barth (1992). The purpose of this paper is to evaluate mangrove and associated plant species for harvest by *Apis mellifera* L. with regard to pollen grains and honeys stored inside the beehives.

MATERIAL AND METHODS

The beehives for study are located in a 2 km-wide mangrove belt in the region of Barão de Iriri, municipality of Magé, state of Rio de Janeiro. This region has suffered the impact of human activities since the first period of land colonization in the sixteenth century. Beside grasses, several herbaceous species and pioneer shrubs occur in the region. A large plantation of *Eucalyptus* trees, exists 3 km from the beehives.

Honey Samples

Honey samples were collected monthly from the same beehive; 11 samples were analyzed. During the month of March no samples were collected due to strong rainfall. In general, the honey was centrifuged, except when it was extracted by hand during times of relative lack of food. Thus, the swarm was not damaged.

Slides of honey samples were prepared in accordance to the standard method (Louveaux at al., 1970). Ten grams of honey were dissolved in 20 ml of distilled water. The acetolysis method was not employed for identification of the origins of the honeys in order to preserve important elements such as pollen grain intine, cytoplasma, adsorbed plant oils, pollen grain color, fungi, and algae (Barth, 1989).

Pollen analyses of the honey samples showed the relative frequency of each pollen type, in accordance to Zander (in Maurizio and Louveaux, 1965). For each honey sample 500 pollen grains were counted. Identification of plant species, genera, or families by pollen grain morphology was based on the pollen types described in Barth (1989). Interpretation of the results was based on dominant pollen types, considering the super- and sub-representation of plant species (Barth, 1989).

Pollen Loads

Samples collected monthly and aleatorichly of standard pollen loads stored inside the hive reach a total of 5 g each month (ca. 20 loads). We analyzed nine samples corresponding to nine months. Pollen loads were dissolved and homogenized in glycerin/water (1:1), centrifuged, and mounted on microscope glass slides with glycerin jelly. Pollen types were identified on five pollen slides for each sample, making a total of 2,000 grains counted for percentage calculation. Slides were examined until no new pollen type was detected.

Luz, Cynthia Fernandes P. da, and Barth, Ortrud Monika
2001 Melissopalynological observations in a mangrove area next to Rio de Janiero, Brazil; *In*: Goodman, D.K., and Clarke, R.T. (eds.), *Proceedings of the IX International Palynological Congress, Houston, Texas, U.S.A., 1996*; American Association of Stratigraphic Palynologists Foundation, p. 489–492.
 ISBN 0–931871–06–9

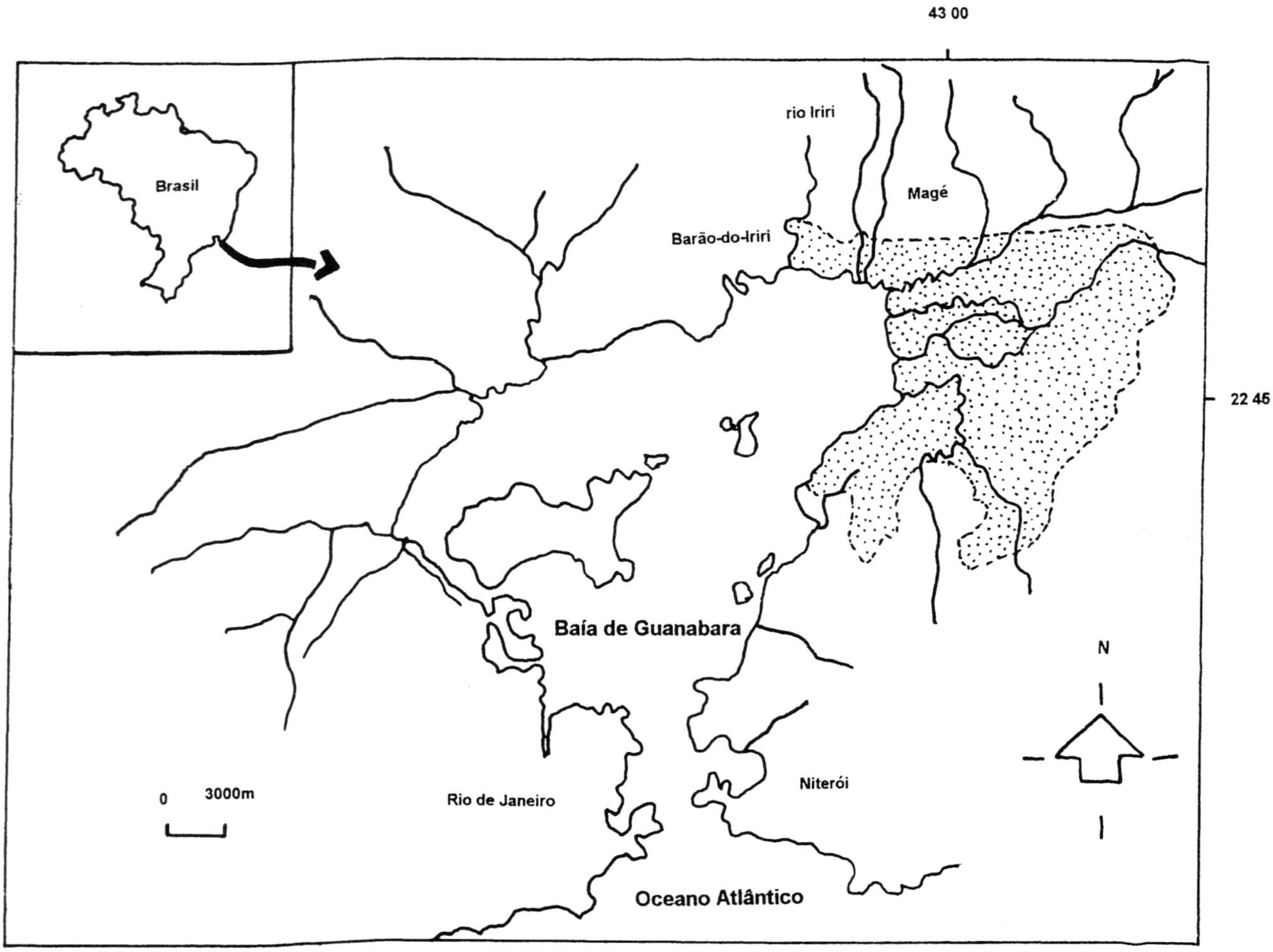

Text-Figure 1. Mangrove area (stippled area) along the Baía de Guanabara; beehives are located at Barão do Iriri.

RESULTS AND DISCUSSION

Plant Species of Importance for Beekeeping

The most frequently visited plant families by the bees for nectar as well as for the pollen loads (Table 1) were the Fabaceae and Asteraceae.

The pollen type *Mimosa scabrella* Benth. (including several species, as *M. pudica* L.), Fabaceae–Mimosoideae, was present during seven months in the honey samples and was the dominant pollen type (more than 45%) in January, February, June and December, in association to the flowering periods. *Mimosa bimucronata (DC.)* O. Ktze. was detected in the honey samples during April, May, June and September; *M. bimucronata* had its highest frequency in May and June. Pollen grains of *Albizzia lebbeck* (L.) Benth. were observed in honey samples during the months of January, February, October and December. In relation to the pollen loads, *Mimosa scabrella* appeared only in January (29.4% of the total pollen amount). *Mimosa bimucronata* was found as a pollen load in April (3.9%). *Piptadenia communis* Benth. (2.1%) occurred in October and November and *Albizzia lebbeck* in October (6.2%).

Nectar from Asteraceae was obtained throughout most of the year. *Gochnatia* sp. is an important type in February and December, and *Eupatorium* pollen is important in April and December. Pollen loads of species of Asteraceae also occured throughout the year, especially during the mid-year period from June to August.

Among the other plant species in the samples examined for their importance in apicultural activities, the following plant taxa were the most representative. Major nectar quantities from *Eucalyptus* were obtained in April. *Croton* sp. ("morrão-de-candeia"), a wild plant species, was visited for nectar in January, February, August, November and December, with a peak in November. Palm species contributed to honey production in April, May, July, August, October and December, with a peak in May. *Spondias* sp. ("cajá") (Anacardiaceae), a native tree of the Atlantic forest that is cultivated near the honeycombs, offered nectar in April and May. Sapindaceae pollen was detected only in honey obtained in November. *Laguncularia racemosa* ("mangue-roxo") offered nectar in April and its pollen grains occur in the samples of April and October.

Qualitative analyses of the honey samples showed the presence of various pollen types characteristic of the state of Rio de Janeiro (Barth, 1970): *Borreria* sp., *Croton* sp., *Cuphea* sp., *Eucalyptus* sp., *Mimosa pudica*, *Hyptis* sp., *Sida* sp., *Trema micrantha* (L.) Blume, *Triumfetta* sp.

TABLE 1. The most important plant species and pollen types detected in honey and pollen load samples during one year of observations.

Plant Species & Pollen Types	Month										
	Jan	Feb	Apr	May	June	July	Aug	Sep	Oct	Nov	Dec
Albizzia lebbeck	X	X							X		X
Croton	X	X					X			X	X
Eupatorium			X								X
Gochnatia		X									X
Laguncularia racemosa			X						X		
Mimosa bimucronata			X	X	X			X			
Mimosa pudica	X	X			X						X
Palmae			X	X		X	X		X		X
Piptadenia communis									X	X	
Sapindaceae										X	
Spondias			X	X							

Nectar and Honeys

Nectar production in the region of Barão de Iriri, occurs from November to June. Nectar is sparse between July and October, and consequently beekeepers must move their honeycombs to other regions that offer better conditions for apicultural activities. Pollen grains from *Mimosa bimucronata*, *Laguncularia racemosa*, Cyperaceae, and Onagraceae are characteristic of the coast or flooded areas and were well-represented in the samples. Periods of nectar secretion provided the following honey types for the region: (1) from December to February there is abundant nectar from "cambará" (*Gochnatia velutina* (Bong.) Cabrera) and "malícia roxa" (*Mimosa pudica*); (2) during April the honey produced was a combination of "salseira" (*Eupatorium maximillianii Schrad.*) and "cajá" (*Spondias* sp.); (3) from May to June a combination of "malícia roxa" and "maricá" (*Mimosa bimucronata*); (4) in July the honey was composed of nectar from species of Asteraceae (*Eupatorium* sp. and *Vernonia* sp.); (5) in August the honey was possibly derived from former loads of "malícia roxa" and "cambará," since these taxa were not observed in flower at this time; (6) in September the honey was obtained from "eucalipto" and "maricá;" (7) in October most commonly from *Myrcia* sp. and *Vernonia* sp.; and (8) in November it comes from "astrapéia rósea" (*Dombeya* sp), "morrão-de-candeia" (*Croton* sp.) and Sapindaceae.

Pollen Loads

During the year, the flowers that produced abundant pollen in the region of Barão do Iriri were *Borreria* sp., *Crotalaria pallida* Ait., *Cuphea* sp., *Laguncularia racemosa*, *Mimosa bimucronata*, *M.pudica*, *Ricinus communis* L., *Triumfetta* sp., Asteraceae, Cyperaceae, Poaceae, Onagraceae, Sapindaceae and several Fabaceae. September and October were the months that presented the greatest diversity of pollen types in the pollen loads.

It was noticed that the bees traveled great distances in search of food, because pollen grains of *Eucalyptus* were found in the honey samples. Those trees grow on a plantation located approximately 3 km from the hives. According to Bastos (1995), bees travel about 10 km in search of nectar and pollen grains from *Eucalyptus*.

Laguncularia racemosa, a species with a sweet perfume, was intensely visited by *Apis*, for nectar and pollen. From field observations, bees stayed about 5 seconds on each flower for nectar collection. As this taxon has a relative long flowering period and its pollen was found in a number of honey samples and pollen loads, this species is of great importance for apiculture in Brazil.

The pollen grains of *Avicennia tomentosa*, another mangrove plant species, were not detected in the honey samples.

ACKNOWLEDGEMENTS

We are grateful to beekeeper Mr. José Lindomar P. de Souza for poviding us with the apiculture material for studies; to Dr. Sérgio Augusto de Miranda Chaves for field excursions. Financial support: Conselho Nacional de Desenvolvimento Científico e Tecnológico (CNPq) and Fundação de Amparo à Pesquisa do Estado do Rio de Janeiro (FAPERJ).

References Cited

BARTH, O.M.

1970 Análise microscópica de algumas amostras de mel. 4. Espectro polínico de algumas amostras de mel do Estado do Rio de Janeiro. *Revista Brasileira de Biologia*, 30: 575–582.

1989 *O Pólem no Mel Brasileiro*. Gráfica Luxor, Rio de Janeiro, Brasil, 150 pgs.

BASTOS, E.M.

1995 Espectro polínico do mel produzido em algumas áreas antrópicas de Minas Gerais. *Revista Brasileira de Biologia*, 55: 789–799.

CARMO, T.M.S., BRITO-ABAUSSE, M.G., MELO, R.M.S., ZANOTTI-XAVIER, S., COSTA, M.B., and HORTA, M.M.M.

1995 Os manguezais da baía norte de Vitória, Espírito Santo: um ecossistema ameaçado. *Revista Brasileira de Biologia*, 55: 801–818.

FEEMA - FUNDAÇÃO ESTADUAL DE ENGENHARIA DO MEIO AMBIENTE
1979 Os manguezais do recôncavo da Baía de Guanabara. Rio de Janeiro, Dicomt (cadernos FEEMA - série técnica, 10/79): 1–60.

LACERDA, L.D., RESENDE, C.E., JOSÉ, D.M.V., and F., M.C.F.
1984 Metallic composition of mangrove leaves from the southeastern Brazilian coast. *Revista Brasileira de Biologia*, 46: 395–400.

LOUVEAUX, J., MAURIZIO, A. and VORWOHL, G.
1970 Les méthodes de la Melisso-Palynologie. *Apidologie*, 1: 211–277.

LUZ, C.F.P., and BARTH, O.M.
1992 Identificação de tipos polínicos em amostras de mel e bolotas de pólen de manguezal na região de Barão de Iriri, Magé, Rio de Janeiro. *XLIII Congresso Nacional de Botânica, Aracaju, Sergipe,* Resumos: 38, (abstract).

MAURIZIO, A., and LOUVEAUX, J.
1965 Pollen de Plantes Mellifères d'Europe. Union des Groupements Apicoles Français, Paris, 148 pgs.

SILVEIRA, F.
1937 Mangrove. *Rodriguésia,* 10: 131–153.

THANIKAIMONI, G.
1987 Mangrove Palynology. UNDP/UNESCO - Regional Project on Training and Research on Mangrove Ecosystems. Institut Français de Pondicherry, 100 pgs.

Authors' addresses

CYNTHIA FERNANDES P. DA LUZ
Laboratório de Palinologia
Departamento de Botânica
Instituto de Biologia
UFRJ, Rio de Janeiro, Brazil

ORTRUD MONIKA BARTH
Laboratório de Palinologia
Departamento de Botânica
Instituto de Biologia
UFRJ, Rio de Janeiro
Brazil
and
Instituto Oswaldo Cruz
FIOCRUZ
Caixa postal 926
Rio de Janeiro
Brazil

PALYNOLOGICAL AND GEOGRAPHICAL CHARACTERIZATION OF UNIFLORAL HONEYS FROM *LOTUS CORNICULATUS* OF URUGUAY

Gloria Daners

Abstract

In a previous examination of 95 honeys from Uruguay, 30% were classified as unifloral from *Lotus corniculatus*. These honeys represent Uruguay's main unifloral honey production and exhibit properties highly appreciated by consumers, separating them from *Lotus* honey from other regions is important. The goal of this study was to palynologically and geographically characterize *L. corniculatus* honey produced in Uruguay. Twenty-nine unifloral honeys from *L. corniculatus* from different beekeeping regions of Uruguay were collected. A total of 89 pollen types were determined, 47% representing native taxa and 53% introduced ones. *Lotus corniculatus* pollen grains in the samples ranged from 45 to 93%. *Eucalyptus* spp., *Trifolium repens*, *Echium plantagineum* and *Baccharis* spp. were found in the secondary pollen class, the latter being the only native taxa. In most samples, pollen from native taxa were classified as minor or present. The pollen spectra revealed a typical association of introduced species in the secondary or minor important classes, and some native taxa in the minor or present. *Trifolium repens*, *T. pratense*, *Eucalyptus* spp. and *Echium plantagineum*, together with *L. corniculatus*, appeared in 100% of the samples. The best represented native taxa were *Eryngium* spp., *Schinus* sp., *Baccharis* spp., *Salix humboldtiana*, *Celtis* sp. and *Phytolacca dioica*. It is proposed that this association could differentiate Uruguay's *Lotus corniculatus* honey from that of other regions. Several features that separate these honeys from those produced in Argentina, Brazil, and Chile are discussed.

INTRODUCTION

Uruguay is located in the Pampean Phytogeographical Province, where the native vegetation is represented by a graminaceous steppe associated to riparian and sierra forests (Cabrera and Willink, 1980). Today, this vegetation type accounts for only 3.5% of the country's surface, and has been mainly substituted by agricultural and forestry areas.

In a previous study to determine the botanical origin of 95 honey samples from Uruguay, 30% were classified as unifloral from *Lotus corniculatus* L. This agrees with the extension of artificial prairies in the country, and with high productivity and relaxed beekeeping management in these botanical communities.

Due to the high proportion of unifloral *Lotus* honeys in Uruguay, and since *Lotus* honeys are also produced in other countries, the main goal of this study was to palynologically and geographically characterize *L. corniculatus* honey from Uruguay. The aim is to establish a base for future botanical, physic-chemical and sensorial classification. This is important because 80% of Uruguay's honey is exported as a standard product. *Lotus corniculatus* honeys exhibit properties preferred by consumers (light color, soft taste). Thus characterization of its pollen spectra, and geographical markers to differentiate it from *Lotus* honeys of other regions, is of primary importance.

Previous references about pollen content of Uruguayan honeys are scarce (Lemarquant, 1982; Daners 1996; Daners and Campá, 1996; Bazzurro et al., 1996). Regional melissopalynological studies developed in Brazil (Barth 1989, 1990; Ramalho et al. 1990, 1991), in Argentina (Tellería 1988, 1992, 1993, 1995, 1996; Gurini and Basilio 1995), and in Chile (Battaglini and Ricciardelli D'Albore 1972), represent the basis for geographical differentiation of the samples analysed in this study.

Daners, Gloria
2001 Palynological and geographical characterization of unifloral honeys from *Lotus corniculatus* of Uruguay; *In*: Goodman, D.K., and Clarke, R.T. (eds.), *Proceedings of the IX International Palynological Congress, Houston, Texas, U.S.A., 1996*; American Association of Stratigraphic Palynologists Foundation, p. 493–500.
 ISBN 0–931871–06–9

MATERIAL AND METHODS

Twenty-nine unifloral honey samples from *Lotus corniculatus* were collected by beekeepers during 1995 in different regions of Uruguay (Text-Figure 1). They were qualitatively analysed following the methodology recommended by the International Commission for Bee Botany (Louveaux et al., 1978), with acetolyzation (Erdtman, 1960). Morphological types were determined using a regional pollen reference collection and pollen atlases (Jones et al., 1995; Markgraf and D'Antoni, 1978; Heusser, 1971; Pire et al., 1992; Roubik and Moreno, 1991; Tellería, 1995). To determine relative frequency of the pollen types, at least 500 grains were counted per sample. An Olympus BH-2 microscope was used for counting and determining the pollen grains.

Pollen spectra of the honeys were analysed using class frequencies (Louveaux et al., 1970): dominant (> 45% of the total grains counted), secondary (16 to 45%), minor important (3 to 15%) and minor (< 3%). In searching for geographical markers, presence or absence of taxa can be more meaningful than their grain count (Vergeron, 1964). Geographical markers are usually observed in small percentages (Jones and Bryant, 1996). Since one objective was to identify geographical markers, a pollen class frequency of those morphological types found in the samples but not counted among the 500 grains, was also included (present).

Polliniferous taxa of Poaceae (83%), *Plantago* sp. (52%), *Ambrosia tenuifolia* Spreng. (31%), Cyperaceae (10%), Chenopodiaceae–Amaranthaceae (10%), *Zea mays* L. (10%), *Casuarina cunninghamiana* Miq. (7%), Moraceae–Urticaceae (3%), *Platanus occidentalis* L. (3%), Juglandaceae (3%) and *Acacia caven* (Mol.) Mol. (3%) were considered contaminants of the honey samples and were not taken into account for the analysis of the pollen spectra. Honeydew elements were scarce in all the samples.

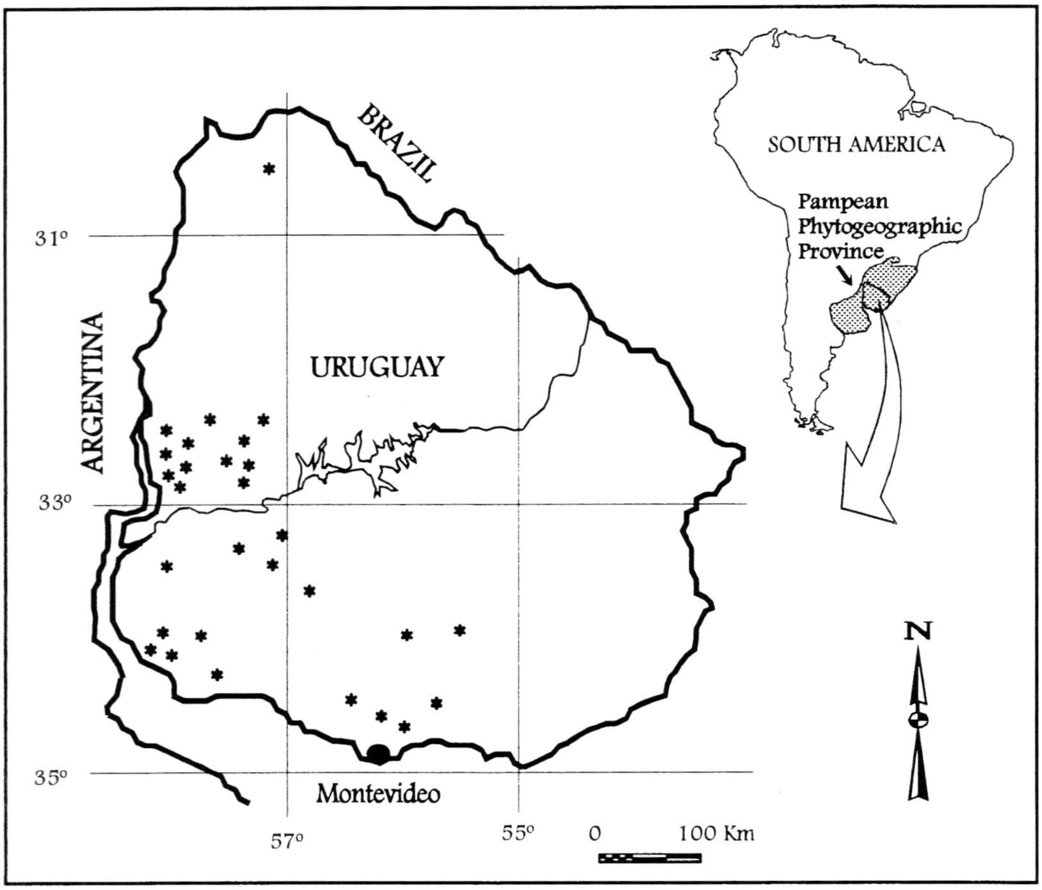

Text-Figure 1. Map of Uruguay showing its location on the Pampean Phytogeographical Province of South America and the sites from where the unifloral *Lotus corniculatus* honeys were obtained.

RESULTS

In 29 honeys from *Lotus corniculatus* from different beekeeping regions of Uruguay, 89 pollen types were determined; 10 to family, 1 to tribe, 26 to genus and 52 to species level. Forty-seven percent represented native taxa, while 53% were introduced.

The pollen of *L. corniculatus* ranged between 45.2 and 93.0%. In addition, only *Eucalyptus* spp. was found in the dominant frequency class (1 sample). *Eucalyptus* spp., *Trifolium repens* L., *Echium plantagineum* L. and *Baccharis* spp. appeared as secondary, the last being the only native genera (Text-Figure 2). In most cases, pollen from native taxa was found as minor or present.

Sample composition exhibited large heterogeneity because the number of morphological types ranged between 10 and 42. However, a high homogeneity in taxa that dominated the spectra was found (Text-Figure 3). Fabaceae and Asteraceae contained the highest morphological diversity of pollen types.

Frequency of pollen types present in percentages higher than 25% are shown in Text-Figure 4. *Trifolium repens, T. pratense* L., *Eucalyptus* spp. and *Echium plantagineum*, together with *Lotus corniculatus*, were observed in all the samples. The most important native taxa were *Baccharis* spp. (84%), *Eryngium* spp. (83%), *Schinus* sp. (83%), *Scutia buxifolia* Reiss. (76%), *Salix humboldtiana* Willd. (66%), *Celtis* sp. (41%) and *Phytolacca dioica* L. (38%).

DISCUSSION

The main features of the *Lotus corniculatus* unifloral honeys of Uruguay are: richness of morphological types, abundance of Fabaceae and Asteraceae, high percentages of *Echium plantagineum, Trifolium repens* and *Eucalyptus* spp., and high frequency of some native taxa. These characteristics indicate that in Uruguay the mean surface area of the *L. corniculatus* fields is smaller than the mean flying radius of *Apis mellifera* L., thus being attracted by native species of surrounding riparian forests.

Regionally, these general features are similar to honeys produced in the pampean region of Buenos Aires (Tellería, 1988, 1992, 1996) and in the Delta of the Paraná River (Gurini and Basilio, 1995). This is expected due to the similarity of dominant vegetation. Nevertheless, the pollen spectra differentiate from a palynological and a geographical standpoint, the honeys.

PALYNOLOGICAL CHARACTERIZATION

Analysis of taxa distribution and the frequency of these unifloral honeys, reveal the presence of a typical

POLLEN TYPE	1	2	3	4	5	6	7	8	9	10	11	12	13	14	15	16	17	18	19	20	21	22	23	24	25	26	27	28	29
Acacia caven																											M		
Acacia sp.																							M					M	M
Acicarpha tribuloides						M								M			p			p		M							
Adesmia bicolor										M				M				M				p							p
Allophyllus edulis																				p									
Alternanthera sp.							M								M													p	
Ambrosia tenuifolia					p			p						M				M			M		M			M	p	M	
Ammi visnaga	M	M	M	M	M	M			M		p	M	p		M	M	M	M	M	M	M	M	M		M	M	M	M	M
Anthemis cotula									M			p																	
Arctium minus								M																					
Baccharis spp.	M	M	M	M	M		M	M	M	M	M	M	M	M	M				M	M	M	M	p	p	p	M	M	S	I
Brassicaceae						M	M		M	M	M								M	I	M				M		M	M	
Casuarina cunninghamiana																											M	p	
Carduus sp./*Cynara* sp.	M	M	M		p							M					M		M		M		M		M	M	M	M	
Caryophyllaceae					p																								
Cedrela cf. *C. lilloi*																									p				
Celtis sp.		M		M	M				M	M	M	M		p				p				M	M				p		
Centaurea sp.								M			M	M	M					M					M		M	M	M	M	
Chenopod.-Amaranthaceae					p																						p	p	
Chorisia speciosa																							M						
Cirsium vulgare	M					M	M									M					M	M			M	M	M		
Citrus sp.																		M					I	M				M	
Cichorium intybus					M	M			M		M				M	M					M		M		M	p	M	M	
Colletia sp.					M				M									M											
Convolvulus arvensis												p																	
Cuphea glutinosa																							p						
Cyperaceae											p	M											M						
Daucus carota											M										M	M							
Dodonaea viscosa											M																		M
Echinodorus sp.					M	M												M									M	M	
Echium plantagineum	I	I	M	M	M	I	M	M	I	I	S	I	M	M	M	I	M	M	M	I	I	I	I	I	I	I	S	I	I
Eryngium spp.	M	M	I	I	M	M			M	M	M	M	p	M	M	M		M	M	M	M	M			p	p	M	M	I
Erythrina crista-galli						M																						M	
Eucalyptus spp.	S	S	S	D	I	I	S	I	I	S	S	I	S	S	I	I	M	I	M	I	M	I	I	I	I	M	I	I	I
Foeniculum vulgare				p			M														p					M			
Gerardia communis					M		M												p		M					M	M		
Helianthus annuus	M					M				M		M			M	p					M	M				M	M		M
Hyptis sp.																	M	M											
Ilex paraguariensis																	I												
Iodina rhombifolia									M																				
Jacaranda mimosifolia																				M									
Juglandaceae																	M												
Ligustrum spp.																				M	M				p				
Liliaceae-Iridaceae					p						M	M						M										M	
Lomatia sp.									p																				

- D: dominant pollen (> 45 %)
- S: secondary pollen (16 - 45 %)
- I: minor important pollen (3 - 15 %)
- M: minor pollen (< 3 %)
- p: present pollen

Text-Figure 2. Distribution of pollen types found in each frequency class for 29 samples of unifloral *Lotus corniculatus* honeys of Uruguay.

POLLEN TYPE	1	2	3	4	5	6	7	8	9	10	11	12	13	14	15	16	17	18	19	20	21	22	23	24	25	26	27	28	29
Lonicera japonica																											M	p	M
Lotus corniculatus	D	D	D	D	D	D	D	D	D	D	D	D	D	D	D	D	D	D	D	D	D	D	D	D	D	D	D	D	D
Ludwigia uruguayensis												M																	
Malpighiaceae						M					p							M											
Matricaria chamomilla	M				M		M										p	M		M	M	M			M			M	M
Medicago sativa																							p						
Melia azedarach																				p									
Mentha pulegium					M		M		M		M	p						M					M				M	M	
Mimosa cf. *M. pudica*														M															
Moraceae-Urticaceae																				M									
Myoporum laetum			p						M																	M	M	M	
Parkinsonia aculeata						M																M							
Phyla canescens																		M				M	M						
Phytolacca dioica		p				M							M	M			p	p	M			M	M			p		M	
t. *Picris* sp.																									p	p			
Plantago sp.		M					M	M	M	M	M	M			M	M	M						p		M	M	M	M	
Platanus occidentalis																													P
Poaceae		M	M	M		M	M		p	p	p	M	M	M	M		p	M	p	M	M	M	M		M	M	M	M	p
Pontederia sp.																				M							M		
Prosopis sp.																		M		M									
Proteaceae																		M	M	M	M						p	M	
Rubus sp.													p																
Sagittaria montevidensis																											M	M	
Salix humboldtiana	M	M	M	M	M	M			M	M	M		p					M	M	M	M	M		M		M	M		M
Salvia sp.																		p										p	
Sapium montevidense					M														M						M			M	
Scutia buxifolia	I	M		M	M	M		M	I	M	M	M	I	p	M		M	M	M	I		M		M			p	M	M
Schinus sp.	M	M		M	M	p		M	M	M	p	M	M	M			M	M	M	I	M	M		I	p	M	M	M	M
Sebastiania brasiliensis																		M							M		M	p	
Sebastiania klotzschiana					p																	p						M	
Senecio sp.	p			M		M			M									M	M	M		M			M			M	p
Solanum spp.									p																				
Spartium junceum		M																											
Struthanthus acuminatus						M								M															
Tamarix sp.																				p								M	M
Taraxacum officinale																		p											
Terminalia australis										M										M								p	
Trifolium pratense	p	M	M	M	I	M	M	I	I	M	M	M	M	M	M	I	M	M	M	I	I	M	I	M	I	M	I	I	I
Trifolium repens	I	M	I	M	I	M	M	S	I	I	S	I	I	I	M	M	M	M	M	I	I	M	S	I	S	I	I	I	I
Verbena sp.					M							p			M			M							p	p	M	M	
Vernonieae				M													M	M											
Wahlenbergia linarioides												M						M											
Xanthium sp.												p									p							M	
Zea mays		p				p																						M	

D	dominant pollen (> 45 %)	I	minor important pollen (3 - 15 %)	p	present pollen
S	secondary pollen (16 - 45 %)	M	minor pollen (< 3 %)		

Text-Figure 2 (continued). Distribution of pollen types found in each frequency class for 29 samples of unifloral *Lotus corniculatus* honeys of Uruguay.

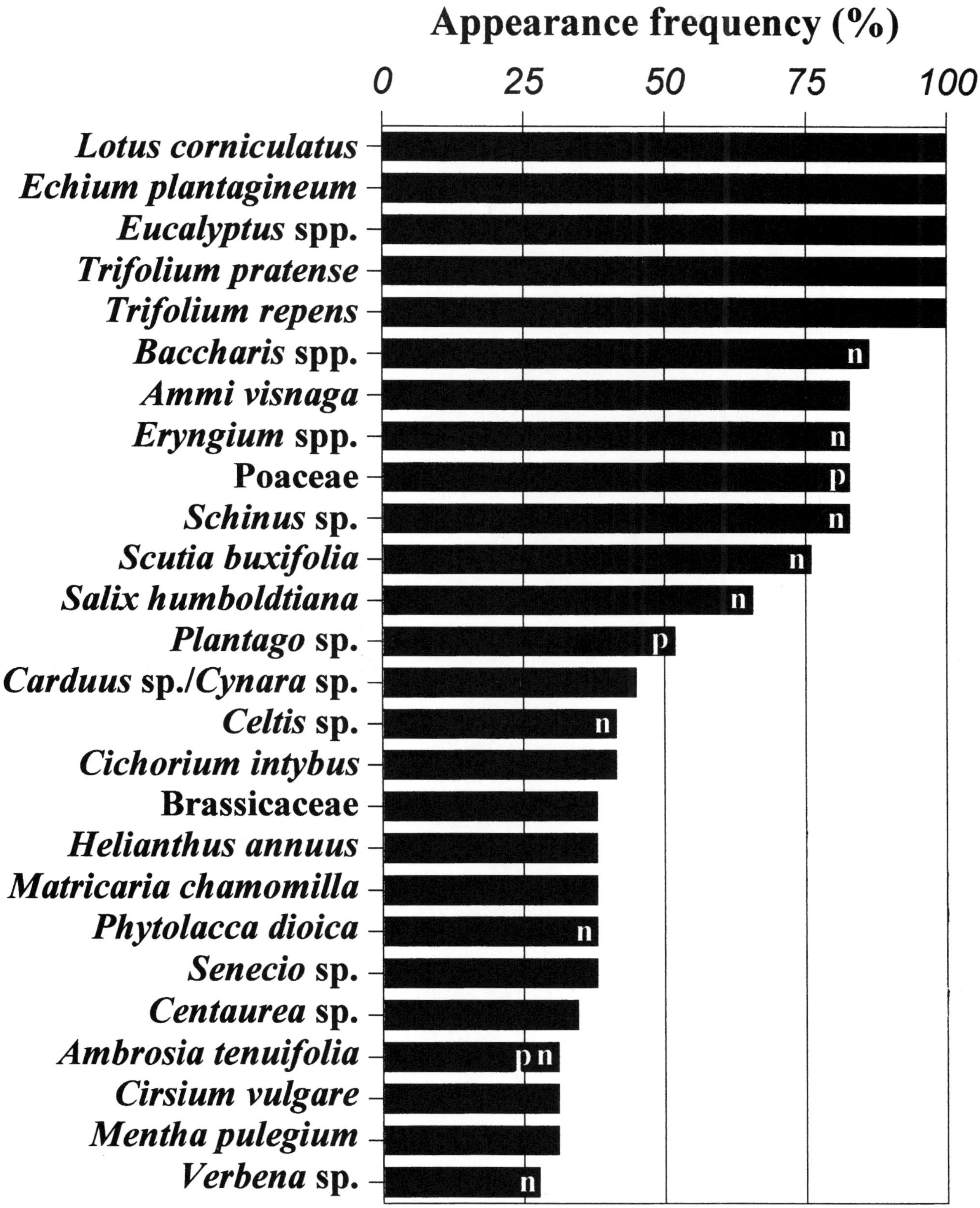

Text-Figue 3. Percentage distribution per sample of the 5 taxa that dominate the pollen spectra of 29 unifloral *Lotus corniculatus* honeys of Uruguay (all other taxa found in each sample are represented in "other taxa," including the 5 dominant ones when appearing as present or minor pollen).

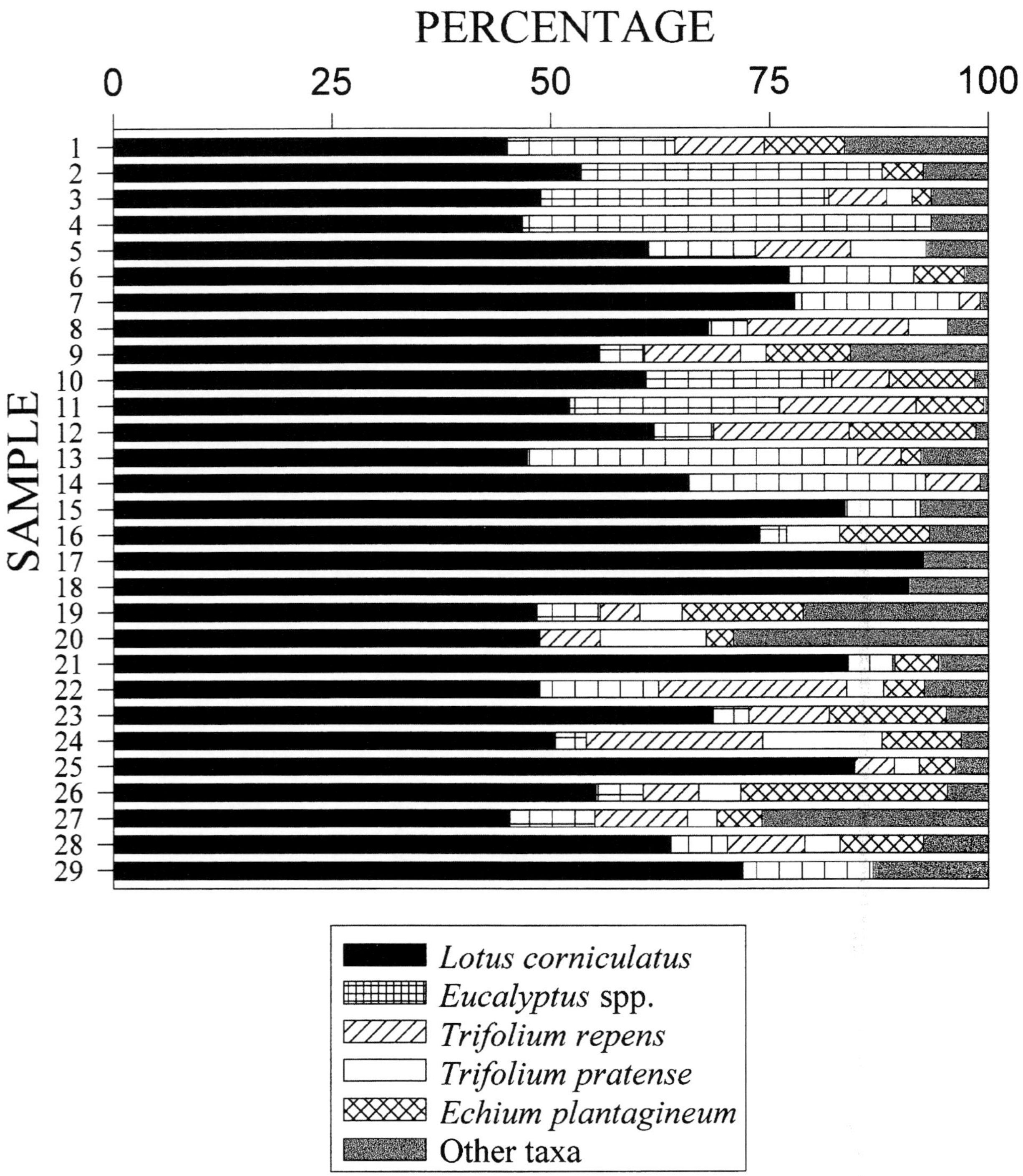

Text-Figure 4. Pollen types with appearance frequency higher than 25 % in 29 unifloral *Lotus corniculatus* honeys of Uruguay (n = native; p = polliniferous).

association composed of introduced and native taxa. The main characteristic of this association is that introduced species were in the secondary and minor important classes, while native species were in the minor and present classes. Introduced taxa *Lotus corniculatus*, *Trifolium repens*, *T. pratense*, *Eucalyptus* spp. and *Echium plantagineum* were found in 100% of the samples, representing an average of 91.8% of the total pollen content, thus dominating the pollen spectra in all the cases (Text-Figure 4). Native taxa *Eryngium* spp., *Scutia buxifolia*, *Schinus* sp., *Baccharis* spp., *Salix humboldtiana*, *Celtis* sp. and *Phytolacca dioica*, mainly tree species, occurred as a whole group or as several subgroups. Although in low percentages, they occurred in high frequency as it is expected for geographical markers (Vergeron, 1964).

GEOGRAPHICAL CHARACTERIZATION

Presence of *Trifolium repens*, *T. pratense*, *Eucalyptus* spp. and *Echium plantagineum* in these *Lotus* honeys, associated to several native taxa (*Eryngium* spp., *Scutia buxifolia*, *Schinus* sp., *Baccharis* spp., *Salix humboldtiana*, *Celtis* sp. and *Phytolacca dioica*), could characterize the honeys geographically. This coincides with Daners (1996), who pointed out *Eryngium* spp., *Scutia buxifolia*, *Schinus* sp., *Baccharis* spp. and *Salix humboldtiana* as potential geographical markers of honeys from Uruguay. Although these taxa are widely distributed, it could be possible to differentiate the *Lotus* honeys produced in Uruguay from those obtained regionally. Although *Lotus* is a taxon of beekeeping importance in the pampean region (Tellería, 1993, 1995), references of unifloral *Lotus* honeys are scarce. Argentinean *Lotus* unifloral honeys (Tellería, 1988, 1992) contained high frequencies of the same introduced taxa as found in Uruguayan honeys. Nevertheless, except for *Celtis* sp., the more frequent native species observed in the unifloral *Lotus* honeys from Uruguay are absent in those in Argentina. Species such as *Centaurea* sp., *Mentha pulegium* and Brassicaceae were more important in Argentinean honeys than in the Uruguayan honeys.

In *Lotus* honeys produced in Chile, Battaglini and Ricciardelli D'Albore (1972) highlight the importance of *Eucryphia* sp. and other characteristic taxa of high latitudes of Los Andes Cordillera. The absence of these taxa in the *Lotus* honeys produced in Uruguay easily distinguishes the two. No references of unifloral *Lotus* honeys from Brazil were found. In addition, no reference to *Lotus* as a bee plant was made in important melliferous plants of Brazil (Ramalho et al., 1990).

Sixty percent of the samples in this study showed percentages of *Lotus corniculatus* pollen higher than 60%, suggesting that this species could be overrepresented. Further studies on floral biology, nectar production and concentration of pollen in nectar of *Lotus corniculatus* are necessary. A quantitative research of the pollen richness in these honeys is required to determine if 45% is the minimum percentage required to consider the honeys from *Lotus* as unifloral.

AKNOWLEDGMENTS

I thank Angeles Beri and Daniel Conde for continuous support and scientific advise during this research, to Pablo Campá for invaluable help in the processing of the samples, and the Botanical Garden of Montevideo for the material for the reference pollen collection. I also thank the referees for helpful comments. This study was financed by the Sectorial Commission for Scientific Research of the University of Uruguay.

References Cited

BARTH, O.M.
1989 *O pólen no mel brasileiro*. Rio de Janeiro, Gráfica Luxor, 150 pgs.
1990 Pollen in monofloral honeys from Brazil. *Journal of Apicultural Research*, 29: 89–94.

BATTAGLINI, M., and RICCIARDELLI D'ALBORE, G.C.
1972 Différentiations entre les miels étrangers sur la base du spectre pollinique. *Simposium International Probleme de flore mellifere et pollinisation, Torino*, 1972. Apimondia, Bucarest, pp. 68–82.

BAZZURRO, D., DÍAZ, R., and SÁNCHEZ, M.
1996 Estudio sobre posibilidades de producción de miel de palma (*Butia capitata*). Actas *V Congreso Ibero Latinoamericano de Apicultura*, Mercedes, Uruguay, 1996: 162–170.

CABRERA, A.L., and WILLINK, A.
1980 *Biogeografía de América Latina*. Serie de Biología, Monografía N° 13, OEA. Chesneau, E. (de.). 122 pgs.

DANERS, G.
1996 Determinación de la flora de importancia melífera de Uruguay mediante el análisis polínico de la miel. *Actas V Congreso Ibero Latinoamericano de Apicultura*, Mercedes, Uruguay, 1996: 184–186.

DANERS, G., and CAMPÁ, P.
1996 Evidencia palinológica de la existencia de miel monofloral de coronilla (*Scutia buxifolia*) en Uruguay. *Actas V Congreso Ibero Latinoamericano de Apicultura*, Mercedes, Uruguay, 1996: 171–173.

ERDTMAN, G.
1960 The acetolysis method. *Svensk Botanisk Tidskrift Utgifuen af Svenska Botaniska föreningen*, 54: 561–564.

GURINI, L.B., and BASILIO, A.
1995 Flora apícola en el Delta del Paraná. *Darwiniana*, 33: 337–346.

HEUSSER, C.
1971 *Pollen and Spores of Chile*. The University of Arizona Press, Tucson, Arizona. 117 pgs.

JONES, G.D., and BRYANT, V.M., Jr.
1996 Melissopalynology. *In*: Jansonius, J., and McGregor, D.C. (eds.), *Palynology: Principles and Applications*, American Association of Stratigraphic Palynologists Foundation. Vol. 3, pp. 933–938.

JONES, G.D., BRYANT, V.M., Jr., LIEUX, M.H., JONES, S.D., and LINGREN, P.D.
1995 *Pollen of the Southeastern United States: with Emphasis on Melissopalynology and Entomology*. American Association of Stratigraphic Palynology, Dallas, TX, Contributions Series Number 30, 76 pgs.

LEMARQUANT, M.
1982 *Origen botánico de la miel para el Uruguay*. Tesis Ingeniero Agrónomo, Montevideo, Uruguay. Facultad de Agronomía, 66 pgs.

LOUVEAUX, J, MAURIZIO, A., and VORWOHL, G.
1970 Methods of Melissopalynology. *Bee World*, 51: 125–131.

LOUVEAUX, M., MAURIZIO, A., and VORWOHL, G.
1978 Methods of Melissopalynology by International Commission of Bee Botany of IUBS. *Bee World*, 59: 139–157.

MARKGRAF, V., and D'ANTONI, H.L.
1978 *Pollen Flora of Argentina*. The University of Arizona Press, Tucson, Arizona, 208 pgs.

PIRE, S.M., ANZÓTEGUI, L.M., and CUADRADO, G.A.
1992 Atlas palinológico del nordeste argentino. I: Amaranthaceae, Anacardiaceae, Apocynaceae, Araliaceae y Sapindaceae. *D'Orbignyana*, N° 7, 75 pgs.

RAMALHO, M., GUIBU, L.S., GIANNINI, T.C. KLEINERT-GIOVANNINI, A., and IMPERATRIZ FONSECA, V.L
1991 Characterization of some southern Brazilian honey and bee plants through pollen analysis. *Journal of Apicultural Research*, 30: 81–86.
RAMALHO, M., KLEINERT-GIOVANNINI, A., and IMPERATRIZ FONSECA, V.L.
1990 Important bee plants for stingless bees (*Melipona* and *Trigonini*) and africanized honeybees (*Apis mellifera*) in neotropical habitats: a review. *Apidologie*, 21: 469–488.
ROUBIK, D.W., and MORENO, J.E.
1991 *Pollen and Spores of Barro Colorado Island*. Monographs in Systematic Botany, Missouri Botanical Garden, St. Louis, Missouri, vol. 36, 268 pgs.
TELLERIA, M.C.
1988 Analyse pollinique des miels du Nord-ouest de la Province de Buenos Aires (République Argentine). *Apidologie*, 19: 275–290.
1992 Caracterización botánica y geográfica de las mieles de la Provincia Fitogeográfica Pampeana (República Argentina) I: Distrito Oriental. *Darwiniana*, 31: 345–350.
1993 Floraison et récolte du pollen par les abeilles doméstiques (*Apis mellifera* L. var. *ligustica*) dans la pampa argentine. *Apidologie*, 2: 109–120.
1995 El polen de las mieles del Noroeste de la Provincia de Buenos Aires, Argentina. *Darwiniana*, 33: 347–364.
1996 Caracterización botánica y geográfica de las mieles de la Provinica Fitogeográfica Pampeana (República Argentina) II. Tandilia. *Boletín de la Sociedad Argentina de Botánica*, 32: 91–96.
VERGERON, Ph.
1964 Interprétation statistique des résultats en matière d'analyse pollinique des miels. *Annales de l'Abeille*, 7: 349–364.

GLORIA DANERS
Sección Palinología
Departamento de Paleontología – Facultad de Ciencias
Iguá 4225
11400–Montevideo, Uruguay
E-mail: glo@fcien.edu.uy

UVALDE, TEXAS, BOLL WEEVILS ARE HUNGRY CRITTERS[1]

Gretchen D. Jones and James R. Coppedge

Abstract

Alternative pollen foraging resources of early season boll weevils, *Anthonomus grandis grandis* Boheman (Coleoptera: Curculionidae), captured in Uvalde, Texas, from April to July 1995 were determined from pollen analyses. One hundred fifty-six (156) pollen types were encountered in the samples: 131 types were identified to family, genus, or species ranks while 25 remained unknown. Pollen representing 44 families, 86 genera, and 14 species occurred in the samples. Boll weevils foraged on the greatest diversity of pollen in April, and the fewest in June. Asteraceae, Fabaceae, Poaceae, and Rhamnaceae pollen were found in all three months. More Fabaceae pollen types (33) were found than any other family, followed by Asteraceae (22). Our research shows that boll weevils captured in Uvalde, Texas forage on a wide range pollen sources.

INTRODUCTION

The boll weevil, *Anthonomus grandis grandis* Boheman (Coleoptera: Curculionidae), invaded the United States about 100 years ago. Today, boll weevils are a major insect pest in all cotton, (*Gossypium hirsutum* C. Linnaeus), growing areas of the United States. During winter and early spring months, boll weevils lie dormant in the soil, under leaf litter, etc. Once temperatures warm, boll weevils begin to emerge. However, amount of boll weevil emergence in spring is low. Since cotton is not in bud until June or July, early emerged boll weevils can not feed on cotton buds nor flowers.

The boll weevil eradication program, a nationwide program, was developed to combat the boll weevil and reduce its numbers and amount of destruction. According to some researchers, this program will eradicate the boll weevil from all cotton-producing areas in the United States. One premise of the eradication program is that boll weevils must enter cotton when cotton buds are immature or boll weevils will die from starvation. This is frequently called suicidal emergence.

In 1978, Cate and Skinner found pollen other than cotton in the boll weevil gut and believed that this "other pollen" might indicate alternative foraging resources. Recent studies have shown that the adult stage forages on a wider range of plants than previously believed. From pollen analyses of boll weevil gut contents, Benedict et al. (1991) and Jones et al. (1992, 1993) found that boll weevils were not as restricted nor specific in their pollen feeding as previously believed. This research was designed to determine if Uvalde, Texas, boll weevils that emerged prior to cotton's blooming fed on pollen of local plants.

METHODS AND PROCEDURES

Early-season emerged boll weevils, captured from April to July, before cotton was in full bloom, were examined for pollen feeding. Boll weevils were trapped in pheromone traps just north of Uvalde, Texas. Five traps were placed about 1 mile (1.6 km) apart along four lines radiating out from three cotton fields. Three boll weevils from traps 2 and 5 of each line about 2 and 5 miles (3.2 and 8 km) away from the cotton fields respectively, and field 3 during the first and third week of each month were examined for pollen feeding. A total of 64 samples per month were examined.

Whole boll weevils were placed in individual 12 ml centrifuged tubes, rinsed several times with distilled water, and acetolyzed. Prior to adding glycerin, samples were strained with a 450 µm screen. Three slides were made from each boll weevil's pollen residue. Light micrographs were taken of all pollen types. Pollen types were identified to the lowest taxonomic rank possible: family, genus, or species. All pollen identification was made utilizing the Stanley D. and Gretchen D. Jones and Areawide Pest Management Research Unit pollen reference collections.

Number of pollen grains, pollen types, identified and unidentified types, percentage of boll weevils with and without pollen were calculated monthly and recalculated for a grand total.

Frequency of occurrence and relative frequency were calculated for the plant families encountered. Frequency of occurrence was calculated by dividing the number of samples in which a taxon occurred by the total number of samples. The resultant was then multiplied by 100. Relative frequency was calculated by dividing the number of samples in which a taxon occurred by the total number of taxa, and multiplying by 100.

RESULTS

A total of 2,271 pollen grains were counted in the boll weevil samples (Table 1). One hundred fifty-six (156) pollen types were found in the samples: 131 were identified to family, genus, or species ranks, and 25 remained unknown. Pollen representing 44 families, 86 genera, and 14 species were identified in the samples (Table 1).

Samples from April not only had the greatest number of pollen grains (1,364), but also the greatest number of pollen

Jones, Gretchen D., and Coppedge, James R.
2001 Uvalde, Texas boll weevils are hungry critters; *In*: Goodman, D.K., and Clarke, R.T. (eds.), *Proceedings of the IX International Palynological Congress, Houston, Texas, U.S.A., 1996*; American Association of Stratigraphic Palynologists Foundation, p. 501–503.
 ISBN 0–931871–06–9

[1] Mention of a commercial or proprietary product in this paper does not constitute an endorsement by the USDA, nor does it imply the recommendation of the product by the USDA to the exclusion of similar products.

TABLE 1. Total number of pollen grains, pollen types, identified and unidentified types, and types classified to family, genus, and species found in Uvalde boll weevils.

Pollen grains	2271
Pollen types	156
Types identified	131
Types unknown	25
Families	44
Genera	86
Species	14

TABLE 3. Frequency of occurrence by month and early season total, April through July (Total), of pollen by family found in Uvalde boll weevils. Only families with a frequency >5 are included.

Family	April	May	June	July	Total
Alismataceae	39				10
Apocynaceae	39				28
Asteraceae	56	44	17	50	42
Cupressaceae	44	39	17		25
Fabaceae	89	72	33	50	61
Fagaceae	94	50			36
Juglandaceae			17	61	20
Moraceae	83	22		11	32
Poaceae	28	6	11	44	22
Rhamnaceae	28	33	39	50	38

types (89, Table 2). Both April and July had the same number of families represented (26). June had the fewest number of pollen grains (253), pollen types (38), and families (17, Table 2). All April and July boll weevils contained pollen (Table 2).

More Fabaceae pollen types were found than from any other family. Thirty-three Fabaceae types were found, including several species of *Acacia*, *Mimosa*, *Neptunia*, and *Trifolium*. Following the Fabaceae in the number of pollen types was Asteraceae (22). Poaceae was represented by five different pollen types. There were four Rhamnaceae types including *Frangula caroliniana* (T. Walter) A. Gray (Sy = *Rhamnus caroliniana* T. Walter), buckthorn. Malvaceae pollen did not occur until June, and was found in both June and July. However, Malvaceae pollen was never a major pollen constituent. Pollen types from the Asteraceae, Fabaceae, Poaceae, and Rhamnaceae were found in each month (Table 3).

Frequency of occurrence (Jones and Coppedge, 1996) indicates how often a pollen type will be encountered within the samples. In April, Fagaceae, Fabaceae, and Moraceae grains were encountered more frequently than other types (Table 3). In the overall total, Fabaceae pollen was encountered most frequently followed by Asteraceae, Rhamnaceae, and Fagaceae (Table 3).

Relative frequency (Jones and Coppedge, 1996) indicates the importance of a pollen type within the sample. Fagaceae pollen was the most important type in April (Table 4). In May, the Fabaceae were the most important, but in June and July it was the Rhamnaceae (Table 4).

DISCUSSION

For years, boll weevils were thought to forage only on cotton pollen, flowers, buds, and fruits. Cate and Skinner (1978) found cotton pollen in the boll weevil midgut up to 18 h after being feeding. Stoner (1968) found adult boll weevils foraging on *Sphaeralcea* spp. (Malvaceae), globe mallow. Rummel et al. (1978) found adults feeding on *Hymenopappus flavescens* A. Gray (Asteraceae), yellow woolly white. From pollen analyses, Rummel (1978) confirmed that boll weevils found on yellow woolly white were feeding on its pollen. Thus, pollen in the gut may identify the species of plant on which boll weevils recently fed (Benedict et al., 1991). From pollen analyses, Benedict et al. (1991) and Jones et al. (1992, 1993) found that boll weevils foraged on pollen from a wide variety of plants.

Our research agrees with Benedict et al. (1991) and Jones et al. (1993) that boll weevils are not restricted in their pollen feeding. Uvalde boll weevils foraged on a large variety of pollen types from acacias to oaks. Based on pollen content, Asteraceae, Fabaceae, Fagaceae, and Rhamnaceae pollen were the most important alternative foraging resources for Uvalde early season boll weevils.

April boll weevils foraged on the greatest variety of pollen. April in Uvalde is flower rich with trees, shrubs, and herbaceous plants in bloom. June, however, is usually dry and is a transition between spring and summer flowering plants. Spring plants have finished flowering and summer plants have not begun. Boll weevils foraged on the least diversity of pollen in June. Once summer rains begin the end of June and beginning of July, the summer flora begins to flower, increasing the number of pollen producing plants as foraging resources for boll weevils. This bimodal blooming season is evident in the number of pollen types encountered from April to July.

There are approximately 6,900 vascular plants in Texas (Jones et al., 1997). Because of this plant diversity, alternative pollen resources in one vegetational zone will be differ-

TABLE 2. Number of pollen grains, pollen types, families and percent boll weevils containing pollen (% with) by month found from pollen analyses of Uvalde boll weevils.

Month	# Grains	# Types	# Families	%with
April	1364	89	26	100
May	363	59	23	83
June	253	38	17	72
July	291	56	26	100

TABLE 4. Relative frequency of the most frequently encountered families in pollen analyses of Uvalde boll weevils.

Family	April	May	June	July
Asteraceae	38	35	18	35
Fabaceae	62	57	35	35
Fagaceae	65	39		
Rhamnaceae	19	26	41	37

ent from those in another. Since pollen feeding provides significant energy and nutrition for dispersing boll weevils, it is important to define the full range of these foraging resources. Therefore, it is important to examine boll weevils from other vegetational zones and during different seasons. At the present time, research is being conducted in Munday and Crockett, Texas, to determine additional boll weevil alternate foraging resources.

References Cited

BENEDICT, J.H., WOLFENBARGER, D.A., BRYANT, V.M., JR., and GEORGE, D.M.

1991 Pollens ingested by boll weevils (Coleoptera: Curculionidae) in southern Texas and northeastern Mexico. *Journal of Economic Entomology*, 84: 126–131.

CATE, J.R., and SKINNER, J.L.

1978 The fate and identification of pollen in the alimentary canal of the boll weevil. *Southwestern Entomologist*, 3: 263–265.

JONES, G.D., and COPPEDGE, J.R.

1996 Pollen feeding by overwintering boll weevils. *In:* Dugger, P., and Richter, D. (eds.), *Proceedings, Beltwide Cotton Production Research Conferences*, National Cotton Council Memphis, TN, p. 976–977.

JONES, S.D., WIPFF, J.K., and MONTGOMERY, P.A.

1997 *Vascular plants of Texas: a comprehensive checklist including synonymy, bibliography, and index.* University of Texas Press, Austin, 404 pgs.

JONES, R.W., CATE, J.R., HERNANDEZ, E.M., and NAVARRO, R.T.

1992 Hosts and seasonal activity of the boll weevil (Coleoptera: Curculionidae) in tropical and subtropical habitats of northeastern Mexico. *Journal of Economic Entomology*, 85: 74–82.

JONES, R.W., CATE, J.R., HERNANDEZ, E.M., and SOSA, E.S.

1993 Pollen feeding and survival of the boll weevil (Coleoptera: Curculionidae) on selected plants species in northeastern Mexico. *Environmental Entomology*, 22: 99–108.

RUMMEL, D.R., WHITE, J.R., and PRUITT, G.R.

1978 A wild feeding host of the boll weevil in West Texas. *Southwestern Entomologist*, 3: 171–175.

STONER, A.

1968 *Sphaeralcea* spp. As hosts of the boll weevil in Arizona. *Journal of Economic Entomology*, 61: 1100–1102.

Authors' addresses

GRETCHEN D. JONES
USDA, ARS, Southern Plains Agricultural Research Center
Areawide Pest Management Research Unit
2771 F&B Road
College Station, TX 77845
U.S.A.
E-mail: g-jones@tamu.edu

JAMES R. COPPEDGE
USDA, ARS, Southern Plains Area Office
7607 Eastmark, Suite 230
College Station, TX 77840
U.S.A.

POLLEN ANALYSIS OF THE CROP OF CORN EARWORM ADULTS[1]

Gretchen D. Jones and Juan D. Lopez, Jr.

Abstract

Pollen found on insect bodies is an important natural marker and is used to determine foraging resources and migration routes. Pollen is ingested along with nectar as insects feed. The possibility of using ingested pollen as a natural marker in moths and butterflies (Lepidoptera) has not been examined. To determine if ingested pollen is usable for adult Lepidopteran pollen analysis, we examined the crop of corn earworm adults for pollen. Seven of eight (88%) adult crops contained pollen. Thirteen pollen types were found. Pollen from nine families was identified from the crop analyses. Three samples contained three different taxa and one sample contained six taxa. *Quercus* spp., oak, pollen was found more frequently than any other pollen type. Pollen from four taxa not previously reported as foraging resources were found in the crop contents. Pollen analyses of the crop is feasible and equal in its results to exterior examination of the proboscis; however, additional research is needed and warranted.

INTRODUCTION

Adults of numerous insect species feed on nectar and other plant exudates associated with flowers. As a result of this feeding activity, these adults become contaminated with pollen. Identification of this pollen is used in adult insect feeding and migratory studies (Courtney et al., 1982; Hendrix et al., 1987; Hendrix and Showers, 1992; Gregg, 1993; Lingren et al., 1993, 1994; Loublier et al., 1994).

One of the first palynological studies of pollinating insects examined the stomach contents of flies that fed on nectar and pollen (Bennett, 1884). Parker (1923) discovered that the source and geographic origin of the nectar being foraged could be identified from pollen in the stomach contents of honeybees. Pollen analyses of the digestive tract and the midgut portion of the alimentary canal is an effective technique in determining foraging resources of the boll weevil (*Anthonomus grandis grandis* Boheman) (Cate and Skinner, 1978; Benedict et al., 1991; Jones et al., 1992, 1993; Jones and Coppedge, 1996). Wratten et al. (1995) found 13 pollen types in the gut contents of hoverflies (*Melanostoma* spp.). Nalepa et al. (1992) examined the stomach contents of 11 lady beetles [*Chilocorus kuwanae* (Silvestri)], and found pollen in all samples.

Historically, only exterior body parts (eyes, proboscis, legs, etc.) of Lepidoptera (moths and butterflies) have been examined for pollen (Wiklund et al., 1979; Courtney et al., 1982; Lazri and Barrows, 1984; Bryant et al., 1991; Gregg, 1993; Lingren et al., 1993, 1994; Berkhousen and Shapiro, 1994). Entomopalynological techniques for Lepidopteran species range from combing the body over a glass slide (Mikkola, 1971), to examining only the proboscis (Gregg, 1993).

Although the digestive system of Lepidopteran has not been examined for pollen, Mikkola (1971) suggested that pollen analyses of the digestive system might be useful in the study of Lepidopteran feeding. Turnock et al. (1978) reported finding pollen in the proboscis tube, but did not examine the digestive system for pollen. The purpose of this research was to determine if pollen analysis of the adult Lepidoptera digestive system is a feasible approach to Lepidopteran pollen analysis.

Jones, Gretchen D., and Lopez, Juan D., Jr.
2001 Pollen analysis of the crop of corn earworm adults; *In*: Goodman, D.K., and Clarke, R.T. (eds.), *Proceedings of the IX International Palynological Congress, Houston, Texas, U.S.A., 1996*; American Association of Stratigraphic Palynologists Foundation, p. 505–509.
 ISBN 0–931871–06–9

METHODS AND PROCEDURES

Eight male corn earworm moths were randomly selected from moths captured in pheromone traps on the night of 16 March 1994, at College Station, TX. The next morning, captured moths were transported to USDA, Areawide Pest Management Research Unit and frozen. The crop (a storage organ for food) of each insect was removed and placed into an individual 1.5 ml Eppendorf™ microcentrifuge tube containing 95% ethanol (**ETOH**). Samples were centrifuged for three minutes and the supernatant decanted. Each sample was mixed for 15 seconds, then acetolyzed for 12 minutes at 100° C (Erdtman, 1963; Lieux, 1980a; Low et al., 1989). After acetolyzation, samples were rinsed four times: twice with one ml of distilled water, once with one ml of 35% **ETOH**, and once with one ml of 95% **ETOH**. Samples were centrifuged for three minutes, the supernatant decanted, and the samples mixed for 15 seconds after each rinse. The pollen residue was stained with two drops of Safranin O. Five drops of glycerin were added to each Eppendorf tube. Samples were placed in a warm Eppendorf hot block (30°C) for 24 hours to evaporate the **ETOH**. Three glass slides were made from each sample's pollen residue. Pollen identification was made to the lowest taxonomic rank (family, genus, species). Percent of moths containing pollen, percent with various taxa, and the number of pollen taxa were calculated.

RESULTS

Seven of the eight crops contained pollen (87.5%, Table 1). Three samples (43%) contained three types, two contained one type (29%), and one contained two types (14%) (Tables 1 and 2). Acetolyzation did not completely dissolve the crop tissue and on two occasions several pollen grains could not be separated from the undissolved tissue.

[1] Mention of a commercial or proprietary product in this paper does not constitute an endorsement by the USDA, nor does it imply the recommendation of the product by the USDA to the exclusion of similar products.

Thirteen pollen types were found in the crops. Three types were classified to family, four to genus, and four to species (Table 2). One pollen grain was unknown and one was deteriorated beyond identification possibly due to enzymes in the saliva of digestive tract.

Pollen from nine families was identified from the crop analyses (Table 2). Two Fabaceae and two Oleaceae types were found in the samples (Table 2).

Quercus spp. (oak) pollen was found in three samples, more frequently than any other pollen (Table 2). *Salix nigra* H. Marshall (black willow), *Fraxinus pennsylvanica* H. Marshall (green ash), *Populus deltoides* W. Barton *ex* H. Marshall (cottonwood) and *Trifolium repens* C. Linnaeus (white clover) were each found in two samples (Table 2). *Magnolia* sp. (magnolia), and *Vicia* sp. (vetch) were found in one sample. Magnolia, white clover, vetch, and cottonwood have not been previously reported as adult corn earworm foraging resources.

DISCUSSION

Although acetolyzation did not completely dissolve the crop tissue, the tissue was cleared enough to identify trapped pollen to family rank. It is possible that a longer acetolyzation time may completely dissolve the crop tissue and allow for more accurate pollen identification.

The percent of moths containing interior pollen (87.5%) was greater that those reported with exterior pollen. Moths containing external pollen ranged from 30% (Turnock and Chong, 1987) to 57% (Lingren et al., 1993). In Missouri, external pollen-contaminated spring-time moths ranged from 30% for female armyworms, *Pseudaletia unipuncta* (Haworth), and 65% for male black cutworms, *Agrostis ipsilon* (Hufnagel) (Hendrix and Showers, 1992).

This higher percentage of moths with internal pollen may be due to several reasons. First, not all plant species produce the same amount of pollen. Todd and Vansell (1942) discovered that equal amounts of nectar from different plant species contained different amounts of pollen. Knowledge that not all taxa produce the same amount of pollen led to the development of pollen coefficient tables for nectar-producing plants (Demianowicz, 1961, 1964; Sawyer, 1988; Jones and Bryant, 1992, 1996). Pollen coefficient tables were intended to enable identification of single floral-sources for honey samples from plants with low pollen yields. Unfortunately, even though used occasionally in melissopalynology (study of pollen in honey), the tables are, at best, a guide to taxa that produce nectar high or low in pollen content (Jones and Bryant, 1992; 1996).

A similar system for Quaternary pollen, "R-values, " was developed by Davis (1963) for a selected group of plant taxa. "R-values" are based on the ratio between the relative pollen percentage of each taxon in a modern sediment sample, and the percentage of area that same plant occupied in the region. Although useful, "R-values," like pollen coefficients, are merely a guide to which taxa are low or high pollen producers. Whether coefficient tables or "R-values" are useful for pollen found in or on insects has not been examined.

TABLE 1. Pollen analysis of the crop contents of 8 corn earworm adults.

	Samples	Percent
with pollen	7	87.5
Quercus spp. pollen	3	37.5
Salix nigra pollen	2	25.0
no pollen	1	12.5
with 6 taxa	1	12.5

A second possible reason internal pollen percentage is higher than external is the various factors influencing pollen retention on or in insects. External pollen retention is partially correlated to a grain's ornamentation. Asteraceae pollen grains with spine-like processes were the most common taxa found on corn earworm adults (Lingren et al., 1993, 1994). Likewise, Asteraceae grains were more frequent on noctuid moths examined by Turnock and Chong (1987).

External pollen retention also depends on pollen size and placement. The alfalfa caterpillar, *Colias eurytheme* (Boisduval), lost 51% of the larger *Phlox* spp. (phlox) pollen grains after coiling and uncoiling its proboscis (Levin and Berube, 1972). Pollen loss occurred more frequently when the grains were on the upper portion of the proboscis because of abrasion by the palps during recoiling.

Todd and Vansell (1942) found that the longer nectar remained in a bee's stomach, the fewer pollen grains could be found. However, pollen loss should be less in the crop, because digestion in the crop is relatively limited.

A third possible reason internal pollen percentage is greater than external may be due to sample size. We analyzed only eight moth crops compared to the external examination of 158 moths in 1988, and 209 in 1990 (Lingren et al., 1993, 1994). However, because our sample was a random sample, the results should be fairly representative. Our purpose for this research was to examine the possibility of pollen occurring in the crop and determine the feasibility of using this technique in examining moths for pollen.

The number of pollen types (13) found internally is similar to the number of types found externally. Six taxa were identified on the proboscis of corn earworm adults collected in Texas, in 1989, and 13 in 1990 (Lingren et al., 1994). In Oklahoma, Lingren et al. (1994) identified 14 taxa on corn earworm adults.

Number of pollen types found internally ranged from zero to six. Three crop samples contained three different taxa, and one contained six taxa (Tables 1 and 2). This pollen diversity per individual is the similar to external pollen diversity of corn earworm adults (Lingren et al., 1993). However, the majority of external contaminated moths had only a single pollen type (Lingren et al., 1993), whereas over 70% of internally contaminated moths contained pollen (calculated from Table 2).

Identification of internal pollen from nine families is in the range of that found externally. Pollen from eight and ten plant families was found in corn earworm adults in 1988 and 1990, respectively (Lingren et al., 1993, 1994). However, in Oklahoma, number of plant families ranged from eight to 12, in 1988 and 1990, respectively (Lingren et al., 1993, 1994).

None of the four newly reported taxa (magnolia, white clover, vetch, and cottonwood) have spine-like processes indicating that crop analyses may yield a different array of foraging resources. White clover, magnolia, and vetch are entomophilous (insect pollinated), but cottonwood is anemophilous (wind pollinated). The most unexpected of the new foraging resources is magnolia.

TABLE 2. Pollen types (taxon) found in the crops contents of pollen containing corn earworm adults and the sample number(s) in which each taxon occurred.

Taxon	Sample Numbers 1	2	3	4	5	6	7
Caryophyllaceae				X			
Fabaceae							
Trifolium repens	X					X	
Vicia sp.						X	
Fagaceae							
Quercus spp.	X	X	X				
Magnoliaceae							
Magnolia sp.						X	
Oleaceae							
Fraxinus pennsylvanica				X			X
Populus deltoides	X		X				
Poaceae					X		
Salicaceae							
Salix nigra	X						X
Ulmaceae							
Ulmus sp.							X
Verbenaceae	X						
Unknown			X				
Indeterminate	X						

Magnolia has large, off-white, bowl-shaped or flat flowers that have a strong fragrance. Magnolia flowers are considered typical for cantharophily (beetle pollinated) and are not expected to be foraged upon by Lepidopteran insects (Faegri and van der Pijl, 1971, 1979). Although not usually mentioned as an important nectar source, magnolia has a heavy nectar flow (Pellett, 1977). Magnolia pollen is found in honey of Texas, Louisiana, and Mississippi (Jones, 1993; Lieux, 1972, 1978, 1981).

Magnolia pollen grains are large, psilate, monocolpate, and boat-shaped in equatorial view (Kapp, 1969; Lieux, 1980b). Their size ranges from 42 to >100 μm in length (Lieux, 1980b). This large pollen grain easily can be dislodged and lost from the proboscis, which may account for it not being previously reported as a foraging resource.

It has been assumed that pollen grains from anemophilous taxa, such as oak and black willow, are accidental contaminants. However, both oak and black willow have been previously reported as foraging resources for noctuid adults (Mikkola, 1971; Turnock and Chong, 1978; Wynne et al., 1989; Bryant et al., 1991; Lingren et al., 1993, 1994; Loublier et al., 1994).

CONCLUSIONS

This research demonstrated that pollen can be found in the crop of adult corn earworms. There is no doubt that this discovery has considerable potential. Less pollen seems to be lost in the crop than on the proboscis. Therefore, a more complete array of natural markers can be identified. In addition, the equipment needed for crop analyses can be found in most palynological laboratory and the technique (acetolyzation) is familiar to palynologists. If future research proves crop analyses to be as good as our preliminary results, pollen analyses of Lepidopteran species can be accomplished by using techniques and equipment used in most palynological laboratories. There is no need for the expense of electron microscopy.

Unfortunately, many questions remain unanswered. Does pollen size or ornamentation influence ingested through the proboscis? How long will pollen remain in the crop or will it be digested? If pollen is digested, does size and/or ornamentation effect digestion rate? Is there a correlation between pollen found internally to that found externally? At the present time, research is being conducted to answer some of these questions, and compare pollen found externally to that found internally.

ACKNOWLEDGMENTS

We are grateful to Doug Foss for his help in preparing crop samples for analyses. We are also indebted to Stanley D. Jones (Botanical Research Center) for his encouragement and reviewing the manuscript, and Vaughn M. Bryant, Jr. (Texas A&M University) for review of the manuscript. We also thank Vaughn M. Bryant, Jr., for access to the Texas A&M Anthropology Palynology Laboratory pollen reference collection.

References Cited

BENEDICT, J.H., WOLFENBARGER, D.A., BRYANT, V.M., JR., and GEORGE, D.M.
1991 Pollens ingested by boll weevils (Coleoptera: Curculionidae) in southern Texas and northeastern Mexico. *Journal of Economic Entomology*, 84: 126–131.

BENNETT, A.W.
1884 On the constancy of insects in their visits to flowers. *Journal of the Linnaean Society of London. Zoology*, 17: 175–185.

BERKHOUSEN, A.E., and SHAPIRO, A.M.
1994 Persistent pollen as a tracer for hibernating butterflies: the case of *Hesperia juba* (Lepidoptera: Hesperiidae). *Great Basin Naturalist*, 54: 71–78.

BRYANT, V.M., PENDLETON, M., MURRAY, R.E., LINGREN, P.D., and RAULSTON, J.R.
1991 Techniques for studying pollen adhering to nectar-feeding corn earworm (Lepidoptera: Noctuidae) moths using scanning electron microscopy. *Journal of Economic Entomology*, 84: 237–240.

CATE, J.R., and SKINNER, J.L.
1978 Fate and identification of pollen in the alimentary canal of the boll weevil, *Anthonomus grandis*. *The Southwestern Entomologist*, 3: 263–452.

COURTNEY, S.P., HILL, C.J., and WESTERMAN, A.
1982 Pollen carried for long distances by butterflies. *Oikos*, 38: 260–263.

DAVIS, M.B.
1963 On the theory of pollen analysis. *American Journal of Science*, 261: 897–912.

DEMIANOWICZ, Z.
1961 Pollenkoeffizienten als Grundlage der quantitativen Pollenanalyse des Honigs. *Pszczelnicze Zeszyty Naukowe*, 5: 95–105.
1964 Characteristik der Einartenhonige. *Annales de l' Abeille*, 7: 273–288.

ERDTMAN, G.
1963 Palynology. *Advances in Botanical Research*, 1: 149–208.

FAEGRI, K., and VAN DER PIJL, L.
1971 *The principles of pollination ecology*. 2nd ed. Pergamon Press, New York. 291 pgs.
1979 *The principles of pollination ecology*. 3rd ed. Pergamon Press, New York. 244 pgs.

GREGG, P.C.
1993 Pollen as a marker for migration of *Helicoverpa armigera* and *H. punctigera* (Lepidoptera: Noctuidae) from western Queensland. *Australian Journal of Ecology*, 18: 209–219.

HENDRIX, W.H., III, MUELLER, T.F., PHILLIPS, J.R., and DAVIS, O.K.
1987 Pollen as an indicator of long-distance movement of *Heliothis zea* (Lepidoptera: Noctuidae). *Environmental Entomology*, 16: 1148–1151.

HENDRIX, W.H., III, and SHOWERS, W.B.
1992 Tracing black cutworm and armyworm (Lepidoptera: Noctuidae) northward migration using *Pithecellobium* and *Calliandra* pollen. *Environmental Entomology*, 21: 1092–1096.

JONES, G.D.
1993 *Melissopalynology of East Texas*. Unpubl. Ph.D. dissertation, Department of Botany, Texas A&M University, College Station, TX, 200 pgs.

JONES, G.D., and BRYANT, V.M., JR.
1992 Melissopalynology in the United States: a review and critique. *Palynology*, 16: 64–71.
1996 Melissopalynology. *In*: J. Jansonius, and D.C. McGregor. (eds.), *Palynology: Principles and applications*. 3. American Association of Stratigraphic Palynologists Foundation, Dallas, TX, pp. 933–938.

JONES, G.D., and COPPEDGE, J.R.
1996 Pollen feeding by overwintering boll weevils. *In*: P. Dugger and D. Richter (eds.), *Proceedings, Beltwide Cotton Production Research Conferences*, National Cotton Council Memphis, TN, pp. 976–977.

JONES, R.W., CATE, J.R., HERNANDEZ, E.M., and NAVARRO, R.T.
1992 Hosts and seasonal activity of the boll weevil (Coleoptera: Curculionidae) in tropical and subtropical habitats of northeastern Mexico. *Journal of Economic Entomology*, 85: 74–82.

JONES, R.W., CATE, J.R., HERNANDEZ, E.M., and SOSA, E.S.
1993 Pollen feeding and survival of the boll weevil (Coleoptera: Curculionidae) on selected plant species in northeastern Mexico. *Population Ecology*, 22: 99–108.

KAPP, R.O.
1969 *How to know the pollen and spores*. Wm. C. Brown Co. Dubuque, Iowa, 249 pgs.

LAZRI, B., and BARROWS, E.M.
1984 Flower visiting and pollen transport by the imported cabbage butterfly (Lepidoptera: Pieridae) in a highly disturbed urban habitat. *Environmental Entomology*, 13: 574–578.

LEVIN, D.A., and BERUBE, D.E.
1972 *Phlox* and *Colias*: the efficiency of a pollination system. *Evolution*, 26: 242–250.

LIEUX, M.H.
1972 A melissopalynological study of 54 Louisiana (U.S.A.) honeys. *Review of Palaeobotany and Palynology*, 13: 95–124.
1978 Minor honeybee plants of Louisiana indicated by pollen analysis. *Economic Botany*, 32: 418–432.
1980a Acetolysis applied to microscopical honey analysis. *Grana*, 19: 95–61.
1980b An atlas of pollen of trees, shrubs, and woody vines of Louisiana and other southeastern states, part II. Platanaceae to Betulaceae. *Pollen et Spores*, 22: 191–243.
1981 An analysis of Mississippi USA honey: pollen, color and moisture. *Apidologie*, 12: 137–158.

LINGREN, P.D., BRYANT, V.M., JR., RAULSTON, J.R., PENDLETON, M., WESTBROOK, J., and JONES, G.D.
1993 Adult feeding host range and migratory activities of corn earworm, cabbage looper and celery looper (Lepidoptera: Noctuidae) moths as evidenced by attached pollen. *Journal of Economic Entomology*, 86: 1429–1439.

LINGREN, P.D., WESTBROOK, J.K., BRYANT, V.M., Jr., RAULSTON, J.R., ESQUIVEL, J.F., and JONES, G.D.
1994 Origin of corn earworm (Lepidoptera: Noctuidae) migrants as determined by *Citrus* pollen markers and synoptic weather systems. *Environmental entomology*, 23: 562–570.

LOUBLIER, Y., DOUAULT, P., CAUSSE, R., BARTHES, J., BUES, R., and POITOUT, S.
1994 Utilisation des spectres polliniques recueillis sur *Agrotis (Scotia) ipsilon* Hufnagel (Noctuidae) comme indicateur des migrations. *Grana*, 33: 276–281.

LOW, N.H., SCHWEGER, C., and SPORNS, P.
1989 Precautions in the use of melissopalynology. *Journal of Apicultural Research*, 28: 50–54.

MIKKOLA, K.
1971 Pollen analysis as a means of studying the migrations of Lepidoptera. *Annales Entomologici Fennici*, 37: 136–139.

NALEPA, C.A., BAMBARA, S.B., and BURROUGHS, A.M.
1992 Pollen and Nectar feeding by *Chilocorus kuwanae* (Silvestri) (Coleoptera: Coccinellidae). *Proceedings of the Entomological Society of Washington*, 94: 596–597.

PARKER, R.L.
1923 Some pollen gathered by bees. *American Bee Journal*, 63: 16–19.

PELLETT, F.C.
1977 *American Honey Plants*. 4th ed. Orange Judd Publishing Co., Inc., Carthage, IL, 467 pgs.

SAWYER, R.
1988 *Honey identification*. Cardiff Academic Press, Wales, UK, 115 pgs.

TODD, F.E., and VANSELL, G.H.
1942 Pollen grains in nectar and honey. *Journal of Economic Entomology*, 35: 728–731.

TURNOCK, W.J., CHONG, J., and LUIT, B.
1978 Scanning electron microscopy: a direct method of identifying pollen grains on moths (Lepidoptera: Noctuidae). *Canadian Journal of Zoology*, 56: 2051–2054.

WIKLUND, C., ERIKSSON, T., and LUNDBERG, H.
1979 The wood white butterfly, *Leptidea sinapis*, and its nectar plants: a case of mutualism or parasitism? *Oikos*, 33: 358–362.

WRATTEN, S.D., WHITE, A.J., BOWIE, M.H., BERRY, N.A., and WEIGMANN., U.
1995 Phenology and ecology of hoverflies (Diptera: Syrphidae) in New Zealand. *Environmental Entomology*, 24: 595–600.

WYNNE, J.W., KEASTER, A.J., GERHART, K.O., and KRAUSE, G.R.
1989 Plant species identified as food sources for adult black cutworm, *Agrotis ipsilon* Hufnagel, and armyworm, *Pseudaletia unipuncta* Haworth, (Lepidoptera: Noctuidae) in Northwest Missouri. *Journal of Kansas Entomological Society*, 64: 381–387.

Authors' addresses

GRETCHEN D. JONES
JUAN D. LOPEZ, JR.
USDA, ARS, Southern Plaine Agricultural Research Center
Areawide Pest Management Research Unit
2771 F&B Road
College Station, TX 77845
U.S.A.
g-jones@tamu.edu

POLLEN ANALYSES OF THE ICEMAN'S COLON CONTENT

Klaus D. Oeggl

Abstract

Pollen and plant macrofossil analyses of feces from the colon were undertaken to discover the environment and the life style of the Iceman. The food residue in his intestinal tract consists mainly of bran from *Triticum monococcum*. Besides these cereal remains, muscle fibres, 30 pollen types, two type of spores, diatoms, mineral particles, a lot of charcoal fragments, and the ova of intestinal parasites were found. The pollen spectrum is complex and may derive from different sources. It is surprising that the AP is represented by 84%, while the NAP is only 16% of the total pollen. Taxa from the deciduous forests (*Querco–Fagetea*) predominate among the AP. Respectively the high amount of *Ostrya*-type indicates the warm loving Hophornbeam (*Ostrya carpinifolia*) and suggests that he came from the valley-bottoms south of the main range of the Eastern Alps. The taxa from acid spruce forests (*Vaccinio–Piceion*) form a minority, but document the characteristic vegetation of the montane regions in the inneralpine area. The six woody species used to construct his tools and equipment are represented by pollen in his colon content. Pollen of the Cerealia-type and of weeds such as *Polygonum persicaria*-type, *Rumex acetosella*-type, *Plantago major*-type, Chenopodiaceae-type and *Urtica* most probably were included in the cereal remains he had eaten. The palynological investigations of his diet confirm the other results of his study. All suggest he came from an environment that was located in the valley-bottoms of the Vinschgau south of the Alps.

INTRODUCTION

In September 1991 a sensational chance discovery was made in the permanent ice of the Tyrolean Alps. At an altitude of 3210 m a naked human body thawed out of the ice in proximity to the Ötztal summit. Within the immediate vicinity of the corpse several weapons, clothes, other personal possessions and different plant remains appeared in the melting ice. The artifacts associated with the body enabled a provisional estimate of the age of the find, and it rapidly became clear that these were the remains of a Neolithic man. During two archaeological excavations of the area a large body of sedimentary, botanical and zoological material was collected (Lippert and Spindler, 1991; Lippert, 1992 a, b; Bagolini et al., 1993). The archaeobotanical investigations of samples taken from the location, from the artifacts and from the body itself have contributed significantly to explain a variety of aspects about the Iceman's everyday existence (Bortenschlager and Oeggl, 2000). One of the most important tasks has been to discover where he started his last journey and where his settlement was located. Somewhat unexpectedly samples from his intestinal tract have provided the most convincing answers to date to these questions, while also giving us new insights into the diet of Neolithic man.

MATERIAL AND METHODS

A 40 mg sample from the transversal section of the colon was removed from the body using an endoscope, in order to limit damage of the body itself. The colon sample was then divided into four equal-sized subsamples A–D, as independent replicates used for verification of the internal consistency of the sampling and the analytical procedures (Text-Figure 1). Three subsamples (A–C) were prepared for light microscopy analysis of pollen- and macroremains and one (subsample D) for ultrastructural investigations by a scanning electron microscope.

Oeggl, Klaus D.
2001 Pollen analysis of the Iceman's colon content; *In*: Goodman, D.K., and Clarke, R.T. (eds.), *Proceedings of the IX International Palynological Congress, Houston, Texas, U.S.A., 1996*; American Association of Stratigraphic Palynologists Foundation, p. 511–516.
 ISBN 0–931871–06–9

Sample Preparation Techniques

Each subsample was rehydrated in a 0.5% solution of trisodium phosphate (Pearsall, 1989; Holden, 1994), and screened through 500µ, 250µ, 125µ and 63µ steel mesh screens; at this stage samples were numbered 2 to 5 noting decreasing mesh size (Text-Figure 1.). The filtered solution (outwash) was also saved for further investigation (spectrum 6). To evaluate the efficiency of the preparation technique and to avoid any problems arising from it, sample screening was carried out initially on subsample A and prior to the preparation of samples B–D. After preparation of subsample A a small random sample was then taken from the 500µ residue to investigate its macrofossil content and to evaluate the state of preservation using light microscopy. Then each subsample was boiled for 10 min in 10% potassium hydroxide to remove the amorphous gel, consisting mainly of starch, protein and fat enclosing the recovered material, washed in distilled water, and then treated with ethanol (Gassner et al., 1989). The pollen analysis samples were stained with fuchsin and mounted in glycerine. The light microscope investigations were carried out under magnifications of x750 and x1200 with the aid of phase contrast and interference contrast.

The results of the pollen analysis are presented in a diagram (Text-Figure 2) which shows the relative pollen quantities of each size fraction from all 4 subsamples (spectra 2–5) and the outwash (spectrum 6) are presented as single spectra. The total colon sample (spectrum 1) was then calculated by adding the values from the spectra 2–6.

RESULTS

Residues in the intestinal tract consist of different types of material. These include primarily cereal remains, muscle fibres, carbon particles, pollen, mineral particles, diatoms and eggs of a human parasite, the whipworm (*Trichuris trichiura*).

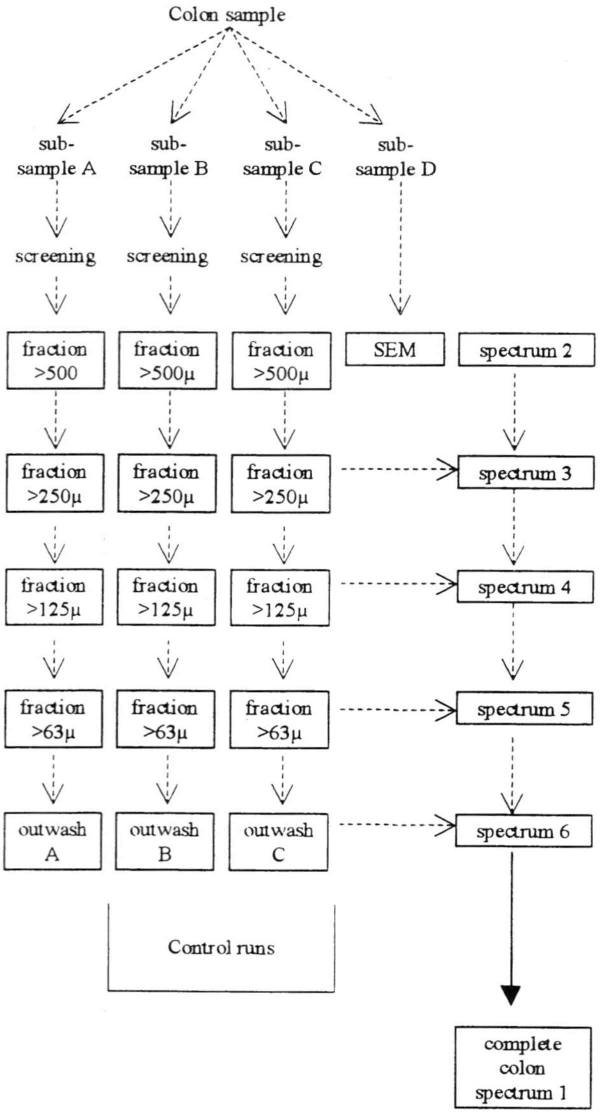

Text-Figure 1. Flowchart of the sampling preparation technique.

Plant Macrofossils

Plant macrofossil contents are derived mainly from bran belonging to the *Triticum/Secale*-type. Small amounts of chaff fragments were also found, most probably belonging to *Triticum*. Given that *Secale* was not cultivated in central Europe during the Neolithic and that several spikelets of Einkorn (*Triticum monococcum*) were found on the body, the bran also belongs to Einkorn (*Triticum monococcum*). Testa remains >500µ make up a small percentage of the total so it is presumed that the wheat used for his last meal was finely ground. Muscle fibres as well as charred bone fragments, were also found and are evidence that meat was also part of his diet.

Pollen

The pollen flora (Text-Figure 2) in the food residues is far more diverse, with 30 different pollen types and two spore types. The AP dominate with 84%, and NAP consists of only 16% of the total pollen. This is of some considerable interest, since an NAP dominance was to have been expected, due to all the edible herbaceous species in the diet.

The pollen content in the different fraction residues (spectra 2–5) varies as a function of the mesh size selection resulting from the adopted recovery technique (Text-Figure 2). The saccate pollen of the conifers *Pinus* and *Picea* dominate in spectra 2 and 3. Overall the samples do not have a high pollen content, with spectra 2, 3 and 4 each containing fewer than one hundered pollen grains.

Spectrum 2 (Text-Figure 2) exhibits the characteristic pollen taxa of the colon sample with a dominance of *Picea*, *Pinus* and *Ostrya*-type. The presence of the *Ostrya*-type pollen was most surprising. Given that the genus *Ostrya* has a restricted distribution in the Alps and its pollen is a good indicator of the Iceman's origin. Some of the *Ostrya*-type pollen were fresh, given the fact that the gametophyte was still visible in the pollen grain. Although sample preparation and analyses were carried out in laboratory conditions outside of the flowering season, the presence of pollen with intact gametophytes led to an initial concern that subsample A had in some way been contaminated, and the small absolute quantities of this pollen type in the fraction subsample (>500µ) gave weight to these concerns. However, a control run of the sampling and the extraction procedures gave the same results indicating that no contamination had taken place and that the *Ostrya* pollen type effectively originated from the colon sample. The control run was also repeated with the subsamples B and C with the same results.

The most exciting thing was that the gametophyte was still visible in what were clearly pollen grains of both *Betula* and in approx. 80% of the *Ostrya*-type pollen. Further investigation was therefore required to highlight those conditions which could guarantee gametophyte preservation in such ancient pollen. It is well known that the cell content of pollen grains decompose after several weeks in water or soil. Reference herbarium material was therefore investigated to see if some indication was forthcoming concerning the survival of gametophytes in pollen grains which had been air-dried. In all the investigated material dating as far back as 1875, gametophytes could clearly be observed in the *Ostrya*-type pollen grains. It therefore seems possible for vegetative cells within the pollen to be preserved for more than 100 years in air-dried conditions, and still be clearly identified under a light microscope.

Diatoms

Beside pollen 16 different diatom species were also recognized in the food residue. The dominant species identified are *Achnanthes minutissima*, *Diatoma ehrenbergii*, *Fragilaria arcus*, *Gomphonema pumilum*, *G. olivaceum* and *Navicula radiosa*. They characterize a freshwater diatom flora typical of mountain rivulets within the investigation area (Rott, 2000).

Pollen Data

The pollen spectrum from the colon (spectrum 1 in Text-Figure 2) reflects a deciduous forest vegetation dominated by *Ostrya* with an admixture of *Betula*, *Corylus*, *Picea* and *Pinus*. The highest *Ostrya*-type concentrations were found in spectra 5 and 6. In the complete pollen sample of the colon

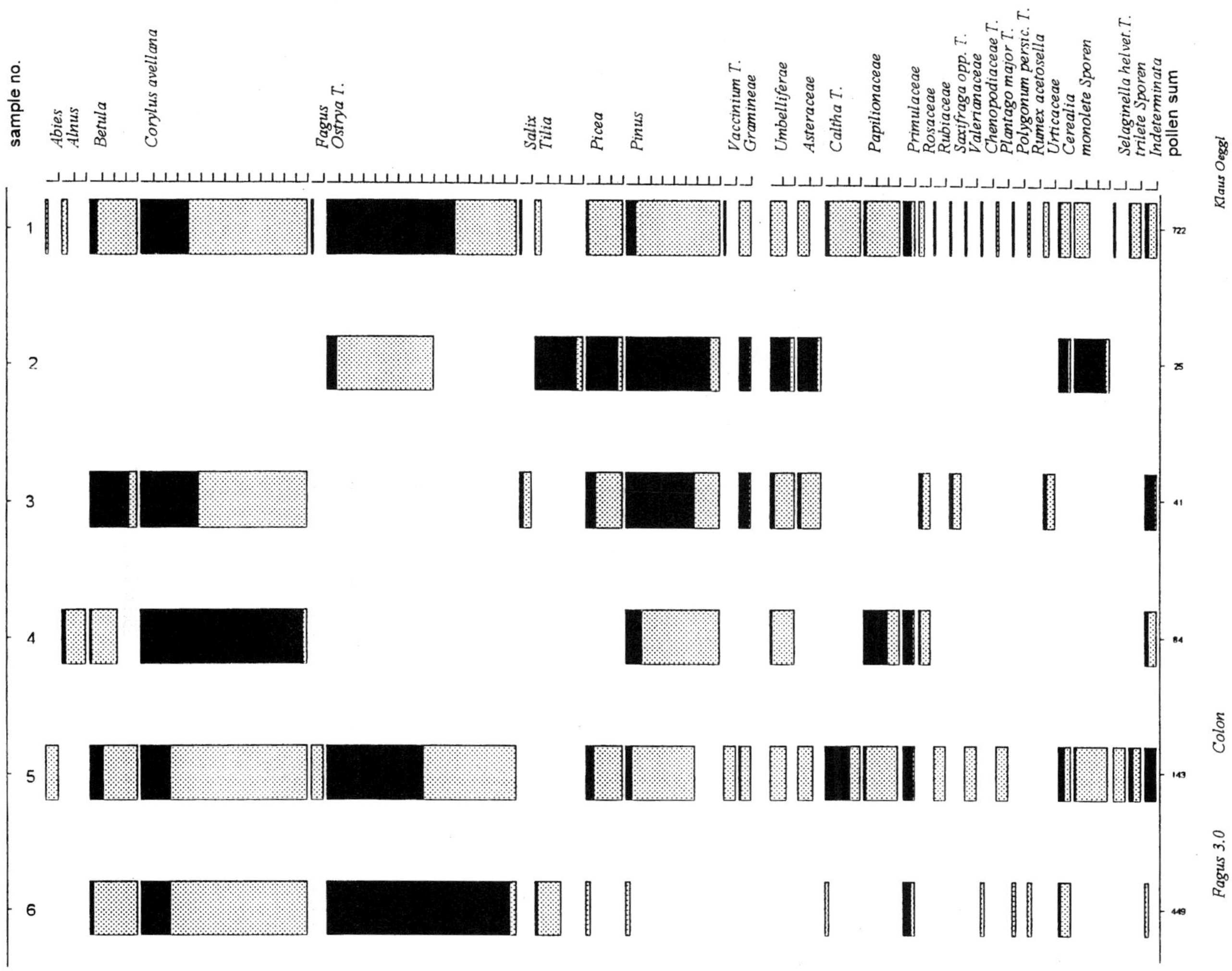

Text-Figure 2. Pollen spectra of the food residue from the colon: 1 = colon complete, 2 = fraction >500μ, 3 = fraction >250μ, 4 = fraction >125μ, 5 = fraction >63μ, 6 = outwash <63μ (scale reading: 1 graduation mark means 5%).

they add up to 54% of the total and represent the dominant tree species with *Corylus* (24%). *Betula, Pinus* and *Picea* are present with values typical for this vegetation type in the investigation area. The NAP represent a mixture of taxa from arable fields, as well as taxa common distributed in meadows and tall forb communities. *Cerealia*-type, *Caltha*-type, Primulaceae and Papilionaceae appear in higher quantities. The *Cerealia*-pollen belongs to the *Triticum*-type and possibly derives from the cereal bran the Iceman had eaten. Associated herb pollen belong to synanthropic species commonly growing in cereal fields, (e.g. Chenopodiaceae, *Plantago major*-type, *Polygonum persicaria*-type, *Rumex acetosella*, and Urticaceae). Most probably they were trapped in the bran during crop processing and were later eaten with the cereal meal. Conspicuously, all the meadow taxa found have an early flowering season while pollen from late flowering species is missing among the NAP spectrum.

Provenience of the Colon Pollen

As briefly outlined above, the pollen taphonomy of the food residue in the colon is complex and may have derived from different sources. There are several ways in which pollen can become incorporated into the intestinal residue: diet, drinking water, breathing and swallowing bronchial mucus, etc. Furthermore it has to be considered whether pollen is incorporated intentionally by consuming plants (economic pollen) or unintentionally as a contamination of food, water or air. Pollen archived by this means in coprolites can be very informative about diet, seasonality of site occupation and palaeoenvironmental conditions (Bryant, 1974; Sobolik, 1988).

To obtain a precise indication of the location of this deciduous vegetation type the colon spectrum was compared with (Text-Figure 3):

- A pollen spectrum developed from material extracted from the ice thought to be contemporaneous with the deposition of the body (sample 2). This one was examined because of the possibility, that the Iceman drank water from the gully in which he was discovered.
- Five pollen spectra developed for sites along a transect through the eastern Alps. Five Burkard-traps collecting airborne pollen over a five-year period were set up at the northern alpine border zone (sample 3), at the northern inneralpine montaine zone (sample 4), at the inneralpine–

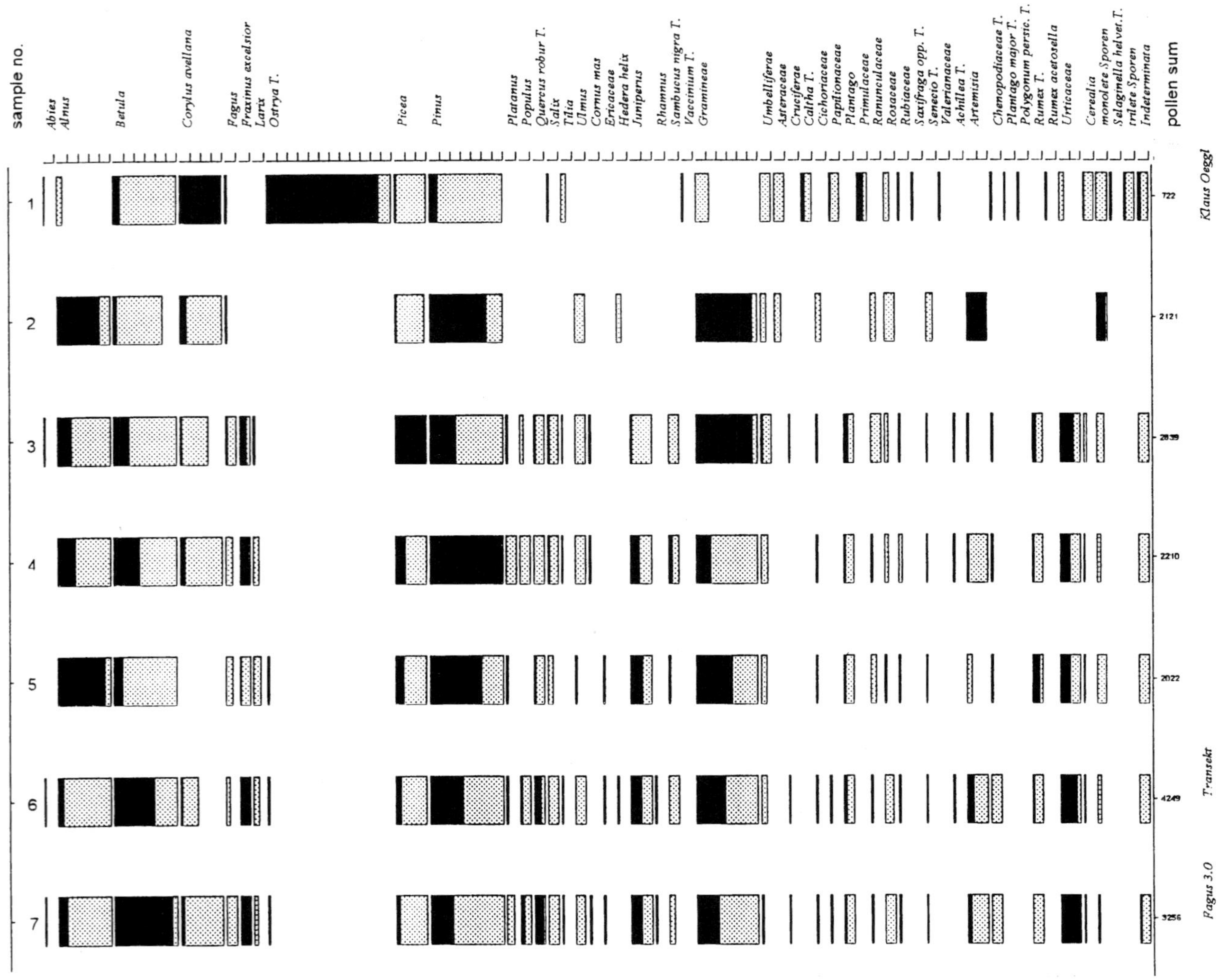

Text-Figure 3. Comparison of the pollen spectrum from the colon with recent pollen spectra along a transect from South to North of the Eastern Alps: 1 = colon complete, 2 = ice sample from the find spot, 3 = northern alpine border zone: montane regions, 4 = northern inneralpine zone: montane regions, 5 = inneralpine zone: subalpine regions, 6 = southern inneralpine zone (Vinschgau): montane regions, 7 = southern inneralpine zone (Bozen): montane regions. μ (scale reading: 1 graduation mark means 5%).

subalpine regions (sample 5), and at two sites (samples 6 and 7) in the southern alpine regions (Vinschgau and Bozen).

Visual comparison, supported by numerical evaluation of the data, clearly indicates that the colon pollen spectrum is very different from those of the transect and ice samples. The pollen spectrum from the ice where the Iceman was found, contains more *Alnus* and *Pinus*, but no *Ostrya*-type, and therefore water from that spot can be ruled out as a pollen source for the colon spectrum. Similar negative results seem to be the case with all the other spectra presented. One underlying reason might be that the pollen spectra from the Burkhard traps represent a regional flora whereas the colon spectrum seems to show specific local features. The only thing that can be deduced from the Burkard pollen spectra is that the *Ostrya* pollen-type is distributed south of the alpine main range. This consideration adds further weight to the evidence that the colon spectrum is representative of a southern alpine flora.

Pollen taxa in the colon spectrum (Text-Figures 2, 3) belonging to the ecologically more demanding deciduous forests (*Querco–Fagetea*) predominate. In particular the numerically predominant *Ostrya*-type indicates the warmth loving Hophornbeam (*Ostrya carpinifolia*), the only indigenous species of this genus in the area, and one which only establishes itself in forests along the valleys south of the main range of the Alps. In the lower Vinschgau and in the Schnalstal this species is a dominant representative in Manna ash–Hophornbeam forests (*Orno–Ostryetum*), which thrive up to 1100 m altitude (Dalla Torre and Sarnthein, 1900; Peer, 1981). Recent stands of *Ostrya carpinifolia* are only *ca* 10 km away from where the Iceman was found. The typical montane vegetation of the inner alpine area, acid spruce forests (*Vaccinio–Piceion*), is also represented by minor amounts of pollen grains belonging to *Pinus*, *Picea*, and *Vaccinium*-type.

In general terms, therefore this pollen spectrum would appear to reflect the transition zone between a warmth demanding vegetation with Hophornbeam-forests (*Orno–Ostryetrum*), and the inneralpine Spruce forests (*Vaccinio–Piceion*) characteristic for the montane vegetation in the valleys of the southern Alpine regions.

Similar conclusions concerning the provenience of the Iceman have also been reached on the basis of dendrological analyses and bryophytes evidence (Dickson et al., 1996). The dendrological analyses carried out on the woody remains, the majority belong to species found in the montane region, with a few from subalpine and alpine species, were also represented. In particular the ecological requirements of the identified woody species point to the transition zone between the thermophilous mixed-oak forest communities (*Quercetalia pubescenti-petreae*) and the montane spruce forest (*Piceetum montanum*). Presence of Norwegian maple (*Acer platanoides*), European yew (*Taxus baccata*), Ash (*Fraxinus excelsior*), Lime (*Tilia* sp.) and Elm (*Ulmus* sp.) allow us to infer a humid habitat with a mineral rich, free-draining soil and a mild winter climate, similar to present-day conditions in woodlands along the slopes and in the gorges of the lower Schnalstal and Vinschgau in South Tyrol (Oeggl and Schoch, 2000).

Alongside this reconstruction of the type of environment in which the Iceman spent the hours preceeding his death, and from the data which emerge from an overview of the botanical materials recovered in the colon, detailed information can be obtained from an evaluation of single components in this complex picture. The most fascinating is the preserved gametophyte within the grains of the *Ostrya*-type pollen. When laboratory contamination of the colon material was ruled out, it was initially assumed that the *Ostrya*–type pollen became associated with the cereals during the preparation of the Einkorn-based (*Triticum monococcum*) food. However, food preparation phases, such as milling, cooking or baking—by heating—would destroy the proteins and damage the cell content irreversibly. No damaged pollen grains were observed in the colon sample, therefore making it improbable that this pollen-type originated through direct association with the cereal remains. Another explanation is that *Ostrya*–type pollen entered the body via drinking water or by breathing. The diatom flora found in the colon residue makes assimilation by drinking water most probable. It would therefore appear highly probable, that the pollen was swallowed when drinking stream water. Regardless of the mechanisms through which the *Ostrya*–type pollen entered the body it is clear that the grains had only recently been released from their inflorescences. *Ostrya*–type pollen was absorbed within a short time after their deposition, otherwise the gametophytes would have been destroyed by oxidizing conditions within a few days or weeks after dispersion. Furthermore, when the flowering season of the other arboreal taxa is also taken into consideration, together with data from the NAP spectrum (Text-Figure 2) in which mainly early flowering taxa are represented and late flowering taxa such as *Artemisia* are missing, this season would appear to be late spring. Depending on the size and shape, pollen can stay within the intestine for up to a month (Gremillion and Sobolik, 1996), thus the Iceman died in early summer at the latest.

Within this reconstruction one further aspect needs discussion, namely why was the pollen content not affected by enzymes and/or digested in the intestinal tract? Food residue in the colon always contains protein and fat, which are only partly digested in this intestinal tract. The exine certainly offers some sort of protection, but this is probably insufficient to enable the preservation of 80% of the pollen in pristine conditions, as is the case of the *Ostrya*-type pollen identified in the colon. Protein and fat content in faeces will increase when the digestive transfer speed is increased. This can be caused, for instance, by a strong endoparasitic intestinal infection, and the material from the Iceman's colon also included a large number of eggs of the whipworm (*Trichuris trichiura*), a human parasite (Aspöck et al., 1995). Furthermore, anatomical investigations have shown that the intestinal epithelium tissue was affected by a strong parasitic infection, which led to a higher transfer speed of the digestive tract, possibly even manifested with diarrhea (Platzer, pers. comm.). The link between the forensic and the microfossil data, which together indicate an increase in the digestive transfer speed and a reduction in the catabolic rate and protein assimilation, explain the presence of remarkably well-preserved *Ostrya*-type pollen in the colon.

CONCLUSIONS

The micro- and macrofossil content of the food residues in the colon give evidence of the Iceman's diet, environment and season of death. His last meal consisted mainly of Einkorn (*Triticum monocccum*) with traces of meat. Pollen was incorporated in a complex way. One part seemed to have come from sticking on plant remains of the cultigens, whereas the diatoms make it plausible that the other part of the pollen contents were swallowed in drinking water.

The occurrence of well preserved *Ostrya*-type pollen in large numbers, together with the other species of the colon pollen spectrum, strongly suggests that in the hours preceeding the Iceman's death, he walked through a warmth-demanding deciduous vegetation dominated by Hophornbeam (*Ostrya carpinifolia*). Regional pollen spectra indicate this vegetation type is typical of the montane valleys of the southern Alps. These results contradict the investigations by Groenman-van Waateringe (1993) on the pollen content of the Iceman's hides and furs. Based on the pollen analyses of his clothes it was deduced that the Iceman lived in the montane zone of the Eastern Central or Northen Alps. Therefore it has to be considered that the pollen in the hides and furs of the Iceman reflect a mixed regional pollen spectrum deriving from all the different habitats he traveled through during his lifetime.

Furthermore the comprehensive occurrence of taxa with early flowering seasons in the colon sample gives a strong indication that the Iceman died in late spring or early summer and not in the autumn as has been suggested (Spindler, 1993). Further confirmation of this reconstruction is expected from additional palynological investigations on the pollen content of the mucose membranes from his bronchial and nasal cavities.

ACKNOWLEDGMENTS

These investigations were supported by the Austrian Science Fund (Project No 10151-SOZ). The author wishes to thank: Dr. Werner Platzer and Dr. Karl-Heinz Künzel (Anatomical Institute of the University of Innsbruck) for helpful discussions and for inclusion in the sample recovery procedures. Dr. Vaughn M. Bryant, Jr. (Texas A & M University),

Dr. Owen K. Davis (University of Arizona) and Dr. James Schoenwetter (Arizona State University) for literature help and for helpful suggestions in the preparation of this manuscript. Dr. James H. Dickson (Institute of Biomedical and Lifescience of the University of Glasgow), Dr. Samuel Piercy Evans (Cranfield University), Dr. Dagfinn Moe (Botanical Institute of Bergen University) and Dr. Vaughn M. Bryant, Jr. (Texas A & M University) reviewed the manuscript and improved the written English.

References Cited

ASPÖCK, H., AUER, H., and PICHLER, O.
1995 The mummy from the Hauslabjoch: A medical parasitology perspective. *Alpe Adria Microbiology Journal*, 2: 105–114.

BAGOLINI, B., DAL RI, L., LIPPERT, A., and NOTHDURFTER, H.
1993 Der Mann im Eis: Die Fundbergung 1992 am Tisenjoch, Gem. Schnals, Südtirol. *In*: Spindler, K., Rastbichler-Zissernigg, E., Wilfling, H., zur Nedden, D. and Nothdurfter, H. (eds.), *Der Mann im Eis: Neue Funde und Ergebnisse*. Wien–New York, p. 3–52.

BORTENSCHLAGER, S., and OEGGL, K.
2000 *The Iceman and his natural environment*. Springer, Wien, 166 pgs.

BRYANT, V.M., Jr.
1974 The Role of Coprolite analysis in Archaeology. *Bulletin of Texas Archaeological Society*, 45: 1–28.

DALLA-TORRE, K.W. von, and SARNTHEIN, L. von
1900 *Flora der gefürsteten Grafschaft von Tirol, des Landes Vorarlberg und des Fürstenthumes Lichtenstein*. 6. Bände. Innsbruck.

DICKSON, J.H., BORTENSCHLAGER, S., OEGGL, K., PORLEY, R., and McMULLEN, A.
1996 Mosses and the Iceman's Provenance. *Proceedings of the Royal Society London, B*, 263: 567–571.

GASSNER, G., HOHMANN, B., and DEUTSCHMANN, F.
1989 *Mikroskopische Untersuchung pflanzlicher Lebensmittel*. Stuttgart: 414 pgs.

GREMILLION, K.J., and SOBOLIK, K.D.
1996 Dietary Variability among Prehistoric Forager-Farmers of Eastern North America. *Current Anthropology*, 37: 529–539.

GROENMAN-VAN WAATERINGE, W.
1993 Analyses of the Hides and Skins from the Hauslabjoch. *In*: Egg, M., Geodecker-Ciolek, R., Groenman-van Waateringe, W., and Spindler, K. (eds.), *Die Gletschermumie vom Ende der Steinzeit aus den Ötztaler Alpen. Jahrbuch des Römisch-Germanischen Zentralmuseums*, 39, p. 114–128.

HOLDEN, T.
1994 Dietary evidence from the intestinal contents of ancient humans with particular reference to desiccated remains from northern Chile. *In*: Hather, J.G. (ed.), *Tropical Archaeobotany: applications and new developments*. London–New York, p. 65–85.

LIPPERT, A.
1992a Topographie der Fundstelle und die weiteren Forschungen. *In*: Payrleitner, A. (ed.), *Der Zeuge aus dem Gletscher. Das Rätsel der frühen Alpen-Europäer*. Wien, p. 35–51.
1992b Die erste archäologische Nachuntersuchung am Tisenjoch. *In*: Höpfel, F., Platzer, W., and Spindler, K. (eds.), *Der Mann im Eis. Band 1. Veröffentlichungen der Universität Innsbruck*, 187: p. 245–253.

LIPPERT, A., and SPINDLER, K.
1991 Die Auffindung einer frühbronzezeitlichen Gletschermumie am Hauslabjoch in den Ötztaler Alpen (Gem. Schnals). *Archäologie Österreichs*, 2/2: 11–17.

OEGGL, K., and SCHOCH, W.
2000 Dendrological analyses of artifacts and other remains. *In*: Bortenschlager, S., and Oeggl, K. (ed.), *The Iceman and his natural environment*. Springer, Wien, p. 29–61.

PEARSALL, D.M.
1989 *Palaeoethnobotany. A Handbook of Procedures*. San Diego: 470 pgs.

PEER, T.
1981 Die aktuellen Vegetationsverhältnisse Südtirols am Beispiel der Vegetationskarte 1:200.000. *Angewandte Pflanzensoziologie*, 26: 151–168.

ROTT, E.
2000 Diatom analyses from the colon content. *In*: Bortenschlager, S., and Oeggl, K. (eds.), *The Iceman and his natural environment*. Springer, Wien, p. 117–125.

SOBOLIK, K.
1988 The importance of pollen concentration values from coprolites: an analysis of southwest Texas samples. *Palynology*, 12: 201–214.

SPINDLER, K.
1993 Die Mumie. *In*: Egg, M., Geodecker-Ciolek, R., Groenman-van Waateringe, W., and Spindler, K. (eds.), *Die Gletschermumie vom Ende der Steinzeit aus den Ötztaler Alpen. Jahrbuch des Römisch-Germanischen Zentralmuseums*, 39, p. 25–35.

Author's address

KLAUS D. OEGGL
Botanical Institute of the University of Innsbruck
Sternwartestrasse 15
A-6020 Innsbruck
Austria

HUMAN INFLUENCE ON HIGH ALTITUDE VEGETATION AT THE TIME OF THE ICEMAN

Sigmar Bortenschlager

Abstract

The discovery of the Iceman in 1991 initiated research work on many peat bogs at high altitude (from 2,200 to 2,800 m a.sl.) in the inner Oetz valley. One of the results was the discovery of a distinct change in NAP around the C14 age of 5,500 BP. From that time onwards the indicators of pastoralism began to increase and so did charcoal fragments. These findings are interpreted as representing the beginning of shepherding and burning of woodland to make new grasslands. The Neolithic people used the area of the Alps for arable farming in the valleys and for pastures on the slopes above the timberline.

INTRODUCTION

On September 19, 1991, at an altitude of 3,200 m near the Tiesenjoch in the Oetztaler Alps a frozen human body was discovered. Because of its importance it will be the subject of study for a large team of scientists for many years to come. After initial archaeological attempts to determine the body's age, AMS (Accelerator-Mass-Spectrometry) dating was used and suggests that the body most probably dates from between 3,350 and 3,300 BC (Bonani et al., 1992). This date was then confirmed by data relating to objects, found associated with the body. Thus for the first time science had access to a person from the end of the Neolithic Age who had frozen to death still wearing his clothes, carrying scraps of food, and who still had food in his gastrointestinal tract.

RESULTS

Initial research results from a number of different fields—medical and natural sciences, archaeology, law—have already been published (Höpfel et al, 1992; Rastbichler-Zissernig, 1995, Wilfling, 1996).

The wooden artifacts found with the frozen man were investigated and the species of wood they came from, were determinated (Oeggl and Schoch, 2000). The area in which these plant species occur provide strong evidence that the iceman was coming from the south. This result is confirmed by the investigation of the charcoal, which again corresponds with the results of the archaeologists. Also his leather clothing was restored and the origin of the hide was determined (Wininger, 1995). The same was done with the grass cape which was made with fibres from *Brachypodium pinnatum* agg. Schrank; the bastefibres of the cape are still being investigated. The question in this region concerns the the influence that humans had on the environment and vice-versa is still the subject of intensive study.

With the coming of the Neolithic, Europe's natural vegetation was, for the first time, subjected to greater and sustained influences by human beings. The mixed oak forests in lower and middle altitudes and the spruce and fir forests at higher altitudes were affected in various ways, as is clearly shown in the pollen profile (Text-Figure 1). In all profiles, the characterization of peat follows Aaby and Berglund (1986).

Arable farming required cleared fields, and with the crops came other weedy flora that accompanies arable farming, as is shown in numerous pollen profiles from valley locations in central Europe (Bieri-Steck, 1990; Bortenschlager and Bortenschlager, 1981; Bortenschlager, 1986; Welten, 1982; Oeggl, unpublished diagrams). The animal husbandry that developed with crop farming is reflected in the pollen profiles by sudden drops in the percentages of pollen from mixed oak forests, in particular ash (*Fraxinus*), elm (*Ulmus*) and lime (*Tilia*) (Text-Figure 1). These trees were cut for their foliage, which was fed to the animals, particulary sheep, as winter fodder. During this early time period, only natural permanent meadows, the treeless areas above the timber line, were used for pasture. Evidence also shows that even at this early stage, attempts were made to create pasture by burning the forests, as is clearly revealed by the increase in charcoal particles (Text-Figure 2) in the sediments in the profile from Gurgler Alm (Vorren et al., 1993).

The geomorphological development of the Central Alps has led to the existence of small and larger peneplanes and flatter areas above the timberline, where dwarf shrubs and alpine meadows now grow. These areas were ideal as pastures, and are still used intensively for this purpose today. Similar but smaller and fewer areas are also found in the limestone regions, such as in the Dachstein area in Upper Austria, where studies have revealed human influence dating back as far as the Neolithic (Kral, 1971a, 1971b, 1994).

THE INFLUENCE OF PASTORALISM ON VEGETATION

In the course of time, intensive grazing alters the natural vegetation. Species that can tolerate regular biting and the compacting of the soil caused by the hooves of large herbivores, gradually dominate, displacing those species that cannot withstand such treatment. This favours species with basal rosettes as for example *Plantago, Arnica, Ligusticum* and species which tolerate dunging such as *Urtica* and *Chenopodium*, as well as those species containing poisonous or bitter substances that protect them against being eaten, such as *Gentiana*. Low bush species, such as *Juniperus* with

Bortenschlager, Sigmar
2001 Human influence on high altitude vegetation at the time of the Iceman; *In*: Goodman, D.K., and Clarke, R.T. (eds.), *Proceedings of the IX International Palynological Congress, Houston, Texas, U.S.A., 1996*; American Association of Stratigraphic Palynologists Foundation, p. 517–525.
 ISBN 0–931871–06–9

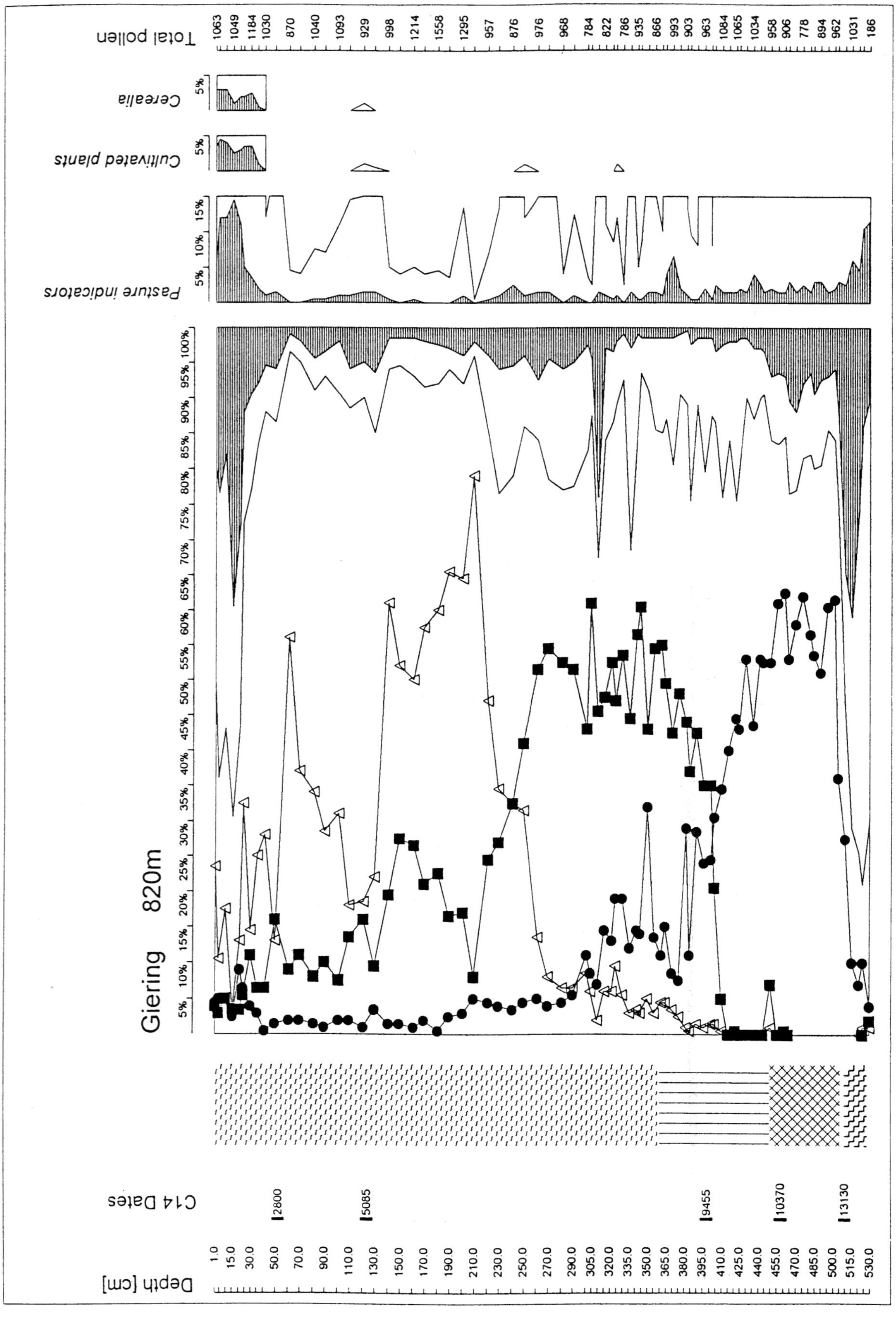

Text-Figure 1. Condensed pollen diagram from the peatbog Giering after Bortenschlager (1976).

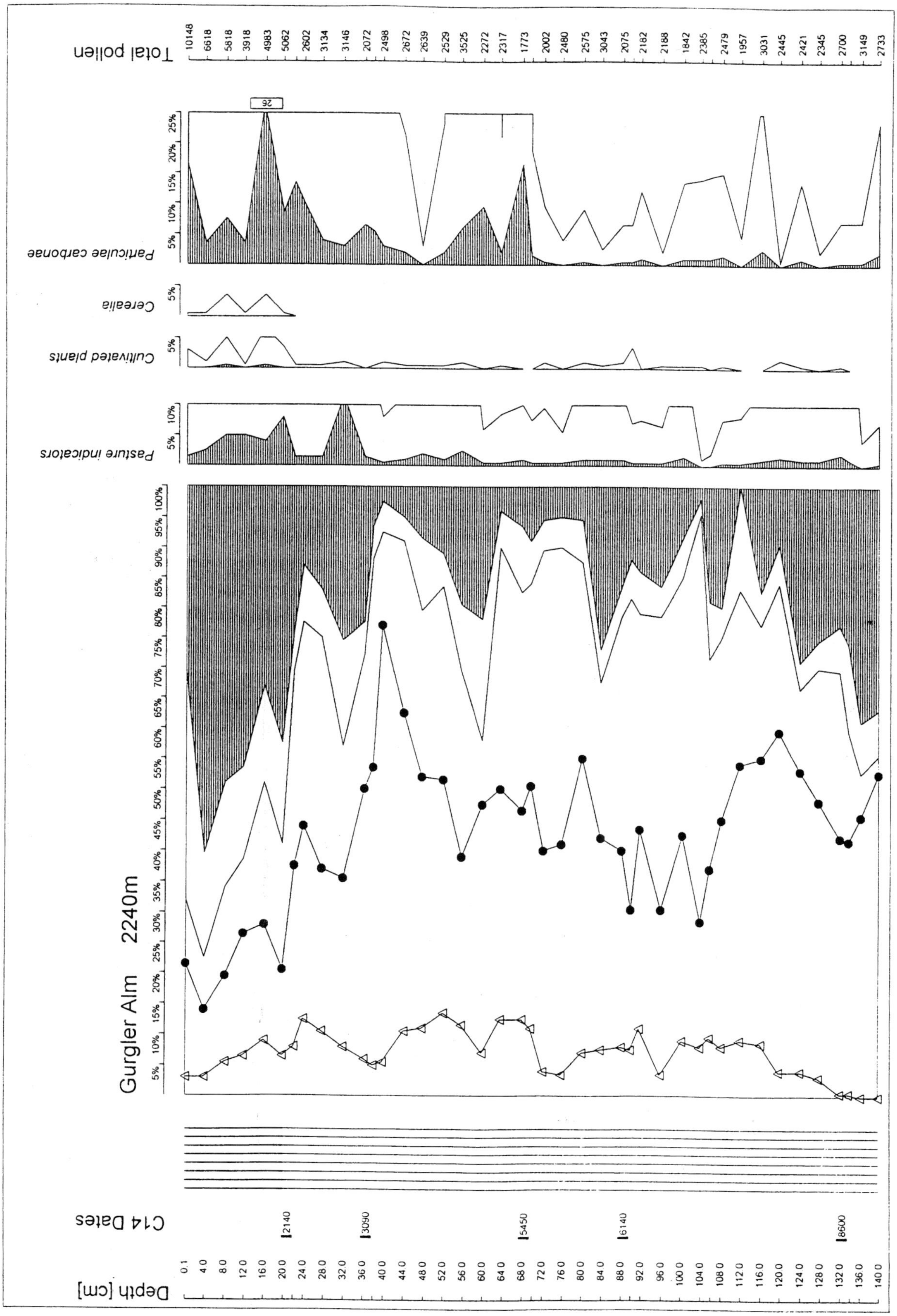

Text-Figure 2. Condensed pollen diagram from the Gurgler Alm peatbog after Vorren et al. (1993).

its needles to protect it against browsing, can also spread under such conditions.

This change in the dominant vegetation can be seen even more clearly in pollen profiles when these individual pasture indicators are taken together. For the Obergurgl–Vent region, the following pollen types have been classified as pasture indicators and are shown together in the graph as "pasture indicators:"

Apiaceae	*Plantago*
Chenopodiaceae	Ranunculaceae
Fabaceae	*Rhinanthus*
Filipendula	Rosaceae
Gentianaceae	*Rumex*
Ligusticum	*Urtica*
Lotus	

This curve shows a significant increase around 5,450 BP that reflects to a certain degree the altitude and distance from the timberline, and to some extent the decline in the tree pollen and an increase in charcoal particles (Text-Figure 2).

The decisive factor, however, is the fact that this increase in pasture indicators took place in the different peat bogs in the Obergurgl–Vent region at almost the same time, as shown below:

Brunnboden	2,660 m	5,460 BP	Text-Figure 5
Am Soom	2,620 m	5,070 BP	Text-Figure 4
Langtalereck	2,425 m	5,610 BP	Text-Figure 3
Gurgler Alm	2,240 m	5,450 BP	Text-Figure 2
Kleinalpl	2,205 m	5,540 BP	Text-Figure 6

These dates are all within the period in which the Iceman died and show that the natural vegetation of the meadows was modified to pasture vegetation during the Neolithic Age. The pollen record also shows that, ever since that time, man has used these high altitudes for pasture.

Of further importance is the fact that the pollen profiles far above the timberline show a more or less uniform tree pollen curve. With decreasing altitude the pollen curves from tree species that form the timber line become more irregular. In the Langtalereck profile (Text-Figure 3), a marked fall in the pine curve coincides with an increase in pasture indicators. In contrast, the Gurgler Alm profile (Text-Figure 2) is already below the timber line and for this reason shows hardly any increase in the pasture indicators, while still revealing a huge increase in charcoal particles around this time. This is an indication that Neolithic man had already begun to try forest clearing at lower elevations and hence extend the pastures, by means of burning down trees. This burning of the forests and the consequent extension of pasture areas was to become particularly intense with the start of the Bronze Age (3,090 BC in the Gurgler Alm profile) (Vorren et al., 1993).

The different sediment thicknesses are to some extent of local origin. The Am Soom site (Text-Figure 4) was on several occasions covered by solifluction deposits, with a particularly clear example at a depth of 120–145 cm. Brunnboden (Text-Figure 5), on the other hand, is a spring fen in a peneplane and hence an ideal pasture; the permanent pasture has meant that only a very thin sediment layer has formed since the beginning of pasturage 5,460 years ago. The Langtalereck profile (Text-Figure 3) is in a protected niche, while the Gurgler Alm profile (Text-Figure 2) was more or less protected in the forest. Both profiles show a sediment formation to the extent usual at this altitude.

The profile Kleinalpl (Text-Figure 6) is a little bit different. It is situated near a glacier and this influence is documented by several layers of sand and gravel. The increase of pasture pollen indicators (arrow) and the decrease of Stone pine (*Pinus cembra* L.) pollen caused by burnig the trees, clearly show the influence of man.

CONCLUSIONS

It can be stated that high-altitude pasturing began in the Alps at about 5,500 y.b.p. ago, alongside with the development of arable farming in the valleys. The Iceman, who came from an agrarian society in the south of the Alps (Dickson et al., 1996; Oeggl, 2000) is evidence from this time on, that humans were using the entire Alps. Whether he was a herdsman or the head of a tribe or something else remains to be determined.

References Cited

AABY, B., and BERGLUND, B.E.
1986 Characterization of peat and lake deposits. *Handbook of Holocene Palaeoecology and Palaeohydrology*, pp. 231–246.

BIERI-STECK, K.
1990 Spät- und postglaziale Vegetationsgeschichte des Lac Du Mont d'Orge im Wallis. *Lizentiatsarbeit am Systematisch-Geobotanischen Institut der Universität Bern*, 88 pgs.

BONANI, G., IVY, S.D., NIKLAUS, T.R., SUTTER, M., HOUSLEY, R.A., BRONK, C.R., JAAP VAN KLINKEN, G., and HEDGES, E.M.
1992 Altersbestimmung von Milligrammproben der Ötztaler Gletscherleiche mit der Beschleunigermassenspektrometrie-Methode (AMS). *In:* Höpfl, F., Platzer, W., and Spindler, K. (eds.), Der Mann im Eis Band 1. *Veröffentlichungen der Universität Innsbruck*, 187, Innsbruck, pp.108–116.

BORTENSCHLAGER, I.
1976 Beiträge zur Vegetationsgeschichte Tirols II: Kufstein – Kitzbühel – Paß Thurn. *Berichte nat.-med. Verein Innsbruck*, 63: 105–137.

BORTENSCHLAGER, I. and BORTENSCHLAGER, S.
1981 Pollenanalytischer Nachweis früher menschlicher Tätigkeit in Tirol. *Veröff. Museum Ferdinandeum*, Innsbruck, 61: 5–12.

BORTENSCHLAGER, S.
1986 Pollenanalytische Untersuchungen zur menschlichen Tätigkeit in den letzten 3000 Jahren im Gebiet des Lindenmooses. *In:* Bachmann H. (ed): *Kundl. Innsbruck*, 32–34.

DICKSON, J.H., BORTENSCHLAGER, S., OEGGL, K., PORLEY, A., and MCMULLEN, A.
1996 Mosses and the Tyrolean Iceman's southern provenance. *Proceedings of the Royal Society London B*, 263: 567–571.

HÖPFEL, F., PLATZER, W., AND SPINDLER, K. (EDS.)
1992 Der Mann im Eis. Band 1. *Veröffentlichungen der Universität Innsbruck*, 187, 463 pgs.

KRAL, F.
1971a Beiträge zur Geschichte der Almwirtschaft im Dachsteinmassiv auf Grund pollenanalytischer Untersuchungen. *Alm und Weide*, 21: 1–7.

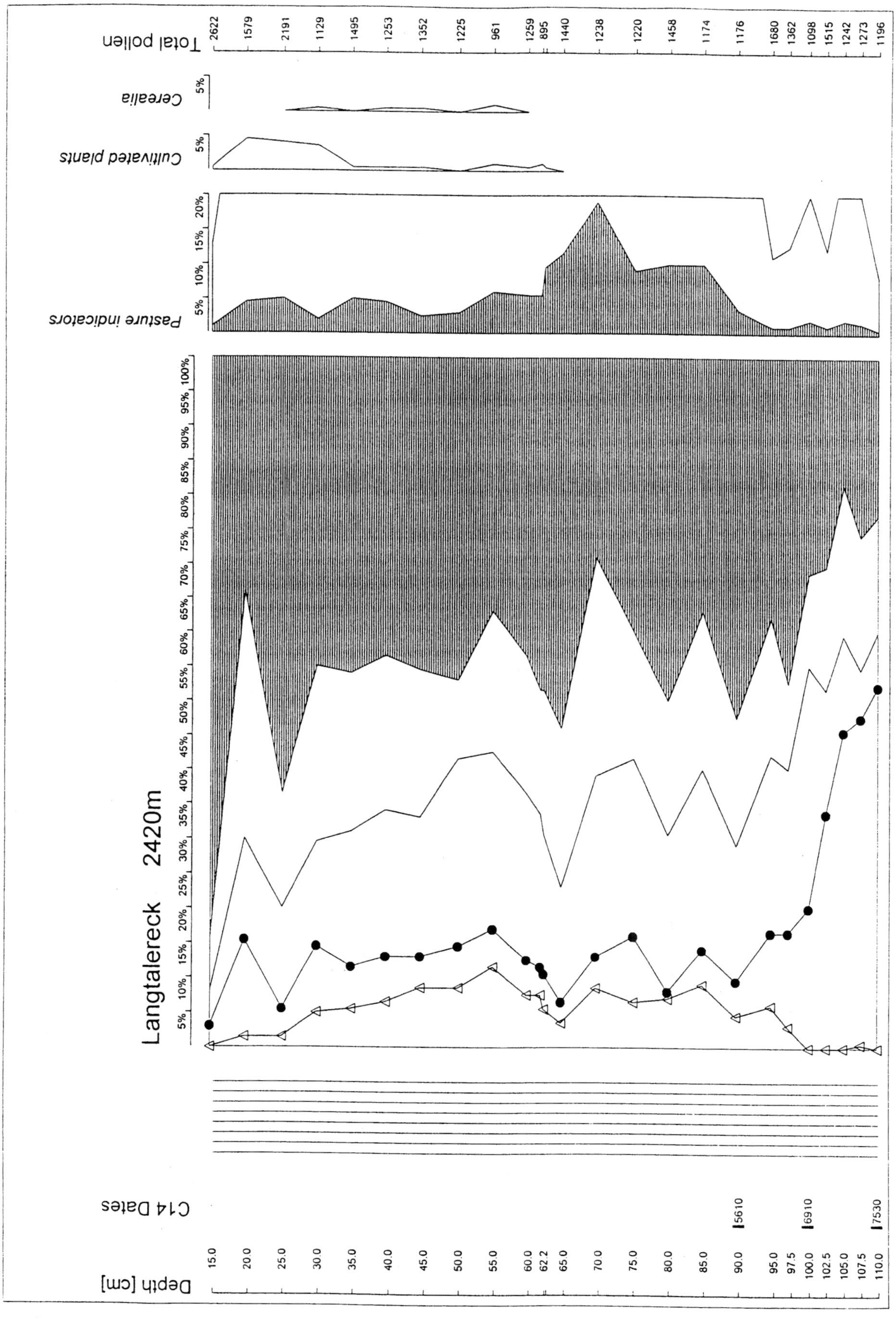

Text-Figure 3. Condensed pollen diagram from the Langtalereck peatbog.

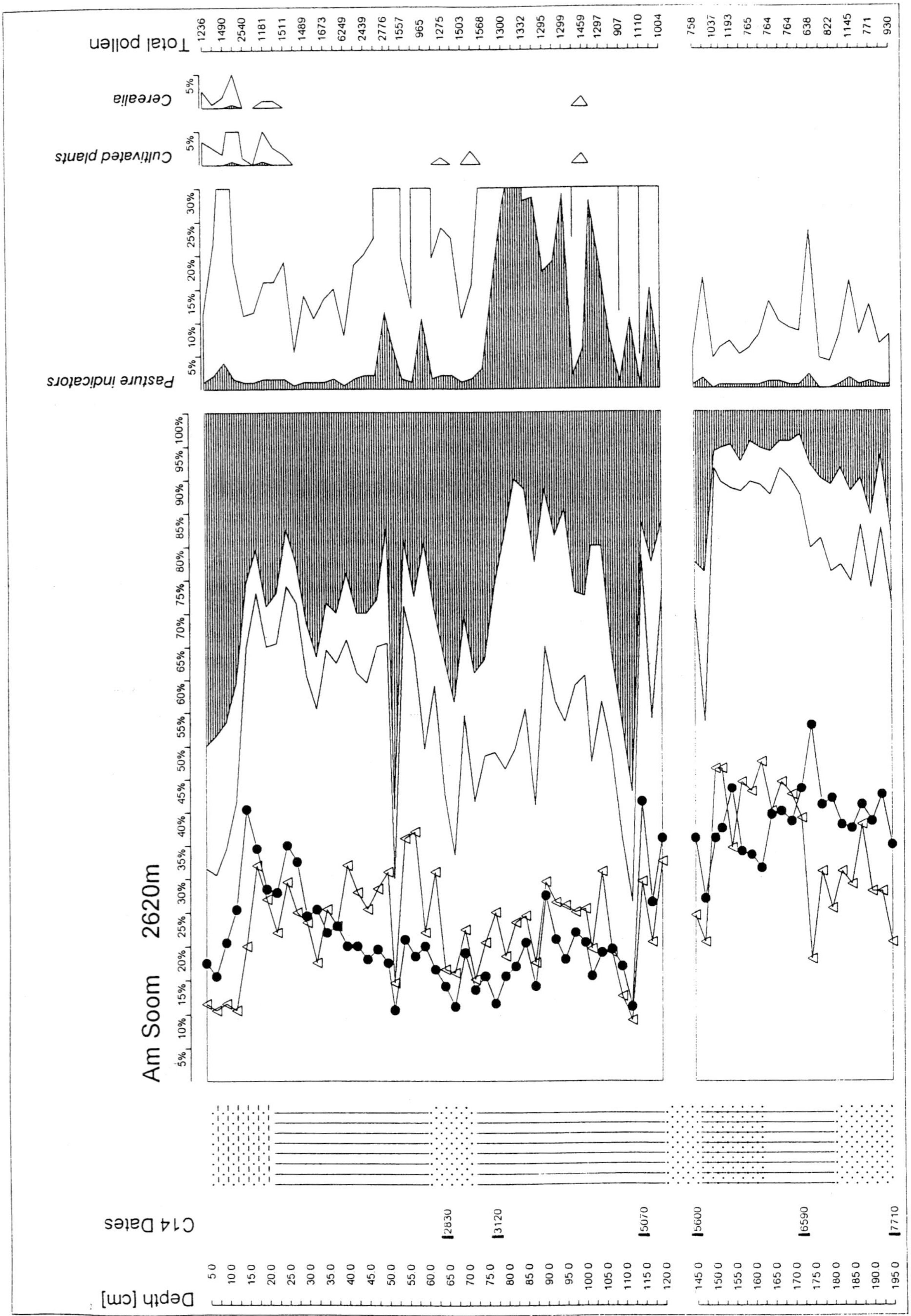

Text-Figure 4. Condensed pollen diagram from the Am Soom peatbog.

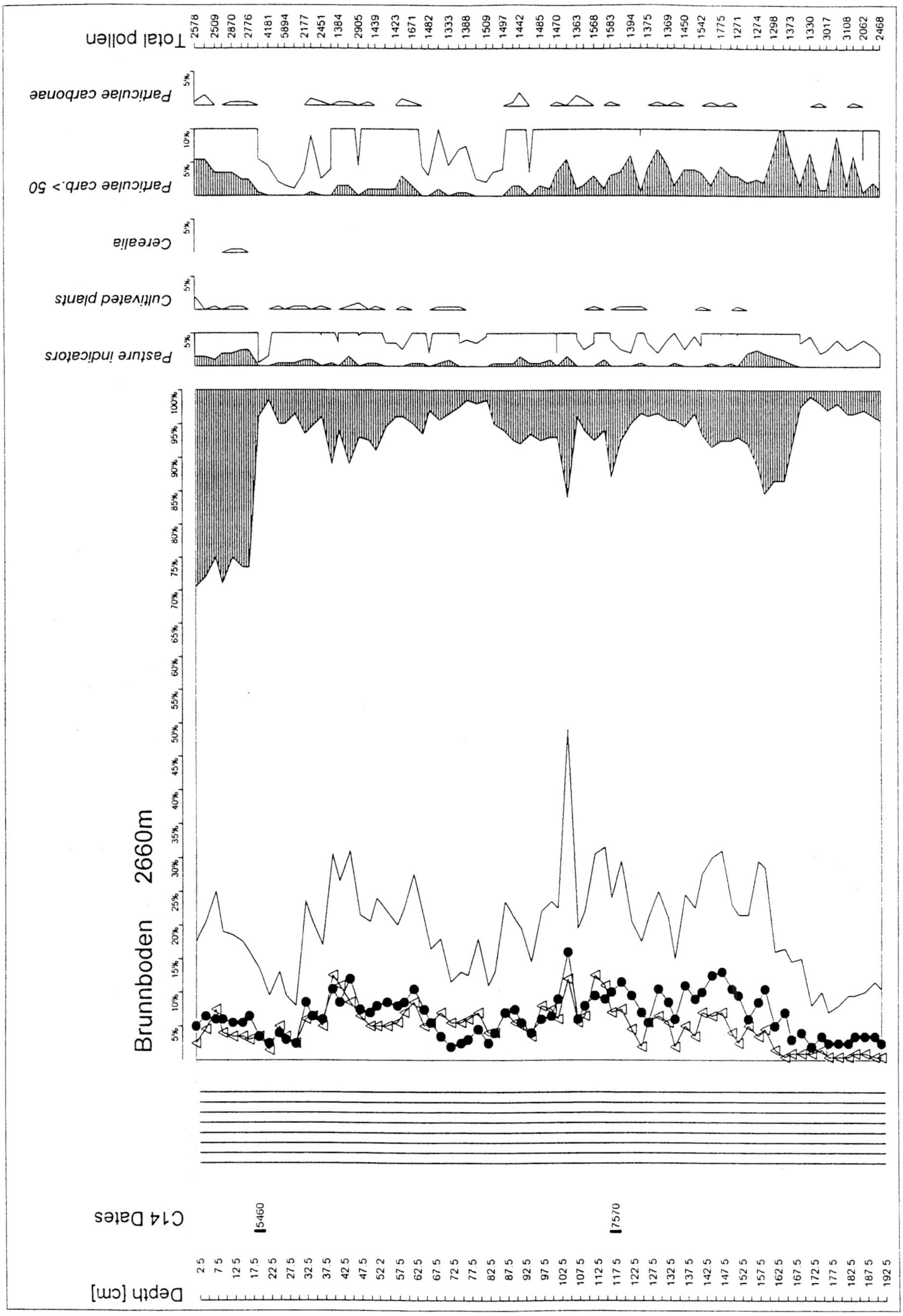

Text-Figure 5. Condensed pollen diagram from the Brunnboden peatbog.

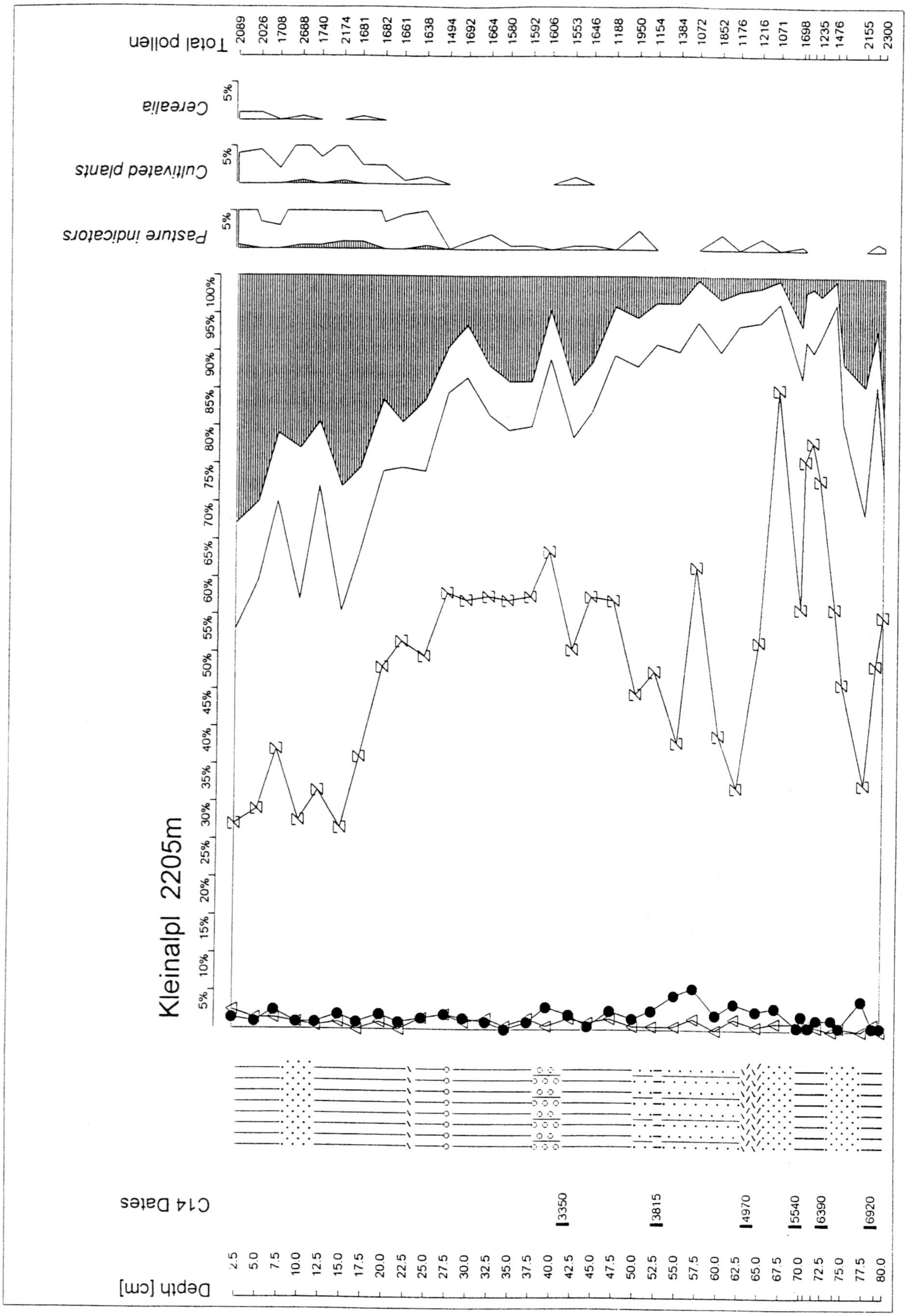

Text-Figure 6. Condensed pollen diagram from the Kleinalpl peatbog.

1971b *Pollenanalytische Untersuchungen zur Waldgeschichte des Dachsteinmassivs. Rekonstruktionsversuch der Waldgrenzdynamik.* Veröffentlichung des Instituts für Waldbau, Hochschule für Bodenkultur Wien. 145 pgs.

1994 Zur Frage früher Almwirtschaft am Östlichen Dachsteinplateau (Pollenprofil Plankenalm). *Mitteilungen der ANISA,* 15: 37–43.

OEGGL, K., AND SCHOCH, W.

2000 Dendrology of the Iceman. *In:* Bortenschlager, S., and Oeggl, K. (eds.). *The Iceman and his Natural Environment.* The Man in the Ice, volume 4. Springer-Verlag, Wien, New York, pp. 29–61.

OEGGL, K.

2000 The Diet of the Iceman. *In*: Bortenschlager S. and Oeggl K. (eds.): *The Iceman and his Natural Environment.* The Man in the Ice, volume 4. Springer-Verlag, Wien, New York, pp. 89–115.

RASTBICHLER-ZISSERNIG, E., WILFLING, H., ZUR NEDDEN, D., and NOTHDURFTER, H.

1995 Der Mann im Eis. Neue Funde und Ergebnisse. Wien, 320 pgs.

VORREN, K.D., MORKVED, B., and BORTENSCHLAGER, S.

1993 Human impact on the Holocene forestline in the central Alps.*Vegetation History and Archaeobotany,* 2: 145–156.

WELTEN, M.

1982 Vegeationsgeschichtliche Untersuchungen in den westlichen Schweizer Alpen: Bern-Wallis. *Denkschriften der Schweizerischen Naturforschenden Gesellschaft,* 95, 104 pgs.

WILFLING, H., RASTBICHLER-ZISSERNIG, E., ZUR NEDDEN, D., and NOTHDURFTER, H.

1996 Human Mummies. *The Man in the Ice, Volume 3,* 294 pgs.

WININGER, J.

1995 Die Bekleidung des Eismannes und die Anfänge der Weberei nördlich der Alpen. *In:* Spindler, K., Rastbichler-Zissernig, E., Wilfing, H., zur Nedden, D., and Nothdurfter, H. (eds.), *Der Mann im Eis Neue Funde und Ergebnisse Band 2.* Springer Verlag Wien New York, pp. 119–187.

Author's address

Sigmar Bortenschlager
Institut für Botanik
Universität Innsbruck
Sternwartestraße 15
A–6020 Innsbruck
Austria

PALYNOSTRATIGRAPHY OF HIMALAYAN LACUSTRINE SEDIMENTS

Chhaya Sharma

Abstract

Pollen analysis of a sedimentary profile from Khechipiri Lake, situated in the temperate zone of Sikkim Himalaya, in conjunction with three ^{14}C dates (2,280 ± 110 yrs. B.P., 2,380 ± 110 yrs. B.P. and 1,680 ± 130 yrs. B.P.) at different depths has unraveled the vegetation history back to Late–Holocene. Around 2,500 yrs. ago, this mountain region is interpreted as covered by dense mixed broad-leaved forests and poorly developed herbaceous vegetation based on exceedingly high values of *Quercus*, followed by *Alnus* and *Pinus*, with other associated arboreal elements such as *Betula, Carpinus, Juglans, Ulmus, Larix, Corylus, Rhododendron, Mallotus, Viburnum*, Rosaceae, and Oleaceae. These reflect warm–temperate and humid climate. Between 1,900 and 1,400 yrs. B.P., a decline in *Quercus, Alnus, Viburnum*, and Oleaceae along with a simultaneous increase in *Larix, Pinus*, and herbaceous vegetation, demonstrates the possibility of a short interval of climatic deterioration, or a phase of cool oscillation. The subsequent restoration of broad-leaved forests between 1,400 and 1,000 yrs. B.P., is indicated by the increased frequencies of *Quercus, Betula, Larix, Rhododendron*, and *Alnus*. Around 600 yrs. B.P., *Rhododendron* and *Alnus* were prominent, whereas *Quercus, Betula*, and *Carpinus* gradually decline. Consistent and considerable increase in the herbaceous elements, along with higher numbers of agricultural pollen, is consistent with the present climatic phase of higher precipitation.

INTRODUCTION

Lake sediments are natural archives which provide important clues reflecting the climatic changes that have taken place in the past. This is especially true during the Quaternary, whose reconstruction is possible using well preserved pollen, spores and other plant remains. The understanding of paleoclimate is the prerequisite for models to predict future climate, especially in view of the present global warming due to the greenhouse effect and alarming environment deterioration.

The Himalayas, with a rich floristic diversity in a variety of fragile ecosystems, are in great danger due to unchecked human interference and development. Nevertheless, the East Himalaya, which includes Upper Assam, Arunachal Pradesh, Sikkim, and the Darjeeling district of West Bengal, still claim to constitute the richest plant diversity center (Rao, 1994). Because no systematic investigations have hitherto been carried out on the Quaternary palynology of the Eastern Himalaya sites at Darjeeling and Sikkim have been selected to reconstruct the paleovegetation from different altitudes in these Himalayan regions. These areas have not been subjected to intense anthropogenic activities and are expected to yield dependable data to decipher the chronology of paleoclimate. Palynostratigraphic studies were recently carried out for two lakes situated in the temperate zone of Darjeeling Himalaya-Mirik (Sharma and Chauhan, 1994) and Jorepokhari (Chauhan and Sharma, 1996), and these sites have portrayed a reasonable vegetation succession since the Last Glacial Maximum.

SIKKIM HIMALAYA

The Indian Himalayan Sikkim state is bounded in the south by the Darjeeling district and in the north, west and east by Tibet, Nepal and Bhutan, respectively. Sikkim-Darjeeling is the most humid region in the Himalayan range, supporting unparalleled floristic diversity under a variety of sensitive ecosystems, derived mainly from the abruptly rising hills that directly confront the moisture-laden winds blowing over the Bay of Bengal.

Sikkim and Darjeeling lie in the same Himalayan region, however, altitude, topographical position, and degree of precipitation have greatly contributed to the variability in the qualitative richness of the vegetation. Both Sikkim and Darjeeling are known for their richness in colourful rhododendrons and orchids, among many plant taxa which are good indicators of specific ecosystems.

Sharma, Chhaya
2001 Palynostratigraphy of Himalayan lacustrine sediments; *In*: Goodman, D.K., and Clarke, R.T. (eds.), *Proceedings of the IX International Palynological Congress, Houston, Texas, U.S.A., 1996*; American Association of Stratigraphic Palynologists Foundation, p. 527–532.

ISBN 0–931871–06–9

KHECHIPIRI LAKE

The profile investigated in this paper from Khechipiri Lake is the first palynostratigraphic study from the regions of Sikkim Himalaya. The lake is situated about 160 km west of Gangtok at an elevation of 1,820 m deep inside the forest, and is most probably the largest and deepest lake of the region (Text-Figure 1). Topograhically, the area is flat but surrounded by steeply rising mountain ridges which are thickly covered with forests (Sharma, in press). This lake has remained undisturbed since the advent of human activity in this remote region because it has always been held highly sacred by the Buddhists and other local inhabitants. The lake has a swampy margin and thus provides a most suitable site for Quaternary palynostratigraphical investigations.

STRATIGRAPHY AND RADIOCARBON DATES

After carrying out a number of trial borings, each only about 3m deep because the corer struck hard substratum beneath, a sedimentary profile was recovered from the swampy margin. Samples from the core were collected at 5 cm intervals for pollen analysis.

Lithologically, the profile is comprised mainly of peaty material except for the lower horizon which had peat mixed with clay (Table 1), and which has been radiocarbon dated at three different depths providing the extrapolated age at the beginning of the profile to about 2,500 years (Table 2).

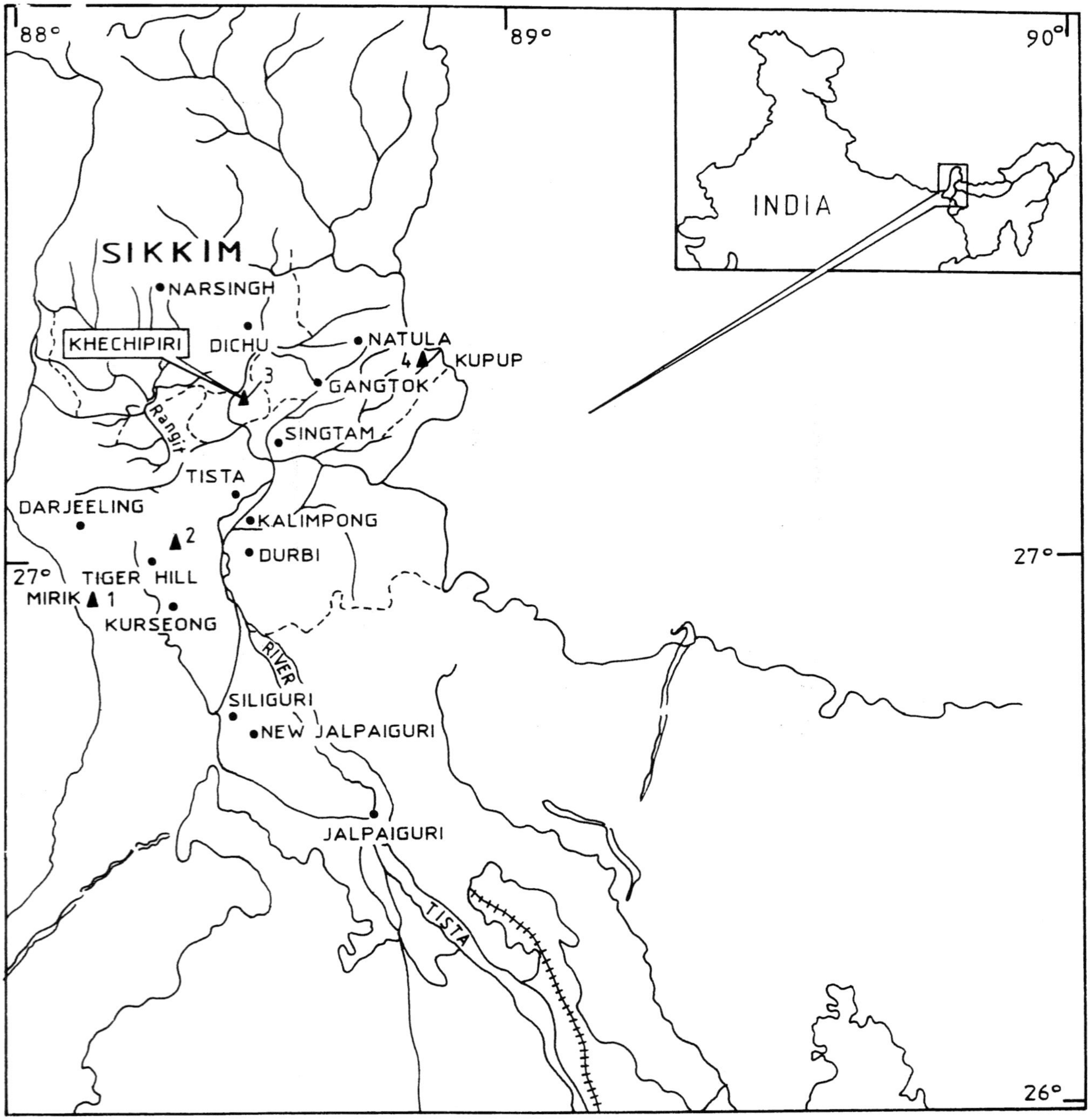

Text-Figure 1. Map of Darjeeling and Sikkim area to show location of lake sites. Index : ▲ lake site: 1. Mirik, 2. Jorepokhari, 3. Khechipiri, 4. Kupup; • Township.

POLLEN ANALYSIS

Erdtman's 1943 techniques were adopted to extract the pollen and spores from the matrix as described earlier for Mirik Lake (Sharma and Chauhan, 1994).

Pollen Diagram

The pollen diagram (Text-Figure 2) shows five pollen assemblage zones on the basis of fluctuations in the frequencies of major vegetation components. These are numbered I through V from the bottom upwards.

Pollen Zone I (305–215 cm): Oak–Alder–Pine–Viburnum–Oleaceae Assemblage

The bottom pollen zone is radiocarbon dated to 2,280 ± 110 yrs. B.P., at 295–305 cm and 2,320 ± 110 yrs. B.P., at 250–275 cm depth, characterized by the high values of *Quercus* followed by *Pinus, Alnus,* and Poaceae. Other broad-leaved

TABLE 1. Lithostratigraphy.

Column Depth	Lithology
0–100 cm	Fibrous-peat, with rootlets
100–150 cm	Fibrous-peat, turning blackish and more compact at the bottom
150–200 cm	Blackish-peat
200–250 cm	Peaty-clay
250–300 cm	Clay, mixed with peat
305–? cm	Compact peat

taxa consistently present are *Betula, Carpinus, Juglans, Ulmus, Corylus, Rhododendron, Mallotus*, Anacardiaceae, and Rutaceae. *Larix*, and *Tsuga*, which are higher altitude taxa (Sharma and Gupta, 1984) in Sikkim mainly north of the lake site, show better representation in this zone than *Cupressus*. The shrubby elements, namely *Viburnum*, Oleaceae, and Rosaceae, occur in low numbers but become quite frequent in the upper half of the zone.

Herbaceous vegetation remains poorly represented except for Poaceae which maintains consistent moderate values throughout the zone. The sporadic occurrence of *Potamogeton* is the only representative of aquatic vegetation. Both monolete and trilete fern spores have fairly moderate values.

Pollen Zone II (215–155 cm): Oak–Pine–Alder–Larch–Oleaceae–Rosaceae Assemblage

This pollen zone with a solitary radiocarbon date 1,680 ± 130 yrs. B.P., at 170–185 cm depth is marked by a slight decline in *Quercus* as well as *Viburnum* at the base, whereas Oleaceae and Rosaceae decline in the upper half. However, increases in the values of both *Pinus* and *Larix* occur in the zone. Other arboreal taxa such as *Alnus, Betula, Ulmus, Rhododendron*, and *Carpinus*, maintain more or less the same values as in the preceding zone.

The herbaceous vegetation is represented by increased frequencies of Poaceae, Cyperaceae, and Hypericaceae, while the aquatic *Potamogeton* registers slightly increased frequency at the base of the zone. *Typha* makes its first appearance but in low numbers. Fern spores are encountered in fluctuating high values.

Pollen Zone III (155–105 cm): Oak–Alder–Birch–Larch–Rhododendron–Rosaceae Assemblage

In this pollen zone, *Quercus, Betula*, and *Alnus* maintain almost their same values but *Larix* and *Rhododendron* register much higher frequencies. On the other hand, *Pinus* declines considerably and *Juglans* makes its first appearance. Among the aquatics, *Potamogeton* increases whereas both *Nymphaea* and *Lemna* appear for the first time but in extremely low frequencies. Fern spores retain their abundance throughout the zone.

Pollen Zone IV (105–70 cm): Oak–Alder–Carpinus–Birch–Rosaceae Assemblage

In this zone, *Quercus* and *Betula* more or less maintain their values, whereas *Alnus, Carpinus*, and *Juglans* increase. *Pinus, Larix*, and Ancardiaceae decline; among the shrubs, Rosaceae increase in the lower half of the zone. Poaceae, Cyperaceae, *Artemisia*, Hypericaceae, and Ranunculaceae are the chief representatives among the herbaceous elements. *Potamogeton* is the only aquatic form but occurs only in low values. Fern spores, both monolete and trilete, increase.

Some fungal spores such as *Tetraploa, Curvularia, Alternaria* and *Helminthosporium* are continuously though sporadically encountered almost throughout the column from the upper part of Pollen Zone-I. *Botryococcus* is recorded in moderate frequencies right from the bottom of the sequence but becomes sporadic within the zone. *Pediastrum* is another fresh water alga continuously represented in low values until it vanishes completely at the commencement of this zone.

Pollen Zone V (70–0 cm)

This uppermost zone is divided into three sub-zones:

1. **Sub-zone V-a (70–40 cm): Oak–Pine–Rosaceae–Artemisia–Ranunculaceae– Hypericaceae Assemblage.** This pollen subzone is delineated by the marked increase in the values of *Alnus* and *Rhododendron* on one hand and reduction in the frequencies of *Quercus, Betula, Carpinus, Pinus, Tsuqa*, and *Larix* on the other. Cyperaceae, *Artemisia*, Cheno/Ams, Ranunculaceae, Urticaceae, *Impatiens*, Lamiaceae, and Rubiaceae are non-arboreal elements with much higher values than below. Hypericaceae alone among the non-arboreals occurs in high values in this subzone and Poaceae, which constantly maintained its dominant position in this subzone declines. Aquatic as well as algal elements disappear at the base of this subzone, whereas trilete fern spores rise further in their values. *Lycopodium* is represented in low but consistent values.

2. **Subzone V-b (40–20 cm): Rhododendron–Birch–Anacardiaceae–Hypericaceae Assemblage**. Subzone V-b is characterized by the sharp decline in *Quercus, Pinus*, Poaceae, Hypericaceae, Ranunculaceae and a

TABLE 2. Radiocarbon dates.

Sample #	Sediment texture	Depth	Lab.No.	^{14}C dates
1	Peaty-clay	170–185 cm	BS-914	1680 ± 130 yrs. B.P.
2	Peaty-clay	250–275 cm	BS-913	2320 ± 110 yrs. B.P.
3	Peaty-clay	295–305 cm	BS-912	2280 ± 110 yrs. B.P.

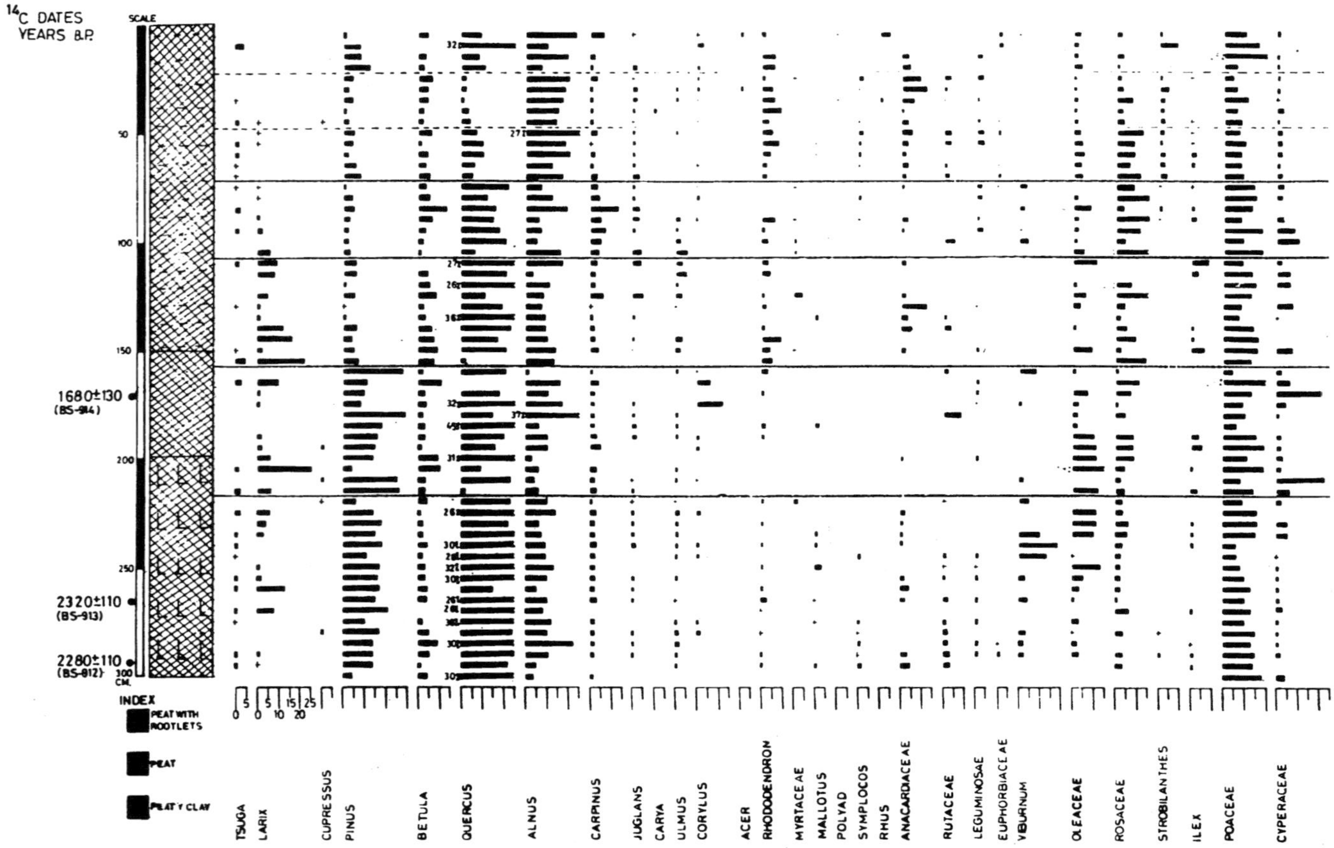

Text-Figure 2. Pollen diagram from Khechipiri Lake, Sikkim, Eastern Himalaya (percentage calculated in terms of total terrestrial plants).

slight increase in the values of *Rhododendron* and Anacardiaceae. *Alnus, Betula,* and *Carpinus* maintain more or less the same values, whereas the two conifers *Tsuga* and *Larix* dwindle to scattered values.

Shrubby elements such as Rosaceae, Oleaceae, and *Viburnum* also decline considerably as none of the aquatics or algal element is represented in this subzone. Fern spores on the whole decline but *Lycopodium* maintains almost the same values as in the preceding subzone.

3. **Subzone V-c (20–0 cm): Oak–Pine–Artemisia–Rhododendron Assemblage.** This subzone is characterized by the regained values of *Quercus, Pinus, Alnus* and *Rhododendron,* although *Tsuga, Juglans, Corylus,* and *Acer* are sporadically represented. However, some shrubby elements such as Oleaceae and Rosaceae do not exhibit any marked change, whereas *Strobilanthes* exhibits improved values only in the upper part of the subzone. Poaceae, Hypericaceae, Urticaceae, Lamiaceae, and *Polygonum* show increased frequencies in the upper part of this subzone.

VEGETATION AND CLIMATE IN SIKKIM REGION SINCE THE LATE–HOLOCENE

The pollen diagram from the Khechipiri Lake profile has three radiocarbon dates, i.e. 2,280 ± 110 yrs. B.P., 2,380 ± 110 yrs. B.P., and 1,680 ± 130 yrs. B.P., at different depths. The extrapolated age of the bottom of the investigated section covers about 2,500 years. The vegetational history begins with the existence of thick, mixed broad-leaved forests and poor herbaceous cover as indicated by very high values of *Quercus,* along with *Alnus,* and *Pinus.* A warm-temperate humid climate is reflected by such forest composition, and is further supported by the preponderance of fern spores. A marked dominance of fern spores in the Recent pollen spectra (Sharma, 2001), as well as the Quaternary pollen diagram, have demonstrated their localized occurrence in the region in wet or shaded humid habitats. Their vertical transportation is limited as indicated through aeropalynological studies in the same region which show very poor representations of these spores in the air catches.

Subsequently, during the period between 1,900 and 1,400 yrs. B.P., a decline in *Quercus* and *Viburnum* occurs that is associated with a simultaneous increase in *Larix* and *Pinus.* An increase in the frequencies of herbaceous elements is also registered during this phase. This change in vegetation pattern indicates a probable short time of climatic deterioration and correlates with the similar change seen in the Darjeeling pollen sequence (Chauhan and Sharma, 1996).

Finally, between 1,400 and 1,000 yrs. B.P., rapid expansion of the broad-leaved forests took place, indicated by an increase in the frequencies of *Quercus, Betula, Larix, Rhodo-*

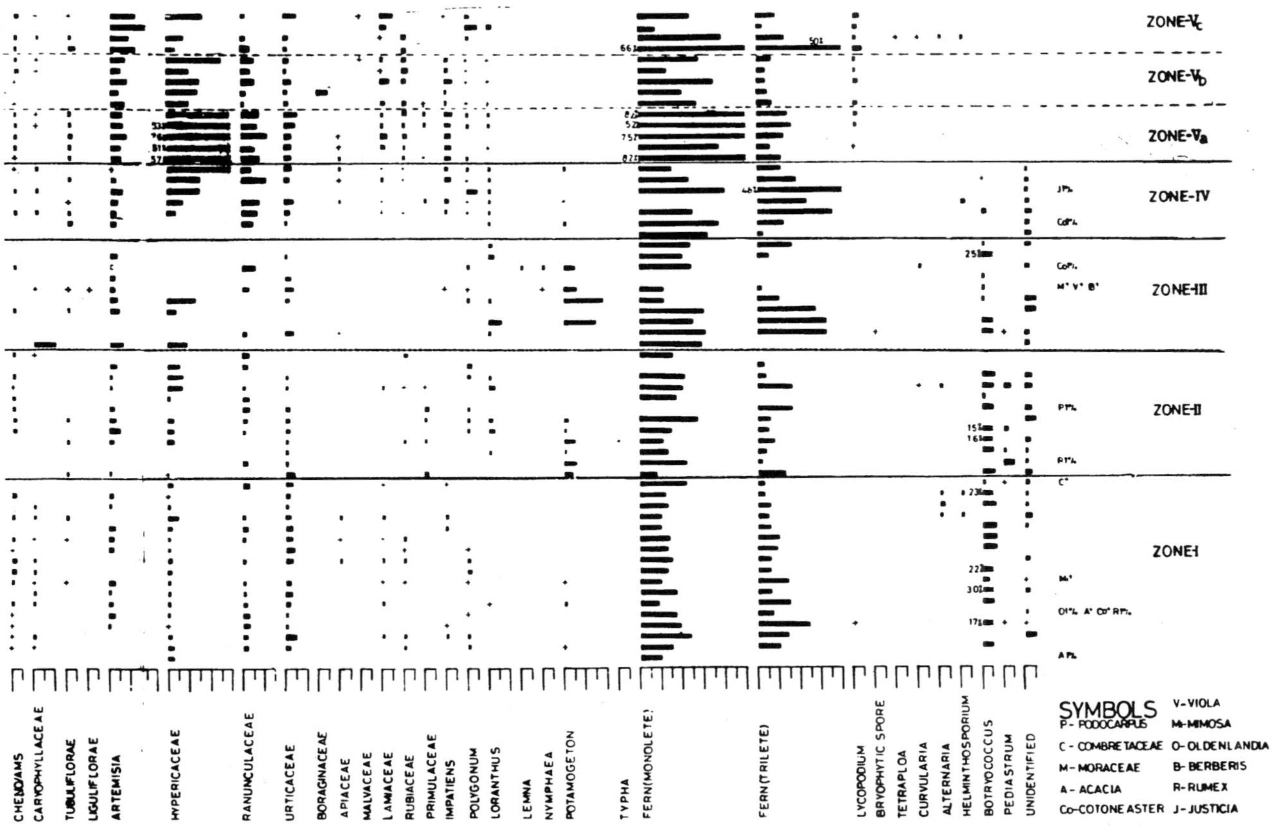

Text-Figure 2 (continued). Pollen diagram from Khechipiri Lake, Sikkim, Eastern Himalaya (percentage calculated in terms of total terrestrial plants).

dendron, and *Alnus*. *Pinus* declined considerably during this phase, whereas aquatic vegetation and ferns register an increasing trend. This reflects the restoration of warm-temperate and humid climate in the region.

During the period between 1,000 and 400 yrs. B.P., more or less similar vegetation occurs but the trend shows a further rise in broad-leaved taxa and a decline in *Larix* and *Pinus*. These trends indicate the climax of the vegetation cycle with well-established forests in the region, that characterize the present floristic zone.

However, about 600 years ago, a phase with marked increase in *Rhododendron* and *Alnus* occurs with a simultaneous reduction in the values of *Quercus*, *Betula*, *Carpinus*, and *Tsuga* along with considerable consistent increase in herbaceous vegetation together with associated agricultural pollen. This is indicative of higher precipitation as well as human effects in the region. The top of the sequence satisfactorily reflects the vegetation scenario existing today.

CORRELATION OF DIAGRAMS CONSTRUCTED FROM EASTERN HIMALAYA

Three pollen diagrams have been constructed to date for the Eastern Himalaya, all of which are from the temperate belt. These are from the lake—Mirik and Jorepokhari in Darjeeling (Sharma and Chauhan, 1994; Chauhan and Sharma, 1996) and Khechipiri in Sikkim. The paleovegetation succession covering the last 20,000 years (since the Last Glacial Maximum) is indicated in a 1.3 m deep profile from Mirik Lake (Sharma and Chauhan, 1994). In contrast, 1.5 and 3 m deep profiles from Jorepokhari (Chauhan and Sharma, 1996) and the Khechipiri sites investigated here respectively portray paleovegetation changes during only the past 2,500 years.

Around 2,500 years. ago, the vegetation in Darjeeling and the Sikkim regions was comprised of mixed broad-leaved forests as is characteristic in the temperate belt today, thus reflecting to the prevalence of a warm-temperate and humid climate in the region.

However, all three areas record a distinct decrease in broad-leaved taxa, with a simultaneous increase in conifers between 2,000 and 1,500 yrs. B.P. This reflects a short period of climatic deterioration, or "Cool Oscillation." However, re-expansion of broad-leaved forests soon took place and has continued thereafter. There is also evidence of enhanced precipitation during the period between 1,000 and 500 yrs. B.P., which is indicated by the prominent increase in *Rhododendron* in the temperate zone. Also, anthropogenic activities are inferred around 600 to 500 years ago which continued untill present time as is evident in Jorepokhari and Khechipiri. Such human interference could not be established at Mirik Lake because its upper profile failed to yield any pollen.

CONCLUDING REMARKS

The Quaternary lacustrine sediments from different regions along the Himalayan range (Sharma, 1993) investigated to date have shown that the Eastern Himalayan sector has a greater potential to decipher the trends in the relative evolution of climate, vis-à-vis the corresponding climate prevailing in the plains of the Indian subcontinent. This, however, depends significantly on the generation of high resolution proxy data through analysis of the deeper stratigraphic profiles from the region. The palynostratigraphic studies carried out to date in Western Rajasthan, the main desert, and in the adjoining state of Gujarat in the region of the western Indian plains, broadly corroborate the paleoclimatic inferences hitherto drawn from the Eastern Himalaya. As evidenced by some investigated lakes in the Indian desert, before 10,000 yrs. B.P. extreme arid conditions prevailed in the region. Strong winds, resulted in the present scenario of stabilized and moving sand dunes, and in the formation of the inland basins, i.e., the table land or the site with the existing lakes. Stabilization of sand dunes and the advent of herbaceous vegetation (composed of typical xeric taxa), coupled with fresh water lake conditions, occurred between 10,000 and 9,000 yrs. B.P. However, the climatic conditions remained stable in the region between 9,500 and 5,000 yrs. B.P., except for increased annual precipitation. Between 5,000 and 3,000 yrs. B.P., the lakes turned saline and arboreal elements simultaneously entered in the area. The climate and resultant floristic characteristics began around 1,200 yrs. B.P. (Singh et al, 1974). However, southwest of this region, in the Gujarat plains, the early Holocene vegetation continued to consist mainly of herbaceous elements. An invasion of few arboreal elements took place around 7,000 yrs. B.P., which then declined between 5,000 and 4,000 yrs. B.P., ultimately resulting in savanna-type vegetational conditions (Vishnu-Mittre and Sharma, 1975, 1979; Sharma and Chauhan, 1991).

ACKNOWLEDGMENTS

The Department of Science and Technology, Government of India, New Delhi, is gratefully acknowledged for the sponsored project "Studies of paleoclimates through the application of paleobotanical methods," under which the present studies were carried out by the author. She is also indebted to Dr. G. Rajagopalan for the radiocarbon dates in this paper.

References Cited

CHAUHAN, M.S., and SHARMA, C.
1996 Late Holocene vegetation of Darjeeling (Jorepokhari) Eastern Himalaya. *Palaeobotanist*, 45: 125–129.

ERDTMAN, G.
1943 *An Introduction to Pollen Analysis*. Ronald Press Company, New York, 239 pgs.

RAO, R.R.
1994 *Biodiversity in India*. New Connaught place, Dehradun, India, 315 pgs.

SHARMA, C.
1993 Palynostratigraphy of Lake Deposits of Himalaya and Palaeoclimate in Quaternary Period. *Current Science*, 64: 930–933.
2001 Modern Pollen Rain compared to vegetation in the Himalayas. *In*: Goodman, D.K., and Clarke, R.T. (eds.), *Proceedings of the IX International Palynological Congress, Houston, Texas, U.S.A., 1996*; American Association of Stratigraphic Palynologists Foundation, p. 557–565.

SHARMA, C., and CHAUHAN, M.S.
1991 Palaeovegetation and palaeoenvironmental inferences for the Quaternary Palynostratigraphy of Western Indian Planes *Man and Environment*, 16: 65–71.
1994 Vegetation and climate since Last Glacial Maximum in Darjeeling (Mirik Lake),Eastern Himalaya. *Proceedings* of 29th *International Geological Congress*, Japan, Part B: pp. 279–288. VSP Publishers, The Netherlands.

SHARMA, C., and GUPTA, H.P.
1984 Past and present distribution of *Larix griffithiana* Hort. ex Carr in the Indian subcontinent as evidenced by palynology. *ZfA Z. Archaeology*. Berlin, 8: 239–246.

SINGH, G., JOSHI, R.D., and SINGH, A.B.
1974 Late Quaternary history of vegetation and climate of the Rajasthan Desert, India. *Philosophical Transactions of the Royal Society*, London (Biological Sciences), 267: 467–501.

VISHNU-MITTRE, and SHARMA, C.
1975 Pollen analysis of the Salt Flat at Malvan, Gujarat. *Palaeobotanist*, 22: 118–123.
1979 Pollen analytical studies at the Nal Lake (Nalsarovar), Gujarat. *Palaeobotanist*, 26: 95–104.

Author's address

CHHAYA SHARMA
Birbal Sahni Institute of Palaeobotany
53 University Road
Lucknow – 226007
India
E-mail: chhaya_sharma@rediffmail.com

RELATIONSHIP BETWEEN POLLEN SPECTRA AND VEGETATION IN FOUR COMMUNITIES OF OTAMENDI NATURAL RESERVE, PARANA RIVER DELTA, BUENOS AIRES, ARGENTINA

Celina A. Fernández and Nora Madanes

Abstract

This work was carried out at the Otamendi Natural Reserve (34° 10′ S - 58° 48′ W) on the Paraná River Delta, in the northeastern section of Buenos Aires Province. Different attributes of vegetation communities, on the community scale were analyzed through their respective pollen spectra. Phytosociological releves of four communities represented by a forest of *Celtis tala* (arboreal), two grasslands of *Distichlis spicata* and *Sporobolus pyramydatus - Xanthium cavanalliesii*, and a marsh of *Spartina densiflora* (herbaceous) were compared with their respective pollen spectra obtained from surface soil samples. The physiognomy in terms of life forms as well as the commmunities' qualitative and quantitative composition were analyzed through pollen spectra. The four communities showed strong physiognomic concordance between the pollen spectra and the original vegetation. Moreover, from the physiognomic standpoint, the forest of *Celtis tala* was separated from grasslands. Regarding taxonomic composition, in every case the dominant types are well represented in the pollen spectra. *Distichlis spicata* and *Spartina densiflora* communities showed a strong distortion in their pollen spectra due to the presence of extra-local types of adjacent vegetation. Furthermore, dominant vegetation taxa present different cases of under and overrepresentation in their pollen values.

INTRODUCTION

Different authors have drawn the attention upon the importance in searching for modern analogues through the study of pollen spectra in relation to its source vegetation (Wright, 1967; Webb, 1974; D'Antoni and Markgraf, 1977; Birks and Birks, 1980; Prentice, 1985). Mancini (1993) described the increase of this type of study during the last decade (particularly their development in Argentina), emphasizing, however, that they have been restricted to localities in Patagonia region (south of Argentina) and Buenos Aires political province (Markgraf et al., 1981; D'Antoni and Schäbitz, 1986; Heusser, 1989; Páez, 1991; Mancini, 1993; Prieto, 1996). No previous work has been published concerning modern analogues in the Paraná River Delta.

This paper focuses on an area included in the Paraná River Delta and constitutes an initial and necessary step for further paleoecological interpretation of fossil pollen spectra of the region, which can later be used to reconstruct its recent vegetation history.

Most of the literature on modern analogues comprises regional scale studies, where phytogeographical units are taken into consideration. However, due to the particular environmental characteristics of the Paraná River Delta area, with unique ecological and biogeographic features in Argentina and South America (Malvárez, 1997), the problem was approached at the community scale (Janssen, 1967). The primary purposes of the present paper are to (a) verify if the pollen spectra from surface pollen samples can be distinguished at the community scale, and (b) compare recent pollen spectra deposited on soil surface with its source vegetation.

Fernández, Celina A., and Madanes, Nora
2001 Relationship between pollen spectra and vegetation in four communities of Otamendi Natural Reserve, Parana River Delta, Buenos Aires, Argentina; *In*: Goodman, D.K., and Clarke, R.T. (eds.), *Proceedings of the IX International Palynological Congress, Houston, Texas, U.S.A., 1996*; American Association of Stratigraphic Palynologists Foundation, p. 533–540.
 ISBN 0–931871–06–9

CHARACTERISTICS OF THE AREA

The Paraná River Delta is an extensive mosaic of wetlands with a surface area of approximately 17,000 km^2, located in two adjacent political provinces of Argentina — Entre Ríos and Buenos Aires. The southern quarter of the delta's surface belongs to the latter province where the Otamendi Natural Reserve is located, on a restricted area lying within 34°10'48"–34°17'37" S Lat and 58°48'49"–58° 53'54" W Long (Text-Figure 1). The climate is humid temperate with annual average temperatures of 16°C. Very low temperatures in winter are infrequent due to the topography and because of the river's regulating effect (García, 1967). The region is subject to temporary and frequent flooding, particularly in the lower depressions, thus generating small ponds surrounded by marshy vegetation. From the floristic standpoint this area presents a mosaic of biogeographic elements coming from three different phytogeographic provinces: Pampean, Espinal and Paranense (Cabrera and Willink, 1980), thus giving rise to a great diversity of arboreal and herbaceous communities (Marchetti et al., 1988; Chichizzola, 1993). According to the latter study, three forest types can be distinguished — *Celtis tala, Salix humboldtiana*, and *Erythrina crista-galli*. Both *Salix humboldtiana* and *Erytrina crista-galli* grow close to the river, and are exposed to flooding. Pristine herbaceous communities of freshwater marshes and grasslands with different floristic composition and vegetation mosaics with co-dominance of species are found (Chichizzola, 1993).

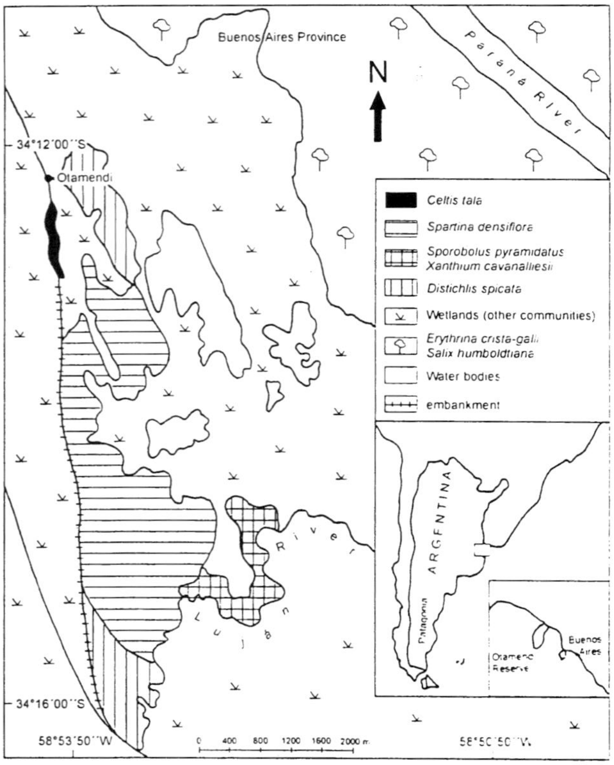

Text-Figure 1. Map showing the location of Otamendi Natural Reserve and the distribution of communities. Adapted from Chichizzola (1993).

MATERIALS AND METHODS

Four communities were selected for sampling from the Otamendi Natural Reserve. A forest of *Celtis tala*, two grasslands of *Sporobolus pyramidatus-Xanthium cavanalliessi* and *Distichlis spicata*, and a marsh community of *Spartina densiflora*. The communities are defined according to physiognomy, species dominance and topography (Text-Figure 1). Phytosociological releves were made in each community following Braun-Blanquet (1979) methods and complemented with descriptions by Chichizzola (1993).

Ten releves of 16 m^2 each were randomly selected in each vegetation type. The vegetation cover was estimated using the Braun-Blanquet cover-abundance scale (Mueller-Dombois and Ellenberg, 1974). For numerical analysis and comparison with pollen percentages, average values were used for + = 0.01; 1 = 2.5; 2 = 15.0; 3 =3 7.5; 4 = 62.5 and 5 = 87.5% respectively.

Within the *Celtis tala* community, three layers were identified: trees, shrubs, and herbs. Cover values for every layer were added and divided by the number of layers in which the species was observed. From the quotient between the sum of these values and the total cover sum of every species in the sampled vegetation, a relative cover value was obtained. Species were classified according to pollen dispersion (Lewis et al., 1983; Faegri and Iversen, 1989)

Surface soil samples for pollen analysis were taken at 0–1 cm depth in each community in sites randomly selected. Laboratory techniques were carried out for silica extraction and pollen concentration following Faegri and Iversen (1989). Residues were mounted in glycerine jelly for observation. Pollen grains were identified to the lowest taxonomic categories possible. Counts greater than 1000 grains per sample were made.

To compare phytosociological releves with their corresponding pollen spectra, vegetation releves were compared to the identified pollen grains (Text-Figure 2). Sorensen's similarity index (Mueller-Dombois and Ellemberg, 1974) was used to evaluate the similarity percentage between vegetation and its respective pollen spectra. A relation between anemophyllous types common to pollen – vegetation (APV) and anemophyllous in vegetation (AV) was calculated and expressed as the percentage APVx100/AV. The same applies for entomophyllous types where the relation is EPVx100/EV.

Species with releves or pollen percentages equal or greater than 2% in at least one of the samples were included in a matrix of 8 x 25 variables, from which pollen diagrams and a dendrogram resulting from cluster analysis were made.

For Q-mode cluster analysis of percentage pollen data and vegetation, the Euclidean distance similarity coefficient and UPGMA method were used.

Pollen-vegetation relationships were evaluated through the R/P value (Braggio-Marucchio et al., 1988), where R is the average cover-percent value for a taxon and P is the pollen percentage in the pollen spectra. Taxa with R/P less than 0.80 were considered over-represented and those with R/P values greater than 1.25 under-represented.

RESULTS

Seventy-seven pollen types, including 108 plant species analogued to 58 taxa (determined to the taxonomic hierarchical level of pollen types) were recorded for each community. In Text-Figure 2, the complete floristic list, pollination mode and analogued vegetation types to pollen types are shown. Average cover-abundance values for each species, summed values of taxa analogued to pollen types and pollen percentage values are shown in the three last columns respectively. Pollen types included in the table are equivalent to variable taxonomic categories, ranging from species to family or genus groups. In turn, these pollen categories were analogued to pollen sources. For example, *Cortadeira selloana* in the vegetation is read as Poaceae in the pollen spectra.

Vegetation

Celtis tala **community.** In Text-Figure 2 we observe that the dominant arboreal species is *Celtis tala*. Other arboreal species present were *Morus alba*, *Melia acedarah* and *Phytolacca dioica*. Epiphitic and bindweed species such as *Tillandsia* and *Clematis denticulata* were found in addition to the trees. The most frequent shrub species were *Pavonia consobrina*, *Lantana camara*, *Lantana entrerriensis* and *Cestrum parquii*. The most important species of the herbaceous layer were *Oxalis corymbosa*, *Viola odorata*, *Solanum glaucophyllum* and *Poa annua*.

Spartina densiflora **community.** The dominant species was *Spartina densiflora* and the following in dominance were: *Salicornia virginica*, *Limonium brasiliensis*, *Distichlis spicata* and *Brassica nigra*.

***Sporobolus pyramidatus-Xanthium cavanalliesii* community.** The cover was recorded in only one layer and the dominant species was *Sporobolus pyramidatus*. This was accompanied by species of secondary importance such as: *Oxalis perdecaria, Limonium brasiliense, Lepidium spicatum* and *Xanthium cavanalliesii* (Chichizzola, 1993).

***Distichlis spicata* community**. The dominant species was *Distichlis spicata* with codominants *Aster squamatus, Scirpus americanus, Atriplex prostata* and *Salicornia virginica*.

Pollen

Cluster analysis separated two principal groups (Text-Figure 3) — the arboreal group and the grasslands group. In the latter, the greatest affinity between pollen spectra were observed in the *Sporobolus pyramidatus-Xanthium cavanallisii* and *Distichlis spicata* communities, not only because of its dominant anemophyllous taxa (Cyperaceae and Poaceae) but also because the entomophyllous taxa are better represented in these two than in the rest of the communities; the communities differ because of the presence-absence of *Xanthiun cavanalliesii* in their pollen spectra. On the other hand, *Spartina* grassland differed from the other two because of the Chenopodiaceae values.

Pollen – Vegetation Relationship

***Celtis tala* community**. Fifty vegetation species were recorded and in turn analogued to 34 pollen taxa, while 22 pollen types were identified in the pollen spectra. The similarity index value was 31.6 %. Nine pollen types (39.1 %) had their sources in the forest, while 13 were extra-local (60.9 %) (Text-Figure 4a). From those 9 taxa common in the pollen spectra and vegetation, 5 were anemophyllous (55.6 %) and 4 were entomophyllous (44.4%). On its turn, from the 34 taxa counted, 9 were anemophyllous (26.5%) and the rest were entomophyllous (73.6 %) (Text-Figure 4b). The relation APVx100/AV was 55% and EPVx100/EV and 16%.

***Spartina densiflora* community**. Forty-one vegetation species were recognized and analogued to 28 vegetation taxa; 25 pollen grains were recorded. The similarity index was 33.4 %. Here, nine species (36.0 %) had their pollen source within the community while 16 (64.0 %) types were extra-local (Text-Figure 4a). From nine taxa recorded and present in vegetation, five (55.0 %) were anemophyllous (Chenopodiaceae, Cyperaceae, *Ambrosia*, Poaceae, *Typha* sp.) and four (44.4 %) were entomophyllous (Astereae, Malvaceae, Apiaceae and Inuleae) (Text-Figure 4b). From 28 vegetation types counted, 40.0% were anemophyllous and 60.0 % entomophyllous. Moreover, only five anemophyllous pollen types were captured from ten anemophyllous types and the same accounts for four pollen types from 18 vegetation types for entomophyllous sources. The relation APVx100/AV was 62% and EPVx100/EV was 31%.

***Sporobolus pyramidatus-Xanthium cavanalliesii* community**. Thirty-four vegetation species were analogued to 24 vegetation taxa while 32 pollen types were recorded. The similarity index value was 38.6 % (Text-Figure 4a). This can be explained because 11 types (33.3%) have their sources in the community while 22 types (66.7 %) were extra-local. Of the six taxa common in the pollen record and vegetation four were anemophyllous (80.0 %) and one was entomophyllous (20.0%). In 24 vegetation types, 20.8 % were anemophyllous and 79.2 % were entomophyllous. The relation APVx100/AV was 80% and EPVx100/EV was 31%.

***Distichlis spicata* community.** Forty-one vegetation species were analogued to 23 taxa and 31 pollen types were recorded. The similarity index indicated 40.7 %. This fact can be explained as seven had their sources in the site (35.4 %) while 20 were extra-local (64.5 %). Furthermore, from seven taxa recorded and present in the vegetation six were anemophyllous (54.5 %) and five were entomophyllous (45.4%). On the other hand nine anemophyllous (39.1 %) and 14 entomophyllous (60.9 %) were recorded from the 23 vegetation types (Text-Figure 4b). The relation APVx100/AV was 55% and EPVx100/EV was 24.

CLUSTER ANALYSIS

Cluster analysis separated two distinctive groups according to the communities physiognomy — one comprising both vegetation censuses and pollen percentage samples of the *Celtis tala* community, and another clustering the other six communities (Text-Figure 5). The affinity shown between vegetation and pollen samples in the first group is generated by *Celtis tala* values (dominant in vegetation) and those of Poaceae. On the other hand, according to R/P values anemophyllous taxa such as Chenopodiaceae, Cyperaceae, Poaceae and *Celtis tala* are over-represented while entomophyllous Apiaceae are under-represented. In addition, four extra-local taxa were recorded, with three of them anemophyllous (*Ambrosia* sp., *Typha* spp. and *Xanthium cavanalliesii*) and one entomophyllous (*Spilanthes* sp.). It has to be emphasized that taxa important for the community description such as Moraceae, *Phytolacca dioica* and Solanaceae, were not recorded in the pollen spectra independently of its dispersion type and abundance values.

Within the grassland group, cluster analysis separated the sample corresponding to the *Spartina densiflora* pollen spectra from the remaining samples. *Sporobolus pyramidatus-Xanthium cavanalliesii* grassland showed the greatest similarity between vegetation and its pollen spectra. The analysis of taxa present in pollen spectra and vegetation in the *Sporobolus-Xanthium* community showed that the anemophyllous type *Ambrosia* was over-represented, while *Xanthium cavanalliesii* and Poaceae were under-represented. Within the entomophyllous types, *Spilanthes* sp. was under represented while Apiaceae and cf. *Medicago* were over represented. On the other hand, four anemophyllous extra-local taxa (*Celtis tala*, *Typha* spp., Cyperaceae and Chenopodiaceae) were recorded in the same community. A vegetation sample of *Spartina densiflora* showed more affinity with samples of the former community than with its own pollen spectra. The similarity was principally due to Poaceae values. Furthermore, *Spartina densiflora* vegetation showed more affinity to the *Sporobolus-Xanthium* pollen spectra because of Chenopodiaceae and Apiaceae values. Vegetation samples of these two grasslands are close due to the presence of Brassicaceae and *Limonium brasiliense*, both important taxa for community description. Both taxa are entomophyllous and together with Apiaceae are absent in

			Vegetation communities								Pollen record			
			Species cover- abundance values (in percentages)				Taxa cover - abundance values (in percentages)				Pollen percentages			
Species	Pollination	Taxa	*C. tala*	*S. den.*	*Spo.-Xan.*	*D.spi.*	*C. tala*	*S. den.*	*Spo.-Xan.*	*D.spi.*	*C. tala*	*S. den.*	*Spo.-Xan.*	*D.spi.*
Acer sp.	A	Aceraceae	0.4				0.4							
Alternanthera philoxeroides	E	*Alternanthera*		0.4	0.1	0.2		0.4	0.1	0.2			0.4	0.2
Zephyrantes minime	E	Amaryllidaceae			0.1				0.1					
Ambrosia scabra	A	*Ambrosia*		0.2	0.1	0.2		0.2	0.1	0.2	1.2	0.4	0.6	3.1
Bowlesia incana	E	Apiaceae	0.8				3.3	2.2	0.1	2.5	0.8		6.2	6.3
Eryngium	E	Apiaceae		0.3										
Eryngium (delta)	E	Apiaceae	0.2											
Eryngium echinatum	E	Apiaceae	0.6	0.2		0.2								
Eryngium sp	E	Apiaceae				0.2								
Apium leptophyllum	E	Apiaceae	0.2	1.3		0.5								
Apium sellowiana	E	Apiaceae	1.0	0.3		0.7								
Hydrocotyle bonaerensis	E	Apiaceae		0.2		0.9								
Torilis nodosa	E	Apiaceae	0.6											
Aster squamatus	E	Astereae		1.3	2.6	2.3		1.3	2.6	2.3				0.2
Berberis ruscifolia	E	Berberidaceae	1.5				1.5							
Heliotropum curassavicum	E	Boraginaceae		0.3				0.3						
Brassica nigra	E	Brassicaceae		2.9		0.5	0.8	3.1	7.6	0.9	0.5		0.2	
Lepidium parodii	E	Brassicaceae				0.5								
Lepidium spicatum	E	Brassicaceae	0.4	0.2	7.5									
Capsella bursa-pastoris	E	Brassicaceae	0.4											
Tillandsia	E	Bromeliaceae	2.1				2.1							
Rhipsalis lumbricoides	E	Cactaceae	0.4		0.1		0.4		0.1					
Acicarpha tribuloides	E	Calyceraceae				0.2				0.2				
Sambucus australis	E	Capprifoliaceae	0.4				0.4							
Carduus acanthoides	E	*Carduus* sp	0.2	0.6			0.2	0.6					0.2	0.4
Cerastium glomeratum	E	Caryophyllaceae	1.2				1.5		0.1		0.1			0.3
Spergula levis	E	Caryophyllaceae			0.1									
Stellaria media	E	Caryophyllaceae	0.4											
Celtis tala	A	*Celtis tala*	29.4				29.4				58.0	0.6	0.6	0.5
Atriplex hastata	A	Chenopodiaceae	0.6	0.9		1.8	0.6	17.4		3.0	2.3	68.4	9.2	30.3
Salicornia virginiaca	A	Chenopodiaceae		16.5		1.1								
Dichondra microcalyx	E	Convolvulaceae			5.0				5.0					
Cayaponia bonaerensis	E	Cucurbitaceae	0.6				0.6							
Cynara cardunculus	E	Cynareae		0.2				0.2						
Carex riparia v. chilensis	A	Cyperaceae	0.2				0.2	3.1		2.1	12.4	11.1	25.5	14.0
Scirpus ameri. v. monophyllus	A	Cyperaceae		3.1		2.1								
Geranium molle	E	Geraniaceae	2.1				2.1							0.1
Flaveria bidentis	E	Helianeae			0.1				0.1					
Blainvillea biaristata	E	Heliantheae	0.6				0.6						0.2	0.3
Hypochoeris microcephala	E	*Hypochoeris*		0.4		0.5		0.4		0.5	0.2		0.1	
Pluchea sagitallis	E	Inuleae		0.3			0.2	0.6						
Conyza bonaeriensis	E	Inuleae	0.2	0.3										
Juncus acutus	A	Juncaceae	0.2				0.2			0.2	0.1	0.8	1.1	0.5
Juncus imbricats	A	Juncaceae				0.2								
Juncus venturianus	A	Juncaceae												
Marrubium vulgare	E	Labiatae	2.7				2.7					0.3	0.1	0.1
Picrosia longuifolia	E	Lactuceae	0.2			0.7	0.2	0.7		1.0	0.2	0.1		0.4
Sonchus asper	E	Lactuceae		0.7		0.5								
Pavonia consobrina	E	Malvaceae	9.3				9.3							
Malvella leprosa	E	*Malvella leprosa*		0.6	2.6			0.6	2.6			0.2		4.8
Medicago sp	E	aff. *Medicago*		0.7		1.4	0.0	0.7		1.4		2.0	6.6	
Melia azedarach	E	Meliaceae	5.8				5.8							
Morus alba	E	Moraceae	15.6				15.6							
Chaptalia arechavaletai	E	Mutisese	1.4				1.4						2.0	
Oxalis corimbosa	E	Oxalidaceae	10.4				10.4		23.8					
Oxalis perdicaria	E	Oxalidaceae			23.8									
Fumaria capreolata	E	Papaveraceae	0.8				0.8							
Melilotus indicus	E	Papilionoideae		0.7		0.7								
Trifolium repens	E	Papilionoideae				0.7								
Passiflora coerulea	E	Passifloraceae	0.8				0.8							
Phyla canescens	E	*Phyla canescens*		0.2	2.5	0.9		0.2	2.5	0.9			1.1	
Phytolacca dioica	A	*Phytolacca dioica*	2.3				2.3							
Plantago tomentosa	A	Plantaginaceae		0.3		0.7		0.3		0.7				
Limonium brasiliense	E	*Limonium* sp		13.8	15.0	0.5		13.8	15.0	0.5				

Text-Figure 2. Table showing the complete floristic list, pollination mode, pollen types identified and vegetation taxa analogued, its pollen types, with its respective percentages. A: anemophyllous; E: entomophyllous.

			Vegetation communities								Pollen record			
			Species cover- abundance values (in percentages)				Taxa cover - abundance values (in percentages)				Pollen percentages			
Species	Pollination	Taxa	*C. tala*	*S. den.*	*Spo.-Xan.*	*D.spi.*	*C. tala*	*S. den.*	*Spo.-Xan.*	*D.spi.*	*C. tala*	*S. den.*	*Spo.-Xan.*	*D.spi.*
Poaceae	A	Poaceae	0.6				6.2	62.1	48.4	98.9	9.6	11.8	33.3	22.2
Cortaderia selloana	A	Poaceae		3.5										
Cynodon dactylon	A	Poaceae			11.3	2.7								
Distichilis spicata	A	Poaceae		4.9		76.2								
Eragrostis curvule	A	Poaceae				0.2								
Poaceae	A	Poaceae				1.6								
Hordeum pusillum	A	Poaceae	0.2	0.2		3.0								
Panicum grumosum	A	Poaceae	1.0											
Paspalum dilatatum	A	Poaceae			2.5	0.9								
Poa annua	A	Poaceae	4.2											
Polypogon chilensis	A	Poaceae				6.4								
Polypogon monspeliensis	A	Poaceae		0.2		0.5								
Setaria geniculata	A	Poaceae	0.2			0.7								
Spartina densiflora	A	Poaceae		53.4		2.7								
Sporobolus indicus	A	Poaceae			2.5									
Sporobolus pyramidatus	A	Poaceae			32.1									
Stenotaphrum secundatum	A	Poaceae				4.1								
Muehlenbachia sagittifolia	A	Polygonaceae		0.4			0.0	0.8	0.1	0.7				
Polygonoum chilensis	A	Polygonaceae		0.2										
Polygonum aviculare	A	Polygonaceae			0.1									
Polygonum punctatum	A	Polygonaceae		0.2										
Samolus valerandi	E	Primulaceae		0.7				0.7						
Clematis denticulata	A/EF	Ranunculaceae	1.9				1.9	0.3						
Ranunculus apiifolius	A/EF	Ranunculaceae		0.3										
Galium aparine	E	Rubiaceae	2.1				2.1						0.1	
Rumex obovatus	A	*Rumex*		0.4		0.7		0.4		0.7				
Salix humboltiana	A	Salicaceae	0.2				0.2							
Gerardia communis	E	Scrophulariaceae		0.3	0.1	0.2		0.3	0.1	0.2				
Senecio	E	*Senecio*		0.2		1.1		0.2		1.1	0.3	0.1		0.2
Senecio bonaerensis	E	*Senecio*				0.2								
Senecio grisebachii	E	*Senecio*		0.2		0.9								
Cestrum parquii	E	Solanaceae	2.9				11.1	0.3	0.1	0.7				
Jaborosa integrifolia	E	Solanaceae			0.1									
Salpicroa origanifolia	E	Solanaceae	0.4											
Solanum laxum	E	Solanaceae	0.4		0.1	0.7								
Solanum glaucophyllum	E	Solanaceae	7.5											
Solanum pilcomayense	E	Solanaceae		0.3										
Sphilanthes stolonifera	E	Sphilanthes			2.6				2.55		0.7	0.1	0.7	3.9
Typha latifolia	A	*Typha spp*		0.9				0.9			10.6	0.9	2.3	0.7
Lantana camara	E	Verbenaceae	4.2				5.4	0.3		0.7				
Lantana entrerreniensis	E	Verbenaceae	1.2											
Verbena gracilescens	E	Verbenaceae		0.3		0.7								
Viola odorata	E	Violaceae	7.7				7.7							
Xanthium canavaliessi	A	*Xanthium*			7.6	0.2			7.6	0.2	0.2	0.4	5.1	0.1
	E	*Acacia*											0.3	
	A	*Alnus*									0.1	0.1	0.1	0.3
	E	Anthemideae									0.1	0.5	0.2	
	E	*Baccharis*									0.4			0.2
	A	*Carya*											0.3	
	A	*Cassuarina*									0.09	0.1	0.08	0.49
	A	*Ephedra*									0.3		0.1	
	E/AF	*Eucalyptus*									0.5	0.3	0.9	1.0
	E	*Helianthus annus*												0.2
	E	Ingae											0.1	
	A	*Junglans*												0.2
	E	Liliaceae										0.1	0.3	0.1
	A	*Nothofagus*										0.1	0.1	
	A	*Pinus*										0.3	0.7	0.4
	E	*Sagittaria*										0.1		
	E	*Sapium*									0.5			0.1
	A	*Schinus*										0.1		
	E	Valerianiaceae										0.1		
	A	*Zea mays*										0.1	0.3	0.2

Text-Figure 2 (continued). Table showing the complete floristic list, pollination mode, pollen types identified and vegetation taxa analogued, its pollen types, with its respective percentages. A: anemophyllous; E: entomophyllous.

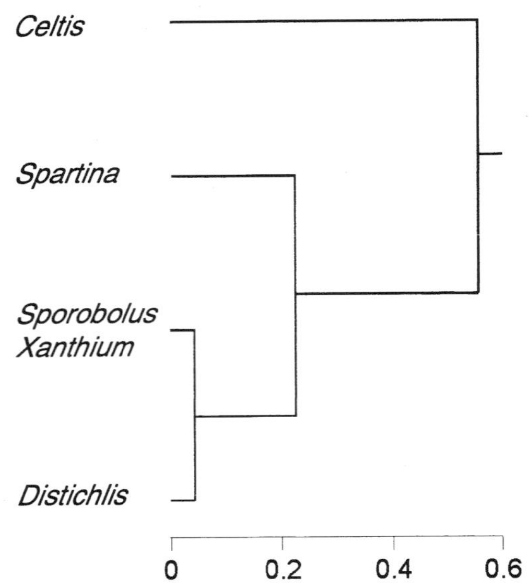

Text-Figure 3. Cluster analysis for pollen samples.

the *Spartina densiflora* pollen spectra. However, these taxa can be found in the *Sporobolus–Xanthium* pollen spectra and in *Spartina* vegetation. In turn, *Spartina* pollen spectra remains isolated due to high percentage values of Chenopodiaceae.

The overall analysis of vegetation and pollen samples of the *Spartina densiflora* community showed that plant types with anemophyllous pollination presented the following characteristics according to its R/P values: Chenopodiaceae, Cyperaceae and *Ambrosia* sp. were over-represented, while Poaceae was under-represented and *Typha* spp was equally represented. Within entomophyllous taxa, *Malvella leprosa* was under-represented while cf. *Medicago* was over-represented. Pollen types such as *Celtis tala*, *Spilanthes* sp. and *Xanthium cavanalliesii* are extra-local and showed low percentage values.

The set of samples corresponding to *Spartina densiflora* vegetation, in addition to both vegetation and pollen spectra of *Sporobolus-Xanthium* showed affinity with *Distichlis spicata* pollen spectra due to Poaceae, Chenopodiaceae, Cyperaceae and Apiaceae values. However, the pollen spectra of the latter community were less similar to their own vegetation source.

In the *Distichlis spicata* grassland, taxa that did not appear in the pollen spectra had little importance regarding the vegetation description. Of the five anemophyllous taxa, Chenopodiaceae, *Ambrosia* sp. and Cyperaceae were over-represented while *Xanthium cavanalliesii* and Poaceae were under represented. Within the entomophyllous types present in pollen spectra and vegetation, Apiaceae was over represented. *Spilanthes* sp., *Typha* spp., Inuleae and *Malvella leprosa* were recorded in the pollen spectra as extra-local elements.

DISCUSSION AND CONCLUSIONS

The Otamendi Natural Reserve comprises a set of communities spatially disposed in a fundamentally anisotropic manner (Eastman, 1993). Pollen spectra from four of

a

Vegetation communities	*Celtis tala*	*Spartina densiflora*	*Sporobolus Xanthium*	*Distichlis spicata*
Vegetation richness	50	41	34	41
Taxonomic richness	34	28	24	23
Pollen richness	22	25	32	30
Taxa common in pollen and vegetation	9	9	11	11
Taxa present in pollen	13	16	22	19
Taxa present in vegetation	25	19	13	12
Sorensen's index (percent)	31.6	35	38.6	40.7

b

Vegetation communities	*Celtis tala*	*Spartina densiflora*	*Sporobolus Xanthium*	*Distichlis spicata*
Anemophyllous taxa present in vegetation	26.5	35.7	20.8	39.1
Entomophyllous taxa present in vegetation	73.5	64.3	79.2	60.9
Anemophyllous taxa common in pollen and vegetation	55.6	55.6	33.4	54.5
Entomophyllous taxa common in pollen and vegetation	44.4	44.4	63.6	45.5

Text-Figure 4. a. Quantitative analysis: number of species, taxa and pollen types are expressed as richness. Number of taxa common to pollen and vegetation, only in pollen and only in vegetation are indicated. b. Pollination mode percentages in vegetation and common in pollen and vegetation.

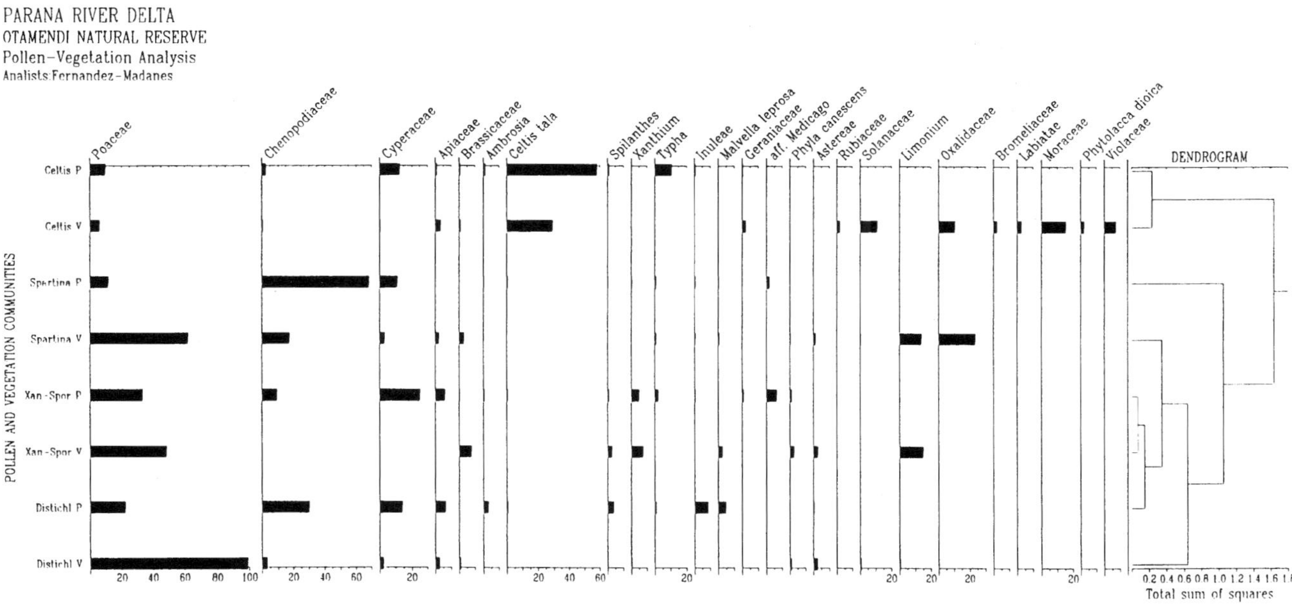

Text Figure 5. Pollen-vegetation diagrams and cluster analysis.

these communities could be distinguished through pollen analysis. This represents a first step for the pollen spectra characterization in the area together with a comparison between pollen-vegetation relations (Janssen, 1966; Wright, 1967).

Comparison between phytosociological censuses and pollen spectra for each community revealed a distortion between present pollen rain and source vegetation (Faegri and Iversen, 1989; D'Antoni and Schäbitz, 1990; Moore et al., 1991). It was therefore found that in every case vegetation community richness was underestimated by its pollen record. Low similarity values resulting from the application of Sorensen's index, are initially due to the absence of taxa in the pollen record but present in the vegetation community, and secondarily because of the presence of extra-local elements. Results obtained from the application of the similarity index showed that the pollen spectra of herbaceous communities were better represented than those of the arboreal community. The greater similarity between pollen and vegetation was given in the following community order: *Distichlis spicata* > *Sporobolus pyramidatus-Xanthium cavanalliesii* > *Spartina densiflora* > *Celtis tala*. Most of the taxa present in the vegetation are entomophyllous higher than 60 % in every case). However, in every pollen spectra a higher proportion of anemophyllous types was found in the pollen spectra. In grasslands, anemophyllous pollen sources important for the community characterization (such as Poaceae) were recorded in the pollen spectra, although according to the R/P values these were under-represented (Hall, 1994). On the other hand, in the *Celtis tala* forest a taxon important for the community description like Moraceae did not appear in the pollen record. None of the entomophyllous types abundant in the vegetation (such as *Limonium brasiliense* and Brassicaceae, among others) was found in the pollen spectra of the *Spartina densiflora* and *Sporobolus-Xanthium* communities. It should be noted that extra-local taxa with these pollination characteristics were recorded in every community. Apiaceae was present in the *Spartina densiflora* vegetation and absent in its pollen record, but was found in the pollen spectra of the *Sporobolus-Xanthium* community.

Multivariate analysis of the vegetation communities and their respective pollen records easily differentiates those communities that were structurally different (forests from grasslands); the analysis, however, was more difficult for those communities that were physiognomically equal. From the three grasslands, the one that showed greater affinity between vegetation and its pollen spectra was that of the *Sporobolus pyramidatus-Xanthium cavanalliesii*. The similarity was given by high values of Poaceae, *Xanthium cavanalliesii* and *Phyla canescens*.

The relationship between present pollen spectra and community through pollen analysis was made on the basis of structural attributes of the source vegetation dispersion characteristics of its pollen taxa. This relationship is, however, considered to be much more complex, involving other variables (D'Antoni, 1990; Moore et al., 1991). This approach allows a first approximation of the problem. However, it is to be expected that incorporation of new variables such as environmental factors will modify this approximation. These studies constitute a first step for the comprehension of recent and fossil pollen spectra in the Lower Delta. The results of this work show that surface samples can be used to characterize communities (Ayyad et al., 1992), although the degree of resolution is limited and these limitations make us think of exercising care when interpreting the fossil record.

ACKNOWLEDGMENTS

We wish to thank Parques Nacionales for authorizing our field work at Otamendi Natural Reserve. We thank Licenciada Marcela Cagnoni for her collaboration in the determination of plants, and Drs. Edgardo Romero and Rubén Quintana for the critical reading of the manuscript. Dr. Raúl Pozner was invaluable for his help in the modification of the map.

References Cited

AYYAD, S.M., MOORE, P.D., and ZAHRAN, M.A.
1992 Modern pollen rain studies of the Nile Delta. Egypt. *New Phytologist*, 121: 663–675.

BIRKS, H.J.B., and BIRKS, H.H.
1980 Quaternary Paleoecology. Edward Arnold. London, 289 pgs.

BRAGGIO MARUCCHIO, G., CORNARA, L., and MARIOTTI, M.
1988 Modern pollen rain and vegetation relationships in microinsular Mediterranean environments. 1 Zannone Island (Pontian Archipielago - Italy). *Pollen et Spores*, 30: 471–484.

BRAUN-BLANQUET, J.
1979 *Fitosociología. Bases para el estudio de las comunidades vegetales*. H. Blume Ediciones, 819 pgs.

CABRERA, A.L., and WILLINK, A.
1980 *Biogeografía de América Latina*. Serie de Biología. Monografía 13. Organización de Estados Americanos, 120 pgs.

CHICHIZZOLA, S.E.
1993 Las comunidades vegetales de la Reserva Estricta Otamendi. *Parodiana*, 8: 227–263.

D'ANTONI, H.L., and MARKGRAF, V.
1977 Dispersión del polen actual en el Oeste árido argentino. *Instituto Argentino de Nivología y Glaciología. CONICET. Anales*, 4: 125–150.

D'ANTONI, H.L., and SCHÄBITZ, F.
1986 Combinación de métodos numéricos para la zonación y distinción de grupos ecológicos en análisis de polen. *Contribuciones al IV Congreso Argentino de Paleontología y Bioestratigrafía. Apéndice de las Actas:* 9–15.

D'ANTONI, H.L., and SHÄBITZ. F.
1990 Pollen analysis for the generation and environmental hypothesis. *Grana*, 29: 295–300.

EASTMAN, J.R.
1993 IDRISI Program. Clarke University. Graduate School of Geography, 209 pgs.

FAEGRI, K., and IVERSEN, J.
1989 *Textbook of Pollen Analysis*. IVth Edition. John Wiley and Sons, 328 pgs.

GARCIA, C.V.
1967 Análisis de las clasificaciones climáticas del territorio argentino. *Buenos Aires. Facultad de Filosofía y Letras. Centro de Estudios Geográficos* A.B.C., 24: 1–80.

HALL, S.A.
1994 Modern pollen influx in tallgrass and shortgrass prairies, southern Great Plains, USA. *Grana*, 33: 321–326.

HEUSSER, C.
1989 Late Quaternary Vegetation and Climate of Southern Tierra del Fiego. *Quaternary Research*, 31: 396–406.

JANSSEN, C.R.
1967 A comparison between the recent regional pollen rain and the subrecent vegetation types in Minnesota (U.S.A). *Revue of Palaeobotany and Palynology*, 2: 331–342.

LEWIS, W.H., VISNAY, P., and ZENGER, V.E.
1993 *Airborne and Allergenic Pollen of North America*. The John Hopkins University Press, 254 pgs.

MALVAREZ, A.I.
1997 La comunidades vegetales del Delta del Río Paraná. Su relación con factores ambientales y patrones de paisaje. Tesis doctoral. Universidad de Buenos Aires, 167 pgs.

MANCINI, M.V.
1993 Recent pollen spectra from forest and steppe of South Argentina. A comparison with vegetation and climate data. *Revue of Palaeobotany and Palynology*, 77: 120–142.

MARCHETTI, B., RUIZ, L., MADANES, N., and CICCHERO, P.
1988 Relevamiento del medio natural y una propuesta del plan de manejo para la futura Area Natural Protejida "Ingeniero Rómulo Otamendi." *Administración de Parques Nacionales. Informe Técnico*, 50 pgs.

MARKGRAF, V., D'ANTONI, H.L., and AGER, T.A.
1981 Modern pollen dispersal in Argentina. *Palynology*, 5: 43–63.

MOORE, P.D., WEBB, J.A., and COLLINSON, M.E.
1991 *Pollen Analysis*. Second Edition. Blackwell Scientific Publications. Oxford, 216 pgs.

MUELLER DOMBOIS, D., and ELLEMBERG, H.
1974 *Aims and Methods of Vegetation Ecology*. John Wiley and Sons (eds), 547 pgs.

PAEZ, M.M.
1991 Palinología del campo Moncada 2 (Chubut). Interpretación paleoecológica para el Holoceno. Tesis doctoral. Universidad Nacional de La Plata, 275 pgs.

PRENTICE, I.C.
1985 Pollen representation, source area and basin size: toward a unified theory of pollen analysis. *Quaternary Research*, 23: 76–86.

PRIETO, A.R.
1996 Late Quaternary vegetation and climatic changes in the Pampa grasslands of Argentina. *Quaternary Research*, 45: 73–88.

WEBB, T., III.
1974 Corresponding patterns of pollen and vegetation in lower Michigan: a comparison of quantitative data. *Ecology*, 55: 17–28.

WRIGHT, H.E., Jr.
1967 The use of surface samples in Quaternary pollen analysis. *Revue of Palaeobotany and Palynology*, 2: 321–330.

Authors' addresses

CELINA A. FERNÁNDEZ
NORA MADANES
Departamento de Ciencias Biológicas
Facultad de Ciencias Exactas y Naturales
Universidad de Buenos Aires
Intendente Güiraldes 2620 (1428)
Buenos Aires
Argentina

POLLEN AND MACROFOSSIL RECORDS FROM AUSTRALIAN STICK-NEST RAT (*LEPORILLUS* SPP.) MIDDENS: DO THEY STAND THE TEST OF TIME?

Lynne McCarthy

Abstract

Uncertainties over temporal resolution currently limit the potential of stick-nest rat *Leporillus* spp. middens to provide a palaeoecological record for Australian semi-arid regions. Concerns over accumulation rates and the possibility of reworking of material must be addressed before interpretations of the macrofossil and pollen records can be refined. Accelerator Mass Spectrometry Radiocarbon (AMS) dating of a midden from the Flinders Ranges, South Australia that displays apparent stratigraphic layering is used as a case study of these issues.

A pollen and macrofossil record from the midden is presented. Leaf fragments and pollen concentrates have been dated to test the assumption that palaeoecological records from within the midden are contemporaneous. The midden has accumulated over approximately 1,000 years and each layer represents a period of 100-200 years. Temporal resolution at this scale will allow the detection of the timing of vegetation changes that have occurred over this time.

The AMS dates on leaf macrofossils are in stratigraphic order and dates on the pollen indicate that the pollen and macrofossil records appear to be contemporaneous. However, the issue of contamination and reworking of pollen can not be ruled out at this stage of the investigation. Analysis of the records indicates taxa represented in pollen and macrofossil assemblages are similar and most likely originate from local sources. For the period from 7,000 to 6,000 yr BP, saltbush (Chenopodiaceae) and shrubs (Apiaceae and Myoporaceae) contribute to a large proportion of the record, similar to present day vegetation communities.

INTRODUCTION

Palaeoecological analysis of *Leporillus* spp. (stick-nest rat) middens is a tool for a non-traditional palynological investigation in environmental conditions that are not always conducive to good pollen preservation. A vegetation history and palaeoclimatic conditions from the Australian semi-arid and arid zones is incomplete (Singh, 1981). Pollen analysis has made a large contribution to Quaternary vegetation history in well studied coastal sites in Australia (Kershaw, 1981) and analysis of stick-nest rat middens will contribute to our understanding of the palaeoecology of the Australian semi-arid zone. However, uncertainty over the temporal resolution of these deposits currently limit their potential to provide a reliable record. Concerns over accumulation rates and the possibility of reworking of material must be addressed before interpretations of the macrofossil and pollen records can be refined.

Native stick-nest rats *Leporillus conditor* and *L. apicalis*, once extant on the Australian mainland, built nests of sticks and twigs collected in a foraging range of approximately 100 m. These constructions have been preserved in caves throughout the semi-arid and arid zones of Australia and are referred to as stick-nest rat middens. Continuing work is focussing on methodological issues concerned with the taphonomy of *Leporillus* midden deposits and attempts to refine the temporal resolution of the records from this material. Stratigraphic and taphonomic issues need to be resolved before the potential of the technique can be fully realised. This is being explored using Accelerator Mass Spectrometry (AMS) radiocarbon dating of macrofossil leaf fragments and pollen concentrates from the midden record in conjunction with field observations.

McCarthy, Lynne
2001 Pollen and macrofossil records from Australian stick-nest rat (*Leporillus* spp.) middens: Do they stand the test of time?; *In*: Goodman, D.K., and Clarke, R.T. (eds.), *Proceedings of the IX International Palynological Congress, Houston, Texas, U.S.A., 1996*; American Association of Stratigraphic Palynologists Foundation, p. 541–548.

 ISBN 0–931871–06–9

The majority of research on *Neotoma* spp. (packrat or woodrat) middens has focussed on using the macrofossil records to reconstruct Late Quaternary plant communities from South West of North America. Researchers in the field have used macrofossil analysis to reconstruct vegetation communities, devise palaeoclimatic interpretations, provide evidence for altitudinal change in species composition, and faunal changes that have taken place in desert environments (Betancourt et al., 1990). More recently, research has focussed on the application of stable isotopic analysis of plant macrofossils (Long et al., 1990; Wigand, 1994) and the investigation of stomatal indicies and density for palaeoclimatic reconstructions (Atchison et al., 2000; Van de Water et al., 1994). Few studies have been concerned with methodologies and assumptions when using middens as a palaeoecological tool (Spaulding et al., 1990; Davis and Anderson, 1987). Factors such as internal variability of midden composition and duration of depositional episodes (Elais, 1990; Spaulding et al., 1990; Webb and Betancourt, 1990; Wells, 1976) may bias the interpretation of midden contents.

AMS radiocarbon dating offers a powerful tool to improve dating resolution and allow different components of the record to be dated separately (Brown et al., 1989, 1992). Of particular interest here is its potential for dating pollen (Long et al., 1992). A range of taphonomic processes can lead to differences between the radiocarbon ages of bulk samples and the proxy climate indicator being studied (eg. pollen, macrofossils). Dating of pollen concentrates will significantly enhance the ability to reconstruct chronologies for these deposits.

This paper will use a case study of one stick-nest rat midden from the Flinders Ranges, to address specific taphonomic and methodological issues concerning the interpretation of pollen and macrofossil records from middens. The aims of this paper are to:

1. Determine a chronology for the *Leporillus* spp. (stick-nest rat) midden using AMS radiocarbon dating.
2. Test the assumption that macrofossil and pollen records from the midden are contemporaneous.
3. Refine interpretations of the palaeceological records preserved in the midden.

This particular case study is one component within a broad regionally based study of a large number of stick-nest rat middens that increase the spatial and temporal coverage of palaeoecological records for the semi-arid zone.

STUDY AREA

The Flinders Ranges are located in South Australia. With an approximate area of 62,000 sq km (Greenwood et al., 1989) they extend from Gulf St. Vincent and Spencer Gulf in the south, through to the Strzelecki Desert and Lake Eyre Basin to the north (Text-Figure 1) . The Ranges receive less than 500 mm annual average rainfall. This precipitation is twice that experienced on the adjacent plains (Greenwood et al., 1989; Singh and Luly, 1991) and therefore, the Flinders Ranges are a relatively well-watered area in an otherwise semi-arid landscape.

Site Description

Brachina Gorge lies approximately 31°20'S and 138°33'E and is part of the Flinders Ranges National Park (Text-Figure 1). Brachina Creek cuts through the Brachina Formation, which is composed of Proterozoic siltstones, sandstones and limestones. The *Leporillus* midden (BR4) was located at the western end of the gorge inside a cave that has been weathered out of the limestone.

Saltbush dominate surrounding plains to the east and west of Brachina Gorge. These semi-succulent shrublands (Chenopodiaceae steppe) include communities with an abundance of herbaceous and perennial plants (Beard, 1994). The chenopods are perennial shrubs (Graetz and Wilson, 1984) and include *Atriplex vesicarius* (saltbush) and *Maireana kochia* (bluebush) These communities may also be accompanied by pasture plants including *Sclerolaena holitana* (burrs) and annual grasses such as *Enneapogan polyphyllus* (limestone bottlewashers), *Stipa scalra* and *Themeda triandra* (porcupine grass) (Sprigg, 1984). Whereas inside the gorge, saltbush, grasses, herbs and ephemeral species colonise the rocky talus slopes on the gorge walls. On lower slopes with more adequate soil development, low semi-arid woodlands dominated by *Casuarina cristata* (black oak), *Heterodendrum oleifolium* (rose wood) and *Callitris columellaris* (white cypress pine) (Stanley, 1983), with and understorey of shrubs such as Myoporaceae spp., *Acacia* spp. and a ground cover of ephemeral grasses and herbs are present. Along the floodplain and creek system, large *Eucalyptus camaldulensis* (river red gum) grow, accompanied by an understorey of trees and shrubs including *Melaleuca* spp. and *Acacia* spp. Cyperaceae spp. and *Typha domingensis* grow in the areas of permanent or semi-permanent water. There is also the presence of exotic plant species as vegetation in the region has been altered by the introduction of domestic stock and feral animals (Beadle, 1981). As a consequence, there has been removal of perennial species due to overgrazing and wide spread colonisation of species that include *Rumex vesicarius* (wild hops), *Echium plantagineum* (Patersons curse) and *Solanum ellipticum* (velvet potato bush).

BRACHINA GORGE STICK-NEST RAT MIDDEN

A stick-nest rat midden displaying stratigraphic layering from top to base , over a section measuring approximately 60 cm has been chosen as a case study for the investigation of the temporal resolution of this type of deposit. The midden was located in a small chamber within a larger cave. This setting has enhanced the preservation of the deposit (Head et al., 1998). Twigs, sticks, leaf fragments, stones, faecal pellets and pollen were sealed within an indurated amberat exterior. A longitudinal section of midden was excavated, to retain the visible stratigraphic layering from top to base (Text-Figure 2).

Laboratory Methods

The deposit was sub-sampled in the laboratory at 10 cm intervals and samples were soaked in distilled water for 24 hr to remove the outer rind thus minimising contamination from modern material. The remainder was soaked in distilled water for up to 5 days, washed through a 2 mm and 500 µm sieve and the residue was then air dried, weighed and placed in a tray to sort and identify macrofossils. Cleaned plant macrofossils were removed for AMS dating. Pollen concentrates to be used for counting and AMS dating were prepared from the residue of fine sediment and decanted supernatent (Mehringer and Wigand, 1990) from each 10 cm layer of midden.

Preparation of the pollen concentrates for counting underwent standard KOH, HF and acetolysis treatments (Moore et al., 1991) and were mounted in silicon oil. Most counts were to a minimum of 200 grains. Where one taxon was dominant, counts continued until 200 grains of other taxa were counted. Pollen taxa recorded in BR4 midden were allocated to structural groupings: trees, shrubs, herbs and halophyptes. The dominant growth form of plant families present at the study site determined what category the pollen taxa was placed in.

Preparation of the pollen concentrates for AMS radiocarbon dating followed techniques described by Long et al. (1992). and Brown et al. (1992). It is a selective sampling technique isolating pollen from organic residues allowing the pollen to be dated directly. The extraction technique is a modification of the standard pollen analysis chemical treatment, combined with additional sieving of the sample to a finer size fraction. The important difference is that chemicals containing carbon are avoided by bleaching less resistant organic material with sodium hypochlorite instead of the traditional acetolysis treatment. The pollen concentrates for AMS dating are composed of pollen grains and small amounts of organic and cellulose material (Brown et al., 1989).

RESULTS

AMS Radiocarbon Dates

Leaf macrofossils from the top layer at a depth of 10 cm dated at 6,290 ± 190 yr BP (OZB295) and AMS dates on leaf macrofossils from successive layers dated at 6,520 ± 50 yr BP (BETA 87299), 6,580 ± 60 yr BP (BETA 87300), 6,880 ± 50 yr BP (BETA 87301) and 6,920 ± 50 yr BP (BETA 87302). The base

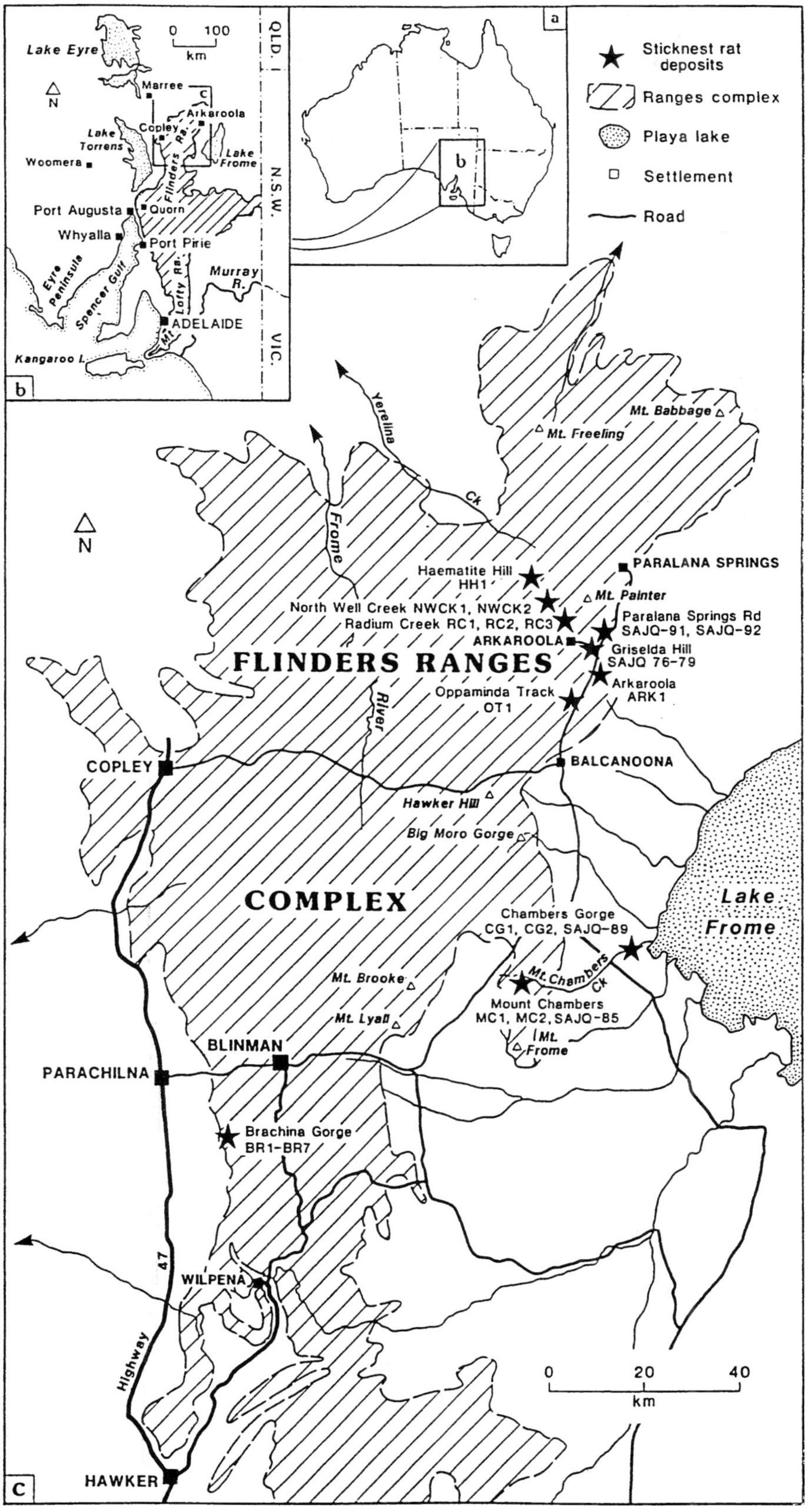

Text-Figure 1. Location map of the Flinders Ranges, South Australia, showing the regionally based location of stick-nest rat midden study sites. BR4 midden at the western end of Brachina Gorge is located in the central region of the Flinders Ranges Complex.

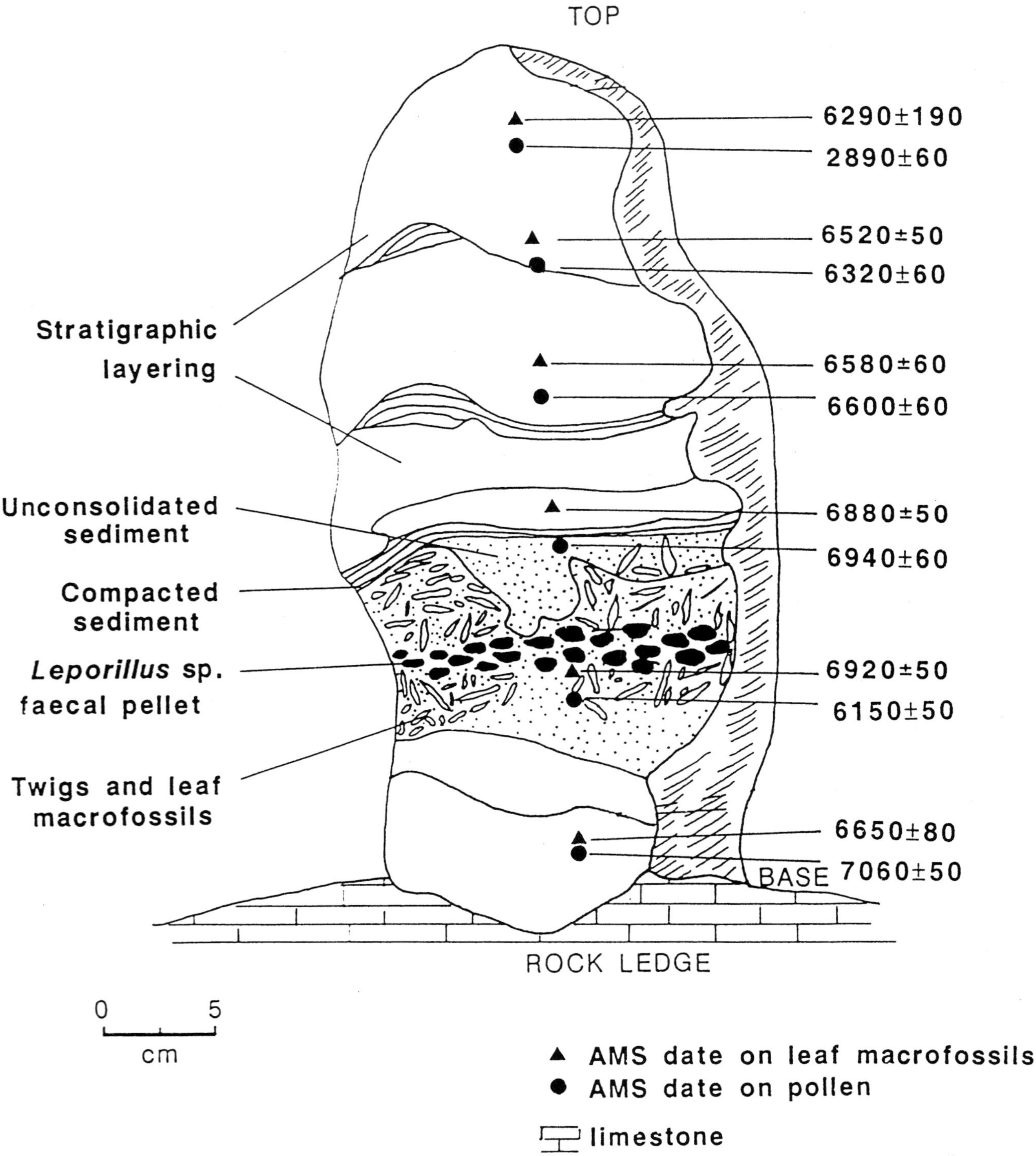

Text-Figure 2. Schematic diagram of Brachina Gorge (BR4) midden. AMS radiocarbon dates on leaf macrofossils and pollen concentrates from the top (10 cm) to base (60 cm) of the midden are shown.

layer at a depth of 60 cm recorded an age of 6,650 ± 80 yr BP (OZB297) (Table 1).The AMS dates on pollen concentrates from corresponding midden layers were 2,890 ± 60 yr BP (BETA 92926), 6,320 ± 60 yr BP (BETA 93952), 6,600 ± 60 yr BP (BETA 93953), 6,940 ± 60 yr BP (BETA 93954), 6,150 ± 50 yr BP (BETA 92927) and 7,060 ± 50 yr BP (BETA 93955) from 10 cm through to 60 cm respectively (Table 1). The AMS dates on leaf macrofossils are in stratigraphic order and dates on the pollen concentrates indicate that both palaeoecological records appear to be contemporary. Assuming that the ages derived from leaf macrofossils from the lower three layers (40 cm, 50 cm and 60 cm) are the same within the error bars reported, the midden has accumulated over a rapid period of time some where between 6,000 and 7,000 yr BP. However, the issue of contamination and reworking of pollen cannot be ruled out at this stage of the investigation as there is an age

TABLE 1. AMS radiocarbon ages of macrofossils and pollen concentrates from Brachina Gorge (BR4) stick-nest rat midden layer. Laboratory codes for dates appear in parentheses ( ).

BR4 midden	LEAF MACROFOSSILS	POLLEN
Layer 1 (10cm)	6 290 ± 190 (OZB295)	2 890 ± 60 (BETA92926)
Layer 2 (20cm)	6 520 ± 50 (BETA87299)	6 320 ± 60 (BETA93952)
Layer 3 (30cm)	6 580 ± 60 (BETA87300)	6 600 ± 60 (BETA93953)
Layer 4 (40cm)	6 880 ± 50 (BETA87301)	6 940 ± 60 (BETA93954)
Layer 5 (50cm)	6 920 ± 50 (BETA87302)	6 150 ± 50 (BETA92927)
Layer 6 (60cm)	6 650 ± 80 (OZB297)	7 060 ± 50 (BETA93955)

reversal in the pollen concentrate at a depth of 50 cm. This issue will be discussed later.

Macrofossil Records

The macrofossil data indicates that plant material used for the construction of the midden and possible diet of the stick-nest rat were of a local nature, within approximately 100 m from the midden site. The presence of *Sida petrophila* (Malvaceae), *Dodonaea viscosa* spp. *angustrissima* (Sapindaceae) and *Ptilotus obovatus* (Amaranthaceae) leaves and leaf fragments suggests a local signal as these shrub and ground cover species are common in rocky slope vegetation communities at Brachina Gorge. These taxa are dominant throughout all midden layers (Table 2). The occurrence of *Casuarina cristata* (black oak) cladodes also reflects local vegetation. Indications of the herbaceous ground cover can be seen with the occurrence of *Daucus glochidiatus* (Apiaceae) seeds and smaller shrubs such as *Cassinia laevis* (Asteraceae) leaves, also a local species in the study area.

Pollen Record

The pollen record preserved within *Leporillus* spp.midden deposits is important as unlike *Neotoma* middens, macrofossil records are usually fragmentary and poorly preserved. Species identifications can be difficult and recovery of a substantial assemblage is not always possible.

Eucalyptus pollen is the dominant tree pollen contributing on average to 10% of the pollen sum in all midden layers (Text-Figure 3). However all tree pollen peaks at a depth of 50 cm after a decline in the signature at 40 cm. There is a large diversity of shrubs in the pollen signal displaying a consistent representation through the midden record but with a slight decrease at 40 cm. Sapindaceae, Myoporaceae Caesalpinaceae and Fabaceae are the most abundant taxa contributing to the shrub pollen sum. The dominant herb pollen taxa include Apiaceae, Solanaceae, Asteraceae and Thymelaceae as these are present throughout all midden layers. There is a variety of other herbaceous taxa such as Halogoraceae, Malvaceae and Cruciferae that contribute to less than 10% of the count and are not consistently represented throughout the pollen record (Text-Figure 3). In the summary diagram (Text-Figure 3), herbs fluctuate with a peak at 30 cm and one at 60 cm. Chenopodiaceae pollen peaks at 40 cm and remains at similar levels throughout the rest of the midden. Poaceae and Cyperaceae pollen is generally more dominant in the record at lower depths of the midden. These taxa peak at 30 cm and 50 cm contributing

TABLE 2. Macrofossil assemblages extracted from Brachina Gorge (BR4) stick-nest rat midden. x indicates the presence in the midden.

	Midden Layers					
Macrofossils	10cm	20cm	30cm	40cm	50cm	60cm
Unidentified leaf fragments	x	x	x	x	x	x
Sida petrophila fragments	x	x	x		x	x
Malvaceae spp. leaf fragments		x	x	x	x	
Ptilotus obovatus fragments		x	x	x		
Myoporaceae spp. leaf fragments					x	
Dodonaea viscosa ssp. *angustrissima*	x	x	x	x	x	
Acacia spp. fragments	x	x		x	x	
Casuarina cristata cladodes	x	x			x	
Cassinia laevis terminalis		x	x			
Daucus glochidiatus fruit	x	x	x	x		
Poaceae spp. seed			x			
Leporillus spp. faecal pellets	x	x	x	x	x	x
Insect fragments	x					x

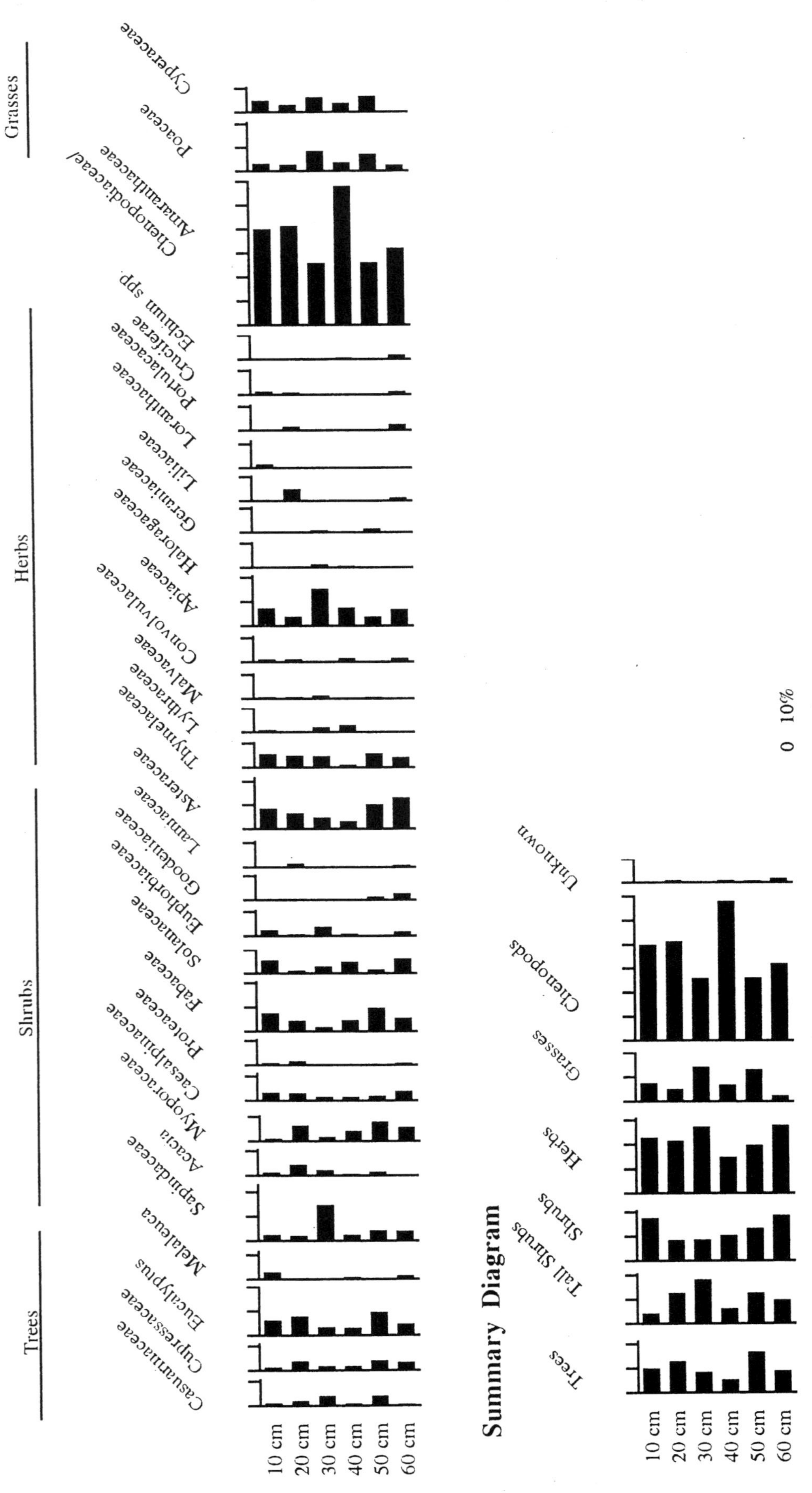

Text-Figure 3. Pollen percentage diagram and summary chart of main pollen taxa grouped into vegetation categories of trees, shrubs, herbs and halophytes for Brachina Gorge (BR4) midden.

greater than 10% of the pollen sum in the midden record at a time when Chenopodiaceae pollen is not as abundant (Text-Figure 3).

DISCUSSION

Chronology of the Midden Deposit

Radiocarbon ages of midden material can be misleading due to the contamination of young middens by the redeposition of older rat material and vice versa (Wells, 1976). This has been demonstrated in this study with two AMS dates from pollen at 10 cm and 50 cm. Given that the macrofossil dates are in stratigraphic order, the 10 cm layer pollen date of 2,890 ± 60 yr BP can be attributed to reworking of younger pollen due to the remobilisation of amberat in more humid periods. The 50 cm layer pollen date of 6,150 ± 50 yr BP may be and indication of reworking of material when the midden was occupied. The rats themselves may have moved through the structure they have built, mixing pollen rich sediment through the layered midden (Head et al., 1998). Changes in the depositional history of the midden has also been dectected by the series of AMS dates on leaf macrofossils. Ages of 40 cm, 50 cm and 60 cm layers may all be contemporary within the reported error bars and therefore indicate a period of rapid accumulation. This complexity in our understanding of the temporal resolution of these deposits illustrates some of the problems encountered when working with midden material. That is, middens can be stratigraphically complex and there is the possibility of breaks in the depositional history or collapse in the structure as the deposit is constructed (Elias, 1990; Spaulding et al., 1990; Webb and Betancourt, 1990).

Modern Pollen Deposition

Pollen collected from traps set up in the Brachina Gorge study site suggest that modern vegetation communities are being consistently represented in the pollen rain. Pollen taxa of local species are dominant, but there is a discernible regional signal. Regional taxa well represented in the pollen rain included Cupressaceae, *Eucalyptus* spp., Poaceae, Sapindaceae and Myoporaceae. Herbaceous taxa such as Asteraceae appeared as a dominant component of the pollen rain in the trap samples whereas other taxa including Convolvulaceae, Solanaceae, Malvaceae and Thymelaceae were less abundant in the pollen sum. Chenopodiaceae pollen was moderately dominant in the pollen rain from traps inside the gorge however, the signal was more abundant in traps located outside the gorge on the surrounding plains. Exotic taxa including *Echium* spp., *Pinus* spp. and Plantaginaceae were also recorded in the pollen traps.

Interpretation of Palaeoecological Records

Generally, the plant families represented in the macrofossil assemblages correspond with the pollen taxa in the midden, suggesting that there is a strong signal from the local vegetation within 100 m of the midden. Pollen indicates that at 6,000 yr BP to 7,000 yr BP the vegetation was dominated by shrubs and chenopod saltbush communities with a less dominant representation of tree pollen. However, there is a change in the pollen at approximately 6,500 yr BP where a decrease in the percentage of tree, grass and shrub pollen occurs with an increase in Chenopodiaceae pollen. It is questionable as to whether this may be indicating drier conditions during this part of the Holocene. Alternatively, it may reflect a period of nest abandonment by the stick-nest rat. Difficulties in the interpretation of midden material can be complicated by the fact that animals other than *Leporillus* may reoccupy a midden after a period of abandonment (Berry, 1991). This situation may be occurring with the Brachina Gorge midden. The change in the pollen signal from 6 500 yr BP may also suggest a change in the depositional history of BR4. Examination of the AMS dates for each layer suggest that 40 cm, 50 cm and 60 cm layers may all be contemporary within reported error bars (Table 1) and therefore indicate a period of rapid accumulation.

CONCLUSIONS

To test the assumption that the age of pollen and macrofossils within the same layer of a midden are contemporary, pollen concentrates from each discrete layer were AMS dated. Dates on pollen concentrates from BR4 reveal the complexity in the temporal resolution of midden deposits. Given that the macrofossil dates are in stratigraphic order, the 10 cm layer pollen date of 2,890 ± 60 BP (BETA92926) can be attributed to possible reworking of younger pollen due to remobilisation of amberat in more humid periods. Alternatively, contamination could have occurred during processing of the sample in the laboratory. The 50cm layer pollen date of 6,150 ± 50 BP (BETA92927) may be an indication of reworking of material when the midden was occupied. The rats themselves may have moved through the nest/midden, mixing pollen rich sediment through the layered midden. The macrofossil assemblages and pollen records are complimentary, with shared taxa occurring in each layer of the midden. Given this and the environmental setting of the midden within a cave, it appears that both palaeoecological records have been derived from local sources.

ACKNOWLEDGMENTS

Funding for this project was provided by the Australian Research Council and the Quaternary Environments Research Centre, University of Wollongong. Thanks to Assoc. Prof. Lesley Head for useful comments on an earlier draft of this manuscript. Figure 1 and 2 drafted by Richard Miller and David Martin.

References Cited

ATCHISON, J., HEAD, L., and McCARTHY, L.
1999 Stomatal parameters and atmospheric change since 7500BP: evidence from *Eremophila deserti* (Myoporaceae) leaves from the Flinders Ranges Region, South Australia. *Australian Journal of Botany*, 48: 223–232.

BEADLE, N.
1981 The vegetation of the arid zone. *In*: Keast, A., (ed.), *Ecological Biogeography of Australia*. Dr. W Junk Publishers, The Hague: p. 51.

BEARD, J.
1984 The vegetation of the Australian arid zone. *In*: Cogger, H., and Cameron, E. (eds.), *Arid Australia*, Surrey Beatty and Sons Pty Ltd, Sydney, pp. 113–117.

BERRY, S.
1991 The potenial of fossil mammal middens as indicators of vegetation history in central Australia. *Australian Journal of Botany*, 39: 305–313.

BETANCOURT, J.L., VAN DEVENDER, T.R., and MARTIN, P.S.
1990 *Packrat Middens: the Last 40 000 years of Biotic Change.* University of Arizona Press, Tucson, 467 pgs.

BOYD, W.
1990 Quaternary pollen analysis in the arid-zone of Australia: Dalhousie Springs, Central Australia. *Review of Palaeobotany and Palynology*, 64: 331–341.

BROWN, T., FARWELL, G., GROOTES, P., and SCHMIDT, F.
1992 Radiocarbon AMS dating of pollen extracted from peat samples. *Radiocarbon*, 34: 550–556.

BROWN, T., NELSON, R., MATHEWES, R., VOGEL, J., and SOUTHON, J.
1989 Radiocarbon dating of pollen by Accelerator Mass Spectrometry. *Quaternary Research*, 32: 205–212

DAVIS, O.K., and ANDERSON, R.S.
1987 Pollen in packrat (*Neotoma*) middens: pollen transport and the relationship of pollen to vegetation. *Palynology*, 11: 185–198.

ELAIS, S.
1990 Observations of the taphonomy of Late Quaternary insect fossil remains in packrat middens of the Chihuahuan Desert. *Palaios*, 5: 356–363.

GRAETZ, R., and WILSON, A.
1984 Saltbush and bluebush. *In*: Harrington, G., Wilson, A., and Young, M. (eds.), *Management of Australia's Rangelands*, CSIRO, Melbourne, pp. 209–222.

GREEN, N., CALDWELL, J., HOPE, J., and LULY, J.
1983 Pollen from an 1800 year old stick-nest rat (*Leporillus* spp.) midden from Gnalta, Western New South Wales. *Quaternary Australasia*, 1: 31–41.

HEAD, L., McCARTHY, L., QUADE, J., WITTER, D., and ALLEN, V.
1998 Classification and radiocarbon dating of *Leporillus* nests in semi-arid Australia, and palaeoclimatic implications. *Palaeoclimates*, 3: 161–177.

KERSHAW, A.
1981 Quaternary vegetation and environments. *In*: Keast, A. (ed.), *Ecological Biogeography of Australia Vol. 1*, Dr. W. Junk Publishers, The Hague, pp. 83–101.

LONG, A., DAVIS, O.K., and DeLANOIS, J.
1992 Separation and ^{14}C dating of pure pollen from lake sediments: nanofossil AMS dating. *Radiocarbon*, 34: 557–560.

LONG, A., WARNEKE, L.A., BETANCOURT, J.L., and THOMPSON, R.S.
1990 Deuterium Variations in Plant Cellulose from Fossil packrat Middens. *In*: Betancourt, J.L., Van Devender, T.R., and Martin. P.S. (eds.), *Packrat Middens: the Last 40,000 years of Biotic Change*, The University of Arizona Press, Tucson, pp. 380–396.

McCARTHY, L., HEAD, L., and QUADE, J.
1996 Holocene palaeoecology of the northern Flinders Ranges, South Australia, based on stick-nest rat (*Leporillus* spp.) middens: a preliminary overview. *Palaeogeography, Palaeoclimatology, Palaeoecology*, 123: 205–218.

MEHRINGER, P., and WIGAND, P.E.
1990 Comparison of Late Holocene environments from woodrat middens and pollen: Diamond Crators Oregon. *In*: Betancourt, J.L., Van Devender, T.R., and Martin. P.S. (eds.), *Packrat Middens: the Last 40,000 years of Biotic Change*, The University of Arizona Press, Tucson, p. 294–325.

MOORE, P., WEBB, J, and COLLINSON, M.
1991 *Pollen Analysis*, 2nd edition, Blackwell Scientific Publications, Oxford, pp. 216.

PEARSON, S., and DODSON, J.
1993 Stick-nest rat middens as sources of palaeoecological data in Australian deserts. *Quaternary Research*, 39: 347–354.

SINGH, G.
1981 Late Quaternary pollen records and seasonal palaeoclimates of Lake Frome, South Australia. *Hydrobiologia*, 82: 419–430.
1988 History of aridland vegetation and climate: a global perspective. *Biological Reviews*, 63: 159–195.

SINGH, G., and LULY, J.
1991 Changes in vegetation and seasonal climate since the last full glacial at Lake Frome, South Australia. *Palaeogeography, Palaeoclimatology, Palaeoecology*. 84: 75–86.

SPAULDING, W., BETANCOURT, J., CROFT, L., and COLE, L.
1990 Packrat middens: their composition and methods of analysis. *In:* Betancourt, J., Van Devender, T., and Martin, P. (eds), *Packrat Middens: The Last 40,000 years of Biotic Change*. The University of Arizona Press, Tucson, pp. 59–84.

SPRIGG, R.
1984 *Arkaroola-Mount Painter in the Northern Flinders Ranges South Australia. The Last Billion Years*. Arkaroola Lutheran, South Australia, p. 298

STANLEY, R.
1983 Soils and vegetation: an assessment of current status. *In*: Messr, J., and Mosley, G. (eds.), *What Future for Australia's Arid Lands? proceedings of the National Arid lands Conference, Broken Hill, New South Wales*, Australian Conservation Foundation, Victoria, pp. 8–18.

VAN de WATER, P.K., LEAVITT, S.W., and BETANCOURT, J.L.
1994 Trends in stomatal density and ^{13}C / ^{12}C ratios of *Pinus flexilis* needles during Last Glacial–Interglacial Cycle. *Science*, 264: 239–243.

WEBB, R., and BETANCOURT, J.
1990 The spatial and temporal distribution of radiocarbon ages from packrat middens *In:* Betancourt, J., Van Devender, T., and Martin, P. (eds.), *Packrat Middens: The Last 40,000 years of Biotic Change*. The University of Arizona Press, Tucson, pp. 85–103.

WELLS, P.
1976 Macrofossil analysis of woodrat (*Neotoma*) middens as a key to the Quaternary vegetational history of arid America. *Quaternary Research*, 6: 223–248.

WIGAND, P.E.
1994 Stable isotopic analyses in palaeoclimatic reconstructions. *In*: Waugh, W.J., Petersen, K.L., Wigand, P.E., Louthan, B., and Walker, R. (eds.), *Climate Change in the Four Corners and Adjacent Regions: Implications for Environmental Restoration and Land-Use Planning, Session-Palaeoclimatic Data—Pleistocene*, Grand Junction, Colorado, pp. 27–38.

Author's address

LYNNE MCCARTHY
National Museum of Australia
GPO Box 1901
Canberra ACT 2601
Australia

QUANTITATIVE CORRELATIONS BETWEEN MYCOFLORA AND LANDSCAPE ECOLOGICAL PARAMETERS FROM A CONTEMPORARY WET HEATHLAND IN THE NETHERLANDS

Christian Mulder

Abstract

In actuopalynology, several problems in the interpretation of pollen data exist in using biological traps, because of the uncertain age of the surface samples. This difficulty in age assignment is reflected in the great variability of vegetation indices, correction coefficients and representation factors available from the literature. The activity of fission-products provides a new tool for a reliable time-standardization of surface samples. All biological traps, whether mosses, lichens or litter, may be selected on the basis of their comparable radiocesium content for further palynological investigation. Several relationships between the plant communities occurring along a Dutch wetland cenocline and the sporae dispersae are discussed. The mycoflora appears closely related to local hydrology. Assemblages of fungal remains have additional ecological implications for a correlation between wetland fungi and plant hosts. The ecological significance of each sporomorph may supply a more complete definition of wetland biomes in the future.

INTRODUCTION

The analysis of the pollen content of moss cushions has become a common tool in both aerobiology and mycology, as mosses are known to be highly receptive biological traps. However, only relative pollen values (percentages) can be calculated from such biological traps, as the time interval during which sporomorphs were captured is usually not known.

In vegetation science it has become important to include soil-inhabiting organisms in the quantitative plant community analysis. In terrestrial ecosystems fungi play for the vegetation an essential role as humus decomposers (Barkman, 1973). According to Sherwood-Pike (1988) the absence of a systematic survey of present-day mycocenoses is particularly evident in paleoenvironmental reconstructions. Numerous studies of the pollen and rhizopod content in recent moss cushions have already been carried out, but records of fungi were often neglected, specially non-mycorrhizal parasitic genera occurring in water films or watersheds (freshwater fungi) or saprophytic forms on dead animals or feces (dung fungi).

Fungal spores can be detected at several thousands of meters from their source in the direction of predominant winds, and the sporomorphs flow can easily be traced by using a specific regional pollen sum (Mulder and Janssen, 1999). Nonetheless the non-linear dispersal of windborne sporomorphs contributes to additional difficulties in a comparison of the mycoflora with the possibly related host plants (Text-Figure 1). In phanerogamic communities, the full implication of the pollen/vegetation relation in terms of dispersal and deposition per unit time becomes clear only after a number of years (Hicks, 1985). However, often the time assumption is vague. The time interval is particularly important in the study of the spore deposition of present-day fungal communities, due to their patchy occurrence if compared to phanerogamic (autotrophic) communities. The sporadic appearance of fungal fruiting bodies may therefore require a collection period of several years.

The aim of this pilot study is to evaluate the percentage fluctuations in the sporomorph deposition along one transect during a uniform time range. In the moss samples, the recovered amount of Cesium originating from the Chernobyl event (April 26th, 1986) can be used to select more or less even-aged samples. The metabolized ^{137}Cs content is indicative of the artificial radionuclide amount taken-up by the cryptogamic layer: all the biological traps with a detectable ^{137}Cs content metabolized the Cesium during the last decade and have been active since 1986 (Mulder, 1996a). Thus, the pollen fixed in the terricolous mosses and humus samples is standardized in time, and can be used for the study of sporomorphs/vegetation relationships within permanent plots.

The palynological significance of surface samples in terms of number of sporomorph types increases with the age of the biological traps. Thus, the Chernobyl radiocesium fallout becomes a selecting criterion for surface samples. All time-standardized samples in Central Europe are marked by a higher radionuclide content due to the Chernobyl plume (Mulder and Janssen, 1998, 1999).

The following questions were addressed:

1. Is the spatial pattern of local vegetation reflected in a spatial pattern of sporomorph deposition? Is it possible to determine the (sub)recent environment of the wetland, and to what extent does bioturbation occur?
2. Which environmental factors are linked to the fungal remains? Is it possible to arrive at a better ecological insight in the mycocenoses through such a palynological study?

Mulder, Christian
2001 Quantitative correlations between mycoflora and landscape ecological parameters from a contemporary wet heathland in The Netherlands; *In:* Goodman, D.K., and Clarke, R.T. (eds.), *Proceedings of the IX International Palynological Congress, Houston, Texas, U.S.A., 1996*; American Association of Stratigraphic Palynologists Foundation, p. 549–555.
 ISBN 0–931871–06–9

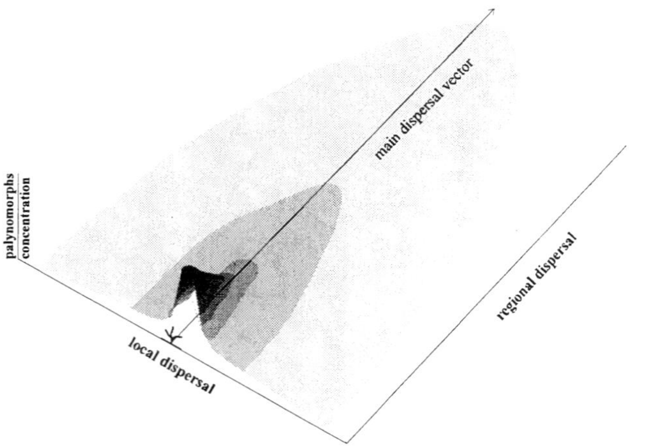

Text-Figure 1. Concentration according to distance: plume fitted with arbitrary values to the main dispersal vector (wind). Model modified after Janssen (1966).

MATERIALS AND METHODS

In order to integrate the comparison of different patches, individual surface samples associated with uniform environmental conditions have to be chosen. For this research the Dwingeloo heath, located in the Dutch Province of Drenthe (52.47° N and 6.22° E), was chosen (Text-Figure 2). A prior screening of the wet communities described by Schaminée et al. (1995) and Stortelder et al. (1998) indicated that the heath of Dwingeloo is extremely suitable for this type of multidisciplinary investigation. The Dwingelose Heide is 1,000 ha large. The heathland has an average precipitation of 822 mm/year with a monthly average temperature that ranges from 1°C to 22°C (Bakker et al., 1986). Winds blow relatively strong, mainly from SW, but in the months of May and June NW and NE winds may also occur (van Heuveln, 1965).

A transect was designed for the application of the negative exponential model of van Raamsdonk and Schouten (1997). Sporomorph dispersal was clearly biased towards known wind directions. The present-day vegetation cover was estimated in the field from February to August 1995. Following Rodwell (1991), plots of 4 x 4 m were used in the heathland, but 10 x 10 m samples were necessary in the pine plantation and the oak forests. The occurrence of each taxon in the vegetation was converted according to a logarithmic scale into the so-called Domin values, estimated from 0 (taxon absent) up to 9 (dense canopy).

Radionuclides Determination

The main fission-product activity comes from the isotope ^{137}Cs. An ecological decay of the artificial long-lived nuclide ^{137}Cs in the moss shoots has not to be taken into account at the scale of a decade (1986–1995), because the dry fallout of radiocesium after the Chernobyl event is negligible (Mulder, 1996b, 1997). According to Andolina and Guillitte (1990), most of the radiocesium deposition is expected to be found in the superficial soil layers. However, also in the deeper soil samples the Chernobyl tracer signal is present. The latter activity has to be distinguished from earlier atmospheric nuclear weapon tests, where a radiocesium peak can be expected in the 1963–64 horizon (cf. Chmura and Kosters, 1994).

The radiocesium content was measured in the laboratory by tracing the 661 KeV γ-peak with a sensitive HP-

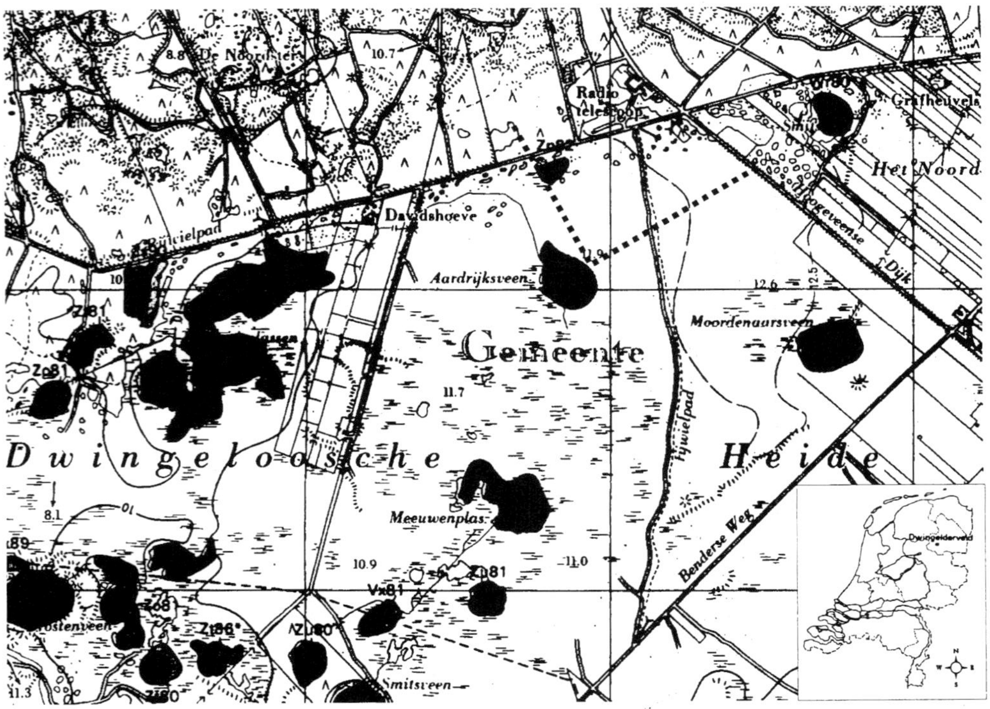

Text-Figure 2. The Dwingelose Heide (Drenthe) at the scale of 1:25,000. The province is known as a whole for its mycological richness (Barkman and Westhoff, 1969). A previous geohydrologic study of the local mires (black areas) by Bakker et al. (1986) was used to recognize the gradients most suitable for the transect approach (dotted line).

Germanium detector. Its specific activity covers a range from 10 Bq/Kg dry matter to 196 Bq/Kg dry matter. It appears that the Chernobyl fallout was extremely variable throughout Europe, but along the Dwingeloo transect the average ^{137}Cs content ranges from 80 Bq/Kg dry matter in the heathlands to 96 Bq/Kg dry matter in the woodlands. Litter and humus samples clearly show much higher radiocesium activity.

Palynological Analysis

The selected samples were boiled in 10% KOH and prepared according to the acetolysis method. The preparations were embedded in silicone-oil 2000 CS, and a known number of *Lycopodium* spores were added to calculate the concentration of sporomorphs. Counting of the sporomorphs was carried out with a Leitz Laborlux 2 microscope at 500x magnification. The works of Barnett and Hunter (1987), Bassett et al. (1978), Moore et al. (1991), Nilsson (1983), Pals et al. (1980), Prescott (1970), van Geel (1978) and van Geel et al. (1983) were used for initial identification. Identifications were checked with the Pollen Reference Collection of the Laboratory of Palaeobotany and Palynology of Utrecht.

Quantitative Approach

The dominant regional pollen types (shown in the synthetic AP/NAP diagram on the left) and a local pollen group (from plant taxa present along the transect, shown as pollen percentage together with the corresponding species average cover) are given in Text-Figure 3. All the percentages of local pollen, spores and fungal remains were calculated on the basis of the regional pollen sum (i.e. pollen produced by taxa away from local pollen sources). The raw concentration matrix was saved as an ASCII file in the Marseille Format of the TILIA program. The lower semimatrix forms were rearranged starting randomly as specified by the SYN-TAX 5.0 package optimizing the diagonal structure of the input matrix with the NMSCAL and DIAGTABL programs. The nonmetric multidimensional scaling of the local cryptogams is shown in Text-Figure 4, while their seriation is given in Text-Figure 5.

RESULTS AND DISCUSSION

The nonmetric multidimensional scaling method measures the stress between the original and the new rank order of distances of the data (Podani, 1993). In this procedure it is essential to define the vegetation unit of interest to avoid background noise. In Text-Figure 4 the iterative two-dimensional solution obtained from the concentration values shows a straight line in the Shepard diagram. This involves a monotonic increase over the original distances, and a high "goodness" of the solution (less scattered local dispersal). The mixing of local and regional information in the sporomorph assemblages makes it difficult to extrapolate the fungal signals from the seriation. The matrix has been redefined by the relocation procedure (178 iterations) in Text-Figure 5 (rows as taxa and columns as samples) in order to optimize the small distances as close to the diagonal as possible (large values far from the diagonal). When comparing these results with Text-Figure 3, it is evident that the highest concentration of fungi occurs under hypertrophic conditions, while the ecotone belt shows the greatest diversity. A continuous data structure does not exist, but the distance shading structure gives an insight in groups of spore assemblages. These groups reflect the structure of the vegetation and its response to disturbance (cattle pressure, human impact, and fire).

Palynoflora

The spatial variation in the pollen deposition allows for a fine-scaled comparison between the mycoflora (representing a clear palynological signal from the mycocenoses) and the palynoflora (as reflecting other plant communities). The trends of sporomorph curves show that only heath and pine/oak forest are reflected in local percentage values, whereas the regional pollen and airborne spores do not show such a distinct pattern. The radiocesium content does not show an age differentiation in the moss samples as they span ages derived from a previous stage in the heath cycle. In the heathlands the information on a (sub) recent succession is rather complete: open landscapes involve a global representation of the regional pollen rain (Bradshaw, 1981), and most of the local vegetation is over-represented. Succession after surface-fire is marked at Dwingeloo landscape by a dense and tall forb layer with luxuriant *Eriophorum vaginatum*. This is common in most Dutch drained bogs (Barkman, 1992), and is also confirmed by the palynological occurrence of spores of *Gelasinospora* (Mulder and Janssen, 1996).

In Text-Figure 3, a comparison of the occurrence and the pollen production of the Ericaceae shows that *Calluna* is much more productive in oligotrophic environments, while *Erica* is not phenologically biased. The past occurrence of *Oxycoccus* and *Empetrum* in a previous stage of heath regeneration is still recorded by the presence of their pollen types. The dispersal of *Andromeda* pollen is extremely local (as these pollen tetrads are large and heavy) and the species is limited to meso/hypertrophic conditions. The mosaic of fen and wet prairies indicates that the plots are accessible to larger mammals, as is evident in the occurrence of coprophilous fungi in the rearranged matrix (Text-Figure 5).

Bioturbation

Soil inhabiting animals can selectively degrade pollen grains (Davis, 1990). The ingestion of Gramineae pollen grains with thin-exine by soil invertebrates caused an under-representation of wild grasses in the JD14 plot (Text-Figure 3). Cereal pollen grains are over-represented in the same plot due to mechanical displacement of pollen: the activity of Collembola and other invertebrates is marked by an abundance of excrement-digesting fungi. In the ectorganic layer of sandy soils under acidic and anaerobic conditions below a pine canopy (sample JD1-5), the presence of dung fungi like *Cercophora* indicates that an increased accumulation of excrement containing pollen can produce a misleading interpretation of the pollen record of the understorey. The palynodebris of Dwingeloo shows that the microfauna became depauperate if compared to the results of past studies conducted in the same area more than sixty years ago by Hoogenraad (1934). The majority of the rhizopods collected in 1995 belong to the Wrightianum Group (Mulder et al., 1995).

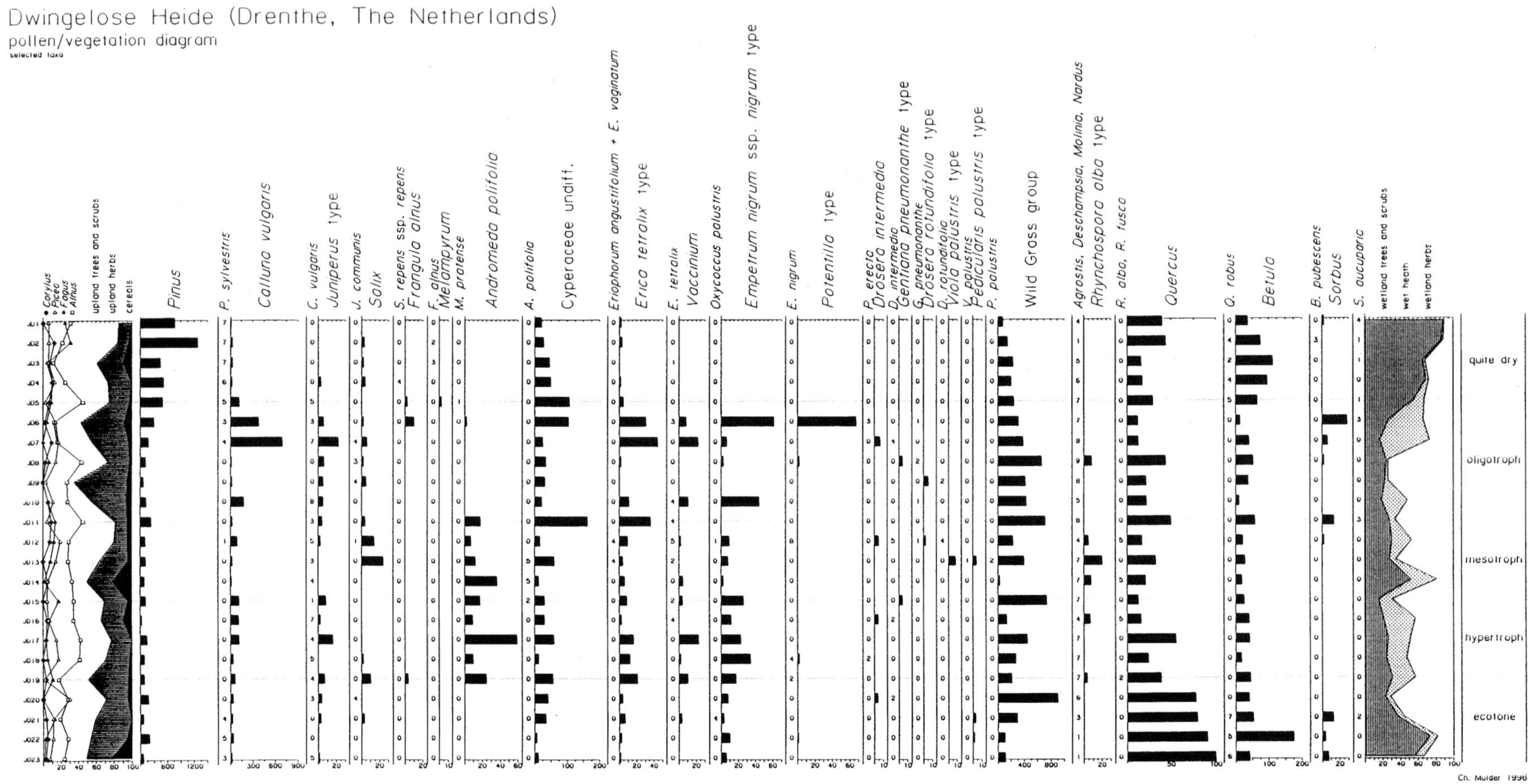

Text-Figure 3. Vegetation abundance and related pollen production/dispersal of selected taxa. The cover for the canopy and the understorey is given in Domin values (0=taxon absent, up to 9=very dense cover, taxon >75%). The percentages are calculated on a regional pollen sum.

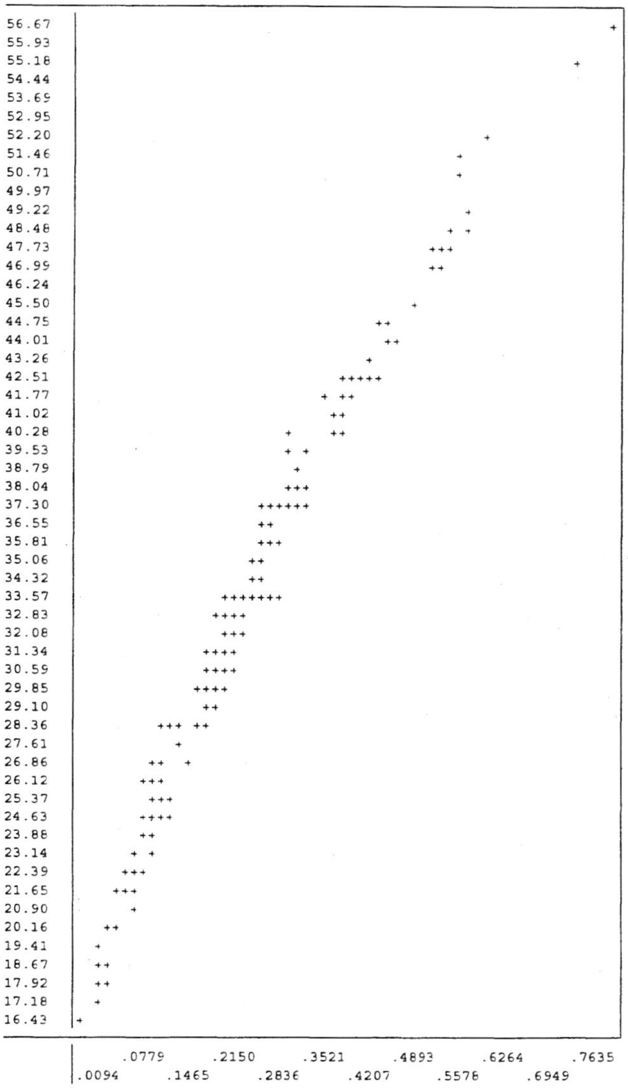

Text-Figure 4. Shepard diagram of the concentration values of all the local taxa recorded at Dwingeloo (original distances vs ordination distances, Nonmetric Multidimensional SCALing program). Random seed value = 9139. This solution narrows an ideal case (a straight line).

Mycoflora

The mycocenoses occurring in the heath are widely represented under hypertrophic conditions (highest concentrations of *Brachysporium*, *Helicoon* and *Sporormiella*). Some fungi appear to be more linked to hydrological conditions, as *Tetraploa aristata* (whose conidia occur only in oligotrophic wetlands), while under mesotrophic conditions other fungi such as *Coniochaeta xylariispora*, *Podospora*, and *Puccinia* spp. are quite rare. The correlation between freshwater genera as *Helicoon* and *Tetraploa* can be assumed as direct (the water film is the primary environment). On the contrary, the occurrence of dung fungi like *Sporormiella* depends on secondary airborne transport (in the forest soils, cf. Davis and Shafer, 1991) or on the primary presence of palatable plants (at Dwingeloo rare under mesotrophic conditions).

The marked presence of fungal remains like the setae of *Beltrania* and the aeciospores of *Puccinia graminis* confirms the close correlation between some mushrooms and the radiocesium availability level in the humus. Andolina and Guillitte (1990) explained the increased ^{137}Cs-availability as a consequence of the lignin-decomposing ability of the fungi (due to the stronger radiocesium retention by lignins in holorganic horizons), but the sporomorphs types here identified are usually more associated with the fulvic and humic acids. Since acidity increases during litter decomposition owing to fulvic and humic acid production (Hemond, 1980), this phenomenon involves a closer relation between fungi and moss carpets.

The mycoflora patterns provide a new possibility to integrate fungal microcommunities in both landscape ecology and the development of recent vegetation. Data from the extant mycoflora should be taken into account when reconstructing terrestrial paleoenvironments.

The inclusion of fungal remains in landscape ecological assessments allows for a more comprehensive view of environmental change both in the course of development of single communities (synepiontology) as in vegetation dynamics as a whole (synchronology).

ACKNOWLEDGMENTS

The author is most grateful to Roel Janssen, Hans Joosten and Bas van Geel for discussions and support. Mario Voltaggio kindly provided the spectrometric analyses at the C.N.R. of Rome. Owen Davis is thanked for the supplied publications. Pierre A. Zippi and David K. Goodman are thanked for comments on the manuscript. Contribution No. 960705 of the Netherlands Research School of Sedimentary Geology.

References Cited

ANDOLINA, J., and GUILLITTE, O.
1990 Radiocesium availability and retention sites in forest humus. *In:* Desmet, G., Nassimbeni, P., and Belli, B. (eds.), *Transfer of radionuclides in natural and semi-natural environments*. Elsevier Applied Science, pp. 135–142.

BAKKER, T.W.M., CASTEL, I.I.Y., EVERTS, F.H., and DE VRIES, N.P.J.
1986 Het Dwingelderveld, een Drents heidelandschap. *Reeks Landschapsstudies*, 8: 198 pgs.

BARKMAN, J.J.
1973 Synusial approach to classification. *In:* Whittaker, R.H. (ed.), *Ordination and Classification of Communities. Handbook of Vegetation Science*, 5: 435–491.
1992 Plant communities and synecology of bogs and heath pools in the Netherlands. *In*: Verhoeven, J.T.A. (ed.), *Fens and Bogs in the Netherlands: Vegetation, History, Nutrient Dynamics and Conservation. Geobotany*, 18: 173–235.

BARKMAN, J.J., and WESTHOFF, V.
1969 Botanical evaluation of the Drenthian district. *Vegetatio*, 19: 330–388.

BARNETT, H.L., and HUNTER, B.B.
1987 *Illustrated Genera of Imperfect Fungi.* Macmillan, New York. 218 pgs.

BASSETT, I.J., CROMPTON, C.W., and PARMELEE, J.A.
1978 *An Atlas of Airborne Pollen Grains and Common Fungus Spores of Canada. Canada Department of Agriculture, Monograph No.* 18: 321 pgs.

BRADSHAW, R.H.W.
1981 Modern pollen-representation factors for woods in south-east England. *Journal of Ecology*, 69: 45–70.

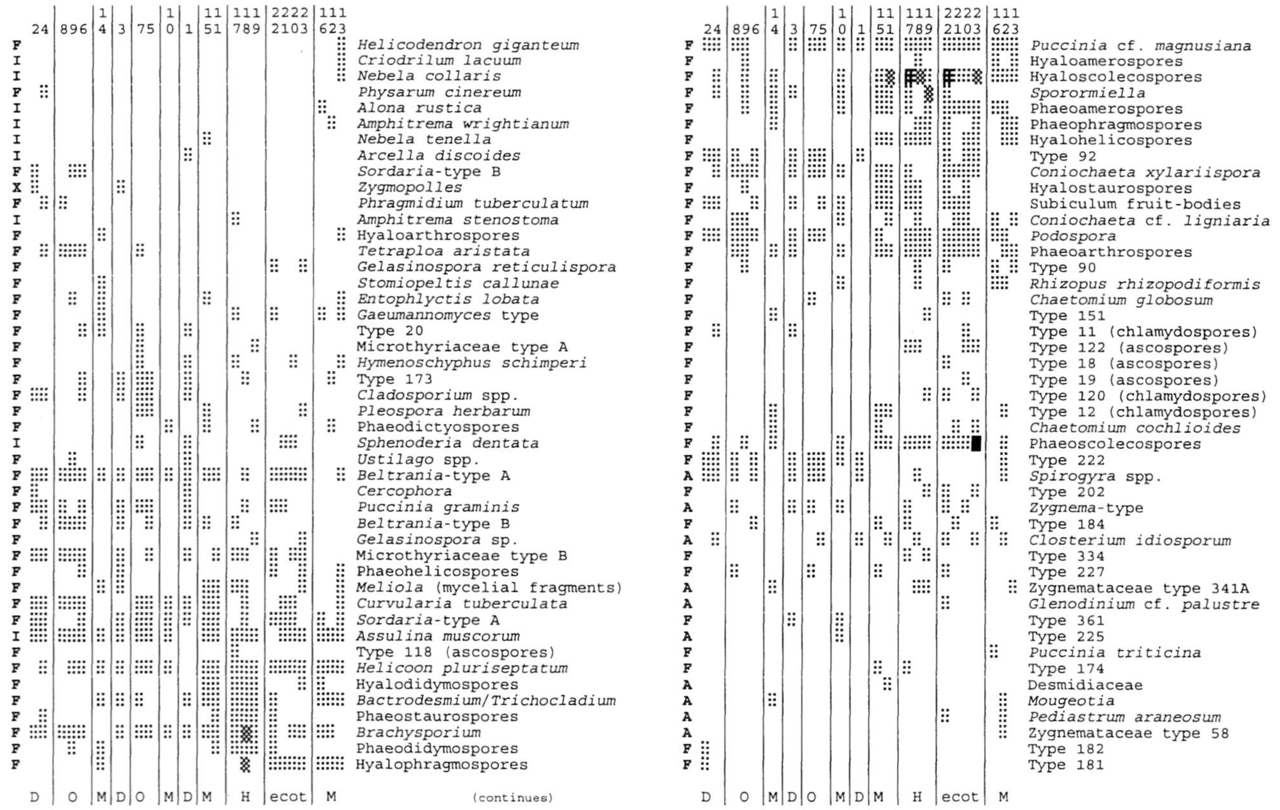

Number of grains/g: = 1.0–138,681; = 138,682–277,362; = 277,363–416,043; = 416,044–554,724.

Text-Figure 5. Modern spore assemblages in the cryptogamic layer of the Dwingelose Heide (DIAGTABL program). Rearranged matrix of the plots along the transect (*x*-axis: JD1-23) and the concentration values of fungal remains (F), algae (A), rhizopods (I) and sporomorphs of unknown origin (X) for each sample (*y*-axis). Landscape ecological parameters of the plots: D=Dry samples (acid litter); O=Oligotrophic, M=Mesotrophic and H=Hypertrophic conditions.

CHMURA, G.L., and KOSTERS, E.C.
1994 Storm deposition and ^{137}Cs accumulation in fine-grained marsh sediments of the Mississippi Delta Plain. *Estuarine, Coastal and Shelf Science*, 39: 33–44.

DAVIS, O.K.
1990 Pollen analysis of archeological sites. *In*: de Barros, P., and Koerper, H.C. (eds.), *Final test investigation report and request for determination of eligibility for 23 sites along the San Joaquin Hills transportation Corridor*. Chamber Group, Inc., pp. J1–J37.

DAVIS, O.K., and SHAFER, D.S.
1991 *Sporormiella*: an indicator of the Clovis-Folsom biomass collapse. *Smithsonian Series in Archaeological Inquiry, 56th Meeting*: 1–17.

HEMOND, H.F.
1980 Biogeochemistry of Thoreau's bog, Concord, Massachusetts. *Ecological Monographs*, 50: 507–526.

HICKS, S.
1985 Modern pollen deposition records from Kuusamo, Finland. I. Seasonal and annual variation. *Grana*, 24: 167–184.

HOOGENRAAD, H.R.
1934 Studien über die sphagnicolen Rhizopoden der Niederländischen Fauna. *Archiv für Protistenkunde*, 84: 100 pgs.

JANSSEN, C.R.
1966 Recent pollen spectra from the deciduous and coniferous-deciduous forests of northwestern Minnesota: a study in pollen dispersal. *Ecology*, 47: 804–825.

MOORE, P.B., WEBB, J.A., and COLLINSON, M.E.
1991 *Pollen Analysis* (2nd ed.). Blackwell Scientific Publications, 216 pgs.

MULDER, Ch.
1996a *Analisi comparata della struttura della vegetazione attuale con lo studio palinologico di campioni di superficie in differenti regioni europee*. PhD Thesis VIII Cycle Bot. Sc. - First University of Rome "La Sapienza" / Lab. Palaeobotany and Palynology, Utrecht – National Libraries of Rome and Florence, 160 pgs.
1996b *Fungal remains and Radiocesium deposition: an actuopalynological approach*. The Linnean Society, Palaeobotany and Palynology: Past, Present and Future. Royal Holloway University of London, 24–25th April, 1996: 9–11.
1997 Actuopalynology and Mediterranean vegetation patterns along the littoral of southern Latium (Italy). *Mededelingen Nederlands Instituut voor Toegepaste Geowetenschappen, TNO*, 58: 285–292.

MULDER, Ch., and JANSSEN, C.R.
1996 *Relations between pollen and fungal spore dispersal and present-day landscape ecological parametres Within Dutch heath communities*. Abstr. Ninth International Palynological Congress, Houston, 23–28th June 1996: 113.
1998 Application of Chernobyl Caesium-137 fallout and naturally occurring Lead-210 for standardization of time in moss samples: recent pollen-flora relationships in the Allgäuer Alpen, Germany. *Review of Palaeobotany and Palynology*, 103: 23–40.

1999 Occurrence of pollen and spores in relation to present-day vegetation in a Dutch heathland area. *Journal of Vegetation Science*, 10: 87–100.

MULDER, Ch., JANSSEN, C.R., and JOOSTEN, J.H.J.
1995 Distributional patterns of palynomorphs and palynodebris along an *Oxycocco-Sphagnetea* transect at Dwingeloo (Drenthe, the Netherlands). *Jahrestreffen des Arbeitskreises für Vegetationsgeschichte der Reinhold-Tüxen-Gesellschaft, Eberswalde*, 24–26. November 1995: 11.

NILSSON, S. (ed.)
1983 *Atlas of Airborne Fungal Spores in Europe*. Springer Verlag, Berlin, 139 pgs.

PALS, J.P., VAN GEEL, B., and DELFOS, A.
1980 Paleoecological studies in the Klokkeweel bog near Hoogkarspel (Prov. of Noord Holland). *Review of Palaeobotany and Palynology*, 30: 371–418.

PODANI, J.
1993 *SYN-TAX-pc version 5.0: Computer Programs for Multivariate Data Analysis in Ecology and Systematics*. Scientia Publishing, Budapest, 104 pgs.

PRESCOTT, G.W.
1970 *How To Know The Freshwater Algae* (2nd ed.). Wm. C. Brown Company Publishers, Dubuque, Iowa, 348 pgs.

RODWELL, J.S. (ed.)
1991 *British Plant Communities – v. 2. Mires and Heaths*. Cambridge University Press, 628 pgs.

SCHAMINÉE, J.H.J, WEEDA, E.J., and WESTHOFF, V. (eds.)
1995 *De vegetatie van Nederland. Deel 2. Plantengemeenschappen van wateren, moerassen en natte heiden*. Opulus Press, Uppsala-Leiden, 357 pgs.

SHERWOOD-PIKE, M.A.
1988 Freshwater fungi: fossil record and paleoecological potential. *Palaeogeography, Palaeoclimatology, Palaeoecology*, 62: 271–285.

STORTELDER, A.H.F., HOMMEL, P.W.F.M., and DE WAAL, R.W.
1998 *Broekbossen. Bosecosystemen van Nederland*, 1. Koninklijke Nederlandse Natuurhistorische Vereniging, Utrecht, 216 pgs.

VAN GEEL, B.
1978 A palaeoecological study of Holocene peat bog sections in Germany and The Netherlands, based on the analysis of pollen, spores and macro- and microscopic remains of fungi, algae, cormophytes and animals. *Review of Palaeobotany and Palynology*, 25: 1–120.

VAN GEEL, B., HALLEWAS, D.P., and PALS, J.P.
1983 A Late Holocene deposit under the Westfriese Zeedijk near Enkhuizen (Prov. of N-Holland, The Netherlands): Palaeoecological and archaeological aspects. *Review of Palaeobotany and Palynology*, 38: 269–335.

VAN HEUVELN, B.
1965 *De bodem van Drente*. Stichting voor Bodemkartering, Wageningen, 135 pgs.

VAN RAAMSDONK, L.W.D., and SCHOUTEN, H.J.
1997 Gene flow and establishment of transgenes in natural plant populations. *Acta Botanica Neerlandica*, 46: 69–84.

Author's address

CHRISTIAN MULDER
Laboratory of Palaeobotany and Palynology
Budapestlaan 4
3584 CD Utrecht
The Netherlands

Present address:
Laboratory for Ecotoxicology
RIVM – P.O. Box 1
3720 BA Bilthoven
The Netherlands
E-mail: christian.mulder@rivm.nl

MODERN POLLEN RAIN COMPARED TO VEGETATION IN THE HIMALAYAS

Chhaya Sharma

Abstract

Pollen analysis of recent moss-cushions in different habitats of the Eastern Himalaya (Darjeeling and Sikkim regions) portrays a reasonably accurate record of the extant vegetation in forested areas. *Alnus* exhibits marked dominance over *Quercus*, *Betula*, *Carpinus*, *Juglans*, *Rhododendron*, *Ulmus*, *Celtis* and *Pinus*, which also occur in reasonable numbers. The presence of *Tsuga* and *Larix* pollen is certainly misleading because neither of these taxa grows in these forests. On the other hand, the pollen of *Shorea robusta*, *Tectona grandis* and *Cryptomeria japonica* is absent even in the spectra derived from the subtropical Sal-Teak forest as well as from thick *Cryptomeria japonica* plantations, thus presenting a distorted picture of the existing vegetation. Similarly, the pollen spectra constructed for deforested areas and for open scrub accurately reflect the dominance of non-arboreals over arboreals as demonstrated by comparatively higher frequencies of grasses, followed by Caryophyllaceae, Asteraceae and Cheno/Ams and the sporadic occurrence of *Quercus*, *Pinus*, *Corylus*, *Alnus* and *Acer*. The preponderance of fern spores, in both forested as well as deforested locations result from the perennial regeneration of ferns in the area.

Comparative analysis of air catches and recent pollen spectra exhibit reasonable coherence in the representation of *Alnus*, *Pinus*, *Betula* and *Carpinus* among arboreals, as well as prominent herbaceous components such as grasses, sedges, Cheno/Ams and Asteraceae among non-arboreals.

INTRODUCTION

The main purpose of Quaternary palynological research is to reconstruct paleovegetation to decipher the past climate. For the correct interpretation of pollen diagrams depicting paleovegetation, it is advisable to know the present vegetation of the area and the local modern pollen rain spectrum or models. The vegetation and environmental conditions are connected so that change in one affects change in the other. Hence, a change in vegetation in a region is a reliable index of environmental change even if it is due to anthropogenic activities. This is particularly true in the Himalaya with its diverse vegetation, rich biodiversity, and highly fluctuating climate, even at a local level. These changes depend primarily on location, its altitude, the monsoon, and relative position within the range.

These lofty mountain ranges support subtropical, temperate and alpine vegetation today. Therefore paleoclimate studies to reveal changes during the Quaternary would be interesting. It is a challenge, however, to find suitable lake deposits which can be accessed for sampling.

This study was carried out to determine the modern pollen/vegetation relationships to help in the resolution and interpretation of pollen diagrams constructed from the Darjeeling and Sikkim Himalayan regions (Sharma, 1993; Sharma, in press; Sharma and Chauhan, 1994; Chauhan and Sharma, 1996). As no such work has been done previously in the Eastern Himalaya, the necessary survey was carried out in forested as well as open areas, ranging in altitude from the plains around Siliguri in the Darjeeling district to the alpine heights of Chhangu in Sikkim. A large number of surface samples was collected (Text-Figure 1) along a transect ascending from the Siliguri foothills (with thick subtropical forests) to the heights of temperate forests in Darjeeling, covering a distance of about 100 km and altitude from 900 to 2700 m (Table 1). Surface samples were also collected in a transect from Gangtok to Khechipiri (Table 2), covering a distance of 160 km west of Gangtok through the temperate belt of Sikkim. However, the samples from the open regions were collected from Gangtok to the alpine localities in Chhangu across a 10 km transect (Table 3).

Sharma, Chhaya
2001 Modern pollen rain compared to vegetation in the Himalayas; *In*: Goodman, D.K., and Clarke, R.T. (eds.), *Proceedings of the IX International Palynological Congress, Houston, Texas, U.S.A., 1996*; American Association of Stratigraphic Palynologists Foundation, p. 557–565.
 ISBN 0–931871–06–9

VEGETATION OF EASTERN HIMALAYA: DARJEELING AND SIKKIM REGIONS

These two regions have a high monsoon intensity and humid climate due to their proximity to the Bay of Bengal and direct exposure to the effects of the moisture-laden southwest monsoons. The lush green vegetation of the Eastern Himalaya has fascinated naturalists since the time the East India Company established its headquarters at Calcutta. Therefore, ahough much has been published on the floristics of this region for the present palynostratigraphical studies, some of the relevant references include Gammie (1893, 1894), Burkill (1907), Chaudhri (1960), Rolla (1963), Mehra and Bir (1964), Hara (1966), Nayar (1977), Sahni (1981, 1982), Mehra et al. (1985), Rao and Hajra (1986), Gangopadhya and Banerjee (1987), Singh and Singh (1987), Chowdhery and Rao (1992), Khoshoo (1992) and Rao (1994). My own observations, as well as data gathered during the field work in the Darjeeling-Sikkim regions, have also been incorporated, chiefly pertaining to the forests around the sites investigated.

The vegetation in Darjeeling and Sikkim is broadly organized by altitude, with tropical forests in the foothills to the alpine meadows beyond the tree line at higher elevations. The vegetation types are discussed below.

Low-Hill Forests

Tropical–Subtropical, to 900 m altitude. These forests occur in the area of Siliguri up to an altitude of 900 m in both Darjeeling and Sikkim. *Shorea robusta* dominates along

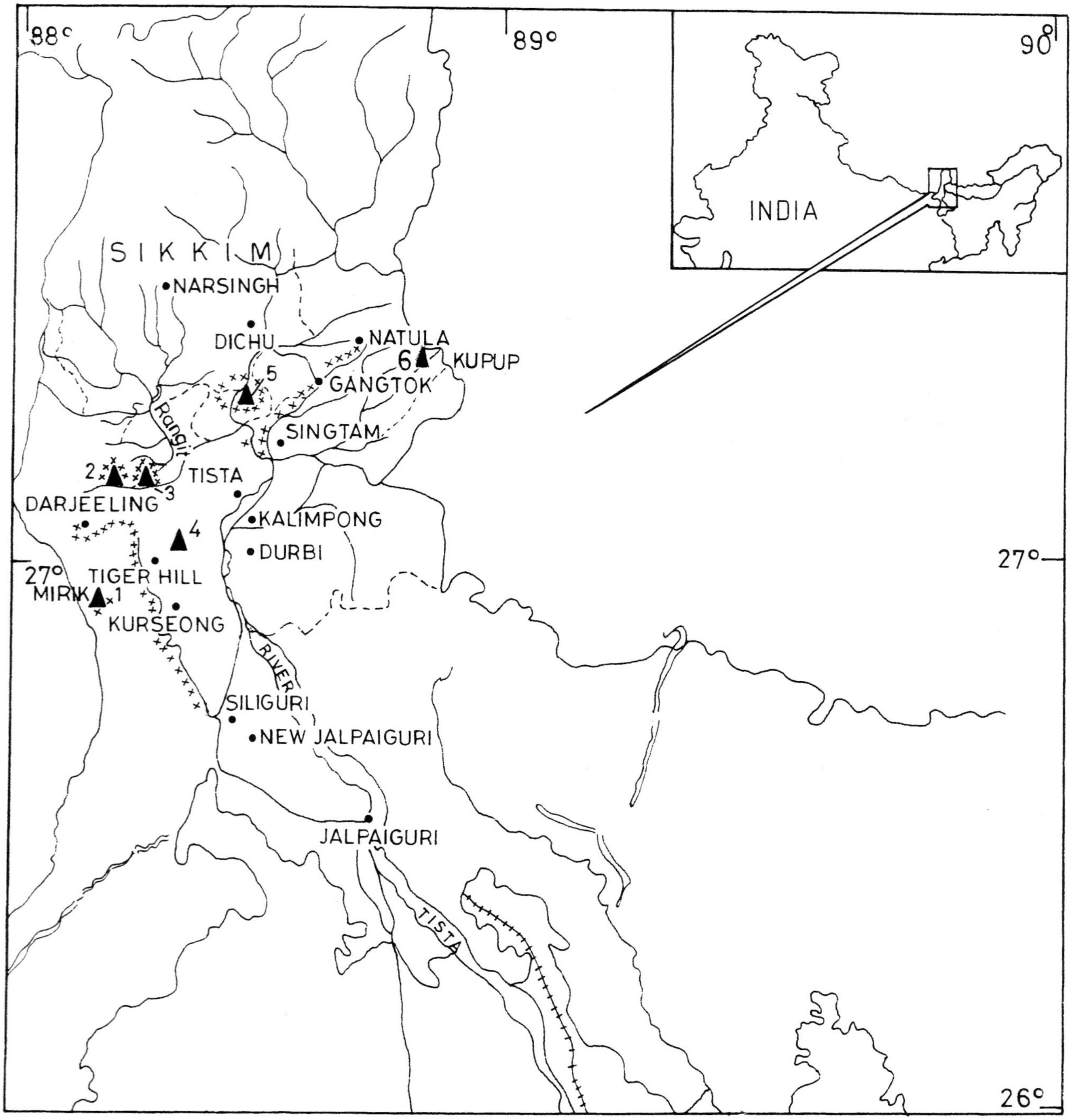

Text-Figure 1. Map of Darjeeling and Sikkim area to show location of lake sites and surface samples. Index: ▲ lake site; 1. Mirik; 2. Sindhap; 3. Sinchar; 4. Jorepokhari; 5. Khechipiri; 6. Kupup; ● Township; xx Surface sample.

the ridges and slopes together with its main associates, *Schima wallichii, Castanopsis, Bauhinia, Cedrela toona, Sterculia, Michelia champaca, Terminalia, Ailanthus* and *Anthocephalus kadamba.* In the Kurseong Division, in addition to these species, *Adina, Terminalia* and *Castanopsis also* occur in abundance. Other common trees in these forests are *Acacia catechu, Dalbergia sisso, Salmalia malabarica, Dillenia, Dipterocarpus, Eugenia, Lagerstroemia, Sterculia, Artocarpus* and *Albizzia.*

Middle-Hill Forests

Subtropical, 750–1500 m altitude. These are largely evergreen forests in which deciduous trees are uncommon. The principal forest constituents are *Alnus nepalensis, Juglans regia, Betula alnoides, Engelhardtia, Cedrela, Castanopsis, Schima wallichii, Eugenia, Cinnamomum, Michelia champaca, Phoebe, Litsea, Ficus roxburghii* and the prominent evergreen arboreals such

TABLE 1. Surface samples collected from Darjeeling Himalaya-forested regions (between Siliguri and Darjeeling).

Sample No.	Locality	Details of surface samples
1–4	Sukna	Moss-cushions from Sal forest
5, 6	Chunbati	Moss-cushions from Sal-Teak forest
7, 8	Tindhara	Moss-cushions from Sal-Teak forest
9, 10	Ghaiyo-Bari	Moss-cushions from mixed forest
11, 12	Mahanadi	Moss-cushions from *Cryptomeria* forest
13, 14	Gidhya-Pahar	Moss-cushions from *Cryptomeria* forest
15, 16	Tong	Moss-cushions from mixed broad-leaved forest
17	Sonada	Moss-cushion from mixed broad-leaved forest
18	Sonada	Soil sample from mixed broad-leaved and conifer forest
19, 20	Ghoom	Moss-cushions from *Cryptomeria* forest
21	Jorepokhri	Moss-cushion from *Cryptomeria* forest

TABLE 2. Surface samples collected from Sikkim Himalaya-forested regions (between Gangtok and Khechipiri).

Sample No.	Locality	Details of Surface samples
1, 2	Deorali,	Moss-cushions from mixed broad-leaved forest
3, 4	Singtam	Moss-cushions from mixed broad-leaved forest
5, 6	Takru	Moss-cushions from mixed broad-leaved forest
7, 8	Babangle	Moss-cushions from mixed broad-leaved forest
1–11	Khechipiri Lake site	Moss-cushions from mixed broad-leaved forest

TABLE 3. Surface samples collected from Sikkim Himalaya-deforested regions (between Gangtok and Chhangu).

Sample No.	Locality	Details of Surface samples
1	1800 m alt.	Moss-cushion from mixed-forest
2	2100 m alt.	Moss-cushion from mixed scrub forest
3, 4	2400 m alt.	Moss-cushions from mixed scrub forest

as *Quercus glauca, Q. spicata, Q. serrata* and *Q. griffithii* which grow in dense formations between 1200 m and 1600 m and higher.

High-Hill Forests

Warm or Wet-Temperate, 1500–2700 m altitude. Most of the mountain slopes between 600 and 2100 m altitude in the Darjeeling district have been cleared of natural forest for tea plantations. However, some remnants of these thick forests still occur around Darjeeling and in the adjoining Sikkim regions at higher elevations. Fortunately, these forests are still undisturbed, with very thick growth supporting rich ground cover.

It should, however, be noted that *Cryptomeria japonica* was introduced in 1891 from Japan as a timber tree and is now extensively planted in and around Darjeeling. This exotic pine now covers large areas, forming dense forests between 2100–2400 m altitude, and thrives in this climate. The existing forests in this zone constitute the characteristic temperate belt whose main elements are oaks and laurels. The latter becomes dominant between 1800–2100 m, *Quercus lamellosa* between 2100–2400 m, and *Q. pachyphylla* between 2400–2700 m; *Michelia* and *Magnolia* occur throughout these zones. Other trees such as *Michelia edulis, Quercus fenestrata* and *Castanopsis hystrix* are quite common in these forests up to 2100 m. *Alnus nepalensis* grows mainly along water courses between 2400 and 2700 m. *Quercus lamellosa, Q. pachyphylla, Castanopsis tribuloides, Acer campbellii, Magnolia cambelli, Symplocos theaefolia* and *Taxus baccata* and are prominent in these forests. *Quercus pachyphylla* occurs in pure formations at about 2700 m altitude and under its shade *Rhododendron griffithii* finds a very favorable habitat together with frequent clumps of *Arundinaria*.

Rhododendron–Conifer Forests

Cold Temperate–Subalpine, 2700–3600 m altitude. Evergreen forests in this altitudinal belt contain a preponderance of *Rhododendron*, alongwith conifers mixed with *Quercus pachyphylla* and *Q. lineata* — the two common broad-leaved arboreals that occur above 2700 m. Higher up, the forests are characterized by *Rhododendron, Betula utilis, Taxus baccata, Tsuga brunoniana* and *Picea spinulosa. Abies densa* grows between 2700 and 3000 m — particularly in the remote Lachen and Lachung valleys. *Larix griffthiana* i.e. Himalayan larch (Sharma and Gupta, 1984), occurs in almost pure formations between Karponang and Chhangu

(East Sikkim) to altitudes up to 3600 m. *Salix wallichiana* often occurs along the streams. *Rhododendron* assumes a scrubby habit at higher altitudes, notably above 3000 m. The vegetation above the tree line in Sikkim is primarily a mosaic of *Rhododendron.*

Alpine Vegetation

4300 m altitude and above. At 3600 m altitude and above, there occurs a good growth of vegetation in the northern and eastern heights of Sikkim — especially in the regions above Thangu (4200 m) and Chhangu (4300 m). *Ephedra saxatilis* and *Rhododendron lepidotum* often cover large areas near the tops of these ridges around Thangu. In the open lush green meadows, a luxuriant growth of *Ranunculus, Anemone, Delphinium, Rhus, Potentilla, Primula, Fragaria, Cassiope* and *Allium* is developed.

LABORATORY PREPARATIONS

Extraction of pollen and spores from the samples was done following the maceration technique by Erdtman (1943, 1969). Ten grams of moss-cushion were boiled in 10% KOH solution to deflocculate the matrix, washed with water, and passed through 200 mesh sieve to obtain residual material for macroscopic remains. The filtrate was centrifuged and then subjected to acetolysis. For clayey samples, the filtrate was treated with hydrofluoric acid for 8 to 10 days to dissolve silica. After decanting the HF, the material was then washed with water, dehydrated with glacial acetic acid, and acetolyzed. The processed samples are stored in 50% glycerine with a few drops of phenol.

Approximately 300 to 500 pollen grains were counted for each sample; the pollen spectra were based on a pollen sum of total terrestrial pollen. The plant taxa are categorized as trees, shrubs, herbs, aquatics, and lower plants. In all, four pollen spectra were constructed: three from the forested regions of Darjeeling and Sikkim, and one from the open region of Sikkim. The aerospora from Darjeeling and Sikkim were also studied using exposed slides during November, 1989.

MODERN POLLEN/VEGETATION RELATIONSHIPS

Recent Pollen Spectra from Darjeeling Himalaya

Forested Regions: Between Siliguri and Darjeeling. Pollen spectra constructed from the Darjeeling Himalaya samples (Text-Figure 2) show the subtropical foothill to temperate high altitude vegetation succession. The subtropical belt of Sal-Teak (*Shorea -Tectona*) forests at lower elevations is characterized by the dominance of non-arboreal taxa over arboreal, the former dominated by grasses and Asteraceae, followed by Cyperaceae, Cheno/Ams and *Polygonum.* In contrast, *Shorea robusta* and *Tectona grandis,* the two chief constituents of these forests, interestingly are not present in the pollen spectra. Other arboreal elements such as *Betula, Alnus, Pinus, Tsuga,* Fabaceae and Rosaceae numbers are recorded in consistently low values.

Pollen spectra from the mixed broad-leaved forests are dominated by *Quercus, Alnus* and *Betula,* followed by *Carpinus, Juglans* and *Pinus.* In the pollen spectra con-

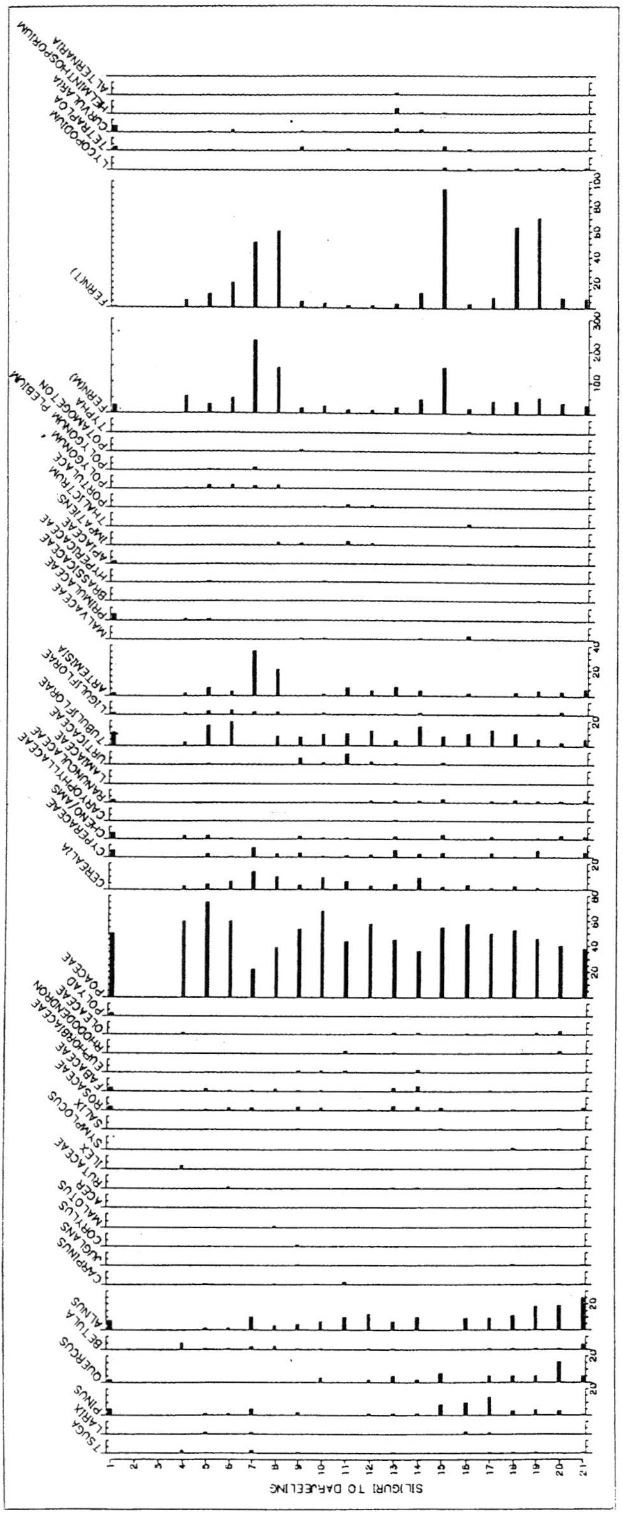

Text-Figure 2. Recent pollen spectra between Siliguri and Darjeeling (percentages calculated in terms of terrestrial plants).

structed from conifer-dominated forests, *Pinus* is represented and attains its highest frequency, however *Tsuga* and *Larix* provide a misleading appearance because they do not grow in the region and are in fact high altitude taxa. The pollen of *Thea sinensis* does not occur in the spectra. The herbaceous vegetation in the mixed broad-leaved forest is dominated by grasses and sedges, followed by Asteraceae, Cheno/Ams and Ranunculaceae. *Lycopodium* spores are also present in low frequencies, and only in the pollen spectra constructed from mixed broad-leaved conifer forests. *Tetraploa* and *Curvularia* fungal spores are more common than *Helminthosporium* and *Alternaria*.

Aerospora of Darjeeling

The aeropalynological investigations carried out at Darjeeling (Text-Figure 3) reveal exceedingly high values of *Alnus*, whereas other taxa such as *Pinus*, *Carpinus* and *Betula* occur in high frequencies compared to *Larix*, *Cryptomeria* and *Syzygium*. Fabaceae and Urticaceae pollen occur sporadically. Among the herbs, Poaceae is most common with *Artemisia*, Tubuliflorae, Cheno/Ams and Ranunculaceae as prominent constituents of modern pollen rain, relative to Brassicaceae, Malvaceae and Cyperaceae. Fern spores are also represented in the air catches. *Helminthosporium*, *Alternaria*, *Tetraploa* and *Curvularia* fungal spores occur in moderate numbers.

Recent Pollen Spectra from the Sikkim Himalaya

Forested Regions: Between Gangtok and Khechipiri. Recent pollen spectra constructed from mixed broad-leaved forests between Gangtok and Khechipiri Lake (Text-Figure 4) show the dominance of arboreal pollen over non-arboreal pollen. *Alnus* is most common, followed by *Betula*, *Quercus*, *Cupressus* and *Pinus*. These pollen taxa accurately reflect the surrounding vegetation. *Tsuga* and *Cryptomeria* are encountered in low numbers only in some spectra. Herbaceous vegetation is represented by moderate numbers of Poaceae, with Cyperaceae, Cheno/Ams, Caryophyllaceae, Asteraceae, Urticaceae and Ranunculaceae in low frequencies. Fern spores are present in all the spectra. Fungal spores such as *Alternaria*, *Helminthosporium*, *Curvularia* and *Tetraploa* occur in low frequencies.

Khechipiri Site. Pollen spectra constructed from Khechipiri Lake (Text-Figure 5), which is in a thick broad-leaved forest, show the dominance of arboreals over non-arboreals. *Alnus* attains maximum frequencies here, followed by *Pinus*, *Quercus*, *Betula*, *Carpinus*, *Juglans* and *Rhododendron*, depicting the existing dense forest in the region. Other arboreal taxa that are represented sporadically are *Ulmus*, *Celtis*, Myrtaceae, *Rhus*, Verbenaceae and *Ilex*. Shrubby elements are represented by low numbers of Oleaceae, Rosaceae, and by scattered occurrences of *Viburnum* and *Berberis*. As is usual among the herbs, Poaceae and *Artemisia* dominate. Other taxa include: Tubuliflorae, Cyperaceae, Cheno/Ams, Urticaceae, Caryophyllaceae, Ranunculaceae and Malvaceae. Fern spores occur in exceedingly high numbers along with many types of fungal spores such as *Alternaria*, *Tetraploa*, *Helminthosporium* and *Curvularia*; these forms occur in low numbers in some samples, reflecting the existing moist ground conditions in the area.

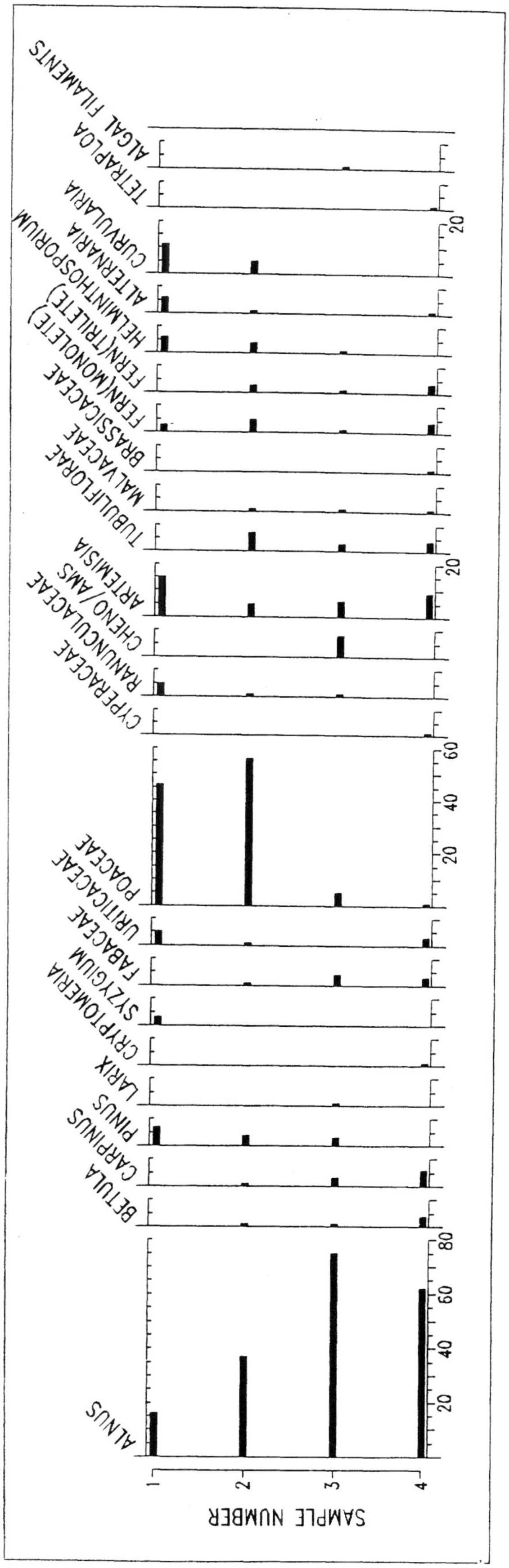

Text-Figure 3. Air catches at Darjeeling, Eastern Himalaya (November 3–6, 1989).

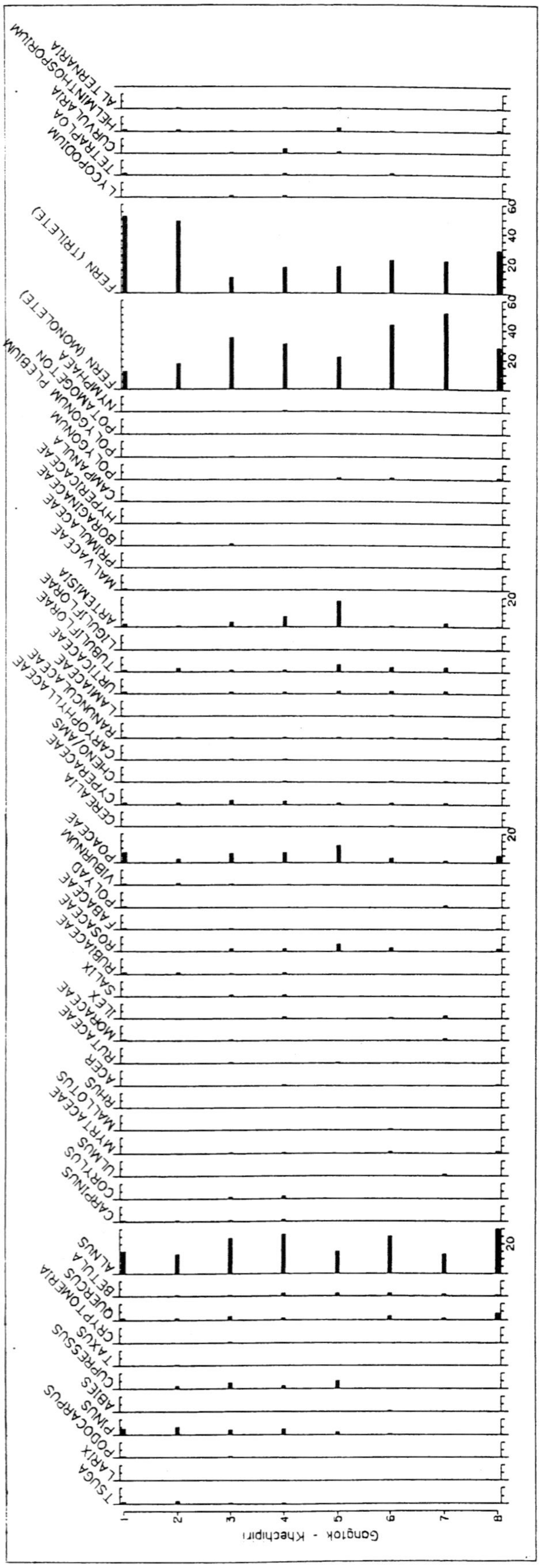

Text-Figure 4. Recent pollen spectra from Gangtok-Khechipiri, Sikkim (percentages calculated in terms of terrestrial plants).

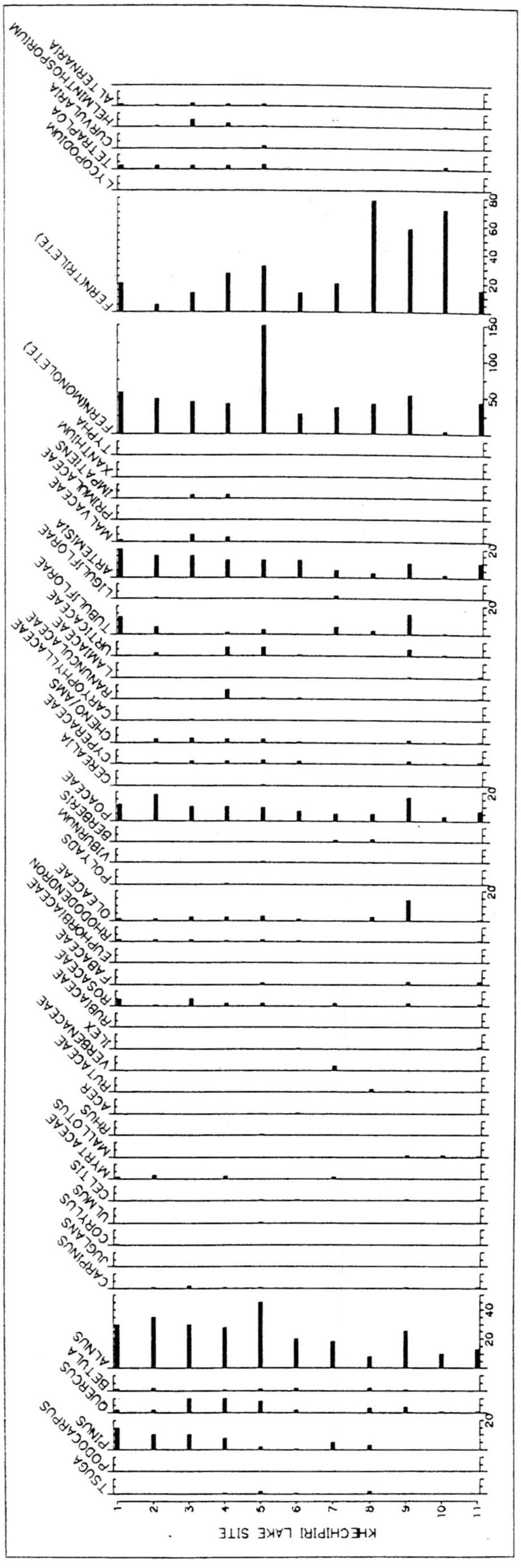

Text-Figure 5. Recent pollen spectra from the Khechipiri lake site, Sikkim (percentages calculated in terms of terrestrial plants).

Deforested Regions: Between Gangtok and Chhangu. Pollen spectra constructed from open areas (Text-Figure 6) truly reflect the depauperate vegetation. Among the arboreals, *Alnus* alone is consistently represented but in low numbers; other arboreal taxa such as *Quercus, Pinus, Corylus, Ulmus, Acer*, Rutaceae and *Mallotus* are also present. Shrubby elements such as *Viburnum*, Rosaceae and Fabaceae are also sporadic, whereas non-arboreals, such as Poaceae and Asteraceae, are more common. However, some other non-arboreal taxa such as the Cheno/Ams, Urticaceae, Caryophyllaceae and Ranunculaceae are poorly represented. Fern spores are reasonably common even in such an open setting, but *Alternaria, Tetraploa* and *Helminthosporium* fungal spores are rare. Aquatics are not present, thus reflecting good drainage in the area of the samples.

Aerospora of Sikkim

Aeropalynological studies carried out at Gangtok (Text-Figure 7) reveal high values of *Alnus*, low variable frequencies of *Betula, Carpinus, Syzygium, Rhus* and Fabaceae, and sporadic numbers of *Juglans, Pinus, Celtis, Acer, Dodonaea*, Rutaceae and *Acacia*. Shrubby elements such as *Viburnum* and Oleaceae occur in low frequencies. Poaceae pollen occurs in high frequencies, followed by Urticaceae, Ranunculaceae, Cheno/Ams, Asteraceae and Malvaceae. Fern and fungal spores are poorly represented.

DISCUSSION AND CONCLUSIONS

Proxy climate data generated by palynological analysis of Quaternary lake sediments depends heavily on the accuracy of the interpreted relationship between vegetation and pollen deposition. Despite the actual floristic diversity present in the Himalayas, the diversity represented in palynologic surface samples is both limited and dominated by arboreal elements. However, the overall picture from the surface samples (moss-cushions, soil samples) and air catches depicts the extant vegetation relatively accurately, both in Darjeeling and in the Sikkim Himalayas.

Pollen spectra from the broad-leaved forests in the Darjeeling Himalaya imply the presence of the majority of broad-leaved taxa (*Quercus, Alnus, Betula, Carpinus* and *Pinus*). The presence of *Tsuga* and *Larix* in the spectra is deceptive because neither of the taxa grows in the vicinity and their pollen is transported from the higher altitudes, particularly from the eastern heights of Sikkim. Sporadic occurrence of only a few grains of these two taxa suggests their transport from distant regions. The pollen spectra from the Sal-Teak (*Shorea–Tectona*) subtropical forest and thick *Cryptomeria* stands (plantations) portray a distorted picture because of the absence of the pollen of *Shorea robusta, Tectona grandis* and *Cryptomeria japonica* in all of the investigated samples. Absence of these three very prominent forest elements in the surface samples can be attributed to poor preservation and microbial degradation of

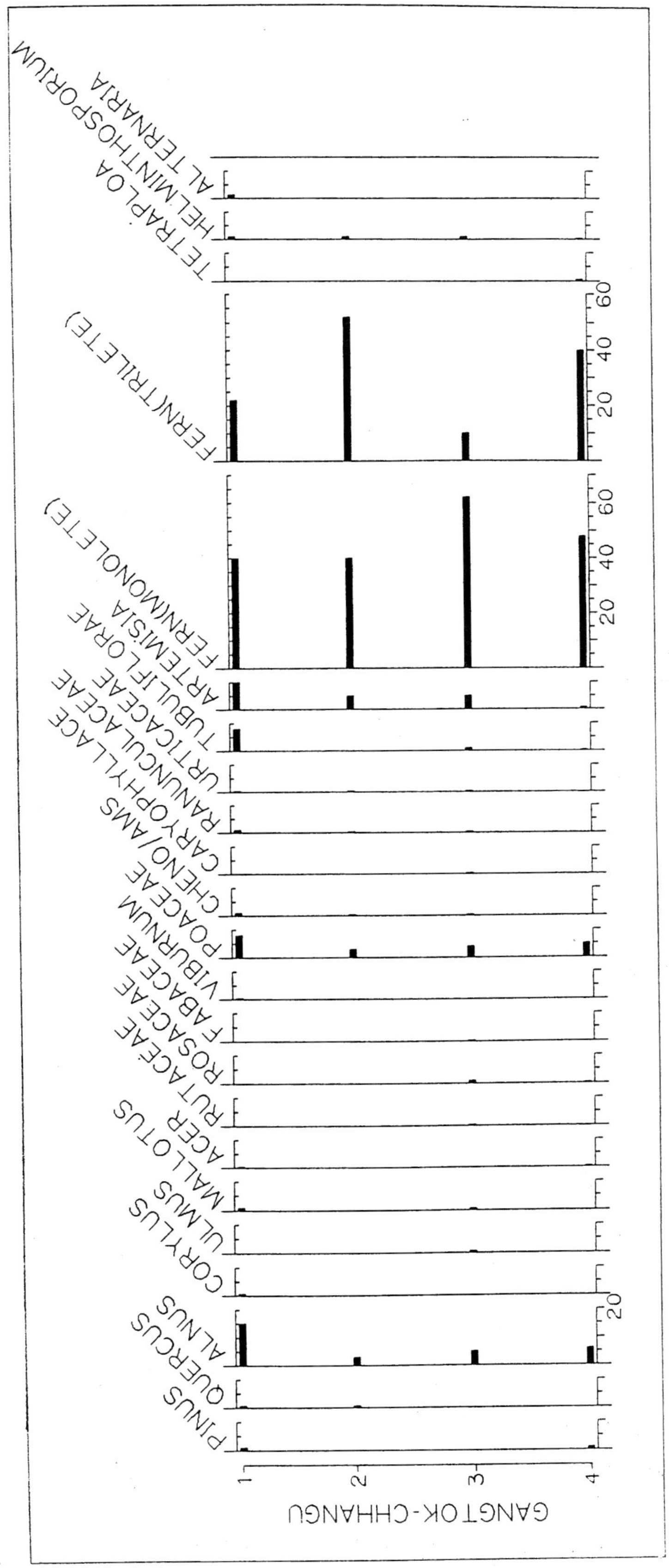

Text-Figure 6. Recent pollen spectra from Gangtok-Chhangu, Sikkim (percentages calculated in terms of terrestrial plants).

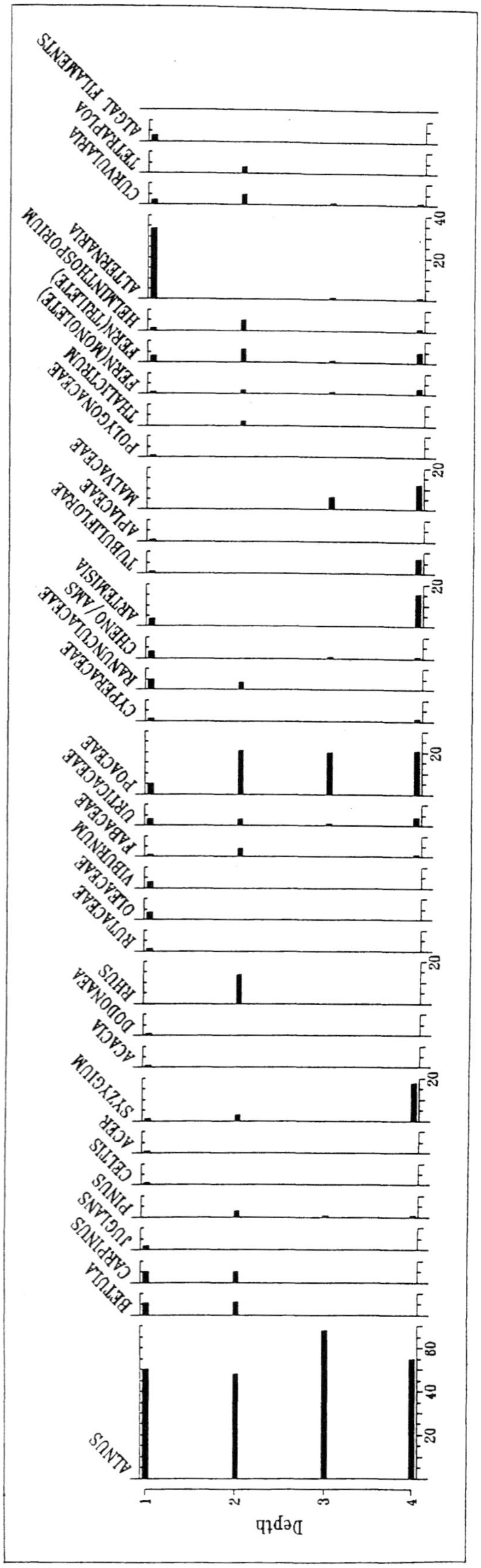

Text-Figure 7. Air catches at Gangtok, Sikkim (November 8–11, 1989).

their pollen, and in the case of *Shorea robusta* and *Tectona grandis* to entomophily. No reasonable explanation is apparent for *Cryptomeria*, which is anemophilous. Comparative studies of air catches, as well as recent pollen spectra, demonstrate that *Alnus*, *Pinus*, *Betula* and *Carpinus* along with prominent herbs such as grasses, sedges, Cheno/Ams, Asteraceae and *Artemisia*, occur. *Quercus* is absent in the air catches, whereas *Alnus* maintains high values, both in the air catches and in the recent spectra, because of its high pollen production. There is no satisfactory explanation for the absence of *Quercus* in the air catches except for the fact that it is a moderate pollen producer.

Fern spores remain sporadic in the air-catches despite their consistent occurrence in the recent spectra, attributable chiefly to their local gregarious growth in wet habitats within the forest and in open areas. The suspension of fern spores in the air may be limited because of their greater mass. Conversely, fungal spores such as *Helmithosporium*, *Tetraploa* and *Curvularia* occur frequently, but in moderate numbers both in the pollen spectra and the air catches.

Pollen spectra for the Sikkim Himalaya from the forested regions around Khechipiri, and across the temperate belt from Gangtok to Khechipiri, accurately reflect the existing broad-leaved forests in the region, as well as overall dominance of arboreal taxa. *Alnus* alone maintains its high frequency throughout, compared to *Quercus*, *Betula*, *Carpinus*, *Juglans*, *Rhododendron*, *Ulmus*, *Celtis*, Myrtaceae and *Rhus*, which are represented by varying frequencies in the spectra. Herbaceous ground vegetation within the forests remains poorly represented as expected, as shown by low values of Poaceae, Asteraceae, Cheno/Ams, Urticaceae, Ranunculaceae, Caryophyllaceae and Malvaceae. The existing humid conditions within the forests is accurately reflected by the exceedingly high values of fern and fungal spores recovered from the surface samples.

Similarly, the pollen spectra constructed from open regions in the Sikkim Himalaya (the open scrubs from Gangtok to Chhangu) reflect the sparse vegetation. These spectra contain low numbers of pollen, particularly tree taxa such as *Quercus*, *Pinus*, *Carpinus*, *Ulmus*, *Acer*, Rutaceae and *Mallotus*. Again, *Alnus* remains comparatively better represented than the other tree taxa. The herbaceous ground cover is well represented, mainly by Poaceae, Asteraceae, Urticaceae and Cheno/Ams. Fern spores recovered from the surface samples are attributed to the frequent presence of fern plants in moist patches within the open habitats.

Thus, the overall model from the four pollen spectra from Darjeeling and the Sikkim Himalaya, based on 45 surface samples from different habitats, indicates that most of the plant taxa present in each habitat also occur in the pollen spectra. Poor recovery of pollen of some very common arboreal taxa such as *Rhododendron*, Rosaceae and *Salix* can be attributed to their entomophilous pollination.

The aerospora study from Darjeeling and Sikkim reveals that the environs of Gangtok have *Alnus* pollen in high values, whereas *Betula*, *Carpinus*, *Syzygium*, *Rhus*, *Juglans*, *Pinus*, *Celtis*, *Acer* and *Dodonaea* are encountered in low numbers. In addition, Poaceae pollen in the Gangtok region dominate, followed by Urticaceae, Ranunculaceae, Cheno/Ams, Asteraceae and Malvaceae. Despite the luxuriant growth of ferns in open areas and in the forests throughout Sikkim, their spores remain poorly represented in the aerospora versus their recorded abundance in recent pollen spectra. This can be attributed to their restricted aerial transportation. Poor recovery of fungal spores is attributed to their limited growing season, which apparently did not coincide with the time when the slides were exposed. *Alnus* pollen in the aerospora from both Darjeeling and Sikkim maintains consistently high values and corresponds very well to its vegetational preponderance in both regions.

ACKNOWLEDGMENTS

The author records her deep sense of gratitude to the Department of Science and Technology, Government of India, New Delhi, for funding the project "Studies of paleoclimates through the application of paleobotanical methods," of which the present contribution is a part.

References Cited

BURKILL, I.H.
1907 Alpine notes from Sikkim. *Kew Bulletin*, 1907: 92–94
CHAUDHRI, A.B.
1960 Common Grasses and Sedges of Kurseong,? Kalimpong and Darjeeling Forest Division, West Bengal. *Indian Forester*, 86: 336–353.
CHAUHAN, M.S., and SHARMA, C.
1996 Late-Holocene vegetation of Darjeeling (Jore-Pokhari) Eastern Himalaya. *Palaeobotanist*, 45: 125–129.
CHOWDHERY, H.J., and RAO, R.R.
1992 Plant life in the Himalayan cold deserts; some adaptive strategies. *Bulletin Botanical Survey India*, 32: 43–56.
ERDTMAN, G.
1943 *An Introduction to Pollen Analysis.* Ronald Press Company, New York, 239 pgs.
1969 *Handbook of Palynology. An Introduction to the Study of Pollen Grains and Spores*, Munksgaard, Copenbagen, 486 pgs.
GAMMI, G.A.
1893 Botanical exploration of Sikkim, Tibet Frontier. *Kew Bulletin, 1893*: 297–314.
1894 Report on Botanical Tour in Sikkim. *Records of Botanical Survey* India, 1: 1–24.
GANGOPADHYA, S.K., and BANERJEE, S.K.
1987 The influrence of Vegetation on the properties of the soils of Sikkim. *Proceedings of the Indian National Science Academy*, B 53: 283–288.
HARA, H.
1966 *The Flora of Eastern Himalaya*, University of Tokyo, Japan, 744 pgs.
HUNTINGTON, E.
1906 Pangong glacial lake in Tibetan plateau. *Journal of Geology*, 15: 599.
KHOSHOO, T.N.
1992 Pandit Govind Ballabh Pant Memorial Lecture: II – Plant Diversity in the Himalaya Conservation and utilization - G.B. Pant Institute of Himalayan Environment and Development, Kosi.
MEHRA, P.N., and BIT, S.S.
1964 Pteridophytic flora of Darjeeling and Sikkim Himalayas. *Research Bulletin of Punjab University*, 15: 69–182
MEHRA P.N., BAWA, K.S., KHOSLA, P.K., and HANS, A.S.
1985 Floristic account of some forest types of the Eastern Himalayas. *Bulletin Botanical Survey of India*, 25: 1–18.
NAYAR, M.P.
1977 Changing patterns of Indian flora. *Bulletin Botanical Survey of India*, 19: 145–155.
RAO, R.R.
1994 Biodiversity in India. New Connaught Place, Dehradun, India, 315 pgs.
RAO, R.R., and HAJRA, P.K.
1986 *Floristic diversity of the Eastern Himalaya* — in a conservation perspective. *Proceedings of the Indian Academy of Science*, pp. 103-125.
ROLLA, S.R.
1963 A Botanical Tour in the Sikkim State, Eastern Himalayas. *Bulletin Botanical Survey of India*, 5: 165–205.
SAHNI, K.C.
1981 Botanical Panorama of the Eastern Himalaya. *In*: Lall, J.S. (ed.), *The Himalaya Aspects of Change*. Oxford University Press, New Delhi, pp. 32–49.
1982 Himalayan flora and physiography. A study in Contrasts. *In*: Paliwal, G.S. (ed.), *The Vegetational Wealth of the Himalayas*, New Delhi, pp. 194–200.
SHARMA, C.
1993 Palynostratigraphy of Lake Deposits of Himalaya and Palaeoclimate in Quaternary Period. *Current Science*, 64: 930–933.
2001 Palynostratigraphy of Himalayan lacustrine sediments. *In*: Goodman, D.K. (ed.), *Proceedings of the IX International Palynological Congress, Houston, Texas, U.S.A., 1996*; American Association of Stratigraphic Palynologists Foundation, p. 527–532.
SHARMA, C., and CHAUHAN, M.S.
1994 Vegetation and climate since Last Glacial Maximum in Darjeeling (Mirik Lake), Eastern Himalaya. *Proceedings of the 29th International Geological Congress.* Part B: 279–828.
SHARMA, C., and GUPTA, H.P.
1984 Past and present distribution of *Larix griffithiana* Hort. ex Carr in the Indian subcontinent as evidenced by palynology. *ZfA Z. Archaeology*, 18: 239–246.
SINGH, J.S., and SINGH, S.P.
1987 Forest Vegetation of the Himalaya. *Botanical Review*, 53: 82–192.

Author's address

CHHAYA SHARMA
Birbal Sahni Institute of Palaeobotany
53 University Road
Lucknow 226 007
India

E-mail: chhaya_sharma@rediffmail.com

AEROPALYNOLOGY AND GLOBAL WARMING

Rewi M. Newnham

Abstract

The onset and peak of the airborne grass pollen season in New Zealand during the summer of 1988/89 varied consistently with latitude and hence temperature, with the season in southernmost localities commencing 8–9 days later than in northernmost localities. These trends were only apparent, however, after sampling localities were classified as being sensitive primarily to rural or to urban pollen sources. The onset of successive pollen seasons for the principal tree species at some localities may also be temperature controlled. As routine, systematic, airborne pollen monitoring is now undertaken extensively in some areas (although not New Zealand), these temperature-controlled spatial variations in pollen production and dispersal indicate the potential of aeropalynology for monitoring biological response to contemporary and future climate change. Under various global warming scenarios, continuous aeropalynological monitoring could provide early signals that pollen production and dispersal is occurring earlier in the season. The north–south lag in onset and peak in grass pollen season in New Zealand means that in this country, as elsewhere, routine monitoring of the onset of the pollen season in lower latitude localities could also assist with the accurate forecasting of the pollinosis season at higher latitudes.

INTRODUCTION

This paper considers the role of airborne pollen monitoring in providing early signals of plant response to climate change. It reports consistent latitudinal variation in the onset and amplitude of airborne grass pollen levels in New Zealand in the summer of 1988–89 that are directly related to temperature variations. It then briefly reviews the results of recent investigations into the relationship between pollen season and climate based on long-term aeropalynological data and discusses the potential scope of future applications.

THE GRASS POLLEN SEASON IN NEW ZEALAND

The physical environment of New Zealand is well suited to examine spatial variations in pollen seasons and their relationship to climate. A broad latitudinal span (34–47°S) coupled with an oceanic climate virtually throughout means that these islands are characterized by an even north - south temperature gradient, with mean annual temperatures in northernmost localities 4–5°C warmer than in southernmost localities (Text-Figure 1). The limited aeropalynological data available for New Zealand suggest that in most, but not all, localities grass pollen forms the major component of the airborne pollen spectrum during summer months (Newnham et al., 1995). Most of the grass pollen derives from pasture grasses introduced since the arrival of Europeans in New Zealand, around 200 years ago. Each species of grass has a distinctive flowering season, usually overlapping with others, and spanning several weeks to months during spring and summer (Newnham et al., 1995).

The onset of the grass pollen season was compared between seven localities monitored during the summer of 1988–89 survey and at an eighth locality, Dunedin (Text-Figure 1), where a survey was conducted using the same methodology, in the summer of 1991–92. Typical annual temperature ranges for these localities are illustrated in Text-Figure 1, with means varying by 4–5°C between northernmost and southernmost localities. Atmospheric pollen samples were collected daily at each sampling locality using the Intermittent Cycling Rotorod sampler (see Newnham et al. (1995) for further details). The start of the grass pollen season is defined as the first day after November 15th when the cumulative daily count reaches 100 grains/m^3 air. November 15th is the first day when continuous data are available for all localities. Grass pollen levels were almost certainly minimal prior to this date.

When comparing data across all localities, no consistent north-to-south lag emerged in the onset of the grass pollen levels. However, if the localities are classified as either large metropolitan centers or smaller rural centers, then a clear latitudinal pattern emerges (Text-Figure 2). In the smaller, rural centers, the grass pollen season began at the northernmost locality, Kaikohe, eight days before the southernmost locality, Gore (Text-Figure 2a). In the large metropolitan centers surveyed, the season commenced at the northernmost city, Auckland nine days before the southernmost city, Christchurch (Text-Figure 2b). The emergence of a consistent latitudinal lag in the onset of the pollen season, once urban and rural sampling localities are separated, suggests that this distinction may be a critical one if pollen season-temperature relationships are to be established. In their investigation into annual variations in concentrations of birch pollen in the London area over a 30 year period, Emberlin et al. (1994) showed that the birch pollen season consistently starts earlier in Central London than at a site 20 km to the south, reflecting the influence of the warmth of the urban climate on birch trees in the city. In contrast, in New Zealand, the smaller rural centers show an earlier onset in the pollen season than urban centers at an approximately equivalent latitude (Text-Figure 2). As the smaller rural centers are often surrounded by vast areas of pastural hinterland this may be a more important factor than urban warmth in determining the onset of pollen seasons in New Zealand.

Airborne grass pollen data collected, using the same sampling instrument and method, from another southern metropolitan center, Dunedin, for the period November to March 1991–92, are consistent with this pattern (Text-Figure 2), although it must be stressed that the Dunedin data are for a different year.

Newnham, Rewi M.
2001 Aeropalynology and global warming; *In*: Goodman, D.K., and Clarke, R.T. (eds.), *Proceedings of the IX International Palynological Congress, Houston, Texas, U.S.A., 1996*; American Association of Stratigraphic Palynologists Foundation, p. 567–570.
 ISBN 0–931871–06–9

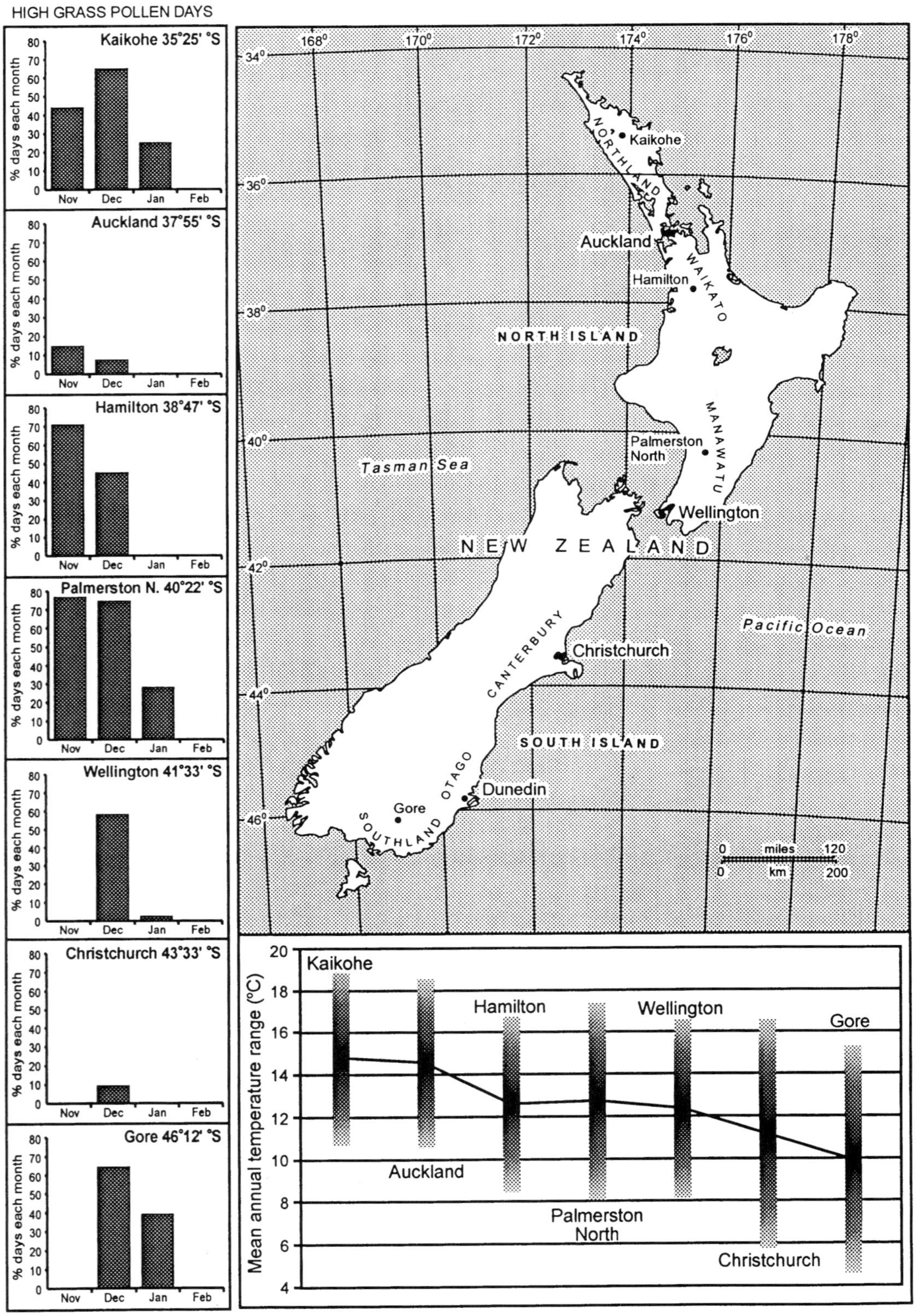

Text-Figure 1. New Zealand pollen sampling localities and regions mentioned in the text. Graphs at side show percentage of days per month (Nov 16th–30th, Dec, Jan, and Feb 1–13) when airborne grass pollen levels exceeded 30 grains/m^3 at New Zealand pollen sampling localities. Graph at base shows mean daily temperatures for July (lower points) and January (upper points) and mean annual temperatures (middle points) for the seven pollen sampling localities. Sources: (Wards, 1976; New Zealand Official Yearbook, 1988/89).

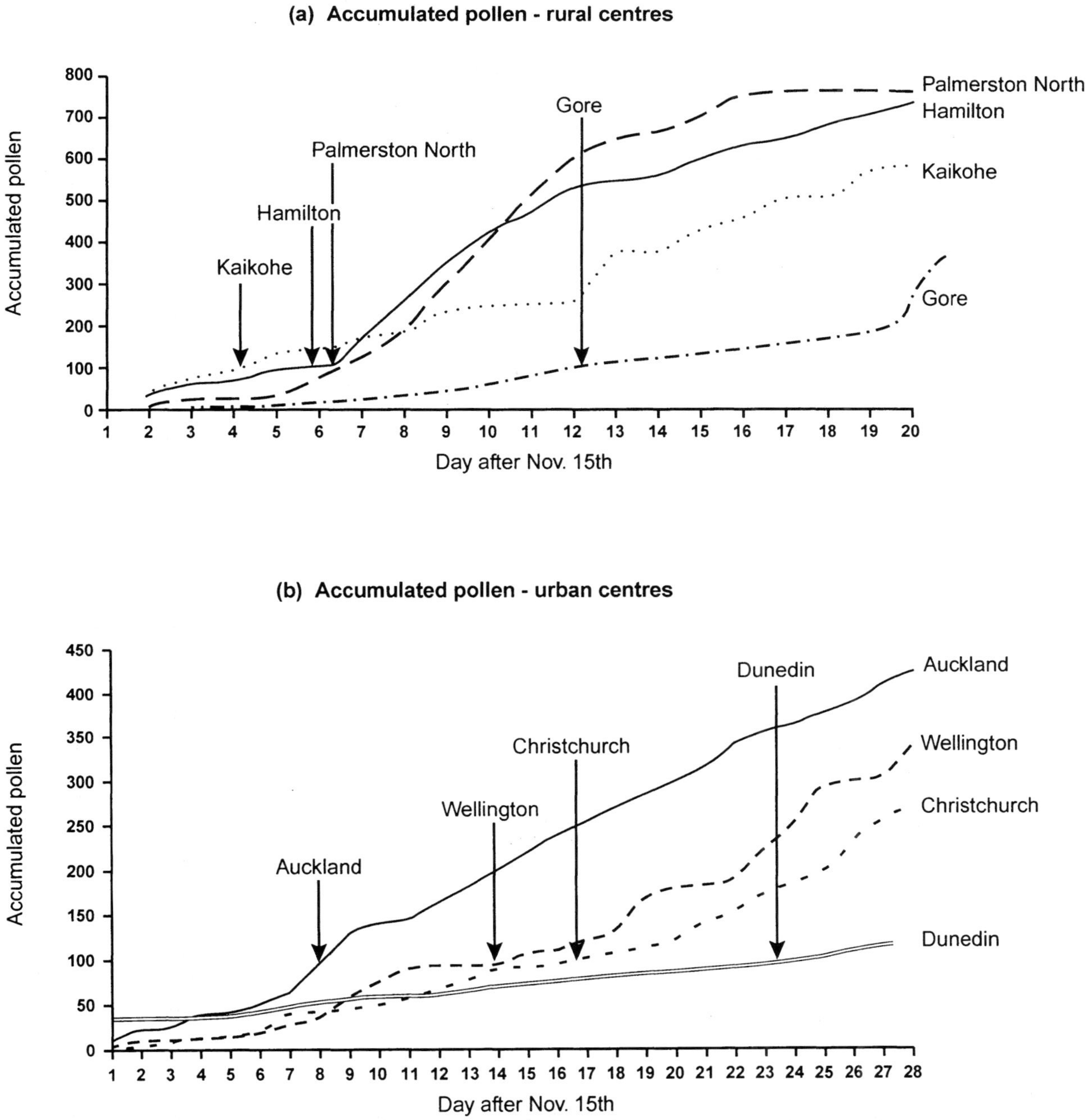

Text-Figure 2. Accumulated airborne grass pollen concentrations after November 1st 1988. Arrows indicate when each locality first exceeded 100 accumulated grass pollen grains/cm^3. (a) Smaller rural centers or cities with greater exposure to rural pollen sources; (b) larger metropolitan centers, including Dunedin, where data are for the 1992/93 season.

FREQUENCY AND TIMING OF 'HIGH' GRASS POLLEN LEVELS

The frequency of 'high' grass pollen days at the seven localities surveyed in 1988/89 is illustrated in Text-Figure 1 as percentage of days in each locality where grass pollen levels exceeded 30 grains/m^3 air. During 1988/89, high grass pollen days occurred at North Island localities (excluding Wellington) during November, December and January, but in South Island localities (including Wellington) only in the latter two months. Generally, more northern localities are likely to experience high pollen levels earlier in the season (November to mid–December) while more southern localities are more likely to experience high pollen levels in mid-season (December to early January). However, as the frequency of peak pollen levels will be influenced by several factors other than temperature, this pattern is likely to vary from season to season. For example, the peak grass pollen season in Auckland in 1988/89 was foreshortened by several weeks due to heavy and persistent rainfall (Newnham et al., 1995). Nevertheless the results presented here suggest that, in a given year, a general north–south temporal trend in timing of the peak pollen season should be discernible, although much inter-annual variability is likely.

ESTABLISHING THE LINK BETWEEN SPATIAL AND TEMPORAL PATTERNS

These results point to a potentially robust relationship between latitude and both onset and peak frequency of pollen levels in New Zealand and, by implication, elsewhere where latitude/temperature gradients are established. As timing of pollen release and pollen production are both strongly related to temperature, which in New Zealand varies consistently with latitude (Text-Figure 1), ongoing monitoring of airborne pollen levels has clear potential to reveal this particular biological response to climate change.

This assertion can be further tested by comparing airborne pollen season parameters with climate variables at sites where monitoring has continued for the past decade or longer. A study of birch pollen measured at five monitoring sites in Europe with records of 18 to 30 years indicated a significant ($p<0.001$) negative correlation at all sites between the onset of the birch pollen season and mean daily temperatures during the preceeding 40 days (Spieksma et al., 1995). As with the New Zealand grass pollen data, a clear latitudinal lag pattern in the onset of the birch pollen season was apparent, with commencement at the northernmost station, Stockholm, on average 29 days after the southernmost station, Basel, for the period 1976–1993. In the United Kingdom between 1987 and 1992, the variation in onset of the grass pollen season at nine widely spaced sites showed a close but complex relationship to cumulative spring temperatures and rainfall in May and June (Emberlin et al., 1994). Earlier onset of birch and alder pollen seasons corresponding with warming in recent years has been reported in Denmark (Anderson, 1991) and in the United Kingdom (Emberlin et al., 1993). The complex relationships between air temperature, bud dormancy and pollen release were investigated by Frenguelli et al. (1991) for alder and poplar at four monitoring stations in Italy and the Netherlands. Both trees require a period of winter chill preceding warmer temperatures in the spring before the pollen season can begin. The close correspondence of these various investigations confirms the complex but consistent relationship between temperature and onset of certain pollen seasons and demonstrates that the potential role of aeropalynology networks extends beyond the medical domain.

IMPLICATIONS FOR ENVIRONMENTAL HEALTH

Besides the implications for monitoring biological response to climate change, the results presented and discussed here indicate that it may be possible to accurately forecast the onset of the grass (or another allergenic) pollen season at higher latitudes by monitoring levels at lower latitudes. This could have important implications for the diagnosis and treatment of allergic rhinitis and other respiratory disorders. In the broader context of forecasting specific impacts of climate change on environmental health, it is vital to determine what changes to the pollen seasons will occur, as these will have a direct bearing on pollinosis and respiratory disorder in the community (Emberlin, 1994). The strong correlation between the onset of pollen season and late winter/early spring, temperatures has been used as a basis for developing empirical regression models to predict timing and severity of *Betula* pollen seasons for the London area (Emberlin et al., 1993) and form the basis of other predictive studies of the onset of pollen seasons elsewhere (Driessen et al., 1989; Frenguelli et al., 1989).

References Cited

ANDERSON, T.B.
1991 A model to predict the start of the pollen season. *Grana*, 30: 269–275.

DRIESSEN, M.N.B., VAN HERPEN, R.M.A., MOELANDS, R.F.M., and SPIEKSMA, F.TH.M.
1989 Prediction of the start of the grass pollen season for the western part of the Netherlands. *Grana*, 28: 37–44.

EMBERLIN, J.
1994 The effects of patterns in climate and pollen abundancy on allergy. *Allergy*, 49, 15–20.

EMBERLIN, J.C., SAVAGE, M., and WOODMAN, R.
1993 Annual variations in the concentration of *Betula* pollen in the London area, 1961–1990. *Grana*, 32: 359–363.

EMBERLIN, J., JONES, S., BAILEY, J., CAULTON, E., CORDEN, J., DUBBELS, S., EVANS, J., MCDONAGH, N., MILLINGTON, W., MULLINS, J., RUSSEL, R., and SPENCER, A.
1994 Variation in the start of the grass pollen season at selected sites in the United Kingdom 1987–1992. *Grana*, 33: 94–99.

FRENGUELLI, G., BRICCHI, E., ROMANO, B., MINCIGRUCCI, G., and SPIEKSMA, F.TH.M.
1989 A predictive study on the beginning of the pollen season for Gramineae and *Olea europea*. L. *Aerobiologia*, 5: 64–70.

FRENGUELLI, G., SPIEKSMA, F.TH.M., BRICCHI, E., ROMANO, B., MINCIGRUCCI, G., NIKKELS, A.H., DANKAART, W., and FERRANTI, F.
1991 The influence of air temperature on the starting dates of the pollen season of *Alnus* and *Populus*. *Grana*, 30: 196–200.

NEW ZEALAND OFFICIAL YEARBOOK
1988/89 Wellington, Govt. Printer.

NEWNHAM, R.M., FOUNTAIN, D.W., CORNFORD, C., and FORDE, M.B.
1995 A national survey of airborne pollen and grass flowering in New Zealand with implications for respiratory disorder. *Aerobiologia*, 11, 239–252.

SPIEKSMA, F.TH.M., EMBERLIN, J.C., HJELMROOS, M., JAGER, S., and LEUSCHNER, R.M.
1995 Atmospheric birch (*Betula*) pollen in Europe: Trends and fluctuations in annual quantities and the starting dates of the seasons. *Grana*, 34: 51–57.

WARDS, I.
1976 *New Zealand Atlas*. Wellington, A.R. Shearer, Govt. Printer.

Author's address

REWI M. NEWNHAM
Department of Geographical Sciences
University of Plymouth
Drake Circus
Plymouth, PL4 8AA
England

AEROPOLLEN OF THE PINGTUNG AREA, SOUTH TAIWAN

Tseng-Chieng Huang and Chia-Chin Liu

Abstract

The diurnal and monthly study of airborne pollen from May, 1993, to April, 1995, at the Pingtung Station, Taiwan, is reported. *Broussonetia* pollen was the most abundant type captured during this time interval. The largest monthly amount captured was in March, 1994. The diurnal collection varies during different months.

INTRODUCTION

A survey conducted by a team of six aeropollen surveyors representing six different localities on Taiwan. The research was carried out by Chen, Su-Hwa (Taipei Station, northern Tainan); Tsou, Tze-Hwa (Taichung Station, central western Taiwan); Kuoh, Chang-Sen (Tainan Station, southwestern Taiwan); Huang, Tseng-Chieng (Pingtung Station, southern Taiwan); Chen, Su-Hui (Hualien Station, northeastern Taiwan); and Chang, Hui-Chu (Taitung Station, southeastern Taiwan) between November, 1991, and October, 1995. The results for the Pingtung Station between May, 1993, and April, 1995, are reported herein. The comprehensive report, including the daily and yearly studies, the pollen calendar of this area and the relationship between weather conditions have been reported by Huang (1998) and Huang et al. (1998).

Locality

A Burkard spore trap was placed on the roof of the Department of Forest Resources, Pingtung Technological University at Pingtung, Taiwan. This location is situated in the southern part of Taiwan, at latitude 22°38'57"N and longitude 120°36'7"E. The trap was located about 15 m above the ground.

Vegetation

The vegetation within a 78.5 km^2 circular area surrounding the site consists of 118 families, 379 genera and 539 species of vascular plants. Of these, Pteridophytes contributed seven families and ten species, Gymnosperms five families and 12 species, Dicotyledons 92 families and 441 species, and Monocotyledons 14 families and 76 species.

METHODS

The Burkard seven-day recording volumetric spore trap was used to collect the airborne pollen for three years between December 1, 1991, and November 10, 1995. Pollen was collected on cellophane tape (Melinex, Burkard) coated with vaseline adhesive. The tape is fixed on a rotating drum that passes an orifice calibrated to expose the tape to 10 liters of air per minute. Exposed tapes were cut into seven 48 mm sections corresponding to each 24 hours of sampling. Sections were mounted on slides and made permanent using Entellan (a xylene and alkylacrytate mixture, made by Merck, Germany). These strips were then covered, and the pollen was scanned using a Leitz compound microscope at x400 or x1,000 magnification. For each section, 24 transverse lines, each 2 mm apart (corresponding to a one-hour sampling period), were scanned and the pollen was recorded to give the amount of pollen (number)/m^3. All pollen grains are reported per m^3 of air or as indicated by hourly (0.6 m^3), daily (14.4 m^3), monthly (432 m^3) or yearly (5184 m^3) values. All slides were recorded for pollen amounts and pollen frequencies every hour. For identification, Huang's (1972) *Pollen Flora of Taiwan* and Huang's (1981) *Spore Flora of Taiwan* were consulted.

Huang, Tseng-Chieng, and Liu, Chia-Chin
2001 Aeropollen of the Pingtung area, south Taiwan; *In:* Goodman, D.K., and Clarke, R.T. (eds.), *Proceedings of the IX International Palynological Congress, Houston, Texas, U.S.A., 1996;* American Association of Stratigraphic Palynologists Foundation, p. 571–577.
 ISBN 0–931871–06–9

RESULTS

Total Aeropollen (Text-Figure 1)

Total pollen grains captured within two years (May 1, 1993, to April 30, 1995) at this station are 151,892 grains. *Broussonetia* was the most abundant with 23,235 grains (15.30 %); other common types included *Trema* 20,018 grains (13.18 %), *Mimosa* – 10.61 %, Gramineae (Poaceae) – 10.19 %, and *Alnus* – 5.48 %. There were 148,788 grains, of which 97.96 % were identifiable at least to the family level; 3,104 grains, or 2.04 %, were unidentifiable.

Monthly Aeropollen Count (Table 1)

The dominant taxa in the assemblages vary in monthly. The top three taxa per month are described.

Diurnal Captured Aeropollen (Text-Figures 2–13)

The diurnal variation of aeropollen per month during these two years are compared. The division of daily periodicity follows that of Reddi et al. (1988). During May to December of 1993 and 1994, the highest peaks in early morning appeared in December 1993 and October 1994; at midday appeared in May, 1994; at pre-midnight appeared in June 1993; at pre-midnight to post-midnight appeared in

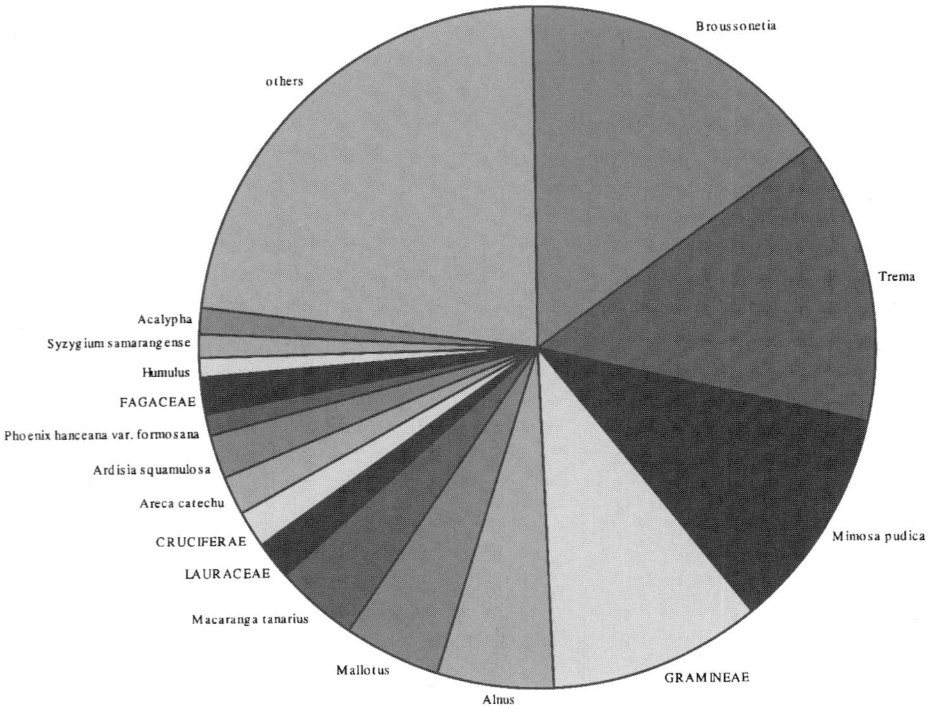

Taxa	Amount	Percent(%)
Broussonetia	23,235	15.30
Trema	20,018	13.18
Mimosa pudica	16,114	10.61
GRAMINEAE	15,484	10.19
Alnus	8,327	5.48
Mallotus	7,101	4.68
Macaranga tanarius	5,767	3.80
LAURACEAE	3,084	2.03
CRUCIFERAE	2,849	1.88
Areca catechu	2,749	1.81
Ardisia squamulosa	2,725	1.79
Phoenix hanceana var. *formosana*	2,025	1.33
FAGACEAE	1,948	1.28
Humulus	1,914	1.26
Syzygium samarangense	1,746	1.15
Acalypha	1,626	1.07
others	35,180	23.16

Text-Figure 1. Total numbers and percentages of the major aeropollen bi-yearly at Pingtung Station during May, 1993, and April, 1995.

TABLE 1. Monthly total, top three taxa, and number of aeropollen at Pingtung from May, 1993 to April, 1995.

Date	Total taxa	Total grains	1	2	3
May-93	63	3594	*Ardisia squamulosa* (18.39)	*Trema spp.*	Gramineae
Jun-93	49	6084	Cruciferae (32.41)	*Trema spp.*	Gramineae
Jul-93	93	5044	*Mimosa* spp. (18.24)	Gramineae	*Areca catechu*
Aug-93	94	8328	*Mimosa* spp. (25.80)	*Macaranga tanarius*	Gramineae
Sep-93	91	13784	*Alnus* spp. (32.41)	*Broussonetia* spp.	Gramineae
Oct-93	76	9180	*Mimosa* spp. (25.80)	*Macaranga tanarius*	*Alnus formosanus*
Nov-93	93	7428	*Mimosa* spp. (41.46)	Gramineae	*Macaranga tanarius*
Dec-93	81	3648	*Mimosa* spp. (41.46)	Gramineae	*Mallotus paniculatus*
Jan-94	63	1940	*Mimosa* spp. (15.26)	Gramineae	*Mallotus paniculatus*
Feb-94	86	3116	*Mallotus paniculatus* (15.53)	*Calocedrus formosanus*	*Mimosa* spp.
Mar-94	158	26832	*Broussonetia* spp. (48.33)	*Trema* spp.	*Mallotus paniculatus*
Apr-94	136	15876	*Trema* spp. (38.55)	*Broussonetia* spp.	*Mallotus paniculatus*
May-94	101	6428	*Trema* spp. (29.87)	Gramineae	*Broussonetia spp.*
Jun-94	105	4705	Gramineae (18.02)	*Mimosa* spp.	*Fraxinus* spp
Jul-94	99	4890	*Fraxinus formosanus* (13.97)	*Mimosa* spp.	*Ardisia squamulosa*
Aug-94	43	1868	Gramineae (25.91)	*Macaranga tanarius*	*Broussonetia* spp.
Sep-94	61	4364	*Alnus* spp. (31.53)	Gramineae	*Mallotus paniculatus*
Oct-94	91	7432	*Alnus formosanus* (21.53)	*Mimosa* spp.	*Mallotus paniculatus*
Nov-94	57	2488	Gramineae (41.46)	*Mallotus paniculatus*	*Mimosa* spp.
Dec-94	47	2088	Gramineae (20.52)	*Mimosa* spp.	Amaranthaceae
Jan-95	55	864	Gramineae (15.28)	Amaranthaceae	*Mallotus paniculatus*
Feb-95	36	661	Amaranthaceae (13.31)	*Broussonetia* spp.	*Mallotus paniculatus*
Mar-95	82	4384	*Broussonetia* spp. (17.61)	*Trema* spp.	Gramineae
Apr-95	110	6940	*Trema* spp. (36.43)	Gramineae	*Euphoria longana*

September 1993; at midnight to post-midnight appeared in August 1993; and at post-midnight appeared in July and November 1993. During January to April of 1994 and 1995, the highest peaks in early morning appeared in April 1994; at fore-noon appeared in February 1994; at post midday appeared in January 1994 and March showed no difference.

DISCUSSION

This study shows great year-to-year variation in pollen concentration. *Ardisia squamulosa* produced the most abundant aeropollen captured in May, 1993, but *Trema* was the most abundant for the month in 1994. The Cruciferae produced the most abundant captured aeropollen in June, 1993, but Gramineae was the most abundant in June of 1994. *Mimosa* produced the most abundant captured aeropollen in July 1993, but *Fraxinus formosanus* did so in 1994. *Mimosa* produced the most abundant captured aeropollen in August 1993, but Gramineae did so in 1994. *Alnus* produced the most abundant captured aeropollen in September 1993 and also in 1994. *Mimosa* produced the most abundant cap-

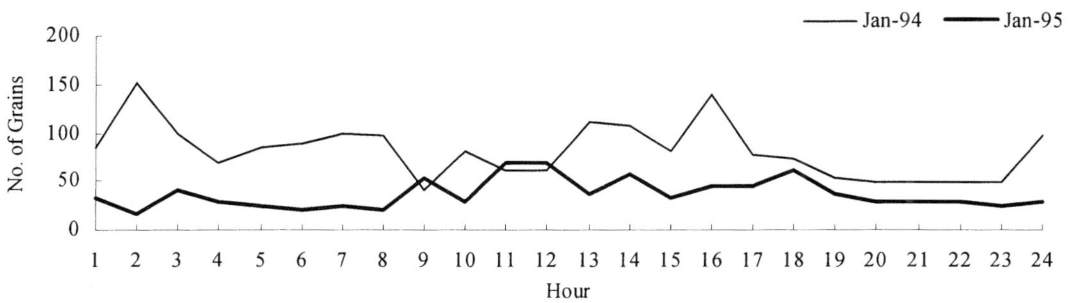

Text-Figure 2. Diurnal variation by total monthly aeropollen in January, 1994 and 1995, at Pingtung Station.

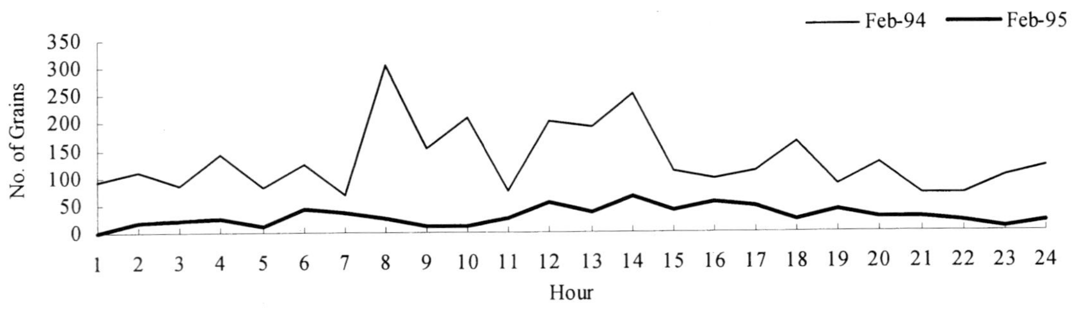

Text-Figure 3. Diurnal variation by total monthly aeropollen in February, 1994 and 1995, at Pingtung Station.

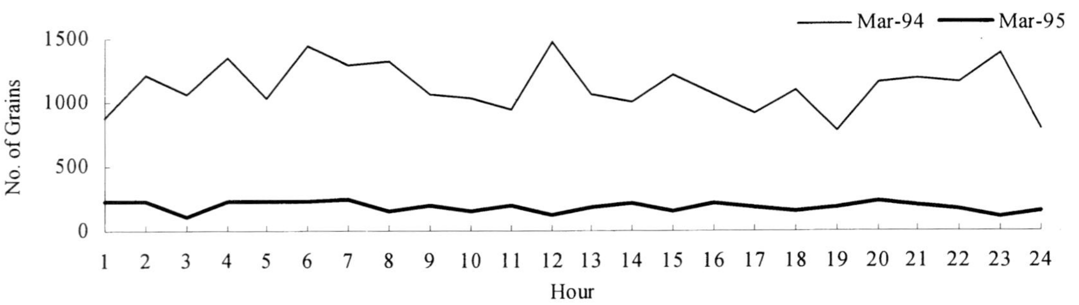

Text-Figure 4. Diurnal variation by total monthly aeropollen in March, 1994 and 1995, at Pingtung Station.

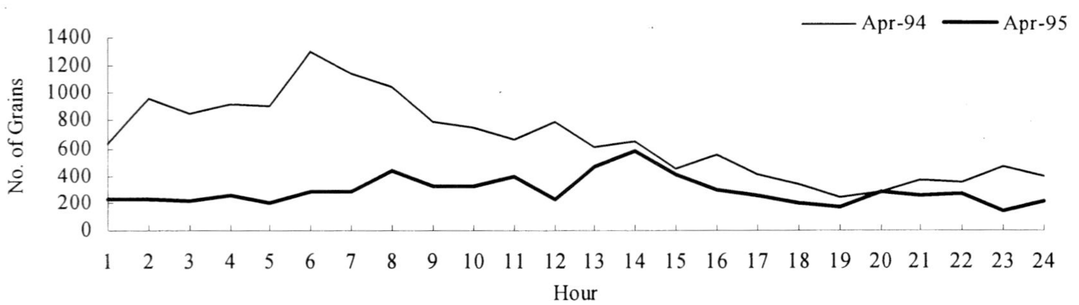

Text-Figure 5. Diurnal variation by total monthly aeropollen in April, 1994 and 1995, at Pingtung Station.

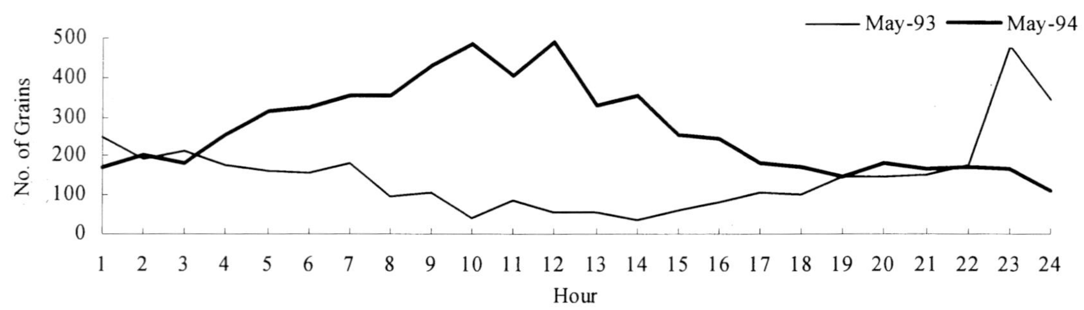

Text-Figure 6. Diurnal variation by total monthly aeropollen in May, 1993 and 1994, at Pingtung Station.

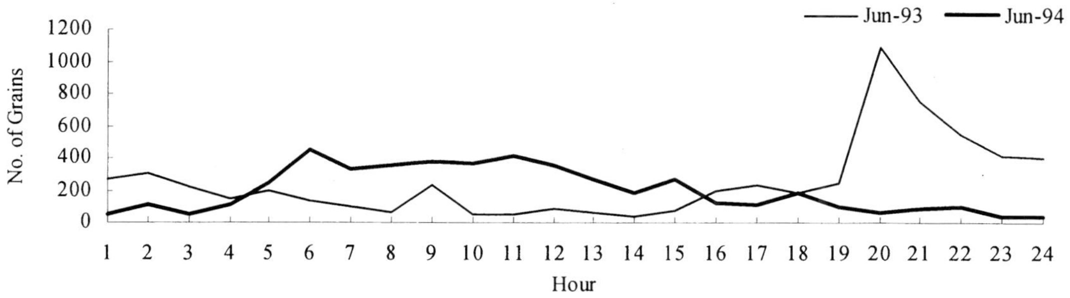

Text-Figure 7. Diurnal variation by total monthly aeropollen in June, 1993 and 1994, at Pingtung Station.

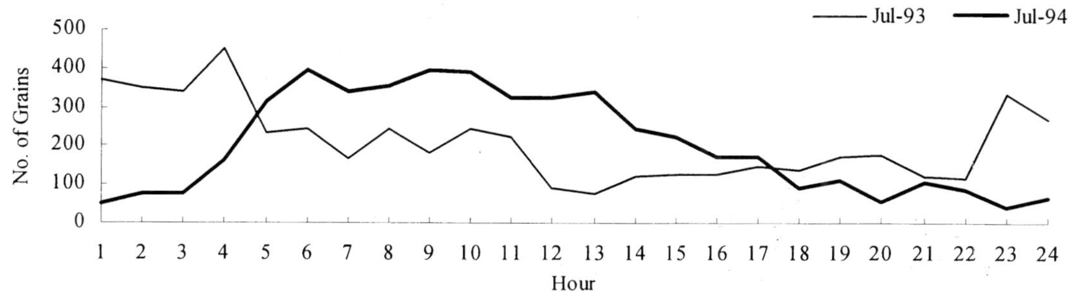

Text-Figure 8. Diurnal variation by total monthly aeropollen in July, 1993 and 1994, at Pingtung Station.

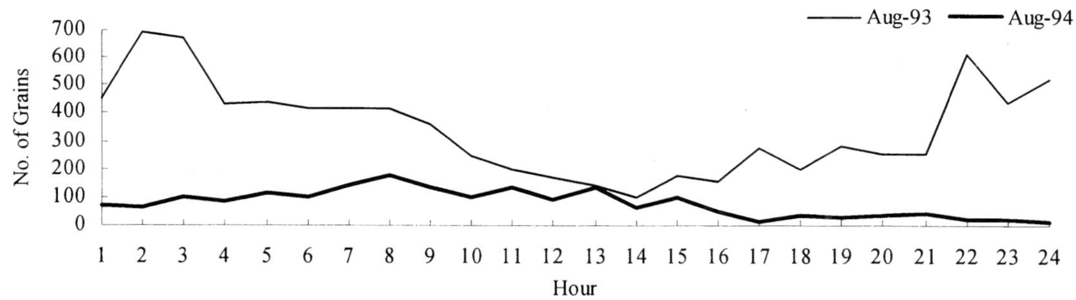

Text-Figure 9. Diurnal variation by total monthly aeropollen in August, 1993 and 1994, at Pingtung Station.

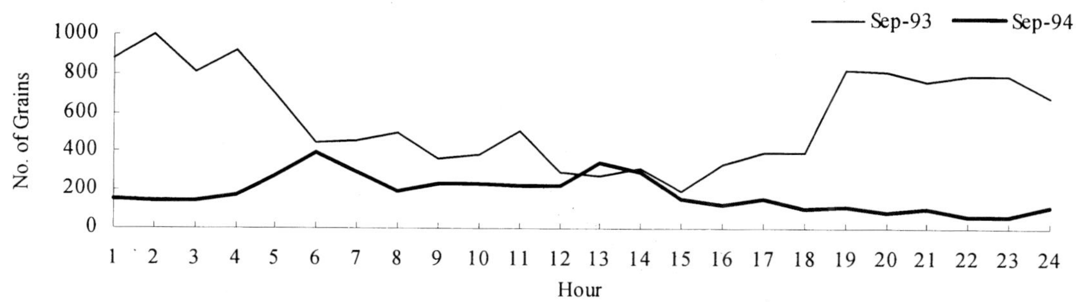

Text-Figure 10. Diurnal variation by total monthly aeropollen in September, 1993 and 1994, at Pingtung Station.

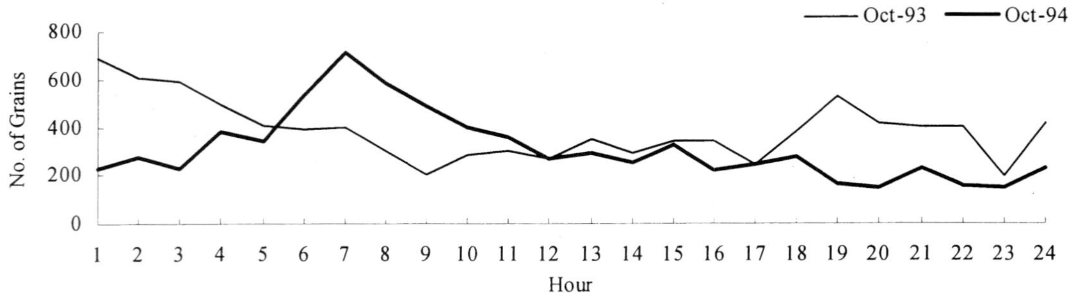

Text-Figure 11. Diurnal variation by total monthly aeropollen in October, 1993 and 1994, at Pingtung Station.

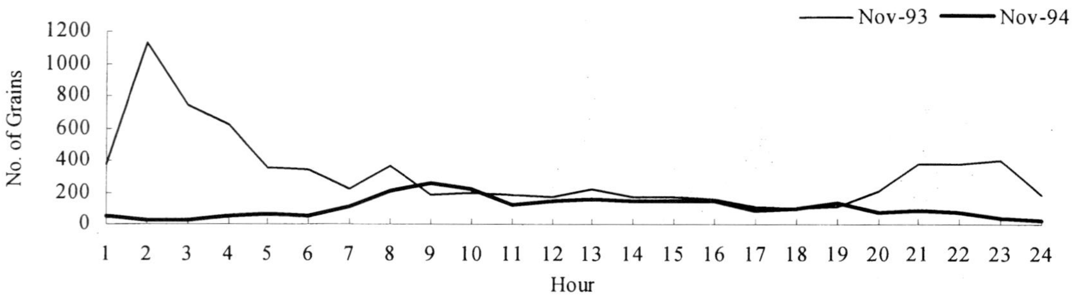

Text-Figure 12. Diurnal variation by total monthly aeropollen in November, 1993 and 1994, at Pingtung Station.

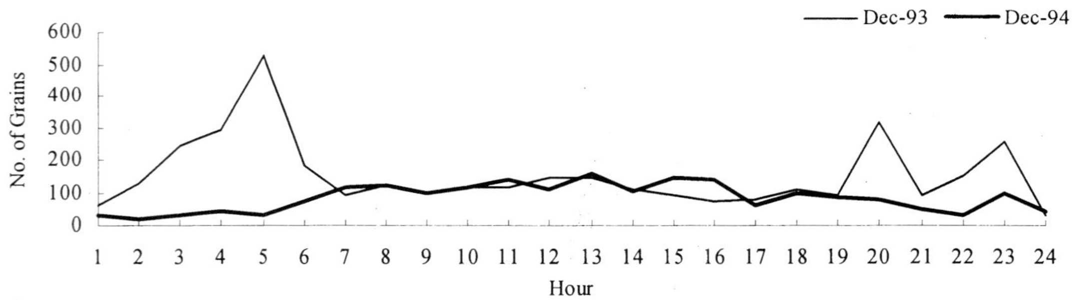

Text-Figure 13. Diurnal variation by total monthly aeropollen in December, 1993 and 1994, at Pingtung Station.

tured aeropollen in October 1993, but *Alnus fomorsana* did so in 1994. *Mimosa* again produced the most abundant captured aeropollen in November 1993, but Gramineae did so in 1994. *Mimosa* again produced the most abundant captured aeropollen in December 1993, but Gramineae did so in 1994. *Mimosa* again produced the most abundant captured aeropollen in January 1994 but Gramineae did so in 1995. *Mallotus paniculatus* produced the most abundant captured aeropollen in February 1994 but Amaranthaceae did so in 1995. *Broussonettia* produced the most abundant captured aeropollen in March 1994 and also in 1995. *Trema* produced the most abundant captured aeropollen in April 1994 and also in 1995.

The daily and diurnal variations were also obvious, but the variation consists of different taxa.

There are still 3,104 grains (or 2.04 %) of captured aeropollen which remain unidentified. The occurrence of this unidentifiable component was higher in January to April, when the grasses and tree pollen were dominant.

ACKNOWLEDGMENTS

This study was supported by the grants of the National Science Council, Republic of China (NSC 82-0211-B 002-370-B08, NSC 83-0409-B002-006-B08, NSC 84-231-B002-

016 and NSC 85-2311-B002-004-B14). We appreciate the airborne collection made by Prof. Ching-Long Yeh of Pingtung Agricultural Technological College, and assistance by Dr. Kuoh-Cheng Yang, Dr. Tsung-Hsin Hsieh, Mr. Chi-Cheng Chou, Miss Lih-Fen Uang, Pei-Chung Li, Yu-Ping Lin and Bei-Yen Chang.

References Cited

HUANG, T.-C.

1972 *Pollen Flora of Taiwania.* Botany Department National. Taiwan University, Taipei, Taiwan, ROC, 1st ed., 297 pgs.

1981 *Spore Flora of Taiwan.* Author's Press. Taipei, Taiwan, Republic of China (privately published), 1st ed., 112 pgs.

1988 Aeropollen of the Pingtung Area, South Taiwan. *Taiwana* 43: 73–100.

HUANG, T.-C., CHEN, S.-H., KUOH, C.-S., CHANG, H.-C., and TSOU, T.-H.

1988 *Airborne Pollen Grains and Spores in Taiwan*. Institute of Botany, National Taiwan University Taipei, 246 pgs. (in Chinese).

REDDI, C. S., REDDI, N.S., and JANAKI, B.A.

1988 Circadian patterns of pollen release in some species of Poaceae. *Review of Palaeobotany and Palynology*, 54: 11–42.

Authors' addresses

TSENG-CHIENG HUANG
CHIA-CHIN LIU
Department of Botany
National Taiwan University
Taipei 106, Taiwan
The Republic of China

THREE YEARS OF POLLEN MONITORING IN MOSCOW: PRELIMINARY POLLEN CALENDAR FOR 1993–1995

E. E. Severova and S. V. Polevova

Abstract

Daily concentration of airborne pollen has been recorded in Moscow during the period of mid-March to mid-September with a Burkard volumetric trap. Pollen grains of about 40 taxa are determined yearly; 19 of them compose 97–99% of the annual pollen spectrum. The concentration of total pollen varies considerably during the season. Three pollen periods are observed. The first and the greatest one (usually about 60–70% of total annual catch) takes place in mid-March through mid-May, and is mainly associated with trees *(Betula, Corylus, Alnus, Salix, Populus, Ulmus, Fraxinus.* The second period is early summer and is characterized by high concentrations of *Pinus* and *Poaceae* pollen (in spite of comparatively low totals during this period (about 2–12% of total annual pollen). The early summer period is regarded as the most dangerous for allergies because of the high grass pollen concentrations. The third period takes place in mid-June through mid-September and consists mainly of Poaceae [at the beginning], *Artemisia,* Chenopodiaceae and *Urtica. Urtica* has the longest pollination period in the Moscow region.

INTRODUCTION

Only a few published measurements of the diurnal variation of air pollen spectra in Moscow are available since 1960. All of them were made with different, mainly gravimetric methods, and the duration of observation did not exceed three years. Regular volumetric sampling of airborne pollen was initiated by Prof. Siwert Nilsson (Palynological Laboratory, Swedish Museum of Natural History) in 1992. The main purpose of these observations was to study the airborne pollen spectrum and factors that influence the local problem of pollinosis. The results of a three-year study — preliminary pollen calendar for 1993–1995 — is presented in this article. We excluded the results of the first year of observation because of some technical problems.

MATERIAL AND METHODS

The pollen grains were sampled with a Burkard Seven-Day Recording Volumetric Spore Trap. The trap was situated on a roof approximately 10 metres above ground in the south-west part of Moscow (55°30'N, 37°30'E), not far from the Moscow University Botanical Garden. The pollen grains were sampled at an air flow rate of 10 cubic liters per minute on tapes, coated with a thin film of Vaseline–Paraffin wax in toluene. The daily tape segments were mounted on slides in glycerine jelly with saffranine for selective staining of pollen grains. For the volumetric evaluation of daily pollen count 12 transversal bands, each of 0.3 mm wide were scanned. A light microscope with an objective lens of 40x and ocular lens of 8x was used for counting and identification.

The observations were made from mid-March to the end of September. Pollen grains of about 40 taxa were determined yearly, 19 of them (the most abundant or the most important from an allergological point of view) were selected for construction a pollen calendar. These include: *Betula, Alnus, Corylus, Salix, Fraxinus, Acer, Populus, Quercus, Ulmus, Pinus, Picea, Plantago, Rumex, Urtica, Tilia,* Poaceae, *Artemisia,* Chenopodiaceae and Rosaceae. Some taxa were identified to species level: *Corylus (C. avellana), Pinus (P. sylvestris), Picea (P. abies), Urtica (U. dioica),* and *Tilia (T. cordata).* Others were identified to a genus level: *Betula, Alnus, Salix, Fraxinus, Acer, Populus, Quercus, Ulmus, Plantago, Rumex, Artemisia;* Poaceae, Chenopodiaceae, Rosaceae were identified to the family level.

The main pollen period (MPP) for each taxa was defined as a period from the date when the sum of concentrations reached 5% of the total annual sum until the time when the sum reached 95%. In other words, the MPP covers the time with 90% of the total annual sum (Nilsson and Persson, 1981). All meteorological data were received from the Meteorological Station of Moscow State University. Phenological data for 1993–1995 were obtained in the Dendrarium of the Botanical Garden of Moscow State University.

Climate

Moscow is situated within a temperate–continental zone that is characterized by long, cold winters and warm summer periods. The annual mean temperature is +3.8°C, with the temperature of July +18.2°C (the warmest month), and the temperature of January being -10.5°C (the coldest month). The annual precipitation of 587 mm falls fairly evenly throughout the year with a summer maximum in July (Text-Figure 1). A snow cover occurs about 115 days a year. The average growing season is about 175–185 days per year.

RESULTS AND DISCUSSION

The airborne pollen spectrum of Moscow consists of about 40 taxa, 19 of them form about 97–99% of total annual pollen rain (Table 1). Pollen grains of exotic or non-typical city plants *(Carpinus, Cupressaceae, Ericaceae, Hipphophae, Juglans,* Typhaceae) do not considerably effect the pollen calendar. Single pollen grains of these taxa occur non-yearly and can be explained by the vicinity of the Botanical Garden.

The pollen season in Moscow lasts from mid-March until mid-September. The total annual pollen count varies

Severova, E. E. and Polevova, S. V.
2001 Three years of pollen monitoring in Moscow: Preliminary pollen calendar for 1993–1995; *In:* Goodman, D.K., and Clarke, R.T. (eds.), *Proceedings of the IX International Palynological Congress, Houston, Texas, U.S.A., 1996;* American Association of Stratigraphic Palynologists Foundation, p. 579–583.
 ISBN 0–931871–06–9

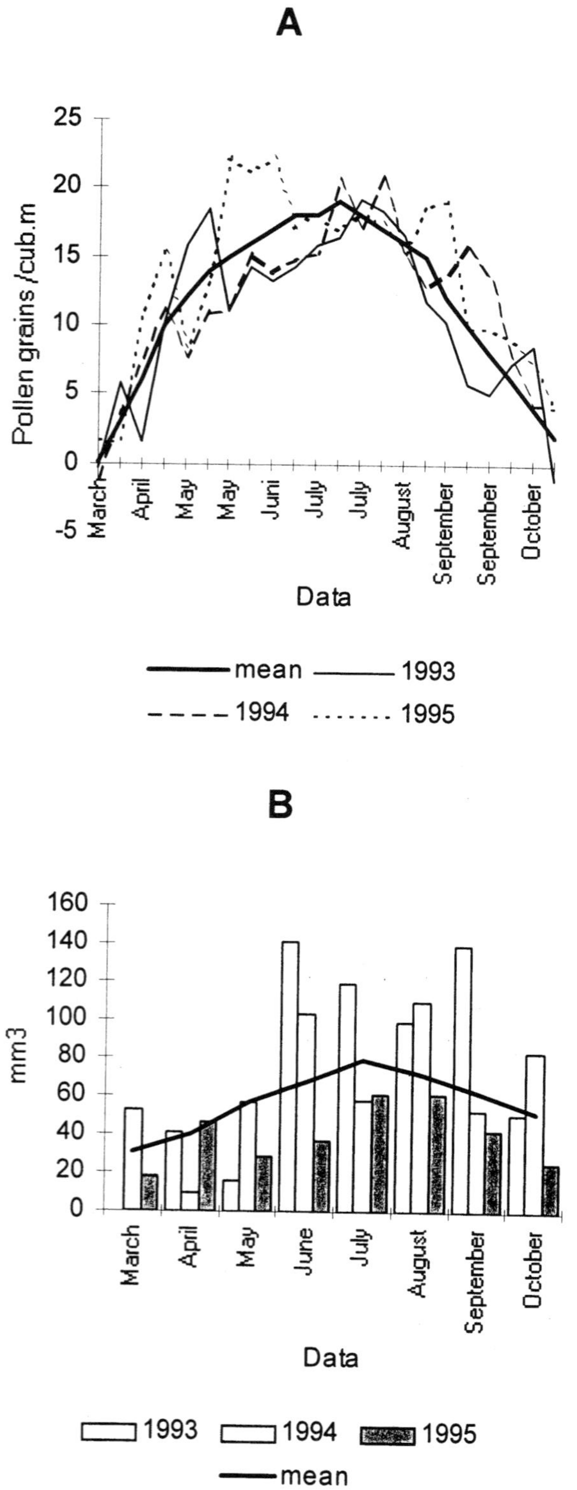

Text-Figure 1. Meteorological parameters for Moscow. A – temperature; B – precipitation.

from 43,610 (1995) to 27,904 (1994). The concentration of total pollen fluctuates considerably during the season. Three periods of pollination can be distinguished by quantitative and qualitative characteristics of the pollen spectrum during each season (Text-Figure 2). The first and the greatest

TABLE 1. Total yearly sum of mean daily concentration of airborne pollen in Moscow, 1993–1995, (pollen grains/m^3).

	Taxon	1993	1994	1995
AC	*Acer*		692	118
AL	*Alnus*	173	3493	5216
AR	*Artemisia*	1383	2636	1555
BE	*Betula*	19919	7040	23965
CH	Chenopodiaceae	127	136	272
CO	*Corylus*	300	116	277
FR	*Fraxinus*		445	569
LA	*Larix*	9	8	16
PC	*Picea*	239	6	77
PI	*Pinus*	897	2165	2704
PL	*Plantago*		177	126
PO	Poaceae	723	2175	1622
PP	*Populus*	5165	700	922
QU	*Quercus*	20	550	270
RO	Rosaceae		168	125
RU	*Rumex*		300	158
SX	*Salix*	591	942	909
TI	*Tilia*	177	34	5
UL	*Ulmus*		165	288
UR	*Urtica*	7387	5430	4185

pollen concentration usually takes place in mid-March to mid-May and consists of about 70% of the annual pollen sum. A low pollen count during the first period of pollination (51% of annual sum) was observed in 1994. However, that season was characterized by a very long, cold spring and early summer. The day temperature did not exceed +10°C until the middle of April and the night temperature exceeded +10°C only after June 25 (Text-Figure 1). The daily pollen catch fluctuation in the first period is very big, in 1993 it was more than 1000 pollen grains/m^3 of air. The second period of pollination lasts about 1–1.5 months, from the end of May utill the beginning of July, and is characterized by the lowest percentage of annual pollen: only 2% (1993) and 12% (1994). The total daily pollen concentration does not fluctuate considerably, daily measurements usually do not exceed 100 pollen grains/m^3. The third pollen period is late summer (July to mid-September) and forms about 20–25% of the total annual pollen sum. In 1993, the total pollen count in this period was 37% of the annual sum. The total daily count is stable during the period and smoothly decreases to the end of August. It rarely exceeds 100–150 pollen grains/m^3.

The taxonomic composition of the pollen spectrum in the first period of pollination consists of tree species: *Betula, Alnus, Corylus, Salix, Populus, Fraxinus, Ulmus, Acer,* and, at the end of the period, *Quercus. Betula* pollen dominates in all years (46–75% of total count during the period). The second most frequent taxon is *Alnus* (15–23%), but in 1993 it was replaced by *Populus* (Text-Figure 3). Except for *Salix,* the duration of the main pollen periods of early spring taxa is considerably short and rarely exceeds 25 days. For *Betula* it varies from 12 to 23 days, and for *Alnus* from 19–24 days

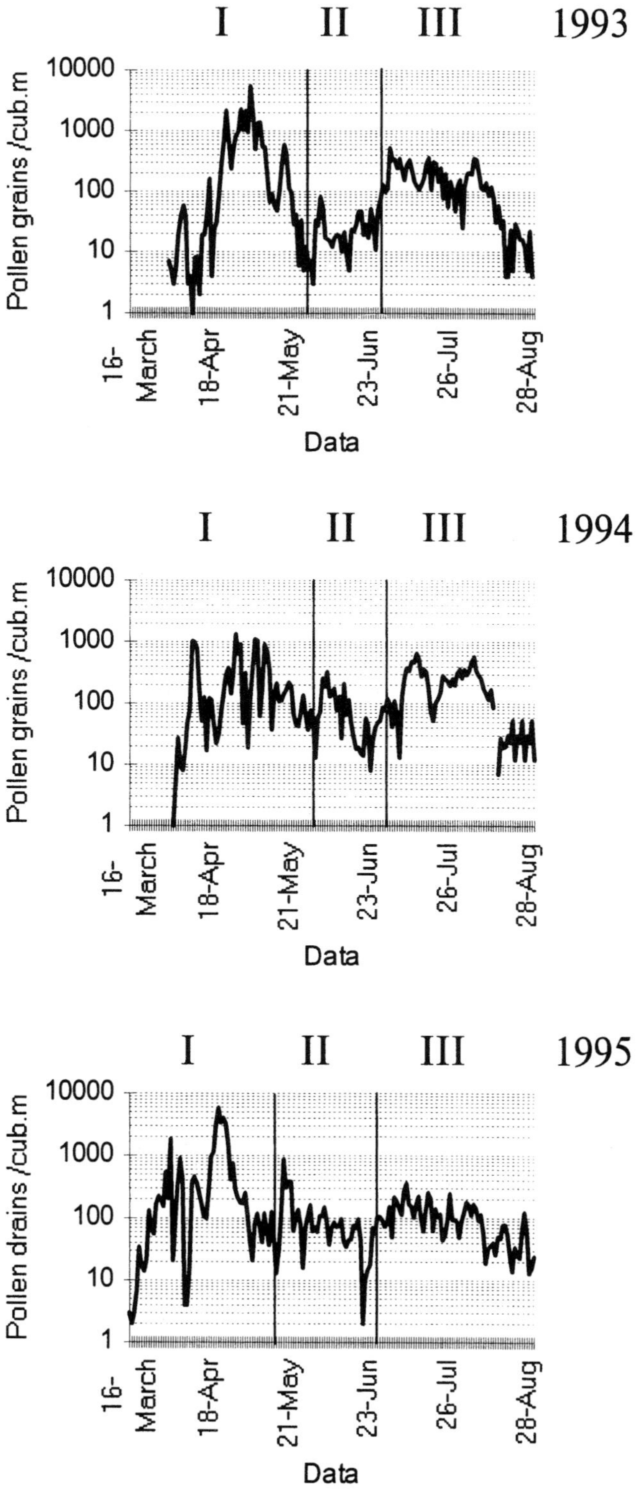

Text-Figure 2. Distribution of airborne pollen sum (1993–1995) during the season. I – first period of pollination; II – second period of pollination; III – third period of pollination.

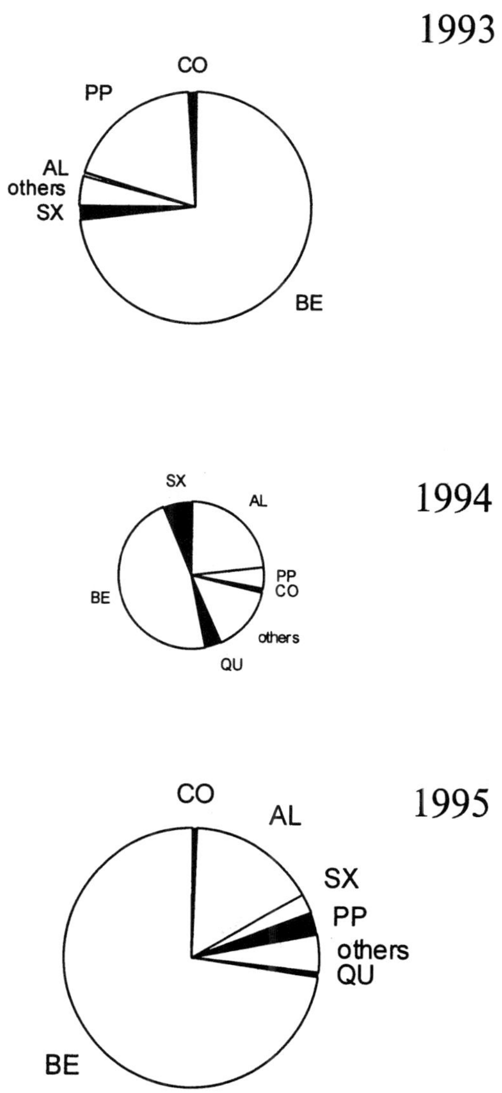

Text-Figure 3. Pollen spectrum in the first period of pollination. Diameters of pies are proportional to the number of pollen grains. CO – *Corylus*, BE – *Betula*, PP – *Populus*, AL – *Alnus*, SX – *Salix*, QU – *Quercus*.

(Table 2). The main pollen period of *Salix* is about 30–35 days. This genus is represented in Moscow by 15 species that are different in their flowering and pollination. We do not identify *Salix* pollen grains to species level and therefore, the observed period of pollination for *Salix* is a combination of species' pollination periods. Other taxa are represented in our region by 1–2 species with one as a dominate. Therefore, the periods of pollination for these taxa focus mainly on the flowering of the dominate taxon. The starting date for the main pollen periods in early spring taxa varies considerably between years and is closely connected with meteorological conditions: *Betula, Alnus, Corylus* and *Salix* started pollination 7 to 10 days earlier in 1995 than other years because of a very warm spring (Table 2).

The taxonomic composition of the second period of pollination is not as well determined as the first one. *Pinus* (16–61% of the total count during the period) and Poaceae (8–25%) pollen grains are the main elements, but their concentration and the time of appearance greatly differ in different seasons (Text-Figure 4). Meteorological conditions mainly influence the Poaceae pollen appearance,

TABLE 2. Pollination periods (1993–1995) for collected taxa. 1 = month and day when pollen from taxon was first trapped in samples; 2 = month and day when taxon began its maximum pollenation period (MPP); 3 = duration in number of days of MPP.

Taxon			1993	1994	1995
		1	First appearance, date		
		2	Start of MPP, data		
		3	Duration of MPP, days		
AC	*Acer*	1		18.04	15.04
		2		29.04	20.04
		3		15	6
AL	*Alnus*	1	5.04	25.03	16.03
		2	6.04	9.04	25.03
		3	24	19	22
AR	*Artemisia*	1	16.07	19.07	30.06
		2	30.07	26.07	23.07
		3	15	20	36
BE	*Betula*	1	19.04	3.04	18.03
		2	29.04	28.04	20.04
		3	16	23	12
CH	Chenopodiaceae	1	25.06	6.06	15.06
		2	9.07	18.07	30.06
		3	41	36	59
CO	*Corylus*	1	1.04	4.04	16.03
		2	4.04	6.04	25.03
		3	23	22	24
FR	*Fraxinus*	1		30.04	19.04
		2		6.05	22.04
		3		14	6
PI	*Pinus*	1	12.05	29.04	29.03
		2	17.05	21.05	19.05
		3	19	39	18
PL	*Plantago*	1		11.06	21.05
		2		27.06	30.05
		3		43	62
PO	Poaceae	1	16.05	6.06	21.05
		2	17.06	19.06	31.05
		3	45	48	61
PP	*Populus*	1	16.04	15.04	24.03
		2	23.04	18.04	17.04
		3	12	15	7
QU	*Quercus*	1	30.04	16.05	27.04
		2	30.04	19.05	13.05
		3	30	12	11
RU	*Rumex*	1		21.05	22.05
		2		6.06	30.05
		3		22	60
SX	*Salix*	1	20.04	15.04	22.03
		2	24.04	24.04	17.04
		3	28	32	35
UL	*Ulmus*	1		9.04	2.05
		2		13.04	16.04
		3		17	12
UR	*Urtica*	1	27.05	10.06	26.05
		2	27.06	6.07	28.06
		3	48	36	45

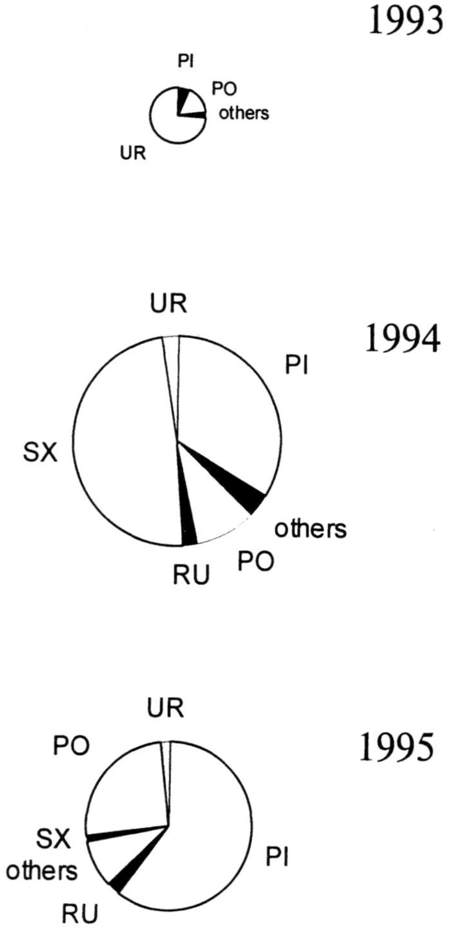

Text-Figure 4. Pollen spectrum in the second period of pollination. Diameters of pies are proportional to the number of pollen grains. PI – *Pinus*; PO – Poaceae; UR – *Urtica*; SX – *Salix*; RU – *Rumex*.

while the pollinating starting date of *Pinus* is fairly constant (May 17–21). In the warm and dry season of 1995, Poaceae started pollination on May 31 and the main pollen periods of these two taxa totally overlapped. In 1993 (wet and cool summer) and in 1994 (very cold early summer) the second period of pollination could be divided into two subperiods: the first one characterized by high *Pinus* pollen concentrations and lasting about a month from mid-May till mid-June, and the second one (the second half of June through the beginning of July) characterized by considerably low *Pinus* and high Poaceae pollen measurements. A similar division was presented in the article by S.G.Gubankova (1981). Besides *Pinus* and Poaceae pollen, grains of *Salix* (extremely high values in 1994), *Quercus, Urtica, Rumex,* and single pollen grains of *Acer, Alnus, Fraxinus,* Cyperaceae, and Brassicaceae noted in early summer pollen spectrum.

The third period of pollination is characterized by the greatest taxonomic diversity of about 30 taxa with 6 of them forming 90% of pollen spectrum. Pollen elements of this period are *Urtica, Artemisia,* Chenopodiaceae, *Rumex, Plantago,* Poaceae (Text-Figure 5). The main taxa are *Urtica* (49–70% of total period sum) and *Artemisia* (20–25%), while

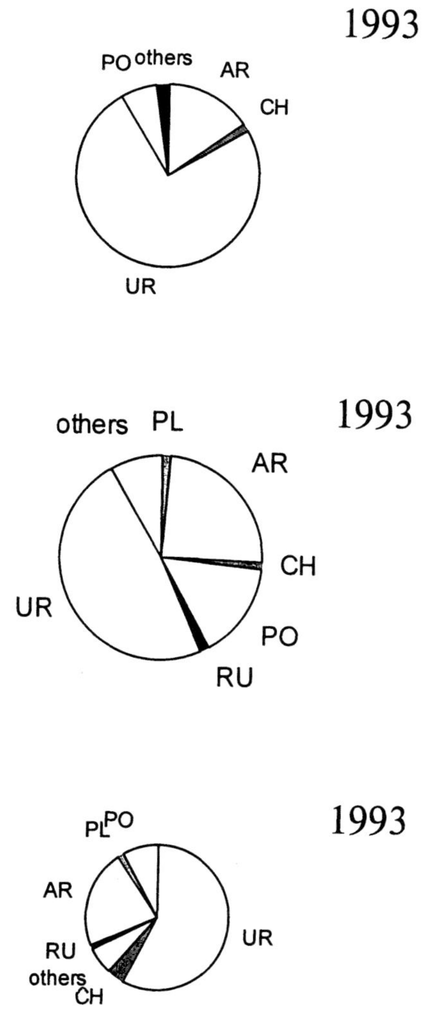

Text-Figure 5. Pollen spectrum in the third period of pollination. Diameters of pies are proportional to the number of pollen grains. PO – Poacea; AR – *Artemisia*; CH – Chenopodiaceae; UR – *Urtica*; RU – *Rumex*; PL – *Plantago*.

pollen grains of Chenopodiaceae (2–5%) play the role of the period marker. In 1994, the pollination of Poaceae was shifted from early to late summer because of cold weather and thus grass pollen formed a significant part of the late summer pollen spectrum (15%).

Except for *Artemisia*, the main summer taxa have considerably long pollination periods: *Urtica* 36–45 days, *Plantago* 43–62 days, *Rumex* 22–60 days, Poaceae 48–60 days, and Chenopodiaceae 36–51 days (Table 2). *Urtica, Plantago* and *Rumex* are weed plants with very long growing and flowering periods, and, therefore, pollination period. *Rumex* is represented in our region by 15 species with different types of flowering. *Urtica* has the longest period of pollination in the Moscow region. Its total duration can exceed 80 days. Poaceae and Chenopodiaceae are very important taxa represented by 67 and 13 genera. Their total pollen counts are the combinations of pollen from individual species.

The comparison of aerobiological and phenological data shows that for all taxa the beginning of the main pollen period coincides with the phenological beginning of flowering and in some cases even forestells it for a few days. The first appearance of pollen in the air is always earlier (sometimes a month earlier) than the phenological flowering because of pollen from long-distance transport.

CONCLUSIONS

The pollen season in Moscow lasts from mid-March until mid-September. Pollen grains of about 40 taxa are determined yearly; 19 of them form 97–99% of the annual pollen spectrum. Three periods of pollination are observed during the season. The first and the greatest one (about 70% of total annual sum) takes place in spring and is mainly associated with tree flowering. The pollen grains of *Betula, Alnus* and *Populus* dominate in that spectrum. Except for *Salix,* early spring taxa have short main pollen periods and their starting date is closely connected to meteorological conditions. The second period (early summer) is characterized by a low pollen concentration (about 2–12% of the total annual sum) that does not fluctuate considerably during the period. *Pinus* and Poaceae pollen grains are the main elements of the spectrum, but their concentration and the time of their first appearance differ in different years. Sometimes the second period can be divided into two subperiods: the first one with mainly *Pinus,* and the second one with high concentrations of Poaceae pollen grains. The third period takes place in July–August and is characterized by *Artemisia,* Chenopodiaceae, *Urtica,* and Poaceae pollen grains. Except for *Artemisia,* the main summer taxa have long main pollen periods without a constant starting date. The comparison of aerobiological and phenological data shows that for all taxa the beginning of the main pollen period coincides with the phenological beginning of flowering. The first appearance of pollen in the air always forestells the phenological flowering because of long-distance transport.

ACKNOWLEDGMENTS

Financial support was provided by The Found of Fundamental Investigations, Russian Academy of Science. The Burkard volumetric trap was rented from Palynological Laboratory of Swedish Museum of Natural History.

References Cited

GUBANKOVA, S.G.
1981 *Aeropalinilogicheskie issledovaniya v Moskve.* Unpubl. Ph.D. thesis. Moskva: 16 pgs.
NILSSON, S, and PERSSON, S.
1981 Tree pollen spectra in the Stockholm region (Sweden), 1973–1980. *Grana,* 20: 179–182.

Author's addresses

E. E. SEVEROVA
S. V. POLEVOVA
Department of Higher Plants
Biological Faculty
Moscow State University
Borob'Evy Gory Moscow 119899
Russia

POLLEN MORPHOLOGY OF *AMORPHOPHALLUS* (ARACEAE)

R. W. J. M. Van der Ham and B. J. Van Heuven

Abstract

The pollen of *Amorphophallus* has been studied with light microscopy, scanning electron microscopy and transmission electron microscopy. The inaperturate pollen shows a large size range, and the exine surface is highly diverse. Eight main surface types have been distinguished, most of which are not clear-cut and show a lot of infratypic variation. The exine consists of a distinct, acetolysis-resistant endexine and a usually non-resistant ectexine. The latter is ultrastructurally diverse by the absence/presence of variously sized, shaped and distributed dark granules of unknown composition. The taxonomic significance of the highly diverse pollen morphology has not been determined to date.

INTRODUCTION

Amorphophallus is a large genus of aroids occurring in Africa, Madagascar, India, continental SE Asia, Malaysia and Australia. The main centre of diversity is continental SE Asia and W Malaysia. The last comprehensive revision was by Engler (1911), who distinguished 78 species in 11 sections. Since then, many new species have been described (Hetterscheid and Ittenbach, 1996). At present the number of species recognized is ca. 170.

Studies dealing with the pollen of *Amorphophallus* are scarce, and most of them treat only one or a few species. The descriptions by Erdtman (1952), Sharma (1968) and Huang (1970, 1972) contain no illustrations, or only drawings without much detail. Thanikaimoni (1969), in his pioneering light microscopic study of Araceae pollen, covered eight species. The descriptions by Bogner (1976), Barthlott and Bogner (1981), Knecht (1983), Straka and Friedrich (1984), Iwanami et al. (1988), Tarasevich (1988), and Bogner and Hetterscheid (1992) contain scanning electron micrographs. Tarasevich (1988) was the first and only worker to provide a transmission electron micrograph and a detailed description of the ultrastructure of the pollen wall.

Grayum (1992), in the framework of an extensive SEM survey of Araceae pollen, treated with 23 species (11 illustrated) belonging to five of Engler's sections. He indicated *Amorphophallus* pollen as the most diverse in the family, and strongly adviced to undertake a more inclusive study of the pollen. The present paper offers the main results of such a study (anno. 1997), which was set up parallel to the taxonomic revisions by Hetterscheid (in prep.) and Ittenbach (1997). A more detailed account, including data of all species and material studied, was published by Van der Ham et al., 1998).

MATERIAL AND METHODS

Two hundred fifty-six collections belonging to 143 species have been studied, of which 110 collections (72 species) were represented by living plants cultivated in the Botanic Garden at Leiden. All other collections were obtainable only as preserved material, nearly always as herbarium specimens. Each collection has been sampled and processed for light (LM) and scanning electron microscopy (SEM); 28 species were selected for transmission electron microscopy (TEM).

Pollen of *Amorphophallus* does not resist the traditional acetolysis method of Erdtman (1960). Usually, the patterned outer layer of the pollen wall is completely removed by this technique (Plate 1, figs. 7–11; see also Tarasevich, 1988). Therefore, the pollen was studied unacetolyzed if fresh, or only boiled in water in case of herbarium material. Except for occasional breakage, boiling did not affect pollen grain size, shape and surface pattern.

Pollen from living plants was suspended in a glycerin/alcohol 96% mixture (2/1). Herbarium material was sampled for complete stamens, which were boiled in water for 10 minutes. Then, the pollen grains could be freed by breaking the anthers, and transferred via alcohol to the glycerin/alcohol mixture.

Fresh and boiled pollen were mounted in glycerin jelly for LM observation. For SEM, the pollen was mounted on stubs and sputter-coated with gold. Preparation for TEM of loose pollen (fresh or boiled; *A. elatus* and *A. gomboczianus* also acetolyzed) included rehydration and fixing in 0.1% glutaraldehyde (1 week), followed by fixation with 1% OsO_4 (1 hour), prestaining with 1% uranylacetate during dehydration, embedding in 3/7 Epon (Luft), poststaining with 5% uranylacetate (10 minutes) and Reynolds' lead citrate (10 minutes). Fresh stamens were prepared with 3% glutaraldehyde (21 hours) and 1% OsO_4 (2 hours), prestained with 1% uranylacetate, and poststained with 3% uranylacetate (10 minutes) and Reynolds' lead citrate (7 minutes).

Pollen grain size (LM) is given as L (= largest diameter) and B (= smallest diameter). The ratio L/B is given as an indication of pollen grain shape. Details of the pollen wall surface were studied and measured with SEM, although nearly all patterns can be recognized with LM as well. The terminology for particular surface types follows Punt et al. (1994). Details relating to thickness and (ultra)structure of the wall were measured and described with TEM.

LIGHT AND SCANNING ELECTRON MICROSCOPY

Pollen grains of *Amorphophallus* are inaperturate monads, often crammed with amyloplasts (Plate 2, figs. 5, 11). Grain size is medium to large, the largest diameter (L) being 28.3–

Van der Ham, R.W.J.M., and Van Heuven, B.J.
2001 Pollen morphology of *Amorphophallus* (Araceae); *In:* Goodman, D.K., and Clarke, R.T. (eds.), *Proceedings of the IX International Palynological Congress, Houston, Texas, U.S.A., 1996;* American Association of Stratigraphic Palynologists Foundation, p. 585–591.

 ISBN 0–931871–06–9

87.8 (av. 53.5) μm. Grain shape is nearly spheroidal to more or less ellipsoidal. In the latter case, one side is often less convex to slightly concave. Grain shape as indicated by L/B is 1.01–1.78 (2.08) (av. 1.25).

The pollen wall surface proved to be highly diverse. Eight surface types have been distinguished, most of which, however, are not clear-cut. In addition, several show a lot of infratypic variation.

1. Echinate Type. With 1–4 μm long spines (Plate 1, fig. 1). Grain size is av. 38.9 μm and grain shape ± spheroidal (av. L/B = 1.14). Variation occurs in the size and shape of the spines: short (1–2 μm) to elongate (2–4 μm), smooth or stacked. This type is found in one Madagascan and four W African species.

2. Verrucate Type. With 1–5 (-10) μm large, ± circular verrucae (Plate 1, figs. 2, 3). The pollen grains are av. 50.8 μm and ± spheroidal (av. L/B = 1.13). Variation exists in the size, shape, prominence and density of the verrucae. This type occurs in five African species and in *A. sylvaticus* (India).

3. Areolate Type. With 1–23 μm large, ± isodiametric, polygonal areolae (Plate 1, figs. 4–6). The pollen size range is fairly wide (37.5–70.8 μm), while shape is ± spheroidal (av. L/B = 1.14). This type is quite heterogeneous with respect to the size and morphology of the areolae: 1–23 μm diameter, 1–2 μm high, smooth to finely fossulate. It occurs in five species from SE Asia.

4. Striate Type. With ± parallel muri < 2 μm wide (Plate 1, figs. 7, 9). Pollen size (28.3–73.5 μm) and shape (L/B = 1.04–1.78) show a wide range, and much diversity exists in the size, shape and direction of the muri, even within species and samples. Moreover, the variation in this type is more or less continuous, which makes it difficult to argue for some subdivision of the type. The distance between the muri varies considerably, and probably depends largely on the degree of expansion (harmomegathic state) of the individual pollen grains. This type contains 63 species, of which only three are African.

5. Striate/scabrate Type. With ± parallel muri with scabrae (Plate 1, fig. 12). Pollen grain size is 32.1 μm. The pattern consists of subparallel muri, which are locally very close to each other or even make up (finely) fossulate areas. The muri themselves are up to 2 μm wide, and mostly irregularly scabrate (locally ± finely serrate). This type is found only in *A. smithsonianus* (India).

6. Fossulate Type. With oblong to irregular elements > 2 μm wide (Plate 1, figs. 13–16). There is a wide size (32.0–86.6 μm) and shape (L/B = 1.04–1.62, av. 1.19) range. This type is clearly heterogeneous, but it is difficult to provide a rigid subdivision without making each species/collection an individual subtype or form. It is found in 18 species (six African or Madagascan), and as a minor form in some psilate species.

7. Psilate Type. Surface ± smooth (Plate 1, figs. 17, 18). Pollen size (41.3–87.8 μm) and shape (L/B = 1.03–1.63, 2.08) show a wide range. Several species show one or a few verrucate to echinate emergences per pollen grain. This type was observed in 49 species (14 African or Madagascan), and as a minor form in some fossulate species.

8. Scabrate Type. Surface finely (< 1 μm) patterned (Plate 1, figs. 19, 20). Pollen size is 34.3 μm. The pattern consists of ± distinct muri of 0.2–0.3 μm wide, mostly < 1 μm long, and often anastomosing into lumps of 1–2 μm in diameter. This type occurs only in *A. galbra*, the only Australian species.

TRANSMISSION ELECTRON MICROSCOPY

The (ultra)structure of the pollen wall has been studied in 28 species, which represent seven surface types; sectioning of the striate/scabrate type failed. In well-developed pollen grains both the intine and the exine are distinct. The 0.3–3.0 μm thick intine is always more or less undulate (Plate 2, figs. 3, 5, 7, 8). Often two or three indistinctly separated sublayers are present. On the whole, intine thickness is fairly constant throughout a grain.

The exine measures 0.4–3.4 μm (spines excluded). It consists of an endexine and an ectexine (see Discussion: Exine Structure, for justification of these terms). The endex-

---

PLATE 1

Pollen of *Amorphophallus*: surface types. SEM. Bar = 10 μm, 5 μm (figs. 10-12, 16, 19) or 1 μm (figs. 4, 20).

1 *A. doryphorus* Ridl. (Eggers s.n., Senegal). Echinate type.
2 *A. abyssinicus* (Rich.) N.E. Br. (H.AM.066B, Nigeria). Verrucate type.
3 *A. sylvaticus* (Roxb.) Kunth (Barnes 880, India). Verrucate type.
4 *A. aberrans* Hett. (Murata et al. T-16977, Thailand). Areolate type (detail).
5 *A. cirrifer* Stapf (H.AM.462, Thailand). Areolate type.
6 *A. tenuistylis* Hett. (van Beusekom et al. 492, Thailand). Areolate type.
7, 8 *A. elatus* Ridl. (H.AM.111, Thailand). Striate type. 7. Unacetolyzed pollen grain. 8. Acetolyzed pollen grain.
9–11 *A. gomboczianus* Pic. Serm. (H.AM.003, Ethiopia). Striate type. 9. Unacetolyzed pollen grain. 10, 11. Acetolyzed pollen grains.
12 *A. smithsonianus* Sivad. (H.AM.216, India). Striate/scabrate type.
13 *A. annulifer* Hett. (H.AM.089, Java). Fossulate type.
14 *A. hohenackeri* (Schott) Engl. & Gehrm. (H.AM.011, India). Fossulate type.
15 *A. konjac* K. Koch (H.AM.168, Japan?). Fossulate type.
16 *A. staurostigma* Ittenb., Hett. & Bogner (Bogner 166, Madagascar). Fossulate type.
17 *A. eichleri* (Engl.) Hook. f. (H.AM.007, central Africa). Psilate type.
18 *A. henryi* N.E. Br. (H.AM.097, Taiwan). Psilate type.
19, 20 *A. galbra* F.M. Bail. (H.AM.174, Australia). Scabrate type.

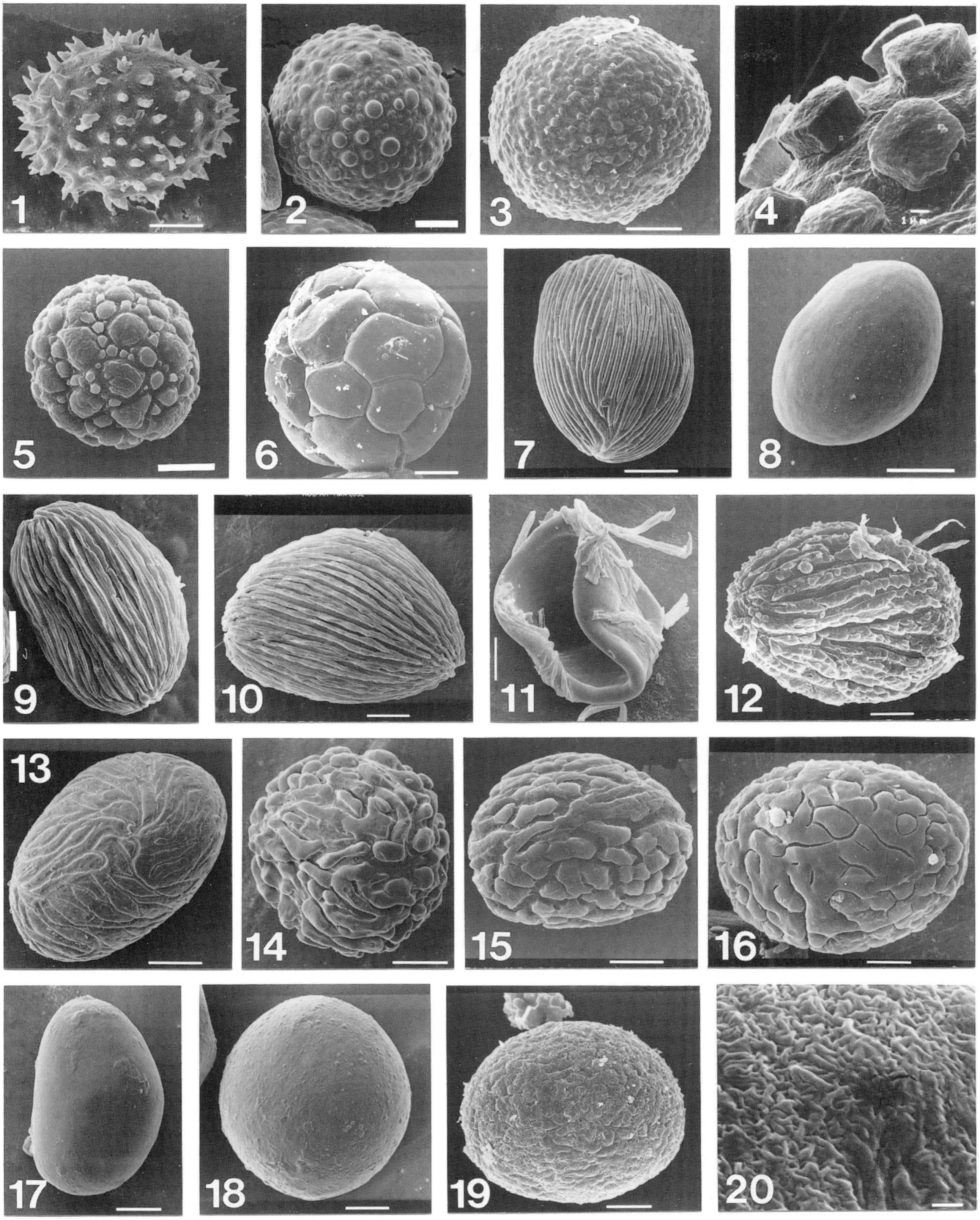
1
2
3
4
5
6
7
8
9
10
11
12
13
14
15
16
17
18
19
20

ine is a uniform, 0.3–1.4 μm thick layer. It is mostly easily recognized by its spongy and/or speckled appearance (Plate 2). Usually, the inner part shows intine intrusions, which in LM appear as irregular endocracks. The outer part is more densely structured than the inner part, and borders regularly on the ectexine. The contact zone is often relatively dark.

Ectexine thickness is 0.1–2.8 μm. All surface patterns observed with SEM reside in this layer. Depending on the surface type, TEM shows a continuous, evenly thick layer, or a series of diversely spaced elements of various sizes and shapes. The individual elements do not stand isolated on the endexine, but are always connected by a mostly thin basal layer. In the striate type the elements (muri) are sometimes also connected peripherally, which results in a seemingly tectate/columellate structure of the ectexine in cross sections (Plate 2, fig. 7). Besides showing a variety of shapes, the surface elements are ultrastructurally diverse, which is mainly determined by the presence, size, shape and distribution of dark granules. Three groups are distinguished.

Group 1. Ectexine without granules or with indistinct granules (Plate 2, figs. 2, 10–12). This group includes 13 species, which represent five surface types: verrucate (2), striate (1), fossulate (4), psilate (5) and scabrate (1).

Group 2. Ectexine with distinct granules (Plate 2, figs. 3, 5–8). This group includes 13 species, which represent four surface types: verrucate (1), areolate (3), striate (9) and fossulate (1). In contrast to group 1, the striate type is common and the fossulate type and psilate type are hardly or not represented here. The granules are more or less densely spaced, and occur ± throughout the ectexine, or in the outer or middle part.

Group 3. Ectexine (ultra)structure deviating. This group includes two rather different species that cannot be easily accommodated in one of the other groups. One species (*A. antsingyensis*) has echinate pollen. The spines stand on a relatively thick continuous ectexine, and consist almost entirely of one to three or four superposed granules (Plate 2, fig. 1). The other species (*A. sylvaticus*) has finely verrucate pollen, and is deviating by small indistinct, densely spaced dark granules in a clearly delimited basal sublayer of the ectexine (Plate 2, fig. 4).

Acetolyzed Pollen

Acetolyzed pollen of *A. elatus* and *A. gomboczianus* was compared with unacetolyzed material. Both species possess striate pollen. LM and SEM showed that in *A. elatus* the striate pattern is completely removed by acetolysis, while a smooth layer is left behind (Plate 1, figs. 7, 8). Using TEM, this smooth layer can be identified as the endexine (Plate 2, figs. 8, 9). *A. gomboczianus* is one of the few species in which the pattern does not disappear by acetolysis, at least not completely (Plate 1, figs. 9–11). TEM of acetolyzed pollen shows, as in *A. elatus*, a distinct endexine. The remaining muri have a lower profile or appear deformed, and seem to have lost the dark granules.

DISCUSSION

Exine Structure

According to Kress and Stone (1982), Zavada (1983) and Stone and Kress (1992), a major difference between dicot and monocot pollen is the rarity of an endexine in monocots. Where present, it is different from the dicotyledonous endexine, being thin, inconspicuous and not thickened in apertural areas. As, moreover, an endexine seems to be absent in the pollen of primitive members of both monocots and dicots (Zavada, 1983), the monocot endexine is possibly not homologous to the dicot endexine.

TEM data of several araceous genera have been published (see also Grayum, 1992: 136), e.g., Lugardon et al. (1988; *Spathiphyllum*), Ohashi et al. (1983; *Arisaema*), Tarasevich (1988; *Amorphophallus, Arisarum, Pistia, Spathiphyllum*), Tarasevich (1989; *Anthurium, Zamioculcas*), Zavada (1983; *Arisaema*). In all genera studied, a usually spongy, thin to rather thick inner exine layer can be observed. In most cases this layer was indicated as the endexine. Zavada (1983) interpreted this layer as a complete exine consisting of an imperforate tectum, a spongy infratectum and an undifferentiated nexine.

PLATE 2

Pollen of *Amorphophallus*: TEM sections of various surface types. Bar = 1 μm; ec = ectexine, en = endexine, in = intine.

1 *A. antsingyensis* Bogner, Hett. & Ittenb. (H.AM.099, Madagascar). Echinate type. Note three superposed large dark granules in spine.

2 *A. abyssinicus* N.E. Br. (H.AM.066B, Nigeria). Verrucate type (see Plate 1, fig. 2). Note endocracks in endexine.

3 *A. preussii* (Engl.) N.E. Br. (Preuss s.n., Cameroon). Verrucate type. Note distinct dark granules.

4 *A. sylvaticus* (Roxb.) Kunth (Barnes 880, India). Verrucate type (see Plate 1, fig. 3). Note small dark granules in basal ectexine.

5, 6 *A. tenuistylis* Hett. (van Beusekom et al. 492, Thailand). Areolate type (see Plate 1, fig. 6). Note densely spaced distinct dark granules.

7 *A. albispathus* Hett. (H.AM.019, Thailand). Striate type. Note peripherally connected muri, distinct dark granules, and endocracks in endexine.

8, 9 *A. elatus* Ridl. (H.AM.111, Thailand). Striate type. 8. Unacetolyzed pollen grain (see Plate 1, fig. 7). 9. Acetolyzed pollen grain (see Plate 1, fig. 8). Note psilate endexine with endocracks.

10 *A. konjac* K.Koch (H.AM.168, Japan?). Fossulate type (see Plate 1, fig. 15), showing indistinct dark granules in basal ectexine.

11 *A. hetterscheidii* Ittenb. & Lobin (H.AM.266, Gabon). Psilate type. Note thin ectexine and amyloplasts.

12 *A. galbra* F.M. Bail. (H.AM.174, Australia). Scabrate type (see Plate 1, figs. 19, 20). Note ectexine covered with dark layer (muri/pollenkitt ?).

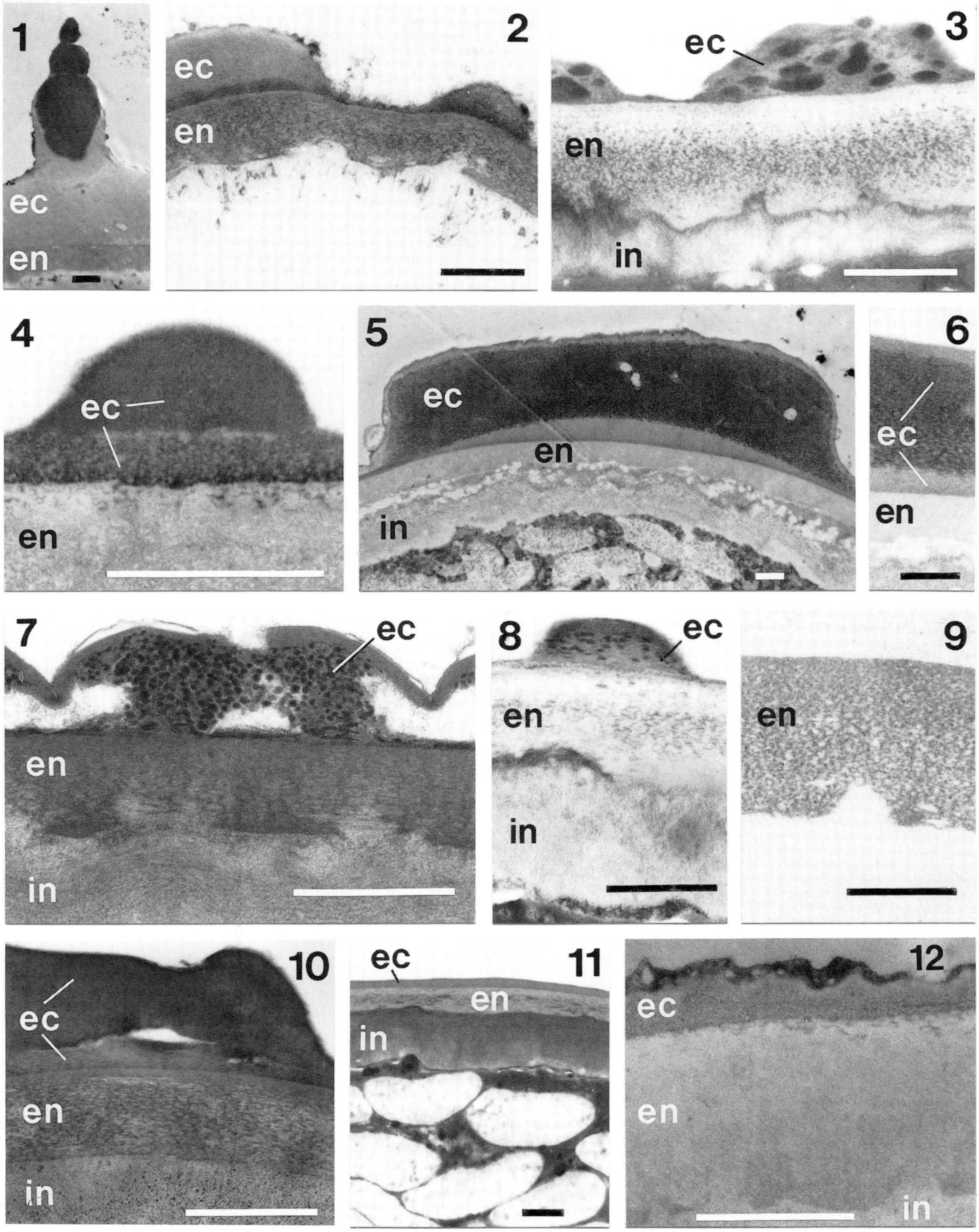
1
ec
en
2
ec
en
3
ec
en
in
4
ec
en
5
ec
en
in
6
ec
en
7
ec
en
in
8
ec
en
in
9
en
10
ec
en
in
11
ec
en
in
12
ec
en
in

Our observations on the structure of the pollen wall in *Spathiphyllum blandum* and *S. cannifolium* show the exine to consist of a striate tectum, a columellate (*S. blandum*) or granular (*S. cannifolium*) infratectum, a foot layer (broken up between the muri in *S. blandum*), and a spongy inner layer. The latter is similar to the endexine reported in the above literature, and to the inner exine layer found in *Amorphophallus*. Because of its resistance to acetolysis (see Descriptions: Acetolyzed Pollen), which indicates it not to be a sublayer of the intine, and its position internal to a foot layer-infratectum-tectum complex in *Spathiphyllum*, the spongy inner layer in *Amorphophallus* and the other Araceae is denoted as the endexine in the present study. Homology within the Araceae is inferred, but no homology with the condition in other monocots or in dicots is presumed.

The outer wall layer in the pollen of *Amorphophallus* is clearly less resistant to acetolysis than the endexine, being completely removed in most cases. Despite its low resistance, the outer layer is denoted here as the ectexine. Possibly, the lower resistance of the ectexine is partly due to the presence of the dark granules. Such granules were first noted by Ohashi et al. (1983) in the echinate pollen of *Arisaema* ("the spinule contains many electron dense, darkly stained inclusions"), and were later, probably independantly, also found by Tarasevich (1988) in *Amorphophallus konjac*, *Arisarum* and *Pistia*. The osmiophilic staining of the granules suggests an oily nature, but further research is needed to elucidate their composition. Comparison of *A. elatus* and *A. gomboczianus* (see Descriptions: Acetolyzed Pollen) shows that their presence alone cannot explain the disappearance of the ectexine. Both have distinct granules in striate ectexine elements (muri), more densely spaced in *A. gomboczianus*, but the ectexine in *A. elatus* is clearly less resistant.

Surface Types

Several surface types can be more or less easily related to each other. This is apparent, either by the presence of intermediate forms or by the occurrence of several types in a single species or sample. The closest relations occur between the psilate type and the fossulate type, and between the fossulate type and the striate type. The areolate type may be connected with the fossulate type and/or the striate type. These relations are only partly reflected by the ultrastructure of the ectexine. The psilate and most fossulate species studied share the absence of distinct dark granules. The striate type nearly always shows distinct dark granules. Thus, the difference between striate and fossulate seems to be more distinct than between psilate and fossulate. The ectexines of the fossulate, striate and areolate species studied with TEM are nearly all thicker than the thickest psilate ectexines. In the fossulate and areolate types the patterns strongly suggest that they originated by some "crack-and-grow" process, and possibly the same pertains to the striate type. This means that besides morphological bonds, there may also exist an ontogenetical relation with the psilate type: cracking and growing is considered to proceed from a ± psilate starting point. Subdivision of the ectexine in more or less separate elements in non-psilate pollen may partly explain its greater thickness, as a cracked ectexine does not have to stretch so much. The psilate and fossulate types occur in Asia as well as in Africa (including Madagascar). The striate type is rare in Africa, while the areolate type is restricted to Asia.

The other four surface types (echinate, verrucate, striate/scabrate, scabrate) are difficult to accommodate. Except may be for the striate/scabrate type, which is possibly a derivative of the striate type, these patterns cannot be easily explained by a "crack-and-grow" mechanism. Ectexine ultrastructure in the four types is deviating too (not studied in the striate/scabrate type), and does not lend support for relations with other types. The echinate type and verrucate type, excluding *A. sylvaticus*, are restricted to Africa (including Madagascar). *A. sylvaticus* and the striate/scabrate type are found in India, and the scabrate type in the only Australian species.

A comparison of the surface types with a preliminary subdivision based on macromorphological data has shown that most of the approximately 23 groups contain two or more surface types. The psilate and striate types occur in many groups. This precludes a straightforward palynological characterization of most groups. A more definite comparison and a phylogenetic analysis of all data are in progress (Van der Ham et al., 1998; Hetterscheid, in prep.).

ACKNOWLEDGMENTS

We thank Wilbert Hetterscheid (Hilversum, The Netherlands), Stephan Ittenbach (Bonn, Germany), Hanneke Jellema (Leiden) and Art Vogel (Leiden) for the pollen samples and/or taxonomic information.

References Cited

BARTHLOTT, W., and BOGNER, J.
1981 Rediscovery of *Amorphophallus staudtii* (Engl.) N.E.Br. in the Tai National Park, Ivory Coast. *Aroideana*, 4: 109–113.
BOGNER, J.
1976 Eine neue *Thomsonia*-Art (Araceae) aus Thailand. *Plant Systematics and Evolution*, 125: 15–20.
BOGNER, J., and HETTERSCHEID, W.L.A.
1992 Notes on the genus *Amorphophallus* (Araceae), 1. Three new species from tropical Asia. *Blumea*, 36: 467–475.
ENGLER, A.
1911 Araceae–Lasioideae. *Das Pflanzenreich*, 48 (IV.23C): 1–130.
ERDTMAN, G.
1952 *Pollen morphology and plant taxonomy. Angiosperms.* Almqvist & Wiksell, Stockholm, 539 pgs.
1960 The acetolysis method. A revised description. *Svensk botanisk Tidskrift*, 54: 561–564.
GRAYUM, M.H.
1992 Comparative external pollen ultrastructure of the Araceae and putatively related taxa. *Monographs in Systematic Botany from the Missouri Botanical Garden*, 43: 1–167.
HETTERSCHEID, W.L.A., and ITTENBACH, S.
1996 Everything you always wanted to know about *Amorphophallus*, but were afraid to stick your nose into. *Aroideana*, 19: 7–131.
HUANG, T.C.
1970 Pollen grains of Formosan plants, 6. *Taiwania*, 15: 73–179.
1972 *Pollen flora of Taiwan*. National Taiwan University Botany Department Press, Taipei, 297 pgs.
ITTENBACH, S.
1997 Revision der afrikanischen Arten der Gattung *Amorphophallus* (Araceae). Thesis, Universität Bonn.
IWANAMI, Y., SASAKUMA, T., and YAMADA, Y.
1988 *Pollen: illustrations and scanning electron micrographs.* Springer/Kodansha, Berlin/Tokyo, 198 pgs.

KNECHT, M.
1983 Comparaison des caractères palynologiques et biologiques d'espèces d'Aracées de Côte d'Ivoire. *Archives des Sciences*, 36: 451–460.
KRESS, W.J., and STONE, D.E.
1982 Nature of the sporoderm in monocotyledons, with special reference to the pollen grains of *Canna* and *Heliconia*. *Grana*, 21: 129–148.
LUGARDON, B., LOBREAU-CALLEN, D., and LE THOMAS, A.
1988 Structures polliniques chez les Araceae, 1. Tribu des Spathiphylleae. *Journal of Palynology*, 23/24: 51–57.
OHASHI, H., MURATA, J., and TAKAHASHI, M.
1983 Pollen morphology of the Japanese *Arisaema* (Araceae). *Science Reports of the Tohoku University*, 4th ser., 38: 219–251.
PUNT, W., BLACKMORE, S., NILSSON, S., and LE THOMAS, A.
1994 *Glossary of pollen and spore terminology*. LPP Contributions Series, 1. LPP Foundation, Laboratory of Palaeobotany and Palynology, Utrecht, 71 pgs.
SHARMA, M.
1968 Pollen morphology of the Indian monocotyledons. *Journal of Palynology*, spec. vol. 1967: 1–98.
STONE, D.E., and KRESS, W.J.
1992 The application of transmission electron microscopy and cytochemistry in palynology. *In*: Nilsson, S., and Praglowski, J. (eds.), *Erdtman's Handbook of Palynology*. Munksgaard, Copenhagen, pp. 432–467.
STRAKA, H., and FRIEDRICH, B.
1984 Palynologia Madagassica et Mascarenica. Familien 19–49. *Tropische und subtropische Pflanzenwelt*, 49: 404–485.
TARASEVICH, V.F.
1988 Peculiarities of morphology in ridged pollen grains in some representatives of Araceae. *In*: Chlonova, A.F. (ed.), *Palynology in the USSR. Papers of the Soviet palynologists to the VII International Palynological Congress, Brisbane, Australia, 1988*. Nauka, Novosibirsk, pp. 58–61.
1989 Pollen grain ultrastructure in the genus *Anthurium* (Araceae) in connection with its systematics. *Botanicheskii Zhurnal*, 74: 314–324.
THANIKAIMONI, G.
1969 Esquisse palynologique des Aracées. *Travaux de la Section Scientifique et Technique, Institut Français de Pondichéry*, 5: 1–31.
VAN DER HAM, R.W.J.M.
1998 Notes on the genus *Amorphophallus* (Araceae) - 8 Pollen morphology of *Amorphophallus* and *Pseudodracontium*. *Review of Palaeobotany and Palynology*, 103: 95–142.
ZAVADA, M.S.
1983 Comparative morphology of monocot pollen and evolutionary trends of apertures and wall structures. *Botanical Review*, 49: 331–379.

Authors' address

R.W.J.M. VAN DER HAM
B.J. VAN HEUVEN
Nationaal Herbarium Nederland
University of Leiden
P. O. Box 9514
2300 RA Leiden
The Netherlands

POLYPORATE POLLEN — ITS PREVALENCE AND PECULIARITY OF EVOLUTION IN THE SUBCLASSES RANUNCULIDAE AND CARYOPHYLLIDAE: A REVIEW

V. D. Savitsky and O. A. Martynyuk

Abstract

Polyporate pollen type is unevenly distributed among angiosperms, evoking different opinions from the point of view of evolutionary status and classification.

We have analyzed this type of pollen grains in representatives of the "Ranunculidae-Caryophyllidae phylogenetic branch" as defined by Takhtajan (1987). In the subclass Ranunculidae, relatively small numbers of species have polyporate pollen in the Ranunculaceae, Papaveraceae, and Menispermaceae families. In the subclass Caryophyllidae, more than half of the families have representatives with polyporate pollen grains. In the families Cactaceae, Phytolaccaceae, Nyctaginaceae, Portulacaceae and Basellaceae (Caryophyllales), and Polygonaceae (Polygonales), half or less of the representatives have polyporate pollen grains. The majority of Caryophyllaceae have polyporate pollen, and all species from the Amaranthaceae, Chenopodiaceae, Achatocarpaceae, and Halophytaceae (Caryophyllales) families have polyporate pollen.

A few species of Papaveraceae (Ranunculidae) have polyporate pollen which differs from that in the Caryophyllales, primarily in disproportional location of pores (e.g., *Papaver* L.), texture, and structure of the pore membrane. Polyporate pollen grains of Ranunculaceae, in general, are more similar with some Caryophyllales, but they are more primitive in structure.

For the most part, polyporate pollen grains appear to be derived from tricolpate pollen. From this point of view, we may suppose that the polyporate pollen type found in the Ranunculidae is ancestral to the Amaranthaceae and Chenopodiaceae.

INTRODUCTION

Panporate pollen grains are considered to be derived from other types of pollen grains. Taxa with panporate grains are usually placed at the end of phylogenetic branches in phylogenetic schemes that have been constructed on the basis of pollen exine structure. Ideas on the evolutionary advance of panporate pollen grains have been presented, for example, by Wodehouse (1935) and Takhtajan (1964).

Panporate pollen originated in parallel in unrelated groups of flowering plants. Representatives of these taxa having this type of pollen are usually considered the most advanced, but, how advanced are the representatives of different taxa with respect to each other?

DISSCUSSION

As model objects for the study, we elected to study pollen from the subclasses Caryophyllidae and Ranunculidae. The two are close relatives according to some investigators (Takhtajan, 1987; for another reference see Grossheim, 1966).

In the subclass Caryophyllidae, we studied the order Caryophyllales (= Centrospermae and Chenopodiales). This order is a natural group in the flowering plants, being separated from other taxa by the presence of betalain pigments instead of anthocyanin, which is normally found in other flowering plants (with the exception of Caryophyllaceae and Molluginaceae). In addition, the taxa possess a special protein ring in their unique P-type sieve-element plastids. It is interesting to precisely determine the possible relationships of this order with closely related plant groups. It is interesting to note the anomalous secondary thickening of stems and roots, C4-type of photosynthesis (with Kranz-structure of leaves) in many representatives of the order. All these points of difference demonstrate the necessity of a multidisciplinary study, but our objective here was to limit our study to that of pollen morphology.

Some systematists consider that the subclass Caryophyllidae is related to the subclass Ranunculidae. For example, Takhtajan (1987) proposed that Caryophyllidae is possibly descended from some ancient Ranunculidae which, for their part, are close to Magnoliidae. However there are other theories; in particular, Volgin (1990) studied the flower structures of Caryophyllidae, and pointed out the distinct isolation of the subclass as well as the impossibility of finding close relatives and possible ancestors among the modern families. Zosimovich (1965) also emphasized that "it is not possible to connect the origin of Centrospermae from Ranales."

Although the boundaries of the subclass are clearly defined, the mutual relationships between different families inside this subclass are still the subject of discussion. Takhtajan (1987) showed that the families Polygonaceae and Plumbaginaceae are in separate orders (Polygonales and Plumbaginales) within the subclass Caryophyllidae, and that the subclass Ranunculidae consists of the orders Ranunculales, Glaucidiales, Paeoniales, and Papaverales (Table 1). One of the latest publications on the systematics of the order Caryophyllales by Bittrich (1993). Almost all of the families within the order have polyporate pollen (see Table 2). We were unable to determine details of the pollen morphology of members of the Stegnospermataceae family.

Savitsky, V. D., and Martynyuk, O. A.
2001 Polyporate pollen — its prevalence and peculiarity of evolution in the subclasses Ranunculidae and Caryophyllidae: A review; *In*: Goodman, D.K., and Clarke, R.T. (eds.), *Proceedings of the IX International Palynological Congress, Houston, Texas, U.S.A., 1996*; American Association of Stratigraphic Palynologists Foundation, p. 593–596.
 ISBN 0–931871–06–9

TABLE 1. List of superorders, orders, suborders and families of the subclasses Caryophyllidae and Ranunculidae (according to Takhtajan, 1987).

Subclass: CARYOPHYLLIDAE
- Superorder: Caryophyllanae
  - Order: Caryophyllales: Phytolaccaceae, Achatocarpaceae, Barbeuiaceae, Nyctaginaceae, Aizoaceae, Tetragoniaceae, Molluginaceae, Stegnospermataceae, Portulacaceae, Hectorellaceae, Basellaceae, Halophytaceae, Cactaceae, Didiereaceae, Caryophyllaceae
    - Suborder Chenopodiinae: Amaranthaceae, Chenopodiaceae
- Superorder: Polygonanae
  - Order: Polygonales: Polygonaceae
- Superorder: Plumbaginanae
  - Order: Plumbaginales: Plumbaginaceae

Subclass: RANUNCULIDAE
- Superorder: Ranunculanae
  - Order: Ranunculales:
    - Suborder: Lardizabalineae: Lardizabalaceae, Sargentodoxaceae, Menispermaceae
    - Suborder: Ranunculinae: Ranunculaceae, Circaeasteraceae
    - Suborder: Berberidineae: Hydrastodaceae, Berberidaceae, Nandinaceae
  - Order: Glaucidiales: Glaucidiaceae
  - Order: Paeoniales: Paeoniaceae
  - Order: Papaverales: Papaveraceae, Hypecoaceae, Fumariaceae

Subclass: MAGNOLIIDAE

TABLE 2. Phylogenetic tree of the order Caryophyllales (according to Bittrich, 1993). Families with polyporate pollen grains are marked with a "plus sign" (+).

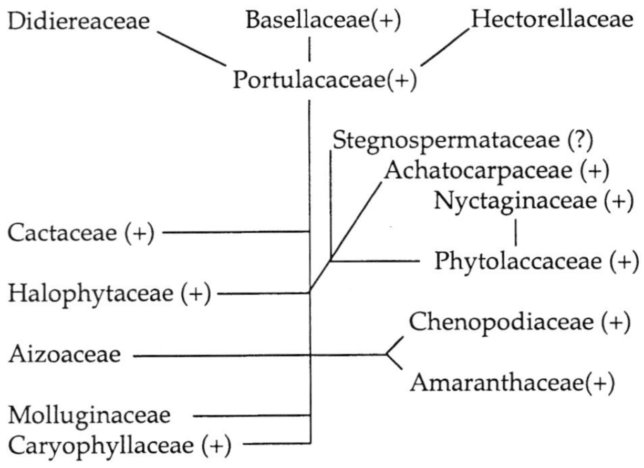

Relationships within the taxa with different types of aperture can be easily traced through comparison of pollen exine morphology. Many families in the order Caryophyllales exhibit a large diversity of pollen morphology, but two comparatively large families of the order, Amaranthaceae and Chenopodiaceae, and the small families Achatocarpaceae (two genera, ten species) and Halophytaceae (1 genus, 1 species), have panporate pollen.

The pollen grains of *Halophytum ameghinoi* Speg. have an unusual cubic form with six pores that are located in the center of planes (Skvarla and Nowicke, 1976; fig. 3). Pollen similar to *Halophytum* is found in two species of *Pseudoplantago* Suess. (Amaranthaceae) Eliasson, 1988; fig.17; Townsend, 1988; fig.2) and in some of the Basellaceae (Nowicke, 1975; figs. 63–65).

Pollen in species of the Achatocarpaceae have an irregular rounded form with large, irregularly spaced pores. The exine is scabrate in contrast to the spinose and perforate exine which is characteristic of the majority of Caryophyllales (Nowicke, 1975; fig.70).

Erdtman (1952) described two distinctive pollen types in the family Amaranthaceae: the *Amaranthus*-type (widespread in different taxa with panporate pollen) and the *Gomphrena*-type (panporate pollen, but with pores located in lacunae). It is assumed that the pollen of the *Gomphrena*-type descended from the *Amaranthus*-type (Martynyuk, 1993). Pore total area results in a reduction in the area of the mesoporium; the exine is very thick in the mesoporium and very thin in the pore region. The thinning of lateral walls associated with deepening (lacunae) sometimes leads to loss of the cover of the collumellae. A columellate layer supports only a thin layer of tectum on its top (for example, in the pollen of genus *Froelichia* Moench.).

Although the Chenopodiaceae family is twice the size of the Amaranthaceae, the pollen is very monomorphic and similar to the pollen of some genera of Amaranthaceae (belonging to the *Amaranthus*-pollen type; in particular, to pollen of the genus *Amaranthus* L.). Based on this observation, some investigators (using pollen analysis) use the term "Chenopodiaceae–Amaranthaceae-complex" (Tsukada, 1967; Lezine and Edorh, 1991). There is no doubt that the pollen morphology data confirm the very close relationship between the two families.

There is some question concerning the evolution of the two families: one proposal states that the two are descended from a single general ancestor, while an alternative states that the family Amaranthaceae is the ancestor of the Chenopodiaceae (Martynyuk, 1993). It is very difficult to imagine that a representative of a monomorphic pollen family such as the Chenopodiaceae could have given rise to a family like the Amaranthaceae, which has morphologically diverse pollen types.

However, because the polyporate pollen did not evolve within the Chenopodiaceae or Amaranthaceae, the ancestor of both families (or perhaps only the Amaranthaceae) has to be a plant which already had developed panporate pollen. Does the ancestor for these families still exist? If so, in which group does it occur? We can suppose that both of these families came from: (1) a family in the order Caryophyllales, (2) a family in the order Polygonales, which have panporate pollen grains, or (3) from an ancestor which is in neither the Caryophyllales nor the Polygonales. The ancestor of all plants in the subclass Caryophyllidae (including the families Amaranthaceae and Chenopodiaceae) should have great variability in both the structure of the pollen exine and pore number. Plumbaginales is the third order within this subclass, it does not, however, have polyporate pollen.

A small number of taxa in the order Polygonales have panporate pollen (for exmple, some species of the genus *Polygonum* L.). These pollen grains have a very distinctive tectum with reticulate ornamentation, which is sharply different from the ornamentation of the panporate pollen of the Caryophyllales (Romanova, 1994).

There are different opinions on the subject of primitive representatives and possible ancestors within the order Caryophyales. Cronquist (1957, 1965, 1968), Soó (1967), Thorne (1968), and Takhtajan (1964) considered the Phytolaccaceae as the most ancient. This is supported by both karyological (Ehrendorfer, 1976; Alexeeva,1977) and cytological investigations (Behnke, 1976).

Similarly, some investigators consider other families to be the most ancient, for example, the Caryophyllaceae (Bessey, 1915; Heintze, 1928), Chenopodiaceae (Wettstein, 1935), Crassulaceae (Hallier, 1912), and Achatocarpaceae (Eichler, 1878).

Volgin (1990) emphasized that one of the modern taxa of the order Caryophyllales could not be considered as an ancestor for this order.

We conducted a comparison of the pollen exine of panporate pollen grains of the Caryophyllidae, and noticed that families which have such pollen types, together with pollen compared with the Chenopodiaceae and Amaranthaceae, have sharply different, specific pollen (typical only for one family). For example, some Portulaccaceae have pollen with very high raised pores, some Nyctaginaceae have panporate pollen with very extended spines on the exine, and the polyporate pollen of the Cactaceae family are often reticulate.

The closest relatives to the Amaranthaceae and Chenopodiaceae are plants with panporate pollen grains in the family Caryophyllaceae (although most of the Caryophyllaceae pollen have specific characteristics). For example, they have a thick exine, large size, low number of pores, and the texture is more obvious under the light microscope. The specific structure of the protein rings of the plastids of both the Amaranthaceae and Chenopodiaceae (dissimilar to other Caryophyllales) and the absence of betalains in Caryophyllaceae force us to look for the ancestor of the Amaranthaceae and Chenopodiaceae (and all Caryophyllidae) among other taxa.

If we agree with Takhtajan's (1987) opinion about the relationships of Caryophyllidae, Ranunculidae, and Magnoliidae, it would be necessary to consider the existence, evolution, and structure of panporate pollen of the latter two subclasses, and compare it with panporate pollen of subclass Caryophyllidae.

Polyporate pollen grains are rare in the subclass Magnoliidae. In addition, there is the exclusion of this type of pollen in the families Chloranthaceae (order Laurales), Aristolochiaceae (Aristolochiales), Rafflesiaceae (Rafflesiales).

In the subclass Ranunculidae, a small number of species in the Ranunculaceae (Ranuncules) and Papaveraceae (Papaverales), as well as the majority of Fumariaceae (Papaverales), have polyporate pollen.

Polyporate pollen grains of Papaveraceae are sharply different from those in the Caryophyllales, differing in both exine thickness and texture. The polyporate pollen of the Fumariaceae have very large pores with high raised borders, and the exine is smooth with small tuberouses (which "undulate" under the light microscope).

Porate pollen from other members of Ranunculaceae (in some *Anemone* L., *Clematis* L., *Ranunculus* L., *Thalictrum* L., etc.) is similar in outline to some pollen of the Caryophyllales. For example, some pollen grains have spines and perforations, but at the same time, there are differences between pollen of the Ranunculaceae and the Caryophyllales, including an uneven border and irregular location of pores on Ranunculaceae pollen.

Therefore, while looking for the ancestor of the Caryophyllidae among the representatives of Ranunculidae, it seems more probable to find it among the Ranunculaceae, at least based on pollen-morphological data.

NOTE

We have used the term "texture" (Kremp, 1965) in a sense of "pattern" (under the light microscope), which assembled from surface and infratectal elements. It often looks like points, nets, streams.

References Cited

ALEXEEVA, T.V.
1977 Karyological survey of the most important taxonomic groups of Centrospermae. *In*: *Materials of the Conference on phyllogeny of Centrospermae*. Moscow University Press, Moscow, p. 3–6. (In Russian).

BEHNKE, H.-D.
1976 Ultrastructure of sieve-element plastids in Caryophyllales (Centrospermae), evidence for the delimitation and classification of the order. *Plant Systematics and Evolution*, 126: 31–54.

BESSEY, C.E.
1915 The phylogenetic taxonomy of flowering plants. *Annals of the Missouri Botanical Garden* II (1–2). (ref. from Grossheim, 1966).

BITTRICH, V.
1993 Introduction to Centrospermae. *In*: *The Families and Genera of Vascular Plants*. Vol. 2. Springer-Verlag, Berlin, p. 13–19.

CRONQUIST, A.
1957 Outline of a new system of families and orders of dicotyledons. *Bull. Jard. Bot. État*, 27: 13–40. (ref. from Behnke, 1976).
1965 The status of the general system of classification of flowering plants. *Ann. Missouri Bot. Gard.*, 52: 281–303. (ref. from Behnke, 1976).
1968 *The evolution and classification of flowering plants*. London; Nelson. (ref. from Behnke, 1976).

ECKARD, T.
1976 Classical morphological features of centrospermous families. *Plant Systematics and Evolution*, 126: 5–25.

EHRENDORFER, F.
1976 Chromosome number and differentiation of Centrospermous families. *Plant Systematics and Evolution*, 126: 27–30.

EICHLER, A.W.
1878 *Blüthendiagramme*. Zweiter Teil. Leipzig, W.Engelmann. (ref. from Behnke, 1976).

ELIASSON, U.H.
1988 Floral morphology and taxonomic relations among the genera of Amaranthaceae in the New World and the Hawaiian Islands. *Botanical Journal of the Linnean Society*, 96: 253–286.

ERDTMAN, G.
1952 *Pollen morphology and plant taxonomy*. Angiosperms. Waltham, Mass., USA, 539 pgs.

GROSSHEIM, A. A.
1966 *Review of modern systems of flowering plants*. (In Russian). Thbilisi, Mizniireba, 198 pgs.

HALLIER, H.
1912 L'origin et le systeme phylétique des Angiospermes exposée a l'aide de leur arbre généalogique. *Archives. Néerlandaises des Sciences. Exactes et Naturelles. Sér. III, B. (Sci. Nat.)* T, 1, Livr, 3. (ref. from Grossheim, 1966).

HEINTZE, A.
1927 Cormofyternas fylogeni (Phylogenie der Cormophyten). Lund. (ref. from Grossheim, 1966).
KREMP, G.O.W
1965 *Morphologic encyclopedia of palynology*. The University of Arizona Press, Tucson, 411 pgs.
LEZINE, A.-M., EDORH, T. M.
1991 Modern pollen deposition in West African Sudanian environments. *Review of Palaeobotany and Palynology*, 67: 41–58.
LIVINGSTONE, D. A., TOMLINSON, M, FRIEDMAN, G., and BROOME, R.
1973 Stellate pore ornamentation in pollen grains of the Amaranthaceae. *Pollen et Spores*, 15: 345–351.
MARTYNYUK, O. A.
1993 Pollen morphology of some genera of Amaranthaceae. (In Ukrainian). *Ukrainian Botanical Journal*, 50: 32–38.
NOWICKE, J.W.
1975 Pollen morphology in the order Centrospermae. *Grana*, 15: 51–77.
ROMANOVA, L.S.
1994 Basic pollen types of Eurasian representatives of family Polygonaceae Juss. (In Ukrainian). *Ukrainian Botanical Journal*, 51: 73–77.
SKVARLA J. J., and NOWICKE, J. W.
1976 Ultrastructure of pollen exine in centrospermous families. *Plant Systematics and Evolution*, 126: 55–78.
SOÓ, R.
1967 Die modernen Systeme der Angiospermen. *Acta Bot. Acad. Sci. Hung.*, 13: 201–233. (ref. from Behnke, 1976).
TAKHTAJAN, A.L.
1964 *Principles of evolutionary morphology of Angiosperms*. (In Russian). Nauka, Moskow, 236 pgs.
1987 *System of Magnoliales*. (In Russian). Nauka, Leningrad, USSR, 439 pgs.
THORNE, R.F.
1968 *Synopsis of a putatively phylogenetic classification of the flowering plants*. Aliso 6: 57–66. (ref. from Behnke, 1976).
TOWNSEND, C.C.
1988 A second species of *Pseudoplantago* from Venezuela. Notes on Amaranthaceae XIX. *Kew Bulletin*, 43: 107–110.
TSUKADA, M.
1967 Chenopod and Amaranth pollen: electron-microscopic identification. *Science*, 157: 80–82.
VOLGIN, S.A.
1990 Comparison morphology of flower and phylogeny of Centrospermae. (Unpubl. doctoral dissertation) Moscow, 48 pgs. (in Russian).
WETTSTEIN, R.
1935 *Handbuch der systematischen Botanik. II. B.* (ref. from Grossheim, 1966).
WODEHOUSE, R. P.
1935 *Pollen Grains*. Hafner, New York, 574 pgs.
ZOSIMOVICH, V.P.
1965 Living forms, polyploidy and evolution of species of the centrospermous families. (In Russian). *In*: *Cytology and genetics*. Naukova dumka, Kiev, Ukraine, 223 pgs.

Authors' addresses

V. D. SAVITSKY
O. A. MARTYNYUK
Institute of Botany of the
National Academy of Sciences of Ukraine
Tereshenkivska str.2
Kiev-MSP-1, 01601
Ukraine

STUDIES ON THE MORPHOLOGY OF MEXICAN PTERIDOPHYTE SPORES

María de la Luz Arreguín-Sánchez, Rodolfo Palacios-Chávez, and David Leonor Quiroz-García

Abstract

The Mexican pteridophytes have been studied by many investigators in the 19th century (e.g. Bentham, 1839–1857; Martens and Galeotii, 1842; Fée (1844–1866); Fournier, 1872; Davenport, 1891), however it was not until 1935 that research was published on Mexican pteridophyte spores. The present work is a review of most of the publications that treat Mexican pteridophytes and an analysis and compilation of those works. Included are a table of species cited per work and an extensive list of pteridophyte species from Mexico whose spore morphology is not described in the literature. The level of study in the various papers is different. Some papers only have photomicrographs and short descriptions of ornamentation, while other papers contain lengthy descriptions; palynological keys are included along with photomicrographs and discussions of taxonomic position and interpretations based on spore morphology. We found that 103 genera and 900 species of pteridophytes have been reported from Mexico, of which 346 species have spore morphology described or illustrated.

REVIEW OF PUBLICATIONS

Mexican pteridophytes have been studied by many investigators during the 19th century (see Bentham, 1839–1857; Martens and Galeotii, 1842; Fée, 1844–1866; Fournier, 1872; Davenport, 1891). However, the first paper to describe pteridophyte spores that grow in Mexico was by R. McVaugh (1935), who studied spores from pteridophytes of northwestern United States. He included palynological keys and drawings of spores. Six of the taxa he described grow in Mexico. A.F. Tryon (1949) studied spores of the genus *Selaginella* in North America and northern Mexico. She included descriptions, short palynological keys, and photomicrographs. Nayar and Devi (1964) illustrated the spore morphology of the families Aspidiaceae, Aspleniaceae and Polypodiaceae of India using photomicrographs, drawings and descriptions, and included nineteen species that grow in Mexico. Using the same type of illustrations, Nayar and Devi (1967) reported on the spore morphology of the Pteridaceae II, the Gymnogrammoid ferns, twenty-five species of which grow in Mexico. The relationship between the geography and evolution of spores of the cheilanthoid ferns was considered by Tryon and Tryon (1973), who illustrated spores using scanning electron microscopy (SEM) and considered general characteristics of perispore morphology; seven of the taxa included in this study grow in Mexico.

More recently, Lloyd (1981) studied the perispore of *Polypodium* and related genera (Polypodiaceae), including 17 Mexican species. The paper contains tables with characters of perispore morphology as well as SEM photomicrographs. Tryon and Tryon (1982) published a book entitled "Ferns and Allied Plants with Special Reference to Tropical America", and included taxonomy, ecology, geography, genetics and palynology; 147 genera and species in this study grow in Mexico. In the palynological section, they included general characters of ornamentation and SEM photomicrographs.

Arreguín-Sánchez, María de la Luz, Palacios-Chavez, Rodolfo, and Quiroz-Garcia, David Leonor
2001 Studies on the morphology of Mexican pteridophyte spores; *In*: Goodman, D.K., and Clarke, R.T. (eds.), *Proceedings of the IX International Palynological Congress, Houston, Texas, U.S.A., 1996*; American Association of Stratigraphic Palynologists Foundation, p. 597–608.
 ISBN 0–931871–06–9

Arreguín-Sánchez (1983, 1985a, b) presented photomicrographs (light microscopy) and palynological keys, and described the relationship between the taxonomic and palynological studies of the cheilanthoid ferns and the subfamilies Dryopteridoideae and Aspleniaceae from Valle de Mexico. The general characters of many pteridophytes, including 148 species distributed in Mexico, were described by Tryon and Lugardon (1990) using both SEM and transmission electron microscope (TEM) photomicrographs. Arreguín-Sánchez and Aguirre-Claverán (1990) studied the morphology of trilete spores of isosporous pteridophytes from Nuevo León and presented descriptions, light microscope and SEM photomicrographs, and palynological keys. They discussed the taxonomic position of some taxa and some interpretations based on spore morphology. Arreguín-Sánchez (1991) studied the spore morphology of 103 species of pteridophytes from Valle de México, and in 1996 studied 119 species of isosporous pteridophytes from Queretaro and presented palynological keys, light microscope and SEM photomicrographs, descriptions, and commentaries on taxonomic and palynological studies.

Other papers which include Mexican taxa and very short descriptions of spores are the floristic studies by Knobloch and Correll (1962) for Chihuahua, Wiggins (1980) for Baja California, Smith (1981) for Chiapas, Mickel and Beitel (1988) for Oaxaca, Aguirre-Claverán and Arreguín-Sánchez (1988) for Nuevo León, Mickel (1992) for Nueva Galicia, and Ameneyro et al. (1995) for Querétaro.

RESULTS

The list in Table 1 is derived from the publications above and shows species that occur in Mexico and which have spore morphology described or illustrated in the 14 keyed publications listed in the bibliography.

Taxa found in Mexico for which spore morphology is not described in the literature are listed below:

Adiantum *amblyopteridium, A. braunii, A. caryotideum, A. clavatum, A. decoratum, A. deflectens, A. dolosum, A. feei, A. fructuosum, A.*

galeottianum, A. latifolium, A. lunulatum, A. oaxacanum, A. obliquum, A. petiolatum, A. princeps, A. raddianum, A. seemannii, A. terminatum, A. trichochlaenum, A. villosum, A. wilesianum.

Alsophilla myosuroides, A. salvinii.

Ananthacorus angustifolium.

Anemia anthriscifolia, A. familiaris, A. karwinskiana, A. munchii, A. paraphyllitidis, A. pastinacaria, A. semihirsuta, A. speciosa, A. tomentosa var. mexicana.

Anogramma chaerophylla.

Antrophyum cajenense, A. chlorosporum, A. ensiforme, A. guayanense, A. lanceolatum, A. lineatum

Asplenium abscissum, A. achilleifolium, A. adiantum-nigrum, A. breedlovei, A. cirrhatum, A. conquisitum, A. dentatum, A. dissectum, A. divaricatum, A. flabellatum var. flabellatum, A. flabellatum var. partitum, A. formosum, A. fournieri, A. hallbergii, A. harpeodes, A. harrisii, A. hesperium, A. heterochroum, A. hoffmannii, A. insolitum, A. juglandifolium, A. laetum, A. lamprocaulon, A. minimum, A. miradorense, A. modestum, A. monodon, A. munchii, A. myriophyllum, A. oligosorum, A. olivaceum, A. otites, A. pyramidatus, A. polyphyllum, A. potosinum, A. pringlei, A. pteropus, A. radicans, A. repandulum, A. rutaceum, A. salicifolium, A. serratum, A. soleirolioides, A. solmsii, A. stolonipes, A. tryonii, A. tuerckheimii, A. uniseriale, A. virillae, A. williamsii, A. yelagagense.

Athyrium arcuatum, A. barnebyanum, A. bourgae.

Azolla mexicana.

Blechnum divergens, B. ensiforme, B. falciforme, B. fragile, B. fraxineum, B. glandulosum, B. gracile, B. lehmannii, B. l'herminieri, B. polypodioides, B. schiedeanum, B. varians, B. wardiae, B. serrulatum.

Bolbitis aliena, B. bernoullii, B. hastata, B. pergamentacea, B. serratifolia.

Bommeria knoblochii.

Botrychium decompositum.

Campyloneurum amphostenon, C. costatum, C. ocultum, C. serpentinum.

Cheilanthes aliena, C aurea (N), C. baitelii, C. bolborrhiza, C. brachypus, C. chaerophylla, C. cinnamonea, C. complanata, C. crassifolia, C. cuneata, C. decomposita, C. feei, C. fendleri, C. fimbriata, C. grayi (N), C. hirsuta, C. hookeri, C. jamaicensis, C. lemmonnii (N), C. lerstenii, C. limitanea (N), var. limitanea, C. longipila, C. lozanii var. lozanii, C. membranacea, C. mickelii, C. pallens, C. pringlei, C. pyramidalis var. arizonica, C. rosei, C. villosa, C. weatherbiana (N), C. wootonii, C. wrightii. (N) before Notholaena.

Cheiloplecton rigidum var. lanceolatum.

Cibotium wendlandii.

Cnemidaria apiculata, C. decurrens.

Cochlidium rostratum, C. serrulatum.

Ctenitis baulensis, C. bullata, C. chiapasensis, C. equestris var. equestris, C. equestris var. erosa, C. excelsa, C. grisebachii, C. hemslayana, C. interjecta, C. lanceolata, C. nigrovenia, C. pansamalensis, C. pulverulenta, C. refulgens, C. salvinii, C. sloanei, C. strigilosa, C. subincisa, C. thelypteroides, C. ursina

Cyathea bicrenata, C. divergens var. tuerckheimii, C. princeps, C. salvinii, C. valdecrenata, C. microdonta.

Cystopteris membranifolia.

Danaea cuspidata, D. elliptica.

Dennstaedtia dissecta.

Dicksonia gigantea

Dicranopteris flexuosa, D. pectinata.

Dictyoxiphium panamense

Diplazium cristatum, D. donnell-smithii, D. drepanolobium, D. entecnum, D. expansum, D. franconis, D. herbaceum, D. hellwigii, D. induratum, D. laciniatum, D. lindbergii, D. lonchophyllum, D. neglectum, D. obscurum, D. plantaginifolium, D. prominulum, D. puberulentum, D. riedelianum, D. striatastrum, D. striatum, D. ternatum, D. urticifolium, D. vera-pax, D. werckleanum.

Doryopteris concolor.

Dryopteris futura, D. karwinskyana, D. munchii, D. nubigena, D. parallelogramma, D. rosea, D. rossii, D. simplicior.

Elaphoglossum acutissimum, E. alan-smithii, E. albomarginatum, E. auricomum, E. chiapense, E. clewellianum, E. decursivum, E. engelii, E. erinaceum, E. glabellum, E. glaucum, E. gratum, E. guatemalense, E. huacssaro, E. ipshookense, E. lanceum, E. leonardii, E. lepidopodum, E. lindenii, E. parduei, E. piloselloides, E. pringlei, E. revolutum, E. rigidum, E. rubescens, E. sartorii, E. seminudum, E. setosum, E. squamipes, E. tambillense, E. tectum, E. tenuifolium, E. vestitum.

Equisetum laevigatum.

Eriosorus hirtus.

Gleichenia palmata, G. underwoodiana.

Grammitis anfractuosa, G. apiculata, G. basiattenuata, G. blepharodes, G. cultrata, G. delicatula, G. deltata, G. firma, G. harrisii, G. hellwigii, G. katasophistes, G. leptostoma, G. margaritata, G. melanosticta, G. mollissima, G. myosuroides, G. oidiophora, G. parietina, G. prionodes, G. pruinosa, G. semihirsuta, G. subtilis, G. trichomanoides, G. xiphopteroides, G. zempoaltepetlensis

Hemionitis elegans.

Hymenophyllum abruptum, H. asplenioides, H. brevifrons, H. brevistipes, H. crassipetiolatum, H. crispum, H. ectocarpon, H. elagantulum, H. fendlerianum, H. fragile, H. hirsutum, H. lanatum, H. microcarpum, H. myriocarpum, H. nigrescens, H. pulchellum, H. radicans, H. sieberi, H. siliquosum, H. trapezoidale, H. undulatum.

Hypolepis blepharochlaena, H. eurychlaena, H. melanochlaena, H. microchlaena, H. pulcherrima, H. thysanochlaena, H. trichochlaena.

Isoëtes montezumae, I. pringlei, I. triangula.

Jamesonia alstonii

Lastreopsis chontalensis

Lindsaea arcuata, L. lancea, L. quadrangularis, L. quadrangularis ssp. *antillensis, L. quadrangularis spp. alata, L. stricta.*

Lomariopsis fendleri, L. mexicana.

Loxosxaphe theciferum

Lycopodium alopecuroides, L. bradeorum, L. crassum, L. cuernavacense, L. dichoromum, L. guatemalense, L. hippurideum, L. mexicanum, L. myrsinites, L. orizabae, L. pithyoides, L. pringlei, L. reflexum, L. serratum, L. skutchii, L. taxifolium, L. thyoides, L. tuerckheimii, L. wilsonii

Marsilea ancylopoda, M. crotophora, M. deflexa, M. mucronata, M. polycarpa.

Microgramma lycopodioides, M. nitida, M. percusa, M. tecta.

Nephrolepis hirsutula, N. pectinata, N. pendula.

Neurodium lanceolatum.

Odontosoria schlechtendalii.

Olenadra guatenalensis.

Olfersia cervina.

Onocleopsis hintonii.

Ophioglossum palmatum, O. crotalophoroides, O. nudicaule, O. petiolatum, O. pycnostichum.

Pecluma consimilis, P. cupreolepis, P. divaricata, P. ferruginea, P. ptilodon.

Pellaea seemannii.

Peltapteris peltata.

Phanerophlebia lindenii.

Phlebodium areolatum, P. decumanum.

Phyllitis lindenii, P. scolopendrium var. serratifolia.

Pityrogramma dealbata.

Plagiogyria pectinata, P. truncata

Pleopeltis angusta var. angusta, P. angusta var. stenoloma, P. conzattii, P. fallax, P. macrocarpa var. trichophora, P. mexicana, P. munchii

Polybotrya aucuparia, P. caudata, P. polybotryoides.

Polypodium adelphum, P. alavae, P. arcanum, P. bolivianum, P. chiapense, P. christensenii, P. collinsii, P. colysoides, P. consimile, P. conterminens, P. diplotrichum, P. eatonii P. echinolepis, P. eperopeutes, P. erythrolepis, P. ferrugineum, P. fissidens, P. furfuraceum, P. fuscopetiolatum, P. glaberulum, P. hispidulum, P. hygrometricum, P. lepidotrychum, P. lindenianum, P. longipinnulatum, P. madrense, P. montigenum, P. oulolepis, P. pleopeltis, P. pleurosorum, P. polypodioides var.

michauxianum, P. polypodioides var. polypodioides, P. puberulum, P. pyrrholepis, P. remotum, P. rhachipterygium, P. sanctae-rosae.
Polystichum drepanoides, P. erythrosorum, P. fournieri, P. furfuraceum, P. hartwegii, P. mickelii, P. munchii, P. ordinatum, P. platyphyllum, P. rachichlaena, P. smithii, P. drepanoides.
Pteris chiapensis, P. erosa, P. muricata, P. muricella, P. paucinervata, P. podophylla, P. pulchra, P. pungens.
Salvinia minima.
Schizaea elegans.
Selaginella arizonica, S. bernoullii, S. chiapensis, S. cladorrhizans, S. corrugis, S. cuneata, S. disticha, S. eurynota, S. finitima, S. flagellata, S. guatemalensis, S. harrisii, S. hoffmannii, S. landii, S. lindenii, S. lineolata, S. macrathera, S. mixteca, S. mollis, S. mosorongensis, S. novae-hollandiae, S. novoleonensis, S. oaxacana, S. pilifera, S. pulcherrima, S. reflexa, S. rhizophora, S. schiedeana, S. schizobasis, S. sertata, S. silvestris, S. stellata, S. steyermarkii, S. subrugosa, S. tarda, S. troodophora, S. underwoodii, S. viridissima, S. wendtii
Stigmatopteris chimalapensis, S. longicaudata, S. sordida.
Tectaria mexicana, T. transiens, T. vivipara.
Thelypteris albicaulis, T. atrovirens, T. balbisii, T. biolleyi, T. blanda, T. blepharis, T. cheilanthoides var. mucosa, T. cinerea, T. cretacea, T. ghiesbreghtii, T. glandulosa var. brachyodus, T. grandis, T. hatchii, T. hispidula, T. imbricata, T. kunthii, T. melanochlaena, T. meniscioides, T. minor, T. munchii, T. nicaraguensis, T. nubigena, T. oaxacana, T. obliterata, T. oligicarpa, T. ovata, T. paucipinnata, T. pilosa var. alabamensis, T. pilosohispida, T. pilosula, T. poiteana, T. reptans, T. resiliens, T. resinifera, T. reticulata, T. rudis, T. scalaris, T. serrata, T. standleyi, T. stolzeana, T. struthiopteroides, T. tablana, T. tetragona, T. thomsonii, T. toganetra, T. tuerckheilii.
Trichomanes angustatum, T. angustifrons, T. bucinatum, T. capillaceum, T. collariatum, T. crispum, T. diversifrons, T. galeotii, T. godmanii, T. hymenoides, T. hymenophylloides, T. krausii, T. lucens, T. membranaceum, T. ovale, T. petersii, T. pinnatum, T. pyxidiferum, T. rigidum, T. tuerckheimii
Vittaria dimorpha, V. flavicosta, V. lineata

We found that 97 genera and 900 species of pteridophytes have been reported from Mexico, of which 359 species have spore morphology described or illustrated (about one third). The number of pteridophytes that grow in Mexico illustrated in the various listed references are indicated on the following list:

McVaugh (1935)	6
Tryon (1949)	5
Nayar and Devi (1964)	17
Nayar and Devi (1967)	25
Tryon and Tryon (1973)	8
Lloyd (1981)	15
Tryon and Tryon (1982)	146
Arreguín-Sánchez (1983)	16
Arreguín-Sánchez (1985) (Dryopteridoideae)	10
Arreguín-Sánchez (1985) (Asplenoideae)	11
Tryon and Lugardon (1990)	153
Arreguín-Sánchez and Aguirre-Claverán (1990)	65
Arreguín-Sánchez (1991)	99
Arreguín-Sánchez at al. (1996)	133

The total number of taxa derived from the table (709) includes descriptions by various authors of the same species; if we eliminate duplication, the total number is 359 species.

The taxonomic level of study in the various papers is different. Some papers only have photomicrographs and short descriptions of ornamentation, while other papers have lengthy descriptions, palynological keys, photomicrographs, and discussions of taxonomic position and interpretations based on spore morphology.

References Cited

Numbers in brackets following the citation (e.g., [#9]) refer to those references listed in Table 1 and serve as a key to Table 1.

AGUIRRE-CLAVERAN, R., and ARREGUÍN-SÁNCHEZ, M.L.
1988 Claves de familias, géneros, especies y variedades de Pteridofitas del estado de Nuevo León, México. *An. Esc. Nac. Cienc. biól. Méx.*, 322: 9–61.
AMENEYRO, G., R., FERNANDEZ-NAVA, R., and ARREGUÍN-SÁNCHEZ, M.L.
1995 Claves para la identificación de géneros, especies y variedades de pteridofitas del estado de Querétaro, México. *An. Esc. Nac. Cienc. biól. Méx.*, 40: 11–82.
ARREGUÍN-SÁNCHEZ, M.L.
1983 Morfología de las esporas de los helehos cheilanthoides del Valle de México. *An. Esc. nac. Cienc. biól. Méx.*, 27: 9–28. [#8]
1985a Morfología de las esporas de la Subfamilia Dryopteridoideae del Valle de México. *An. Esc. Nac. Cienc. biól. Méx.*, 29: 29–41. [#9]
1985b Morfología de las esporas de la Subfamilia Asplenioideae del Valle de México *In*: Estudios palinológicos y paleoetnobotánicos. *Colección Científica*, 147: 7–22. [#10]
1991 Claves para la separación de los taxa de pteridofitas del Valle de México por medio de sus esporas. *Palynol. et Paleobot.*, 3: 5–54. [#13]
ARREGUÍN-SÁNCHEZ, M.L., and AGUIRRE-CALVERAN, R.
1990 Morfología de las esporas isospóreas triletes de Pteridofitas de Nuevo León, México. *Palynol. et Paleobot.*, 2: 5–52. [#12]
ARREGUÍN-SÁNCHEZ, M.L., FERNANDEZ-NAVA, R. PALACIOS-CHÁVEZ, R., and QUIROZ-GARCÍA, D.L.
1996 Morfología de las esporas de pteridófitas isospóreas del Estado de Querétaro, México. *Polibotánica*, 1: 10–60. [#14]
BENTHAM, G.
1839-1857 *Plantae Hartwegianae*, 393 pgs., London.
DAVENPORT, G.E.
1891 Filices mexicanae. *Gard. and For.*, 4: 448–450; 483–484; 519–520; 555, 556.
FEE, A.L.A.
1844-1866 *Mémoires sur la famille des fougères.* Strasbourg, Berger-Leurault, 1547 pgs.
FOURNIER, E.
1872 Mexicanas Plantas nuper a collectoribus expeditions scientificae allatus aut longis ab annis in herbario Musei Parisiensis depositas, Pars Prima, Cryptogamia. Enumeradas curavit Eug. Fournier. Filices, pp. 59–149. Paris.
KNOBLOCH, I.W., and CORRELL, J.S.
1962 *Fern and Fern Allies of Chihuahua, México.* Texas Research Foundation, Renner, Texas, 198 pgs.
LLOYD, R.M
1981 The perispore in *Polypodium* and related genera (Polypodiaceae). *Canadian Journal of Botany*, 59: 175–187. [#6]
MARTENS, M., and GALEOTTI, H.
1842 Memoire sur les fougères du Mexique et considérations sur la géographie botanique de cette contrée. *Mem. Acad. Sci. Belg.*, 15: 1–99.

McVAUGH, R.
1935 Spores of some Northeastern Ferns. *American Fern Journal*, 25:73–85. [#1]
MICKEL, J.
1992 Pteridophytes Flora Novo-Galiciana. *The University of Michigan Herbarium*, 17: 120–455.
MICKEL, J., and BEITEL, J.
1988 Pteridophyte Flora of Oaxaca, México. *Memoirs of the New York Botanical* Gardens, 46: 1–568.
NAYAR, B.K., and DEVI, D.
1964 Spore morphology of Indian ferns II and III, Aspleniaceae, Blechnaceae, and Polypodiaceae, *Grana Palynologica*, 5(2, 3): 222–246 and 242–395. [#3]
1967 Spore morphology of the Pteridaceae II. The gymnogrammoid ferns. *Grana Palynologica*, 7: 568–600. [#4]
SMITH, A.R.
1981 *Flora of Chiapas Part. 2. Pteridophytes.* The California Academy of Sciences, 370 pgs.
TRYON, A. F.
1949 Spores of the genus *Selaginella* in North America north of Mexico. *Annals of the Missouri Botanical Gardens*, 36: 413–432. [#2]
TRYON, A.F., and LUGARDON, B.
1990 *Spores of Pteridophyta.* Springer-Verlag, New York, Berlin, Heidelberg, 648 pgs. [#11]
TRYON, R.M., and TRYON, A.F.
1973 Geography, spores and evolutionary relations in the cheilanthoid ferns I Jermy, A.C., Crabbe, J.A., and Thomas, B.A., The phylogeny and classification of the ferns. *Linnean Society of London*, p. 145–153. [#5]
1982 *Ferns and allied plants with special reference to Tropical America.* Springer-Verlag, New York, Heidelberg, Berlin, 857 pgs. [#7]
WIGGINS, I.L.
1980 *Flora of Baja California.* Stanford University Press, pp. 51–71.

Authors' addresses

MARÍA DE LA LUZ ARREGUÍN-SÁNCHEZ
RODOLFO PALACIOS-CHÁVEZ
DAVID LEONOR QUIROZ-GARCÍA
Instituto Politécnico Nacional
Escuela Nacional de Ciencias Biológicas
Departamento de Botánica
Plan de Ayala y Carpio
Col. Santo Tomás, México D.F. 11340
Mexico

arregin1@df1.telmex.net.mx
rpalacios@alquimia.encb.ipn.mx
lquiroz@alquimia.encb.ipn.mx

Scholarship of COFAA, Instituto Politécnico Nacional

TABLE 1. Species that occur in Mexico and which have spore morphology described in the 14 publications listed in the bibliography. Numbers in the top row indicate principal literature references (refer to References Cited for key).

	1	2	3	4	5	6	7	8	9	10	11	12	13	14
Acrostichum *aureum*							X				X			
A. danaeifolium							X				X			
Adiantopsis *chlorophylla*							X				X			
A. radiata				X	X		X				X			
Adiantum andicola													X	X
A. capillus-veneris				X							X	X	X	X
A. concinnum				X			X				X	X	X	X
A. macrophyllum				X										
A. patens											X			
A. pedatum	X										X	X		
A. poiretii				X								X		X
A. pulverulentum											X			
A. tenerum				X								X		X
A. tetraphyllum											X			
A. trapeziforme							X				X			X
A. tricholepis												X		X
Anemia *adiantifolia*											X	X		X
A. hirsuta											X			
A. mexicana											X	X		X
A. oblongifolia							X				X			
A. phyllitidis							X				X	X		X
Anitum *citrifolium*							X							
Anogramma *leptophylla*							X				X		X	
Aspidotis *meifolia*					X									
Arachnioides *denticulata*											X			
Asplenium *auriculatum*														X
A. auritum											X			X
A. castaneum							X			X	X		X	
A. commutatum										X			X	
A. cristatum														X
A. cuspidatum														X
A. delitescens							X							X
A. exiguum										X			X	
A. falcinellum											X		X	
A. feei							X							
A. fibrillosum										X			X	
A. lacerum														X
A. monanthes			X							X	X		X	X
A. palmeri										X			X	X
A. praemorsum										X			X	X
A. pumilum										X				X
A. resiliens			X							X			X	X
A. serra							X				X			
A. sessilifolium			X								X		X	X
A. sphaerosporum														X
A. trichomanes											X			

TABLE 1 (continued).

	1	2	3	4	5	6	7	8	9	10	11	12	13	14
Athyrium *dombei*							X				X			
A. filix-femina			X							X	X		X	X
A. palmense											X			
A. skinneri							X							
Azolla *coroliniana*											X			
A. filiculoides							X				X		X	
Blechnum *occidentale*							X				X		X	X
B. serrulatum											X			
B. stoloniferum													X	
Bolbitis *hemiotis*							X							
B. portoricensis							X							
Bommeria *ehrenbergiana*							X				X	X		
B. hispida				X			X				X			
B. pedata							X	X					X	
B. subpalacea							X							
Botrychium *cicutarium*											X			
B. dissectum											X			
B. schaffneri											X			
B. virginianum	X						X				X	X	X	X
Campyloneurum *angustifolium*			X				X				X		X	X
C. phyllitidis			X				X				X			X
C. repens			X											
C. tenuipes														X
C. xalapense														X
Cheilanthes *aemula*												X		
C. affinis														X
C. alabamensis												X		
C. allosuroides(P)								X			X		X	X
C. angustifolia								X				X		X
C. aschenborniana (N)												X		
C. bonariensis				X	X		X	X			X	X	X	X
C. bryopoda (N)												X		
C. candida (N)							X				X	X		X
C. castanea			X									X		
C. cochicensis											X			
C. cucullans								X					X	X
C. delicatula												X		
C. eatonii											X			X
C. farinosa				X			X	X			X	X	X	X
C. formosa (N)												X		
C. galeottii														X
C. greggii (N)							X				X	X		
C. horridula							X					X		
C. incana (N)			X	X	X		X				X	X	X	X
C. integerrima												X	X	X
C. kaulfussii								X				X	X	X
C. lendigera					X			X						
C. leonina (N)												X		

TABLE 1 (continued).

	1	2	3	4	5	6	7	8	9	10	11	12	13	14
C. leucopoda							X				X	X		X
C. limitanea var. *mexicana* (N)												X		
C. lindheimeri														X
C. lozanii var. *seemannii*														X
C. marginata												X	X	X
C. marsupianthes													X	
C. meifolia												X		
C. mexicana													X	
C. microphylla var. *microphylla*								X				X	X	X
C. myriophylla				X			X	X					X	X
C. neglecta (N)												X		
C. notholaenoides												X	X	X
C. rigida												X		
C. parvifola (N)				X							X	X		
C. pyramidalis var. *pyramidalis*								X				X	X	
C. schaffneri (N)												X	X	
C. sinuata var. *sinuata* (N)				X				X			X	X		X
C. sinuata var. *cochisensis* (N)												X		
C. skinneri							X							
C. standleyi (N)												X		
C. sulphurea											X	X		X
C. tomentosa												X		
C. villosa												X		
(N) *before Notholaena;* (P) *before Pellaea*														
Cheiloplecton *rigidum* var. *rigidum*					X							X		X
Cibotium *regale*							X				X			
Ctenitis *melanosticta*											X			X
C. mexicana														X
C. submarginalis							X				X			
Culcita *coniifolia*							X							
Cyathea *costaricensis*							X				X			
C. divergens var. *divergens*														X
C. fulva											X			
C. mexicana							X				X	X		
C. myosuroides							X				X			
C. schiedeana											X			
Cyclopeltis *semicordata*							X							
Cyrtomium see *Phanerophlebia*														
Cystopteris *fragilis* var. *fragilis*	X						X			X	X		X	X
Danaea *nodosa*							X				X			
Dennstaedtia *arborescens*											X			
D. bipinnata							X							X
D. cicutaria											X			X
D. distenta														X
D. globulifera												X		X
D. obtusifolia							X							
Diplazium *hians*							X							
Diplopterygium (see *Gleichenia*)							X							

TABLE 1 (continued).

	1	2	3	4	5	6	7	8	9	10	11	12	13	14
Doryopteris *pedata*				X			X							
Dryopteris *cinnamonea*							X		X			X	X	X
D. paleacea									X			X	X	X
D. patula							X		X			X	X	
D. pseudo-filix-mas									X			X	X	X
D. wallichiana									X			X	X	X
Elaphoglossum *affine*													X	
E. crinipes											X			
E. eximium											X			
E. latifolium							X							
E. lindenii											X		X	
E. mathewsii													X	
E. muelleri														X
E. muscosum													X	
E. paleaceum											X		X	
E. petiolatum														X
E. pilosum													X	
Equisetum *giganteum*							X				X			
E. hyemale var. *affine*											X		X	X
E. myriochaetum														X
Eriosorus *flexuosus*											X			
Gleichenia *bancroftii*											X			
G. bifida							X				X			
Grammitis *asplenifolia*							X				X			
G. delitescens							X				X			
G. heteromorpha													X	
G. jungermannioides							X				X			
G. moniliformis							X				X			
G. pilosissima													X	
Gymnopteris (see *Hemionitis*)											X			
Hecistopteris *pumila*											X			
Hemidictyum *marginatum*							X				X			
H. levyi							X							
H. palmata				X			X					X		X
H. pinnatifida							X				X			
H. rufa							X							
H. subcordata							X				X			
Histiopteris *incisa*							X				X			
Holodictyum *ghiesbreghtii*							X				X			
Hymenophyllum *fucoides*							X							
H. polyanthos											X			
H. tunbrigense							X				X			
Hypolepis *bogotensis*							X							
H. nigrescens							X							
H. repens							X							
Isoetes *cubana*							X				X			
I. mexicana											X		X	
Lastreopsis *effusa*							X				X			

TABLE 1 (continued).

	1	2	3	4	5	6	7	8	9	10	11	12	13	14
L. exculta							X							
Lindsaea *portoricensis*							X							
Llavea *cordifolia*							X				X	X		X
Lomariopsis *recurvata*							X							
Lonchitis *hirsuta*											X			X
Lophosoria *quadripinnata*							X				X			X
Loxogramme *mexicana*							X				X			
Lycopodium *cernuum*							X							
L. clavatum							X				X			X
L. linifolium							X							
Lygodium *heterodoxum*							X							
L. venustum							X							X
Marattia *excavata*							X				X			
M. laxa							X							
M. weinmannifolia							X				X			X
Marsilea *mollis*													X	
Metaxya *rostrata*											X			
Microgramma *piloselloides*							X				X			
M. reptans							X							
Mildella *intramarginalis* var. *intramarginalis*					X									X
M. intramarginalis var. *serratifolia*								X				X	X	X
Nephelea (see *Cyathea*)														
Nephrolepis *biserrata*							X				X			
N. cordifolia											X			
N. exaltata							X				X			
N. multiflora											X			
N. occidentalis							X							
N. rivularis											X			
Niphidium *crassifolium*							X				X			X
Notholaena (see *Cheilanthes*)														
Odontosoria *guatemalensis*							X							
Oleandra *articulata*							X							
Ophioglossum *crotalophoroides*											X			
O. engelmannii											X		X	
O. petiolatum													X	
O. reticulatum											X		X	
O. vulgatum	X										X			
Osmunda *cinnamomea*											X			
O. regalis var. *spectabilis*	X						X				X			
Pecluma *alfredii*														X
P. atra														X
P. dispersa							X							
P. plumula			X											X
Pellaea *cordata*												X		
Pellaea *atropurpurea*	X			X								X		X
P. cordifolia											X		X	X
P. intermedia											X	X		
P. longimucronata				X										

TABLE 1 (continued).

	1	2	3	4	5	6	7	8	9	10	11	12	13	14
P. notabilis											X	X		
P. ovata				X				X				X	X	X
P. sagittata				X			X	X			X		X	X
P. skinneri					X						X	X		
P. ternifolia var. *ternifolia*				X			X	X			X	X	X	X
P. ternifolia var. *wrightiana*					X									
Phanerophlebia *auriculatum*							X				X			
P. juglandifolia							X				X			
P. macrosora							X							X
P. nobilis							X		X		X		X	X
P. remotispora											X			X
P. umbonata							X				X			
Phlebodium *areolatum*														X
P. aureum			X										X	
Pityrogramma *colomelanos*				X			X				X			X
P. tartarea				X							X		X	X
P. triangularis				X			X				X			
P. trifoliata							X				X			
Plagiogyra *semicordata*							X				X			
Plecosorus *speciosissimus*									X				X	
Pleopeltis *astrolepis*														X
** *P. crassinervata*														X
** *P. interjecta*														X
P. macrocarpa var. *macrocarpa*			X				X				X		X	X
** *as P. macrocarpa* var. *crassinervata or* var. *interjecta*							X							
P. polylepis											X		X	X
P. revoluta							X							
Polybotrya *cervina*											X			
Polypodium *californicum*						X								
P. cryptocarpum														X
P. cupreolepis													X	
P. dissimile														X
P. fallax						X								
P. fraternum			X											X
P. fraxinifolium			X				X							
P. guttatum													X	X
P. hartwegianum						X							X	
P. loriceum						X	X							
P. madrense						X								
P. martensii						X							X	X
P. oulolepis													X	X
P. platylepis													X	
P. plebeium						X							X	X
P. plesiosorum							X				X			X
P. polypodioides						X							X	
P. polypodioides Var. *aciculare*														X
P. ptilodon						X								
P. rhodopleuron						X								X

TABLE 1 (continued).

	1	2	3	4	5	6	7	8	9	10	11	12	13	14
P. ptilodon						X								
P. rhodopleuron						X								X
P. rosei						X							X	X
P. subpetiolatum													X	X
P. thyssanolepis						X							X	X
P. triseriale			X			X								
P. vulgare			X			X								
Polystichum *distans*									X				X	X
P. muricatum											X			
Psilotum *complanatum*						X					X		X	
P. nudum											X			
Pteridium *arachnoideum* **							X				X			X
*P. caudatum***							X				X	X		X
*P. feei***							X							
P. pubescens **													X	X
** *as Pteridium aquilinum* var. *arachnoideum, caudatum, feei, pubescens*														
Pteris *altissima*							X				X			
P. cretica												X	X	X
P. biaurita											X			
P. grandifolia							X				X			X
P. longifolia							X				X	X		X
P. orizabae													X	X
P. orizabae													X	X
P. quadriaurita							X							X
Saccoloma *elegans*							X							
S. inaequale							X							
Salvinia *auriculata*							X				X			
Schaffneria *nigripes*							X				X			X
Selaginella *apoda*		X									X			
S. arsenei													X	
S. delicatissima							X				X		X	
S. galeottii							X				X			
S. lepidophylla		X					X						X	
S. martensii											X			
S. microdendron							X							
S. minima							X				X			
S. pallescens											X		X	
S. peruviana		X											X	
S. porphyrospora							X							
S. rupincola		X									X		X	
S. sartorii							X				X			
S. sellowii														X
S. tarapotensis							X							
S. wrightii		X											X	
Sphaeropteris (see *Cyathea*)														
Sticherus (see *Gleichenia*)														
Tectaria *heracleifolia*							X						X	

TABLE 1 (continued).

	1	**2**	**3**	**4**	**5**	**6**	**7**	**8**	**9**	**10**	**11**	**12**	**13**	**14**
T. incisa							X							
Thelypteris *cheilanthoides* var. *Cheilanthoides*							X							
T. concinna														X
T. deflexa							X							
T. dentata							X							
T. falcata							X							
T. interrupta							X							
T. ovata var. *lindheimeri*														X
T. linkiana							X							
T. patens									X				X	
T. pilosa var. *pilosa*													X	X
T. pilosa var. *major*									X				X	
T. puberula var. *puberula*														X
T. torresiana														X
Trichipteris (see *Cyathea*)														
Trichomanes *polypodioides*							X				X			
T. radicans														X
T. reptans							X							
Trismeria *trifoliata*				X										
Vittaria *graminifolia*							X				X			X
Woodsia *mexicana*													X	X
W. mollis							X						X	X
W. plummerae							X							
Woodwardia *fimbriata*											X		X	
W. martinezii							X							
W. spinulosa														X

FORENSIC PALYNOLOGY: THE RUIDOSO PLANE CRASH

Alan Graham

Abstract

The crash of a private airplane in New Mexico in 1989 was blamed on a pellet of biological material found in the fuel control unit. The material was composed of pollen, trichomes, and insect remains. Study revealed that it was a nest of the bee *Ashmeadiella* built after the crash and while the wreckage was in an outdoor storage facility.

INTRODUCTION

A Beechcraft Super Kingair F90 crashed on December 2, 1989, while on its way from San Diego, California, to the Sierra Blanca Regional Airport in Ruidoso, New Mexico (Brunk, 1997). The airplane also had landed a few weeks before at a remote airport at Sand Springs, Montana. The pilot and his wife were on their way to a vacation home near Ruidoso. There was blowing snow, a low 800 ft ceiling, and visibility was less than 1.5 miles. The instrument approach to the Ruidoso airport is difficult, and the NDB (Non-Directional Beacon) instrumentation on the aircraft is an older type of navigational aid. The pilot had logged less than 100 hours of flight time in this plane. The plane lost the NDB and was heard flying off course above the clouds. It was then seen emerging from the clouds, heading straight down at full power, to a fiery crash in the mountains.

The airplane had two Pratt and Whitney turboprop engines with a FCU (fuel control unit) made by Allied Signal. The wreckage was recovered by the NTSB (National Transportation and Safety Board) and sent to Pratt and Whitney in Canada for examination, No defects in the engine were found. The wreckage was trucked back to the Sierra Blanca Regional Airport in the spring of 1990 and dumped in an outdoor storage yard. It remained there uncovered through the spring and summer except for some soil which had been shoveled onto the wreckage and which had been blown in by wind.

In the fall of 1990, a group of attorneys from Philadelphia visited the site. They found a mass of biological material ca. 4 mm in diameter in a metal tube, later designated as the B2 elbow, that was part of the FCU. A lawsuit was filed in San Diego on behalf of the two adult children of the dead couple. It was the contention of the plaintiffs' attorneys that the engine had ingested biological contaminants from the air, and that these materials had blocked the fuel line and caused the plane to crash.

There was some preliminary evidence suggesting it was unlikely the biological material could have gotten into the line and remained there during the operation of the engines. It would have had to pass through compression blades turning at tens of thousands of RPMs; the average temperature in the line is ca. 260°C (500°F), which likely would have incinerated any biological material; and air pressure through the line is ca. 300 PSI, or ca. 10x the pressure in a fully inflated automobile tire. It was assumed that the biological material could not have remained in the B2 elbow at this pressure. However, an elaborate apparatus was designed at the Garrett Fluid Systems Division in Tempe, Arizona, to remove the mass with a jet of forced air so it could be divided between plaintiffs and defendants for further study. The jet of air was admittedly at pressures less than those found in the B2 elbow during normal operations, but the mass did not budge.

In contrast to the presumption of innocence in criminal trials, in civil trials it often is not enough just to counter a plaintiff's theory. It is usually necessary to offer the jury a plausible alternative explanation. Thus, a team of scientists was assembled and given the tasks of (1) identifying the various components of the biological material, (2) determining how it got into the fuel line, and (3) assessing whether it was a pre-crash or post-crash accumulation.

THE POLLEN EVIDENCE

Within the mass of biological material there was embedded a clump of pollen ca. 2 mm in diameter. Lewis (1997) identified the following pollen types: *Grindelia* (curlycup gumweed, Asteraceae, 62.24%), *Melilotus* (yellow sweet clover, Fabaceae, 36.56%), *Sphaeralcea* (globemallow, Malvaceae, 0.82%), and trace amounts of a few other types. TEM (transmission electron microscopy) revealed that some of the pollen still retained the intine and protoplast. Germination experiments using a 10% sucrose solution were conducted on pollen from the B2 elbow, and pollen tubes emerged from the apertures of *Grindelia* and *Melilotus*. The plants all represent insect-pollinated types that flower between May and October, and all were found growing in or near the outdoor storage facility. Recall that the wreckage had been stored there through the spring and summer of 1990.

By coincidence, an airborne pollen survey had been conducted at Hobbs, New Mexico, southeast of Ruidoso (Lewis et al., 1990). Pollen in the atmosphere consisted of wind-pollinated trees and shrubs (*Juniperus*, 13.7%; *Pinus*, 7.1%; *Morus*, 41.2%; *Quercus*, 3.2%) and herbs (Poaceae, 11.1%; Amaranthaceae/Chenopodiaceae, 14.7%; *Ambrosia*, 3.0%). No pollen of entomophilous species were present. There is no known mechanism whereby pollen of insect-pollinated plants, which is rare to absent in the atmosphere, could selectively be filtered out to the exclusion of the vastly more abundant wind-pollinated types. Furthermore, the

Graham, Alan
2001 Forensic palynology: The Ruidoso plane crash; *In*: Goodman, D.K., and Clarke, R.T. (eds.), *Proceedings of the IX International Palynological Congress, Houston, Texas, U.S.A., 1996*; American Association of Stratigraphic Palynologists Foundation, p. 609–612.

ISBN 0–931871–06–9

plane was flying in the winter, during snow, when production of any pollen would have been nil.

When I first examined the pollen pellet in the biological mass at Seal Laboratories in Los Angeles in April, 1992, it had just been dislodged from the B2 elbow using needles and forceps; it was to be divided between plaintiffs and defendants. Under the dissecting microscope two features of the pollen were immediately apparent—it was glistening and golden yellow in color, and there were strands of a sticky substance connecting the grains (Graham, 1997). It is well known that pollen grains and spores darken upon heating and, indeed, this is a commonly used measure of thermal maturity as shown in the diagrams of Pearson (1984) and Sengupta (1978; see also Traverse, 1988, p. 417). Pollen of *Hibiscus* (Malvaceae, the same family to which *Sphaeralcea* belongs) was heated to 80°C for four hours. This is a much lower temperature than that found in the line of the B2 elbow (260°C). The heated pollen was noticeably darker than fresh *Hibiscus* pollen, and the sticky strands connecting the grains had disappeared. Furthermore, the post-crash fire was so hot that it melted fiberglass (400°C) and aluminum (500°–600°C) parts of the plane. It was clear from the palynological evidence that the pollen was a post-crash accumulation and was unrelated to the crash of the plane. As noted earlier, however, in this civil case it was necessary to clearly explain the origin of the biological mass in the B2 elbow.

OTHER BIOLOGICAL EVIDENCE

In addition to pollen, the mass consisted of microscopic, finely branched structures and was underlain by a dark amorphous substance. Scanning electron microscopy (SEM) revealed the branched structures to be trichomes of *Sphaeralcea* (Bates et al., 1997).

Two other members of the defense team (Rozen and Eikwort, 1997) were entomologists. They were familiar with the fact that bees of the genus *Ashmeadiella* build nests in tubular structures; in fact, one way to collect the bees is to set soda straws in the ground where they will construct the nests. These nests consist of trichomes that, in this region, are often gathered from *Sphaeralcea*. A pellet of pollen is placed in the mass as food for the developing insects, and the eggs are laid. In other wreckage from the storage yard, remains of the bee were recovered, and a hair of *Ashmeadiella* with pollen attached was found in the B2 elbow. A convincing case developed that the biological mass was a nest of *Ashmeadiella* which, like the pollen, probably would have been altered by the heat in the line, and certainly would have been destroyed by the post-crash fire.

THE CHEMICAL EVIDENCE

Two other components of the nest still required identification and explanation. One was the dark amorphous mass underlying the trichomes and pollen. Spectral analyses performed at Seal Laboratories revealed that the mass was similar to leaves of *Sphaeralcea* (Liddell, 1997). The bees often use masticated leaves of this plant to cement the nest in place. As shown by the experiment at Garrett Fluid Systems Division, the substance is very effective as an adhesive agent. Liddell (1997) also demonstrated by ft-Infrared (FTIR) spectroscopic analysis done at the Seal Laboratories that spectra from the strands connecting pollen grains in the B2 elbow, and spectra from commercial honey were virtually identical (Text-Figures 1 and 2).

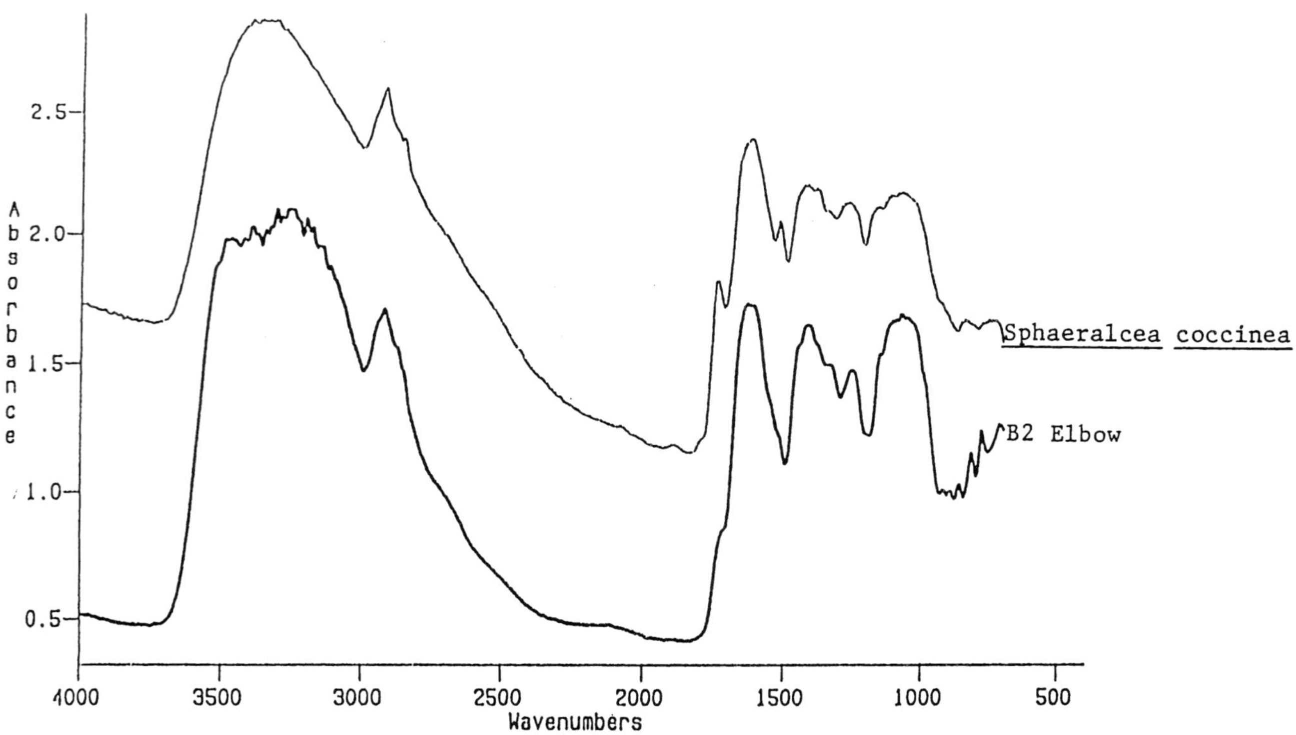

Text-Figure 1. FTIR spectrogram of *Sphaeralcea coccinea* leaves and the dark amorphous mass from the B2 elbow. From Liddell (1997).

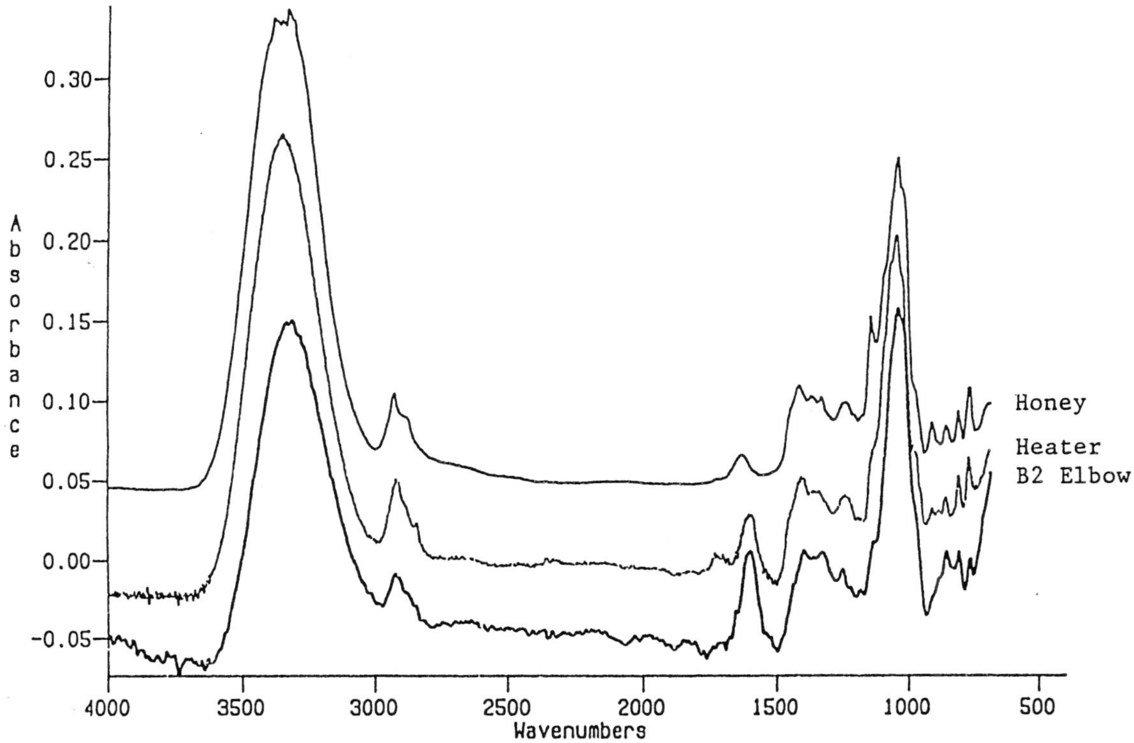

Text-Figure 2. FTIR spectrogram of honey, and the glistening strands connecting the pollen in the heater and the B2 elbow. From Liddell (1997).

THE SOIL ALTERNATIVE

During the course of the trial the Plaintiffs proposed a second explanation for the crash. It was suggested that (1) because soil was found in the wreckage of the engine, it had been ingested and caused the engine to malfunction, and (2) that the soil came from the vicinity of the previous landing site in Montana, and was not soil blown in from the vicinity of the open storage yard during several months of exposure, or soil that had been used to cover the engine. A soil expert on the defense team used EDX x-ray (energy dispersive x-ray) maps and spectra of soil from the wreckage, the storage yard, and the Montana site to assess this theory (Daugherty, 1997). The generalized silicate-based soil was similar at all three sites including, in particular, the soil from the engine and the storage yard. The soils were subjected to heat of up to 1,000°C in the laboratory to simulate the post-crash fire, and the iron components oxidized to a red color. Soil from inside the engine wreckage was not red.

THE VERDICT

As recounted by Brunk (written commun., 1996), late in the trial, after part of the biological evidence had been presented, the judge called the attorneys to sidebar and wanted to know 'how many torpedoes we wanted to put in a sinking ship?' The case was turned over to the jury and a verdict was rendered after eight days of deliberation. When the jury was impaneled the defense was delighted to get an analytical chemist and a retired McDonnell Douglas employee on the jury, who supposedly would understand the science and be sympathetic to the aviation industry. The analytical chemist was elected foreman. The verdict was 10–2 for the defense (a 9–3 vote is required for civil verdicts). The two holdouts were the analytical chemist and the McDonnell Douglas employee. Two of the strongest supporters were a body shop owner and a car mechanic who believed it impossible that contamination could accumulate at the air pressures operating in the system.

The Ruidoso plane crash case is another example in the growing list of instances where, under certain circumstances, palynology can provide useful evidence in court proceedings. The information was particularly important in this case because of the large amounts of money involved, the potential awards, insurance ramifications, the establishment of case law, decisions of the multinational corporations involved regarding the manufacturing of small aircraft, and the future of the private plane industry.

References Cited

BATES, D.M., ANDERSON, G.J., and LEE, R.D.
1997 Forensic botany: trichome evidence. *Journal of Forensic Sciences,* 42: 380–386.

BRUNK, S.K.
1997 The Ruidoso plane crash—The background and the trial verdict. *Journal of Forensic Sciences,* 42: 378–379.

DAUGHERTY, L.A.
1997 Soil science contribution to an airplane crash investigation, Ruidoso, New Mexico. *Journal of Forensic Sciences,* 42: 401–405.

GRAHAM, A.K.
1997 Forensic palynology and the Ruidoso, New Mexico plane crash—the pollen evidence II. *Journal of Forensic Sciences,* 42: 391–393.

LEWIS, W.H.
1997 Pollen composition in a crashed plane's engine. *Journal of Forensic Sciences,* 42: 387–390.
LEWIS, W.H., DIXIT, A.B., and WARD, W.A.
1990 Distribution and incidence of North American pollen aeroallergens. *American Journal of Otolaryngology,* 12: 205–226.
LIDDELL, C.M.
1997 Field sampling and chemical analysis. *Journal of Forensic Sciences,* 42: 398–400.
PEARSON, D.L.
1984 Pollen/spore color "standard." Version 2. Phillips Petroleum Company. Privately circulated.
ROSEN, J.C., and EIKWORT, G.C.
1997 The entomological evidence. *Journal of Forensic Sciences,* 42: 394–397.
SENGUPTA, S.
1978 Experimentally matured *Lycopodium.* Privately circulated.
TRAVERSE, A.
1988 *Paleopalynology.* Unwin Hyman, Boston. 600 pp.

Author's address

ALAN GRAHAM
Department of Biological Sciences
Kent State University
Kent, OH 44242
U.S.A.

FORENSIC PALYNOLOGY: ASSESSING THE WEIGHT OF THE EVIDENCE

Mark Horrocks and Kevan A. J. Walsh

Abstract

The main forensic application of palynology is in determining the possibility of associative evidence. Forensic palynologists should assess the significance of their evidence to assist the court in determining the guilt or innocence of the defendant. They also need to present this assessment in language which can be readily understood by judges and juries. The purpose of this paper is to establish a framework for making such assessments using the likelihood ratio (LR). The LR forms part of the assessment of forensic evidence using Bayes' Theorem and considers the quality of the pollen "match" and how common the pollen assemblage is. A pollen case is used as an example. A more detailed version of this approach will is given in the *Review of Palaeobotany and Palynology, Special Issue: New Frontiers and Applications in Palynology* (Horrocks and Walsh, 1998).

INTRODUCTION

Various methods and examples of forensic palynology have been described by Bryant et al. (1990), Mildenhall (1982, 1988, 1990), Stanley (1992, 1993), and Horrocks et al. (1997, 1998). Its main application is in providing associative evidence, i.e., assisting to prove or disprove a link between people and objects with places or with other people. For example, soil on a suspect's clothing can be analyzed for pollen and compared with control soil samples from the crime scene.

There are two very important considerations when presenting pollen evidence. Firstly, all information needs to be presented in plain, non-scientific language which is easily understandable by lay people. Secondly, scientists should assist the judge or jury to assess the weight of the evidence for or against the defendant. Palynologists finding a "match" between pollen samples may conclude that the samples are "palynologically identical" or "came from the same type of environment." Such conclusions meet neither requirement, and therefore do little to assist the judge or jury.

The aim of this paper is to establish a framework for assessing the strength of pollen evidence, using a case as an example. Forensic palynologists need to clearly state *how* strong their evidence is in simple language so that non-scientists can fully appreciate its significance.

DISCUSSION

Most types of forensic evidence, including pollen, involve a two-step process when comparing samples:

1. do the samples **match**? (Yes/no/inconclusive)
2. how **significant** is this evidence?

With some evidence types, if two samples match, then the evidence may prove conclusive association between the samples. If the samples are so different that they clearly do not match, then this evidence may exclude any association between the samples. Some types of evidence, however, involve so many variables that a perfect match is rarely or never found. Also, although there may be a match between two samples, this does not mean that the samples are unique to any particular source. Pollen evidence has these problems. As pollen samples may contain up to fifty or more different variables (pollen types), even samples known to be from the same environment will not be identical in terms of pollen amounts and types because of random sampling variation. For example, a pollen type of low abundance in an environment may be found in one sample but not in another sample taken from that environment.

When the pollen evidence is exclusionary, the assessment of the evidence is straightforward. For example, if a soil sample from the suspect's clothing does not contain pine pollen (a widely dispersed, abundantly produced pollen type) and the crime scene was in a pine forest, we can say with certainty that the suspect (or strictly, the suspect's clothing) had not been at the crime scene. If the evidence is not exclusionary, then the assessment of the significance of the evidence is difficult.

A very useful approach to assessing the significance of the evidence is to consider two questions and their significance:

1. what is the **quality** of the pollen match? (i.e., how many pollen types are common to both samples and how close were their relative abundance?)
2. how **common** is the pollen assemblage? (i.e., how common would it be to find such a pollen assemblage in other samples unrelated to the crime?)

These two questions are essentially the primary components of the likelihood ratio (LR). The LR forms part of the assessment of forensic evidence using Bayes' Theorem. This is used for many evidence types, such as glass, fibres and DNA (Evett, 1986; Evett and Buckleton, 1989, 1990; Buckleton et al., 1991; Lindley, 1977; Robertson and Vignaux, 1995). The LR is simply a ratio of probabilities of the evidence given alternative hypotheses and is formally written in statistical nomenclature as:

$$\frac{p(E|C)}{p(E|\overline{C})}$$

Horrocks, M. and Walsh, K.A.J.
2001 Forensic palynology: Assessing the weight of the evidence; *In*: Goodman, D.K., and Clarke, R.T. (eds.), *Proceedings of the IX International Palynological Congress*; American Association of Stratigraphic Palynologists Foundation, p. 613–615.
 ISBN 0–931871–06–9

where p denotes a probability; E is the scientific evidence; the vertical bar separates the event to the left of the bar (in this case, the scientific evidence) from those things that the event is conditional upon; C denotes the condition where, for example, the samples are from the same source; C denotes the condition where the samples are not from the same source. Therefore, the numerator considers the probability of obtaining this evidence if the samples are indeed from the same source. The denominator considers the probability of obtaining this evidence if the samples are not from the same source.

Each of the probability terms of the LR can have a value between 0 and 1. The value of 0 occurs if there is no chance that the event could occur, while the value of 1 occurs if the event would definitely occur.

For the numerator of the LR, a high value indicates a better match. For the denominator, a very low value indicates a very uncommon assemblage. In most cases these values will be subjective estimates based on the palynologist's knowledge and experience.

Although it is possible to objectively determine values for the numerator and denominator of the LR, it would require a vast surveying effort. The variability between pollen assemblages from different sources is enormous, and therefore very large surveys would be required to estimate probabilities for the denominator (the probability of obtaining at random a sample with the pollen assemblage that was observed in the "recovered" sample) and the numerator (the probability of obtaining the assemblage that was observed, in light of what assemblages might be expected to be obtained from the "control" or "scene" sample as a result of sampling variation). The authors are considering ways to tackle this problem. At the moment, however, the assessment is subjective but within the appropriate framework.

Non-Bayesian approaches to evidential assessment typically involve a "frequentist" approach which looks only at the denominator of the likelihood ratio. For example, the frequentist approach might involve asking "how common is the pollen assemblage that was observed in the soil from the shoes of the suspect?" Without the survey data, the frequentist approach is also subjective. In the opinion of the authors, the advantage of the Bayesian approach over the frequentist approach is that, with the Bayesian approach, consideration is given to the quality of the match. Fienberg and Schervish (1986) have discussed in detail the different approaches considered for legal decision making.

A very strong match and a very uncommon assemblage results in a very large number for the LR which indicates that the evidence is very strong, i.e., highly significant. On the other hand, there may be no pollen match and the numerator is 0, in which case the pollen assemblage becomes irrelevant because the resulting value for the LR is 0 regardless, in which case the evidence is exclusionary. These are the two extremes generally found with pollen evidence. Between these, a range of descriptions is required when the evidence is, for example, only of weak or moderate significance. A proposal for this is shown in Text-Figure 1 (after Evett and Buckleton, 1989).

In Text-Figure 1, the axis is the value obtained for the LR. A value of 1 is neutral or inconclusive (i.e., the evidence neither supports nor negates the contention that the suspect was at the crime scene). The higher the value of the LR, the greater the support for the contention that the suspect was at the crime scene. The categories within this range are described in language considered more appropriate for non-scientists (Evett and Buckleton, 1989):

LR = 1–10: the pollen evidence **weakly supports** the contention that the suspect was at the crime scene.

LR = 10–100: the pollen evidence **supports** the contention that ...

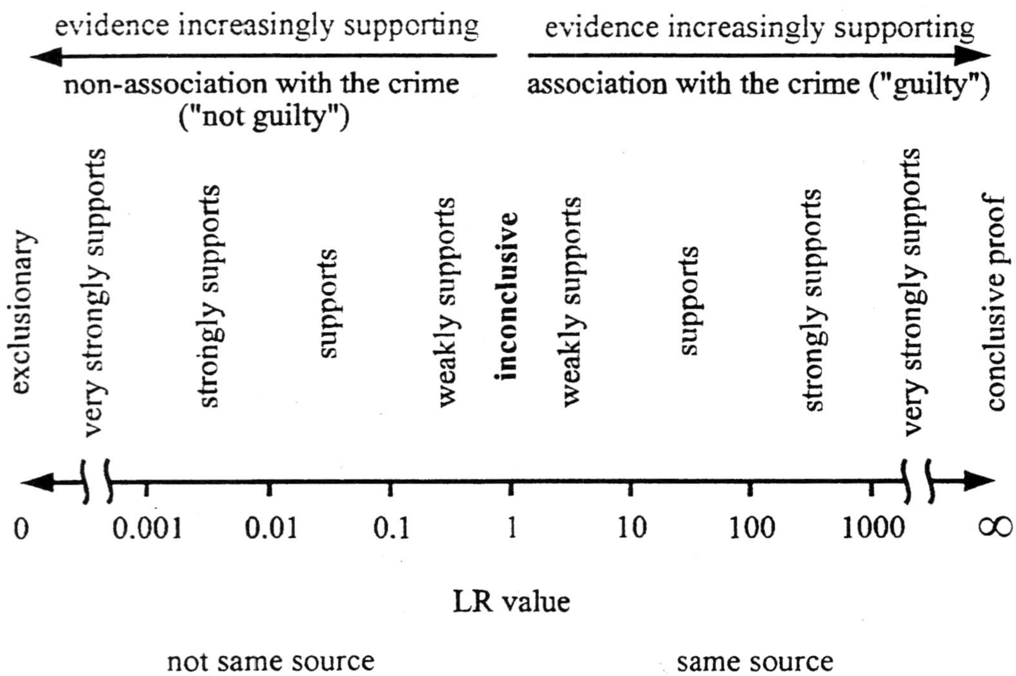

Text-Figure 1. Diagram showing a range of descriptions of the significance of numbers obtained from the Likelihood Ratio (LR).

LR = 100–1000: the pollen evidence **strongly supports** the contention that ...

LR >1000: the pollen evidence **very strongly supports** the contention that ...

An exactly equivalent but inverse set of categories and conclusions are formed for evidence which supports the contention that the suspect was not at the crime scene (i.e., LR <1):

The case example involves a car chase. The pursued car failed to take a bend and crashed into a tree next to a driveway. The driver struggled on the ground with pursuing police before escaping through kiwifruit (*Actinidia deliciosa*) vines growing alongside the driveway. A suspect was found later who denied being the driver. Pollen in soil from the suspect's jeans was compared with pollen in a soil sample from the crime scene. Both samples contained nine of the same uncommon pollen types of similar abundance. Four of these were not differentiated but the remaining five were identified as pollen of *Eucalyptus*, poplar (*Populus*), kiwifruit (*Actinidia deliciosa*), English ivy (*Hedera helix*) and Chinese fan palm (*Trachycarpus fortunei*), all of which were growing at the scene. Most, if not all of these pollen types are not well dispersed by wind. Direct contact with the parent plants or with soil containing these pollen types would be necessary to transfer them to clothing.

The occurrence of nine significant pollen types (even though four are undifferentiated) shared between these samples, and of similar abundance, makes this a very strong match. The presence of similar pollen assemblages of similar abundance in the two samples makes the probability of this evidence high if they are from the same source. Also, the pollen assemblage would be rarely encountered in soil taken from the clothing of people unrelated to this crime, so a figure very close to 0 is attributed to the denominator, i.e., the probability of this evidence is very low if they are not from the same source. The resulting value of the LR is certainly greater than 1000 which falls in the highest category on the likelihood scale (Text-Figure 1). Therefore, the conclusion reached after this analysis was that "the pollen evidence **very strongly** supports the contention that the soil from Mr. Brown's jeans came from the driveway at ..."

CONCLUSIONS

The use of Bayes' Theorem for the assessment of forensic palynological evidence allows for a well structured and considered framework in which to assess the evidence. Therefore, the court is assisted by having the scientist assess the evidence and present it in a manner which is more readily understood by lay people.

References Cited

BRYANT, V.M. JR., JONES, J.G., and MILDENHALL, D.C.
1990 Forensic palynology in the United States of America. *Palynology*, 14: 193–208.

BUCKLETON, J., WALSH, K.J., and TRIGGS, C.M.
1991 A continuous model for interpreting the positions of bands in DNA locus-specific work. *Journal of the Forensic Science Society*, 31: 353–363.

EVETT, I.W.
1986 A Bayesian approach to the problem of interpreting glass evidence in forensic science casework. *Journal of the Forensic Science Society*, 26: 3–18.

EVETT, I.W., and BUCKLETON, J.S.
1989 Some aspects of the Bayesian approach to evidence evaluation. *Journal of the Forensic Science Society*, 29: 317–324.
1990 The interpretation of glass evidence. A practical approach. *Journal of the Forensic Science Society*, 30: 215–223.

FIENBERG, S., and SCHERVISH, M.J.
1986 The relevance of Bayesian inference for the presentation of statistical evidence and for legal decision making. *Boston University Law Review*, 66: 771–798.

LINDLEY, D.
1977 A problem in forensic science. *Biometrika*, 64: 207.

HORROCKS, M., BEDFORD, K.R., and MORGAN-SMITH, R.K.
1997 The filtering effects of various household fabrics on the pollen content of hash oil (cannabis extract). *Journal of Forensic Sciences*, 42: 255–258.

HORROCKS, M., COULSON, S.A., and WALSH, K.A.J.
1998 Forensic palynology: variation in the pollen content of soil surface samples. *Journal of Forensic Sciences*, 43: 320–323.

HORROCKS, M., and WALSH, K.A.J.
1998 Forensic palynology: assessing the value of the evidence. *In*: Boyd, W.E., and Hall, J.A. (eds.), *Review of Palaeobotany and Palynology*, Special Issue: New Frontiers and Applications in Palynology, 103: 69–74.

MILDENHALL, D.C.
1882 Forensic palynology. *Geological Society of New Zealand Newsletter*, 58: 25.
1988 Deer velvet and palynology: an example of the use of forensic palynology in New Zealand. *Tuatara*, 30: 1–11.
1990 Forensic palynology in New Zealand. *Review of Palaeobotany and Palynology*, 64: 227–234.

ROBERTSON, B., and VIGNAUX, G.A.
1995 *Interpreting evidence*. John Wiley & Sons, Chichester, 240 pgs.

STANLEY, E.A.
1992 Application of palynology to establish the provenance and travel history of illicit drugs. *Microscope*, 40: 149–152.
1993 Forensic palynology. *In*: *Federal Bureau of Investigation International Symposium on Trace Evidence*. US Government Printing Office, Washington DC, p. 17–30.

Authors' addresses

MARK HORROCKS
Centre for Archaeological Research
University of Auckland
Private Bag 92–019
Auckland
New Zealand

KEVAN A. J. WALSH
Institute of Environmental Science & Research Ltd.
Mt. Albert Science Centre
Private Bag 92–021
Auckland
New Zealand

Author Index